Hydraulics/Mechanics

CIVIL ENGINEERING PRACTICE

2/Hydraulics/Mechanics

EDITED BY

PAUL N. CHEREMISINOFF

NICHOLAS P. CHEREMISINOFF

SU LING CHENG

IN COLLABORATION WITH

B. S. Bhatt
M. G. Bos
R. P. Chhabra
A. J. Clemmens
G. Echávez
R. C. Gupta
W. R. Gwinn
M. I. Haque
H. M. Hathoot
E. J. Hickin
J. L. Humar
B. Hunt
S. Ikeda
N. Jnanasekaran
H. Kobayashi
B. L. Krayterman
E. Levi
S.-Y. Liong
J. A. McCorquodale
H. Majcherek
J. Marsalek
L. W. Mays
B. W. Melville
E. Nonveiller
M. Patarapanich
N. Rajaratnam
J. A. Replogle
N. Rodríguez-Cuevas
N. C. Sacheti
D. S. Sandegren
M. Sathyamoorthy
H. T. Shen
M. Shimada
K. Sonoda
B. Stanić
Y.-K. Tung
L. C. van Rijn
L. S. Willardson
P. D. Yapa

TECHNOMIC PUBLISHING CO., INC.

LANCASTER · BASEL

Published in the Western Hemisphere by
Technomic Publishing Company, Inc.
851 New Holland Avenue
Box 3535
Lancaster, Pennsylvania 17604 U.S.A.

Distributed in the Rest of the World by
Technomic Publishing AG

Printed in the United States of America

10 9 8 7 6 5 4 3 2 1

Main entry under title:
Civil Engineering Practice 2—Hydraulics/Mechanics

A Technomic Publishing Company book
Bibliography: p.
Includes index p. 777

Library of Congress Card No. 87-50629
ISBN No. 87762-546-8

TABLE OF CONTENTS

PREFACE

While the designation civil engineering dates back only two centuries, the profession of civil engineering is as old as civilized life. Through ancient times it formed a broader profession, best described as master builder, which included what is now known as architecture and both civil and military engineering. The field of civil engineering was once defined as including all branches of engineering and has come to include established aspects of construction, structures, and emerging and newer sub-disciplines (e.g., environmental, water resources, etc.). The civil engineer is engaged in planning, design of works connected with transportation, water and air pollution, as well as canals, rivers, piers, harbors, etc. The hydraulic field covers water supply/power, flood control, drainage and irrigation, as well as sewerage and waste disposal.

The civil engineer may also specialize in various stages of projects such as investigation, design, construction, operation, etc. Civil engineers today, as well as engineers in all branches, have become highly specialized, as well as requiring a multiplicity of skills in methods and procedures. Various civil engineering specialties have led to the requirement of a wide array of knowledge.

Civil engineers today find themselves in a broad range of applications, and it was to this end that the concept of putting this series of volumes together was made. The tremendous increase of information and knowledge all over the world has resulted in proliferation of new ideas and concepts, as well as a large increase in available information and data in civil engineering. The treatises presented are divided into five volumes for the convenience of reference and the reader:

VOLUME 1 Structures
VOLUME 2 Hydraulics/Mechanics
VOLUME 3 Geotechnical/Ocean Engineering
VOLUME 4 Surveying/Transportation/Energy/Economics & Government/Computers
VOLUME 5 Water Resources/Environmental

A serious effort has been made by each of the contributing specialists to this series to present information that will have enduring value. The intent is to supply the practitioner with an authoritative reference work in the field of civil engineering. References and citations are given to the extensive literature as well as comprehensive, detailed, up-to-date coverage.

To insure the highest degree of reliability in the selected subject matter presented, the collaboration of a large number of specialists was enlisted, and this book presents their efforts. Heartfelt thanks go to these contributors, each of whom has endeavored to present an up-to-date section in their area of expertise and has given willingly of valuable time and knowledge.

PAUL N. CHEREMISINOFF
NICHOLAS P. CHEREMISINOFF
SU LING CHENG

CONTRIBUTORS TO VOLUME 2

B. S. Bhatt, University of the West Indies, St. Augustine, Trinidad (W.I.)

M. G. Bos, International Institute for Land Reclamation and Improvement/ILRI, Wageningen, The Netherlands

R. P. Chhabra, Indian Institute of Technology, Kanpur, India

A. J. Clemmens, U.S. Water Conservation Laboratory, USDA-ARS, Phoenix, AZ

G. Echávez, Universidad Nacional Autónoma de México

R. C. Gupta, National University of Singapore, Singapore

W. R. Gwinn, retired, U.S. Department of Agriculture, Stillwater, OK

M. I. Haque, The George Washington University, Washington, DC

H. M. Hathoot, Alexandria University, Alexandria, Egypt

E. J. Hickin, Simon Fraser University, British Columbia, Canada

J. L. Humar, Carleton University, Ottawa, Canada

B. Hunt, University of Canterbury, Christchurch, New Zealand

S. Ikeda, Saitama University, Saitama, Japan

N. Jnanasekaran, A.C. College of Engineering and Technology, Karaikudi, India

H. Kobayashi, Osaka City University, Osaka, Japan

B. L. Krayterman, University of Maryland, College Park, MD

E. Levi, Instituto Mexicano de Technología del Agua

S.-Y. Liong, National University of Singapore, Singapore

J. A. McCorquodale, University of Windsor, Windsor, Ontario, Canada

H. Majcherek, Technical University of Poznan, Poznan, Poland

J. Marsalek, National Water Research Institute, Burlington, Ontario, Canada

L. W. Mays, University of Texas, Austin, TX

B. W. Melville, University of Auckland, Auckland, New Zealand

E. Nonveiller, University of Zagreb, Yugoslavia

M. Patarapanich, National University of Singapore, Singapore

N. Rajaratnam, University of Alberta, Edmonton, Alberta, Canada

J. A. Replogle, U.S. Water Conservation Laboratory, USDA-ARS, Phoenix, AZ

N. Rodríguez-Cuevas, Universidad Nacional Autónoma de México

N. C. Sacheti, University of Jodhpur, Jodhpur, India

D. S. Sandegren, Anna University, Madras, India

M. Sathyamoorthy, Clarkson University, Potsdam, NY

H. T. Shen, Clarkson University, Potsdam, NY

M. Shimada, National Research Institute of Agricultural Engineering, Ibaraki, Japan

K. Sonoda, Osaka City University, Osaka, Japan

B. Stanić, University of Zagreb, Yugoslavia

Y.-K. Tung, University of Wyoming, Laramie, WY

L. C. van Rijn, Delft Hydraulics, The Netherlands

L. S. Willardson, Utah State University, Logan, UT

P. D. Yapa, Clarkson University, Potsdam, NY

CIVIL ENGINEERING PRACTICE

SECTION ONE

Hydraulics/Open Channel Flow

An Asymptotic Solution for Dam-Break Floods in Sloping Channels

BRUCE HUNT*

INTRODUCTION

Floods resulting from sudden dam breaks have been responsible for causing large losses in human life and property. Thus, it is of some importance for a civil engineer to be able to estimate flood depths and flow rates downstream from a ruptured dam so that risks can be evaluated and reduced to an acceptable level. In mathematical terms, however, this problem requires the simultaneous solution of two nonlinear partial differential equations which have time and distance as independent variables. The result is that a complete, closed-form solution to this problem has never been obtained, and singularities that appear in one of these two equations when flow depths vanish make it extremely difficult to even obtain satisfactory numerical solutions. As a result, approximate solutions have played a prominent role in the history of this problem.

Ritter [6] obtained the first dam-break solution over 90 years ago. This remarkable closed-form solution was for an instantaneous rupture of a dam holding a reservoir of semi-infinite length upon a horizontal dry channel. Bed resistance was also neglected. These assumptions meant that Ritter's solution gave a satisfactory approximation away from the leading edge of the flood wave and during the initial period after rupture before the flood had a chance to advance very far downstream. About 60 years later Dressler [3] and Whitham [11] obtained approximate solutions that included resistance effects. These solutions used the Ritter solution together with a correction for the effects of bed resistance near the leading edge of the flood wave. The results gave more realistic approximations over the entire length of the flood wave, but only during the initial time period following the dam rupture. After this time period, however, the effects of a finite reservoir length, non-zero channel slope and bed resistance near the trailing edge of the flood wave become too large to ignore.

In more recent years, emphasis has shifted toward numerical solutions such as those given by Chen [1], Chen and Armbruster [2], Sakkas and Strelkoff [7,8], and Strelkoff, Schamber and Katopodes [9]. These solutions have the advantage of providing a relatively large degree of flexibility when specifying variables such as channel geometry and roughness. On the other hand, difficulties with these numerical solutions are encountered near both the leading and trailing edges of the flood wave as water depths approach zero. In addition, numerical solutions are not as useful as analytical solutions for obtaining an understanding of physical processes.

The approximate analytical solution given herein follows an earlier solution published by Hunt [5] and is obtained by applying the method of matched asymptotic expansions described by Van Dyke [10]. This solution is for the complete and instantaneous collapse of a dam in a wide, infinitely-long sloping channel, and order of magnitude arguments suggest that the result is an asymptotic solution that only becomes valid after the leading edge of the flood wave advances three or four reservoir lengths downstream from the dam. After the flood wave reaches this point, the free surface becomes nearly parallel to the channel bottom, which means that the depth-gradient and acceleration terms in the momentum equation become negligibly small everywhere except near the leading edge of the flood wave. Thus, the momentum equation reduces to the Chézy equation of open-channel flow throughout this region, a result which is known as the "kinematic-wave approximation." Near the leading edge of the flood wave, where the curvature of the free surface is relatively large, a second solution is obtained that includes the bed-slope, bed-resistance, and depth-gradient terms in the momentum equation. This inner

*Reader, Department of Civil Engineering, University of Canterbury, Christchurch, New Zealand

boundary-layer solution is then combined with the outer kinematic-wave solution to form a composite solution that gives a valid description of the physics over the entire length of the flood wave.

PROBLEM FORMULATION

A reservoir with a length L and a maximum depth H lies in an infinitely-long, wide channel that has a slope of H/L. The x coordinate is measured from the upstream end of the reservoir. At $t = 0$ the dam is suddenly and completely removed, allowing the water within the reservoir to rush downstream over a dry channel bed. The leading edge of the flood wave is called a shock front and has the coordinate $x = x_s(t)$. A definition sketch for this problem is shown in Figure 1.

The system of equations describing this problem consists of a momentum equation, continuity equation, and initial condition:

$$g\left(\frac{\partial h}{\partial x} + S_f - S_o\right) + u\frac{\partial u}{\partial x} + \frac{\partial u}{\partial t} = 0 \qquad (1)$$

$$\frac{\partial (uh)}{\partial x} + \frac{\partial h}{\partial t} = 0 \qquad (2)$$

$$h(x, 0) = H\frac{x}{L} \text{ for } 0 \le x < L \qquad (3)$$

$$= 0 \qquad \text{for all other } x$$

in which h = flow depth; u = velocity; x = distance along the sloping channel; t = time; g = gravitational constant; S_f = bed resistance term; S_O = bed slope; H = maximum reservoir depth; and L = reservoir length.

It has already been noted that the channel slope can be written in terms of the reservoir dimensions:

$$S_0 = \frac{H}{L} \qquad (4)$$

One traditional way of approximating the bed resistance

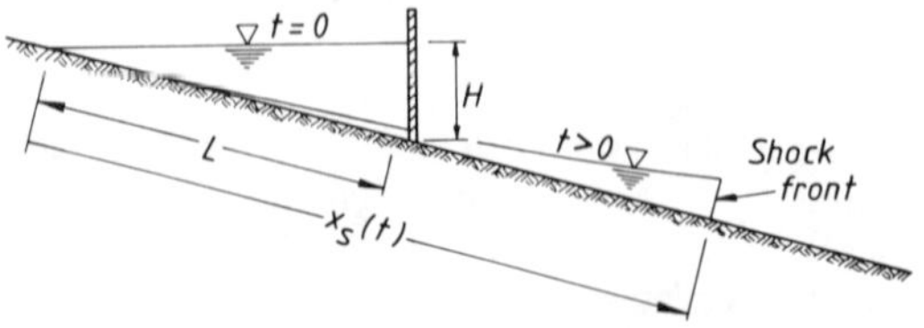

FIGURE 1. A definition sketch for the dam-break problem.

term is by using the Chézy equation of uniform open-channel flow:

$$u = C\sqrt{h\, S_f} \qquad (5)$$

in which C = Chézy coefficient. Thus, if u_0 and h_0 are a corresponding uniform flow velocity and depth when $S_f = S_0$, Equation (5) becomes

$$u_0 = C\sqrt{h_0\, S_0} \qquad (6)$$

and eliminating C from Equations (5–6) gives

$$S_f = S_o\frac{(u/u_0)^2}{h/h_0} \qquad (7)$$

Finally, using Equations (4) and (7) to eliminate S_0 and S_f from Equation (1) gives an alternative form of the momentum equation:

$$g\frac{\partial h}{\partial x} + g\frac{H}{L}\left[\frac{(u/u_0)^2}{h/h_0} - 1\right] + u\frac{\partial u}{\partial x} + \frac{\partial u}{\partial t} = 0 \qquad (8)$$

An approximate solution of Equations (2), (3), and (8) will be obtained by using one of the techniques of singular perturbation theory.

Perturbation techniques basically consist of scaling terms in a set of governing equations and then dropping terms that are relatively small. Thus, it is necessary to introduce the following scaled dimensionless terms into Equations (2), (3), and (8):

$$(h^*, u^*, x^*, t^*, H^*, F) = \left[\frac{h}{h_0}, \frac{u}{u_0}, \frac{x}{\ell}, \frac{u_0 t}{\ell}, \frac{H}{h_0}, \frac{u_0}{\sqrt{g h_0}}\right] \qquad (9)$$

in which F = Froude number. Since the scaled terms must have an order of magnitude that does not exceed unity, we will choose h_0 = maximum value of h at the instant that the solution is being computed, u_0 = uniform flow velocity corresponding to the depth h_0 and $\ell = x_s(t)$ at the instant under consideration. Introducing Equation (9) into Equations (2), (3), and (8) and dropping the asterisk superscript for notational convenience gives

$$1 - \frac{u^2}{h} = \epsilon\left[\frac{\partial h}{\partial x} + F^2\left(u\frac{\partial u}{\partial x} + \frac{\partial u}{\partial t}\right)\right] \qquad (*10)$$

$$\frac{\partial (uh)}{\partial x} + \frac{\partial h}{\partial t} = 0 \qquad (*11)$$

$$h(x, 0) = \frac{x}{\epsilon} \text{ for } 0 \le x < \epsilon H \qquad (*12)$$

$$= 0 \text{ for all other } x$$

in which an asterisk superscript appears in front of the equation numbers to show that the equations are written in dimensionless variables. The small perturbation parameter, ϵ, is given by

$$\epsilon = \frac{L}{\ell}\frac{h_0}{H} \qquad (13)$$

Since L/ℓ and h_0/H both become smaller as ℓ increases, it is seen that a solution for small values of ϵ will be an asymptotic solution that applies only after the shock front has advanced far enough downstream. In this case, Equation (12) is seen to model the reservoir as a point source of mass as $\epsilon \to 0$.

THE OUTER SOLUTION

The outer solution, which applies over the major portion of the solution domain, is obtained by letting $\epsilon \to 0$ in Equations (10)–(12). This procedure reduces Equation (10) to the kinematic-wave approximation

$$1 - \frac{u^2}{h} = 0 \qquad (*14)$$

Eliminating u from Equations (11) and (14) gives

$$\frac{3}{2}\sqrt{h}\,\frac{\partial h}{\partial x} + \frac{\partial h}{\partial t} = 0 \qquad (*15)$$

which must be solved subject to the initial condition given by Equation 12.

Equation (15) can be written as

$$\frac{dh}{dt} \equiv \frac{dx}{dt}\frac{\partial h}{\partial x} + \frac{\partial h}{\partial t} = 0 \qquad (*16)$$

provided that the following choice is made for dx/dt:

$$\frac{dx}{dt} = \frac{3}{2}\sqrt{h} \qquad (*17)$$

Equations (16)–(17) are the characteristic form of Equation 15, and their simultaneous integration gives

$$h = C_1 \qquad (*18)$$

along the straight-line characteristic curves

$$x - \frac{3}{2}t\sqrt{h} = C_2 \qquad (*19)$$

in which the integration constants C_1 and C_2 must be determined from the initial condition given by Equation (12).

In the limit as $\epsilon \to 0$, Equation (12) shows that $h(x, 0)$ vanishes along the x axis at all points except the origin. Thus, an application of Equations (18–19) along a characteristic joining the two points $(\xi, 0)$ and (x, t) gives

$$h = 0 \text{ along } x = \xi \text{ for } 0 < |\xi| < \infty \qquad (*20)$$

Equation (20) gives a parametric solution for h in a certain portion of the (x, t) plane. Since h takes on all values between 0 and infinity at the origin, as $\epsilon \to 0$ in Equation (12), an application of Equations (18–19) along one of an infinite number of straight-line characteristics joining the two points (0, 0) to (x, t) gives

$$h = \eta \text{ along } x - \frac{3}{2}t\sqrt{\eta} = 0 \text{ for } 0 \le \eta < \infty \qquad (*21)$$

Equation (21) also gives a parametric solution for h in a certain portion of the (x, t) plane. Alternative forms for Equations (20–21) can be found by eliminating the parameters ξ and η:

$$h = 0 \text{ for } 0 < |x| < \infty \qquad (*22)$$

$$h = \left(\frac{2x}{3t}\right)^2 \text{ for } 0 \le h < \infty \qquad (*23)$$

Either Equations (20–21) or Equations (22–23) can be used to calculate the solution.

A more detailed examination of the characteristic curves in Equations (20–21) shows that both families of characteristics are superimposed upon each other in the upper right quadrant of the (x, t) plane. Since each characteristic carries a different value of h, the solution will be double-valued in this region unless a curve passing through the origin is inserted within the first quadrant to prevent characteristics from crossing. Thus curve is called a shock, and Equations (22) and (23) give the solution for h downstream and upstream, respectively, from the shock. Mass conservation requires that the shock be located so that

$$\int_0^{x_s(t)} h(x,t)\,dx = \int_0^{\epsilon H} \frac{x}{\epsilon}\,dx \equiv \frac{1}{2}\epsilon H^2 \text{ for } 0 \le t < \infty \qquad (*24)$$

and the use of Equation (23) in Equation (24) gives the

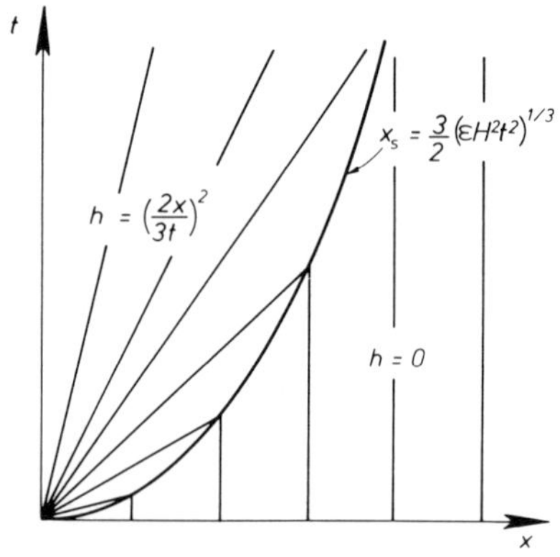

FIGURE 2. A sketch showing the characteristics, the shock location and the outer (kinematic-wave) solution in the (x, t) plane.

shock coordinate, $x_s(t)$:

$$x_s(t) = \frac{3}{2}(\epsilon H^2 t^2)^{1/3} \tag{*25}$$

A sketch of the characteristics and shock location is shown in Figure 2, and it is important to note that the outer solution for h is discontinuous along the shock.

THE INNER SOLUTION

One of the distinguishing characteristics of a singular perturbation solution is that order of magnitude estimates used to scale equations for the outer solution become invalid within some small region of the solution domain. In this case, it is seen that the depth-gradient term in Equation (10) is ignored in the outer solution but becomes infinite along the location of the shock, which is given by Equation (25) and plotted in Figure 2. Thus, the next step is to rescale the governing equations to obtain an approximate solution in the neighborhood of this shock. This can be done by introducing the following inner variables:

$$(v, \phi, y, \tau) = \left[u, h, \frac{x - x_s(t)}{\Delta(\epsilon)}, t\right] \tag{*26}$$

in which $\Delta(\epsilon) \to 0$ as $\epsilon \to 0$. Substituting Equation (26) into Equations (10–11) gives

$$1 - \frac{v^2}{\phi} = \frac{\epsilon}{\Delta(\epsilon)}\left[\frac{\partial \phi}{\partial y} + F^2(v - x_s')\frac{\partial v}{\partial y}\right] + \epsilon F^2 \frac{\partial v}{\partial \tau} \tag{*27}$$

$$\frac{\partial[(v - x_s')\phi]}{\partial y} + \Delta(\epsilon)\frac{\partial \phi}{\partial \tau} = 0 \tag{*28}$$

in which x_s' denotes the derivative of $x_s(t)$. Since the depth-gradient term must be retained in the limit as $\epsilon \to 0$ in the inner equations, Equation (27) shows that

$$\Delta(\epsilon) = \epsilon \tag{*29}$$

Using Equation (29) and taking the limit $\epsilon \to 0$ in Equations (27–28) gives the first-order inner equations.

$$1 - \frac{v^2}{\phi} = \frac{\partial \phi}{\partial y} + F^2(v - x_s')\frac{\partial v}{\partial y} \tag{*30}$$

$$\frac{\partial[(v - x_s')\phi]}{\partial y} = 0 \tag{*31}$$

Equation (31) can be integrated immediately to give

$$(v - x_s')\phi = K(\tau) \tag{*32}$$

in which $K(\tau)$ is an unknown function of τ. But $v - x_s' = 0$ at the shock front for all values of τ. Thus, $K(\tau)$ vanishes for all τ, and Equation (32) reduces to

$$v = x_s'(\tau) \tag{*33}$$

Equation (33) states that velocities near the shock vary with τ but not y and, therefore, are equal to the shock velocity. This is an approximation that was first used by Whitham [11] with less formal justification.

Substituting Equation (33) into Equation (30) gives an equation that can be integrated to find ϕ.

$$1 - \frac{x_s'^2}{\phi} = \frac{\partial \phi}{\partial y} \tag{*34}$$

Since acceleration terms do not appear in Equation (34), it can be concluded that velocities are nearly constant, that accelerations are negligible and that the bed slope, bed resistance and depth-gradient terms are the only important terms in the momentum equation near the shock front. It is convenient to rewrite $x_s'(\tau)$ in Equation (34) in terms of the shock depth, $h_s(\tau) \equiv \phi_s(\tau)$. This can be done by substituting Equation (25) into Equation (23) to obtain an expression for the shock depth:

$$h_s(t) \equiv \phi_s(t) = \left(\frac{\epsilon H^2}{t}\right)^{2/3} \tag{*35}$$

Next, differentiating Equation (25) gives

$$x_s'(t) = \left(\frac{\epsilon H^2}{t}\right)^{1/3} \tag{*36}$$

and eliminating t from Equations (35–36) gives

$$x_s'(\tau) = \sqrt{\phi_s(\tau)} \qquad (*37)$$

in which $\phi_s(\tau)$ is the shock depth written in inner variables. Thus, Equation (34) can be written in the form

$$1 - \frac{\phi_s(\tau)}{\phi(y,\tau)} = \frac{\partial \phi(y,\tau)}{\partial y} \qquad (*38)$$

Integration of Equation (38) gives

$$y = \phi + \phi_s(\tau) \ln \left[1 - \frac{\phi}{\phi_s(\tau)} \right] + y_0(\tau) \qquad (*39)$$

in which $y_0(\tau)$ is an unknown function of τ. Matching of the inner and outer solution, which consists of rewriting the outer solution in inner variables, the inner solution in outer variables and taking the limit $\epsilon \to 0$, gives

$$\phi(-\infty, \tau) = \phi_s(\tau) \qquad (*40)$$

However, Equation (39) satisfies Equation (40) for all values of $y_0(\tau)$, and some other condition must be used to evaluate $y_0(\tau)$. Since the kinematic shock front has the location $y = 0$ in the outer solution, and since the shock front in the inner solution extends past this point by an amount $y_0(\tau)$, the function $y_0(\tau)$ can be determined from the mass conservation requirement

$$\int_{-\infty}^{0} [\phi_s(\tau) - \phi]\, dy = \int_{0}^{y_0(\tau)} \phi\, dy \qquad (*41)$$

But Equation (38) gives

$$dy = \frac{\phi}{\phi - \phi_s}\, d\phi \qquad (*42)$$

which can be substituted into Equation (41) to give

$$-\int_{\phi_s}^{\phi_0} \phi\, d\phi = \int_{\phi_0}^{0} \frac{\phi^2}{\phi - \phi_s}\, d\phi \qquad (*43)$$

The value for ϕ_0 is found by setting $y = 0$ in Equation (39) to obtain

$$0 = \frac{\phi_0}{\phi_s} + \ln \left(1 - \frac{\phi_0}{\phi_s} \right) + \frac{y_0}{\phi_s} \qquad (*44)$$

Calculating the integrals in Equation (43) leads to a second equation involving ϕ_0 and ϕ_s.

$$\frac{\phi_0}{\phi_s} + \ln \left(1 - \frac{\phi_0}{\phi_s} \right) = -\frac{1}{2} \qquad (*45)$$

Finally, eliminating ϕ_0 from Equations (44–45) gives

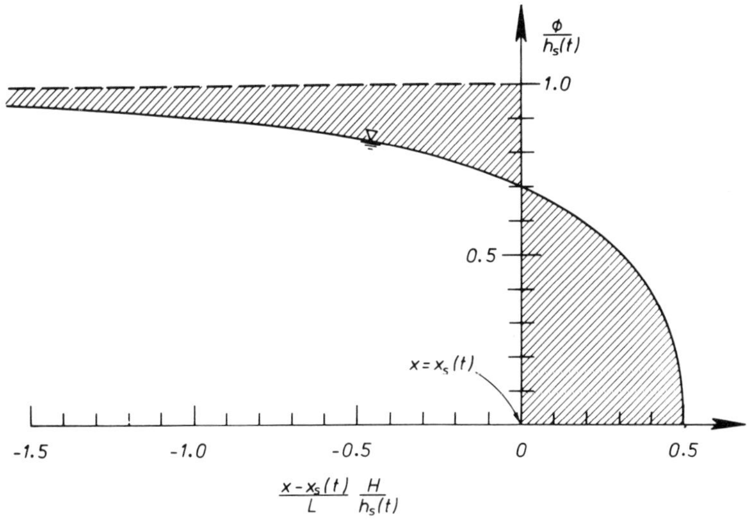

FIGURE 3. A plot of the inner solution in the neighborhood of the kinematic shock.

$$y_0(\tau) = \frac{1}{2}\,\phi_s(\tau) \qquad (*46)$$

and the inner solution is obtained from Equations (39) and (46).

$$y = \phi + \phi_s(\tau)\ln\left[1 - \frac{\phi}{\phi_s(\tau)}\right] + \frac{1}{2}\,\phi_s(\tau) \qquad (*47)$$

A plot of Equation (47) is shown in Figure 3, and it is worth noting that the cross-hatched areas on each side of the origin in this figure are equal because of the mass conservation statement given by Equation (41).

A COMPOSITE SOLUTION

The physical significance of the inner solution given by Equation (47) is that the discontinuity in h that appears in the outer kinematic-wave solution has been eliminated by decreasing peak depths behind the shock and by placing a rounded nose in front of the kinematic shock. Mass has been conserved during this process by making the area of water contained between the new shock nose and the kinematic shock equal to the area that was lost behind the kinematic shock when the kinematic-wave solution was replaced with the inner solution. Since Equation (40) shows that the common part of the inner and outer solution behind the shock is given by $\phi_s(\tau)$, this common part can be subtracted from the sum of the inner and outer solutions behind the kinematic shock to obtain a composite solution, h_c.

$$h_c = h(x, t) + \phi(y, \tau) - \phi_s(\tau) \text{ for } 0 \le x \le x_s(\tau) \qquad (*48)$$

The outer solution vanishes in front of the kinematic shock, which means that the composite solution in front of the kinematic shock is given by

$$h_c = \phi(y, \tau) \text{ for } x_s(\tau) \le x \qquad (*49)$$

The composite solution, h_c, is seen from Equations (48–49) to be continuous at the shock since $h = h_s = \phi_s$ at $x = x_s$.

DIMENSIONAL VARIABLES

The complete first-order solution is given in dimensionless variables by Equations (22), (23), (25), (47), (48), and (49). One way to carry out numerical calculations with these equations is to first convert dimensional variables into dimensionless variables by using Equations (9) and (26). In this case, it is necessary to choose representative values for h_0, u_0 and ℓ. These were initially chosen as the maximum value of h, the corresponding value for u under conditions of uniform flow and the value of $x_s(t)$, respectively, at the instant for which calculations are being carried out. This choice was made initially for the purpose of determining orders of magnitude for the various terms in the equations of motion. However, since ℓ does not appear in the dimensional form of the governing equations [Equations (2), (3), and (8)], one finds that the variable ℓ cancels out and does not appear in the solution when it is rewritten in dimensional variables. Thus, when calculating numerical values for the dimensionless variables, it is possible to choose an arbitrary value for ℓ. Likewise, h_0 may also be chosen arbitrarily, and u_0 must be the uniform flow velocity calculated from the Chézy equation, or from field measurements, that corresponds to the flow depth h_0.

Many engineers prefer to work with equations in dimensional form. Inserting dimensional variables in the solution gives the following result for the location, speed and depth of the kinematic shock:

$$\frac{x_s(t)}{L} = \frac{3}{2}\left(\frac{u_0 t}{L}\sqrt{\frac{H}{h_0}}\right)^{2/3} \qquad (50)$$

$$\frac{x_s'(t)}{u_0} = \left(\frac{H}{h_0}\frac{L}{u_0 t}\right)^{1/3} \qquad (51)$$

$$\frac{h_s(t)}{h_0} = \left(\frac{H}{h_0}\frac{L}{u_0 t}\right)^{2/3} \qquad (52)$$

The outer solution is given by

$$\frac{h(x, t)}{h_0} = \left(\frac{2}{3}\frac{x}{u_0 t}\right)^2 \text{ for } 0 \le x < x_s(t)$$
$$= 0 \text{ for } -\infty < x \le 0 \text{ and } x_s(t) < x < \infty \qquad (53)$$

and the inner solution is given by

$$\frac{x - x_s(t)}{L}\,\frac{H}{h_s(t)} = \frac{\phi}{h_s(t)} + \ln\left[1 - \frac{\phi}{h_s(t)}\right] + \frac{1}{2} \qquad (54)$$

A plot of Equation (54) is shown in Figure 3. Finally, the composite solution is given in dimensional variables by

$$h_c = h(x, t) + \phi - h_s(t) \text{ for } 0 \le x \le x_s(t)$$
$$= \phi \text{ for } x_s(t) \le x \qquad (55)$$

Inspection of Equation (55) and Figure 3 shows that the composite solution is the difference between the kinematic-wave (outer) solution and the cross-hatched region for

$x \leq x_s(t)$ and becomes identical with the cross-hatched region for $x \geq x_s(t)$.

SOLUTION VALIDITY

The solution just calculated is shown from the problem formulation to be an asymptotic solution that becomes valid only after $x_s(t)$ becomes large enough. The obvious question, of course, is how large is "large enough." The only indisputable way to answer this question is to compare the asymptotic approximation with a more exact solution obtained by experimental, numerical, or mathematical methods. Unfortunately, however, more exact solutions are not obtained easily, and a successful comparison of this nature has not yet been carried out.

Letting $\epsilon \rightarrow 0$ in Equations (10–12) has two effects: first, depth-gradient and acceleration terms vanish from the momentum equation, and, second, the initial condition giving the distribution of water along the river channel at $t = 0$ is replaced with an initial condition that models the reservoir as a point source of mass. One way to examine the effect of neglecting the depth-gradient and acceleration terms is to obtain second-order corrections by substituting the first-order solution into the right side of Equation (10) and solving Equations (10–12). This implies that second-order corrections will have the same order of magnitude as the approximated terms on the right side of Equation (10). Hunt [4] substituted the kinematic-wave solution given by Equations (14) and (23) into the right side of Equation (10) and calculated a magnitude of 10 percent for these approximated terms after the kinematic shock travels about four reservoir lengths downstream from the ruptured dam. A similar estimate can be made by calculating a value for ϵ in Equation (10), since all other terms have been scaled to have a unit order of magnitude. Thus, eliminating t from Equations (50) and (52) gives

$$\frac{h_s(t)}{H} = \frac{3}{2}\frac{L}{x_s(t)} \tag{56}$$

In scaling the equations, it was assumed that h_0 was the maximum value of h at the instant of interest. Since this maximum value of h occurs at $x = x_s(t)$ in the outer solution, setting $h_s(t) \equiv h_0$ and $x_s(t) \equiv \ell$ in Equation (56) allows ϵ to be estimated from Equation (13) as

$$\epsilon = \frac{3}{2}\left(\frac{L}{\ell}\right)^2 \tag{57}$$

Thus, Equation (57) shows that $\epsilon \leq 1/10$ when

$$\frac{\ell}{L} \geq \sqrt{15} \cong 4 \tag{58}$$

Equation (58) suggests that a second-order correction would contribute less than 10 percent to the first-order solution after the kinematic-shock travels about three reservoir lengths downstream from the ruptured dam. The difference of one reservoir length between Equation (58) and the result obtained earlier by Hunt [4], which is insignificant from the viewpoint of an order of magnitude analysis, is undoubtedly the result of the relatively steep depth-gradient that occurs in Equation (23) on the upstream side of the kinematic shock, as shown in Figure 4. On the other hand, the order

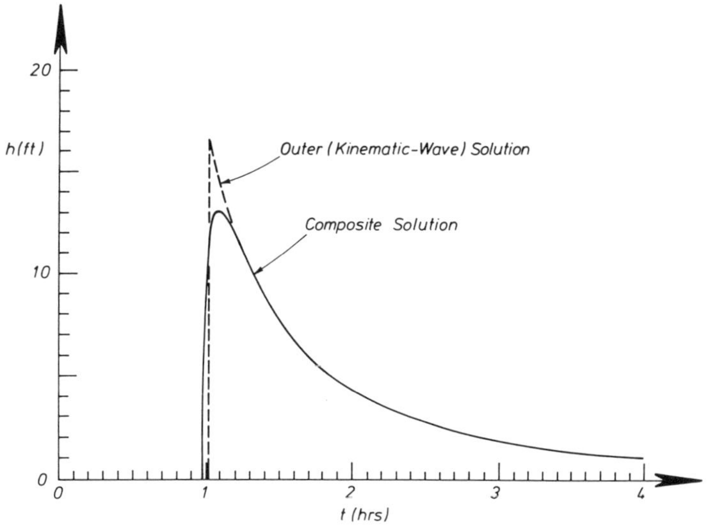

FIGURE 4. The depth hydrograph for the numerical example.

of magnitude argument used to obtain Equation (58) essentially estimates the depth-gradient magnitude by approximating the free surface with a straight line that passes through the points $h = 0$ and $h = h_s(t)$ at $x = 0$ and $x = x_s(t)$, respectively.

The effect of approximating the reservoir as a point source of mass can be investigated by letting $\epsilon \to 0$ in Equation (10), which gives the kinematic-wave approximation, but retaining a finite value for ϵ in Equation (12). In other words, the kinematic-wave solutions obtained by modeling the reservoir with a point source and a distributed source will be compared. Applying Equations (18–19) to the initial condition given by Equation (12) for $\epsilon > 0$ gives

$$h = \frac{\xi}{\epsilon} \text{ for } 0 \leq \xi < \epsilon H \qquad (*59)$$

along the characteristic

$$x - \frac{3}{2} t\sqrt{h} = \xi \text{ for } 0 \leq \xi < \epsilon H \qquad (*60)$$

For all other values of ξ, it is found that h vanishes. Eliminating the parameter ξ gives the following solution in nonparametric form:

$$x = \frac{3}{2} t\sqrt{h} + \epsilon h \text{ for } 0 \leq x < x_s(t) \qquad (*61)$$

The equation of the shock is found by inserting Equation (61) into Equation (24) after calculating dx from Equation (61):

$$dx = \left(\frac{3}{4}\frac{t}{\sqrt{h}} + \epsilon\right) dh \qquad (*62)$$

Putting Equation (62) into Equation (24) and integrating gives

$$th_s^{3/2} + \epsilon h_s^2 = \epsilon H^2 \qquad (*63)$$

in which $h_s(t)$ is the flow depth just upstream from the shock. A second relationship is obtained by evaluating Equation (61) on the upstream side of the shock.

$$x_s = \frac{3}{2} t\sqrt{h} + \epsilon h_s \qquad (*64)$$

Eliminating h_s between Equations (63) and (64) would give the shock location in the form $x_s = x_s(t)$.

The result of approximating the reservoir as a point source of mass can now be determined by substituting the following equations into Equation (61), (63), and (64) and equating powers of ϵ:

$$h = \left(\frac{2x}{3t}\right)^2 + \epsilon h_1 + \ldots \qquad (*65)$$

$$h_s = \left(\frac{\epsilon H^2}{t}\right)^{2/3} + \epsilon h_{s_1} + \ldots \qquad (*66)$$

$$x_s = \frac{3}{2} (\epsilon H^2 t^2)^{1/3} + \epsilon x_{s_1} + \ldots \qquad (*67)$$

This leads to the following result:

$$h = \left(\frac{2x}{3t}\right)^2 - \epsilon \frac{32}{81}\frac{x^3}{t^4} + \ldots \qquad (*68)$$

$$h_s = \left(\frac{\epsilon H^2}{t}\right)^{2/3} - \epsilon \frac{2}{3}\frac{\epsilon H^2}{t^2} + \ldots \qquad (*69)$$

$$x_s = \frac{3}{2} (\epsilon H^2 t^2)^{1/3} + \frac{\epsilon}{2}\left(\frac{\epsilon H^2}{t}\right)^{2/3} + \ldots \qquad (*70)$$

Equations (65–70) show clearly that the second-order correction is of order ϵ and that this correction decreases as t increases. Thus, approximating the reservoir with a point source of mass is entirely consistent with the first-order approximation calculated herein.

A NUMERICAL EXAMPLE

It will be assumed for a numerical example that a reservoir with a maximum depth of 60 ft (18.3 m) is located in a valley that has an average slope of 1:200. It is estimated, either by using the Chézy equation, the Manning equation or field measurements, that a uniform flow velocity of 5 ft/s (1.52 m/s) occurs at a flow depth of 3 ft (0.914 m) within the valley. The object of the exercise is to compute the depth hydrograph at a point 10 miles (16.1 km) downstream from the dam if the dam is completely and instantaneously removed.

The following variables are given:

$$L/H = 200 \qquad (71)$$

$$H = 60 \text{ ft } (18.3 \text{ m}) \qquad (72)$$

$$u_0 = 5 \text{ ft/s } (1.52 \text{ m/s}) \qquad (73)$$

$$h_0 = 3 \text{ ft } (0.914 \text{ m/s}) \qquad (74)$$

The reservoir length is calculated from Equations (71–72).

$$L = 12{,}000 \text{ ft} = 2.27 \text{ mi } (3.66 \text{ km}) \qquad (75)$$

Thus, the depth hydrograph is to be computed at a point 4.4

reservoir lengths downstream from the dam, which means that the asymptotic solution given herein should be applicable.

In the ft-s system of units, Equations (50–54) become

$$x_s(t) = 273\,t^{2/3} \tag{76}$$

$$x_s'(t) = \frac{182}{t^{1/3}} \tag{77}$$

$$h_s(t) = \frac{3{,}962}{t^{2/3}} \tag{78}$$

$$h = \frac{2.239(10)^8}{t^2} \text{ for } x_s(t) > 64{,}800 \text{ ft}$$
$$= 0 \text{ for } x_s(t) < 64{,}800 \text{ ft} \tag{79}$$

$$\frac{64{,}800 - x_s(t)}{12{,}000}\,\frac{60}{h_s(t)} = \frac{\phi}{h_s(t)} + \ln\left[1 - \frac{\phi}{h_s(t)}\right] + \frac{1}{2} \tag{80}$$

Putting $x_s(t) = 64{,}800$ ft (19,751 m) in the left side of Equation (76) allows the arrival time of the kinematic shock to be calculated as 1.02 hrs after the dam has been ruptured, and Equations (77–78) show that the kinematic shock will have a speed and depth of 11.8 ft/s (3.64 m/s) and 16.7 ft (5.07 m) at this instant. Next, putting different values for t in Equations (76), (78), and (79) allows values to be computed for x_s, h_s, and h, and these values, in turn, are inserted in Equation (80) to compute ϕ. Finally, values of h and ϕ for the same value of t are inserted in Equation (55) to compute the composite solution for the depth hydrograph, which is plotted in Figure 4. In this plot it is seen that a maximum depth of 13.1 ft (3.99 m) will occur at a time of 1.08 hrs after the dam rupture.

NOTATION

F = Froude number
g = gravitational constant
H = maximum reservoir depth
h = water depth in outer solution
h_c = water depth in composite solution
h_0 = uniform flow depth with velocity u_0
h_s = shock depth in outer solution
L = reservoir length
ℓ = value of x_s at time of interest
S_f = bed-resistance term
S_0 = bed slope
t = time
u = velocity in outer solution
u_0 = uniform flow velocity at depth h_0
v = velocity in inner solution
x = distance along channel bed
x_s = shock coordinate in outer solution
y = magnified inner coordinate
ϵ = small dimensionless parameter
ξ = parameter
τ = time in inner solution
ϕ = water depth in inner solution
ϕ_s = shock depth in inner solution

REFERENCES

1. Chen, C., "Laboratory Verification of a Dam-Break Flood Model," *Journal of the Hydraulics Division, ASCE,* Vol. 106 (No. HY4), Paper 15324, pp. 535–556 (Apr. 1980).
2. Chen, C. and J. T. Armbruster, "Dam-Break Wave Model: Formulation and Verification," *Journal of the Hydraulics Division, ASCE,* Vol. 106 (No. HY5), Paper 15395, pp. 747–767 (May 1980).
3. Dressler, R. F., "Hydraulic Resistance Effect Upon the Dam-Break Functions," *Journal of Research,* National Bureau of Standards, Vol. 49 (No. 3), pp. 217–225 (Sept. 1952).
4. Hunt, B., "Asymptotic Solution for Dam-Break Problem," *Journal of the Hydraulics Division, ASCE,* Vol. 108 (No. HY1), Paper 16801, pp. 115–126 (Jan. 1982).
5. Hunt, B., "Perturbation Solution for Dam-Break Floods," *Journal of Hydraulic Engineering, ASCE,* Vol. 110 (No. 8), Paper 19056, pp. 1058–1071 (Aug. 1984).
6. Ritter, A., "Die Fortpflanzung der Wasserwellen," *Zeitschrift des Vereines Deutscher Ingenieure,* Vol. 36 (No. 33), pp. 947–954 (Aug. 1982).
7. Sakkas, J. G. and T. Stelkoff, "Dam-Break Flood in a Prismatic Dry Channel," *Journal of the Hydraulics Division, ASCE,* Vol. 99 (No. HY12), Paper 10233, pp. 2195–2216 (Dec. 1973).
8. Sakkas, J. G. and T. Strelkoff, "Dimensionless Solution of Dam-Break Flood Waves," *Journal of the Hydraulics Division, ASCE,* Vol. 102 (No. HY2), Paper 11910, pp. 171–184 (Feb. 1976).
9. Strelkoff, T., D. Schamber and N. Katopodes, "Comparative Analysis of Routing Techniques for the Floodwave from a Ruptured Dam," *Proceedings of Dam-Break Routing Workshop,* Bethesda, Md., pp. 228–291 (1977).
10. Van Dyke, M., *Perturbation Methods in Fluid Mechanics,* annotated ed., The Parabolic Press, Stanford, Calif. (1975).
11. Whitham, G. B., "The Effects of Hydraulic Resistance in the Dam-Break Problem," *Proceedings, Royal Society of London, Series A,* Vol. 227, pp. 399–407 (1955).

Chute Entrances for HS, H, and HL Flumes

WENDELL R. GWINN*

ABSTRACT

In tests of the HL flume with a flat floor approach the discharge ranged from 6.5% below published table values for low flows to almost no difference at the maximum head. Generalized discharge equations for HS, H, and HL flumes with chute entrances are given. A direct solution for the effect of submergence on HL flumes is presented in equation form. The hydraulic jump would form on the sloping floor of the chute for all flows. The turbulence in the jump will help keep sediment in suspension and provide mixing for water quality measurements. The depth of ponding over the approach is reduced to half of what it would be if no chute were provided. Sand in the flow will deposit at the lower end of the chute and in the flume. However, satisfactory ratings are provided for flows with concentrations of sediment less than 10,000 ppm.

INTRODUCTION

The HS, H, and HL flumes have a proven performance for measuring runoff from watersheds and flows in open channels. They have wide applicability because they are simple in form and relatively easy to construct; also, and most important, they can be quite accurate. Ratings can be calculated (see Appendix) from the dimensions of the flume [2]. Accuracy of small flow measurement is obtained by the narrowness of the flume control section at the bottom. However, shaping the flume for the desired sensitivity throughout the flow range requires a relatively large head. [A 3-ft (0.9 m) H flume requires 3 ft (0.9 m) of head to measure a 31.1-cfs (0.881 m^3/s) flow.] A flume installed with its invert at watershed outlet elevation results in ponding over the watershed, which can be quite undesirable from the standpoint of hydrologic research. Also, sediment deposition can extend into the flume and affect its accuracy [3]. Hydrologists have suggested a drop box approach [1] to correct this problem. The floor of the box is sloped 2% in the direction of flow, and a transverse slope of 1-on-8 is suggested "where silting is a serious problem." However, ratings are not available for flumes equipped with such a drop box, so the published rating tables [1] obtained from tranquil approach calibrations are used. This practice can lead to error because unpublished data (V. D. Young, National Hydraulic Laboratory) on the $0.7D$ drop showed a maximum increase of 28% in discharge (head = $0.2D$) when compared with the tranquil flow calibration.

In field studies of this earlier drop box sometimes accumulated high loads of deposit, an action which left the accuracy of the flow measurement in doubt. Also, laboratory data [6] showed that the drop box with a side gutter entry (a nonsymmetrical approach to the box) could produce a large eddy in the flume and cause discharge errors in excess of 50%. It was reasoned that some of these undesirable performance characteristics of the old box could be eliminated by sloping the floor uniformly between the entrance to the box and the entrance to the measuring flume. If this slope were proportioned correctly, a hydraulic jump could be created on the sloping floor for all flows. The turbulence in the jump would help keep sediment in suspension and provide mixing for the water quality measurements. Investigations were made to evaluate the performance characteristics of the chute entrance, to determine its most effective dimensions, and obtain its rating.

The calibration of the HL flume with a flat floor and tranquil flow was done to provide the rating needed for comparison purposes. Since calibration data for the HL flume were not available, the results of these calibrations will be reported in detail. These calibrations were done first; and since the model was available, the development of the chute entrance was started with the HL flume. Five different chutes were tested in the laboratory to determine the re-

*Research Hydraulic Engineer (Retired), Agricultural Research Service, U.S. Department of Agriculture, Stillwater, OK quired dimensions. Data from these tests, along with data on hydraulic jumps in sloping channels obtained by Kindsvater [5] and Hickox [4], were used to design the new chute entrances for HS, H, and HL flumes. The flumes were calibrated with the new chute entrance in place for both free and submerged outfall conditions. Sediment was fed into the flow to evaluate chute entrance and flume performance when conveying sediment-laden flows.

CHUTE ENTRANCE DESIGN

Three dimensions of a chute entrance were considered in the design. The width was to be the width of the flume to avoid contraction or separation effects at the junction of the chute and the flume. To insure formation of the hydraulic jump, the drop in the chute needed to be sufficient that backwater from the flume when flowing at maximum capacity would not reach an elevation at the chute entrance greater than 0.6 of the head on the entrance. The chute entrance needed to be long enough so that flume outlet would normally be above ground level. Preliminary tests indicated $5D$ to be a suitable length, and this length was adopted. The slope of the floor of the chute is critical because it controls the location of the hydraulic jump; therefore, chute length could vary with floor slope. Also, if possible, the chute dimensions were to be those shown in the Handbook [1] in the event that older boxes were to be changed according to the sloping floor design. A rounded entrance to the box was selected to provide the most efficient opening and to reduce the possibility of undesirable wave formation in the upper part of the chute.

An overriding consideration in the design of the chute was to provide for a hydraulic jump on the sloping floor of the chute and not in the flume itself. Kindsvater's formula [5] for a hydraulic jump in a sloping channel was used to predict the jump characteristics. The formula is

$$\frac{d_2}{d_1} = \frac{1}{2 \cos a}\left[\left(\frac{8\, F_1'^2 \cos^3 a}{1 - 2\, \varnothing \tan a} + 1\right)^{0.5} - 1\right] \tag{1}$$

in which d_1 = depth of flow (normal to slope) upstream of jump; d_2 = depth of flow downstream of jump; a = angle of the sloping floor with horizontal; $F_i' = v_1/(gd_1)^{0.5}$ = Froude number; and $\varnothing$ = experimentally evaluated dimensionless pressure coefficient. In examining Kindsvater's paper [5], Hickox [4] presented the relationship between $\varnothing$ and $F_1'^2$ for 1-on-3 slopes (p. 1142, Figure 18b). A limited portion of this relationship is expressed in the formula

$$\varnothing = 5.25 - 13.9 \tan a + (-0.404 + 1.34 \tan a)F_1' \tag{2}$$

as developed from Hickox's [4] data in which $0.1 \Leftarrow \tan a \Leftarrow 0.2$; and $2.236 \Leftarrow F_1' \Leftarrow 7.746$.

CALIBRATION OF HL FLUME WITH FLAT FLOOR

The discharge equations for the HL flume published earlier [2] were obtained by fitting equations to the head-discharge relationship of published rating tables [1] because no calibration data were found for the HL flume. Flume depths, D, of 0.5 ft (0.2 m), 1 ft (0.3 m), and 2 ft (0.6 m) were chosen for calibration.

Main Flows

For the main flow range [2], the heads are greater than 0.2 ft (0.06 m). The main flow discharge [2] is

$$\frac{Q}{(2g)^{0.5}(H + v^2/2g)^{1.5}\, B_0} = C_0 + C_1\, B_1\, (H + v^2/2g)/B_0 \tag{3}$$

$$C_0 = E_0 + E_1\, D \tag{4}$$

$$C_1 = F_0 + F_1\, D \tag{5}$$

in which Q = total discharge through flume; g = acceleration by gravity; H = depth of flow at the head-measurement section; v = average velocity in the head-measurement section; B_0 = one-half the bottom width of the flume at its outlet; B_1 = vertical projected slope of control edge (front view); C_0 and C_1 = experimentally evaluated discharge coefficients; and E_0 and E_1, F_0, and F_1 = constants for $D < 1$ ft (0.305 m). (The velocity distribution coefficient is assumed to be 1.0.) Values of C_0 and C_1 [Equation (3)] were obtained by analyzing the data by the least-squares method. B_0 and B_1 values are obtained from flume dimensions of the individual flumes. The results are shown in Figure 1, where $C_0 = 0.746$ and $C_1 = 0.355$ for $D < 1$ ft (0.305 m) and where $C_0 = 0.856$ and $C_1 = 0.322$ for $D = 0.5$ ft (0.152 m). Substituting the values of C_0 in Equation (4) results in $E_0 = 0.967$ and $E_1 = -0.221$ ($E_m = 0.725$). Substituting the values of C_1 in Equation (5) results in $F_0 = 0.289$ and $F_1 = 0.0659$ ($F_m = 0.216$). The main flow discharge equation for HL flume with flat floor is

$$Q = [(0.967 - 0.221\, D)B_0 + (0.289 + 0.0659\, D)B_1\, (H + v^2/2g)]\; (2g)^{0.5}\; (H + v^2/2g)^{1.5} \tag{6}$$

Equation (6) applies to all flumes whether larger or smaller than 1 ft (0.305 m) deep if D is set equal to 1.0 ft (0.305 m) for all flumes with $D > 1.0$ ft (0.305 m). The values of v must be determined by successive approximations.

Low Flows

For low flows, heads are less than 0.1 ft (0.03 m). The flow is zero for heads less than 0.01 ft (0.003 m) because of sur-

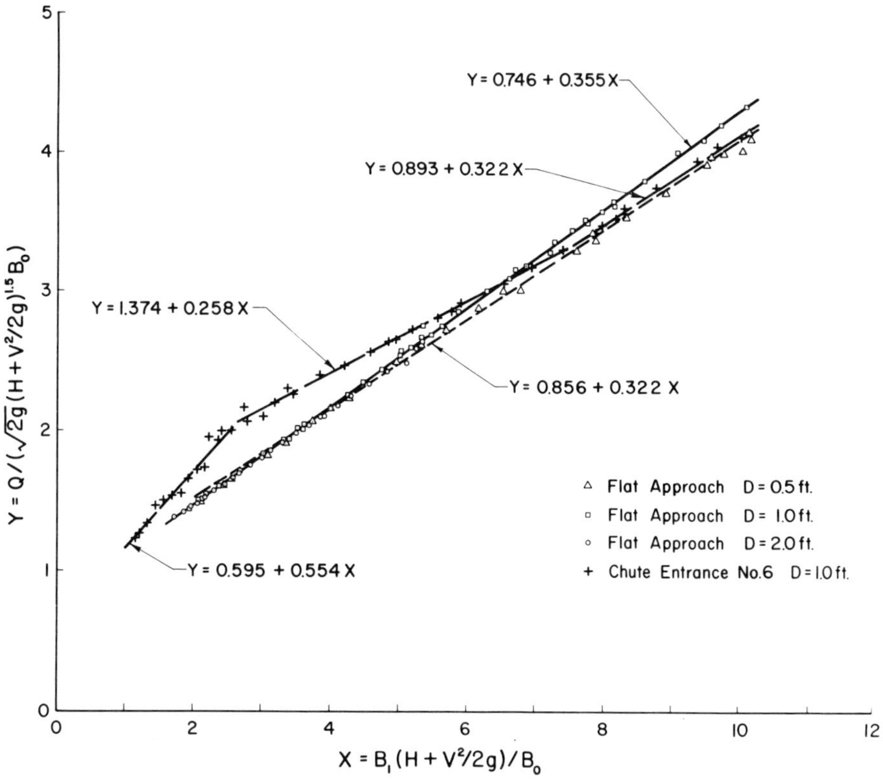

FIGURE 1. Dimensionless rating curves for the HL flume (design B_0 = 0.1 D, design B_1 = 1).

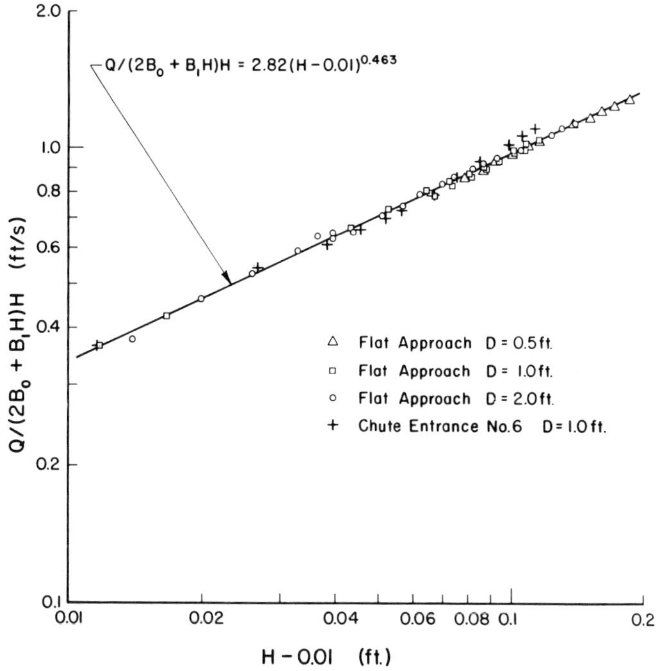

FIGURE 2. Rating curve of low flow range for HL flume.

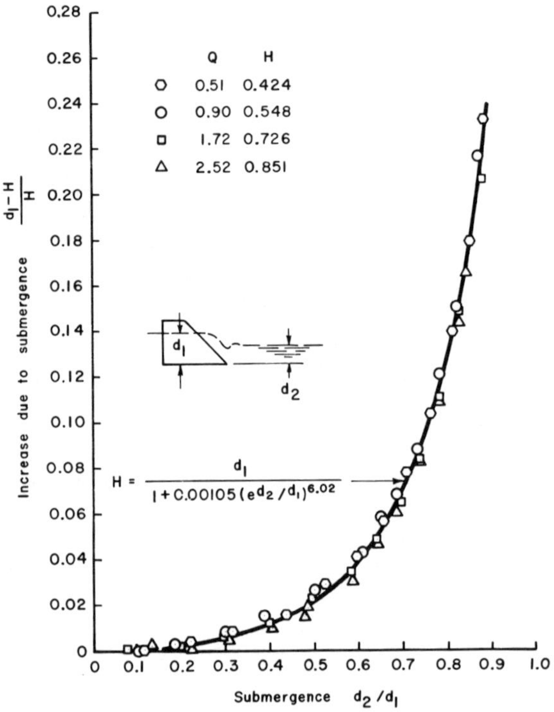

FIGURE 3. Effect of submergence on calibration of HL flumes with floor.

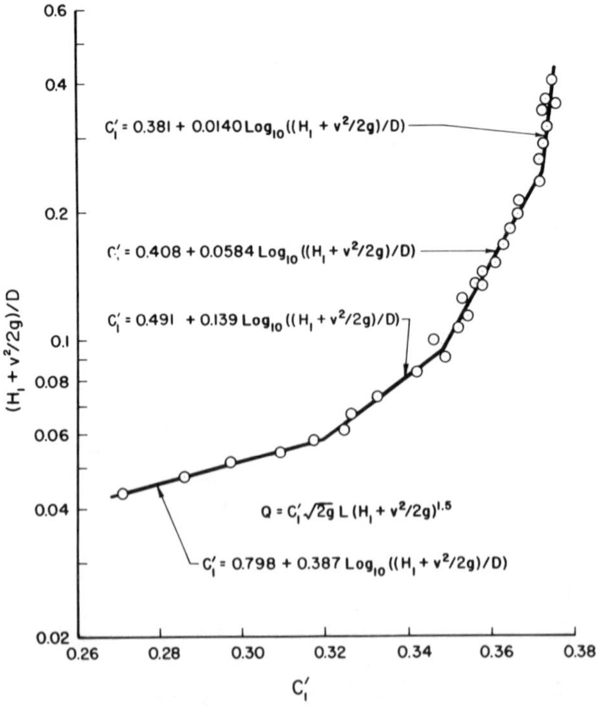

FIGURE 4. Discharge equations for the chute entrance (data from No. 5).

face tension. The generalized discharge equation for low flows is not dimensionless because of the 0.01 ft constant. The low flow Equation (2) is

$$Q = A_0 (2 B_0 + B_1 H) (H) (H - 0.01)^{A_1} \quad (7)$$

in which Q = discharge in cfs, A_0 = constant in $ft^{(1-A_1)}/s$, A_1 = constant, H = head in ft, and 0.01 = zero flow constant in ft (0.003 m). The A_0 and A_1 were obtained from low flow data by the method of least squares. These are shown in Figure 2, in which $A_0 = 2.82\ ft^{(1-A_1)}/s$ ($1.49\ m^{(1-A_1)}/s$) and $A_1 = 0.463$.

Transition Between Low and Main Flows

A computed Equation (2) is used to define the head-discharge relationship in transition range from a nondimensionless equation to the dimensionless equation for main flows. The equation is

$$Q = (K_0 B_0 + K_1 B_1 H) (2g)^{0.5} H^{1.5} \quad (8)$$

Values of K_0 and K_1 are determined by a simultaneous solution of a pair of equations similar to Equation (8). One equation will have 0.1 ft (0.03 m) for H and corresponding value of Q obtained from a solution of the low flow Equation (7). The other equation will have 0.2 ft (0.06 m) for H and the corresponding value of Q obtained from Equation (6).

Four-Foot HL Flume Comparison

The published 4-ft HL flume rating table [1] was compared with a computed rating by using ideal design dimensions and the new data. The difference between the rating ranged from −6.5% to about +1% at minimum and maximum head, respectively.

Effect of Submergence

Submergence of the flume outlet increases the water depth at the head-measuring section without a corresponding increase in the flow rate. If this submergence effect is ignored, the discharge estimate will be too large. Therefore, it is desirable for flumes to discharge freely. However, some installations do not allow free outfall. The correction for submergence effects in HL flumes is shown in Figure 3.

A direct solution for the free-flow head, H (the head that would have occurred if there had been no submergence), can be approximated with

$$H = \frac{d_1}{1 + 0.00105[(e)^{d_2/d_1}]^{6.02}} \quad (9)$$

in which H = free flow head; d_1 = actual head with submergence; d_2 = tailwater depth above flume floor elevation; e = 2.71828, the base of natural logarithms; and $0.15 < d_2/d_1 < 0.90$. The discharge is then computed by using H in Equations (6), (7), or (8).

CHUTE ENTRANCE TESTS

Chute Entrance Rating

Tests were made for the 0.5D rounded entrance to determine the dimensionless coefficient C_1' in the rating formula

$$Q = C_1' (2g)^{0.5} L (H_1 + v^2/2g)^{1.5} \quad (10)$$

in which L = length of crest (for the HL flume, L = 3.2 D; for the H flume, L = 1.9 D; and for the HS flume, L = 1.05 D), H_1 = the head above the crest (2 D upstream of crest) of the chute, v_1 = velocity in the approach (2 D upstream) to the chute entrance, and C_1' = experimental evaluated discharge coefficient. Values of C_1' (data from No. 5) are shown in Figure 4.

HL FLUME CHUTE ENTRANCE

Chute Performance

Six different chute entrance designs (Table 1) were tested for the HL flume with slopes of 1-on-10, 1-on-5, and 1-on-7.5. The floor of all designs sloped uniformly from the entrance of the chute to the floor of the HL flume except No. 3, which had a 0.25D vertical drop at the chute entrance. Chute Nos. 3 and 4 reflect attempts to shorten the box length. However, in certain flow ranges the rating curve was not unique, so the short chute was dropped from further consideration.

Chute No. 6 required only three equations to define the main flow, and produced a hydraulic jump on the 1-on-7.5 sloping floor of the chute for the complete flow range. The final design selected (Chute No. 6) is shown in Figure 5.

Discharge Equations

Rating tests were run on the HL flume with Chute No. 6 in place. The low flow rating was the same as the low flow rating for tranquil approach shown in Figure 2.

The equation is

$$Q = 2.82 (2 B_0 + B_1 H) (H - 0.01)^{0.463} \quad (11)$$

in which Q = discharge in cfs; H = head in ft, or

$$Q_m = 1.49 (2 B_0 + B_1 H_m) (H_m) (H_m - 0.003)^{0.463} \quad (12)$$

in which Q_m = discharge in m^3/s, and H_m = head in m.

The main flow range required three equations (Table 2) to

TABLE 1. HL Flume Chute Entrance Designs.

Chute number (1)	Length of chute (2)	Floor slope (3)	Sill height at entrance[a] (4)	Chute side wall height[a] (5)	Entrance radius (6)	Equations required to define main flow (7)	Remarks (8)
1	5D	1 on 10	0.5D	0.6D	—	6	Flow over side wall produces sudden change in rating
2	5D	1 on 10	0.5D	1.1D	0.5D	6	Only minor changes in rating
3	2.5D	1 on 10	0.5D	1.1D	0.5D	10	Rating not unique for rising and falling flows
4	2.5D	1 on 5	0.5D	1.1D	0.5D	6	Rating not unique for rising and falling flows
5	3.75D	1 on 7.5	0.5D	1.1D	0.5D	4	Submergence of jump for H > 0.7D
6	5D	1 on 7.5	0.67D	1.17D	0.5D	3	Uniform rating

[a]Referenced to flume floor (Zero H) in which D = flume maximum depth.
Note: Width of Chute = 3.2D.

define the rating as contrasted to the flume with the tranquil approach, which requires only one equation (see Figure 1). The discontinuities in the rating occur at $X = 2.63$ and 7.52, in which $X = B_1\,(h + v^2/2g)/B_0$. The departure from the tranquil rating in the plus direction (greatest at $X = 2.63$) was attributed to higher velocities in the flume because of incomplete energy dissipation at the hydraulic jump. The chute entrance rating fell below the flat floor rating for the higher flows ($X > 7.52$) because stagnation along the sidewalls of the flume increased the head measured in the gage well. The dimensionless coefficients (Table 2) can be used for $D > 1$ ft (0.305 m).

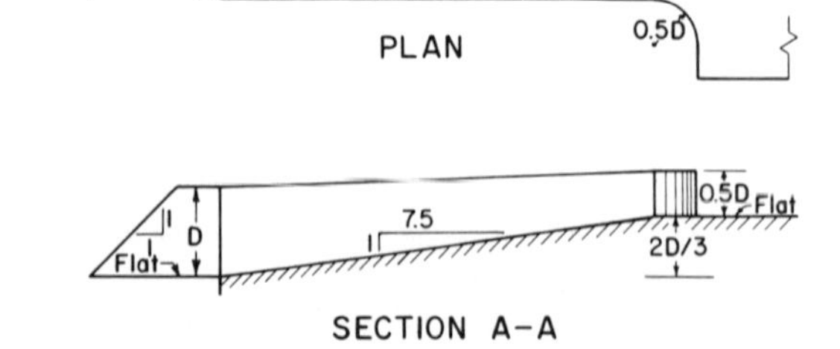

FIGURE 5. Chute entrance for HL flume (Chute No. 6).

Submergence Effect

The submergence data (Figure 6) for the HL flume with chute entrance scatter a little more than the submergence data (Figure 3) for the HL flume with flat floor, but the equations for submergence correction are very similar. The equation for the HL flume with chute entrance is

$$H = \frac{d_1}{1 + 0.00082[(e)^{d_2/d_1}]^{6.13}} \tag{13}$$

in which d_1 = head at the measurement section;

TABLE 2. Main Flow Discharge Coefficient (Equation 3) for HL Flume with Chute Entrance No. 6.

Range $B_1\,(H + v^2/2g)/B_0$ (1)	C_0 (2)	C_1 (3)
2 – 2.63	0.595	0.554
2.64 – 7.51	1.374	0.258
7.52 – 10.2	0.893	0.322

d_2 = tailwater depth above the flume floor elevation; and
H = free flow head.

H FLUME CHUTE ENTRANCE

Chute Entrance Performance

Experience with the HL flume chute entrance indicated that the 5D length and drop of 0.7D would be satisfactory. These dimensions were used in the final design shown in Figure 7.

The performance of the chute was predicted by preliminary analysis and then verified by tests.

To illustrate the design method, the chute entrance for a 1 ft (0.305 m) H flume (Figure 7) will be used. Substituting the maximum discharge of 1.99 cfs (0.056 m³/s) into Equation (10) and using the appropriate C_1' equation from Figure 5 gives an $(H_1 + v^2/2g)/D$ value of 0.49. The required drop of the chute is equal to $1D - (0.49 \times 0.6)D$ or 0.7D. The 0.7D provided by the chute therefore meets the required drop.

The hydraulic jump characteristics were predicted for the four flow rates shown in column 1, Table 3, computed values. The specific energy $(H_1 + v^2/2g)$ at the box entrance was computed by using Equation (10) and the appropriate C_1' equation from Figure 5 for $L - 1.9D$. The values are given in column 2. Total head shown in column 3 is referenced to the flume floor (zero H). The distance upstream of flume selected in column 4 had little effect on the computed d_2 result in column 11. Columns 7 and 8 were computed by using a standard step energy balance method with energy loss neglected and velocity coefficient assumed to be 1.0. The parameter d_1 in column 9 represents a measurement that is perpendicular to the sloping floor, and it is computed from values in columns 5, 6, and 7. The parameter d_2 in column 11 was computed by using Equations (1) and (2) and values from columns 5, 9, and 10. These computed values of d_2 approximate the head-discharge relationship of the 1-ft H flume. Therefore, the preliminary design was assumed satisfactory.

The chute entrance was built as shown in Figure 7, with D = 1 ft (0.305 m), and tested. The hydraulic jump formed on the sloping approach floor for all flows. Average depths of flow on the floor as close as possible to the point of the hydraulic jump were measured for four design discharges. These results are shown at the bottom of Table 3. The computed d_2 values were very close to the design values even though the experimental total head was less than the design. The d_2 value for the maximum discharge was computed even though the F_1' value was below the range of Equation (2). The actual head for the flume, H, is shown in column 12. H for the maximum discharge was slightly less than the computed d_2 required to maintain the jump on the sloping

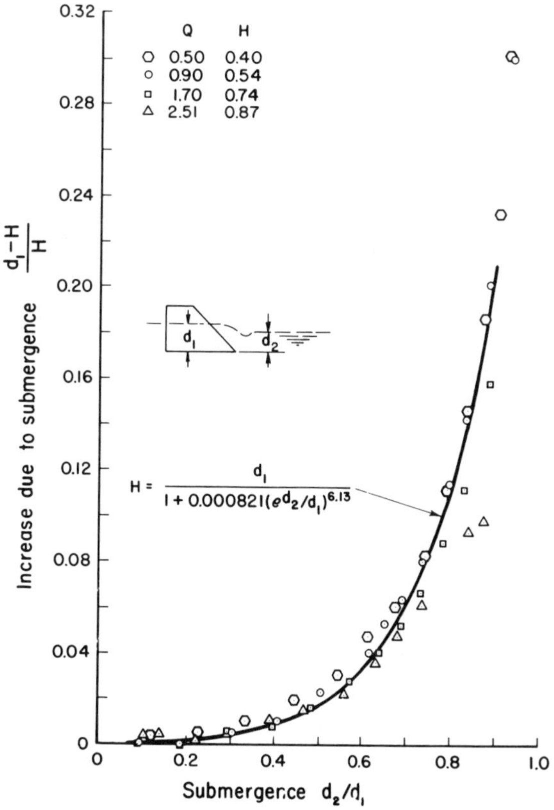

FIGURE 6. Effect of submergence on calibration of HL flumes with chute entrance (data are for Chute No. 6).

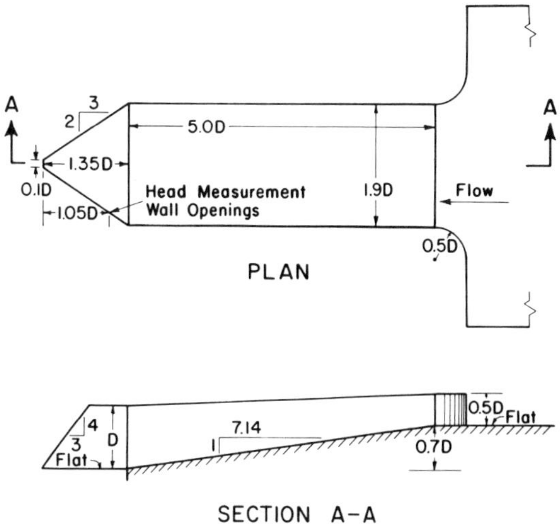

FIGURE 7. Chute entrance for H flume.

TABLE 3. Chute Entrance Design Values for 1-Foot H Flume.

Discharge Q in cubic feet per sec. (1)	$H_1 + v^2/2g$ in feet (2)	Elevation total head in feet (3)	Distance upstream of flume in feet (4)	Approach floor slope tan a (5)	Average bottom elevation in feet (6)	Average water surface elevation, ft (7)	Velocity head, $v^2/2g$, in ft. (8)	d_1 in feet (9)	Froude number, $F_1' = v_1/(gd_1)^{0.5}$ (10)	Computed d_2 in ft. (11)	Computed h in ft. (12)
(a) Computed using Equations (8), (9) and (10)											
1.9	0.488	1.188	4	0.14	0.56	0.753	0.435	0.189	2.15	1.169	
1.0	0.313	1.013	4	0.14	0.56	0.675	0.338	0.113	2.44	0.768	
0.5	0.200	0.900	4	0.14	0.56	0.623	0.277	0.062	2.98	0.487	
0.5	0.200	0.900	3	0.14	0.42	0.471	0.429	0.050	4.14	0.487	
0.25	0.130	0.830	4	0.14	0.56	0.595	0.235	0.034	3.69	0.311	
0.25	0.130	0.830	3	0.14	0.42	0.448	0.382	0.027	5.32	0.308	
0.25	0.130	0.830	2	0.14	0.28	0.303	0.527	0.023	6.76	0.303	
(b) Experimental data											
1.894		1.118	3.83	0.139	0.531	0.736	0.382	0.201	1.95	1.150	0.989
1.000		0.963	3.21	0.139	0.445	0.549	0.414	0.102	2.85	0.768	0.749
0.500		0.771	2.49	0.139	0.346	0.401	0.370	0.054	3.70	0.485	0.555
0.254		0.671	1.79	0.139	0.248	0.275	0.396	0.026	5.47	0.306	0.407

Note: 1 ft = 0.305 m; 1 cfs = 0.028 m^3/s.

floor. The design slope is correct in that the H equals or exceeds d_2 for all flows except the maximum.

Discharge Equations

The discharge ratings for the H flume with chute entrance were almost the same (Q within $\pm 2\%$) as the ratings for the H flume with flat floor [2] (Figures 8, 9, and 10).

Submergence Effect

The submerged flow data (Figure 10) were almost identical to the data [2] for the H flume with flat approach floor. Therefore, no new equation was computed for the effect of submergence on the H flume with chute entrance.

HS Flume Chute Entrance With Collector Trough

The HS flume is used to measure runoff from small plots. A collector trough that acts as a weir at the downstream end of the plot (Figure 4.5 of Reference 1) is used to convey the water to the flume. The required drop of the chute is computed by using a 1-ft HS flume. Substituting the maximum discharge of 0.82 cfs (0.023 m^3/s) into Equation (10) and using appropriate C_1' equation from Figure 4 gives $(H_1 + v^2/2g)/D$ value of 0.41. The drop of the chute is equal to $1D - (0.41 \times 0.6)D$ or $0.75D$.

Two experimental chute entrances with collector troughs were built as shown in Figure 11 with $D = 1$ ft (0.305 m) and $D = 0.6$ ft (0.183 m). The hydraulic jump formed on the sloping approach floor for all flows in each size flume.

Discharge Equations

The discharge ratings for the 0.6-ft and 1.0-ft HS flume with chute entrance and collector trough were almost the same for heads less than 0.1 ft (0.03 m). The data are shown in Figure 12, where $A_0 = 3.54$ ft$^{(1-A_1)}$/s (1.857 m$^{(1-A_1)}$/s) and $A_1 = 0.457$ in Equation (7).

For the main flow range [$H > 0.2$ ft, (0.06 m)], values of C_0 and C_1 in Equation (3) were almost the same. The results are shown in Figure 13, where $C_0 = 0.941$ and $C_1 = 0.457$ for $D = 1.0$ ft (0.305 m) and where $C_0 = 0.971$ and $C_1 = 0.449$ for $D = 0.6$ ft (0.183 m). Substituting the values of C_0 in Equation (4) results in $E_0 = 1.016$ and $E_1 = -0.075$ ($E_m = -0.246$). Substituting the values of C_1 in Equation (5) results in $F_0 = 0.437$ and $F_1 = 0.02$ ($F_m = 0.0656$). The main flow discharge equation for HS flume with chute entrance and collector trough is

$$Q = [(1.016 - 0.075D)\,B_0 + (0.437 + 0.02D)\,B_1\,(H + v^2/2g)]\,(2g)^{0.5}\,(H + v^2/2g)^{1.5} \quad (14)$$

If $D > 1$ ft (0.305 m) use $D = 1$ ft ($Dm = 0.305$ m).

Equation (8) is used to define the transition range between 0.1 ft (0.03 m) and 0.2 ft (0.06 m) by the method previously described.

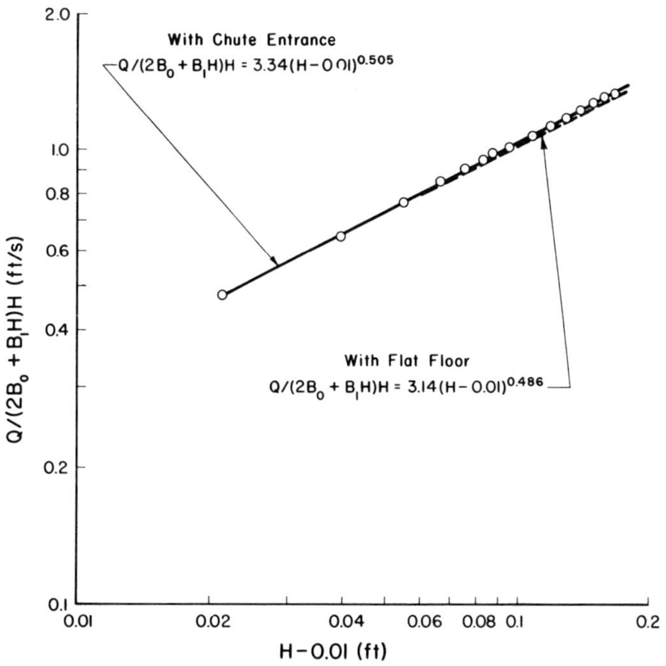

FIGURE 8. Rating curve of low flow range for H flume with chute entrance (dashed line is for H flume with flat floor).

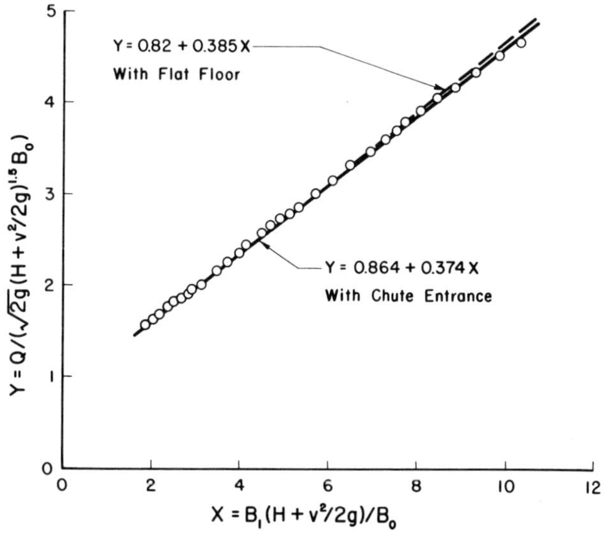

FIGURE 9. Dimensionless rating curve for H flume with chute entrance (dashed line is for H flume with flat floor).

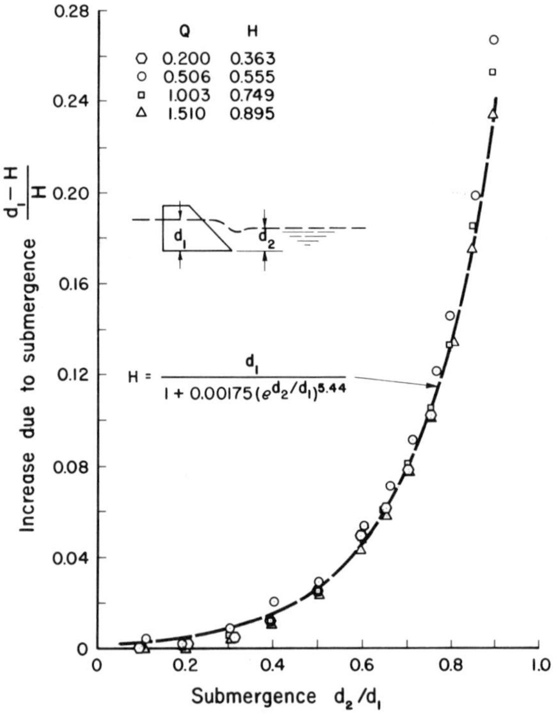

FIGURE 10. Effect of submergence on calibration of H flumes with chute entrance (dashed line is for H flume with flat floor).

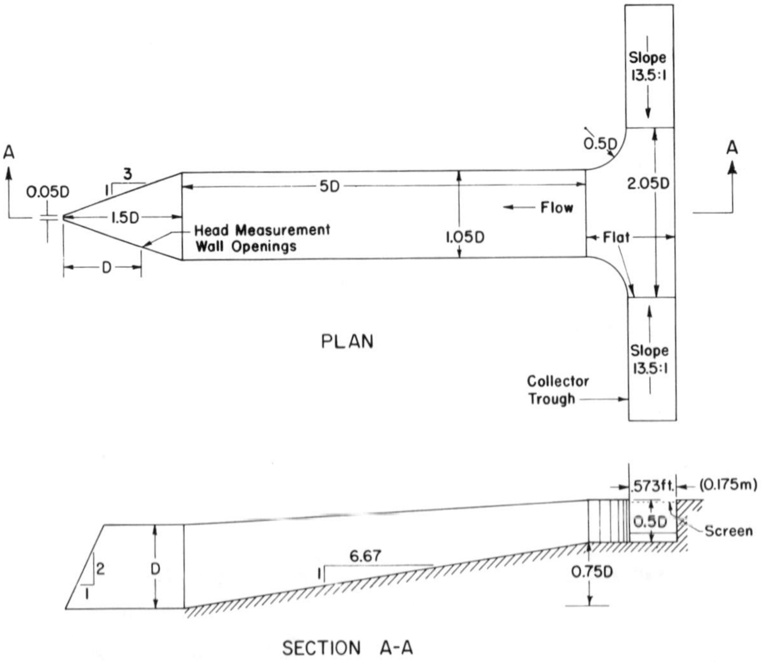

FIGURE 11. Chute entrance and collector trough for HS flume.

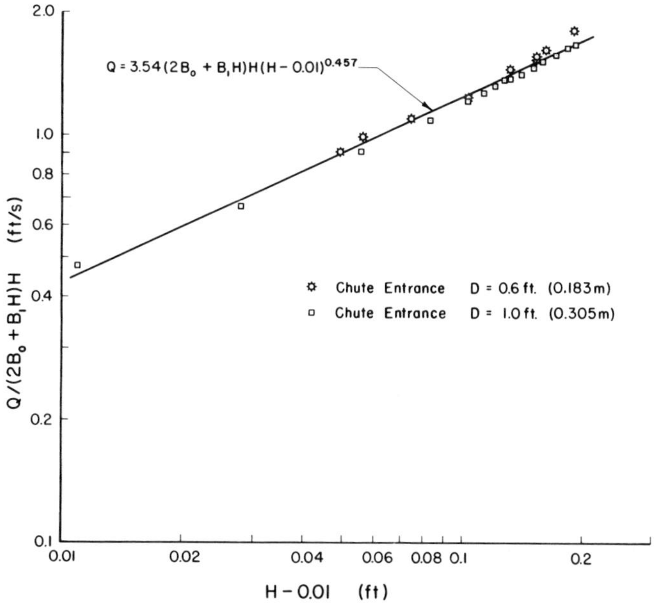

FIGURE 12. Rating curve of low flow range for HS flume with chute entrance and collector trough.

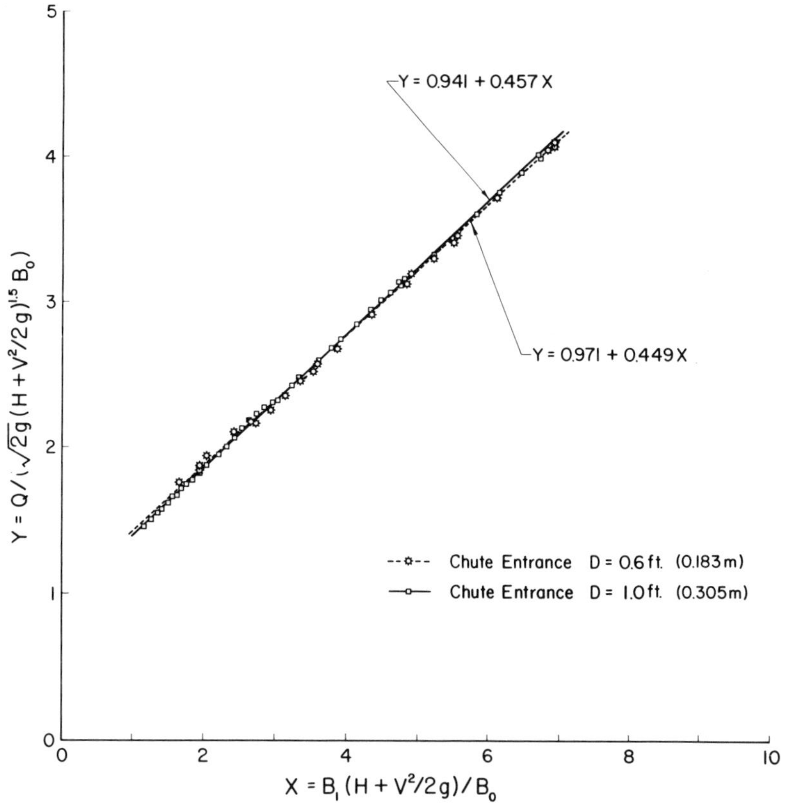

FIGURE 13. Dimensionless rating curves for HS flume with chute entrance and collector trough.

TABLE 4. Flows with Sediment.

$Q/(2g)^{0.5}D^{2.5}$ (1)	H/D with clear water (2)	H/D with sediment (3)	Maximum depth of deposit (4)	Average depth gagewell (5)	Average sediment concentration (6)	Time of feeding in hours (7)
(a) 1-ft HL flume, d_{50} = 0.46 mm						
0.092	0.490	0.491	0.23D	0.09D	737	4
0.125	0.570	0.570	0.12D	0.03D	555	4
(b) 1-ft H flume, d_{50} = 0.46 mm						
0.0125	0.260	0.261	0.25D	0.19D	6,300	2
0.025	0.362	0.356	0.32D	0.22D	2,950	2
0.062	0.551	0.552	0.43D	0.21D	1,240	2
0.125	0.748	0.747	0.24D	0.04D	609	2
0.237	0.988	0.987	0.04D	0.01D	298	2
(c) 0.6-ft HS flume, d_{50} = 0.46 mm						
0.0089	0.327	0.293	0.32D	0.12D	13,200	2
0.023	0.507	0.467	0.39D	0.35D	7,820	2
0.023	0.506	0.439	0.44D	0.34D	13,800	2
0.023	0.509	0.426	0.48D	0.38D	29,600	2
0.045	0.690	0.660	0.48D	0.36D	4,120	2
0.045	0.690	0.642	0.53D	0.38D	7,710	2
0.045	0.689	0.637	0.53D	0.39D	9,170	2
0.096	0.965	0.946	0.59D	0.30D	1,840	2
0.095	0.963	0.933	0.57D	0.35D	3,230	2
0.096	0.965	0.912	0.68D	0.39D	6,160	2
(d) 1-ft HS flume, d_{50} = 0.46 mm						
0.0025	0.169	0.165	0.21D	0.12D	14,100	2
0.0062	0.272	0.248	0.27D	0.20D	12,600	2
0.0063	0.274	0.256	0.30D	0.21D	20,900	2
0.0062	0.271	0.255	0.31D	0.22D	30,900	2
0.012	0.381	0.354	0.35D	0.26D	5,430	2
0.012	0.383	0.337	0.36D	0.26D	10,100	2
0.012	0.383	0.329	0.36D	0.26D	15,800	2
0.025	0.529	0.498	0.42D	0.30D	3,280	2
0.025	0.531	0.495	0.44D	0.33D	5,400	2
0.025	0.533	0.488	0.44D	0.33D	7,540	2
0.043	0.767	0.663	0.46D	0.29D	1,700	2
0.042	0.677	0.658	0.48D	0.35D	2,790	2
0.043	0.674	0.648	0.48D	0.35D	4,420	2
0.042	0.673	0.643	0.48D	0.36D	4,440	2
0.093	0.952	0.952	0.14D	0.05D	474	2
0.093	0.955	0.952	0.16D	0.06D	541	2
0.093	0.955	0.952	0.28D	0.16D	1,040	2
0.093	0.955	0.947	0.48D	0.23D	2,010	2
(e) 1-ft HS flume, d_{50} = 0.3 mm						
0.0025	0.169	0.155	0.22D	0.13D	15,200	2
0.0063	0.273	0.246	0.28D	0.20D	8,930	2
0.0063	0.272	0.242	0.29D	0.20D	19,500	2
0.0062	0.273	0.251	0.32D	0.22D	31,700	2
0.012	0.381	0.336	0.33D	0.25D	10,700	2
0.012	0.383	0.332	0.37D	0.27D	15,600	2

FLOWS WITH SEDIMENT

The HL flume, H flume, and HS flume with chute entrances were tested for performance with sediment-laden flows.

Washed river sand with a size distribution of $d_{84} = 0.69$ mm, $d_{50} = 0.46$ mm, and $d_{16} = 0.27$ mm was used as sediment for all flumes. Also, a river sand with a size distribution of $d_{84} = 0.45$ mm, $d_{50} = 0.3$ mm, and $d_{16} = 0.19$ mm was used as sediment for the 2-ft HS flume. This material was fed by using a vibrator feeder at the entrance to the drop box. Feeding rates ranged from equivalent concentrations of 298 to 31,700 ppm. The sand moved down the drop box with the flow. After passing through the jump, the sand started to deposit at or near the flume entrance. In less than 2 hr after the start of the constant sand inflow, further deposition ceased and an equilibrium condition was established. At the end of the test, the sand feeder and the flow were shut off simultaneously, and at the same time the sand deposit was kept submerged by raising the tailwater gate on the basin. The water was then slowly drained to keep the established deposit of sand in the flume unchanged. Data for all tests are listed in Table 4. The sand deposits had little effect on the measured head on the H and HL flume except for the one with a sediment concentration of 2,950 mg/L in the 1-ft H flume. The maximum depth of deposit was 90% of the measured head and occurred upstream of the head measuring section on the sloping floor. However, the change in head represented only a 3.6% error in discharge.

The HS flume was expected to have higher concentrations of sediment in the field. With the smaller flows in the HS flume, the concentration could be increased by setting the plot slope to 3% above the collector trough and concentrating the flow under the sand feeder. For sediment concentration S_c less than 500 mg/L the sand deposits had no effect on the measured head. For S_c between 500 mg/L and 11,000 mg/L the discharge Q can be obtained by using the equation

$$Q/Q_{ND} = 0.984 + 0.0000296\, S_c \tag{15}$$

in which Q_{ND} = discharge for no deposit; and S_c = sediment concentration in mg/L. If the average depth of sediment at the head measuring section is known, the discharge (±5%) can be obtained by applying the average $v^2/2g$ factor in Equation (14). Equation (15) will overestimate the discharge for sediment concentrations greater than 11,000 mg/L.

CONCLUSIONS

Calibration of the HL flume with a flat floor approach had discharges which ranged from 6.5% below published table values [1] for low flows to almost no difference at the maximum head. Equations for the ratings are given. An equation to correct for submergence effect is also provided.

Rating tests of HS, H, and HL flumes with chute entrance approaches were made, and equations for the head-discharge relationships were developed. Comparison of ratings for a chute-entrance-equipped H flume with ratings for a flat floor approach showed differences to be within ±2%. Comparisons for the HL flume showed low flow ratings for the chute entrance and the flat floor approaches to be the same. For the main flow range, however, a difference of nearly 23% at one head was found between the ratings for the two approach conditions. Three equations are required for the chute entrance flume to express the head-discharge relationship in the main flow range for the HL flume.

The hydraulic jump would form on the sloping floor of the chute entrance for all flows. The depth required for the hydraulic jump agreed with the depth predicted by hydraulic analysis.

The entrance to the chute controls the depth of flow in the approach to the H and HL flumes. With the proposed design the depth of ponding over the approach to the H and HL volumes is reduced to half of what it would be if no chute entrance were provided. With the chute entrance and collector trough design for the HS flume, no ponding will occur.

Sand in the flow will deposit at the lower end of the chute and in the flume. Discharge corrections are provided for sediment concentrations less than 11,000 mg/L (1.1%) in the HS flume with chute entrance and collector trough.

NOTATION

The following symbols are used in this paper.

A_0 = constant in low-flow equation, in $ft^{(1-A_1)}$/sec
A_1 = constant in low-flow equation
a = angle of sloping floor with horizontal
B_0 = one-half bottom width of flume at outlet
B_1 = vertical projected slope of control edge of flume (front view)
C_0 = experimental discharge coefficient for flume
C_1 = experimental discharge coefficient for flume
C_1' = experimental discharge coefficient for drop box entrance
D = flume maximum depth
d_1 = head at the flume measurement section with submergence and depth of flow upstream of hydraulic jump
d_2 = tailwater depth above flume floor elevation and depth of flow downstream of hydraulic jump
d_{16} = diameter of sediment with 16% passing
d_{50} = median diameter of sediment
d_{84} = diameter of sediment with 84% passing
E_m = constant in main flow equation, in $meters^{-1}$

E_0 = constant in main flow equation
E_1 = constant in main flow equation, in feet^{-1}
e = 2.71828, base of natural logarithm
F_m = constant in main flow equation, in meters^{-1}
F_0 = constant in main flow equation
F_1 = constant in main flow equation, in feet^{-1}
F_1' = $v_1/(gd_1)$ = Froude number
g = acceleration by gravity [value used 32.16 f/s^2 (9.802 m/s^2)]
H = head (free flow) at head measurement section of flume, in feet
H_m = head (free flow) at head measurement section of flume, in meters
H_1 = head above crest of box
K_0 = computed experimental discharge coefficient, transition flow
K_1 = computed experimental discharge coefficient, transition flow
L = length of crest
Q = total discharge through flume
Q_m = discharge in m^3/s
Q_1 = 4-ft HL rating table discharge [4]
S_c = sediment concentration (mg/L)
v = average velocity at head measurement section and average velocity in the approach to the drop box
v_1 = average velocity upstream of hydraulic jump
x = abscissa
y = ordinate
$\varnothing$ = dimensionless pressure coefficient for a hydraulic jump

REFERENCES

1. Brakensiek, D. L., H. B. Osborn, and W. J. Rawls, Coordinators, "Field Manual for Research in Agricultural Hydrology," Agriculture Handbook No. 224, U.S. Dept. of Agriculture, Agricultural Research Service, Washington, D.C. (Revised Feb. 1979).
2. Gwinn, W. R. and D. A. Parsons, "Discharge Equations for HS, H, and HL Flumes," *Journal of the Hydraulics Division, ASCE, Vol. 102* (No. HY1), Proc. Paper 1187, pp. 73–88 (Jan. 1976).
3. Hermsmeier, L. F. and R. A. Young, "Effect of Sediment Load on the Rating Curves of 0.6-Foot HS Flumes," ARS 41-142, U.S. Department of Agriculture, Agricultural Research Service, Washington, D.C. (June 1968).
4. Hickox, G. H., "Discussion of The Hydraulic Jump in Sloping Channels," by C. E. Kingsvater, *Transactions, ASCE, Vol. 109*, pp. 1141–1146 (1944).
5. Kindsvater, C. E., "The Hydraulic Jump in Sloping Channels," *Transactions, ASCE, Vol. 109*, pp. 1107–1154 (1944).
6. Rice, C. E. and W. R. Gwinn, "The Performance of a Modified Coshocton Type Runoff Sampler," *Transactions, American Society of Agricultural Engineers, Vol. 20* (No. 1), pp. 134–138 (1981).

APPENDIX

BASIC Programs

```
10 REM                  ****************************************************
20 REM                  *       RATING TABLE FOR HS, H, AND HL FLUMES      *
30 REM                  *              By Wendell R. Gwinn                 *
40 REM                  *         Dimensions in feet, Table in cfs         *
50 REM                  *                  April, 1985                     *
60 REM                  ****************************************************
70 DIM Q(10),NQ%(10),M%(10),QI(10),MI%(10),CI(10),C2(10),CH(8),DH(8),
       EH(10),FH(10)
80 CH(1)=3.54:CH(2)=3.93:CH(3)=3.14:CH(4)=3.3:CH(5)=2.82:CH(6)=0:
   CH(7)=3.34:CH(8)=2.82
90 DH(1)=.457:DH(2)=.526:DH(3)=.486:DH(4)=.5:DH(5)=.463:DH(6)=0:
   DH(7)=.505:DH(8)=.463
100 C1(1)=1.016:C1(2)=.861:C1(3)=.612:C1(4)=.63:C1(5)=.967:C1(6)=0:
    C1(7)=.655:C1(8)=.595:C1(9)=1.374:C1(10)=.893
110 C2(1)=.437:C2(2)=.479:C2(3)=.409:C2(4)=.4:C2(5)=.289:C2(6)=0:
    C2(7)=.398:C2(8)=.554:C2(9)-.258:C2(10)=.322
120 EH(1)=-.075:EH(2)=.112:EH(3)=.209:EH(4)=.174:EH(5)=-.221:EH(6)=0:
    EH(7)=.209:EH(8)=0:EH(9)=0:EH(10)=0
130 FH(1)=+.02:FH(2)=-.035:FH(3)=-.024:FH(4)=-.018:FH(5)=.0659:
    FH(6)=0:FH(7)=-.024:FH(8)=-0:FH(9)=-0:FH(10)=-0
140 LINE INPUT;"Line one of table identification,      []   centered
    on screen                          ";IDENT1$
150 PRINT
```

```
160 LINE INPUT;"LINE TWO OF TABLE IDENTIFICATION,        []    CENTERED
    ON SCREEN                        ";IDENT2$
170 PRINT
180 LINE INPUT"Date using 18 spaces   ";DATE$
190 INPUT"Change in head between each value in table, .01 or .001";
    DELTYH
200 PRINT "CODE FOR FLUME TYPE:":PRINT"      1=HS Flume with sloping
    drop box entrance and collector trough"
210 PRINT "      2=HS Flume, flat floor":
    PRINT "      3=H Flume with flat floor":
    PRINT "      4=H Flume with 1 on 8 sloping floor"
220 PRINT"      5=HL Flume, flat floor":
    PRINT"      6=Laboratory rating, read in constants for the
    equation of discharge"
230 PRINT"      7=H Flume with sloping drop box entrance":
    PRINT"      8=HL Flume with sloping drop box entrance"
240 INPUT"CODE FOR FLUME TYPE";NFT%
250 INPUT "ELEVATION OF ZERO HEAD";ZERO
260 INPUT "SIZE OF FLUME (MAXIMUM HEAD OR DEPTH)";DEPTH
270 PRINT "                                              HS          H
       H          HL"
280 PRINT "                                              Flume       Flume
     Flume      Flume"
290 PRINT "                  STANDARD                    Flat        Flat
     1 on 8     Flat"
300 PRINT "                  DIMENSION                   Floor       Floor     S
    loping Floor Floor"
310 PRINT "                   (feet)"
320 PRINT USING  "B0, Half Width of Throat                 #.####     #.#
    ###     #.####       #.####";DEPTH*.025,DEPTH*.05,DEPTH*.05,DEPTH*.1
330 PRINT        "B1, Slope of Entrance Control            0.1667     0.5
         0.5          1.0"
340 PRINT USING  "Width of 'flat floor' at Gage Section   #.####     #.#
    ###     #.####       #.####";DEPTH*.7167,DEPTH*1.5,DEPTH*.3,DEPTH*2.7
350 PRINT        "Slope of floor at Gage Section              0
    0          8            0"
360 PRINT USING  "Total Width at Gage Section              #.####     #.#
    ###     #.####       #.####";DEPTH*.7167,DEPTH*1.5,DEPTH*1.5,
    DEPTH*2.7
370 N=NFT%
380 PRINT            "       ********* USE ACTUAL DIMENSIONS WHEN AVAILABLE
    ********"
390 INPUT"B0, HALF WIDTH OF THROAT";B0
400 INPUT"B1, SLOPE OF ENTRANCE";B1
410 INPUT"WIDTH OF FLAT AT GAGE SECTION";E
420 INPUT"SLOPE OF FLOOR AT GAGE SECTION";F
430 INPUT"TOTAL WIDTH AT GAGE SECTION";W
440 GI=.2
450 IF N=6 THEN GOTO 470
460 GOTO 500
470 PRINT"You have selected to read in the coefficients for the equatio
    ns of discharge":PRINT" ":PRINT"      Q=A0*(2*B0+B1*H)*H*(H-0.01)^A1"
480 PRINT"      Q=[(E0+E1*D)*B0+(F0+F1*D)*B1*(H+V^2/2g)]*2g^.5*
    (H+V^2/2g)^1.5":PRINT" ":PRINT"NOTE: E1 AND F1=0 FOR MOST DATA"
490 INPUT "A0,A1,E0,E1,F0,F1";CH(6),DH(6),C1(6),EH(6),C2(6),FH(6)
500 NOL=45
510 PAGE%=1:J=1:L=1
520 INPUT"AVERAGE SEDIMENT DEPTH AT HEAD MEASURING SECTION";SD
530 INPUT"DO YOU WANT THE INPUT DATA (Yes input 1, No press RETURN)";
    COPY
```

```
540 TQ=.02
550 D=DEPTH
560 IF D>1 THEN D=1
570 IZERO%=INT(ZERO/DELTYH)
580 IZER1%=INT(ZERO/(10*DELTYH))*10
590 J=IZERO%-IZER1%+1
600 FOR I=1 TO J
610 Q(I)=0:QI(I)=0
620 NEXT I
630 HEAD=IZER1%*DELTYH-10*DELTYH
640 H=IZERO%*DELTYH-ZERO
650 NEXT1%=1
660 IF DELTYH=.01 THEN LAST%=10
670 IF DELTYH=.001 THEN LAST%=100
680 TEND%=1
690 GOTO 1160
700 HA=.1
710 IF SD>=.09 THEN 970
720 QA=CH(N)*((2*B0+B1*HA)*HA-(2*B0+SD*B1)*SD)*(HA-.01-SD)^DH(N)
730 TBQ=4*TQ
740 IF E>=W THEN 810
750 IF E+F*GI/2>=W THEN 780
760 AREA=GI*(E+F*GI/2)-SD*(E+F*SD/2)
770 GOTO 820
780 IF E+F*SD>=W THEN 810
790 AREA=GI*W-(W-E)*(W-E)/(2*F)-SD*(E+F*SD/2)
800 GOTO 820
810 AREA=(GI-SD)*W
820 V2G=TBQ*TBQ/(AREA*AREA*64.31)
830 QB=((C1(N)+EH(N)*D)*B0+(C2(N)+FH(N)*D)*B1*(GI+V2G))*
    8.02*(GI+V2G)^1.5
840 IF V2G=0 THEN GOTO 910
850 DELTAQ=ABS(TBQ-QB)
860 IF DELTAQ<=.000001 THEN GOTO 910
870 TBQ=QB
880 IF TBQ<10000 THEN GOTO 820
890 LAST%=INT((GI-.001)/DELTYH+1.01)
900 GOTO 1010
910 Y1=QA/(8.02*HA^1.5*B0)
920 X1=B1*HA/B0
930 Y2=QB/(8.02*GI^1.5*B0)
940 X2=B1*GI/B0
950 CI2=(Y2-Y1)/(X2-X1)
960 CI1=Y2-CI2*X2
970 LAST%=INT((GI-.001)/DELTYH+1.01)
980 TEND%=2
990 IF SD>.09 THEN GOTO 1010
1000 GOTO 1160
1010 TEND%=7
1020 GOTO 1160
1030 IF N<8 THEN GOTO 1140
1040 CROSS1=B0*2.58786/B1
1050 LAST%=INT%((CROSS1-.001)/DELTYH+1.01)
1060 TEND%=3
1070 GOTO 1160
1080 N=9
1090 CROSS2=B0*7.51563/B1
1100 LAST%=INT((CROSS2-.001)/DELTYH+1.01)
1110 TEND%=4
1120 GOTO 1160
```

```
1130 N=10
1140 LAST%=INT(DEPTH/DELTYH+1.01)
1150 TEND%=5
1160 INDEX%=NEXT1%
1170 NEXT1%=LAST%+1
1180 V2G=0
1190 FOR I=INDEX% TO LAST%
1200 IF H-.01-SD<.0001 THEN GOTO 3040
1210 IF TEND%=1 THEN Q(J)=CH(N)*((2*B0+B1*H)*H-(2*B0+SD*B1)*SD)*
     (H-.01-SD)^DH(N)
1220 IF TEND%=2 THEN Q(J)=(CI1*B0+CI2*B1*H)*8.02*H^1.5
1230 IF TEND%<=2 THEN GOTO 1400
1240 IF E>=W THEN GOTO 1310
1250 IF E+F*H>=W THEN GOTO 1280
1260 AREA=H*(E+F*H/2)-SD*(E+F*SD/2)
1270 GOTO 1320
1280 IF E+F*SD>=W THEN GOTO 1310
1290 AREA=H*W-(W-E)*(W-E)/(2*F)-SD*(E+F*SD/2)
1300 GOTO 1320
1310 AREA=(H-SD)*W
1320 V2G=TQ*TQ/(AREA*AREA*64.31)
1330 Q(J)=((C1(N)+EH(N)*D)*B0+(C2(N)+FH(N)*D)*B1*(H+V2G))*
     8.02*(H+V2G)^1.5
1340 IF V2G=0 THEN GOTO 1400
1350 DELTAQ=ABS(TQ-Q(J))
1360 IF DELTAQ<=.00005 THEN GOTO 1400
1370 TQ=Q(J)
1380 IF TQ>10000 THEN GOTO 3040
1390 GOTO 1320
1400 H=H+DELTYH
1410 IF Q(J)<.00005 THEN GOTO 1440
1420 TQ=Q(J)
1430 GOTO 1450
1440 Q(J)=0
1450 IF J=10 THEN GOTO 1530
1460 IF TEND%<>5 THEN GOTO 3070
1470 IF I<>LAST% THEN GOTO 3070
1480 IX2%=J+1
1490 FOR IX=IX2% TO 10
1500 Q(IX)=0
1510 NEXT IX
1520 TEND%=6
1530 FOR K=1 TO 10
1540 M%(K)=1
1550 IF Q(K)>.1 THEN M%(K)=2
1560 IF Q(K)>1 THEN M%(K)=3
1570 IF Q(K)>10 THEN M%(K)=4
1580 IF Q(K)>100 THEN M%(K)=5
1590 IF Q(K)>1000 THEN M%(K)=6
1600 NEXT K
1610 HEAD=HEAD+10*DELTYH
1620 IF NOL<>45 THEN GOTO 1800
1630 LPRINT "                                        RATING TABLE
                Page ";
1640 LPRINT USING"##";PAGE%
1650 LPRINT IDENT1$
1660 LPRINT IDENT2$
1670 LPRINT"                           Discharge in cubic feet per second"
1680 IF SD=0 THEN GOTO 1720
1690 LPRINT " "
```

```
1700 LPRINT"           Average Sediment Depth at Head Measuring Section
     = ";SD
1710 LPRINT:LPRINT:LPRINT
1720 IF ZERO=0 THEN LPRINT "  HEAD"
1730 IF ZERO<>0 THEN LPRINT "  ELEV."
1740 IF DELTYH=.001 THEN LPRINT" (Feet)   .000   .001   .002   .003   .
     004   .005   .006   .007   .008   .009"
1750 IF DELTYH=.01 THEN LPRINT " (Feet)   .00    .01    .02    .03    .
     04    .05    .06    .07    .08    .09"
1760 LPRINT
1770 PAGE%=PAGE%+1
1780 NOL=0
1790 IF SD<>0 THEN NOL=5
1800 IF Q(1)<>0 THEN MX%=M%(1)
1810 IF Q(1)=0 THEN MX%=M%(10)
1820 IF TEND%=6 THEN MX%=M%(1)
1830 IF DELTYH=.001 THEN LPRINT USING "###.##  ";HEAD;
1840 IF DELTYH=.01 THEN LPRINT USING "####.#  ";HEAD;
1850 ON MX% GOTO 1860,1860,2080,2300,2520,2740
1860 IF TEND%=6 THEN GOTO 1890
1870 LPRINT USING " #.####";Q(1),Q(2),Q(3),Q(4),Q(5),Q(6),Q(7),Q(8),
     Q(9),Q(10)
1880 GOTO 2960
1890 ON IX2% GOTO 1900,1900,1920,1940,1960,1980,2000,2020,2040,2060
1900 LPRINT USING" #.####";Q(1)
1910 GOTO 2960
1920 LPRINT USING" #.####";Q(1),Q(2)
1930 GOTO 2960
1940 LPRINT USING" #.####";Q(1),Q(2),Q(3)
1950 GOTO 2960
1960 LPRINT USING" #.####";Q(1),Q(2),Q(3),Q(4)
1970 GOTO 2960
1980 LPRINT USING" #.####";Q(1),Q(2),Q(3),Q(4),Q(5)
1990 GOTO 2960
2000 LPRINT USING" #.####";Q(1),Q(2),Q(3),Q(4),Q(5),Q(6)
2010 GOTO 2960
2020 LPRINT USING" #.####";Q(1),Q(2),Q(3),Q(4),Q(5),Q(6),Q(7)
2030 GOTO 2960
2040 LPRINT USING" #.####";Q(1),Q(2),Q(3),Q(4),Q(5),Q(6),Q(7),Q(8)
2050 GOTO 2960
2060 LPRINT USING" #.####";Q(1),Q(2),Q(3),Q(4),Q(5),Q(6),Q(7),Q(8),Q(9)
2070 GOTO 2960
2080 IF TEND%=6 THEN GOTO 2110
2090 LPRINT USING" ##.###";Q(1),Q(2),Q(3),Q(4),Q(5),Q(6),Q(7),Q(8),Q(9),
     Q(10)
2100 GOTO 2960
2110 ON IX2% GOTO 2120,2120,2140,2160,2180,2200,2220,2240,2260,2280
2120 LPRINT USING" ##.###";Q(1)
2130 GOTO 2960
2140 LPRINT USING" ##.###";Q(1),Q(2)
2150 GOTO 2960
2160 LPRINT USING" ##.###";Q(1),Q(2),Q(3)
2170 GOTO 2960
2180 LPRINT USING" ##.###";Q(1),Q(2),Q(3),Q(4)
2190 GOTO 2960
2200 LPRINT USING" ##.###";Q(1),Q(2),Q(3),Q(4),Q(5)
2210 GOTO 2960
2220 LPRINT USING" ##.###";Q(1),Q(2),Q(3),Q(4),Q(5),Q(6)
2230 GOTO 2960
2240 LPRINT USING" ##.###";Q(1),Q(2),Q(3),Q(4),Q(5),Q(6),Q(7)
```

```
2250 GOTO 2960
2260 LPRINT USING" ##.###";Q(1),Q(2),Q(3),Q(4),Q(5),Q(6),Q(7),Q(8)
2270 GOTO 2960
2280 LPRINT USING" ##.###";Q(1),Q(2),Q(3),Q(4),Q(5),Q(6),Q(7),Q(8),Q(9)
2290 GOTO 2960
2300 IF TEND%=6 THEN GOTO 2330
2310 LPRINT USING" ###.##";Q(1),Q(2),Q(3),Q(4),Q(5),Q(6),Q(7),Q(8),Q(9),
     Q(10)
2320 GOTO 2960
2330 ON IX2% GOTO 2340,2340,2360,2380,2400,2420,2440,2460,2480,2500
2340 LPRINT USING" ###.##";Q(1)
2350 GOTO 2960
2360 LPRINT USING" ###.##";Q(1),Q(2)
2370 GOTO 2960
2380 LPRINT USING" ###.##";Q(1),Q(2),Q(3)
2390 GOTO 2960
2400 LPRINT USING" ###.##";Q(1),Q(2),Q(3),Q(4)
2410 GOTO 2960
2420 LPRINT USING" ###.##";Q(1),Q(2),Q(3),Q(4),Q(5)
2430 GOTO 2960
2440 LPRINT USING" ###.##";Q(1),Q(2),Q(3),Q(4),Q(5),Q(6)
2450 GOTO 2960
2460 LPRINT USING" ###.##";Q(1),Q(2),Q(3),Q(4),Q(5),Q(6),Q(7)
2470 GOTO 2960
2480 LPRINT USING" ###.##";Q(1),Q(2),Q(3),Q(4),Q(5),Q(6),Q(7),Q(8)
2490 GOTO 2960
2500 LPRINT USING" ###.##";Q(1),Q(2),Q(3),Q(4),Q(5),Q(6),Q(7),Q(8),Q(9)
2510 GOTO 2960
2520 IF TEND%=6 THEN GOTO 2550
2530 LPRINT USING" ####.#";Q(1),Q(2),Q(3),Q(4),Q(5),Q(6),Q(7),Q(8),Q(9),
     Q(10)
2540 GOTO 2960
2550 ON IX2% GOTO 2560,2560,2580,2600,2620,2640,2660,2680,2700,2720
2560 LPRINT USING" ####.#";Q(1)
2570 GOTO 2960
2580 LPRINT USING" ####.#";Q(1),Q(2)
2590 GOTO 2960
2600 LPRINT USING" ####.#";Q(1),Q(2),Q(3)
2610 GOTO 2960
2620 LPRINT USING" ####.#";Q(1),Q(2),Q(3),Q(4)
2630 GOTO 2960
2640 LPRINT USING" ####.#";Q(1),Q(2),Q(3),Q(4),Q(5)
2650 GOTO 2960
2660 LPRINT USING" ####.#";Q(1),Q(2),Q(3),Q(4),Q(5),Q(6)
2670 GOTO 2960
2680 LPRINT USING" ####.#";Q(1),Q(2),Q(3),Q(4),Q(5),Q(6),Q(7)
2690 GOTO 2960
2700 LPRINT USING" ####.#";Q(1),Q(2),Q(3),Q(4),Q(5),Q(6),Q(7),Q(8)
2710 GOTO 2960
2720 LPRINT USING" ####.#";Q(1),Q(2),Q(3),Q(4),Q(5),Q(6),Q(7),Q(8),Q(9)
2730 GOTO 2960
2740 IF TEND%=6 THEN GOTO 2770
2750 LPRINT USING" ######";Q(1),Q(2),Q(3),Q(4),Q(5),Q(6),Q(7),Q(8),Q(9),
     Q(10)
2760 GOTO 2960
2770 ON IX2% GOTO 2780,2780,2800,2820,2840,2860,2880,2900,2920,2940
2780 LPRINT USING" ######";Q(1)
2790 GOTO 2960
2800 LPRINT USING" ######";Q(1),Q(2)
2810 GOTO 2960
```

```
2820 LPRINT USING" ######";Q(1),Q(2),Q(3)
2830 GOTO 2960
2840 LPRINT USING" ######";Q(1),Q(2),Q(3),Q(4)
2850 GOTO 2960
2860 LPRINT USING" ######";Q(1),Q(2),Q(3),Q(4),Q(5)
2870 GOTO 2960
2880 LPRINT USING" ######";Q(1),Q(2),Q(3),Q(4),Q(5),Q(6)
2890 GOTO 2960
2900 LPRINT USING" ######";Q(1),Q(2),Q(3),Q(4),Q(5),Q(6),Q(7)
2910 GOTO 2960
2920 LPRINT USING" ######";Q(1),Q(2),Q(3),Q(4),Q(5),Q(6),Q(7),Q(8)
2930 GOTO 2960
2940 LPRINT USING" ######";Q(1),Q(2),Q(3),Q(4),Q(5),Q(6),Q(7),Q(8),Q(9)
2950 GOTO 2960
2960 NOL=NOL+1
2970 C=NOL MOD 5
2980 IF C=0 THEN LPRINT
2990 J=0
3000 IF NOL<>45 THEN GOTO 3070
3010 LPRINT"  Wendell R. Gwinn, PE PhD
     ";
3020 LPRINT USING "&";DATE$:LPRINT :LPRINT:LPRINT:LPRINT
3030 GOTO 3070
3040 Q(J)=QI(J)=0
3050 TQ=.02
3060 GOTO 1400
3070 J=J+1
3080 NEXT I
3090 IF TEND%=1 THEN GOTO 700
3100 IF TEND%=2 THEN GOTO 1030
3110 IF TEND%=3 THEN GOTO 1080
3120 IF TEND%=4 THEN GOTO 1130
3130 IF TEND%=7 THEN GOTO 1030
3140 IF C<>0 THEN PRINT " "
3150   IF   NOL<>45   THEN   LPRINT"   Wendell   R.   Gwinn,   PE   PhD
";
3160 IF NOL<>45 THEN LPRINT USING"&";DATE$:LPRINT:LPRINT:LPRINT:LPRINT
3170 FOR P%=1 TO 56
3180 LPRINT" "
3190 NEXT P%
3200 HAH=HA+ZERO
3210 GIH=GI+ZERO
3220 IF COPY=1 THEN 3240
3230 IF COPY=0 THEN 3620
3240 LPRINT "            INPUT DATA":LPRINT
3250 LPRINT IDENT1$
3260 LPRINT IDENT2$
3270 IF SD<>0 THEN LPRINT"    Average Sediment Depth at Head Measuring
     Section = ";:PRINT SD:PRINT " "
3280 LPRINT "   DATE ";DATE$;"  DELTYH = ";DELTYH;"     FLUME TYPE = ";
     NFT%
3290 LPRINT "   ZERO = ";ZERO;"    DEPTH = ";DEPTH
3300 LPRINT "   B0 = ";B0;"    B1 = ";B1
3310 LPRINT"  Width of Flat, Slope of Floor, Total Width,   AT GAGE SEC
     TION"
3320 LPRINT "    E =  ";E;"    F = ";F;"    W = ";W:LPRINT " "
3330 LPRINT "    COMPUTED BY PROGRAM ":LPRINT
3340 IF N=10 THEN GOTO 3440
3350 IF ZERO>0 THEN GOTO 3400
3360 LPRINT "    A0 = ";CH(N);"    A1 = ";DH(N):LPRINT " ":
     LPRINT "            HEAD = ";HAH:LPRINT
```

```
3370 LPRINT "     KO = ";CI1;"     K1 = ";CI2:LPRINT " ":
     LPRINT      "          HEAD = ";GIH:LPRINT
3380 LPRINT "     EO = ";C1(N);"     E1 = ";EH(N):
     LPRINT " ":LPRINT "     FO = ";C2(N);"     F1 = ";FH(N)
3390 GOTO 3620
3400 LPRINT "     AO = ";CH(N);"     A1 = ";DH(N):LPRINT:
     LPRINT "          ELEV. = ";HAH:LPRINT
3410 LPRINT "     KO = ";CI1;"     K1 = ";CI2:LPRINT:
     LPRINT "          ELEV. = ";GIH:LPRINT
3420 LPRINT "     EO = ";C1(N);"     E1 = ";EH(N):LPRINT:
     LPRINT "     FO = ";C2(N);"     F1 = ";FH(N)
3430 GOTO 3620
3440 CROS1H=CROSS1+ZERO
3450 CROS2H=CROS2H+ZERO
3460 IF ZERO>0 THEN GOTO 3550
3470 LPRINT "     AO = ";CH(8);"     A1 = ";DH(8):LPRINT:
     LPRINT "          HEAD = ";HAH:LPRINT " "
3480 LPRINT "     KO = ";CI1;"     K1 = ";CI2:LPRINT:
     LPRINT "          HEAD = ";GIH:LPRINT " "
3490 LPRINT "     EO = ";C1(8);"     E1 = ";EH(8):LPRINT:
     LPRINT "     FO = ";C2(8);"     F1 = ";FH(8):LPRINT " "
3500 LPRINT "          HEAD = ";CROSS1H:LPRINT:
     LPRINT "     EO = ";C1(9);"     E1 = ";EH(9)
3510 LPRINT "     FO = ";C2(9);"     F1 = ";FH(9):LPRINT
3520 LPRINT "          HEAD = ";CROSS2H:LPRINT:
     LPRINT "     EO = ";C1(10);"     E1 = ";EH(10)
3530 LPRINT "     FO = ";C2(10);"     F1 = ";FH(10)
3540 GOTO 3620
3550 LPRINT "     AO = ";CH(8);"     A1 = ";DH(8):LPRINT:
     LPRINT "          ELEV. = ";HAH:LPRINT " "
3560 LPRINT "     KO = ";CI1;"     K1 = ";CI2:LPRINT:
     LPRINT "          ELEV. = ";GIH
3570 LPRINT "     EO = ";C1(8);"     E1 = ";EH(8):LPRINT:
     LPRINT "     FO = ";C2(8);"     F1 = ";FH(8)
3580 LPRINT "          ELEV. = ";CROS1H:LPRINT:
     LPRINT "     EO = ";C1(9);"     E1 = ";EH(9)
3590 LPRINT "     FO = ";C2(9);"     F1 = ";FH(9)
3600 LPRINT "          ELEV. = ";CROS2H:LPRINT:
     LPRINT "     EO = ";C1(10);"     E1 = ";EH(10)
3610 LPRINT "     FO = ";C2(10);"     F1 = ";FH(10)
3620 INPUT "  DO YOU WANT ANOTHER COPY (Yes input 1, No press RETURN)";
     COPY
3630 IF COPY=1 THEN GOTO 500
3640 STOP
3650 END
```

```
10  REM      ***********************************************************
20  REM      *           RATING TABLE FOR HS, H, AND HL FLUMES         *
30  REM      *                  By Wendell R. Gwinn                    *
40  REM      *       Dimension in meters, Table in m^3/s or L/s        *
50  REM      *                     April, 1985                         *
60  REM      ***********************************************************
70 DIM Q(10),NQ%(10),M%(10),QI(10),MI%(10),CI(10),C2(10),CH(8),DH(8),
   EH(10),FH(10)
80 CH(1)=1.857:CH(2)=2.238:CH(3)=1.705:CH(4)=1.822:CH(5)=1.49:CH(6)=0:
   CH(7)=1.855:CH(8)=1.49
90 DH(1)=.457:DH(2)=.526:DH(3)=.486:DH(4)=.5:DH(5)=.463:DH(6)=0:
   DH(7)=.505:DH(8)=.463
```

```
100 C1(1)=1.016:C1(2)=.861:C1(3)=.612:C1(4)=.63:C1(5)=.967:C1(6)=0:
    C1(7)=.655:C1(8)=.595:C1(9)=1.374:C1(10)=.893
110 C2(1)=.437:C2(2)=.479:C2(3)=.409:C2(4)=.4:C2(5)=.289:C2(6)=0:
    C2(7)=.398:C2(8)=.554:C2(9)=.258:C2(10)=.322
120 EH(1)=-.246:EH(2)=.367:EH(3)=.686:EH(4)=.571:EH(5)=-.725:EH(6)=0:
    EH(7)=.686:EH(8)=0:EH(9)=0:EH(10)=0
130 FH(1)=+.0656:FH(2)=-.115:FH(3)=-.079:FH(4)=-.059:FH(5)=.216:
    FH(6)=0:FH(7)=-.079:FH(8)=-0:FH(9)=-0:FH(10)=-0
140 LINE INPUT;"Line one of table identification,      []   centered on
     screen                       ";IDENT1$
150 PRINT
160 LINE INPUT;"LINE TWO OF TABLE IDENTIFICATION,      []   CENTERED ON
    SCREEN                        ";IDENT2$
170 PRINT
180 LINE INPUT"Date using 18 spaces right addjusted ";DATE$
190 INPUT"Change in head between each value in table, .01 or .001";
    DELTYH
200 PRINT "CODE FOR FLUME TYPE:":PRINT "      1=HS Flume with sloping
    drop box entrance and collector trough"
210 PRINT "      2=HS Flume, flat floor":PRINT "      3=H Flume with flat
    floor":PRINT "      4=H Flume with 1 on 8 sloping floor"
220 PRINT"      5=HL Flume, flat floor":PRINT"      6=Laboratory rating,
    read in constants for the equation of discharge"
230 PRINT"      7=H Flume with sloping drop box entrance":PRINT"      8=HL
     Flume with sloping drop box entrance"
240 INPUT"CODE FOR FLUME TYPE";NFT%
250 N=NFT%
260 INPUT"ELEVATION OF ZERO HEAD (METERS)";ZERO
270 INPUT "SIZE OF FLUME (MAXIMUM HEAD OR DEPTH)";DEPTH
280 PRINT "                                               HS          H
      H          HL"
290 PRINT "                                              Flume       Flume
      Flume      Flume"
300 PRINT "                  STANDARD                    Flat        Flat
      1 on 8     Flat"
310 PRINT "                  DIMENSION                   Floor       Floor    Sl
    oping Floor Floor"
320 PRINT "                  (meters)"
330 PRINT USING "BO, Half Width of Throat                  #.####      #.##
    ##      #.####    #.####";DEPTH*.025,DEPTH*.05,DEPTH*.05,DEPTH*.1
340 PRINT       "B1, Slope of Entrance Control             0.1667      0.5
         0.5       1.0"
350 PRINT USING "Width of 'flat floor' at Gage Section     #.####      #.##
    ##      #.####    #.####";DEPTH*.7167,DEPTH*1.5,DEPTH*.3,DEPTH*2.7
360 PRINT       "Slope of Floor at Gage Section              0            0
           8         0"
370 PRINT USING "Total Width at Gage Section               #.####      #.##
    ##      #.####    #.####";DEPTH*.7167,DEPTH*1.5,DEPTH*1.5,DEPTH*2.7
380 PRINT "   ****************** INPUT ACTUAL DIMENSIONS IF AVAILABLE *
    ***********"
390 INPUT"BO, HALF WIDTH OF THROAT IN METERS";BO
400 INPUT"B1, SLOPE OF ENTRANCE";B1
410 INPUT"WIDTH OF FLAT AT GAGE SECTION IN METERS";E
420 INPUT"SLOPE OF FLOOR AT GAGE SECTION";F
430 INPUT"TOTAL WIDTH AT GAGE SECTION IN METERS";W
440 INPUT"CODE FOR UNITS OF RATING TABLE [0=m^3/s or 1=l/s(0.001m^3/s)]"
    ;LITER
450 GI=.06096
460 IF N=6 THEN GOTO 480
470 GOTO 510
```

```
480 PRINT"You have selected to read in the coefficients for the equation
    s of discharge":PRINT" ":PRINT"      Q=A0*(2*B0+B1*H)*H*(H-0.003)^A1"
490 PRINT"      Q=[(E0+E1*D)*B0+(F0+F1*D)*B1*(H+V^2/2g)]*2g^.5*
    (H+V^2/2g)^1.5":PRINT" ":PRINT"NOTE: E1 AND F1=0 FOR MOST DATA"
500 INPUT "A0,A1,E0,E1,F0,F1";CH(6),DH(6),C1(6),EH(6),C2(6),FH(6)
510 NOL=45
520 PAGE%=1:J=1:L=1
530 INPUT"AVERAGE SEDIMENT DEPTH AT HEAD MEASURING SECTION IN METERS";SD
540 INPUT " DO YOU WANT THE INPUT DATA (Yes Input 1, No press RETURN) ";
    COPY
550 TQ=.02
560 D=DEPTH
570 IF D>.3048 THEN D=.3048
580 IZERO%=INT(ZERO/DELTYH)
590 IZER1%=INT(ZERO/(10*DELTYH))*10
600 J=IZERO%-IZER1%+1
610 FOR I=1 TO J
620 Q(I)=0:QI(I)=0
630 NEXT I
640 HEAD=IZER1%*DELTYH-10*DELTYH
650 H=IZERO%*DELTYH-ZERO
660 NEXT1%=1
670 IF DELTYH=.01 THEN LAST%=3
680 IF DELTYH=.001 THEN LAST%=30
690 TEND%=1
700 GOTO 1170
710 HA=.03048
720 IF SD>=.03 THEN 980
730 QA=CH(N)*((2*B0+B1*HA)*HA-(2*B0+SD*B1)*SD)*(HA-.003-SD)^DH(N)
740 TBQ=4*TQ
750 IF E>=W THEN 820
760 IF E+F*GI/2>=W THEN 790
770 AREA=GI*(E+F*GI/2)-SD*(E+F*SD/2)
780 GOTO 830
790 IF E+F*SD>=W THEN 820
800 AREA=GI*W-(W-E)*(W-E)/(2*F)-SD*(E+F*SD/2)
810 GOTO 830
820 AREA=(GI-SD)*W
830 V2G=TBQ*TBQ/(AREA*AREA*19.605)
840 QB=((C1(N)+EH(N)*D)*B0+(C2(N)+FH(N)*D)*B1*(GI+V2G))*4.4277*
    (GI+V2G)^1.5
850 IF V2G=0 THEN GOTO 920
860 DELTAQ=ABS(TBQ-QB)
870 IF DELTAQ<=.000001 THEN GOTO 920
880 TBQ=QB
890 IF TBQ<10000 THEN GOTO 830
900 LAST%=INT((GI-.001)/DELTYH+1.01)
910 GOTO 1020
920 Y1=QA/(4.4277*HA^1.5*B0)
930 X1=B1*HA/B0
940 Y2=QB/(4.4277*GI^1.5*B0)
950 X2=B1*GI/B0
960 CI2=(Y2-Y1)/(X2-X1)
970 CI1=Y2-CI2*X2
980 LAST%=INT((GI-.001)/DELTYH+1.01)
990 TEND%=2
1000 IF SD>.03 THEN GOTO 1020
1010 GOTO 1170
1020 TEND%=7
1030 GOTO 1170
```

```
1040 IF N<8 THEN GOTO 1150
1050 CROSS1=B0*2.58786/B1
1060 LAST%=INT((CROSS1-.001)/DELTYH+1.01)
1070 TEND%=3
1080 GOTO 1170
1090 N=9
1100 CROSS2=B0*7.51563/B1
1110 LAST%=INT((CROSS2-.001)/DELTYH+1.01)
1120 TEND%=4
1130 GOTO 1170
1140 N=10
1150 LAST%=INT(DEPTH/DELTYH+1.01)
1160 TEND%=5
1170 INDEX%=NEXT1%
1180 NEXT1%=LAST%+1
1190 V2G=0
1200 FOR I=INDEX% TO LAST%
1210 IF H-.003048-SD<.0001 THEN GOTO 3070
1220 IF TEND%=1 THEN Q(J)=CH(N)*((2*B0+B1*H)*H-(2*B0+SD*B1)*SD)*
     (H-.003048-SD)^DH(N)
1230 IF TEND%=2 THEN Q(J)=(CI1*B0+CI2*B1*H)*4.4277*H^1.5
1240 IF TEND%<=2 THEN GOTO 1410
1250 IF E>=W THEN GOTO 1320
1260 IF E+F*H>=W THEN GOTO 1290
1270 AREA=H*(E+F*H/2)-SD*(E+F*SD/2)
1280 GOTO 1330
1290 IF E+F*SD>=W THEN GOTO 1320
1300 AREA=H*W-(W-E)*(W-E)/(2*F)-SD*(E+F*SD/2)
1310 GOTO 1330
1320 AREA=(H-SD)*W
1330 V2G=TQ*TQ/(AREA*AREA*19.605)
1340 Q(J)=((C1(N)+EH(N)*D)*B0+(C2(N)+FH(N)*D)*B1*(H+V2G))*4.4277*
     (H+V2G)^1.5
1350 IF V2G=0 THEN GOTO 1410
1360 DELTAQ=ABS(TQ-Q(J))
1370 IF DELTAQ<=.00005 THEN GOTO 1410
1380 TQ=Q(J)
1390 IF TQ>10000 THEN GOTO 3070
1400 GOTO 1330
1410 H=H+DELTYH
1420 IF Q(J)<.00005 THEN GOTO 1450
1430 TQ=Q(J)
1440 GOTO 1460
1450 Q(J)=0
1460 IF J=10 THEN GOTO 1540
1470 IF TEND%<>5 THEN GOTO 3100
1480 IF I<>LAST% THEN GOTO 3100
1490 IX2%=J+1
1500 FOR IX=IX2% TO 10
1510 Q(IX)=0
1520 NEXT IX
1530 TEND%=6
1540 FOR K=1 TO 10
1550 IF LITER=1 THEN Q(K)=Q(K)*1000
1560 M%(K)=1
1570 IF Q(K)>.1 THEN M%(K)=2
1580 IF Q(K)>1 THEN M%(K)=3
1590 IF Q(K)>10 THEN M%(K)=4
1600 IF Q(K)>100 THEN M%(K)=5
1610 IF Q(K)>1000 THEN M%(K)=6
```

```
1620 NEXT K
1630 HEAD=HEAD+10*DELTYH
1640 IF NOL<>45 THEN GOTO 1830
1650 LPRINT "                                    RATING TABLE
            Page ";
1660 LPRINT USING"##";PAGE%
1670 LPRINT IDENT1$
1680 LPRINT IDENT2$
1690 IF LITER=0 THEN LPRINT"                         Discharge in cubic met
     ers per second"
1700 IF LITER=1 THEN LPRINT"                            Discharge in liters
      per second"
1710 IF SD=0 THEN GOTO 1750
1720 LPRINT " "
1730 LPRINT"          Average Sediment Depth at Head Measuring Section
     = ";SD
1740 LPRINT:LPRINT:LPRINT
1750 IF ZERO=0 THEN LPRINT "  HEAD"
1760 IF ZERO<>0 THEN LPRINT "  ELEV."
1770 IF DELTYH=.001 THEN LPRINT"(Meters)  .000   .001   .002   .003   .0
     04   .005   .006   .007   .008   .009"
1780 IF DELTYH=.01 THEN LPRINT "(Meters)  .00    .01    .02    .03    .0
     4    .05    .06    .07    .08    .09"
1790 LPRINT
1800 PAGE%=PAGE%+1
1810 NOL=0
1820 IF SD<>0 THEN NOL=5
1830 IF Q(1)<>0 THEN MX%=M%(1)
1840 IF Q(1)=0 THEN MX%=M%(10)
1850 IF TEND%=6 THEN MX%=M%(1)
1860 IF DELTYH=.001 THEN LPRINT USING "###.##  ";HEAD;
1870 IF DELTYH=.01 THEN LPRINT USING "####.#  ";HEAD;
1880 ON MX% GOTO 1890,1890,2110,2330,2550,2770
1890 IF TEND%=6 THEN GOTO 1920
1900 LPRINT USING " #.####";Q(1),Q(2),Q(3),Q(4),Q(5),Q(6),Q(7),Q(8),
     Q(9),Q(10)
1910 GOTO 2990
1920 ON IX2% GOTO 1930,1930,1950,1970,1990,2010,2030,2050,2070,2090
1930 LPRINT USING" #.####";Q(1)
1940 GOTO 2990
1950 LPRINT USING" #.####";Q(1),Q(2)
1960 GOTO 2990
1970 LPRINT USING" #.####";Q(1),Q(2),Q(3)
1980 GOTO 2990
1990 LPRINT USING" #.####";Q(1),Q(2),Q(3),Q(4)
2000 GOTO 2990
2010 LPRINT USING" #.####";Q(1),Q(2),Q(3),Q(4),Q(5)
2020 GOTO 2990
2030 LPRINT USING" #.####";Q(1),Q(2),Q(3),Q(4),Q(5),Q(6)
2040 GOTO 2990
2050 LPRINT USING" #.####";Q(1),Q(2),Q(3),Q(4),Q(5),Q(6),Q(7)
2060 GOTO 2990
2070 LPRINT USING" #.####";Q(1),Q(2),Q(3),Q(4),Q(5),Q(6),Q(7),Q(8)
2080 GOTO 2990
2090 LPRINT USING" #.####";Q(1),Q(2),Q(3),Q(4),Q(5),Q(6),Q(7),Q(8),Q(9)
2100 GOTO 2990
2110 IF TEND%=6 THEN GOTO 2140
2120 LPRINT USING" ##.###";Q(1),Q(2),Q(3),Q(4),Q(5),Q(6),Q(7),Q(8),Q(9),
     Q(10)
2130 GOTO 2990
```

```
2140 ON IX2% GOTO 2150,2150,2170,2190,2210,2230,2250,2270,2290,2310
2150 LPRINT USING" ##.###";Q(1)
2160 GOTO 2990
2170 LPRINT USING" ##.###";Q(1),Q(2)
2180 GOTO 2990
2190 LPRINT USING" ##.###";Q(1),Q(2),Q(3)
2200 GOTO 2990
2210 LPRINT USING" ##.###";Q(1),Q(2),Q(3),Q(4)
2220 GOTO 2990
2230 LPRINT USING" ##.###";Q(1),Q(2),Q(3),Q(4),Q(5)
2240 GOTO 2990
2250 LPRINT USING" ##.###";Q(1),Q(2),Q(3),Q(4),Q(5),Q(6)
2260 GOTO 2990
2270 LPRINT USING" ##.###";Q(1),Q(2),Q(3),Q(4),Q(5),Q(6),Q(7)
2280 GOTO 2990
2290 LPRINT USING" ##.###";Q(1),Q(2),Q(3),Q(4),Q(5),Q(6),Q(7),Q(8)
2300 GOTO 2990
2310 LPRINT USING" ##.###";Q(1),Q(2),Q(3),Q(4),Q(5),Q(6),Q(7),Q(8),Q(9)
2320 GOTO 2990
2330 IF TEND%=6 THEN GOTO 2360
2340 LPRINT USING" ###.##";Q(1),Q(2),Q(3),Q(4),Q(5),Q(6),Q(7),Q(8),Q(9),
     Q(10)
2350 GOTO 2990
2360 ON IX2% GOTO 2370,2370,2390,2410,2430,2450,2470,2490,2510,2530
2370 LPRINT USING" ###.##";Q(1)
2380 GOTO 2990
2390 LPRINT USING" ###.##";Q(1),Q(2)
2400 GOTO 2990
2410 LPRINT USING" ###.##";Q(1),Q(2),Q(3)
2420 GOTO 2990
2430 LPRINT USING" ###.##";Q(1),Q(2),Q(3),Q(4)
2440 GOTO 2990
2450 LPRINT USING" ###.##";Q(1),Q(2),Q(3),Q(4),Q(5)
2460 GOTO 2990
2470 LPRINT USING" ###.##";Q(1),Q(2),Q(3),Q(4),Q(5),Q(6)
2480 GOTO 2990
2490 LPRINT USING" ###.##";Q(1),Q(2),Q(3),Q(4),Q(5),Q(6),Q(7)
2500 GOTO 2990
2510 LPRINT USING" ###.##";Q(1),Q(2),Q(3),Q(4),Q(5),Q(6),Q(7),Q(8)
2520 GOTO 2990
2530 LPRINT USING" ###.##";Q(1),Q(2),Q(3),Q(4),Q(5),Q(6),Q(7),Q(8),Q(9)
2540 GOTO 2990
2550 IF TEND%=6 THEN GOTO 2580
2560 LPRINT USING" ####.#";Q(1),Q(2),Q(3),Q(4),Q(5),Q(6),Q(7),Q(8),Q(9),
     Q(10)
2570 GOTO 2990
2580 ON IX2% GOTO 2590,2590,2610,2630,2650,2670,2690,2710,2730,2750
2590 LPRINT USING" ####.#";Q(1)
2600 GOTO 2990
2610 LPRINT USING" ####.#";Q(1),Q(2)
2620 GOTO 2990
2630 LPRINT USING" ####.#";Q(1),Q(2),Q(3)
2640 GOTO 2990
2650 LPRINT USING" ####.#";Q(1),Q(2),Q(3),Q(4)
2660 GOTO 2990
2670 LPRINT USING" ####.#";Q(1),Q(2),Q(3),Q(4),Q(5)
2680 GOTO 2990
2690 LPRINT USING" ####.#";Q(1),Q(2),Q(3),Q(4),Q(5),Q(6)
2700 GOTO 2990
2710 LPRINT USING" ####.#";Q(1),Q(2),Q(3),Q(4),Q(5),Q(6),Q(7)
```

```
2720 GOTO 2990
2730 LPRINT USING" ####.#";Q(1),Q(2),Q(3),Q(4),Q(5),Q(6),Q(7),Q(8)
2740 GOTO 2990
2750 LPRINT USING" ####.#";Q(1),Q(2),Q(3),Q(4),Q(5),Q(6),Q(7),Q(8),Q(9)
2760 GOTO 2990
2770 IF TEND%=6 THEN GOTO 2800
2780 LPRINT USING" ######";Q(1),Q(2),Q(3),Q(4),Q(5),Q(6),Q(7),Q(8),Q(9),
     Q(10)
2790 GOTO 2990
2800 ON IX2% GOTO 2810,2810,2830,2850,2870,2890,2910,2930,2950,2970
2810 LPRINT USING" ######";Q(1)
2820 GOTO 2990
2830 LPRINT USING" ######";Q(1),Q(2)
2840 GOTO 2990
2850 LPRINT USING" ######";Q(1),Q(2),Q(3)
2860 GOTO 2990
2870 LPRINT USING" ######";Q(1),Q(2),Q(3),Q(4)
2880 GOTO 2990
2890 LPRINT USING" ######";Q(1),Q(2),Q(3),Q(4),Q(5)
2900 GOTO 2990
2910 LPRINT USING" ######";Q(1),Q(2),Q(3),Q(4),Q(5),Q(6)
2920 GOTO 2990
2930 LPRINT USING" ######";Q(1),Q(2),Q(3),Q(4),Q(5),Q(6),Q(7)
2940 GOTO 2990
2950 LPRINT USING" ######";Q(1),Q(2),Q(3),Q(4),Q(5),Q(6),Q(7),Q(8)
2960 GOTO 2990
2970 LPRINT USING" ######";Q(1),Q(2),Q(3),Q(4),Q(5),Q(6),Q(7),Q(8),Q(9)
2980 GOTO 2990
2990 NOL=NOL+1
3000 C=NOL MOD 5
3010 IF C=0 THEN LPRINT " "
3020 J=0
3030 IF NOL<>45 THEN GOTO 3100
3040 LPRINT "  Wendell R. Gwinn, PE PhD
      ";
3050 LPRINT USING "&";DATE$:LPRINT :LPRINT:LPRINT:LPRINT
3060 GOTO 3100
3070 Q(J)=QI(J)=0
3080 TQ=.02
3090 GOTO 1410
3100 J=J+1
3110 NEXT I
3120 IF TEND%=1 THEN GOTO 710
3130 IF TEND%=2 THEN GOTO 1040
3140 IF TEND%=3 THEN GOTO 1090
3150 IF TEND%=4 THEN GOTO 1140
3160 IF TEND%=7 THEN GOTO 1040
3170 IF C<>0 THEN PRINT " "
3180   IF   NOL<>45   THEN   LPRINT"    Wendell   R.   Gwinn,   PE   PhD
";
3190 IF NOL<>45 THEN LPRINT USING"&";DATE$:LPRINT:LPRINT:LPRINT:LPRINT
3200 FOR P%=1 TO 55
3210 LPRINT" "
3220 NEXT P%
3230 HAH=HA+ZERO
3240 GIH=GI+ZERO
3250 IF COPY=1 THEN 3270
3260 IF COPY=0 THEN 3650
3270 LPRINT "            INPUT DATA":LPRINT
3280 LPRINT IDENT1$
```

```
3290 LPRINT IDENT2$
3300 IF SD<>0 THEN LPRINT"    Average Sediment Depth at Head Measuring
     Section = ";:LPRINT SD:LPRINT
3310 LPRINT "   DATE ";DATE$;"  DELTYH = ";DELTYH;"      FLUME TYPE = ";
     NFT%
3320 LPRINT "   ZERO = ";ZERO;"     DEPTH = ";DEPTH
3330 LPRINT "   B0 = ";B0;"     B1 = ";B1
3340 LPRINT " Width of Flat, Slope of Floor, Total Width     AT GAGE
     SECTION"
3350 LPRINT "    E =   ";E;"     F = ";F;"     W = ";W:LPRINT
3360 LPRINT "    COMPUTED BY PROGRAM ":LPRINT
3370 IF N=10 THEN GOTO 3470
3380 IF ZERO>0 THEN GOTO 3430
3390 LPRINT "    A0 = ";CH(N);"     A1 = ";DH(N):LPRINT:
     LPRINT "            HEAD = ";HAH:LPRINT " "
3400 LPRINT "    K0 = ";CI1;"     K1 = ";CI2:LPRINT:
     LPRINT      "          HEAD = ";GIH:LPRINT " "
3410 LPRINT "    E0 = ";C1(N);"     E1 = ";EH(N):LPRINT:
     LPRINT "    F0 = ";C2(N);"     F1 = ";FH(N)
3420 GOTO 3650
3430 LPRINT "    A0 = ";CH(N);"     A1 = ";DH(N):LPRINT:
     LPRINT "            ELEV. = ";HAH:LPRINT " "
3440 LPRINT "    K0 = ";CI1;"     K1 = ";CI2:LPRINT:
     LPRINT "            ELEV. = ";GIH:LPRINT " "
3450 LPRINT "    E0 = ";C1(N);"      E1 = ";EH(N):LPRINT:
     LPRINT "    F0 = ";C2(N);"     F1 = ";FH(N)
3460 GOTO 3650
3470 CROS1H=CROSS1+ZERO
3480 CROS2H=CROS2H+ZERO
3490 IF ZERO>0 THEN GOTO 3580
3500 LPRINT "    A0 = ";CH(8);"     A1 = ";DH(8):LPRINT:
     LPRINT "            HEAD = ";HAH:LPRINT
3510 LPRINT "    K0 = ";CI1;"    K1 = ";CI2:LPRINT:
     LPRINT "            HEAD = ";GIH:LPRINT
3520 LPRINT "    E0 = ";C1(8);"     E1 = ";EH(8):LPRINT:
     LPRINT "    F0 = ";C2(8);"     F1 = ";FH(8):LPRINT
3530 LPRINT "            HEAD = ";CROSS1H:LPRINT:
     LPRINT "    E0 = ";C1(9);"     E1 = ";EH(9)
3540 LPRINT "    F0 = ";C2(9);"     F1 = ";FH(9):LPRINT
3550 LPRINT "            HEAD = ";CROSS2H:LPRINT:
     LPRINT "    E0 = ";C1(10);"     E1 = ";EH(10)
3560 LPRINT "    F0 = ";C2(10);"     F1 = ";FH(10)
3570 GOTO 3650
3580 LPRINT "    A0 = ";CH(8);"     A1 = ";DH(8):LPRINT:
     LPRINT "            ELEV. = ";HAH:LPRINT
3590 LPRINT "    K0 = ";CI1;"     K1 = ";CI2:LPRINT:
     LPRINT "            ELEV. = ";GIH
3600 LPRINT "    E0 = ";C1(8);"     E1 = ";EH(8):LPRINT:
     LPRINT "    F0 = ";C2(8);"     F1 = ";FH(8)
3610 LPRINT "            ELEV. = ";CROS1H:LPRINT:
     LPRINT "    E0 = ";C1(9);"     E1 = ";EH(9)
3620 LPRINT "    F0 = ";C2(9);"     F1 = ";FH(9)
3630 LPRINT "            ELEV. = ";CROS2H:LPRINT:
     LPRINT "    E0 = ";C1(10);"     E1 = ";EH(10)
3640 LPRINT "    F0 = ";C2(10);"     F1 = ";FH(10)
3650 INPUT "  DO YOU WANT ANOTHER COPY (Yes enter 1, No press RETURN)";
     COPY
3660 IF COPY=1 THEN GOTO 510
3670 STOP
3680 END
```

RATING TABLE Page 1
1-Ft. H Flume with Drop Box Entrance
Using Standard Dimensions
Discharge in cubic feet per second

HEAD (Feet)	.00	.01	.02	.03	.04	.05	.06	.07	.08	.09
0.0	0.0000	0.0000	0.0007	0.0016	0.0027	0.0041	0.0057	0.0076	0.0098	0.0122
0.1	0.0149	0.0178	0.0210	0.0245	0.0284	0.0325	0.0370	0.0418	0.0469	0.0523
0.2	0.0581	0.0640	0.0702	0.0768	0.0836	0.0908	0.0984	0.1063	0.1145	0.1231
0.3	0.1321	0.1414	0.1511	0.1612	0.1716	0.1825	0.1937	0.2053	0.2174	0.2298
0.4	0.2427	0.2559	0.2696	0.2838	0.2983	0.3133	0.3287	0.3446	0.3609	0.3777
0.5	0.3950	0.4127	0.4308	0.4495	0.4686	0.4883	0.5084	0.5290	0.5501	0.5717
0.6	0.5939	0.6165	0.6397	0.6634	0.6876	0.7124	0.7377	0.7635	0.7900	0.8169
0.7	0.8445	0.8726	0.9012	0.9305	0.9604	0.9908	1.0218	1.0535	1.0857	1.1186
0.8	1.152	1.186	1.221	1.256	1.292	1.329	1.366	1.404	1.443	1.482
0.9	1.522	1.563	1.605	1.647	1.690	1.733	1.777	1.822	1.868	1.915
1.0	1.962									

Wendell R. Gwinn, PE PhD April 5, 1985

INPUT DATA

1-Ft. H Flume with Drop Box Entrance
Using Standard Dimensions
DATE April 5, 1985 DELTYH = .01 FLUME TYPE = 7
ZERO = 0 DEPTH = 1
B0 = .05 B1 = .5
Width of Flat, Slope of Floor, Total Width, AT GAGE SECTION
E = 1.5 F = 0 W = 1.5

COMPUTED BY PROGRAM

A0 = 3.34 A1 = .505

HEAD = .1

K0 = .720894 K1 = .450184

HEAD = .2

E0 = .655 E1 = .209

F0 = .398 F1 = -.024

RATING TABLE Page 1
0.3048-m H Flume with Drop Box Entrance
Using Standard Dimensions
Discharge in liters per second

HEAD (Meters)	.00	.01	.02	.03	.04	.05	.06	.07	.08	.09
0.0	0.000	0.053	0.191	0.408	0.707	1.100	1.591	2.164	2.836	3.614
0.1	4.503	5.507	6.633	7.885	9.267	10.786	12.447	14.253	16.209	18.321
0.2	20.59	23.03	25.64	28.43	31.40	34.56	37.92	41.47	45.24	49.21
0.3	53.41									

Wendell R. Gwinn, PE PhD April 5, 1985

```
          INPUT DATA

                     0.3048-m H Flume with Drop Box Entrance
                            Using Standard Dimensions
       DATE       April 5, 1985 DELTYH =  .01   FLUME TYPE =  7
       ZERO =  0      DEPTH =  .3048
       BO =  .0152      B1 =  .5
Width of Flat, Slope of Floor, Total Width       AT GAGE SECTION
   E =   .4572      F =  0            W =  .4572

   COMPUTED BY PROGRAM

   AO =  1.855     A1 =  .505

         HEAD =  .03048

   KO =  .723207     K1 =  .44879

        HEAD =  .06096

   EO =  .655     E1 =  .686

   FO =  .398     F1 = -.079
```

Design Charts of Channel Transition for Flow Without Choke

SHIE-YUI LIONG* AND MANA PATARAPANICH*

INTRODUCTION

Most long channels have channel transitions. Devices used for measuring discharge through channels, such as venturi flume and Parshall flume, are based on the principle of channel transition. To avoid scouring, velocities in the irrigation channels have to be reduced; to prevent shoaling in the navigation channels velocities have to be increased. Such reduction or increase of velocities may be made possible through channel transition—expansion or contraction of channel section. Also, as flow passes below a bridge, a channel transition in the form of contraction and subsequent expansion is involved. To dissipate the tremendous amount of energy of flow over a spillway the channel bottom is usually lowered suddenly. Many more examples of channel transition may be cited.

In designing a channel transition, such as expansion or contraction, or a change in bed elevation, or both, it is always desirable to prevent the transition from acting as a "choke" [2] influencing the upstream flow. Choke is a process in which the transition causes a temporary or permanent reduction of discharge passing through sections upstream of the transition zone.

Choke prevention requires criteria on the maximum allowable changes on width and bed elevation. These limits can be determined by applying the energy and continuity equations to the cross sections immediately before and after the transition zone. Allens [1] presented design charts for flow of no-choke condition solely for a change in bed elevation of a rectangular channel. Liong [3] expanded Allens' design charts to accommodate changes in both bed elevation and channel width of a rectangular channel. This chapter presents the analysis and design charts of channel transition for flow without choke in channel of rectangular, triangular, parabolic, or trapezoidal cross section.

*Department of Civil Engineering, National University of Singapore, Singapore

ASSUMPTIONS

In cases of changing width and cases of changing width and bed elevation together, the following assumptions concerning the channel shape and flow are made: (1) The transition zone connects one uniform channel to another which is of the same cross-sectional form; (2) the transition is gradual and smooth, and thus energy loss in the transition zone is negligibly small; (3) the channel is horizontal; and (4) the flow is one-dimensional, steady and uniform.

ANALYSIS

Figure 1 schematically displays one-dimensional, steady, uniform flow in a channel of rectangular, triangular, parabolic, or trapezoidal cross section with changing width and channel bed elevation. Bernoulli's equation applied to sections 1 and 2 is

$$h_1 + \frac{V_1^2}{2g} = h_2 + \frac{V_2^2}{2g} + \Delta Z \tag{1}$$

in which h_1, V_1 and h_2, V_2 = the flow depth and velocity in sections 1 and 2, respectively; and ΔZ = the elevation difference between channel bed at sections 1 and 2. With Froude number defined by

$$F = \frac{V}{\sqrt{g \frac{A}{B}}} \tag{2a}$$

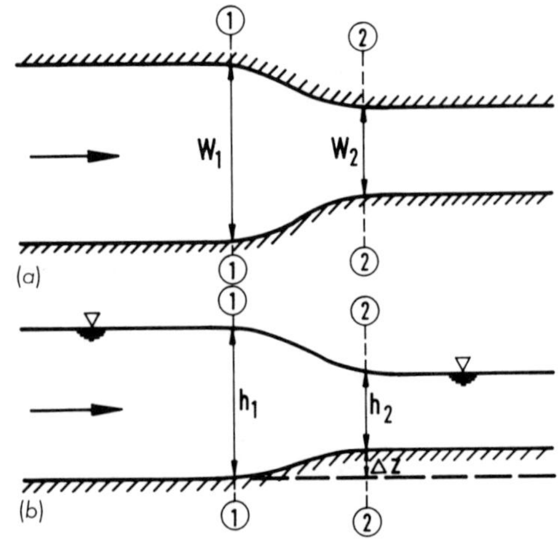

FIGURE 1. Gradually changing width and depth in channel transition: (a) plan view; (b) elevation view.

or

$$F = \frac{Q}{\sqrt{g\dfrac{A^3}{B}}} \tag{2b}$$

in which Q = discharge; A = flow cross-sectional area; and B = water surface width, Equation (1) can then be rewritten as

$$h_1 + \frac{1}{2}\left(\frac{A_1}{B_1}\right)F_1^2 = h_2 + \frac{1}{2}\left(\frac{A_2}{B_2}\right)F_2^2 + \Delta Z \tag{3}$$

Dividing Equation (3) throughout with h_1 an explicit equation for the dimensionless term ($\Delta Z/h_1$) expressed as

$$\frac{\Delta Z}{h_1} = 1 + \frac{1}{2}\left(\frac{A_1}{B_1 h_1}\right)F_1^2\left[1 - \left(\frac{A_2}{A_1}\right)\left(\frac{B_1}{B_2}\right)\left(\frac{F_2}{F_1}\right)^2\right] - \frac{h_2}{h_1} \tag{4}$$

is then obtained.

Continuity equation applied to sections 1 and 2 has the following form

$$V_1A_1 = V_2A_2 \tag{5}$$

which after substitution of Equation (2a) into Equation (5) yields an expression for (B_1/B_2) as

$$\left(\frac{B_1}{B_2}\right) = \left(\frac{A_1}{A_2}\right)^3\left(\frac{F_1}{F_2}\right)^2 \tag{6}$$

Substitution of Equation (6) into Equation (4) yields

$$\frac{\Delta Z}{h_1} = 1 + \frac{1}{2}\left(\frac{A_1}{B_1h_1}\right)F_1^2\left[1 - \left(\frac{A_1}{A_2}\right)^2\right] - \frac{h_2}{h_1} \tag{7}$$

Equation (7) was derived for channel of any cross-sectional shape. However, for channel of rectangular, triangular or parabolic cross section Equation (7) will be transformed and expressed in different form. For rectangular channel, for example, an expression for the ratio (A_1/A_2) is obtained from Equation (6) and is given as

$$\left(\frac{A_1}{A_2}\right) = \left(\frac{b_1}{b_2}\right)^{1/3}\left(\frac{F_2}{F_1}\right)^{2/3} \tag{8}$$

where b_1 and b_2 are the channel width at sections 1 and 2, respectively. The ratio (h_2/h_1) in Equation (7) can be written as the product of (A_2/A_1) and (b_1/b_2) with (A_2/A_1) as expressed in Equation (8), viz.

$$\left(\frac{h_2}{h_1}\right) = \left(\frac{b_1}{b_2}\right)^{2/3}\left(\frac{F_1}{F_2}\right)^{2/3} \tag{9}$$

After substituting Equations (8) and (9) into Equation (7) the transformed Equation (7) for a rectangular channel has the following form

$$\frac{\Delta Z}{h_1} = \left(1 + \frac{1}{2}F_1^2\right) - \left(\frac{b_1}{b_2}\right)^{2/3}\left(\frac{F_1}{F_2}\right)^{2/3}\left(1 + \frac{1}{2}F_2^2\right) \tag{10}$$

Similarly, the following expressions

$$\frac{\Delta Z}{h_1} = \left(1 + \frac{1}{4}F_1^2\right) - \left(\frac{m_1}{m_2}\right)^{2/5}\left(\frac{F_1}{F_2}\right)^{2/5}\left(1 + \frac{1}{4}F_2^2\right) \tag{11}$$

for a triangular channel and

$$\frac{\Delta Z}{h_1} = \left(1 + \frac{1}{3}F_1^2\right) - \left(\frac{W_1}{W_2}\right)^{1/2}\left(\frac{H_2}{H_1}\right)^{1/4}\left(\frac{F_1}{F_2}\right)^{1/2}\left(1 + \frac{1}{3}F_2^2\right) \tag{12}$$

for a parabolic channel are obtained. Note that m_1 and m_2 are the side slope of a triangular channel at sections 1 and 2, respectively; W_1, H_1 and W_2, H_2 are the maximum channel width and its corresponding channel depth of the parabolic channel at sections 1 and 2, respectively.

For flow without choke the flow regime downstream of the transition zone must be the same as that upstream. Hence, the limiting value of F_2 is one. With $F_2 = 1$, Equation (10) becomes

$$\frac{\Delta Z}{h_1} = \left(1 + \frac{1}{2}F_1^2\right) - \frac{3}{2}\left[\left(\frac{b_1}{b_2}\right)F_1\right]^{2/3} \tag{13}$$

TABLE 1. Computed ΔZ and Design ΔZ.

Channel Cross Section	Formulae to Compute $(\Delta Z)_{Chart}$	$(\Delta Z)_{Design}$
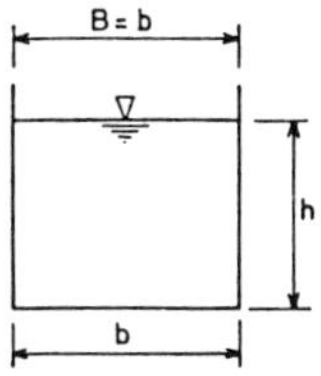 	$A = bh$ $B = b$ $F = \left[\frac{Q^2}{g\,b^2h^3}\right]^{1/2}$ $\frac{\Delta Z}{h_1} = \left(1 + \frac{1}{2}F_1^2\right) - \frac{3}{2}\left[\left(\frac{b_1}{b_2}\right)F_1\right]^{2/3}$	For a hump $[(\Delta Z)_{Chart} > 0]$: $(\Delta Z)_{Design} \le (\Delta Z)_{Chart}$ For a drop $[(\Delta Z)_{Chart} < 0]$: $(\Delta Z)_{Design} \ge \lvert(\Delta Z)_{Chart}\rvert$
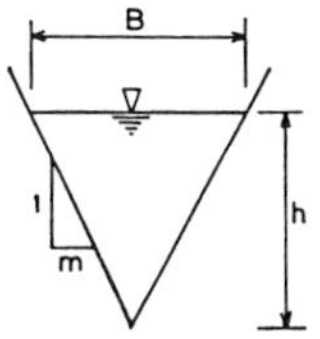 	$A = mh^2$ $B = 2mh$ $F = \left[\left(\frac{Q^2}{g\,h^5}\right)\left(\frac{2}{m^2}\right)\right]^{1/2}$ $\frac{\Delta Z}{h_1} = \left(1 + \frac{1}{4}F_1^2\right) - \frac{5}{4}\left[\left(\frac{m_1}{m_2}\right)F_1\right]^{2/5}$	
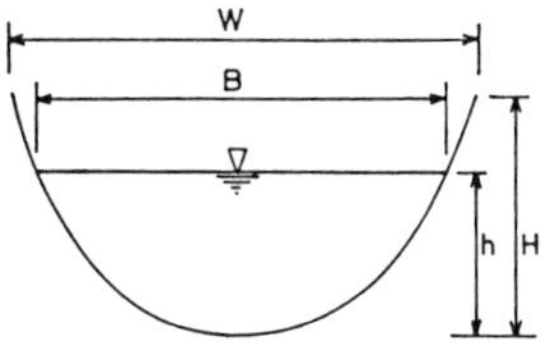 	$A = \frac{2}{3}\,\frac{W\,h^{3/2}}{H^{1/2}}$ $B = W\left(\frac{h}{H}\right)^{1/2}$ $F = \left[3.375\left(\frac{H}{h}\right)^4\left(\frac{Q^2}{g\,W^2H^3}\right)\right]^{1/2}$ $\frac{\Delta Z}{h_1} = \left(1 + \frac{1}{3}F_1^2\right) - \frac{4}{3}\left[\left(\frac{W_1}{W_2}\right)\left(\frac{H_2}{H_1}\right)^{1/2}F_1\right]^{1/2}$	
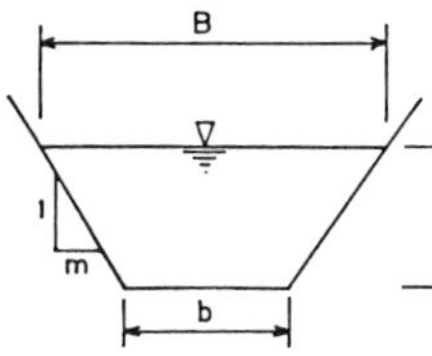 	$A = h\,(b + mh)$ $B = b + 2mh$ $F = \left\{\left(\frac{m^3Q^2}{g\,b^5}\right)\frac{\left[1 + 2\left(\frac{mh}{b}\right)\right]}{\left[\left(\frac{mh}{b}\right)\left(1 + \frac{mh}{b}\right)\right]^3}\right\}^{1/2}$ $\frac{\Delta Z}{h_1} = 1 + \frac{1}{2}\left(\frac{A_1}{B_1h_1}\right)(F_1)^2\left[1 - \left(\frac{A_1}{(A_2)_c}\right)^2\right] - \frac{(h_2)_c}{h_1}$	

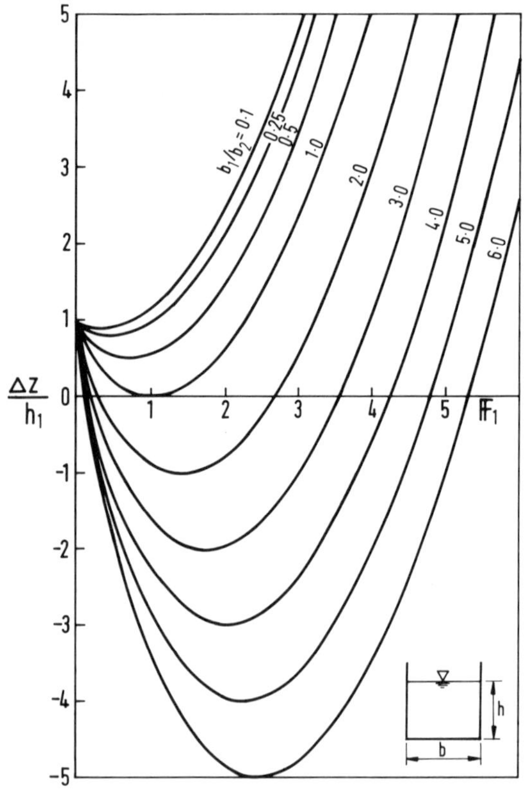

FIGURE 2. Design chart for flow without choke in rectangular channels.

for a rectangular channel; Equation (11) becomes

$$\frac{\Delta Z}{h_1} = \left(1 + \frac{1}{4}F_1^2\right) - \frac{5}{4}\left[\left(\frac{m_1}{m_2}\right)F_1\right]^{2/5} \quad (14)$$

for a triangular channel; and Equation (12) becomes

$$\frac{\Delta Z}{h_1} = \left(1 + \frac{1}{3}F_1^2\right) - \frac{4}{3}\left[\left(\frac{W_1}{W_2}\right)\left(\frac{H_2}{H_1}\right)^{1/2}F_1\right]^{1/2} \quad (15)$$

For trapezoidal channel Equation (7) will be used. With $F_2 = 1$ the flow cross-sectional area and the flow depth at section 2 will be at their critical values. Equation (7) now becomes

$$\frac{\Delta Z}{h_1} = 1 + \frac{1}{2}\left(\frac{A_1}{B_1h_1}\right)F_1^2\left\{1 - \left[\frac{A_1}{(A_2)_c}\right]^2\right\} - \frac{(h_2)_c}{h_1} \quad (16)$$

DESIGN CHARTS

In Table 1 the formulae used to compute the limiting values of ΔZ, $(\Delta Z)_{chart}$, for channel of four different cross-sectional shapes are summarized.

For rectangular channel Equation (13) is plotted and shown in Figure 2. $(\Delta Z/h_1)$ is plotted against F_1 with various values of (b_1/b_2). Note that the curve for $(b_1/b_2) = 1$ represents a prismatic rectangular channel; those for $(b_1/b_2) > 1$ and $(b_1/b_2) < 1$ represent a channel with contraction and a channel with expansion, respectively.

Figure 3 shows the plot of Equation (14) which was derived for channel of triangular cross section. In Figure 3, $(\Delta Z/h_1)$ is plotted against F_1 with various values of (m_1/m_2). $(m_1/m_2) > 1$ represents a contraction; $(m_1/m_2) < 1$ an expansion; and $(m_1/m_2) = 1$ no change in cross section size.

Equation (15) of parabolic channel is plotted in Figure 4. Similar to Figures 2 and 3 $(\Delta Z/h_1)$ is plotted against F_1 with various values of $[(W_1/\sqrt{H_1})/(W_2/\sqrt{H_2})]$. $[(W_1/\sqrt{H_1})/(W_2/\sqrt{H_2})] > 1$ indicates a contraction; $[(W_1/\sqrt{H_1})/(W_2/\sqrt{H_2})] < 1$ an expansion; and $[(W_1/\sqrt{H_1})/(W_2/\sqrt{H_2})] = 1$ no change in cross section size.

To obtain $(\Delta Z)_{chart}$ for trapezoidal channel from Equation (16) there are several other parameters we first have to determine. To illustrate the determination of $(\Delta Z)_{chart}$ the following procedure is given:

1. With known h, b, B, m at section 1 and discharge Q, compute $A_1 = h_1 (b_1 + m_1h_1)$; (A_1/B_1h_1); and $F_1 = [(Q^2 \cdot B_1)/gA_1^3]^{1/2}$.
2. Compute $[(m_2^3Q^2)/g\, b_2^5]^{1/2}$ from known values of m, b at section 2 and from discharge Q. Use Figure 5a to obtain $[m_2\,(h_2)_c/b_2]$ and then compute $(h_2)_c$.
3. Use Figure 5(b) to obtain $[A_2/B_2(h_2)_c]$ with the value of $[m_2(h_2)_c/b_2]$ obtained in step 2. With $(B_2)_c = b_2 + 2m_2\,(h_2)_c$ then compute · $(A_2)_c$.
4. With the computed values of F_1 and $[A_1/B_1h_1)]$ in step 1 use Figures 5(c) or 5(e) to obtain the function $[\frac{1}{2}(A_1/B_1h_1)\,F_1^2]$. Figure 5(c) is for $F_1 \geq 1$ while Figure 5(e) is for $F_1 \leq 1$. For trapezoidal cross section the values of (A/Bh) is between 0.5 and 1.0. It can be easily verified that the limiting value of (A/Bh) of 1.0 is for rectangular channel, and that of 0.5 is for the triangular section.
5. Form $[A_1/(A_2)_c]$. Use Figures 5(d) or 5(f) to obtain $\{[(h_2)_c + \Delta z]/h_1\}$ with the value of $[\frac{1}{2}(A_1/B_1h_1)F_1^2]$ obtained in step 4. Figure 5(d) is for $[\frac{1}{2}(A_1/B_1h_1)F_1^2] > 0.5$ while Figure 5(f) for $[\frac{1}{2}(A_1/B_1h_1)F_1^2] \leq 0.5$.
6. Compute $(\Delta Z)_{chart}$ from $\{[(h_2)_c + \Delta Z]/h_1\}$ obtained in step 5.

With Figures 2, 3, 4, and 5 we can easily determine $(\Delta Z)_{chart}$ for no-choke flow in channel of rectangular, triangular, parabolic, or trapezoidal cross section, respectively.

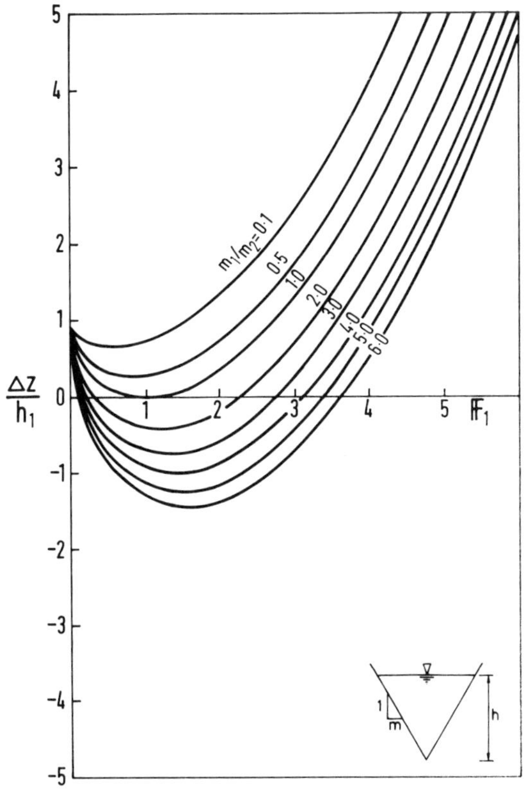

FIGURE 3. Design chart for flow without choke in triangular channels.

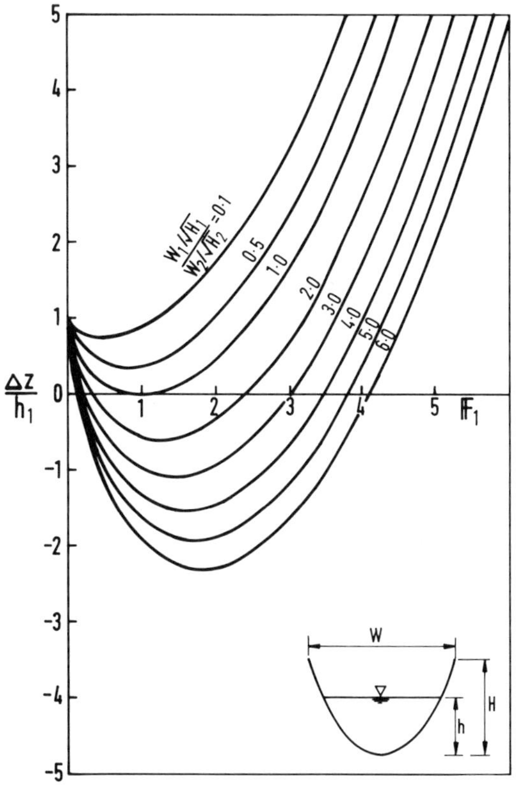

FIGURE 4. Design chart for flow without choke in parabolic channels.

Should $(\Delta Z)_{chart}$ be a positive value, i.e., a hump, for a particular flow problem it implies that the designed ΔZ, $(\Delta Z)_{design}$, must not exceed $(\Delta Z)_{chart}$ in order to prevent choke from occurring. This can be explained by considering the energy equation, Equation (1), once again. Writing Equation (1) in different form as

$$E_2 = E_1 - \Delta Z \quad (17)$$

where E = specific energy = $h + V^2/2g$, shows that there exists a maximum positive ΔZ at which a minimum energy at section 2 is required to carry the discharge passing through section 1.

Should E_2 be lower than the minimum required value choke will immediately occur.

For a drop, negative $(\Delta Z)_{chart}$, Equation (17) has the following form

$$E_2 = E_1 + |\Delta Z| \quad (18)$$

Equation (18) implies that there exists a minimum ΔZ at which a minimum energy at section 2 is required to carry the discharge passing through section 1. Hence, the designed ΔZ, $(\Delta Z)_{design}$, should be at least equal to $(\Delta Z)_{chart}$ or the structure at section 2 should be further lowered in order to prevent choke from occurring.

Thus, the relationship between the design value for $(\Delta Z)_{design}$ and the values obtained from charts for $(\Delta Z)_{chart}$ may be summarized as follows:

1. For a hump (positive ΔZ):

$$(\Delta Z)_{design} \leq (\Delta Z)_{chart} \quad (19)$$

2. For a drop (negative ΔZ):

$$(\Delta Z)_{design} \geq |(\Delta Z)_{chart}| \quad (20)$$

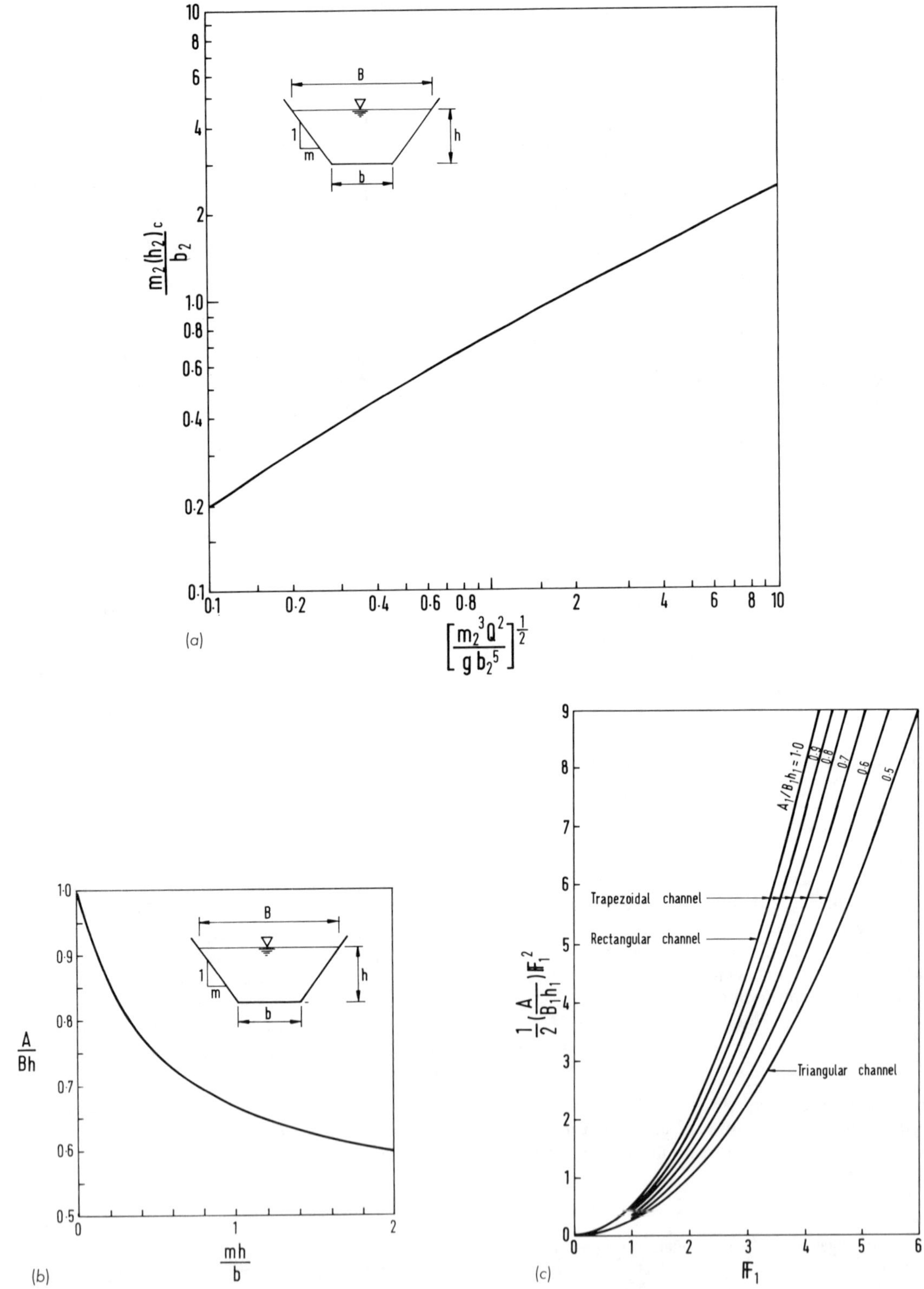

FIGURE 5. Design chart for flow without choke in trapezoidal channels: (a) to determine the critical flow depth at section 2; (b) to determine the dimensionless cross-sectional areas at sections 1 and 2; (c) to determine the function $\frac{1}{2}(A_1/B_1h_1)F_1^2$ in supercritical flow condition (continued).

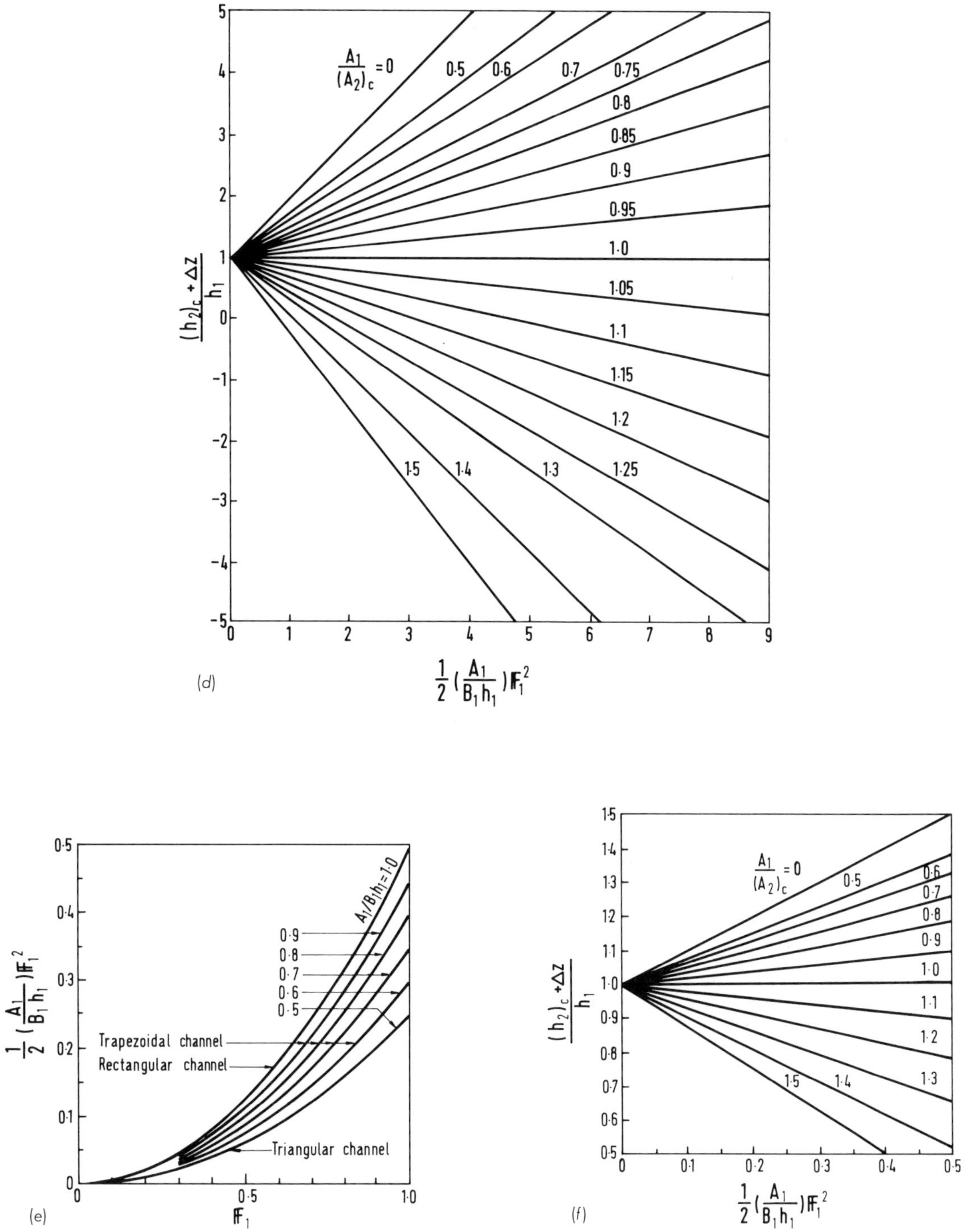

FIGURE 5 (continued). Design chart for flow without choke in trapezoidal channels: (d) to determine ΔZ for $[\frac{1}{2}(A_1/B_1h_1)F_1^2] > 0.5$; (e) to determine the function $\frac{1}{2}(A_1/B_1h_1)F_1^2$ in subcritical flow condition; (f) to determine ΔZ for $[\frac{1}{2}(A_1/B_1h_1)F_1^2] \leq 0.5$.

SUMMARY

Design charts for one-dimensional, steady, uniform and no-choke flow in channel of rectangular, triangular, parabolic, or trapezoidal cross section are presented. With the design charts a designer can select the maximum allowable hump or the minimum required drop in a channel transition with or without a change in cross section size.

NOTATION

A = flow cross-sectional area
B = water surface width
b = channel bottom width
E = specific energy
g = acceleration of gravity
h = flow depth
H = maximum channel depth
m = channel side slope
Q = channel discharge
V = flow velocity
W = maximum channel width
ΔZ = elevation difference between upstream and downstream section
F = Froude number

Subscripts

1 upstream section of the transition zone
2 downstream section of the transition zone
c critical

REFERENCES

1. Allen, R. F., "Steady Solution for River Flow," *Journal of the Hydraulic Division,* ASCE, Vol. 106, No. HY4, Proc. Paper 15305, pp. 608–611 (Apr. 1980).
2. Henderson, F. M., *Open Channel Flow,* McMillan Book Co., New York, NY, p. 49 (1966).
3. Liong, S. Y., "Channel Design and Flow Operation Without Choke," *Journal of Irrigation and Drainage Engineering,* Vol. 110, No. 4, Paper No. 19318, pp. 403–407 (Dec. 1984).

Internal Flow in the Hydraulic Jump

JOHN A. MCCORQUODALE*

INTRODUCTION

General

PURPOSE

This chapter reviews the available experimental and theoretical information on the internal flow in hydraulic jumps. An attempt is made to summarize the literature in a form that will be useful for the practicing engineer. The relationship between the internal flow and the external or macroscopic characteristics of the jump is also considered.

Most of the research work on hydraulic jumps has been concerned with their macroscopic behaviour. The important parameters in these studies included the sequent depth ratio and the jump length which are required for stilling basin design. The internal flow has received much less attention than it deserves.

There are several reasons for studying the internal flow in hydraulic jumps. Foremost is the problem of scaling-up physical model data to predict prototype behaviour. Stilling basins are often tested in small-scale models and the prototype data are determined by applying the Froude Law. The Froude Law assumes that inertia and gravity are the dominant independent forces acting on the stilling basin while viscous forces and surface tension are assumed to be negligible. A complete mathematical model of the internal flow would enable the modeller to assess the possible scale effects in a physical model. Other effects such as the behaviour of the air phase and the effect of the entrance boundary layer are generally not identical in the physical model and in the prototype. Other reasons for studying the internal flow are:

(a) To assess the cavitation potential within the hydraulic jumps which depends on the internal flow and turbulence pattern
(b) To estimate the forces on walls, floors and appurtenances in a stilling basin which depend on the internal flow
(c) To assess the mixing and re-aeration characteristics of a hydraulic jump
(d) To determine the size and extent of riprap protection downstream from a stilling basin

DEFINITIONS AND CLASSIFICATIONS

The hydraulic jump which is illustrated in Figure 1 refers to the phenomenon that occurs when there is an abrupt transition from supercritical (inertia dominanted) flow to subcritical (gravity dominated) flow. The most important parameter affecting the flow in a hydraulic jump is the initial Froude number defined by

$$F_1 = \frac{U_1}{\sqrt{g'D}} \qquad (1)$$

where

$g' = g(\Delta\varrho/\varrho)$
g = acceleration of gravity
ϱ = density of active fluid
$\Delta\varrho$ = difference in density between the active and ambient fluid
u = longitudinal fluid velocity relative to a coordinate system attached to the hydraulic jump
D = hydraulic mean depth
$= A/B$
A = flow area

*Department of Civil Engineering, University of Windsor, Windsor, Ontario, Canada

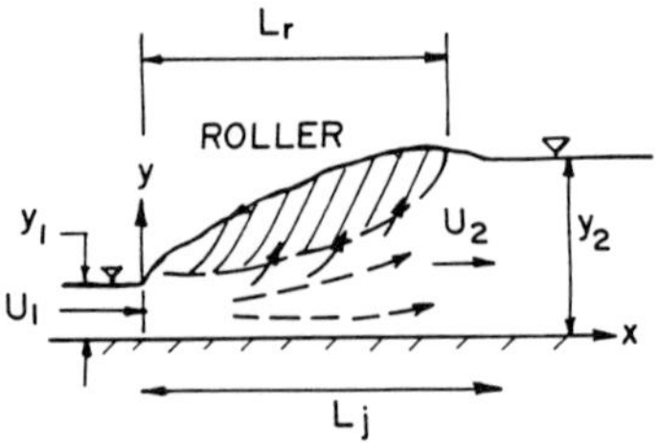

FIGURE 1. Defining diagram of the classical hydraulic jump.

B = contact length between the active and ambient fluids perpendicular to u
U_1 = mean value of u at the initial section

Figure 2 shows schematically the effects of Froude number on the hydraulic jump (Chow 1957). The ambient fluid may be a liquid or a gas. If the $F_1 > 1$, the flow is supercritical and a hydraulic jump is possible. If $F_1 < 1$, the flow is subcritical. A characteristic of hydraulic jumps is that active fluid tends to entrain the ambient fluid.

Another important condition affecting the hydraulic jump is the boundary geometry. Figure 2 illustrates some of the many geometrices that are used in hydraulic jump stilling basins.

Figure 2 shows other possible classifications of the hydraulic jump. A jump involving two liquids or two gases of different densities is termed an internal jump while a jump involving a liquid as the active (entraining) fluid with a gas as the entrained fluid is referred to as a "free" jump if the inflow is unsubmerged and if there are no drag inducing appurtenaces. If the inflow is submerged, the jump does not tend to entrain the gas and is called a submerged jump. A forced jump occurs where baffles and/or sills are placed on the stilling basin boundary.

A hydraulic jump may be stationary (steady mean flow) or travelling (unsteady) as in the case of a hydraulic bore. This chapter is primarily concerned with the stationary case although some of the analyses can be applied to both cases.

MACROSCOPIC CONSIDERATIONS

A definition sketch of a typical hydraulic jump is shown in Figure 1. The important macroscopic parameters of the jump are:

y_1 = the initial depth
y_2 = the sequent depth
U_1 = initial mean velocity
U_2 = mean velocity at the exit of the jump
L_j = jump length
L_r = length of the roller

The internal equations of motion for a hydraulic jump can be integrated in two ways depending on whether the control volume is defined on a microscopic or macroscopic scale. The latter will be dealt with in this section while the internal problem will be considered in detail in subsequent sections.

Since the hydraulic jump takes place over a relatively short distance, of the order of five sequent depths, the transition is dominated by the initial momentum flux and pressure force due to the sequent depth. Boundary shear forces are secondary. Figure 3 shows an unsubmerged forced hydraulic jump in a radially diverging sloping channel. This will be used to illustrate the macroscopic approach.

Rouse, Siao and Nagaratnam (1958) have applied the Reynolds equation for turbulent flow in the form of

$$\frac{\partial(\bar{u}_i\bar{u}_j)}{\partial x_j} + \frac{\partial(\overline{u_i'u_j'})}{\partial x_j} = -\frac{1}{\varrho}\frac{\partial \bar{p}}{\partial x_j} + X_i + \frac{\mu}{\varrho}\frac{\partial^2 \bar{u}_i}{\partial x_j \partial x_i} \qquad (2)$$

to the control volume $\mathcal{V}$ of a hydraulic jump, in which

$\bar{u}_i$ = local mean velocity in direction i
ϱ = fluid density
$\bar{p}$ = local mean pressure
X_i = body force
μ = dynamic viscosity
u_i' = random component of u_i

Using Green's theorem, Equation (2) can be integrated over the volume, $\mathcal{V}$, to obtain the following macroscopic equation:

$$\underbrace{\int_\xi \varrho\bar{u}_i\bar{u}j\frac{\partial x_j}{\partial n}\,d\xi}_{\text{I}} + \underbrace{\int_\xi \varrho\overline{u_i'u_j'}\frac{\partial x_j}{\partial n}\,d\xi}_{\text{II}} \qquad (3)$$

$$= -\underbrace{\int_\xi \bar{p}\frac{\partial x_i}{\partial n}\,d\xi}_{\text{III}} + \underbrace{\int_{\mathcal{V}} \varrho X_i d\mathcal{V}}_{\text{IV}} + \underbrace{\int_\xi \mu\frac{\partial \bar{u}_i}{\partial x_j}\frac{\partial x_j}{\partial n}\,d\xi}_{\text{V}}$$

where n is the outwardly directed normal.

The terms represent the following:

1. The net flux of momentum through the boundary, ξ, due to the mean flow
2. The net momentum transfer through the boundary, ξ, due to turbulence
3. The resultant mean normal (pressure) force exerted on the fluid boundary, ξ
4. The net weight of the fluid within the control volume, $\mathcal{V}$
5. The mean tangential force exerted on the boundary, ξ

A macroscopic momentum equation is obtained if Equation (3) is applied to the control volume, $\mathcal{V}$, shown in Figure 3. Considering the streamwise, x-momentum, the

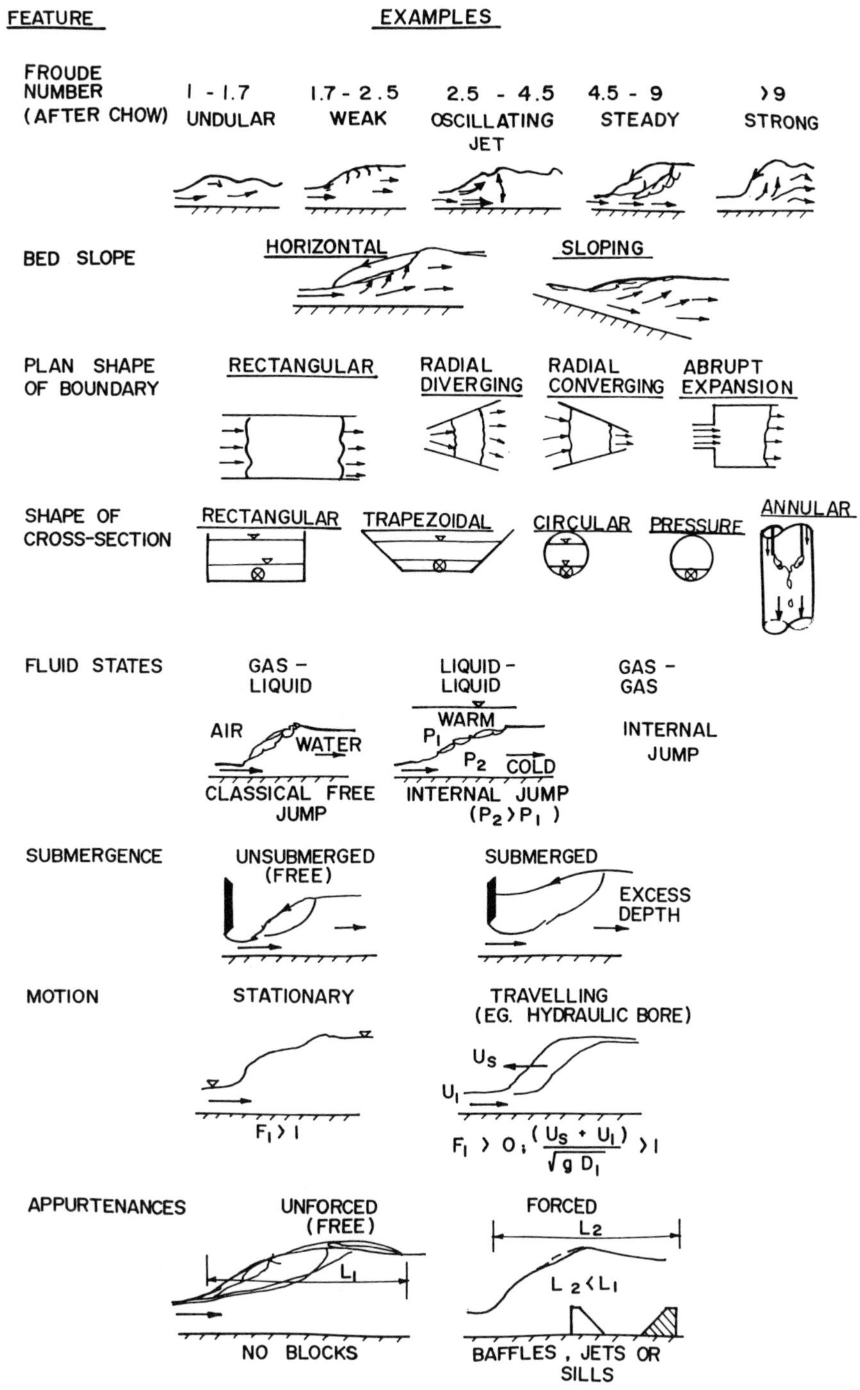

FIGURE 2. Classification of the hydraulic jump.

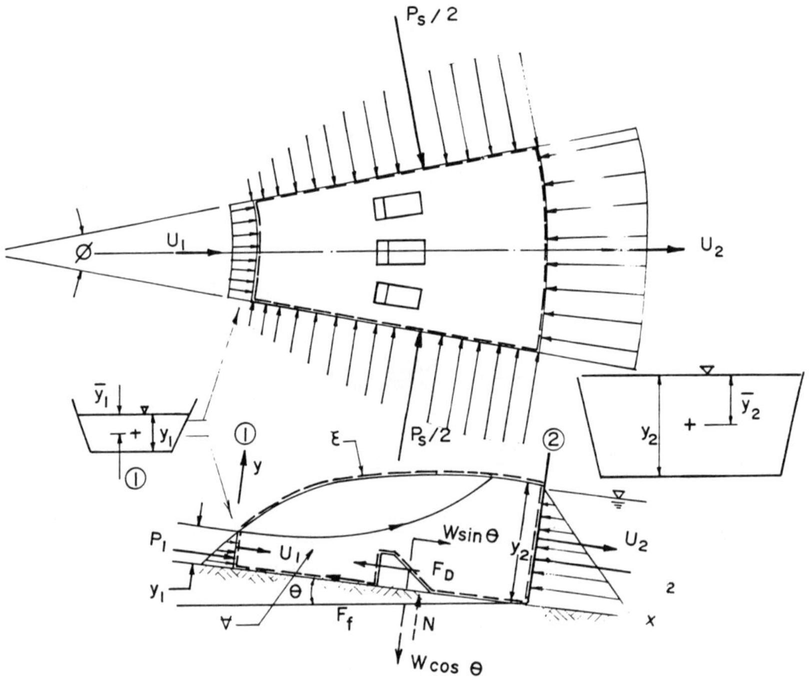

FIGURE 3. General defining diagram of a hydraulic jump in an open channel.

following equation results:

$$\underbrace{\{\beta_2 \varrho U_2 Q - \beta_1 \varrho U_1 Q\}}_{\text{I}} + \underbrace{\{\varrho I_2 - \varrho I_1\}}_{\text{II}}$$

$$= P_1 - P_2 + P_s \sin(\phi/2) - F_D - W \sin\theta - F_f \tag{4}$$

in which

$\beta = \int_A \bar{u}^2 dA/(QU)$
U = average velocity over A
Q = total flow
A = flow area
$I = \int_A \bar{u}' dA$
P = pressure force = $\beta' g \varrho A \bar{y} \cos\theta$
$\bar{y}$ = centroidal depth below the water surface measure perpendicular to the bed
P_s = total side force
F_D = form drag
W = weight of the water in volume $\forall$
F_f = boundary shear force
θ = bed slope angle
ϕ = side slope angle
β' = pressure correction factor

The mass conservation equation (the continuity equation) is

$$\frac{\partial \bar{u}_i}{\partial x_i} = 0 \tag{5}$$

or in the integrated form for the problem in Figure 3,

$$Q = U_1 A_1 = U_2 A_2 \tag{6}$$

Equations (4) and (6) can be solved simultaneously to determine the sequent conditions (U_2, y_2) for given initial conditions (U_1, y_1).

Several terms in Equation (4) can be dropped for a hydraulic jump in a horizontal rectangular channel with no baffles. These are:

(a) The side force $P_s \sin(\phi/2)$
(b) The drag force F_D
(c) The body force component $W \sin\theta$

Furthermore, the turbulent intensity term $\varrho(I_2 - I_1)$ and the bed shear force F_f are relatively small. The momentum and pressure correction factors β and β' are assumed to be close to unity. With these simplifications Equation (4) reduces to

$$\frac{U_2^2 y_2}{g} + \frac{1}{2} y_2^2 = \frac{U_1^2 y_1}{g} + \frac{1}{2} y_1^2 \tag{7}$$

while the continuity equation becomes

$$y_2 U_2 = y_1 U_1 \tag{8}$$

The simultaneous solution of Equations (7) and (8) gives the well known cubic equation in the sequent depth ratio ($y_o = y_2/y_1$):

$$y_o^3 - (2F_1^2 + 1)y_o + 2F_1^2 = 0 \quad (9)$$

which yields the classical Belanger solution:

$$y_o = \frac{1}{2}(\sqrt{1 + 8F_1^2} - 1) \quad (10)$$

Rajaratnam (1967) showed that the effect of bed friction on y_o increases with increasing F_1, reaching a reduction of about 4% at $F_1 = 10$.

It is important to note that the above macroscopic approach yields only the sequent depth and gives no information about the surface profile or the jump length. This information can be found in the process of solving the internal flow. In many cases, such as radial stilling basins, sloping stilling basins and forced hydraulic jumps, even the sequent depth depends on the internal flow. Traditionally the additional information on the hydraulic jump has been obtained through the use of physical models.

INTERNAL FLOWS

The Classical Hydraulic Jump

THE INTERNAL MECHANICS

Experimental studies should be the basis for the development of a numerical model for complex phenomena like the hydraulic jump. A number of excellent experimental studies can be found in the literature.

Early investigation of the internal flow in hydraulic jumps was hampered by the problem of making accurate velocity measurements in the presence of entrained air. Studies by Rouse and his co-workers (1958, 1954, 1957) used air models with solid boundaries that conformed to the mean configurations of actual hydraulic jumps. They applied the hot-wire technology and obtained information about the mean flow, the turbulence intensities and the Reynolds stresses within the domain of the simulated jumps. The air model could not simulate all aspects of the hydraulic jump, e.g., the surface gravity waves, the entrainment of the ambient fluid and certain aspects of the actual pressure field. In addition, the solid boundary at the surface introduced a shear stress which would be higher than the corresponding stress for the air–water interface of an actual jump. Nevertheless, their work was a breakthrough in the understanding of the internal mechanics of the hydraulic jump and has provided a framework for future research.

Their analytical treatment of the hydraulic jump was based on the Reynolds and continuity equations for the fluid [Equations (2) and (5)]. They integrated these equations over the flow domain to obtain the macroscopic Equations (3) and (6).

Furthermore, they developed a differential equation of work and energy in the form:

$$\varrho\bar{u}_j \frac{\partial}{\partial x_j}\left(\frac{\bar{U}_T^2}{2} + \frac{\bar{u}_T'^2}{2}\right) + \overline{\varrho u_j' \frac{\partial}{\partial x_j}\left(\frac{u_T'^2}{2}\right)} + \varrho \frac{\partial}{\partial x_j}(\bar{u}_i\overline{u_i'u_j'})$$

$$= -\bar{u}_i \frac{\partial \bar{p}}{\partial x_i} - \overline{u_i' \frac{\partial p'}{\partial x_i}} + \varrho\bar{u}_i\bar{X}_i + \mu\bar{u}_i \frac{\partial^2 \bar{u}_i}{\partial x_j \partial x_i} \quad (11)$$

$$+ \overline{\mu u_i' \frac{\partial^2 u_i'}{\partial x_i \partial x_j}}$$

in which

$\bar{u}_T$ = magnitude of the local mean velocity vector
u_T' = random component of u_T
p' = random component of the pressure p

This equation was segregated into two parts relating to the work energy relationships for the mean flow and turbulence, respectively, as follows:

$$\bar{u}_j \frac{\partial(\varrho\bar{u}_T^2/2)}{\partial x_j} + \varrho\bar{u}_i \frac{\partial\overline{u_i'u_j'}}{\partial x_j} = -\bar{u}_i \frac{\partial \bar{p}}{\partial x_i} + \varrho\bar{u}_i\bar{X}_i + \mu\bar{u}_i \frac{\partial^2\bar{u}_i}{\partial x_j \partial x_j} \quad (12)$$

$$\bar{u}_j \frac{\partial(\varrho\overline{u_T'^2}/2)}{\partial x_j} + \overline{u_j' \frac{\partial(\varrho u_T'^2/2)}{\partial x_j}} + \varrho\overline{u_i'u_j'} \frac{\partial \bar{u}_i}{\partial x_j} = -\overline{u_i' \frac{\partial p'}{\partial x_i}} + \overline{\mu u_i' \frac{\partial^2 u_i'}{\partial x_j \partial x_j}} \quad (13)$$

Neglecting the initial turbulence and the free surface stresses and assuming hydrostatic conditions throughout, the following integrated energy equation was derived for transport of the mean flow kinetic energy:

$$\int_o^y \frac{\bar{u}_T^2}{2g}\bar{u}\,dy - \int_o^{y_1} \frac{u_T^3}{2g}\,dy + \int_o^y \frac{\bar{u}\,\overline{u'^2} + \bar{v}\,\overline{u'v'}}{g}\,dy$$

Net flux of K.E. due to mean motion — Work of Reynolds stresses at surfaces (turb. energy flux)

$$-\frac{1}{g}\int_o^y\int_o^x \left[\overline{u'v'}\left(\frac{\partial\bar{u}}{\partial y} + \frac{\partial\bar{v}}{\partial x}\right) + (\overline{u'^2} - \overline{v'^2})\frac{\partial\bar{u}}{\partial x}\right] dydx$$

= Net work by Reynolds stresses from o to x (turb. energy flux)

$$qd_1 - qd + \frac{\mu}{\gamma}\int_o^y \left[2\bar{u}\,\frac{\partial \bar{u}}{\partial x} + \bar{v}\left(\frac{\partial \bar{u}}{\partial y} + \frac{\partial \bar{v}}{\partial x}\right)\right]dy$$

Rate of work by pressure forces — Rate of work by viscous forces (conservative)

$$-\frac{\mu}{\varrho}\int_o^y\int_o^x \left[4\left(\frac{\partial \bar{u}}{\partial x}\right)^2 + \left(\frac{\partial \bar{u}}{\partial y} + \frac{\partial \bar{v}}{\partial x}\right)^2\right] dydx \qquad (14)$$

Dissipation rate of mean K.E. due to viscous stresses

A similar equation was obtained for the turbulent kinetic energy transport:

$$\int_o^y \frac{\overline{u_T'^2}}{2g}\,\bar{u}\,dy + \int_o^y \frac{\overline{u_T'^2 u'}}{2g}\,dy$$

(mean flow) (turb. diff.)

Flux of turbulent K. E.

$$+\frac{1}{g}\int_o^y\int_o^x \left[\overline{u'v'}\left(\frac{\partial \bar{u}}{\partial y} + \frac{\partial \bar{v}}{\partial x}\right) + (\overline{u'^2} - \overline{v'^2})\,\frac{\partial \bar{u}}{\partial x}\right]dydx$$

Production rate of turbulent K.E.

$$= -\int_o^y \frac{\overline{p'u'}}{\gamma}\,dy \qquad (15)$$

Rate of work by fluctuating pressure on the D/s surface

$$+\frac{\mu}{\gamma}\int_o^y \left[\frac{\partial}{x}\left(\frac{\overline{u_T'^2}}{2} + \overline{u'^2}\right) + \frac{\partial \overline{u'v'}}{\partial y}\right]dy$$

Rate of work by viscous stresses of turbulence on the D/s surface

$$-\int_o^y\int_o^x \left[K\left(\frac{\partial \overline{u'}}{\partial x}\right)^2\right] dydx$$

Rate of dissipation of turbulent K.E.

DISTRIBUTION OF VELOCITY AND TURBULENCE IN THE HYDRAULIC JUMP

Rouse *et al.*, based on their air model (1958), presented experimental results for three values of F_1 (2, 4 and 6). Figure 4(a) shows the vertical distributions of mean velocity, turbulence intensity and turbulent shear stress for the simulation of a jump at $F_1 = 6$. Figure 4(b) shows the longitudinal variation in the maximum velocity, turbulent shear stresses and turbulence intensities in dimensionless form. Using Equation (2) integrated over intermediate ranges of x, a momentum balance was obtained as shown in Figure 5(a). Similarly, Rouse investigated the mean energy balance as represented by Equation (14). Some of his experimental data are shown in Figure 5(c). Rouse also presented experimental data on the rates of production, dissipation and convection of turbulent kinetic energy. His results for $F_1 = 6$ are shown in Figure 5(b).

Leutheusser and his co-workers (1971, 1972, 1974, 1976, 1977) used a hot-film anemometer to measure the Reynolds stresses in actual hydraulic jumps. They investigated the effect of the initial boundary layer development on the internal and macroscopic flow of the jump. Their results for an undeveloped initial boundary layer are in general agreement with the air model data of Rouse *et al.* (1958). However, the results for the developed initial boundary layer are significantly different. Figure 6 compares the mean flow patterns for a developed and undeveloped initial boundary layer at $F_1 = 6$. Figure 6(a) indicates that the position of maximum velocity for developed flow is shifted upward near the start of the jump, resulting in a surface wave. Similar flow patterns were obtained for $F_1 = 2.85$. Leutheusser *et al.* (1979) also found that the tendency for flow separation on the bed and walls of the stilling basin is affected by the inflow boundary layer development. For example, with developed flow the range of potential separation was found to be $5 < x/y_2 < 7$ for $F_1 = 4$ while for undeveloped flow this range was about $3 < x/y_2 < 7$. A similar delay in possible separation was noted at other values of F_1. The tendency for boundary layer separation increased with increasing F_1 in undeveloped flow.

Leutheusser and Kartha (1972) studied the effect of initial boundary layer development on the boundary shear stress. Figure 7(a) compares the boundary stresses for developed and undeveloped inflows for $F_1 = 2.85$ and 14.39. The value L_{j*} is the influence length of the jump as indicated in Figure 7(b); this is considerably greater than the traditional jump length L_j. It is also evident from Figure 7 that the extent of influence of the jump is greater for developed flow.

Turbulence data obtained by Rouse *et al.* (1958) and Resch *et al.* (1971, 1976) are compared in Table 1.

Rajaratnam (1965, 1967) has studied the velocity distribution, bed shear stress and air entrainment for various hydraulic jumps. He showed that the hydraulic jump could be studied as a wall jet under a strong adverse pressure gradient. He presented his data (1967) in a dimensionless plot of $f(\eta) = \bar{u}/U_m$ versus $\eta = y/\delta_1$ where U_m is the maximum velocity and δ_1 is the depth to $\bar{u} = U_m/2$ in the outer layer (see Figure 8). The required values of δ_1 and U_m can be obtained from Figures 9 and 10.

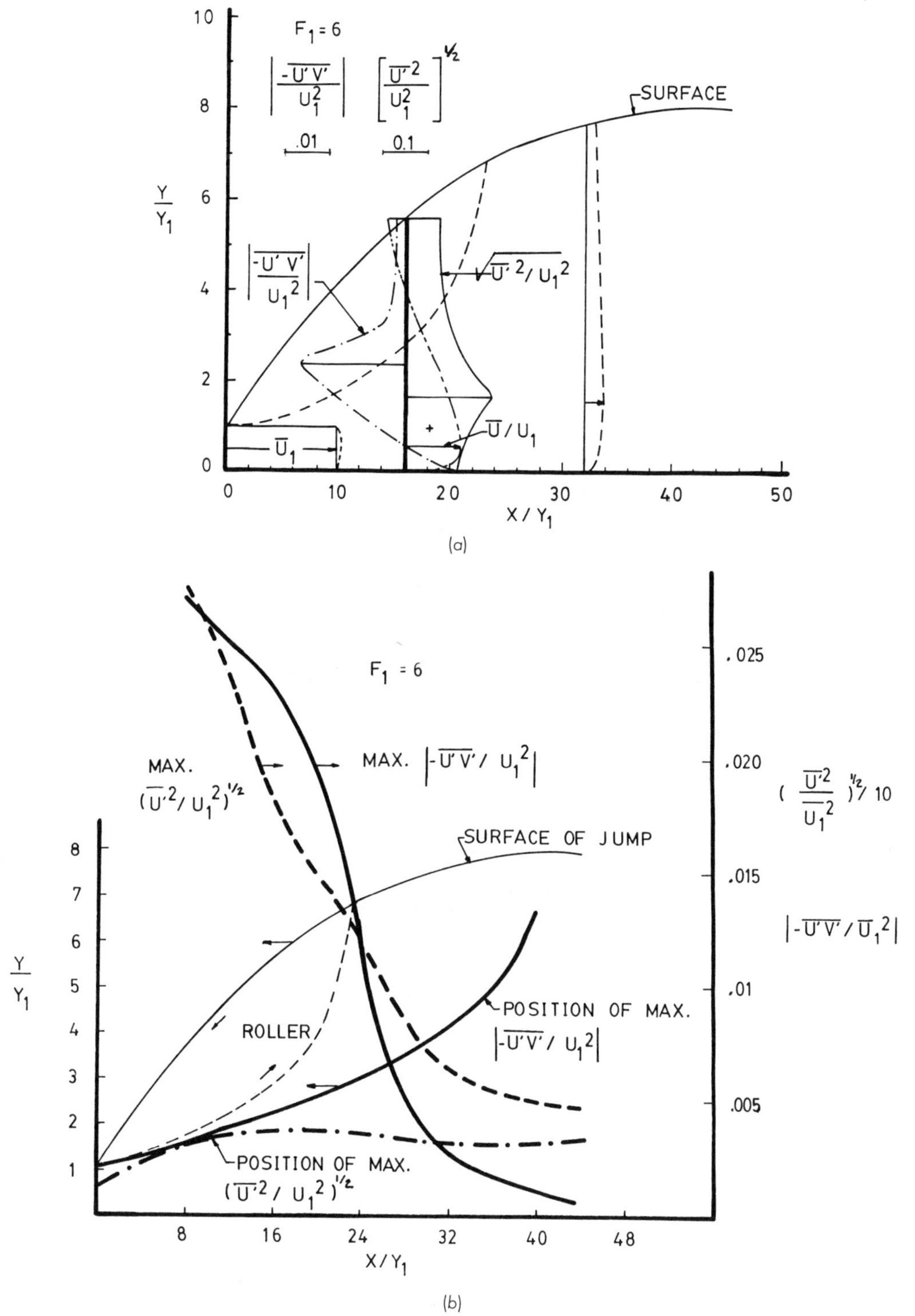

FIGURE 4. (a) Vertical distribution of velocity, turbulence intensity and turbulence shear stress at an $F_1 = 6$ [after Rouse et al. (1958)]; (b) longitudinal variation of the maximum velocity, turbulence intensity and turbulence shear stress at $F_1 = 6$ [after Rouse (1958)].

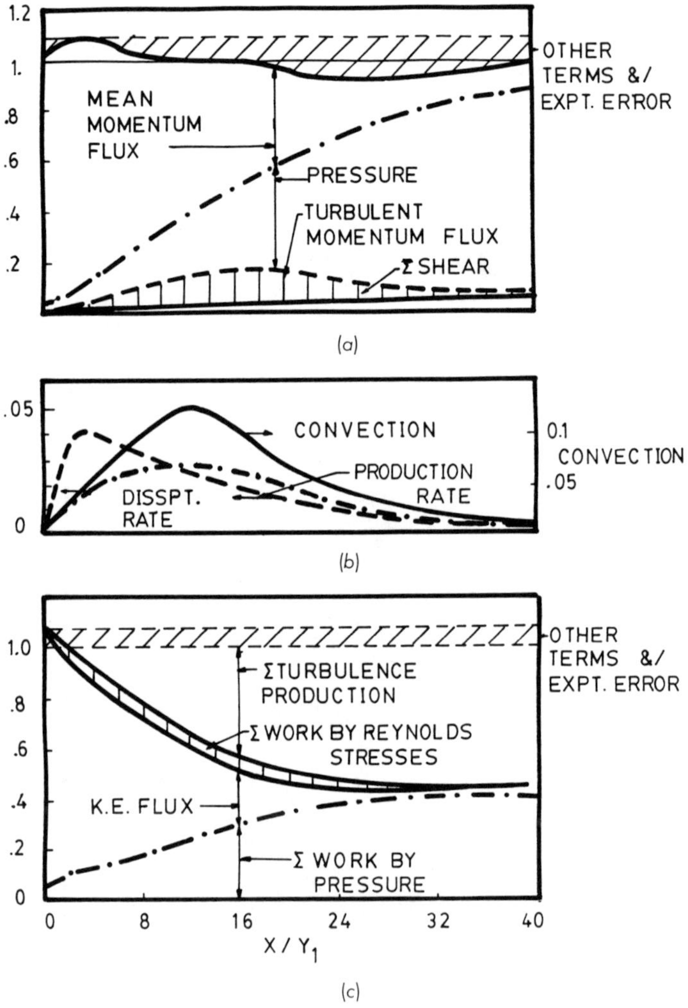

FIGURE 5. (a) Momentum balance for hydraulic jump at $F_1 = 6$; (b) convection, production and dissipation rates for turbulent kinetic energy; (c) the energy balance in a hydraulic jump at $F_1 = 6$ [after Rouse (1958)].

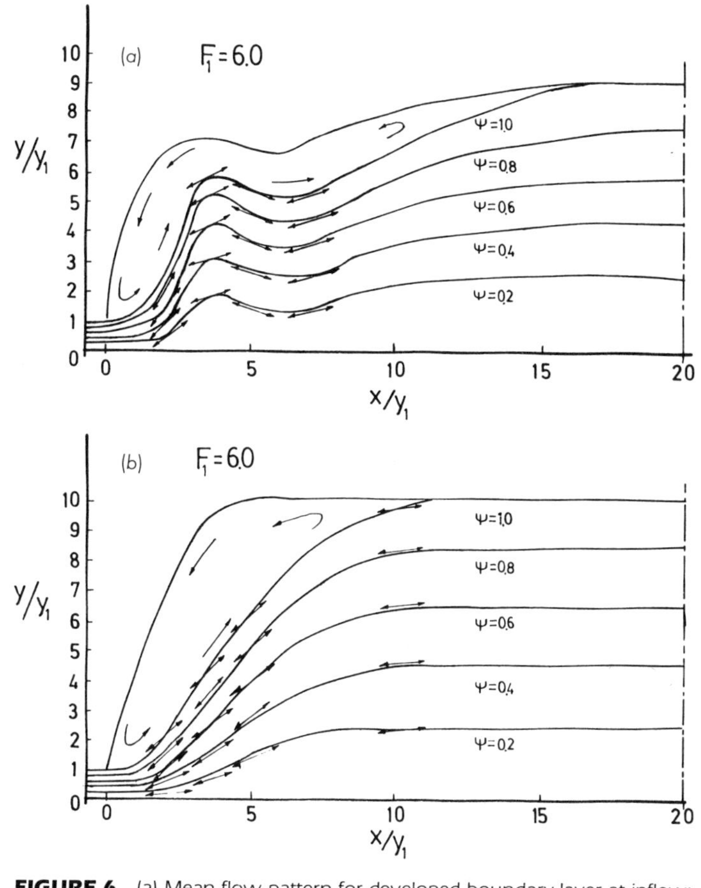

FIGURE 6. (a) Mean flow pattern for developed boundary layer at inflow; (b) mean flow pattern for undeveloped boundary layer at inflow [after Leutheusser and Kartha (1972)].

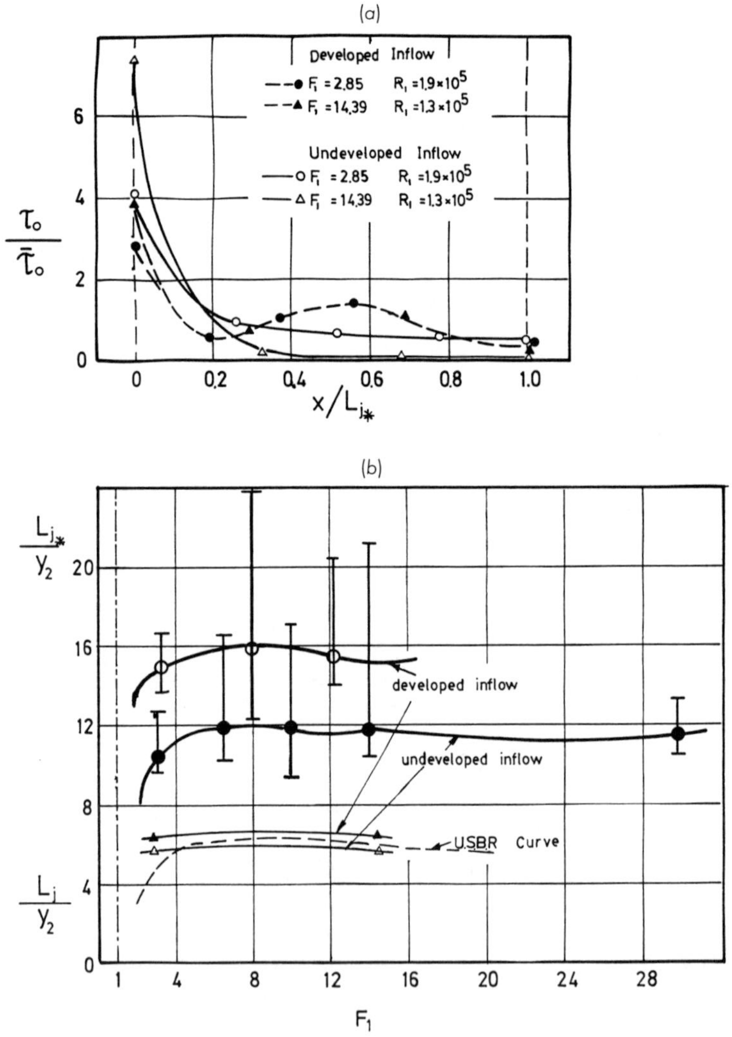

FIGURE 7. Effect of inflow boundary layer development on (a) boundary shear stress and (b) jump length (after Leutheusser and Kartha).

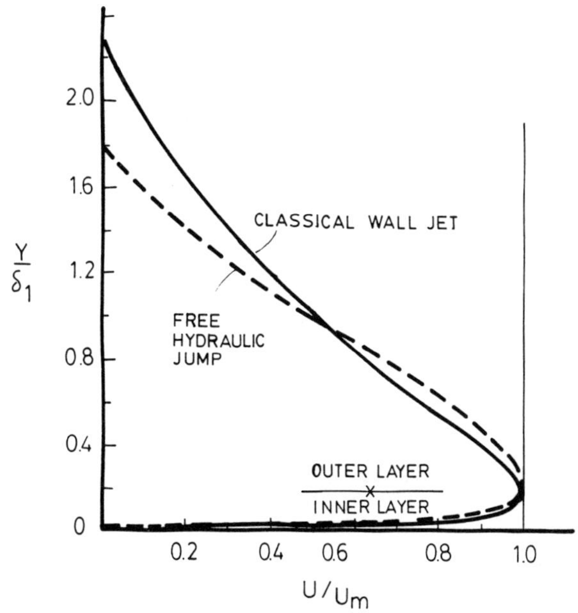

FIGURE 8. Comparison of the free jump velocity distribution with the wall jet distribution [after Rajaratnam (1967)].

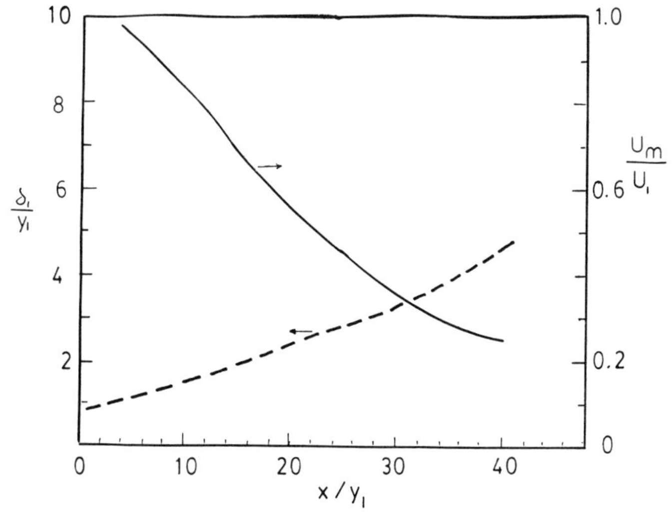

FIGURE 9. Longitudinal variation of the maximum velocity and boundary layer thickness of a hydraulic jump at $F_1 = 6$ [after Rajaratnam (1967)].

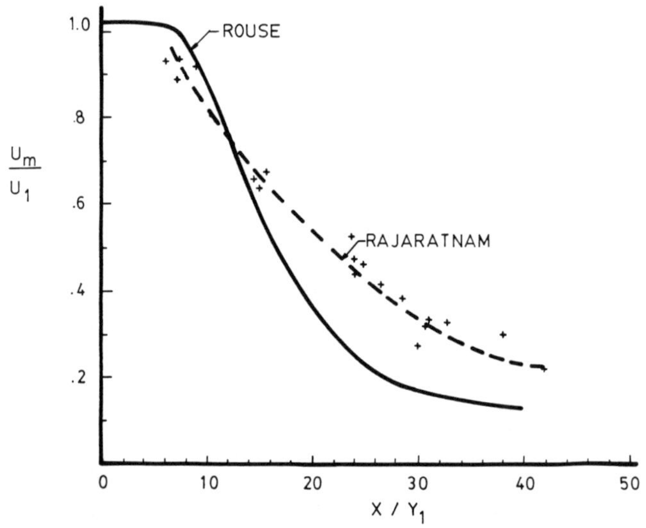

FIGURE 10. Longitudinal variation of the maximum velocity in a hydraulic jump at F_1 = 6 [after Rouse et al. (1958) and Rajaratnam (1967)].

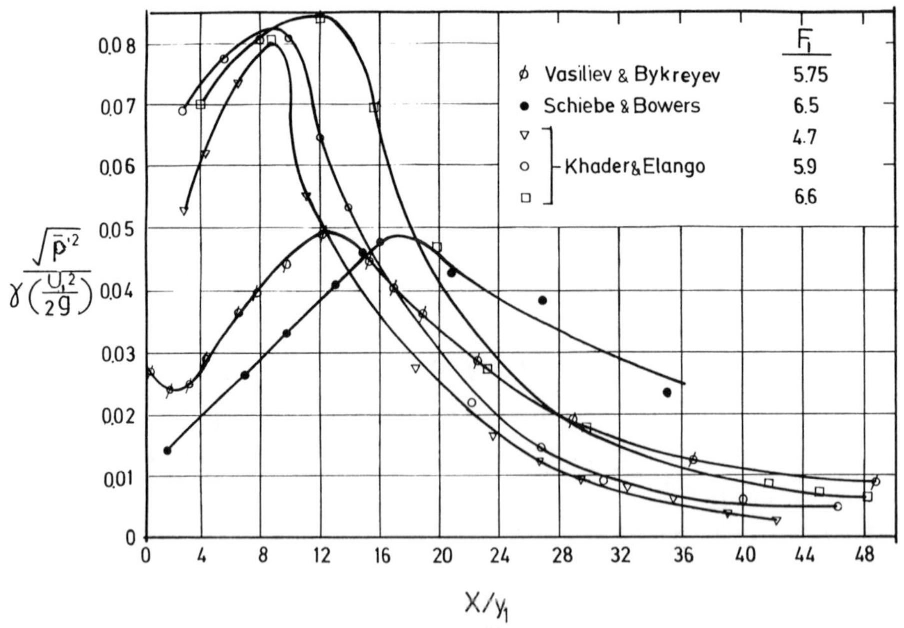

FIGURE 11. The rms pressure fluctuation for various F_1 along the bed of a free hydraulic jump.

TABLE 1. Turbulence Characteristics in Hydraulic Jumps Characteristic.

Reference and Condition	F_1	$\max\left(\frac{\overline{U'^2}}{U_1^2}\right)^{1/2}$	$\max\left\lvert\frac{\overline{u'v'}}{U_1^2}\right\rvert$	Location $\frac{x}{y_2} \cong$	Location $\frac{y}{y_2} \cong$
Rouse (air)	2	0.20		1	0.6
			0.024	1	0.6
	4	0.24		1	0.25
			0.023	1	0.25
	6	0.28		1	0.2
			0.028	1	0.2
Leutheusser (water undeveloped)	2.85	0.23		1	0.3
			0.016		0.36
	6	0.32		1	0.15
			0.026	1	0.15
Leutheusser (water developed)	2.85	0.38		1	0.8
			0.1	1	0.85
	6	0.29		>1	0.6
			0.035	1	0.28

Rajaratnam's (1967) velocity decay curve is compared with that given by Rouse *et al.* (1958) in Figure 10. The differences between the results cannot be explained by experimental error alone. Judging from the apparently longer potential core lengths in the Rouse study, it would appear that a more developed turbulent flow existed at the initial section in Rajaratnam's study.

DECAY OF TURBULENCE DOWNSTREAM FROM A STILLING BASIN

The decay of turbulence downstream from a stilling basin is important in the design of erosion protection. Lipay and Pustovoit (1967) studied the decay of turbulence downstream from a stilling basin. They summarized their results in an equation for u_x^*, the maximum instantaneous velocity near the bed:

$$\frac{u_x^*}{U_2} = 1.2 + \frac{0.2F_1}{1 + 0.07\left(\frac{x}{y_2}\right)^2} \tag{16}$$

It is noted that between five and ten sequent depths are required for substantial decay of the excess turbulent velocity components in the outflow from a stilling basin.

PRESSURE FLUCTUATIONS IN THE HYDRAULIC JUMP

Pressure fluctuations in the hydraulic jump have been studied by many researchers, for example, Vasiliev and Byrkreyeu (1967), Wisner (1967), Bourkov (1965), King (1967), Narayanan (1978, 1980), Narasimhan and Bharagava (1967) and Khader and Elango (1974). Many of these studies used pressure transducers mounted on the stilling basin floor. The fluctuating pressure can be characterized by its rms value. Figure 11 compares dimensionless rms pressure fluctuations along the bed for various values of F_1 as obtained by Vasiliev and Bykrevey (1967) Schiebe and Bowers (1971) and Khader and Elango (1974). Khader and Elango indicated that the reason for the large difference between their peak rms values and those of the other researchers was probably the fact that the inflow in their case had a more developed boundary layer. Figure 12 compares the dimensionless rms values for several researchers Khader and Elango (1974), Bourkov (1965), Vasiliev and Bykreyev (1967) and Wisner (1967) as a function of the distance relative to the end of the roller (L_r) where $L_r \cong 0.7L_j$. The peak dimensionless rms pressures of 0.05 to 0.082 occur at 30 to 40% of the distance to the end of the roller or at about 25% of the jump length.

Khader and Elango (1974) also studied the autocorrelation function for the bed pressure fluctuations at the location of the maximum rms pressure. Figure 13 shows the mean curve for their results and those of Vasiliev and Bykreyev (1967); the plot has been normalized as $R(\tau_p U_1/y_1)$ versus $\tau_p U_1/y_1$ where τ_p is the time lag. The $V_e \cong 0.1\ U_1$ was obtained.

Rajaratnam (1967) presented an experimental study on air entrainment in a classical hydraulic jump. His results in terms of depth averaged air fractions $\bar{\alpha}_a$ are shown in Figure 19. A peak air fraction $\bar{\alpha}_a$ of almost 20% occurred at $x \cong 0.17L_j \cong y_2$. The location of the maximum agrees with the peak horizontal air flux obtained by Babb *et al.* [Figure 18(b)]. The dimensionless air flux and the depth average air fractions cannot be directly compared since Babb *et al.* could not measure near the surface and the reverse flow

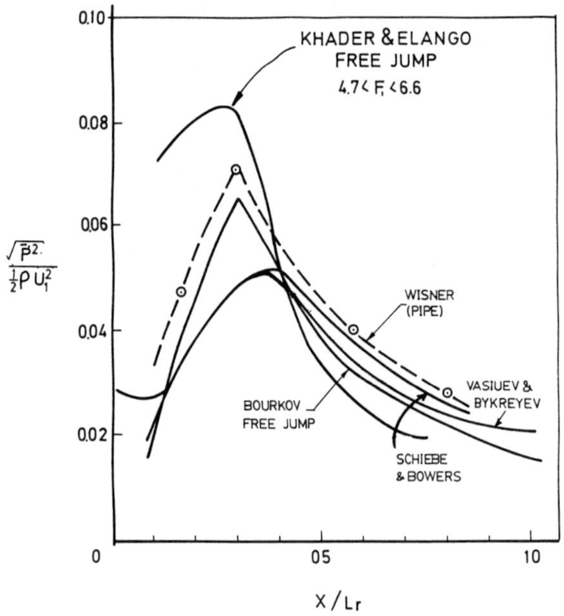

FIGURE 12. Typical rms pressures versus X/L_r at bed of hydraulic jumps.

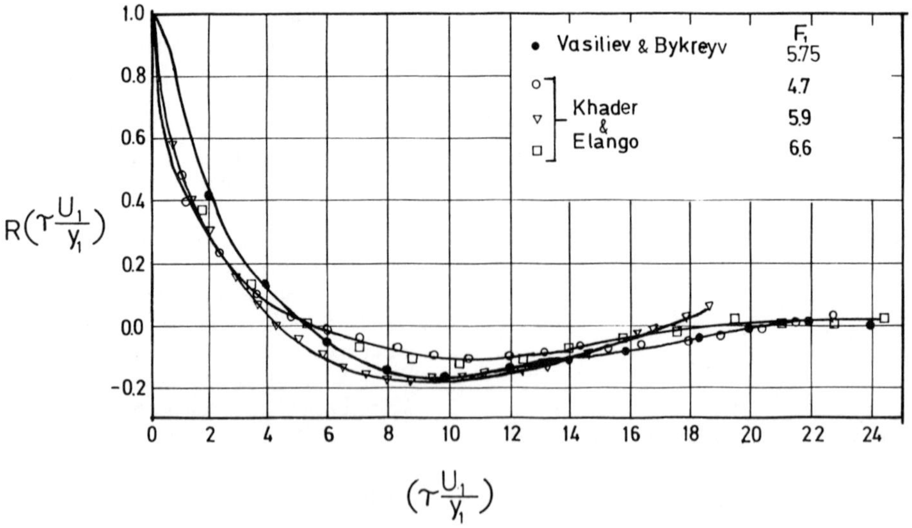

FIGURE 13. Auto-correlations of pressure fluctuations on the bed of a free hydraulic jump [after Khader and Elango (1974)].

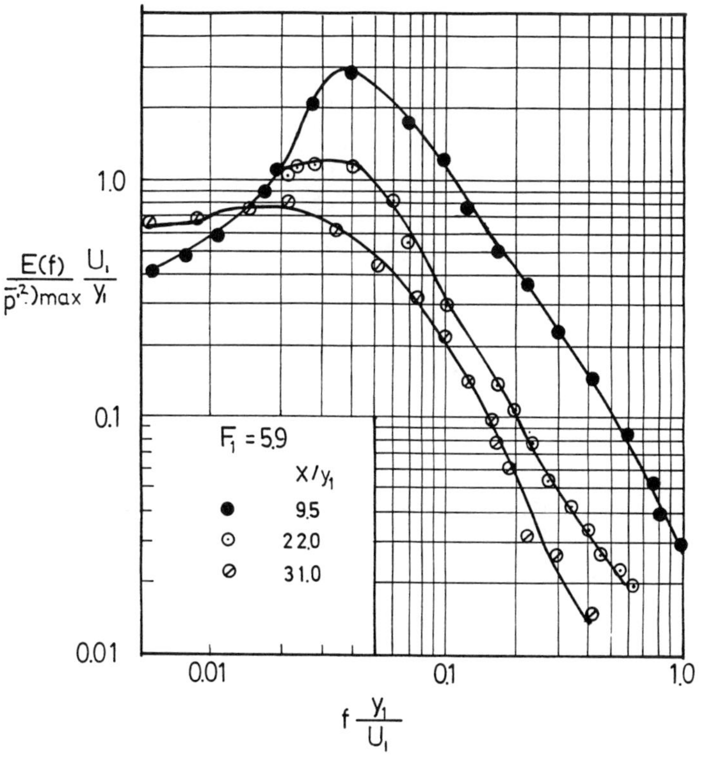

FIGURE 14. Spectra for fluctuating pressure for various position along the bed of a free hydraulic jump (after Khader and Elango).

would make the dimensionless flux less than the air fraction $\bar{\alpha}_a$.

Resch, Leutheussen and Alemu (1974) studied the entrainment of air in hydraulic jumps with undeveloped and fully developed boundary layers at the inflow. Their air fraction distribution for developed and undeveloped inflows at $F_1 = 2.85$ and 6 are shown in Figure 20(a) to (d). Their results for fully developed flow at $F_1 = 6$ are in good agreement with those of Babb *et al.* (Figure 16). The undeveloped inflow gave a bimodal vertical air distribution in the vicinity of $x = 2y_2$. It appeared that the undeveloped flow had a higher initial entrainment rate than the fully developed case; however, the air was released more quickly in the undeveloped case. It is possible that the bimodel air distribution causes an instability that results in the rapid upward air movement. Based on the data presented by Resch *et al.* the maximum depth averaged air concentration would be about 18% for undeveloped flow and 14% for developed flow.

Zero R value increases with x/y_1 as

$$(\tau U_1/y_1)_{R=0} \cong 0.6\ (x/y_1) \qquad (17)$$

Figure 14 shows the spectral density function $[E(f)]$ for pressure fluctuations at various x/y_1 locations and for $F_1 = 5.9$ (Khader and Elango) ordinate of Figure 14 has been made non-dimensional in the form $[E(f)/(\bar{p}'^2)_{max}]$ (U_1/y_1) while the frequency, f, has been reduced to the dimensionless form $f\,y_1/U_1$ where $(\bar{p}')^2_{max}$ is the maximum mean square value of the pressure fluctuations. Figure 15 shows the effect of F_1 on the dimensionless spectral density function at the maximum rms location. There is a tendency for the spectral density function to increase with increasing F_1. A peak value of $[E(f)/\bar{p}'^2](U_1/y_1) \cong 5$ occurs at $F_1 = 6.6$ and $f\,y_1/U_1 \cong 0.045$.

AIR ENTRAINMENT

Both free and forced unsubmerged hydraulic jumps entrain air. There is no reliable mathematical model for the prediction of air distribution in a hydraulic jump. However, several experimental studies have been reported in which air distribution was measured.

Babb and Aus (1981) studied the air entrainment effect in a free rectangular jump at $F_1 = 6$. They gave the air frac-

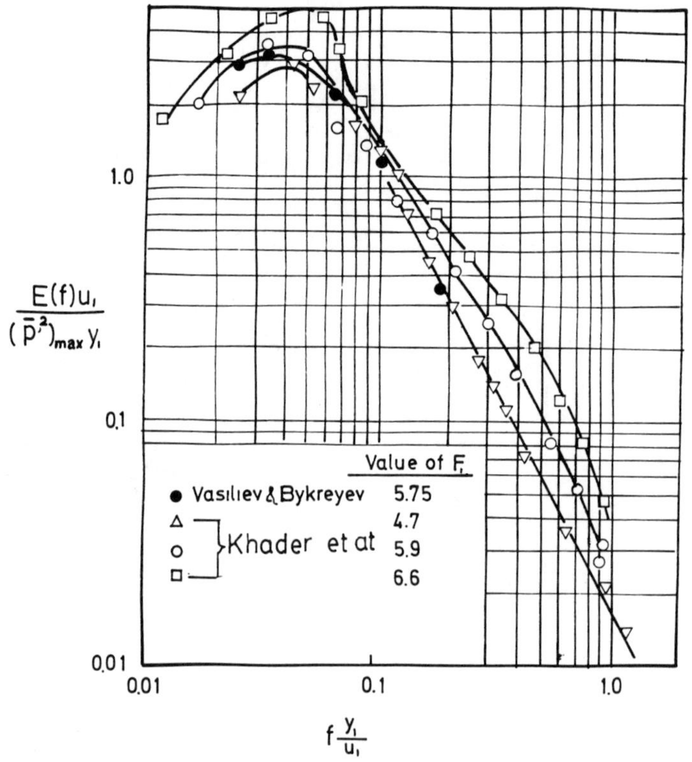

FIGURE 15. Spectra for fluctuating pressure at maximum rms location at the bed of a free hydraulic jump (after Khader and Elango).

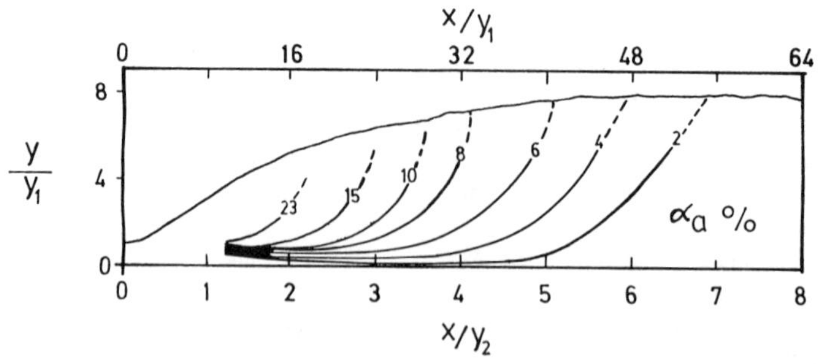

FIGURE 16. Distribution of air fraction, α_a, in the classical hydraulic jump [after Babb and Aus (1981)].

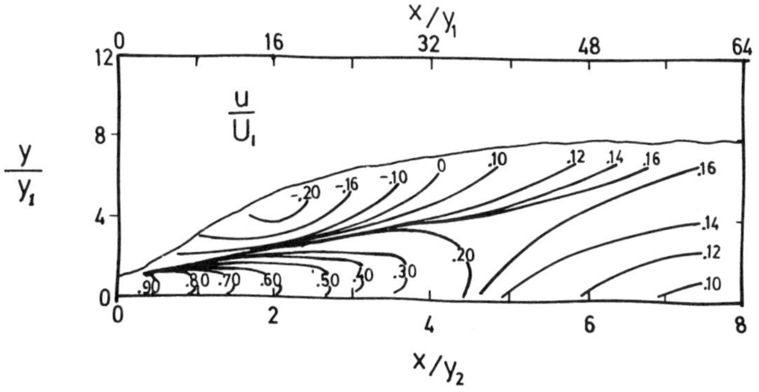

FIGURE 17. Distribution of velocity in the hydraulic jump as U/U_1 (after Babb and Aus).

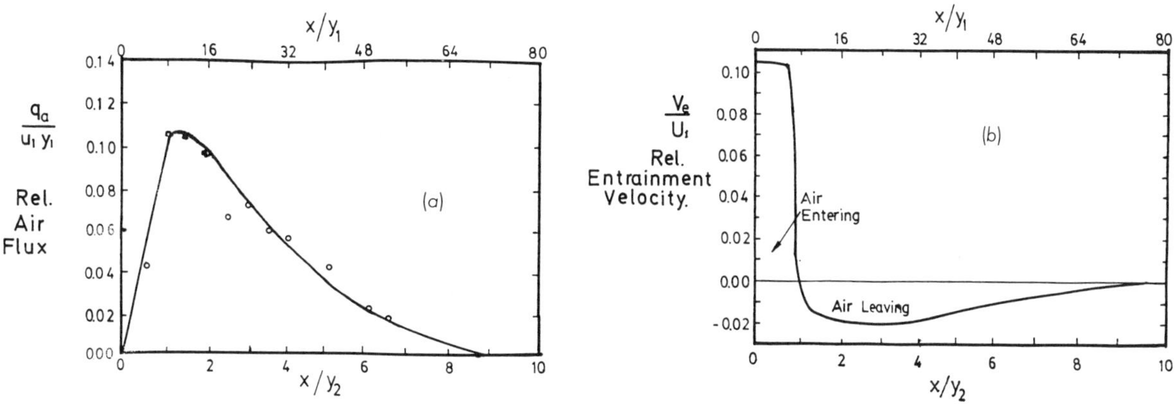

FIGURE 18. (a) Dimensionless air flux along the jump; (b) dimensionless entrainment velocity Ve/U_1 (after Babb and Aus).

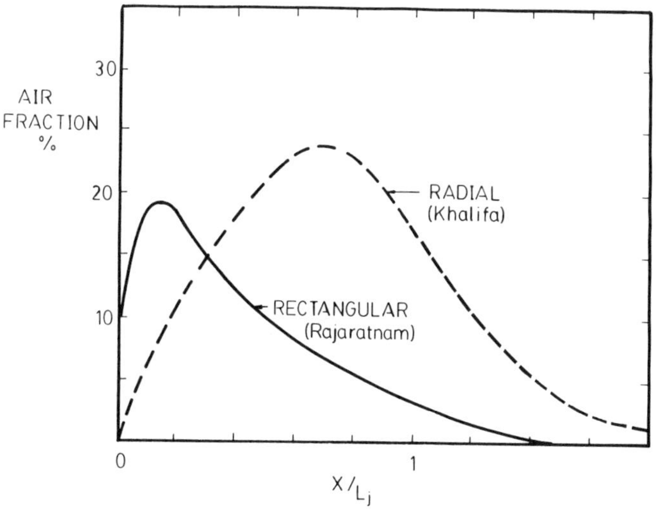

FIGURE 19. Longitudinal variation of air fraction in radial and rectangular hydraulic jumps [after Khalifa (1980) and Rajaratnam (1967)].

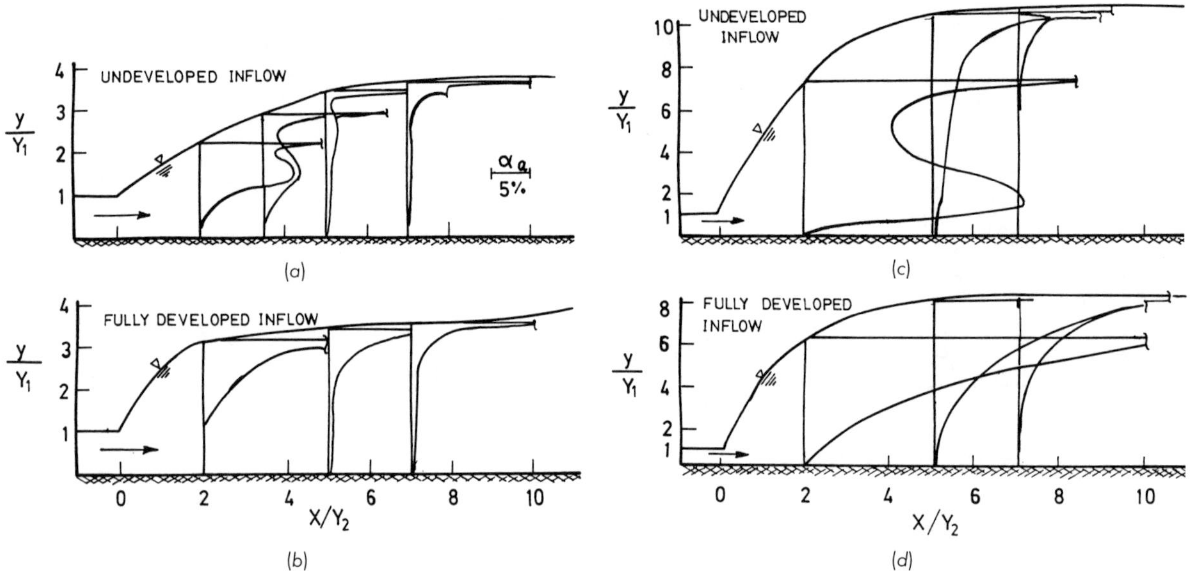

FIGURE 20. Air fraction distribution in the free hydraulic jump for: (a) F_1 – 2.85 with developed inflow; (b) F_1 – 2.85 with undeveloped inflow; (c) F_1 = 6 with developed inflow; (d) F_1 = 6 with undeveloped inflow [after Leutheusser et al. (1974)].

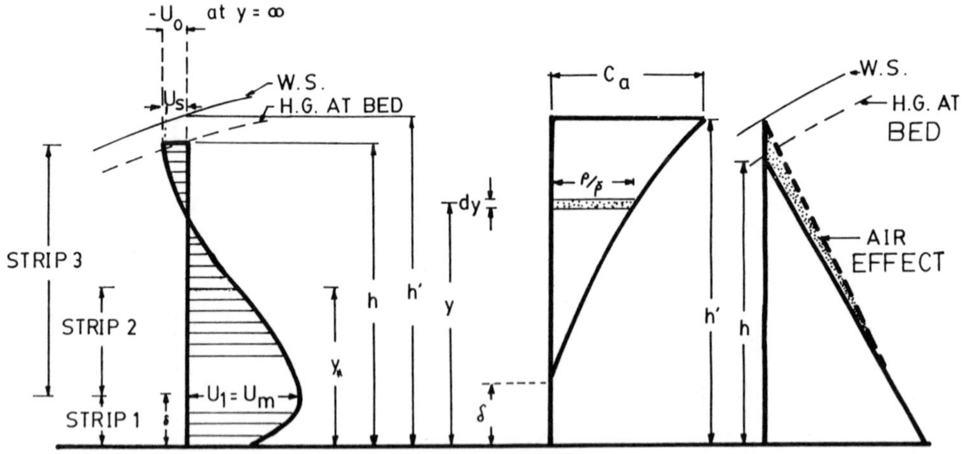

FIGURE 21. Definition sketch for strip integral method [after McCorquodale and Khalifa (1983)].

tion distribution shown in Figure 16. They also measured the velocity distribution in the jump (Figure 17) and computed the dimensionless air flux along the jump. Figure 18(a) shows the depth average horizontal air flux while Figure 18(b) shows the derived air entrance velocity for $x < y_2$ an entrainment velocity.

Caution should be used in applying air entrainment data based on small-scale models. These models are assumed to operate by the Froude Law but the air release mechanism is not properly modelled by this law. It can be expected that entrained air will persist for relatively greater longitudinal distances in the prototype stilling basins. Another complication in prototype structures is the greater potential for entrainment of air on the spillway chute. Uppal, Gulati and Sharma (1967) presented a comparison of the prototype and model jump profiles for the Bhakra Dam Spillway and Stilling Basin. The prototype depth at the middle of the hydraulic jump was increased by about 50% above the model prediction. This increase is probably caused by pre-entrained air.

An important characteristic of closed conduit flows is their capacity to remove air from the conduit. Kalinske and Robertson (1943) suggested the following air demand equation:

$$\frac{Q_a}{Q_w} = 0.0066(F_1 - 1)^{1.4} \tag{18}$$

where

Q_a = air discharge
Q_w = water discharge
$F_1 = U_1/\sqrt{gy_1}$

The U.S. Army Corps of Engineers (1964) suggested a design equation in the form

$$\frac{Q_a}{Q_w} = 0.03(F_1 - 1)^{1.06} \tag{19}$$

as the upper envelope to the empirical data.

Wisner (1965) suggested the following equations for prototype air demand:

$$\frac{Q_a}{Q_w} = .014(F_1 - 1)^{1.4} \text{ for } F_1 < 8 \tag{20a}$$

and

$$= 0.04(F_1 - 1)^{0.85} \text{ for } F_1 > 8 \tag{20b}$$

Sharma (1976) reviewed the existing research and after completing additional studies recommended that the F_1 in Equation (18) be replaced by the Froude number at the vena contracta of the gate in the tunnel.

A number of studies have been made to determine the oxygen transfer in a hydraulic jump. Avery and Novak (1978) cite a transfer efficiency of 0.91 to 0.16 kg O_2/kwh for a prototype hydraulic jump with an initial deficiency of 50%.

ANALYTICAL APPROACHES TO INTERNAL FLOW

The Strip Integral Method was applied by Narayanan (1975) to compute the flow patterns for the classical hydraulic jump. He used a power law relationship to represent the velocity distribution in the inner layer and a cosine function for the outer layer. McCorquodale and Khalifa (1983) refined the work of Narayanan by: (1) using the Gaussian velocity distribution suggested by Rajaratnam (1976); (2) incorporating the entrained air and its effect on the hydrostatic force and on the turbulent shear; (3) using the kinematic boundary condition at the water surface; (4) including the turbulence pressure; and (5) the centrifugal effects. A brief account of this method is presented here.

The Strip Integral Method is well established as a method of solving boundary layer problems (Moses, 1968). It has also been applied to other hydrodynamics and sedimentation problems in which a dominant flow direction can be identified. The method involves the selection of velocity shape functions with parameters which are functions of x (the dominant flow direction). These shape functions are inserted in the differential continuity and momentum equations which can then be partially integrated, over selected strips, to obtain a set of ordinary differential equations in the unknown parameters.

Based on the experimental observations of Rajaratnam (1965) and Nagaratnam (1957), the mean velocity distributions (Figure 21) in the hydraulic jump can be represented by:

$$\bar{u} = U_m\left(\frac{y}{\delta}\right)^{1/7} \qquad 0 \le y \le \delta \tag{21}$$

and

$$\bar{u} = u_o + u_t \exp\left(-4c[(y - \delta)/(h - \delta)]^2\right) \qquad \delta < y \le h \tag{22}$$

where $\bar{u}$ = horizontal velocity; $u_t = (\ddot{u}_m - u_o)$; $\ddot{u}_m$ = maximum horizontal velocity at x; δ = ordinate of the maximum velocity; u_o = horizontal velocity as $y \to \infty$; h = depth of flow; c = 0.693. The problem is to find U_m, δ, u_o and h which are functions of x.

The momentum equation [Equation (2)] is simplified to

$$\bar{u}\frac{\partial \bar{u}}{\partial x} + v\frac{\partial \bar{u}}{\partial y} = -\frac{1}{\varrho}\frac{\partial p_*}{\partial x} + \frac{1}{\varrho}\frac{\partial}{\partial y}(\tau_b + \tau_t) \tag{23}$$

and

$$p_* = \gamma'(h' - y) + \varrho(\overline{u'^2} - \overline{v'^2} + \overline{v_s'^2}) + \gamma c \quad (24)$$

where

γ' = specific weight of air–water mixture above y
h' = depth to the surface of the air–water mixture
τ_b = laminar shear
τ_t = turbulent shear
$\overline{v_s'^2} = \overline{v'^2}$ at $y = h$
c = pressure head due to curvilinear flow
$\cong [U_m^2(h - y)/g\bar{R}$ (25)
$\bar{R} \cong 200y_1$ (26)

The value of $(\overline{u'^2} - \overline{v'^2} + \overline{v_s'^2})$ is approximated by $\overline{u'^2}$.

The continuity equation[1]

$$\frac{\partial u}{\partial x} + \frac{\partial v}{\partial y} = 0 \quad (27)$$

is also required for the solution.

The Strip Integral Method (Narayanan, 1975) is used to reduce Equations (23) and (27) to a set of ordinary differential equations which describe the longitudinal variation of the selected characteristics, $\ddot{u}_m$, δ, h, u_o, of the hydraulic jump. Four differential equations are required to obtain these unknowns.

The integral continuity equation is

$$\int_o^h \frac{\partial v}{\partial y}\,dy = -\int_o^h \frac{\partial y}{\partial x}\,dy = u_s\frac{dh}{dx} \quad (28)$$

in which u_s = horizontal component of the surface velocity; $u_s(dh/dx)$ is an approximation to the vertical velocity component at the effective jump surface, $y = h$.

Three other equations are obtained by integrating Equation (23) over the three strips shown on Figure 21:

$$\int_o^\delta u\frac{\partial u}{\partial x}\,dy + \int_o^\delta v\frac{\partial u}{\partial y}\,dy = -\frac{1}{\varrho}\int_o^\delta \frac{\partial p_*}{\partial x}\,dy + \frac{1}{\varrho}\int_o^\delta \frac{\partial \tau}{\partial y}\,dy \quad (29)$$

$$\int_\delta^{y_*} u\frac{\partial u}{\partial x}\,dy + \int_\delta^{y_*} v\frac{\partial u}{\partial y}\,dy = -\frac{1}{\varrho}\int_\delta^{y_*} \frac{\partial p*}{\partial x}\,dy + \frac{1}{\varrho}\int_\delta^{y_*} \frac{\partial \tau}{\partial y}\,dy \quad (30)$$

$$\int_\delta^h u\frac{\partial u}{\partial x}\,dy + \int_\delta^h v\frac{\partial u}{\partial y}\,dy = -\frac{1}{\varrho}\int_\delta^h \frac{\partial p*}{\partial x}\,dy + \frac{1}{\varrho}\int_\delta^h \frac{\partial \tau}{\partial y}\,dy \quad (31)$$

in which

$$y* = \frac{1}{\sqrt{8C}}(h - \delta) + \delta \quad (32)$$

and

$$\tau = \tau_b + \tau_t \quad (33)$$

$y*$ is the ordinate at which maximum turbulent shear occurs.

Equations (29) to (31) introduce bed and turbulent shear at $y*$. Zero shear is assumed at $y = h$. An expression suggested by Rajaratnam (1965) was used for shear stress at the bed:

$$\tau_b = \frac{.0424}{\left(\frac{u_m\delta}{\nu}\right)^{.25}}\varrho\frac{U_m^2}{2} \qquad y = 0 \quad (34)$$

The Prandtl mixing length concept along with the free jet theory and Equation (22) were used to develop the following expression for the maximum turbulent shear:

$$\tau_t = \varrho D_*^2 C\frac{u_t^2}{e} \qquad y = y* \quad (35)$$

where $D* = 0.11$ which was obtained from the experimental air model data of Rouse (1958) and agrees with the free jet theory6 (Rajaratnam, 1976).

The values of the average air fraction C_o along the jump can be estimated from the experimental data of Rajaratnam (1967) which can be approximated by

$$C_o = 0.066F_1^{1.35}\frac{x}{L_j} \qquad x < 8y_1 \quad (36)$$

$$C_o = 0.0115F_1^{1.35}\left(1 - \frac{x}{1.6}\right) \qquad x \geq 8y_1 \quad (37)$$

[1]The bar on u and v have been omitted hereafter for simplicity.

The experimental work of Rouse (1958), Nagaratnam (1957), Narayanan (1978), Narasimhan and Bharagava (1976) and Resch *et al.* (1971, 1972) shows that $\overline{u'^2}$ increase rapidly to a maximum at about $x \cong 12y_1$ and thereafter decays. The following relationships are suggested for the turbulence "pressure" effect [Le Méhauté (1976)]:

$$\varrho \overline{u'^2} = \varrho\, \alpha \frac{u_1^2}{2}\left[K_1 + \frac{x(K_3 - K_1)}{12y_1}\right] \quad \text{for } x \le 12y_1 \tag{38}$$

$$\varrho \overline{u'^2} = \varrho\, \alpha \frac{u_1^2}{2}\left[\frac{12y_1}{x}\right]^n K_1 \text{ for } x > 12y_1 \tag{39}$$

in which $K_1 = .0067$; $K_3 = .0174$; $\alpha = 1.0$; $n = 1.0$.

A set of four differential equations are obtained by substituting the velocities from Equations (21) and (22) and the shear functions from Equations (34) and (35) into the integral continuity and momentum equations, Equations (28) to (31); these equations can be expressed in the form of:

$$A_i \frac{dh}{dx} + B_i \frac{d\delta}{dx} + C_i \frac{du_o}{dx} + D_i \frac{dU_m}{dx} = E_i \tag{40}$$

$$(i = 1,2,3,4)$$

Inverting Equation (40) gives:

$$\begin{Bmatrix} \frac{dh}{dx} \\ \frac{d\delta}{dx} \\ \frac{du_o}{dx} \\ \frac{du_m}{dx} \end{Bmatrix} = \begin{bmatrix} A_1 & B_1 & C_1 & D_1 \\ A_2 & B_2 & C_2 & D_2 \\ A_3 & B_3 & C_3 & D_3 \\ A_4 & B_4 & C_4 & D_4 \end{bmatrix}^{-1} \begin{Bmatrix} E_1 \\ E_2 \\ E_3 \\ E_4 \end{Bmatrix} \tag{41}$$

The coefficient matrix $[A]$ and vector $\{E\}$ are functions of h, δ, u_o and U_m (McCorquodale and Khalifa, 1983).

A correction to the mixing length was made for the density gradient to the entrained air, viz.

$$\ell_p = \frac{\ell_m}{\sqrt{1 + \beta_* R_i}} \tag{42}$$

where R_i = the Richardson Number and $\beta_* \lesssim 160$ (Odd and Rodger, 1978).

The bulking effect of the entrained air was also included where the air concentrations and distributions were estimated from the work of Rajaratnam (1967, 1965) and Resch *et al.* (1974).

The solution was started with initial conditions at the end of the potential flow, $x \cong 4y_1$. The macroscopic continuity and momentum equations were applied with the assumption that $\delta \cong y_1/2$ and $U_m \approx U_1$ at $x = 4y_1$ and the requirement that the energy loss > 0.

Figure 22 shows that the jump lengths of the SIM model and those given by Rajaratnam are very close; however, the predicted lengths are slightly lower than the USBR (1964) for $F_1 = 3$ to 5 and slightly higher for $F_1 > 5$.

The model was run with and without the air entrainment effect. Except for the bulking effect of the air, the model indicated that air plays a minor role in determining the shape of the jump.

Figure 23 illustrates the effect of the turbulence "pressure" on the hydraulic jump at F_1 of 4 and 9.05. The model was run with and without the turbulence "pressure" term. Turbulence pressure had a much greater effect at $F_1 = 9.05$ than at $F_1 = 4$. The general effect was to lower the water surface near the beginning of the jump, although the length of the jump was not significantly changed. Equation (38) shows that the turbulence "pressure" relative to the hydrostatic pressure depends on F_1^2.

The relative importance of the centrifugal force also increases as F_1^2.

Figure 24 shows a reasonable agreement between the theoretical velocity distribution and the measured velocity distribution of Rouse *et al.* (1958). Figure 25 shows good agreement between predicted surface velocity and that of Rouse *et al.* while predicted maximum velocity falls between the measured value of Rouse *et al.* (1958) and Rajaratnam (1967, 1965).

Non-Classical Hydraulic Jumps

Most of the internal flow studies have been restricted to the classical hydraulic jump, i.e., the free jump with parallel side walls and a horizontal bed. This section summarized a number of studies that have been presented on other types of jumps, namely the internal jump, the submerged jump, the radial jump, jumps at abrupt drops, forced hydraulic jumps and jumps on sloping beds.

THE INTERNAL HYDRAULIC JUMP

Internal hydraulic jumps involve two fluids in the same state, e.g., hot and cold water. This type of jump has been noted in several environmental engineering problems. The writer has observed this phenomenon in hydraulic models of primary clarifiers that have been subjected to diurnal variations in influent temperatures; alternating unsteady bottom and surface jets and jumps have been observed. The internal jump may also occur at wastewater diffusers operating under stagnant ambient conditions [Koh (1971)].

Many experimental studies have been carried out on both bottom and surface internal jumps and jets, e.g., Chu

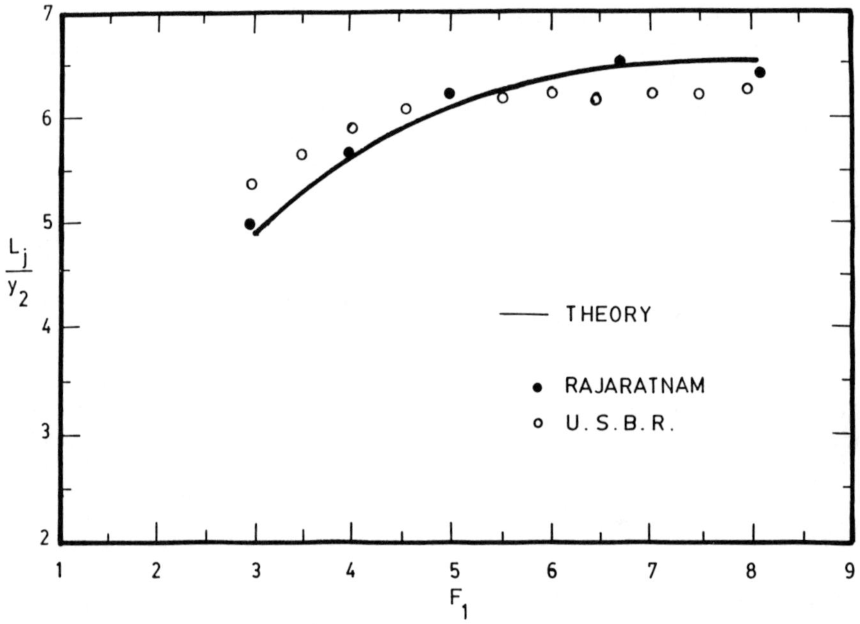

FIGURE 22. Comparison of computed and experimental jump lengths [after McCorquodale (1985)].

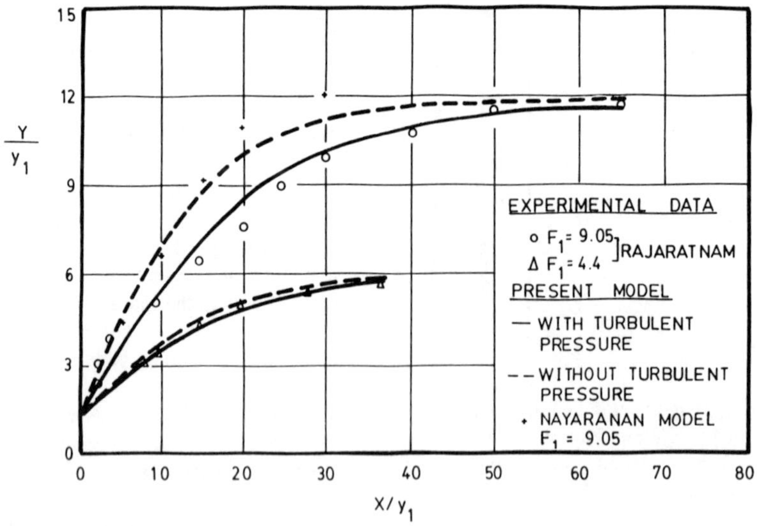

FIGURE 23. Effect of turbulence pressure on the profile of a hydraulic jump [after Mc Corquodale et al. (1983)].

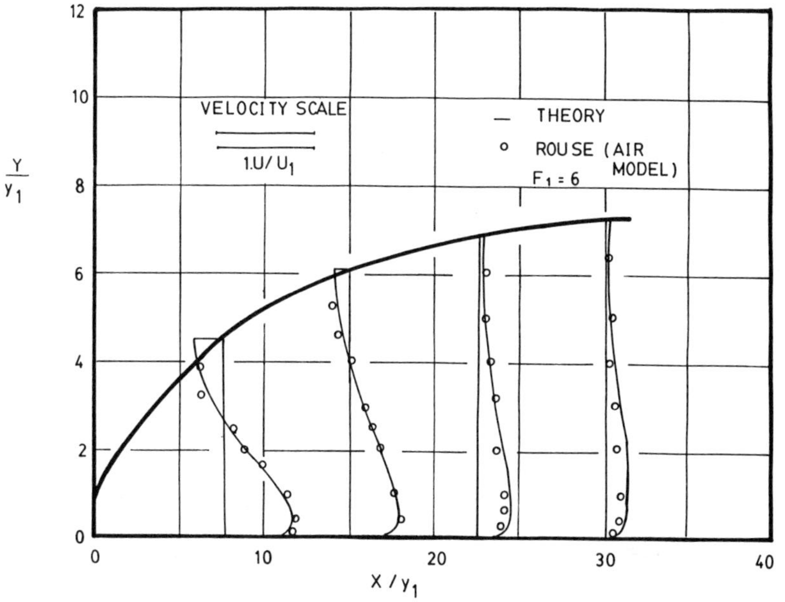

FIGURE 24. Comparison of the SIM computed velocity distributions with those measured by Rouse et al. [after McCorquodale et al. (1983)].

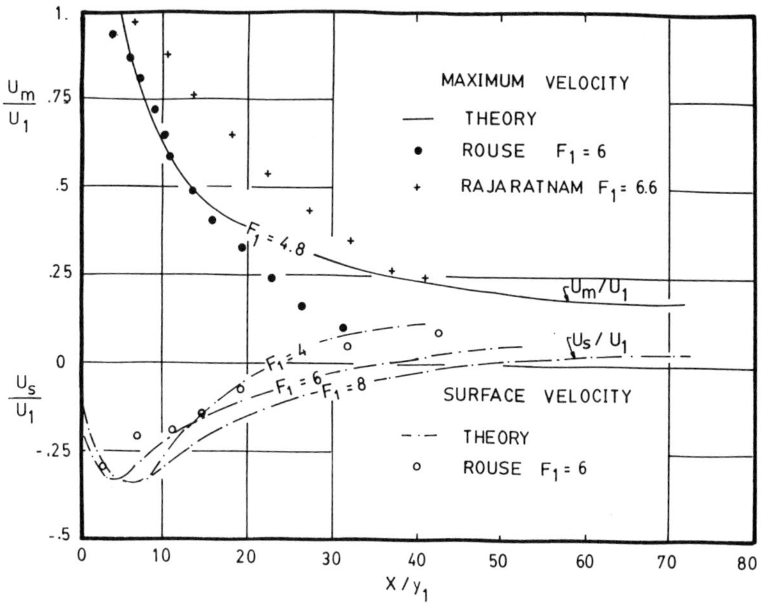

FIGURE 25. Comparison of predicted and measured surface and maximum velocities along the classical hydraulic jump [after McCorquodale and Khalifa (1983)].

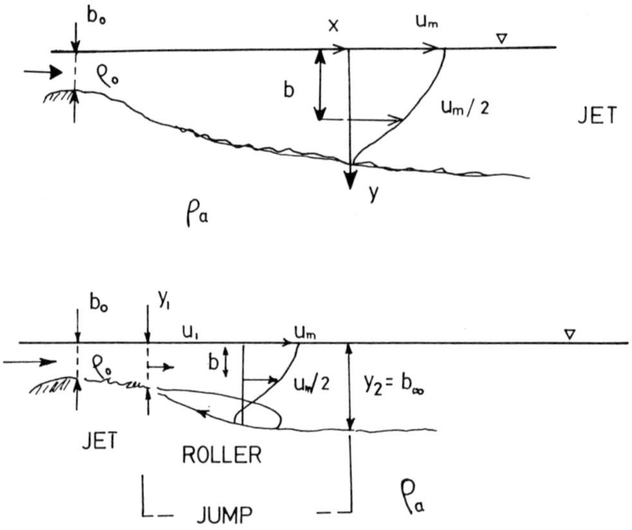

FIGURE 26. Definition of interval hydraulic jumps and buoyant surface jets.

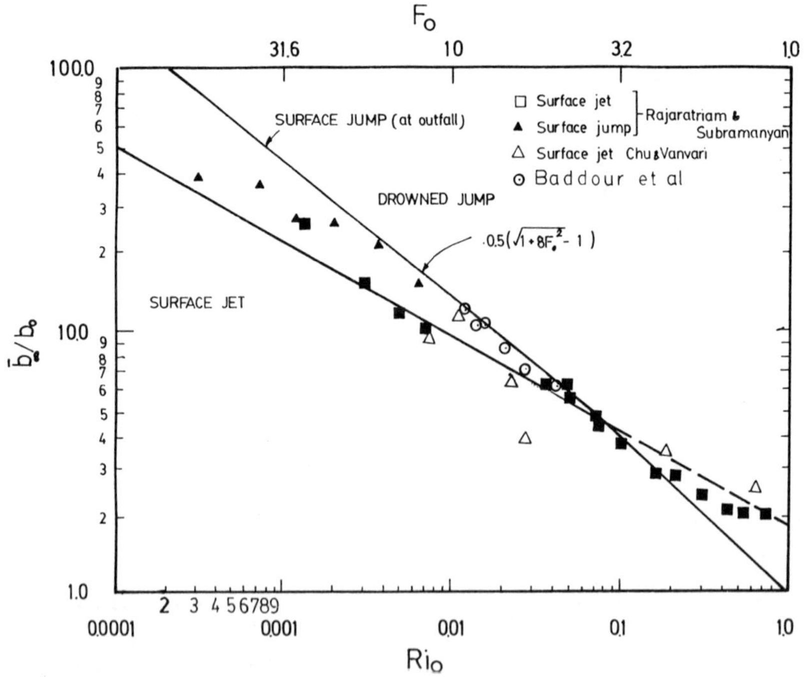

FIGURE 27. Segment depth ratio for the internal hydraulic jump [after Rajaratnam (1985)].

and Baddour (1984), Baddour and Abbink (1983), Hayakawa (1971), Yih and Guha (1955) and Rajaratnam and Subramanyan (1985). Most of the studies have dealt with the macroscopic characteristics of the internal jump.

Figure 26 defines some of the important variables in internal jumps and buoyant jets. One feature that distinguishes the jet from the jump is the reserve flow (roller) associated with the jump. The jet has approximately constant entrainment throughout the length while the jump has an initial zone of entrainment followed by a large zone of low or no entrainment.

For turbulent flows it has been shown that the classical sequent depth ratio equation is applicable to internal jumps, i.e.,

$$y_o = \frac{1}{2}(\sqrt{1 + F_1^2} - 1) \tag{44}$$

where

$$F_o = 1/\sqrt{R_{io}} = U_1/\sqrt{gy_1|\Delta\varrho_o/\varrho_o|} \tag{45}$$

= source densimetric

R = Froude number

R_{io} = source Richardson number

ϱ_o = source density

$\Delta\varrho_o = (\varrho_o - \varrho_a)$

ϱ_a = density of the stagnant fluid

In Figure 27 the data of Rajaratnam *et al.* (1985), Chu and Vanvari (1976) and Baddour and Abbink (1983) indicate the validity of Equation (44).

Rajaratnam *et al.* (1985) studied the velocity distribuiton in a surface internal jump. They found that the vertical distribution of the forward horizontal velocity, u, can be approximated by

$$\frac{u}{U_m} = \exp\left[-0.693\left(\frac{y}{b}\right)^2\right] \tag{46}$$

where

U_m = the maximum horizontal velocity as shown in Figure 28(a)

b = velocity length scale as shown in Figure 28(b)

Equation (46) applies beyond the potential core which typically extends from $x = 0$ to $4\ b_o$ and for $y \leq 2b$. Chu and Baddour (1984) obtained similar distributions for surface buoyant jets.

THE SUBMERGED HYDRAULIC JUMP

The submerged jump is usually encountered downstream from a control gate. The mean and turbulent nature of the submerged jump is important in determining the discharge from the gate as well as the hydraulic loading on the gate. This section will first consider submerged jumps in stilling basins with parallel side wall (rectangular) and then basins with diverging side walls (radial).

A typical submerged jump is illustrated in Figure 29. Chow (1959) used the momentum and continuity equations to obtain an equation for the depth, y_s, at the control gate:

$$y_s/y_{2s} = [1 + 2F_{2s}^2(1 - y_{2s}/y_1)]^{1/2} \tag{47}$$

where

y_1 = initial jet depth

y_{2s} = tailwater depth

$F_{2s} = U_{2s}/\sqrt{gy_{2s}}$

Govinda and Rao *et al.* (1963) obtained the following empirical equation for the submerged jump length:

$$L_j/y_2 = 4.9(y_{2s}/y_2 - 1) + 6.1 \tag{48}$$

where y_2 is the free jump depth corresponding to y_1 and F_1.

Rajaratnam (1967) showed that the velocity distribution in the rectangular submerged jump has the form of a wall jet in the inner layer and a Gaussian form similar to the free jump in the outer layer. Figure 30 compares the longitudinal variation of U_m and the velocity scale δ_1 for the free and submerged rectangular jumps as well as the classical wall jet.

An extensive study of the submerged jump was made by Narasimhan and Bhargava (1976); their velocity decay and velocity length scales are shown in Figures 31(a) and 31(b), respectively. They found δ_1 nearly independent of F_1 and submergence while U_m/U_1 for a given x/y_1 increased with increasing submergence for y_{2s}/y_1 up to 12. Figure 32 compares the results of Rajaratnam (1967), Narasimhan *et al.* (1976) and Narayanan (1978) for $y_{2s}/y_1 \geq 12$.

Narayanan (1978) and Narasimhan *et al.* (1976) measured the vertical distribution of the horizontal velocity, $\bar{u}$, at various positions along the submerged jump. Their results are compared in Figure 33. At high submergence the submerged jump distribution approaches that of the classical wall jet.

Narasimhan and Bhargava (1976) measured the rms pressure fluctuations along the bed of a submerged jump as shown in Figure 34. Narayanan (1978) carried out a similar study; his results are compared with the free jump results of Vaseliev (1967) in Figure 35. Narayanan also presented the dimensionless frequency spectra for pressure fluctuations at three x/y_1 values.

Narayanan (1978) used the linear (inviscid) Rayleigh instability theory to predict the dominant frequency of the pressure fluctuations. His measurements and theory are compared in Figure 36.

The bed shear in the free and submerged jump [Rajaratnam (1965)] can be estimated by

$$C_f = \tau_o/\varrho U_m^2 = 0.055/(U_m\delta/\nu)^{0.25} \tag{49}$$

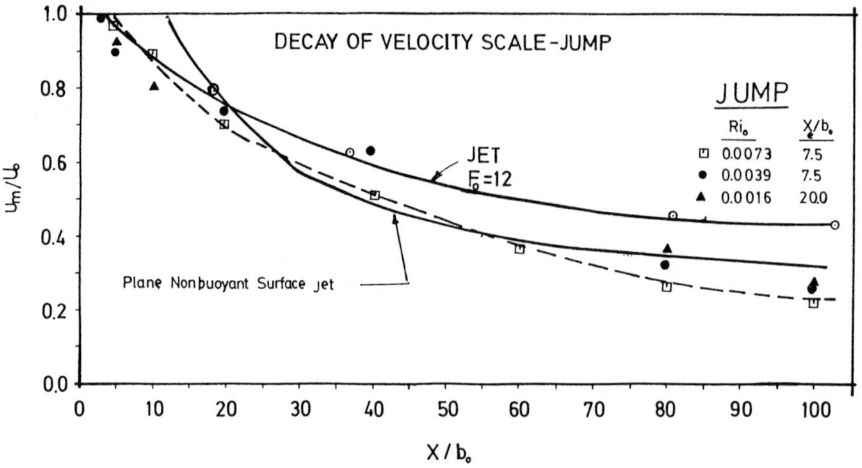

(a)

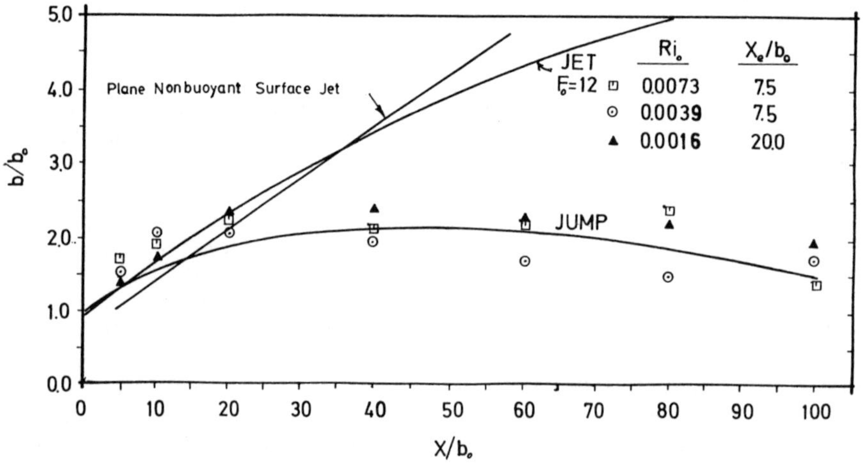

(b)

FIGURE 28. (a) Decay of U_m in the internal jump and the buoyant surface jet; (b) velocity length scales, b, in the internal jump and surface buoyant jet.

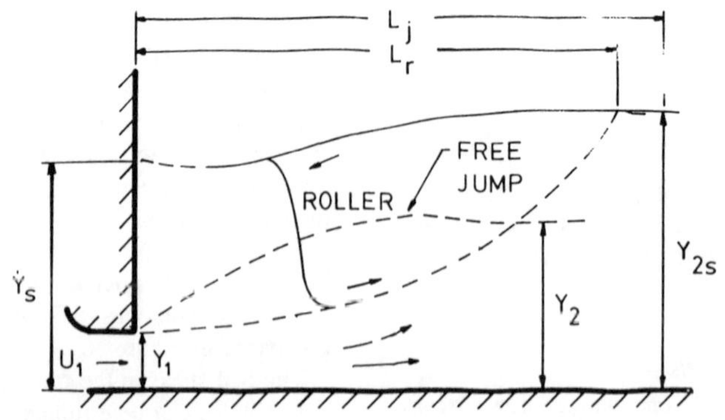

FIGURE 29. Definition sketch of submerged hydraulic jump.

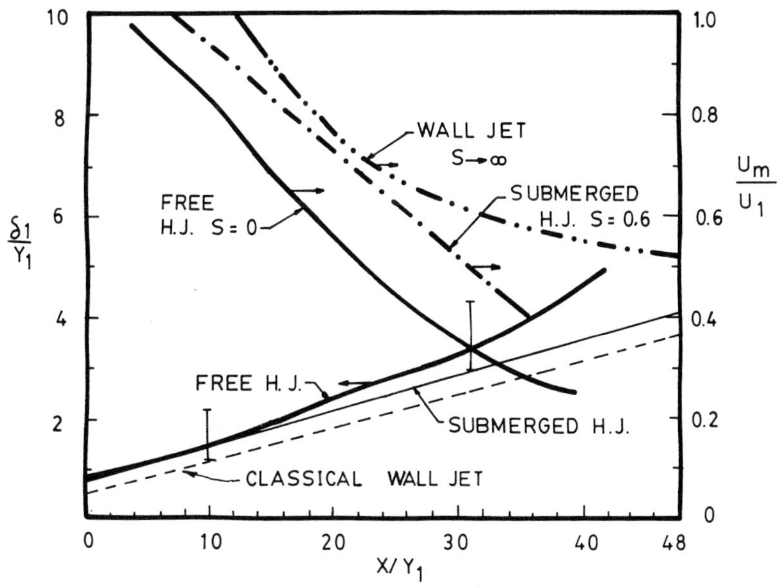

FIGURE 30. Longitudinal variation of U_m and δ_1 for the submerged jump. Note: $S = (Y_{2s}/Y_2 - 1)$ [after Rajaratnam (1967)].

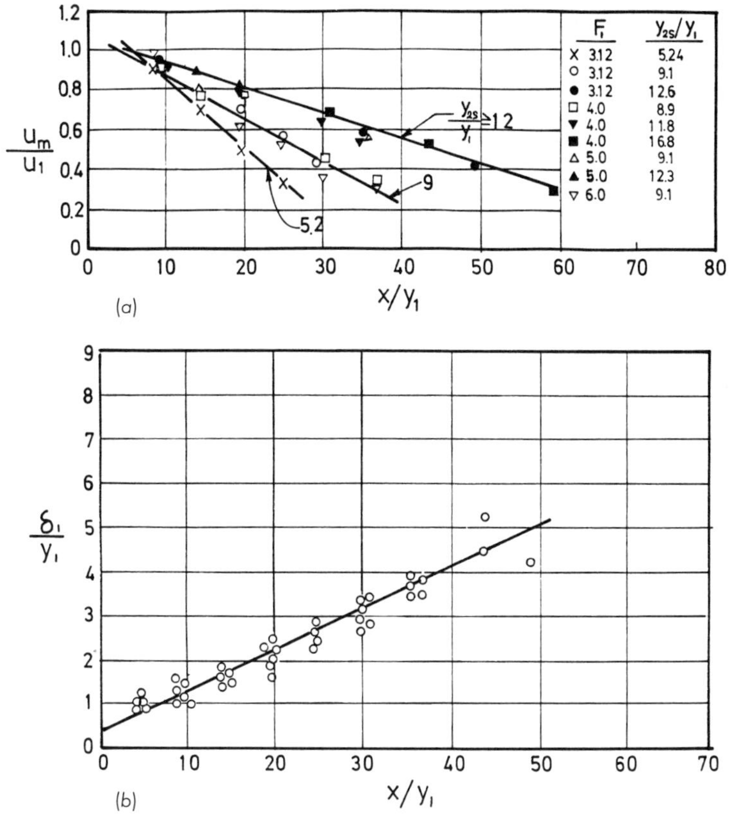

FIGURE 31. Longitudinal variation of (a) U_m and (b) δ_1 for a submerged jump [after Narasimhan and Bhargava (1976)].

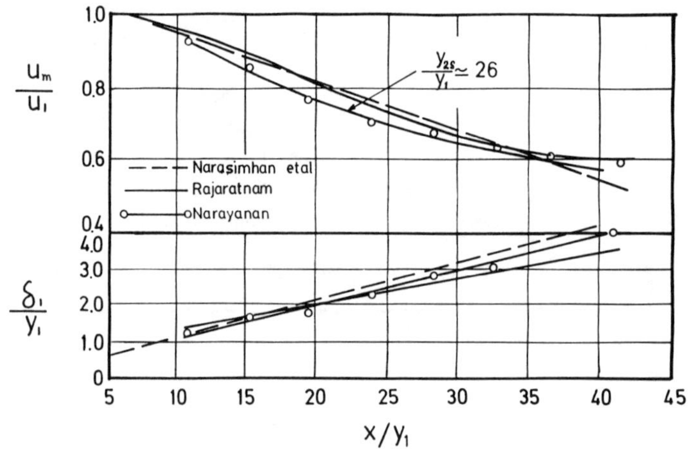

FIGURE 32. Comparison of the velocity decay and scale data of Rajaratnam (1967), Narasimham et al. (1976) for the rectangular submerged jump for $Y_{2s}/y_1 > 12$.

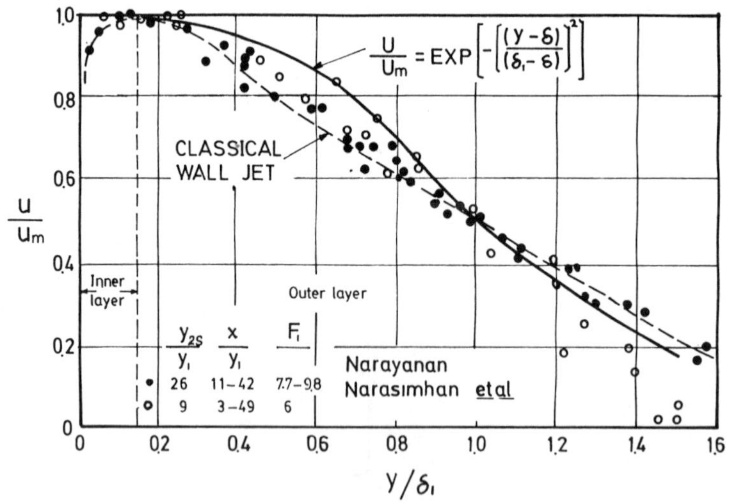

FIGURE 33. Vertical distribution of the horizontal velocity in the submerged jump.

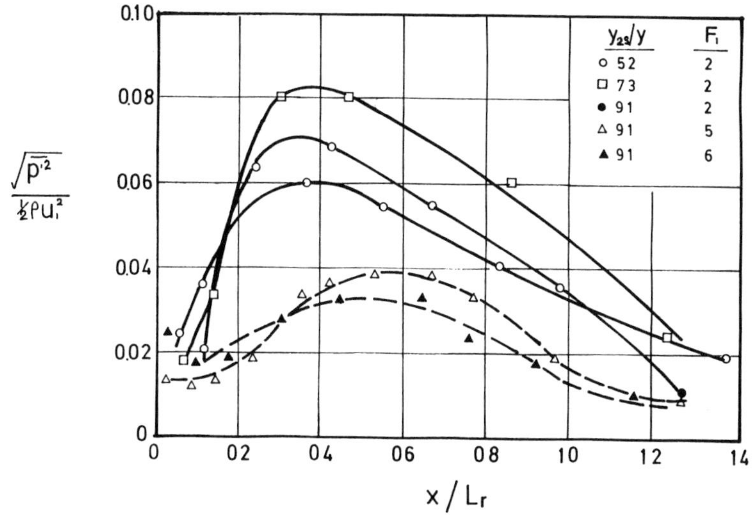

FIGURE 34. The rms pressures along the bed of a submerged hydraulic jump [after Narasimhan et al. (1976)].

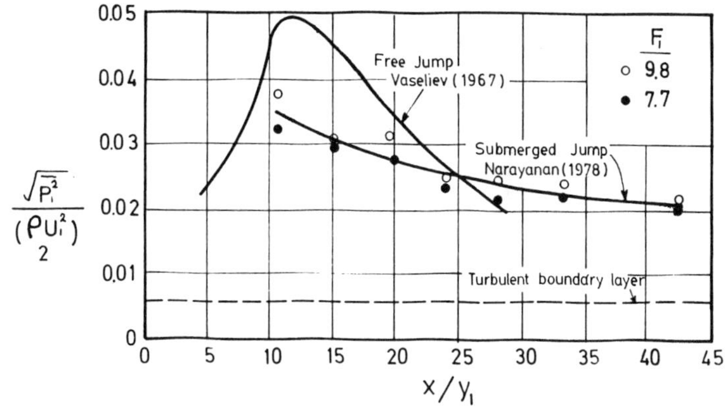

FIGURE 35. Comparison of the rms pressures along the bed of submerged and free jumps [after Narayanan (1978)].

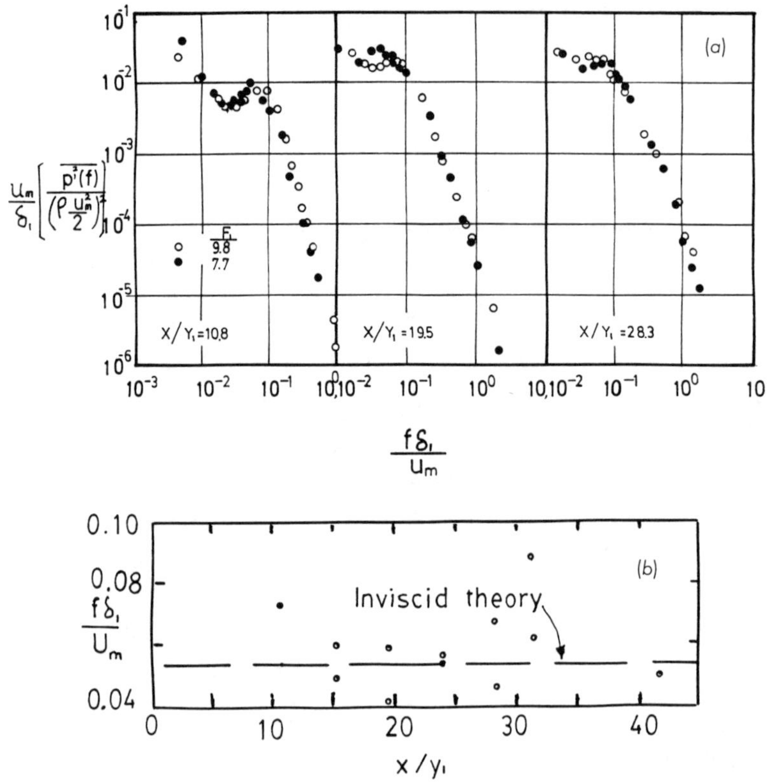

FIGURE 36. Dominant frequency of pressure fluctuations in the submerged jump (a) spectra for Y_{2s}/y_1 = 26; (b) comparison of theory and experiment [after Narayanan (1978)].

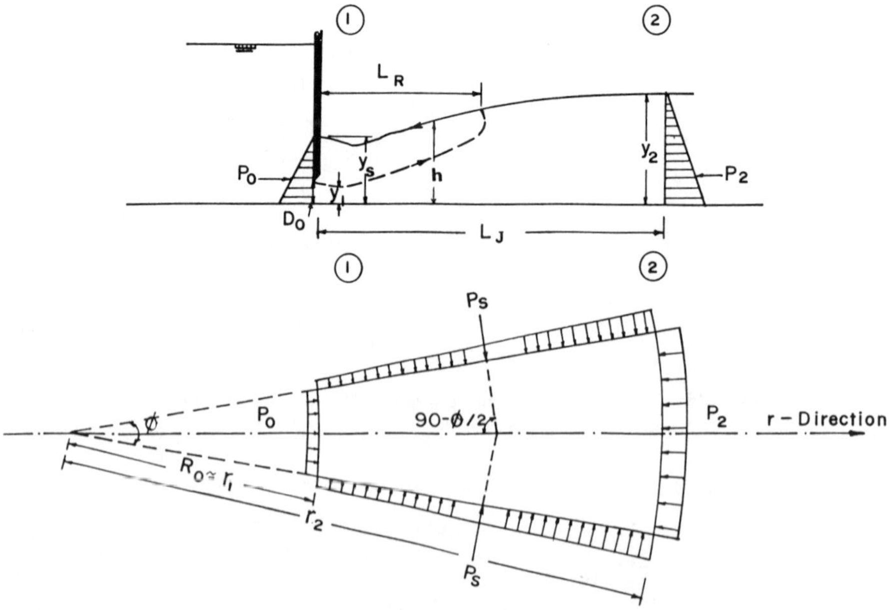

FIGURE 37. Defining sketch for submerged radial hydraulic jump [after Khalifa and McCorquodale (1980)].

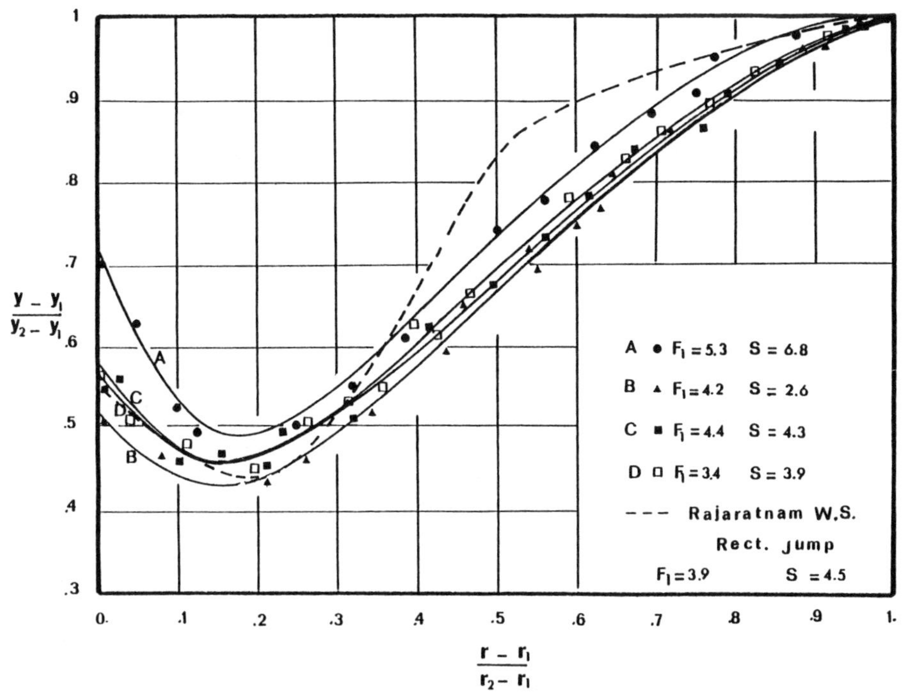

FIGURE 38. Surface profile of a submerged radial jump [after McCorquodale and Khalifa (1980)].

where

τ_o = bed shear

δ = turbulent boundary layer thickness, i.e., y at $u = U_m$

Narayanan (1978) found generally lower values of C_f in his experiments on the submerged jump.

The radial submerged jump was studied experimentally by Khalifa (1980) and Khalifa and McCorquodale (1980) and analytically and experimentally by Abdel-Gawad and McCorquodale (1985). The defining diagram for the submerged radial jump is given in Figure 37.

Khalifa *et al.* (1980) using the momentum principle obtained the following equation for the depth at the control gate:

$$S = y_s/y_1 = -A_s + [A_s^2 + r_o y_o^2 + 2F_1^2 (1/r_o y_o - 1)]^{1/2} \quad (50)$$

in which $r_o = R_2/R_o$; $y_o = y_{2s}/y_1$; $F_1 = u_1/\sqrt{gy_1}$; $A_s = 1/2\ y_o(r_o - 1)\ B$; B = side force correction factor $\cong 1.04 - 0.074\sqrt{F_1}$. The surface profiles of the submerged radial jump are compared with Rajaratnam's submerged rectangular jump profile in Figure 38. Khalifa (1980) gave the jump length as

$$L_j = 3.75\ (F_1 + S) - 0.002\ (F_1 + S)^2 - 5.89 \quad (51)$$

The mean bed pressure along the submerged radial jump is shown in Figure 39.

Abdel-Gawad and McCorquodale (1985) used the strip integral method to model the submerged radial jump. Figure 40 shows typical measured and predicted internal velocity distributions. Figure 41 shows the experimental and theoretical velocity decay curves for various values of F_1 and $\psi = y_s/y_1 = S$.

THE FREE RADIAL HYDRAULIC JUMP

Khalifa (1979, 1980) made an experimental and theoretical study of the radial hydraulic jump. Khalifa obtained the empirical and theoretical sequent depth ratios, y_o, shown in Figure 42. He also determined that the minimum acceptable jump length for the radial jump is approximately

$$L_{jr_o} \cong 0.72 L_j \quad (52)$$

in which L_{jr_o} = radial jump length and L_j = free rectanusing jump length.

Using the Strip Integral Method he simulated the internal flow in the radial jump. Figure 43 shows typical experimental and theoretical velocity distributions for a radial jump. In Figure 44 the surface velocity, U_s, and the maximum velocity, U_m, are shown as a function of x/y_1. Figure 45 indicates that the growth of the turbulent boundary layer is slower in the radial jump than in the free rectangular jump.

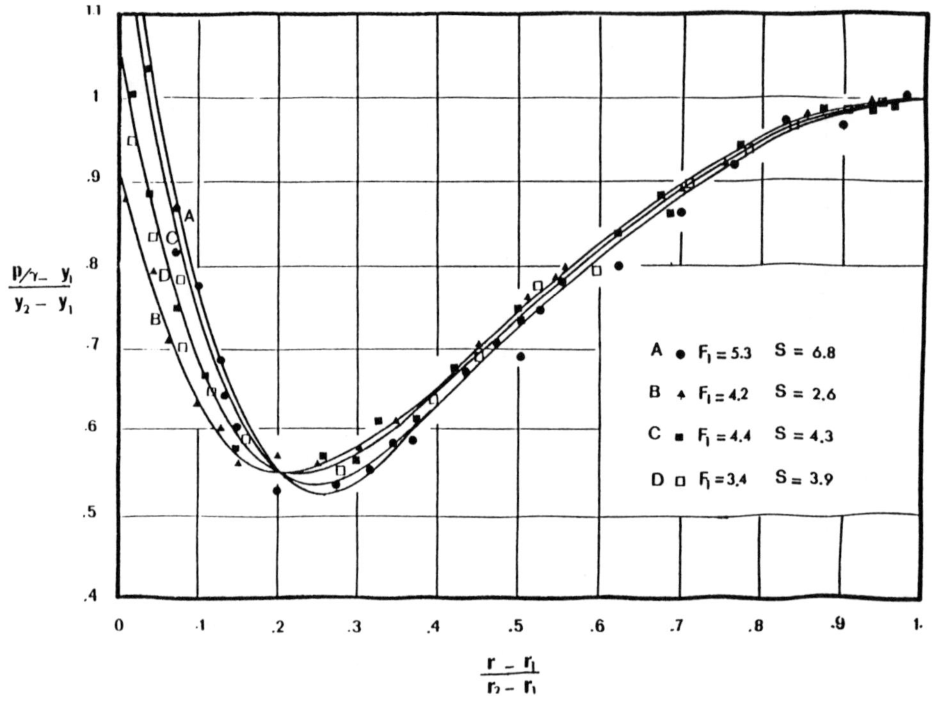

FIGURE 39. Bed pressures for a submerged radial jump [after McCorquodale and Khalifa (1980)].

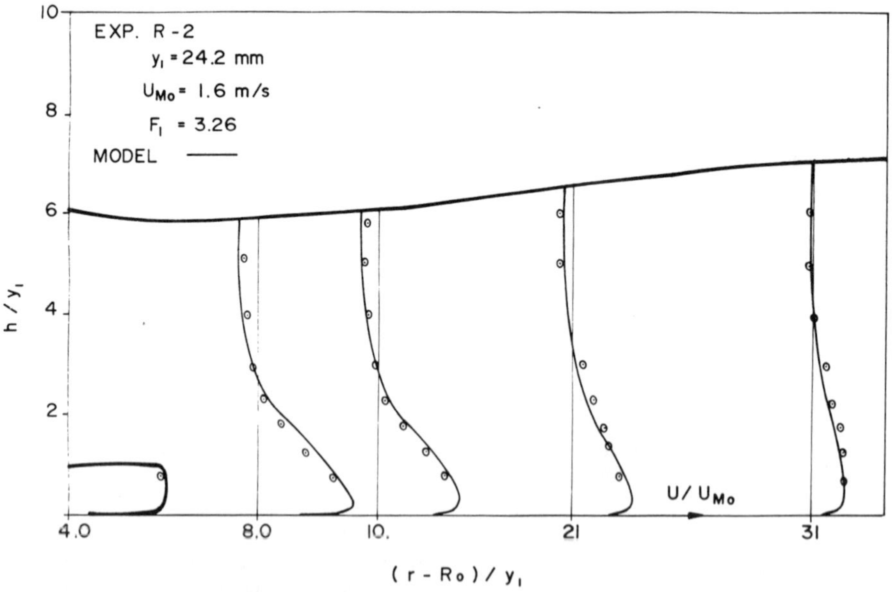

FIGURE 40. Typical pressured and SIM predicted velocity distributions in a submerged radial jump [after Abdel-Gawad and McCorquodale (1985)].

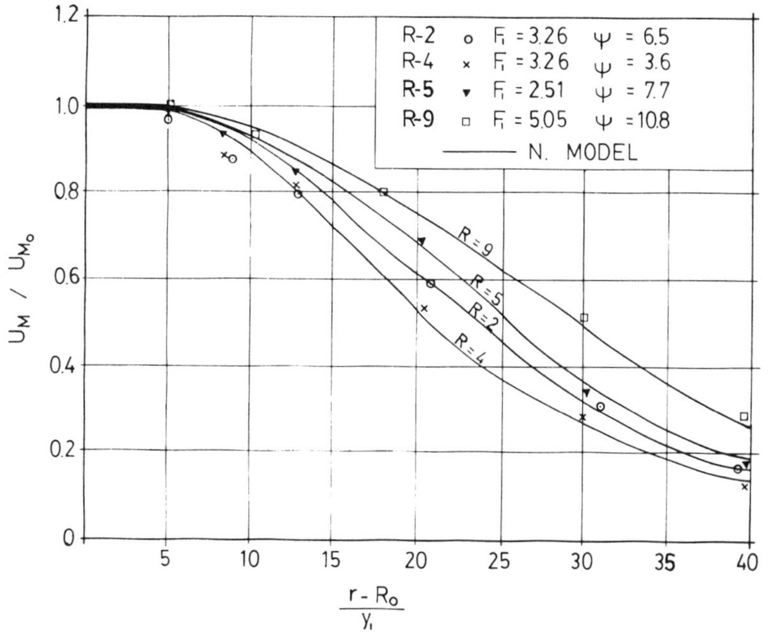

FIGURE 41. Experimental and SIM predicted velocity decay curves for the submerged radial hydraulic jump [after Abdel-Gawad and McCorquodale (1985)].

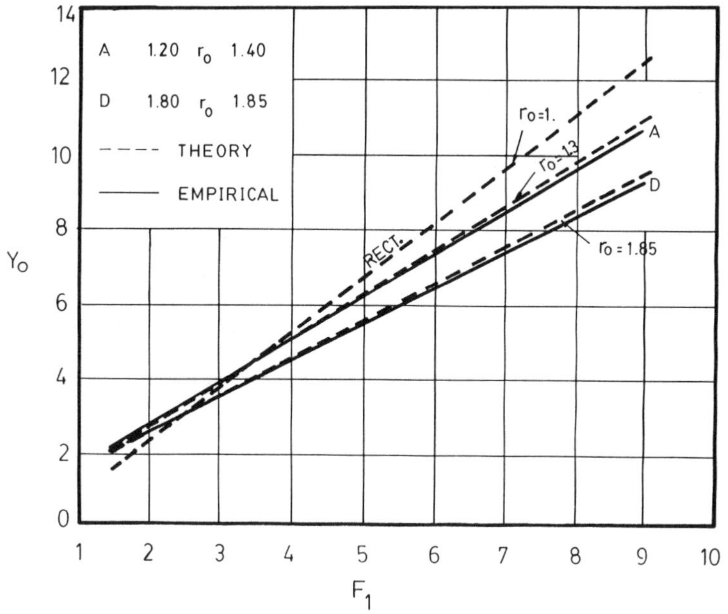

FIGURE 42. Sequent depth ratio for radial hydraulic jump—comparison of experimental and theoretical curves [after Khalifa (1980)].

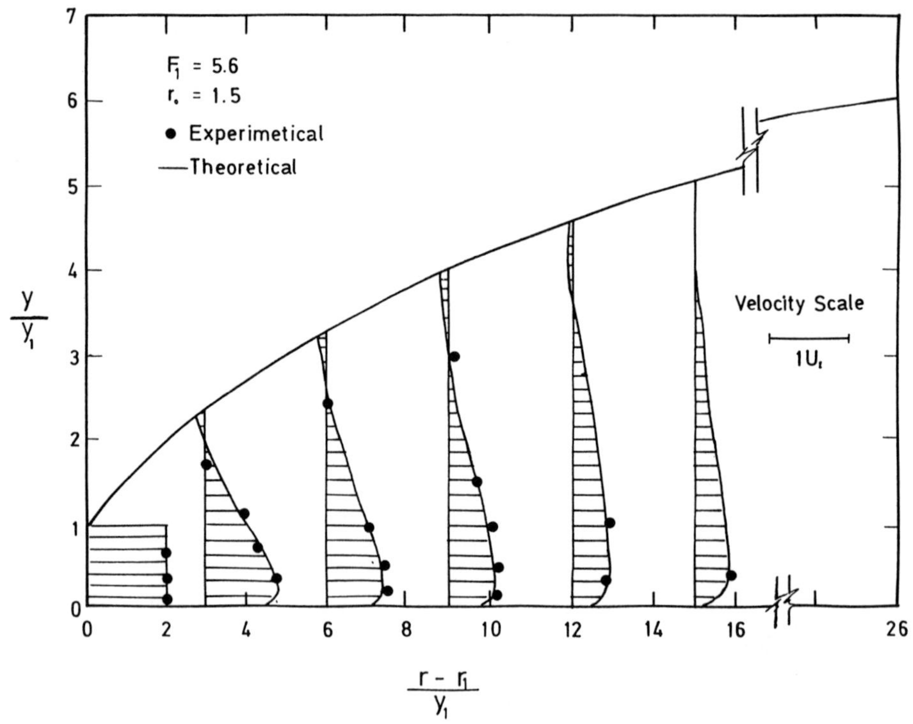

FIGURE 43. Comparison of experimental and SIM predicted velocity distributions in the radial hydraulic jump [after Khalifa (1980)].

Khalifa (1980) also estimated the depth averaged air fraction in the force radial jump (see Figure 19).

An important consideration in a radial jump stilling basin is the bistability of the jump. In general the flare angle

$$\phi/2 \cong \tan^{-1}(1/C_1F_1) \quad (53)$$

where $C_1 \cong 3$ for centrally stable effluent. Khalifa found that a cylindrically curved inflow gate gave $C_1 \cong 2$.

HYDRAULIC JUMPS AND WAVES AT ABRUPT DROPS

Rajaratnam and Ortiz (1977) investigated the internal flow in jumps downstream of abrupt drops. These jumps can be classified as indicated in Figure 46. Their velocity profiles are reproduced in dimensionless form for the B-jump [Figure 47(a)] and the wave-jump [Figure 47(b)]. The decay curves for the maximum velocity for the abrupt drop jumps and the classical jump are shown on Figure 48. Figure 49(a) gives the free mixing zone length scale $(\delta_1 - \delta)$ for the B-jump while Figure 49(b) shows the turbulent boundary layer δ. The free mixing velocity distribution is close to the classical wall jet. The jump length L_j in the B-jump varies from about $7y_2$ to $5y_2$ for $F_1 = 3$ to 9.

The readers are referred to the work of Talatt (1983) for a detailed study of the transition of the B-jump (plunging nappe) to the wave-jump (undular jump). Tallat explains the hysteresis' phenomenon in this transition; i.e., if the tailwater was raised with plunging flow the transition depth to wavy flow was greater compared to the transition depth when there was wavy flow and the depth decreased until plunging flow started.

THE FORCED HYDRAULIC JUMP

Many stilling basins contain appurtenances, such as baffle blocks, which are intended to stabilize the jump position and shorten the length of the basin. The U.S. Bureau of Reclamation (1955) has given several options. A standard stilling basin design has also been suggested by the St. Anthony Falls Hydraulic Laboratory (Blaisdell, 1949). In the following discussion the rectangular forced jump will be considered first and then a brief note will be added on the diverging forced jump.

The Forced Rectangular Jump

Another important case of Equation (4) is that involving appurtenances on the floor of the stilling basin. With the appropriate simplifications for a horizontal bed, the combination of Equations (4) and (6) gives

$$y_o^3 + \left[G\left(\frac{u_B}{U_1}\right)^2 F_1^2 - 2F_1^2 - 1 \right] y_o + 2F_1^2 = 0 \quad (54)$$

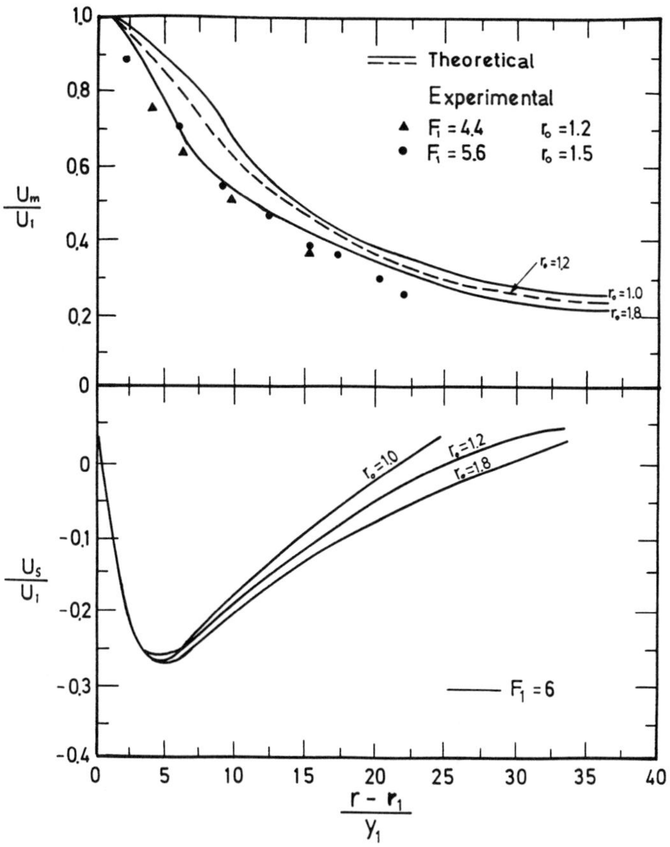

FIGURE 44. Variation of the surface and maximum velocities along the radial jump [after Khalifa (1985)].

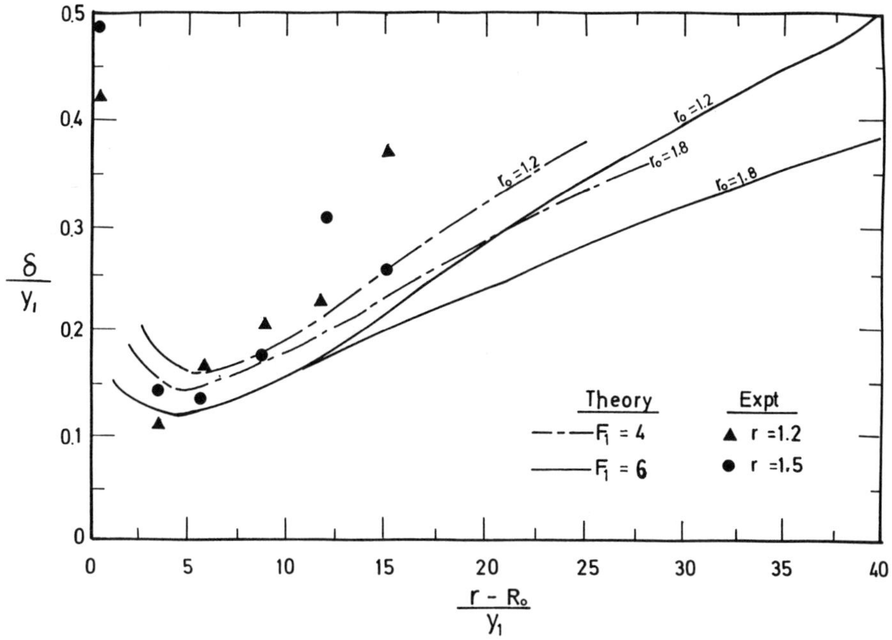

FIGURE 45. Growth of the boundary layer in a radial hydraulic jump [after Khalifa (1980)].

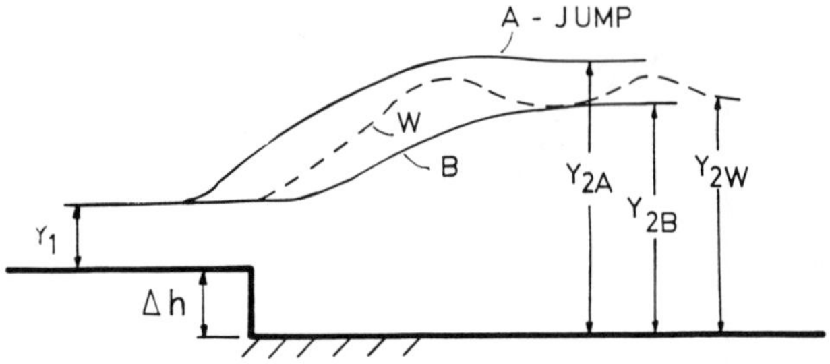

FIGURE 46. Forms of jumps at an abrupt drop [adapted from Rajaratnam (1967) and Chow (1959)]

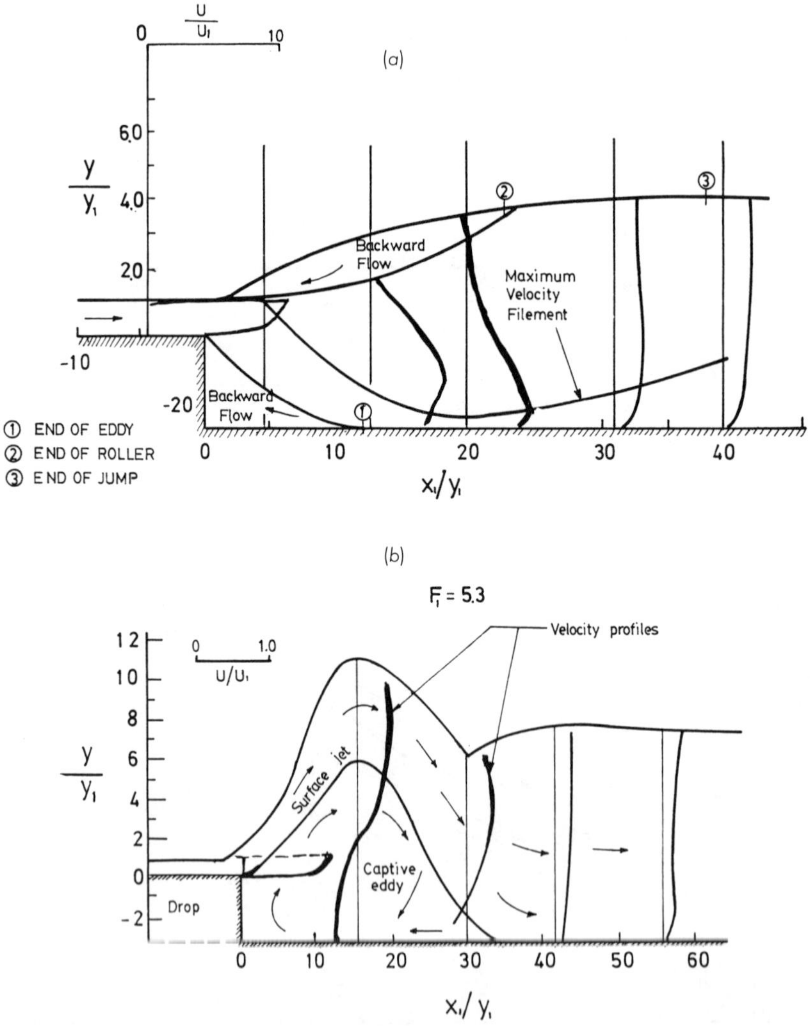

FIGURE 47. Velocity distributions in (a) B-jumps and (b) wave jumps [after Rajaratnam and Ortiz (1977)].

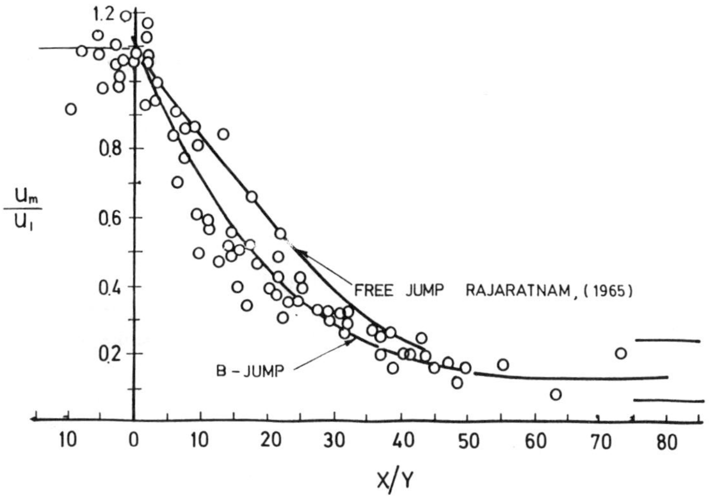

FIGURE 48. Comparison of velocity decay data for abrupt drops and the classical jump [after Rajaratnam and Ortiz (1977)].

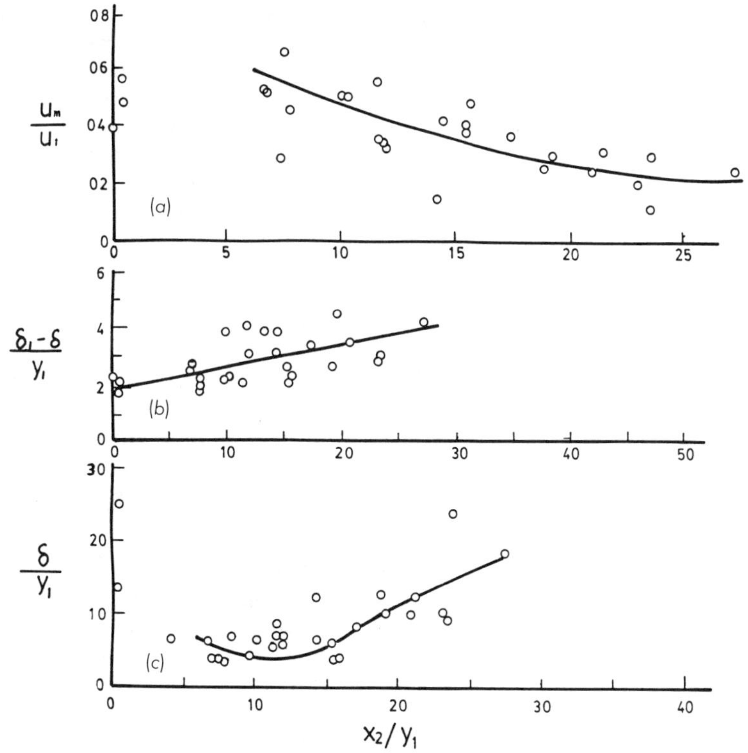

FIGURE 49. Velocity and length scales for the B-jump: (a) U_m/U_1; (b) free mixing zone $(\delta_1 - \delta)$; (c) turbulent boundary layer, δ [after Rajaratnam and Ortiz (1977)].

where

$$G = S_B \frac{h_B^*}{y_1} C_D \tag{55}$$

$$S_B = \frac{w_B}{s_B + w_B} = \text{blockage ratio} \tag{56}$$

h_B^* = y_B or h_B, whichever is less (57)
C_D = drag coefficient
s_B = baffle spacing
w_B = baffle width
h_B = baffle height
u_B = jet velocity at the baffle
y_B = jet depth at baffle

The jet velocity at the baffle varies from U_1 to U_2 from the beginning to the end of the jump. McCorquodale and Regts (1968) applied the momentum and continuity equations to estimate the expansion of the initial jet under the adverse pressure gradient of the forced hydraulic jump; the jet depth is

$$y_B = q/u_B \tag{58}$$

where

$$\frac{u_B}{U_1} = 1 - \frac{y_o(x_B/y_1)}{\left(\frac{x_B}{y_1} + y_o\right) F_1^2} \left\{ 1 + \frac{1}{2} \left[\frac{y_o}{\frac{x_B}{y_1} + y_o} \right] \frac{x_B}{y_1} \right\} \tag{59}$$

x_B = distance from the initial section to the baffle

The determination of the drag coefficient on baffle blocks and sills in hydraulic jumps has been studied by Rajaratnam (1964, 1971), Harleman (1955), Rand (1965), Weide (1951), Pillai and Unny (1964), McCorquodale and Giratella (1972), Narayanan (1980), Tyagi *et al.* (1978) and Karki (1976). Rajaratnam (1967) represented the drag coefficient on a sill in a hydraulic jump as a function of the portion of the wall from the start of the jump. He represented the drag force as

$$F_D = \frac{1}{2} C_d \varrho U_1^2 h_B \tag{60}$$

where h_B = baffle height and

$$C_d = f(x/L_j) \tag{61}$$

He found that C_d varied from about 0.6 at the start of the jump to about 0 at $x/L_j \cong 0.8$; C_d then increased to about 0.12 for $x/L_j \geq 1.3$.

McCorquodale *et al.* (1968, 1972) attempted to define the drag coefficient, C_D, in terms of the baffle geometry. Thus, the drag force was defined, as in Equation (15), by

$$F_D = \frac{1}{2} C_d \varrho u_B^2 A_B^* \tag{62}$$

where A_B^* = (area jet)Ω(area baffle) (63)

In order to estimate an upper limit for C_D, McCorquodale *et al.* studied the deflection of a supercritical jet by a single row of ventilated baffles. The measured drag forces were the highest possible values since the wake was aerated.

Figure 50 shows the C_D values obtained by Giratella. The author also carried out some experiments where a back pressure was allowed to develop. Figure 50 also shows the correction for this back pressure at a Froude number of approximately 5. The experimental values of C_D obtained by Karki (1976) for an unventilated baffle are also shown in Figure 50. The experimental conditions of Karki are probably closer to the hydraulic jump than those of the author. It was also noted by Giratella (1969) that the upstream separation and associated vortex are stronger for a continuous wall than for a row of equally spaced baffle blocks. It is suggested that the Karki curve be used for C_D in the central portion of the jump while the higher values of C_D should be used to estimate the highest mean force on the baffles as well as for determining sweep-out conditions. In the subcritical flow in the downstream portion of the jump, the work of Rajaratnam (1964) indicates that C_D varies from 0.6 to more than 2.

Two special cases of Equation (54) are of interest, i.e.,

1. $x_B = 0$ and $u_B = U_1$; $y_B = y_1$
2. $x_B \cong L_j$ and $u_B = U_2$; $y_B = y_2$

Case 1 corresponds to impending sweep-out and probably has a C_D very close to the value for a deflected jet. Figure 51 shows the sequent depth ratio, y_o, as a function of G and F_1 along with the specific example of: $h_B = 1.25y_1$; $h_B^*/y_1 = 1$; $s_B = 0.5$ and $C_D = 0.9$ which gives $G \cong 0.45$.

In case 2 the baffles are near the end of the jump and exert a relatively low drag force on th flow. Figure 51 shows the sequent depth ratio for $S_B = 0.5$, $C_D = 1$ and $(u_B/U_1) = (y_1/y_2)$ which gives $G \cong 1/(2y_o)$. It is evident that baffle blocks or even a sill at the end of the jump are ineffective in substantially reducing the sequent depth; however, these appurtenances may be useful in deflecting the outflowing jet away from the bed. A large sill, such as that used in the USBR Stilling Basin III (1974) may act as a weir to limit the minimum sequent depth to the sill height plus the critical depth at the sill. Forster and Skrinde (1950) presented an analysis of the hydraulic jump with an abrupt rise at the end of the jump.

Figure 51 shows the Waterways Experimental Station, WES (Basco, 1969) data for stilling basins with baffles. The validity of Equation (54) and the magnitude of the C_D values

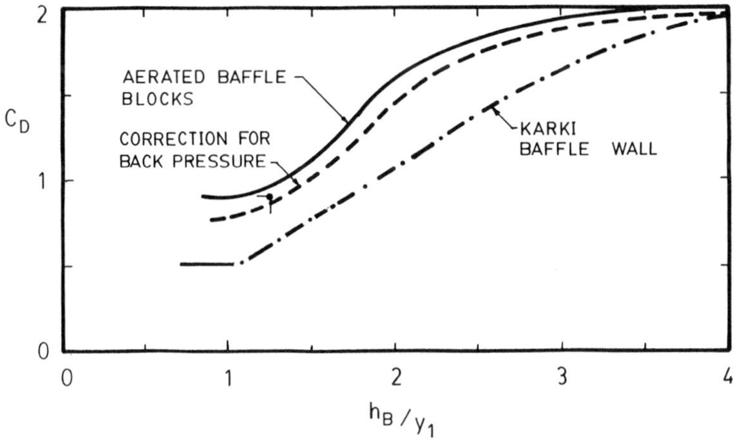

FIGURE 50. Drag coefficients on baffles in supercritical flow.

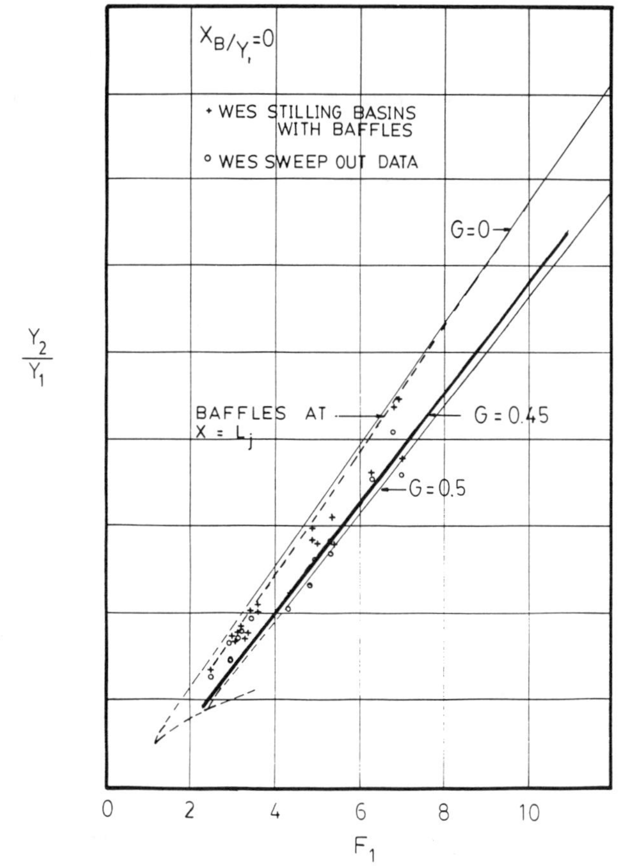

FIGURE 51. Sequent depth relationships for stilling basins with baffles [after McCorquodale and Regts (1968) and WES (1969)].

in Figure 50 are supported by the agreement between the theoretical and WES experimental data.

Regts (1967) used a dynamometer to measure forces in a forced hydraulic jump. He represented the mean drag force in dimensionless form as

$$\frac{F_D}{F_j} = \frac{F_b}{S_B Q W y_1 U_1^2} \tag{64}$$

where

$W = W_B + S_B$
F_b = drag on one baffle

His results are compared with the theoretical solution (McCorquodale and Regts, 1968) in Figure 52.

Narayanan and Schizas (1978, 1980) studied the drag force on a baffle wall as well as the fluctuation in this force. They used the same definition for C_d as Rajaratnam; maximum values of C_d were about 0.5. They found the rms value of the random component of C_d to be approximately 0.05 (C_d'); however, instantaneous fluctuations could be as high as seven times this value. Figure 53, presented by Narayanan *et al.* (1980), shows the observed variation of C_d with F_1 and the tailwater depth ratio for a baffle wall at $x/y_1 = 34.6$ and $h_B/y_1 = 1$. Figure 54 indicates the variation of C_d' (the rms value of C_d) with y_{2s}/y_1 for various values of h_b/y_1. Narayanan *et al.* (1980) present force frequency spectra for near sweep-out, normal jump and submerged jump; they found a dominant frequency at the transition from sweep-out to a normal jump. Figure 55 shows their spectra. Tyagi *et al.* (1978) obtained similar results to Narayanan *et al.* and Rajaratnam (1964); however, they found the rms value of the random component of the fluctuating force to be about three times higher if the baffle wall was placed at $x/y_1 < 10$. They considered the force on individual baffles of the USBR standard design as well as a baffle wall. In their study it was found that the force fluctuation, F_B', was normally distributed as indicated in Figure 56. For baffle walls with $h_B/y_1 > 1.4$, they found the average baffle force per unit with $\bar{F}_{BW}$ was

$$\bar{F}_{BW} \cong C_x\, \gamma\, y_2^2\, (h_B/y_1)^{-1.5} \tag{65}$$

where $C_x \cong 0.1$ for $5 < X_B/y_1 < 10$ and $C_x \cong 0.05$ for $X_B/y_1 \cong 20$. Also y_2 is obtained from the free jump. The rms value of the force equation fluctuations for $h_b/y_1 > 1.4$ varies from 0.03 to 0.5 with an average of about 0.15. For USBR type baffles they found that the average force per unit width was nearly proportional to the blockage ration; however, the force fluctuation on the individual baffles are very large, e.g.:

$$\sqrt{\overline{F_B'^2}}/\bar{F}_B \cong 0.75 \tag{66}$$

where

$$\bar{F}_B \cong S_B\, \bar{F}_{BW} \tag{67}$$

and S_B = blockage ratio.

The results of Basco and Adams (1971) indicate that C_x in Equation (67) should be increased by 100% to 150% for $h_B/y_1 \lesssim 1$. Ranga Raju *et al.* (1980) reviewed the work of Basco and Adams (1971), Gomasta *et al.* (1977), and Rajaratnam and Murahari (1971) and proposed a design procedure for stilling basins with baffles. They stress that the baffles should be designed for the maximum force which, if the baffles are near the start of the jump, can be 10% to 50% greater than the average force. For baffles near the end of the jump the maximum force can be double the average.

In addition to studies on the C_d for baffle walls, Rajaratnam and Murahari (1971) measured the velocity distribution in forced jumps under three conditions: (a) $K = 0$, near sweep out; (b) $K = 0.4$, baffles in the mid-portion of the jump; and (c) $K = 1.0$, baffles near the end of the jump. Figures 57 to 59 show the jump profiles and the corresponding dimensionless velocity distributions. The velocity profiles are analysed in two regions, i.e., $X < X_B$ and $X > X_B$ where X_B distance from the start of the jump to the baffle wall. Figure 60 compares the growth of the boundary layer to that of the classical jump. A power law

$$u/U_m = (y/\delta)^n \tag{68}$$

was suggested for the boundary layer. The value of δ is given in Figure 60, while n varies from 5 to 8.65. Figure 61(a) shows the velocity decay data for the outer layer for $X < X_B$ and the corresponding length scale is given in Figure 61(b). The vertical distribution of u in the outer layer is close to the classical wall jet as can be seen in Figure 62. The downstream zone, $X > X_B$, has a curved jet with the decay curve shown in Figure 63, in which $U_{m_j} = U_m$ at $X = X_B$. The positive velocity in this jet is close to the Tolimien solution for a plane turbulent jet. The value for the velocity length scale in this zone is

$$\delta_1 \cong 0.23\,(X - X_B) \tag{69}$$

The maximum reverse velocity behind the baffle wall was found to be

$$U_{mr} \cong 0.25\, U_1 \text{ for } K = 0 \tag{70}$$

and

$$U_{mr} \cong 0.1\, U_1 \text{ for } K = 1 \tag{71}$$

The eddy length was 4 to 12 times the baffle height.

The standard baffle blocks (USBR, 1974) are subject to cavitation damage if the inflow velocity exceeds about 50 ft/sec (15 m/s). In order to overcome this problem, the U.S. Army Corps of Engineers (Berryhill, 1957) designed a streamlined baffle block with elliptical upstream curves. Harleman (1955) studied these baffles and recorded similar drag forces to stepped baffles.

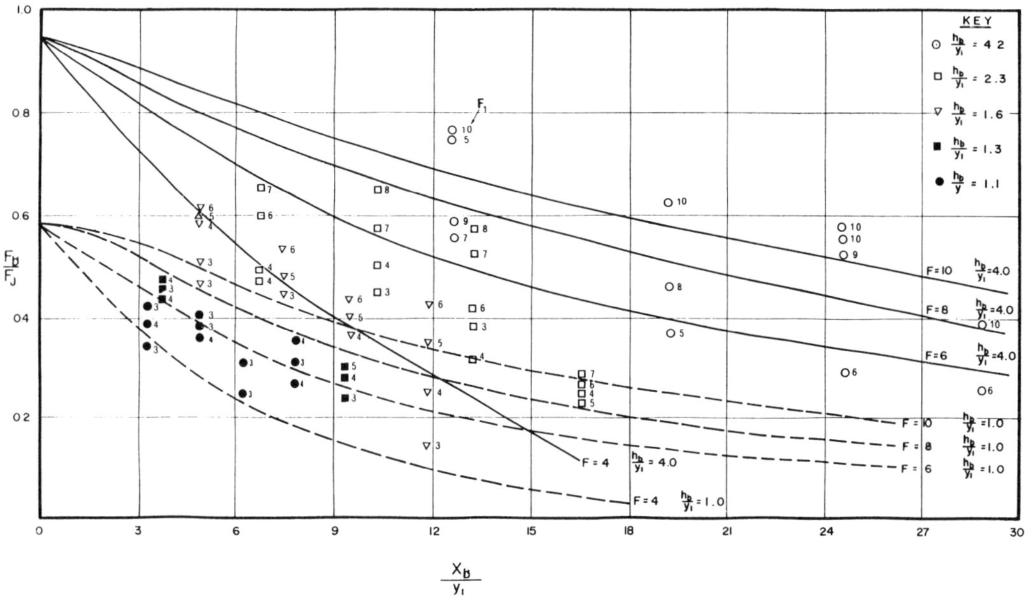

FIGURE 52. Comparison of predicted and measured mean baffle forces in a hydraulic jump [after Regts (1967) and McCorquodale and Regts (1968)].

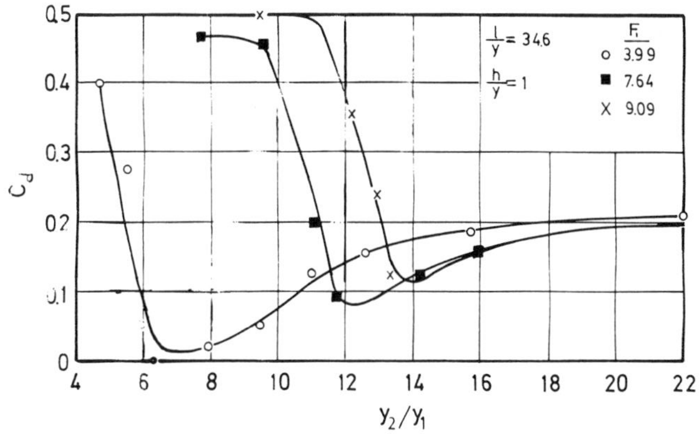

FIGURE 53. Drag coefficient, C_d for a baffle wall in a forced hydraulic jump [after Narayanan et al. (1980)].

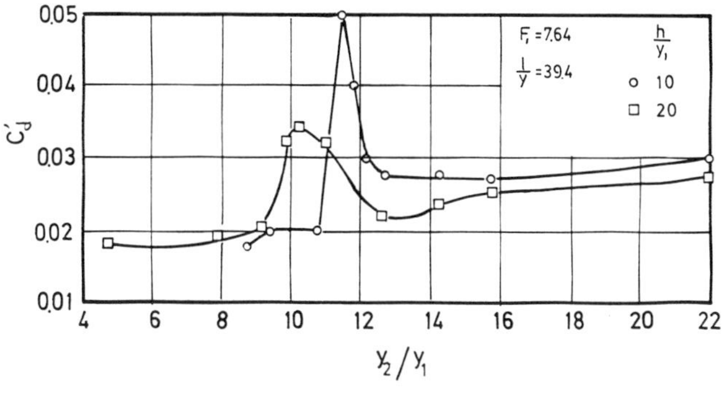

FIGURE 54. The rms value of the drag coefficient for various submergences and baffle wall heights [after Narayana et al. (1980)].

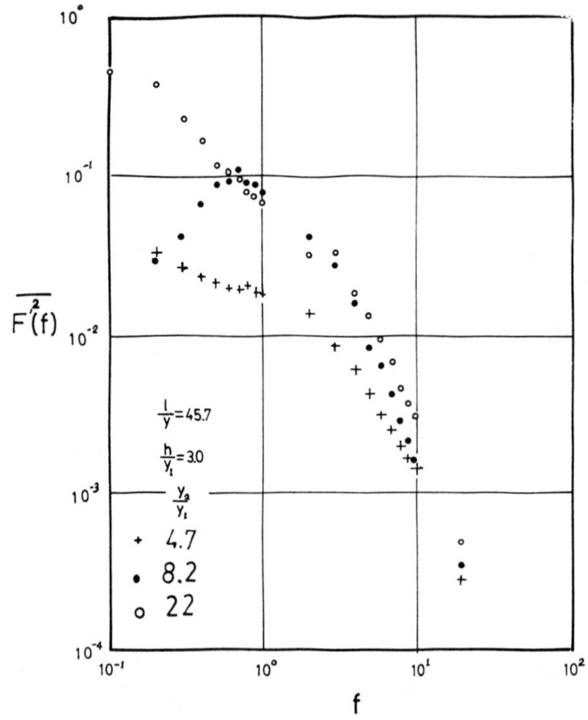

FIGURE 55. Force-frequency spectra for a baffle wall in a forced jump (after Narayana et al.).

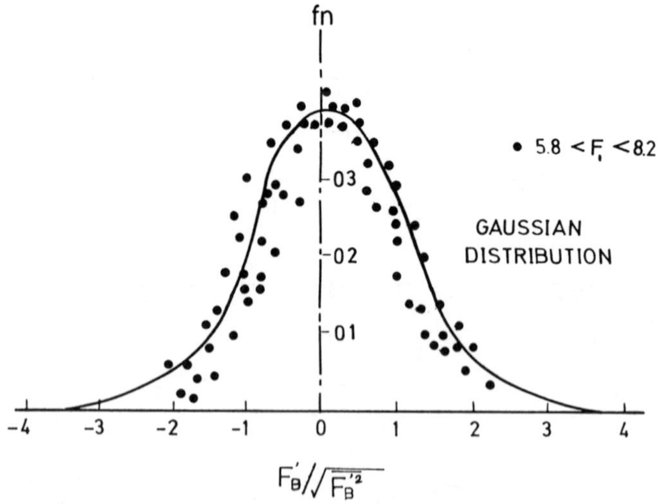

FIGURE 56. Probability distribution of force fluctuations on a baffle wall [after Tyagi et al. (1978)].

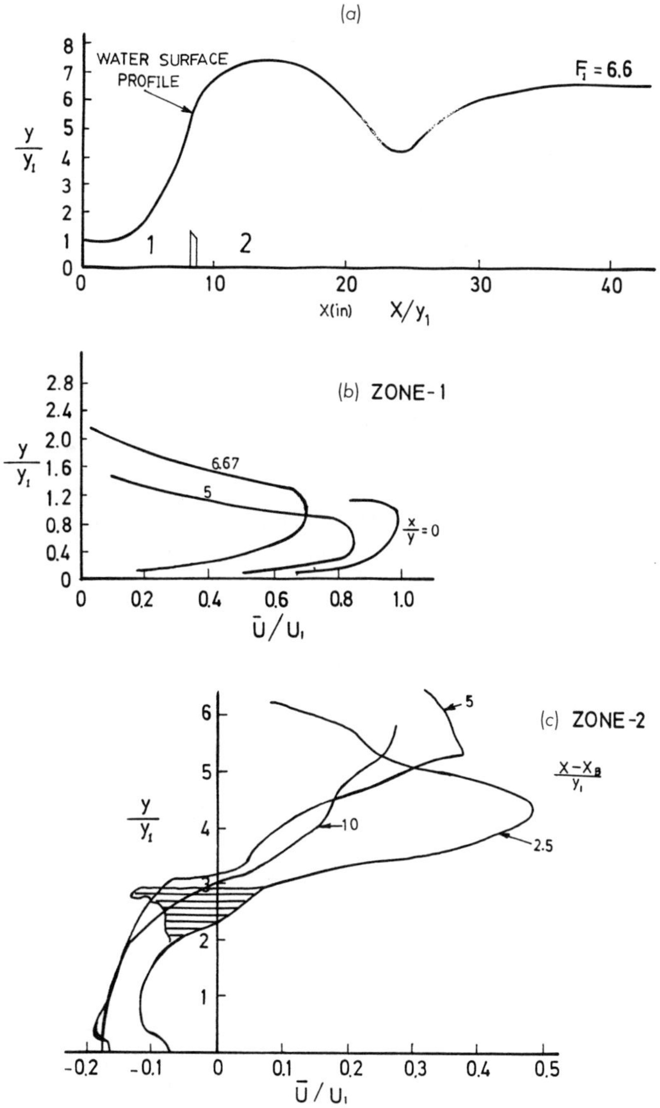

FIGURE 57. Velocity distributions in the forced jump with a baffle wall under near sweep-out conditions [after Rajaratnam and Murahari (1971)].

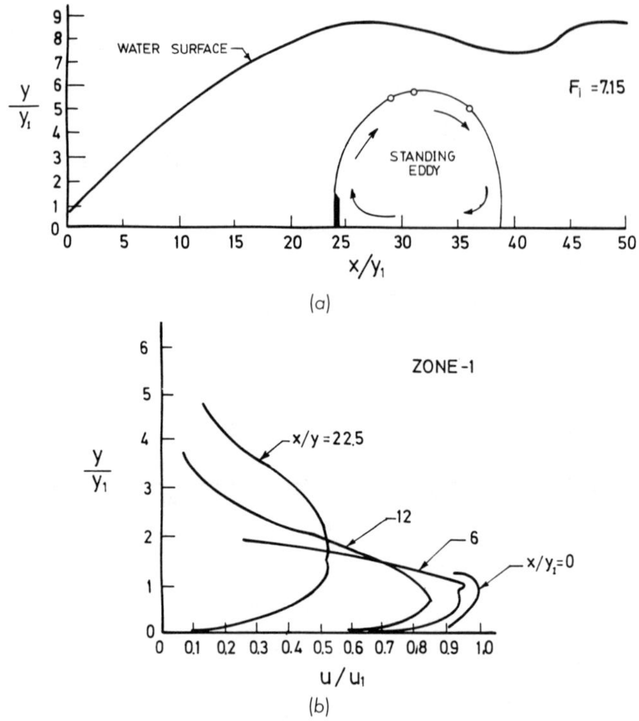

FIGURE 58. Velocity distributions in the forced jump with a baffle wall near the middle of the basin [after Rajaratnam et al. (1971)].

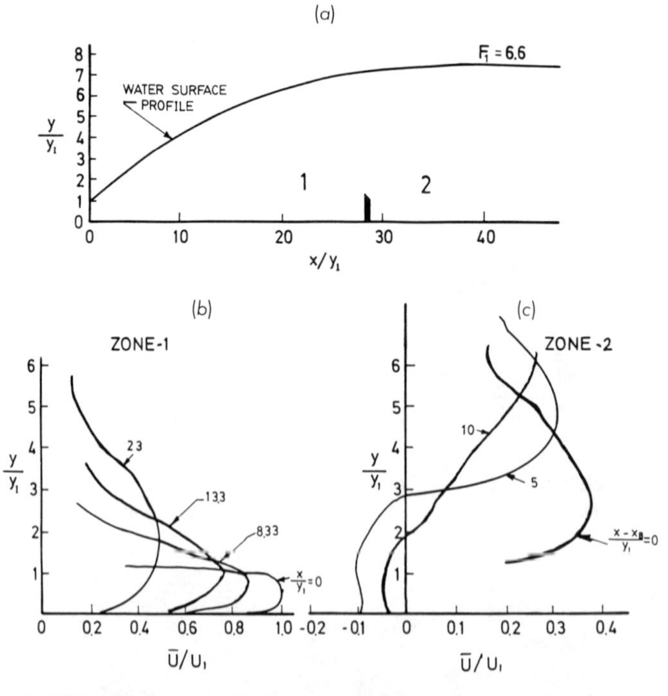

FIGURE 59. Velocity distributions in the forced jump with a baffle wall near the end of the basin [after Rajaratnam et al. (1971)].

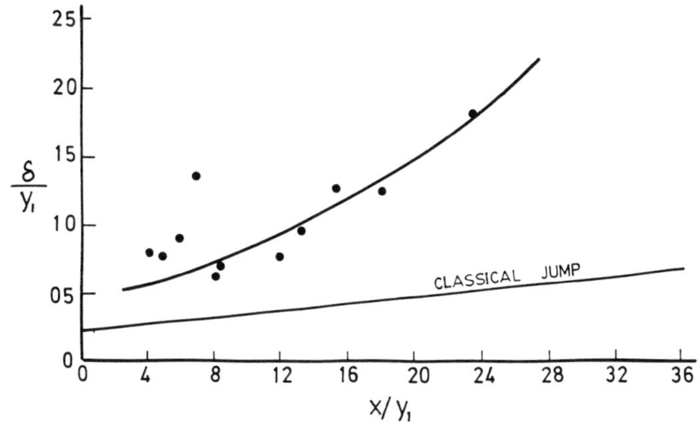

FIGURE 60. Boundary layer growth in the forced and free rectangular hydraulic jump [after Rajaratnam et al. (1971)].

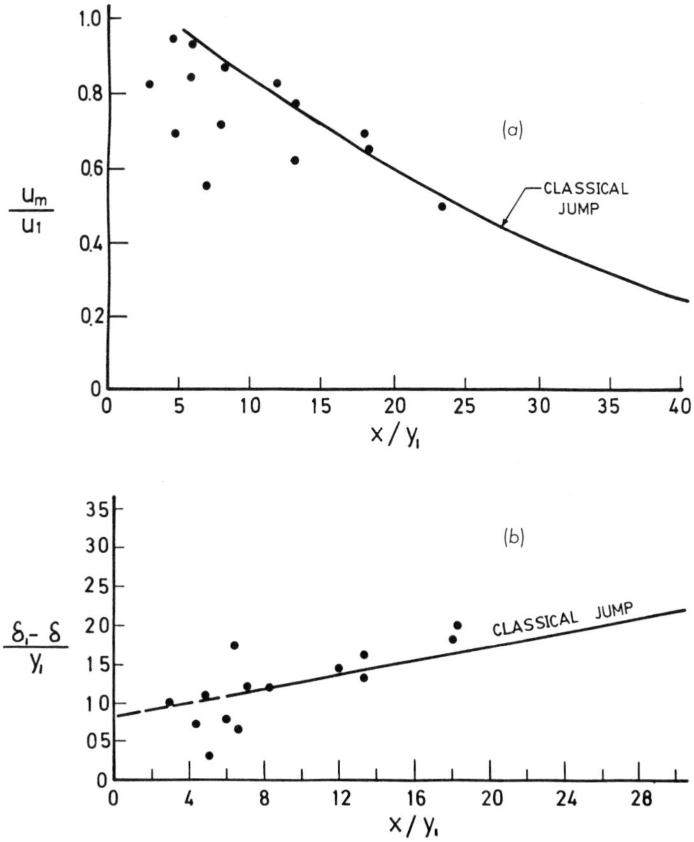

FIGURE 61. (a) Velocity decay for $X < X_B$; (b) velocity length scale for $X < X_B$ [after Rajaratnam et al. (1971)].

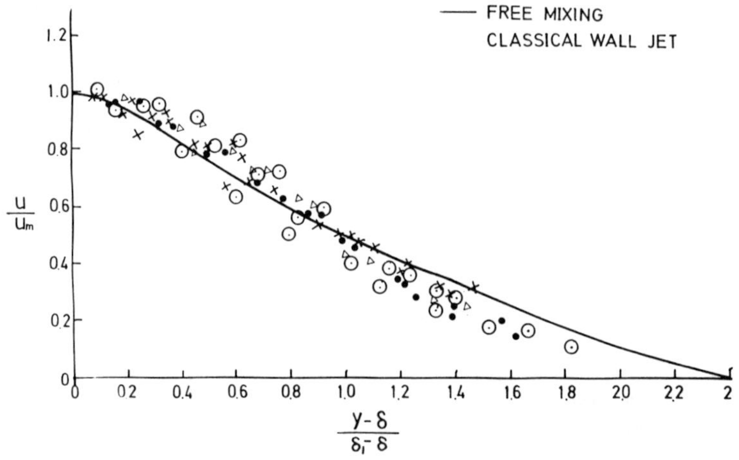

FIGURE 62. Velocity distribution in the outer layer for X < X_B [after Rajaratnam et al. (1971)].

Diverging Forced Jump

An analysis of the radial stilling basin with baffles was presented by Nettleton and McCorquodale (1983). An attempt was made to apply the macroscopic energy balance in order to determine the sequent depth; however, this approach was not totally successful because of the difficulty in determining the side force in the presence of the hump, produced by the baffles. Figures 64(a) to (e) illustrate the variability of the free surface profile with baffle position and Froude number. Figures 65 and 66 indicate typical theoretical sequent depth ratios. Nettleton (1983) found a large degree of scatter in his measurements of the length of a forced radial jump. One of the problems in measuring the length of the jump involved the estimation of where the flow reaches the sequent depth because of secondary waves caused by the baffles. Figure 67 shows the scatter in the experimentally determined jump lengths. The work of Nettleton shows that the addition of baffles did not substantially decrease the jump length, as has been observed when baffles are added to rectangular stilling basins.

Nettleton found that the outflow from a forced radial hydraulic jump became unstable for flare angle:

$$\frac{\phi}{2} \geq \tan^{-1}\left(\frac{1}{1.5F_1}\right) \tag{72}$$

Typical lateral velocity profiles are shown in Figure 68. Lateral instability is evident at $F_1 = 7$.

Chong, Fann and Sadeli (1985) extended the research of Nettleton on forced radial stilling basins. They developed the upper and lower bounds for y_2/y_1 and L_j/y_2 shown in Figure 69. They also determined the bounds on the surface wave amplitudes and dominant periods of the surface waves as shown in Figure 70. The writer used the data surface profile and bed pressure data of Chong *et al.* to estimate the air concentration in the forced radial jump (Figure 71).

Thomas and Lean (1963) reported on the dynamic loading on a baffle wall in an impact stilling basin at the outlet of the Mangla Dam diversion tunnels. They installed pressure transducers on the impact wall. The rms pressure head fluctuations $\sqrt{\overline{p'^2}}/\gamma$ varied from 0.08 to 0.23 ($U_1^2/2g$) with

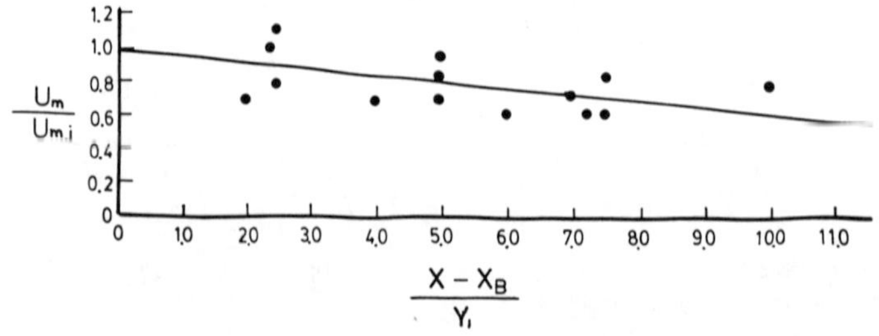

FIGURE 63. Velocity decay curve for X < X_B [after Rajaratnam et al. (1971)].

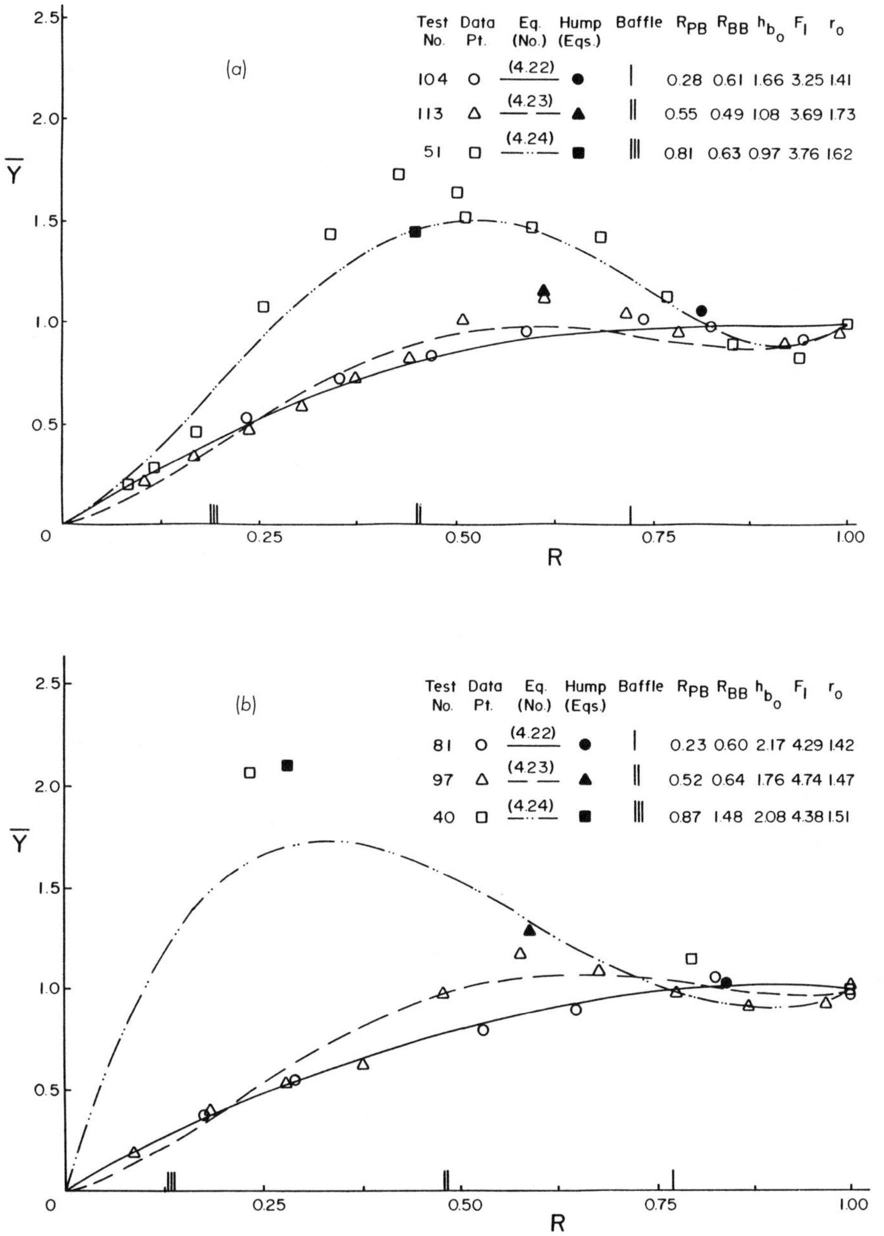

FIGURE 64. Water surface profiles for the forced radial hydraulic jump for various baffle positions and Froude numbers. $\bar{y} = (y - y_1)/(y_2 - y_1)$; $R = (r - r_1)/(r_2 - r_1)$ [after Nettleton (1983)]. (a) $3 < F_1 < 4$; (b) $4 < F_1 < 5$; (c) $5 < F_1 < 6$; (d) $6 < F_1 < 7$; (e) $7 < F_1 < 8$.

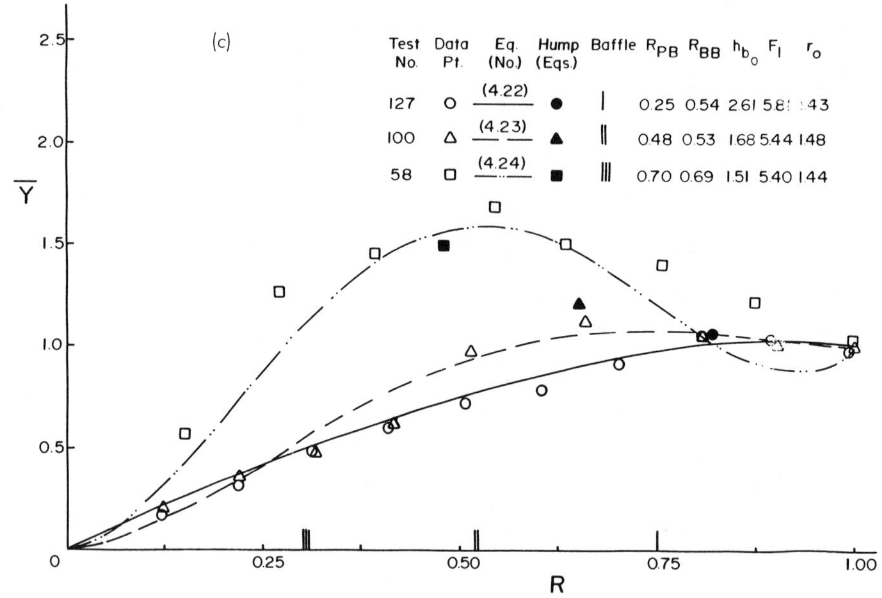

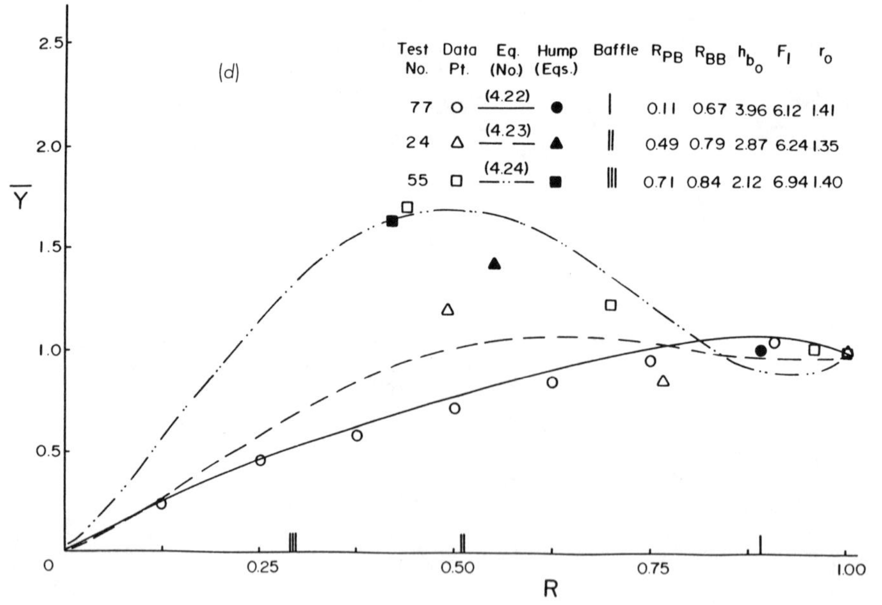

FIGURE 64 (continued). Water surface profiles for the forced radial hydraulic jump for various baffle positions and Froude numbers. $\bar{y} = (y - y_1)/(y_2 - y_1)$; $R = (r - r_1)/(r_2 - r_1)$ [after Nettleton (1983)]. (a) $3 < F_1 < 4$; (b) $4 < F_1 < 5$; (c) $5 < F_1 < 6$; (d) $6 < F_1 < 7$; (e) $7 < F_1 < 8$.

Test No.	Data Pt.	Eq. (No.)	Hump (Eqs.)	Baffle	R_{PB}	R_{BB}	h_{b_0}	F_1	r_0
123	○	(4.22)	●	|	0.18	0.58	3.93	7.68	1.35
103	△	(4.23)	▲	||	0.55	0.53	1.83	7.23	1.60
35	□	(4.24)	■	|||	0.84	2.27	4.31	7.49	1.37

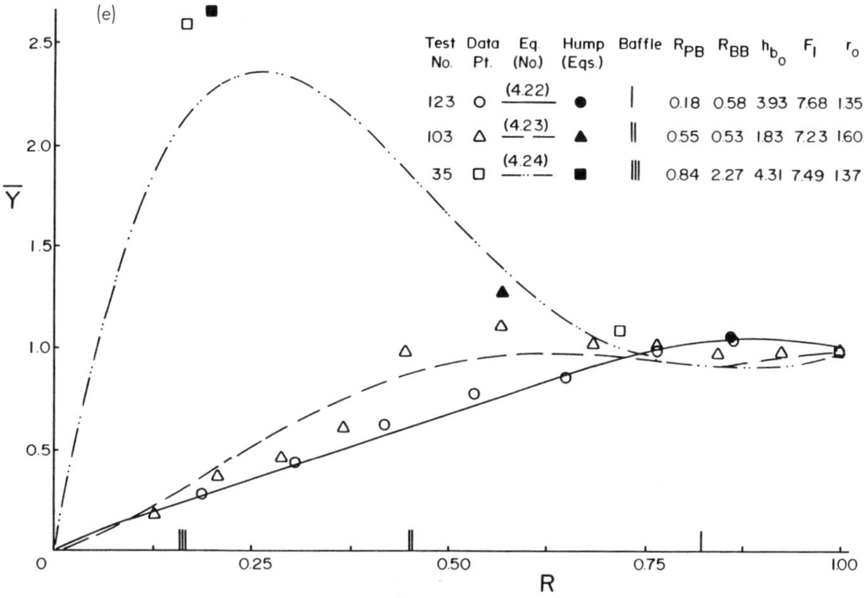

FIGURE 64 (continued). Water surface profiles for the forced radial hydraulic jump for various baffle positions and Froude numbers. $\bar{y} = (y - y_1)/(y_2 - y_1)$; $R = (r - r_1)/(r_2 - r_1)$ [after Nettleton (1983)]. (a) $3 < F_1 < 4$; (b) $4 < F_1 < 5$; (c) $5 < F_1 < 6$; (d) $6 < F_1 < 7$; (e) $7 < F_1 < 8$.

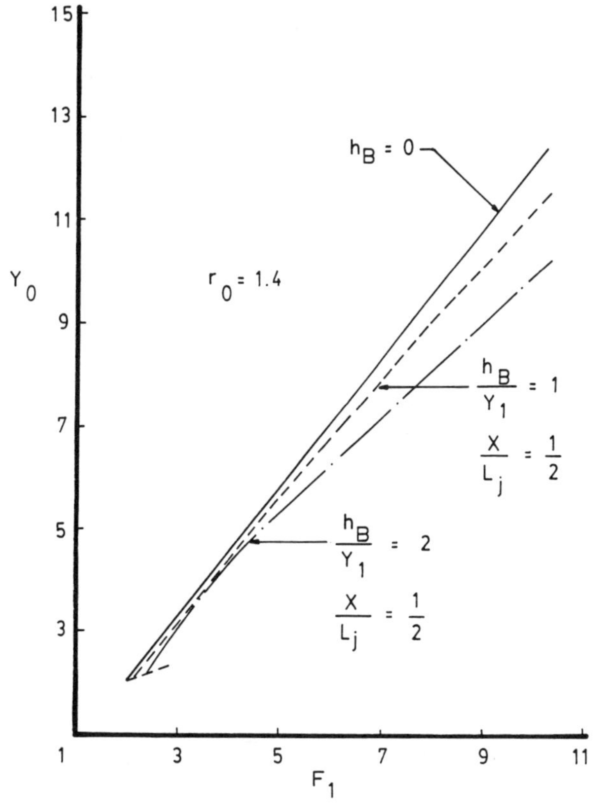

FIGURE 65. The suggested sequent depth ratio as a function of F_1 and h_B for $r_0 = 1.4$.

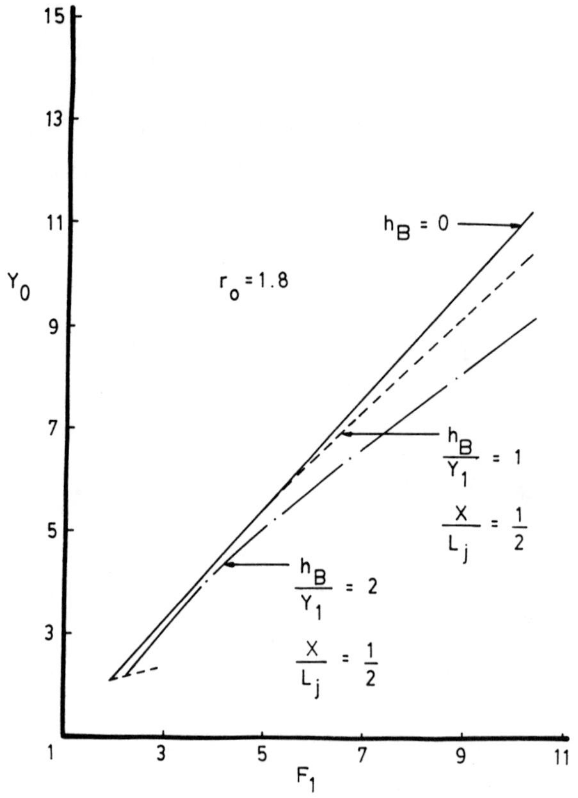

FIGURE 66. The suggested sequent depth ratio as a function of F_1 and h_B for $r_o = 1.8$.

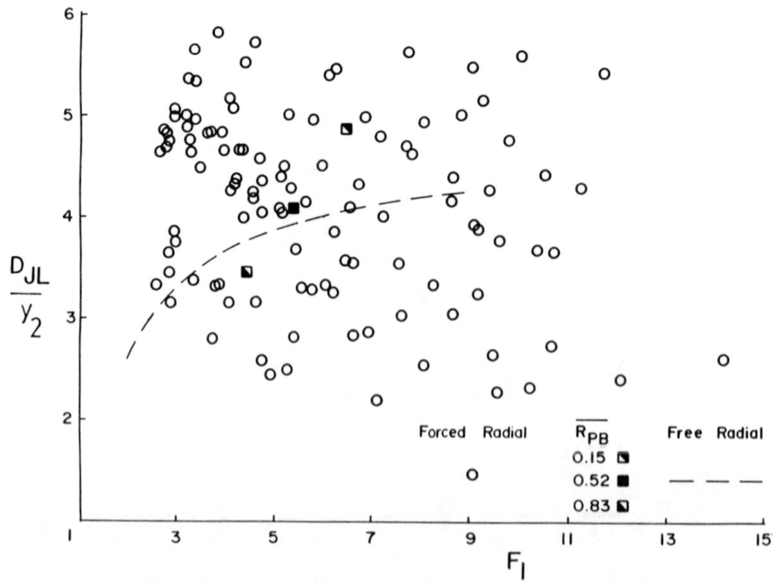

FIGURE 67. Estimated jump length as a function of F_1 [after Nettleton (1983)]. (Note D_{JL} = length to the second standing wave.)

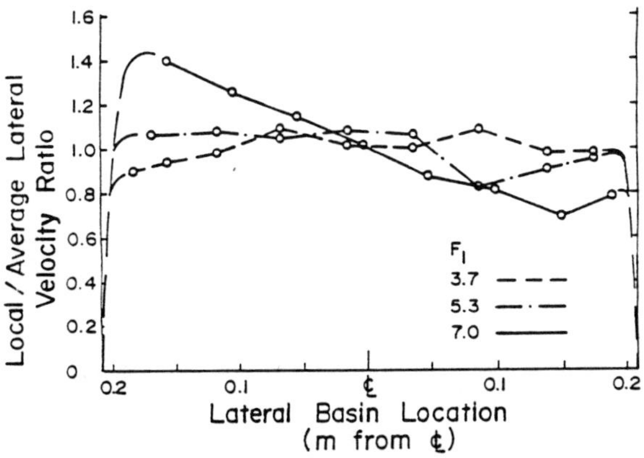

FIGURE 68. Lateral velocity profiles at the downstream end of a forced radial jump [after Nettleton (1983)].

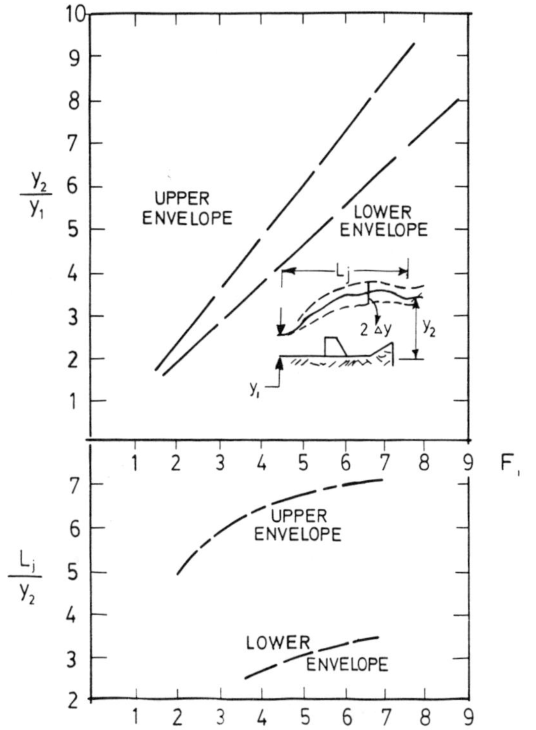

FIGURE 69. Sequent depth ratio and jump length for the forced radial hydraulics jump [after Chong et al. (1985)].

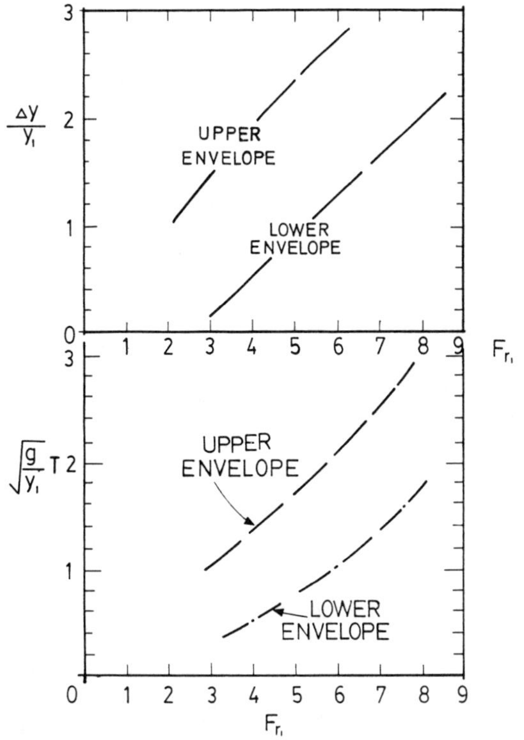

FIGURE 70. Surface wave amplitudes and dominant periods for forced radial jump [after Chong et al. (1985)].

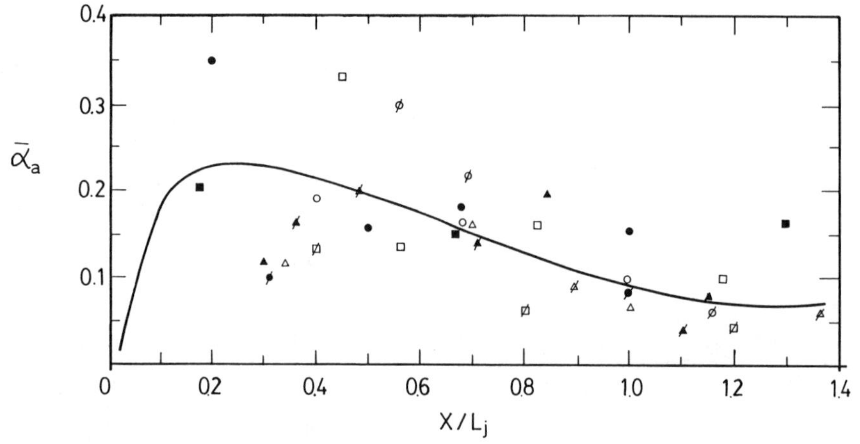

FIGURE 71. *Air concentration in the forced radial jump.*

$fy_1/U_1 \cong 0.5$. The maximum pressure fluctuation

$$\Delta p_{max}/(\varrho\ U_1^2/2) \cong \pm 0.5 \qquad (73)$$

at

$$fy_1/U_1 \cong .03 \qquad (74)$$

where y_1 = tunnel area/basin width.

The study showed that the point pressures could be assumed to act over approximately the 1/36 of the tunnel area while the rest of the wall would experience random pressure fluctuations.

HYDRAULIC JUMPS ON SLOPING BEDS

If the bed of the hydraulic jump is sloped, the longitudinal component of the body force term in Equation (5) must be retained. Even if the appurtenances are excluded, the macroscopic momentum and continuity equations, by themselves, do not yield the sequent depth ratio since the body force depends on the length and surface profile of the jump. These are determined by the internal mechanics of the jump.

Chow (1959) and Kindsvater (1944) represented the body forces in the rectangular case by

$$W = \frac{1}{2} K\gamma L_j(y_1 + y_2) \qquad (75)$$

where K is a factor to account for the shape of the jump. Using the assumptions that $\beta = \beta' = 1$ and $F_f \cong 0$ and substituting Equation (6) into Equation (4) leads to

$$y_o^3 - K_o\, y_o^2 - \left(\frac{2F_1^2}{\cos\theta} + K + 1\right) y_o + \frac{2F_1^2}{\cos\theta} = 0 \qquad (76)$$

where $K_o = (KL_j/y_1) \tan\theta = f(\theta\ F_1)$.

An approximate solution to Equation (76) was given by Rajaratnam (1967).

$$y_o = \frac{1}{2\cos\theta} (\sqrt{1 + 8G_1^2} - 1) \qquad (77)$$

where

$$G_1 = F_1\Gamma_1 \qquad (78)$$

and $\Gamma_1 = 10^{1.55\theta^{rad}}$

Figure 72 compares the experimentally derived lengths and sequent depth ratios for horizontal and sloping hydraulic jumps. It is evident that the sequent depth increases sharply with increasing bed slope; however, the jump length in terms of y_1 is not greatly affected by θ.

A stilling basin with a sloping bed is sometimes used to accommodate uncertain or variable tailwater rating curves. In such cases, the stilling basin is designed to prevent sweep-out under the lowest tailwater levels. If higher tailwater levels are encountered, the start of the hydraulic jump will move upstream on the sloping apron. This arrangement is thought to give a more rapid reduction in the maximum velocity than would occur with a submerged hydraulic jump.

An important case of a sloping jump is encountered when high tailwater levels cause the normal jump to shift upstream onto the chute of the spillway. Since the chutes are generally very steep (e.g., $\theta \cong 45°$) the nature of these jumps is distinctly different from those with small bed slopes. The jumps with large slopes behave like submerged jumps; this is to be expected since the jet in these jumps is subjected to a very small adverse piezometric gradient in comparison to the classical jump. McCorquodale and Smith (1977) studied hydraulic jumps on a 3H:1V chute that was followed by a standard USBR type stilling basin. It was found that the vertical distribution of the longitudinal

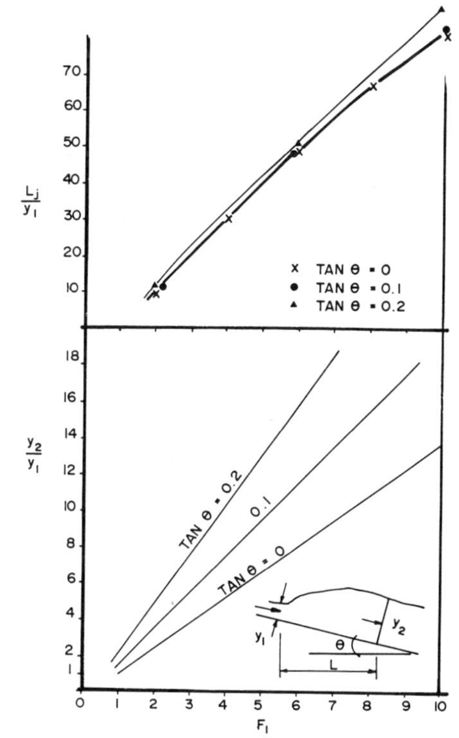

FIGURE 72. Jump length and sequent depth ratios for sloping rectangular channels [adapted from Chow (1959)].

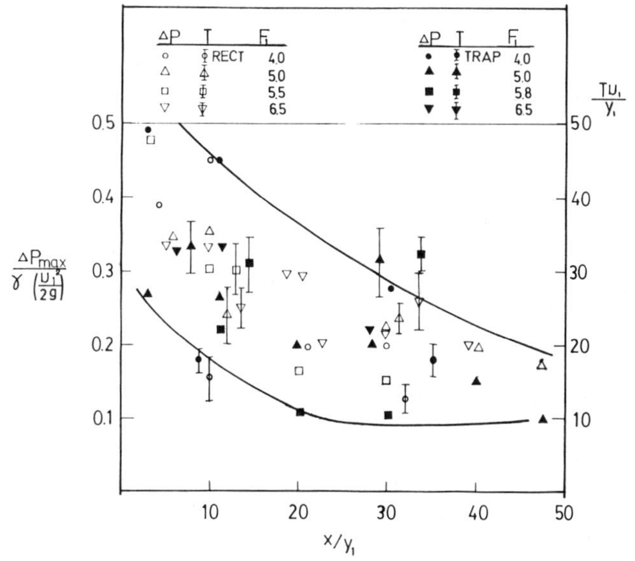

FIGURE 73. Maximum peak-to-trough pressure fluctuations and the corresponding periods in sloping jumps (chute slope 1V:3H) [after Siu and Ip (1977)].

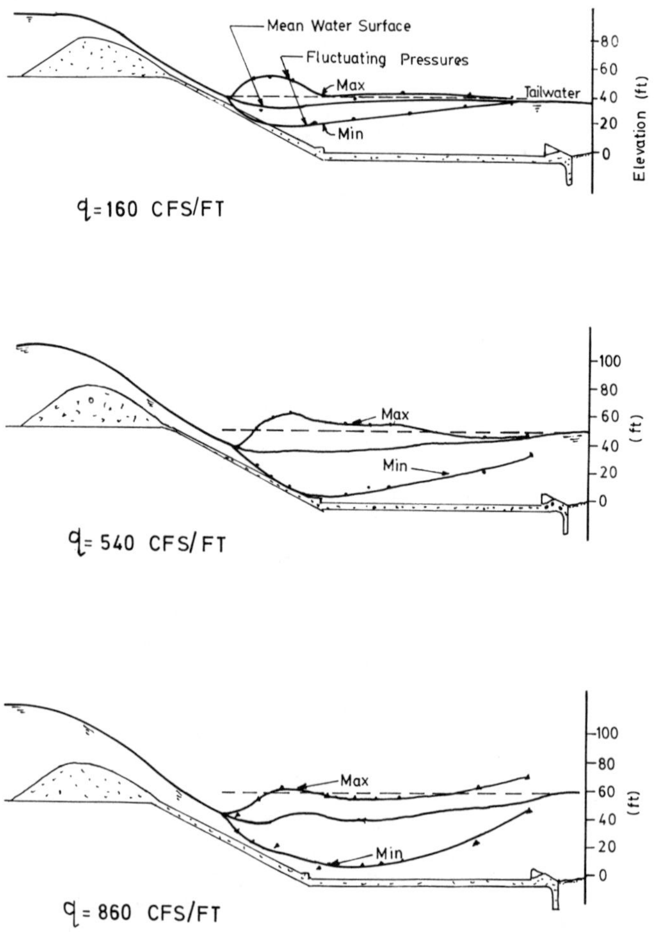

FIGURE 74. Typical maximum and minimum instantaneous pressure envelopes for 3 flows [after Bowers and Tsai (1969)].

velocity resembled the classical wall jet described by Rajaratnam (1976).

Siu and Ip (1977) studied the pressure fluctuations of a hydraulic jump on a 3H:1V chute. They placed pressure transducers near the bed on a splitter wall mid-way across the jump. They considered rectangular and trapezoidal cross sections. Their records were analyzed for maximum peak-to-trough pressure fluctuations (Δp^*_{max}) and the corresponding periods (T). Their results are summarized in Figure 73. Since the fluctuations are often asymmetrical the maximum positive or negative fluctuations can be close to Δp^*_{max}.

Bowers and Tsai (1969) point out the importance of considering the possibility of a jump on the chute and need to include the pressure fluctuations in the design of the chute and stilling basin. They noted that the pressure fluctuations can be transmitted through the drainage system and can results in large uplift forces. Figure 74 shows maximum and minimum instantaneous pressures for a jump originating on a spillway chute followed by a horizontal stilling basin (typical prototype dimensions have been given). They suggest that maximum pressure fluctuations of

$$\Delta p_{max}/(\varrho U_1^2/2) \cong \pm 0.4 \tag{79}$$

are possible with prototype periods of about 5 seconds or $fy_1/U_1 \cong .03$ in their example. An instantaneous pressure difference between two locations on the chute (potential uplift pressure head) of 50% of the initial velocity head was indicated in their study.

Uppal, Gulati and Sharma (1967), in connection with their study of the failure of a training wall at the Bhakra Dam Spillway, used a model to investigate the pressure fluctuation on the wall. In the prototype the velocity was of the order of 100 ft/sec (30 m/s) and this flow was highly aerated. The prototype jump occurred in the region where the chute was curved, i.e., the start of the bucket. The maximum differential pressure head between one side of the training wall and the other was 20% of the initial velocity head with $fy_1/U_1 \cong .01$ to .03 at $F_1 \cong 6$.

From the above studies it would appear that fluctuations in pressure head for a hydraulic jump starting on the chute of a spillway can be $\pm 40\%$ of the initial velocity head with fy_1/U_1 between 0.01 and 0.1.

CONCLUDING REMARKS

This article has attempted to summarize the published data on the internal flow in hydraulic jumps. The survey was limited to the internal flow in the most common hydraulic jumps, i.e., the classical jump (free rectangular), internal hydraulic jumps, submerged jumps, forced hydraulics, diverging flow jumps, depressed stilling basins, and hydraulic jumps on sloping beds. In addition to the sequent depth ratio and jump length characteristics, the mean velocities, the velocity length scales, turbulence intensities, turbulent shears and the pressure fluctuation data are discussed. The reader is referred to Rajaratnam (1967), ASCE Task Force (1964), Bradley *et al.* (1957), Elevatorski (1959) and McCorquodale (1985) for a summary of the information on other types of hydraulic jumps.

ACKNOWLEDGEMENTS

Some of the data presented in this chapter was developed with the help of my present and former graduate students. I acknowledge the assistance of E. Regts, M. Giratella, C. Y. Li, A. Hannoura, A. Khalifa, M. Hamam, P. Nettleton and S. Abdel-Gawad.

Some of the author's research reported herein was supported by operating grants from the Natural Sciences and Engineering Research Council of Canada.

NOTATION

A = flow area
A_B^* = effective area of baffle
b = velocity length scale
B = top width
c = coefficient
C = coefficient
C_D = baffle drag coefficient
C_d = baffle drag coefficient related to x
C_x = coefficient in baffle force equation
D = hydraulic mean depth
D = diameter of pipe
$E_{(f)}$ = spectral density function
f = frequency in hz
F = force
F_1 = number on the inflow to the jump
F_{B_1} = baffle force per unit width of jump
F_B' = random or fluctuating component of F_B'
$\bar{F}_B$ = mean value of F_B
F_{BW} = baffle wall force per unit width of jump
g = acceleration due to gravity
G = drag parameter for forced jump
h = depth in hydraulic jump
h_B = baffle height
h_B^* = effective baffle height
I = integral
K = coefficient
ℓ = mixing length
L = length
L_j = jump length
L = roller length
L_{j*} = influence length for a hydraulic jump
L_{jr_o} = length in a radial hydraulic jump
p = pressure
p' = random component of p
$\bar{p}$ = local average pressure
p_x = effective pressure at a point
P = force due to pressure
q = flow per unit width
Q = Volumetric flow of fluid
Q_a = air flow
Q_w = water flow
r_o = radius ratio of hydraulic jump = r_2/r_1
r_1 = initial radius of hydraulic jump
r_2 = radius to end of hydraulic jump
R_i = Richardson number
R_o = radius to start of submerged radial jump
R_2 = radius to end of submerged radial jump
s_B = baffle spacing
S = inlet submergence factor = y_s/y_1 = ψ
S' = Rajaratnam's submergence factor ($y_{2s}/y_2 - 1$)
S_B = blockage ratio
t = time
T = period of fluctuation
u = instantaneous longitudinal velocity (x-direction) (*Note:* for convenience w is also used for local mean value of u)
u' = random component of u
$\bar{u}$ = local mean value of u
u_s = surface velocity
U = depth average of $\bar{u}$
U_1 = initial velocity
U_2 = velocity at downstream section
U_m = maximum value of $\bar{u}$ in a cross section
v = velocity in the y-direction
w_B = baffle width
w = width or weight
x,y,z = Castesian coordinates
X_Bq = distance from start of jump to the baffle

y_1 = initial depth
y_2 = sequent for hydraulic jump
y_o = sequent depth ratio = y_2/y_1
y_s = depth at the gate in a submerged jump
y_{2s} = tailwater depth in a submerged jump
α = kinetic energy correction factor
α_a = air fraction
$\bar{\alpha}_a$ = depth average air fraction (also C_o)
β = momentum correction factor
β' = pressure correction factor
γ = specific weight
δ = boundary layer thickness
δ_1 = depth to $\bar{u} = U_m/2$
θ = bed angle
μ = dynamic viscosity
ν = kinematic viscosity
ϱ = density
τ = shear
ϕ = angle of divergence of side walls in radial jump

REFERENCES

Abdel-Gawad, S. M. and J. A. McCorquodale, "Modelling Submerged Radial Hydraulic Jumps," *Proceedings Annual Conference of Canadian Society for Civil Engineering,* Halifax, Nova Scotia, Canada (1984).

Abdel-Gawad, S. M. and J. A. McCorquodale, "Analysis of the Submerged Radial Jump," *Canadian Journal of Civil Engineering,* CSCE., Vol. 12 (Sept., 1985).

Arbhabhirama, A. and A. Abella, "Hydraulic Jump Within Gradually Expanding Channel," *Journal of the Hydraulics Division, ASCE,* 97:HY1, Proc. Paper 7831, pp. 31–41 (1971).

ASCE Task Force, "Energy Dissipators for Spillways and Outlet Works," *Journal of the Hydraulics Division, ASCE,* 90:HY1, Proc. Paper 3762 (January, 1964).

Avery, S. T. and P. Novak, "Oxygen Transfer at Hydraulic Structures," *Journal of Hydraulics Division, Proceedings, ASCE,* Vol. 104, No. HY11, pp. 1521–1540 (Nov., 1978).

Babb, A. F. and H. C. Aus, "Measurement of Air in Flow Water," *Journal of Hydraulics Division, Proceedings, ASCE,* Vol. 107, No. HY12, pp. 1615–1630 (Dec., 1981).

Baddour, R. E. and V. H. Chu, "Buoyant Surface Discharge on a Step and on a Sloping Bottom," Technical Report 75-2 (FML), Department of Civil Engineering and Applied Mechanics, McGill University, Montreal, Canada (1975).

Baddour, R. E. and H. Abbink, "Turbulent Underflow in a Short Channel of Limited Depth," *Journal of the Hydraulics Division, ASCE,* 109:6, pp. 722–740 (1983).

Basco, D. R., "Trends in Baffled, Hydraulic Jump Stilling Basin Designs of the Corps of Engineers Since 1947," Misc. Paper H-69-1, Waterway Experimental Station, Vicksburg, Mississippi (1969).

Basco, D. R. and J. R. Adams, "Drag Forces on Baffle Blocks in Hydraulic Jumps," *Journal of Hydraulics Division, Proceedings, ASCE,* Vol. 97, No. HY12, pp. 2023–2035 (Dec., 1971).

Berryhill, R. H., "Stilling Basin Experiences of Engineers," *Journal of the Hydraulics Division, ASCE,* 83, HY3 (June, 1957).

Blaisdell, F. W., "Development and Hydraulic Design, Saint Anthony Falls Stilling Basin," *Proceedings, ASCE,* 73:2 (February, 1947).

Blaisdell, F. W., "The St. Anthony Falls Stilling Basin," U.S. Conservation Service, Report SCS-TP-79 (1949).

Bourkov, V. J., "Experimental Studies of Pressure Pulsations in the Hydraulic Jump," (Russian), V28, Gosenergoizet Leningrad and Moscow (January, 1965).

Bowers, C. E. and F. Y. Tsai, "Fluctuating Pressure in Spillway Stilling Basins," *Journal of the Hydraulics Division, ASCE,* 95, HY6 (November, 1969).

Bowers, C. E. and F. Y. Tsai, "Fluctuating Pressures in Spillway Stilling Basins," *Journal of the Hydraulics Division, ASCE,* Vol. 95, No. HY6, pp. 2071–2079 (Nov., 1969).

Bradley, J. N. and A. J. Peterka, "The Hydraulic Design of Stilling Basins: Hydraulic Jumps on a Horizontal Apron (Basin I), *Journal of the Hydraulics Division, ASCE,* 83, HY5, pp. 1–24 (1957).

Bradley, J. N. and A. J. Peterka, "Hydraulic Design of Stilling Basins: Stilling Basin with Sloping Apron (Basin V), *Journal of the Hydraulics Division, ASCE,* 85, HY5, pp. 1–32 (1957).

Chong, C. L., W. B. Fann, and B. Sadeli, "Experimental Study of Radial Stilling Basins," B.A.Sc. Thesis, Department of Civil Engineering, University of Windsor, Windsor, Ontario, Canada (1985).

Chow, V. T., *Open Channel Hydraulics,* McGraw-Hill Book Co.. Inc., New York (1959).

Chu, V. H. and M. R. Vanvari, "Experimental Study of Turbulent Stratified Shearing Flow," *Journal of the Hydraulics Division, ASCE,* 102, HY6, pp. 691–706 (1976).

Chu, V. H. and R. E. Baddour, "Turbulent Gravity Stratified Flows," *Journal of Fluid Mechanics,* Vol. 138, pp. 353–378 (1984).

Elevatorski, E. A., *Hydraulic Energy Dissipators,* McGraw-Hill Book Co., Inc., New York (1959).

Forster, J. W. and R. A. Skrinde, "Control of the Hydraulic Jump by Sills," *Transactions, ASCE,* 115, pp. 973–1022 (1950).

Gomasta, S. K., M. K. Mittal, and P. K. Pande, "Hydrodynamic Forces on Baffle Blocks in Hydraulic Jump," XVII Congress IAHR, Vol. 4, pp. 453–460 (Aug., 1977).

Govinda Rao, N. S. and N. Rajaratnam, "The Submerged Hydraulic Jump," *Journal of the Hydraulics Division, ASCE,* 89, HY1, Proc. Paper 3404 (January, 1963).

Harleman, D. R. F., "Effect of Baffle Piers on Stilling Basin Performance," *Boston Society of Civil Engineers Journal,* 42:2 (April, 1955).

Hayakawa, N., "Internal Hydraulic Jump in Co-Current Stratified Flow," *Journal of the Engineering Mechanics Division, ASCE,* 96, EM5, pp. 797–800 (1970).

Kalinske, A. A. and J. W. Robertson, "Entrainment of air in Flowing Water-Closed Conduct Flow," *Transactions, ASCE,* 108, pp. 1435–1447 (1943).

Karki, K. S., S. Chander, and R. C. Malhotra, "Supercritical Flow Over Sills at Incipient Conditions," *Journal of the Hydraulics Division, ASCE,* 98, HY10, pp. 1753–1764 (October, 1972).

Karki, K. S., "Supercritical Flow Over Sills," *Journal of the Hydraulics Division, ASCE,* 102, HY10, pp. 1449–1459 (1976).

Khalifa, A. and J. A. McCorquodale, "Radial Hydraulic Jump," *Journal of the Hydraulics Division, ASCE,* 105, HY9, pp. 1065–1078 (September, 1979).

Khalifa, A., "Theoretical and Experimental Study of the Radial Hydraulic Jump," Ph.D. Thesis, University of Windsor, Windsor, Canada (1980).

Kindsvater, C. E., "The Hydraulic Jump in Sloping Channels," *Transactions, ASCE,* 109, pp. 1107–1154 (1944).

King, D. L., "Analysis of Random Pressure Fluctuations in Stilling Basin," *Proceedings 12th Congress IAHR,* 2 (1967).

Koh, R. C. Y., "Two-Dimensional Surface Warm Jets," *Journal of the Hydraulics Division, ASCE,* 97, HY6, pp. 819–836 (1971).

Koloseus, H. J. and D. Ahmad, "Circular Hydraulic Jump," *Journal of the Hydraulics Division, ASCE,* 95, HY1, Proc. Paper 6367, pp. 409–422 (January, 1969).

Le Mehaute, B., *An Introduction to Hydrodynamics and Water Waves,* Springer-Verlag, New York (1976).

Leutheusser, H. J. and Kartha, V. C., "Effects of Inflow Condition on Hydraulic Jump," *Journal of the Hydraulics Division, ASCE,* 98, HY8 (1972).

Leutheusser, H. J. and S. Alemu, "Flow Separation under Hydraulic Jump," *Journal of Hydraulic Research, IAHR,* 17:3, pp. 193–206 (1979).

Lipay, I. E. and V. F. Pustovoit, *Proceedings 12th Congress, IAHR,* Fort Collins, Colorado, pp. 362–369 (1967).

McCorquodale, J. A. and A. Khalifa, "Internal Flow in a Hydraulic Jump," *Journal of the Hydraulics Division, ASCE,* 106, HY3, pp. 355–367 (1983).

McCorquodale, J. A. and E. H. Regts, "A Theory for the Forced Hydraulic Jump," *Transactions of the Engineering Institute of Canada,* 11:C-1 (May, 1968).

McCorquodale, J. A. and M. Giratella, "Supercritical Flow over Sills," *Journal of the Hydraulics Division, ASCE,* 98, HY4, pp. 667–679 (1972).

McCorquodale, J. A., "Hydraulic Study of the Circular Clarifiers at the West Windsor Pollution Control Plant," for Lafontaine, Cowie, Buratto and Associates, Windsor, Canada (1976).

McCorquodale, J. A. and A. F. Smith, "Hydraulic Model Study of the Wilkesport Drop-Structure and Trapezoidal Stilling Basin," Report to M. M. Dillon Limited, Windsor, Canada (1977).

McCorquodale, J. A., "Temperature Measurements in Circular Clarifiers," for City of Windsor, Canada (1977).

McCorquodale, J. A., "Model Studies of the W. Darcy McKeough Dam and Diversion Control Studies," IRI-11-3, for M. M. Dillon Limited, Windsor, Canada (1978).

McCorquodale, J. A., "Report on the Hydraulics of the Expanded West Windsor Pollution Control Plant," for Lafontaine, Cowie, Buratto and Associates, Windsor, Canada (1979).

McCorquodale, J. A. and A. Khalifa, "Submerged Radial Hydraulic Jump," *Journal of the Hydraulics Division, ASCE,* 106, HY3, pp. 355–367 (March, 1980).

McCorquodale, J. A., "Hydraulic Jumps and Internal Flows," *Encyclopedia of Fluid Mechanics,* Gulf Publishing Co., West Orange, N. J. (1985).

Mehotra, S. C., "Circular Jumps," *Journal of the Hydraulics Division, ASCE,* 100, HY8, pp. 1133–1140 (August, 1974).

Mehotra, C., "Length of Hydraulic Jump," *Journal of the Hydraulics Division, ASCE,* 162, HY7, pp. 1027–1033 (July, 1976).

Moore, W. L. and K. Meshgin, "Adaptation of the Radial Energy Dissipator for Use with Circular or Box Culverts," Research Report 116-1, University of Texas at Austin (1970).

Moses, H. L., "A Strip-Integral Method for Predicting the Behaviour of the Turbulent Boundary Layers," *Proceedings, Computation of Turbulent Boundary Layers, AFOSR-IFP-Stanford Conference* (1968).

Nagaratnam, S., "The Mechanism of Energy Dissipation," M.S. Thesis, University of Iowa (June, 1957).

Narasimhan, S. and P. Bharagava, "Pressure Fluctuations in Submerged Jump," *Journal of the Hydraulics Division, ASCE,* 102, HY3, pp. 339–350 (March, 1976).

Narayanan, R., "Wall Jet Analogy to Hydraulic Jump," *Journal of the Hydraulics Division, ASCE,* 101, HY3, Proc. Paper 11172, pp. 347–360 (March, 1975).

Narayanan, R., "Pressure Fluctuations Beneath Submerged Jump," *Journal of the Hydraulics Division, ASCE,* 104, HY9, Proc. Paper 14039, pp. 1331–1342 (September 1978).

Narayanan, R. and L. S. Schizas, "Force Fluctuations on Sill of Hydraulic Jump," *Journal of the Hydraulics Division, ASCE,* 106, HY4, pp. 589–599 (April, 1980).

Narayanan, R. and L. S. Schizas, "Force on Sill of Forced Jump," *Journal of Hydraulic Division, ASCE,* Vol. 106, No. HY4, pp. 1159–1172 (1980).

Nettleton, P. C. and J. A. McCorquodale, "Radial Stilling Basins with Baffles," *Proceedings 6th Canadian Hydrotechnical Conference,* Ottawa, Canada, pp. 651–670 (1983).

Nettleton, P. C., "The Forced Radial Hydraulic Jump," M.A.Sc. Thesis, Department of Civil Engineering, University of Windsor, Windsor, Canada (1983).

Odd, N. and J. Rodger, "Vertical Mixing in Stratified Tidal Flows," *Journal of the Hydraulics Division, ASCE,* 104, HY3, Proc. Paper 13599, pp. 337–351 (March, 1978).

Pillai, N. N. and T. E. Unny, "Shapes for Appurtenances in Stilling Basins," *Journal of the Hydraulics Division, ASCE* (November, 1964).

Rajaratnam, N., "Profile Equation of the Hydraulic Jump," *Water Power,* 14, pp. 324–327 (1962).

Rajaratnam, N., "The Forced Hydraulic Jump," *Water Power,* 16, pp. 14–19, 61–65 (1964).

Rajaratnam, N., "The Hydraulic Jump as a Wall Jet," *Journal of the Hydraulics Division, ASCE,* 91, HY5, Proc. Paper 4482, pp. 107–132 (September, 1965).

Rajaratnam, N., "Submerged Hydraulic Jump," *Journal of the Hydraulics Division, ASCE,* 91, pp. 71–96 (1965).

Rajaratnam, N., "Hydraulic Jumps," in *Advances in Hydroscience,* V. T. Chow, ed. Academic Press, New York, NY, pp. 198–280 (1967).

Rajaratnam, N. and K. Subramanya, "Profile of the Hydraulic Jump," *Journal of the Hydraulics Division, ASCE,* 94, HY3, Proc. Paper 5931, pp. 663–673 (May, 1968).

Rajaratnam, N. and K. Subramanya, "Hydraulic Jump Below Abrupt Symmetrical Expansions," *Journal of the Hydraulics Division, ASCE,* 94, HY2, pp. 481-503 (1968).

Rajaratnam, N. and V. Murahari, "A Contribution to Forced Hydraulic Jumps," *Journal of Hydraulic Research, IAHR,* 9:2 (1971).

Rajaratnam, N., *Turbulent Jets,* Elsevier Scientific Co., New York (1976).

Rajaratnam, N. and N. Ortiz, "Hydraulic Jumps and Waves at Abrupt Drops," *Journal of the Hydraulics Division, ASCE,* 103: 4, pp. 381-394 (1977).

Rajaratnam, N. and S. Subramanyan, "Plane Turbulent Buoyant Surface Jets and Jumps," *Journal of Hydraulics Division, IAHR,* Vol. 23, No. 2, pp. 131-146 (1985).

Rand, W., "Flow Over a Vertical Sill in an Open Channel," *Journal of the Hydraulics Division, ASCE,* 91, HY4, pp. 97-121 (1965).

Ranga Raju, K. G., Kitaal, M. S. Verma, and V. R. Ganeshan, "Analysis of Flow Over Baffle Blocks and End Sills," *Journal of Hydraulic Research, IAHR,* Vol. 18, No. 3, pp. 227-241 (1980).

Regts, E., "A Theoretical and Experimental Study of the Forced Hydraulic Jump," M.S. Thesis, University of Windsor, Windsor, Canada (1967).

Resch, F. J. and H. J. Leutheusser, "Research Studies on Stilling Basins, Energy Dissipators, and Associated Appurtenances," Hydraulic Laboratory Report No. Hyd-399, United States Bureau of Reclamation, Denver, Colorado (June 1, 1955).

Resch, F. J. and H. J. Leutheusser, "Mesures de Turbulence dans le Ressault Hydraulique," *La Houille Blanche,* No. 1, pp. 279-294 (1971).

Resch, F. J. and H. J. Leutheusser, "Reynolds Stress Measurements in Hydraulic Jumps," *Journal of Hydraulic Research, IAHR,* 10, 4, pp. 409-430 (1972).

Resch, F., H. Leutheusser, and S. Alemu, "Bubbly Two-Phase Flow in a Hydraulic Jump," *Journal of the Hydraulics Division, ASCE,* 100, HY1, Proc Paper 10297, pp. 137-150 (1974).

Resch, F. J., H. J. Leutheusser, and M. Coantic, "Study of the Kinematic and Dynamic Structures of the Hydraulic Jump," *Journal of Hydraulic Research, IAHR,* Vol. 14, No. 4, pp. 293-319 (1976).

Rouse, H., *Elementary Mechanics of Fluids,* Dover Publications Inc., New York, NY (1946).

Rouse, H., *Hydraulic Engineering,* John Wiley and Sons, New York, NY (1949).

Rouse, H. and I. Simon, *History of Hydraulics,* Dover Publications, Inc., New York, NY (1957).

Rouse, H., T. T. Siao, and R. Nagaratnam, "Turbulence Characteristics of the Hydraulic Jump," *Journal of the Hydraulics Division, ASCE,* 84, HY1, Proc. Paper 1528 (February, 1958).

Schiebe, F. R. and C. E. Bowers, "Boundary Fluctuations Due to Macroturbulence in Hydraulic Jumps," Laboratory Report, St. Anthony Falls Hydraulic Laboratory, University of Minnesota (1971).

Scott-Moncrieff, A., "Behaviour of Water Jet in a Diverging Shallow Open Channel," Fifth Australian Conference on Hydraulics and Fluid Mechanics, Christ Church, New Zealand, pp. 42-48 (December, 1974).

Sharma, Hari R., "Air Entrainment in High Head Gated Conduits," *Journal of the Hydraulics Division, ASCE,* 102, 11, pp. 1629-1646 (November, 1976).

Siao, T. T., "Characteristics of Turbulence in an Air-Flow Model of the Hydraulic Jump," Ph.D. Dissertation, State University of Iowa (1954).

Siu, F. P. and R. K. C. Ip, "Experimental Study of Trapezoidal Stilling Basins," B.A.Sc. Thesis, Department of Civil Engineering, University of Windsor, Windsor, Ontario, Canada (1977).

Talatt, A. M., "Abfluss-verhaltnissen an Sohlabsturzen," Dr. Ing Dissertation, Technical University Carolo-Wilhelmina at Braunschweig, West Germany (1983).

Thomas, A. R. and G. H. Lean, "The Vibration of a Submerged Wall Exposed to a Jet," *Proceedings, IAHR Congress, London,* pp. 89-100 (1963).

Tyagi, D. M., P. K. Prande, and M. K. Mittal, "Drag on Baffle Walls in Hydraulic Jump," *Journal of the Hydraulics Division, ASCE,* 104, HY4, pp. 515-525 (1978).

Unny, T. E., "The Spiral Hydraulic Jump," *Proceedings International Association of Hydraulics Research,* Dubrovnik, Yugoslavia, pp. 32-42 (1961).

Uppal, H. L, T. D. Gulati, and B. A. D. Sharma, "A Study of Damage to the Central Training Wall of Bhakra Dam Spillway," *Journal of Hydraulic Research, IAHR,* Vol. 5, No. 3, pp. 209-224 (1967).

U.S. Army Corps of Engineers, *Hydraulic Design Criteria,* Hydraulic Design Charts (1964).

U.S. Bureau of Reclamation, "Research Studies on Stilling Basins, Energy Dissipators and Associated Appurtenances," Hydraulic Laboratory Report No. Hyd-399 (1955).

U.S. Department of Interior Bureau of Reclamation, *Design of Small Dams,* Second Edition, pp. 442-446 (1974).

Vasiliev, O. F. and V. I. Bukreyev, "Statistical Characteristics of Pressure Fluctuations in the Region of Hydraulic Jump," *Proceedings 12th Congress IAHR,* 2 (1967).

Weide, L., "The Effect of the Size and Spacing of Floor Blocks in the Control of the Hydraulic Jump," M.Sc. Thesis, Colorado A and M (1951).

Wilson, E. and A. Turner, "Boundary Layer Effects on Hydraulic Jump Location," *Journal of the Hydraulics Division, ASCE,* 98, HY7, pp. 1127-1142 (July, 1972).

Wisner, P., "Bottom Pressure Pulsations of the Closed Conduit and Open Channhel Hydraulic Jump," *Proceedings 12th Congress, IAHR,* 2 (1967).

Yih, C. S. and C. R. Guha, "Hydraulic Jump in a Fluid System of Two Layers," *Tellus,* 7, pp. 358-366 (1955).

Parabolic Canal Design and Analysis

LYMAN S. WILLARDSON*

INTRODUCTION

Unlined channels carrying water tend to assume a parabolic shape. River beds, unlined canals and irrigation furrows all tend to approximate a stable parabolic shape more closely than the triangular or trapezoidal shape in which they are initially constructed. When channels are lined, they can be made hydraulically stable in any shape. Unlined channels could be made hydraulically more stable by constructing them initially in a parabolic shape. There is also the possibility that they may be economically and hydraulically more efficient.

In the Soviet Union, canals with parabolic cross sections are more common than they are in the United States. Prefabricated irrigation water distribution channels of the form shown in Figure la are used extensively in irrigation systems of the Middle Asian Republic, such as Uzbekistan and Tadjekistan. The standard size of the prefabricated sections is 8 meters long (Figure la). The heights vary from 0.5 to 1.5 meters Figure lb shows methods of supporting the flume sections on or above ground. The joints between flume sections are sealed with rubber gaskets and asphalt mastic. The flume sections are reinforced in such a way that they only need support at the joints.

Large irrigation canals such as the Kahovsky Canal east of Odessa that delivers water for irrigation from the Dniepr River to the Crimea are constructed with a parabolic cross section. The Kahovsky Canal carries a maximum of 600 m^3/sec (21,000 cfs). The top width is 200 meters and the flow depth is 12 meters. The canal has a 2 mm polyethylene film seepage barrier under a one meter protective layer of sand. There is a longitudinal concrete lining panel that protects the canal bank at the water line (Figure 2).

In the United States, parabolic channels are relatively uncommon and are limited to small irrigation laterals being paved by hand (Figure 3) with prefabricated blocks.

One of the factors preventing the acceptance of parabolic channels is the assumed difficulty in computing the cross section and other design parameters. The advent of small programmable calculators has made these computations relatively simple. Another negative factor is the difficulty of construction. Slipforms can be easily adapted to a parabolic shape, but forming an accurate shape of the subgrade is difficult. Large parabolic canals are more simple to construct than trapezoidal channels. A simple bulldozer is adequate.

PARABOLIC CROSS SECTIONS

The general form of the equation of a parabola that opens upward with its center at the origin is given by Equation (1):

$$x^2 = 2\,Py \qquad (1)$$

Figure 4 shows the configuration of a parabolic canal and equations for the area and wetted perimeter. The factor P defines the shape of the parabola. P is related to the top width and depth of the canal. The value of P can be found in terms of the limiting side slope, m, of the canal at the maximum water surface. The m value is the allowable soil side slope in the canal.

Beginning with the equation of Chezy:

$$Q = AC\sqrt{RS} \qquad (2)$$

where Q is the flow in m^3/sec, A is the cross-sectional area

*Department of Agricultural and Irrigation Engineering, Utah State University, Logan, UT

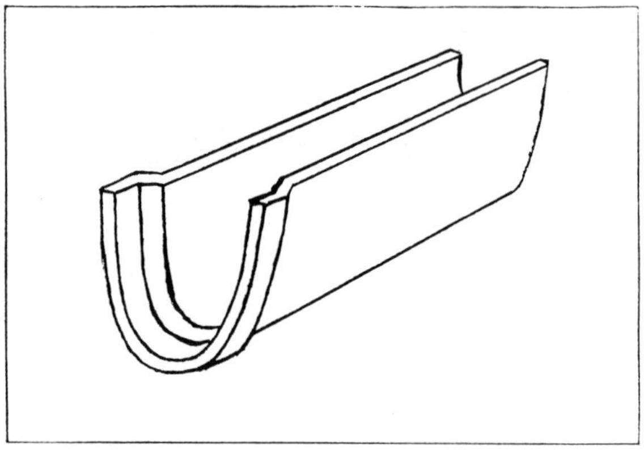

FIGURE 1a. The shape of the prefabricated flume section.

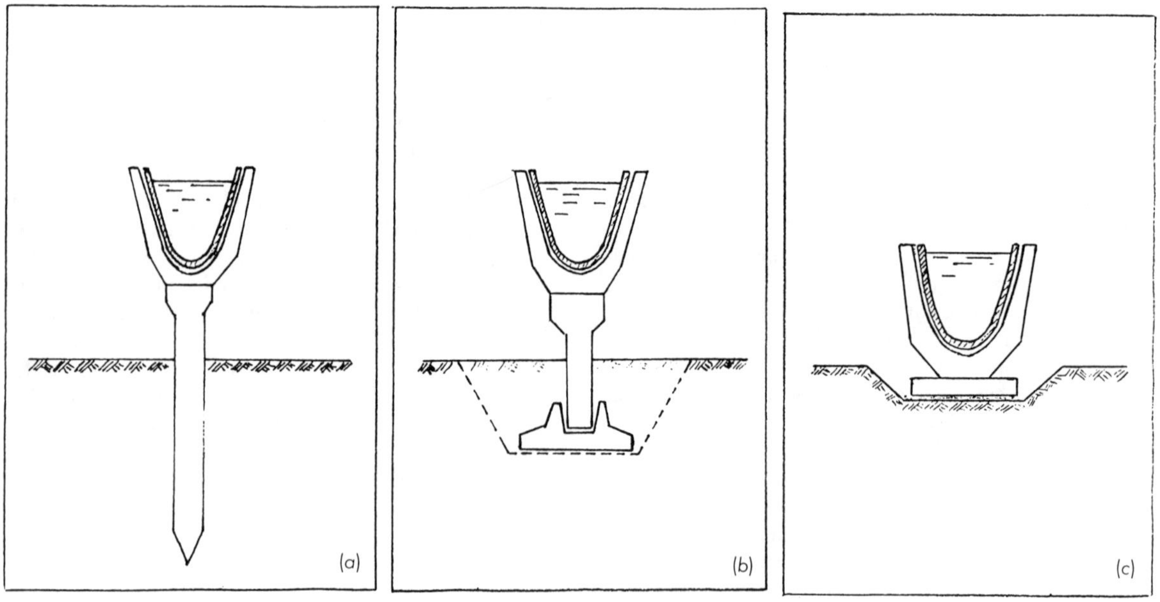

FIGURE 1b. Methods of supporting the flume sections.

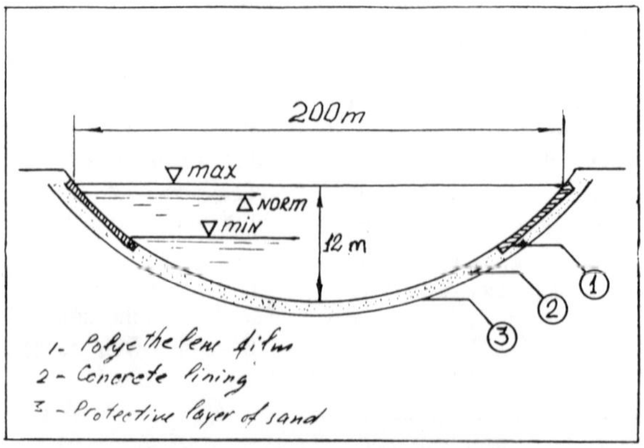

FIGURE 2. The cross section of the Kahovsky Canal.

FIGURE 3. Placing precast concrete slabs with tongue and groove joints.

of flow in m², C is the Chezy coefficient, R is the hydraulic radius and S is the slope of the canal.

The derivative of Equation (1) will give the side slope of the parabolic canal section at any distance from the center. The slope of the canal bank at $B/2$ and depth equal to h is:

$$m = \frac{2P}{B} \tag{3}$$

The equation for wetted perimeter for the parabola is shown in Figure 4. If Equation (3) is solved for P and is substituted in the equation for wetted perimeter, an equation for wetted perimeter in terms of m is obtained:

$$WP = \frac{B}{2}\left[\sqrt{1 + \frac{1}{m^2}} + m \ln\left(\frac{1}{m} + \sqrt{1 + \frac{1}{m^2}}\right)\right] \tag{4}$$

The function within the brackets will be designated N for the discussion that follows. Substituting Equation (4) and the equation for area from Figure 4 in Chezy's equation

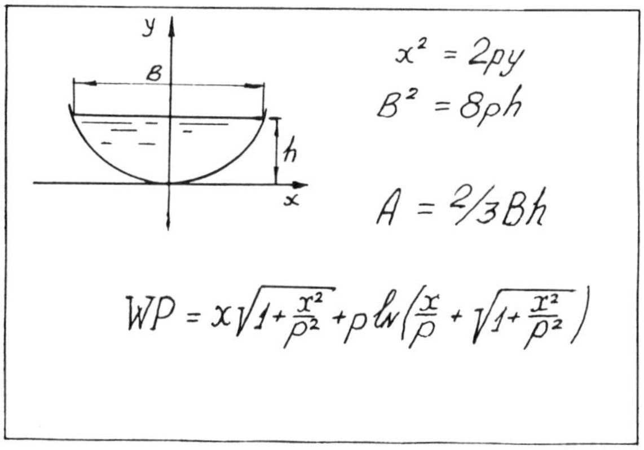

FIGURE 4. The main parameters of a parabolic cross section.

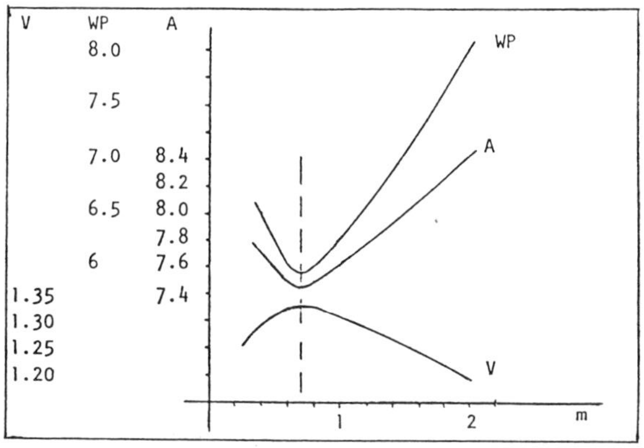

FIGURE 5. The changes in velocity, area and wetted perimeter.

TABLE 1. Comparison between Parabolic and Trapezoidal Canals.

	m	Q = 1.0 m³/S			Q = 10 m³/S			Q = 75 m³/S		
		Trap	Par	±	Trap	Par	±	Trap	Par	±
WP,m	0.7	3.19	2.54	−0.65	7.56	6.02	−1.54	16.09	12.83	−3.26
	1.0	3.29	2.64	−0.65	7.79	6.27	−1.52	16.59	13.35	−3.24
	1.5	3.59	2.99	−0.60	8.51	7.09	−1.42	18.12	15.09	−3.03
A,m²	0.7	1.46	1.33	−0.13	8.20	7.50	−0.70	37.17	33.98	−3.19
	1.0	1.48	1.35	−0.13	8.30	7.62	−0.68	37.63	34.52	−3.11
	1.5	1.53	1.42	−0.11	8.60	8.00	−0.60	38.98	26.26	−2.72
V,m/s	0.7	0.68	0.75	+0.07	1.22	1.33	+0.11	2.02	2.21	+0.19
	1.0	0.68	0.74	+0.06	1.20	1.31	+0.11	1.99	2.17	+0.18
	1.5	0.65	0.70	+0.05	1.16	1.25	+0.09	1.92	2.07	+0.15

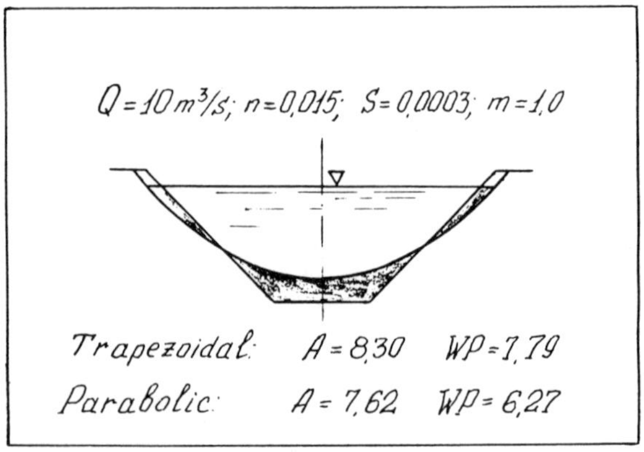

FIGURE 6. The shapes of comparable trapezoidal and parabolic canals.

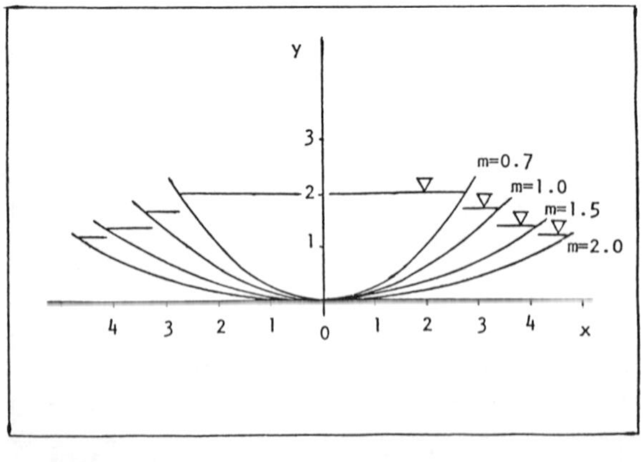

FIGURE 7. The effect of the value m on the shape of a parabolic cross section.

[Equation (2)] and solving for h gives:

$$h = \left(\frac{0.31\ Q \cdot n \cdot N^{2/3}}{S \cdot m}\right)^{3/8} \tag{5}$$

Equation (5) is only a function of the known values of Q, n, s and m where n is the Manning Roughness Coefficient. When the depth h is known, the top width can be calculated from

$$B = 4m\, h \tag{6}$$

and the value of P can be obtained from

$$P = \frac{B^2}{8h} \tag{7}$$

and the wetted perimeter can be obtained from

$$WP = \frac{BN}{2} \tag{8}$$

A parameter that is useful in comparing parabolic canal cross sections is the side slope m at the water line.

Chugayev [1] developed an analysis to find the most efficient hydraulic section for trapezoidal canals for a given side slope and suggested that it could be applied to parabolic channels. King [2] mentioned parabolic channels, but did not provide a design procedure. The design procedure presented here is related to selection of the maximum allowable side slope at the water surface.

The hydraulically optimum cross section for a parabolic channel occurs when $m = 0.7$. An example for a given flow rate is shown in Figure 5. The velocity is maximum and both the area and wetted perimeter are minimum for $m = 0.7$. For different canal discharges, other combinations of canal design parameters will give the same result. The most efficient hydraulic section occurs for $m = 0.7$.

Table 1 is a comparison between parabolic and trapezoidal channels for different flow rates. The trapezoidal channel was designed for 1:1 side slopes. The slope of the parabolic channel at the water line is also 1:1. The wetted perimeter of the parabolic channel is 19.5% less than the wetted perimeter of the trapezoidal channel. The smaller wetted perimeter would represent an equal percent savings of lining material. The shapes of comparable trapezoidal and parabolic canals are shown in Figure 6. Even though the m value is greater than the optimum value of 0.7, the parabolic section is more hydraulically efficient, as evidenced by the smaller cross section of the parabolic channel. There is an 8% difference in cross-sectional area in favor of the parabolic section.

Figure 7 shows the effect of the value of m on the shape of a parabolic cross section. The shape for $m = 0.7$ is hydraulically optimum, but soil stability may dictate the use of flatter side slopes. Figure 5 shows the changes in velocity, area and wetted perimeter that can be expected from changes in m.

The most complicated equation in the design of a parabolic channel is the one for wetted perimeter [Equation (4)]. A programmable calculator makes solution of this equation relatively simple. For a programmable calculator having Reverse Polish Notation, Q, N, s and m can be entered and the values of h, B, P, Area, and WP, R and velocity can be calculated in less than 90 program steps. An example program for an HP-67 calculator is shown in the Appendix. The execution of the program as written requires only nine addressable memories. The program can be written for any other calculator or computer in a straightforward manner from the equations given.

CONCLUSIONS

Parabolic canal sections are advantageous in that they are close to the natural shape of open channels. They are also hydraulically efficient as is indicated by the comparison with the trapezoidal channel in Figure 6. Parabolic channels may be more difficult to construct precisely, but the savings in excavation and lining material due to the smaller cross section may make them more economical. A lined parabolic channel has no sharp angles of stress concentration where cracks may occur, and can be prefabricated in molded sections. Large canals can be easily constructed by bulldozers, tractors and scrapers because only one point in the cross section is at maximum side slope. The structural shape of a prefabricated parabolic shape makes it possible to use the section as a suspended flume with support only at the joints.

The development shown in Figure 5 can serve as a guide to efficient design of parabolic canal sections for maximum stability and hydraulic efficiency.

NOTATION

The following symbols are used in this chapter:

A = cross-sectional area
B = top-width of a channel
C = Chezy coefficient
h = maximum depth of water in a channel
m = casual side slope at the water surface
n = Manning roughness coefficient
N = function
P = parameter of a parabola
Q = flow rate
R = hydraulic radius
S = hydraulic slope of a channel
WP = wetted perimeter
x,y = horizontal and vertical coordinates

REFERENCES

1. Chugayev, R. R., *Hydraulics (Technical Fluid Mechanics) Energiia.*, Leningrad, U.S.S.R. (1960); Translated from Russian, U.S. Department of the Interior, WPRS, Washington, DC (1977).
2. King, H. W., *Handbook of Hydraulics,* McGraw-Hill Book Co., Inc., NY and London (1939).

APPENDIX

The Program for a Parabolic Channel Parameter (for HP-67 Calculator)

001	fLBLA	÷		3		2
	R/S→Q	RCL 4		enter		×
	STO 0	+		8		3
	R/S→n	STO 5		÷	070	÷
	STO 1	2		hy^x		f-x-(A)
	R/S→S	enter	050	f-x-(h)		STO 9
	STO 2	3		STO 6		RCL 7

(continued)

The Program for a Parabolic Channel Parameter (for HP-67 Calculator) (continued)

008	R/S→m	030	÷		RCL 3		RCL 5
	STO 3		hy^x		×		×
010	gx^2		RCL0		4		2
	h 1/x		×		×		÷
	1		RCL 1		f-x-(B)		f-x-(WP)
	+		×		STO 7		STO A
	f√		·		RCL 3	080	h 1/x
	STO 4		3		×		RCL 9
	RCL 3		1	060	2		×
	h 1/x		×		÷		f-x-(R)
	+	040	RCL 2		f-x-(p)		RCL 0
	fLN		f√		STO 8		RCL 9
020	RCL 3		÷		RCL 6		÷
	×		RCL 3		RCL 7		f-x-(v)
	2		÷		×	088	GTO A

Submerged Logarithmic Weirs

Hanna Majcherek*

ABSTRACT

Theoretical and experimental studies of the effect of weir submersion on the logarithmic discharge characteristic were performed. The problem was solved theoretically by accepting the simplifying assumption that the coefficient of discharge has a constant value. Strict formulae which determine the relative accuracy, ϵ, of conformity with the logarithmic characteristic for the weir operating under the submersion condition were derived as a function of the submersion ratio, h/H, and the filling ratio, a/H. The function $\epsilon(h/H,\ a/H)$ was derived for three particular cases: for the weir with rectangular and triangular bottom part and also for the weir with the shape having no angle points. Because of the complicated structure of this function, an approximate formula which simplifies computation was derived. Experimental studies of the weir with rectangular bottom part showed that the coefficient of discharge is a function of nappe head; however, the acceptance of a constant value for this coefficient results in a discrepancy between theoretically computed and experimentally derived values of discharge does not exceed 3.5%.

KEY WORDS

Hydraulics, water flow, weirs.

INTRODUCTION

Logarithmic weirs are characterized by high sensitivity, i.e., small variations in flow rate cause great variations of nappe head and are therefore used especially to measure the flow rate. The problem of determining the shape of free falling, sharpcrested weirs has been presented in previous works [1,2,5,6]. The aim of this study is to determine how far the submersion of weirs effects the discharge characteristic of free falling weirs. Theoretical and experimental examinations will be conducted for three types of weirs with different shapes of lower weir section. Equations determining the submersion correction will be derived with the simplifying assumption consisting in the acceptance of constant coefficient of discharge. This coefficient accounts for the effect of many factors on the liquid flow through the weirs, such as liquid viscosity, surface tension, velocity and direction of flow approach to the weir. The coefficient of discharge is a function of nappe head and also of submersion head in the case of weir submersion. The effect of the accepted simplifying assumption on the precision of derived equations was examined experimentally.

THEORETICAL CONSIDERATIONS

The analysis of the weir operation under submerged flow conditions was accomplished for the weirs with the shape composed of two sections, described with the function $f_1(y)$ for $y \le a$ and $f_2(y)$ for $y \ge a$ (Figure 1). In free falling conditions the above-mentioned weirs satisfy the discharge characteristic described with a power function $Q_1\ (H)$ for $H \le a$ and logarithmic function $Q_2\ (H)$ for $H \ge a$. The effect of weir submersion on the characteristic $Q_2(H)$ alteration can be calculated from the following functions:

$$\varepsilon\left(\frac{h}{H}, \frac{a}{H}\right) = \frac{Q_2\left(\frac{a}{H}\right) - \bar{Q}_2\left(\frac{h}{H}, \frac{a}{H}\right)}{Q_2\left(\frac{a}{H}\right)} \quad \text{for } \frac{h}{H} \leqslant \frac{a}{H} \qquad (1)$$

$$\varepsilon\left(\frac{h}{H}, \frac{a}{H}\right) = \frac{Q_2\left(\frac{a}{H}\right) - \bar{\bar{Q}}_2\left(\frac{h}{H}, \frac{a}{H}\right)}{Q_2\left(\frac{a}{H}\right)} \quad \text{for } \frac{h}{H} \geqslant \frac{a}{H} \qquad (2)$$

*Institute of Environmental Engineering, Technical University of Poznan, Poznan, Poland

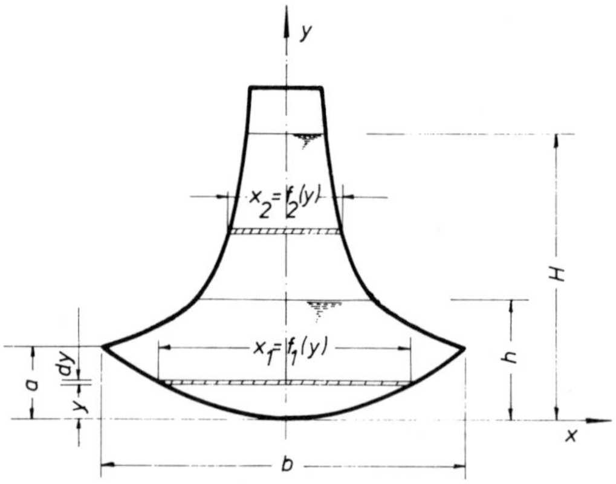

FIGURE 1. Weir described by function $x_1 = f_1(y)$ for $y \leq a$ and $x_2 = f_2(y)$ for $y \geq a$.

in which H = nappe head; h = submersion head; a = height of the lower weir section, described with a function $f_1(y)$; $Q_2(a/H)$ = discharge characteristic through the free falling weir, for $H \geq a$; $\overline{Q}_2(h/H, a/H)$ = discharge characteristic through the submerged weir, for $H > a$ and $h \leq a$; $\overline{\overline{Q}}_2(h/H, a/H)$ = discharge characteristic through the submerged weir for $H > a$ and $h \geq a$. Equations determining the discharge characteristic for the submerged weir have the form:

$$\overline{Q}_2(H, h) = C_d\sqrt{2g}\left[\sqrt{H-h}\int_0^h f_1(y)dy + \int_h^a \sqrt{H-y}\, f_1(y)dy + \int_a^H \sqrt{H-y}\, f_2(y)dy\right] \tag{3}$$

for $H > a$; $h \leqslant a$

$$\overline{\overline{Q}}_2(H, h) = C_d\sqrt{2g}\left[\sqrt{H-h}\left(\int_0^a f_1(y)dy + \int_a^h f_2(y)dy\right) + \int_h^H \sqrt{H-y}\, f_2(y)dy\right] \tag{4}$$

for $H > a$; $h \geqslant a$

in which C_d is a coefficient of discharge.

General characteristic equations for free falling weirs with the shape having angle points at the height $y = a$ (that is at the intersection points between the lower and upper sections of the weir) are accepted in the form:

$$Q_1(H) = k_1 H^{m_1} \tag{5}$$

$$Q_2(H) = k_2 \ln\left(1 + \frac{H - \lambda a}{a}\right) \tag{6}$$

in which k_1, k_2 = proportionality factors; λ = coefficient determining datum level for nappe head; and m_1 = power function exponent (its value should be taken from the range $3/2 \leq m_1 \leq 5/2$). Basic equations describing weir shape have been previously derived [1,2]:

$$f_1(y) = \frac{2k_1 m_1(m_1 - 1)}{C_d\pi\sqrt{2g}} \times B\left(m_1 - 1, \frac{1}{2}\right) y^{m_1 - (3/2)} \tag{7}$$

$$f_2(y) = f_1(y) - \frac{2}{C_d\pi\sqrt{2g}}\left\{\int_a^y \frac{1}{\sqrt{y - H}} \times \left[k_1 m_1(m_1 - 1)H^{m_1 - 2} + \frac{k_2}{[H + a(1 - \lambda)]^2}\right] dH\right\} \tag{8}$$

in which $B(m_1 - 1, 1/2)$ = beta function. The coefficients k_1, k_2 and λ can be obtained from two boundary conditions for $Q_1(H)$ and $Q_2(H)$ when $H = a$:

$$Q_1(a) = Q_2(a) \tag{9}$$

$$\left[\frac{d}{dH} Q_1(H)\right]_{H=a} = \left[\frac{d}{dH} Q_2(H)\right]_{H=a} \tag{10}$$

and from the third condition that the width of the weir at the height $y = a$ is equal to b:

$$f_1(a) = b \tag{11}$$

If conditions 9–10 are satisfied then $f_1(a) = f_2(a)$. Coefficients k_1, k_2 and λ are defined as:

$$k_1 = \frac{C_d\pi\sqrt{2g}\, a^{3/2 - m_1} b}{2m_1(m_1 - 1)B\left(m_1 - 1, \frac{1}{2}\right)} \tag{12}$$

$$k_2 = \frac{C_d\pi\sqrt{2g}a^{3/2}b(2 - \lambda)}{2(m_1 - 1)B\left(m_1 - 1, \frac{1}{2}\right)} \tag{13}$$

$$(2 - \lambda) \ln (2 - \lambda) = \frac{1}{m_1} \tag{14}$$

The characteristics for free falling weirs with the shape having no angle points at the height $y = a$ are described by the following discharge functions:

$$Q_1(H) = k_{10}H^{m_{10}} + k_{11}H^{m_{11}} \tag{15}$$

$$Q_2(H) = k_2 \ln \left(1 + \frac{H - \lambda a}{a}\right) \tag{16}$$

in which k_{10}, k_{11}, and k_2 = proportionality factors; and m_{10} and m_{11} = exponents. The shape of the weir is described by equations:

$$f_1(y) = \frac{2}{C_d\pi\sqrt{2g}}\left[k_{10}m_{10}(m_{10} - 1) \times B\left(m_{10} - 1, \frac{1}{2}\right) y^{m_{10}-3/2} + k_{11}m_{11}(m_{11} - 1) \times B\left(m_{11} - 1, \frac{1}{2}\right) y^{m_{11}-3/2}\right] \tag{17}$$

$$f_2(y) = f_1(y) - \frac{2}{C_d\pi\sqrt{2g}}\left\{\frac{1}{\sqrt{y - a}} \times \left(k_{10}m_{10}a^{m_{10}-1} + k_{11}m_{11}a^{m_{11}-1} - \frac{k_2}{a(2 - \lambda)}\right) + \int_a^y \frac{1}{\sqrt{y - H}} \times \left[k_{10}m_{10}(m_{10} - 1)H^{m_{10}-2} + k_{11}m_{11}(m_{11} - 1)H^{m_{11}-2} + \frac{k_2}{[H + a(1 - \lambda)]^2}\right] dH\right\} \tag{18}$$

The coefficients k_{10}, k_{11}, k_2, and λ are determined from Equations (9–11) and from the fourth boundary condition for $Q_1(H)$ and $Q_2(H)$ when $H = a$:

$$\left[\frac{d^2}{dH^2} Q_1(H)\right]_{H=a} = \left[\frac{d^2}{dH^2} Q_2(H)\right]_{H=a} \tag{19}$$

When three conditions 9, 10, and 19 are satisfied, $f_1(a)$ and $f_2(a)$ have a common tangent (i.e., the shape of the weir is smooth at $y = a$). In this case the factors k_{10}, k_{11}, k_2, and λ can be expressed as:

$$k_{10} = k_2 \frac{m_{11}(2 - \lambda)\ln(2 - \lambda) - 1}{a^{m_{10}}(m_{11} - m_{10})(2 - \lambda)} \tag{20}$$

$$k_{11} = -k_2 \frac{m_{10}(2 - \lambda)\ln(2 - \lambda) - 1}{a^{m_{11}}(m_{11} - m_{10})(2 - \lambda)} \tag{21}$$

$$k_2 = \frac{C_d\pi\sqrt{2g}a^{3/2}b(2 - \lambda)(m_{11} - m_{10})}{2m_{10}(m_{10} - 1)B\left(m_{10} - 1, \frac{1}{2}\right)[m_{11}(2 - \lambda)\ln(2 - \lambda) - 1] - 2m_{11}(m_{11} - 1)B\left(m_{11} - 1, \frac{1}{2}\right)[m_{10}(2 - \lambda)\ln(2 - \lambda) - 1]} \tag{22}$$

$$m_{10}m_{11}(2 - \lambda)^2 \ln(2 - \lambda) + \frac{m_{11}(m_{11} - 1) - m_{10}(m_{10} - 1)}{(m_{10} - m_{11})} \times (2 - \lambda) - 1 = 0 \tag{23}$$

The above equations show that the coefficient k_2 has a positive value if $m_{11} > m_{10}$.

A detailed study will be carried out for weirs with rectangular and triangular bottom parts and also for the weir with no angle points at the height $y = a$.

Weir with Rectangular Bottom Part

In this case the discharge characteristics are described with Equations (5) and (6) in which $m_1 = 3/2$. The coefficients k_1, k_2, and λ, according to Equations (12–14), are expressed by the following functions:

$$k_1 = \frac{2}{3} C_d\sqrt{2g}b \tag{24}$$

$$k_2 = C_d\sqrt{2g}(2 - \lambda)a^{3/2}b \tag{25}$$

$$\lambda = 0.4588 \tag{26}$$

The characteristics of the weir described by Equations (5) and (6) for $m_1 = 3/2$, taking into account the aforementioned equations for the factors k_1 and k_2, take the form:

$$Q_1 = \frac{2}{3} C_d \sqrt{2g} b H^{3/2} \tag{27}$$

$$Q_2 = C_d(2 - \lambda)\sqrt{2g} a^{3/2} b \ln\left(1 + \frac{H - \lambda a}{a}\right) \tag{28}$$

The shape of the weir, according to Equations (7) and (8), is expressed by:

$$f_1(y) = b \tag{29}$$

$$f_2(y) = b\left[1 - \frac{2}{\pi} \operatorname{arc\,tg} \sqrt{\frac{y - a}{a}} - \frac{2\sqrt{a(y - a)}}{\pi[(1 - \lambda)a + y]} - \frac{(2 - \lambda)a^{3/2}}{\pi[(1 - \lambda)a + y]^{3/2}} \times \ln \frac{\sqrt{(1 - \lambda)a + y} + \sqrt{y - a}}{\sqrt{(1 - \lambda)a + y} - \sqrt{y - a}}\right] \tag{30}$$

The rate of flow through a submerged weir may be calculated from Equations (3) and (4) taking into account the shape functions Equations (29) and (30):

$$\overline{Q}_2 = \frac{1}{3} C_d \sqrt{2g} b \left[3(2 - \lambda)a^{3/2} \ln \times \left(1 + \frac{H - \lambda a}{a}\right) - 2H^{3/2} + \sqrt{H - h}(2H + h)\right] \tag{31}$$

$$\overline{\overline{Q}}_2 = \frac{2}{3\pi} C_d \sqrt{2g} b \left\{\sqrt{H - h} \times \left[(2H + h) \operatorname{arc\,ctg} \sqrt{\frac{h - a}{a}} - \sqrt{a(h - a)}\right] + 2a^{3/2} \operatorname{arc\,tg} \sqrt{\frac{H - h}{h - a}} - 2H^{3/2} \operatorname{arc\,tg} \sqrt{\frac{a(H - h)}{H(h - a)}} + 6(2 - \lambda)a^{3/2} \times \left[\frac{\pi}{2} \ln \frac{\sqrt{(1 - \lambda)a + H} + \sqrt{H - a}}{\sqrt{(2 - \lambda)a}} - \ln \frac{\sqrt{(1 - \lambda)a + h} + \sqrt{h - a}}{\sqrt{(2 - \lambda)a}} \times \operatorname{arc\,sin} \sqrt{\frac{(1 - \lambda)a + h}{(1 - \lambda)a + H}} - S_H \sqrt{\frac{H - a}{a}} + S_h \sqrt{\frac{h - a}{a}}\right]\right\} \tag{32}$$

in which $S_H = S$ for $y = H$; and $S_h = S$ for $y = h$. The symbol S is described as follows:

$$S = H_1^{-1/2} + H_1^{-3/2}\left(\frac{2 - \lambda}{3^2} + \frac{1}{2 \cdot 3^2} y_1\right) + H_1^{-5/2}\left(\frac{(2 - \lambda)^2}{5^2} + \frac{2 - \lambda}{2 \cdot 5^2} y_1 + \frac{1 \cdot 3}{2 \cdot 4 \cdot 5^2} y_1^2\right) + H_1^{-7/2}\left(\frac{(2 - \lambda)^3}{7^2} + \frac{(2 - \lambda)^2}{2 \cdot 7^2} y_1 + \frac{1 \cdot 3(2 - \lambda)}{2 \cdot 4 \cdot 7^2} y_1^2 + \frac{1 \cdot 3 \cdot 5}{2 \cdot 4 \cdot 6 \cdot 7^2} y_1^3\right) \tag{33}$$

in which $y_1 = 1 - \lambda + y/a$; and $H_1 = 1 - \lambda + H/a$.

The submersion correction may be calculated in accordance with Equations (1) and (2) and then accepting the Q_2, $\overline{Q}_2$ and $\overline{\overline{Q}}_2$ functions correspondingly to the derived Equations (28), (31) and (32):

$$\varepsilon = \frac{2 - (2 + h_1)\sqrt{1 - h_1}}{3(2 - \lambda)a_1^{3/2} \ln\left(\frac{1}{a_1} + 1 - \lambda\right)} \quad \text{for } h_1 \leqslant a_1 \tag{34}$$

$$\varepsilon = 1 - \frac{2}{3\pi(2 - \lambda)a_1^{3/2} \ln\left(\frac{1}{a_1} + 1 - \lambda\right)} \times \left\{\sqrt{1 - h_1}\left[(2 + h_1) \operatorname{arc\,ctg} \sqrt{\frac{h_1 - a_1}{a_1}}\right.\right.$$

$$
\begin{aligned}
&- \sqrt{a_1(h_1 - a_1)}\Big] \\
&+ 2a_1^{3/2} \operatorname{arc\,tg} \sqrt{\frac{1 - h_1}{h_1 - a_1}} \\
&- 2 \operatorname{arc\,tg} \sqrt{\frac{a_1(1 - h_1)}{h_1 - a_1}} \\
&+ 6(2 - \lambda)a_1^{3/2} \\
&\times \left[\frac{\pi}{2} \ln \frac{\sqrt{(1 - \lambda)a_1 + 1} + \sqrt{1 - a_1}}{\sqrt{(2 - \lambda)a_1}}\right. \\
&- \ln \frac{\sqrt{(1 - \lambda)a_1 + h_1} + \sqrt{h_1 - a_1}}{\sqrt{(2 - \lambda)a_1}} \\
&\times \arcsin \sqrt{\frac{(1 - \lambda)a_1 + h_1}{(1 - \lambda)a_1 + 1}} - S_H \sqrt{\frac{1 - a_1}{a_1}} \\
&\left.+ S_h \sqrt{\frac{h_1 - a_1}{a_1}}\right]\Bigg\} \quad \text{for } h_1 \geq a_1
\end{aligned} \tag{35}
$$

in which $h_1 = h/H$; and $a_1 = a/H$. Functions 34 and 35 are shown in Figure 2. The regular shape of the curves makes it possible to replace derived Equations (34) and (35), which have complicated structures, with a simple approximative formula:

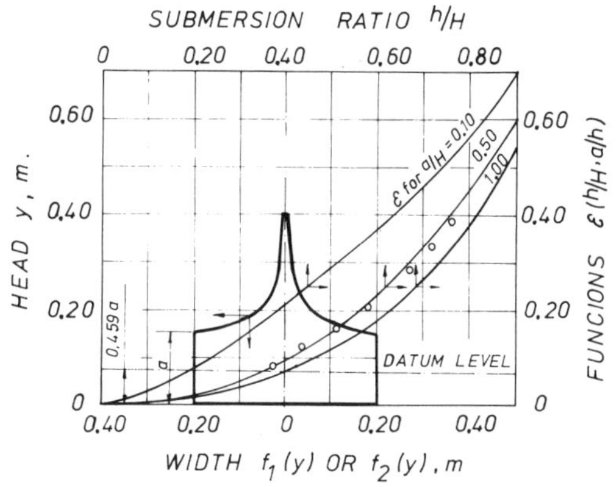

FIGURE 2. Functions ϵ (h/H, a/H) for the weir with rectangular bottom part.

$$\epsilon = s_1 h_1^{(s_2 a_1 + s_3 a_1^{s} + s_4)} a_1^{s_5} e^{s_6 a_1} \tag{36}$$

in which the coefficients are: $s_1 = 0.8687$; $s_2 = -1.0846$; $s_3 = 3.5343$; $s_4 = -0.2865$; $s_5 = -0.0238$; $s_6 = -0.4753$; and $s = 0.2500$.

The general form of the approximate Equation (36) can be employed in calculating the submersion correction for all further logarithmic weirs being analyzed, as well as for weirs with linear [3] and "square" [4] discharge characteristics.

Weir with Triangular Bottom Part

Assuming the triangular bottom part of the weir and solving the entire problem in the similar manner to the weir with rectangular bottom part, one obtains the equations determining the submersion correction.

The discharge characteristics are described with Equations (5) and (6) in which $m_1 = 5/2$. The coefficients k_1, k_2 and λ, according to Equations (12–14), are defined as:

$$k_1 = \frac{4}{15a} C_d \sqrt{2g} b \tag{37}$$

$$k_2 = \frac{2}{3} C_d \sqrt{2g} (2 - \lambda) a^{3/2} b \tag{38}$$

$$\lambda = 0.6540 \tag{39}$$

Substituting Equations (37–38) and $m_1 = 5/2$ into Equations (5) and (6) yields:

$$Q_1 = \frac{4}{15a} C_d \sqrt{2g} b H^{5/2} \tag{40}$$

$$Q_2 = \frac{2}{3} C_d \sqrt{2g} (2 - \lambda) a^{3/2} b \ln \left(1 + \frac{H - \lambda a}{a}\right) \tag{41}$$

The shape of the weir can be determined using Equations (7) and (8):

$$f_1(y) = \frac{b}{a} y \tag{42}$$

$$
\begin{aligned}
f_2(y) = \frac{b}{a} y \Bigg\{1 - \frac{2}{\pi} \Bigg[&\operatorname{arc\,tg} \sqrt{\frac{y - a}{a}} \\
&- \frac{\sqrt{a(y - a)} \left[\left(\frac{5}{3} - \lambda\right) a + y\right]}{y[(1 - \lambda)a + y]}
\end{aligned} \tag{43}
$$

$$- \frac{(2 - \lambda)a^{5/2}}{3y[(1 - \lambda)a + y]^{3/2}}$$

$$\times \ln \frac{\sqrt{(1 - \lambda)a + y} + \sqrt{y - a}}{\sqrt{(1 - \lambda)a + y} - \sqrt{y - a}}\Bigg]\Bigg\}$$

The rate of flow through a submerged weir is determined using Equations (3) and (4) in conjunction with Equations (42) and (43):

$$\overline{Q}_2 = \frac{2}{15a} C_d \sqrt{2g} b \Bigg[5a^{5/2}(2 - \lambda)$$

$$\times \ln \left(1 + \frac{H - \lambda a}{a}\right) - 2H^{5/2} \tag{44}$$

$$+ \frac{1}{4} \sqrt{H - h}(8H^2 + 4Hh + 3h^2) \Bigg]$$

$$\overline{\overline{Q}} = \frac{2}{15\pi a} C_d \sqrt{2g} b \left\{ \frac{1}{2} \sqrt{H - h} \right.$$

$$\left[(8H^2 + 4Hh + 3h^2) \operatorname{arc\,ctg} \sqrt{\frac{h - a}{a}} \right.$$

$$\left. - (4H + 3h + 2a) \sqrt{a(h - a)} \right] + \frac{1}{4} \sqrt{a}$$

$$\times (15H^2 - 30Ha + 31a^2) \operatorname{arc\,tg} \sqrt{\frac{H - h}{h - a}}$$

$$- 4H^{5/2} \operatorname{arc\,tg} \sqrt{\frac{a(H - a)}{H(h - a)}} - \frac{15}{8} \sqrt{a}(H - a)^2$$

$$\times \left(\frac{\pi}{2} - \arcsin \frac{2h - H - a}{H - a}\right) + 20(2 - \lambda)a^{5/2} \tag{45}$$

$$\times \left[\frac{\pi}{2} \ln \frac{\sqrt{(1 - \lambda)a + H} + \sqrt{H - a}}{\sqrt{(2 - \lambda)a}} \right.$$

$$- \arcsin \sqrt{\frac{(1 - \lambda)a + h}{(1 - \lambda)a + H}}$$

$$\times \ln \frac{\sqrt{(1 - \lambda)a + h} + \sqrt{h - a}}{\sqrt{(2 - \lambda)a}}$$

$$\left.\left. - S_H \sqrt{\frac{H - a}{a}} + S_h \sqrt{\frac{h - a}{a}} \right] \right\}$$

in which S is defined by Equation (33).

The submersion correction can be calculated based on Equations (1) and (2) and using Equations (41), (44) and (45):

$$\varepsilon = \frac{8 - \sqrt{1 - h_1}(3h_1^2 + 4h_1 + 8)}{20(2 - \lambda)a_1^{5/2} \ln \left(\frac{1}{a_1} + 1 - \lambda\right)} \quad \text{for } h_1 \leq a_1 \tag{46}$$

$$\varepsilon = 1 - \frac{1}{5\pi(2 - \lambda)a_1^{5/2} \ln \left(\frac{1}{a_1} + 1 - \lambda\right)}$$

$$\times \left\{ \frac{1}{2} \sqrt{1 - h_1} \left[(3h_1^2 + 4h_1 + 8) \right. \right.$$

$$\times \operatorname{arc\,ctg} \sqrt{\frac{h_1 - a_1}{a_1}}$$

$$\left. - (3h_1 + 2a_1 + 4) \sqrt{a_1(h_1 - a_1)} \right]$$

$$+ \frac{1}{4} \sqrt{a_1} (31a_1^2 - 30a_1 + 15)$$

$$\times \operatorname{arc\,tg} \sqrt{\frac{1 - h_1}{h_1 - a_1}}$$

$$- 4 \operatorname{arc\,tg} \sqrt{\frac{a_1(1 - a_1)}{h_1 - a_1}} \tag{47}$$

$$- \frac{15}{8} \sqrt{a_1} (1 - a_1)^2$$

$$\times \left(\frac{\pi}{2} - \arcsin \frac{2h_1 - a_1 - 1}{1 - a_1}\right)$$

$$+ 20(2 - \lambda)a_1^{5/2} \left[\frac{\pi}{2} \right.$$

$$\times \ln \frac{\sqrt{(1 - \lambda)a_1 + 1} + \sqrt{1 - a_1}}{\sqrt{(2 - \lambda)a_1}}$$

$$- \arcsin \sqrt{\frac{(1 - \lambda)a_1 + h_1}{(1 - \lambda)a_1 + 1}}$$

$$\times \ln \frac{\sqrt{(1 - \lambda)a_1 + h_1} + \sqrt{h_1 - a_1}}{\sqrt{(2 - \lambda)a_1}}$$

$$- S_H \sqrt{\frac{1 - a_1}{a_1}} + S_h \sqrt{\frac{h_1 - a_1}{a_1}}\Bigg]\Bigg\}$$

for $h_1 \geqslant a_1$

in which $h_1 = h/H$; and $a_1 = a/H$. The above functions are shown in Figure 3. Equations (46) and (47) have been replaced with a common approximate Equation (36) in which the coefficients become: $s_1 = 1.9955$; $s_2 = -2.1721$; $s_3 = 7.3842$; $s_4 = -2.0258$; $s_5 = 0.2137$; $s_6 = -1.4782$; and $s = 0.2500$.

Weir with the Shape Having No Angle Points

The discharge characteristics for weirs with the shape having no angle points at the height $y = a$ are described with Equations (15) and (16). Assume the exponent values: $m_{10} = 3/2$ and $m_{11} = 2$. In this case the coefficients k_{10}, k_{11}, k_2 and λ, according to Equations (20–23) are defined as:

$$k_{10} = A_{10}C_d\pi\sqrt{2g}b \tag{48}$$

$$k_{11} = -A_{11}C_d\pi\sqrt{2g}a^{-1/2}b \tag{49}$$

$$k_2 = A_2C_d\pi\sqrt{2g}a^{3/2}b \tag{50}$$

$$\lambda = 0.2230 \tag{51}$$

in which

$$A_{10} = \frac{2[2(2 - \lambda)\ln(2 - \lambda) - 1]}{3\pi[2(2 - \lambda)\ln(2 - \lambda) - 1] - 8[3(2 - \lambda)\ln(2 - \lambda) - 2]}$$

$$A_{11} = \frac{3(2 - \lambda)\ln(2 - \lambda) - 2}{3\pi[2(2 - \lambda)\ln(2 - \lambda) - 1] - 8[3(2 - \lambda)\ln(2 - \lambda) - 2]}$$

$$A_2 = \frac{2 - \lambda}{3\pi[2(2 - \lambda)\ln(2 - \lambda) - 1] - 8[3(2 - \lambda)\ln(2 - \lambda) - 2]}$$

Functions Equations (15) and (16) for $m_{10} = 3/2$, $m_{11} = 2$, taking into account the aforementioned equations for the factors k_{10}, k_{11} and k_2, take the form:

$$Q_1 = C_d\pi\sqrt{2g}b(A_{10}H^{3/2} - A_{11}a^{-1/2}H^2) \tag{52}$$

$$Q_2 = A_2C_d\pi\sqrt{2g}a^{3/2}b\ln\left(1 + \frac{H - \lambda a}{a}\right) \tag{53}$$

The shape of the weir according to Equations (17) and (18) is described by the following functions:

$$f_1(y) = b\left(\frac{3}{2}\pi A_{10} - 8A_{11}\sqrt{\frac{y}{a}}\right) \tag{54}$$

$$f_2(y) = b\Bigg[\frac{3}{2}\pi A_{10} + 8A_{11}\left(\sqrt{\frac{y - a}{a}} - \sqrt{\frac{y}{a}}\right) - 3A_{10}\,\text{arc tg}\sqrt{\frac{y - a}{a}} - \frac{2}{2 - \lambda}A_2\frac{\sqrt{a(y - a)}}{y + (1 - \lambda)a} - A_2\left(\frac{a}{y + (1 - \lambda)a}\right)^{3/2} \times \ln\frac{\sqrt{y + (1 - \lambda)a} + \sqrt{y - a}}{\sqrt{y + (1 - \lambda)a} - \sqrt{y - a}}\Bigg] \tag{55}$$

The rate of flow through the submerged weir of the shape described with Equations (54) and (55) can be calculated using Equations (3) and (4). Thus,

$$\overline{Q}_2 = C_d\sqrt{2g}b\Bigg\{\pi A_2a^{3/2}\ln\left(1 + \frac{H - \lambda a}{a}\right) + \sqrt{H - h}\left[\frac{\pi}{2}A_{10}(2h + h) - \frac{2}{3}A_{11}\sqrt{\frac{h}{a}}(3H + 2h)\right] - A_{10}\pi H^{3/2} + A_{11}\frac{H^2}{\sqrt{a}}\left(\frac{\pi}{2} + \text{arc sin}\frac{2h - H}{H}\right)\Bigg\} \tag{56}$$

$$\overline{\overline{Q}}_2 = C_d\sqrt{2g}b\Bigg\{\sqrt{H - h}\Bigg[A_{10}(2H + h) \times \text{arc ctg}\sqrt{\frac{h - a}{a}} + \frac{2}{3}A_{11}(3H + 2h) \times \left(\sqrt{\frac{h - a}{a}} - \sqrt{\frac{h}{a}}\right) - 2\left(\frac{5}{3}A_{11} - A_{10}\right.$$

$$+ \frac{1}{2-\lambda} A_2\Big) \sqrt{a(h-a)}\Big]$$

$$+ \sqrt{a}\left[\left(3A_{10} - \frac{2}{2-\lambda}A_2\right) H\right.$$

$$- \left.\left(A_{10} - \frac{2}{2-\lambda}A_2\right) a\right] \text{arc tg} \sqrt{\frac{H-h}{h-a}}$$

$$- 2A_{10}H^{3/2} \text{arc tg} \sqrt{\frac{(H-h)a}{H(h-a)}} \tag{57}$$

$$+ A_{11}\frac{1}{\sqrt{a}}\left[(H-a)^2\left(\frac{\pi}{2}\right.\right.$$

$$\left. - \text{arc sin} \frac{2h-H-a}{H-a}\right)$$

$$\left.-H^2\left(\frac{\pi}{2} - \text{arc sin} \frac{2h-H}{H}\right)\right]$$

$$+ 4A_2a^{3/2}\left[\frac{\pi}{2} \ln \frac{\sqrt{H+(1-\gamma)a} + \sqrt{H-a}}{\sqrt{(2-\lambda)a}}\right.$$

$$- \text{arc sin} \sqrt{\frac{(1-\lambda)a+h}{H+(1-\lambda)a}}$$

$$\times \ln \frac{\sqrt{h+(1-\lambda)a} + \sqrt{h-a}}{\sqrt{(2-\lambda)a}}$$

$$\left.\left. - S_H\sqrt{\frac{H-a}{a}} + S_h\sqrt{\frac{h-a}{a}}\right]\right\}$$

The submersion correction factor can be derived in an analogous manner to that previously presented for the weirs with rectangular and triangular bottom part. Substituting Equations (53), (56) and (57) into Equations (1) and (2) yields:

$$\varepsilon = \frac{1}{A_2\pi a_1^{3/2} \ln\left(\frac{1}{a_1} + 1 - \lambda\right)}\left\{\sqrt{1-h_1}\left[\frac{2}{3}\right.\right.$$

$$\left.\times A_{11}\sqrt{\frac{h_1}{a_1}}(3+2h_1) - \frac{\pi}{2}A_{10}(2+h_1)\right] \tag{58}$$

$$+ A_{10}\pi - A_{11}\frac{1}{\sqrt{a_1}}\left[\frac{\pi}{2}\right.$$

$$\left.\left. + \text{arc sin}(2h_1 - 1)\right]\right\} \text{ for } h_1 \leqslant a_1$$

$$\varepsilon = 1 - \frac{1}{A_2\pi a_1^{3/2} \ln\left(\frac{1}{a_1} + 1 - \lambda\right)}\left\{\sqrt{1-h_1}\right.$$

$$\times \left[A_{10}(2+h_1) \text{arc ctg} \sqrt{\frac{h_1-a_1}{a_1}}\right.$$

$$+ \frac{2}{3}A_{11}(3+2h_1)\left(\sqrt{\frac{h_1-a_1}{a_1}}\right.$$

$$\left.- \sqrt{\frac{h_1}{a_1}}\right) - 2\left(\frac{5}{3}A_{11} - A_{10}\right.$$

$$\left. + \frac{1}{2-\lambda}A_2\right) \sqrt{a_1(h_1-a_1)}\Big]$$

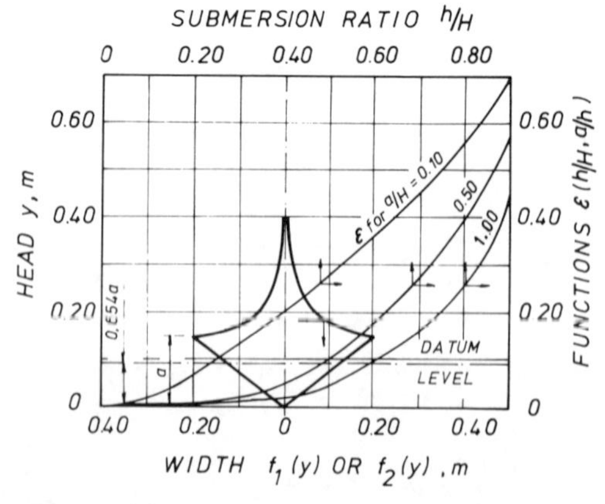

FIGURE 3. Functions ε (h/H, a/H) for the weir with triangular bottom part.

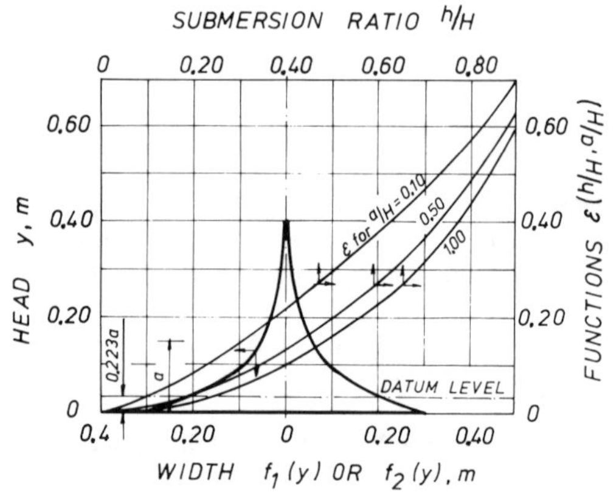

FIGURE 4. Functions ε (h/H, a/H) for the weir with the shape having no angle points.

$$+ \sqrt{a_1}\left[3A_{10} - \frac{2}{2-\lambda}A_2 - \left(A_{10} - \frac{2}{2-\lambda}A_2\right)a_1\right] \times \text{arc tg}\sqrt{\frac{1-h_1}{h_1-a_1}} - 2A_{10}\,\text{arc tg}\sqrt{\frac{(1-h_1)a_1}{h_1-a_1}} + A_{11}\frac{1}{\sqrt{a_1}}\left[(1-a_1)^2\left(\frac{\pi}{2} - \text{arc sin}\,\frac{2h_1-1-a_1}{1-a_1}\right) - \frac{\pi}{2} + \text{arc sin}\,(2h_1-1)\right] + 4A_2a_1^{3/2} \times \left[\frac{\pi}{2}\ln\frac{\sqrt{1+(1-\lambda)a_1}+\sqrt{1-a_1}}{\sqrt{(2-\lambda)a_1}} - \text{arc sin}\sqrt{\frac{(1-\lambda)a_1+h_1}{1+(1-\lambda)a_1}}\ln \times \frac{\sqrt{h_1+(1-\lambda)a_1}+\sqrt{h_1-a_1}}{\sqrt{(2-\lambda)a_1}} - S_H\sqrt{\frac{1-a_1}{a_1}} + S_h\sqrt{\frac{h_1-a_1}{a_1}}\right]\Bigg\} \tag{59}$$

for $h_1 \geqslant a_1$

in which $h_1 = h/H$; and $a_1 = a/H$. The function $\epsilon\ (h/H, a/H)$ is presented in Figure 4. The approximate formula for this function is taken in accordance with Equation (36) in which $s_1 = 0.8647$; $s_2 = -0.5115$; $s_3 = 2.3942$; $s_4 = 0.1125$; $s_5 = 0.0232$; $s_6 = -0.2830$; and $s = 0.2500$.

EXPERIMENTAL STUDIES

The studies incorporated: determination of the datum line for the free falling weir; establishment of the coefficient of discharge curve for the free falling weir; and calculation of the average value of this coefficient for a constant argument value of $a/H = 0.50$. The results of these studies formed the basis for calculating the submersion correction factor and for checking the conformity of theoretical solutions with the measured phenomenon.

Investigations were performed in a hydraulic channel having the dimensions 0.60 m width, 1.00 m depth and 23.80 m length. The water supplied to the flume was pumped from a lower tank to the upper surge tank, from which the water flowed by gravity to the hydraulic channel through the triangular measuring weir and calming barrier. The examined weir was installed at a distance of 7.5 m from the calming chamber. A valve was located at the outlet of the channel, thus making it possible to control the filling of the channel downstream from the weir. To measure the water levels in the channel, point gages were installed. Readings were taken with a precision of ± 0.1 mm. A weir having a crest width of $b = 0.40$ m and a height of the lower rectangular section of $a = 0.15$ m was examined. The shape of the weir notch was determined in accordance with Equations (29) and (30).

Using the values of $a = 0.15$ m, $b = 0.40$ m, and $g = 9.81$ msec^{-2} in the discharge characteristic Equation (28), the result for $H \geq a$ is

$$Q_2 = 0.159C_d \ln\left(\frac{H}{0.15} + 0.541\right),\ m^3s^{-1} \tag{60}$$

in which H = the height of a nappe, in meters. Solving Equation (60) with respect to C_d one obtains

$$C_d = \frac{Q_2}{0.159\ln\left(\frac{H}{0.15} + 0.541\right)} \tag{61}$$

After computing the coefficient of discharge value according to Equation (61) (using experimentally determined Q_2 and H values) this coefficient was plotted as a function of nappe height in Figure 5. Based on the shape of this curve, an approximate formula was accepted in the form

$$C_d = a + \frac{b}{H} + \frac{c}{H^2} \tag{62}$$

in which the coefficients assume the values: $a = 0.619$; $b = -0.040$; and $c = 0.006$.

The diagram of Equation (62) and values of the coefficient of discharge (computed from Equation (61) for experimentally determined values of Q_2 and H) are shown in Figure 5. The relative error of the function $C_d\ (H)$ values, computed from Equations (61) and (62), does not exceed 0.4%. The average value for the examined range of nappe head from 0.1825–0.3136 m is $C_d = 0.556$. Using this value ($C_d = 0.556$) in Equation (60) the discharge equation becomes

$$Q_2 = 0.088\ln\left(\frac{H}{0.15} + 0.541\right),\ m^3s^{-1} \tag{63}$$

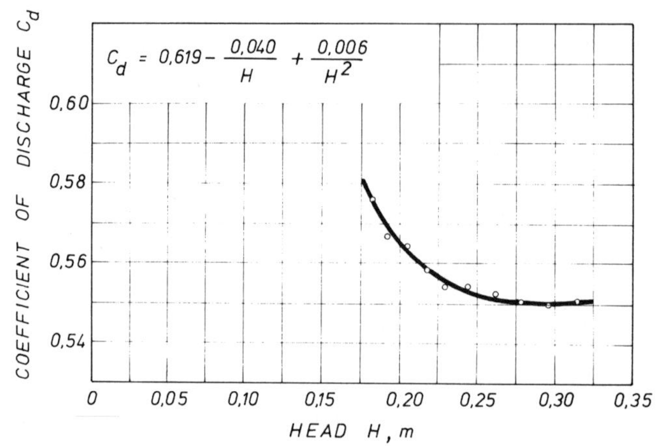

FIGURE 5. Coefficient of discharge for weir with rectangular bottom part.

The relative error of Q_2 (H) function values, determined experimentally and computed from Equation (63) does not exceed 3.5%. The plot of Equation (63) and the results of the head measurements for variable flow rates are presented in Figure 6. The diagram of the ϵ $(h/H, a/H)$ function for $a/H = 0.50$ made in accordance with the strict Equations (34) and (35) and the results of experimental measurements are shown in Figure 2.

CONCLUSIONS

From the theoretical and experimental investigations, it has been established that:

1. The submersion correction factor ϵ $(h/H, a/H)$, that characterizes the effect of weir submersion on the change in its characteristic, is a monotonically increasing function of the argument h/H and a monotonically decreasing one of the argument a/H (Figures 2–4). This function makes it possible to replace Equations (34), (35), (46), (47), (58) and (59), which have a complicated structure with the approximate Equation (36), thereby simplifying the computation of the discharge values.
2. The coefficient of discharge for the free falling weirs under investigation is a function of nappe head. As a result of analyzing theoretical solutions and experimental measurements, it has been shown that the acceptance of a constant value for this coefficient does not cause essential variations in the predetermined discharge characteristic.
3. The formulas derived for the function ϵ $(h/H, a/H)$ render it possible to use logarithmic submerged weirs for determining discharge.

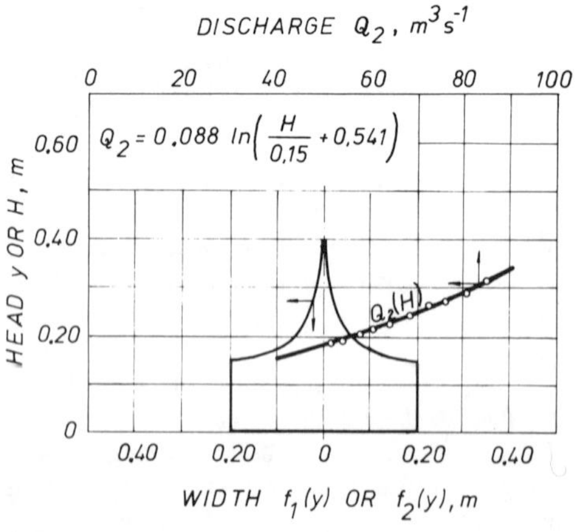

FIGURE 6. Discharge characteristic for weir with rectangular bottom part.

NOTATION

The following symbols are used in this paper:

a = height of the lower weir section, described with a function $f_1(y)$
B = beta function
b = width of the weir for $y = a$
C_d = coefficient of discharge
$f_1(y)$ = function describing the shape of the lower part of the weir, for $y \leq a$
$f_2(y)$ = function describing the shape of the upper part of the weir, for $y \geq a$
g = acceleration due to gravity
H = nappe head
h = submersion head

k_1, k_{10}, k_{11}, k_2 = proportionality factors
m_1, m_{10}, m_{11} = exponents
Q_1 = discharge characteristic for the emerged weir for $H \leq a$
Q_2 = discharge characteristic for the emerged weir for $H \geq a$
$\bar{Q}_2$ = discharge characteristic for the submerged wier for $H > a$ and $h \leq a$
$\bar{\bar{Q}}_2$ = discharge characteristic for the submerged weir for $H > a$ and $h \geq a$
ϵ = submersion correction
λ = coefficient determining datum level for nappe head

REFERENCES

1. Grabarczyk, C. and H. Majcherek, "Teoretyczne podstawy wyznaczania ksztaltu niezatopionych przelewów ostrobrzeznych o zadanej charakterystyce przeplywu," *Archiwum Hydrotechniki, Vol. 20,* No. 1, pp. 7-19 (1973).
2. Grabarczyk, C. and H. Majcherek, "Przelewy o charakterystyce przeplywu opisanej funkcją potęgową i logarytmiczną," *Archiwum Hydrotechniki,* Vol. 20, No. 2, pp. 145-170 (1973).
3. Majcherek, H., "Analiza pracy przelewów o liniowej charakterystyce przeplywu w warunkach zatopienia," *Archiwum Hydrotechniki,* Vol. 24, No. 3, pp. 289-311 (1977).
4. Majcherek, H., "Analiza pracy przelewów o kwadratowej charakterystyce przeplywu w warunkach zatopienia," *Archiwum Hydrotechniki,* Vol. 24, No. 4, pp. 471-488 (1977).
5. Rao, N. S. G. and K. K. Murthy, "On the Design of Logarithmic Weirs," *Journal of Hydraulic Research,* Vol. 4, No. 1, pp. 51-59 (1966).
6. Di Ricco, G., "Stramazzi con Data Equazione di Portata," *L'Acqua Nell' Agricoltura, Nell'Igiene e Nell'Industria,* No. 1, pp. 2-6 (1963) and No. 2, pp. 31-40 (1963).

Developing Laminar Power-Law Fluid Flow in a Straight Channel

Ramesh C. Gupta*

1. INTRODUCTION

Consider a viscous fluid entering a channel with uniform velocity. A boundary layer of zero initial thickness is formed at the channel walls. This leaves a practically unsheared core of the fluid which shrinks with the progress of the flow as the growing boundary layers gradually envelope the core flow. The velocity profile, therefore, consists of two boundary layer velocity profiles near the channel walls joined by a line of constant core velocity. As the fluid moves downstream, the velocity pattern undergoes a change and finally attains a fully developed form which does not vary with the distance from the entrance plane. The region of the channel where this development takes place is designated as the "developing" or "inlet" region.

An accurate description of the flow in the developing region is not only of significance in polymer processing and in the measurement of basic flow characteristics of materials but it is also of extreme importance in the proper design of any flow intake device.

Since the flow flux must be the same at every cross section, the flow retardation near the walls is compensated by a corresponding acceleration in the core region and this results in a fall in pressure.

The fully developed state is attained only asymptotically; the "inlet length" is usually taken as the channel length required for the core velocity to attain 98% or 99% of its fully developed value.

The main problems are to know (1) the inlet length, (2) the pressure-drop, and (3) the velocity profiles in the developing region.

Exact analytic solutions are not known. The difficulty is due to the presence of non-linear inertia terms in the governing differential equations. Various attempts have, therefore, been made to obtain the approximate solution. Most of these involve some form of the Prandtl's boundary layer equations. Four methods of approach may be discerned in the literature: 1) linearisation techniques, 2) matching methods, 3) integral solutions, and 4) numerical solutions. These treatments have been reviewed by Shah and London [17] and Boger [1].

Few of these methods have been extended to study the non-Newtonian flow behaviour in the developing region. In this chapter we shall discuss various solutions for the developing laminar non-Newtonian Power-law fluid flow in a straight channel. Power-law fluids are characterised by the rheological relation

$$\tau_{ij} = k \left| \left(\sum_{\ell=1}^{3} \sum_{m=1}^{3} e_{\ell m}\, e_{m\ell} \right)^{1/2} \right|^{n-1} e_{ij} \tag{1}$$

between the components of deviatoric stress tensor τ_{ij} and those of rate of strain tensor e_{ij}; k and n being the consistency and flow behaviour index, respectively.

2. FORMULATION OF THE PROBLEM

The differential equations governing the steady laminar two-dimensional flow of an incompressible constant property Power-law fluid, which is described by the relation (1), entering a channel formed by two semi-infinite horizontal plates separated by a distance $2h$ with uniform velocity U_0 are [11]

$$\frac{\partial U}{\partial X} + \frac{\partial V}{\partial Y} = 0 \tag{2}$$

$$U \frac{\partial U}{\partial X} + V \frac{\partial U}{\partial Y} = -\frac{1}{\varrho}\frac{dP}{dX} + \gamma \frac{\partial}{\partial Y}\left(\left| \frac{\partial U}{\partial Y} \right|^{n-1} \frac{\partial U}{\partial Y} \right) \tag{3}$$

*Department of Mathematics, National University of Singapore, Singapore with

$$\gamma = \frac{1}{\varrho} k\, 2^{(n-1)/2} \tag{4}$$

U and V are the axial (X-) and transverse (Y-) components of the velocity in the boundary layer; P and ϱ respectively represent pressure and the fluid density. The axis of X has been chosen to coincide with the lower wall of the channel and Y is measured in a direction perpendicular to it. Origin is taken at the upstream end of the lower plate.

Introduction of dimensionless variables given by

$$X = xh^{n+1}\, U_0^{2-n}/\gamma,$$

$$Y = hy,\ U = uU_0,\ v = Vh^n\, U_0^{1-n}/\gamma,\ P = p\varrho\, U_0^2$$

transforms Equations (2) and (3) to the form

$$\frac{\partial u}{\partial x} + \frac{\partial v}{\partial y} = 0 \tag{5}$$

$$u\frac{\partial u}{\partial x} + v\frac{\partial u}{\partial y} = -\frac{dp}{dx} + \frac{\partial}{\partial y}\left(\left|\frac{\partial u}{\partial y}\right|^{n-1}\frac{\partial u}{\partial y}\right) \tag{6}$$

Symmetry of the flow field about the axis of the channel leads to the boundary conditions:

at the entry plane $x = 0$, $u = 1,\ v = 0$ for all y (7a)

at the channel wall $y = 0,\ u = 0,\ v = 0$ for all x (7b)

and at the axis of the channel $y = 1$, $\dfrac{\partial u}{\partial y} = 0$,

$v = 0$ for all x (7c)

and, therefore, complete flow picture can be obtained by solving (5) and (6) subject to the boundary conditions (7).

3. VARIOUS APPROXIMATE SOLUTIONS

We discuss below various solutions to the problem under consideration which have been devised to obtain more and more accurate flow behaviour in the developing region.

3.1 Momentum Integral Solution

In this approach integral forms of the equations of continuity and momentum are applied to a model consisting of an inviscid core and a developing boundary layer. The axial velocity distribution in the boundary layer is expressed as a polynomial function of the boundary layer thickness δ, as in the standard Kármán-Pohlhausen procedure. In addition to the integrated versions of equations of continuity and momentum, a third relation is needed to determine the three unknown quantities δ, u, and p.

Schiller [15] initiated this approach. He considered the pressure "to be impressed" on the boundary layer and determined the same by an application of Bernoulli's equation to the inviscid core flow. The assumption of a parabolic velocity distribution in the boundary layer rendered the problem determinate.

Kapur and Gupta [12,13] employed this approach to examine developing Power-law fluid flow problem. They used cubic [12] and quartic boundary layer velocity profiles [13]. These profiles do not appear to represent adequately pseudoplastic ($n < 1$) flow behaviour [4], and impose additional restrictions on the analysis [7].

3.2 Solution by Matching Technique

This technique, first used by Schlichting [16], consists in subdividing the developing flow region into two zones. Near the entry, it is treated as a boundary layer problem. The boundary layer development does not yield similarity type velocity profiles and an approximate solution is obtained in terms of a perturbation of the Blausius external flow solution. Far downstream, where the flow is nearly fully developed the series solution is obtained in terms of small perturbation of the fully developed flow, this part of the analysis proceeding in the upstream direction. The solution is completed by matching these solutions at some appropriate location. Collins and Schowalter [3] used this procedure to study the developing flow of Power-law fluids in a channel. It involves cumbersome numerical calculations. Furthermore, the results for dilatant ($n > 1$) fluids are not included in Reference [3].

In these analyses [3,12,13], the Bernoulli's equation (mechanical energy balance) has been applied only to the inviscid core, so that the viscous dissipation in the boundary layer was ignored. The loss of energy due to viscous dissipation in the boundary layer is of increasing importance because, as the flow proceeds, the boundary layer grows thicker. Thus, to have a better flow picture we present momentum energy integral solution to the problem. This accounts for the loss of energy due to viscous dissipation in the boundary layer.

3.3 Momentum Energy Integral Solution

Elimination of v between (5) and (6) followed by integration over half the channel width leads to the integrated version of the principle of conservation of momentum as

$$\frac{d}{dx}\int_0^1 u^2 dy + \frac{dp}{dx} + \left(\left|\frac{\partial u}{\partial y}\right|^{n-1}\frac{\partial u}{\partial y}\right)\Bigg|_0 = 0 \tag{8}$$

Using (5) the principle of conservation of mass can be restated as

$$\int_0^1 u dy = 1 \tag{9}$$

Following Campbell and Slattery [2] and Gupta [6] dp/dx is determined by an application of the mechanical energy balance to the entire flow field. To accomplish this we obtain the equation describing the rate of change of kinetic energy per unit mass by multiplying each term of (6) by $2u$ and integrate the same over the flow cross section to arrive at

$$\frac{d}{dx}\int_0^1 u^3 dy + 2\frac{dp}{dx} + 2\int_0^1 \left|\frac{\partial u}{\partial y}\right|^{n-1}\left(\frac{\partial u}{\partial y}\right)^2 dy = 0 \tag{10}$$

We now assume the axial velocity distribution to be given by

$$u = \begin{cases} \phi & , \delta \le y \le 1, \text{ for all } x \\ \phi\left[1 - \left(1 - \frac{y}{\delta}\right)^{(n+1)/n}\right] & , 0 \le y \le \delta, \text{ for all } x \end{cases} \tag{11}$$

in which δ denotes the dimensionless boundary layer thickness, and ϕ is the dimensionless axial velocity in the inviscid core. This distribution reduces to the fully developed velocity profile [18] when the boundary layer merges with the axis of the channel. Combination of this distribution with overall continuity Equation (9) gives nondimensional axial core velocity in the form

$$\phi = \frac{1}{1 - \frac{n}{2n+1}\delta} \tag{12}$$

Finally, elimination of u and ϕ between (8), (10), (11), and (12) and subsequent rearrangement leads to

$$x^* = \frac{X/h}{R_n} = \frac{1}{24}\int_0^\delta \left(\frac{\delta}{C_1}\right)^n \times \frac{B_3 + B_1 - 2B_2 + 2B_1(B_3 - 2B_1)\delta + 2B_1^2 B_2 \delta^2}{(1 - B_1\delta)^{3-n}[C_1 - (1 - B_1\delta)]} d\delta \tag{13}$$

in which

$$B_1 = n/(2n + 1) \tag{14}$$

$$B_2 = n(4n + 3)/(2n + 1)(3n + 2) \tag{15}$$

$$B_3 = n(18n^2 + 28n + 11)/(2n + 1)(3n + 2)(4n + 3) \tag{16}$$

$$C_1 = (n + 1)/(2n + 1)$$

and R_n is the contraction for the Reynolds number $(= 12[nh/(2n + 1)]^n U_0^{2-n}/\gamma)$.

This relation gives the location where a given boundary layer thickness is attained.

Either of the Equations (8) or (10) can be used to obtain the relation between pressure and the boundary layer thickness. If P_0 is the pressure at the entry, then

$$p^* = \frac{P_0 - P}{\frac{1}{2}\varrho U_0^2} = \int_0^\delta \frac{B_3 - 3B_1 + 2C_1(2B_1 + B_2) - 2B_1(B_3 - C_1B_2)\delta}{(1 - B_1\delta)^3[C_1 - (1 - B_1\delta)]} d\delta \tag{17}$$

This relation coupled with (13) gives inlet region pressure distribution.

These relations provide the complete flow picture in the developing region.

The solution reduces to that in [6] for Newtonian ($n = 1$) fluids.

3.4 Another Integral Solution

Here we present another integral solution which utilises both the differential and integral momentum equations. The method was first suggested and used by Fargie and Martin [5] for the developing laminar flow in a circular pipe.

Elimination of u between (8), (11) and (12) leads to

$$\frac{d\delta}{dx} + \frac{(1 - B_1\delta)^3}{2B_1 - B_2 - B_1B_2\delta} \times \left[\frac{dp}{dx} + \left(\frac{n+1}{n(1 - B_1\delta)\delta}\right)^n\right] = 0 \tag{18}$$

As x increases, δ continuously increases from zero at the entry plane to 1 when the flow becomes fully developed. Thus, the fully developed axial pressure gradient can be obtained from (18) by substituting $\delta = 1$. This gives

$$\frac{dp}{dx} = -\left(\frac{2n+1}{n}\right)^n \tag{19}$$

Since both p and δ are functions of x, a relation additional to (18) is required to complete the solution; this, following Fargie and Martin [5], is determined by applying the differential momentum Equation (6) at the wall where $y = 0$ and $u = 0 = v$. This leads to the explicit expression for the pressure gradient

$$\frac{dp}{dx} = \left[\frac{\partial}{\partial y}\left(\left|\frac{\partial u}{\partial y}\right|^{n-1}\frac{\partial u}{\partial y}\right)\right]_0 \tag{20}$$

Substitution of u from (11) into this equation gives

$$\frac{dp}{dx} = -\left(\frac{n+1}{n}\phi\right)^n \frac{1}{\delta^{n+1}} \tag{21}$$

In principle Equations (18) and (21) give x as a function of δ and this completes the solution. It is, however, necessary to examine any irregularities in the relationship [21]. Normalising the pressure gradient under the developing flow conditions [Equation (21)] with that under fully developed conditions [Equation (19)] and defining its reciprocal by λ, we arrive at

$$\lambda = \frac{(dp/dx)_{\delta = 1}}{dp/dx} = \left(\frac{2n + 1 - n\delta}{n+1}\right)^n \delta^{n+1} \tag{22}$$

As expected this relation gives correct values of λ when $\delta = 0$ and 1. Moreover, it does not have any point of inflection in $0 \le \delta \le 1$ for pseudoplastic ($n < 1$) and Newtonian fluids, whereas for dilatant ($n > 1$) fluids it shows a point of inflection in $0 \le \delta \le 1$.

Therefore, for pseudoplastic ($n < 1$) fluids one can use (21) in conjunction with (18) to obtain

$$x^* = \frac{1}{12}\int_0^\delta \frac{(2B_1 - B_2 - B_1B_2\delta)\,\delta^{n+1}}{C_1^n(1 - B_1\delta)^{3-n}(1-\delta)}\,d\delta \tag{23}$$

This relation shows that the fully developed state is attained asymptotically.

And the pressure distribution is found to be given by

$$p^* = 2\int_0^\delta \frac{2B_1 - B_2 - B_1B_2\,\delta}{(1 - B_1\delta)^3(1-\delta)}\,d\delta \tag{24}$$

These relations can be used to obtain the developing flow of Newtonian ($n = 1$) fluids in a straight channel [9].

On the other hand, for dilatant fluids, following Fargie and Martin [5], we assume the normalized inverse pressure gradient to be given by

$$\lambda = a_0 + a_1\delta + a_2\delta^2 \tag{25}$$

and determine the constants a_0, a_1 and a_2 using the conditions $\lambda(0) = 0$ and $\lambda(1) = 1$, and the third condition is obtained by assuming that (22) provides correct gradient of λ at $\delta = 1$. This provides

$$\lambda = \frac{1}{n+1}\,\delta(1 + n\delta) \tag{26}$$

The corrected pressure gradient under developing flow conditions is, therefore, obtained from (22) as

$$\frac{dp}{dx} = -\left(\frac{2n+1}{n}\right)^n \frac{n+1}{\delta(1 + n\delta)} \tag{27}$$

This relation combined with (18) gives

$$x^* = \frac{1}{12}\int_0^\delta \frac{(2B_1 - B_2 - B_1B_2\delta)\delta^n\,d\delta}{(1 - B_1\delta)^3[(n+1)\delta^{n-1}(1 + n\delta)^{-1} - C_1^n(1 - B_1\delta)^{-n}]} \tag{28}$$

To complete the solution, Equations (27) and (28) can be combined to obtain the pressure distribution for dilatant fluids.

3.5 An Improved Momentum Energy Integral Solution

Integral solutions presented in 3.3 and 3.4 use momentum integral Equation (8). This contains the shear stress term expressed in terms of the derivative of the velocity evaluated at the channel wall. When this term is replaced with the help of an assumed axial velocity distribution, an appreciable error might have been introduced in the analysis. Probably for this reason Kármán-Pohlhausen procedure leads to less satisfactory results for the retarded flows [14]. Volkov [19] suggested that the method can be improved if the wall shear stress term is expressed in terms of integrals involving unknown velocity distribution. He [19] showed that this refinement yields better results. Thus, to have a more accurate description of the Power-law fluid flow behaviour in the developing region of a channel we incorporate Volkov's suggestion into the momentum energy integral solution of the problem.

To begin with, let us eliminate v between (5) and (6) and integrate from $y = 0$ to (some) y to obtain

$$\left(\left|\frac{\partial u}{\partial y}\right|^{n-1}\frac{\partial u}{\partial y}\right)\Bigg|_y - \left(\left|\frac{\partial u}{\partial y}\right|^{n-1}\frac{\partial u}{\partial y}\right)\Bigg|_0 \tag{29}$$

$$= \int_0^y \frac{\partial u^2}{\partial x}\,dy - u\int_0^y \frac{\partial u}{\partial x}\,dy + \frac{dp}{dx}\,y$$

When $y = 1$, this reduces to the Equation (8). To determine $(|\partial u/\partial y|^{n-1}\partial u/\partial y)|_0$ in terms of integrals involving u, we integrate each term of (29) over flow cross section to obtain

$$-\left(\left|\frac{\partial u}{\partial y}\right|^{n-1}\frac{\partial u}{\partial y}\right)\Bigg|_0 = -\int_0^1 \left|\frac{\partial u}{\partial y}\right|^{n-1}\frac{\partial u}{\partial y}\,dy$$

$$+\int_0^1\int_0^y \frac{\partial u^2}{\partial x}\,dy\,dy \tag{30}$$

$$-\int_0^1 u\int_0^y \frac{\partial u}{\partial x}\,dy\,dy + \frac{1}{2}\frac{dp}{dx}$$

This equation when coupled with (8) gives

$$-\frac{1}{2}\frac{dp}{dx} = \frac{d}{dx}\int_0^1 u^2 dy + \int_0^1 \left|\frac{\partial u}{\partial y}\right|^{n-1}\frac{\partial u}{\partial y}\,dy \tag{31}$$

$$-\int_0^1\int_0^y \frac{\partial u^2}{\partial x}\,dy\,dy + \int_0^1 u\int_0^y \frac{\partial u}{\partial x}\,dy\,dy$$

Elimination of dp/dx between this modified momentum integral equation and the mechanical energy balance (10) leads to

$$\frac{d}{dx}\int_0^1 u^3 dy + 2\int_0^1 \left|\frac{\partial u}{\partial y}\right|^{n-1}\left(\frac{\partial u}{\partial y}\right)^2 dy$$

$$= 4\frac{d}{dx}\int_0^1 u^2 dy + 4\int_0^1 \left|\frac{\partial u}{\partial y}\right|^{n-1}\frac{\partial u}{\partial y}\,dy \tag{32}$$

$$-4\int_0^1\int_0^y \frac{\partial u^2}{\partial x}\,dy\,dy + 4\int_0^1 u\int_0^y \frac{\partial u}{\partial x}\,dy\,dy$$

Now u and ϕ may be eliminated between Equations (10), (11) and (32) to give after simplification

$$x^* = \int_0^\delta \frac{(1 - B_1\delta)\{\tfrac{1}{2}B_1 + \tfrac{1}{4}B_3 + (A_1 - C_2)\delta - B_1(A_1 - \tfrac{1}{2}B_1^2)\delta^2\} - \tfrac{3}{4}B_1(1 - B_3\delta)}{6C_1^n(1 - B_1\delta)^{3-n}(C_1 - \delta + B_1\delta^2)}\,\delta^n d\delta \tag{33}$$

in which

$$A_1 = n^2(27n^2 + 32n + 9)/2(2n + 1)^2(3n + 1)(3n + 2) \tag{34}$$

$$C_2 = n^2(9n + 7)/(2n + 1)(3n + 1)(3n + 2) \tag{35}$$

and the remaining symbols have been described in the preceding section.

Either of the Equations (31) or (10) can be used in conjunction with (33) to obtain the pressure drop as a function of δ. This is given by

$$p^* = \frac{1 - B_3\delta}{(1 - B_1\delta)^3} + 24\,C_1^{n+1}\int_0^\delta \frac{\dfrac{dx}{d\delta}\,d\delta}{(1 - B_1\delta)^{n+1}\delta^n} - 1 \tag{36}$$

Equations (33) and (36) provide p^* at any location in the developing region.

The slope of $p^* - 24x^*$ curve at the beginning of the fully developed state can be computed from these relations and is found to be given by

$$\tan\theta = \frac{C_1^n}{(1 - B_1)^{n+1}}\left[C_1 + \frac{(3B_1 - B_3 - 2B_1B_3)(C_1 - 1 + B_1)}{4(1 - B_1)\{\tfrac{1}{2}B_1 + \tfrac{1}{4}B_3 + (A_1 - C_2) - B_1(A_1 - \tfrac{1}{2}B_1^2)\} - 3B_1(1 - B_3)}\right]$$

$$= 1 \text{ for all } n \tag{37}$$

For $n = 1$, this solution simplifies to that in [8].

4. RESULTS

In 3.3 we utilised integrated versions of principles of conservation of mass, momentum, and energy along with an appropriate form of the boundary layer velocity profile (11) to

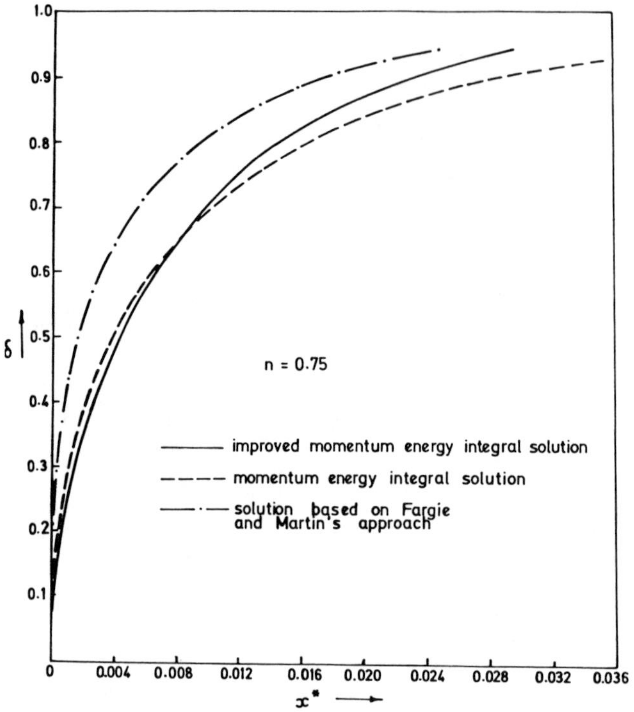

(a)

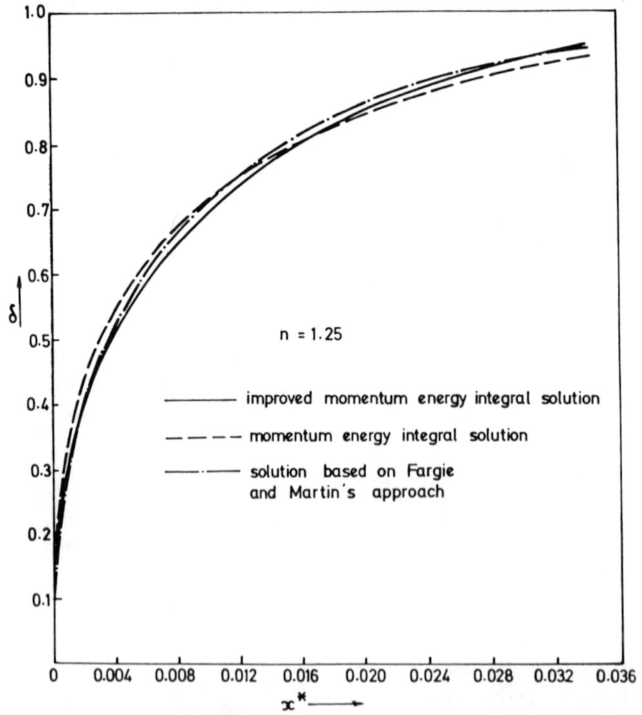

(b)

FIGURE 1. Boundary layer development in the inlet region: (a) pseudoplastic fluids; (b) dilatant fluids.

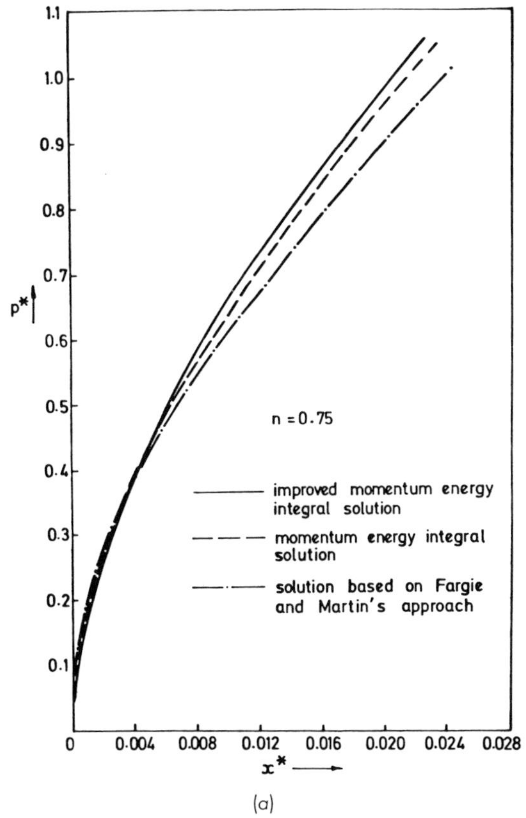

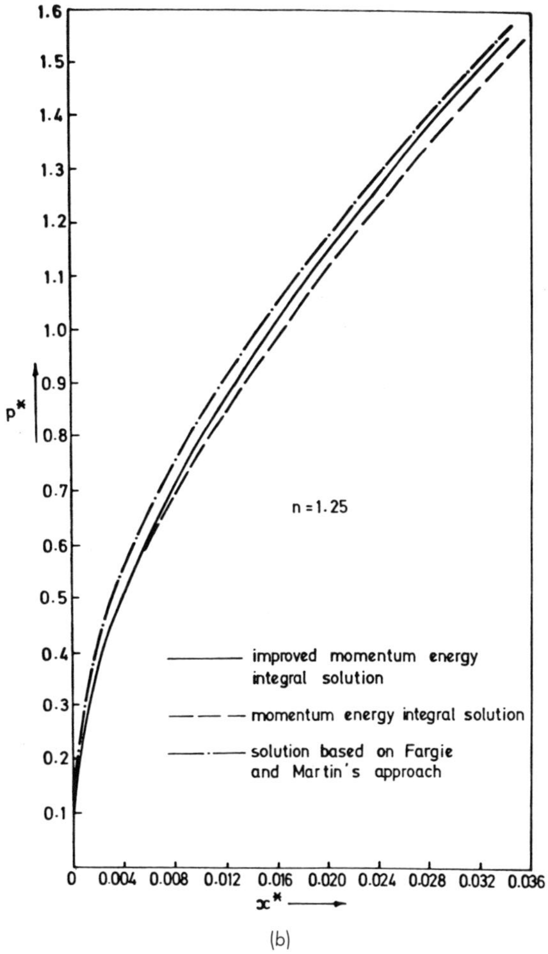

FIGURE 2. Pressure distribution in the developing region: (a) pseudoplastic fluids; (b) dilatant fluids.

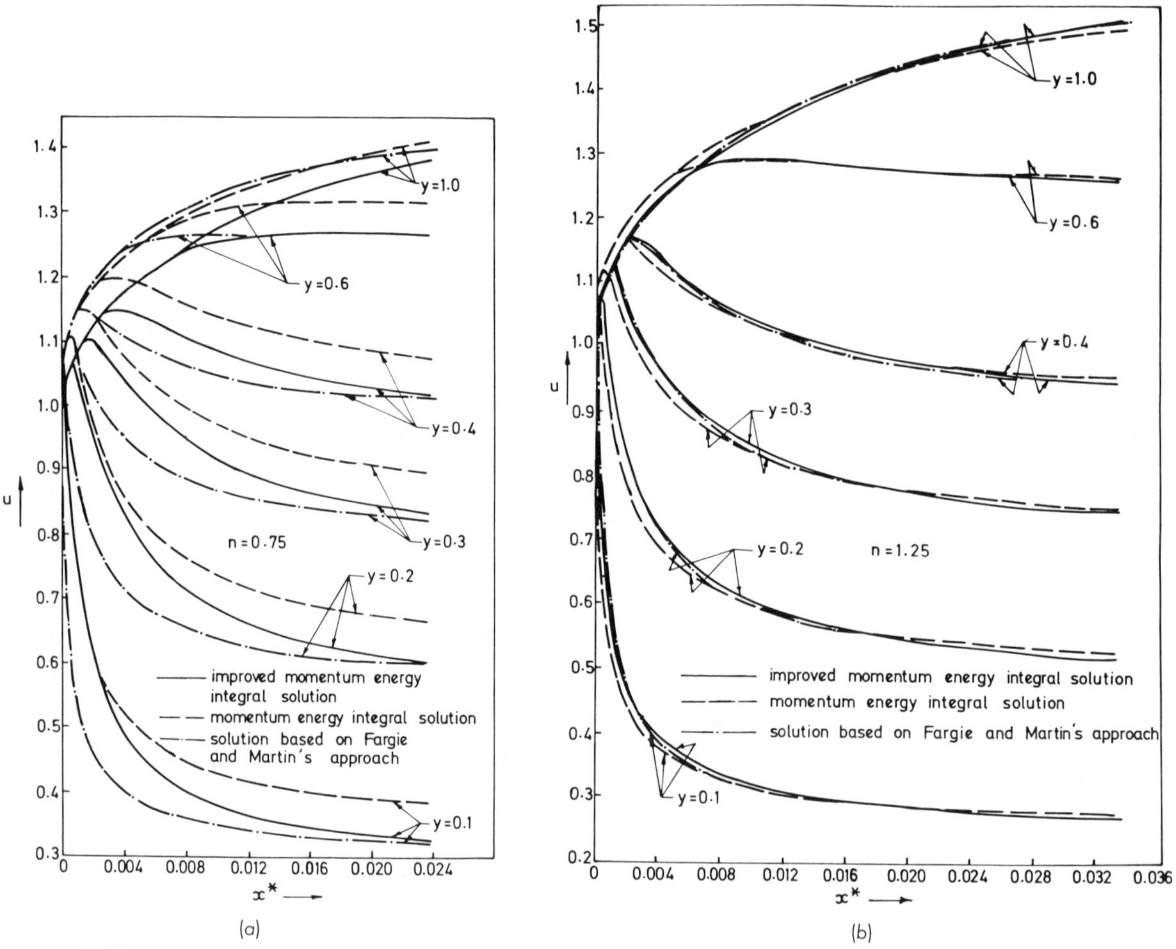

FIGURE 3. Axial velocity as a function of distance from the entry plane: (a) pseudoplastic fluids; (b) dilatant fluids.

obtain the developing Power-law fluid flow in a straight channel. It was noted that the cubic and quartic profiles used earlier [7] do not represent well the flatter profiles for pseudoplastic fluids. The choice (11) for the velocity profiles in the boundary layer not only accounts for this inadequacy but also leads exactly to the fully established distribution for all flow behaviour indices. This in turn takes care of the objections raised by Defrawi and Finlayson [4] against polynomial profiles.

Fargie and Martin [5] used a simple and elegant approach for developing flow in a circular pipe in which they determined the pressure gradient by an application of differential momentum Equation (6) at the pipe wall. Complete solution has been obtained by combining this with the momentum integral equation and choosing the boundary layer velocity profile appropriately. This approach has been extended to solve the title problem [9]. A simplified version of [9] is contained in 3.4.

The effectiveness of integral solutions may be attributed to the fact that the values of integrals do not differ much from their actual values if integrands vary slightly from their exact values provided the boundary conditions are satisfied. Thus, in these analyses (3.3 and 3.4) the only source of error appears to be the skin friction term appearing in the momentum integral Equation (8). This is in terms of derivative of the velocity evaluated at the lower wall of the channel. When the boundary layer velocity distribution is assumed, the error in its derivative at the wall may be considerable.

To get rid of this possible source of error, in 3.5 skin friction has been expressed in terms of integral involving the unknown velocity distribution. This, following Volkov [19], has been accomplished by integrating the equation of momentum twice over the flow cross section, which leads to the modified momentum integral Equation (31). This gives rise to an improved momentum energy integral solution [10] which is presented in 3.5.

All three solutions are displayed in Figures 1–3.

Boundary layer thickness as a function of location is displayed in Figure 1. For pseudoplastic ($n = 0.75$) fluids [Figure 1(a)], except near the entry, the location at which a given boundary layer thickness is attained as predicted by the solution in 3.5 lies betwen those predicted on the basis of solutions in 3.4 and 3.3. But solution in the former shows a relatively rapid rate of boundary layer growth and the station for a given thickness is predicted to be the closest to the entry plane. However, near the entry, for the given δ, the distances as predicted by the solutions in 3.4, 3.3 and 3.5 are in increasing order. Figure 1(b) displays the boundary layer development for the dilatant ($n = 1.25$) fluids.

Inlet length may be defined as the distance from the entry plane at which the centre line has reached, say 98% of its fully established value. Therefore, for a particular n, the inlet length can be computed as the value of x^* from the relevant relation for δ for which $\phi = 0.98[(2n+1)/(n+1)]$.

Inlet region pressure distribution is shown in Figure 2. For pseudoplastic ($n = 0.75$) fluids the solution in 3.5 predicts most rapid rate of growth of pressure drop with distance, whereas very near the entry the rate of growth is least amongst the three solutions.

For dilatant ($n = 1.25$) fluids the location at which a given pressure drop occurs is least inward for the solution in 3.3, whereas the prediction of 3.5 lies between those based on 3.3 and 3.4.

In the fully established state, the slope of the $p^* - 24x^*$ curve is unity. In the inlet region the slope of this curve is continuously changing, and that at the onset of the fully established state the solution in 3.5 predicts its value (from the left) to be unity. The solutions in 3.3 and 3.4 fail in this respect.

Figure 3 compares the axial velocity distributions for pseudoplastic ($n = 0.75$) and dilatant ($n = 1.25$) fluids at various distances from the lower wall of the channel.

From the experimental standpoint the confirmation of these theoretical predictions should be difficult because of the bounding end walls and the need for very large flow rates to even approximate a channel of infinite extent.

5. CONCLUDING REMARKS

Volkov's refinement of Kármán-Pohlhausen procedure has been incorporated in the solution presented in 3.5. In this analysis the pressure gradient has been obtained by an application of mechanical energy balance to the entire flow field and the assumed axial velocity distribution reduces to the one in the fully developed state when the boundary layer finally merges with the axis of the channel for all flow behaviour indices. Unlike the solutions in 3.3 and 3.4, this solution predicts the continuity of the slope of pressure drop-distance curve at the beginning of the fully established state. It is, therefore, believed that the improved momentum energy integral solution in 3.5 provides the best over-all Power-law fluid flow picture in the developing region of a straight channel presently available.

NOTATION

A_1, B_1, B_2, C_1, C_2 = Constants depending on n, defined in the text
e_{ij} = Components of rate of strain tensor
$2h$ = Channel width
k = consistency of the fluid
n = flow behaviour index
p = dimensionless pressure ($= P/\varrho U_0^2$)
p^* = dimensionless pressure drop ($= (P_0 - P)/\frac{1}{2}\varrho U_0^2$)
P = pressure at any point
P_0 = pressure at the entrance plane

R_n = Reynolds number ($= 12[nh/(2n + 1)]^n U_0^{2-n}/\gamma$)
u = dimensionless axial velocity ($= U/U_0$)
U = axial velocity
U_0 = uniform entry velocity
v = dimensionless transverse velocity ($= Vh^n U_0^{1-n}/\gamma$)
V = transverse velocity
x = dimensionless distance from the entrance plane ($= X\gamma/h^{n+1}U_0^{2-n}$)
x^* = dimensionless distance from the entrance plane ($= X/h\ R_n$)
X = axial distance measured from the entrance plane
y = dimensionless distance from the lower wall of the channel ($= Y/h$)
Y = distance measured from the lower wall of the channel
γ = fluid kinematic viscosity ($= k2^{(n-1)/2}/\varrho$)
δ = dimensionless boundary layer thickness
λ = reciprocal of normalised pressure gradient
ϱ = fluid density
τ_{ij} = components of stress tensor
ϕ = dimensionless core velocity

REFERENCES

1. Boger, D. V., "Circular Entry Flows of Inelastic and Viscoelastic Fluids," in *Advances in Transport Processes*, Vol. II, Wiley Eastern Ltd., New Delhi, pp. 43–104 (1982).
2. Campbell, W. D. and J. C. Slattery, "Flow in the Entrance of a Tube," *Jour. Basic Eng., Trans. ASME,* 85: 41–46 (1963).
3. Collins, M. and W. R. Schowalter, "Behaviour of Non-Newtonian Fluids in the Inlet Region of a Channel," *Amer. Instn. Chem. Engrs. Jour.*, 9: 98–102 (1963); also *Amer. Instn. Chem. Engrs. Jour.*, 10: 597 (1964).
4. Defrawi, M. E. and B. A. Finlayson, "Entrance Region Flows of Non-Newtonian Fluids," *Amer. Instn. Chem. Engrs. Jour.*, 18: 251–254 (1972).
5. Fargie, D. and B. W. Martin, "Developing Laminar Flow in a Pipe of Circular Cross-Section," *Proc. Roy. Soc. Lond. A*, 321: 461–476 (1971).
6. Gupta, R. C., "Flow Development in the Hydrodynamic Entrance Region of a Flat Duct," *Amer. Instn. Chem. Engrs. Jour.*, 11: 1149–1151 (1965).
7. Gupta, R. C., "Behaviour of Power-law Fluids in the Inlet Region of a Straight Channel," *Int. Jour. Nonlinear Mech.*, 5: 325–334 (1970).
8. Gupta, R. C., "Laminar Flow in the Hydrodynamic Inlet Region of a Flat Duct," *Arch. Hydrotechniki*, 23: 173–178 (1976).
9. Gupta, R. C., "Power-law Fluid Flow in the Hydrodynamic Inlet Region of a Straight Channel," *Jour. Phys. Soc. Japan*, 53: 585–591 (1984).
10. Gupta, R. C., "Developing Laminar Power-law Fluid Flow in Flat Duct," *Jour. Engg. Mech.*, 110: 752–760 (1984).
11. Kapur, J. N., "A Note on Boundary Layer Equations for Power-law Fluids," *Journ. Phys. Soc. Japan*, 18: 144 (1963).
12. Kapur, J. N. and R. C. Gupta, "Two Dimensional Flow of Power-law Fluids in the Inlet Length of a Straight Channel," *Zeit. Angew. Math. Mech.*, 43: 135–141 (1963).
13. Kapur, J. N. and R. C. Gupta, "Two Dimensional Flow of Power-law Fluids in the Inlet Length of a Straight Channel II," *Zeit. Angew. Math. Mech.*, 44: 277–284 (1964).
14. Meksyn, D. *New Methods in Laminar Boundary Layer Theory.* Pergamon Press Ltd. (1961).
15. Schiller, L., "Die Entwiklung der laminaren Geshwindigkeitverteilung und ihre Bedeutung für Zähigkeitmessungen," *Zeit. Angew. Math. Mech.*, 2: 96–106 (1922).
16. Schlichting, H., "Laminare Kanaleinlaufstromung," *Zeit. Angew. Math. Mech.*, 14: 368–373 (1934).
17. Shah, R. K. and A. L. London. *Laminar Flow Forced Convection in Ducts.* Supplement to *Advances in Heat Transfer.* Academic Press, New York (1978).
18. Skelland, A. H. P. *Non-Newtonian Flow and Heat Transfer.* John Wiley & Sons, Inc. New York (1967).
19. Volkov, V. N., "A Refinement of the Kármán-Pohlhausen Integral Method in Boundary Layer Theory," *Jour. Engg. Phy.*, 9: 371–374 (1965).

Broad-Crested Weirs and Long-Throated Flumes for Open Channel Flow Measurement

ALBERT J. CLEMMENS,* JOHN A. REPLOGLE* AND MARINUS G. BOS**

INTRODUCTION

Description and Definition

Most structures built for the purpose of measuring or regulating the rate of flow (Figure 1) consist of a converging transition where subcritical flowing water is accelerated and guided into the throat without flow separation, a throat where it accelerates through critical flow so that the discharge is controlled; and a diverging transition where the flow velocity is gradually reduced to an acceptable subcritical velocity and potential energy is recovered. Thus, a free-flowing, critical-flow flume is essentially a streamlined contraction built into an open channel where there must be sufficient water surface fall through the contraction so that critical flow occurs in the throat of the flume. The channel contraction may be formed from the sides only, from the bottom only, or from both the side and the bottom.

An approach channel of sufficient length upstream from the structure is necessary for the development of parallel flow and a stable water surface, the elevation of which can be determined accurately. Downstream from the structure is a tailwater channel that is fundamentally important to the design of the structure. The range of tailwater levels resulting from varying flow rates determines the elevation of the crest of the throat with respect to the tailwater channel bottom (or amount of contraction in flow needed).

Several commonly accepted structures eliminate or ignore one or more of these parts. As a result, the function of that part is not fulfilled, making the applicability of the structure less general. Of most practical significance is straight-lined parallel flow both in the approach channel and within the contraction (or throat) itself where critical flow occurs. When this parallel critical flow exists, classical hydraulic principles can be used to approximate the stage(head)-discharge relationship with a minimum of observed data. For curvilinear flow the rate of streamline curvature must be known at particular control points. This information is usually difficult to incorporate into practical hydraulic models and is even less practical to obtain in field situations. Consequently, flow-measuring devices based on curvilinear flow are usually laboratory rated, are sensitive to downstream water levels, and must be reproduced in careful detail to obtain reasonably accurate measurements. This theoretical predictability is important not only for designing structures for matching specific conditions but also for determining the acceptability of construction errors and modifications. Short (or nonexistent) throats cause three-dimensional flow for which no valid theory is available and thus limit the predictability of hydraulic behavior.

In this chapter we deal with "long-throated flumes" and "broad-crested weirs," which from a hydraulic point of view are similar. The name "weir" is used if the control section is formed by a raised channel floor or sill. By custom, the "flume" is formed by side contractions in the channel. If the control section is formed by both side contractions and bottom sill it also is usually called a "flume." Classified under the above family of structures are those weirs or flumes over which the streamlines are nearly parallel in the flume throat, at least over a short distance, and the upstream approach channel near the gauging station where the sill referenced head is to be observed (Figure 1). The sill referenced head is the water level in the upstream channel relative to the flume throat bottom or weir crest. To obtain parallel flow at the control section, the throat length L in the direction of flow must be of a certain minimum length equal to one to two times the total upstream sill-referenced energy head H_1. In the following sections the limitations on the ratio H_1/L will be specified in more detail.

*U.S. Water Conservation Laboratory, USDA-ARS Phoenix, AZ
**International Institute for Land Reclamation and Improvement/ILRI, Wageningen, The Netherlands

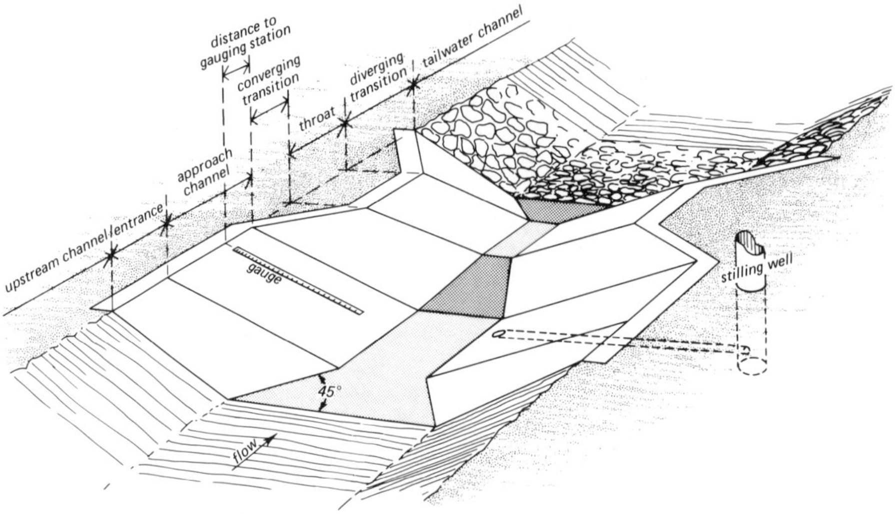

FIGURE 1. General layout of a flow-measuring structure.

FIGURE 2. Broad-crested weirs require only a small drop in water surface elevation (Arizona).

Advantages

The use of long throated flumes and the related broad-crested weir (with gradual approach ramp) is recommended for measuring flows in open channels whenever the water surface at the control section must remain free. They have the following major advantages over other styles of weirs and flumes (Parshall flume, cutthroat flume, H-flume, sharp-crested weirs, broad-crested weirs with abrupt entrance, etc):

a) Provided that critical flow occurs in the throat, a rating table can be calculated with an error of less than 2% in the listed discharge. This can be done for any combination of a prismatic throat and an arbitrarily shaped approach channel. Some examples of the wide variety of sizes and shapes of structures are given in later sections.
b) The throat, perpendicular to the direction of flow, can be shaped in such a way that the complete desired range of discharge can be measured accurately.
c) The headloss over the weir or flume required to have a unique relationship between the upstream sill-referenced head and the discharge is minimal (see Figure 2).
d) This headloss requirement can be estimated with sufficient accuracy for any of these structures placed in an arbitrarily shaped channel.
e) Because of their gradual converging transition, these structures have few problems with floating debris.
f) Field observations have shown that the structure can be designed to pass sediment transported by channels with subcritical flow.
g) Provided that the throat is horizontal in the direction of flow, a rating table can be produced which is based upon post-construction dimensions. This allows an accurate rating table to be made to compensate for deviations from designed dimensions. It also allows reshaping of the throat if required.
h) Under similar hydraulic and other boundary conditions, these weirs/flumes are usually the most economical of all structures for accurately measuring flow.

For a brief history on the development of flumes and weirs, see Bos (1985).

THEORY

Head Discharge Equations

IDEAL FLUID FLOWS

The head-discharge relationship for a long-throated flume is determined by balancing the energy head across the flume. For an ideal fluid, the energy or head loss is zero. It is convenient to work with the sill (or weir crest) referenced energy head, defined as

$$H = h + \frac{v^2}{2g} = h + \frac{Q^2}{2gA^2} \tag{1}$$

in which h = the sill-referenced head, v = average flow velocity, g = acceleration due to gravity, $Q = Av$ = flow rate, and A = flow area. A definition sketch for the terms to be used is given in Figure 3, in which y = actual water depth; and ΔH = the energy or head loss across the flume. The subscript, 1, refers to the conditions upstream from the flume at the gauging station where the water depth, h_1, is to be measured. The subscript, c, refers to the conditions at the control section, which is the approximate location of critical flow. And the subscript, 2, refers to the conditions in the tailwater channel downstream from the flume.

If flow is critical and Q is constant, there is only one value of y_c for each value $H = H_c$, which can be calculated by the

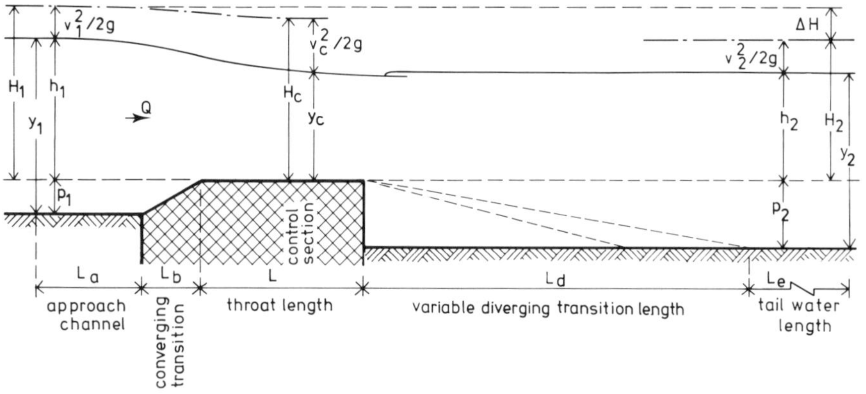

FIGURE 3. Definition sketch for flow through long-throated flume (profile).

following equation for critical flow:

$$H_1 = H_c = y_c + \frac{A_c}{2B_c} \qquad (2)$$

in which A_c = wetted area at the control section if the depth of flow equals y_c; and B_c = water surface width at the control section.

For a discharge measurement structure, the depth of flow at the control section must be critical to avoid the need to measure the depth there.

Since, by definition, $h_c = y_c$, Equation (1) then can be written as

$$Q = A_c \{2\ g\ (H_1 - y_c)\}^{0.5} \qquad (3)$$

in which, according to Equation (2),

$$y_c = H_1 - \frac{A_c}{2B_c} \qquad (4)$$

Combining these two equations gives

$$Q = \left\{\frac{gA_c^{\ 3}}{B_c}\right\}^{0.5} \qquad (5)$$

This general discharge equation is valid for all arbitrarily shaped control sections. It is easily applicable if simple equations exist for both A_c and B_c (Jameson, 1925).

For a rectangular control section in which flow is critical, we may write $A_c = b_c y_c$ and $B_c = b_c$ where b_c is the bottom width, so that Equation (4) may be written as

$$y_c = \frac{2}{3} H_1 \qquad (6)$$

Substitution of this relation and $A_c = by_c$ into Equation (3) gives, after simplification,

$$Q = \frac{2}{3}\left(\frac{2}{3}\, g\right)^{0.5} b_c H_1^{\ 1.5} \qquad (7)$$

REAL FLUID FLOW

In reality the above-mentioned assumptions are not entirely correct because there are frictional losses, nonuniform velocity distributions, and deviations from straight parallel flow. Therefore, these deviations from the idealized flow assumptions must be compensated for by the introduction of a discharge coefficient C_d. Equation (7) rearranged then reads

$$Q = C_d 2/3(2/3g)^{0.5} bH_1^{\ 1.5} \qquad (8)$$

In an open channel it is not convenient to measure the energy head H_1 directly. It is therefore common practice to relate the flow rate to the upstream sill-referenced water level h_1 (or head) in the following way:

$$Q = C_v C_d 2/3(2/3g)^{0.5} bh_1^{\ 1.5} \qquad (9)$$

where C_v is the approach-velocity coefficient which accounts for neglecting the velocity head at the gauging location, $v_1^2/2g$.

The calibration for any flume cross-sectional shape can be approximated by a generalized form of Equation (9), expressed as a power function:

$$Q = C_v C_d K(h_1)^u \qquad (10)$$

where K is a constant that includes flow area and other factors and u is a function of the flume shape (1.5 for rectangular flumes and 2.5 for triangular flumes).

HEAD-DISCHARGE EQUATIONS FOR OTHER CONTROL SHAPES

Following the same procedure that led to Equation (7) and making the same assumptions that underlie the C_d and C_v coefficients in Equation (9), Bos (1978) derived head-discharge equations for commonly used shapes of the control section. The results (which have been verified by laboratory studies) are shown in Figure 4. Tables 1 and 2 are included to calculate the critical depth y_c for trapezoidal and circular controls, respectively. Tables 2 and 3 provide useful values for computing the head-discharge relationships for circular controls and controls made from placing a bottom sill in a circular section (Clemmens, Bos, Replogle, 1984). In Table 2, the distance along one side of a circular section (1/2 of wetted perimeter, *WP*) as a function of pipe diameter is given to aid in the construction of gauges for pipe walls.

Computing Actual Discharge

There are basically three methods of determining the discharge of a flume or weir as a function of head: 1) by direct laboratory or field determination of head versus discharge; 2) by statistical approximation of discharge coefficient from prior studies (Bos, 1985) and the calculation of an approach velocity coefficient, and 3) by use of a mathematical model developed by Replogle (1975).

DIRECT CALIBRATION

One of the purposes of this chapter is to show that there is no need to actually calibrate the flume or weir in the laboratory. The best calibration procedure is one which uses a weight tank or volume tank. These can give accuracies on the order of ±0.1%. Likewise, laboratory (e.g., weight tank) calibrated meters can repeat such accuracy. Unfortunately, field methods for calibrating flumes can rarely be relied upon to give accuracies of ±5%. Velocity-area method can often exceed ± 10% in error. Commonly used

SHAPE OF CONTROL SECTION	HEAD–DISCHARGE EQ. TO BE USED	HOW TO FIND THE y_c - VALUE	SOURCE OF BASIC EQUATION
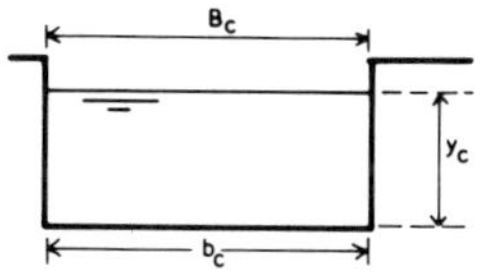	$Q = C_d C_v \frac{2}{3} (\frac{2}{3} g)^{1/2} b_c h_1^{3/2}$	$y_c = \frac{2}{3} H_1$	Belanger, 1849
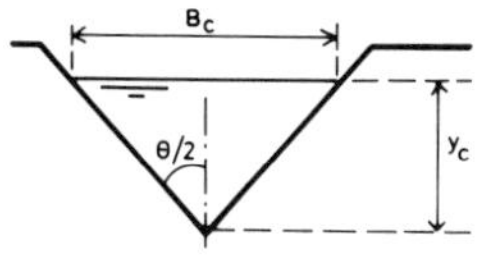	$Q = C_d C_v \frac{16}{25} (\frac{2}{5} g)^{1/2} \tan\frac{\theta}{2}\, h_1^{5/2}$	$y_c = \frac{4}{5} H_1$	Jameson, 1925
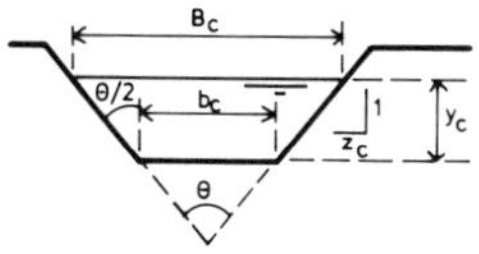	$Q = C_d [b_c y_c + z_c y_c^2][2g(H_1 - y_c)]^{1/2}$	Use Table 1	Bos, 1976
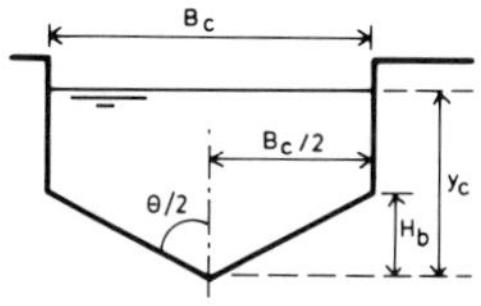	If $H_1 \leqslant 1.25\, H_b$ $Q = C_d C_v \frac{16}{25} (\frac{2}{5} g)^{1/2} \tan\frac{\theta}{2}\, h_1^{5/2}$	$y_c = \frac{4}{5} H_1$	Jameson, 1925
	If $H_1 \geqslant 1.25\, H_b$ $Q = C_d C_v \frac{2}{3} (\frac{2}{3} g)^{1/2} B_c (h_1 - \frac{1}{2} H_b)^{3/2}$	$y_c = \frac{2}{3} H_1 + \frac{1}{6} H_b$	Bos, 1976
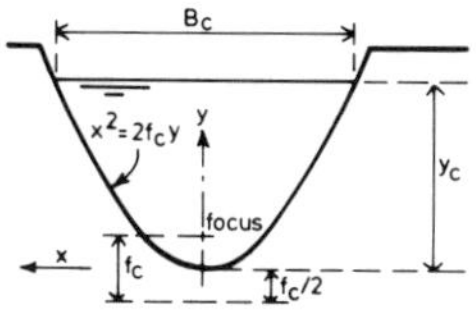	$Q = C_d C_v (\frac{3}{4} f_c g)^{1/2} h_1^2$	$y_c = \frac{3}{4} H_1$	Jameson, 1925
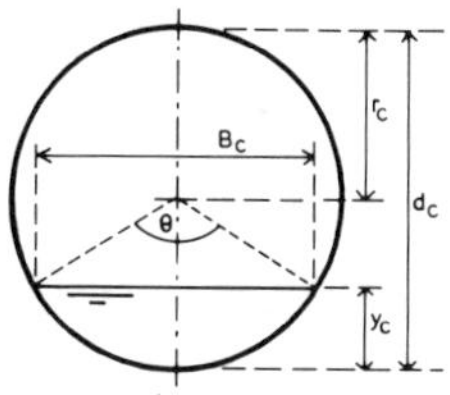	$Q = C_d d_c^{5/2} \sqrt{g}\, [f(\theta)]$ use table 2 to find $f(\theta)$	Use Table 2	Bos, 1976
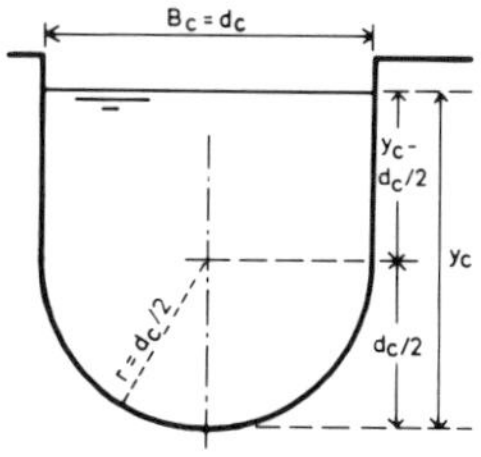	If $H_1 \leqslant 0.70\, d_c$ $Q = C_d d_c^{5/2} \sqrt{g}\, [f(\theta)]$ use table 2 to find $f(\theta)$	Use Table 2	Bos, 1976
	If $H_1 \geqslant 0.70\, d_c$ $Q = C_d C_v \frac{2}{3} (\frac{2}{3} g)^{1/2} d_c (h_1 - 0.1073 d_c)^{3/2}$	$y_c = \frac{2}{3} H_1 + 0.0358 d_c$	Bos, 1977
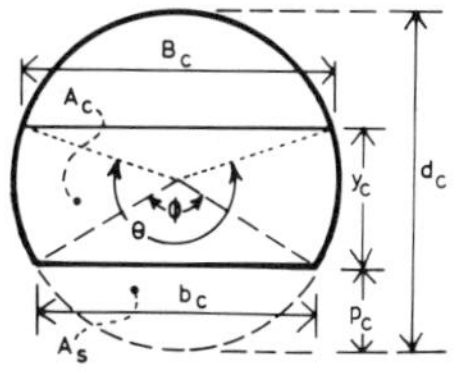	$Q = C_d d_c^{5/2} \sqrt{g}\, [f(\phi, \theta)]$ use table 3 to find $f(\phi, \theta)$	y_c is variable	Clemmens, Bos and Replogle, 1984

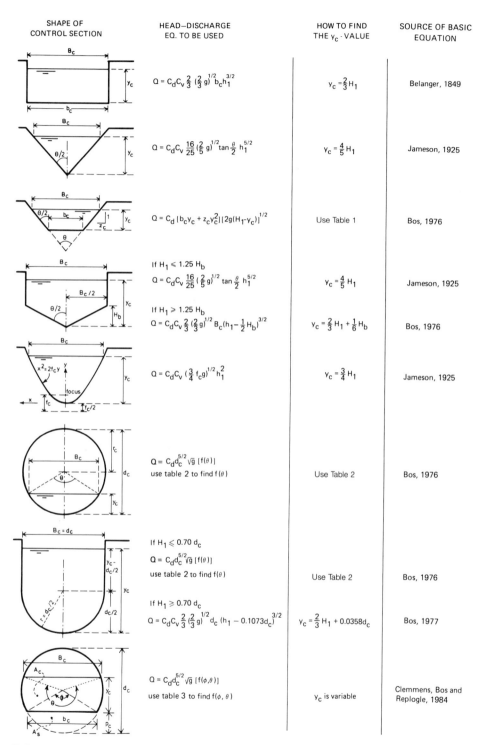

FIGURE 4. Head-discharge relationship for long-throated flumes (from Bos, Replogle and Clemmens, 1984).

TABLE 1. Values of the Ratio y_c/H_1 as a Function of z_c and H_1/b_c for Trapezoidal Control Sections.

	Side slopes of channel, ratio of horizontal to vertical (z_c)									
H_1/b_c	Vertical	0.25:1	0.50:1	0.75:1	1:1	1.5:1	2:1	2.5:1	3:1	4:1
.00	.667	.667	.667	.667	.667	.667	.667	.667	.667	.667
.01	.667	.667	.667	.668	.668	.669	.670	.670	.671	.672
.02	.667	.667	.668	.669	.670	.671	.672	.674	.675	.678
.03	.667	.668	.669	.670	.671	.673	.675	.677	.679	.683
.04	.667	.668	.670	.671	.672	.675	.677	.680	.683	.687
.05	.667	.668	.670	.672	.674	.677	.680	.683	.686	.692
.06	.667	.669	.671	.673	.675	.679	.683	.686	.690	.696
.07	.667	.669	.672	.674	.676	.681	.685	.689	.693	.699
.08	.667	.670	.672	.675	.678	.683	.687	.692	.696	.703
.09	.667	.670	.673	.676	.679	.684	.690	.695	.698	.706
.10	.667	.670	.674	.677	.680	.686	.692	.697	.701	.709
.12	.667	.671	.675	.679	.684	.690	.692	.701	.706	.715
.14	.667	.672	.676	.681	.686	.693	.699	.705	.711	.720
.16	.667	.672	.678	.683	.678	.696	.703	.709	.715	.725
.18	.667	.673	.679	.684	.690	.698	.706	.713	.719	.729
.20	.667	.674	.680	.686	.692	.701	.709	.717	.723	.733
.22	.667	.674	.681	.688	.694	.704	.712	.720	.726	.736
.24	.667	.675	.683	.689	.696	.706	.715	.723	.729	.739
.26	.667	.676	.684	.691	.698	.709	.718	.725	.732	.742
.28	.667	.676	.685	.693	.699	.711	.720	.728	.734	.744
.30	.667	.677	.686	.694	.701	.713	.723	.730	.737	.747
.32	.667	.678	.687	.696	.703	.715	.725	.733	.739	.749
.34	.667	.678	.689	.697	.705	.717	.727	.735	.741	.751
.36	.667	.679	.690	.699	.706	.719	.729	.737	.743	.752
.38	.667	.680	.691	.700	.708	.721	.731	.738	.745	.754
.40	.667	.680	.692	.701	.709	.723	.733	.740	.747	.756
.42	.667	.681	.693	.703	.711	.725	.734	.742	.748	.757
.44	.667	.681	.694	.704	.712	.727	.736	.744	.750	.759
.46	.667	.682	.695	.705	.714	.728	.737	.745	.751	.760
.48	.667	.683	.696	.706	.715	.729	.739	.747	.752	.761
.5	.667	.683	.697	.708	.717	.730	.740	.748	.754	.762
.6	.667	.686	.701	.713	.723	.737	.747	.754	.759	.767
.7	.667	.688	.706	.718	.728	.742	.752	.758	.764	.771
.8	.667	.692	.709	.723	.732	.746	.756	.762	.767	.774
.9	.667	.694	.713	.727	.737	.750	.759	.766	.770	.776

(continued)

TABLE 1 (continued).

H_1/b_c	Side slopes of channel, ratio of horizontal to vertical (z_c) Vertical	0.25:1	0.50:1	0.75:1	1:1	1.5:1	2:1	2.5:1	3:1	4:1
1.0	.667	.697	.717	.730	.740	.754	.762	.768	.773	.778
1.2	.667	.701	.723	.737	.747	.759	.767	.772	.776	.782
1.4	.667	.706	.729	.742	.752	.764	.771	.776	.779	.784
1.6	.667	.709	.733	.747	.756	.767	.774	.778	.781	.786
1.8	.667	.713	.737	.750	.759	.770	.776	.781	.783	.787
2	.667	.717	.740	.754	.762	.773	.778	.782	.785	.788
3	.667	.730	.753	.766	.773	.781	.785	.787	.790	.792
4	.667	.740	.762	.773	.778	.785	.788	.790	.792	.794
5	.667	.748	.768	.777	.782	.788	.791	.792	.794	.795
10	.667	.768	.782	.788	.791	.794	.795	.796	.797	.798
∞		.800	.800	.800	.800	.800	.800	.800	.800	.800

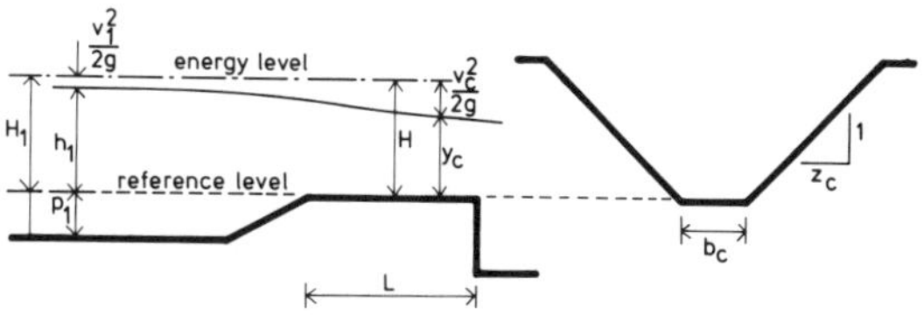

TABLE 2. Ratios for Determining the Discharge Q of a Broad-Crested Weir and Long-Throated Flume with Circular Control Section.

y_c/d_c	H_1/d_c	A_c/d_c^2	$W_p/2_d$	$f(\Theta)$	y_c/d_c	H_1/d_c	A_c/d_c^2	$W_p/2_d$	$f(\Theta)$
.01	.0133	.0013	0.1002	0.0001	.51	.7114	.4027	0.7954	0.2556
.02	.0267	.0037	0.1419	0.0004	.52	.7265	.4127	0.8054	0.2652
.03	.0401	.0069	0.1741	0.0010	.53	.7417	.4227	0.8154	0.2750
.04	.0534	.0105	0.2014	0.0017	.54	.7570	.4327	0.8254	0.2851
.05	.0668	.0147	0.2255	0.0027	.55	.7724	.4426	0.8355	0.2952
.06	.0803	.0192	0.2475	0.0039	.56	.7879	.4526	0.8455	0.3056
.07	.0937	.0242	0.2678	0.0053	.57	.8035	.4625	0.8556	0.3161
.08	.1071	.0294	0.2868	0.0068	.58	.8193	.4724	0.8657	0.3268
.09	.1206	.0350	0.3047	0.0087	.59	.8351	.4822	0.8759	0.3376
.10	.1341	.0409	0.3218	0.0107	.60	.8511	.4920	0.8861	0.3487
.11	.1476	.0470	0.3381	0.0129	.61	.8672	.5018	0.8963	0.3599
.12	.1611	.0534	0.3537	0.0153	.62	.8835	.5115	0.9066	0.3713
.13	.1746	.0600	0.3689	0.0179	.63	.8999	.5212	0.9169	0.3829
.14	.1882	.0688	0.3835	0.0214	.64	.9165	.5308	0.9273	0.3947
.15	.2017	.0739	0.3977	0.0238	.65	.9333	.5404	0.9377	0.4068

(continued)

TABLE 2 (continued).

y_c/d_c	H_1/d_c	A_c/d_c^2	$Wp/2_d$	$f(\Theta)$	y_c/d_c	H_1/d_c	A_c/d_c^2	$Wp/2_d$	$f(\Theta)$
.16	.2153	.0811	0.4115	0.0270	.66	.9502	.5499	0.9483	0.4189
.17	.2289	.0885	0.4250	0.0304	.67	.9674	.5594	0.9589	0.4314
.18	.2426	.0961	0.4381	0.0340	.68	.9848	.5687	0.9695	0.4440
.19	.2562	.1039	0.4510	0.0378	.69	1.0025	.5780	0.9803	0.4569
.20	.2699	.1118	0.4636	0.0418	.70	1.0204	.5872	0.9912	0.4701
.21	.2836	.1199	0.4760	0.0460	.71	1.0386	.5964	1.0021	0.4835
.22	.2973	.1281	0.4882	0.0504	.72	1.0571	.6054	1.0132	0.4971
.23	.3111	.1365	0.5002	0.0550	.73	1.0759	.6143	1.0244	0.5109
.24	.3248	.1449	0.5120	0.0597	.74	1.0952	.6231	1.0357	0.5252
.25	.3387	.1535	0.5236	0.0647	.75	1.1148	.6319	1.0472	0.5397
.26	.3525	.1623	0.5351	0.0698	.76	1.1349	.6405	1.0588	0.5546
.27	.3663	.1711	0.5464	0.0751	.77	1.1555	.6489	1.0706	0.5698
.28	.3802	.1800	0.5576	0.0806	.78	1.1767	.6573	1.0826	0.5855
.29	.3942	.1890	0.5687	0.0863	.79	1.1985	.6655	1.0948	0.6015
.30	.4081	.1982	0.5796	0.0922	.80	1.2210	.6735	1.1071	0.6180
.31	.4221	.2074	0.5905	0.0982	.81	1.2443	.6815	1.1198	0.6351
.32	.4361	.2167	0.6013	0.1044	.82	1.2685	.6893	1.1326	0.6528
.33	.4502	.2260	0.6119	0.1108	.83	1.2938	.6969	1.1458	0.6712
.34	.4643	.2355	0.6225	0.1174	.84	1.3203	.7043	1.1593	0.6903
.35	.4784	.2450	0.6331	0.1289	.85	1.3482	.7115	1.1731	0.7102
.36	.4926	.2546	0.6435	0.1311	.86	1.3777	.7186	1.1873	0.7312
.37	.5068	.2642	0.6539	0.1382	.87	1.4092	.7254	1.2019	0.7533
.38	.5211	.2739	0.6642	0.1455	.88	1.4432	.7320	1.2171	0.7769
.39	.5354	.2836	0.6745	0.1529	.89	1.4800	.7384	1.2327	0.8021
.40	.5497	.2934	0.6847	0.1605	.90	1.5204	.7445	1.2490	0.8293
.41	.5641	.3032	0.6949	0.1683	.91	1.5655	.7504	1.2661	0.8592
.42	.5786	.3130	0.7051	0.1763	.92	1.6166	.7560	1.2840	0.8923
.43	.5931	.3229	0.7152	0.1844	.93	1.6759	.7612	1.3030	0.9297
.44	.6076	.3328	0.7253	0.1927	.94	1.7465	.7662	1.3233	0.9731
.45	.6223	.3428	0.7353	0.2012	.95	1.8341	.7707	1.3453	1.0248
.46	.6369	.3527	0.7454	0.2098					
.47	.6517	.3627	0.7554	0.2186					
.48	.6665	.3727	0.7654	0.2276					
.49	.6814	.3827	0.7754	0.2368					
.50	.6964	.3927	0.7854	0.2461					

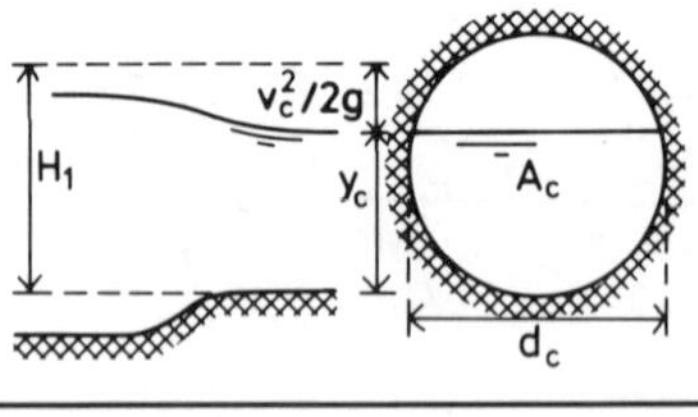

Note: $f(\Theta) = \frac{A_c}{d_c^2}\{2(\frac{H_1}{d_c} - \frac{y_c}{d_c})\}^{0.5} = \frac{(\Theta - \sin\Theta)^{1.5}}{8(8\sin 1/2\Theta)^{0.5}}$

TABLE 3. Ratios for Determining the Discharge of a Broad-Crested Weir in a Circular Pipe. $C_d = 1.0$, $\alpha_c = 1.0$, $H_1 = H_c$.

$$f(\phi, \Theta) = \frac{(\Theta - \phi + \sin\phi - \sin\Theta)^{1.5}}{8(8\sin\frac{1}{2}\Theta)^{0.5}}$$

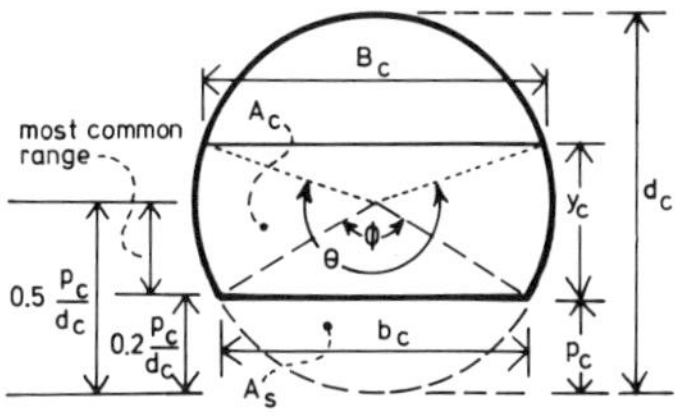

$\frac{P_c + H_1}{d_c}$	P_c/d_c = 0.15	0.20	0.25	0.30	0.35	0.40	0.45	0.50
0.16	0.0004							
0.17	0.0011							
0.18	0.0021							
0.19	0.0032							
0.20	0.0045							
0.21	0.0060	0.0004						
0.22	0.0076	0.0012						
0.23	0.0094	0.0023						
0.24	0.0113	0.0036						
0.25	0.0133	0.0050						
0.26	0.0155	0.0066	0.0005					
0.27	0.0177	0.0084	0.0013					
0.28	0.0201	0.0103	0.0025					
0.29	0.0226	0.0124	0.0038					
0.30	0.0252	0.0145	0.0054					
0.31	0.0280	0.0169	0.0071	0.0005				
0.32	0.0308	0.0193	0.0090	0.0014				
0.33	0.0337	0.0219	0.0110	0.0026				
0.34	0.0368	0.0245	0.0132	0.0040				
0.35	0.0399	0.0273	0.0155	0.0057				
0.36	0.0432	0.0302	0.0179	0.0075	0.0005			
0.37	0.0465	0.0332	0.0205	0.0094	0.0015			
0.38	0.0500	0.0363	0.0232	0.0115	0.0027			
0.39	0.0535	0.0396	0.0260	0.0138	0.0042			
0.40	0.0571	0.0429	0.0289	0.0162	0.0059			
0.41	0.0609	0.0463	0.0320	0.0187	0.0077	0.0005		
0.42	0.0647	0.0498	0.0351	0.0214	0.0097	0.0015		
0.43	0.0686	0.0534	0.0383	0.0242	0.0119	0.0028		
0.44	0.0726	0.0571	0.0417	0.0271	0.0143	0.0043		
0.45	0.0767	0.0609	0.0451	0.0301	0.0167	0.0060		
0.46	0.0809	0.0648	0.0487	0.0332	0.0193	0.0079	0.0005	
0.47	0.0851	0.0688	0.0523	0.0365	0.0220	0.0100	0.0015	
0.48	0.0895	0.0729	0.0561	0.0398	0.0249	0.0122	0.0028	
0.49	0.0939	0.0770	0.0599	0.0432	0.0279	0.0145	0.0043	
0.50	0.0984	0.0813	0.0638	0.0468	0.0309	0.0170	0.0061	
0.51	0.1030	0.0856	0.0678	0.0504	0.0341	0.0197	0.0080	0.0005
0.52	0.1076	0.0900	0.0719	0.0541	0.0374	0.0224	0.0101	0.0015
0.53	0.1124	0.0945	0.0761	0.0579	0.0408	0.0253	0.0123	0.0028
0.54	0.1172	0.0990	0.0803	0.0618	0.0443	0.0283	0.0147	0.0044
0.55	0.1221	0.1037	0.0847	0.0658	0.0479	0.0314	0.0172	0.0061
0.56	0.1270	0.1084	0.0891	0.0699	0.0515	0.0346	0.0198	0.0080
0.57	0.1320	0.1132	0.0936	0.0741	0.0553	0.0379	0.0226	0.0101
0.58	0.1372	0.1180	0.0981	0.0783	0.0592	0.0413	0.0255	0.0123
0.59	0.1423	0.1230	0.1028	0.0826	0.0631	0.0448	0.0285	0.0147
0.60	0.1476	0.1280	0.1075	0.0870	0.0671	0.0484	0.0316	0.0172
0.62		0.1382	0.1172	0.0960	0.0754	0.0559	0.0381	0.0225
0.64		0.1486	0.1271	0.1053	0.0840	0.0637	0.0449	0.0283
0.66		0.1593	0.1373	0.1149	0.0929	0.0718	0.0522	0.0346
0.68		0.1703	0.1477	0.1247	0.1020	0.0802	0.0597	0.0412
0.70		0.1815	0.1584	0.1348	0.1114	0.0888	0.0676	0.0481

(continued)

TABLE 3 (continued).

$$f(\phi,\theta) = \frac{(\theta - \phi + \sin\phi - \sin\theta)^{1.5}}{8(8\sin\frac{1}{2}\theta)^{0.5}}$$

$\frac{p_c + H_1}{d_c}$	$p_c/d_c = 0.15$	0.20	0.25	0.30	0.35	0.40	0.45	0.50
0.72		0.1929	0.1692	0.1451	0.1211	0.0978	0.0757	0.0554
0.74		0.2045	0.1804	0.1556	0.1310	0.1070	0.0841	0.0629
0.76		0.2163	0.1917	0.1663	0.1411	0.1164	0.0928	0.0707
0.78		0.2283	0.2031	0.1773	0.1514	0.1260	0.1016	0.0788
0.80		0.2405	0.2148	0.1884	0.1618	0.1358	0.1107	0.0870
0.82		0.2528	0.2267	0.1997	0.1725	0.1458	0.1200	0.0955
0.84		0.2653	0.2386	0.2111	0.1833	0.1559	0.1294	0.1042
0.86		0.2780	0.2508	0.2227	0.1943	0.1662	0.1390	0.1130
0.88		0.2907	0.2630	0.2344	0.2054	0.1767	0.1487	0.1220
0.90		0.3036	0.2754	0.2462	0.2166	0.1872	0.1586	0.1311
0.92		0.3166	0.2879	0.2581	0.2279	0.1979	0.1686	0.1404
0.94		0.3297	0.3005	0.2701	0.2394	0.2087		
0.96		0.3428	0.3131	0.2823	0.2509			
0.98		0.3561	0.3259	0.2944				
1.00		0.3694	0.3387					
1.02		0.3827						
1.04		0.3961						

flumes and weirs which require laboratory calibrations are subject to significant errors when placed in a setting which differs from the laboratory conditions. For these reasons field calibrations are generally not as reliable as direct calculation of discharge. However, they do offer a check against gross errors.

APPROXIMATIONS FOR C_d AND C_v

Discharge Coefficient C_d.

The discharge coefficient corrects for such phenomena as energy loss between gauging and control sections, nonuniformity of velocity distribution, and streamline curvature in the control section. These factors are closely related to the value of the ratio H_1/L, where L is the length of the contraction section in the direction of flow. Thus, the discharge coefficient for a weir or flume is not constant, but varies with the head H_1. When the flow is relatively shallow over a particular throat of length L (for example, an H_1/L ratio of 0.1), the frictional effects amount to a relatively large proportion of the value for H_1, approaching perhaps a 5% reduction in idealized discharge, or a C_d value of about 0.95. For deeper relative flow depth, say $H_1/L = 0.5$, the frictional and other effects are a smaller fraction of H_1 and may require a C_d value of perhaps 0.99. Finally, as the flow approaches an H_1/L value of 1, the flow streamlines become so curved through the region of critical flow that there are significant pressure reductions. The discharge then starts to exceed the computed idealized flow in spite of frictional losses and velocity distribution effects, and C_d may slightly exceed 1.

Bos (1978, 1985) compiled a comprehensive collection of data on the calibration of long-throated flumes and modified broad-crested weirs (i.e., weirs with a gradual approach ramp). Reasonably reliable predictions of C_d were obtained. The 95% confidence interval on C_d was within ±5% of the average C_d value within this range which could be approximately described by

$$C_d = 0.93 + 0.1\ H_1/L \text{ for } 0.1 \leq \frac{H_1}{L} \leq 1.0 \qquad (11)$$

Thus with this procedure, the value of C_d can be approximated to within ± 5% (95% of the time).

Approach Velocity Coefficient, C_v

The approach velocity coefficient corrects for the use of h_1 instead of H_1 in the head-discharge equation and thus for neglecting $\alpha_1 v_1^2/2g$. The exact value of C_v equals the ratio of Equation (8) divided by Equation (9). In general, C_v thus equals

$$C_v = \left\{\frac{H_1}{h_1}\right\}^u = \left\{1 + \frac{\alpha_1 v_1^2}{2gh_1}\right\}^u \qquad (12)$$

where u equals the power of h_1 in the head-discharge equation, being $u = 1.50$ for a rectangular control section, and α_1 is the velocity distribution coefficient (typically 1.04).

The power u of h_1 also determines the range of discharges that can be measured by a structure and the sensitivity of the structure.

If the approach velocity v_1 is small, that is, if the cross-sectional area of flow at the gauging station, A_1, is large in comparison with the control section, the velocity head $\alpha_1 v_1^2/2g$ is small with respect to H_1. Hence, H_1 and h_1 are almost equal to each other, in which case the C_v value is just slightly more than 1.0.

Because the discharge is mainly determined by the area of flow at the control section [Equation (5)] and the related approach velocity by the area of flow at the gauging station, it was found to be convenient to correlate C_v to the area ratio $\sqrt{\alpha_1}\, C_d\, A^*/A_1$ (Bos, 1978). In the latter area ratio, the value A^* equals the imaginary area of flow at the control section if the water depth would equal h_1. For a rectangular section

$$A^* = b_c h_1 \tag{13}$$

and for the trapezoidal approach channel

$$A_1 = b_1 y_1 + z_1 y_1^2 = y_1 (b_1 + z_1 y_1) \tag{14}$$

Values of C_v as a function of the area ratio $\sqrt{\alpha_1}\, C_d A^*/A_1$ are shown in Figure 5 for various control shapes. Because of the use of A^* in the area ratio, the C_v value is almost the same for all control shapes. Discharges are then computed for each value of h_1 with Equation (9) and C_d and C_v as computed here. Note that C_v and C_d both vary with head and thus also discharge.

MATHEMATICAL MODELING OF DISCHARGE

Replogle (1975) developed a mathematical model for the determination of actual discharges based on the calculation of head losses with a boundary layer drag method and on a procedure to account for velocity distributions (see Bos, Replogle, Clemmens 1984). It has been shown (Replogle, 1978) that this model can reliably predict discharges (and thus C_d) to within $\pm 2\%$ of actual flow over the range 0.05 to 0.5 for H_1/L. Here C_d is computed after the fact. Since permanent flumes change in roughness over time, predictions down to $H_1/L = 0.05$ may become unreliable. Also, deviations from predictions vary gradually for $H_1/L > 0.5$, often still within the $\pm 2\%$. Owing to these later conditions, the range of application for the rating tables presented in this chapter is $0.075 \leq H_1/L \leq 0.75$. The tables presented here were all developed with the model discussed above.

The model uses an iterative procedure to solve for Q given a value of h_1. Thus a C_v correction is not necessary. Also, the procedure starts by computing the ideal discharge, then adds friction and other effects and determines actual discharge. Thus C_d and C_v are calculated after the actual discharge is determined. Details of the model can be found in Bos, Replogle, Clemmens (1984).

Determining Acceptable Tailwater Levels

SUBMERGENCE

For long-throated devices some submergence is allowed, and thus the entire upstream energy head is not lost as for a sharp-crested weir. The submergence ratio is defined as the ratio between the downstream and upstream sill referenced energy heads, H_2/H_1. For low submergence ratios, critical flow occurs in the flume throat and the tailwater conditions have little or no effect on the upstream head; this flow is then called modular flow. At very high submergence ratios, critical flow no longer exists in the flume throat. This nonmodular flow condition requires that two heads be measured and

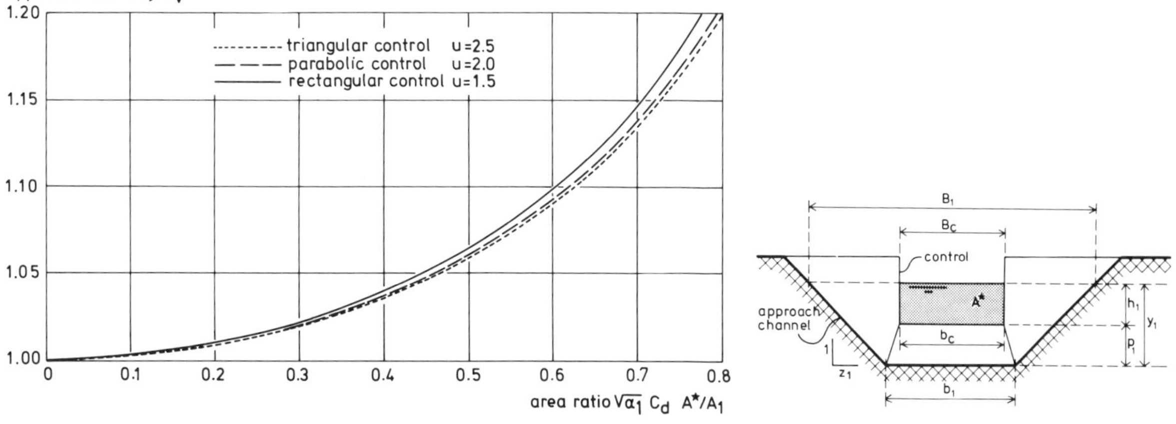

FIGURE 5. The approach velocity coefficient C_v as a function of the area ratio $\sqrt{\alpha_1}\, C_d A^*/A_1$.

leads to considerably greater error in the discharge measurement. While this is a common practice, the writers do not recommend it. The limiting submergence ratio between modular and nonmodular flow is called the modular limit and is defined as

$$ML = \frac{H_2}{H_1} = 1 - \frac{\Delta H}{H_1} \quad (15)$$

in which H_2 = is the downstream energy head; and ΔH = the change in energy head ($H_2 - H_1$) at the highest tailwater level, such that the flow through the flume is not changed by more than 1% for a given upstream head. The modified broad-crested weirs can have very high modular limits. For these reasons, the writers do not recommend the use of and provide no rating tables for nonmodular flow conditions.

REQUIRED HEAD LOSS

All critical-depth measuring devices require some energy or head loss for their operation. The required head loss is a result of wall friction and internal energy losses. Bos and Reinink (1981) developed a procedure for estimating the required head loss and modular limit for long-throated flumes. This procedure was incorporated into the mathematical model of Replogle and is presented in the computer program of Bos, Replogle, and Clemmens (1984). The frictional losses are computed from boundary layer theory as for the flume calibrations. The internal losses are primarily from the rapid expansion of flow on the downstream side of the structure. These can be estimated with an expansion coefficient, ξ. The energy loss due to the expansion is found from

$$\Delta H_k = \xi \frac{(v_c - v_2)^2}{2g} \quad (16)$$

Thus the velocity in the exit channel, v_2, is important to the calculated head loss. Discharging into a lake, $v_2 = 0$, will give the highest value for head loss and the lowest modular limit. Maintaining a high velocity in the exit channel will minimize the required head loss. The expansion coefficient, ξ, is a function of the downstream expansion ratio, m (horizontal/vertical).

The values of modular limit for various expansion ratios are given in Table 4. It is interesting that these values are significantly different than 0.67 (or 2/3), which corresponds to critical depth in the flume throat [$y_c = (2/3)\, H_1$], Equation (6). For a rapid expansion into a pool, there is considerable loss that requires a drop in water level across the expansion ($y_c > h_2$).

Also shown in Table 4 are the recommended modular limit values for channel flow from Bos, Replogle, and Clemmens (1984). For long expansions, i.e., $m = 10$, frictional losses over the expansion are very high, thus defeating the purpose of the expansion. Very short expansions have a minor effect on the modular limit and produce undesirable flow patterns. Therefore, either a rapid expansion ($m = 0$) or a gradual expansion of 6:1, is recommended. If short expansions are desired, the 6:1 expansion should be truncated abruptly and not curved downward toward the channel bottom.

The value of H_1/L appears to have a significant influence on the required head loss for maintaining modular flow. Canal shape also has an influence. Procedures developed for these calculations are generally conservative for $H_1/L < 0.5$, and under most conditions for $H_1/L < 0.75$. The values of required head loss given for the flumes and weirs presented here should be reasonably conservative. Generally, estimations of downstream tailwater levels are less precise. Once constructed, visual observations are an easy way to detect flume conditions. As long as the water surface continues to drop from the approach channel to the end of the flume throat (start of diverging transition), flow is modular. Once the water surface profile begins to turn up-

TABLE 4. Approximate Modular Limits for Rectangular Flumes.

Expansion ratio (horizontal/vertical) m	Expansion coefficient ξ[1]	Modular limit for discharge into standing water[2]	Recommended values of modular limit for broad-crested weirs	
			Rectangular control	Trapezoidal control
0:1	1.20	0.60	0.70[3]	0.75
1:1	1.06	0.65	0.72	0.77
2:1	0.94	0.69	0.74	0.80
4:1	0.77	0.74	0.77	0.83
6:1	0.66	0.78	0.79	0.85
10:1	0.51	0.83	0.80	0.87

[1]From Bos, Replogle and Clemmens 1984, adapted from Bos and Reinink 1981.
[2]Assuming no frictional energy losses.
[3]Minimum values based on experience from using estimation procedures.

ward at the end of the throat, or a standing wave (roller) exists over the flume crest, the calibration is no longer reliable. Deviations from predicted discharges are minor at first, then change rather rapidly. This along with problems with accurate head detection make calibrations for nonmodular flow impractical.

DESIGN

There are three major steps in the process of designing a flume or weir: 1) site selection, 2) selection of head measurement technique and 3) selecting an appropriate structure. Design is an iterative process between these steps. The order and importance of these depend on the specific conditions encountered. If a structure to measure or regulate the flow rate is to function well, it should be selected properly. All demands that will be made on the structure should be listed and matched with the properties of the known structures.

Broadly speaking, these demands or operational requirements originate from four sources:

1. The hydraulic performance
2. The construction and/or installation cost
3. The ease with which the structure can be operated
4. The cost of maintenance

To aid in the selection of a structure, the imposed demands will be discussed in more detail.

Operational Requirements

BASIC FUNCTIONS

While in most cases the purpose of flumes and weirs is to measure flow rates, there are also situations where the structure is used to regulate flow or water levels. This may require a weir with a vertically movable crest as shown in Figure 6 (see Bos, Replogle, Clemmens 1984). For flow measurement, the frequency and duration of measurements may determine whether to use a portable/reusable structure (Figure 7), a temporary structure (Figure 8) or a permanent structure (Figure 9). This chapter will primarily deal with the latter two. A variety of portable structures are discussed in Bos, Replogle, and Clemmens (1984).

RANGE OF DISCHARGES

The flow rate in an open channel tends to vary with time. The range between Q_{min} and Q_{max} through which the flow should be measured strongly depends on the nature of the channel in which the structure is placed. Irrigation canals, for example, require a considerably narrower range of discharges than do natural streams. The anticipated range of discharges to be measured may be classified by the ratio

$$\gamma = \frac{Q_{max}}{Q_{min}} \tag{17}$$

Table 5 shows approximate γ values for a variety of control shapes (see Bos, Replogle, and Clemmens 1984).

The listed γ values illustrate that whenever the anticipated ratio, $\gamma = Q_{max}/Q_{min}$, will exceed about 35, the rectangular control cannot be used. If γ exceeds about 55, we may use a (wide) semicircular, parabolic, a (wide) truncated triangular, a narrow-bottom trapezoidal, or a triangular shaped control. In irrigation canals, the ratio $\gamma = Q_{max}/Q_{min}$ rarely exceeds 35, so that all shapes of the control can be used. In natural drains, however, the range of flows to be measured usually will determine the shape of the control.

SENSITIVITY OF THE METERING STRUCTURE

The accuracy to which a flow rate can be measured depends on:

1. The accuracy with which a rating table for a structure can be made. The percentage error in the flow rating tables in this chapter is less than 2%.
2. The accuracy with which the upstream sill-referenced head, h_1, can be determined. This will be discussed in detail later.

The error in the flow rate which originates from an incorrect determination of h_1 upstream of a given structure can be evaluated by the sensitivity, S, of the structure. For modular flow it reads

$$S = \frac{100\Delta Q}{Q} = 100u\frac{\Delta h_1}{h_1}\% \tag{18}$$

where Δh_1 = the difference between the determined and true value of h_1, and ΔQ = the error in discharge measurement caused by Δh_1. The value of Δh_1 can refer to an unnoticed change in water level, head reading error on scale or paper, mislocation of gauging station, error in zero-setting of the scale or recorder, internal resistance in recorder, and so on. The total value of Δh_1 rapidly increases if insufficient care is given to the determination of h_1. The sensitivity is an important consideration when weirs are used for regulation at canal bifurcations.

SEDIMENTATION AND DEBRIS

Open channels, especially those that pass through forested or populated areas, transport all kinds of floating or suspended debris. If this debris is trapped by the gauge or the structure, the approach channel and control section become clogged, impairing the ability of the structure to measure discharges, and causing overtopping of the upstream channel banks. To avoid trapping of debris, the staff-gauge or recorder housing should not interfere with the flow pattern. All weirs and flumes described in this chapter are

FIGURE 6. Movable weirs combine the functions of flow regulation and measurement into one structure. Indonesia (Courtesy DHV, Amersfoort, The Netherlands).

FIGURE 7. Small portable flume in a natural stream (The Netherlands).

FIGURE 8. Temporary wooden weir in an irrigation canal (Arizona).

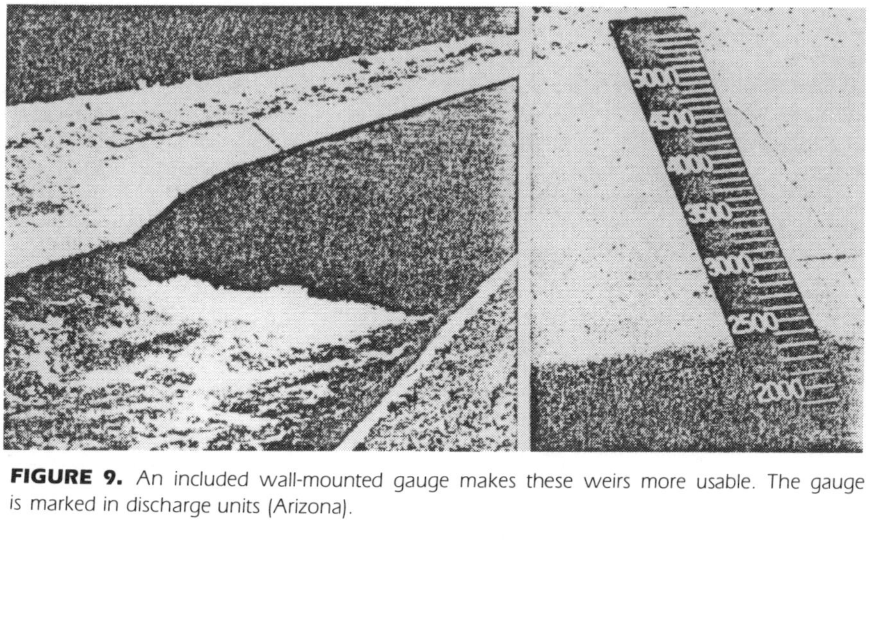

FIGURE 9. An included wall-mounted gauge makes these weirs more usable. The gauge is marked in discharge units (Arizona).

TABLE 5. Values of u and γ as a Function of the Control Shape.

Shape of the control			
basic form	width with respect to h_1	u-value dimensionless	γ-value dimensionless
rectangular	all	1.5	35
triangular	all	2.5	335
trapezoidal	large	1.7	55
	small	2.3	210
truncated V	large	2.4	265
	small	1.7	55
parabolic	all	2.0	105
semi-circle	large	2.0	105
	small	1.6	40

sufficiently streamlined to avoid debris trapping provided that the debris size does not exceed the size of the control. If two or more weirs are installed side-by-side, the intermediate piers should be at least 0.30 m wide and have rounded noses. Sharp-nosed and narrow piers tend to trap debris.

Most open channels carry some amount of sediment, either as washload or bed material load. If the placement of a weir or flume causes a decrease in channel velocity upstream from the structure, some sediment deposition may take place. Flumes and weirs with gradual approach ramps will pass reasonably high sediment loads. Where sedimentation is a problem, flumes can be designed to maintain high relative flow velocities. To avoid changes in the upstream water level after flume or weir installation, the structure can be placed at a drop in the channel. Then the stage-discharge relationship of the flume can be closely matched to that of the channel. Moving sediment may require relatively high Froude numbers which may not be desirable for accurate flow measurement.

REQUIRED HEAD LOSS

The water level in the downstream channel is important to the design of a structure. Sometimes this level is directly predictable by channel discharge equations, but frequently this level is influenced by downstream confluences with other channels, operation of gates and reservoirs, etc. The approximate channel water levels are necessary to determine the required vertical location of the long-throated flume or the broad-crested weir to achieve modular flow. When the level is influenced by downstream backwater effects, modular flow for the flume must be checked at both the high and low flow rate for which a measurement is desired. When no downstream backwater effects are present, modular flow need only be checked at the high flow. An exception to this are flat bottomed flumes where the exponent u of the control section is less than that of the channel.

REQUIRED ACCURACY

The accuracy with which a discharge can be measured with a particular structure is limited by the accuracy with which a measurement can be reproduced. If, independently from each other, two identical structures are constructed and operate with exactly the same upstream sill-referenced head, both flow rates usually are not the same. For the flumes and weirs in this chapter, the difference between these flow rates and the true flow rate will be less than 2% if the flow rate is calculated with the computer program of Replogle (Bos, Replogle, and Clemmens, 1984).

If the Q versus h_1 rating of a custom-built structure is calculated by Equation (9) this error, X_c, will be within $\pm 5\%$ (95% confidence level). Besides the error X_c, the most important error in the measured flow rate is the error inherent to the determination of h_1. Three types of errors can be distinguished: 1) systematic errors which occur at each head reading and can be corrected if discovered, 2) random errors which are normal reading errors, and 3) spurious errors which negate the entire measurement (e.g., writing down wrong number, recorder failure, restrictions to normal flow). Sources of these errors include: 1) zero-setting errors, 2) algal growth which changes the effective crest level, 3) stilling-well lag errors, 4) construction related errors, 5) head-reading errors, and 6) zero shift (e.g., float density changes, nonuniform settling of structures).

As discussed earlier, the measurement of flow rate is subject to two errors:

X_c = error in the used rating table
X_{h1} = error in the upstream sill-referenced head

The value of X_{h_1} is a combination of all known random errors of h_1 evaluated by the equation

$$X_{h_1} = \frac{100}{h_1}\sqrt{\delta_{h_i}^2 + \delta_{h_{ii}}^2 + \text{------} + \delta_{h\infty}^2} \qquad (19)$$

where δ_{hi}, δ_{hii}, and so on are the various random errors in the head measurement. Note that the systematic errors in h_1 are then added algebraically to the measured h_1 value. The total error X_Q in the measured flow rate can then be calculated by the equation

$$X_Q = \sqrt{X_c^2 + (u\, X_{h1})^2} \tag{20}$$

Again, this is for 95% level of confidence, if X_c and Xh_i represent the 95% confidence level.

Site Conditions

The selection of the measuring site and the selection of the structure to be used are usually closely related. Some structures are more appropriate for certain sites, and some sites require specific structures. Even when the general site and structure have been chosen, some consideration must be given to the exact location of the structure and the flow conditions both upstream and downstream.

All structures for measuring or regulating the rate of flow should be located in a channel reach where an accurate value of h_1 can be measured and where sufficient head loss can be created to obtain a unique Q versus h_1 relation (modular flow).

The survey of a channel to find a suitable location for a structure should also provide information on the following relevant factors that influence the performance of a future structure.

1. Upstream of the potential site, the channel should be straight and have a reasonably uniform cross section for a length equal to approximately ten times its average width. If there is a bend closer to the structure, the water elevations at the two sides of the channel become different. Reasonably accurate measurements can be made (added error about 3%) if the upstream straight channel has a length equal to about 2 times its width. In this case, the water level should be measured at the inner bend of the channel.
2. The channel reach should have a stable bottom elevation. In some channel reaches, sedimentation occurs in dry seasons or periods. These sediments may be eroded again during the wet season. Such sedimentation changes the approach velocity towards the structure or may even bury the structure, while the erosion may undercut the foundation of the structure.
3. Whether the water level in the channel is directly predictable by the channel discharge or whether it is influenced by downstream confluences with other channels, operation of gates, reservoir operation, and so on must be determined. The channel water levels greatly influence the sill height necessary to obtain modular flow.
4. Based on the channel water levels and the required sill height in combination with the Q versus h_1 relation of the structure, the possible inundation of upstream surroundings or overtopping of canals should be studied. These inundations usually cause sedimentation because of the subsequent change in the approach flow conditions.
5. For the full range of anticipated discharges the Froude number, Fr_1, at the gauging station should be calculated by use of the equation

$$Fr_1 = \frac{v_1}{\sqrt{g\,\dfrac{A_1}{B_1}}} \tag{21}$$

   where

   A_1 = cross-sectional area perpendicular to the flow
   B_1 = water surface width at the gauging station

   To obtain a reasonably smooth water surface for which the elevation can be determined accurately, the Froude number Fr_1 should not exceed 0.45 over a distance of at least 30 times h_1 upstream from the structure. If feasible, the Froude number should be reduced to 0.2. Some care should be taken when locating a flume immediately downstream from a structure with a large head drop.
6. Subsoil conditions: At the site of the structure, leakage around and beneath it due to the head loss over the structure must be cut off at reasonable costs. Also, a stable foundation, without significant settling, must be secured.
7. To avoid sedimentation upstream of the structure, sufficient head must be available in the selected channel reach.

An example form for data collection for site selection is given in Table 6.

Measurement of Head

GENERAL CONSIDERATIONS

As discussed above, the accuracy of a flow measurement depends strongly on the true determination of the upstream, sill-referenced head. The success of a measuring structure often depends entirely on the effectiveness of the gauge or recorder used.

The sill-reference head refers to the effective control section, which is located on the weir crest, or in the flume throat, at a distance of about $L/3$ from the downstream edge of the sill. In the direction of the flow, the top of the sill (weir crest, or invert of flume throat) must be truly level. If minor undulations in level occur, it is recommended that the average level at the effective control section be used rather than the average level of the entire sill.

LOCATION FOR HEAD MEASUREMENT

The gauging or head-measuring station should be located sufficiently far upstream to avoid detectable water surface

TABLE 6. Form for Recording Information About a Potential Measuring Site.

NAME OF SITE: ______________________________ DATE: __________

HYDRAULIC DEMANDS

Range of flow to be measured, Q	Present water depth in channel, y_2	Maximum permitted error in measurement, X_q
Q_{min} = ________ m^3/s	y_{2min} = ________ m	X_{qmin} = ________ ± %
Q_{max} = ________ m^3/s	y_{2max} = ________ m	X_{qmax} = ________ %

HYDRAULIC DESCRIPTION

Channel bottom width b_1 = ________ m
Channel side slope z = ________
Channel depth d = ________ m
Maximum allowable water depth, y_{1max} = ________ m
Manning's n n = ________
Hydraulic gradient s = ________
Available drop in water surface at site Δh = ________ m
Drop in channel bottom at site Δp = ________ m

Sketch of channel cross-section.

Concrete lined ☐
Earthen channel ☐

Profile along bottom of channel over length of 100 b_1

FUNCTION OF STRUCTURE

Measurement only ☐
Regulation and measurement of flow rate ☐

SERVICE PERIOD OF STRUCTURE

daily ☐ seasonally ☐
monthly ☐ permanently ☐

DESCRIPTION OF ENVIRONMENT

Irrigation system		Drainage System	
Main	☐	From irrigated area	☐
Lateral	☐	Artificial drain	☐
Farm ditch	☐	Natural stream	☐
In field	☐		

Plan of site:

FURTHER DESCRIPTION: (attach photo)

drawdown, yet close enough for the energy losses between the gauging station and approach section to be negligible. This is particularly true if the ratings are derived using the C_d and C_v curves. In the computer derived ratings, the gauge location drawdown and friction losses are an integral part of the calculation, and therefore the gauge should be located as indicated in the appropriate tables.

For the ratings derived using C_d and C_v, the gauging station should be located at a distance between two and three times the maximum h_1 reading from the beginning of the throat section (sill) or at a distance equal to h_1 from the beginning of the converging transition, whichever is greater.

HEAD MEASUREMENT METHODS

Basically, the head is sensed either in the channel itself or in a stilling well located to one side of the channel and connected to the channel by means of a small pipe. Many methods can be used to detect the water surface. Some methods exploit the electromagnetic properties of the water and of the water-air interface. Other methods depend on reflecting a sonic wave from the air–water interface. Still other methods detect water depth with a variety of pressure sensing devices and infer the head from that information.

The most frequently encountered methods, however, are the vertical and side-wall-mounted staff gauges in the canal or in a stilling well, and the float-operated recorder placed in a stilling well.

Stilling Wells

Stilling wells serve two purposes, (1) to facilitate the accurate reading of a water level at a gauging station where the water surface is disturbed by wave action, and (2) to house the float of a recorder system or other surface detecting equipment.

The cross-sectional dimensions of the well depend on the method by which the head is to be measured, ranging from a recommended minimum size of 0.1 m for hand-inserted dipsticks to over 0.5 m to accommodate large-diameter floats. The pipe connecting the stilling well to the canal should be large enough to allow the stilling well to respond quickly to water level changes. Usually this pipe diameter is about one-tenth the diameter of the stilling well.

Further details on stilling wells can be found in Bos (1978), Brakensick, Osborn and Rawls (1979) and Bos, Replogle and Clemmens (1984).

Staff Gauges

Where no continuous information on the flow rate is needed, as in canals where the fluctuation of flow is gradual, periodic readings on a calibrated staff gauge may provide adequate information. The gauge should be placed in such a manner that the water level can be read from the canal bank and so that its surface can be cleaned by the observer (see Figure 9).

For concrete-lined canals, the gauge can be mounted directly on the canal wall. In the trapezoidal-shaped canals, appropriate adjustments to the value for h_1 as read on the canal wall must be made to convert the reading to the vertical h_1 value before entering the discharge tables. For unlined canals, a vertical support is required.

Most permanent gauges are enameled steel, cast aluminum, or polyester. Enameled linear scales marked in metric or English units are available from commercial sources. Important flow rates can be noted on these scales by separate markings to avoid the need for tables to be always at hand. Within an irrigation district it is frequently desirable to use a limited number of standardized structures. It is convenient to mark the gauges of these structures directly in discharge units rather than in head units.

SELECTION OF HEAD-MEASUREMENT DEVICE

The selection of a suitable head-measurement device contributes greatly to the success or failure of the structure and to the value of the collected data. The three most important factors that influence the choice of a device are:

1. Frequency of discharge measurement
2. Allowable error in the head detection
3. Type of structure over which the head must be measured

In Table 7, common reading errors Δh_1 in the sill-referenced head are listed for the most common head-measurement devices. The listed errors are somewhat higher than the random errors to accommodate the effect of several sources of systematic error such as zero-setting, instrument lag, reading error, temperature, and stilling well leakage.

If no device with sufficient accuracy is found from this procedure, there are two choices:

(a) Allow a greater error in the measured discharge for the minimum upstream head, h_{1min}.
(b) Redesign the structure with a narrower bottom width resulting in a higher value of h_{1min}.

GAUGE PLACEMENT AND ZERO-SETTING

The accurate determination of the sill-referenced head h_1 is the most important factor in obtaining accurate discharge measurements. The upstream sill-referenced head h_1 can be measured by a gauge or recorder only if the observed water level is known with respect to the weir sill (or flume crest) level at the control section. The method by which the relative setting of gauge, recorder, and so on is determined depends on factors such as the canal size in which the structure is located, the flow rate in the channel during the setting procedure, and available equipment. Surveying is the most practical method for setting a wall or staff gauge (see Figure 10).

Often the canal side slopes are not constructed to the intended slope. If this occurs, it is recommended that the

TABLE 7. Common Reading Errors in Sill-Referenced Head.

Device	Reading error Δh_1 in h_1 if head detection is in: open channel	stilling well	Remarks
Point gauge	n.a.	0.1 mm	Commonly used for research
Dipstick	n.a.	1.0 mm	Good for research and field use
Staff gauge	4.0 mm	4.0 mm	$Fr_1 \le 0.1$
	7.0 mm	5.0 mm	$Fr_1 = 0.2$
	>15.0 mm	7.0 mm	$Fr_1 = 0.5$
Pressure bulb + recorder	up to 20 mm	n.r.	Very suitable for temporary installations. (Error is 2% of h_{1max})
Bubble gauge + recorder	10.0 m	n.r.	Stilling well is not required but can be used
Float-operated recorder	n.a.	5.0 mm	Stilling well is required
Flow totalizer tied to recorder	—	—	Some additional random and systematic error is possible.

n.a. = not applicable; n.r. = not required

gauge be mounted so that the correct reading occurs at the correct elevation for a point in the most frequently used flow range (Figure 11). Thus, the greatest reading errors will occur in the flow ranges that are seldom used. If the zero setting is displaced more than about 5 mm, or if accuracy for the full flow range is required, then the actual side slope should be separately determined for adjustments to the gauge.

There are several methods for zeroing a water level recorder, three of which are particularly suitable. The recorder can be set when the canal is dry, when water is ponded over the flume or when water is flowing through the canal (Figure 12). These zero-setting methods assume that the sill-reference elevation can be measured during the procedure. This is not always practical, especially on wide structures. A stable bench mark (bronze cap poured in concrete) should be added to such structures, the elevation of which is known with respect to the sill-reference point. For detailed information on zero-setting procedures see Bos, Replogle, and Clemmens (1984).

Determining Structure Dimensions

Long-throated flumes and broad-crested weirs operate by using a channel contraction to cause critical flow. If not enough contraction exists, critical flow will not occur, flow is nonmodular and an accurate measurement is not feasible. If too much contraction exists, the water surface upstream is raised too much causing canal overtopping or other problems. The problem facing the designer is to select the shape of the control section or throat so as to produce critical flow throughout the full range of discharges to be measured with a sensitivity and accuracy that are acceptable while not causing too much disruption in upstream flow conditions (e.g., sediment deposition, canal overtopping). While this appears to be a difficult task, design aids and rating tables have been produced and are presented here to make this task more manageable. Based on practical experience and theory an infinite number of possible designs have been sorted to a few structures from which the designer may choose.

Direct stage-discharge or rating tables for four types of channel conditions are presented in this chapter: 1) lined trapezoidal channels, 2) earthen canal or lined rectangular channels, 3) circular pipes and conduits, and 4) natural streams. For the lined trapezoidal channels, the broad-crested weir (i.e., a bottom contraction only) was selected for the development of rating tables. A design aid was developed to provide sufficient sensitivity, limit the Froude number and aid in selection. The designer need only select a weir width with its corresponding sill height. Trapezoidal flumes with side contractions are more difficult to construct and require more head loss. For unlined canals, broad-crested weirs were again chosen. However, this time the designer must select both a channel (and throat) width and a sill height. In this case, the designer must be more aware of the other design considerations. For lined rectangular channels, only the sill height must be determined. These tables can also be used to determine rating for side contracted flumes with rectangular controls by appropriate adjustments for C_v. For circular pipes, again the broad-crested weirs were chosen. These resemble several earlier versions of Palmer-Bowlus flumes. However, these structures with a single horizontal crest are far easier to construct than the commercially available Palmer-Bowlus flumes which have a trapezoidal throat. Here the designer simply selects a sill height. For natural streams, V-shaped flumes were chosen because of the wide range of discharges which must be measured. The only design choice here is the throat sideslopes. Of course, the designer always has the option of designing a flume shape or size not present here by using theory or the computer program.

Sediment carrying capabilities are sometimes an important design characteristic. V-shaped flumes do not pass

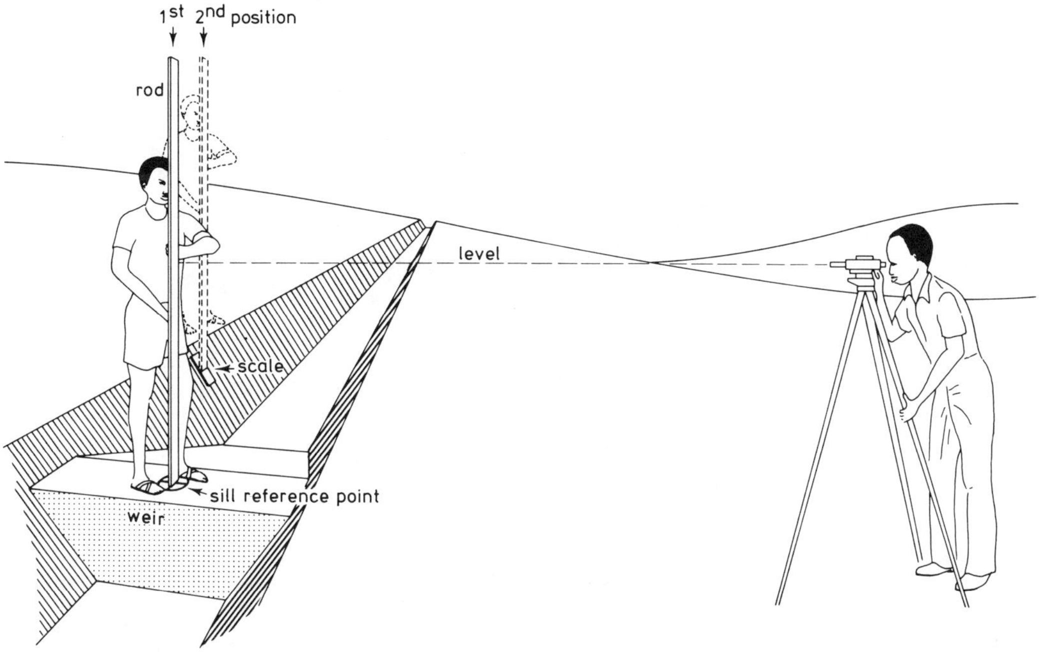

FIGURE 10. Stages of setting of a gauge on a lined canal slope.

FIGURE 11. Setting the rod on the inclined gauge (Arizona). Note that gauge is registered to most common discharge rather than gauge zero.

FIGURE 12. Equipment for setting recorder in a canal with flowing water. This equipment is also used with temporary weirs (California).

bedload sediment well. The wide discharge range of V-shaped flumes often necessary in watershed measurements is in opposition to the ability to pass sediment. Sediment movement is also a consideration in sewer flow. The weirs for pipes can be designed for reasonable good sediment passage (Clemmens, Bos, Replogle 1984). These weirs have the advantage of a wide crest, which appears to be better for sediment transport than a narrow bottom with a lower crest height.

The rating tables provided here each assume a particular known approach channel cross section (or flow area). However, any particular control section size and shape can be used with any approach section size and shape. The discharge can be adjusted by the use of the approach velocity coefficient C_v. This procedure is described in detail in the section on as-built calibrations. The rating tables given here automatically limit the Froude number. If smaller approach areas are used, the designer is responsible for assuring that the Froude number is below about 0.45.

In many cases, the site conditions will call for flumes the dimensions of which fall outside the ranges provided by the ratings given in this chapter. Hopefully, we have provided sufficient information to allow users to extend beyond these limits. Further information can be found in Bos, Replogle, Clemmens (1984), Clemmens, Replogle, Bos (1984), Bos (1978), Bos (1984) and Ackers, White, Perkins, Harrison (1978).

Scaling Flume Ratings by Froude Modeling

It is a common practice in the study of fluid flow to test a scale model of a large structure in a hydraulics laboratory. The flow properties of the large structure are inferred from the model. This is referred to as hydraulic similitude. For the study of open channel flow, the ratio between the forces of inertia and gravity must be the same for both the model and the prototype. The square root of this ratio is known as the Froude number; thus, this type of hydraulic similitude is often referred to as Froude modeling.

By this hydraulic similtude concept, if the properties of one structure are known, we can determine the properties of a "similar" structure. That is, a weir that is 1 m high, 1 m wide, with a head of 1 m is similar to a weir 2 m high, 2 m wide, with a head of 2 m. The discharge, however, is not doubled, but is found from the following equation:

$$\frac{Q_{prototype}}{Q_{model}} = \left\{\frac{L_{prototype}}{L_{model}}\right\}^{(2.5)} \tag{22}$$

This equation is valid when all the physical structure dimensions and the heads are of the same ratio. With this equation, the ratings for a number of flume sizes can be determined from a given rating.

The procedures in this chapter are consistent with the

concept of Froude modeling; that is, all of the head-discharge equations have Q on one side and dimensions of length to the exponent 2.5 on the other. (The units are balanced by $g^{0.5}$) The effects of frictional and turbulent energy loss resulting from fluid viscosity are accounted for by the Reynolds number which is the ratio of the inertial and viscous forces. The coefficient, C_d, is used to account for these and other effects. However, in Equation (11), C_d is strictly a function of a length ratio. This is an average curve for a wide range of flume shapes and sizes. Thus, the use of this curve is an approximate adjustment for friction within the Froude modeling concept. The mathematical model of Replogle accounts for the effects of friction through application of Reynolds number. The effects of scale from this model due to friction are minor until the relative roughness gets large. Thus, reasonably accurate ratings usually can be made with the Froude model extensions to other sizes, particularly if the scale ratio is 4 or less. Thus, if ratings are not available for a particular flume size, it may be possible to determine them by use of Equation (22).

RATINGS AND DESIGN AIDS

Structures for Lined Trapezoidal Canals

Standard weir sizes were selected and precomputed for use in selected slipformed canals of convenient metric dimensions. The weir is simply placed in an existing channel as shown in Figure 13 (see also Figures 2, 6, and 9). In selecting standard-sized canals and the related flow rates, consideration was given to proposals by the International Commission on Irrigation and Drainage (ICID, 1979), to the construction practices of the United States Bureau of Reclamation, and to design criteria for small canals used by the United States Soil Conservation Service.

Present practice is toward sideslopes of 1:1 for small, monolithic, concrete-lined canals with bottom widths less than about 0.8 m, and depths less than about 1 m. Deeper and wider canals tend toward side slopes of 1.5 horizontal to 1 vertical. When the widths and depths are greater than about 3 m, the trend is more toward 2:1 side slopes, particularly if canal operating procedures may allow rapid dewatering of the canal, which, in some soil conditions, can cause hydrostatic pressures on the underside of the canal walls that lead to wall failure. Most of the lined canals used in a tertiary irrigation unit or on large farms are of the smaller size: they have 0.3- to 0.6-m bottom widths, 1:1 side slopes, and capacities below 1 m^3/s (35 ft^3/s).

We attempt to accommodate subdivisions in metric units that can be anticipated for usual selections in metric dimensions and simultaneously cover tendencies to match equipment designed to English unit dimensions. In Table 8, precomputed broad-crested weir selections are given for canals with bottom widths at quarter-meter increments, with special insertions for 0.3 m (approximately 1 ft) and 0.6 m (ap-

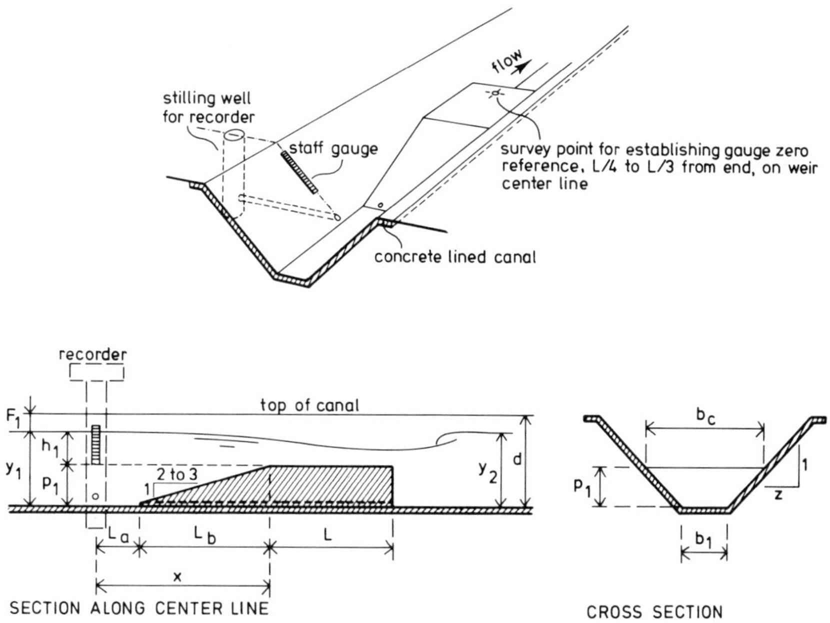

FIGURE 13. Layout of broad-crested weir in trapezoidal concrete-lined canal.

TABLE 8. Choices of Weir Sizes and Rating Tables for Trapezoidal Lined Canals.

Canal Shape		Maximum canal depth 1/	Range of canal capacities		Weir selections	Weir Shape		Minimum head loss
side-slopes	bottom width					Crest width	Sill height	
z_1 -	b_1 m	d m	lower 2/ m^3/s	upper m^3/s	see rating Table 3.3	b_c m	p_1 m	ΔH m
(1)	(2)	(3)	(4)	(5)	(6)	(7)	(8)	(9)
1.0	0.25	0.70	0.08	0.14+	A_m	0.5	0.125	0.015
			0.09	0.24+	B_m	0.6	0.175	0.018
			0.10	0.38+	C_m	0.7	0.225	0.022
			0.11	0.43+	D_{m1}	0.8	0.275	0.026
			0.12	0.37	E_{m1}	0.9	0.325	0.030
			0.13	0.32	F_{m1}	1.0	0.375	0.033
1.0	0.30	0.75	0.09	0.21+	B_m	0.6	0.15	0.017
			0.10	0.34+	C_m	0.7	0.20	0.021
			0.11	0.52+	D_{m1}	0.8	0.25	0.025
			0.12	0.52	E_{m1}	0.9	0.30	0.029
			0.13	0.44	F_{m1}	1.0	0.35	0.033
			0.16	0.31	G_{m1}	1.2	0.45	0.039
1.0	0.50	0.80	0.11	0.33+	D_{m2}	0.8	0.15	0.019
			0.12	0.52+	E_{m2} or E_{m1}	0.9	0.20	0.024
			0.12	0.68+	F_{m1} or F_{m2}	1.0	0.25	0.029
			0.16	0.64	G_{m1}	1.2	0.35	0.037
			0.18	0.46	H_m	1.4	0.45	0.043
			0.20	0.29	I_m	1.6	0.55	0.048
1.0	0.60	0.90	0.12	0.39+	E_{m2}	0.9	0.15	0.021
			0.13	0.62+	F_{m2}	1.0	0.20	0.025
			0.16	1.09	G_{m1}	1.2	0.30	0.035
			0.18	0.86	H_m	1.4	0.40	0.043
			0.20	0.64	I_m	1.6	0.50	0.050
			0.22	0.43	J_m	1.8	0.60	0.049
1.0	0.75	1.0	0.16	0.91+	G_{m2}	1.2	0.225	0.030
			0.18	1.51	H_m	1.4	0.325	0.038
			0.20	1.22	I_m	1.6	0.425	0.047
			0.22	0.94	J_m	1.8	0.525	0.053
1.5	0.60	1.2	0.20	1.3 +	K_m	1.50	0.300	0.031
			0.24	2.1 +	L_m	1.75	0.383	0.038
			0.27	2.5	M_m	2.00	0.467	0.044
			0.29	2.2	N_m	2.25	0.550	0.050

(continued)

TABLE 8 (continued).

Canal Shape: side-slopes z_1 -	Canal Shape: bottom width b_1 m	Maximum canal depth [1/] d m	Range of canal capacities: lower [2/] m^3/s	Range of canal capacities: upper m^3/s	Weir selections: see rating Table 3.3	Weir Shape: Crest width b_c m	Weir Shape: Sill height p_1 m	Minimum head loss ΔH m
(1)	(2)	(3)	(4)	(5)	(6)	(7)	(8)	(9)
			0.32	1.8	P_m	2.50	0.633	0.056
			0.35	1.4	Q_m	2.75	0.717	0.059
1.5	0.75	1.4	0.24	1.8 +	L_m	1.75	0.333	0.036
			0.27	2.8 +	M_m	2.00	0.417	0.042
			0.29	3.9 +	N_m	2.25	0.500	0.049
			0.32	3.5	P_m	2.50	0.583	0.055
			0.35	3.1	Q_m	2.75	0.667	0.062
			0.38	2.6	R_m	3.00	0.750	0.066
1.5	1.00	1.6	0.29	3.4 +	N_m	2.25	0.417	0.046
			0.32	4.7	P_m	2.50	0.500	0.052
			0.35	5.7	Q_m	2.75	0.583	0.059
			0.38	5.1	R_m	3.00	0.667	0.065
			0.43	3.9	S_m	3.50	0.833	0.081
1.5	1.25	1.7	0.32	4.1 +	P_m	2.50	0.417	0.048
			0.35	5.6 +	Q_m	2.75	0.500	0.055
			0.38	7.2	R_m	3.00	0.583	0.061
			0.43	5.9	S_m	3.50	0.750	0.074
			0.49	4.5	T_m	4.00	0.917	0.084
			0.55	3.3	U_m	4.50	1.083	0.089
1.5	1.50	1.8	0.35	4.8 +	Q_m	2.75	0.417	0.051
			0.38	6.5 +	R_m	3.00	0.500	0.058
			0.43	8.1	S_m	3.50	0.667	0.071
			0.49	6.6	T_m	4.00	0.833	0.083
			0.55	5.1	U_m	4.50	1.000	0.092

1/ Maximum recommended canal depth
2/ Limited by sensitivity
\+ Limited by Froude number; otherwise limited by canal depth

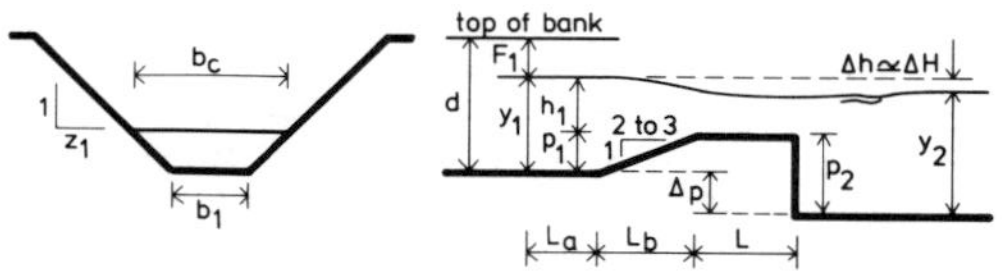

$L_a > H_{1max}$
$L_b = 2$ to $3\ p_1$
$x = L_a + l_b > 2$ to $3\ H_{1max}$
$L > 1.5\ H_{1max}$, but within range given in Rating Table
$d > 1.2\ h_{1max} + p_1$
$\Delta H > 0.1\ H_1$

proximately 2 feet). It is hoped that the offering of so many precomputed sizes will aid in retrofitting older canal systems and yet not prevent the adoption of standard sized canals as proposed by ICID (International Commission on Irrigation and Drainage). Canal sizes with bottom widths in excess of 1.5 m or 5 ft, respectively, are avoided in the precomputed tables on the assumption that these sizes deserve special design consideration.

Table 8 shows a number of precomputed weirs that may be used for the various combinations of bottom widths and sidewall slopes as given in the first two columns. The third column gives recommended values of maximum canal depth, d, for each side-slope, bottom-width combination. For each canal size, a number of standard weirs can be used (Column 6). Columns 4 and 5 give the limits on canal capacity for each canal-weir combination. These limits on canal capacity originate from three sources:

1. The Froude number in the approach channel is limited to 0.45 to assure water surface stability, where

$$Fr_1 = \frac{v_1}{\sqrt{\frac{gA_1}{B_1}}}$$

2. The canal freeboard F_1 upstream from the weir should be greater than 20% of the upstream sill-referenced head, h_1. In terms of canal depth this limit becomes $d \geq 1.2\, h_1 + p_1$.
3. The sensitivity of the weir at maximum flow should be such that a 0.01-m change in the value of the sill-referenced head h_1 causes less than 10% change in discharge.

Although Table 8 is primarily intended for the selection of these standard weirs, they are also useful for the selection of canal sizes. The Froude number in the canal is automatically limited to 0.45, and selecting the smallest canal for a given capacity will give a reasonably efficient section. For instance, if the design capacity of the canal is to be 1.0 m³/s, the smallest canal that can be incorporated with a measuring structure has $b_1 = 0.60$ m, $z_1 = 1.0$, and $d = 0.90$ m. Larger canals can also be used. The hydraulic grade line of the channel should be checked to assure an adequate design.

Each standard weir can be used for several different bottom widths. This is possible because the change in flow area upstream from the weir causes only a small change in velocity of approach and thus energy head. We have limited the error in discharge caused by the change in flow area to about 1%. This is a systematic error for any particular approach area, and the value of this error varies with discharge. If a weir can be used for several bottom-width canals, it can also be used for any intermediate width. For example, in Table 8, weir G_{ml} can be used in canals with bottom widths of 0.30, 0.50, and 0.60 m, or any width in between, say $b_1 = 0.40$ m. The user, however, will need to calculate the sill height, head loss, and upper limit on design discharge. The rating tables for each weir are given in Table 9 and were computed using the following criteria (Figure 13):

1. Each weir has a constant bottom width b_c and a sill height p_1 that varies with the canal dimensions.
2. The ramp length can be chosen such that it is between 2 and 3 times the sill height. The 3:1 ramp slope is preferable.
3. The gauge is located a distance at least H_{1max} upstream from the start of the ramp. In addition, it should be located a distance of roughly two to three times H_{1max} from the entrance to the throat.
4. The throat length should be 1.5 times the maximum expected sill-referenced depth h_{1max}, but should be within the limits indicated in Table 8.
5. The canal depth must be greater than the sum of $p_1 + h_{1max} + F_1$, where F_1 is the freeboard requirement, or roughly 0.2 times h_{1max}.

Table 9 contains the specific calibrations for the canal-weir combination specified in Table 8. The ratings in Table 9 are prepared with discharge rate expressed as the independent variable in the first column and the upstream head h_1 as the dependent variable in the second column. This is the reverse of the usual style for tables for flumes and weirs. This method allows simple calculations of values for marking sidewall gauges in direct discharge units for the canals by multiplying the listed h_1 depth for each selected unit of discharge by a function of the side slope ratio.

Also indicated in the last column of Table 8 is a minimum head loss ΔH that the weir must provide. Excessive downstream water levels may prevent this minimum head loss, which means that the weir exceeds its modular limit and no longer functions as an accurate measuring device. The required head losses for the various broad-crested weirs were evaluated by the method described earlier and, for design purposes, listed for each weir size with the restriction that the computed modular limit shall not exceed 0.90. Thus the design head loss is either 0.1 H_1 or the listed value for ΔH, whichever is greater. For these calculations, it was assumed that the weir was placed in a continuous channel with a constant cross section (e.g., $p_1 = p_2$, $b_1 = b_2$, and $z_1 = z_2$) and that the diverging transition was omitted (rapid expansion, $m = 0$). Technically, the modular limit is based on the drop in total energy head through the weir, (i.e., including velocity head), but in the above continuous channels, the velocity head component is usually of the same order of magnitude upstream and downstream of the structure when $p_1 \cong p_2$, so that Δh may be satisfactorily substituted for ΔH.

TABLE 9. Rating Tables for Weirs in Trapezoidal Lined Canals in Metric Units.
(See Table 8 for Details on Weir Dimensions and Head Loss Values.)

Weir A_m b_c = 0.50 m $0.23<L<0.34$ m		Weir B_m b_c = 0.60 m $0.30<L<0.42$ m		Weir C_m b_c = 0.70 m $0.35<L<0.51$ m		Weir D_{m1} b_c = 0.80 m $0.40<L<0.58$ m		Weir D_{m2} b_c = 0.80 m $0.30<L<0.45$ m		Weir E_{m1} b_c = 0.90 m $0.38<L<0.56$ m	
Q m^3/s	h_1 m	Q m^3/s	h_1 m	Q m^3/s	h_1 m	Q m^3/s	h_1 m	Q m^3/s	h_1 m	Q m^3/s	h_1 m
.005	.032	.005	.029	.010	.041	.010	.038	.010	.038	.010	.036
.010	.049	.010	.045	.020	.063	.020	.059	.020	.058	.020	.055
.015	.063	.015	.057	.030	.081	.030	.076	.030	.074	.030	.071
.020	.075	.020	.069	.040	.096	.040	.090	.040	.089	.040	.085
.025	.085	.025	.078	.050	.110	.050	.103	.050	.102	.050	.097
.030	.095	.030	.088	.060	.123	.060	.116	.060	.113	.060	.109
.035	.104	.035	.096	.070	.135	.070	.127	.070	.124	.070	.119
.040	.112	.040	.104	.080	.146	.080	.137	.080	.134	.080	.129
.045	.120	.045	.111	.090	.156	.090	.147	.090	.144	.090	.139
.050	.127	.050	.118	.100	.165	.100	.156	.100	.153	.100	.148
.055	.134	.055	.125	.110	.175	.110	.165	.110	.162	.110	.156
.060	.141	.060	.131	.120	.183	.120	.174	.120	.170	.120	.165
.065	.148	.065	.138	.130	.192	.130	.182	.130	.178	.130	.172
.070	.154	.070	.144	.140	.200	.140	.190	.140	.186	.140	.180
.075	.160	.075	.149	.150	.208	.150	.197	.150	.193	.150	.187
.080	.165	.080	.155	.160	.216	.160	.205	.160	.200	.160	.195
.085	.171	.085	.160	.170	.223	.170	.212	.170	.207	.170	.202
.090	.176	.090	.165	.180	.230	.180	.219	.180	.214	.180	.208
.095	.182	.095	.170	.190	.237	.190	.226	.190	.220	.190	.215
.100	.187	.100	.175	.200	.243	.200	.233	.200	.227	.200	.221
.105	.192	.105	.181	.210	.250	.210	.239	.210	.233	.210	.227
.110	.197	.110	.185	.220	.256	.220	.245	.220	.239	.220	.233
.115	.201	.115	.190	.230	.262	.230	.251	.230	.245	.230	.239
.120	.206	.120	.194	.240	.268	.240	.257	.240	.251	.240	.245
.125	.211	.125	.199	.250	.274	.250	.263	.250	.256	.250	.251
.130	.215	.130	.203	.260	.280	.260	.269	.260	.262	.260	.257
.135	.219	.135	.207	.270	.286	.270	.275	.270	.267	.270	.262
.140	.224	.140	.211	.280	.292	.280	.280	.280	.273	.280	.267
		.145	.215	.290	.297	.290	.285	.290	.278	.290	.273
		.150	.219	.300	.303	.300	.291	.300	.283	.300	.278
		.155	.223	.310	.308	.310	.296	.310	.288	.310	.283
		.160	.226	.320	.313	.320	.301	.320	.293	.320	.288
		.165	.230	.330	.318	.330	.306	.330	.298	.330	.293
		.170	.234	.340	.323	.340	.311			.340	.298
		.175	.238	.350	.329	.350	.316			.350	.302
		.180	.241	.360	.334	.360	.321			.360	.307
		.185	.245	.370	.339	.370	.326			.370	.312
		.190	.248	.380	.344	.380	.330			.380	.316
		.195	.252			.390	.335			.390	.321
		.200	.255			.400	.339			.400	.325
		.205	.259			.410	.344			.410	.330
		.210	.261			.420	.348			.420	.334
		.215	.266			.430	.353			.430	.338
		.220	.270			.440	.357			.440	.343
		.225	.273			.450	.361			.450	.347
		.230	.276			.460	.365			.460	.351
		.240	.282			.470	.370			.470	.355
						.480	.374			.480	.359
						.490	.378			.490	.363
						.500	.382			.500	.367
						.510	.386			.510	.371
						.520	.390			.520	.375

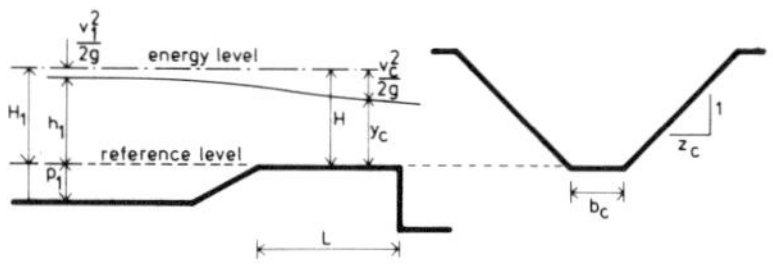

(continued)

TABLE 9 (continued).

Weir E_{m2}, b_c = 0.90 m, $0.38<L<0.56$ m: Q m^3/s	h_1 m	Weir F_{m1}, b_c = 1.0 m, $0.42<L<0.61$ m: Q m^3/s	h_1 m	Weir F_{m2}, b_c = 1.0 m, $0.42<L<0.61$ m: Q m^3/s	h_1 m	Weir G_{m1}, b_c = 1.2 m, $0.50<L<0.75$ m: Q m^3/s	h_1 m	Weir G_{m2}, b_c = 1.2 m, $0.45<L<0.68$ m: Q m^3/s	h_1 m	Weir H_m, b_c = 1.4 m, $0.56<L<0.84$ m: Q m^3/s	h_1 m
.010	.035	.010	.033	.010	.033	.02	.046	.02	.046	.02	.042
.020	.055	.020	.051	.020	.051	.04	.072	.04	.071	.04	.066
.030	.070	.030	.066	.030	.066	.06	.093	.06	.092	.06	.085
.040	.084	.040	.080	.040	.079	.08	.111	.08	.110	.08	.102
.050	.096	.050	.091	.050	.091	.10	.127	.10	.126	.10	.117
.060	.108	.060	.102	.060	.102	.12	.142	.12	.141	.12	.131
.070	.118	.070	.112	.070	.112	.14	.156	.14	.154	.14	.143
.080	.127	.080	.122	.080	.122	.16	.169	.16	.167	.16	.156
.090	.136	.090	.131	.090	.131	.18	.181	.18	.179	.18	.167
.100	.145	.100	.140	.100	.139	.20	.193	.20	.191	.20	.178
.110	.153	.110	.148	.110	.147	.22	.204	.22	.202	.22	.189
.120	.161	.120	.155	.120	.155	.24	.215	.24	.212	.24	.199
.130	.169	.130	.163	.130	.162	.26	.226	.26	.222	.26	.208
.140	.177	.140	.170	.140	.169	.28	.236	.28	.232	.28	.218
.150	.184	.150	.178	.150	.176	.30	.245	.30	.241	.30	.227
.160	.191	.160	.185	.160	.183	.32	.254	.32	.251	.32	.236
.170	.198	.170	.191	.170	.189	.34	.263	.34	.259	.34	.244
.180	.204	.180	.197	.180	.196	.36	.272	.36	.268	.36	.253
.190	.210	.190	.204	.190	.203	.38	.281	.38	.276	.38	.261
.200	.217	.200	.210	.200	.209	.40	.289	.40	.285	.40	.269
.210	.223	.210	.216	.210	.215	.42	.296	.42	.292	.42	.277
.220	.228	.220	.222	.220	.220	.44	.304	.44	.300	.44	.285
.230	.234	.230	.228	.230	.226	.46	.312	.46	.308	.46	.292
.240	.240	.240	.234	.240	.231	.48	.319	.48	.315	.48	.299
.250	.245	.250	.239	.250	.237	.50	.327	.50	.323	.50	.306
.260	.251	.260	.244	.260	.242	.52	.334	.52	.330	.52	.313
.270	.256	.270	.250	.270	.248	.54	.341	.54	.337	.54	.320
.280	.261	.280	.255	.280	.253	.56	.348	.56	.344	.56	.327
.290	.266	.290	.260	.290	.258	.58	.355	.58	.350	.58	.333
.300	.272	.300	.265	.300	.263	.60	.362	.60	.357	.60	.340
.310	.277	.310	.270	.310	.267	.62	.368	.62	.363	.62	.346
.320	.282	.320	.275	.320	.272	.64	.375	.64	.370	.64	.352
.330	.286	.330	.280	.330	.277	.66	.381	.66	.376	.66	.359
.340	.290	.340	.284	.340	.281	.68	.387	.68	.382	.68	.365
.350	.296	.350	.289	.350	.286	.70	.394	.70	.388	.70	.371
.360	.300	.360	.294	.360	.291	.72	.400	.72	.394	.72	.377
.370	.305	.370	.299	.370	.295	.74	.406	.74	.400	.74	.382
.380	.309	.380	.303	.380	.300	.76	.412	.76	.406	.76	.388
.390	.314	.390	.307	.390	.304	.78	.418	.78	.412	.78	.394
.400	.319	.400	.312	.400	.309	.80	.423	.80	.418	.80	.399
.410	.324	.410	.316	.410	.313	.82	.429	.82	.423	.82	.405
.420	.328	.420	.320	.420	.317	.84	.435	.84	.429	.84	.410
.430	.333	.430	.324	.430	.321	.86	.440	.86	.434	.86	.416
.440	.337	.440	.328	.440	.325	.88	.446	.88	.440	.88	.420
.450	.341	.450	.332	.450	.329	.90	.451	.90	.445	.90	.425
.460	.345	.460	.336	.460	.333	.92	.457	.92	.450	.92	.430
.470	.349	.470	.340	.470	.337	.94	.462			.94	.436
.480	.353	.480	.344	.480	.341	.96	.467			.96	.441
.490	.357	.490	.348	.490	.345	.98	.472			.98	.446
.500	.361	.500	.352	.500	.349	1.00	.477			1.00	.451
.510	.365	.520*	.359	.520*	.356	1.02	.482			1.05*	.463
.520	.369	.540	.367	.540	.363	1.04	.488			1.10	.475
		.560	.374	.560	.370	1.06	.492			1.15	.486
		.580	.381	.580	.377	1.08	.497			1.20	.498
		.600	.388	.600	.384	1.10	.502			1.25	.509
		.620	.395	.620	.391					1.30	.520
		.640	.402	.640	.398					1.35	.530
		.660	.409	.660	.405					1.40	.541
		.680	.416	.680	.412					1.45	.551
										1.50	.561

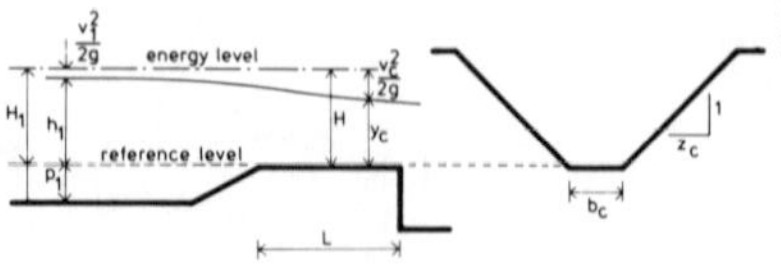

* Change in discharge increment

(continued)

TABLE 9 (continued).

Weir I_m b_c = 1.6 m 0.48<L<0.71 m Q m^3/s	h_1 m	Weir J_m b_c = 1.8 m 0.40<L<0.60 m Q m^3/s	h_1 m	Weir K_m b_c = 1.50 m 0.48<L<0.72 m Q m^3/s	h_1 m	Weir L_m b_c = 1.75 m 0.58<L<0.87 m Q m^3/s	h_1 m	Weir M_m b_c = 2.00 m 0.65<L<0.97 m Q m^3/s	h_1 m	Weir N_m b_c = 2.25 m 0.75<L<1.10 m Q m^3/s	h_1 m
.02	.039	.02	.036			.05	.065	.05	.061	.10	.088
.04	.060	.04	.056	.04	.062	.10	.100	.10	.094	.20	.136
.06	.078	.06	.072	.06	.080	.15	.129	.15	.121	.30	.174
.08	.094	.08	.087	.08	.095	.20	.154	.20	.144	.40	.207
.10	.108	.10	.100	.10	.109	.25	.176	.25	.165	.50	.236
.12	.121	.12	.113	.12	.122	.30	.196	.30	.184	.60	.263
.14	.133	.14	.124	.14	.134	.35	.216	.35	.202	.70	.288
.16	.145	.16	.135	.16	.146	.40	.233	.40	.219	.80	.311
.18	.156	.18	.146	.18	.156	.45	.250	.45	.234	.90	.333
.20	.166	.20	.155	.20	.166	.50	.265	.50	.249	1.00	.354
.22	.176	.22	.165	.22	.176	.55	.280	.55	.264	1.10	.374
.24	.186	.24	.174	.24	.185	.60	.294	.60	.277	1.20	.392
.26	.195	.26	.183	.26	.194	.65	.307	.65	.290	1.30	.411
.28	.204	.28	.192	.28	.202	.70	.320	.70	.303	1.40	.428
.30	.213	.30	.200	.30	.211	.75	.333	.75	.315	1.50	.444
.32	.221	.32	.208	.32	.219	.80	.345	.80	.327	1.60	.460
.34	.229	.34	.216	.34	.226	.85	.357	.85	.339	1.70	.476
.36	.237	.36	.223	.36	.234	.90	.368	.90	.350	1.80	.491
.38	.245	.38	.231	.38	.241	.95	.379	.95	.360	1.90	.506
.40	.253	.40	.238	.40	.248	1.00	.390	1.00	.371	2.00	.520
.42	.260	.42	.245	.42	.255	1.05	.401	1.05	.381	2.10	.534
.44	.267	.44	.252	.44	.262	1.10	.411	1.10	.391	2.20	.548
.46	.274	.46	.259	.46	.269	1.15	.421	1.15	.401	2.30	.561
.48	.281	.48	.266	.48	.275	1.20	.431	1.20	.411	2.40	.573
.50	.288	.50	.272	.50	.282	1.25	.441	1.25	.420	2.50	.586
.52	.295	.52	.279	.52	.288	1.30	.450	1.30	.429	2.60	.599
.54	.301	.54	.285	.54	.294	1.35	.459	1.35	.438	2.70	.611
.56	.308	.56	.291	.56	.300	1.40	.468	1.40	.447	2.80	.623
.58	.314	.58	.298	.58	.306	1.45	.477	1.45	.456	2.90	.635
.60	.320	.60	.304	.60	.312	1.50	.486	1.50	.464	3.00	.646
.62	.327	.62	.310	.62	.318	1.55	.495	1.55	.473	3.10	.658
.64	.333	.64	.315	.64	.323	1.60	.503	1.60	.481	3.20	.669
.66	.339	.66	.321	.66	.329	1.65	.511	1.65	.489	3.30	.680
.68	.345	.68	.327	.68	.334	1.70	.519	1.70	.497	3.40	.691
.70	.350	.70	.333	.70	.339	1.75	.527	1.75	.505	3.50	.702
.72	.356	.72	.338	.72	.345	1.80	.535	1.80	.513	3.60	.713
.74	.362	.74	.344	.74	.350	1.85	.544	1.85	.521	3.70	.724
.76	.367	.76	.349	.76	.355	1.90	.552	1.90	.528	3.80	.734
.78	.373	.78	.354	.78	.360	1.95	.559	1.95	.536	3.90	.744
.80	.378	.80	.360	.80	.365	2.00	.567	2.00	.542		
.82	.384	.82	.365	.82	.370	2.05	.574	2.05	.550		
.84	.389	.84	.370	.84	.375	2.10	.581	2.10	.557		
.86	.394	.86	.375	.86	.380			2.15	.564		
.88	.399	.88	.380	.88	.384			2.20	.571		
.90	.404	.90	.385	.90	.389			2.25	.578		
.92	.409	.92	.390	.92	.394			2.30	.585		
.94	.414	.94	.395	.94	.398			2.35	.592		
.96	.419			.96	.403			2.40	.598		
.98	.424			.98	.407			2.45	.605		
1.00	.429			1.00	.412			2.50	.611		
1.02	.434			1.05*	.422			2.55	.617		
1.04	.439			1.10	.433			2.60	.623		
1.06	.443			1.15	.443			2.65	.629		
1.08	.448			1.20	.453			2.70	.635		
1.10	.453			1.25	.463			2.75	.642		
1.12	.457			1.30	.473			2.80	.648		
1.14	.462										
1.16	.466										
1.18	.471										
1.20	.475										
1.22	.480										

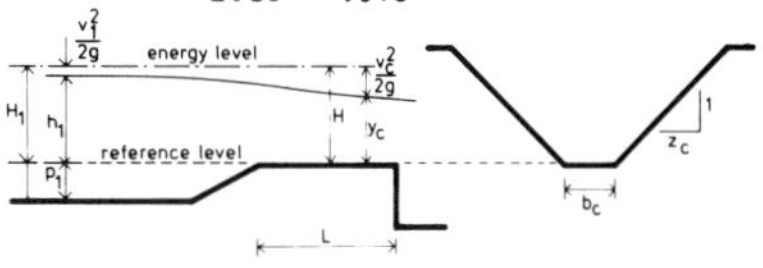

* Change in discharge increment

(continued)

TABLE 9 (continued).

Weir P_m b_c = 2.50 m $0.80<L<1.20$ m Q m^3/s	h_1 m	Weir Q_m b_c = 2.75 m $0.85<L<1.28$ m Q m^3/s	h_1 m	Weir R_m b_c = 3.00 m $0.95<L<1.40$ m Q m^3/s	h_1 m	Weir S_m b_c = 3.50 m $0.95<L<1.40$ m Q m^3/s	h_1 m	Weir T_m b_c = 4.00 m $0.85<L<1.20$ m Q m^3/s	h_1 m	Weir U_m b_c = 4.50 m $0.68<L<1.00$ m Q m^3/s	h_1 m
.10	.082	.10	.077	.10	.075	.10	.067	.10	.062	.10	.058
.20	.127	.20	.120	.20	.115	.20	.104	.20	.097	.20	.090
.30	.163	.30	.155	.30	.149	.30	.135	.30	.125	.30	.116
.40	.195	.40	.185	.40	.177	.40	.162	.40	.150	.40	.139
.50	.224	.50	.213	.50	.203	.50	.186	.50	.173	.50	.160
.60	.249	.60	.237	.60	.227	.60	.209	.60	.193	.60	.180
.70	.273	.70	.260	.70	.250	.70	.230	.70	.213	.70	.198
.80	.295	.80	.282	.80	.270	.80	.249	.80	.231	.80	.216
.90	.316	.90	.301	.90	.290	.90	.268	.90	.249	.90	.232
1.00	.336	1.00	.322	1.00	.309	1.00	.286	1.00	.266	1.00	.248
1.10	.355	1.10	.340	1.10	.327	1.10	.302	1.10	.282	1.10	.263
1.20	.373	1.20	.357	1.20	.344	1.20	.319	1.20	.297	1.20	.278
1.30	.391	1.30	.374	1.30	.361	1.30	.334	1.30	.312	1.30	.292
1.40	.408	1.40	.390	1.40	.377	1.40	.350	1.40	.326	1.40	.306
1.50	.424	1.50	.407	1.50	.392	1.50	.364	1.50	.340	1.50	.319
1.60	.440	1.60	.422	1.60	.407	1.60	.379	1.60	.354	1.60	.332
1.70	.455	1.70	.437	1.70	.422	1.70	.393	1.70	.367	1.70	.345
1.80	.470	1.80	.451	1.80	.436	1.80	.407	1.80	.380	1.80	.357
1.90	.484	1.90	.465	1.90	.450	1.90	.420	1.90	.392	1.90	.369
2.00	.498	2.00	.479	2.00	.463	2.00	.433	2.00	.405	2.00	.381
2.10	.511	2.10	.492	2.10	.476	2.10	.445	2.10	.417	2.10	.392
2.20	.524	2.20	.505	2.20	.489	2.20	.457	2.20	.428	2.20	.403
2.30	.537	2.30	.518	2.30	.501	2.30	.469	2.30	.440	2.30	.414
2.40	.550	2.40	.531	2.40	.514	2.40	.481	2.40	.451	2.40	.425
2.50	.563	2.50	.543	2.50	.525	2.50	.492	2.50	.462	2.50	.436
2.60	.575	2.60	.554	2.60	.537	2.60	.504	2.60	.473	2.60	.446
2.70	.587	2.70	.566	2.70	.549	2.70	.515	2.70	.483	2.70	.456
2.80	.599	2.80	.578	2.80	.560	2.80	.526	2.80	.494	2.80	.466
2.90	.610	2.90	.589	2.90	.571	2.90	.537	2.90	.504	2.90	.476
3.00	.621	3.00	.600	3.00	.582	3.00	.547	3.00	.514	3.00	.486
3.10	.633	3.10	.611	3.10	.593	3.20*	.568	3.10	.524	3.10	.496
3.20	.643	3.20	.622	3.20	.602	3.40	.588	3.20	.534	3.20	.505
3.30	.654	3.30	.632	3.30	.613	3.60	.606	3.30	.544	3.30	.514
3.40	.665	3.40	.643	3.40	.623	3.80	.625	3.40	.553	3.40	.523
3.50	.675	3.50	.653	3.50	.633	4.00	.642	3.50	.562	3.50	.532
3.60	.685	3.60	.663	3.60	.643	4.20	.660	3.60	.572	3.60	.541
3.70	.696	3.70	.672	3.70	.652	4.40	.677	3.70	.581	3.70	.550
3.80	.706	3.80	.682	3.80	.662	4.60	.694	3.80	.590	3.80	.559
3.90	.715	3.90	.692	3.90	.672	4.80	.711	3.90	.599	3.90	.568
4.00	.725	4.00	.701	4.00	.681	5.00	.727	4.00	.607	4.00	.576
4.10	.735	4.10	.711	4.20*	.699	5.20	.743	4.10	.616	4.10	.584
4.20	.745	4.20	.720	4.40	.717	5.40	.758	4.20	.625	4.20	.593
4.30	.754	4.30	.729	4.60	.735	5.60	.773	4.30	.633	4.30	.601
4.40	.764	4.40	.738	4.80	.752	5.80	.788	4.40	.642	4.40	.609
4.50	.774	4.50	.747	5.00	.769	6.00	.803	4.50	.650	4.50	.617
4.60	.783	4.60	.756	5.20	.785	6.20	.818	4.60	.658	4.60	.625
4.70	.791	4.70	.765	5.40	.800	6.40	.832	4.70	.666	4.70	.633
		4.80	.773	5.60	.816	6.60	.846	4.80	.674	4.80	.641
		4.90	.782	5.80	.831	6.80	.860	4.90	.682	4.90	.648
		5.00	.790	6.00	.846	7.00	.873	5.00	.690	5.00	.656
		5.10	.799	6.20	.861	7.20	.887	5.20*	.705	5.10	.663
		5.20	.807	6.40	.876	7.40	.900	5.40	.720		
		5.30	.815	6.60	.892	7.60	.913	5.60	.735		
		5.40	.823	6.80	.906	7.80	.926	5.80	.750		
		5.50	.831	7.00	.920	8.00	.938	6.00	.764		
		5.60	.839	7.20	.934	8.20	.951	6.20	.778		
		5.70	.847					6.40	.792		
								6.60	.806		

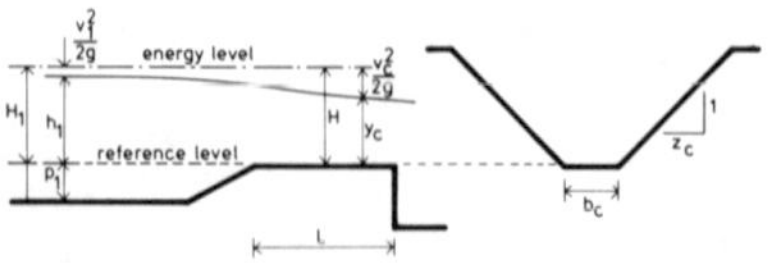

* Change in discharge increment

In some cases, a weir that will work satisfactorily cannot be found from these tables. Some judgment is necessary at this point, and many options are available. For example:

1. Find a new site for the flume where more head loss is available.
2. Add to the canal wall height upstream from the site, so that more backwater effect can be created.
3. Try one of the other weir shapes.
4. Use the tables to interpolate and get a rating for an intermediate width, probably with some sacrifice in accuracy.
5. Produce a special design from theory.

Rectangular Structures

Weirs and flumes for earthen (unlined) channels require a structure that contains the following basic parts: entrance to approach channel, approach channel, converging transition, throat, diverging transition, stilling basin, and riprap protection. As illustrated in Figure 14, the discharge measurement structure for an earthen channel is longer, and thus more expensive, than a structure in a concrete-lined canal (Figure 13) because, in the latter, the approach channel and sides of the control section are already available and the riprap is not needed. An example of an actual rectangular structure is shown in Figure 15.

The purpose of the approach canal of Figure 14 is to provide a known flow area and velocity of approach. The rating tables for the rectangular weirs given in this chapter assume that the approach section is rectangular and of the same width as the throat. The application of these tables to a structure for which the upstream sill-referenced head is not measured in a rectangular approach canal but in the wider earthen or trapezoidal upstream channel causes an error in Q because of a wide variation in the approach velocity and thus, in $v_1^2/2g$ and the related approach velocity coefficient, C_v. Consequently, the Q values for a weir without the rectangular approach channel of Figure 14 must be corrected by adjustments to C_v given in a later section.

Further shortening of the full-length structure of Figure 14 can be obtained by deleting the diverging transition or the rectangular tailwater channel. The diverging transition may be deleted if the available head loss over the structure exceeds $0.4H_1$, so that no velocity head needs to be recovered. The rectangular tailwater channel may be deleted if at maximum flow the Froude number at the entrance of this channel is less than 1.7.

A rectangular broad-crested weir discharges nearly equal quantities of water over equal widths. The major differences are associated with the friction along the walls, which reduces the discharge per unit width next to the walls. Thus the flow is nearly two-dimensional over the weir, so that rating tables can be made which give the flow rate, q, in m^3/s per meter width of sill for each value of h_1. This allows a wide variety of sizes for rectangular broad-crested weirs

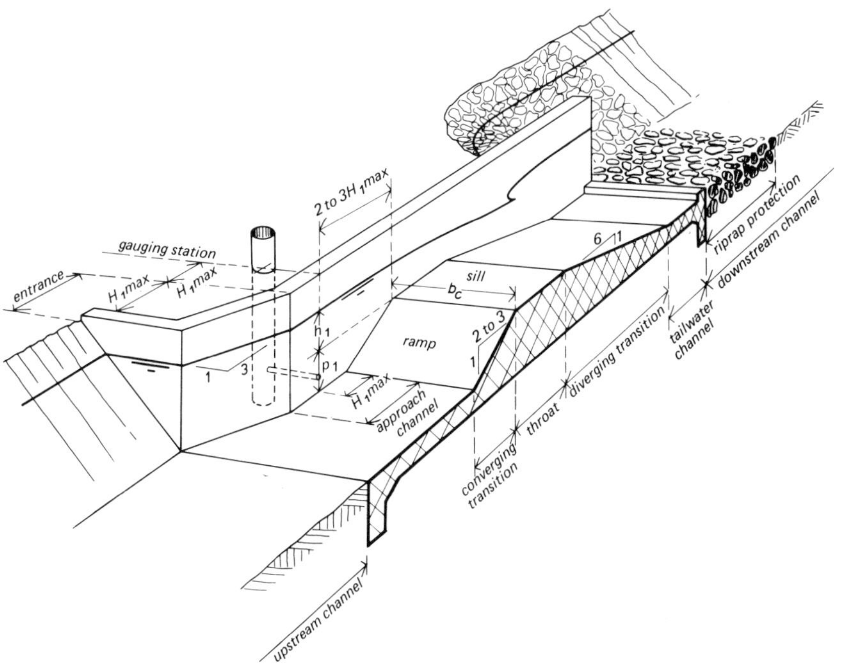

FIGURE 14. Flow-measuring structure for earthen channel with rectangular control section.

FIGURE 15. Measuring structure with a rectangular control section in an earthen canal (India).

since, for each width, b_c, of the weir, an accurate rating table can be made by multiplying the table discharges by b_c. Thus:

$$Q = b_c q \quad (17)$$

Table 10 gives a series of rating tables for rectangular broad-crested weirs that were developed from the computer model (Bos, Replogle, and Clemmens, 1984). The groupings of weir width were selected to keep the error due to the effects of the side walls to less than 1%. Ratings are given for a number of sill heights, p_1, to aid in design. Interpolation between sill heights will give reasonable results. If the approach area, A_1, is larger than that used to develop these rating tables either because of a higher sill or a wider approach, the ratings must be adjusted for C_v. To simplify this process, the discharge over the weir for a C_v value of 1.0 is given in the far right column of each grouping. This discharge column is labeled as $p_1 = \infty$ since for $C_v = 1.0$ the velocity of approach is zero, as would be the case if the weir were the outlet of a reservoir or lake. Under this circumstance, the weir has the lowest discharge for a given upstream head. Note that at the very low heads, the discharge for the weirs with rectangular approach channels approaches Q at $p_1 = \infty$ since the approach velocities are small.

For lined rectangular canals, design is relatively straightforward. It consists of selecting a sill height, p_1, such that modular flow exists throughout the flow range, and sufficient freeboard is maintained at the maximum discharge. For unlined canals, some judgment is needed in choosing an appropriate width. There are probably a number of widths that will work equally well. However, extremely wide, shallow flows are subject to measurement errors due to poor head detection sensitivity, and extremely narrow, deep flows require long structures and large head losses.

Because of the wide variety of shapes that can be encountered in earthen channels and in the range of discharges to be measured, it is rather complicated to determine the interrelated values of h_{1max}, p_1, and b_c of the structure. While this makes the design process somewhat more complicated, it allows the designer greater flexibility and expands the applicability of the weirs. The following criteria should be considered by the designer:

1. The discharges to be measured (per meter width) must be within the range of discharges shown in the rating table for the selected weir (i.e., if these flume dimensions and rating tables are to be used).
2. The allowable measurement error should not be exceeded. This allowable error may be different at different flow rates. [See Bos (1978) or Bos, Replogle, Clemmens (1984).]
3. Flow over the weir should be modular at all flow rates to be measured.
4. Placing a weir in the canal should not cause overtopping upstream.
5. The measuring structure should be placed in a straight section with a relatively uniform cross section for a distance of about 10 times the width of the channel.
6. The Froude number should not exceed 0.45 for a distance of at least 30 times h_1 upstream from the weir.

If these criteria are followed, the designer should obtain a satisfactory structure which will operate as intended.

TABLE 10. Rating Tables for Rectangular Weirs with Discharge per Meter Width. L_b = 2 to 3 Times p_1, $L_a = H_{1max}$, $L_a + L_b$ = 2 to 3 Times H_{1max}.

$0.10 \leq b_c \leq 0.20$ m, L = 0.2 m

h_1 (m)	q (m³/s per m width), p_1 = 0.05m	q (m³/s per m width), p_1 = ∞
.014	.0026	.0026
.016	.0032	.0032
.018	.0039	.0038
.020	.0046	.0045
.022	.0054	.0053
.024	.0062	.0060
.026	.0070	.0068
.028	.0079	.0076
.030	.0088	.0085
.032	.0097	.0094
.034	.0107	.0103
.036	.0117	.0112
.038	.0128	.0122
.040	.0138	.0132
.042	.0150	.0142
.044	.0161	.0153
.046	.0173	.0164
.048	.0185	.0175
.050	.0197	.0186
.052	.0210	.0197
.054	.0223	.0209
.056	.0236	.0221
.058	.0250	.0233
.060	.0264	.0245
.062	.0278	.0257
.064	.0293	.0270
.066	.0307	.0283
.068	.0322	.0296
.070	.0338	.0309
.072	.0353	.0323
.074	.0369	.0337
.076	.0385	.0350
.078	.0402	.0365
.080	.0419	.0379
.082	.0436	.0393
.084	.0453	.0408
.086	.0470	.0423
.088	.0488	.0438
.090	.0506	.0453
.092	.0524	.0468
.094	.0543	.0484
.096	.0562	.0499
.098	.0581	.0515
.100*	.0600	.0531
.105	.0649	.0571
.110	.0700	.0613
.115	.0753	.0656
.120	.0806	.0699
.125	.0861	.0744
.130	.0918	.0789

ΔH = 0.012m or $0.1H_1$

$0.20 \leq b_c \leq 0.30$ m, L = 0.35 m

h_1 (m)	q (m³/s per m width), p_1 = 0.1m	q (m³/s per m width), p_1 = ∞
.025	.0064	.0063
.030	.0085	.0084
.035	.0108	.0107
.040	.0133	.0131
.045	.0160	.0157
.050	.0189	.0184
.055	.0220	.0213
.060	.0252	.0244
.065	.0285	.0275
.070	.0321	.0308
.075	.0357	.0342
.080	.0396	.0377
.085	.0435	.0414
.090	.0476	.0451
.095	.0519	.0490
.100	.0563	.0529
.105	.0608	.0570
.110	.0655	.0611
.115	.0702	.0654
.120	.0752	.0697
.125	.0802	.0741
.130	.0854	.0787
.135	.0907	.0833
.140	.0961	.0880
.145	.1017	.0928
.150	.1074	.0977
.155	.1132	.1026
.160	.1191	.1077
.165	.1251	.1128
.170	.1312	.1180
.175	.1375	.1233
.180	.1439	.1286
.185	.1504	.1340
.190	.1567	.1396
.195	.1633	.1451
.200	.1701	.1508
.205	.1770	.1565
.210	.1840	.1623
.215	.1911	.1681
.220	.1983	.1741
.225	.2056	.1801
.230	.2130	.1861
.235	.2205	.1923

ΔH = 0.025m or $0.1H_1$

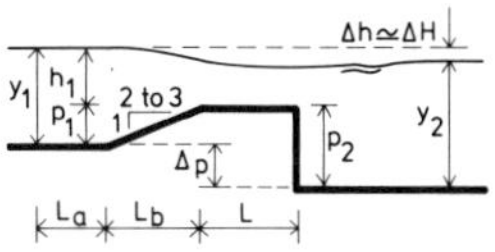

$0.30 \leq b_c \leq 0.50$ m, L = 0.5 m

h_1 (m)	q (m³/s per m width), p_1 = 0.1m	q (m³/s per m width), p_1 = 0.2m	q (m³/s per m width), p_1 = ∞
.035	.0108	.0106	.0106
.040	.0133	.0131	.0130
.045	.0160	.0157	.0156
.050	.0188	.0185	.0183
.055	.0219	.0214	.0212
.060	.0251	.0245	.0242
.065	.0285	.0278	.0274
.070	.0320	.0312	.0307
.075	.0357	.0347	.0341
.080	.0395	.0383	.0376
.085	.0435	.0421	.0412
.090	.0476	.0460	.0450
.095	.0519	.0500	.0488
.100	.0561	.0540	.0528
.105	.0606	.0583	.0567
.110	.0652	.0626	.0608
.115	.0700	.0671	.0651
.120	.0748	.0717	.0694
.125	.0798	.0764	.0738
.130	.0850	.0812	.0783
.135	.0902	.0861	.0828
.140	.0956	.0911	.0875
.145	.1011	.0962	.0923
.150	.1067	.1014	.0971
.155	.1125	.1068	.1020
.160	.1183	.1122	.1070
.165	.1243	.1177	.1121
.170	.1304	.1234	.1173
.175	.1366	.1291	.1225
.180	.1429	.1349	.1278
.185	.1493	.1409	.1332
.190	.1559	.1469	.1387
.195	.1625	.1530	.1442
.200	.1693	.1593	.1498
.205	.1762	.1656	.1555
.210	.1831	.1720	.1612
.215	.1902	.1786	.1671
.220	.1974	.1852	.1730
.225	.2047	.1919	.1789
.230	.2121	.1987	.1849
.235	.2196	.2056	.1910
.240	.2272	.2125	.1972
.245	.2349	.2196	.2034
.250*	.2427	.2268	.2097
.260	.2587	.2414	.2225
.270	.2750	.2563	.2355
.280	.2917	.2716	.2488
.290	.3088	.2872	.2623
.300	.3262	.3032	.2760
.310	.3441	.3195	.2900
.320	.3623	.3361	.3042
.330	.3808	.3531	.3186

ΔH = 0.027m or $0.1H_1$ (for p_1 = 0.1m); 0.044m (for p_1 = 0.2m)

* Change in head increment

(continued)

TABLE 10 (continued).

$0.5 \leq b_c \leq 1.0$ m
L = 0.75 m

h_1 m	q m³/s per m width, p_1 = 0.1m	p_1 = 0.2m	p_1 = 0.3m	p_1 = ∞
.050	.0186	.0183	.0182	.0181
.055	.0216	.0212	.0210	.0209
.060	.0248	.0242	.0240	.0239
.065	.0281	.0274	.0272	.0270
.070	.0316	.0308	.0305	.0303
.075	.0352	.0342	.0339	.0336
.080	.0390	.0378	.0374	.0371
.085	.0429	.0416	.0411	.0407
.090	.0470	.0454	.0449	.0444
.095	.0512	.0494	.0488	.0482
.100	.0555	.0535	.0528	.0521
.105	.0600	.0577	.0570	.0561
.110	.0646	.0621	.0612	.0602
.115	.0693	.0665	.0656	.0644
.120	.0742	.0711	.0700	.0688
.125	.0792	.0758	.0746	.0732
.130	.0843	.0806	.0793	.0776
.135	.0896	.0855	.0840	.0822
.140	.0949	.0905	.0889	.0869
.145	.1004	.0956	.0939	.0916
.150	.1061	.1009	.0989	.0965
.155	.1118	.1062	.1041	.1014
.160	.1176	.1116	.1094	.1064
.165	.1236	.1172	.1147	.1115
.170	.1297	.1228	.1202	.1166
.175	.1359	.1285	.1257	.1219
.180	.1422	.1344	.1314	.1272
.185	.1486	.1403	.1371	.1325
.190	.1552	.1464	.1430	.1380
.195	.1618	.1525	.1489	.1435
.200	.1686	.1587	.1549	.1492
.210*	.1824	.1715	.1671	.1606
.220	.1697	.1846	.1798	.1723
.230	.2113	.1981	.1927	.1843
.240	.2264	.2119	.2060	.1965
.250	.2419	.2262	.2197	.2090
.260	.2578	.2407	.2336	.2217
.270	.2741	.2557	.2479	.2348
.280	.2908	.2709	.2625	.2480
.290	.3078	.2866	.2775	.2610
.300	.3253	.3025	.2927	.2752
.310	.3431	.3188	.3083	.2892
.320	.3613	.3355	.3242	.3034
.330	.3799	.3524	.3404	.3178
.340	.3988	.3697	.3568	.3325
.350	.4181	.3873	.3736	.3473
.360	.4378	.4053	.3907	.3624
.370		.4235	.4081	.3777
.380		.4421	.4258	.3932
.390		.4610	.4438	.4089
.400		.4802	.4620	.4248
.410		.4998	.4806	.4409
.420		.5196	.4994	.4573
.430		.5397	.5185	.4738
.440		.5601	.5379	.4905
.450		.5809	.5576	.5074
.460		.6019	.5776	.5245
.470		.6232	.5978	.5418
.480		.6448	.6183	.5593
.490		.6667	.6391	.5769
.500		.6888	.6601	.5948
ΔH =	.028m or $0.1H_1$	.048m or $0.1H_1$	.063m	

$1.0 \leq b_c \leq 2.0$ m
L = 1.0 m

h_1 m	q m³/s per m width, p_1 = 0.2m	p_1 = 0.3m	p_1 = 0.4m	p_1 = ∞
.070	.0304	.0301	.0300	.0298
.080	.0374	.0370	.0369	.0298
.090	.0450	.0445	.0442	.0439
.100	.0531	.0524	.0521	.0516
.110	.0616	.0608	.0604	.0597
.120	.0706	.0696	.0691	.0683
.130	.0801	.0788	.0782	.0771
.140	.0900	.0885	.0877	.0864
.150	.1004	.0985	.0976	.0960
.160	.1112	.1090	.1079	.1059
.170	.1224	.1198	.1185	.1161
.180	.1339	.1319	.1295	.1267
.190	.1459	.1426	.1408	.1375
.200	.1583	.1545	.1525	.1487
.210	.1711	.1668	.1646	.1601
.220	.1842	.1794	.1769	.1718
.230	.1977	.1924	.1896	.1838
.240	.2116	.2058	.2027	.1961
.250	.2259	.2194	.2160	.2086
.260	.2405	.2334	.2297	.2214
.270	.2555	.2477	.2436	.2344
.280	.2708	.2624	.2579	.2477
.290	.2864	.2774	.2725	.2612
.300	.3024	.2927	.2873	.2749
.310	.3188	.3083	.3025	.2889
.320	.3355	.3242	.3180	.3032
.330	.3525	.3404	.3337	.3176
.340	.3698	.3569	.3498	.3323
.350	.3875	.3738	.3661	.3472
.360	.4055	.3909	.3828	.3623
.370	.4238	.4083	.3997	.3776
.380	.4424	.4261	.4168	.3931
.390	.4614	.4441	.4343	.4088
.400	.4806	.4624	.4520	.4248
.410	.5002	.4810	.4701	.4409
.420	.5200	.4999	.4883	.4573
.430	.5401	.5190	.5069	.4738
.440	.5607	.5385	.5257	.4905
.450	.5815	.5582	.5447	.5075
.460	.6025	.5782	.5641	.5246
.470	.6238	.5984	.5837	.5419
.480	.6455	.6189	.6035	.5594
.490	.6674	.6398	.6236	.5771
.500	.6896	.6608	.6440	.5950
.510	.7122	.6822	.6646	.6130
.520	.7350	.7038	.6855	.6312
.530	.7580	.7257	.7065	.6496
.540	.7814	.7478	.7279	.6682
.550	.8050	.7702	.7495	.6869
.560	.8290	.7929	.7715	.7059
.570	.8532	.8158	.7936	.7249
.580	.8776	.8390	.8159	.7442
.590	.9024	.8624	.8385	.7636
.600	.9274	.8861	.8613	.7832
.610	.9527	.9102	.8844	.8029
.620	.9782	.9343	.9077	.8228
.630	1.004	.9588	.9312	.8429
.640	1.030	.9835	.9550	.8632
.650	1.056	1.008	.9790	.8836
.660	1.083	1.034	1.003	.9041
.670	1.110	1.059	1.020	.9249
ΔH =	.046m or $0.1H_1$	.066m or $0.1H_1$	.086m	

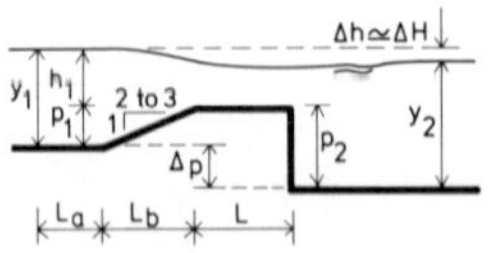

* Change in head increment

(continued)

TABLE 10 (continued).

$b_c \geq 2.0$ m
$L = 1.0$ m

h_1 m	q m^3/s per m width			
	P_1 = 0.2 m	P_1 = 0.4 m	P_1 = 0.6 m	P_1 = ∞
.100	.0521	.0511	.0508	.0506
.120	.0695	.0680	.0675	.0671
.140	.0889	.0866	.0858	.0852
.160	.1099	.1067	.1056	.1046
.180	.1326	.1283	.1268	.1253
.200	.1569	.1513	.1493	.1473
.220	.1827	.1756	.1732	.1704
.240	.2101	.2013	.1982	.1946
.260	.2389	.2283	.2245	.2199
.280	.2691	.2565	.2519	.2461
.300	.3008	.2859	.2805	.2733
.320	.3337	.3165	.3101	.3015
.340	.3681	.3483	.3409	.3306
.360	.4037	.3812	.3727	.3606
.380	.4406	.4153	.4056	.3914
.400	.4788	.4505	.4395	.4231
.420	.5182	.4868	.4744	.4556
.440	.5588	.5241	.5103	.4889
.460	.6007	.5626	.5472	.5229
.480	.6437	.6020	.5851	.5577
.500	.6878	.6425	.6239	.5932
.520	.7331	.6840	.6636	.6295
.540	.7796	.7265	.7042	.6664
.560	.8271	.7699	.7458	.7041
.580	.8758	.8144	.7884	.7425
.600	.9257	.8600	.8319	.7815
.620	.9765	.9063	.8762	.8212
.640	1.028	.9537	.9214	.8615
.660	1.081	1.002	.9674	.9025
.680	1.135	1.051	1.014	.9441
.700	1.191	1.101	1.062	.9864
.720		1.153	1.111	1.029
.740		1.205	1.160	1.073
.760		1.257	1.210	1.117
.780		1.311	1.262	1.161
.800		1.366	1.314	1.207
.820		1.422	1.367	1.252
.840		1.478	1.420	1.299
.860		1.535	1.474	1.346
.880		1.593	1.530	1.393
.900		1.652	1.586	1.441
.920		1.712	1.642	1.490
.940		1.773	1.700	1.539
.960		1.834	1.758	1.588
.980		1.897	1.817	1.638
1.000		1.960	1.877	1.689
ΔH =	0.047m or $0.1H_1$	0.087m or $0.1H_1$	.124m or	

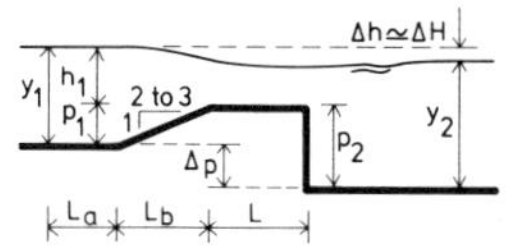

For a rectangular structure in an earthen canal, the rectangular section need not extend 10 times its width upstream from the flume if a gradual taper is used to guide the flow into the rectangular section. For the flumes given here, it is recommended that the rectangular section extend upstream from the head measurement location (gauging station), as shown in Figure 14. It is also recommended that riprap be placed downstream from the structure for a distance of 4 times y_{2max}. A step should be provided at the transition between the rectangular section and the riprap section to avoid local erosion from floor jets. Sizing of riprap and filters is discussed by Bos (1978) and Bos, Replogle, Clemmens, et al. (1984).

An analysis of head measurement errors is presented by Bos (1978) and Bos, Replogle and Clemmens (1984) and will not be repeated here. For lined channels, a freeboard criteria of 0.2 h_{1max} has been used satisfactorily. For unlined channels it may be more appropriate to specify a maximum water depth, y_{1max}.

Submergence or modular flow should be checked at both minimum and maximum expected discharges. Usually, for lined channels where flow depth is controlled by channel friction, designing for the maximum discharge is sufficient.

If the channel is rectangular, or the length of the rectangular throated structure downstream from the weir sill is as in Figure 14, then we can use the lower value of $\Delta H = 0.1H_1$ or the ΔH-value given at the bottom of Table 10. If a shorter length in an earthen channel is used, and the tailwater channel is significantly larger than the stilling basin would be, then considerably more head loss will probably be required. The designer could use the head loss value for the discharge into a lake or pool, $\Delta H = 0.4H_1$. This may represent a drastic difference in the value of head loss. The designer may decide to use the shortened structure and calculate the actual modular limit by use of the computer model (Bos, Replogle, and Clemmens, 1984), or the procedure of Bos and Reinink (1981). Another alternative is to actually build a prototype in the field and set the crest to the appropriate level by trial and error.

Structures for Circular Channels

The broad-crested weir was chosen for flow measurements in pipes with a free water surface because of its simplified construction. Other shapes can be and have been used; however, no direct ratings are presented here. There are two situations of interest for designing a weir in a pipe. The weir can be placed somewhere in the middle of a straight pipe, or at the end of a pipe, such as the entrance to a deep manhole. The latter is probably more common because the site is more accessible (Figure 16). When the weir is placed in the interior of a pipe, the presence of the weir must cause a rise in the upstream water surface due to the required head loss. This increase in flow area upstream causes a proportional decrease in velocity and subsequent

FIGURE 16. Broad-crested weir with horizontal sill in a drainage pipe outlet (Arizona).

sediment deposition. Where this is a problem, a downstream ramp should be considered to reduce the head loss as much as possible. Because of the shape of these weirs, sediment problems are further aggravated at low flows since the free flowing (normal depth) water surface drops proportionally faster than the water surface upstream from the weir. Thus the velocity difference becomes greater and greater with decreasing discharge. For situations where wide fluctuations in flow rate exist and sedimentation is a problem, an alternative weir shape or location should be considered.

An alternative location for a measuring site is the end of a pipe, particularly where a drop in water surface exists. For long pipes where flow depth is controlled by channel friction, a weir can often be designed so that the water level upstream from the weir either matches or is below the normal water level in the pipe. In this way, we can considerably reduce the effects of the weir on sediment deposition.

If the weir is located in a section of pipe, the normal depth, y_n, equals y_2. Hence, at maximum flow:

$$y_2 + \Delta h \leq p_1 + h_1 \leq 0.9d \qquad (24)$$

Equation (24) gives the limits on design to provide for modular flow $y_1 \geq y_2 + \Delta h$ and to keep the pipe from flowing full $y_1 \leq 0.9d$ at maximum flow. If flow in the pipe is caused to occur by downstream backwater effects (in excess of normal depth), these criteria should be checked at low flows.

With a weir at the end of a pipe and sufficient overfall, the weir can be lower for better sediment transport. Preferably, now the normal depth, y_n, should be larger than or equal to y_1. Hence at maximum flow:

$$y_1 = p_1 + h_1 \leq 0.9d \qquad (25)$$

Since all weir and pipe combinations can be considered as Froude models of each other, rating tables were developed for a diameter $d = 1.0$ m and for sill heights p_c, ranging from 0.05 to 0.50 m.

There is a direct relationship between $Q/(d^{2.5}\, g^{0.5})$ and h_1/d. Because of the complex geometries involved in the circular cross sections it is convenient to use an empirical equation for this relationship. A good fit of $Q/(d^{2.5}\, g^{0.5})$ vs. h_1/d can be represented as a polynomial or as a modified power function of the form:

$$Q/(d^{2.5}\, g^{0.5}) = C_e\,(h_1/d + K_h/d)^u \qquad (26)$$

where C_e, K_h, and u are constants in the empirical equation. The above ratings were fit to Equation (26) with a curve-fit routine for complex functions. This routine uses an iterative procedure to search for the optimum values of the coefficients of Equation (26).

The resulting values of C_e, K_h/d, and u are given in Figure 17. The equations reproduce the values computed by the mathematical model to within less than $\pm 0.25\%$ and most differences were on the order of $\pm 0.1\%$.

However, since relative roughnesses will be different for small and large pipes, additional frictional effects will cause slight differences in discharge. For pipe diameter of 0.3–2.0 m, these differences are generally less than 0.5% for most of the flow range. At low relative heads (e.g., $H_1/L < 0.2$), these differences were slightly greater (e.g., 1 or 2%).

The required head loss for maintaining modular flow over this type of structure in a pipe with a rapid expansion as a function of upstream sill-referenced depth for several values of sill height is plotted in Figure 18. Because of some uncertainty in modular limit calculations and uncertainty about actual flow conditions, we recommend that the modular limit not exceed 0.9 (equal to a head loss of $0.1H_1$) for truncated weirs, i.e., weirs without a downstream transition. As shown in Figure 18, the computed head loss is used until the modular limit reaches 0.9. Then a head loss value of $0.1H_1$ is used.

If the pipe is discharging into a reservoir or pool, the head loss will be greater since there is less energy recovery. The modular limit for this case should be taken as 0.60 or $\Delta h_1 = 0.4H_1$. The addition of a downstream ramp can

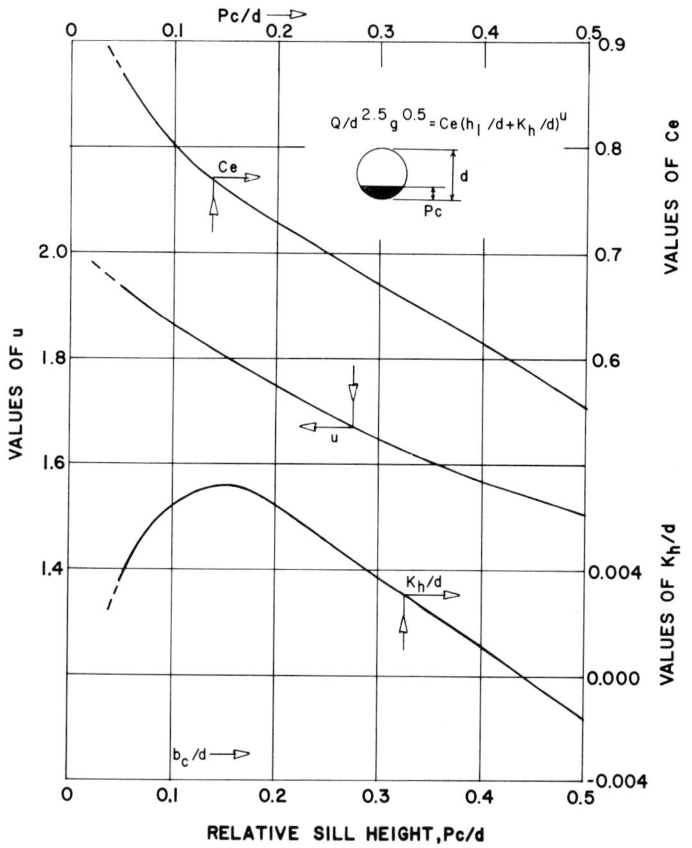

FIGURE 17. Values of discharge equation coefficients C_e, u and k_h/d for broad-crested weirs (horizontal sills) in circular channels.

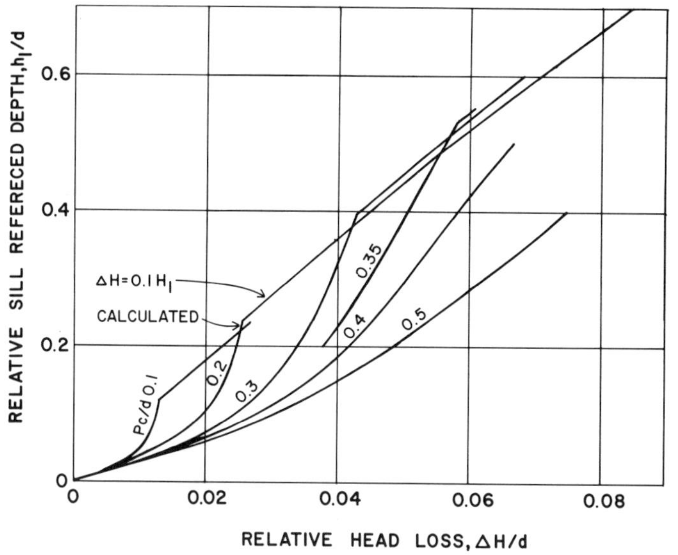

FIGURE 18. Head loss required for maintaining modular flow over RBC broad-crested weirs with sudden expansion (for a weir placed in a continuous channel of constant cross section with no sudden drop).

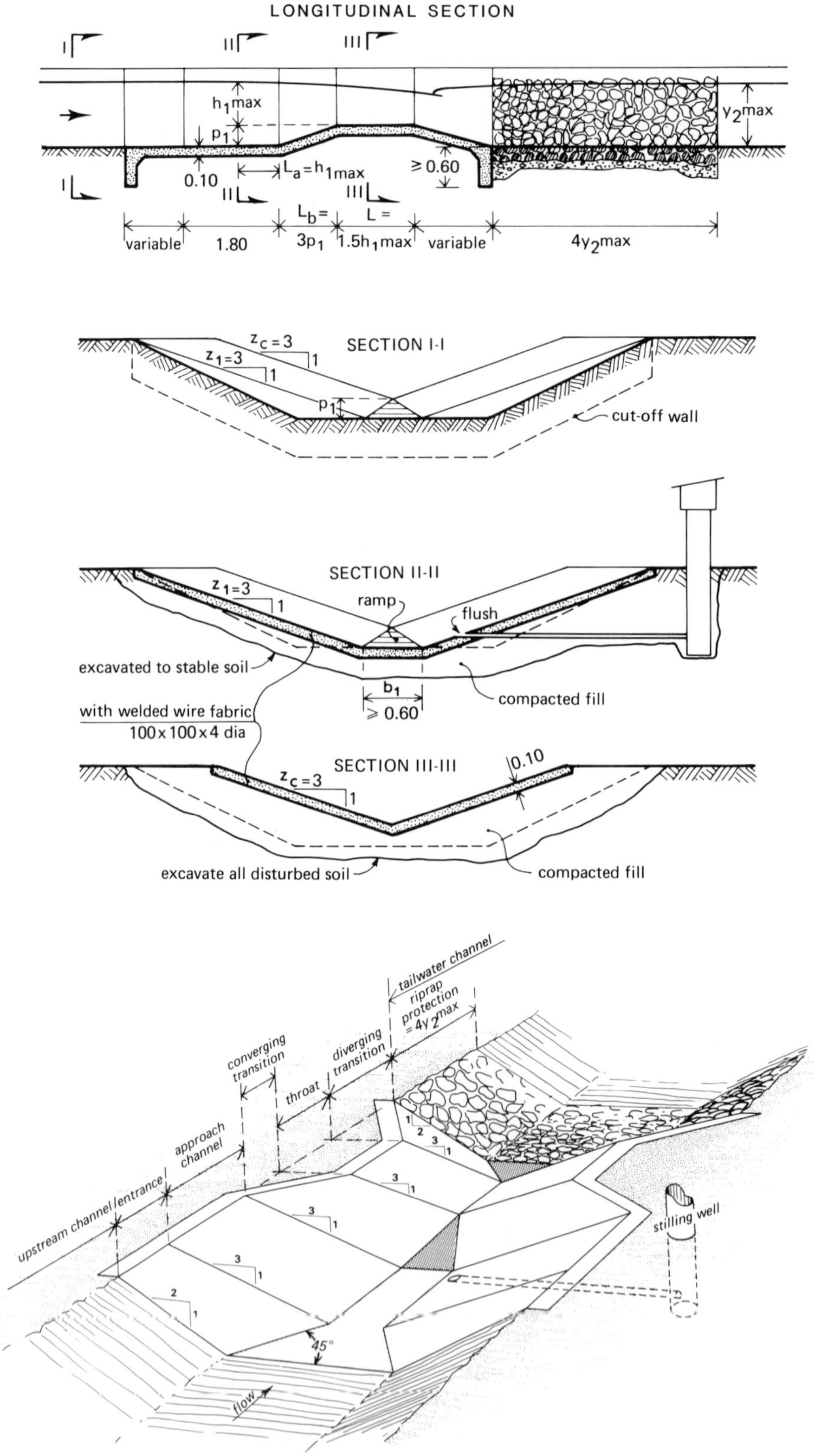

FIGURE 19. Layout of triangular-throated flume.

reduce the required head loss. A method for calculating the head loss for different combinations of tailwater conditions and ramp slopes is given in Bos and Reinink (1981).

Structures for Natural Channels

For monitoring return flows and operational spillages from irrigation systems or for measuring flows in natural streams, a structure is needed that can measure a wide range of flows. A structure with triangular control section is very suitable for this purpose because $\gamma \cong 335$ (see Table 5). Here again a wide variety of options exist for side slopes, sill height, and approach channel geometry. Figure 19 shows the general layout of such a flume. Rating tables for several sideslopes and a fixed sill height and approach bottom width are given in Table 11. These structure dimensions have proven to be very useful in a wide variety of applications (Figure 20). The head-loss values given in Table 11 are for a rapid expansion into a tailwater channel of the same size as the approach channel. A gradual transition into a wide channel is more common. For a triangular control section, the differences are not too significant, particularly since the design head loss should not exceed $0.1H_1$, while the theoretical head loss without velocity head recovery is $0.2H_1$.

CONSTRUCTION AND AS-BUILT CALIBRATIONS

Construction Techniques and Materials

To summarize the foregoing discussion as it relates to construction and as-built calibrations: flow measurements,

TABLE 11. Rating Tables for Triangular Throated Flumes with $b_1 = b_2 = 0.60$ m, $b_c = 0$, $p_1 = p_2 = 0.15$ m, $L_a = 0.90$ m, $L_b = 1.0$ m, $L = 1.2$ m, $z_1 = z_c = z_2$.

Metric Units			
h_1	Q in m³/s		
in m	z_c=1.0	z_c=2.0	z_c=3.0
0.08	.0020	.0042	.0063
0.10	.0036	.0074	.0111
0.12	.0057	.0118	.0178
0.14	.0085	.0175	.0264
0.16	.0121	.0247	.0372
0.18	.0163	.0333	.0503
0.20	.0214	.0436	.0658
0.22	.0273	.0557	.0840
0.24	.0341	.0695	.1049
0.26	.0419	.0853	.1288
0.28	.0506	.1031	.1557
0.30	.0604	.1230	.1859
0.32	.0712	.1451	.2193
0.34	.0831	.1694	.2562
0.36	.0962	.1962	.2968
0.38	.1104	.2253	.3410
0.40	.1259	.2570	.3891
0.42	.1427	.2913	.4411
0.44	.1607	.3282	.4973
0.46	.1800	.3679	.5577
0.48	.2008	.4105	.6225
0.50	.2229	.4560	.6917
0.52	.2464	.5044	.7655
0.54	.2715	.5560	.8440
0.56	.2980	.6106	.9273
0.58	.3260	.6685	1.015
0.60	.3556	.7296	1.109
0.62	.3868	.7942	1.207
0.64	.4197	.8621	1.311
0.66	.4542	.9335	1.419
0.68	.4904	1.008	1.534
0.70	.5284	1.087	1.654
0.72	.5681	1.169	1.779
0.74	.6096	1.255	1.911
0.76	.6529	1.345	2.048
0.78	.6981	1.439	2.191
0.80	.7451	1.536	2.340
ΔH =	0.09 m	0.07 m or $0.1H_1$	0.06 m or $0.1H_1$

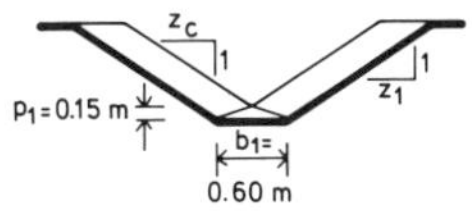

FIGURE 20. Triangular flume flowing at approximately 0.1 m^3/s. A toe-wall constructed down to a firm clay layer control undercutting of the flume (Florida).

even in lined canals, have usually been costly, frequently of questionable accuracy, and otherwise difficult to apply in field situations. Major problems include the requirement of reshaping the canals to accommodate a limited assortment of calibrated devices, as is the case for Parshall Flumes, and the requirement for relatively large water surface drops that is associated with sharp-crested weirs. Not the least of these problems is the general inability to control installation errors and readout errors for an accurate, convenient, and reliable head reading. Many of these problems are significantly reduced by the use of long-throated flumes and the hydraulically related broad-crested weirs.

Theoretical and applied research on flow measuring structures has singled out the broad-crested weir as the most promising discharge measurement structure for accurately measuring flow rate in lined canals. Error control is incorporated in several ways:

1. The width dimensions of the weir crest are chosen to be large enough to easily absorb normal concrete construction errors in the width of the control section. Therefore, existing concrete-lined canals can be used for most of the measuring structures.
2. The weir-crest length in the direction of flow and a sloping approach ramp provide flow situations that can be accurately modeled mathematically ($\pm 2\%$) and solved by a mathematical model for nearly any cross-sectional channel shape.
3. Surveying methods are used to accurately determine wall-gauge mounting positions, or stilling-well reference levels, above the elevation of the constructed sill crest of the weir. This vertical gauge positioning, or stilling-well reference, which is upstream from the weir, is the most critical of the field determinations, and special care with the surveying methods should be observed.

Construction of the illustrated broad-crested weir in a lined canal is simple. This style of broad-crested weir requires only one carefully constructed surface: the weir sill crest (which should be horizontal in both directions). All other surfaces can be approximately dimensioned and rough-finished to about $\pm 10\%$ without affecting the calibration beyond about 1% (see Table 12). For large canals and unlined canals, the control section walls should also be reasonably plane surfaces, such that the cross section is reasonably uniform in the direction of flow. Because of these simplified construction requirements, construction costs are about 10% of previous, more complicated flumes (such as the Cutthroat and Parshall flumes) for small canals (e.g., < 3m 3/5). This cost ratio in very large sizes is about 25%.

BROAD-CRESTED WEIRS IN SMALL LINED CANALS

Once a suitable broad-crested weir has been selected, construction is relatively easy and straightforward. The weir has two sections: the control section made by the sill of length L and height p_1 (Figure 13) and a sloping ramp of length $3p_1$ that forms the transition from the channel shape to that of the sill. The sill top should be level with no large irregularities. The constructed width of the weir, b_c, should be as close to the value in the appropriate tables as the design accuracy of the measurement, because a 1% error in b_c will produce about a 1% error in discharge.

While the sill height p_1 is important for selecting the weir, because it controls the water surface elevation and thus the modular limit and freeboard relationships, its precise vertical dimension is not critical to the weir calibration relation. Also, the length L can be adjusted $\pm 10\%$ without noticeable effect. The ramp length of $3p_1$ is also approximate and its purpose is to convey the water smoothly to the weir crest. Thus, the ramp may be straight or gently curved, but the straight kind is usually more readily constructed. These liberal tolerances should facilitate construction, but should not be an excuse for sloppy construction.

A drain tube should be installed to allow the upstream canal to drain for control of mosquitoes, particularly if the canal is for intermittent use. Flow through the drain tube is negligible if its diameter is about 30 mm or less and the canal flow is near capacity for the weir. If the tubes are larger, they should be at least partly blocked during use as a measuring structure. This can be done with valves, bricks, or rags. The latter can be quickly pushed out of the tube with a stick or rod when drain-down of the canal is desired. Complete blockage is not required. Sediments, which tend to accumulate at the base of the ramp, may plug the drain pipe and should be cleaned out periodically. Sediments will usually not significantly accumulate in the canal upstream from the weir if they did not normally accumulate in the canal before the weir was installed. Canals that normally accumulate significant depths of sediments will continue to accumulate similar depths upstream from the weir.

If the canal has construction joints, the sill should be placed so that the gauge location is at least 0.5 m downstream from the joint. If that is difficult, the gauge should be located slightly upstream of the joint. If possible, construction joints should be avoided in the area of the sill although they are acceptable in the ramp section. Care should be exercised to assure that if a construction joint is located between weir sill and gauge, no vertical movement occurs, or else the zero settings may not remain reliable.

Cast-in-Place Construction

Cast-in-place construction is usually very simple. There are a wide variety of techniques possible. The main requirement is that the edges of the crest be properly formed so that the crest is level in both directions and rectangular, and so that the concrete can be properly screeded. The structure can be poured as a monolithic volume of concrete extending to the canal bottom, or as a thin shell. When cast as monolithic concrete, the edges of the sill can be made with two sheets of plywood, two pieces of angle iron or one of each. With two pieces of angle iron, the sill and both ramps are

TABLE 12. Percentage Error in the Discharge From That Indicated in the Rating Tables Due to Change in Actually Constructed Dimensions.

1% change in dimension of:	Causes shown percentage error in discharge	Remarks
Upstream ramp length	0.01%	Slope of ramp may vary from 1:2 to 1:3
Sill height, p_1	0.03%	Influences approach velocity
Sill or throat length, L	0.1%	Depending on H_1/L value
Bottom width control, b_c	up to 1%	Depends upon percentage change in wetted flow area at control
Wetted flow area at control	1%	
1 degree change of:		
Cross slope of sill	0.1%	Has minor effect on area of flow
Sill slope in direction of flow	up to 3%	Is most difficult factor to correct for
Side slope of control	0.5%	Depends on change of wetted flow area

poured at once. The angles can be tied together or bolted to the canal wall. When two sheets of plywood are used, the sill is poured alone, and one or more ramps poured after the sill has set and the plywood is removed. The plywood sheets are cut to match the canal cross section and tied together at the desired spacing. When only an approach ramp is poured, an angle iron is used for the upstream edge, and both ramp and sill are poured at the same time (Figure 21).

These flumes can be poured as a thin shell by using earth fill to form the bottom of the structure. Angle iron is again used for screed edges along the crest; however, it also adds considerable strength. The crest is typically poured 60 to 150 mm thick depending on length of span, amount of reinforcing steel used, and whether or not fill material below the crest will remain or wash out. If no downstream ramp is poured, the downstream angle iron can act as a form as well. However, the fill material will wash out. If both ramps are poured, coarse grained fill material should be used since it is less likely to wash out through cracks and holes. For larger structures, canal walls may not be able to handle the large forces developed. Here most of the structure weight should be transmitted to the canal floor, e.g., through coarse fill material overlaid by a concrete shell.

Both the ramp and the sill may be broom finished. The upper edge of the ramp should join the sill edge so that there is no abrupt rise or drop that might cause unpredictable flow separation. Precision edge matching of the two sections or slight rounding of the joint corner is desirable. This is easily done with the concrete trowel. The ramp should be fairly uniform with no large humps or depressions. Its surface should approximate a slope of 3 horizontal to 1 vertical as seen in profile (Figure 13). For sediment passage, it is best that the ramp be slightly sunken or concave rather than rounded upward, or convex. The ramp need not taper to zero thickness but can be ended abruptly when it becomes about 50 mm thick.

Pre-Cast Concrete Weirs

For the smaller canals, precast versions of the permanent weirs can be made (Figure 22). Reinforcing wire and two 15-mm-diameter reinforcing rods are used in thin shell copies of the weirs. The concrete thickness is usually about 5 cm with concrete crossbeams cast into the sill piece. A groove is formed on the upstream edge of the sill piece to support the pre-cast ramp section. For safety in handling, simple wire reinforcing in the flat portions of the sill piece and in the ramp piece is recommended.

Temporary and Portable Weirs

Temporary weirs for lined canals can be made out of wood, sheet metal or any such material. These generally require stiffening to limit deflection. Structures of this type that are to be left unattended for long periods of time should

FIGURE 21. When constructing a cast-in-place broad-crested weir, the upstream form can be made of angle iron and is left in weir.

FIGURE 22. Small precast weir being placed in lined canal.

FIGURE 23. Portable weir for use in concrete-lined canals (Arizona).

be anchored to the canal wall, even though water pressure will hold them in place. These structures are particularly useful for determining flow conditions for design of a permanent structure, for flow survey work, or for temporary periods. Head measurements can be made with a staff or wall gauge surveyed to the crest, or with a point gauge mounted on a board spanning the channel. Here the water level in the upstream channel is siphoned into a cup suspended above the crest. The point gauge is used to measure the difference between the crest level and the water level in the cup. Care should be taken to assure proper upstream head sensing. This can be accomplished with a 25 mm diameter pipe 0.6 m long with a battery of 3 mm holes drilled radially around the middle of the pipe and with one end plugged or welded shut and the other end connected to the siphon tube. Such a device can also be used for permanent flumes and for zeroing recorders. For smaller sized portable weirs in lined canals, the point gauge, cup, siphon tube and sensing pipe can all be connected to the flume via an exterior framework (Figure 23). This makes for convenient flow monitoring.

Other portable weirs are available for unlined canals. Small rectangular flumes including the approach, converging and control sections have been constructed of sheet metal and fiberglass for discharges in the range of 10 to 100 l/s. Smaller trapezoidal flumes (Figure 7) have been constructed to measure flows on the order of 0.1 to 10 l/s. In this case, stilling wells are located on the side of the throat, or immediately downstream from the throat to minimize leveling errors. For more details, see Bos, Replogle and Clemmens (1984).

BROAD-CRESTED WEIRS IN LARGE CANALS

The construction of large weirs in main or lateral irrigation canals (Figure 24) may require special attention because of both hydraulic and foundation problems. Adding excessive concentrated loads to previously lined sections of canal, which are likely to be on nearly saturated soils, invites settling and concrete cracking. In unlined canals, the addition of fill material to make a large sill, frequently in excess of 1 meter high, also causes large concentrated loadings. On previously constructed and operated canals, these bottom soils are likely to be unstable. A solution that appears feasible is to remove much of the unstable material and replace it with easily drainable coarse material. The sill with ramps at both ends is then constructed as a compacted fill much like a low earth dam. The requirement for impervious backfill is less important than the requirement for minimum differential settlement. Therefore, coarse-textured, easily drained fill is desirable. The fill is compacted and shaped to the rough profile of the finished sill, and where time permits, allowed to settle for a time period consistent with local soil conditions. Some sites may require an extensive foundation design. The compacted sill material is then covered with 100 to 150 mm of reinforced concrete (Figure 25). All slopes are on flat enough grades that flat-slab techniques can be used for economical concrete placement. Canal lining upstream and downstream for several meters is usually reinforced, which serves to limit cracking and thus seepage flow through the pervious fill materials and reduces the chances for destructive removal of the fill through cracks that otherwise could form. Minor settling, including differential settling, can be periodically compensated for by adjusting the gauge zero and/or recomputing the calibration (see Construction-Related Errors).

STRUCTURES IN EARTHEN CANALS AND CHANNELS

In constructing a rectangular weir or flume as in Figure 14, the designer may select any locally available construction material. For example, the wing and sidewalls can be brickwork containing a mortar-plastered sill (Figure 26); the entire structure can be made of reinforced concrete, or a (wooden) sheet piling can be driven across the channel in which a steel or aluminum control section is bolted on seals (Figure 27). The latter method of construction is very suitable on soils with a low bearing capacity.

The structure shown in Figure 14 can be considerably shortened if an earth approach section is used (with appropriate adjustment with C_v) and if no diverging transition and stilling basin are required. However, cutoff walls and rock riprap should still be used. A structure of this type is commonly used in drains or irrigation canals.

A structure can be built that resembles a short section of lined canal with a sill placed in the bottom, as given in the section on structures for lined trapezoidal canals (Figure 28). The structural requirements are the same as for the rectangular structures. Shortcutting these design requirements will ultimately lead to structural failure. The walls should be placed on well-compacted fill and can be either formed or hand-plastered.

The channel section needs to be accurate in the region of the sill to about $\pm 1\%$. A template cut to the correct shape is useful in controlling this for plastering operations. The wall slope in the region where the gauge is to be attached should also be accurate. The width dimensions of the channel at this location, however, need be only $\pm 10\%$. These liberal tolerances on construction allow the use of stone and mortar for these structures or almost any convenient rigid material. The dimensions of the structure must follow the same criteria as the rectangular weirs as shown in Figure 14. If an earthen approach channel is used, which is of a significantly different size from that used in the rating, then the rating tables must be corrected for C_v. Similarly, if the tailwater channel is of significantly different size, then the head-loss values may also need to be recalculated (see Bos, Replogle, Clemmens 1984).

The triangular throated structures (Figure 20) can be constructed with flat slab techniques. The simplest method for

FIGURE 24. Large weir on Arizona canal at approximately 34 m^3/s.

FIGURE 25. Large weir on Arizona canal during construction.

FIGURE 26. Construction of a rectangular measuring structure in an earthen canal with stones and mortar (India).

FIGURE 27. Metal broad-crested weir in wooden sheet piling (The Netherlands).

FIGURE 28. Trapezoidal weir flowing roughly 7 m^3/s (250 ft^3/s) (Idaho). Weir was placed in a short lined section.

constructing this type of flume is: (1) excavate unstable soil, (2) place compacted fill, (3) install stilling well and pipe and compact fill around pipes, (4) excavate two toe-wall trenches of at least 0.60-m depth, (5) install 4-mm-diameter welded wire fabric of 100- to 150-mm spacing, (6) place forms to mark edges of concrete slabs, (7) pour concrete for both toe walls plus related entrance and diverging transitions and for the converging transition (Figure 29), (8) upon hardening of this concrete, remove forms and pour concrete for approach channel and throat, (9) place riprap protection, and (10) finish stilling-well inlet pipe flush with concrete and install recorder.

Construction-Related Errors

To justify the use of the rating tables given here, the constructed dimensions of the structure must be sufficiently close to those given in the related drawing or table heading. A change in these dimensions will influence the "error" between the true flow and that indicated by the tables. The relative order of magnitude of these added systematic errors is illustrated in Table 12. As shown, the most important errors in the rating table flow rates are caused by the changes in the wetted flow area at the control perpendicular to the direction of flow. The given rating tables may be corrected by the percentages shown in Table 12 provided that the sum of all errors does not exceed 5%. For larger deviations, we recommend generating a new rating table by use of the theory or by use of the computer program. If the sill or throat is not level, but slopes in the direction of flow, this influences both the flow rating and the modular limit of the structure. Because it is also difficult to correct for slopes larger than 2 degrees, it is easier to level the sill than to correct for larger slopes.

TABLE 13. Approach Velocity Coefficients for Rectangular Weirs and Flumes (u = 1.5).

$\frac{\sqrt{\alpha_1} C_d A^*}{A_1}$	.00	.01	.02	.03	.04[a]	.05	.06	.07	.08	.09
0.0	1.000	1.000	1.000	1.000	1.000	1.001	1.001	1.001	1.001	1.002
0.1	1.002	1.003	1.003	1.004	1.004	1.005	1.006	1.007	1.007	1.008
0.2	1.009	1.010	1.011	1.012	1.013	1.014	1.016	1.017	1.018	1.019
0.3[a]	1.021	1.022	1.024	1.026	1.027[a]	1.029	1.031	1.033	1.035	1.037
0.4	1.039	1.041	1.043	1.045	1.048	1.050	1.053	1.055	1.058	1.061
0.5	1.063	1.066	1.070	1.073	1.076	1.079	1.083	1.086	1.090	1.094
0.6	1.098	1.102	1.106	1.111	1.115	1.120	1.125	1.130	1.135	1.141
0.7	1.146	1.152	1.159	1.165	1.172	1.179	1.186	1.193	1.201	1.210

[a]Example: If $\sqrt{\alpha_1} C_d A^*/A_1 = 0.3 + 0.04 = 0.34$, then $C_v = 1.027$

FIGURE 29. Concrete is poured for the downstream toe-wall and diverging transition of this flume with a triangular crest (Arizona).

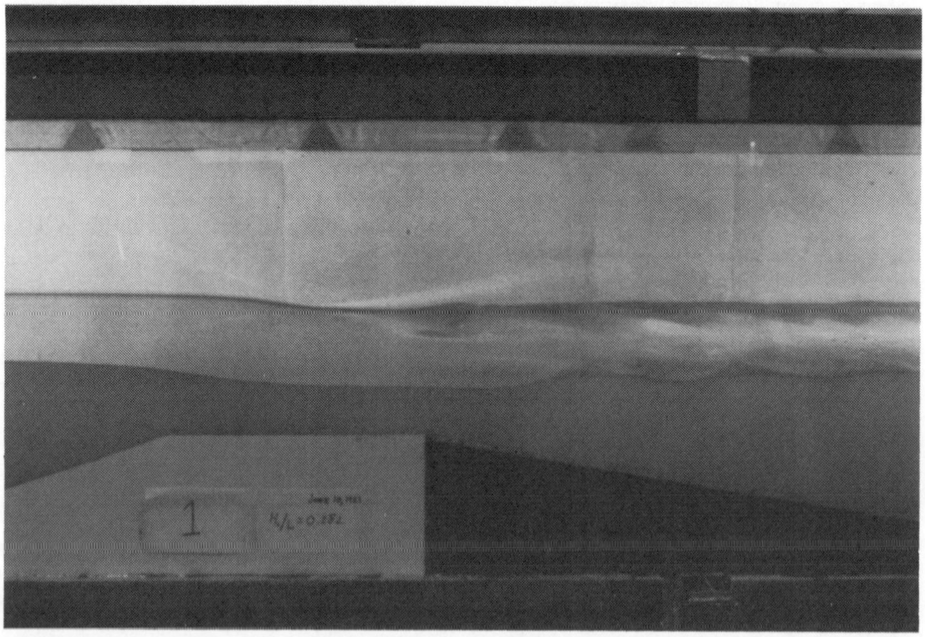

FIGURE 30. Rectangular laboratory weir at $H_1/L = 0.282$. Tailwater level is just below that required for modular flow. Note slight drop in water level downstream from throat or crest.

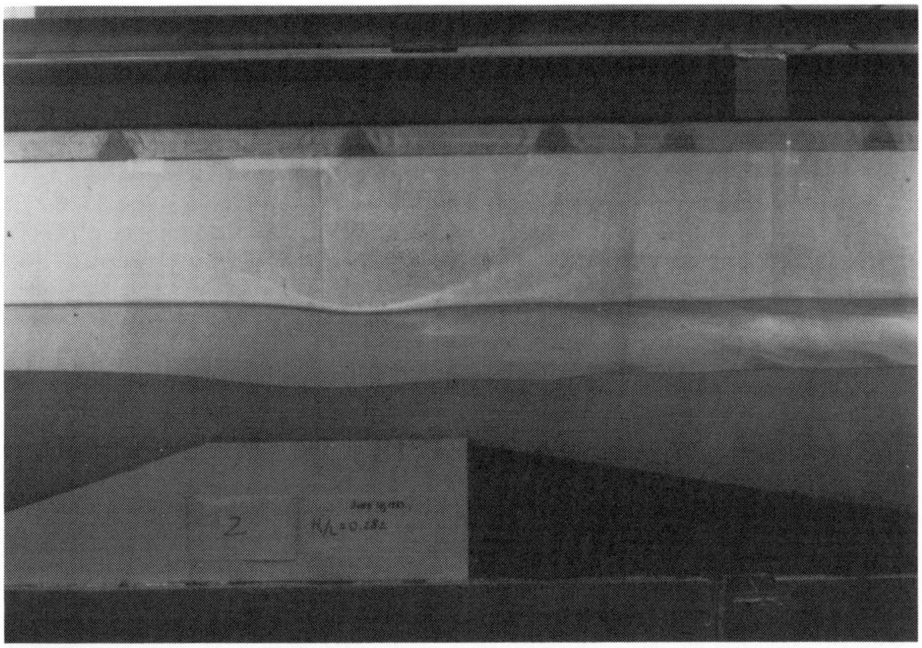

FIGURE 31. Same weir as above with tailwater level raised just past modular limit. Note rise in water of last half of flume throat. Here the upstream water level is influenced by the downstream tailwater level.

FIGURE 32. Large flume on Arizona Canal. Rollers have moved back onto crest. The flow profile here is similar (i.e., approximate Froude scale model) to that shown in previous figure. Whitewater is present in large flume due to scale effects not accounted for by Froude modeling.

Adjustments to Rating Tables with C_v

There may be situations when site conditions call for use of an approach channel section other than those assumed in the development of the rating tables presented here. For example, a trapezoidal (or rough-form earthen) approach to a rectangular weir. In most cases, these alterations can be accommodated through adjustments in C_v. The rating tables give values for Q versus h_1. This relationship is described by Equation (9) or similar equations in Figure 4. For a given weir and h_1 value, only C_d and C_v can change. It can be assumed that C_d does not change with minor changes in velocity head since H_1/L changes very little. Thus, we need only evaluate C_v to adjust the rating table values. The new discharge for a given value of h_1 can be found from

$$Q_{new} = Q_{rat} \frac{C_{v\ new}}{C_{v\ rat}} \tag{27}$$

where Q_{rat} is the value obtained from the printed rating table, $C_{v\ rat}$ is the velocity coefficient for the approach section assumed in developing these rating tables, and $C_{v\ new}$ is the velocity coefficient for the actual or design approach section. Values of C_v can be obtained from Figure 5 or Table 13 for each value of h_1. The value of C_d used in Figure 5 can be estimated from Equation (11). Although the rating tables were not calculated by this method, it is a reliable adjustment procedure.

Visual Observation of Modular Flow

After construction, it is important to be able to judge whether or not the flume or weir is operating properly. Actual tailwater levels often differ from those used in design. Also, required head loss calculations are not exact. As a result, properly designed flumes may sometimes exhibit nonmodular flow, negating the flow measurement. It is easy to test for modular flow if one can control the downstream water level. With a constant inflow, simply raise the downstream water level until the upstream head increases enough to cause a 1% shift in discharge. Flumes with a rapid expansion may exhibit a slight ($< 1\%$) shift in discharge between unsubmerged and partially submerged flow (i.e., tailwater level below to above crest). This is not at the modular limit.

In most cases, control of tailwater levels is not feasible. Fortunately, it is possible to determine flow conditions visually. The presence of modular or nonmodular flow can be determined by observing the water surface profile. In a normal water surface profile, drawdown begins in the approach channel and continues through the converging transition. As flow enters the throat there is a slight inflection as the long crest tries to level out the profile. The profile continues to drop across the entire crest. At the end of the crest, the profile again drops off, and then goes through a hydraulic jump on the diverging transition or tailwater channel, and rises up (Figure 30). Now as the tailwater channel backs water higher on the weir, the jump changes to rollers and the rollers begin to impinge on the throat. As soon as the water surface profile at the end of the throat is level or slightly upward, flow can no longer be considered modular (Figures 31 and 32). At still higher tailwater levels, a roller exists on the crest, with a downward sloping water surface at the end of the crest. Do not be fooled. Here flow is nonmodular, since the profile does not continually drop from the approach channel to the end of the crest.

REFERENCES

1. Ackers, P., W. R. White, J. A. Perkins, and A. J. M. Harrison. *Weirs and Flumes for Flow Measurement,* John Wiley and Sons, New York, 327 pp. (1978).
2. Bos, M. G. (Editor), *Discharge Measurement Structures,* 2nd ed. Publication No. 20, International Institute for Land Reclamation and Improvement/ILRI, Wageningen, The Netherlands, 464 pp. (1978).
3. Bos, M. G. *Long-Throated Flumes and Broad-Crested Weirs,* Martinus Nijhoff/Dr. W. Junk Publishers, Dordrecht, The Netherlands, 141 pp. (1985).
4. Bos, M. G., and Y. Reinink, "Required Head Loss over Long-Throated Flumes," *Journal of the Irrigation and Drainage Division,* American Society of Civil Engineers, Vol. 107, No. IRI, pp. 87–102 (1981).
5. Bos, M. G., J. A. Replogle, and A. J. Clemmens, *Flow Measuring Flumes for Open Channel Systems,* John Wiley and Sons, New York, 321 pp. (1984).
6. Brakensiek, D. L., H. B. Osborn, and W. J. Rawls, Coordinators, Field manual for research in agricultural hydrology, Agricultural Handbook No. 224, U.S. Department of Agriculture, U.S. Government Printing Office, Washington, DC, pp. 550 (1979).
7. Chow, V. T., *Open-Channel Hydraulics,* McGraw-Hill Book Co., New York, 680 pp. (1959).
8. Clemmens, A. J., M. G. Bos, and J. A. Replogle, "RBC Broad-Crested Weirs for Circular Sewers and Pipes," in: G. E. Stout and G. H. Davis, Eds., *Global Water: Science and Engineering—The Ven Te Chow Memorial Volume, Journal of Hydrology,* Vol. 68, pp. 349–368 (1984).
9. Clemmens, A. J., J. A. Replogle, and M. G. Bos, "Rectangular Measuring Flumes for Lined and Earthen Channels," *Jrl. of Irrig. and Drainage Engineering Am. Soc. of Civil Engineers,* 110(2): 121–137 (June 1984).
10. ICID, Recommendation for design criteria and specifications for machine-made lined irrigation canals. Committee on Irriga-

tion and Drainage Techniques, International Commission on Irrigation and Drainage Bulletin, Vol. 28, No. 2, pp. 43–55 (1979).

11. Jameson, A. H., "The Venturi Flume and the Effects of Contractions in Open Channels," *Trans. Inst. of Water Engin.*, London, England, Vol. 30, pp. 19–24 (June 1925).
12. Replogle, J. A., "Critical Flow Flumes with Complex Cross Section," *Proc., Spec. Conf. Irrig. Drain. Age Competition Resour.*, Am. Soc. Civ. Eng., pp. 366–388 (1975).
13. Replogle, J. A., "Compensating for Construction Errors in Critical-Flow Flumes and Broad-Crested Weirs," NBS Spec. Publ. (U.S.) 484 (1), pp. 201–218 (1977).
14. Replogle, J. A., "Flumes and Broad-Crested Weirs—Mathematical Modeling and Laboratory Ratings," in *Flow Measurement of Fluids*, (H. H. Dijstelbergen and E. A. Spencer, eds.), pp. 321–328, North-Holland Pub., Amsterdam (1978).
15. Replogle, J. A. and M. G. Bos, "Flow Measurement Flumes: Application to Irrigation Water Management," in: *Advances in Irrigation*, D. I. Hillel, (ed.), Academic Press, pp. 147–217 (1982).

SECTION TWO

Flow in Pipes

Head Losses at Sewer Junctions

JIRI MARSALEK*

INTRODUCTION

Stormwater, sanitary sewage and combined sewage collection systems consist of sewer pipes and various appurtenances and special structures among which sewer junction manholes are the most common. An example of a layout of a storm sewer system with various types of junction manholes is shown in Figure 1.

A properly designed sewer system must convey the maximum design flow, transport suspended solids, minimize odour nuisance and meet restrictions on hydraulic grade line elevations in the case of a surcharged system. To meet such design objectives, the sewer network has to be designed as a system in which capacities of individual sewers depend not only on sewer characteristics, but also on flow conditions at manholes and other structures.

The hydraulic design of sewer networks is based on equations of mass continuity and energy conservation. The latter equation requires consideration of two types of head losses – skin friction losses in sewer pipes and form losses at various appurtenances and special structures, such as manholes. While skin friction losses are caused primarily by viscous and turbulent shears along the conduit boundary, form losses may be caused by shear as well as pressure differentials caused by flow separation, changes in flow alignment, and drag on flow obstructions. Friction head losses have been studied extensively in the past and can be adequately characterized for design purposes. On the other hand, form losses at junction manholes are less well understood and the available information on such head losses is fairly limited. Yet in some cases, form losses at junctions may be fairly large, in comparison to friction losses, and junctions then act as bottlenecks which seriously limit the capacity of the sewer system. Under such circumstances, the sewer system becomes surcharged and this condition may lead to basement flooding or sewage overflows. Consequently, relief facilities may be required or new development halted in order to protect adjoining property. Such problems can be often avoided by minimizing form head losses in new as well as existing sewer systems.

Although junction head losses need to be considered in sewer design regardless of the design approach taken, the importance of such considerations increased in recent years with the introduction of sophisticated computerized design methods. In the traditional approach, sewer systems are designed as open-channel networks in which the hydraulic grade line does not exceed crown elevations and form head losses are not excessive. Under such circumstances, even crude approximations of junction head losses may be adequate, particularly when dealing with subcritical flows of low velocities (less than 1.5 m/s). There are, however, cases where sewer systems surcharge and increased head losses at junctions and hydraulic grade line elevations are of primary importance.

The surcharging of sewers occurs for various reasons. For example, in combined and storm sewers, surcharging is caused by the occurrence of rare storms which produce higher-than-design peak flows. During wet weather, surcharging may occur also in sanitary sewers with high infiltration and inflow. Finally, it is sometimes economical to design deep sanitary sewers (without service connections) for surcharge, or to design storm sewers for surcharge at peak design flow [17]. The surcharging of sewer systems is not necessarily harmful as long as the hydraulic grade line does not exceed the critical elevation above which flood damages or overflows occur. A proper design of surcharged systems is based on computerized pressure flow routing through the sewer network and on computations of the hydraulic grade line elevations. The sophistication and ac-

*National Water Research Institute, Burlington, Ontario, Canada

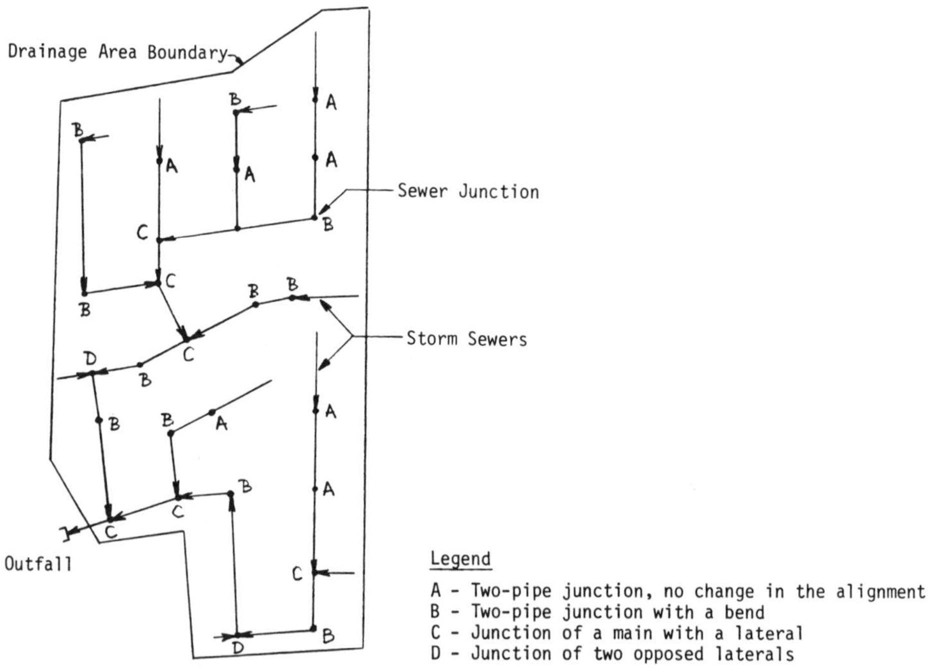

FIGURE 1. Storm sewer system layout with various types of sewer junctions.

curacy of such calculations is defeated by neglect or improper consideration of junction head losses.

The design of junctions of combined or sanitary sewers should not be limited to hydraulic computations only, but it should also consider sulphide gas releases at junctions. Dissolved sulphide in wastewater tends to pass into air from exposed surfaces. The rate of sulphide gas releases is proportional to the degree of flow turbulence [17] which may be particularly high at sewer junctions. Escaping sulphide, usually in the form of hydrogen sulphide, causes odour nuisances and produces lethal atmosphere in sewers. To minimize such problems, the junction's susceptibility to sulphide releases should be also considered in junction design.

It follows from the preceding discussion that head losses at sewer junctions warrant full consideration in sewer design, particularly when dealing with surcharged systems. Guidance for determination of junction head losses is given in the following sections. The presentation of material starts with a general discussion of the hydraulics of sewer junctions followed by design data for individual types of junctions.

HYDRAULICS OF SEWER JUNCTIONS

General problems of the flow through channel junctions were discussed by Chow [4] who concluded that the flow through junctions was a rather complicated problem whose generalization by analytical means was not possible and the best solutions would be found by experimental studies of individual junction designs. This suggestion agrees well with the literature survey findings which indicate that experimental investigations, usually done in scale models, are the most common approach to the study of junction hydraulics. Another approach is the application of the momentum equation. Both approaches are discussed later in this section.

As mentioned in the Introduction, sewer junction head losses are particularly important in surcharged sewer systems with pressure flow in sewers but a free water surface at junction manholes. For this reason, the majority of experimental studies dealt only with surcharged sewers and, consequently, the discussion in this chapter also concentrates on problems of junctions of surcharged sewers. Whenever available, additional information on junctions with open-channel flow in all branches is also presented.

The presentation of material in this section starts with basic definitions, followed by application of the momentum equation to sewer junctions, and basic considerations for experimental studies of sewer junctions.

Basic Definitions

A sketch of flow conditions at a common junction of a main with a perpendicular lateral is shown in Figure 2. This example is used for a general discussion of flow problems at junctions and introduction of basic definitions.

Flows entering the junction are subject to an energy head loss which comprises various loss components depending on flow conditions. In both pressure and open-channel flows such loss components may include losses at the junction entrance (a sudden expansion), losses due to turbulence inside the junction, losses due to the flow direction change, losses at the outlet (a sudden contraction), and losses due to increased turbulence downstream of the junction. In open-channel flow, additional losses include those caused by flow deceleration upstream of the junction, and flow acceleration and surface waves downstream of the junction. Because it is impractical and often even impossible to separate and evaluate the individual loss components, they are lumped together and referred to as the junction energy head loss, or simply the junction head loss. Such a loss is plotted as a sudden drop in the projected upstream energy grade line above the centre of the junction (see Figure 2).

In design of surcharged sewer systems, it is not sufficient to make only the energy calculations, but it is also required to calculate the hydraulic grade line elevations for the entire system in order to check whether they meet design restrictions. Consequently, it is necessary to establish pressure changes caused by flow conditions at junctions. For this reason, both head losses and pressure changes are discussed in this chapter.

It is customary to express both the junction head loss, ΔE, and pressure change, ΔP, in terms of the outfall velocity head in the following form:

$$\Delta E = K \frac{v_o^2}{2g} \tag{1}$$

$$\Delta P = K_p \frac{v_o^2}{2g} \tag{2}$$

where K is the head loss coefficient, K_p is the pressure

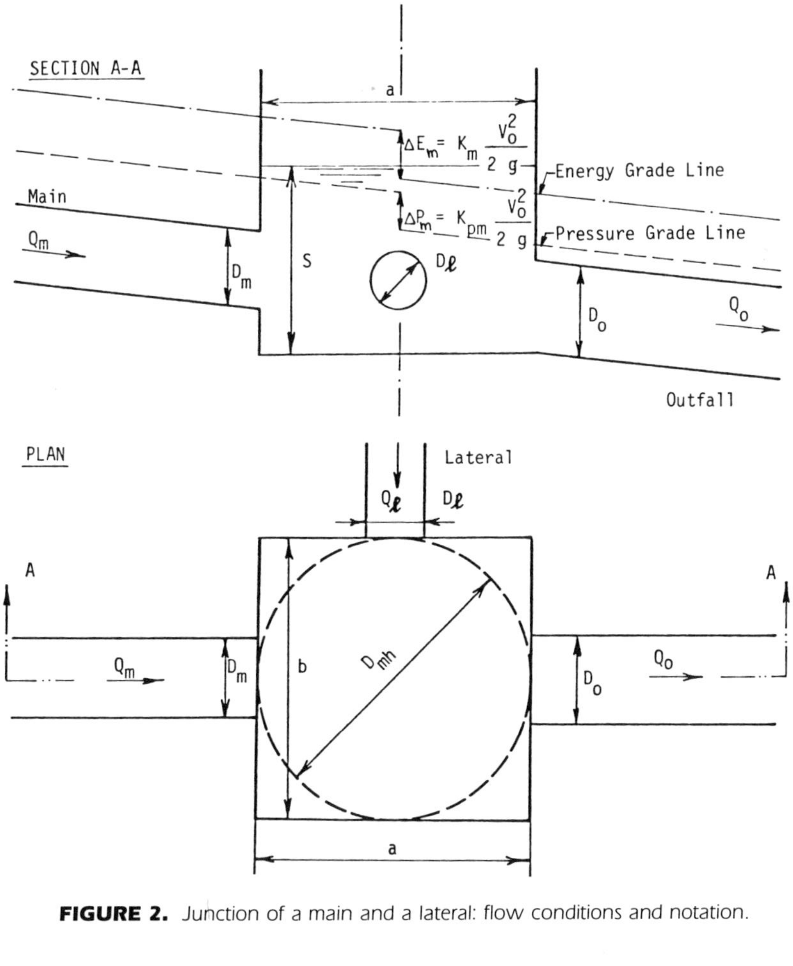

FIGURE 2. Junction of a main and a lateral: flow conditions and notation.

change coefficient, v_o is the outfall mean flow velocity, and g is the acceleration due to gravity. Because both head losses and pressure changes are considered for all inflow pipes, Equations (1) and (2) are applied to all inflow pipes and appropriate subscripts are introduced for ΔE, ΔP, K and K_p. Most of the discussion in the following sections then concentrates on establishing coefficients K and K_p for various junction designs.

Applications of the Momentum Equation to Sewer Junctions

Theoretically, the impact head loss at a junction may be computed by application of the momentum principle. This principle states that, for a particular direction x, the sum of external forces (ΣF_x) acting on the junction water body is equal to the change in momentum flux through the junction [4]:

$$\Sigma F_x = K_o \varrho Q_o v_o - \sum_{i=1}^{n} K_i \varrho Q_i v_i \qquad (3)$$

where K is the momentum flux correction coefficient, ϱ is the mass density, Q is the branch discharge, v is the mean branch velocity, and subscripts o and i refer to the outfall and inlet branches ($i = 1$ to n), respectively. External forces include pressure forces at the inlets and outfall, forces transmitted to the fluid from the boundary which confines the flow, and friction forces. Expressions analogous to Equation (3) could be written for other directions.

The pressure head change is introduced into Equation (3) as the difference in pressure heads between the inflow and outfall sections.

In practice, the application of Equation (3) was further simplified by derivation of the so-called Thompson's formula which states that the summation of all pressure forces acting on the junction water body, ignoring friction, is equal to the average cross-sectional area through the junction, multiplied by the change in the hydraulic gradient through the junction [14]. The product of the average cross-sectional area and the pressure change then equals the change in the momentum flux.

The accuracy of the momentum equation applications to junction problems may be questioned. In particular, the pressure components, in the direction of flow, along the walls and floor of channels cannot be determined accurately, because of the effects of the impact of streams and the curvature of channels [17]. Problems with applications of the momentum equation are further demonstrated below for a surcharged straight-flow-through junction shown in Figure 3.

Considering the junction shown in Figure 3, the momentum equation can be written as follows:

$$P_m + P_{uw} - P_{dw} - P_o - F_f = K_o \varrho Q_o v_o - K_m \varrho Q_m v_m \qquad (4)$$

where P_m is the hydrostatic force, F_f is the boundary drag force, K is the momentum correction coefficient, ϱ is the mass density, Q is the discharge, v is the mean branch velocity, and subscripts m, o, uw and dw refer to the main, outfall, upstream junction wall and downstream junction wall, respectively. Equation (4) can be further simplified by the following considerations:

1. $P_{uw} = P_{dw}$, thus $P_{uw} - P_{dw} = 0$.
2. The drag forces exerted by junction walls are usually neglected in comparison to pressure forces, thus $F_f \sim 0$.
3. It follows from the junction geometry that $Q_m = Q_o = v_o \pi D^2/4$, and $v_m = v_o$.

Using the above simplifications and substitutions, Equation (4) can be written as:

$$\gamma\,(h - D/2)\frac{\pi D^2}{4} - \gamma\,(h - \Delta P - D/2)\frac{\pi D^2}{4} = (K_o - K_m)\varrho \frac{\pi D^2}{4} v_o^2 \qquad (5)$$

where γ is the specific weight, h is the hydrostatic grade line elevation above the junction floor and ΔP is the pressure change at the junction. After solving Equation (5) for ΔP, the following final expression is obtained:

$$\Delta P = 2(K_o - K_m)\frac{v_o^2}{2g} = K_p \frac{v_o^2}{2g} \qquad (6)$$

where $K_p = 2(K_o - K_m)$.

It is obvious from Equation (6) that if both momentum flux correction coefficients were equal, or neglected as commonly done in engineering applications, Equation (6) would indicate no pressure change at the junction. Yet experimental observations presented later for the junction shown in Figure 3 clearly indicate a pressure drop at the junction.

The evaluation of the pressure change coefficient K_p by means of Equation (6) would require the knowledge of velocity distributions in the inflow and outfall sections of the junction. From such distributions, K_m and K_o can be determined. In turbulent flow, K typically varies from 1.01 to 1.07 [4]. Assuming the lower value for the inflow section and the higher value for the outfall section, the pressure change coefficient would be calculated as $K_p = 2(1.07 - 1.01) = 0.12$. Such a calculated value has a correct order of magnitude in relation to the experimental values.

It appears from the preceding discussion that although the application of the momentum equation offers some insight into the hydraulics of junctions, it does not provide a universal solution because of inherent inaccuracies. Such difficulties with analytical approaches then contributed to the popularity of experimental studies as the most common approach to investigations of junction hydraulics.

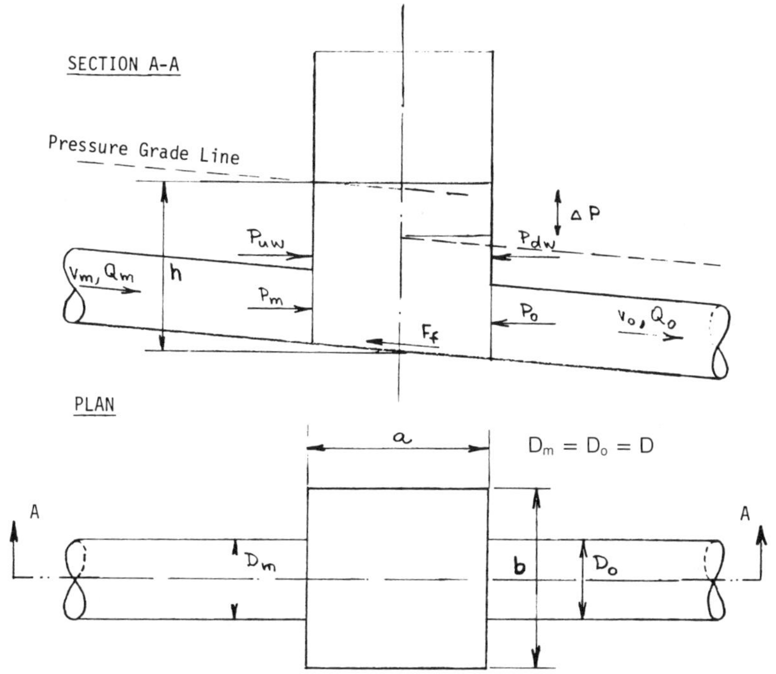

FIGURE 3. Application of the momentum equation to a straight-flow-through junction.

Experimental Studies of Sewer Junctions

Head loss and pressure change coefficients used in sewer design have been typically determined from observations at actual or model sewer junctions. Because of difficulties with observations at actual sewer installations [1], almost all experimental observations have been done in laboratory scale models of sewer junctions. Before presenting the results of such studies, it is desirable to examine experimental variables, their practical range of values, and scaling of model data to the prototype scale.

EXPERIMENTAL VARIABLES

In experimental studies of sewer junctions, it is useful to start with a dimensional analysis of the problem. Considering the surcharged sewer junction shown in Figure 2, the head loss coefficient K can be expressed as a function of the following variables:

$$K = f_1 (\varrho, \mu, g, Q_o, Q_\ell, S, a, b, D_m, D_\ell, D_o) \quad (7)$$

where f_1 is a function, ϱ is the fluid density, μ is the fluid viscosity, g is the acceleration due to gravity, Q_o is the outlet pipe discharge, Q_ℓ is the lateral pipe discharge (the main pipe discharge $Q_m = Q_o - Q_\ell$), S is the water depth at the junction, a is the junction length, b is the junction width, and D_m, D_ℓ, D_o are the diameters of the main, lateral and outlet pipes, respectively. Dimensional analysis then yields the following expression for the head loss coefficient:

$$K = f_2 \left(\frac{Q_o}{g^{1/2} D_o^{5/2}}, \frac{\varrho Q_o}{\mu D_o}, \frac{Q_\ell}{Q_o}, \frac{S}{D_o}, \frac{a}{D_o}, \frac{b}{D_o}, \frac{D_m}{D_o}, \frac{D_\ell}{D_o} \right) \quad (8)$$

Equation (8) can be further modified by substituting $v_o = 4Q_o/\pi D_o^2$ and $\nu = \mu/\varrho$ and by adding the junction benching to the list of independent variables. For simplicity, the benching geometry is described by a dimensionless factor B to avoid the introduction of several geometrical parameters needed to describe various benching shapes. After these modifications, Equation (8) can be written as

$$K = f \left(\frac{v_o}{\sqrt{gD_o}}, \frac{v_o D_o}{\nu}, \frac{Q_\ell}{Q_o}, \frac{S}{D_o}, \frac{a}{D_o}, \frac{b}{D_o}, \frac{D_m}{D_o}, \frac{D_\ell}{D_o}, B \right) \quad (9)$$

Note that Equation (9) applies to junctions with a rectangular base. For junctions with a round base, the terms a/D_o

and b/D_o would be replaced by a single term D_{mh}/D_o, where D_{mh} is the manhole diameter.

Similar procedures would be used to derive expressions for a more general case with three inflow pipes for which no experimental data were found in the literature survey. Note also that an expression analogous to Equation (9) would be obtained for the pressure change coefficient K_p.

A discussion of independent variables listed in Equation (9) follows. This discussion focuses on the importance of individual variables in sewer junction design.

The first term, $v_o/\sqrt{gD_o}$ is the Froude number written for the outfall. Earlier studies indicate that for surcharged sewers K does not depend on the Froude number [13,16]. On the other hand, the operation of junctions with open-channel flow is affected by the Froude number of branch flows [15]. Thus, the Froude number may be neglected in design of surcharged sewer junctions, but it should be considered in design of junctions with open-channel flow in branch pipes.

The second term, $R = v_o D_o/\nu$, is the Reynolds number written for the outfall. Experimental studies indicate that K does not depend on the Reynolds number if R is greater than 10^4 [3,5]. This condition is always met in practice, because specifications of the minimum pipe diameter (0.3 m) and the minimum flow velocity (0.61 m/s) lead to Reynolds numbers greater than 10^5. In order to eliminate R from further considerations as an independent variable, it is necessary to undertake scale model studies for R greater than 10^4.

The third term is the relative lateral inflow. This term defines implicitly the relative main inflow as $Q_m/Q_o = 1 - Q_l/Q_o$. In junctions with more than one lateral, more than one lateral inflow would need to be considered. Experimental studies indicate that relative inflows are very important variables affecting junction head losses and pressure changes [15,16].

The fourth term is the relative junction submergence indicating the depth of water at the surcharged junction. This term is important for operation of certain junctions, such as the straight-flow-through junction manholes with a round base [9,10]. For all other junction designs discussed later, no evidence of the effects of submergence on K or K_p was found in the literature [16].

The remaining five independent variables in Equation (9) are geometrical terms. The first two, a/D_o and b/D_o describe the junction width and length. For round manholes, these two terms can be replaced by a single term D_{mh}/D_o. The manhole width and length have some influence on junction head losses. Such influences were observed in some laboratory studies in which both these parameters were varied over a very wide range of values [10,15,16]. From the practical point of view, the variations in relative manhole sizes are somewhat limited. Considering the most commonly used standard manhole, which is a 1.22 m (4 ft) prefabricated round-base manhole, the pipe sizes applied with this manhole range from 0.2 m to 0.61 m and the corresponding D_m/D_o ratio varies from 2 to 6. Experimental investigations indicate that the effects of the relative manhole size on K are limited for D_{mh}/D_o or b/D_o greater than 2.0 [16].

The next two terms, D_m/D_o and D_l/D_o, indicate the relative sizes of inflow pipes. These terms are important in considerations of head losses because, in conjunction with relative inflows, they determine the momentum fluxes into the junction. Such fluxes then strongly affect head losses and pressure changes at the junction [16]. For junctions with more than one lateral, additional relative lateral sizes would be introduced.

Finally, the last term, which is denoted as benching, refers to channels inside the junction. The main function of such channels is to guide flows smoothly through the junction and thereby to reduce head losses. Horizontal benches incorporated in such designs are helpful for maintenance operations. Benchings were found very important in considerations of head losses at sewer junctions [13]. Although details of junction benchings are given later, a simple classification of practical benching designs is given here. For better understanding of descriptions of individual designs and their applications to various junction designs, reference is made to Figures 4, 5, 10 and 16.

Benching *B*1 represents the simplest design in which no benching or flow guidance is provided at the junction. This design is expected to produce the largest head losses and for that reason should be avoided in practical design.

Benching *B*2 is formed by extending the lower half of inflow pipes through the junction and adding horizontal or slightly sloping benches at the top of the semicircular channel. This design is used in municipal design practice and should yield lower losses than *B*1.

Benching *B*3 is an improved version of design *B*2. This improvement is achieved by extending the semicircular channels vertically to the pipe crown, thus forming an U-channel, and placing the benches at that level. This design should provide a better flow guidance and lower head losses than *B*2.

The last design considered, *B*4, is a special design attempting to further reduce head losses. In principle, it is the *B*3 design with an expanded flow cross section through the junction. For this purpose, pipe expanders and reducers are installed upstream and downstream of the junction. Because of increased costs of this design, its use would be limited to critical cases where large reductions of junction head losses are needed [13].

In summary, the eleven independent variables listed in Equation (9) affect junction head losses to various degrees. For surcharged junctions, the most important hydraulic variables seem to be the relative inflows. Among the geometrical parameters, junction benchings and branch pipe diameters are particularly important. The effects of the remaining variables on junction head losses or pressure changes are relatively limited. For open channel flow junctions, the considerations of independent variables would be

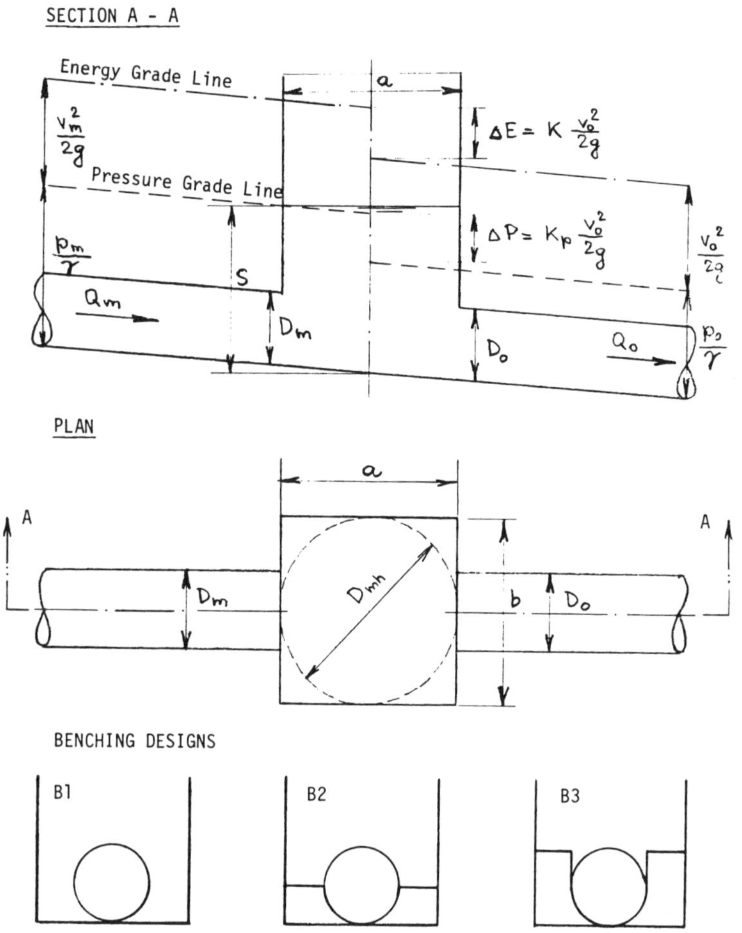

FIGURE 4. Straight-flow-through junction.

even more complicated. In Equation (9), the relative submergence would be irrelevant, but additional terms, such as the depths of flow in branches and Froude numbers for branches, would have to be added.

SCALING OF MODEL RESULTS

As mentioned earlier, almost all experimental data on head losses at sewer junctions were observed in scale models. Consequently, it is necessary to establish whether such observations are affected by the scale of the model and whether they need to be scaled up for applications to the prototype.

In general, head loss or pressure change coefficients measured in model junctions are directly transferable to the prototype, if the dimensionless parameters listed in Equation (9) are identical for both the model and the prototype. Although the identity of all parameters listed in Equation (9) is not feasible in all cases, an approximate identity can be attained for special cases discussed below.

For many cases, Equation (9) may be substantially simplified. For example, for surcharged junctions with high Reynolds numbers (greater than 10^4) and higher surcharges ($S/D_o > 1.3$), the list of independent variables in Equation (9) may be reduced to relative inflows and junction geometric parameters. In such cases, if the relative inflows (Q_l/Q_o's) in the model and prototype are identical and the model is geometrically similar to the prototype, the head or pressure change coefficients observed in the model are directly transferable to the prototype.

Scaling effects were studied for manholes with a 90° bend using two different size models. On the average, the head losses coefficients observed in both models differed by less than 4% [13]. Considering uncertainties involved in such observations, no scaling effects were detected.

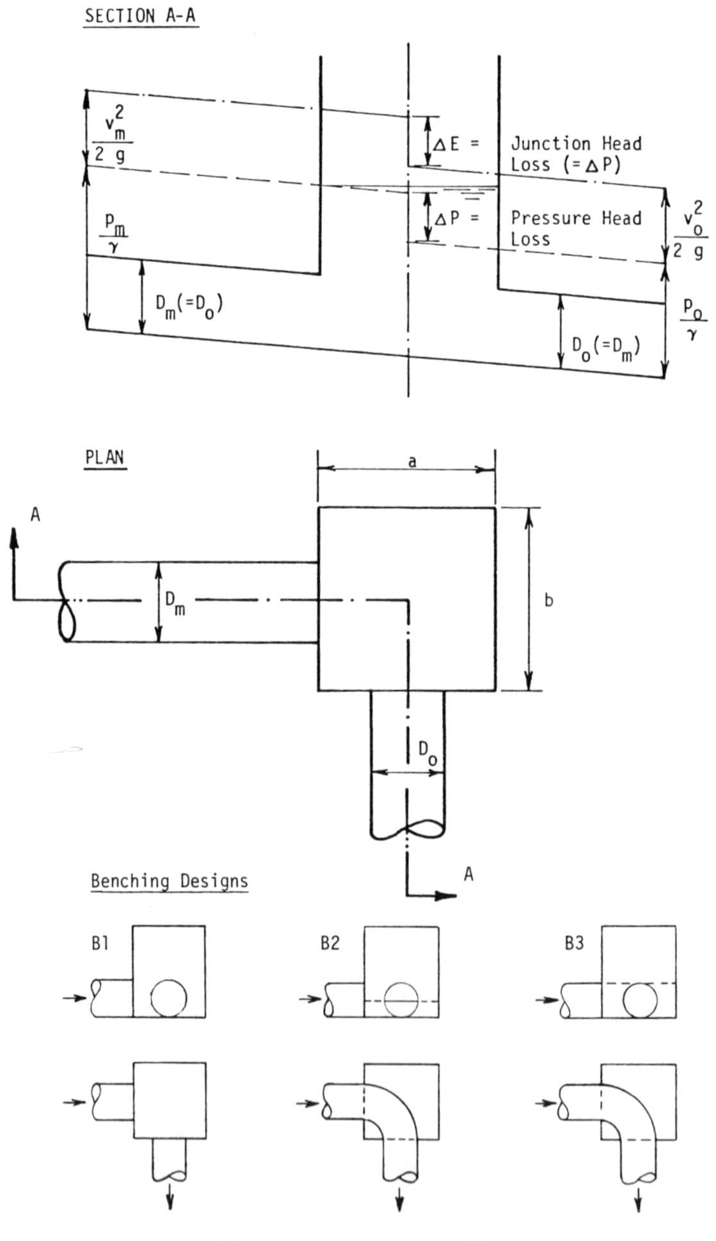

FIGURE 5. Manhole with a 90° bend.

Considerations of model similarity for open-channel flow junctions are more complicated than in the former case, because of additional variables affecting the operation of such junctions. The findings reported in the literature [15] indicate that the gravity forces tend to dominate the junction flow processes and that the viscous forces may be neglected. Consequently, the Froude similarity would apply. For this similarity, the head and pressure change coefficients are directly transferable from the model to the prototype for identical Froude numbers in the model and prototype.

DESIGN DATA

Considering the preceding discussion of the hydraulics of sewer junctions, experimental data from junction models represent the best source of information for evaluation of junction head losses and pressure changes. Such data have been compiled from numerous sources and presented in this section. Wherever required, the literature data are further interpreted and extrapolated to enhance their usefulness for practical design.

Sewer junctions are classified here into the following four categories:

1. Two-Pipe Junctions
2. Three-Pipe Junctions
3. Four-Pipe Junctions
4. Special Junction Structures

Each of the above categories has a number of subcategories which are discussed later. The presentation of data starts with the description of data sources, followed by data listings and the listing of recommended design values.

Two-Pipe Junctions

Two-pipe junctions are the simplest junctions characterized by a single inflow pipe, referred to as the main, and a single outfall pipe. These junctions can be further divided into straight-flow-through junctions without any change in the pipe alignment and junctions with a change in flow alignment. Further discussion of both junction types follows.

STRAIGHT-FLOW-THROUGH JUNCTIONS

Straight-flow-through junctions are shown schematically in Figure 4. In the simplest form, sewer pipes upstream and downstream of the junction manhole maintain the same alignment and pipe diameter. Such manholes are installed for maintenance purposes, or where the pipe slope changes.

Head losses and pressure changes at straight-flow-through junctions are primarily caused by changes in the flow channel geometry at the junction. For surcharged junctions, the head loss coefficient can be expressed as

$$K = f(S/D_o,\ b/D_o(\text{or } D_{mh}/D_o),\ B) \tag{10}$$

For open-channel flow junctions, the functional expression would be similar to Equation (10), but the head losses are so small that they hardly warrant a detailed consideration.

The literature survey identified six sources of data on straight-flow-through junctions. Some of the reported data were observed in junctions of a main and lateral as a limiting case with no lateral inflow. A brief description of data sources is given in Table 1.

It appears from the above sources that data for straight-flow-through junctions are fairly extensive and consistent. In general, the reported head loss coefficients ($K = K_p$, for $D_m = D_o$) vary from 0.05 to 0.35 for junctions without special flow phenomena present at the junction. The findings regarding various variables affecting head losses and pressure changes at straight-flow-through junctions are summarized below.

In pressurized flows, the head loss or pressure change coefficient varies with the surcharge depth, base shape, manhole width, and benching. The effects of the surcharge depth are perhaps the least understood. For smaller surcharge depths ($S/D_o < 2$), these depths seem to affect head losses and pressure changes by inducing the formation of special flow patterns at the junction. The formation of these patterns is further affected by the manhole width base shape, and benching. In particular, for round-base manholes with a half-pipe benching (*B*2) and certain ranges of relative manhole widths and depths of surcharge, the formation of strong swirls or periodic sloshing inside the junction was reported [7,9,10]. Such phenomena increased the head loss coefficient values to levels as high as 0.9. The highest values were observed for $S/D_o = 1.5$ (9) and they dropped to normal values (<0.35) for higher depths of surcharge ($S/D_o > 2$). Similar phenomena were observed even for higher benchings placed asymmetrically inside wide junction manholes [7].

TABLE 1. Sources of Data on Head Losses at Straight-Flow-Through Junctions.

Author	Type of Flow	Benching	Reference
Archer et al.	P[1]	B2	2
Howarth and Saul	P	B2, special design	7
Liebmann	P, OC[2]	special design	8
Lindvall	P	B2	10
Marsalek	P,OC	B1, B2 and B3	11
Sangster et al.	P	B1	16

[1]Pressurized flow.
[2]Open-channel flow.

From the practical point of view, sudden increases in head losses at straight-flow-through junctions can be prevented by using a full pipe benching ($B3$) placed symmetrically inside the junction manhole. In any case, the sudden increases in head losses represent a transient phenomenon which would increase the junction surcharge depth which then leads to the breakdown of the junction swirl or sloshing and concomitant reduction in head losses.

The effects of the manhole base shape are not significant except for the fact that the round base manholes are clearly more susceptible to the formation of junction swirls or sloshing [7].

Head losses seem to increase slightly with the relative manhole width [10,16]. Larger manholes are more susceptible to the formation of swirls or sloshing motion.

The benching has a strong influence on head losses at sewer junctions [11]. Junctions without any benching exhibit the highest head losses. Such losses can be substantially reduced by installing a full pipe benching ($B3$) at the junction. This design also seems to avoid the formation of swirls or sloshing motion at the junction.

Head losses in open-channel flow junctions are generally much lower than those at surcharged junctions. Limited data were found in the literature for subcritical flows and average depths of flow. When dealing with supercritical flows, it is required to check whether a hydraulic jump forms at the junction. Standard formulas derived from the momentum Equation (4) can be used for this purpose.

Head loss and pressure change coefficients for design of straight-flow-through junctions are presented in Table 2.

Finally, the straight-flow-through junctions with a change in the pipe diameter need to be also considered. Since many design criteria do not allow reductions in the pipe diameter in the downstream direction, only the case of pipe expansion ($D_o > D_m$) is of interest. For this case, the pressure change coefficient can be derived from the momentum equation in the following form [16]:

$$K_p = 2[1 - (D_o/D_m)^2] \qquad (11)$$

Experimental verification of Equation (11) produced an acceptable agreement between the observed and calculated values for D_m/D_o greater than 0.53 and smaller than 1.0.

For $D_m \neq D_o$, the head loss and pressure change coefficients are not equal. The relationship between both coefficients can be derived as

$$K = K_p - 1 + (D_o/D_m)^4 \qquad (12)$$

and after substituting from Equation (11), the final expression for K was obtained in the form

$$K = (D_o/D_m)^4 - 2(D_o/D_m)^2 + 1 \qquad (13)$$

Note that for $D_o > D_m$, Equation (11) yields negative K_p's and the pressure grade line elevation increases downstream of the junction. For $D_o \sim D_m$, Equation (13) yields $K \sim O$ and obviously underestimates head losses. This is caused by simplifying assumptions used in the derivation of Equation (11).

TWO-PIPE JUNCTION WITH A CHANGE IN ALIGNMENT

Although the pipe alignment can be changed gradually by using curved sewers, it is much more practical to change the alignment abruptly at a junction manhole. Such changes are described by the deflection angle θ which typically varies from 22.5° to 90°. An example of a two-pipe junction with a 90° bend is shown in Figure 5.

Head losses at surcharged junctions with a bend are caused not only by changes in the flow cross section at the junction, but also by changes in the flow direction. For the limiting case of θ = 90°, it appears that the head loss

TABLE 2. Head Loss and Pressure Change Coefficients for Straight-Flow-Through Junctions ($D_o = D_m$).

Junction Manhold				K
Base Shape	Relative Size b/D_o, D_{mh}/D_o	Benching	Flow	(= K_p for Pressurized Flow)
Square, Round	2–5	B1	Pressurized	0.15–0.30[1]
Square, Round	2–5	B2	Pressurized	0.15–0.25[1,2]
Square, Round	2 5	B3	Pressurized	0.10–0.15[1]
Square, Round	2–5	B1	Open-Channel	0.15[3]
Square, Round	2–5	B2	Open-Channel	0.10[3]
Square, Round	2–5	B3	Open-Channel	0.05[3]

[1]Lower values should be used for small b/D_o's or D_{mh}/D_o's and vice versa.
[2]For round-base manholes, this value may increase up to 0.9 for the range of submergence depths from 1.2 to 2.0.
[3]Approximate values obtained as averages of observations for various depths and a subcritical flow.

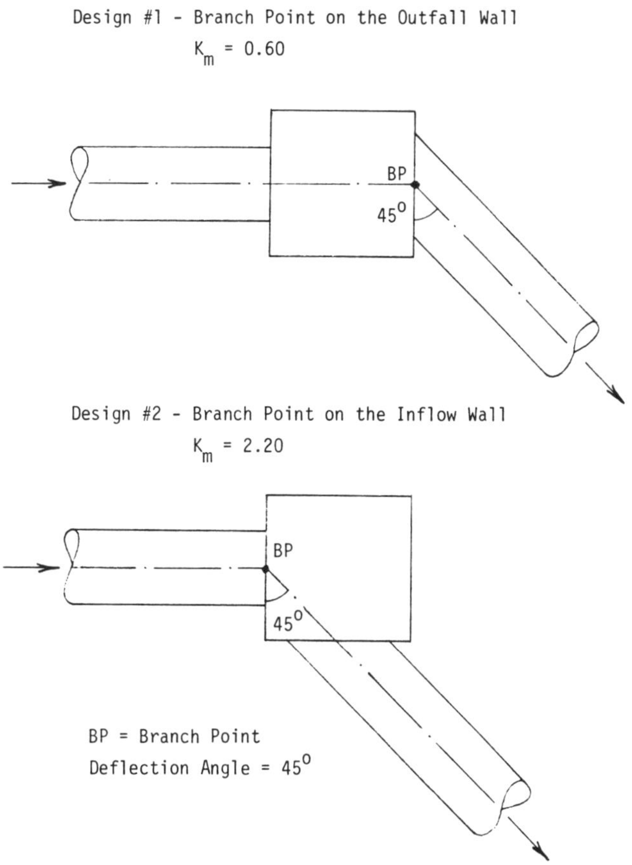

FIGURE 6. Manhole layouts with a 45° bend and different locations of branch points [6].

coefficient depends primarily on the junction benching [13]. Other factors which were listed in Equation (10) for straight-flow-through junctions, such as S/D_o, b/D_o (or D_{mh}/D_o), and the base shape, have only minor influence.

Another factor affecting head losses at manholes with a bend is the location of the branch point at which the inflow and outfall pipe axes intersect. The lowest losses were found for branch points on the outfall wall [6] as shown in Figure 6 for two manhole layouts. Both junctions in Figure 6 have the same deflection angle of 45°, but significantly different pressure change coefficients. The design with the branch point located on the inflow wall has $K_p = 2.20$ ($D_m = D_o$) and the design with the branch point on the outfall wall has $K_p = 0.60$ ($D_m = D_o$). The hydraulic effectiveness of the latter design is obvious.

In summary, it appears that head loss and pressure change coefficients for surcharged two-pipe junctions with a bend depend primarily on the deflection angle, junction benching, and location of the branch point. Junctions with open-channel flow would behave similarly and their head loss would be also affected by the Froude numbers of the inflow and outfall.

A literature survey identified three sources of information on head or pressure changes at junctions with a bend [1,6,13]. The basic data from these sources are summarized in Table 3.

It is obvious from Table 3 that none of the existing data sets covers a full range of both major variables, the deflection angle and benching. Consequently, it is necessary to use the existing data for interpolations and extrapolations of missing values. Toward this end, the appropriate data from the references listed in Table 3 were plotted in Figure 7 and used in interpolations and extrapolations. Such procedures increase uncertainties in the final design data which are presented in Table 4 for various deflection angles and benchings.

It appears from the data in Table 4 that head losses and pressure changes depend strongly on both the deflection an-

TABLE 3. Sources of Data on Head Losses and Pressure Changes at Junctions with Bend (D_m = D_p).

	Junction Characteristics						
Author	Type of Flow	θ	Base	b/D_o (D_{mh}/D_o)	Benching	$K = K_p$	Reference
Archer et al.	P[1]	30°	RC[3]	5	B3	0.40	2
	P	60°	RC	5	B3	0.85	2
	P	30°	Rd[4]	5	B3	0.50	2
	P	60°	Rd	5	B3	0.95	2
Hare	P	22.5°	Sq[5]	2	B1	0.30	6
	P	45°	Sq	2	B1	0.60	6
	P	45°	Sq	2	B1	2.20	6
	P	67.5°	Sq	2	B1	2.00	6
	P	90°	Sq	2	B1	1.85	6
Marsalek	P	90°	Sq,Rd	2.3,4.6	B1	1.75	13
	P	90°	Sq,Rd	2.3,4.6	B2	1.65	13
	P	90°	Sq,Rd	2.3,4.6	B3	1.10	13
	P	90°	Sq,Rd	2.3,4.6	B4	0.65	13
	OC[2]	90°	Sq,Rd	2.3,4.6	B1	1.10	13
	OC	90°	Sq,Rd	2.3,4.6	B2	0.60	13
	OC	90°	Sq,Rd	2.3,4.6	B3	0.30	13

[1]Pressurized flow.
[2]Open-channel flow.
[3]Rectangular.
[4]Round.
[5]Square.

TABLE 4. Head Loss and Pressure Change Coefficients for Surcharged Junctions with a Bend (D_m = D_o).

	Deflection Angle θ								
	30°			60°			90°		
Benching	B1	B2	B3	B1	B2	B3	B1	B2	B3
$K = K_p$	0.90	0.80	0.50	1.35	1.25	0.85	1.85	1.65	1.10

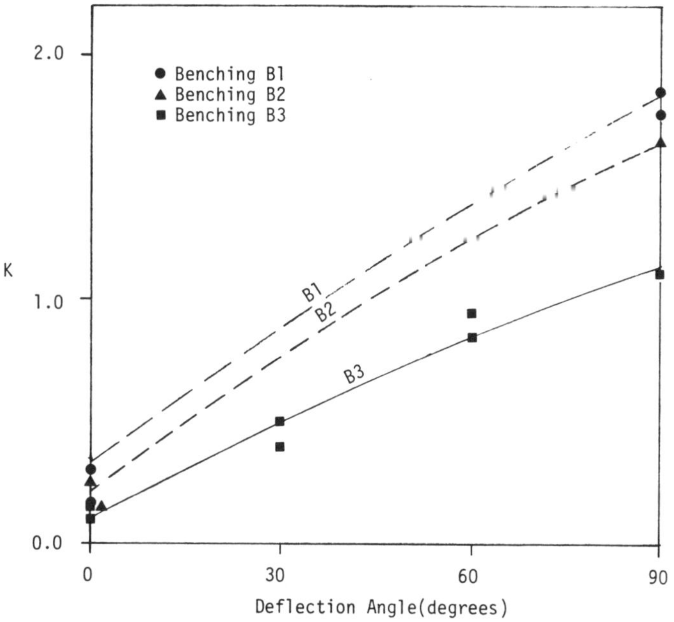

FIGURE 7. Head loss coefficients for various deflection angles and benchings.

gle and benching. The finding about the benching effects is particulary important because it can be used to develop a special low head loss design [13]. Such a design, characterized by the full depth benching and expanded flow cross sections at the junction, is shown in Figure 8 for $\theta = 90°$. This design produced a low head loss coefficient of 0.65.

The above data were produced for identical pipe diameters upstream and downstream of the junction. The junctions with changes in both pipe alignment and diameter were also studied by some researchers [6,16] and the pertinent data are listed in Table 5.

The data in Table 5 indicate that for high deflection angles and branch points upstream of the outfall wall, there is little variation observed in K_p's. In a conservative approach, it is possible to adopt a constant K_p (for $D_m = D_o$) and apply it to all values of D_m/D_o. Another possibility is to reduce K_p for lower values of D_m/D_0 as indicated by trends in Table 5. The corresponding K would be calculated from Equation (13).

The available experimental data for head losses at junctions with bends and open-channel flow conditions are very limited. The only case studied was a 90° bend and three junction benchings [13]. For subcritical flows and full benching, there are no changes in flow areas at the junction and the head losses arise only from changes in the flow direction.

In comparison to surcharged junctions, losses at junctions with open-channel flow are significantly smaller. Consequently, even crude approximations of the head loss coefficient obtained from interpolations of data for $\theta = 0°$ and $\theta = 90°$ may be acceptable. Both observed and interpolated K's are given in Table 6 and recommended for use in junction design. The coefficients given in Table 6 represent averages for various depths of flow.

Three-Pipe Junctions

Among three-pipe junctions, the most common are T-junctions of a main and lateral, or junctions of two opposed laterals. Less common are Y-junctions.

A general analysis of surcharged three-pipe junctions indicates that the head loss or pressure change coefficients can be described in the following form:

$$K, K_p = f(Q_\ell/Q_o, \theta, b/D_o \text{ or } D_{mh}/D_o, D_m/D_o, D_\ell/D_o, \text{base shape}, B) \tag{14}$$

where θ is the lateral branch angle.

Thus for pressurized flow the junction head loss or pressure change coefficients depend only on one hydraulic variable, the relative lateral inflow Q_ℓ/Q_o, and several geometric characteristics of the junction and sewer pipes. Note the absence of the junction surcharge depth which had some

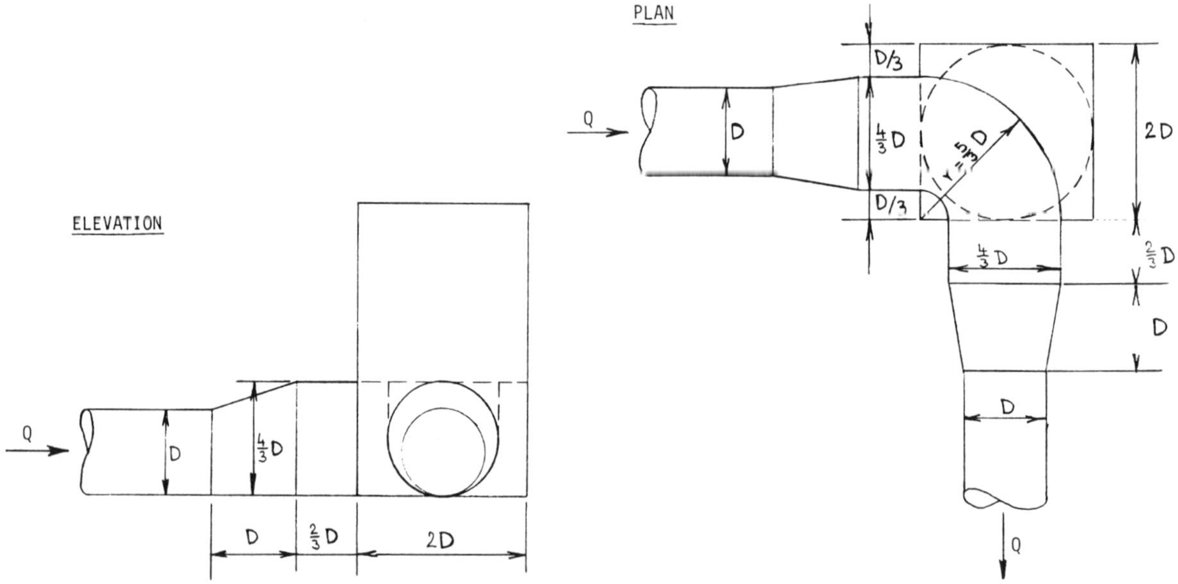

FIGURE 8. Low head loss design for manholes with a 90° bend (K = 0.65; [13]).

effects on operation of certain two-pipe junctions. Among the geometric parameters, the most important ones are the benching [13], relative pipe sizes [16], and possibly the lateral branch angle θ. Other geometrical parameters, such as the relative manhole size and base shape, are barely significant.

The discussion of T-junctions starts with junctions of a main with perpendicular lateral followed by junctions with two opposed laterals.

TABLE 5. Pressure Change Coefficients for Junctions with Changes in Pipe Alignment and Diameter [6,16].

	K_p						
	Deflection Angle						
D_m/D_o	22.5°	45°[1]	45°[2]	67.5°	90°	90°[3]	90°[3]
0.52						1.71	1.91
0.65						2.20	1.87
0.70	−1.6	−0.9	2.05	1.70	1.50		
0.80	−0.6	0.0	2.10	1.80	1.65		
0.83						2.31	2.20
0.90	0.0	0.45	2.15	1.90	1.75		
1.00	0.30	0.60	2.20	2.00	1.85	2.00	2.00

[1]Branch point on the outfall wall.
[2]Branch point on the inflow wall.
[3]Data from Reference [16] for very narrow manholes. All other data were adopted from Reference [6].

JUNCTIONS OF A MAIN WITH LATERAL

Four sources of data for junctions of a main with perpendicular lateral are listed in Table 7. It appears that such data are fairly extensive and provide a good basis for junction design.

It can be inferred from Table 7 that different studies focused on different aspects of the junction problem and none of them can serve as a sole source of design data. Consequently, two design aids were prepared: a graph for junctions with various pipe diameters and tables of K_p's for junctions with comparable pipe sizes.

The earlier published design graphs [16] present K_p as a function of the relative lateral discharge Q_ℓ/Q_o and the relative main pipe diameter D_m/D_o. These graphs were prepared for surcharged junctions of a main with perpendicular lateral, and no benching at the junction. The basic graph from Reference [16] is replotted in Figure 9. This graph was prepared for junctions with comparable D_m and D_ℓ, and in practical applications, identical pressure change coefficients are assumed for both the main and lateral ($K_{pm} = K_{p\ell}$). It appears from Figure 9 that K_p increases with D_m/D_o and Q_ℓ/Q_o.

For $D_m/D_o = 1$, the data in Figure 9 were verified against some recent published data [13]. Graphs for junction benchings $B2$ and $B3$ (see Figures 10–12) were produced by extrapolation. Such extrapolated graphs are only approximate, because they are based on observations for $D_m/D_o = 1$ and the general shape of curves shown in Figure 9. The above extrapolations were done for a limited range of D_m/D_o from 0.7 to 1.0.

TABLE 6. Head Loss Coefficients for Junctions With Bends and Subcritical Open-Channel Flow ($D_m = D_o$).

	K											
	Deflection Angle θ											
	30°			60°			90°					
Benching	B1	B2	B3	B1	B2	B3	B1	B2	B3	B1	B2	B3
K	0.15	0.10	0.05	0.47	0.27	0.13	0.79	0.44	0.21	1.10	0.60	0.30

The other case of interest is characterized by comparable sizes of the main and outfall, and a similar size or smaller lateral. Such a case was studied for two D_ℓ/D_o's and three benchings [13]. Observed K_{pm}'s and $K_{p\ell}$'s were plotted in Figures 11 and 12 for $D_\ell/D_o = 0.5$ and 1.0, and extrapolated for other values.

The data in Figures 11 and 12 indicate differences in K_{pm} and $K_{p\ell}$ for smaller lateral pipe diameters. For $D_\ell/D_o = 1$, both K_{pm} and $K_{p\ell}$ are practically identical. For equal lateral discharges, smaller laterals are characterized by a higher flow momentum which leads to a greater disruption and pressure changes in the main pipe. For small laterals and low Q_ℓ/Q_o's, the pressure change coefficient for the lateral pipe is negative.

Among other parameters listed in Equation (14), the manhole base shape did not affect K_p's at all. The relative manhole size had some effect on K_p's: the coefficient values slightly increased with the manhole size. Head losses seem to increase with the lateral angle θ. This parameter, however, cannot be fully evaluated because of the lack of supporting experimental data.

Open-channel flow junctions of a main and lateral were investigated in two studies [13,15]. Both studies indicated the importance of benching for reduction of head losses. Full pipe depth benchings (*B*3) resulted in the lowest head losses. It was further found that the hydraulically effective junctions should be as short as practical and the drop between the inflow and outfall inverts should be gradual [15]. In junctions without benching, some reduction in head losses was achieved by installing hinged deflector plates [15].

Generally, head losses in open-channel flow junctions are fairly small with most *K*'s being smaller than 0.5. Some guidance for selection of *K* values for junctions with various benchings can be obtained from Figure 13. Head loss coefficients in Figure 13 represent average values obtained for various subcritical flow depths. It should be further noted in Figure 13 that for low Q_ℓ/Q_o's, the K_ℓ values are negative and the lateral flow experiences an apparent energy gain. Under such circumstances, the water is drawn from the lateral in a manner resembling ejector action. Note, however, that the energy gain is indicated by calculations in which the downstream channel energy is based on the mean flow velocity in this channel. Such a gain is only apparent because the water from lateral enters into the main flow region where the velocity is below the mean value. In any case, negative head losses, or apparent energy gains, are neglected in practical design.

JUNCTIONS OF TWO OPPOSED LATERALS

Junctions of two opposed laterals are characterized by large head losses and pressure changes and, whenever possible, their use in sewer network layouts should be avoided.

Experimental studies indicate that the head loss and pressure change coefficients for surcharged junctions of two opposed laterals are primarily dependent on the relative lateral inflow $Q_{\ell 1}/Q_o$ ($Q_{\ell 2}/D_o = 1 - Q_{\ell 1}/Q_o$), the relative

TABLE 7. Sources of Data on Head Losses and Pressure Changes at Junctions of a Main with Perpendicular Lateral.

Author	Junction Characteristics				Reference
	Flow	Benching	(D_m/D_o)	(D_ℓ/D_o)	
Lindvall	P[1]	B2,B3	1.0	0.4–1.0	10
Marsalek	P,OC[2]	B1–B3	1.0	0.5,1.0	13
Prins	OC	B1–B3	1.0	0.67	15
Sangster et al.	P	B1	0.5–1.3	0.5–1.0	16

[1]Pressurized flow.
[2]Open-channel flow.

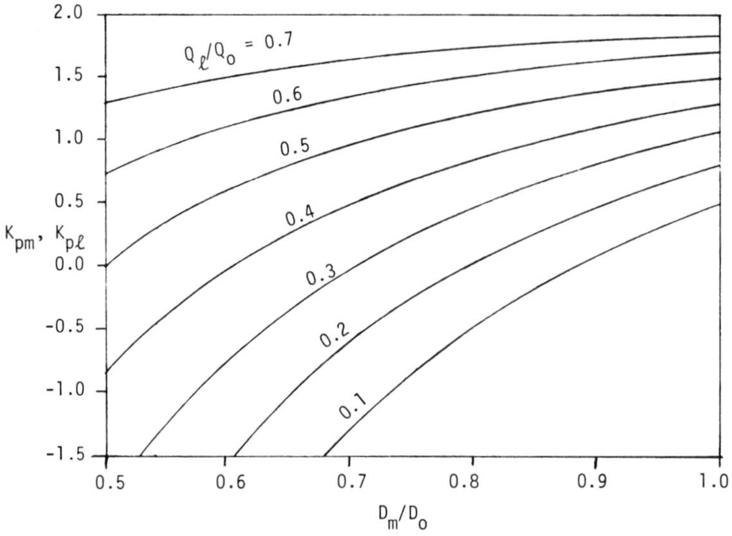

FIGURE 9. Pressure change coefficients for junctions of a main and a lateral without benching ($D_m \sim D_\ell$; [16]).

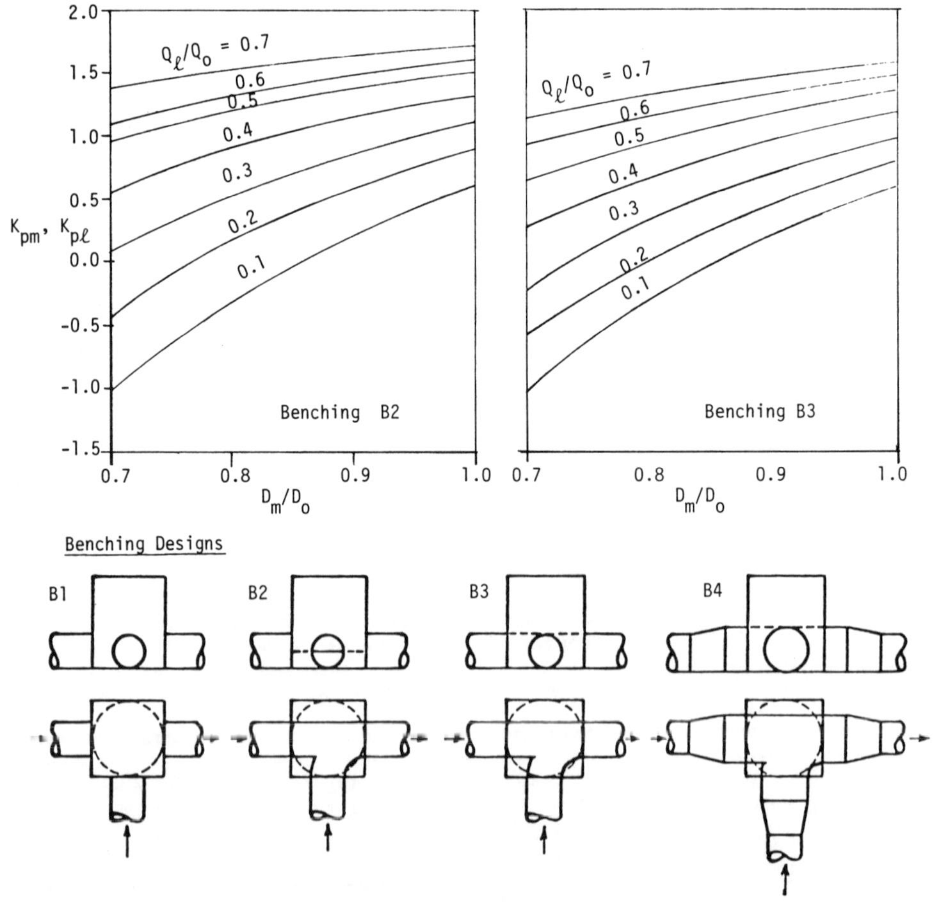

FIGURE 10. Pressure change coefficients for junctions of a main and a lateral with benchings ($D_m \sim D_\ell$; for $D_m/D_o < 1$, all data extrapolated).

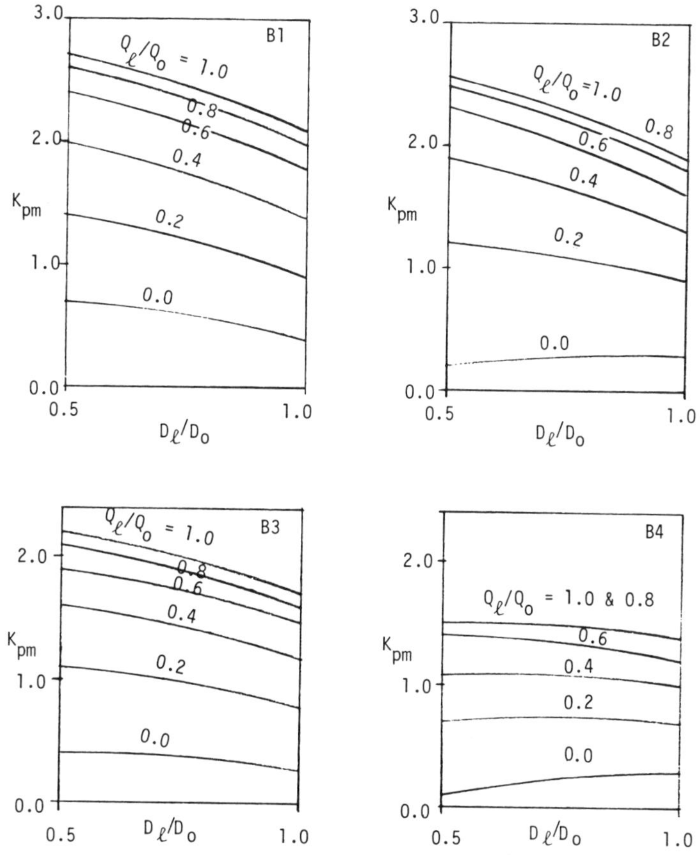

FIGURE 11. Main pressure change coefficient for junctions of a main and a lateral with benchings ($D_m = D_o$; $D_\ell \leq D_o$; for $0.5 < D_\ell/D_o < 1$, data interpolated).

lateral sizes $D_{\ell 1}/D_o$ and $D_{\ell 2}/D_o$, benching, and lateral alignment [13,16]. In the preceding notation, $_{\ell 1}$ and $_{\ell 2}$ denote the first and second lateral, as shown in Figure 14.

The pressure change coefficients of two opposed laterals, joined at a junction without benching, depend on the relation of flow velocities in both laterals [16]. K_p for the higher flow velocity lateral, K_{phv}, is approximately constant throughout a wide range of flows and attains a value of 1.8. For the lower velocity lateral, the pressure change coefficient $K_{p\ell v}$ can be determined as the difference of two pressure factors, H and L, plotted in Figure 14 [16]. Thus, both lateral pressure change coefficients can be determined from the following expressions:

$$\text{Higher velocity lateral } K_{phv} = 1.8 \qquad (15)$$

$$\text{Lower velocity lateral } K_{p\ell v} = H - L - 0.2 \qquad (16)$$

Investigations of junctions of two opposed laterals with benchings indicate that for identical pipe diameters ($D_{\ell 1} = D_{\ell 2} = D_o$) and intermediate flow divisions ($Q_{\ell 1}/Q_o = 0.3 - 0.7$), the benching barely contributes to the reduction of pressure change coefficients [13]. For benchings *B*2 and *B*3 shown in Figure 14, K_p's were only 8% and 17%, respectively, smaller than those corresponding to the junction without benching. Thus it appears that the earlier presented data for junctions without benching may be transposed to junctions with benching without introduction of significant inaccuracies.

Experimental studies further indicate that the lowest head losses and pressure changes at surcharged junctions of two opposed laterals are found for junctions with comparable flow velocities in both laterals and full pipe depth benching [13]. Another means of reducing pressure changes is by offsetting laterals at the junction [16]. For design of junction with opposed offset laterals, the design chart in Figure 15 can be used.

Limited experimental data are available for junctions of

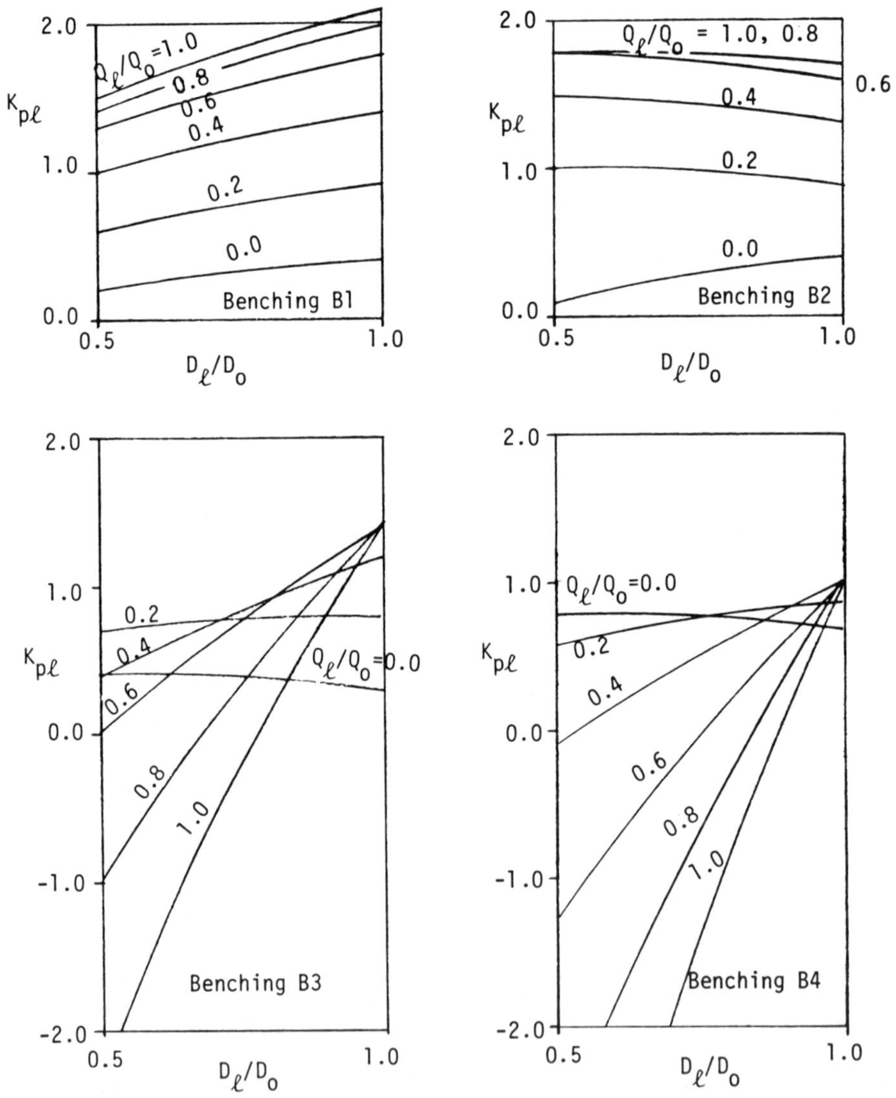

FIGURE 12. Lateral pressure change coefficient for junctions of a main and a lateral with benchings ($D_m = D_o$; $D_\ell \leq D_o$; for $0.5 < D_\ell/D_o < 1$, data interpolated).

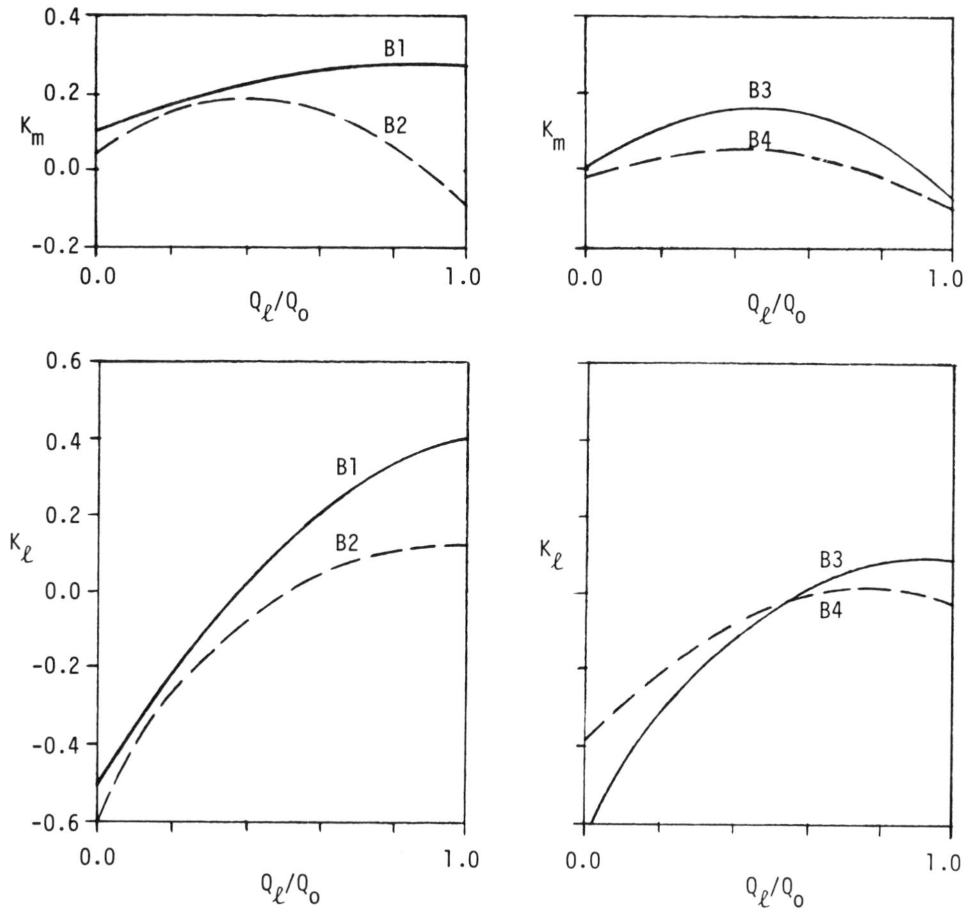

FIGURE 13. Junctions of a main and a lateral: head loss coefficients for subcritical open-channel flow [13].

two opposed laterals with open-channel flow. Such data were produced for subcritical flow with Froude numbers ranging from 0.1 to 0.6, identical pipe sizes $D_{\ell 1} = D_{\ell 2} = D_o$, the depths of flow from 0.4 D_o to 0.9 D_o, and three junction benchings $B1$–$B3$. For such conditions, approximate head loss coefficients were calculated as averages for all depths and plotted in Figure 16. The uncertainties associated with data curves in Figure 16 are estimated as ± 0.1.

It can be inferred from Figure 16 that the relative inflow and benching are again fairly important parameters. The highest losses were observed at junctions without benching. Such losses were substantially reduced for benching $B2$ and even more for $B3$.

Y-JUNCTIONS

Although T-junctions are the most common among three-pipe junctions, other layouts, such as Y-junctions are also used in some sewer networks. In these junctions, flows from two laterals combine at the junction and leave through the outfall. In comparison to junctions of opposed laterals, Y-junctions should be hydraulically more effective, because some momentum from both laterals should be preserved at the junction and contributed to the outflow. This would be particularly true when both lateral inflows are comparable in terms of flow rates and velocities.

No data on pressure changes at Y-junctions were found in the literature. It was reported, however, that the junction design can be improved by installing a vertical divider separating both inflows as shown in Figure 17 [15]. This divider should split the flow area in the same ratio as that of the two lateral peak inflows.

Four-Pipe Junctions

Four-pipe junctions shown schematically in Figure 18 are rarely used in design practice because of high losses taking

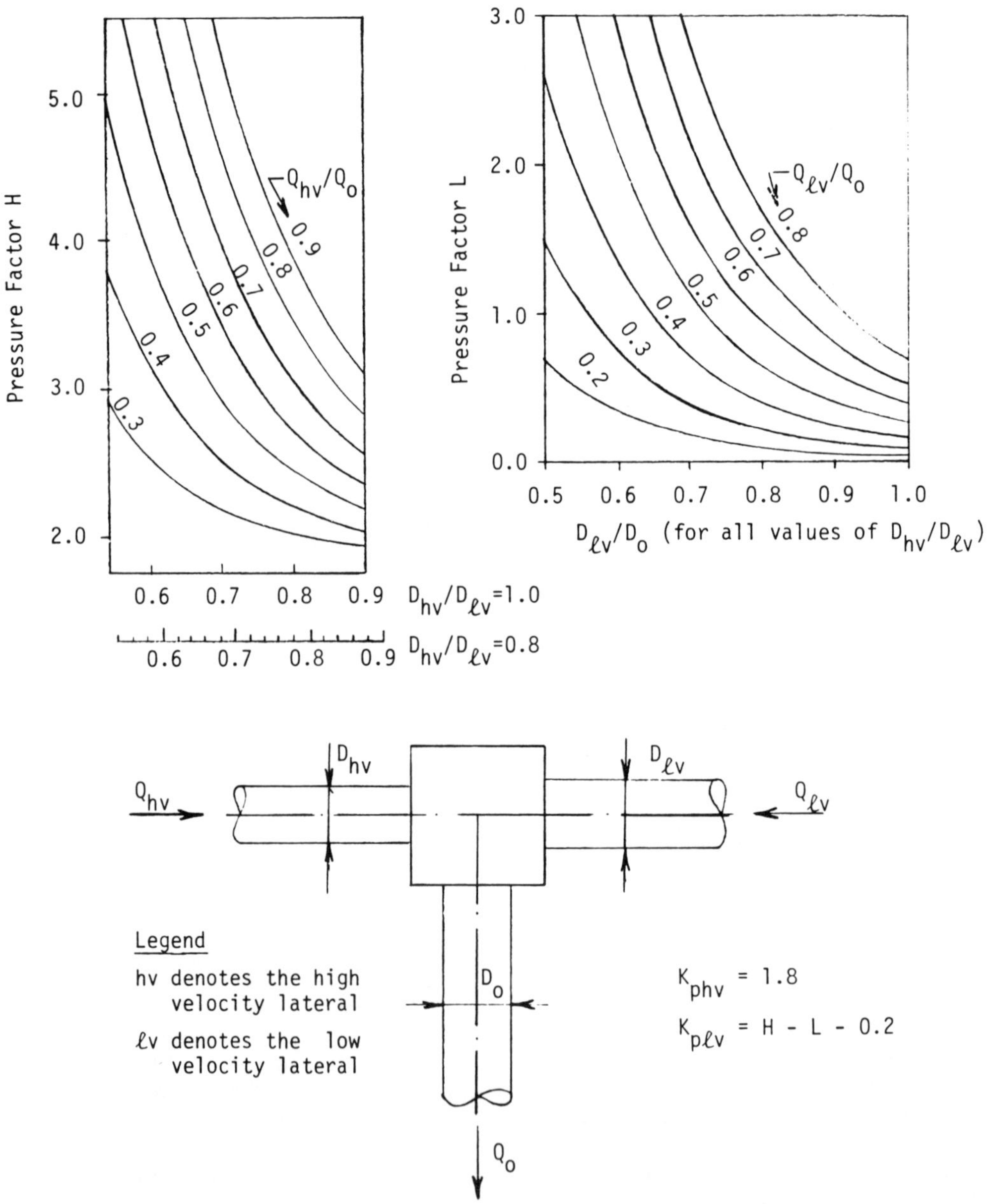

FIGURE 14. Junctions of two in-line opposed laterals: pressure change coefficients (no benching; [16]).

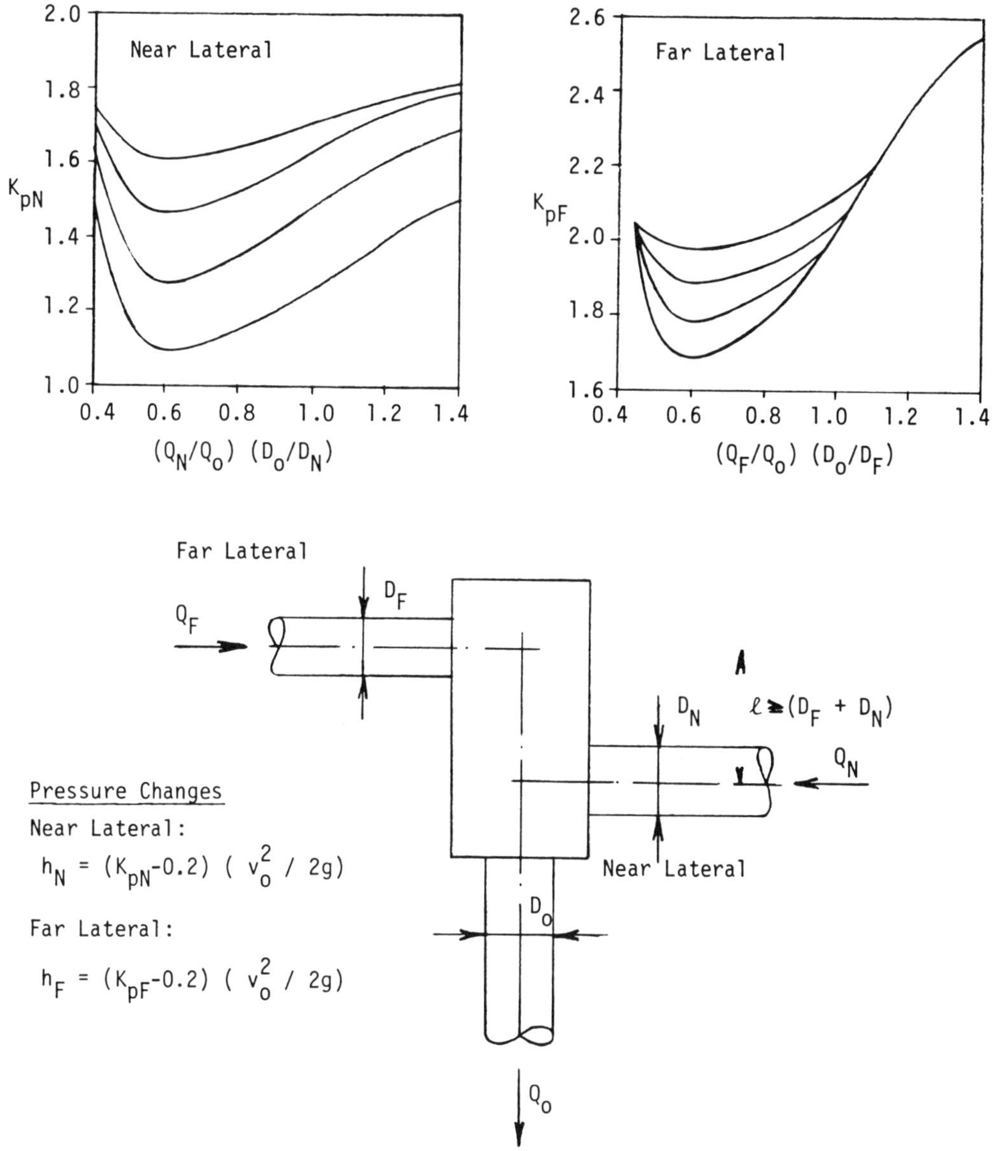

FIGURE 15. Junctions of two offset opposed laterals: pressure change coefficients (no benching; [16]).

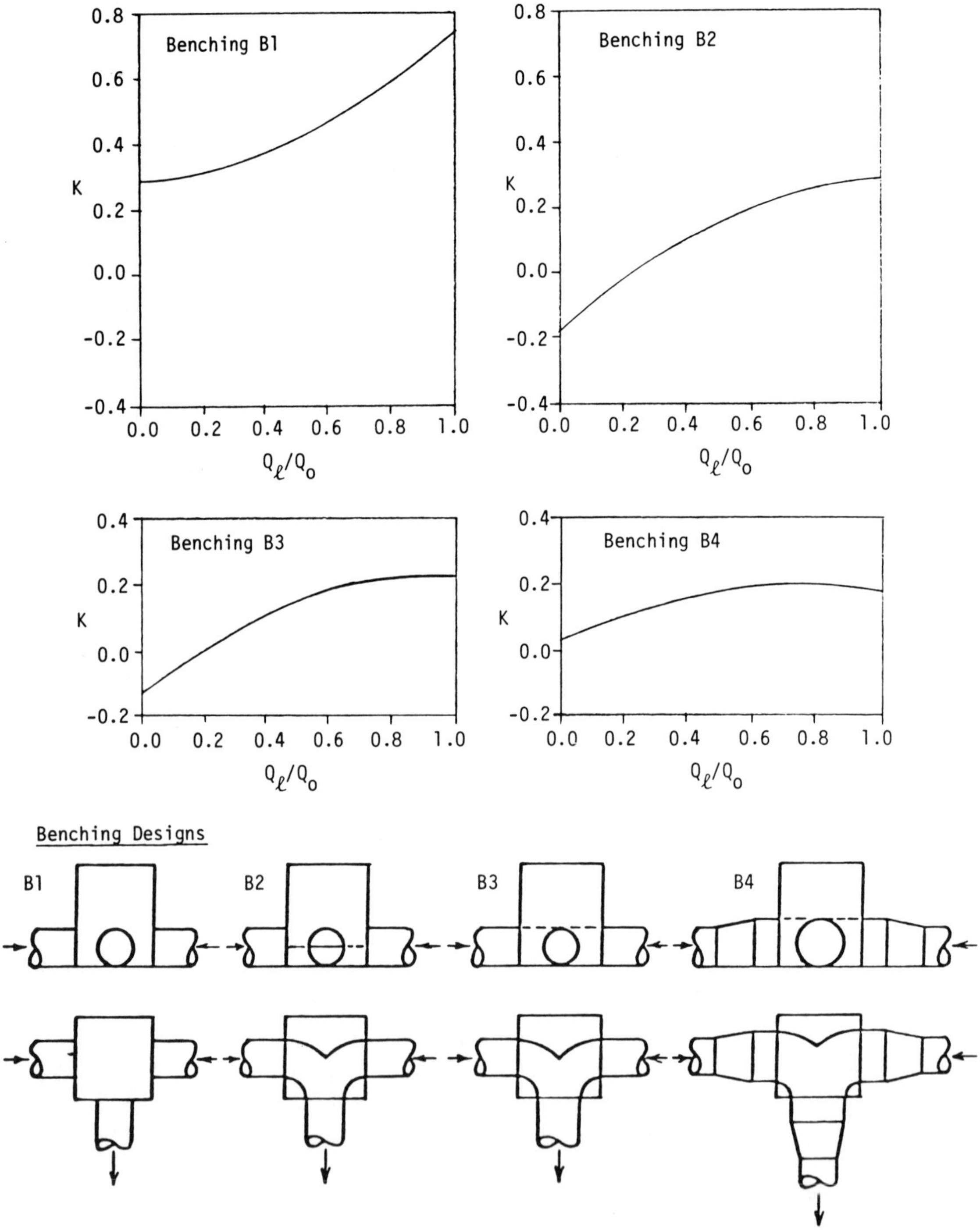

FIGURE 16. Junctions of two in-line opposed laterals: head loss coefficients for subcritical open-channel flow [13].

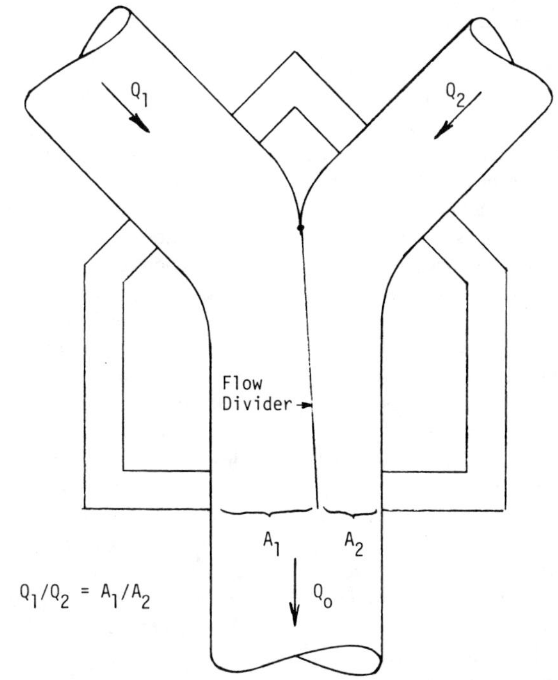

FIGURE 17. Y-Junction with a flow divider [15].

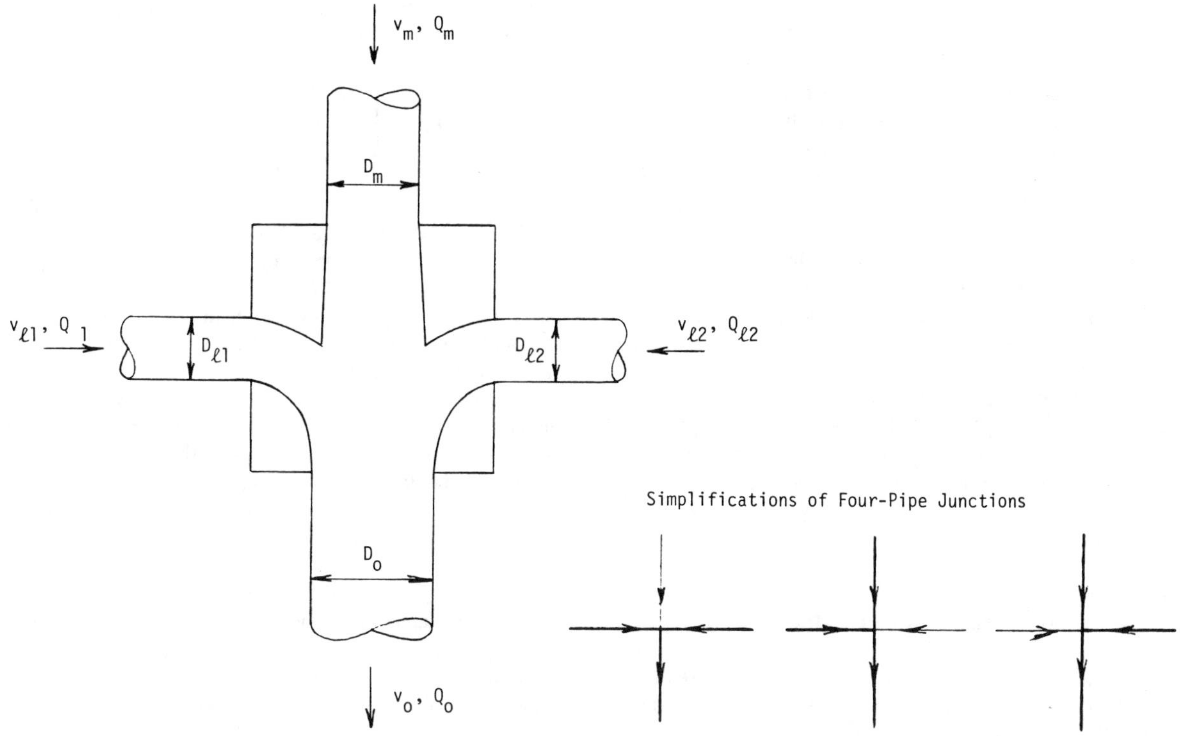

FIGURE 18. Four-pipe junctions.

place at such junctions. The literature survey did not reveal any published information on these junctions. In the absence of specific design data, the designer has to approach four-pipe junctions using the information available for other junction types. Toward this end, it is suggested to calculate momentum fluxes (ρQv) for all inflow pipes. After comparing magnitudes of such fluxes, it will be possible in many cases to consider the four-pipe junction as one of three-pipe junctions discussed earlier.

For a relatively low momentum flux in the main pipe, the four-pipe junction may be considered as a junction of two opposed laterals discussed in the preceding section. The higher of the two lateral pressure grade line elevations would be assumed for the main pipe as well. When either of the two laterals has a low momentum flux, the four-pipe junction may be considered as a T-junction of a main with perpendicular lateral. Only when all three inflow momentum fluxes are comparable, the above simplifications are not realistic, although they may represent the only solution available to the designer.

Inferences from the earlier discussed studies of T-junctions indicate that it would be desirable to avoid high head losses and pressure changes expected at four-pipe junctions with in-line laterals. This could be done by offsetting the laterals, either within the junction manholes, or by inserting another junction manhole and replacing the four-pipe junction by two T-junctions of a main and lateral.

Special Junction Structures

The previous discussion dealt primarily with common sewer junctions and junction manholes which are often prefabricated and can be characterized by relatively small dimensions and flows. In large metropolitan sewer networks, there may be cases where several large trunk sewers are joined at special junction chambers. Other special structures comprise various drop manholes which again can have large dimensions. Considering the high costs and importance of such special structures, it is desirable to investigate their operation and design in detail. Such investigations include a computational analysis and possibly a scale model study designed to develop hydraulically effective layouts of special structures. Because the flow conditions at such structures are dominated by gravity effects, the scale models would be typically designed and operated according to the Froude similarity.

APPLICATIONS OF JUNCTION HEAD LOSSES IN PRESSURIZED SEWER FLOW ROUTING

As stated in the introduction, junction head losses and pressure changes are particularly large and important in surcharged sewer systems. The analysis of surcharged sewer networks differs from that of fully pressurized water distribution networks, because there is a free water surface at junctions and possible overflows (flow losses) at various structures.

Flow routing in surcharged sewer networks requires calculation of flow velocities in individual pipes and calculations of pressure grade line elevations throughout the network. For such calculations, it is important to consider head losses and pressure changes at junctions. Since flow routing calculations are usually computerized, head loss or pressure change coefficients can be used as inputs to the flow routing model. These coefficients may depend on certain flow characteristics and their variations can be described using the design data from the preceding section.

If the routing model does not consider explicitly junction head losses or pressure changes, such factors can be still considered implicitly by using the equivalent roughness concept [12]. In this concept, the pipe roughness is increased in calculations to compensate for the form losses at the adjacent junction. This condition can be expressed as

$$H_{eq} = H_j + H_p \tag{17}$$

where H_{eq} is the equivalent pipe head loss, H_j is the junction head loss, and H_p is the actual pipe head loss.

After substituting for $H_j = Kv^2/2g$ and $H_p = Lv^2n^2 (D/4)^{-4/3}$ from the Manning equation, the following expression is obtained for the equivalent pipe roughness:

$$n_{eq} = \left[n^2 + \frac{K}{2gL} \left(\frac{D}{4} \right)^{4/3} \right]^{1/2} \tag{18}$$

where n_{eq} is the equivalent (increased) Manning's roughness coefficient, K is the junction head loss coefficient, D is the pipe diameter, L is the pipe length, and n is the actual pipe roughness coefficient. Note that the junction head loss could be also compensated for by lengthening the pipe, but this may lead to problems with numerical solutions employed in some routing models [12].

Approximations of junction head losses by equivalent pipe losses are quite appropriate in cases where the junction coefficient K does not vary strongly with such flow characteristics as the depth of surcharge, or the relative lateral inflows. These characteristics will vary during the passage of surcharge waves through the sewer network and the corresponding changes in K cannot be reflected by the equivalent pipe roughness which remains constant during computations. Under such circumstances, it may be necessary to use iterations in order to establish the flow characteristics corresponding to the peak flow and then use the appropriate values of K and equivalent pipe roughness in final computations.

TABLE 8. Susceptibility of Selected Junctions to Sulphide Gas Releases.

Junction Type	Susceptibility to Sulphide Releases According to Junction Benching		
	Low	Medium	High
90° Bend	B3	—	B1,B2
Main and Lateral	B3	B1,B2	B1
Two Opposed Laterals	B3	B2	B2

SULPHIDE GAS RELEASES AT JUNCTIONS

A proper design of sewer systems receiving discharges of sanitary sewage must consider sulphide gas releases which are closely related to flow conditions in the system. Although sulphide gases may escape from any exposed surfaces of the sewage flow, such releases are strongly enhanced by flow turbulence which is particularly strong at sewer junctions. In order to minimize sulphide gas releases, which create odour nuisances and safety problems for personnel entering sewers, it is desirable to use hydraulically effective junction designs with low levels of turbulence.

Ratings of susceptibility of selected junction designs to sulphide gas releases were reported and presented in Table 8 [13].

It is obvious from Table 8 that where sulphide releases are of concern, the use of hydraulically effective junction designs with full pipe depth benching is required. It was further noted that lower turbulence and sulphide releases can be expected for higher depths of surcharge and comparable magnitudes of inflows [13].

SUMMARY

Head losses and pressure changes at sewer junctions are significantly large to affect the system discharge capacity. Junction flow phenomena are particularly important in surcharged sewer systems with higher flow velocities as well as higher head losses, and design requirements for calculation of pressure grade line elevations. Therefore, it is a good design practice to evaluate head losses at sewer junctions and to compensate for these losses by equivalent invert drops at the junction. A gradual lowering of the invert elevation is more effective than sudden drops.

Head losses and pressure changes at junctions are affected by a number of hydraulic and geometric variables. The designer's control over flow variables, such as discharges in branch lines, is usually limited and, consequently, it is important to concentrate on the geometrical design of junctions. The most important parameter in this regard is the junction benching. Hydraulically effective junctions should be designed with benching extending to the pipe crown (*B*3). In the plan, such channels should provide for a gradual change in the flow direction, merging of incoming flows, and deflection into the outfall. In critical cases, a special design incorporating an enlarged pipe immediately upstream of the junction and a correspondingly wide and deep U-channel benching at the junction can be used. Such installations will be more costly.

In existing sewer systems with surcharge problems, the retrofitting of junctions with proper benching should be considered as one of remedial measures. For this purpose, fibreglass inserts or concrete benchings can be used.

Other geometrical parameters of junction manholes are less important. In terms of head losses, no significant differences were found between square- or round-base manholes. Round-base manholes may be more susceptible, in special cases, to formation of swirls or lateral water surface oscillations at the junction which then contribute to increased head losses.

The relative size of manholes has a small influence on head losses. Larger manholes, compared to the outfall diameter, produce somewhat higher head losses. For maintenance purposes, it is desirable to maintain a certain minimum manhole size, in relation to adjacent sewers, to allow some working space. In practice, the minimum value of D_{mh}/D_o is about two.

In the vertical arrangement, it is common to match pipe crown levels at the junction. The junction invert should be sloping in the flow direction to compensate for the maximum head loss. For pipes with smaller losses, or for lower flows, there may be an energy surplus at the junction.

The list of flow variables affecting junction operations includes the relative surcharge, relative inflows, and the Froude number for open-channel flow. For open-channel flow, junctions behave somewhat differently than in pressurized flow. If the flow is supercritical, hydraulically ineffective junctions may cause hydraulic jumps with concomitant increases in the flow depth and possible sewer surcharging. Such disruptions should be avoided. In subcritical flow, head losses at junctions are fairly small because of relative small changes in the flow area at the junction.

The relative surcharge depth may affect head losses at some sewer junctions. Among the junction types discussed in Section 3, only the straight-flow-through junctions were affected by the surcharge depth for S/D_o smaller than two. Such junctions, with a round base and without benching, were susceptible to formation of swirls and water oscillations at the junction and this resulted in a sudden increase of head losses.

The relative junction inflows are usually given by the overall project layout and their control may not be within designer's power. It is possible, however, to affect the momentum flux of incoming flows by varying the pipe diameter. In this regard, it is important to reduce the velocity of the stream disturbing the main flow through the junction. For example, in a T-junction of a main with lateral, it is desirable to reduce the lateral velocity by using a larger lateral pipe. Similarly, it may be sometimes advantageous to increase the pipe diameter upstream of the junction. Although the head loss coefficient is generally not affected by the pipe diameter, the loss magnitude expressed as $Kv^2/2g$ will change substantially with the pipe diameter and the resulting change in flow velocity. For example, by enlarging the pipe diameter 1.2 times, the velocity head and the head loss are reduced in half.

Hydraulically effective junction designs with full pipe depth benching are also effective in reducing sulphide releases caused by high turbulence at junctions. Such considerations are important in sewer systems receiving sanitary sewage discharges.

REFERENCES

1. Ackers, P., "An Investigation of Head Losses at Sewer Manholes," *Civil Engineering and Public Works Review,* Vol. 54, No. 637, pp. 882–884 and 1033–1036 (1959).
2. Archer, B., F. Bettes, and P. J. Colyer, "Head Losses and Air Entrainment at Surcharged Manholes," Report No. IT185, Hydraulics Research Station, Wallingford (1978).
3. Black, R. G. and T. L. Pigott, "Head Losses at Two Pipe Stormwater Junction Chambers," *Proc. Second National Conf. on Local Government Engineering,* Brisbane, pp. 219–223 (September 19–22, 1983).
4. Chow, V. T. *Open Channel Hydraulics.* McGraw-Hill, New York (1959).
5. de Groot, C. F. and M. J. Boyd, "Experimental Determination of Head Losses in Stormwater Systems," *Proc. Second National Conference on Local Government Engineering,* Brisbane (September 19–22, 1983).
6. Hare, C. M., "Magnitude of Hydraulic Losses at Junctions in Piped Drainage Systems," *Civil Engineering Transactions, Institution of Civil Engineers,* pp. 71–77 (1983).
7. Howarth, D. A. and A. J. Saul, "Energy Loss Coefficients at Manholes," *Proc. 3rd Int. Conf. on Urban Storm Drainage,* Goteborg, pp. 127–136 (June 4–8, 1984).
8. Jensen, M., "Hydraulic Energy Relations for Combined Flow and Their Application to the Ninety Degree Junction," *Series Paper No. 11,* The Royal Veterinary and Agricultural University, Copenhagen (1981).
9. Liebman, H., "Der Einfluss von Einsteigschachten auf den Abflussforgang in Abwasserkanalen" (Effects of Manholes on Flow Processes in Storm Drains), *Wasser und Abwasser in Forschung und Praxis,* Band 2, Erich Schmid Verlag, Bielefeld, FRG (1970).
10. Lindvall, G., "Head Losses at Surcharged Manholes With a Main Pipe and a 90° Lateral," *Proc. 3rd Int. Conf. on Urban Storm Drainage,* Goteborg, pp. 137–146 (June 4–8, 1984).
11. Marsalek, J., "Head Losses at Sewer Junction Manholes," *Journal of Hydraulic Engineering, ASCE,* Vol. 110, pp. 1150–1154 (1984).
12. Marsalek, J., "Urban Runoff Peak Frequency Curves," *Nordic Hydrology, Vol. 15,* pp. 85–101 (1984).
13. Marsalek, J., "Head Losses at Selected Sewer Manholes," A report submitted to APWA, National Water Research Institute, Burlington, Ontario, Report No. 85–15 (July, 1985).
14. Pardee, L. A., "Hydraulic Analysis of Junctions," *Office Standard No. 115, Storm Drain Design Division,* City of Los Angeles (1968).
15. Prins, R. J., "Storm Sewer Junction Geometry and Related Energy Losses," M.A.Sc. Thesis, University of Ottawa, Ottawa (December, 1975).
16. Sangster, W. M., H. W. Wood, E. T. Snerdon, and H. G. Bossy, "Pressure Changes at Storm Drain Junctions," *Engineering Series Bull. 41, Engng. Exp. Station,* University of Missouri, Columbia, Missouri (1958).
17. Water Pollution Control Federation, "Design and Construction of Sanitary and Storm Sewers," *WPCF Manual of Practice No. 9,* Water Pollution Control Federation, Washington, DC (1970).

Numerical Analysis for Water Hammer

MASASHI SHIMADA*

INTRODUCTION

Transients in a pipeline caused by a change in boundary condition such as operations of pumps and valves and pump failures are defined as the unsteady flow varying from a steady state toward another steady state finally established. In general, transients are divided into two phenomena: surging and water hammer [16]. Surging, governed by slowly varying quasi-static forces, appears in pipelines including some tanks having the water free surfaces. Surging can be dealt with by the rigid column theory or the assumption of imcompressible fluid, which is described by ordinary differential equations. On the other hand, water hammer is the dynamic process that the kinetic and elastic energies are transformed into each other due to the elasticities of liquid and pipe material, in which abnormally higher and lower pressure heads may be generated, since the small change in volume of a liquid produces much greater change in pressure of the liquid. It is very important to accurately and economically predict and control the transients in pipelines so that it should prevent the pipeline from collapsing due to the abnormal pressure heads and the liquid should be safely transported and distributed.

Classification of Transients

Transients in pipeline are classified into three different regions [3], depending on the amount of free gas and liquid vapor.

WATER HAMMER

In water hammer region the effect of free gas and vapor on the transients is negligible when liquid contains little gas, that is, the void fraction defined by Equation (1) is extremely small,

$$\alpha = \frac{\text{¥}_g}{\text{¥}_g + \text{¥}_\ell} \tag{1}$$

in which ¥_g = volume of the gas (m³) and ¥_ℓ = liquid volume (m³). Not depending on the local pressure, the wavespeed of pressure pulse, a(m/s), for stiff thin-wall pipes becomes [1]

$$a = \sqrt{\frac{\frac{K_\ell}{\rho_\ell}}{1 + \frac{DK_\ell}{Ee} c_1}} \tag{2}$$

in which ρ_ℓ = density of a liquid (kg/m), e = pipe wall thickness (m), D = pipe diameter (m), E = Young modulus of elasticity (Pa), K_ℓ = bulk modulus of liquid (Pa), and c_1 = constant depending on how to sustain the pipe.

Halliwell gives more general expressions for the wavespeed [15]. The head rise due to water hammer in instantaneous stoppage of flow at a downstream valve is theoretically estimated by the Joukowsky's equation,

$$\Delta H = \frac{aV_0}{g} \tag{3}$$

in which V_0 = the initial flow velocity (m/s) and g = the gravitational acceleration (m/s²).

CAVITATION

When the pressure in pipelines due to rarefaction waves becomes sufficiently less than the saturation pressure of dissolved gas, cavitation nuclei grow to small gas bubbles, at the vapor pressure of the liquid the liquid vaporizes, and

*National Research Institute of Agricultural Engineering, Kannondai 2-1-2, Tsukuba, Ibaraki, 305 Japan

cavities are formed. The void fraction becomes much less than unity for small velocity gradients. The flow is called "the cavitating flow" of the bubbly flow type, in which the cavities generated are dispersed through the pipe cross-section. When the effect of the pipe wall elasticity becomes negligible, the increase in the void fraction reduces the wavespeed,

$$a = \sqrt{K/\rho}$$

$$K = \frac{K_\ell}{1 + \alpha \left(\dfrac{K_\ell}{K_g} - 1 \right)} \qquad (4)$$

$$\rho = \rho_g \alpha + \rho_\ell (1 - \alpha)$$

in which the subscripts g and ℓ denote the quantities corresponding to the gas and liquid, respectively. If the wall effect is significant, then the wavespeed becomes

$$a = \sqrt{\frac{\dfrac{K}{\rho}}{1 + \dfrac{DK}{Ee}}} \qquad (5)$$

in which K and ρ are the same as those in Equation (4). Equation (5) is in good agreement with the experiment [20,30].

To be important, even small amount of air or gas greatly influences on the wavespeed. The impulse due to the collapse of cavities caused by the reflected compressive waves may burst the pipeline.

COLUMN SEPARATION

The void fraction can increase to values comparable to unity for large velocity gradients. When the cavities grow and become so large to fill the whole cross section, the liquid flow separates into two columns. This is called "the column separation". The rejoining of the two columns due to the positive waves produces very high pressures, which may damage the pipeline. The magnitude of the head rise due to the rejoining is estimated [51] by

$$\Delta H = \frac{a \cdot \Delta V}{2g} \qquad (6)$$

in which $\Delta V =$ the difference in velocity at instant of collapse of the cavity. The presence of air cavities causes the damping of pressure waves produced by the column separations and their rejoining, since the expansion and compression of cavities causes the additional energy dissipation due to the heat transfer to the liquid [49].

In this article, water hammer is dealt with as in the elastic model, which is able to generally describe the transients, and some methods numerically integrating governing equations for water hammer are addressed. The reader should refer to the papers and textbooks [3,21,44,52,55] about the numerical analysis on cavitating flow and column separation. Due to numerical predictions to the transients some kinds of effective devices have to be taken for eliminating the undesirable transients, such as excessively low and high pressures, and column separation:

1. Installations of valves, surge tanks, and air-chamber at proper positions
2. Changing the pipeline profile
3. Increasing the pipe diameters
4. Reducing the wavespeed
5. Increasing moment of inertia of rotating parts of turbomachines such as a flywheel

BASIC EQUATIONS FOR WATER HAMMER

Assumptions of Mathematical Model

1. The liquid and pipe wall are linearly elastic (small deformation) and the changes in the liquid density, ρ, and the cross-section area in a uniform pipe, $A(m^2)$, are also negligible during the transients.
2. The pipe flow is dealt with as in one-dimensional, in which the velocity averaged over the cross-section, $V(m/s)$, and the piezometric head above a datum, $H(m)$, become two dependent variables.
3. The void fraction is very small and the wavespeed is constant if otherwise specified.
4. The resistant law valid for steady flows in turbulent or laminar flow regimes is also used during the transients, though the shear stress depends on the rate of change of velocity [57].

Governing Equations [51]

EQUATION OF MOTION

The equation of motion for a pipe of uniform cross-section is derived by directly applying the Newton's second law to the control volume having the length, Δx, as shown in Figure 1.

Force Balance Toward x Axis

inertia force	pressure acted on cross section
$\rho A \dot{V} \Delta x$	$= pA - \left(pA + \dfrac{\partial pA}{\partial x} \Delta x \right)$
$\rho A \left(\dfrac{\partial V}{\partial t} + V \dfrac{\partial V}{\partial x} \right) \Delta x$	frictional forces due to shear stress on wall $\qquad (7)$ $- \tau_0 \Pi D \Delta x$
	component force of gravitation $- \rho g A \Delta x \sin \theta$

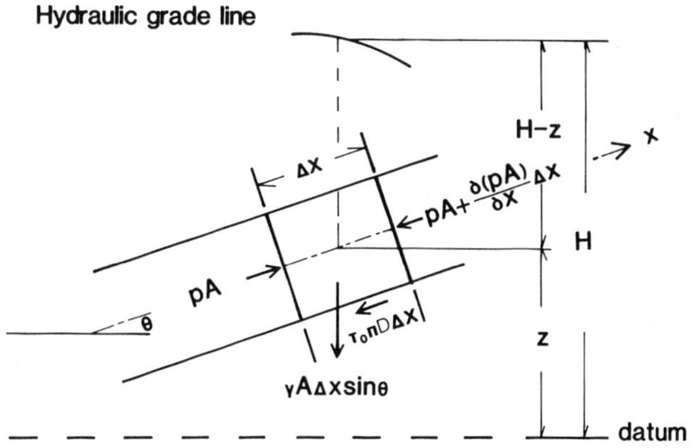

FIGURE 1. Control volume (cross section area, A, and length, Δx).

in which x = distance coordinate along pipe axis (m), p = the pressure (Pa/m^2), τ_0 = the wall shear stress (N/m^2), θ = the angle between the pipe axis and horizon, and a dot over a dependent variable denotes the total derivative with respect to time.

Using the assumption (4), the shear stress is expressed by

$$\tau_0 = \varrho f V|V|/8 \tag{8}$$

in which f = the Darcy-Weisbach's friction factor being a function of Reynold number and/or the wall roughness.

The friction factor may be given by the implicit formulas such as the Colebrook equation [4], which is the basis for the Moody diagram [28].

$$\frac{1}{\sqrt{f}} = \begin{cases} -2\log_{10}\left(\dfrac{2.51}{Re\sqrt{f}}\right) & \text{(smooth)} \\ 1.74 - 2\log_{10}\left(\dfrac{2k}{D} + \dfrac{18.7}{Re\sqrt{f}}\right) & \text{(transition)} \\ -2\log_{10}\left(0.266\,\dfrac{k}{D}\right) & \text{(rough)} \end{cases} \tag{9}$$

in which Re = the Reynold's number, VD/v, v = the kinematic viscosity (m^2/s), and k = the wall roughness (m).

To numerically deal with Equation (9), it requires a subroutine program in which the friction factor is calculated from the values of V, D, and k at each sections. Instead of the awkward way, some kinds of explicit formulas valid within the range defined from experimental data are developed [22]. The Hazen-William's formula being one of the explicit formulas [43] is:

$$f = \frac{1014.2}{C^{1.852}D^{0.0184}} Re^{-0.148} \tag{10}$$

in which C = the Hazen-Williams coefficient.

For laminar flow, the friction factor becomes

$$f = \frac{64}{Re}$$

Substituting Equation (8) into Equation (7), we have

$$\frac{\partial V}{\partial t} + V\frac{\partial V}{\partial x} + \frac{1}{\rho}\frac{\partial p}{\partial x} + g\sin\theta + \frac{f}{2D}V|V| = 0 \tag{11}$$

Giving the local gradient of the pressure with respect to the piezo-metric head and the height of pipe axis, z, we have

$$\frac{1}{\rho}\frac{\partial p}{\partial x} = g\left(\frac{\partial H}{\partial x} - \sin\theta\right) = g\left(\frac{\partial H}{\partial x} - \frac{\partial z}{\partial x}\right) \tag{12}$$

Equation (11) becomes

$$\frac{\partial V}{\partial t} + V\frac{\partial V}{\partial x} + g\frac{\partial H}{\partial x} + \frac{f}{2D}V|V| = 0 \tag{13}$$

If the empirical formulas are used, then Equation (13) can be generally expressed as

$$\frac{\partial V}{\partial t} + V\frac{\partial V}{\partial x} + g\frac{\partial H}{\partial x} + FV|V|^m = 0 \tag{14}$$

in which F = the constant consisting of friction factor, pipe diameter, and so on, and m = the exponent.

Hazen-William's formula

$$F = \frac{507.1}{C^{1.852}} \frac{\nu^{0.148}}{D^{1.1664}}$$

$$m = 0.852$$

Darcy-Weisbach's formula

$$F = f/(2D)$$

$$m = 1$$

Due to the assumption (1), the pressure is directly given by integrating Equation (12),

$$p = \rho g(H - z) \tag{15}$$

CONTINUITY EQUATION

Applying the mass conservation's law to the conduit with a uniform cross-section, we have

$$\frac{\dot{\rho}}{\rho} + \frac{\dot{A}}{A} + \frac{\partial V}{\partial x} = 0 \tag{16}$$

The first term in the left-hand side of Equation (16) is related to the change of the pressure

$$\frac{\dot{\rho}}{\rho} = \frac{\dot{p}}{K} \tag{17}$$

The static-balance between the pressure and the stress acting on the pipe wall and the perturbation from the balancing condition give the relation, (18), for a stiff thin wall pipe:

$$\frac{\dot{A}}{A} = C_1 \frac{D}{Ee} \dot{p} \tag{18}$$

Equations (15) to (18) give

$$\frac{\partial H}{\partial t} + \frac{a^2}{g}\frac{\partial V}{\partial x} + V\frac{\partial H}{\partial x} - V \sin\theta = 0 \tag{19}$$

in which a = the wavespeed of pressure pulse given by Equation (2).

Characteristic Equations

The method of characteristics in mathematical theory offers the way to solve the basic equations by covering the integral surface with a net of characteristic curves or lines and by understanding that the transients are interaction between advancing and receding pressure waves, i.e., characteristic curves. With the method of characteristics the partial differential Equations, (14) and (19), are transformed into the four ordinary differential equations:

$$C^+: \quad dx = (V + a)dt \tag{20}$$

$$dH + (a/g)dV - V \sin\theta dt + \frac{FaV}{g}|V|^m dt = 0 \tag{21}$$

$$C^-: \quad dx = (V - a)dt \tag{22}$$

$$-dH + (a/g)dV + V \sin\theta dt + \frac{FaV}{g}|V|^m dt = 0 \tag{23}$$

Equations (20) and (22) are the equations defining the trajectories of advancing and receding pressure waves, respectively. Equations (21) and (23), called "the compatibility equations", give the relationships between the head, the velocity, the gravitational forces, and the frictional effect, which are valid for on the corresponding pressure waves, c^+ and c^-, respectively.

When the wavespeed, a, is much greater than the flow velocity, Equations (20) to (23) are simplified,

$$C^+: \quad dx = adt \tag{24}$$

$$dH + (a/g)dV + \frac{Fa}{g} V|V|dt = 0 \tag{25}$$

$$C^-: \quad dx = -adt \tag{26}$$

$$-dH + (a/g)dV + \frac{Fa}{g} V|V|dt = 0 \tag{27}$$

The assumption that $a \gg |V|$ is valid for stiff pipes such as metal, concrete, and so on. The characteristics become straight lines.

NUMERICAL ANALYSIS

The arithmetic method [1] and the graphical method [2,29] are known as the solution procedures for water hammer problems but they are not suitable for complex and complicated pipeline systems with more than about ten pipes and other facilities. The characteristic methods numerically integrating Equations (20) to (23) or Equations (24) to (27) are the most widely-used solution procedure for water hammer, since the solution procedures are relatively easy to implement.

Characteristic Methods

CHARACTERISTIC GRID METHOD (CGM)

A transient is schematically shown in Figure 2 ($x - t$ plane) in which travelling characteristic curves irregularly

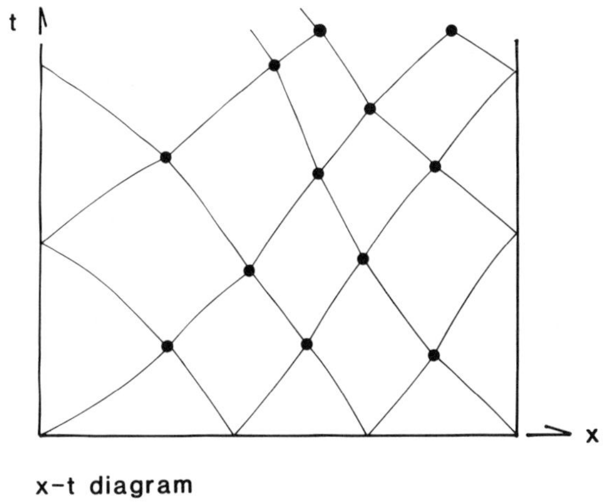

FIGURE 2. Characteristic grids.

cross each other. The cross points are called "the characteristic grids", between which Equations (20) to (23) are integrated to get finite difference equations. But it is not useful to directly apply the characteristic grid method except for special purposes such as for transient analysis in highly deformable pipes or in low wavespeed pipelines.

METHOD OF SPECIFIED TIME INTERVAL (MSTI)

For a system of the simplified equations, (24) to (27), we can carry out the orderly computations by using the regular time-space grid with a specified time interval as shown in Figure 3, in which the ratio, $\Delta x/\Delta t$, is equal to the wavespeed, a.

Therefore, MSTI does not require any interpolation, since characteristic lines justly pass the regular grids as shown in Figure 4a. The finite difference equations with MSTI without any interpolations are derived by approximately integrating Equations (24) to (27) between two regular grid points, A and P or C and P. Depending on how to approximate the non-linear frictional term, a first-order approximation model and three second-order approximation models are known as the solution procedures. At present the solution method, based on the first-order approximation model (FI), is organized and the most widely used [3,11,39,48,51]. Due to the numerical unstability of FI some second-order approximation models are also used for analysing the transients in long oil pipe-lines [40]. In addition, it is discussed to use second-order approximation models for efficient and accurate computations saving time [33,35,53,54].

First-Order Approximation Model (FI)

Assuming H_A, H_B, H_C, V_A, V_B, and V_C at point A, B, and C to be known while H and V at P to be unknown, the integration of Equation (25) or (27) with respect to time is approximated by

$$\int_t^{t+\Delta t} V(t')|V(t')|^m dt' \cong V_A|V_A|^m \cdot \Delta t$$

or (28)

$$V_C|V_C|^m \cdot \Delta t$$

in which t' = the parameter of time integration.

Equation (28) gives the finite difference equations valid for on C^+ and C^-, respectively:

$$C^+: \quad H - H_A + B(V - V_A) + RV_A|V_A|^m = 0 \quad (29)$$

$$C^-: \quad -(H - H_C) + B(V - V_C) + RV_C|V_C|^m = 0 \quad (30)$$

in which $B = (a/g)$ and $R = Fa(\Delta t)/g$.

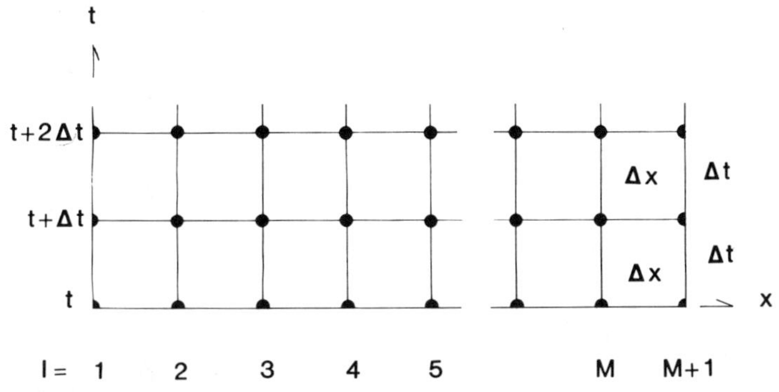

FIGURE 3. Regular time-space grids.

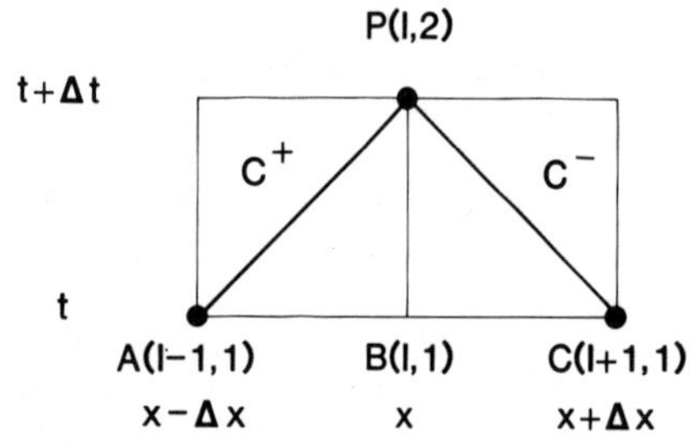

FIGURE 4a. Calculation at an interior point with MSTI.

Equations (29) and (30) are simplified,

$$C^+: \quad H + BV = C_P \tag{31}$$

$$C^-: \quad -H + BV = C_M \tag{32}$$

in which $C_P = H_A + BV_A - RV_A|V_A|^m$ and $C_M = -H_C + BV_C - RV_C|V_C|^m$. The solution of Equations (31) and (32) becomes

$$V = (C_P + C_M)/(2B) \tag{33}$$

The head is given by substituting V into Equations (31) or (32).

It is defined that the symbols, B_1, C_{P1} and B_2, C_{P2}, and so on are used when the corresponding compatibility equations are applied to pipe 1, 2, and so on, respectively. The subscripts, 1 and 2, denote pipe 1 and 2, respectively.

Second-Order Approximation Models

LINEAR MODEL (LSAM) [53,54]

Instead of Equation (28), Equation (34) is used

$$\int_t^{t+\Delta t} V|V|^m dt' \cong V|V_A|^m \Delta t \quad \text{or} \quad V|V_C|^m \Delta t \tag{34}$$

The finite difference equations become

$$C^+: \quad H - H_A + B(V - V_A) + RV|V_A|^m = 0 \tag{35}$$

$$C^-: \quad -(H - H_C) + B(V - V_C) + RV|V_C|^m = 0 \tag{36}$$

or

$$C^+: \quad H + B_P V = C_P \tag{37}$$

$$C^-: \quad -H + B_M V = C_M \tag{38}$$

in which $B_P = B + R|V_A|^m$, $B_M = B + R|V_C|^m$, $C_P = H_A + BV_A$, and $C_M = -H_C + BV_C$.

The solution of Equations (37) and (38) becomes

$$V = \frac{C_P + C_M}{B_P + B_M} \tag{39}$$

The head is given by substituting V into Equations (37) or (38).

It is defined that the symbols, C_{P1}, B_{P1} and C_{M1}, B_{M1}, and so on are used when the corresponding compatibility equations are applied to pipe 1. The subscript 1 denotes pipe number.

NON-LINEAR MODELS

Propson-Newton-Raphson Model (PNR) [31,33]

The second-order model giving the best accuracy uses the approximation:

$$\int_t^{t+\Delta t} V|V|^m dt' \cong \frac{V_A + V}{2}\left|\frac{V_A + V}{2}\right|^m \Delta t$$

or (40)

$$\frac{V_C + V}{2}\left|\frac{V_C + V}{2}\right|^m \Delta t$$

The finite difference equations are

$$C^+: \quad H + BV + FP(V) = C_P \tag{41}$$

$$C^-: \quad -H + BV + FM(V) = C_M \tag{42}$$

in which $FP(V) = S(V + V_A)|V + V_A|^m$, $FM(V) = S(V + V_C)|V + V_C|^m$, and $S = R/2$, $C_P = H_A + BV_A$, and $C_M = -H_C + BV_C$.

Through eliminating the head from Equations (41) and (42), the unknown velocity is obtained by solving the following equation by the Newton-Raphson iterative technique.

$$E = 2BV + FP(V) + FM(V) - (C_P + C_M) = 0$$

$$\frac{dE}{dV} = YP(V) + YM(V)$$

in which $YP(V) = B + DP(V)$, $DP(V) = S(m + 1)|V + V_A|^m$, $YM(V) = B + DM(V)$, and $DM(V) = S(m + 1)|V + V_C|^m$.

It is shown that any iteration is not required to get the solution with the same accuracy as the truncation error due to the discretization if the non-viscous solution is used as the first estimate in the iterative process [33]. This is also valid for boundary calculations.

Calculation Steps

1. Estimate the non-viscous solution for the velocity,

$$V^* = (C_P + C_M)/(2B) \tag{43}$$

2. Calculate the correction by

$$\Delta V = -\frac{E^*}{\left(\dfrac{dE}{dV}\right)^*} \tag{44}$$

in which $E^* = FP(V^*) + FM(V^*)$, $(dE/dV^*) = YP(V^*) + YM(V^*)$,

3. Set the solution by

$$V = V^* + \Delta V \tag{45}$$

4. And calculate the head by substituting V into Equations (41) or (42). Or use the average of the heads obtained from two compatibility equations improves the accuracy [35].

It should be noted that the first estimate, Equation (43), for the velocity, i.e., the non-viscous solution obtained by setting that $S = 0$ in Equations (41) and (42) is formally identical with the solution, Equation (33), obtained by *FI*. That is also valid for computations of boundary conditions. The functions of the velocity, *FP*, *FM*, *YP*, *YM*, *DP*, and *DM*, are used such as $FP2(V_2)$, $FM2(V_2)$, $YP2(V_2)$, and so on when the corresponding compatibility equation is valid for pipe 2. The subscript 2 denotes pipe number.

Predictor-Corrector Scheme (PCS) [9]

If the integration is approximated by

$$\int_t^{t+\Delta t} V|V|^m dt' \cong \frac{\Delta t}{2}(V|V|^m + V_A|V_A|^m)$$

or (46)

$$\frac{\Delta t}{2}(V|V| + V_C|V_C|^m)$$

then we have

$$C^+:\ H - H_A + B(V - V_A) + (R/2)(V|V|^m + V_A|V_A|^m) = 0 \tag{47}$$

$$C^-:\ -H + H_C + B(V - V_C) + (R/2)(V|V|^m + V_C|V_C|^m) = 0 \tag{48}$$

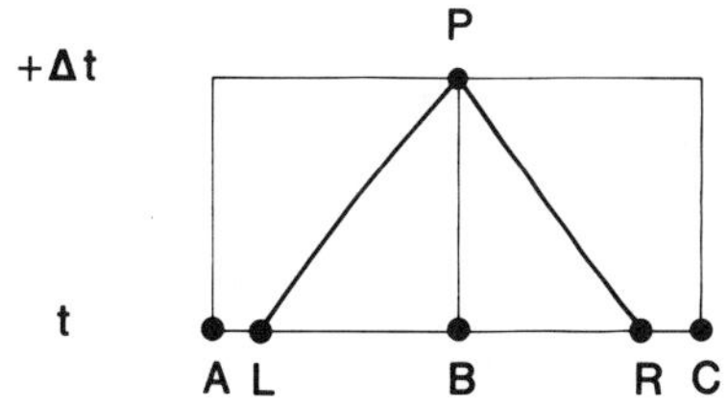

FIGURE 4b. Calculation at an interior point with IMSTI.

The non-linear equations are approximately solved by substituting the velocity, Equation (33), obtained by the first approximation model, into the unknown velocity in the frictional terms. The solution becomes

$$V = (C_P + C_M)/(2B) \tag{49}$$

in which $C_P = H_A + BV_A - (R/2)(V^*|V^*|^m + V_A|V_A|^m)$, $C_M = -H_C + BV_C - (R/2)(V^*|V^*|^m + V_C|V_C|^m)$, and V^* = the velocity, Equation (33), obtained by *FI*. The head is given by substituting V into Equations (47) or (48).

INTERPOLATED METHOD OF SPECIFIED TIME INTERVAL (IMSTI) [36,37]

The basic equations in their entirety must be used when the wavespeed is about the same as the velocity and/or greatly varies with the void fraction or the pressure.

As shown in Figure 4b, characteristic curves do not always pass the regular time-space grids. A linear interpolation is usually made to give V_L, V_R, H_L, and H_R at points L(left) and R(right) by using the knowns, V_A, V_B, V_C, H_A, H_B, and H_C [23].

$$\frac{X_B - X_L}{X_B - X_A} = \frac{(V_L + a)\Delta t}{\Delta x} = \frac{V_B - V_L}{V_B - V_A} = \frac{H_B - H_L}{H_B - H_A}$$

$$\frac{X_B - X_R}{X_C - X_B} = \frac{(V_R - a)\Delta t}{\Delta x} = \frac{V_B - V_R}{V_C - V_B} = \frac{H_B - H_R}{H_C - H_B}$$

$$V_L = \frac{V_B - \theta a(V_B - V_A)}{1 + \theta(V_B - V_A)} \tag{50}$$

$$H_L = H_B - \theta(V_L + a)(H_B - H_A)$$

$$V_R = \frac{V_B + \theta a(V_C - V_B)}{1 + \theta(V_C - V_B)} \qquad (51)$$

$$H_R = H_B + \theta(V_R - a)(H_B - H_C)$$

in which $\theta = \Delta t/\Delta x$.

Equations (20) to (23) are applied to LP and RP, and the finite difference equations become

$$C^+: \quad H + BV = C_P \qquad (52)$$

$$C^-: \quad -H + BV = C_M \qquad (53)$$

in which $C_p = H_L + (BV_L + V_L\Delta t \sin\theta - RV_L|V_L|^m)$, $C_M = -H_R + (BV_R - V_R\Delta t \sin\theta - RV_R|V_R|^m)$. Under the condition that the wavespeed is much greater than the velocity, Equations (24) to (27) give

$$V_L = V_B - a\theta(V_B - V_A)$$
$$H_L = H_B - a\theta(H_B - H_A) \qquad (54)$$

$$V_r = V_B + a\theta(V_C - V_B)$$
$$H_r = H_B + a\theta(H_C - H_B) \qquad (55)$$

$$C^+: \quad H + BV = C_P \qquad (56)$$

$$C^-: \quad -H + BV = C_M \qquad (57)$$

in which $C_P = H_L + BV_L - RV_L|V_L|^m$ and $C_M = -H_R + BV_R - RV_R|V_R|^m$ [18].

Under the condition that the wavespeed is constant in the pipe, Equations (52) and (53) or Equations (56) and (57) have formally the same expression as that with *FI*. The interpolation causes the numerical damping and dispersion due to the error, especially for sharp transients [39].

To avoid this inaccuracy, three other characteristics methods are known: 1. the reachback time-line interpolations method with the characteristics extending back in time [12,52], 2. the method with spatial interpolations over more than a single reach [46,50], and 3. combined implicit method with characteristics method [33]. The former two methods are useful, especially in transient analysis where the wavespeed in a pipeline varies within a wide range. The time step size with IMSTI is restricted to small one by the Courant-Lewy-Friedrichs condition for stability, and to overcome that difficulty, the implicit time-line interpolations method [12] has been proposed to give a larger time step size for computational efficiency, which corpolates the advantages of other implicit numerical schemes.

COMBINED IMPLICIT METHOD WITH CHARACTERISTIC METHOD (CIWC) AND IMPLICIT METHOD

It is effective to use *CIWC* in dealing with very short pipes relative to other pipes and the numerical damping caused is less than those with the lumped model and *IMSTI*. The centered-implicit finite difference is applied to the short reach or the pipe with length, Δx, which satisfies the condition that $\Delta x < a \cdot \Delta t$ as shown in Figure 5. $\bar{V}_U$, $\bar{H}_U$, $\bar{V}_D$, and $\bar{H}_D$ are knowns while V_U, H_U, V_D and H_D are unknowns.

$$\begin{aligned}
\frac{\partial H}{\partial x} &= \frac{H_U + \bar{H}_U - (H_D + \bar{H}_D)}{2 \cdot \Delta x} \\
\frac{\partial H}{\partial t} &= \frac{H_D + H_U - (\bar{H}_D + \bar{H}_U)}{2 \cdot \Delta t} \\
\frac{\partial V}{\partial x} &= \frac{V_U + \bar{V}_U - (V_D + \bar{V}_D)}{2 \cdot \Delta x} \\
\frac{\partial V}{\partial t} &= \frac{V_D + V_U - (\bar{V}_D + \bar{V}_U)}{2 \cdot \Delta t} \\
V &= \frac{\bar{V}_D + \bar{V}_U}{2}
\end{aligned} \qquad (58)$$

Substituting the Equations (58) into the simplified basic equations, we have

$$\begin{aligned}
V_D + V_U + C_1H_U - C_1H_D &= C_2 \\
-V_D + V_U + C_3H_U + C_3H_D &= C_4
\end{aligned} \qquad (59)$$

in which $C_1 = g \cdot \Delta t/\Delta x$, $C_2 = (\bar{V}_D + \bar{V}_U) - C_1(\bar{H}_U - \bar{H}_D) - \Delta t F(\bar{V}_D + \bar{V}_U)|(\bar{V}_D + \bar{V}_U)/2|^m$, $C_3 = (g/a^2)(\Delta x/\Delta t)$, and $C_4 = C_3(\bar{H}_D + \bar{H}_U) - (\bar{V}_U - \bar{V}_D)$. Two other equations are required for solving four unknowns. They are

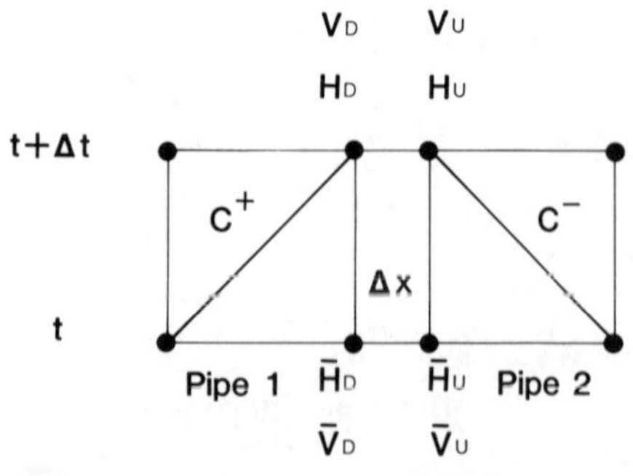

FIGURE 5. Combined implicit method with characteristic method.

generally the boundary conditions or the compatibility equations valid for the characteristic lines travelling the reach considered. In the case as shown in Figure 5, two compatibility equations are applied,

$$C^+: \quad H_D + B_1 V_D = C_{P1}$$

$$C^-: \quad -H_U + B_2 V_U = C_{M2}$$

$$(FI \text{ and } IMSTI)$$

in which the subscripts below the variables 1 and 2, denote the upstream and downstream pipes, respectively.

The method above described is called "the combined implicit method with the characteristics method" [40]. The model to be combined should be the first approximation model (FI). Giving the velocity in the frictional term with the average of four values, i.e., V_U, $\overline{V}_U$, V_D, and $\overline{V}_D$,

$$V = \frac{V_U + V_D + \overline{V}_U + \overline{V}_D}{4} \ (PNR) \qquad (60)$$

the higher accuracy may be attained when PNR with the second order accuracy is combined [42].

And it is possible to compose the implicit method by using only the centered implicit method [42]. In this case, Δt can be changed in each time step.

Selection For Δx and Δt

In general, any implicit method has no restrictions on the ratio of the distance and time step sizes, $\Delta x/\Delta t$. On the other hand, when any explicit numerical model is employed, the ratio, $\Delta x_i/\Delta t_i$, for ith pipe in a pipeline with multi-pipes must always satisfy the Courant stability condition [6],

$$\frac{\Delta x_i}{\Delta t_i} \geqq \max|V_i + a_i| \qquad (61)$$

or

$$\frac{\Delta x_i}{\Delta t_i} \geqq a_i \quad \text{for the simplified equations} \qquad (62)$$

In addition, it is convenient to use the time step size common with all the pipes in computations except for the implicit method. Let be the number of reaches M_i, the pipe length, L_i, the distance and time step size, Δx_i and Δt_i, the wavespeed, a_i, the velocity, V_i, for ith pipe, and the common time step, Δt. The Courant stability conditions are

$$\Delta t \leqq \Delta t_i \leqq \frac{L_i}{M_i} \frac{1}{\max|V_i + a_i|} \quad (i = 1, Np) \qquad (63)$$

in which Np = the total number of pipes. For the simplified euqations, we have

$$\Delta t \leqq \Delta t_i \leqq \frac{L_i}{M_i a_i} \quad (i = 1, Np) \qquad (64)$$

as well as Equation (63). With the method of specified time interval, the equality must be valid but it is difficult to determine a set of integers, M_i, giving the common time step size. To do this, the pipe lengths and the wavespeeds are slightly modified, that is,

$$L_i' = L_i + \Delta L_i$$
$$a_i' = a_i + \Delta a_i \qquad (65)$$

in which L_i' and a_i' are the length and wavespeed adjusted. When much shorter pipes are included in a pipeline with multi pipes, one is forced to select a very small time step size, which reduces in computational efficiency. To avoid that situation, there are three devices [40]: 1. the lumped model due to the rigid column theory, 2. the interpolated method of specified time interval, and 3. the combined implicit method with characteristic methods.

Subscript Notations of Grid Points

For computational purposes, two integer subscripts, I and J, may be used. Let M be the number of reaches in a pipe, and then the total number of grid points, MS, is equal to $(M + 1)$. The subscript I, denotes the position of grid points, that is, $I = 1$ to MS corresponds to the grid at the upstream end and to the grid at the downstream end, respectively. The subscript, J, denotes the time of grid points, that is, $J = 1$ to 2 means $t = t$ and $t = t + \Delta t$, respectively. For computations at an interior grid, V_A, H_A, V_C, H_C, V, and H at the points A, B, and P as shown in Figures 4a and 4b are expressed as

$$V_A = V(I-1,1),\ H_A = H(I-1,1),\ V_C = V(I+1,1)$$

$$H_C = H(I+1,1),\ V = V(I,2), \text{ and } H = H(I,2)$$

For computations at boundary grids as shown in Figure 6, V_A, H_A or V_C, H_C are expressed as

$$V_A = V(M,1) \text{ and } H_A = H(M,1) \quad \text{on } C^+ \text{ (downstream end of pipe)}$$

or

$$V_C = V(2,1) \text{ and } H_C = H(2,1) \quad \text{on } C^- \text{ (upstream end of pipe)}$$

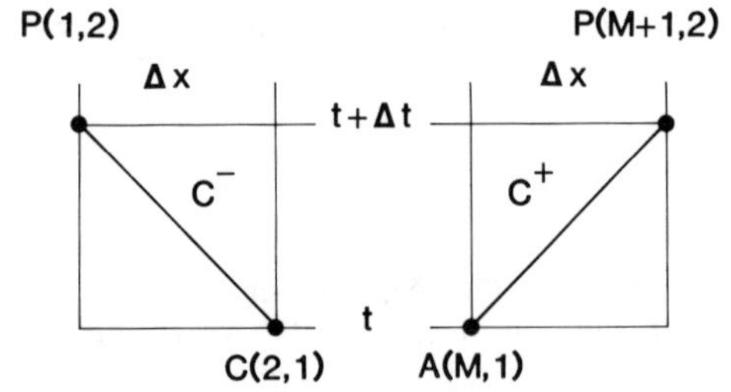

FIGURE 6. Calculations at boundary points.

when C^+ or C^- compatibility equation is applied to the downstream or upstream end, respectively.

Since it is convenient, in a pipeline with a number of pipes, to use the subscript, I, as the subscript for one-dimensional array expressing all grid points in a series, the following data are required:

$M(N_i)$ = the number of reaches

$MS(N_i)$ = the number of grid points, $M(N_i) + 1$

for the N_ith pipe and $NGP(N_i)$ = the sum of all grid points for pipes 1 to N_i,

$$NGP(N_i) = \sum_{L=1}^{N_i} MS(L)$$

From the above data, computations at the interior points of the Nth pipe is formulated as

$DO\ 100\ I = NGP(N_i - 1) + 2, NGP(N_i) - 1$

Calculation algorithm at interior points

100 CONTINUE

in which we define that $NGP(0) = 0$.

For example, Figure 7 shows two branching pipes composed of pipes 1, 2, and 3 ($MS(1) = 3$, $MS(2) = 4$, $MS(3) = 3$, $NGP(1) = 3$, $NGP(2) = 7$, and $NGP(3) = 10$). After the computations at $t = t + \Delta t$ are finished and before it proceeds to the next time step, it is necessary to transfer the values stored in the array $(I,2)$ into the array $(I,1)(I = 1$ to $NGP(N_i))$.

For computations at boundary points, the incidence matrix, defined in the graph theoretic formulation, may be useful to give how many pipes and how they connect to the boundary grids as well as in the steady flow analysis based on the basic equation of steady flow [7,19]. In addition, another data to distinguish the types of boundary such as two branching, valve inline, surge tank, and so on should be offered by some indexes [38].

Initial Conditions and Steady Flow Calculations

Numerical analysis for water hammer, which is formulated as in the initial and boundary values problems of partial differential equations, requires the initial and boundary conditions. The initial condition is usually given by a steady-state condition before the transients start. The steady flow analysis, in which the proper systematic calculations are organized to solve the simultaneous non-linear equations composed of the head loss and continuity equations [19,32,48], is used for preparing the initial steady condition. An alternative is to use the transient analysis: When we run a program of transient analysis under the time invarying boundary conditions, the computational results converge a steady flow solution if the numerical stability is guaranteed. But, the transient analysis consumes much more time than the steady flow analysis. Therefore, special devices to reduce the computational time are made in the transient analysis such that the whole pipe length is taken as a reach [10] or the wavespeeds were changed to much lower ones within the extent that the Courant condition for stability is satisfied [47]. It is also possible to check the program for the transient analysis through investigating whether the converged steady flow solution is essentially identical with that obtained by the steady flow analysis or not. It should be noted that the converged solution is not always identical with the steady flow condition obtained by the steady flow analysis [34,54,56]. The discrepancy is mainly due to the two reasons: 1. the concept of steady flow due to the rigid column theory is also used in the analysis in the elastic theory and 2. the finite difference equations is not exactly valid for steady flows.

In the rigid column theory the steady flow for a uniform

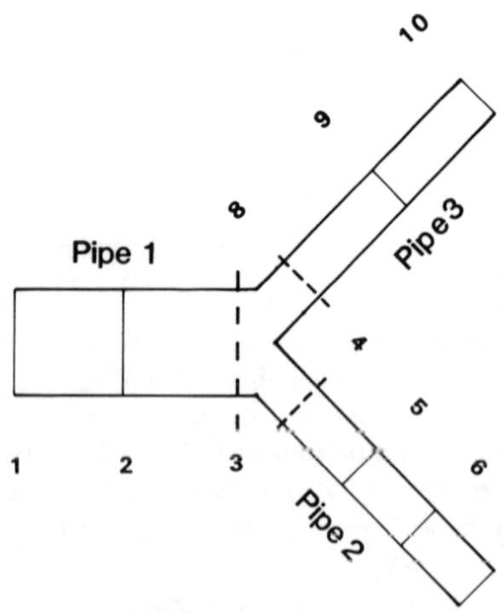

FIGURE 7. Indexing to grid points for complex piping systems.

pipe is usually described by

$$\frac{dH}{dx} = -FV|V|^m \qquad (66)$$

$$\frac{dV}{dx} = 0 \quad \text{or} \quad \frac{dQ}{dx} = 0$$

while the basic equations for the steady flow in the elastic theory are given by setting the partial differential terms with respect to time in Equations (14) and (19) to be zero,

$$\begin{pmatrix} V, & (a^2/g) \\ g, & V \end{pmatrix} \begin{pmatrix} \dfrac{dH}{dx} \\ \dfrac{dV}{dx} \end{pmatrix} = \begin{pmatrix} V \sin\theta \\ -FV|V|^m \end{pmatrix} \qquad (67)$$

that is,

$$\frac{dH}{dx} = \frac{-V}{a^2 - V^2}(V \sin\theta + (a^2/g)FV)$$

$$\frac{dV}{dx} = \frac{V}{a^2 - V^2}(g \sin\theta + FV|V|^m) \qquad (68)$$

or

$$\frac{dQ}{dx} = \frac{a^2}{(K/\rho)} A \frac{dV}{dx}$$

Though Equation (66) is usually very good approximation to Equation (68) for practical purposes, strictly speaking, they are not identical to each other.

For the simplified equations *FI*, *LSAM*, and *PCS* hold for steady flows while *PNR* produces the error due to neglecting the smaller terms during the transients. The magnitude of the maximum error, δ_v, in velocity is estimated by discussing the valve closure problem in the simple piping system with a reservoir at the upstream and a valve at the downstream [35]:

$$\delta_v \cong M \cdot \Delta$$

$$\Delta \cong |V_S^* - V_S|/V_0 = \frac{\varepsilon^3}{B_A^3} \times \left(16 - 64\frac{\varepsilon}{B_A} v_S + 128\frac{\varepsilon^2}{B_A^2} v_S^2\right) v_S \qquad (69)$$

in which V_s = the steady flow velocity specified at the downstream (m/s), v_S = the dimensionless velocity, $v_s = \bar{V}_s/\bar{V}_0$ = the representative velocity (m/s), H_0 = the constant head at the reservoir (m), V_S^* = the finally converged value of velocity (m/s), $\varepsilon = \sigma/(4M)$, $\sigma = fV_0^2L/(2gDH_0)$, $B_A = (aV_0)/(gH_0)$, M = the number of reaches, and the frictional term is assumed to be proportional to the velocity square. The computational results are shown in Table 1. To reduce the error, it is effective to use the average of the heads obtained by the two compatibility equations as shown in Table 2. With this the maximum error in velocity is estimated by

$$\delta_v \cong \Delta \qquad (70)$$

We find that the error can be neglected for the usual number of reaches.

TABLE 1.

Error of Steady Flow Calculations

M	PNR		
(1)	(2) $v_s = 1.0$	(3) $v_s = 0.5$	(4) $v_s = 0.25$
5	0.05587	0.00328	0.00020 (Numeric)
	0.04095	0.00305	0.00021 (Eq. 69)
10	0.01405	0.00087	0.00005 (Numeric)
	0.01219	0.00083	0.00005 (Eq. 69)
20	0.00357	0.00022	0.00001 (Numeric)
	0.00333	0.00022	0.00001 (Eq. 69)

$B_A = 0.5 \quad \sigma = 0.9 \quad m = 1$

Boundary Conditions

The computations different from those at the interior grid points are necessary at boundary grid points where a cross section area discontinuously varies to another such as at

TABLE 2. (The head is calculated by the average of values obtained by the two compatibility equations).

Error of Steady Flow Calculations

M	PNR		
(1)	(2) $v_s = 1.0$	(3) $v_s = 0.5$	(4) $v_s = 0.25$
5	0.00553	0.00036	0.00002 (Numeric)
	0.00819	0.00061	0.00004 (Eq. 70)
10	0.00072	0.00005	0.00000 (Numeric)
	0.00122	0.00008	0.00001 (Eq. 70)
20	0.00009	0.00001	0.00000 (Numeric)
	0.00017	0.00001	0.00000 (Eq. 70)

$B_A = 0.5 \quad \sigma = 0.9 \quad m = 1$

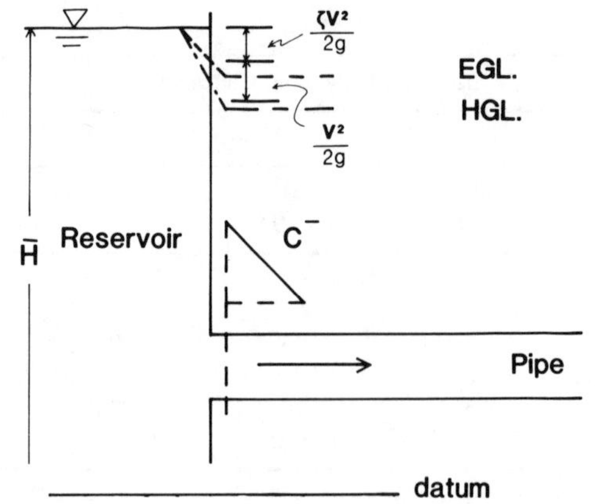

FIGURE 8a. Reservoir with the specified head at the upstream end of pipe.

valves, junctions at compound pipes, branching pipes, and surge tank, pumps, inlet or exit at a reservoir, and so on. Some new relationships related to the continuity of mass and energy are required to obtain the boundary conditions, in addition to the compatibility equations valid for on the characteristics travelling at the boundary point considered. From a mathematical point of view, the number of unknowns must be equal to the number of the relations, and the unknowns are obtained by solving simultaneous nonlinear or linear equations.

In this section, the expressions for some representative boundary conditions are offered for some kinds of the solution procedures:

1. The boundary condition for the first-order approximation model combined with the method of specified time interval is first given.
2. The boundary conditions with LSAM are formally obtained by replacing B in C^+ and B in C^- with *FI* by B_P and B_M in Equations (37) and (38), respectively.
3. For the boundary conditions with *PNR*, the initial estimates given by the non-viscous solution are formally identical with the expressions with *FI*, though C_P 'and C_M with *FI* are different from those with *PNR* in content. The correction to the initial estimates is calculated only one time by the Newton-Raphson algorithm.
4. The boundary condition with *IMSTI* is formally the same expressions as that with *FI* when the wavespeed is assumed to be constant.

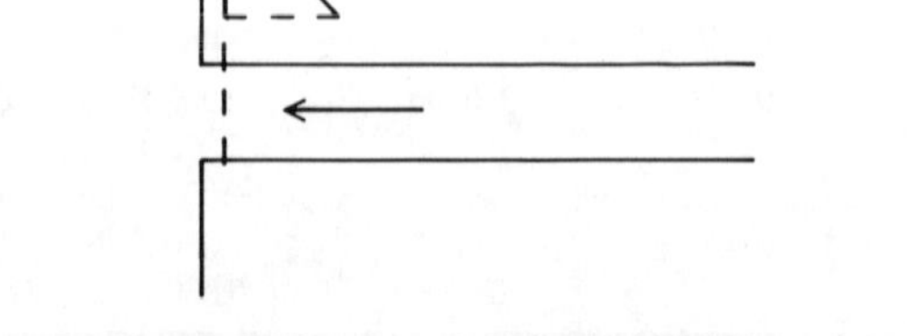

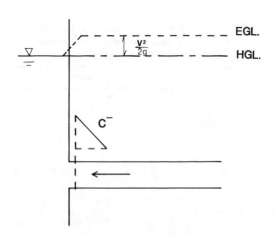

FIGURE 8b. Reverse flow as well as Figure 8a.

RESERVOIR WITH A SPECIFIED HEAD, $\overline{H}$

Pipe Next to a Reservoir at the Upstream End
Figures 8a and 8b

$H = \overline{H}$ when the minor losses are neglected.

$$FI \quad C^-: \quad -H + BV = C_M$$
$$V = (\overline{H} + C_M)/B \tag{71}$$

$$LSAM \quad V = (\overline{H} + C_M)/B_M \tag{72}$$

$$PNR \quad V = V^* - \frac{FM(V^*)}{YM(V^*)} \tag{73}$$

in which V^* is formally given by Equation (71).

When the minor losses are significant, the energy equation is added [5].

$$H = \overline{H} - (1 + \zeta)\frac{V^2}{2g}\zeta \tag{74}$$

in which ζ = the pipe entrance loss coefficient.

$$FI \quad V = \frac{2(C_M + \overline{H})}{B + \sqrt{B^2 + \dfrac{2(1+\zeta)}{g}(C_M + \overline{H})}} \tag{75}$$

LSAM B in Equation (75) is replaced by B_M in Equation (38).

$$PNR\ V = V^* - \frac{FM(V^*)}{YM(V^*) + \dfrac{(1+\zeta)}{g}V^*} \tag{76}$$

in which V^* is formally given by Equation (75).

For reverse flow, Equations (71) to (73) are applied or

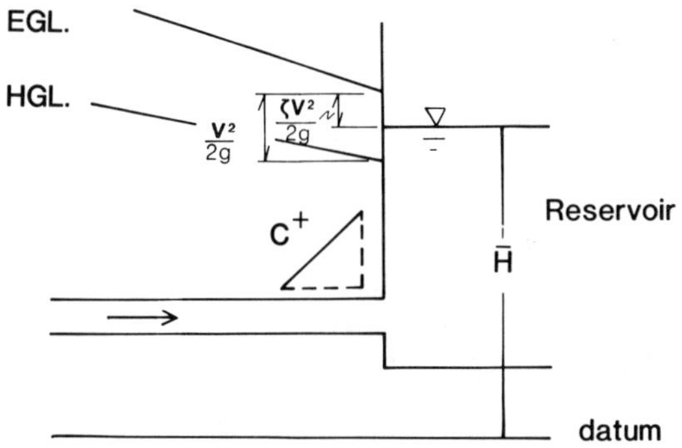

FIGURE 9. Reservoir at the downstream end.

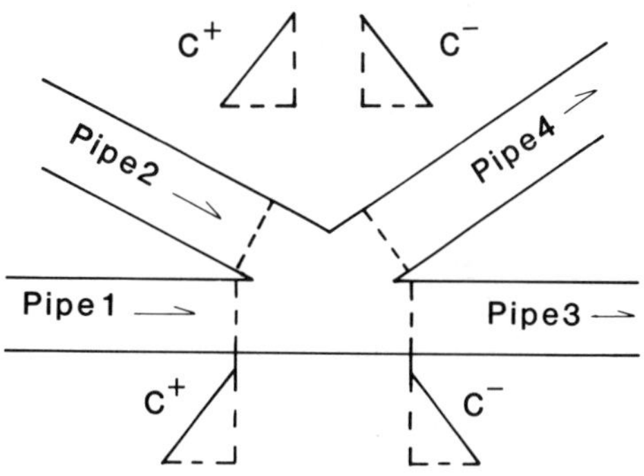

FIGURE 10. Four pipes connection.

replace ζ in Equations (75) to (76) by -1 when the kinetic energy is all lost.

Pipe Next to a Reservoir at the Downstream End (Figure 9)

$$\text{energy equation} \quad H = \bar{H} - (1 - \zeta)\frac{V^2}{2g} \tag{77}$$

FI C^+: $H + BV = C_P$

$$V = \frac{2(C_P - \bar{H})}{B + \sqrt{B^2 - \dfrac{2(1 - \zeta)}{g}(C_P - \bar{H})}} \tag{78}$$

LSAM B is replaced by B_P in Equation (37).

$$PNR \; V = V^* - \frac{FP(V^*)}{YP(V^*) - \dfrac{(1 - \zeta)}{g} V^*} \tag{79}$$

in which V^* is formally given by Equation (78).

COMPOUND PIPES

Branching Junction Composed of Four Pipes as Shown in Figure 10

Continuity Equation

$$A_1V_1 + A_2V_2 = A_3V_3 + A_4V_4$$

Heads Continuity

$$H = H_1 = H_2 = H_3 = H_4$$

in which the subscripts 1, 2, 3, and 4 denote pipe 1, 2, 3, and 4, respectively.

$$\begin{aligned} FI \quad C_1^+&: \quad H_1 + B_1V_1 = C_{P1} \\ C_2^+&: \quad H_2 + B_2V_2 = C_{P2} \\ C_3^-&: \quad -H_3 + B_3V_3 = C_{M3} \\ C_4^-&: \quad -H_4 + B_4V_4 = C_{M4} \end{aligned} \tag{80}$$

The unknown head is solved as follows,

$$H = \frac{\left(\dfrac{A_1}{B_1}\right) C_{P1} + \left(\dfrac{A_2}{B_2}\right) C_{P2} - \left(\dfrac{A_3}{B_3}\right) C_{M3} - \left(\dfrac{A_4}{B_4}\right) C_{M4}}{\sum_{i=1}^{4}\left(\dfrac{A_i}{B_i}\right)} \tag{81}$$

The velocities, V_1, V_2, V_3, and V_4 are obtained by substituting H into the corresponding compatibility equations, respectively.

$$\begin{aligned} V_1 &= (C_{P1} - H)/B_1 \\ V_2 &= (C_{P2} - H)/B_2 \\ V_3 &= (C_{M3} + H)/B_3 \end{aligned} \tag{82}$$

and

$$V_4 = (C_{M4} + H)/B_4$$

LSAM B_1, B_2, B_3, and B_4 in Equations (81) and (82) are replaced by B_{P1}, B_{P2}, B_{M3}, and B_{M4} respectively.

PNR The compatibility equations are

$$C_1^+: \quad H_1 + B_1V_1 + FP1(V_1) = C_{P1}$$

$$C_2^+: \quad H_2 + B_2V_2 + FP2(V_2) = C_{P2}$$

$$C_3^-: \quad -H_3 + B_3V_3 + FM3(V_3) = C_{M3}$$

$$C_4^-: \quad -H_4 + B_4V_4 + FM4(V_4) = C_{M4}$$

The non-linear equations for V_2, V_3, and V_4 are

$$E_2 = B_1V_1 + FP1(V_1) - C_{P1} - B_2V_2 - FP2(V_2) + C_{P2} = 0$$

$$E_3 = B_1V_1 + FP1(V_1) - C_{P1} + B_3V_3 + FM3(V_3) - C_{M3} = 0$$

$$E_4 = B_1V_1 + FP1(V_1) - C_{P1} + B_4V_4 + FM4(V_4) - C_{M4} = 0$$

in which $V_1 = (-A_2V_2 + A_3V_3 + A_4V_4)/A_1$.

The initial estimates, H^*, V_1^*, V_2^*, V_3^*, and V_4^* are formally given by Equations (81) and (82).

The corrections to V_2^*, V_3^*, and V_4^* are obtained by solving

$$\begin{pmatrix} -(\alpha_2Y_1 + Y_2), & \alpha_3Y_1, & \alpha_4Y_1 \\ -\alpha_2Y_1, & \alpha_3Y_1 + Y_3, & \alpha_4Y_1 \\ -\alpha_2Y_1, & \alpha_3Y_1, & \alpha_4Y_1 + Y_4 \end{pmatrix} \begin{pmatrix} \Delta V_2 \\ \Delta V_3 \\ \Delta V_4 \end{pmatrix} = -\begin{pmatrix} E_2^* \\ E_3^* \\ E_4^* \end{pmatrix} \tag{83}$$

in which $\alpha_k = A_k/A_1 (k = 2, 3, \text{and } 4)$, $Y_1 = YP1(V_1^*)$, $Y_2 = YP2(V_2^*)$, $Y_3 = YM3(V_3^*)$, $Y_4 = YM4(V_4^*)$, $E_2^* = FP1(V_1^*) - FP2(V_2^*)$, $E_3^* = FP1(V_1^*) + FM3(V_3^*)$, and $E_4^* = FP1(V_1^*) + FM4(V_4^*)$.

Equation (83) gives the solution for V_2, V_3, and V_4 with the required accuracy, though any iteration in the Newton-Raphson algorithm is not made.

Two Branching Junction (Figure 11)

$$A_1V_1 = A_2V_2 + A_3V_3 \quad \text{and} \quad H = H_1 = H_2 = H_3$$

$$FI \quad H = \frac{\left(\frac{A_1}{B_1}\right) C_{P1} - \left(\frac{A_2}{B_2}\right) C_{M2} - \left(\frac{A_3}{B_3}\right) C_{M3}}{\sum_{k=1}^{3} \frac{A_k}{B_k}} \tag{84}$$

$$V_1 = (C_{P1} - H)/B_1$$

$$V_2 = (C_{M2} + H)/B_2$$

$$V_3 = (C_{M3} + H)/B_3$$

LSAM B_1, B_2, and B_3 in Equation (84) are replaced by B_{P1}, B_{M2}, and B_{M3}, respectively.

PNR The compatibility equations are

$$C_1^+: \quad H_1 + B_1V_1 + FP1(V_1) = C_{P1}$$

$$C_2^-: \quad -H_2 + B_2V_2 + FM2(V_2) = C_{M2}$$

$$C_3^-: \quad -H_3 + B_3V_3 + FM3(V_3) = C_{M3},$$

and the non-linear equations are

$$E_2 = B_1V_1 + FP1(V_1) - C_{P1} + B_2V_2 + FM2(V_2) - C_{M2} = 0$$

$$E_3 = B_1V_1 + FP1(V_1) - C_{P1} + B_3V_3 + FM3(V_3) - C_{M3} = 0$$

in which $V_1 = (A_2V_2 + A_3V_3)/A_1$.

The first estimates are formally given by Equation (84). The corrections to V_2 and V_3, are obtained by solving

$$\begin{pmatrix} \alpha_2Y_1 + Y_2, & \alpha_3Y_1 \\ \alpha_2Y_1, & \alpha_3Y_1 + Y_3 \end{pmatrix} \begin{pmatrix} \Delta V_2 \\ \Delta V_3 \end{pmatrix} = -\begin{pmatrix} E_2^* \\ E_3^* \end{pmatrix} \tag{85}$$

in which $Y_1 = YP1(V_1^*)$, $Y_2 = YM2(V_2^*)$, $Y_3 = YM3(V_3^*)$,

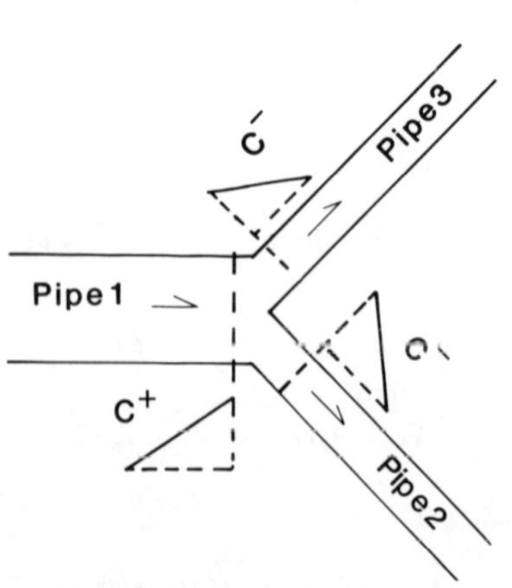

FIGURE 11. Two branching pipes.

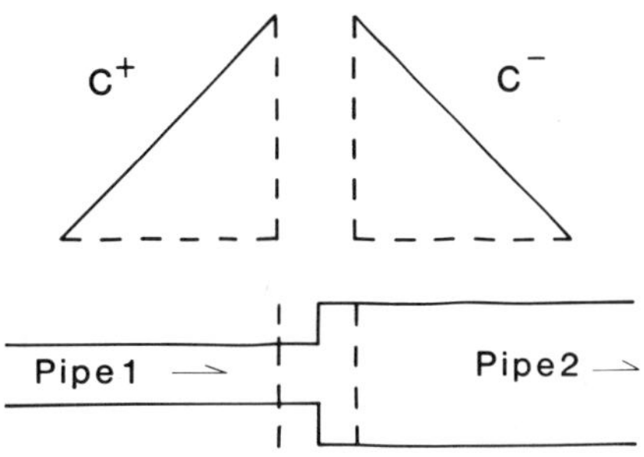

FIGURE 12. Series pipes.

$E_2^* = FP1(V_1^*) + FM2(V_2^*)$, and $E_3^* = FP1(V_1^*) + FM3(V_3^*)$.

V_1^* and V_2^*, and V_3^* are formally given by Equation (84). The head is given by substituting V_1, V_2, or V_3 into the corresponding compatibility equations, respectively.

Series Pipes (Figure 12)

$$A_1V_1 = A_2V_2 \quad \text{and} \quad H = H_1 = H_2$$

$$FI \quad H = \frac{\left(\dfrac{A_1}{B_1}\right) C_{P1} - \left(\dfrac{A_2}{B_2}\right) C_{M2}}{\sum_{k=1}^{2} \dfrac{A_k}{B_k}} \tag{86}$$

$$V_1 = (C_{P1} - H)/B_1$$

$$V_2 = (A_1/A_2)V_1$$

LSAM B_1 and B_2 are replaced by B_{P1} and B_{M2}, respectively.
PNR The compatibility equations are

$$C_1^+: \quad H_1 + B_1V_1 + FP1(V_1) = C_{P1}$$

$$C_2^-: \quad -H_2 + B_2V_2 + FM2(V_2) = C_{M2},$$

and the non-linear equation for V is

$$E = B_1V_1 + FP1(V_1) - C_{P1} + B_2V_2 + FM2(V_2) - C_{M2} = 0 \tag{87}$$

in which $V_1 = (A_2V_2)/A_1$.

The correction to V_2 is

$$\Delta V_2 = -\frac{E^*}{\alpha_2 Y_1 + Y_2}$$

in which $Y1 = YP1(V_1^*)$, $Y2 = YM2(V_2^*)$, $E^* = FP1(V_1^*) + FM2(V_2^*)$ and V_1^* and V_2^* = the initial estimate formally given by Equation (86).

VALVES

In dealing with the boundary conditions including a valve, test data for the head loss characteristics of the valve has to be known. The expressions of the head loss characteristics are generally given in three different ways [48]:

1. The loss coefficient ζ

The following equations describe the head loss through the valve,

$$\Delta H = \zeta \frac{V^2}{2g} \tag{88}$$

$$\Delta H_0 = \zeta_0 \frac{V_0^2}{2g} \tag{89}$$

in which the subscript 0 denotes the head loss in the full opening of the valve.

2. The flow coefficient C_v

C_V is related to the pipe diameter, $D(m)$, and the loss coefficient in SI unit,

$$C_V = 46093\ D^2/\sqrt{\zeta} \tag{90}$$

3. Tau-curves

The discharge through the valve is described by

$$Q = AV = (C_dA_V)\sqrt{2g\Delta H}$$
$$Q_0 = AV_0 = (C_dA_V)_0\sqrt{2g\Delta H_0} \tag{91}$$

in which C_dA_V = the area of the opening, A_V, times the discharge coefficient, C_d, and the subscript 0 denotes the quantity in full opening of valve.

$$\tau = \sqrt{\frac{\zeta_0}{\zeta}} = \sqrt{\frac{(C_dA_V)}{(C_dA_V)_0}} \tag{92}$$

ζ, C_V, or τ must be specified as a function of valve stem movement with respect to time in transient computations. Interpolated approximations in linear or parabolic equations are required to numerically deal with the head loss characteristics generally offered in tabular form.

Valve in line (Figure 13)

$$H_1 = H_2 \pm \zeta \frac{V^2}{2g} \quad \text{and} \quad V_1 = V_2$$

(The minus sign is valid for reverse flow.)

$$FI \quad C_1^+: \quad H_1 + B_1V_1 = C_{P1}$$
$$C_2^-: \quad -H_2 + B_2V_2 = C_{M2} \tag{93}$$

$$V = V_1 = V_2$$

$$= \frac{2(C_{P1} + C_{M2})}{B_1 + B_2 + \sqrt{(B_1 + B_2)^2 \pm \frac{2\zeta}{g}(C_{P1} + C_{M2})}} \tag{94}$$

$$(V \geqslant 0)$$

LSAM B_1 and B_2 are replaced by B_{P1}, B_{M2}, respectively, in Equations (93) and (94).

$$PNR \quad V = V^* - \frac{FP1(V^*) + FM2(V^*)}{YP1(V^*) + YM2(V^*) \pm \left(\frac{\zeta}{g}\right) V^*} \tag{96}$$

in which the first estimates, V_1^* is formally given by Equation (94) and plus or minus sign is valid for normal or reverse flow, respectively.

Pressure Reducing Valve (PRV) [17,48]

PRV is designed to control an excessive pressure and maintain a constant downstream pressure, H_S, not depending on the upstream pressure. *PRV* with a spring actuator and no-damping is assumed to instantaneously change in flow conditions. *PRV* has three different operational models depending on the upstream and downstream heads, H_1 and H_2, respectively.

$$1. \; H_1 - H_2 > \Delta H_m \tag{97}$$

ΔH_m is the minimum pressure difference for *PRV* to normally operate. Then we have

$$H_2 = H_S \tag{98}$$

$$FI \quad V_1 = V_2 = (C_{M2} + H_S)/B_2 \quad \text{and}$$
$$H_1 = C_{P1} - B_1V_1 \tag{99}$$

LASM B_1 and B_2 are replaced by B_{P1} and B_{M2}, respectively in Equation (99).

$$PNR \quad V_2 = V_2^* - \frac{FM2(V_2^*)}{YM2(V_2^*)} \tag{100}$$

in which V_2^* is the non-viscous solution given formally by Equation (99).

$$2. \; H_2 < H_1 < H_2 + \Delta H_m \tag{101}$$

PRV does not work, and so if the cross section areas on both sides of *PRV* are the same, then the equations valid for "valve in line" are applied.

$$3. \; H_2 > H_1 \tag{102}$$

PRV operates as a check valve to prevent reverse flow situations.

$$V_1 = V_2 = 0 \tag{103}$$

To obtain the heads, both sides of the valve are considered as the dead ends:

$$FI \quad H_1 = C_{P1} \quad \text{and} \quad H_2 = -C_{M2}$$

$$PNR \quad H_1 = C_{P1} - FP1(0) \quad \text{and}$$
$$H_2 = FM2(0) - C_{M2}.$$

ONEWAY SURGING TANK (FIGURE 14)

Oneway surging tanks are often installed downstream near pumps to pour liquid from the tanks into the pipeline and prevent the pressure in the pipe from dropping below the liquid level in the tank, H_S. If $H_S + z > H$, then the check valve opens while if not so, the valve closes:

$$A_1V_1 + Q_S = A_2V_2 \quad \text{and}$$
$$Q_S = C_dA_V\sqrt{2g(H_S + z - H)} \tag{104}$$

$$H = H_1 = H_2$$

$$A_S\frac{dHs}{dt} = -Q_S \tag{105}$$

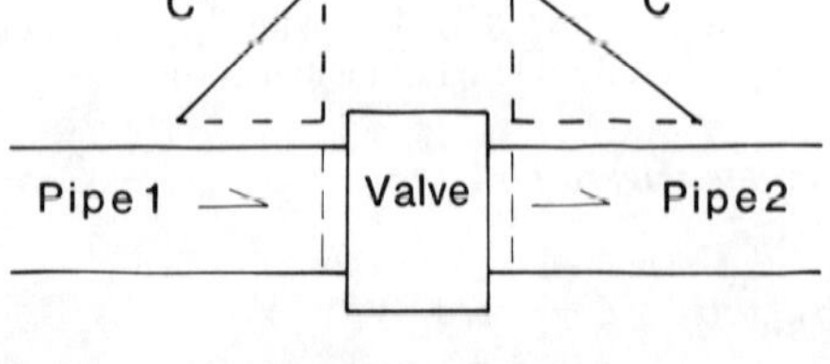

FIGURE 13. Valve in-line.

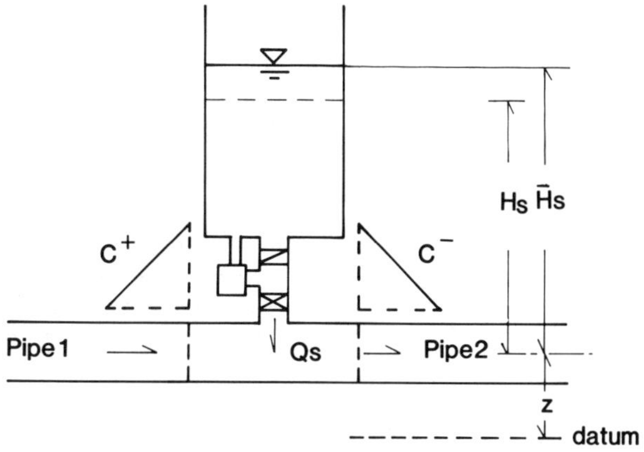

FIGURE 14. One-way surging tank.

in which Q_S = the discharge out of the tank, C_d = the discharge coefficient, A_V = the valve opening area, H_S = the water level in the tank from the pipe axis, z = the height of the pipe axis above a datum, and A_S = the cross section area of the tank.

Integration of Equation (105) is given by

$$H_S = \overline{H}_S - \frac{\Delta t}{2A_S}(Q_S + \overline{Q}_S) \tag{106}$$

in which $\overline{H}_S$ and $\overline{Q}_S$ = H_S and Q_S at the beginning of Δt, respectively.

$$\begin{aligned} FI \quad C_1^+&: \quad H_1 + B_1 V_1 = C_{P1} \\ C_2^-&: \quad -H_2 + B_1 V_2 = C_{M2} \end{aligned} \tag{107}$$

$$Q_S = \frac{2C_5}{\sqrt{C_4^2 + 4C_5} + C_4} \tag{108}$$

in which

$$C_4 = C_6\left(1 + \left(\frac{A_1}{B_1} + \frac{A_2}{B_2}\right)\frac{\Delta t}{2 \cdot A_S}\right)$$

$$C_5 = C_6\left[\left(\overline{H}_S + z - \frac{\overline{Q}_S \cdot \Delta t}{2A_S}\right)\right.$$

$$\left.\times \left(\frac{A_1}{B_1} + \frac{A_2}{B_2}\right) + \left(\frac{A_2}{B_2}\right) C_{M2} - \left(\frac{A_1}{B_1}\right) C_{P1}\right]$$

and

$$C_6 = \frac{2^1_g(C_dA_v)^2}{\dfrac{A_1}{B_1} + \dfrac{A_2}{B_2}}$$

H_S, H_1, V_1, and V_2 are obtained by substituting the known values into Equations (106), (105), and (107) in turns.

LSAM B_1 and B_2 are replaced by B_{P1} and B_{M2}, respectively.

PNR The corrections to V_1^* and V_2^* are obtained by solving

$$\begin{pmatrix} A_1C_7 - C_8 YP1(V_1^*), & -A_2C_7 \\ A_1C_7, & -A_2C_7 + C_8 YM2(V_2^*) \end{pmatrix} \times \begin{pmatrix} \Delta V_1 \\ \Delta V_2 \end{pmatrix} = -\begin{pmatrix} E_1^* \\ E_2^* \end{pmatrix} \tag{109}$$

in which the first estimates, V_1^* and V_2^*, are formally identical with the solution with *FI*, $C_7 = 2(gC_9(C_dA_V)^2 + A_1V_1^* - A_2V_2^*)$, $C_9 = -\Delta t/(2A_S)$, $C_8 = 2g(C_dA_V)^2$, $E_1^* = -C_8FP1(V_1^*)$, and $E_2^* = C_8FM2(V_2^*)$.

CENTRIFUGAL PUMP WITH A CONSTANT SPEED AT UPSTREAM END (FIGURE 15)

In two normally operating states, i.e., a pump start up and pump-stopping, the pump speed remains almost constant during the transients. As for pump start up the discharge valve is gradually operated as the speed of pump becomes the rated speed. In a pump-stopping the discharge valve is closed slowly, and then the power supply to the motor of pump is switched off. For a centrifugal pump

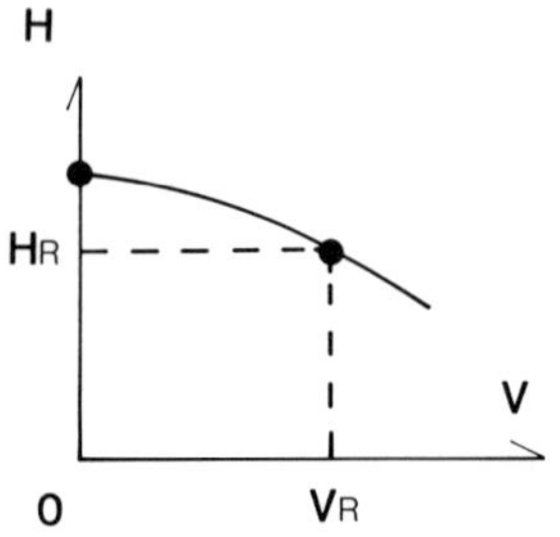

FIGURE 15. Pump characteristics at the constant speed.

running at a constant speed, the head-discharge is approximated by a parabolic equation:

$$H = H_S + C_1V - C_2V^2 \tag{110}$$

in which H_S = the shut-off head and C_1, C_2 = constants.

FI C^-: $-H + BV = C_M$

$$V = \frac{2(H_S + C_M)}{B - C_1 + \sqrt{(B - C_1)^2 + 4C_2(H_S + C_M)}} \tag{111}$$

LSAM B in Equation (111) is replaced by B_M.

PNR C^-: $-H + BV + FM(V) = C_M$

$$V = V^* - \frac{FM(V^*)}{YM(V^*) + 2C_2V^* - C_1} \tag{112}$$

in which V^* is formally given by Equation (111) with *FI*.

AIR CHAMBER (FIGURE 16)

An air chamber is installed near pumps to prevent extremely low pressures and column separation by allowing inflow into the outflow out of the chamber. The compressed air at its top contracts and expands depending on the inflow and outflow, respectively. A differential orifice is usually provided at the connection of the chamber and the pipeline so that the outflow from the chamber can become free rather than the inflow. The polytropic equation is assumed for air:

$$H_a \mathcal{V}_a^m = H_{a0} \mathcal{V}_{a0}^m = C \tag{113}$$

in which H_a and $\mathcal{V}_a$ = the absolute pressure head (m) and the volume (m^3) of the enclosed air at $t = t + \Delta t$, H_{a0} and $\mathcal{V}_{a0}$ = those in the rated condition (or initial steady condition), and m = the exponent in the polytropic gas equation.

The averaged value between $m = 1$ (isothermal) and $m = 1.4$ (adiabatic) may be assumed in the computations, though the polytropic equation is not always valid for modelling the behavior of air volume [13,14].

$$A_1V_1 = Q_a + A_2V_2 \quad \text{and} \quad H = H_1 = H_2$$

(Continuity of mass and head)

$$\Delta H_{or} = C_{or}Q_a|Q_a|$$

(head loss across the orifice)

$$H_a = H + H_b - z - \Delta H_{or} \tag{114}$$

(absolute pressure head of air)

$$\mathcal{V}_a = \overline{\mathcal{V}}_a - A_C(z - \bar{z})$$

(change in liquid level or air volume)

$$z = \bar{z} + \frac{\Delta t}{2 \cdot A_c}(Q_a + \overline{Q}_a)$$

in which H_b = the barometric pressure head, Q_a = the

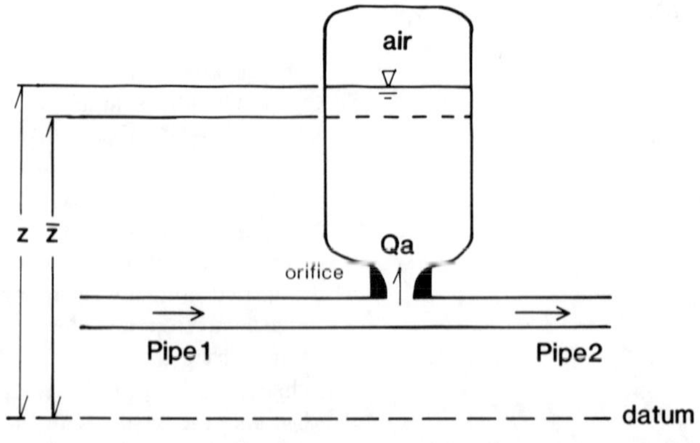

FIGURE 16. Air chamber.

discharge across the orifice, C_{or} = the coefficient related to the orifice, z = the height of liquid level in the chamber above a datum, ΔH_{or} = the head loss across the orifice, A_C = the cross section area of the chamber, H = the head at the junction between the pipeline and the chamber, and the bar over the variables denotes the known quantity at the beginning of Δt.

$$FI \quad C_1^+: \quad H_1 + B_1 V_1 = C_{P1}$$
$$C_2^-: \; -H_2 + B_2 V_2 = C_{M2} \tag{115}$$

The above equations give the non-linear equation with respect to Q_a, which is solved by the Newton-Raphson method:

$$E = (C_1 - C_2 Q_a - C_3 Q_a |Q_a|) \times (C_4 - C_5 Q_a)^m - C = 0$$

$$\frac{dE}{dQ_a} = -(C_2 + 2C_3|Q_a|)(C_4 - C_5 Q_a)^m - mC_5(C_1 - C_2 Q_a - C_3 Q_a|Q_a|) \times (C_4 - C_5 Q_a)^{m-1}$$

in which

$$C_1 = \frac{\left(\frac{A_1}{B_1}\right) C_{P1} - \left(\frac{A_2}{B_2}\right) C_{M2}}{\frac{A_1}{B_1} + \frac{A_2}{B_2}} + H_b - \left(\bar{z} + \frac{\Delta t}{2A_C}\bar{Q}_a\right)$$

$$C_2 = \frac{\Delta t}{2 \cdot A_C} + \frac{1}{\frac{A_1}{B_1} + \frac{A_2}{B_2}}, \; C_3 = C_{or}$$

$$C_4 = \bar{\forall}_a - \frac{\bar{Q}_a}{2}\Delta t, \quad \text{and} \quad C_5 = \Delta t/2$$

$$Q_a^{(k)} = Q_a^{(k-1)} - \left.\frac{E}{\frac{dE}{dQ_a}}\right|_{Q_a = Q_a^{(k-1)}} \tag{116}$$

The iteration is repeated until the tolerance is satisfied:

$$|Q_a^{(k)} - Q_a^{(k-1)}| < \varepsilon$$

LSAM B_1 and B_2 are replaced by B_{P1} and B_{M2}, respectively.

$$PNR \quad C_1^+: \quad H_1 + B_1 V_1 + FP1(V_1) = C_{P1}$$
$$C_2^-: \quad -H_2 + B_2 V_2 + FM2(V_2) = C_{M2}$$

$$E_1 = (C_1 - C_2 Q_a - C_3 Q_a|Q_a| - C_6 FP1(V_1) + C_7 FM2(V_2))(C_4 - C_5 Q_a)^m - C = 0 \tag{117}$$

$$E_2 = B_1 V_1 + B_2 V_2 + FP1(V_1) + FM2(V_2) - (C_{P1} + C_{M2}) = 0 \tag{118}$$

in which

$$V_2 = (A_1 V_1 - Q_a)/A_2, \; C_6 = \frac{\frac{A_1}{B_1}}{\frac{A_1}{B_1} + \frac{A_2}{B_2}}$$

and

$$C_7 = \frac{\frac{A_2}{B_2}}{\frac{A_1}{B_1} + \frac{A_2}{B_2}}$$

Therefore, the non-linear equations for V_1 and Q_a are solved by the Newton-Raphson method,

$$\begin{pmatrix} \frac{\partial E_1}{\partial V_1}, & \frac{\partial E_1}{\partial Q_a} \\ \frac{\partial E_2}{\partial V_1}, & \frac{\partial E_2}{\partial Q_a} \end{pmatrix} \begin{pmatrix} \Delta V_1 \\ \Delta Q_a \end{pmatrix} = - \begin{pmatrix} E_1 \\ E_2 \end{pmatrix} \tag{119}$$

in which

$$\frac{\partial E_1}{\partial V_1} = \left(-C_6 DP1(V_1) + C_7 DM2(V_2)\frac{A_1}{A_2}\right) \forall_a^m,$$

$$\frac{\partial E_1}{\partial Q_a} = (-C_2 - 2C_3|Q_a| - C_7 DM2(V_2)/A_2)\forall_a^m - H_a m C_5 \forall_a^{m-1}$$

$$\frac{\partial E_2}{\partial V_1} = YP1(V_1) + YM2(V_2)A_1/A_2$$

and

$$\frac{\partial E_2}{\partial Q_a} = -YM2(V_2)/A_2$$

The initial estimates for V_1, V_2, $\forall_a$, H_a, and Q_a can be formally given by the solution with *FI*.

Boundary Conditions For Pump Failure [3,51]

Emergency pump operations cause transients in pumping systems, and cavitation or column separation may occur at the place where the transient-state hydraulic line becomes below the elevation of the pipeline. The pipeline should be designed to bear the highest or relative lower pressure, and if necessary, effective devices such as one-way surging tank, air chamber, and/or a flywheel must be taken. For more accurate analysis except for pump normal operations, we must know the relationships, called "the complete pump characteristics, between the discharge, $Q(m^3/s)$, the pumping head, $H(m)$, the rotational speed, N(rpm), and the shaft torque acting on the fluid, T(Nm). The pump characteristics must be given in at least three zones (1, 2, and 3) of the four operations: 1. pump (normal) zone ($V \geqslant 0$ and $N \geqslant 0$), 2. energy dissipation zone ($V < 0$ and $N > 0$), 3. turbine zone ($V \leqslant 0$ and $N \leqslant 0$), and 4. turbine energy dissipation zone ($V > 0$ and $N < 0$). If the complete pump characteristics data are not available, then that is approximated by the pump characteristics for almost the same specific speed, N_S, (data for $N_S = 25$, 147, and 261 are shown in [8]).

HOMOLOGOUS RELATIONSHIPS AND COMPLETE PUMP CHARACTERISTICS [43]

The similarity law for turbomachines is used to predict the performance characteristics of a pump under some different rotational speeds or to give the pump characteristics of a series of pumps, which are geometrically similar to each other. The necessary condition for homologous pumps that two geometrically similar units have similar velocity vector diagrams gives the following equations:

$$\frac{H}{N^2D^2} = const \tag{120}$$

$$\frac{Q}{ND^3} = const \tag{121}$$

in which D = diameter of impeller (m).

Eliminating D from Equations (120) and (121), the specific speed is defined by

$$N_S = \frac{N_R \cdot Q_R^{1/2}}{H_R^{3/4}} \tag{122}$$

in which N_S = the specific speed and N_R, Q_R, and H_R denote the speed, the discharge, and the head at the rated condition. The rated values are generally specified at the point giving the best efficiency of the pump. For a particular unit, Equations (120) to (121) become

$$\frac{H}{N^2} = const \quad \text{and} \quad \frac{Q}{N} = const \tag{123}$$

since the diameter D can be included in the constants.

The equations related to the torque, T,

$$\frac{T}{Q^2} = const \quad \text{and} \quad \frac{T}{N^2} = const \tag{124}$$

are obtained by using that $T \cdot N$ is proportional to $Q \cdot H$. Q, H, N, and T are non-dimensionalized by Q_R, H_R, N_R, and T_R, respectively as follows,

$$v = Q/Q_R,\ h = H/H_R,\qquad \alpha = N/N_R, \quad \text{and} \quad \beta = T/T_R \tag{125}$$

Therefore, the dimensionless homologous relation is possible to be expressed as

$$f_1(h/\alpha^2, v/\alpha) = 0 \quad \text{and} \quad f_2(\beta/\alpha^2, v/\alpha) = 0 \tag{126}$$

or

$$g_1(h/v^2, \alpha/v) = 0 \quad \text{and} \quad g_2(\beta/v^2, \alpha/v) = 0 \tag{127}$$

in which f_1, f_2, g_1, and g_2 are functions to be known from test data. For example, the references [26,27,45] are available as the data in the form of Equations (126) and (127). But, since α and v become zero in analysing the transients due to a pump failure, it is convenient to use

$$f\left(\frac{h}{\alpha^2 + v^2}, \theta\right) = 0 \quad \text{and} \quad g\left(\frac{\beta}{\alpha^2 + v^2}, \theta\right) = 0 \tag{128}$$

in which f and g = functions of θ and $(h/\alpha^2 + v^2)$ or $(\beta/\alpha^2 + v^2)$, respectively, and $\theta = \pi + \tan^{-1}(v/\alpha)$ $(0 \leqslant \theta \leqslant 2\pi)$ [24].

To numerically deal with the curves described by Equation (128), discrete points on that curve at equal intervals of $\Delta\theta$, are stored in a digital computer [8,45]. Each segment corresponding to the curve between θ and $\theta + \Delta\theta$ is approximated by a straight line:

$$\frac{h}{\alpha^2 + v^2} = a_1 + a_2(\pi + \tan^{-1}(v/\alpha)) \tag{129}$$

in which a_1 and a_2 are constants giving the straight line, taking values different from those on other straight lines.

The dimensionless torque, β, is described, as well as Equation (129), by

$$\frac{\beta}{\alpha^2 + v^2} = b_1 + b_2(\pi + \tan^{-1}(v/\alpha)) \tag{130}$$

in which b_1 and b_2 are constants.

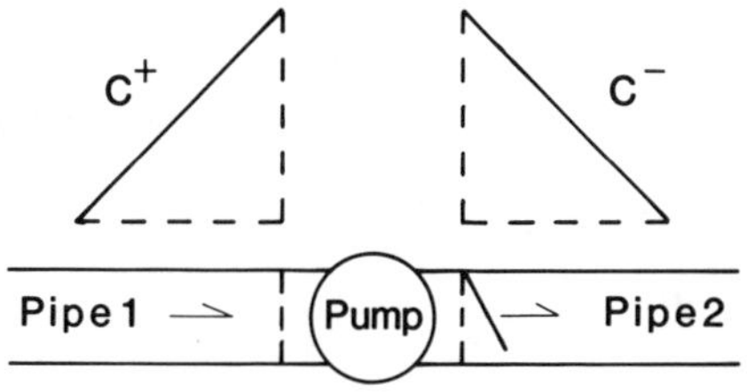

FIGURE 17. Pump in-line.

SINGLE PUMP IN PIPELINE

To develop the boundary condition at a pump as shown in Figure 17, in addition to the pump characteristics and two compatibility equations, two basic equations are required, i.e., the head balance equation through the pump and the equation related to the change in rotational speed of the pump.

Head Balance Equation

$$H_1 + H_{tdh} - \Delta H_v = H_2 \tag{131}$$

$$H_{tdh} = hH_R = \frac{V_d^2}{2g} + \frac{P_d}{\rho g} + z_d - \left(\frac{V_1^2}{2g} + \frac{P_1}{\rho g} + z_1\right) \tag{132}$$

$$\Delta H_v = \Delta H_0 v|v|/\tau^2 \tag{133}$$

$$v = Q/Q_R = A_1V_1/Q_R = A_2V_2/Q_R \tag{134}$$

in which H_{tdh} = the total dynamic head (energy increase per unit weight of liquid pumped) (m), the subscripts, d and 1, denote the discharge and suction flange, ΔH_v = the valve head loss, ΔH_0 = the head loss across the valve for the rated flow Q_R when $\tau = 1$, and V_2 and H_2 = the velocity and head downstream the valve.

$$FI \quad C_1^+: \quad H_1 + B_1V_1 = C_{P1} \quad \text{(on the suction-side)} \tag{135}$$

$$C_2^-: \quad -H_2 + B_2V_2 = C_{M2} \quad \text{(downstream the valve)} \tag{136}$$

The head balance equation becomes

$$C_1 + C_2v + H_R(\alpha^2 + v^2) \times (a_1 + a_2(\pi + \tan^{-1}(v/\alpha))) - \frac{\Delta H_0 \cdot v|v|}{\tau^2} = E_1(=0) \tag{137}$$

in which

$$C_1 = C_{P1} + C_{M2},\ C_2 = -Q_R\left(\frac{B_1}{A_1} + \frac{B_2}{A_2}\right)$$

and τ must be given at each time step.

Change in Rotational Speed of the Pump

In addition to Equation (137), another equation including only v and α is given by the relation of the unbalanced torque. After a pump failure, the power is switched off, and the change in speed is governed by

$$T = -\frac{W \cdot R_g^2}{g}\frac{d\omega}{dt} = -\frac{W \cdot R_g^2}{g}\frac{2\pi N_R}{60}\frac{d\alpha}{dt} \tag{138}$$

in which W = the weight of rotating parts plus entrained liquid, R_g = the radius of gyration of the rotating mass, ω = the angular velocity (rad/s).

Equation (138) is integrated, and the dimensionless expression becomes

$$\beta + \beta_0 + C_3(\alpha - \alpha_0) = 0 \tag{139}$$

in which $\beta_0 = T_0/T_R$, $\alpha_0 = N_0/N_R$, the subscript 0 denotes the corresponding quantities at the beginning of the time step, Δt, and

$$C_3 = \frac{W \cdot R_g^2 N_R \pi}{15 g T_R \Delta t}$$

Combining Equation (139) with Equation (130), we have

$$(\alpha^2 + v^2)(b_1 + b_2 \tan^{-1}(v/\alpha)) + C_4 + C_3\alpha = E_2(=0) \tag{140}$$

in which $C_4 = \beta_0 - C_3\alpha_0$.

The simultaneous non-linear equations, (137) and (140), are solved by the Newton-Raphson iterative technique:

$$\begin{pmatrix} \dfrac{\partial E_1}{\partial v}, & \dfrac{\partial E_1}{\partial \alpha} \\ \dfrac{\partial E_2}{\partial v}, & \dfrac{\partial E_2}{\partial \alpha} \end{pmatrix} \begin{pmatrix} \Delta v^{(k)} \\ \Delta \alpha^{(k)} \end{pmatrix} = \begin{pmatrix} -E_1 \\ -E_2 \end{pmatrix} \tag{141}$$

in which $\Delta v^{(k)}$ and $\Delta\alpha^{(k)}$ = the corrections to the assumed kth values, $v^{(k)}$ and $\alpha^{(k)}$

$$\frac{\partial E_1}{\partial v} = C_2 + 2H_Rv(a_1 + a_2\theta) + a_2H_R\alpha - \frac{2\Delta H_0}{\tau^2}|v|$$

$$\frac{\partial E_1}{\partial \alpha} = 2H_R\alpha(a_1 + a_2\theta) - a_2H_Rv$$

$$\frac{\partial E_2}{\partial v} = 2v(b_1 + b_2\theta) + b_2\alpha$$

and

$$\frac{\partial E_2}{\partial \alpha} = 2\alpha(b_1 + b_2\theta) - b_2v + C_3$$

The first estimate of v and α, $v^{(1)}$ and $\alpha^{(1)}$, is approximated by

$$\begin{aligned} v^{(1)} &= v_0 + \Delta v_0 \\ \alpha^{(1)} &= \alpha_0 + \Delta\alpha_0 \\ \Delta v_0 &= v_0 - v_{-1} \\ \Delta\alpha_0 &= \alpha_0 - \alpha_{-1} \end{aligned} \tag{142}$$

in which v_0 and α_0 = the known values at $t = t$ and v_{-1} and α_{-1} = the known values at $t = t - \Delta t$. In general, the corrections, $\Delta v^{(k)}$ and $\Delta\alpha^{(k)}$ are obtained by solving Equation (140).

$$\begin{aligned} v^{(k+1)} &= v^{(k)} + \Delta v^{(k)} \\ \alpha^{(k+1)} &= \alpha^{(k)} + \Delta\alpha^{(k)} \end{aligned} \tag{143}$$

The iterative process must be continued until the tolerance is satisfied:

$$|\Delta v^{(k)}| + |\Delta\alpha^{(k)}| < \varepsilon \tag{144}$$

in which ε = a small parameter of the tolerance.

The solution finally obtained should be checked whether the solution, v and α, is within the assumed range on the straight line.

LSAM B_1 and B_2 in the above derivation are replaced by B_{P1} and B_{M2}, respectively.

PNR Two compatibility equations are replaced by

$$C_1^+: \quad H_1 + B_1V_1 + FP1(V_1) = C_{P1}$$

$$C_2^-: \quad -H_2 + B_2V_2 + FM2(V_2) = C_{M2}$$

and Equation (137) is modified to

$$E_1' = E_1 - FP1\left(\frac{Q_R}{A_1}v\right) - FM2\left(\frac{Q_R}{A_2}v\right) = 0 \tag{145}$$

PARALLEL PUMP IN PIPELINE [25,51]

Two parallel pumps are in line as shown in Figure 18, which are dealt with as well as in the case of the single pump.

Two total dynamic head equations

$$H_1 + H_{tdh1} - \Delta H_1 v|v|/\tau_1^2 - H_2 = 0 \tag{146}$$

$$H_1 + H_{tdh2} - \Delta H_2 v|v|/\tau_2^2 - H_2 = 0 \tag{147}$$

$$H_{tdh1} = H_{R1}(\alpha_1^2 + v_1^2)(C_{11} + C_{12}\theta_1) \tag{148}$$

$$H_{tdh2} = H_{R2}(\alpha_2^2 + v_2^2)(C_{21} + C_{22}\theta_2) \tag{149}$$

in which $\theta_1 = \pi + \tan^{-1}(v_1/\alpha_1)$ and $\theta_2 = \pi + \tan^{-1}(v_2/\alpha_2)$.

$$A_1V_1 = v_1Q_{R1} + v_2Q_{R2} = A_2V_2 \quad \text{(Continuity equation)} \tag{150}$$

The torque unbalance gives the relations for two pumps:

$$E_1 = (\alpha_1^2 + v_1^2)(C_{31} + C_{32}\theta_1) + C_{33}\alpha_1 + C_{34} = 0 \tag{151}$$

$$E_2 = (\alpha_2^2 + v_2^2)(C_{41} + C_{42}\theta_2) + C_{43}\alpha_2 + C_{44} = 0 \tag{152}$$

$$FI \quad C_1^+: \quad H_1 + B_1V_1 = C_{P1} \tag{153}$$

$$C_2^-: -H_2 + B_2V_2 = C_{M2} \tag{154}$$

From Equations (148), (149), (150), (153), and (154), we have

$$\begin{aligned} E_3 = {} & (C_{P1} + C_{M2}) - (B_1/A_1 + B_2/A_2) \\ & \times (v_1Q_{R1} + v_2Q_{R2}) + H_{R1}(\alpha_1^2 + v_1^2) \\ & \times (C_{11} + C_{12}\theta_1) - \Delta H_1 v_1|v_1|/\tau_1^2 \\ = {} & 0 \end{aligned} \tag{155}$$

$$\begin{aligned} E_4 = {} & (C_{P1} + C_{M2}) - (B_1/A_1 + B_2/A_2) \\ & \times (v_1Q_{R1} + v_2Q_{R2}) + H_{R2}(\alpha_2^2 + v_2^2) \\ & \times (C_{21} + C_{22}\theta_2) - \Delta H_2 v_2|v_2|/\tau_2^2 \\ = {} & 0 \end{aligned} \tag{156}$$

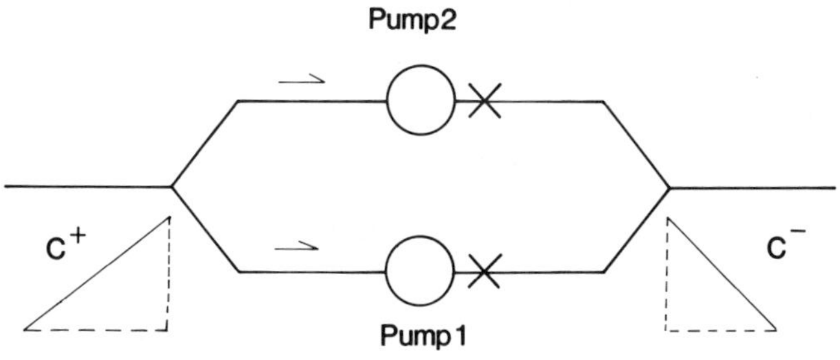

FIGURE 18. Parallel pumps.

The four variables, v_1, v_2, α_1, and α_2, are obtained by solving Equations (151), (152), (155), and (156) with the Newton-Raphson method as well as in the case of the single pump. E_1 to E_4 become zero for the true solution.

LSAM B_1 and B_2 are formally replaced by B_{P1} and B_{M2}, respectively.

PNR The compatibility equations become

$$C_1^+: \quad H_1 + B_1V_1 + FP1(V_1) = C_{P1} \tag{157}$$

$$C_2^-: \quad -H_2 + B_2V_2 + FM2(V_2) = C_{M2} \tag{158}$$

and Equations (155) and (156) are replaced by

$$E_3' = E_3 - FP1((v_1Q_{R1} + v_2Q_{R2})/A_1) - FM2((v_1Q_{R1} + v_2Q_{R2})/A_2) \tag{159}$$

$$E_4' = E_4 - FP1((v_1Q_{R1} + v_2Q_{R2})/A_1) - FM2((v_1Q_{R1} + v_2Q_{R2})/A_2) \tag{160}$$

and Equations (151), (152), (159), and (160) have to be solved.

It should be noted that the calculation steps reduce to the same one for the single pump in pipeline when one of the check valves is completely closed.

NOTATION

The following symbols are used in this article:

A = cross-section area
a = wavespeed
a_1, a_2 = constants
B, B_P, and B_M = constants in compatibility equations
B_A = Allievi's constant
$b_{1,2}$ = constants
C_k = constants; (k = 1, 2, . . .)
C = Hazen-Williams coefficient
C_v = flow coefficient of valve
C_d = discharge coefficient of valve
C^+ and C^- = advancing and receding characteristics
C_P and C_M = constants in compatibility equations
D = pipe diameter
E = Young modulus of elasticity: error of algebraic equation to be solved by Newton-Raphson
e = pipe wall thickness
F = generalized friction factor
f = friction factor: general function expressing pump characteristics
g = gravitational acceleration: function expressing pump characteristics
H = piezometric head
h = dimensionless head
I = integer variable related to section of pipes
J = integer variable related to time in computations
k = wall roughness
K = bulk modulus of liquid or mixture of liquid and gas
L = pipe length
m = exponent in frictional term or polytropic equation
M = number of reaches
MS = number of grid points
N = rotational speed of pump shaft
N_P = total number of pipes in pipeline
N_S = specific speed of pumps
NGP = sum of grid points for pipes
p = pressure
Q = discharge
R_g = radius of gyration
R = $Fa\Delta t/g$
Re = Reynolds number
S = friction factor times time step size in compatibility equation
T = torque
t = time

t' = parameter of time integration
V = velocity
$\forall$ = volume
V^* = initial estimates for velocity with *PNR* (non-viscous solution)
v = dimensionless velocity or discharge
W = weight of rotating parts plus entrained liquid
x = distance coordinate taken toward downstream
Y = function used with *PNR* method
z = elevation of pipe axis or water level in air chamber
α = void fraction, dimensionless rotational speed of pump shaft
β = dimensionless torque
γ = unit weight of fluid
δv = magnitude of maximum error in steady flow calculations with *PNR*
ε = control parameter for convergence in Newton-Raphson iterative calculations
ζ = loss coefficient
θ = angle between pipe axis and horizontal, ratio of time or $\pi + \tan^{-1}(\nu/\alpha)$
ν = kinetic viscosity
ρ = density
σ = dimensionless friction factor
τ_0 = shear stress acting on pipe wall
ω = angular velocity

Subscripts

a = air chamber
b = barometric pressure
A, B, and C = position in regular grids
D = downstream
d = discharge line
g = gas: gyration
i = pipe number
L = left side point of P with *IMSTI*
l = liquid
M = minus characteristics curves or lines
m = minimum
or = orifice
P = plus characteristics curves or lines
R = rated condition: right side point of P with IMSTI
s = suction side of pump: sustained head downstream valves: surge tank: and shut off of check-valve downstream pump
tdh = total dynamic head
U = upstream
v = valve
0 = representative quantities: initial value: corresponding values at the beginning of Δt
1, 2, and so on = pipes 1, 2, and so on, respectively

Superscripts

$*$ = initial estimates with *PNR* (non-viscous solution)
$-$ = known values, or known values at former time step and
$\cdot$ = total derivative with respect to time

REFERENCES

1. Allievi, L., ''Theory of Water Hammer, (translated by E. E. Halmos), Riccard Garoni, Rome, 1925.
2. Angus, R. W., ''Simple Graphical Solution for Pressure Rise in Pipes Pump Discharge Lines,'' *J. of Eng. Inst.*, Canada, 1935.
3. Chaudhry, M. H., ''Applied Hydraulic Transients,'' Van Nostrand Reinhold Co., 1979.
4. Colebrook, C. F., ''Turbulent Flow in Pipes with Particular Reference to the Transition Region between the Smooth and Rough Pipe Laws,'' *J. of the Inst. of Civil Engineers*, Vol. 11, 1938–1939, London, pp. 133–156.
5. Contractor, D. N., ''The Reflection of Waterhammer Pressure Waves From Minor Losses,'' *J. of Basic Engrg., ASME*, Vol. 87, Ser. D, 1965, pp. 445–452.
6. Courant, R., K. Friedrichs, and H. Lewy, ''Über die Partiellen Differenzen-Gleichungen der Mathematischen Physik,'' *Math. Ann.*, 100, 1928, p. 32.
7. Desor, C. A. and E. S. Kuh, ''Basic Circuit Theory,'' McGraw-Hill Book Co., 1969.
8. Donsky, B., ''Complete Pump Characteristics and the Effects of Specific Speeds on Hydraulic Transients,'' *J. of Basic Engrg., ASME*, 1961, pp. 685–699.
9. Evangelisti, G., ''Waterhammer Analysis by the Method of Characteristics,'' *L-Energ. Elec.*, Vol. XLVI, Nos. 10, 11, and 12, Milan, 1969.
10. Fox, J. A. and A. E. Keech, ''Pipe Network Analysis—A Novel Steady State Technique,'' *J. of Inst. Water Engrs. and Sci.*, 29, 1975, pp. 183–194.
11. Fox, J. A., ''Hydraulic Analysis of Unsteady Flow in Pipe Networks,'' the Macmillan Press Ltd., 1977.
12. Goldberg, D. E. and E. B. Wylie, ''Characteristics Method Using Time-Line Interpolations,'' *J. of Hydraulic Engrg.*, ASCE, Vol. 109, No. 5, 1983, pp. 670–683.
13. Graze, H. R., ''The Importance of Temperature in Air Chamber Operations,'' presented at *1st Int. Conf. on Pressure Surges*, Paper F2, BHRA Fluid Engrg., Cranfield, Bedford, England, 1972.
14. Graze, H. R., J. Schubert, and J. A. Forrest, ''Analysis of

Field Measurements of Air Chamber Installations,'' presented at *2nd Int. Conf. on Pressure Surges*, Paper K2, BHRA.

15. Haliwell, A. R., ''Velocity of Waterhammer Wave in an Elastic Pipe,'' *J. of the Hyd. Div., ASCE*, Vol. 89, No. HY4, 1963, pp. 1–21.
16. Jaeger, C., ''Fluid Transients in Hydro-Electric Engineering Practice,'' Blackie, 1977.
17. Jeppson, R. W. and A. L. Davis, ''Pressure Reducing Valves in Pipe Network Analyses,'' *J. of the Hyd. Engrg., ASCE*, Vol. 102, HY7, 1976, pp. 987–1001.
18. Kaplan, M., V. L. Streeter, and E. B. Wylie, ''Computation of Oil Pipeline Transients,'' *J. of Pipeline Div., ASCE*, Vol. 93, PL3, 1967, pp. 59–72.
19. Kesavan, H. K. and M. Chadrashekar, ''Graph-theoretic Models for Pipe Network Analysis,'' *J. of Hyd. Div., ASCE*, Vol. 98, No. HY2, 1972, pp. 345–364.
20. Kobori, T., S. Yokoyama, and H. Miyashiro, ''Propagation Velocity of Pressure Wave in Pipe Line,'' *Hitachi Hyoron*, Vol. 37, No. 10, 1955.
21. Kranenburg, C., ''Gas Release During Transient Cavitation in Pipes,'' *J. of Hyd. Div., ASCE*, Vol. 100, No. HY10, 1974, pp. 1383–1398.
22. Lamont, P. A., ''Common Pipe Flow Formulas Compared with the Theory of Roughness,'' *J. of AWWA*, Vol. 73, No. 5, 1981, pp. 274–280.
23. Lister, M., ''The Numerical Solution of Hyperbolic Partial Differential Equations by the Method of Characteristics,'' in Antony Ralston and H. S. Wilf (edits), *Numerical Methods for Digital Computers*, John Wiley and Sons, Inc., New York, 1960.
24. Marchal, M., G. Flesh, and P. Suter, ''The Calculation of Waterhammer Problems by Means of Digital Computer,'' *Proc. Int. Symp. Waterhammer Pumped Storage Projects, ASME*, Chicago, Nov. 1965.
25. Miyashiro, H., ''Waterhammer Analysis of Pumps in Parallel Operation,'' *Bull. of JSME*, Vol. 5, No. 19, 1962, pp. 479–484.
26. Miyashiro, H., ''Waterhammer Analysis of Pump Discharge Line with Several One-way Tanks,'' *J. of Engrg. for Power, ASME*, 1967.
27. Miyashiro, H., ''Water Hammer Analysis of Pump System,'' *Bull. of JSME*, Vol. 10, No. 42, 1967, pp. 952–958.
28. Moody, L. F., ''Friction Factors for Pipe Flow,'' *Trans. ASME*, Vol. 66, 1944, p. 671.
29. Parmakian, J., ''Water Hammer Analysis,'' Dover Publications, Inc., New York, 1963.
30. Pearsall, I. S., ''The Velocity of Water Hammer Waves,'' *Symp. on Surges in Pipelines*, Institution of Mech. Engrs., Vol. 180, part 3E, Nov. 1965, pp. 12–20.
31. Propson, T. P., ''Discussion of Unsteady Flow Calculations by Numerical Methods,'' by V. L. Streeter, *J. of Basic Engrg., Trans. ASME*, 1972, pp. 465–466.
32. Shamir, U. and C. D. D. Howard, ''Water Distribution System Analysis,'' *J. of Hyd. Div., ASCE*, Vol. 94, No. HY1, 1968, pp. 219–234.
33. Shimada, M. and S. Okushima, ''New Numerical Model and Technique,'' *J. of Hyd. Engrg., ASCE*, Vol. 110, No. 6, 1983, pp. 736–748.
34. Shimada, M., ''Pipe Steady Flow in Elastic Theory,'' (submitted).
35. Shimada, M., ''Second Model for Water Hammer,'' (in preparation).
36. Streeter, V. L., ''Waterhammer Analysis of Pipelines,'' *J. of Hyd. Div., ASCE*, Vol. 90, No. HY4, 1964, pp. 151–172.
37. Streeter, V. L. and C. Lai, ''Water Hammer Analysis Including Fluid Friction,'' *J. of Hyd. Civ., ASCE*, Vol. 88, No. HY3, 1962, pp. 79–112.
38. Streeter, V. L., ''Water Hammer Analysis of Distribution Systems,'' *J. of Hyd. Div., ASCE*, Vol. 93, No. HY5, 1967, pp. 185–200.
39. Streeter, V. L. and E. B. Wylie, ''Hydraulic Transients,'' McGraw-Hill Book Co., New York, 1967.
40. Streeter, V. L., ''Waterhammer Analysis,'' *J. of Hyd. Div., ASCE*, Vol. 95, No. HY6, 1969, pp. 1959–1972.
41. Streeter, V. L., ''Transients in Pipelines Carrying Liquids or Gases,'' *J. of Transportating Engrg., ASCE*, Vol. 97, No. TE1, 1971, pp. 15–29.
42. Streeter, V. L., ''Unsteady Flow Calculations by Numerical Methods,'' *J. of Basic Engrg., ASME*, 1972, pp. 457–466.
43. Streeter, V. L. and E. B. Wylie, ''Fluid Mechanics,'' McGraw-Hill Co., New York, 1983.
44. Streeter, V. L., ''Transient Cavitating Pipe Flow,'' *J. of Hyd. Engrg., ASCE*, Vol. 109, No. 11, 1983, pp. 1408–1423.
45. Thomas, G., ''Determination of Pump Characteristics for Computerized Transient Analysis,'' presented at *Int. Conf. on Pressure Surges*, Paper A3, Univ. of Kent, Canterbury, BHRA, England, 1972, pp. 21–32.
46. Vardy, A. E., ''On the Use of Method of Characteristics for the Solution of Unsteady Flows in Networks,'' presented at *2nd Int. Conf. on Pressure Surges*, Paper H2, Bedford, England, BHRA, 1976.
47. Vardy, A. E., ''Rapidly Attenuated Water Hammer and Steel Hammer,'' presented at *4th Int. Conf. on Pressure Surges*, Paper A1, Bath, England, organized and sponsored by BHRA, 1983, pp. 1–12.
48. Watters, G. Z., ''Analysis and Control of Unsteady Flow in Pipelines,'' Butterworth, 1984.
49. Weyler, M. E., V. L. Streeter, and P. S. Larsen, ''An Investigation of the Effect of Cavitation Bubbles on the *M* Mentum Loss in Transient Pipe Flow,'' *J. of Basic Engrg., ASME*, 1971, pp. 1–10.
50. Wiggert, D. C. and M. J. Sundquist, ''Fixed-Grid Characteristics for Pipeline Transients,'' *J. of Hyd. Div., ASCE*, Vol. 103, No. HY12, 1977, pp. 1403–1416.
51. Wylie, E. B. and V. L. Streeter, ''Fluid Transients,'' McGraw-Hill Book Co., New York, 1978.
52. Wylie, E. B., ''Free Air in Liquid Transient Flow,'' presented

at *3rd Int. Conf. on Pressure Surges*, Paper B1, Canterbury, England, organized and sponsored by BHRA, 1980, pp. 27–42.

53. Wylie, E. B., ''The Computer and Pipeline Transients,'' *J. of Hyd. Engrg., ASCE*, Vol. 109, No. 2, 1982, pp. 1723–1739.
54. Wylie, E. B., ''Advances in the Use of MOC in Unsteady Pipeline Flow,'' presented in *4th Int. Conf. on Pressure Surges*, Paper A3, Bath, England, organized and sponsored by BHRA, 1983, pp. 27–37.
55. Wylie, E. B., ''Simulation of Vaporous and Gaseous Cavitation,'' *J. of Fluids Engrg., ASME*, Vol. 106, Sept., 1984, pp. 307–311.
56. Wylie, E. G., "Fundamental Equations of Waterhammer," *J. of Hyd. Engrg., ASCE*, Vol. 110, No. 4, 1984, pp. 539–542.
57. Zielke, W., ''Frequently-Dependent Friction in Transient Pipe Flow,'' *J. of Basic Engrg., ASME*, Vol. 90, Ser. D, 1968, pp. 109–115.

Minimum Cost Design of Horizontal Pipelines

HELMI M. HATHOOT*

INTRODUCTION

In this section the problem of designing a horizontal pipeline with equally spaced similar pumping units, Figure 1, is taken into account [4].

Among the factors to be considered in designing a horizontal pipeline of minimum cost are the initial investment cost of pipes and pumps, the annual operating and maintenance costs for the life of the pipes and pumps, and the salvage value of the pipeline [1].

In general, a pipe of large diameter produces a small friction head loss against which the pump should act. On the other hand, though a pipe of a smaller diameter is cheaper, it produces a greater friction head loss [10]. Therefore, an optimum diameter exists for which the total annual cost of pipes and energy is a minimum.

COST OF PIPE

The levelized net annual cost of pipe per unit pipe length [6,8] may be given by

$$K = \pi D t \gamma_p C_1 \tag{1}$$

in which D is the pipe diameter; t is the pipe wall thickness; γ_p is the specific weight of the pipe material; and C_1 is the levelized net annual cost of pipes per unit weight of pipe material. In general the thickness t may be assumed to be roughly proportional to the pipe diameter [5,11], so that

$$t = CD \tag{2}$$

*Department of Civil Engineering, Faculty of Engineering, Alexandria University, Alexandria, Egypt

in which C is a constant. Substituting Equation (2) for t into Equation (1) and rearranging, it follows that

$$K = CC_1\gamma_p\pi D^2 \tag{3}$$

COST OF ENERGY

For a pipeline with similar equidistant pumping units, the power required per pump can be given by

$$P = \frac{\gamma Q h_L}{E} \tag{4}$$

in which γ is the specific weight of the liquid to be pumped; Q is the discharge; h_L is the friction head loss between two successive pumping units; and E is a constant representing the overall efficiency of each pumping unit. The friction head loss can be given by

$$h_L = \frac{8fLQ^2}{\pi^2 g D^5} \tag{5}$$

in which f is the friction factor, L is the pipe length between two successive pumping units; and g is the acceleration due to gravity. From Equations (4) and (5), the power required per unit length of pipe may be written as

$$W = \frac{8f\gamma Q^3}{gE\pi^2 D^5} \tag{6}$$

and hence the levelized net annual cost of pumping energy per unit pipe length becomes

$$K_{en} = \frac{8fC_2\gamma Q^3}{gE\pi^2 D^5} \tag{7}$$

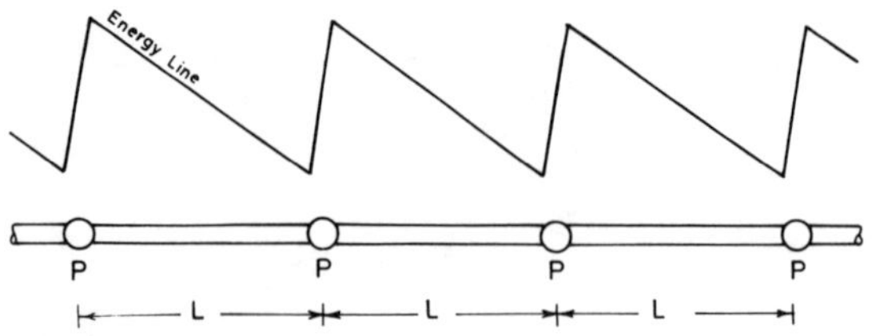

FIGURE 1. Horizontal pipeline.

in which C_2 is the levelized net annual cost of pumping energy per watt.

The total annual cost of the pipeline is given by the sum of the annual costs of the pipes and energy. From Equations (3) and (7) the levelized total annual cost of the pipeline per unit pipe length can be written as

$$K_{tu} = CC_1\gamma_p\pi D^2 + \frac{8fC_2\gamma Q^3}{gE\pi^2 D^5} \tag{8}$$

The friction factor f is generally a function of both Reynolds number and the relative roughness of the pipe [9] and each zone on Moody diagram can be represented by a corresponding friction factor formula [2,3,12,14].

FRICTION FACTOR

Laminar Flow

In laminar flow the friction factor is given by

$$f = \frac{16\pi D\nu}{Q} \tag{9}$$

in which ν is the kinematic viscosity of the liquid to be pumped.

Turbulent Flow

In most cases of pipe flow, the flow is turbulent. A significant portion of the turbulent region on the Moody diagram could be represented [13] by the following formula

$$f = \frac{1.325}{\ln\left(\dfrac{\varepsilon}{3.7D} + \dfrac{5.74}{R^{0.9}}\right)^2} \tag{10}$$

in which ε is the absolute roughness of pipe, and R is the Reynolds number. The above formula yields errors less than 1 percent in the range $10^{-6} \leq \varepsilon/D \leq 10^{-2}$ and $5 \times 10^3 \leq R \leq 10^8$.

Though the upper portion of the turbulent region on Moody diagram ($\varepsilon/D > 10^{-2}$) is not represented by Equation (10), yet this portion is represented by the following equation [14] for completely rough pipe flow

$$f = \frac{1}{\left(1.14 + 2.0 \log \dfrac{D}{\varepsilon}\right)^2} \tag{11}$$

Equations (9), (10) and (11) mostly cover Moody diagram.

MINIMUM-COST DESIGN FORMULAS

Laminar Flow

The levelized total annual cost of pipeline per unit pipe length in the case of laminar flow can be written as

$$K_{tu} = CC_1\gamma_p\pi D^2 + \frac{128C_2\gamma\nu Q^2}{gE\pi D^4} \tag{12}$$

Equation (12) is obtained by substituting the value of f as given by Equation (9) into Equation (8).

Differentiating both sides of Equation (12) with respect to D, equating to zero and solving for D:

$$D = 1.72\left(\frac{C_2\gamma\nu Q^2}{CC_1\gamma_p gE}\right)^{1/6} \tag{13}$$

By means of Equation (13) optimum pipe diameter for laminar flow may be evaluated.

NUMERICAL EXAMPLE 1

It is required to pump 0.2 m^3/S of a liquid ($\nu = 1.19 \times 10^{-3}\ m^2/S$, $\gamma = 12337\ N/m^3$). Find the optimum pipe diameter for the following data:

$C = 0.01$; $C_1 = 0.01$ \$/N $C_2 = 2.0$ \$/W, $E = 0.66$,

$g = 9.81\ m/S^2$ $\gamma_p = 73575\ N/m^3$

and

$\varepsilon = 0.25 \times 10^{-3}\ m.$

Solution

Since the viscosity of the liquid is high, the flow is expected to be laminar, applying Equation (13):

$$D = 1.72\left(\frac{2.0(12337)1.19 \times 10^{-3}(0.2)^2}{0.01(0.01)73575(9.81)0.66}\right)^{1/6} = 0.93\ m$$

check

$$R = \frac{4Q}{\pi D \nu} = \frac{4(0.2)}{\pi(0.93)1.19 \times 10^{-3}} = 230$$

which means that the assumption of laminar flow is correct.

Turbulent Flow

TURBULENT FLOW IN GENERAL [8]

Substituting the value of f from Equation (10) into Equation (8) we get

$$K_{tu} = CC_1\gamma_p\pi D^2 + \frac{10.6C_2\gamma Q^3}{gE\pi^2 D^5\left\{\ln\left(\dfrac{\varepsilon}{3.7D} + \dfrac{4.618}{\left(\dfrac{Q}{\nu D}\right)^{0.9}}\right)\right\}^2} \tag{14}$$

For minimum-cost design, differentiation of K_{tu} with respect to the diameter D should equal zero. Differentiating both sides of Equation (14), equating to zero and simplifying

$$CC_1\gamma_p\pi D = \frac{10.6C_2\gamma Q^3}{gE\pi^2 D^6} \times \left\{\frac{-1 + \dfrac{8.774}{\left(\dfrac{Q}{\nu D}\right)^{0.9}\left(\dfrac{\varepsilon}{3.7D} + \dfrac{4.618}{\left(\dfrac{Q}{\nu D}\right)^{0.9}}\right)} + 2.5\ln\left(\dfrac{\varepsilon}{3.7D} + \dfrac{4.618}{\left(\dfrac{Q}{\nu D}\right)^{0.9}}\right)}{\left\{\ln\left(\dfrac{\varepsilon}{3.7D} + \dfrac{4.618}{\left(\dfrac{Q}{\nu D}\right)^{0.9}}\right)\right\}^3}\right\}$$

Solving for D

$$D = \left(\frac{10.6C_2\gamma Q^3}{gCEC_1\gamma_p\pi^3}\right)^{1/7} \times \left\{\frac{-1 + \dfrac{8.774}{\left(\dfrac{Q}{\nu D}\right)^{0.9}\left(\dfrac{\varepsilon}{3.7D} + \dfrac{4.618}{\left(\dfrac{Q}{\nu D}\right)^{0.9}}\right)} + 2.5\ln\left(\dfrac{\varepsilon}{3.7D} + \dfrac{4.618}{\left(\dfrac{Q}{\nu D}\right)^{0.9}}\right)}{\left\{\ln\left(\dfrac{\varepsilon}{3.7D} + \dfrac{4.618}{\left(\dfrac{Q}{\nu D}\right)^{0.9}}\right)\right\}^3}\right\}^{1/7} \tag{15}$$

For convenience, Equation (15) may be put in the form

$$D = \left\{F\left(\frac{-1 + \dfrac{10.905}{R^{0.9}\phi(D)} + 2.5\ln\phi(D)}{(\ln\phi(D))^3}\right)\right\}^{1/7} \tag{16}$$

in which

$$F = \frac{10.6C_2\gamma Q^3}{gCEC_1\gamma_p\pi^3} \tag{17}$$

and

$$\phi(D) = \left(\frac{\varepsilon}{3.7D}\right) + \left(\frac{5.739}{R^{0.9}}\right) \tag{18}$$

By means of Equation (16) the minimum-cost diameter of a pipeline may be evaluated.

The practical value as well as the convergence characteristics of Equation (16) may be illustrated by the following numerical example.

NUMERICAL EXAMPLE 2

For a horizontal pipeline having similar equidistant pumping units, it is required to design the most economical pipe diameter if the following data are given: $C = 0.01$; $C_1 = 0.01$ $/N; $C_2 = 2.0$ $/W; $E = 0.65$; $g = 9.81\ m/S^2$; $Q = 2.3\ m^3/S$; $\gamma = 9{,}810\ N/m^3$, $\nu = 1 \times 10^{-6}\ m^2/S$; $\gamma_p = 73{,}575\ N/m^3$, and $\varepsilon = 0.25 \times 10^{-3}\ m$.

Solution

$$F = \frac{10.6(2.0)9810(2.3)^3}{9.81(0.01)0.65(0.01)73575\pi^3} = 1739.51$$

To apply Equation (16), a first trial value $D = 1.0\ m$ is assumed.

$$R = \frac{4(2.3)}{\pi(10)^{-6} \times 1.0} = 2928450.9$$

$$\phi(D) = \frac{0.25 \times 10^{-3}}{3.7(1.0)} + \frac{5.739}{R^{0.9}} = 7.62544 \times 10^{-5}$$

$$D = \left\{ 1739.51 \times \left(\frac{-1 + \dfrac{10.6}{(2928450.9)^{0.9}(7.62544 \times 10^{-5})} + 2.5 \ln(7.62544 \times 10^{-5})}{(\ln (7.62544 \times 10^{-5}))^3} \right) \right\}^{1/7}$$

$$= 1.75\ m$$

For a second trial let us assume $D = 1.75\ m$.

$$R = \frac{4(2.3)}{\pi(10)^{-6} \times 1.75} = 1673400.5$$

$$\phi(D) = \frac{0.25 \times 10^{-3}}{3.7(1.75)} + \frac{5.739}{(1673400.5)^{0.9}} = 5.29846 \times 10^{-5}$$

$$D = \left\{ 1739.51 \times \left(\frac{-1 + \dfrac{10.6}{(1673400.5)^{0.9}(5.29846 \times 10^{-5})} + 2.5 \ln(5.29846 \times 10^{-5})}{(\ln (5.29846 \times 10^{-5}))^3} \right) \right\}^{1/7}$$

$$= 1.73\ m$$

It is evident that a third trial is not necessary.

Since the final diameter is computed according to Equation (16), which is based on Equation (10) [13], both Reynolds number and the relative roughness should be verified. Hence, $R = 1.67 \times 10^6$ (between 5000 and 10^8, O.K.); and $\varepsilon/D = 0.25 \times 10^{-3}/1.73 = 1.45 \times 10^{-4}$ (between 10^{-2} and 10^{-6}, O.K.).

The extremely fast convergence characteristics of the design formula are quite apparent.

GRAPHICAL DESIGN

It is of practical importance for pipeline designers to have graphical solutions of Equation (16). The graph of Figure 2 is non-dimensionally plotted on a log-log graph paper, covering a wide range of the expected values of both $(C_2\gamma Q^3/(C_1\gamma_p gCE))^{1/7}/\varepsilon$ and D/ε. By virtue of Figure 2 no trials are needed and as the term $(C_2\gamma Q^3/(C_1\gamma_p gCE))^{1/7}/\varepsilon$ is evaluated and the term $Q/\nu\varepsilon$ is known, the diameter ratio D/ε can be found directly. Curves of Figure 2 are observed to be nearly parallel straight lines each corresponding to an indicated $Q/\nu\varepsilon$ value. In designing the graph of Figure 2, the limits $5000 \le R \le 10^8$ and $10^{-2} \le \varepsilon/D \le 10^{-6}$ are taken into account. The use of Figure 2 is illustrated by the following examples.

NUMERICAL EXAMPLE 3

It is required to design the pipeline of Example 2 using the chart of Figure 2.

Solution

$$\frac{\left(\dfrac{C_2\gamma Q^3}{C_1\gamma_p gCE}\right)^{1/7}}{\varepsilon} = \frac{\left(\dfrac{2.0(9810)(2.3)^3}{0.01(73575)9.81(0.01)0.65}\right)^{1/7}}{(0.25 \times 10^{-3})}$$

$$= 1.354 \times 10^4$$

$$\frac{Q}{\nu\varepsilon} = \frac{2.3}{10^{-6} \times 0.25 \times 10^{-3}} = 9.2 \times 10^{-9}$$

From the chart, $D/\varepsilon = 7 \times 10^3$, from which $D = 7 \times 10^3 \times 0.25 \times 10^{-3} = 1.75\ m$.

NUMERICAL EXAMPLE 4

Using the design chart of Figure 2 find the optimum pipe diameter of the pipeline with the following data: $C = 0.01$;

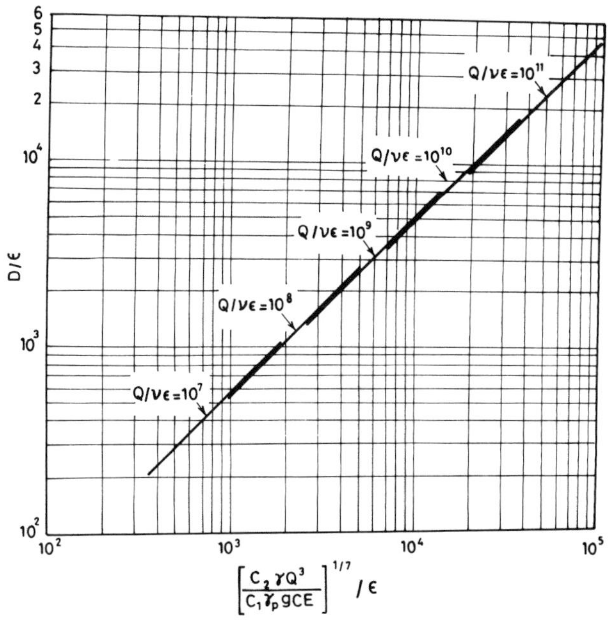

FIGURE 2. Pipe diameter design chart.

$C_1 = 0.012$ \$/N; $C_2 = 2.1$ \$/W; $E = 0.63$; $g = 9.81\ m/S^2$; $Q = 1.0\ m^3/S$; $\gamma = 8{,}436.6\ N/m^3$; $\nu = 1 \times 10^{-5}\ m^2/S$; $\gamma_p = 76{,}518\ N/m^3$; and $\varepsilon = 0.046 \times 10^{-3}\ m$.

Solution

$$\frac{\left[\dfrac{C_2\gamma Q^3}{C_1\gamma_p gCE}\right]^{1/7}}{\varepsilon} = \frac{\left(\dfrac{2.1(8436.6)1.0^3}{0.012(76518)9.81(0.01)0.63}\right)^{1/7}}{0.046 \times 10^{-3}}$$

$$= 4.9384 \times 10^4$$

$$\frac{Q}{\nu\varepsilon} = \frac{1.0}{10^{-5} \times 0.046 \times 10^{-3}} = 2.17 \times 10^9$$

From the design chart, $D/\varepsilon = 2.6 \times 10^4$, thus $D = 2.6 \times 10^4 \times 0.046 \times 10^{-3} = 1.196\ m$.

COMPLETELY ROUGH TURBULENT FLOW [7]

Combining Equations (8) and (11), the levelized total annual cost of the pipeline per unit pipe length for completely rough flow can be written as

$$K_{tu} = CC_1\gamma_p\pi D^2 + \frac{8C_2\gamma Q^3}{gE\pi^2 D^5\left(1.14 + 0.87\ \ln\dfrac{D}{\varepsilon}\right)^2} \quad (19)$$

Differentiating with respect to D, equating to zero and rearranging:

$$D = \left\{\frac{0.129C_2\gamma Q^3}{CC_1\gamma_p gE\left(1.14 + 0.87\ \ln\dfrac{D}{\varepsilon}\right)^2} \times \left(\frac{1.74}{1.14 + 0.87\ \ln\dfrac{D}{\varepsilon}} + 5.0\right)\right\}^{1/7} \quad (20)$$

The optimum pipe diameter for the case of completely rough flow is given by Equation (20), which is implicit and to be solved by a trial and error procedure.

NUMERICAL EXAMPLE 5

It is required to design the most economical diameter of a horizontal pipeline with equidistant pumping units for the following data:

$C = 0.01$; $C_1 = 0.015$ \$/N; $C_2 = 2.0$ \$/W;

$E = 0.68$; $g = 9.81\ m/S^2$;

$Q = 0.05\ m^3/S$; $\gamma = 6673.7\ N/m^3$,

$\nu = 4.26 \times 10^{-7}\ m^2/S$, $\gamma_p = 76518\ N/m^3$

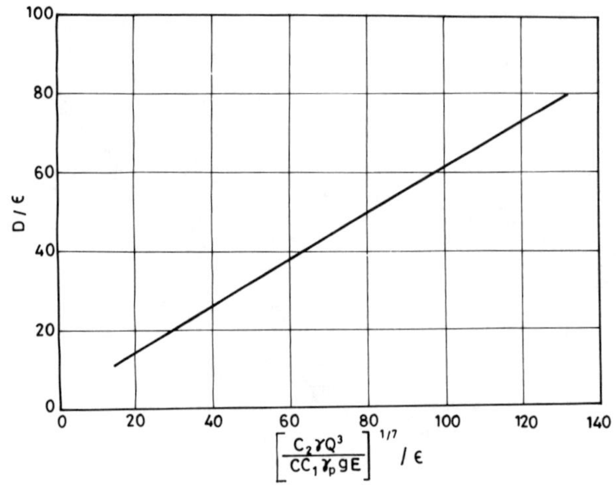

FIGURE 3. Design chart for wholly rough flow.

and

$$\varepsilon = 5 \times 10^{-3}\, m.$$

Solution

It is expected that the flow in this case is beyond the range of Figure 2 since ε is large.

As a first trial assume $D = 0.5\ m$. Applying Equation (20):

$$D = \left\{ \frac{0.129(2.0)6673.7(0.05)^3}{0.01(0.015)76518(9.81)0.68\left(1.14 + 0.87 \ln \dfrac{0.5}{0.005}\right)^2} \left(\frac{1.74}{1.14 + 0.87 \ln \dfrac{0.5}{0.005}} + 5.0 \right) \right\}^{1/7}$$

$$= 0.344\ m$$

Now inserting this new value of the diameter into Equation (20)

$$D = \left\{ \frac{0.129(2.0)6673.7(0.05)^3}{0.01(0.015)76518(9.81)0.68\left(1.14 + 0.87 \ln \dfrac{0.344}{0.005}\right)^2} \left(\frac{1.74}{1.14 + 0.87 \ln \dfrac{0.344}{0.005}} + 5.0 \right) \right\}^{1/7}$$

$$= 0.35\ m$$

This means that a third trial is not necessary.

$$\text{check: } \frac{\varepsilon}{D} = \frac{0.005}{0.35} = 0.0143 > 10^{-2}$$

$$R = \frac{4(0.05)}{\pi(0.35)4.26 \times 10^{-7}} = 4.27 \times 10^5$$

(completely rough flow zone).

The above value of (ε/D) is beyond the range of validity of Equation (16).

GRAPHICAL DESIGN

Since Equation (20) is implicit, it is of practical importance to have a graphical design chart for the optimum pipe diameter. Introducing natural logarithms into Equation (20), dividing throughout by ε and simplifying:

$$\frac{D}{\varepsilon} = 0.8575 \left\{ \left(\frac{C_2\gamma Q^3}{CC_1\gamma_p g E} \right)^{1/7} \Big/ \varepsilon \right\} \times \left(\frac{1.0}{\left(1.31 + \ln \dfrac{D}{\varepsilon}\right)^3} + \frac{2.5}{\left(1.31 + \ln \dfrac{D}{\varepsilon}\right)^2} \right)^{1/7} \tag{21}$$

Rearranging Equation (21):

$$\left(\frac{C_2\gamma Q^3}{CC_1\gamma_p gE}\right)^{1/7} \Big/ \varepsilon = \frac{1.1662\left(\frac{D}{\varepsilon}\right)}{\left(\frac{1.0}{\left(1.31 + \ln\frac{D}{\varepsilon}\right)^3} + \frac{2.5}{\left(1.31 + \ln\frac{D}{\varepsilon}\right)^2}\right)^{1/7}} \quad (22)$$

In Figure 3, D/ε versus $(C_2\gamma Q^3/CC_1\gamma_p gE)^{1/7}/\varepsilon$ is shown plotted. In fact Figure 3 is the pipeline design chart for completely rough flow.

NUMERICAL EXAMPLE 6

It is required to design the most economical diameter of the pipeline of Example 5 using the chart of Figure 3.

Solution

$$\text{put } G = \frac{C_2\gamma Q^3}{C_1\gamma_p gCE}$$

$$G = \frac{2.0(6673.7)(0.05)^3}{0.015(76518)9.81(0.01)0.68} = 0.021791$$

$$G^{1/7}/\varepsilon = (0.021791)^{1/7}/0.005 = 115.8$$

From Figure 3 we can get

$$\frac{D}{\varepsilon} = 70$$

from which

$$D = 0.35\ m$$

NOTATION

C = constant
C_1 = levelized net annual cost of pipes per unit weight of pipe material
C_2 = levelized net annual cost of pumping energy per watt
D = pipe diameter
E = overall efficiency of pumping unit
f = coefficient of friction
g = acceleration due to gravity
h_L = friction head loss between two pumping units
K = levelized net annual cost of pipes per unit pipe length
K_{en} = levelized net annual cost of pumping energy per unit pipe length
K_{tu} = levelized total annual cost of pipeline per unit pipe length
L = spacing between two successive pumping units
P = power required per pumping unit
Q = pipeline discharge
R = Reynolds number
t = pipe wall thickness
W = power required per unit pipe length
γ = specific weight of liquid
γ_p = specific weight of pipe material
ε = absolute roughness of pipe and
ν = kinematic viscosity of liquid

REFERENCES

1. Albertson, M., Barton, J., and Simons, D., *Fluid Mechanics for Engineers*, Prentice Hall, New York, 1960.
2. Blasius, H., "Das Ähnlichkeitsgezetz bei Reibungsvorgängen in Flüssigteiten," *Forschungsarbeiten auf dem Gebiete des Ingenieurwesens*, No. 131, Berlin, 1913.
3. Colebrook, C., "Turbulent Flow in Pipes, with Particular Reference to the Transition Region Between the Smooth and Rough Pipe Laws," *Journal of the Institution of Engineering*, Vol. 11, London, 1939, pp. 133–156.
4. Daugherty, R. and Ingersoll, A., "Steady Flow of Incompressible Fluids," *Fluid Mechanics with Engineering Applications*, 5th ed., McGraw-Hill Co., New York, 1954.
5. Davis, C. and Sorensen, K., *Handbook of Applied Hydraulics*, 3rd ed., McGraw-Hill Co., New York, 1969.
6. Hathoot, H., "Optimum Design of Irrigation Pipelines," *Bulletin of the International Commission on Irrigation and Drainage (ICID)*, Vol. 29, No. 2, July, 1980, pp. 73–76.
7. Hathoot, H., "Optimum Design of Horizontal Circular Pipelines," *Bul. Faculty of Engrg.*, Alexandria Univ., Vol. XXII, 1983.
8. Hathoot, H., "Minimum-Cost Design of Horizontal Pipelines," *Journal of Transportation Engrg., ASCE*, May 1984.
9. Moody, L., "Friction Factors for Pipe Flow," *Transactions of the American Society of Mechanical Engineers*, 1944, pp. 671–684.
10. Russel, G., *Hydraulics*, 5th ed., Holt, Rinehart & Winston, New York, 1963.
11. Standard Handbook for Mechanical Engineering, McGraw-Hill Co., New York, 1966.
12. Streeter, V. L. and Wylie, B. E., "Fluid Resistance," *Fluid Mechanics*, 7th ed., McGraw-Hill Co., New York, 1979, pp. 182–261.
13. Swamee, P. and Jain, A., "Explicit Equations for Pipe Flow Problems," *Journal of the Hydraulics Division, ASCE*, May, 1976, pp. 657–664.
14. Vennard, J. and Street, R., "Fluid Flow in Pipes," *Elementary Fluid Mechanics*, 5th ed., John Wiley and Sons, Inc., New York, 1976, pp. 379–463.

Hydraulic Transport of Solids in Horizontal Pipes

R. P. Chhabra*

INTRODUCTION

The use of slurry pipelines for the conveyance of particulate solids is neither a new concept nor a recent innovation. It is safe to assume that one of the earliest slurry pipeline system in operation is within ourselves. The arterial system carries a heterogeneous mixture of solids in suspension throughout our bodies for a life time. Nature abounds with further examples of slurry transport. For instance, it is the phenomenon of slurry transport that has been building river deltas and scouring canyons throughout geologic history. It works on the principle that a liquid sufficiently turbulent and fast moving can transport a particulate solid an indefinite distance. One of the great natural slurry transportation systems is the Mississippi River in the spring: soil is fed into the muddy brown flood from headwaters and tributaries and is carried to the Mississippi delta more than 1000 miles away.

From a practical point of view, solid conveyance through pipes is now an industrial reality. Indeed the slurry pipelines have come of an age. There are many installations around the globe which testify to the practicality, and in most cases to the economic advantages of this mode of transportation [1]. Typical examples are listed in Table 1. Although the earliest industrial solid line dates back to 1884—when in Pennsylvania the anthracite culm (fine waste material) was pumped through pipes into the mines for extinguishing a mine fire [4], the systematic investigation of this subject began only about thirty to forty years ago. In recent years, slurry pipelines have received further attention for conveying coal. Indeed the movement of coal and other solids by slurry pipelines is not only economical but is also more acceptable to the public than the conventional modes of rail/road transport. Wasp [5,6], Cabrera [7,8], and others [9] have dealt with the environmental, legal and aesthetic aspects, and other public issues of slurry pipelines.

During the last three decades or so, considerable efforts have been expended in improving our understanding of the basic physics of the phenomenon of slurry transport. From a theoretical point of view, the governing equations are highly nonlinear and complex in form, and are amenable to rigorous analysis only under highly idealized conditions. Such solutions though are enlightening and do further our understanding but unfortunately are of little practical utility. Therefore, most of the progress in this area has been made through the use of empirical formulations. Most of the resulting information is readily available in a number of books [10–14] and in the voluminous proceedings of several conferences [15,16]. There is no question that the major research efforts for slurry flows have been directed towards an elucidation of the following two aspects:

1. The identification and prediction of various flow regimes encountered during the slurry flow
2. The prediction of pressure gradients (or frictional head loss) in straight lengths of pipe and pipe fittings

Other important aspects of the slurry transportation systems include erosion and wear characteristics of pipes and pumps, design of solid feeders, selection and evaluation of pumps etc. In this chapter, however, the attention is focused only on the two above-mentioned aspects, namely, flow regimes and estimation of frictional losses. Throughout the discussion, the flow is assumed to be steady, fully developed (no entrance effects) and isothermal in nature. Furthermore, since the literature is vast on the subject, only a selection of methods which have been adequately tested is presented herein.

*Department of Chemical Engineering, Indian Institute of Technology, Kanpur 208016, India

TABLE 1. Summary of Industrial Slurry Pipelines in Operation (Modified after Gandhi [2]).

System	Material	Maximum Particle Size	Length (km)	Pipe Diameter (mm)	Annual Capacity (MMT)	Type of Pump	Operation
Consolidation, USA	Coal	8 mesh	194	225	1.3	Reciprocating slurry pump	1957
Black Mesa, USA	Coal	14 mesh	490	450	4.8	Reciprocating slurry pump	1970
Calaveras, USA	Limestone	28 mesh	31	175	1.5	Reciprocating slurry pump	1971
Rugby, England	Limestone	35 mesh	103	250	1.7	Reciprocating slurry pump	1964
Trinidad	Limestone	48 mesh	11	200	—	Reciprocating slurry pump	—
Columbia	Limestone	—	10	200	—	Reciprocating slurry pump	—
Japan	Slime	—	68	300	—	Reciprocating slurry pump	—
Samarco, Brazil	Hematite	100 mesh	396	500	12	Plunger pump	1977
Valep, Brazil	Phosphate	65 mesh	120	225	2	Plunger pump	1979
Savage River, Australia	Magnetite	100 mesh	95	225	2.5	Plunger pump	1967
Sierra Grande, Argentina	Magnetite	100 mesh	36	200	2.1	Plunger pump	—
Las Truchas, Mexico	Magnetite	150 mesh	31	250	1.5	Plunger pump	1976
Bougainville, Papua	Copper	65 mesh	31	150	1	Plunger pump	1972
West Irian, Indonesia	Copper	100 mesh	124	100	0.3	Plunger pump	1972
Pinto Valley, USA	Copper	65 mesh	20	100	0.4	Plunger pump	1974
Vaal Reefs, USA	Gold ore fines	65 mesh	—	—	—	Marspumps	—
Oppu Mining	Milled ore	—	8	27	—	Marspumps	—
Hokuroku, Japan	Mill tailings	—	72	300	—	Marspumps	—
Cuban American	Iron ore	80 mesh	—	—	—	Diaphragm pump	—
W. Germany	Bauxite	3 mm	—	—	—	Diaphragm pump	—
Wippipi, New Zealand	Iron sand	—	3.3	1465	1	Centrifugal pump	1971
Kudremukh, India	Magnetite	—	67	1115	7.5	Centrifugal pump	1980
Shirasu, Japan	Ash	—	6.8	600	—	Centrifugal pump	—
English Clays, England	China clay	30 μm	13	200/250	—	Centrifugal pump	—
ETSI	Coal	—	1660	950	25	—	1979
Alton	Coal	—	290	600	10	—	1981
Chandrapur, India	Coal	—	35	—	—	—	In design stage (3)

TYPES OF SLURRY FLOW

The flow of mixtures of a solid and a liquid in pipes differs from the flow of a single liquid in a number of ways. In the case of single homogeneous liquids, the complete range of velocities is possible, and the state of flow is classified as laminar or turbulent depending upon the value of Reynolds number. Characterisation of slurry flow, on the other hand, is not as simple as for liquid for two reasons. Firstly, there are two distinct components present, and therefore any discussion should incorporate the properties of both the species. Secondly, a range of slurry behaviour is possible depending upon the particular conditions prevailing e.g. physico-chemical properties of the solid-liquid system, velocity, particle size, density etc.

Broadly speaking, the slurry behaviour can conveniently be classified into two types:

1. The type in which the presence of small colloidal particles alters the rheological behaviour of the carrier liquid. The mixture may be treated as a pseudohomogeneous liquid exhibiting complex non-Newtonian behaviour. Examples of such behaviour are encountered during the pipe-line flow of aqueous suspensions of kaolin/clay [17–21,55], of fine anthracite coal [22], of titanium dioxide [23,24], of thoria [25,26], of red mud [27], sewage sludges

[28] and drilling muds [29,182] etc. This type of slurry flow is also known as "homogeneous" and "non-settling type slurry" flow. In this work the term "*pseudohomogeneous non-Newtonian suspensions*" will be used.

2. The second type of slurry behaviour is characterised by the fact that the particles are too large (and noncolloidal in nature) for the mixture to be treated as pseudohomogeneous. The two components namely the vehicle and the solids retain their identity. Obviously, any discussion on the hydrodynamics of such flows must consider the two phases and their relevant properties. This type of behaviour is exemplified by the flow of water and coarse sand, limestone, coal, phosphate rocks etc. Since the particles are large and have a tendency to settle down, this type of behaviour is also characterised as "settling type of suspension." In this work we shall simply call this type of flow as flow of *two phase solid liquid mixtures*.

It should be emphasized here that the foregoing classification is not unique and is in fact quite arbitrary. Nor is one able to predict a priori as to what kind of slurry flow behaviour will be encountered. But nonetheless, this classification procedure does facilitate the presentation of information in the ensuing sections; hence it has been adopted here.

FLOW OF PSEUDOHOMOGENEOUS NON-NEWTONIAN SUSPENSIONS

Due to the interaction between hydrodynamic and electrochemical forces, a pseudohomogeneous mixture of fine colloidal particles and water (or any other solvent) often displays a variety of non-Newtonian characteristics in pipeline flow. The extent and type of non-Newtonian behaviour is influenced by a large number of factors such as particle size (or size distribution), particle density, viscosity and density of the liquid medium, concentration, and pH etc. Clearly a detailed discussion on how all these variables govern the rheological behaviour of suspensions is beyond the scope of this chapter, attention may, however, be drawn to a number of excellent review articles and books which are available on this subject [30–36]. Here we will simply accept that such suspensions are non-Newtonian in behaviour.

Rheological Models

Before any prediction about flow behaviour in a pipe line can be made, the constitutive relationship must be established. Most suspensions of practical interest exhibit a wide range of non-Newtonian characteristics including shear thinning, shear thickening [37], elasticity [38,39], time dependency [27] and a combination of two or more of these.

The most common inelastic time independent non-Newtonian behaviour observed is shearthinning (or pseudoplasticity) and in simple shear it is characterised by an apparent viscosity (shear stress divided by shear rate) which decreases with increasing shear rate. Admittedly the other non-Newtonian characteristics (such as elasticity etc.) do affect the kinematics of flow in pipes, it is now generally agreed that the pressure gradient is largely governed by the shearthinning viscosity. Therefore the ensuing discussion is limited only to the flow of time independent shearthinning fluids. Pipe flows of other types of non-Newtonian fluids have been adequately discussed by Govier and Aziz [2] and others [40].

Empirical expressions of varying complexity have been proposed and generalized to model the shearthinning characteristics [41]. The most common (and particularly suitable for suspensions) constitutive relations in use are the power law model which in simple shear is written as

$$\tau = k(\dot{\gamma})^n \tag{1}$$

and the ideal Bingham plastic model which in pure shearing flow may be expressed as

$$\tau = \tau_0 + \mu_P\dot{\gamma} \tag{2}$$

where τ is shear stress corresponding to shear rate $\dot{\gamma}$; k and n are known as the fluid consistency and the flow behaviour index respectively; τ_0 is the yield stress and μ_P is the plastic viscosity. Clearly Equation (1) displays shearthinning viscosity for $n < 1$ and reduces to the Newton's law of viscosity when $n = 1$. Likewise the Bingham plastic model reduces to the Newtonian case when $\tau_0 = 0$.

Cheng [42] has suggested using a generalized Bingham fluid as a universal model for time independent fluids with the constitutive relation as

$$\tau = \tau_0 + k(\dot{\gamma})^n \tag{3}$$

Equation (3) contains both the power law as well as the Bingham plastic fluid as limiting cases. This model is also known as the Herschel-Bulkley model. The rheological parameters appearing in Equations (1–3) for a given material are evaluated from $\tau - \dot{\gamma}$ data which are obtained using one or a combination of viscometers. Typical comparisons between experimental data and each of the three models referred to above are depicted in Figures 1–3 where they seem to give satisfactory fit of experimental data.

It is important to bear in mind that all these constitutive relations are entirely empirical and usually it is possible that all of them may fit a given set of $\tau - \dot{\gamma}$ data equally well over a limited range of conditions. No single set of values of (k, n) or (τ_0, k, n), however, can adequately describe the shear stress-shear rate data over the entire range. Therefore the extrapolation of the values of the rheological parameters

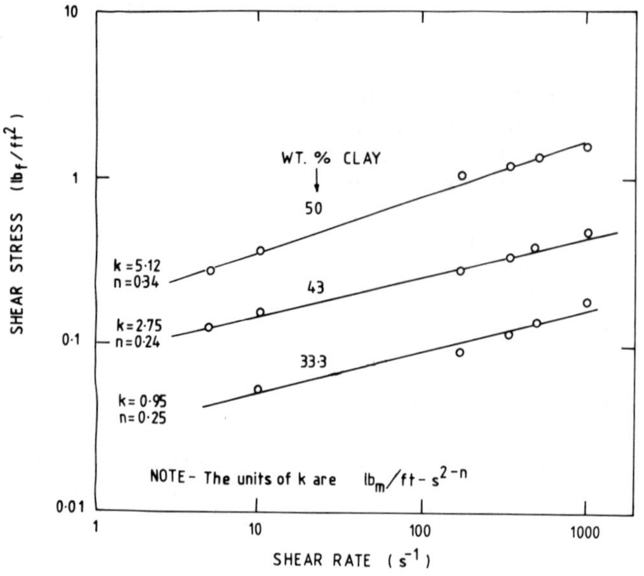

FIGURE 1. Comparison of experimental data with the power law model (replotted from [43]).

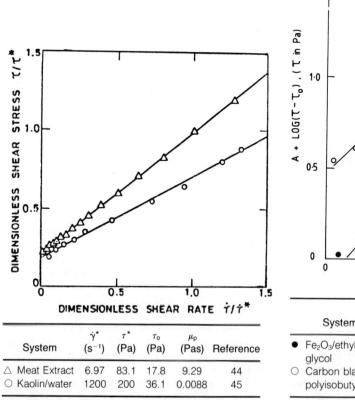

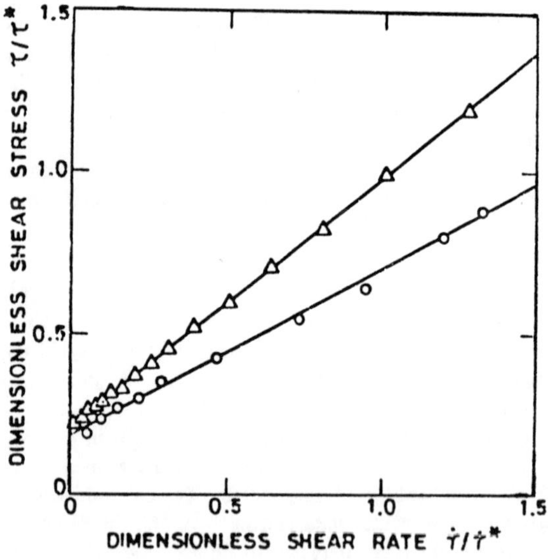

System	$\dot{\gamma}^*$ (s^{-1})	τ^* (Pa)	τ_0 (Pa)	μ_p (Pas)	Reference
△ Meat Extract	6.97	83.1	17.8	9.29	44
○ Kaolin/water	1200	200	36.1	0.0088	45

FIGURE 2. Comparison of experimental data with the Bingham plastic model (replotted from [94]).

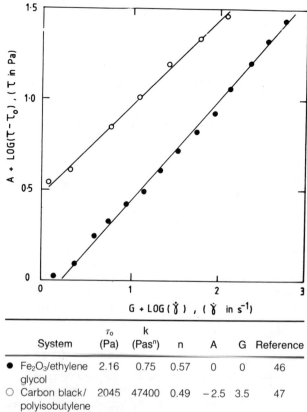

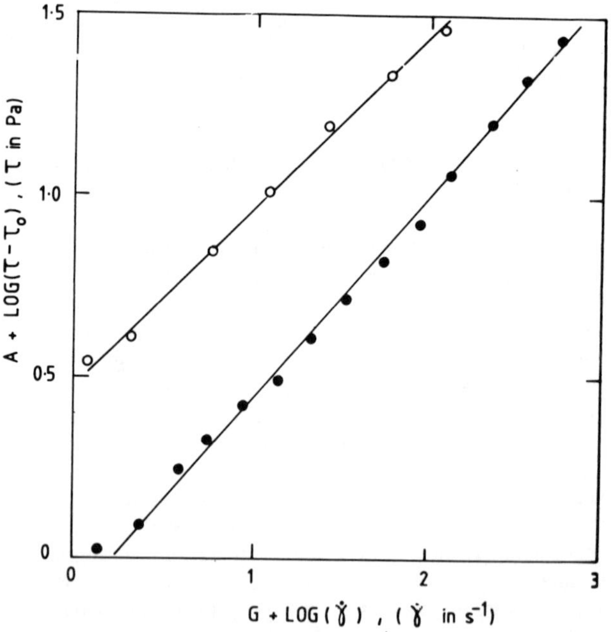

System	τ_0 (Pa)	k (Pas^n)	n	A	G	Reference
● Fe_2O_3/ethylene glycol	2.16	0.75	0.57	0	0	46
○ Carbon black/ polyisobutylene	2045	47400	0.49	−2.5	3.5	47

FIGURE 3. Comparison of experimental data with the Herschel-Bulkley model (replotted from [94]).

from one range of conditions of shear to another completely new range of shear rates must be done with caution.

Furthermore, unlike the viscosity of a Newtonian fluid which can readily be found from reference hand books or by using various estimation methods available in the literature e.g. see Reid, Prausnitz and Sherwood [48], the rheological characterisation of non-Newtonian fluids is not so simple. First, there is no data source book or estimation method to use, because non-Newtonian properties are very sensitive to the exact composition, and different batch of the same fluid can have quite different properties. Direct measurement is necessary and often a number of different viscometers have to be used to cover a sufficiently wide range of interest. Additional complications arise in dealing with fluids that are pseudohomogeneous in nature such as suspensions. These fluids can give rise to $\tau - \dot{\gamma}$ data that are widely scattered and hence the uncertainty associated with the estimation of constants appearing in Equations (1–3). Detailed discussions on the rheological measurement techniques are available in the literature [49,50].

Consideration will now be given to the pipe line flow of time independent non-Newtonian suspensions whose steady shearthinning characteristics can be adequately described by either of the three above mentioned rheological models. Like in the mechanics of Newtonian fluids, the pipe flow of non-Newtonian fluids also undergoes laminar to turbulent transition depending upon the relative importance of inertial and viscous forces. Our primary aim here is the calculation of pressure drop for a given flow and rheology or the maximum achievable flow for a fixed pressure drop. In the treatment to follow, it is assumed that the flow is fully developed and that isothermal conditions prevail.

Laminar Flow

Consider the laminar flow of a power law fluid in a circular pipe of radius R (or diameter D) as shown in Figure 4. By writing momentum balance over a thin shell, one can show:

$$\frac{\tau_{rz}}{\tau_w} = \frac{r}{R} \tag{4}$$

where τ_{rz} is the shear stress at radial location r and τ_w is the shear stress at the pipe wall. One can express the wall shear stress in terms of pressure drop gradient and rewrite Equation (4) as

$$\tau_{rz} = \frac{r}{2}\left(\frac{\Delta P}{L}\right) \tag{5}$$

Since the fluid is undergoing simple shearing motion, the power law model is written as

$$\tau_{rz} = k\left(-\frac{dV_z}{dr}\right)^n \tag{6}$$

Combining Equations (5) and (6) and integrating between the limits $r = 0$ to $r = R$, one gets:

$$V_z = \left(\frac{1}{2k}\frac{\Delta P}{L}\right)^{1/n}\left(\frac{n}{n+1}\right)R^{(n+1/n)} \times \left\langle 1 - \left(\frac{r}{R}\right)^{(n+1/n)}\right\rangle \tag{7}$$

As expected when $n = 1$ i.e. the fluid is Newtonian, Equation (7) reduces to the parabolic velocity profile. One can further relate volumetric flow rate to pressure gradient as

$$Q = \pi\left(\frac{1}{2k}\frac{\Delta P}{L}\right)^{1/n}\left(\frac{n}{1+3n}\right)R^{(1+3n/n)} \tag{8}$$

This equation is of particular interest as it allows the calculation of pressure gradient or flow rate for a material of given rheology (i.e. n and k) flowing through a pipe of radius R. Equation (8) is what might be called the power law equivalent of the Hagen-Poiseuille formula for Newtonian liquids.

Following the same methodology as used above, one can derive $Q - \Delta P$ relationships for any time independent fluid model, and a summary of these is available in a number of books [12,40,51], and the relevant expressions are reproduced below for the other two models of interest here.

For a Bingham model fluid, the $Q - \Delta P$ relationship is:

$$\frac{8Q}{\pi D^3} = \frac{1}{4\mu_P}\left(\frac{D\Delta P}{4L}\right)\left(1 - \frac{4b}{3} + \frac{b^4}{3}\right) \tag{9}$$

Similarly for the flow of a generalized Bingham model fluid

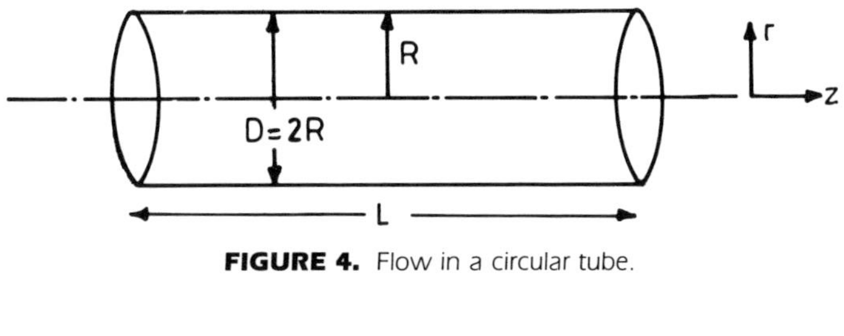

FIGURE 4. Flow in a circular tube.

(or the Herschel-Bulkley model fluid)

$$\frac{32Q}{\pi D^3} = \left(\frac{4n}{1 + 3n}\right)\left(\frac{\tau_w}{k}\right)^{1/n} (1 - b) \times \left\{1 - \frac{b}{(1 + 2n)} 1 + \frac{2nb(1 + nb)}{(1 + n)}\right\} \tag{10}$$

The new parameter b introduced in Equations (9) and (10) is defined as:

$$b = \frac{\tau_0}{\tau_w} = \tau_0\left(\frac{4L}{D\Delta P}\right) \tag{11}$$

For a given pressure gradient, the calculation of attainable velocity or volumetric flow rate is rather straightforward due to the explicit form of Equations (9) and (10); however, the reverse calculation requires an iterative procedure.

It is customary to express the flow rate-pressure drop relations in the form of dimensionless groups. For this purpose, a friction factor is defined as:

$$f = \frac{D\Delta P}{2\rho LV^2} \tag{12}$$

Combining Equation (8) with (12) for a power law fluid one gets

$$f = \frac{16}{Re} \tag{13}$$

where the Reynolds number Re is defined as

$$Re = \frac{D^n V^{2-n}\rho}{k} \cdot 8\left(\frac{n}{2 + 6n}\right)^n \tag{14}$$

Note this cumbersome looking expression reduces to the usual definition of Reynolds number for Newtonian liquids when $n = 1$ is substituted in Equation (14).

In the same manner, Equation (9) may be expressed as given below:

$$\frac{1}{Re_B} = \frac{f}{16} - \frac{Y}{6Re_B} + \frac{Y^4}{3f^3 Re_B^4} \tag{15}$$

where $Re_B = (\rho VD/\mu_P)$ and $Y = (D\tau_0/V\mu_P)$. The new dimensionless number Y is known as the yield number. The values of friction factor calculated using Equation (15) are shown in Figure 5. An alternative form of Equation (15) is also sometimes used which is written as:

$$\frac{1}{Re_B} = \frac{f}{16} - \frac{He}{6Re_B^2} + \frac{He^4}{3f^3 Re_B^8} \tag{16}$$

where the new dimensionless group He is known as Hedstrom number and is defined as $(D^2\tau_0\rho/\mu_P^2)$; it is, of course, obvious that $He = YRe_B$. The predictions of Equation (16) are plotted in Figure 6. A typical comparison between the predictions of Equation (16) and the experimental data [52] for the flow of a coal/water pseudohomogeneous suspensions in a 52 mm pipe is shown in Figure 7. Finally, one can also express the corresponding $Q - \Delta P$ relation for the Herschel-Bulkley model fluid in the following dimensionless form [53]:

$$f = \frac{16}{\phi Re} \tag{17}$$

where

$$\phi = (1 + 3n)^n (1 - b)^{1+n}\left[\frac{(1 - b)^2}{(1 + 3n)} + \frac{2b(1 - b)}{(1 + 2n)} + \frac{b^2}{(1 + n)}\right]^n \tag{18}$$

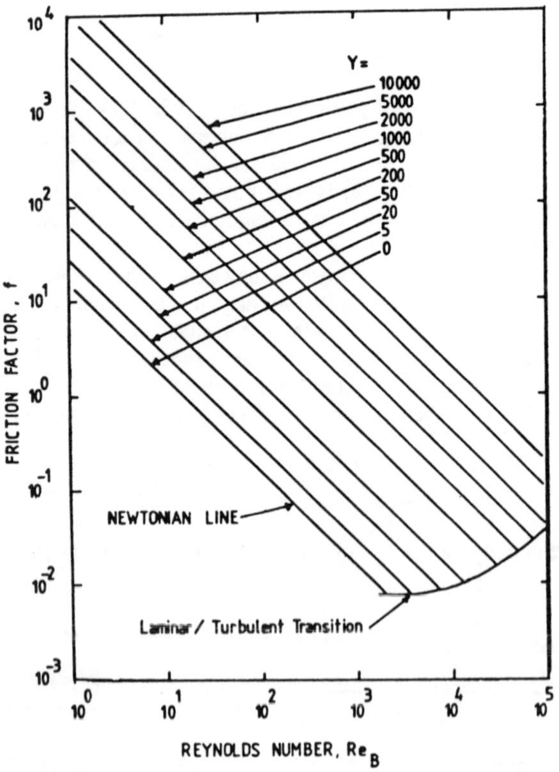

FIGURE 5. Friction factor for the laminar flow of Bingham plastic fluids (from [12]).

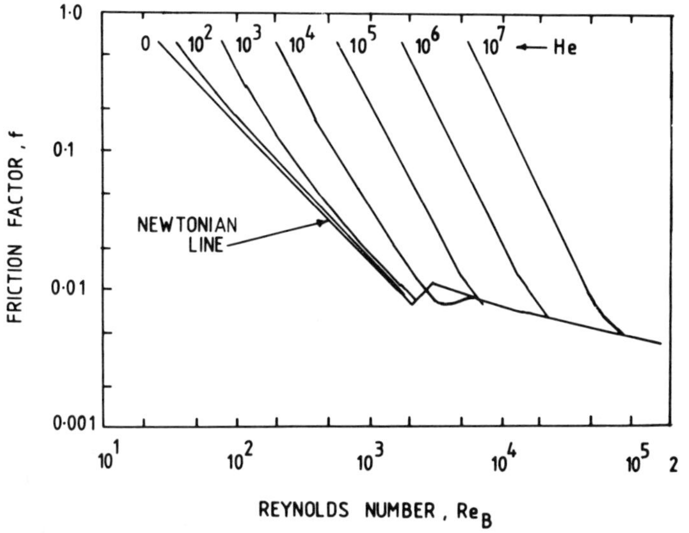

FIGURE 6. Friction factor for the laminar flow of Bingham plastic fluids (from [12]).

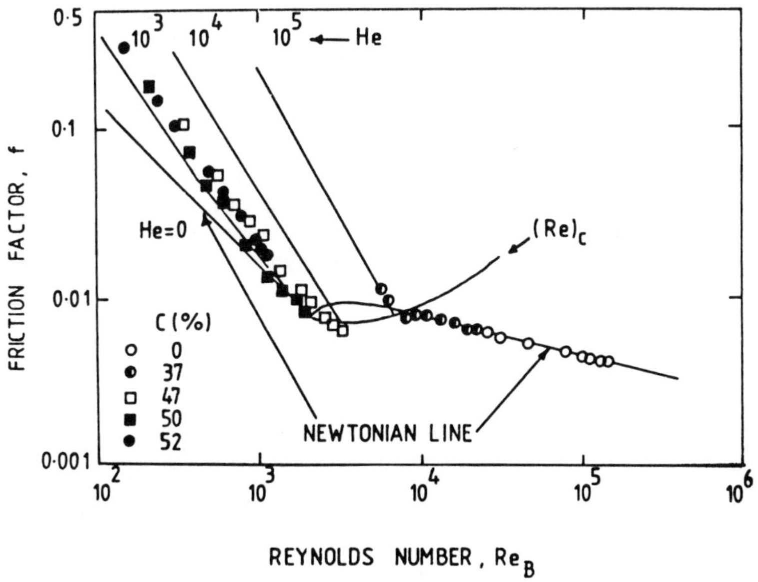

FIGURE 7. Typical comparison between the predictions of Equation (16) and the experimental values of Shook et al. [52].

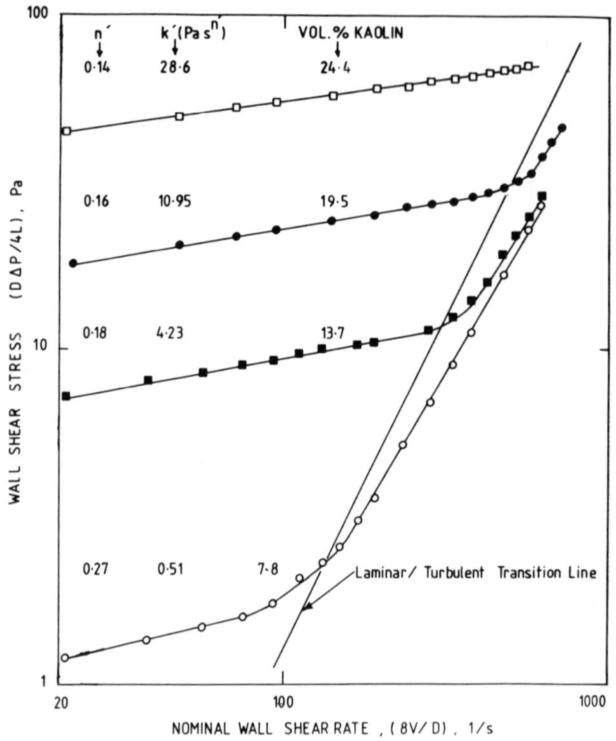

FIGURE 8. Typical wall shear stress–nominal wall shear rate data for the flow of kaolin suspensions in a 42 mm diameter pipe (replotted from [21,176]).

All the above mentioned equations are valid for smooth as well as rough pipes.

Hence the pressure loss due to friction can easily be estimated by using one of the aforementioned equations depending upon the form of relationship between shear stress-shear rate data. However, it is usually the case that the precise nature of the $\tau - \dot{\gamma}$ relationship is not known a priori; this difficulty is obviated by following the general approach suggested by Metzner and Reed [54], who showed that the Fanning friction factor (defined by Equation (12)) can be rigorously related to a generalized Reynolds number known as Metzner-Reed Reynolds number, and defined as,

$$Re_{MR} = \frac{D^{n'}V^{2-n'}\rho}{8^{n'-1}k'} \tag{19}$$

in the following way:

$$f = \frac{16}{Re_{MR}} \tag{20}$$

where k' and n' are apparent fluid consistency index, and flow behaviour index respectively, and are evaluated from $\Delta P - V$ data obtained in a circular pipe under laminar flow conditions as follows:

$$\left(\frac{D\Delta P}{4L}\right) = k'\left(\frac{8V}{D}\right)^{n'} \tag{21}$$

In general the values of n' and k' are shear rate dependent, but for most materials of practical interest these values can be treated constant over the shear rate range encountered in the envisaged application. This type of flow behaviour is exemplified by the data shown in Figure 8 where $(D\Delta P/4L)$ is plotted against $(8V/D)$ for a range of kaolin suspensions flowing in a pipe of 42 mm diameter [21,176]. It is clear that over the limited range of shear rate a constant value of n' is sufficient to describe each of the lines in laminar flow; whereas in turbulent flow they all have slopes in the range of 1.75 to 2.0. Metzner [56] has also derived relationships between the true rheological parameters such as k and n, μ_P, τ_0 etc. and k' and n'. For the rheological models of interest here these are:

Newtonian Fluid

$$n' = 1, \; k' = \mu \tag{22}$$

Power Law Fluid

$$n' = n, \; k' = k\left(\frac{3n' + 1}{4n'}\right)^{n'} \tag{23}$$

Bingham Plastic Fluid

$$n' = \frac{1 - \frac{4b}{3} + \frac{b^4}{3}}{1 - b^4}$$

$$k' = \left(\frac{D\Delta P}{4L}\right)\left[\frac{\mu_P}{\left(\frac{D\Delta P}{4L}\right)\left(1 - \frac{4b}{3} + \frac{b^4}{3}\right)}\right]^{n'} \tag{24}$$

The generalizations of Metzner and Reed are of considerable help in instances where the exact relationship between shear stress and shear rate is not known a priori. It is obvious that once the rheological characteristics are known either in the terms of a rheological model or k' and n', head loss can be estimated for a given situation under laminar flow conditions which is found to occur for $Re_{MR} < 2000$–2100.

Typical results for the laminar flow of kaolin suspensions in a 207 mm diameter pipe [57] are shown in Figure 9.

Transition from Laminar to Turbulent Flow

Regardless of the type of non-Newtonian fluid, laminar flow gives way to turbulent flow at sufficiently high values of Reynolds number. Ryan and Johnson [58] established that for a power law fluid, the transition from laminar to turbulent flow occurs at a critical value of Reynolds number (defined by Equation (14)) given by:

$$(Re)_c = \frac{6464n}{(1 + 3n)^2(2 + n)^{-(2+n/1+n)}} \tag{25}$$

In the range $0.1 \leqslant n \leqslant 1$, the value of $(Re)_c$ varies from 1880 to 2100; thus for all practical purposes it is safe to assume that the transition takes place at $Re = 2100$ which also coincides with the value generally accepted for the transition of Newtonian fluid. The experimental measurements of Hanks and Christiansen [59] support the theoretical analysis of Ryan and Johnson [58]. In the case of Bingham plastic fluids, Hanks [60] demonstrated that the transition takes place at a critical value of the relevant Reynolds number (Re_B) given by the expression:

$$(Re_B)_c = \frac{He}{8b_c}\left(1 - \frac{4b_c}{3} + \frac{b_c^4}{3}\right) \tag{26}$$

where b_c is obtained from

$$\frac{b_c}{(1 - b_c)^3} = \frac{He}{16800} \tag{27}$$

The predictions of Equations (26) and (27) have been experimentally confirmed by the data of Caldwell and Babbitt [61], Wilhelm *et al.* [62], and Alves *et al.* [63].

More recently, Froishteter and Vinogradov [64] have given the following criterion for the transition from laminar to turbulent flow for a Herschel-Bulkley model fluid:

$$\left.\begin{aligned}
(Re_{MR})_c &= \frac{2100}{\alpha} \\
\alpha &= \frac{1}{W^3}\left(\frac{b^2}{2} + (1 - b^2)A \right. \\
&\qquad \left. + \, b(1 - b)B\right) \\
W &= b^2 + 2b(1 - b)\left(\frac{1 + n}{1 + 2n}\right) \\
&\qquad + (1 - b^2)\left(\frac{1 + n}{1 + 3n}\right) \\
A &= \frac{3(1 + n)^3}{2(1 + 3n)(1 + 2n)(3 + 5n)} \\
B &= \frac{6(1 + n)^3}{(1 + 2n)(2 + 3n)(3 + 4n)}
\end{aligned}\right] \tag{28}$$

Note that when $n = 1$ and $b = 0$, Equation (28) reduces to its proper Newtonian limit. It does not, however, approach the result given by Ryan and Johnson [58] when $b = 0$ is substituted in Equation (28). Nor does it approach the limiting case of Bingham plastic fluid as n goes to unity.

Finally, Dodge and Metzner [65] found experimentally that the critical value of Re_{MR} denoting the transition from laminar to turbulent conditions depends on the value of n'. However, the dependence is so weak that it is generally accepted that the transition occurs at or around $Re_{MR} = 2000$–2100 over the complete range of n'. Over the past three decades, experimental results obtained with a wide variety of test fluids exhibiting a range of rheological properties have shown that this is not a bad engineering approximation. All the aforementioned criteria seem to suggest an abrupt transition while in practice it is gradual and there is a range

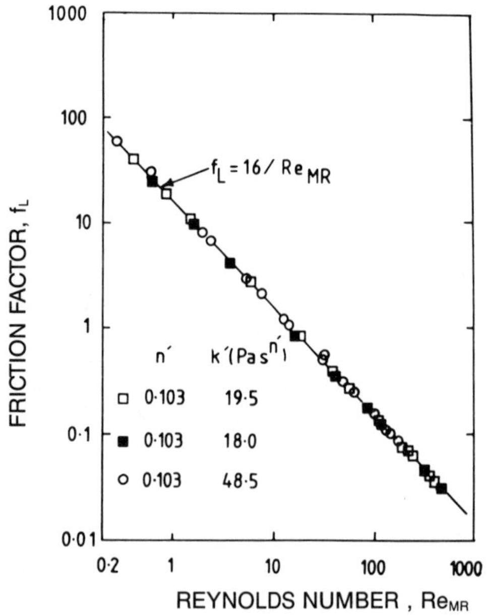

FIGURE 9. Experimental verification of the validity of Equation (20). Points are for the flow of kaolin suspensions in a 207 mm diameter pipe [57].

of velocity for each liquid (depending upon the rheological properties and pipe diameter etc.) in which the flow can be neither classified as laminar nor turbulent. This intermediate region is known as transition region. In view of this, therefore, the critical values of Reynolds number calculated from Equations (25–28) should be interpreted as the upper limits for steamline or laminar flow to exist.

Turbulent Flow ($Re_{MR} > 2000$)

Just as there are many alternative equations for predicting friction factor for turbulent Newtonian flow, so there are numerous available for turbulent non-Newtonian flow. Most of them are empirical in nature, as pointed out by Heywood and Cheng [53], and are based on the use of the power law model [65–71]. These require modifications before they can be used for the Herschel-Bulkley model fluids while others have been specifically derived for use with the Herschel-Bulkley model fluids e.g. Torrance [71], Hanks [72]. All assume the pipe wall to be hydrodynamically smooth. For rough-walled pipes, Szilas *et al.* [70] and Torrance [71] have introduced small corrections in their equations. A critical review of all these equations has been recently presented by Heywood and Cheng [53]. Here only a selection of proven methods is given.

POWER LAW FLUIDS

Dodge and Metzner [65] carried out a semi-theoretical analysis of turbulent flow of power law fluids in smooth walled pipes. Following the dimensional arguments, originally advanced by Millikan [73] for Newtonian fluids, they derived the following expression for friction factor:

$$\frac{1}{\sqrt{f}} = \frac{4}{(n')^{0.75}} \log\left[Re_{MR} f^{(2-n'/2)}\right] - \frac{0.4}{(n')^{1.2}} \tag{29}$$

Note that when $n' = 1$, Equation (29) reduces to the well known Nikuradse equation for the Newtonian turbulent flow. Equation (29) is plotted in Figure 10 with n' as a parameter.

Dodge and Metzner [65] reported excellent agreement between calculated and experimental values of friction factor over values of n' from 0.36 to 1 and Re_{MR} from 2900 to 36000. For materials that do not conform to the power law behaviour over the relevant shear stress range, Dodge and Metzner [65] suggest that their approach may be used provided n' is evaluated from laminar flow data in the wall shear stress range applicable to the turbulent flow conditions occurring in the pipe of interest. Although this method has proved to be quite useful, Heywood and Richardson [20] and Heywood [74] have, however, pointed out some practical difficulties in using the method of Dodge and Metzner [65]. Usually in practice n' is obtained from laminar flow data at wall shear stresses substantially below the relevant stress levels in turbulent flow. The reason for this is either because turbulent flow predictions are required for the same sized pipe (used for laminar flow tests) or because the material is so markedly shearthinning that even tests using small bore tubes do not result in sufficiently high wall shear stresses. Consequently, the value of n' (and the corresponding value of f) is often underestimated. Metzner [75] has, however, refuted this conclusion. For the turbulent flow of aqueous flocculated kaolin suspensions whose laminar flow behaviour could be approximated by the power law fluid model, Kemblowski and Kolodziejski [69] have reported the following explicit formula for friction factor:

$$f = \frac{(E/4)\phi^{1/Re_{MR}}}{(Re_{MR})^{m'}}$$

where

$$E = 8.9 \times 10^{-3} \exp(3.57n^2) \tag{30}$$

$$\phi = \left\langle \exp\left(0.572\,\frac{1 - n^{4.2}}{n^{0.435}}\right)\right\rangle^{1000}$$

$$m' = 0.314n^{2.3} - 0.064$$

This correlation is valid for $Re_{MR} < \dfrac{31600}{(n')^{0.435}}$ and above this critical value of Reynolds number, the usual Blasius equation can be used.

Admittedly Equation (30) is completely empirical in nature, its use requires the knowledge of rheological proper-

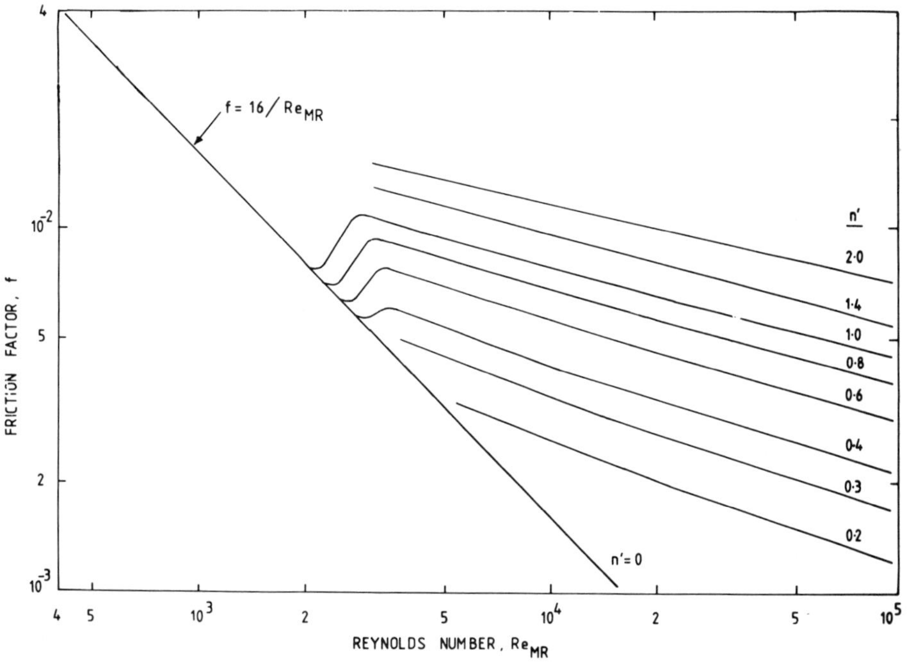

FIGURE 10. Friction factor–Reynolds number relation for non-Newtonian fluids [65].

ties in laminar flow regime which is a distinct advantage over the method of Dodge and Metzner [65]. Besides, Equation (30) does not require an iterative procedure. Though the utility of Equation (30) for the flow of kaolin suspensions is well established [55,76], its general applicability remains yet to be demonstrated.

A typical comparison between the correlations of Dodge and Metzner [65] and the predictions of Equation (30) and the experimental data is shown in Figure 11.

BINGHAM PLASTIC FLUIDS

Tomita [66] has applied the concept of Prandtl mixing length to the turbulent flow of Bingham plastic fluids in smooth pipes, and obtained the following relationship:

$$\sqrt{\frac{1}{f_{BT}}} = 4 \log(Re_{BT}\sqrt{f_{BT}}) - 0.4 \qquad (31)$$

where the modified Reynolds number and friction factor are defined as:

$$\left.\begin{aligned} Re_{BT} &= \frac{DV\rho}{\mu_P}\left[\frac{(1-b)(b^2-4b+3)}{3}\right] \\ f_{BT} &= f/(1-b) \end{aligned}\right\} \qquad (32)$$

In these relations the value of b to be inserted is calculated as if the flow were laminar. Tomita's equation has been experimentally validated over the range $2000 < Re_{BT} < 100{,}000$ by twenty data points covering a good range of pipe diameter, a sevenfold variation of τ_0 but only a slight variation in plastic viscosity (μ_P).

Thomas [45,77–80] undertook a comprehensive review of the information available in the area of non-Newtonian suspensions. In one of his papers [80], he reported the turbulent friction factors for flocculated suspensions of thoria, kaolin, and titania (whose behaviour could be well described by the Bingham plastic model). In all cases the friction factors followed the Newtonian curve except that the data fell below the line by 20–30%. Thomas explained this decrease in friction factor by postulating the thickening of the laminar sublayer due to non-Newtonian properties. For yield stresses less than 2.44 Pa, the friction factor values approached the Newtonian behaviour. Both sets of data were correlated with the Blasius type relation as:

$$f = B(Re_B)^{-\alpha}$$

where

$$\left.\begin{aligned} B &= 0.079\left[\left(\frac{\mu}{\mu_P}\right)^{0.48} + \left(\frac{\rho\tau_0\kappa^2}{\mu_2}\right)^2\right] \\ \alpha &= 0.25\left[\left(\frac{\mu}{\mu_P}\right)^{0.15} + \left(\frac{\tau_0\kappa^2\rho}{\mu_2}\right)^2\right] \end{aligned}\right\} \qquad (33)$$

with $\kappa = 1.83 \times 10^{-6}\ m$.

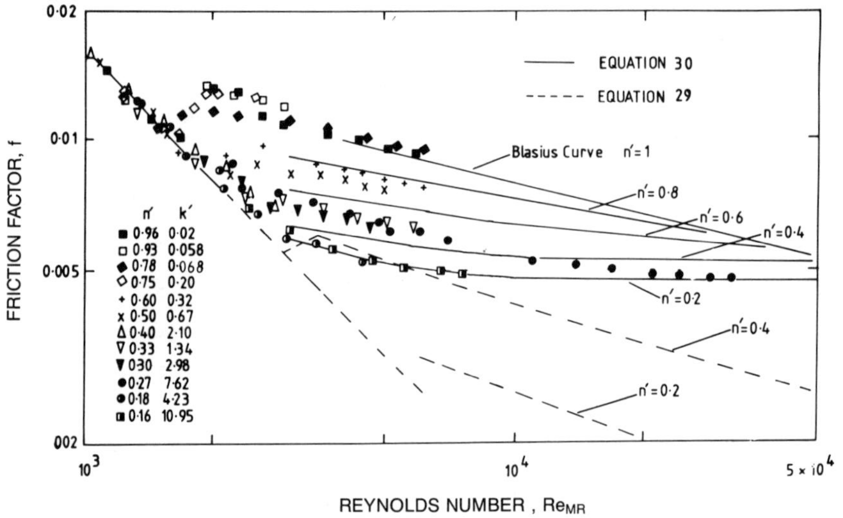

FIGURE 11. Typical comparisons between the predictions of Equations (29) and (30), and experimental data [76,176].

This correlation has, however, not gained wide acceptance due to its complex form.

Other correlations for the turbulent flow of Bingham plastic fluids have been reviewed by Govier and Aziz [12].

HERSCHEL-BULKLEY MODEL FLUIDS

Not much is known about the turbulent flow of Herschel-Bulkley model fluids. Torrance [71] has investigated the turbulent flow in smooth walled pipes and has proposed the following relationship for calculating friction factor:

$$\sqrt{\frac{1}{f}} = \left(0.45 - \frac{2.75}{n}\right) + \frac{1.97}{n}\ln(1 - b) + \frac{1.97}{n}\ln(Ref^{(1-n/2)}f(n)) \tag{34}$$

where

$$f(n) = \left(\frac{1 + 3n}{4n}\right)^n$$

Although most investigators claim general validity for their correlation/equation, only a few have been adequately tested. A notable exception to this is the celebrated correlation of Dodge and Metzner [65]. In a recent paper, Heywood and Cheng [53] have conducted a detailed comparative study of fourteen different methods available for the estimation of friction factors, and they have also discussed the sensitivity of friction factors to the uncertainty associated with the evaluation of rheological model parameters. They concluded that unlike in the case of Newtonian turbulent flow where most of the existing correlations predict values of f to within $\pm 4\%$ of each other, the predictions for the non-Newtonian turbulent flow can differ by as much as a factor of three. They go on to say that, "the choice of the best correlation equations to use cannot be made until predictions have been thoroughly compared with experimental data covering a wide range of fluids." The best strategy for the designer to adopt therefore is to use as many correlations as possible to establish the upper and lower bounds of friction factor, with proper attention being paid to the uncertainty in friction factor owing to uncertainty in evaluating the rheological parameters.

EFFECT OF PIPE ROUGHNESS ON PRESSURE DROP

It is well established that the magnitude of irregularities on the inside wall of a pipe has no effect on the pressure drop of either Newtonian nor non-Newtonian materials in laminar flow. Such irregularities, on the other hand, have appreciable effect on pressure drop in turbulent flow. In general, the larger the pipe surface irregularities for a given pipe diameter and Reynolds number, the higher is the pressure drop. Despite this observation for Newtonian materials, the effect of pipe roughness on the flow of non-Newtonian materials has received no attention at all. One of the few references to consider pipe roughness is the text by Govier and Aziz [12]. In their recommended design methods, Govier and Aziz [12] suggest that if the pipe is rough, the pressure gradient calculated by the various methods for smooth pipes should be multiplied by the ratio of the friction factor for rough pipe to the friction factor for smooth pipe as if the fluid were Newtonian. If the relative pipe roughness is such that fully rough pipe wall turbulence (i.e. where f is independent of Re but is dependent upon relative roughness ε/D) may be expected, the equation due to Torrance [71] may be used:

$$\frac{1}{\sqrt{f}} = \frac{1.77}{n} \ln\left(\frac{D}{2\varepsilon}\right) + \left(6.0 - \frac{2.65}{n}\right) \qquad (35)$$

This equation has not been confirmed experimentally so the predictions must be treated with caution.

Pressure Drop in Pipe Fittings

Consideration has so far been given to the estimation of pressure losses in straight pipe lengths. To determine the total pressure drop in a pipe network, one must add the additional pressure drop arising from various fittings such as bends, expansions, contractions, valves etc. For the flow of Newtonian fluids the magnitude of these losses are well documented and the relevant data may be found in standard reference books [81], whereas only scant information regarding the flow of non-Newtonian materials is available. Skelland [40] cites the work of Weltmann and Keller [23] who reported some experimental work on such flows with Bingham plastics and shearthinning suspensions. Presumably turbulent conditions prevailed in their work, and the pressure losses were found to be comparable to those for Newtonian liquids. This may be so partly due to the fact that non-Newtonian materials behave not too differently from the Newtonian fluids in turbulent conditions. Therefore, the conventional friction loss factors in fittings that are available for Newtonian liquids may be used as first approximations for designing non-Newtonian flow systems. However, at low Reynolds number, losses for non-Newtonian fluids may be as much as ten times greater than for Newtonian liquids [74,82]. Only a limited amount of data is available, and thus it is not possible to formulate general predictive methods for head losses through process fittings.

The preceding section has been concerned with the flow of pseudohomogeneous suspensions displaying time independent shearthinning characteristics in circular pipes. The available methods for estimation of friction factors (or head losses) in laminar flow regime are based on sound theory whereas the situation is not so simple in the case of turbulent flow. Based on the existing information, the best possible methods have been identified for turbulent flow conditions. There is a real paucity of information with regard to the effect of pipe roughness on friction losses and the pressure drop incurred in flow through pipe fittings.

FLOW OF TWO-PHASE SOLID-LIQUID MIXTURES

A salient characteristic of the heterogeneous suspension or mixtures is the fact that the solid and liquid phases (usually water) remain identifiable; there is not, for instance, any increase in the viscosity of the liquid phase due to the presence of solid particles. In cases where some of the solids are ultrafine in size and colloidal in nature and are held in homogeneous suspension by water, the heavy medium so formed must be re-evaluated in terms of its density and viscosity (as described in the previous section). The new heavy carrier medium so formed and the coarse particles which are not constituents of the medium still behave respectively as liquid and solid. The increased buoyancy of the solid phase due to the increased density of the vehicle can be of great help in reducing the pipeline pressure loss. In the treatment to follow it is assumed that there is no interaction between the solids and liquid such as the dissolution of solids or the adsorption of the liquid on the solids or through electrostatic charges. Furthermore, the information given here is for suspensions of mono or closely sized and reasonably regular shaped particles.

It is now well known from observations that the behaviour of solids in a moving liquid is largely determined by the particle size and density, but the manner in which the fluid causes particle movement is not fully understood. The nature of the mechanism of energy transfer between the two phases varies from one flow regime to another, while in any given regime, the relative importance of the various forces causing particle movement is a function of the relative velocity between the two phases. Currently held views on the nature of the fluid-particle interactions have been largely gleaned from deductions based on experimental work and empirical correlations derived there from. Although a number of hypotheses have been put forward which shed some light on certain aspects of two phase flows, they are not yet refined to the extent to be used as a basis for design calculations. Most of the design methods available are therefore empirical in nature, and lack universality. Theoretical developments in this area have been summarised in several references [12,83,84,183].

Flow Regimes

A wide variety of flow patterns or regimes has been observed and reported for the two phase flow of liquid-solid heterogeneous mixtures. It is now generally agreed that all patterns can be described in terms of the four major and easily identifiable types; these are defined below and sketched in Figure 12.

HOMOGENEOUS FLOW

This type of flow is characterised by uniform distribution of solids throughout the liquid medium. Homogeneous flow, or a close approximation to it, is achieved in the flow of slurries of fine particles and high concentrations flowing at moderately high mixture velocities. Under certain conditions the concentration profile approaches the velocity profile [85,86]. Under these conditions the concentration of solids in the discharged mixture is almost equal to the *in-*

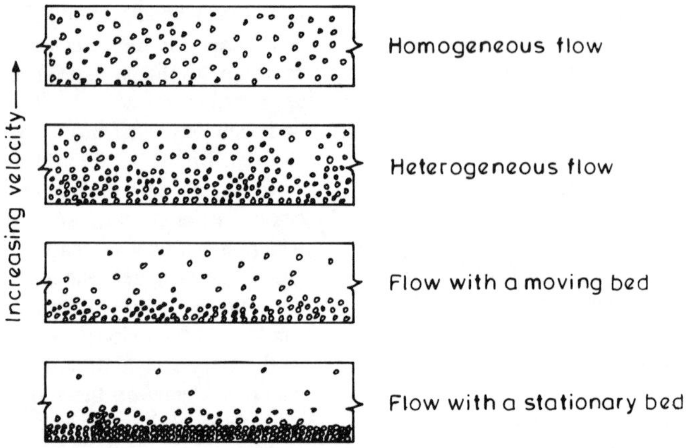

FIGURE 12. Schematic representation of the four major flow patterns (or regimes).

situ concentration. The term "symmetric suspension flow" is also used in the literature to describe this kind of flow [12].

HETEROGENEOUS FLOW

As the mixture velocity and hence the intensity of turbulence and lift forces are reduced, the resultant of the weakened forces tending to suspend the particles coupled with the settling tendency of the particles causes a distortion of the concentration and velocity profile [86] with more of the solids being transported in the lower half of the pipe. This type of flow is encountered when the solids are coarse and of large density. Govier and Aziz [12] have used the term "asymmetric suspension flow" to define this flow regime.

MOVING BED OR DENSE PHASE FLOW

If the mixture velocity is reduced further to an extent such that the gravitational forces become more dominant than the hydrodynamic (turbulent eddies) forces, particles tend to accumulate on the bottom of the pipe, first in the form of dunes [87] and finally as a moving bed. The dunes or the bed move along the bottom of the pipe at a lower velocity than the carrier fluid (and fine particles) in the top clear path.

SALTATION AND/OR FLOW WITH A STATIONARY BED

As the mean value of mixture velocity is further decreased, the lower most particles of the bed become nearly stationary, the bed thickens and bed motion is caused by the uppermost particles tumbling over one another i.e. saltation. Such a flow occurs in the case of appreciably large and heavy particles which cannot be held in suspension by turbulence. Due to the presence of a stationary bed, the free area available for flow is decreased. Eventually, with continued reduction in the mixture velocity and gradual buildup of the bed, pressure gradient needed to sustain the flow increases rapidly, and in the absence of an abnormally high applied pressure, one comes across the most undesirable conditions of the clogging of pipe.

The foregoing description is for reasonably regular shaped and closely sized solid particles in a liquid such as water which is the most commonly employed transporting medium. For particles of mixed size and a wide range of settling velocities, the fastest settling particles will form a bed in the bottom of the pipe, the ones with intermediate settling velocities will be conveyed as a heterogeneous (or asymmetric) suspension and the slowest settling particles may be uniformly distributed across the pipe diameter. Besides, if there are ultrafine colloidal particles, the carrier medium can exhibit non-Newtonian shearthinning viscosity. Hence, one may indeed encounter all possible regimes and the resulting complexities in one single application.

It is also important to bear in mind that the distinction between any two flow patterns is usually ill-defined and the transition from one to another regime occurs over a range of conditions. Since the mechanism of the flow and energy transfer (that is, friction losses) is different for each flow pattern, it is essential to know which flow pattern is occurring in a given situation before attempting an analysis or interpretation of experimental data. In the next section, an appraisal of some of the widely used methods to delineate flow regimes is presented.

PREDICTION OF FLOW REGIMES

Figure 13 shows a typical pressure drop-mixture velocity relationship for the flow of two phase solid-liquid mixtures. Four major flow patterns (or regimes) identified above are also schematically indicated in this figure. Transition from one to another flow pattern is reflected in the visible change in the shape of the curve. Govier and Aziz [12] and Turian and Yuan [88] have defined four transition velocities to cover the complete range of behaviour observed, and these velocities, namely V_1, V_2, V_3, V_4 are also marked in Figure 13. For all velocities greater than V_1, the liquid-solid mixture behaves much as a true fluid, although nonuniform but symmetric distribution of the particles across the cross-section results in a holdup. Not withstanding these effects, the pressure gradient line nearly parallels that for the conveying liquid, and may often be approximated as that for a pure liquid having the properties of the input mixture. At very high velocities, the additional pressure drop due to the presence of solids is negligible and all the lines (corresponding to different solid concentration) approach that for the flow of carrier liquid.

The transition velocity V_2 is perhaps the most important one as it marks the onset of bed formation on the bottom of the pipe. This transition velocity has been variously referred to as the limit deposit velocity, the critical velocity and the critical deposit velocity. Much confusion exists in the literature regarding the interpretation of V_2 [89,90] e.g. a number of researchers [78,91,92,93] have assumed that this velocity corresponds to the minimum in the pressure gradient-velocity curve (Figure 13). Round [90] has, however, pointed out that it was necessarily not so and the minimum in ΔP–V curve may occur at or just below V_2 [128].

The transition velocity V_3, which demarcates the flow with

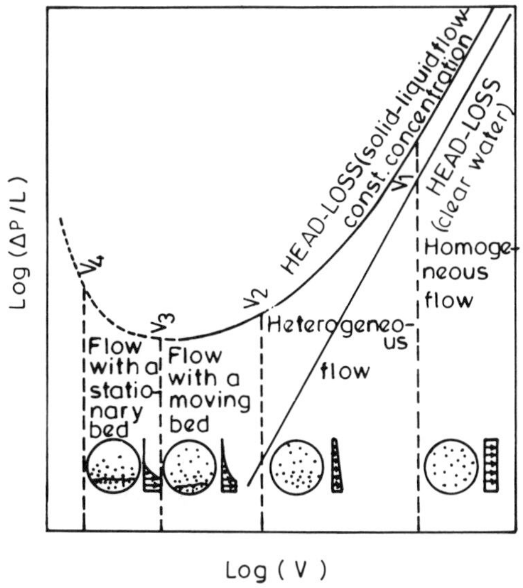

FIGURE 13. Typical pressure drop–mixture velocity curve for the flow of solid–liquid mixtures.

a moving bed from that with a stationary bed, is not very well reflected in the pressure gradient-velocity curve, and V_4 is seldom reached in actual experiments, and is of little significance.

TRANSITION VELOCITY V_1

The transition from homogeneous (symmetric) to heterogeneous (asymmetric) flow regime takes place gradually with reduction in mixture velocity, and due to the increasing relative importance of gravitational forces. In 1955, Spells [95], using a combination of dimensional considerations and empirical approach, presented correlations for what he called the standard velocity and the minimum velocity. Govier and Aziz [12] and Round [90] have interpreted his standard velocity as V_1 as defined in the present context. Spells' equation is:

$$V_1^2 = 0.075 \left(\frac{D V_1 \rho_m}{\mu_m} \right)^{0.775} g d_{85}(s - 1) \tag{36}$$

where

d_{85} = particle diameter such that 85% (by weight) of the particles are finer than d_{85}
ϱ_m = mixture density
μ_m = mixture viscosity
s = specific gravity of solids
D = pipe diameter

Govier and Aziz [12] have combined Equation (36) with Newton's law of settling to obtain:

$$V_1 = 9.1 C_D^{0.816} D^{0.633} V_0^{1.63} \tag{37}$$

where D is now in meters; V_0 is the free settling velocity of a single particle in ms^{-1}. Equation (37) further assumes that the kinematic viscosity of the mixture is same as that of the conveying liquid.

Based on semi-empirical considerations, Newitt *et al.* [96] have also given an equation which enables the calculation of V_1 as:

$$V_1 = (1800 g D V_0)^{1/3} \tag{38}$$

In a comprehensive review, Turian and Yuan [88] culled data from various sources, and created a data bank containing 2848 entries covering wide ranges of conditions. Based on carefully collated experimental results, they proposed the following empirical formula for calculating the value of transition velocity V_1;

$$V_1^2 = 0.2859 C^{1.075} f_L^{-0.67} C_D^{-0.9375} g D(s - 1) \tag{39}$$

Equation (39) was claimed to yield much better results than both Equations (37) and (38). This is hardly surprising in view of the fact that Equation (39) contains four adjustable parameters which have been evaluated using the least square analysis method. It is certainly cumbersome to use. Furthermore, since not much data is available about this transition velocity, detailed comparisons of Equations (37–39) with experimental observations as well as with each other have not been carried out to date. In absence of any definite information, Zandi [1] and Wani *et al.* [97] have suggested the use of Equation (38) only as a guideline rather than a fast rule. Wani *et al.* [97] have also extended the applicability of Equation (38) to mixed size particles.

TRANSITION VELOCITY V_2

As mentioned earlier, perhaps this is the most important transition in the whole of the slurry transport technology; consequently it has received the greatest amount of attention in the literature resulting in a large number of correlations. Most of these methods, along with their merits and demerits, have been reviewed by Carleton and Cheng [98] and by Oroskar and Turian [99] while Parzonka *et al.* [89] have re-evaluated most of the available experimental results. One of the earliest, and probably the most widely used correlation is that of Durand and Condolios [100], as summarised by Gilbert [101] is:

$$V_2 = F_L(2gD(s - 1))^{1/2} \tag{40}$$

where F_L is a dimensionless parameter and is a function of both particle size and concentration; this dependence is shown graphically in Figure 14. This figure is based on data obtained with sand/water and coal/water mixtures ($C \leqslant 15\%$ by vol.) flowing in pipes of 40–700 mm diameter. It is of interest to note that the particle size (d) does not appear to have appreciable effect on F_L for $d > 2$ mm. Wasp *et al.* [102] and Parzonka *et al.* [89] have critically examined the validity of Equation (40). The latter workers divided all the data relating to the flow of heterogeneous mixtures into five categories according to the type of solid and particle size range, as follows:

1. Small size sand particles (100–280 μm)
2. Medium and coarse size sand particles (400–850 μm)
3. Coarse size sand and gravel (1.15–19 mm)
4. Small size high density materials (50–300 μm)
5. Coal particles (1–2.26 mm)

Although different types of representations have been used in the literature, Parzonka *et al.* [89] have re-expressed all results in the form of F_L as defined by Equation (40). The variation of F_L with concentration is shown in Figures 15–19 relating to each of the categories mentioned above. An examination of Figures 15–19 reveals that the value of F_L (thus the value of V_2) generally goes through a maximum in each case at about $C \approx 15\%$. For the values of concentration greater than 15%, the parameter F_L in fact decreases. This observation is also consistent with the findings of Toda *et al.* [125]. This type of dependence was neither envisaged by Durand and Condolios [100] nor is borne out by their experimental work in which the maximum concentration

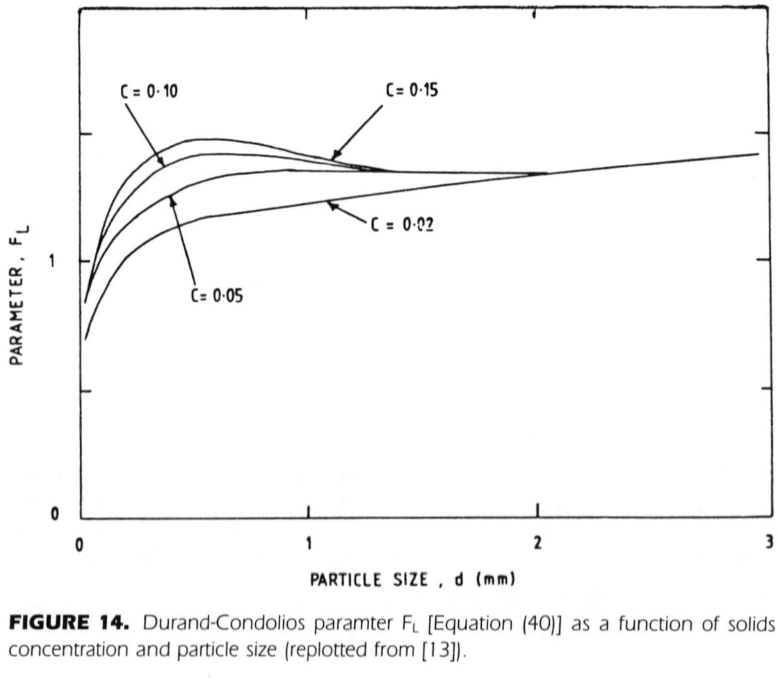

FIGURE 14. Durand-Condolios paramter F_L [Equation (40)] as a function of solids concentration and particle size (replotted from [13]).

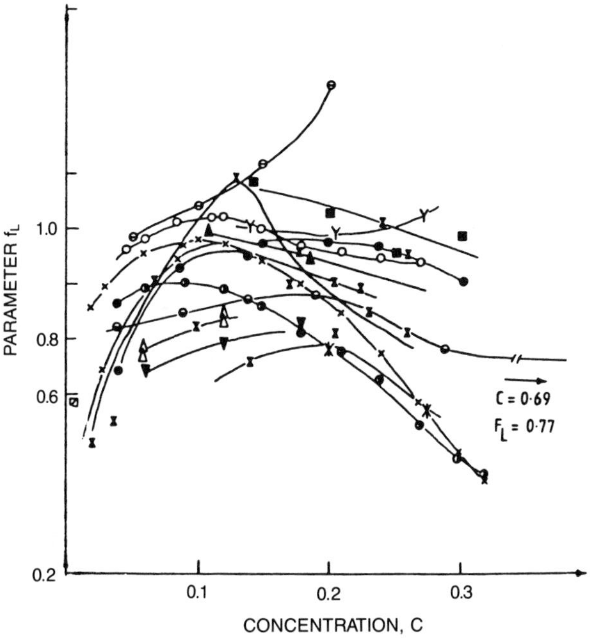

Symbol	d (mm)	Presence of small particles	D (mm)	References
○	0.25	?	103	
△	0.28	?	614,800	103–105
▼	0.25	?	410	
⊠	0.25	none	27,127	120
⧄	0.25	none	52	121
⊗	0.10	none	51	106
⊖	0.23	none	108	107
⧗	0.19	none	76	108
X	0.25	3%, d = 0.06 mm	103	103–105
◑	0.25	3%, d = 0.006 mm	103	
*	0.10	2.5%, d < 0.074 mm	26	92
Y	0.23	2%, d < 0.074 mm	26	
◒	0.20	none	202	93
●	0.17	32%, d < 0.074 mm	108	
■	0.17	32%, d < 0.074 mm	159	111
▲	0.17	32%, d < 0.074 mm	263	

FIGURE 15. Variation of F_L with solids concentration for small sized sand being conveyed in water (replotted from [89]).

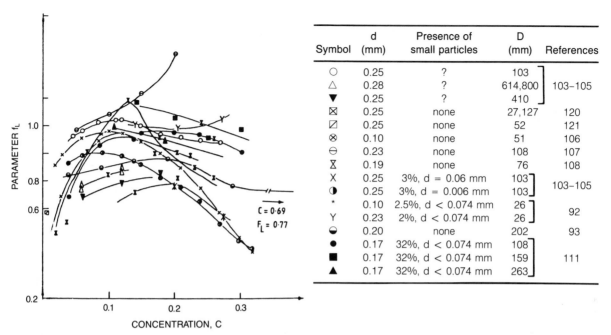

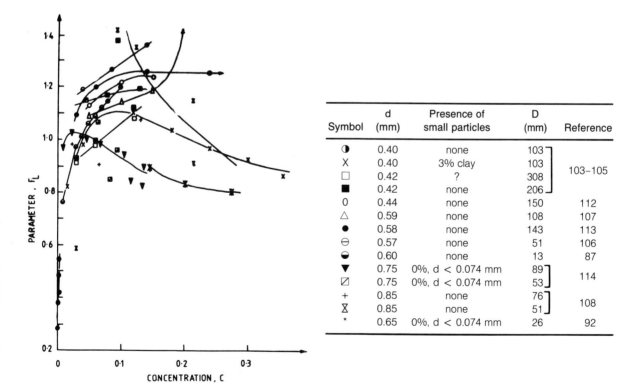

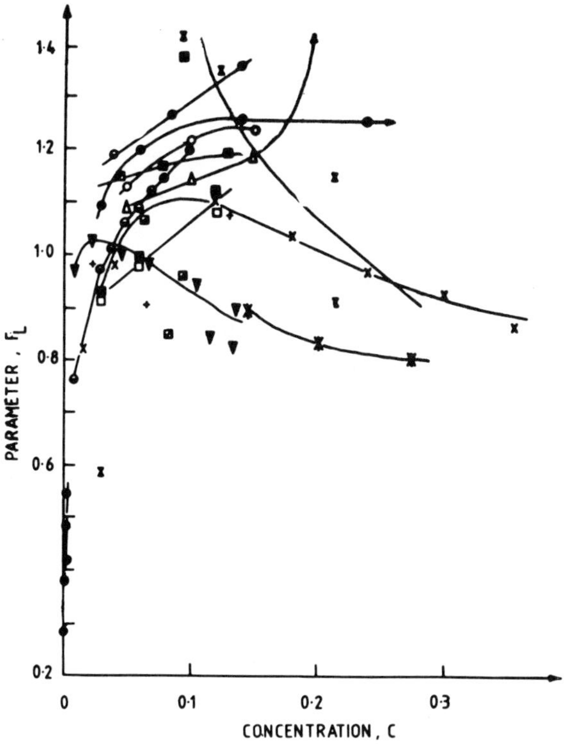

Symbol	d (mm)	Presence of small particles	D (mm)	Reference
◑	0.40	none	103	
X	0.40	3% clay	103	103–105
□	0.42	?	308	
■	0.42	none	206	
0	0.44	none	150	112
△	0.59	none	108	107
●	0.58	none	143	113
⊖	0.57	none	51	106
◒	0.60	none	13	87
▼	0.75	0%, d < 0.074 mm	89	114
⧄	0.75	0%, d < 0.074 mm	53	
+	0.85	none	76	108
⧗	0.85	none	51	
*	0.65	0%, d < 0.074 mm	26	92

FIGURE 16. Dependence of F_L on solids concentration for medium and coarse sized sand in water (replotted from [89]).

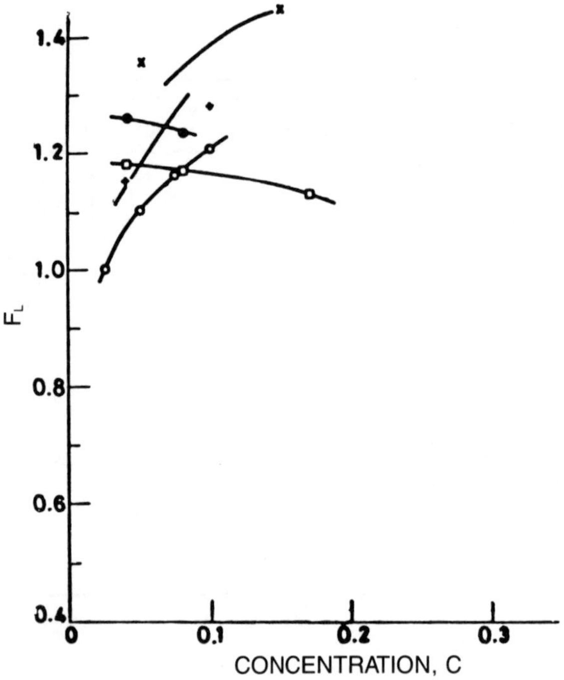

Symbol	d (mm)	D (mm)	Reference
0	2.04	150	112
X	1.15	108	107
●	1.35	407	117
□	19.0	407	
+	3.7	51	106

FIGURE 17. Variation of F_L with solids concnetration for coarse sized sand and gravel being conveyed in water (replotted from [89]).

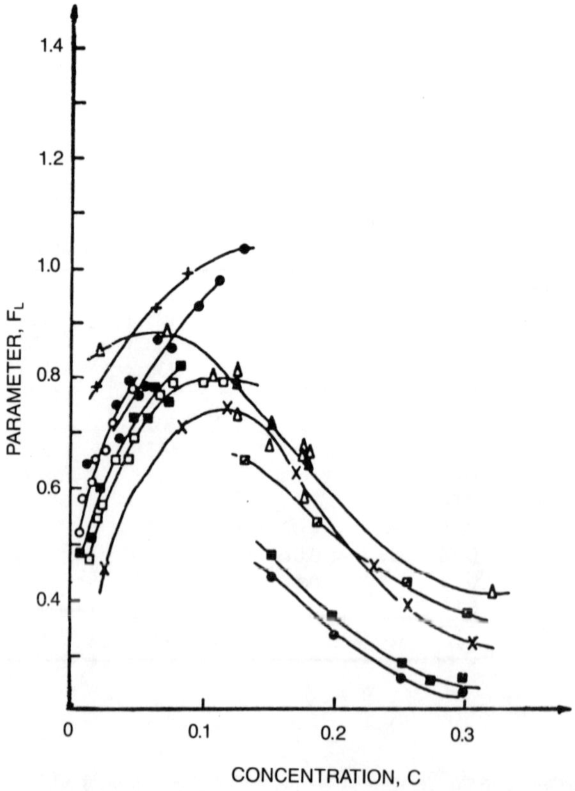

Symbol	d (mm)	% of Particles of d < 0.075 mm	ϱ_s (kgm^{-3})	D (mm)	Reference
+	0.30	16	3360	103	103–105
X	0.20	60	3360	103	
△	0.14	16	2690	207	116
▲	0.14	16	2690	100	
○	0.075	50	4000	207	117
■	0.07	55	4000	149	
●	0.09	45	3000	207	
□	0.06	80	3100	149	
⧄	0.05	72	5250	52	118
⌧	0.05	77	5250	209	
⊖	0.05	77	5250	263	

FIGURE 18. Variation of F_L with solids concentration for small sized, high density materials in water (replotted from [89]).

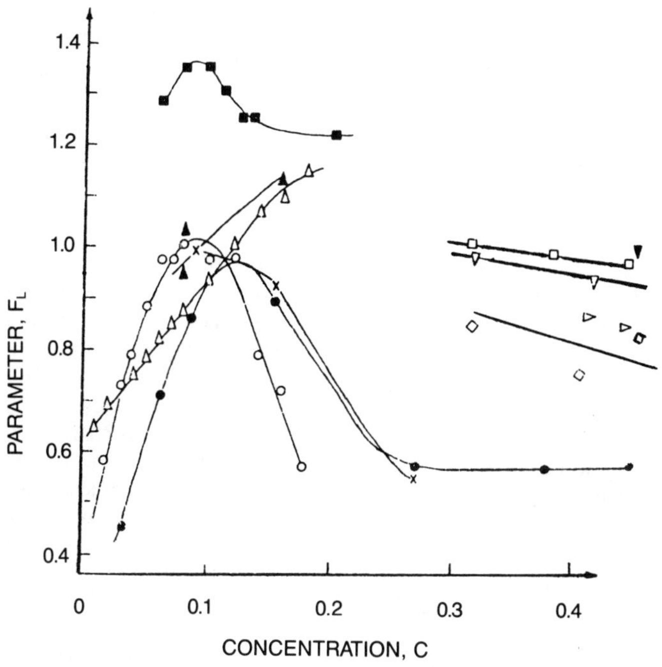

Symbol	d (mm)	Presence of small particles	ϱ_s (kgm^{-3})	D (mm)	Reference
▲	2	?	1630	103	103–105
X	2	?	1630	203	103–105
●	2	?	1630	255	103–105
■	1	?	1900	206	103–105
○	2.26	?	1400	13	87
△	2.20	?	1400	25	87
▽	0.21	33%, d < 0.074 mm	1390	158	119
□	0.21	33%, d < 0.074 mm	1390	208	119
▷	0.21	33%, d < 0.074 mm	1390	263	119
◇	0.21	33%, d < 0.074 mm	1390	315	119

FIGURE 19. Variation of F_L with solids concentration for aqueous coal slurries (replotted from [89]).

was limited to about 15%. The data of Sinclair [87] on the flow of iron ore/water mixtures further show a solids density dependence on the parameter F_L. Therefore, the exact nature of functional dependence of F_L on the pertinent variables (d, C, ρ_s, etc.) is not yet fully known; this deficiency, obviously, raises serious doubts about the usefulness of Equation (40) in making a priori prediction of V_2.

In 1967, Zandi and Govatos [122] re-appraised the information available in this area, and proposed a new flow pattern index to distinguish the heterogeneous flow from flow with a moving bed as follows:

$$N_1 = \frac{V^2\sqrt{C_D}}{gCD(s - 1)} \qquad (41)$$

and the authors indicated that the flow in the form of a heterogeneous suspension would cease to exist at $N_1 = 40$. This results in the following expression for V_2:

$$V_2 = \left(\frac{40gCD(s - 1)}{\sqrt{C_D}}\right)^{1/2} \qquad (42)$$

Babcock [123] has, however, suggested that the heterogeneous flow would continue till $N_1 = 10$. Note that the predictions of V_2 based on these two values of N_1 differ by a factor of two. In a recent work concerning the flow of gravel/water mixtures in a 42 mm diameter pipe, Chhabra and Richardson [124] found that their experimental findings were in line with the predictions of Equation (42). Turian and Yuan [88] also gave an empirical relation for this transition velocity as:

$$V_2^2 = 2.411C^{0.2263}f_L^{-0.2334}C_D^{-0.384}Dg(s - 1) \qquad (43)$$

To say the least, once again, this equation is tedious to use as compared with the criterion of Zandi and Govatos [122].

In 1980, Oroskar and Turian [99] re-visited this problem and carried out a semi-theoretical analysis. The essence of their analysis lies in balancing the energy required to suspend the particles with that derived from dissipation of an appropriate fraction of the turbulent eddies. Though the new analysis is elegant and useful in furthering our understanding of the problem, it does not lead to completely predictive criterion. By combining the theory with least square analysis approach, Oroskar and Turian [99] derived the following semi-empirical relation for V_2:

$$\frac{V_2}{\sqrt{gd(s - 1)}} = 1.85C^{0.1536}(1 - C)^{0.3564}(d/D)^{-0.378}\psi^{0.30} \times \left(\frac{D\rho\sqrt{gd(s - 1)}}{\mu}\right)^{0.09} \qquad (44)$$

where ψ is a parameter and is a function of the ratio (V_0/V_2); the functional dependence is available in the original publication [99]. In the same paper, the authors also assessed the performance of seven correlations against the available data, and concluded that Equation (44) resulted in the smallest average deviation of 22%, whilst all other equations including Equations (40) and (42) showed average deviations ranging from 50% to 75%. There is certainly a definite advantage in using Equation (44), but from a user's point of view the new equation is rather inconvenient. Besides, keeping in mind the poor reproducibility and accuracy of data in this field, it is believed that the use of the simple criterion due to Zandi and Govatos [122] should suffice for all practical purposes.

TRANSITION VELOCITY V_3

A number of theoretical attempts with widely differing approaches to predict the transition velocity between the moving bed flow and flow with a stationary bed, namely V_3, have been made by Thomas [126,127], Wicks [120], and Wilson [128–130]. Thomas's work was concerned with the flow of fine flocculated thoria slurries and whence is not of much relevance in the present context. Wicks' work is based on the flow of lean sand/water mixtures and his final correlation does not contain a concentration term in it. Wilson's approach is radically different in that he determined the pressure gradient necessary to initiate sliding of a layer of solids. His equation for the required pressure gradient is:

$$\frac{1}{\rho}\left(\frac{\Delta P}{L}\right)_3\left[\frac{\theta - \sin\theta\cos\theta}{4} + \frac{R'}{D}\left(\sin\theta - \frac{\theta\eta_s}{\tan\phi}\right)\right] = \frac{\eta_s(s - 1)C_b(\sin\theta - \theta\cos\theta)g}{2} \qquad (45)$$

where

$\left(\frac{\Delta P}{L}\right)_3$ = pressure gradient at velocity V_3
θ = one half the angle subtended at the pipe centre by the surface of the bed of solids
η_s = solid/wall friction coefficient
R' = cross-sectional area of bed divided by bed width
ϕ = angle of repose of solids
C_b = volume fraction of solids in the bed

From tests with 0.7 mm sand/water mixtures flowing in a 3.5 inch diameter pipe, Wilson confirmed the constancy of η_s and from that concluded that his analysis and the underlying assumptions were correct. In principle at least one should be able to calculate velocity corresponding to the pressure gradient given by Equation (45). Unfortunately, due to its complex form, this method has not been used very often.

Like for other transitions, Turian and Yuan [88] have also derived a criterion to predict the transition velocity V_3 as:

$$V_3^2 = 31.93C^{1.083}f_L^{1.064}C_D^{-0.0616}gD(s - 1) \qquad (46)$$

Intuitively it appears that the drag coefficient C_D is unlikely to be a significant variable both in flow with a moving as well as with a stationary bed, yet is is included in Equation (46); the functional dependence is very tenuous, however. Only very scant data is available on this transition velocity; hence neither of the procedures mentioned above has been adequately tested. Other methods which require a prior knowledge of pressure gradient have been dealt with by Govier and Aziz [12].

TRANSITION VELOCITY V_4

This transition velocity denotes the conditions when the motion of solids as well as liquid ceases and the clogging of pipe occurs. One would rarely want to operate or design an installation for these conditions to occur. Therefore, the lack of proven methods for estimating V_4 is not a serious deficiency from a practical point of view.

From the foregoing discussion it is evident that satisfactory schemes are available for predicting the homogeneous/heterogeneous flow and heterogeneous/flow with moving bed transitions; whereas very little is known about the transition of flow with a moving bed to one with a stationary bed which represents an important transition from operational point of view. This is so partly due to the lack of suitable experimental data.

Pressure Drop in Pipeline Flows

This variable is perhaps the most important design parameter for slurry pipelines. Considerable amounts of effort, therefore, have been expended in devising universal predictive methods. Unfortunately the occurrence of a wide variety of possible flow patterns has hindered the development of generalized procedures for the estimation of pressure gradient for a given duty. From a theoretical point of view, the governing equations are hopelessly complex and are not amenable to rigorous analysis; hence most of the progress in this area has been made using empirical formulations which inevitably lack universality.

Generally, the pressure loss incurred during the flow of a two phase solid/liquid heterogeneous mixture is expressed as a sum of two terms, viz.:

$$i = i_L + i_s \qquad (47)$$

where i is the total pressure gradient; i_L is the pressure gradient due to the flow of conveying liquid alone flowing at the same velocity, and i_s represents the contribution due to the presence of solids. The latter is also known as "additional" or "excess" pressure drop due to the presence of solids. The estimation of i_L is rather straightforward for all types of fluids—Newtonian as well as non-Newtonian—under all conditions of practical interest (e.g. see [12]). The situation, however, is not so simple for the estimation of i_s.

Complications arise from the fact that the mechanisms responsible for momentum transfer and pressure losses vary from one flow pattern to another. For example, in the case of homogeneous suspension flow, it is the increased density and viscosity which result in the excess pressure drop (i_s) while in the case of flow with a moving bed, the friction between the moving bed and pipe wall contributes appreciably to i_s. These complexities indeed preclude the possibility of a single correlation applicable to all flow patterns ever becoming available. In the following, therefore, estimation methods (for i or i_s) are presented separately for each flow pattern.

HOMOGENEOUS FLOW ($V > V_1$)

This type of flow occurs when the particles are fully suspended by the turbulent eddies and the concentration profile is symmetrical about the pipe axis: Generally the assumption of no hold up is a good approximation i.e. the *in-situ* volume fraction of solids is the same as the input or discharged fraction of solids.

There have been two independent approaches assuming pseudohomogeneity and make use of the Newtonian fluid mechanics and involving the concept of dimensionless excess pressure gradient; the second approach developed by Julian and Dukler [131], and subsequently employed by Shook and Daniel [132]. The first approach has proved to be quite useful and has yielded reliable equations for the estimation of i as described below.

With the assumption of pseudohomogeneity and no hold up, one can conveniently express the dimensionless excess pressure drop as:

$$\frac{i - i_L}{i_L} = \frac{\dfrac{2f_M\rho_M V^2}{D} - \dfrac{2f_L\rho_L V^2}{D}}{\dfrac{2f_L\rho_L V^2}{D}}$$

or

$$\frac{i - i_L}{i_L} = \frac{f_M\rho_M - f_L\rho_L}{f_L\rho_L} \qquad (48)$$

where f_M and f_L are, respectively, the friction factors for the mixture and the transporting fluid, both evaluated at the same velocity V. Both the density and the viscosity of the mixture will usually exceed those of the conveying liquid. The density of the mixture can be expressed as:

$$\rho_M = (1 + (s - 1)C)\rho \tag{49}$$

where C is the volumetric fractional concentration of solids in the mixture.

The viscosity of the mixture is given by [12]:

$$\mu_M = (1 + 2.5C + 14.1C^2)\mu \tag{50}$$

or

$$\mu_M = \mu(1 + 2.5C + 10.05C^2 + 0.00273 \exp(16.6C)) \tag{51}$$

Using Equations (49) and (50), one can write

$$\frac{Re_M}{Re_L} = \frac{1 + (s - 1)C}{(1 + 2.5C + 14.1C^2)} \tag{52}$$

Generally the conveying liquid is maintained in highly turbulent conditions, and if it is assumed that the presence of particles does not appreciably alter the level of turbulence, one would expect a Blasius type of equation to apply to both the mixture flow as well as the flow of liquid alone, that is,

$$\frac{f_M}{f_L} = \left(\frac{Re_M}{Re_L}\right)^{-0.25} = \left[\frac{1 + 2.5C + \cdots}{1 + (s - 1)C}\right]^{0.25} \tag{53}$$

Depending on the values of C and s, the ratio (f_M/f_L) may slightly be greater or smaller than one. Thus if it is further assumed that this ratio is equal to unity. Equation (48) takes on a particularly simple form as:

$$\frac{i - i_L}{Ci_L} = (s - 1) \tag{54}$$

Newitt *et al.* [96] also obtained a similar result by arguing that since the highly turbulent conditions generally prevail, so the increase in the viscosity (above that of the liquid) is of little consequence, and they attributed the increased pressure drop solely to the increased density of the mixture.

In practice [86,96,133,134], Equation (54) has proved to be a useful approximation, but significant deviations have also been observed. There are mainly two reasons for departures from Equation (54): firstly, at relatively high velocities and with low to moderate concentrations, the solids tend to concentrate near the pipe axis leaving the wall region relatively free of particles, and one finds that

$$i - i_L \approx 0$$

Secondly, it is now well established [135–137] that water carrying a small amount of fine (noncolloidal) particles produces less pressure gradient than that for water flowing on its own, all other conditions being identical. This suppression effect is clearly shown in Figure 20. Zandi [137] observed the same phenomenon when low concentration of either fine coal, fine charcoal, or fine ash is added to the flowing water. A number of investigators [136,137] have attributed the decrease in pressure gradient to the damping effect of solid particles on the level of turbulence. Other mechanisms [138] have also been postulated. At the present time, there is no method available to enable the prediction of this suppressive action quantitatively. This suppressive phenomenon is included in the correlation of Newitt *et al.* [96] as:

$$\frac{i - i_L}{Ci_L} = K(s - 1) \tag{55}$$

where Newitt *et al.* reported a value of 0.6 for K. Govier and Aziz [12] have suggested that the value of K generally ranges from 0.6 to 1.0. Since there is no way to differentiate, a priori, between those suspensions which may exhibit suppression in pressure gradient and those which would not, Zandi [1] has suggested the use of Equation (55) with $K = 1$. In the case of the suppressing type of flow, the calculation would be conservative.

A typical comparison between experimental values and those calculated using Equations (55) is shown in Figure 21.

Rose and Duckworth [139] and Molerus and Wellmann [140] have both applied dimensional considerations and have presented master plots which can be used to predict the additional pressure drop in homogeneous as well as heterogeneous flow regimes. The graphical method of Rose and Duckworth [139] has been discussed in detail by Govier and Aziz [12] and by Duckworth [141] himself; however, the method of Molerus and Wellmann is briefly presented here. Molerus and Wellmann defined the following four dimensionless variables:

Particle Froude number, Fr*

$$= \frac{V}{\sqrt{gd(s - 1)}} \tag{56}$$

Pipe Froude number, Fr

$$= \frac{V_0}{\sqrt{gD(s - 1)}} \tag{57}$$

Dimensionless relative slip velocity,

$$\frac{V_{rel}}{V} = \left(\frac{x}{2}\sqrt{(1 + 4x)} - 1\right) \tag{58}$$

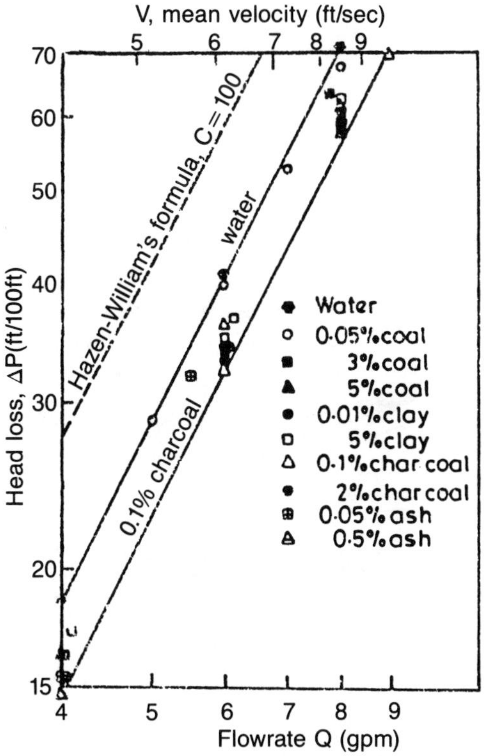

FIGURE 20. Pressure drop–mean velocity relationship on suppression type of suspensions (replotted from [1]).

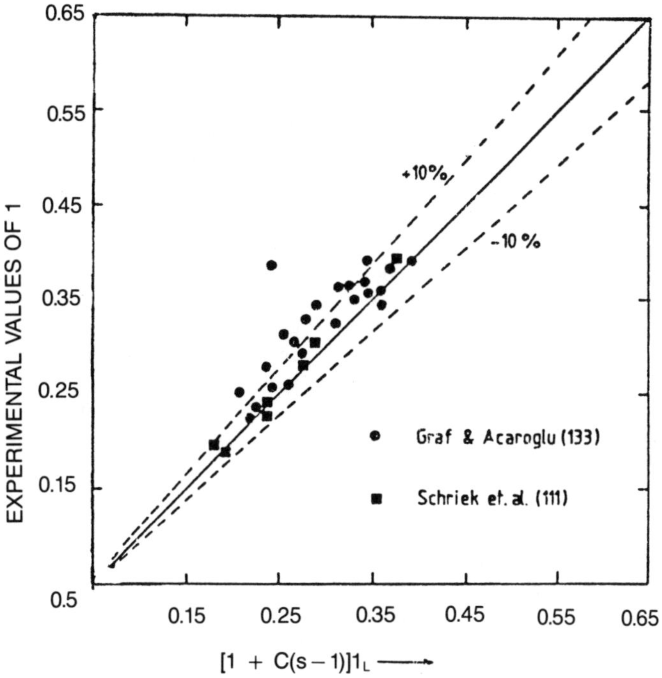

FIGURE 21. Typical comparison between the predictions of Equation (55) (with K = 1) and experimental measurements.

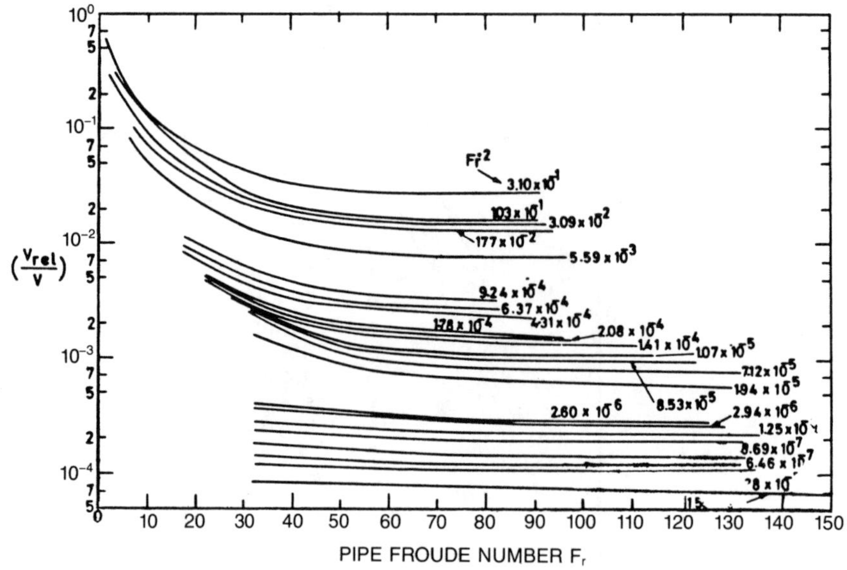

FIGURE 22. Empirical correlation of Molerus and Wellman [140].

where V_{rel} is the difference between the actual in-situ velocities of the liquid and solids.

$$\text{Finally, } x = \frac{\Delta P_s}{C(\rho_s - \rho)Lg}\left(\frac{V_0}{V}\right)^2 \tag{59}$$

Using these four parameters, Molerus and Wellmann reconciled most of the data from literature in the form of a master plot which is reproduced herein Figure 22.

Molerus and Wellmann [140] also asserted in their paper that their analysis was based on a sound theory. This is, however, not so as demonstrated by Chhabra and Richardson [142] who compared the values of slip velocities calculated using the approach of Molerus and Wellmann with the experimental measurements [143–145] and large discrepancies were found. Despite this internal inconsistency, the utility and validity of the final correlation of Molerus and Wellmann [140] is not hampered.

Several other correlations and extensions of those mentioned above are available and these have been reviewed by Govier and Aziz [12] and by Turian and Yuan [88].

HETERGENEOUS FLOW ($V_2 < V < V_1$)

This type of flow occurs when the solid particles are coarse, of large density and the mean velocity of the flow is such that it allows partial separation of the solid particles from the liquid. The particles are conveyed by being in suspension but a concentration gradient exists. Durand [91] defines the maximum particle size for this type of flow as being 2 mm, while Govier and Charles [146] suggest free settling velocity below 1.5 mm/s as a criterion. Examples of heterogeneous flow can be found in the transport of coarse sand [147], nickel suspensions [148] and coal [100]. Due to the great importance and wide occurrence of this non-deposit flow regime in industry, it has been the subject of many review studies [92,100,149,150,151]. Despite all these efforts, the main useful body of available knowledge is still empirical and very little is known about the basic physics of the flow. Furthermore, laboratory scale studies are usually limited to mono (or closely) sized solid particles which represent an over-simplification of the actual situation.

There is no scarcity of empirical relations which have been proposed in the literature for the estimation of friction losses in this mode of flow. A summary of some of these along with their limitations is given in Table 2.

Any discussion of prediction of pressure gradient in heterogeneous flow regime must inevitably begin with the first definite and comprehensive detailed experimental study reported by Durand and Condolios [100,112]. They obtained extensive data on the flow of sand/water and gravel/water mixtures in horizontal pipes up to 580 mm in diameter with solids up to 25 mm size and up to concentrations of 22%. Their results, as expressed by Govier and Aziz [12], are given by the following relation:

$$\frac{i - i_L}{Ci_L} = 121 \times \left[\left(\frac{gD(s-1)}{V^2}\right)\left(\frac{V_0}{\sqrt{gd(s-1)}}\right)\right]^{1.5} \tag{60}$$

TABLE 2. Correlations for Heterogeneous Flow (Modified After Aandi [1]).

Investigator	Equation	Basis of Correlation	Range of Application	Remarks
Durand and Condolios [112] Durand [91]	(a) $\phi = K\left[\frac{V^2\sqrt{C_D}}{gD}\right]^{-3/2}$ (b) $\phi = K\psi^{-3/2}$	Experiments with coal and sand	$V > F_L\sqrt{2gD(s-1)}$	1. K is not specified 2. Data are derived for $s = 2.65$ 3. F_L is given in graphical form (see Figures 14–19) 4. Equation (a) is extended arbitrarily to produce Equation (b).
Worster [152]	$\frac{(i)^{1/2} - (i_L)^{1/2}}{(i)^{1/2}} = \frac{4V_0^{0.173}(CgD(s-1))^{0.143}}{V - \frac{2g\sqrt{Di_L}}{V}}$	Reanalysis of data from Ref. 160	$\frac{\frac{2V}{(Di_Lg)^{1/2}} - \frac{2V}{(Dig)^{1/2}}}{\sqrt{g}} < 13$	This expression was obtained by Zandi and Govatos [122]
Newitt *et al.* [96]	$\phi = 1100\frac{gDV_0(s-1)}{V^3}$	Experiment with sand & gravel	$(1800gDV_0)^{1/3} > V > 17V_0$	Semitheoretical analysis
Bonnington [151]	$\phi = 150\psi^{-3/2}$	Reporting	Same as Durand	No data for heterogeneous flow is given
Koch [153]	$\phi = 81\psi^{-3/2}$	Just presented	Same as Durand	Review of the existing literature. No new data or analysis
Wayment *et al.* [147]	$i = A_1 + A_2(\rho_s) + A_3(V) + A_4(\rho_s^2) + A_5(V^2) + A_6(\rho_sV)$	—	None given	Coefficients A_1–A_6 to be evaluated experimentally and are function of pipe diameter
Condolios and Chapus [92]	$\phi = 180\left(\frac{V^2\sqrt{C_D}}{gD}\right)^{-3/2}$	Experiment same as Durand	None stated	No new data
Ellis *et al.* [154]	$\phi = 85\psi^{-3/2}$	Just reported	None stated	Review of the existing work. No new work or analysis
Ellis and Round [148]	$\phi = 385\psi^{-3/2}$	Experiments with nickel/water mixtures	None given	—
Babcock [149]	$\phi = 81\psi^{-3/2}$	—	None	Review of the existing work
Zandi and Govatos [122]	$\phi = 6.3\psi^{-0.35}$ for $\psi > 10$ $= 280\psi^{-1.93}$ for $\psi < 10$	Analysis of most published data	$\frac{V^2\sqrt{C_D}}{gDC(s-1)} > 40$	—
Hayden and Stelson [106]	$\phi = 100\left(\frac{gD(s-1)V_0}{V^2\sqrt{gd(s-1)}}\right)^{1.3}$	Experiments with sand	None given	Modifications of Durand–Condolios relation
Turian and Yuan [88]	$f - f_L = 0.9857C^{0.8687}f_L^{1.2}C_D^{-0.1677}\left(\frac{V^2}{Dg(s-1)}\right)^{-0.6938}$	Analysis of most published data and some new experiments with sand	Mean velocity bounded by Equations (39) and (43)	Marginally advantageous over the other simpler equation

$\phi = \frac{i - i_L}{Ci_L}$; $\psi = \frac{V^2\sqrt{C_D}}{gD(s-1)}$

Alternatively, one can re-write Equation (60) in terms of the drag coefficient C_D as:

$$\frac{i - i_L}{Ci_L} = 150 \left(\frac{gD(s-1)}{V^2} \cdot \frac{1}{\sqrt{C_D}} \right)^{1.5} \tag{61}$$

Durand and Condolios [112] and Durand [91] did not report the numerical values of constants, these have been incorrectly inferred by some investigators as pointed out by Govier and Aziz [12].

Although the numerical values of the constants appearing in Equations (60) and (61) have been a matter of disagreement among several workers, the basic functional form has been used almost unaltered for thirty years now. For example, Zandi and Govatos [122], modified Equation (61) as follows:

$$\frac{i - i_L}{Ci_L} = 6.3 \left(\frac{V^2\sqrt{C_D}}{gD(s-1)} \right)^{-0.354} \quad \text{for } (CN_1) > 10 \tag{62}$$

and

$$\frac{i - i_L}{Ci_L} = 280 \left(\frac{V^2\sqrt{C_D}}{gD(s-1)} \right)^{-1.93} \quad \text{for } (CN_1) < 10 \tag{63}$$

where N_1 is defined by Equation (41). Admittedly Equations (62) and (63) were reported to give better representation of data than Equation (61); however, the reasons for the abrupt change in the values of the two constants in Equations (63) and (64) are not obvious. It remains to be seen whether this shift corresponds to a change in flow mechanism or is it merely the scatter in the data used to derive the correlations.

Likewise, Hayden and Stelson [106] collected experimental data on the flow of sand-water mixtures in pipes of different diameters, and adjusted the values of the constants in Equation (60) as:

$$\frac{i - i_L}{Ci_L} = 100 \left(\frac{gD(s-1)}{V^2} \cdot \frac{V_0}{\sqrt{gd(s-1)}} \right)^{1.3} \tag{64}$$

A comparison between their experimental data and the predictions of Equation (64) is shown in Figure 23. Included in the same figure are the values calculated using Equations (60), (62), and (63). Admittedly a significant amount of scatter is seen in this figure, the agreement between experiments and the two equations of Zandi and Govatos [122] is not particularly very good. Indeed one is tempted to conclude that the original Durand–Condolios correlations or the Hayden–Stelson relationship are about as good as may be expected, especially when the ordinate variable has an uncertainty of $\pm 40\%$.

Newitt *et al.* [96] have carried out a semi-theoretical analysis of the heterogeneous suspension flow. Their treatment hinges on the fact that the work done in keeping particles in suspension is proportional to their effective weight

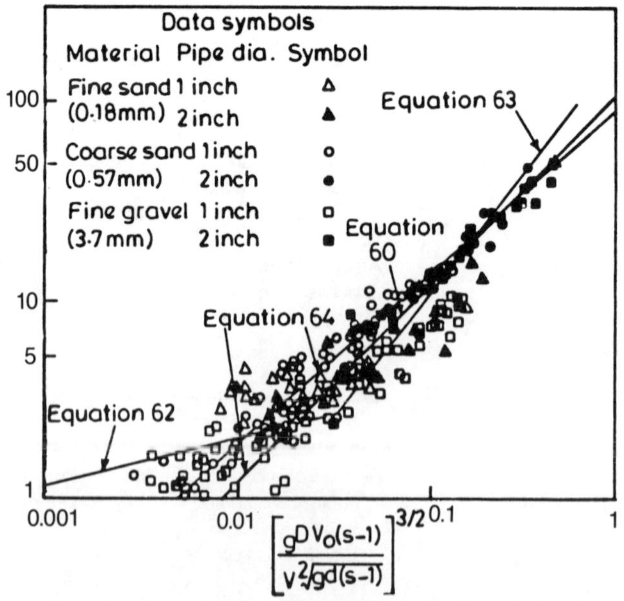

FIGURE 23. Comparison of predictions of Equations (60,62–64) with the experimental data of Hayden and Stelson [106].

times the settling velocity, and, in turn, this is equal to the energy dissipated due to the presence of solids. Based on these considerations, Newitt *et al.* [96] derived the following relation:

$$\frac{i - i_L}{Ci_L} = K_1 \frac{gDV_0}{V^3}(s - 1) \tag{65}$$

where K_1 is a system constant involving the concentration effect on V_0 and the friction factor for the flow of liquid alone. Intuitively one would expect K_1 to depend upon concentration and pipe Reynolds number etc., but Newitt *et al.* [96] found that a single value of 1100 fitted well their experimental data covering wide ranges of solid size, density and concentration in one inch diameter pipe. Typical results from Newitt *et al.* [96] are shown in Figure 24 where a satisfactory correspondence is seen to exist. Newitt *et al.* [96] also observed that when the solids, with a broad side distribution, are being transported, the presence of finer solids aids in carrying coarse particles. They have suggested that Equation (65) could also be used for mixed sizes by using a weighted mean diameter.

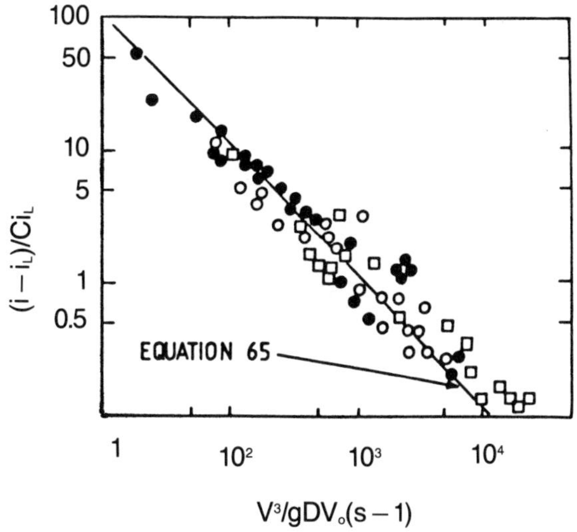

FIGURE 24. Comparison between the predictions of Equation (65) (with K_1 = 1100) and the experimental data of Turtle [160] and of Abbott [161]. (○) Sand; (●) Coal; (□) perspex.

Unfortunately, conflicting conclusions have been reached and reported in the literature regarding the validity and reliability of Equation (65). For example, Hayden and Stelson [106] found it completely unsatisfactory for their experimental results, whereas Zandi and Govatos [122] have concluded it to be as reliable as the Durand–Condolios relation. In fact, Turian and Yuan [88] have demonstrated that the correlation of Newitt *et al.* [96] leads to better predictions than both of the Zandi and Govatos [122] and Durand–Condolios relations. Further proof for the usefulness of Equation (65) is provided in Figure 25 where data drawn from different sources are included. Equations (38) and (42) were used to ascertain the flow regime in each case, and only those data which relate to heterogeneous mode of flow are included in Figure 25. Certainly, there is appreciable scatter present but nonetheless majority of points are seen to be bounded by ±40% bands which is a conservative estimate of experimental error in this field [122]. Thus the comparison shown in Figure 25 is regarded to be satisfactory and acceptable.

It is interesting to note, as pointed out by Round [90], that the assumption of the linear dependence of the total pressure gradient on concentration is central to the formulation of all correlations listed in Table 2 (a notable exception is the correlation of Turian and Yuan), yet it has never been verified independently. This is so partly because of the inherent difficulties in performing experiments where the mean mixture velocity is kept constant and the solids concentration is gradually varied.

Yen and Zandi [155] have presented a detailed analysis of the heterogeneous flow of suspensions in a rectangular duct, while Graf and Acaroglu [133] and Shih [156] have studied the heterogeneous flow in inclined pipes.

It is appropriate to add here that Zandi [1] has also identified a mixed or intermediate flow regime between homogeneous and heterogeneous flow patterns. Essentially this mode of flow corresponds to the transition band between the two well defined modes of flow. A simultaneous occurrence of these two conditions is possible when the fine particles constitute a homogeneous mixture (which may or may not exhibit non-Newtonian behaviour). The coarse particles will result in a density gradient. Since in most industrial applications the material to be transported usually contains particles of different sizes, mixed or intermediate regime thus may be encountered quite frequently. Unfortunately, not much is known about this mode of flow. Wasp *et al.* [157,158] have suggested a methodology for analyzing this situation. They envisioned that a mixed type slurry flow consists of two components: a vehicle, which acts as the carrier fluid, and an ensemble of coarse particles which concentrates near the bottom of the pipe, and moves more slowly than the carrier fluid. The overall pressure gradient is thus a sum of two components arising from the above-mentioned processes. There are several complicating features of this approach that need to be resolved in order that the technique may be used for a priori calculations. The following information is needed in order to predict the head loss without resorting to pilot plant studies:

1. The properties (density and viscosity) of the carrier fluid
2. The suitable value of V_0 and/or C_D to be used in head loss correlations

The only other work relevant to this flow regime is that of Charles [159] who suggested a linear combination of the

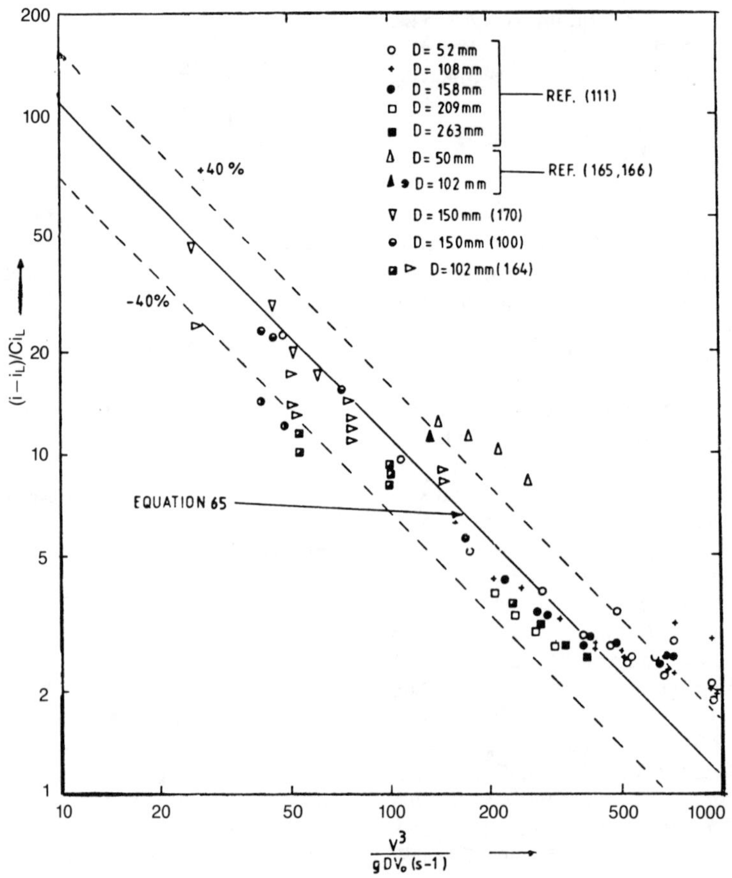

FIGURE 25. Comparison between the predictions of Equation (65) (with $K_1 = 1100$) and the experimental data for large diameter pipes.

two contributions, that is

$$\frac{i - i_L}{Ci_L} = (s - 1) + K_1 \left(\frac{V^2 \sqrt{C_D}}{gD(s - 1)} \right)^{-1.5} \tag{66}$$

The second term on the right hand side may be replaced by any other equation such as those listed in Table 2. For high density materials [148], this approach leads to values of the pressure gradient which are higher than the experimentally observed values.

MOVING BED FLOW ($V_3 < V < V_2$)

Under certain conditions, it is inevitable that particles will not be conveyed either in the heterogeneous or in the pseudohomogeneous or in mixed regimes, or conversely the velocities required to achieve suspended flow will be uneconomically high. Normally, the reason for this is that the particles are either too large or too heavy or both. The gravitational forces far outweigh the forces produced because of turbulence, and the particles will form a bed on the bottom of pipe which slides along the wall. Though Zandi [1] regards this mode of flow to be highly undesirable but for coarse particles this is the only possible mode for transportation in pipes. Newitt *et al.* [96] were the first to recognize the inherently different character of this type of flow. Based on the considerations of work done on the solids and the energy dissipation due to the friction between the bed and pipe wall, they derived the following expression:

$$i_s = K_2 C(s - 1) \tag{67}$$

Equation (67) can be rearranged to be written as:

$$\frac{i - i_L}{Ci_L} = \frac{K_2 gD(s - 1)}{f_L V^2} \tag{68}$$

where K_2 is a constant incorporating solid/pipe wall friction etc. Newitt *et al.* [96] argued that, in most instances, water is used as the carrier liquid and invaribly turbulent conditions are maintained. Therefore, the friction factor is relatively constant, and thus they lumped it with the system constant to simplify Equation (68) as:

$$\frac{i - i_L}{Ci_L} = \frac{K_3 gD(s - 1)}{V^2} \tag{69}$$

Based on their own data relating to the flow of sand/water, gravel/water etc. mixtures in a one inch diameter pipe, they proposed a value of 66 for K_3. A comparison between their data [160,161] and Equation (69) is displayed in Figure 26.

Wilson *et al.* [162] have developed a mechanistic model for this mode (also known as dense phase) of transport. Their analysis is useful in improving our understanding of the fluid mechanical aspects of the problem, but is not yet refined to the extent of being useful for a priori calculations. Shook [163] and others [168] have extensively demonstrated the applicability of Wilson's approach.

Like in the case of correlations for heterogeneous flow, there is an implicit assumption in deriving Equation (69) that the total pressure gradient is a linear function of concentration in the discharged mixture. Babcock [123] and Chhabra and Richardson [124] have experimentally investigated this aspect and concluded that the assumption of linear concentration dependence was only valid when there was a well defined bed of solids formed on the bottom of the pipe. The latter authors studied the pressure gradient characteristics of the flow of gravel/water mixtures in a 42 mm diameter pipe. In this series of tests, the mean mixture velocity was kept constant and the solids concentration was gradually increased.

In the past, the frictional head loss has usually been presented as the dimensionless excess pressure drop ($i - i_L/Ci_L$), and thus the raw experimental data have been subjected to several arithmetic operations which have a tendency to mask any real trends, and to exaggerate the experimental scatter. The latter point has been unequivocally demonstrated by Babcock [123]. Chhabra and Richardson [124] thus plotted their results in the form of total pressure gradient versus discharged concentration at constant value of the mean velocity as shown in Figures 27 to 30. An examination of these figures shows that the particle size is not a significant variable which is consistent with the findings of James and Broad [164]. However, the form of functional relationship between the total pressure gradient and the concentration changes from linear to nonlinear as the mean velocity is increased from 1.22 ms^{-1} to 3.92 ms^{-1}. At the lowest velocity a straight line may be drawn through the data points and the intercept corresponds to the gradient for the flow of water alone. From visual observations through a glass section, the conditions in the pipe appeared to cor-

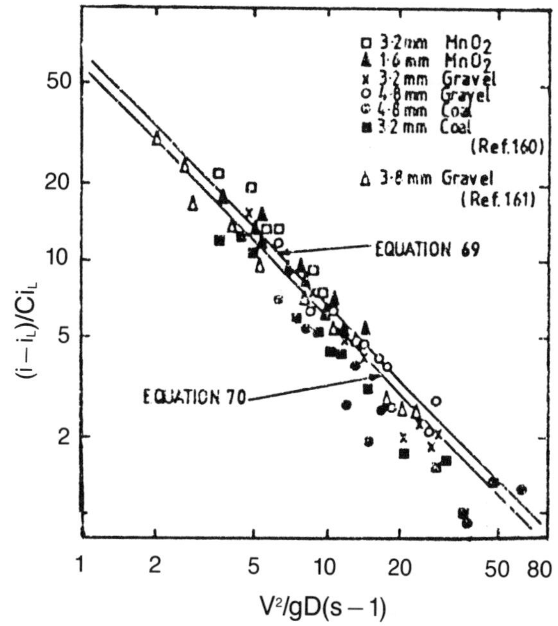

FIGURE 26. Comparison between the predictions of Equations (69) (with K_3 = 66) and (70) and the data of Turtle [160] and Abbott [161] (replotted from [124]).

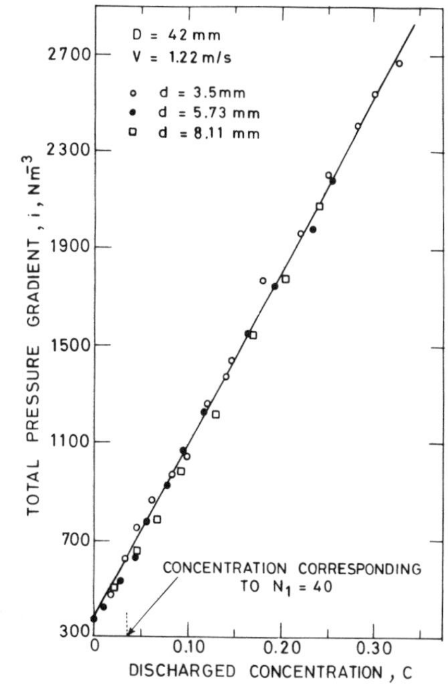

FIGURE 27. Total pressure gradient as a function of discharged concentration for V = 1.22 ms^{-1} (replotted from [124]).

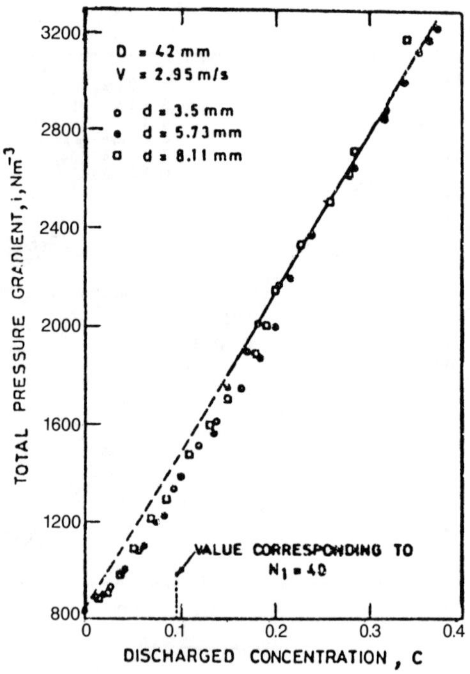

FIGURE 28. Total pressure gradient as a function of discharged concentration at V = 1.95 ms^{-1} (replotted from [124]).

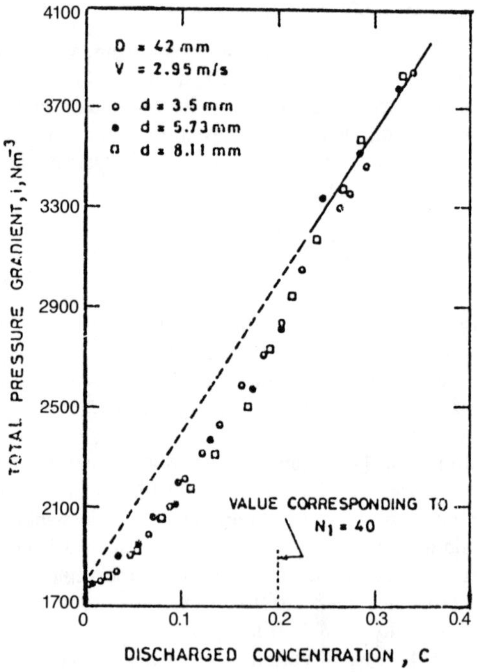

FIGURE 29. Total pressure gradient as a function of discharged concentration at V = 2.95 ms^{-1} (replotted from [124]).

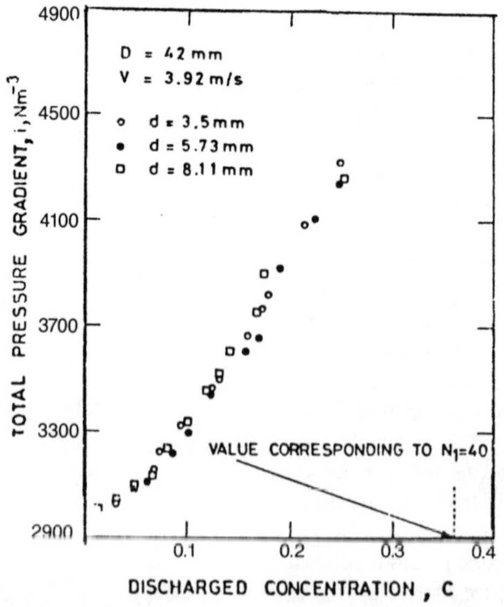

FIGURE 30. Total pressure gradient as a function of discharged concentration at V = 3.92 ms^{-1} (replotted from [124]).

TABLE 3. Summary of Results Shown in Figure 31.

Investigator	*D* (mm)	*d* (mm)	*s*	Symbol Used in Figure 31
Howard [165,166]	51	0.45	2.65	○
	102	0.45	2.65	●
	102	2.50	2.71	△
Boothroyde *et al.*	200	4.30	2.55	+
[169]	200	16	2.80	⊡
James and Broad	102	5	2.70	◩
[164]		10	2.70	⊖
		20	2.70	▲
	156	5	2.70	■
		10	2.70	◪
		20	2.70	▽
		40	2.70	▼
	207	5	2.70	◓
		10	2.70	⬒
		20	2.70	⬓
		40	2.70	*
Chhabra and Richardson [124]	42	3.5, 5.7 and 8.11	2.65	▷
Duckworth [141]	32.2	2.4	11.14	◐
Durand and	104	3.78	2.65	◑
Condolios [100]		7.50	2.65	▶
		12	2.65	☒
		18	2.65	◁
		22	2.65	◀
	150	4.37	2.65	*x*
	580	0.44	2.65	*V*
Worster [170]	150	2	2.65	*a*
Abbott [161]	25	3.1	2.65	*b*
Worster and	40.6	2.04	1.50	*S*
Denny [171]		2.04	2.65	*C*
		2.04	3.65	*Z*

respond to a moving bed over the whole range of concentration. As the mixture velocity is raised, bed formation becomes progressively less marked and the flow moves towards the regime of heterogeneous flow. This shift of flow regime manifests itself in the form of a nonlinearity in *i-C* curve at low values; however, the relationship again becomes linear at concentrations high enough for there still to be evidence of bed formation (see Figure 29). At the highest velocity used (3.92 ms^{-1}), the relationship was nonlinear over the whole range of concentration studied (Figure 30). Indicated in these Figures 27–30 are also the predictions of Equation (42) which seem to correspond roughly to the transitions from nonlinear to linear dependence in *i-C* curves. Thus, this form of data presentation can also be useful in identifying flow regimes. Using the criterion of Zandi and Govatos [122], Chhabra and Richardson [124], re-analysed the data of Abbott [161] and of Turtle [160] and found that a value of 59 for K_3 in Equation (69) gave a better representation of data than the originally suggested value of 66. Besides the new value of K_3 is also close to the value of 60.6 proposed by Babcock [123]. Therefore Equation (69) becomes:

$$\frac{i - i_L}{Ci_L} = \frac{59gD(s - 1)}{V^2} \tag{70}$$

The predictions of Equation (70) have also been plotted in Figure 26.

Owing to its simple form, and proven reliability the correlation of Newitt *et al.* [Equations (69) or (70)] has gained considerable acceptance for calculating pressure losses in small diameter pipes. Much confusion, however, exists regarding the effect of pipe diameter. In spite of the fact that the pipe diameter appears explicitly in Equation (69) or (70), many workers have reported that the value of constant depends upon pipe diameter [123,134,164]. For example, Coulson and Richardson [167] cite the work of James and Broad [164] to suggest that the value of constant increases with pipe diameter. Babcock [123], on the other hand, reported an inverse relationship between the value of constant K_3 and pipe diameter. An analysis of the data of Maruyama *et al.* [134] further shows particle size dependence on K_3. Thus conflicting conclusions have been reached and reported in this regard.

To resolve this uncertainty, data have been culled from as many different sources as possible covering wide range of pipe diameter and materials. A brief summary of data and related information is presented in Table 3 which encompasses more than an order of magnitude variation in pipe diameter. The criterion of Zandi–Govatos [122], that is, $N_1 = 40$ was used to ascertain the flow regime. Only those points satisfyng the condition $N_1 < 40$ were subsequently used to assess the effect of pipe diameter on K_3. In this regard, a comment about the work of Boothroyde *et al.* [169] is in order. They investigated the flow of gravel/water and granite chips/water mixtures in a 200 mm diameter pipe, and the mean mixture velocity ranged from 2 to 7 ms^{-1}. The value of N_1 exceeds 40 for the mean mixture velocity greater than 6 ms^{-1}. However, a plot of total pressure gradient against the discharged concentration (similar to Figure 27) was found to be highly linear over the entire range of conditions studied. Besides, Boothroyde *et al.* [169] wrote in their article:

> The initial aim of this test was to observe the critical deposit velocity i.e. that velocity at which particles drop out of suspension and start to move as a continual sliding bed. However, it soon became obvious that with this material (gravel) even at velocities of 7 ms^{-1} the great majority of solid was moving as a sliding bed.

Boothroyde *et al.* [169] recorded similar observations for the case of granite chips/water mixtures. An additional complication arises in the case of granite chips due to their highly irregular shape, whereas the criterion of Zandi and Govates [122] is derived from data obtained with reasonably regular shape particles. Therefore, in spite of the fact that the data of Boothroyde *et al.* [169] relate to conditions $N_1 > 40$, the sliding bed flow was assumed to occur. The predictions of Equation (70) are plotted in Figure 31 along with the data drawn from various sources as identified in Table 3. Albeit, the wide scatter present in this figure is discomforting and disturbing but there is certainly no discernible trend with regard to the pipe diameter. We hesitate to perform a rigorous statistical analysis due to the different degrees of uncertainty associated with each set of data, but a preliminary examination does suggest that 95% of the points are enclosed by +60% line which is about the accuracy of different correlations. One can argue that Equation (70), on the whole, underpredicts the values of pressure gradient, but we feel, however, that any firm conclusions regarding the effect of pipe diameter on the value of K_3 must await further work.

Until now the discussion has been restricted to the studies wherein the turbulently flowing water was the vehicle. The advantages of heavy media in the conveyance of coarse particles have been recognised for quite some time now [43,96]. Yet very little is known about the influence of physical properties (density and viscosity) of the vehicle on the transport phenomenon [43,134,169,172–175]. Kenchington [174] and Charles and Charles [43] reported some preliminary results on the conveyance of 750 μm and 216 μm size sand respectively in aqueous kaolin suspensions. Both suggested that the existing correlations (based on data with water as the carrier) were unsatisfactory for the representation of their results. However no definite suggestion was made in either of the two papers.

More recently, Chhabra and Richardson [175] conducted a comprehensive study on the transport of coarse gravel

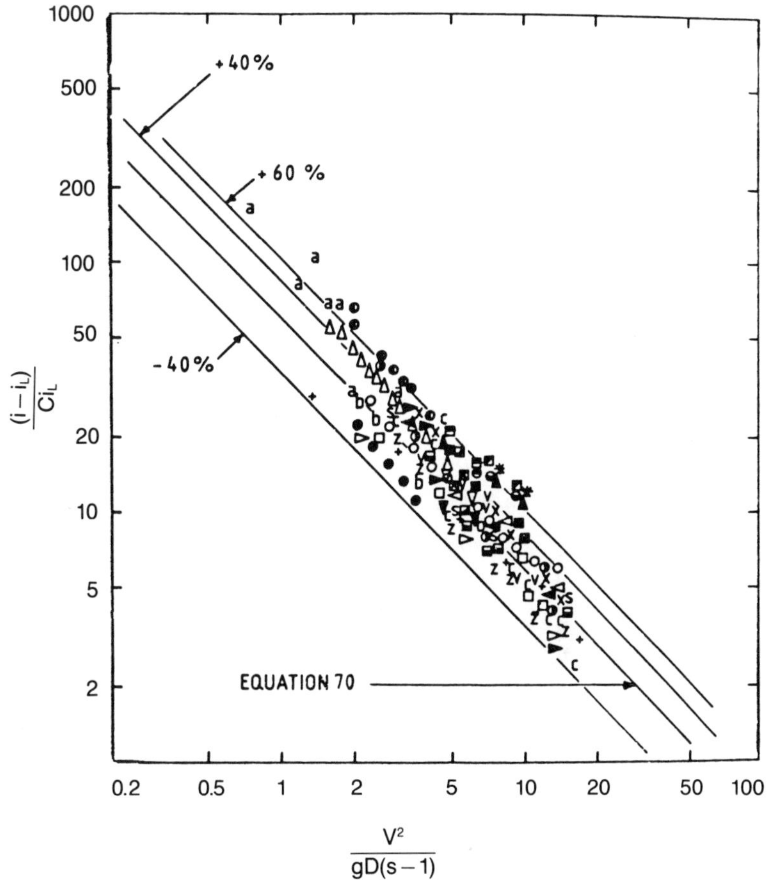

FIGURE 31. Effect of pipe diameter on the pressure drop in the flow with a moving bed regime. Symbols are identified in Table 3.

particles of three different sizes (3.5, 5.7, 8.1 mm) in a 42 mm diameter pipe using water/glycerol mixtures, aqueous kaolin suspensions and aqueous solutions of high molecular weight polymers. The initial objective of this study was to exploit the advantages of increased density and of higher viscosity (than water) of carrier both of which are favourable for achieving suspended flow. Furthermore, due to the non-Newtonian viscosity of kaolin suspensions and of polymer solutions, the pressure gradient will not rise too rapidly with velocity in laminar flow conditions. Tests were performed by keeping the mean mixture velocity constant at a number of values and gradually increasing the solids concentration. This type of experiments are useful in elucidating the nature of dependence of two phase pressure drop on solids concentration under otherwise constant conditions (as explained in the preceding section; also see Figures 27–30).

The rheological properties of the aqueous polymer solutions and kaolin suspensions were adequately represented by the apparent power law model [Equation (21)] and the resulting values of n' and k' along with the operating conditions are summarised in Table 4 where wide ranges of k' and n' values are seen to be covered; however, the variation in the carrier density is limited to only about 20%.

Figures 32 to 36 show the experimental results plotted in the form of a dimensionless pressure gradient i^* ($= i/i_L$) versus the solids concentration in the discharged mixture for a range of mixture velocities and for a number of vehicles identified by a run number designated in Table 4. Though the chief objective of this study [175] was to investigate the flow characteristics in the suspended flow regime, but it proved to be impossible to achieve suspended flow even for the smallest particle used with the available experimental facility; hence almost all data gleaned by Chhabra and Richardson [175] relate to the flow with moving bed regime.

TABLE 4. Summary of Properties of Liquids and the Operating Conditions Used by Chhabra and Richardson [175].

Run No.	Type of Liquid	n'	K' Pas$^{n'}$	kg m^{-3}	d (mm)	V (ms^{-1})	Maximum Value of C	Re_{MR}	No. of Data Points
1	80% Glycerol	1	0.067	1210	3.50	0.98	0.35	735	15
2	76% Glycerol	1	0.046	1208	5.73	0.98	0.42	1070	14
3	76% Glycerol	1	0.045	1208	8.11	0.98	0.365	1090	12
4	74% Glycerol	1	0.038	1201	8.11	0.98	0.34	1270	12
5	64% Glycerol	1	0.013	1164	3.50	0.98	0.33	3510	12
6	63% Glycerol	1	0.012	1164	5.73	0.98	0.30	3790	10
7	42% Glycerol	1	0.004	1103	3.50	0.98	0.26	1.17×10^4	8
8	1.5% CMC	0.56	2.11	1000	5.73	0.73	0.28	128	11
9	1.0% CMC	0.62	1.09	1000	8.11	0.73	0.31	184	11
10	1.0% CMC	0.62	1.05	1000	3.50	0.73	0.31	190	13
11	0.75% CMC	0.71	0.34	1000	3.50	1.46	0.28	918	13
12	0.60% CMC	0.80	0.20	1000	8.11	1.46	0.22	940	8
13	0.60% CMC	0.80	0.16	1000	5.73	1.46	0.28	1170	10
14	0.50% CMC	0.85	0.085	1000	5.73	0.73	0.20	755	6
15	0.50% CMC	0.88	0.078	1000	8.11	0.73	0.28	710	8
16	0.30% CMC	1.0	0.032	1000	3.50	0.73	0.37	940	15
17	0.30% CMC	1.0	0.026	1000	3.50	1.46	0.36	2320	13
18	0.30% CMC	1.0	0.025	1000	8.11	1.46	0.31	2420	11
19	0.30% CMC	1.0	0.024	1000	5.73	1.46	0.33	2550	11
20	0.30% CMC	1.0	0.024	1000	5.73	2.20	0.33	3840	11
21	0.30% CMC	1.0	0.022	1000	8.11	2.20	0.30	4080	10
22	0.30% CMC	1.0	0.028	1000	3.50	2.20	0.31	3340	11
23	0.15% CMC	1.0	0.014	1000	3.50	2.68	0.19	7980	7
24	0.15% CMC	1.0	0.014	1000	5.73	2.68	0.16	7980	6
25	0.15% CMC	1.0	0.014	1000	8.11	2.68	0.11	7980	4
26	1.2% Meyprogat	0.75	0.21	1000	3.50	0.73	0.14	500	5
27	1.2% Meyprogat	0.75	0.21	1000	3.50	1.46	0.33	1190	9
28	1.0% Meyprogat	0.75	0.137	1000	8.11	0.733	0.076	770	3
29	1.0% Meyprogat	0.75	0.137	1000	8.11	1.46	0.031	1820	9
30	0.7% Meyprogat	0.90	0.050	1000	5.73	0.733	0.068	1000	3
31	0.7% Meyprogat	0.90	0.050	1000	5.73	1.46	0.032	2140	10
32	0.5% Meyprogat	0.90	0.017	1000	5.73	2.92	0.24	1.35×10^4	7
33	0.5% Meyprogat	0.90	0.017	1000	3.50	2.92	0.27	1.35×10^4	8
34	0.5% Meyprogat	0.90	0.017	1000	8.11	2.92	0.20	1.35×10^4	6
35	1% CMC + 0.2% Separan	0.37	4.81	1000	3.50	0.976	0.22	230	6
36	0.75% CMC + 0.15% Separan	0.45	2.40	1000	8.11	1.71	0.19	720	7
37	0.75% CMC + 0.15% Separan	0.40	3.57	1000	8.11	0.976	0.27	263	8
38	0.4% CMC + 0.08% Separan	0.58	0.88	1000	5.73	0.976	0.31	416	11
39	0.4% CMC + 0.08% Separan	0.58	0.88	1000	5.73	1.71	0.29	922	10
40	0.2% CMC + 0.04% Separan	0.65	0.244	1000	3.50	1.71	0.32	2200	9
41	0.2% CMC + 0.04% Separan	0.65	0.244	1000	3.50	2.44	0.22	3600	7
42	0.2% CMC + 0.04% Separan	0.65	0.244	1000	5.73	2.44	0.23	3600	7
43	0.15% CMC + 0.03% Separan	0.72	0.138	1000	8.11	2.44	0.24	4120	7
44	11.5% (vol) kaolin	0.20	5.55	1190	3.50	1.35	0.12	1030	9
45	12.8% (vol) kaolin	0.19	10.40	1210	3.50	1.57	0.11	770	5
46	12.8% (vol) kaolin	0.19	9.16	1210	8.11	1.57	0.13	880	9

Note: 1. Tap water was used as solvent in each case. 2. The quoted concentration of polymer solutions are only approximate. 3. CMC stands for Carboxymethyl cellulose.

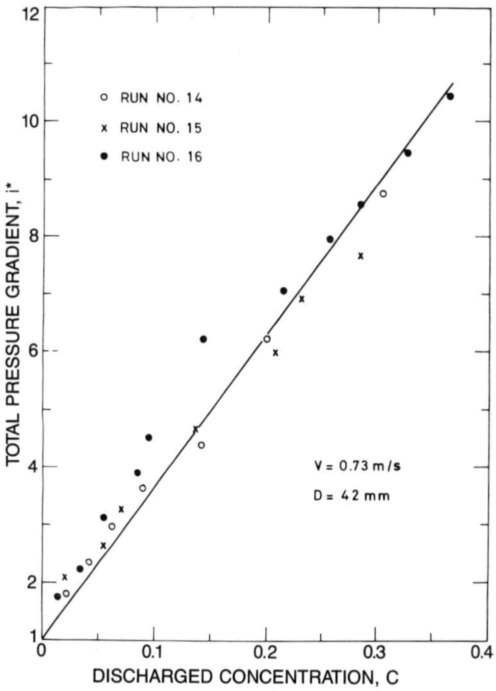

FIGURE 32. Dimensionless total pressure gradient as a function of discharged concentration (replotted from [175]).

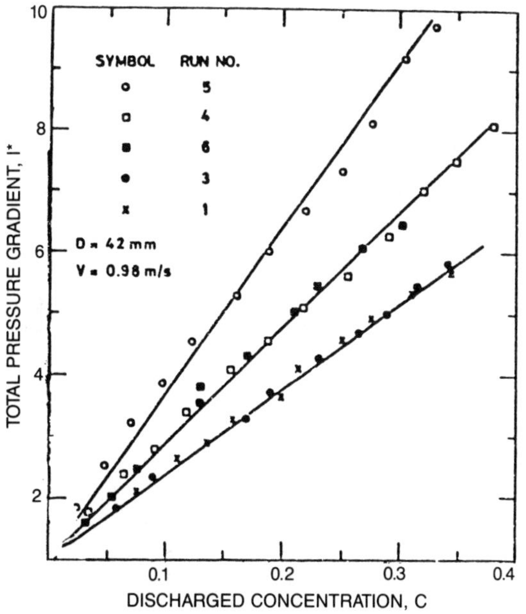

FIGURE 33. Dimensionless total pressure gradient as a function of discharged concentration (replotted from [175]).

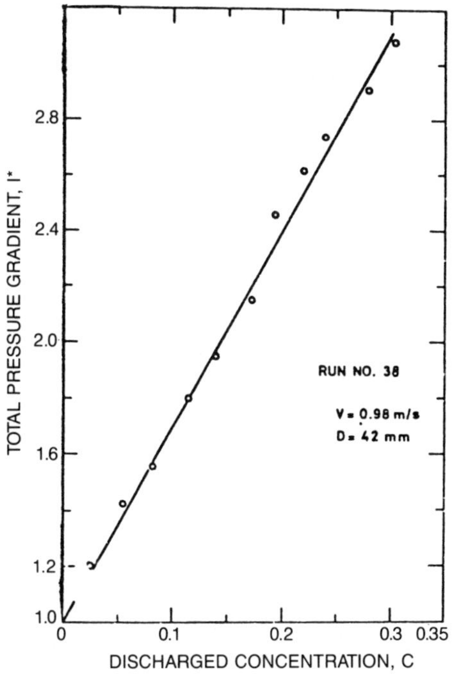

FIGURE 34. Dimensionless total pressure gradient as a function of discharged concentration (replotted from [175]).

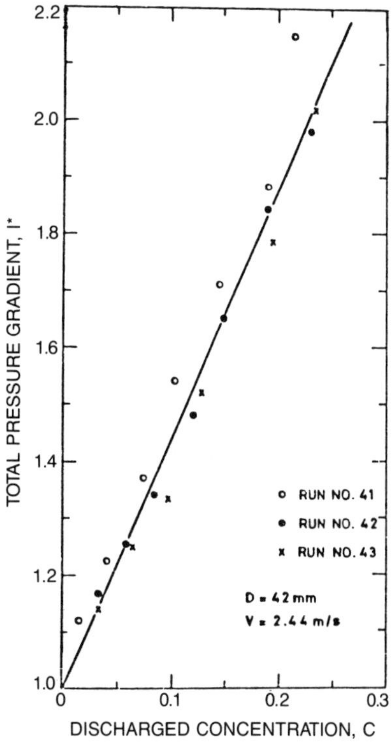

FIGURE 35. Dimensionless total pressure gradient as a function of discharged concentration (replotted from [175]).

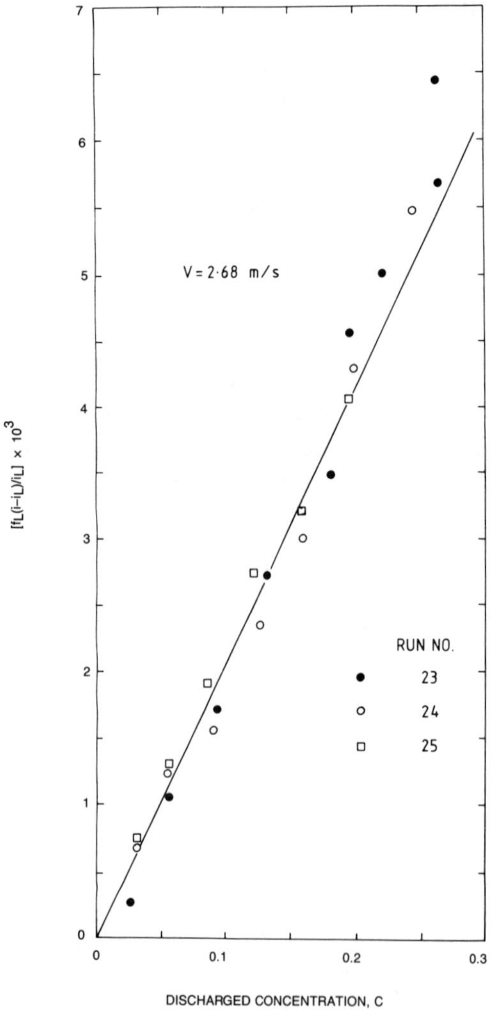

FIGURE 36. Dimensionless total pressure gradient as a function of discharged concentration (replotted from [175]).

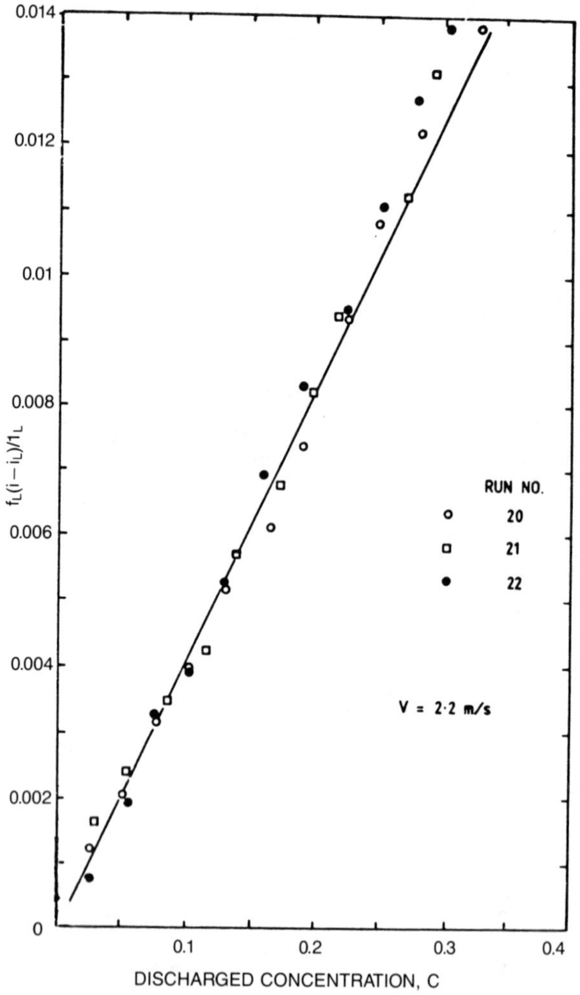

FIGURE 37. Plot of $[(i - i_L)/i_L]f_L$ versus discharged concentration (replotted from [175]).

This is also borne out by the linear relationship between the total pressure gradient and the solids concentration. However, the flow regime cannot be ascertained by using the criterion of Zandi and Govatos [122] simply because the value of C_D cannot be estimated. From visual observations through a glass section, a well defined moving bed was seen to exist under all conditions except at the highest velocity of 2.93 ms^{-1} where the bed formation was less marked. As remarked earlier, since in the derivation of Equations (67) and (68) no assumption has been made regarding the nature of carrier fluid or mode of flow (i.e. laminar or turbulent), in principle they should therefore be applicable to this series of tests [175]. However, since the value of Reynolds number [Equation (19)] varied from 128 to 8×10^4, the assumption of near constancy of friction factor is no longer valid; hence no further simplification of Equation (68) [such as to Equation (69)] is now possible. This point was also noted by Masuyama *et al.* [173] and by Kenchington [174].

One can rearrange Equation (68) as:

$$\frac{f_L(i - i_L)}{i_L} = \frac{k_2 g D(s - 1)}{V^2} C \qquad (71)$$

which suggests a linear relationship between the quantity $(f_L(i - i_L)/i_L)$ and solids concentration in the discharged mixtures, everything else being constant. This behaviour is established in Figures 37–39 for a range of polymer solutions and mixture velocities. An inspection of Figures 37–39 reveals that most of the points are well within ±25% bands which is of the order of experimental error.

Finally, Chhabra and Richardson [175] transformed the results (shown in Figures 32–39) to a master plot as shown in Figure 40. The twelve data points plotted in Figure 40 represent nearly 425 actual experimental tests. Included in the same figure is a line of slope −1 as implied by Equation (71) which does not seem to give satisfactory correlation of the results. Instead by using the nonlinear regression approach, the following relation correlated the results rather well

$$\frac{(i - i_L)f_L}{Ci_L} = 0.55 \left(\frac{gD(s - 1)}{V^2}\right)^{-1.25} \qquad (72)$$

Average and maximum deviations between the predictions of Equations (72) and experiments are 14% and 29% respectively. Included in the same figure are the scant results reported by Kenchington [174]. Admittedly his results show some scatter, there are no discernible trends, however. Furthermore keeping in mind the fluctuating nature of the two phase flow, and the difficulties of sampling [174], the agreement is about as satisfactory as may be expected.

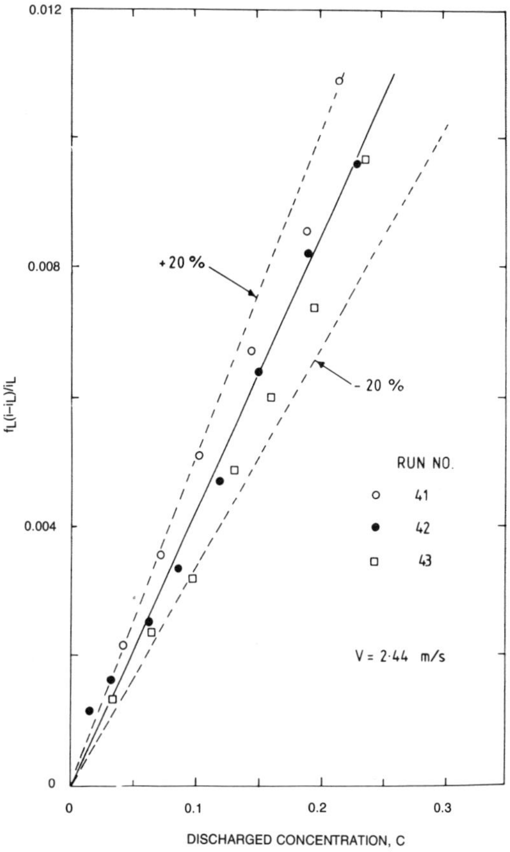

FIGURE 38. Plot of $[(i - i_L)/i_L]f_L$ versus discharged concentration (replotted from [175]).

FLOW WITH A STATIONARY BED ($V < V_3$)

When the mean mixture velocity is decreased below the transition velocity V_3, the lower part of the moving bed ceases to move, and stays stationary. The motion occurs by saltation. As the bed thickens, the free area available for flow gradually decreases, and any further reduction in the mean velocity leads to the clogging of pipe. Experimental work in this area is extremely difficult, and the reproducibility and accuracy of results is much poorer than in the other modes of flow. From a practical point of view, this represents an undesirable mode of flow, and it is therefore not surprising that this flow regime has received very little attention.

Wicks [177] has carried out a semi-theoretical analysis by considering the rolling and lifting of particles on the bed surface. Carstens [178] has also studied the flow of a heterogeneous mixture above a stationary bed. Undoubtedly both of these analyses are useful in shedding some light on

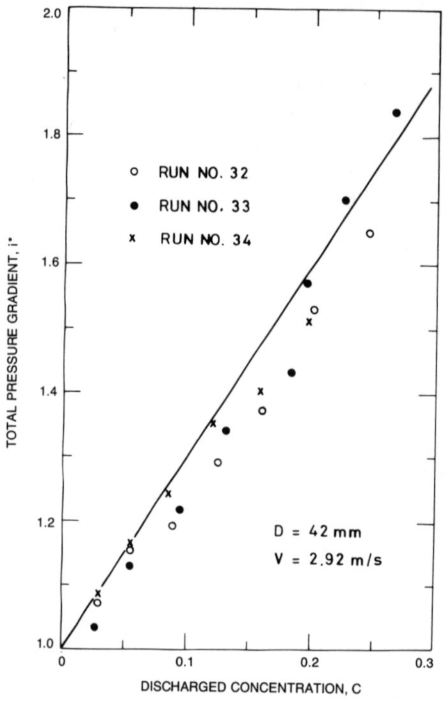

FIGURE 39. Plot of $[(i - i_L)/i_L]f_L$ versus discharged concentration (replotted from [175]).

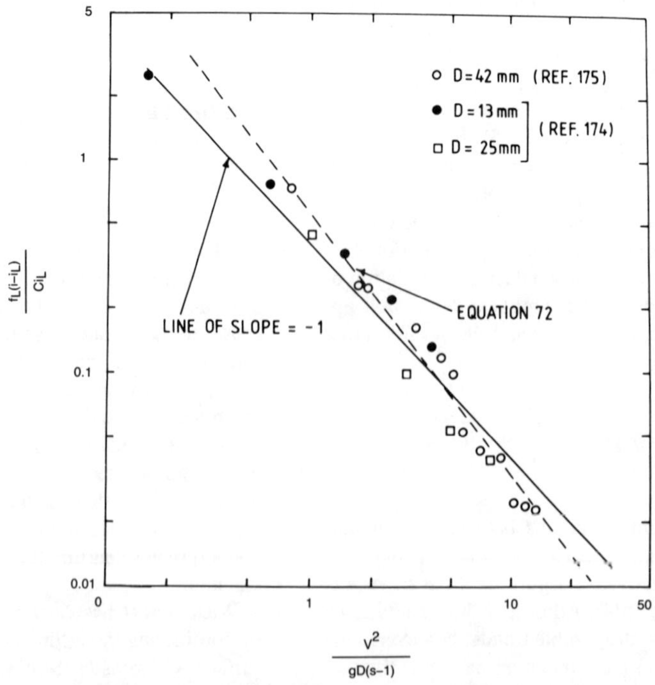

FIGURE 40. Overall representation of results in the form of $f_L(i - i_L)/C\ i_L$ and $V^2/gD\ (s - 1)$ (replotted from [175]).

the mechanism of flow, but the final equations include variables like bed height, porosity of bed, actual velocity in the free flow area etc. which are neither always known nor are there any methods available for their estimation in a new application. Thus the usefulness of these results, which are available in graphical form, is somewhat limited.

In their comprehensive study, Turian and Yuan [88] also developed a correlation for this flow regime. This is:

$$f - f_L = 0.4036C^{0.7389}f_L^{0.7717}C_D^{-0.4054} \times \left(\frac{V^2}{gD(s-1)}\right)^{-1.096} \tag{73}$$

An examination of Equation (73) suggests the drag coefficient to be a significant variable, though on physical grounds it appears that drag coefficient (or free settling velocity) is unlikely to be important in this mode of transport. But nevertheless Turian and Yuan concluded that their correlation reproduced about 400 experimental tests with an average error 29%. This equation has not yet been verified by independent experimental data. Perhaps the lack of data and/or correlations in this area is not too serious from an engineering applications point of view.

Flow of Solid-Liquid Mixtures Through Pipe Fittings

Much has been written in the foregoing pages about the head losses and their estimation in straight pipes. In a pipe network, however, additional pressure drop is incurred due to the presence of pipe fittings (e.g. bends, elbows, tees, etc.). Despite the pragmatic importance of slurry transport, the flow through pipe fittings remains hitherto an unexplored terrain. The only pertinent works, as far as we know, are those of Toda *et al.* [144,179], of Gregory [180] and of Turian *et al.* [181]. Both Toda *et al.* [179] and Gregory [180] have studied the flow of solid-liquid mixtures through 90° bends, whereas Turian *et al.* [181] have measured pressure drops for the flow of glass beads/water mixtures through a range of fittings including bends (45°, 90°, 180°), gate and globe valves etc. The pressure drop measurements in homogeneous flow pattern expressed in the form of loss coefficients are well correlated by the available Newtonian values provided the density of suspension is used instead of that of water. No effect of particle size was reported. Since very little information is available, it is not yet possible to outline generalized methods for the estimation of head losses through fittings.

In this chapter, the treatment has been limited to hydraulic transport of solids in horizontal pipes. However, in practice flow may take place in inclined and vertical pipes. Furthermore, the discussion is applicable only to mono or closely sized particle suspensions, whereas in practice one often deals with a spectrum of particle sizes. Some of the methods described in this chapter have been shown to be applicable to the conveyance of mixed size particles; however, this aspect of slurry transport has not been studied systematically. Finally, there is one more equally important and well established mode of solid conveyance, namely capsule flow, which has not been included in this review. The work in this area up to 1970 has been summarised by Govier and Aziz [12].

CONCLUDING REMARKS

This chapter has been concerned with the hydrodynamics of slurry—both pseudohomogeneous non-Newtonian as well as two-phase solid/liquid mixtures types—flows in straight horizontal pipes and in pipe fittings. In particular, the discussion has been focussed on the development of methods for the prediction of flow patterns and pressure losses due to friction. The state-of-art may be summarised as follows:

1. For the laminar flow of particulate suspensions displaying shearthinning characteristics, methods based on sound theory are available for the estimation of pressure losses in straight lengths of pipe. However, for the turbulent flow, the situation is not so simple as different methods yield widely divergent values of pressure gradient. Until the available methods have been thoroughly compared with experimental results, no single correlation can be recommended as *the equation* to be used. It is, therefore, suggested that for a given situation as many methods should be employed as possible to establish upper and lower bounds on the head loss. Very little is known about the turbulent head losses in rough walled pipes, and virtually nothing is available in terms of correlation as far as the flow through pipe fittings is concerned. This remains an area for future research activity.
2. For the flow of two phase solid-liquid mixtures, a wide variety of flow patterns are observed. Existing methods which purport to predict the transition conditions from one flow pattern to another have been reviewed. The transition velocities corresponding to the onset of heterogeneous flow and, eventually, to the flow with a moving bed can be predicted reasonably well using the criteria of Newitt *et al.* [Equation (38)] and of Zandi and Govatos [Equation (42)], respectively. However, the phenomenon of transition from flow with a moving bed to that with a stationary bed is poorly understood; this is so partly due the lack of appropriate data on this flow situation.
3. For the flow of mono or closely sized particle suspensions in straight pipes, satisfactory means are available for the estimation of head losses in homogeneous flow [Equation (55)], in heterogeneous flow [Equation (65)], and in flow with a moving bed [Equation (70)]. Once again little is

known about the flow with a stationary bed. Also, most correlations have been shown to perform well in relatively small diameter pipes (typically less than 50 mm), and at present, there appears to be some uncertainty with regard to the effect of pipe diameter. There is a real paucity of data on the pressure losses in pipe fittings; currently available information is limited to the homogeneous flow conditions.
4. Although the advantages in the use of heavy media for transporting coarse particles are readily acknowledged, it has received only scant attention during the last 10–15 years. This subject, we feel, merits much more attention than it has received in the past.

Once again it is reiterated that the foregoing discussion is generally applicable to the suspensions of mono (or closely) sized particles of reasonably regular shape. Thus future studies should be planned to study the hydrodynamics of mixed size particle suspensions. Evidently, the empirical formulations presented herein have enabled the calculations of pressure gradient with reasonable levels of accuracy and reliability, our basic understanding of the mechanics of flow is rather limited. Until this situation is rectified, the scale up to large sizes cannot be done without performing pilot plant studies which are expensive both in terms of time and money. Therefore, at least a part of future research efforts should be diverted towards gleaning fundamental information such as particle velocities, effect of solids on turbulence etc. which will, hopefully, assist in building some sort of mathematical models. Roco and Shook [184–186] have already embarked upon the modelling of slurry flows. Considerable scope therefore exists both for theoreticians as well as for experimentalists since much remains yet to be done in this field.

ACKNOWLEDGEMENTS

It would not be fitting for me to end this chapter without expressing gratitude to my mentor and a very special friend, Professor J. F. Richardson, who has never failed to stimulate and inspire all those who have worked with him.

NOTATION

b = dimensionless ratio (τ_0/τ_w)
C = volumetric fractional concentration of solids in the discharged mixture
C_D = drag coefficient of a single particle
d = particle size
D = pipe diameter
f = Fanning friction factor, defined by Equation (12)
f_L = Fanning friction factor for the flow of liquid alone
f_{BT} = modified Fanning friction factor, defined by Equation (32)
F_L = dimensionless parameter, defined by Equation (40)
g = acceleration due to gravity
He = Hedstrom number, defined as $(D^2\tau_0\varrho/\mu_p^2)$
i = total pressure gradient, (m/m)
i_L = pressure gradient for the flow of liquid alone, (m/m)
i_s = additional pressure gradient due to the presence of solids (m/m)
i^* = dimensionless total pressure gradient $(= i/i_L)$
k = Power law fluid consistency, Equation (1)
k' = fluid consistency index, Equation (21)
L = Length of pipe
n = Power law flow behaviour index, Equation (1)
n' = flow behaviour index, Equation (21)
N_1 = Parameter defined by Equation (41)
ΔP = pressure drop
Q = volumetric flow rate
r = radial coordinate
R = radius of pipe
Re = Reynolds number, Equation (14)
Re_B = Reynolds number for Bingham plastic fluids defined as $(\varrho VD/\mu_P)$
Re_{BT} = modified Reynolds number, Equation (32)
Re_{MR} = Metzner–Reed Reynolds number, Equation (20)
s = specific gravity of solids (ρ_s/ρ)
V = mean velocity
V_0 = free fall velocity of a single particle
V_1–V_4 = transition velocities, defined in Figure 13
V_z = average velocity in z-direction
Y = yield number, defined as $(D\tau_0/V\mu_P)$

Greek Symbols

$\dot{\gamma}$ = shear rate
μ = Newtonian viscosity
μ_p = plastic viscosity, Equation (2)
ρ = fluid density
ρ_s = density of solids
τ = shear stress
τ_0 = yield stress
τ_w = wall shear stress

Subscripts

c = refers to the critical value corresponding to laminar/turbulent transition
L = refers to pure liquid
M = mixture

REFERENCES

1. Zandi, I., "Hydraulic Transport of Bulky Materials," in *Adv. in Solid-Liquid Flow in Pipes and its Applications*, Pergamon Press, 1971, pp. 1–34.

2. Gandhi, R. L., ''An Overview of Slurry Pipelines,'' *J. Pipelines*, 3, 1982, 1–12.
3. *The Times of India*, New Delhi, April 12, 1985.
4. Nardi, J., ''Pumping Solids Through a Pipeline,'' *Chem. Eng.*, 66, July 27, 1959, 119–122.
5. Wasp, E. J., ''Coal Slurry Pipeline,'' *Mech. Eng.*, 101, No. 12, 1979, 38–45.
6. Wasp, E. J., ''Slurry Pipelines,'' *Sci. Amer.*, 249, 1983, 42–49.
7. Cabrera, V. R., ''Slurry Pipelines: Theory, Design, and Equipment,'' *World Mining*, 32, No. 1, 1979, 56–64; also see 32, No. 3, 1979, 59–65.
8. Cabrera, V. R., ''Slurry Pipelines of the Future,'' *Mech. Eng.*, 103, No. 4, 1981, 30–34.
9. Stripling, T. E. and R. G. Holter, ''Long Distance, High-Volume Transport of Coal by Slurry Pipelines,'' *J. Eng. Resour. Tech.*, *(ASME)*, 103, 1981, 322–329.
10. Zandi, I. (Ed.), ''Adv. in Solid-Liquid Flow in Pipes and its Applications,'' Pergamon Press, 1971.
11. Bain, A. G. and S. T. Bonnington, ''The Hydraulic Transport of Solids by Pipeline,'' Pergamon Press, 1970.
12. Govier, G. W. and K. Aziz, ''The Flow of Complex Mixtures in Pipes,'' R. E. Krieger Pub. Co., Florida, 1982.
13. Wasp, E. J., J. P. Kenney, and R. L. Gandhi, ''Solid-Liquid Flow: Slurry Pipeline Transportation,'' Trans. Tech. Pub., 1977.
14. Round, G. F., ''Solid-Liquid Flow Abstracts,'' Vol. 1–3, Gordon and Breach, NY, 1969.
15. *International Technical Conference on Solid-Liquid Transportation*. Slurry Transportation Association, U.S.A. (held every year since 1976).
16. *Proceedings of Hydrotransport 1–8*, British Hydromechanics Research Association, Cranfield, England.
17. Cheng, D. C-H., D. J. Ray, and F. H. Valentin, ''The Flow of Thixotropic Bentonite Suspension Through Pipes and Pipe Fittings,'' *Trans. Inst. Chem. Engrs.*, 43, 1965, 176–186.
18. Gabrysh, W. F., H. Eyring, L. S. Pan, and A. F. Gabrysh, ''Rheological Factors for Bentonite Suspensions,'' *J. Am. Ceramic. Soc.*, 46, 1963, 523–529.
19. Michaels, A. S. and J. C. Bolger, ''The Plastic Flow Behaviour of Flocculated Kaolin Suspensions,'' *Ind. Eng. Chem. Fundam.*, 1, 1962, 153–162.
20. Heywood, N. I. and J. F. Richardson, ''Rheological Behaviour of Flocculated and Dispersed Aqueous Kaolin Suspensions in Pipe Flow,'' *J. Rheo.*, 22, 1978, 599–613.
21. Farooqi, S. I. and J. F. Richardson, ''Rheological Behaviour of Kaolin Suspensions in Water and Water-Glycerol Mixtures,'' *Trans. Inst. Chem. Engrs.*, 58, 1980, 116–124.
22. Farooqi, S. I., N. I. Heywood, and J. F. Richardson, ''Drag Reduction by Air Injection in Suspension Flow in a Horizontal Pipeline,'' *Trans. Inst. Chem. Engrs.*, 58, 1980, 16–27.
23. Weltmann, R. N. and T. A. Keller, ''Pressure Losses of Titania and Magnesium Slurries in Pipes and Pipeline Transition,'' NASA Tech. Note #3889, Jan. 1957.
24. Quader, A. K. M. A. and W. L. Wilkinson, ''Turbulent Heat Transfer Characteristics to Dilute Aqueous Suspensions of Titanium Dioxide in Pipes,'' *Int. J. Multiphase Flow*, 7, 1981, 545–554; also see *ibid.*, 6, 1980, 553–561.
25. Eissenberg, D. M., ''Measurement of the Turbulent Velocity Profile of Thoria Suspensions in Pipe Flow,'' Oak Ridge National Laboratory, ORNL-TM-701, 1963.
26. Kearsey, H. A., ''The Rheology of Dispersed Suspensions of Thoria and its Relation to the Non-Newtonian Behaviour of Flocculated Suspensions,'' *Trans. Inst. Chem. Engrs.*, 40, 1962, 140–145.
27. Nguyen, Q. D., ''Rheology of Concentrated Bauxite Residue Suspensions,'' Ph.D. Thesis, Monash University, Australia, 1983.
28. Manoliadis, O. and P. L. Bishop, ''Temperature Effect on Rheology of Sludges,'' *J. Environ. Div., Proc. ASCE*, 110, 1984, 286–290.
29. Metzner, A. B., ''Pipe-Line Design for Non-Newtonian Fluids,'' *Chem. Eng. Prog.*, 50, 1954, 27–34.
30. Jeffrey, D. J. and A. Acrivos, ''The Rheological Properties of Suspensions of Rigid Particles,'' *A.I.Ch.E.J.*, 22, 1976, 417–432.
31. Russell, W. B., ''Review of the Role of Colloidal Forces in the Rheology of Suspensions,'' *J. Rheo.*, 24, 1980, 287–317.
32. Chaffey, C. E., ''Mechanism and Equations for Shearthinning and Thickening in Dispersions,'' *Colloid Polym. Sci.*, 255, 1977, 691–698.
33. Schowalter, W. R., ''Mechanics of Non-Newtonian Fluids,'' Pergamon Press, 1978, pp. 264–289.
34. Barnes, H. A., ''Dispersion Rheology 1980—A Survey of Industrial Problems and Academic Progress,'' Report prepared for the Process Tech. Group of the Royal Society of Chemistry, 1981.
35. Mill, C. C. (Ed.), ''Rheology of Disperse Systems,'' Pergamon, 1959.
36. Kao, D. T. Y., ''Rheology of Suspensions,'' in *Handbook of Fluids in Motion*, N. P. Cheremisinoff and R. Gupta (Eds.), Ann Arbor Sci., 1983.
37. Bullivant, S. A. and T. E. R. Jones, ''Elasticoviscous Properties of Deflocculated China Clay Suspensions—Concentration Effects,'' *Rheo. Acta*, 20, 1981, 64–77.
38. Van de Ven, T. G. M. and R. J. Hunter, ''The Energy Dissipation in Sheared Coagulated Sols,'' *Rheo. Acta.*, 16, 1977, 534–543.
39. Buscall, R. *et al.*, ''Viscoelastic Properties of Concentrated Lattices,'' *J. Chem. Soc.*, Faraday Trans. I., 78, 1982, 2873–2899.
40. Skelland, A. H. P., ''Non-Newtonian Flow and Heat Transfer,'' John Wiley, 1967.
41. Cramer, S. D. and J. M. Marchello, ''Procedure for Fitting Non-Newtonian Viscosity Data,'' *A.I.Ch.E.J.*, 14, 1968, 814–815; *ibid.*, 1968, 980–983.
42. Cheng, D. C-H., ''A Design Procedure for Pipeline Flow of Non-Newtonian Dispersed Systems,'' *Hydrotransport 1*, Paper J5, 1971.
43. Charles, M. E. and R. A. Charles, ''The Use of Heavy Media in the Pipe Line Transport of Particulate Solids,'' in *Adv. in*

Solid-Liquid Flow in Pipes and its Applications, Ed. I. Zandi, Pergamon, 1971, 187–197.
44. Denn, M. M., "Process Fluid Dynamics," Prentice Hall, Englewood Cliffs, 1977, p. 20.
45. Thomas, D. G., "Non-Newtonian Suspensions," *Ind. Eng. Chem.*, 55, No. 11, 1963, 19–29.
46. Smith, T. L. and C. A. Bruce, "Intrinsic Viscosities and Other Rheological Properties of Flocculated Suspensions of Nonmagnetic and Magnetic Ferric Oxides," *J. Coll. and Intfac. Sci.*, 72, 1979, 13–26.
47. Vinogradov, G. V. *et al.*, "Viscoelastic Properties of Filled Polymers," *Int. J. Polymeric Materials*, 2, 1972, 1–27.
48. Reid, R. C., J. M. Prausnitz, and T. K. Sherwood, "The Properties of Gases and Liquids," McGraw Hill, 1977; 3rd edition.
49. Walters, K., "Rheometry," Chapman and Hall, London, 1975.
50. Van Wazer, J. R. *et al.*, "Viscosity and Flow Measurement," Interscience, 1963.
51. Metzner, A. B., "Non-Newtonian Technology: Fluid Mechanics, Mixing and Heat Transfer," *Adv. Chem. Eng.*, 1, 1956, 79–150.
52. Shook, C. A. *et al.*, "Flow of a Coal Slurry with a Yield Stress," *J. Pipelines*, 4, 1984, 289–297.
53. Heywood, N. I. and D. C-H. Cheng, "Comparison of Methods for Predicting Head Loss in Turbulent Pipe Flow of Non-Newtonian Fluids," *Trans. Inst. Meas. Control*, 6, 1984, 33–45.
54. Metzner, A. B. and J. C. Reed, "Flow of Non-Newtonian Fluids—Correlation of Laminar, Transition, and Turbulent Flow Regions," *A.I.Ch.E.J.*, 1, 1955, 434–440.
55. Heywood, N. I., "Air Injection into Suspensions Flowing in Horizontal Pipelines," Ph.D. thesis, University College of Swansea, U.K., 1976.
56. Metzner, A. B., "Non-Newtonian Flow," *Ind. Eng. Chem.*, 49, 1957, 1429–1432.
57. Chhabra, R. P. *et al.*, "Cocurrent Flow of Air and China Clay Suspension in Large Diameter Pipes," *Chem. Eng. Res. Des.*, 61, 1983, 56–61.
58. Ryan, N. W. and M. M. Johnson, "Transition from Laminar to Turbulent Flow in Pipes," *A.I.Ch.E.J.*, 5, 1959, 433–435.
59. Hanks, R. W. and E. B. Christiansen, "The Laminar Turbulent Transition in Non-Isothermal Flow of Pseudoplastic Fluids in Tubes," *A.I.Ch.E.J.*, 8, 1962, 467–477.
60. Hanks, R. W., "The Laminar-Turbulent Transition for Fluids with a Yield Stress," *A.I.Ch.E.J.*, 9, 1963, 306–309; also see *ibid.*, 14, 1968, 691–695.
61. Caldwell, D. H. and H. E. Babbitt, "The Flow of Muds, Sludges and Suspensions in Circular Pipes," *Trans. A.I.Ch.E.*, 37, 1941, 237–266.
62. Wilhelm, R. H., D. M. Wroughton, and W. F. Loeffel, "Flow of Suspensions Through Pipes," *Ind. Eng. Chem.*, 31, 1939, 622–629.
63. Alves, G. E., D. F. Boucher, and R. L. Pigford, "Pipeline Design for Non-Newtonian Solutions and Suspensions," *Chem. Eng. Prog.*, 48, 1952, 385–393.
64. Froishteter, G. B. and G. V. Vinogradov, "The Loss of Laminar Flow Stability in Non-Isothermal Flow of Plastic Disperse Systems Through Circular Pipes," *Rheo. Acta*, 16, 1977, 620–627.
65. Dodge, D. W. and A. B. Metzner, "Turbulent Flow of Non-Newtonian Systems," *A.I.Ch.E.J.*, 5, 1959, 189–204; Corrigenda in *A.I.Ch.E.J.*, 8, 1962, 143.
66. Tomita, Y., "A Study of Non-Newtonian Flow in Pipelines," *Bull. JSME*, 2, 1959, 10–16; *ibid.*, 4, 1961, 77.
67. Shaver, R. G. and E. W. Merrill, "Turbulent Flow of Pseudoplastic Polymer Solutions in Straight Cylindrical Tubes," *A.I.Ch.E.J.*, 5, 1959, 181–188.
68. Thomas, G., Ph.D. Thesis, Uni. College of Swansea, U.K., 1960.
69. Kemblowski, Z. and J. Kolodziejski, "Flow Resistances of Non-Newtonian Fluids in Transitional and Turbulent Flow," *Int. Chem. Eng.*, 13, 1973, 265–279.
70. Szilas, A. P., E. Bobok, and L. Navratil, "Determination of Turbulent Pressure Loss of Non-Newtonian Oil Flow in Rough Pipes," *Rheo. Acta*, 20, 1981, 487–496.
71. Torrance, B. McK., "Friction Factors for Turbulent Non-Newtonian Fluid Flow in Circular Pipes," *South African Mech. Engr.*, 13, 1963, 89–91.
72. Hanks, R. W., "Low Reynolds Number Turbulent Flow of Pseudohomogeneous Slurries," *Hydrotransport*, 5, Paper C2, 1978.
73. Millikan, C. B., "A Critical Discussion of Turbulent Flows in Channels and Circular Tubes," *Proc. 5th Int. Cong. App. Mech.*, John Wiley & Sons, NY, 1939, pp. 386–390.
74. Heywood, N. I., "Pipeline Design for Non-Newtonian Fluids," *Proc. Interflow 80, I. Chem. E. Sym. Ser. 60*, 1980, 33–52.
75. Metzner, A. B., "Criticism of Rheological Behaviour of Flocculated and Dispersed Aqueous Kaolin Suspensions in Pipe Flow by Heywood and Richardson," *J. Rheo.*, 24, 1980, 115–119; Reply by J. F. Richardson, *ibid.*, 24, 1980, 119–121.
76. Farooqi, S. I. and J. F. Richardson, "Horizontal Flow of Air and Liquid (Newtonian and Non-Newtonian) in a Smooth Pipe, Part I: A Correlation for Average Liquid Hold Up," *Trans. Inst. Chem. Engrs.*, 60, 1982, 292–305.
77. Thomas, D. G., "Transport Characteristics of Suspensions, Part IV: Friction Loss of Concentrated Flocculated Suspensions in Turbulent Flow," *A.I.Ch.E.J.*, 8, 1962, 266–271.
78. Thomas, D. G., "Transport Characteristics of Suspensions, Part IX," *A.I.Ch.E.J.*, 10, 1964, 303–308.
79. Thomas, D. G., "Heat and Momentum Transport Characteristics of Non-Newtonian Aqueous Thorium Oxide Suspensions," *A.I.Ch.E.J.*, 6, 1960, 631–639.
80. Thomas, D. G., "Non-Newtonian Suspensions, Part II: Turbulent Transport Characteristics," *Ind. Eng. Chem.*, 55, No. 12, 1963, 27–35.

81. Perry, J. H., "Chemical Engineers Handbook," 4th edition, McGraw Hill, 1963.
82. Edwards, M. F. *et al.*, "Head Losses in Pipe Fittings at Low Reynolds Number," *Chem. Eng. Res. Des.*, 63, 1985, 43–50.
83. Soo, S. L., "Fluid Dynamics of Multiphase Systems," Blaisdell (Ginn), 1967.
84. Soo, S. L., "Development of Theories on Liquid-Solid Flows," *J. Pipelines*, 4, 1984, 137–142.
85. Soo, S. L. and J. A. Regalbuto, "Concentration Distribution in Twophase Pipe Flow," *Can. J. Chem. Eng.*, 38, 1960, 160–166.
86. Newitt, D. M., J. F. Richardson, and C. A. Shook, "Hydraulic Conveying of Solids in Horizontal Pipes, Part II: Distribution of Particles and Slip Velocities," *Proc. Sym. Int. between Fluids and Particles, I. Chem. E.*, London, 1962.
87. Sinclair, C. G., "The Limit Deposit Velocity of Heterogeneous Suspensions," *Proc. Symp. Int. between Fluids and Particles, Inst. Chem. Engrs.*, London, 1962. pp. 68–76.
88. Turian, R. M. and T. F. Yuan, "Flow of Slurries in Pipelines," *A.I.Ch.E.J.*, 23, 1977, 232–243.
89. Parzonka, W., J. M. Kenchington, and M. E. Charles, "Hydrotransport of Solids in Horizontal Pipes: Effects of Solids Concentration and Particle Size on the Deposit Velocity," *Can. J. Chem. Eng.*, 59, 1981, 291–295.
90. Round, G. F., "Hydraulic Transport of Solids in Pipelines," *App. Mech. Rev.*, 35, 1982, 1349–1355.
91. Durand, R., "Basic Relationships of the Transportation of Solids in Pipes—Experimental Research," *Proc. Int. Assoc. Hyd. Res.*, Minneapolis, Minn., Sept. 1–4, 1953, p. 89.
92. Condolios, E. and E. E. Chapus, "Transporting Solid Materials in Pipes," *Chem. Eng.*, 70, June 24, 1963, 93–98.
93. Babcock, H. A., "Heterogeneous Flow of Heterogeneous Solids," Paper presented at the *Int. Sym. on Solid-Liquid Flow in Pipes*, Univ. of Pennsylvania, Philadelphia, Pa., March, 1968.
94. Bird, R. B., G. C. Dai, and B. J. Yarusso, "The Rheology and Flow of Viscoplastic Materials," *Rev. Chem Eng.*, 1(1983) 1–70.
95. Spells, K. E., "Correlations for Use in Transport of Aqueous Suspensions of Fine Solids Through Pipes," *Trans. Inst. Chem. Engrs.*, 33, 1955, 79–84.
96. Newitt, D. M., J. F. Richardson, M. Abbott, and R. B. Turtle, "Hydraulic Conveying of Solids in Horizontal Pipes," *Trans. Inst. Chem. Engrs.*, 33, 1955, 93–110.
97. Wani, G. A., M. K. Sarkar, and B. Pitchumani, "Pressure Drop Prediction in Multisize Particle Transportation Through Horizontal Pipes," *J. Pipelines*, 3, 1982, 23–33.
98. Carleton, A. J. and D. C-H. Cheng, "Design Velocities for Hydraulic Conveying of Settling Suspensions," *Hydrotransport*, 3, E.57–74, 1974.
99. Oroskar, A. R. and R. M. Turian, "The Critical Velocity in Pipeline Flow of Slurries," *A.I.Ch.E.J.*, 26, 1980, 550–558.
100. Durand, R. and E. Condolios, "The Hydraulic Transportation of Coal and Solid Materials in Pipes," in *Colloquium on Hydraulic Transportation of Coal*, London, Nov. 1952.
101. Gilbert, R., "Transport Hydraulique et Refoulement des Mixtures en Conduites," *Ann. des Ponts et Chaussees*, 130, 1960, 307–373; 437–486.
102. Wasp, E. J. *et al.*, "Deposition Velocities, Transition Velocities and Spatial Distribution of Solids in Slurry Pipelines," *Hydrotransport*, 1, Paper H4, 1971.
103. Silin, N. A. *et al.*, "Fluid Mechanics," *Soviet Res.*, 2, Nov.–Dec. 1973, 1.
104. Silin, N. A. *et al.*, "Hydrotransport," p. 104, *Naukova Dumka*, Kiev, 1971 (in Russian).
105. Silin, N. A. *et al.*, "Operational Mechanics of Large Dredgers and Pipelines," p. 48, *Ukr. Acad. Sci.*, Kiev, 1962 (in Russian).
106. Hayden, J. W. and Stelson, T. E., "Hydraulic Conveyance of Solids in Pipes," in *Adv. Solid-Liquid Flow in Pipes and its Applications*, Zandi, I. (Ed.), Pergamon, 1971, pp. 149–163.
107. Yotsukura, N., *U.S. Geol. Survey*, Prof. Paper 450-E, 1963, pp. 172–173.
108. Smith, R. A., "Experiments on the Flow of Sand-Water Slurries in Horizontal Pipes," *Trans. Inst. Chem. Engrs.*, 33, 1955, 85–92.
109. Stevens, G. S. and M. E. Charles, "The Pipeline Flow of Slurries—Transition Velocities," *Hydrotransport*, 2, Paper E3, 1972.
110. Pokrovskaya, V. N., "Ways of Improving the Efficiency of Transport," p. 19, Nedra, Moscow, 1972 (in Russian).
111. Schriek, W. *et al.*, "Experimental Studies on Solids Pipelining of Canadian Commodities, Report VII: Experimental Studies on the Transport of Two Different Sands in Water in 2, 4, 6, 8, 10 and 12 Inch Pipelines," Report No. E73-21, Nov. 1973.
112. Durand, R. and E. Condolios, "Etude Experimentale du Refoulement des Materiaux en Conduits," *2 emes Journees de l'Hydraulique*, Societe Hydrotechnique de France, p. 29, June 1952.
113. Ambrose, H. H., "The Transportation of Uniform Sand in a Smooth Pipe," Ph.D. Thesis, Uni. Iowa, Ames, 1952.
114. Wilson, K. C., "Proc. Ist. Int. Sym. Dredging Technol.," Paper C3, *BHRA Fluid Eng.*, 1975.
115. Bielova, N. T., "Movement of Uniform and Non-Uniform Fluids," *Moskovski Inzhynierno-Stroitelnyi Institut in Kuybycheva*, Moscow, p. 112, 1968 (in Russian).
116. Wolanski, Z., Ph.D. Thesis, *Acad. Agriculture*, Wroclaw, Poland, 1972.
117. Vodolazski, V. I., V. P. Gruchko, and E. D. Telechkin, "An Investigation of the Critical Velocity for Hydraulic Transport of Suspensions from Gas Scrubbing of Blast Furnace Plants in Metallurgical Factories," p. 53, *Vodosnabchenie Kanalizatsia, Gidrotechnicheskie, Sooruzenia*, Kiev, 1971.
118. Schriek, W. *et al.*, "Experimental Studies of Solids Pipelining of Canadian Commodities, Report III: Experimental

Studies on the Hydraulic Transport of Iron Ore,'' *Saskatchewan Res. Council*, Canada, July 1973.

119. Schriek, W. *et al.*, ''Experimental Studies of Solids Pipelining of Canadian Commodities, Report V: Experimental Studies on the Hydraulic Transport of Coal,'' *Saskatchewan Res. Council*, Canada, Oct. 1973.
120. Wicks, M., ''Transport of Solids at Low Concentration in Horizontal Pipes,'' in *Adv. Solid-Liquid Flow in Pipes and its Applications*, I. Zandi (Ed.), Pergamon Press, 1971, pp. 101–124.
121. Craven, J. P., ''A Study of the Transportation of Sand in Pipes,'' Ph.D. Thesis, State Uni. Iowa, 1951.
122. Zandi, I. and G. Govatos, ''Heterogeneous Flow of Solids in Pipeline,'' *J. Hyd. Div., Proc. ASCE*, 93, 1967, 145–159.
123. Babcock, H. A., ''Heterogeneous Flow of Heterogeneous Solids,'' in *Adv. Solid-Liquid Flow in Pipes and its Applications*, Zandi, I. (Ed.), Pergamon Press, 1971, pp. 125–148.
124. Chhabra, R. P. and J. F. Richardson, ''Hydraulic Transport of Coarse Gravel Particles in a Smooth Horizontal Pipe,'' *Chem. Eng. Res. Des.*, 61, 1983, 313–317.
125. Toda, M., H. Konno, and S. Saito, ''Simulation of Limit Deposit Velocity in Horizontal Liquid Solid Flow,'' *J. Chem. Eng.*, Japan, 13, 1980, 439–444.
126. Thomas, D. G., ''Transport Characteristics of Suspensions: II. Minimum Transport Velocity for Flocculated Suspensions in Horizontal Pipes,'' *A.I.Ch.E.J.*, 7, 1961, 423–430.
127. Thomas, D. G., ''Transport Characteristics of Suspensions: VI. Minimum Transport Velocity for Large Particle Size Suspensions in Round Horizontal Pipes,'' *A.I.Ch.E.J.*, 8, 1962, 373–378.
128. Wilson, K. C., ''Slip Point of Beds in Solid-Liquid Pipeline Flow,'' *J. Hyd. Div., Proc. ASCE*, 96, 1970, 1–12.
129. Wilson, K. C., ''Parameters for Bed Slip Point in Two-Phase Flow,'' *J. Hyd. Div., Proc. ASCE*, 97, 1971, 1665–1679.
130. Wilson, K. C., ''Co-ordinates for the Limit of Deposition in Pipeline Flow,'' in *Int. Conf. on the Hydraulic Transport of Solids in Pipes*, 3rd, Golden, Colo., 1974; Also *Hydrotransport*, 3, Paper E1, 1974.
131. Julian, F. M. and A. E. Dukler, ''An Eddy Viscosity Model for Friction in Gas-Solid Flows,'' *A.I.Ch.E.J.*, 11, 1965, 853–858.
132. Shook, C. A. and S. M. Daniel, ''A Variable Density Model of the Pipeline Flow of Suspensions,'' *Can. J. Chem. Eng.*, 47, 1969, 196–200.
133. Graf, W. H. and E. R. Acaroglu, ''Homogeneous Suspensions in Circular Conduits,'' *J. Pipeline Div., Proc. ASCE*, 93, 1967, 63–69.
134. Maruyama, T., J. Ando, and T. Mizushina, ''Flow of Settling Slurries in Horizontal Pipes,'' *J. Chem. Eng.*, Japan, 13, 1980, 269–274.
135. Vanoni, V. A., ''Transportation of Suspended Sediment by Water,'' *Trans. ASCE*, Publication 2267, 111, 1964, 67–133.
136. Babkowicz, A. J. and W. H. Gauvin, ''The Turbulent Flow Characteristics of Model Fiber Suspensions,'' Pulp and Paper Research Institute of Canada, TR 357, 1964.
137. Zandi, I., ''Decreased Head Losses in Raw Water Conduits,'' *J. Am. Waterworks Ass.*, 59, Feb. 1967, 213–226.
138. Daily, J. W. and G. Bugliarella, ''Basic Data for Dilute Fiber Suspensions in Uniform Flow with Shear,'' *TAPPI*, 44, 1961, 497.
139. Rose, H. E. and R. A. Duckworth, ''Transport of Solid Particles in Liquids and Gases,'' *The Engineer*, 227, 1969, 392–396; 227, 1969, 430–433; 227, 1969, 478–483.
140. Molerus, O. and P. Wellmann, ''A New Concept for the Calculation of Pressure Drop with Hydraulic Transport of Solids in Horizontal Pipes,'' *Chem. Eng. Sci.*, 36, 1981, 1623–1632.
141. Duckworth, R. A., ''The Hydraulic Transport of Materials by Pipeline,'' *South Afr. Mech. Engr.*, 28, 1978, 291–306.
142. Chhabra, R. P. and J. F. Richardson, ''Comments on a New Concept for the Calculation of Pressure Drop with Hydraulic Transport of Solids in Horizontal Pipes,'' *Chem. Eng. Sci.*, 37, 1982, 1575–1578.
143. Shook, C. A., ''The Hydraulic Conveying of Solids in a One Inch Diameter Pipe,'' Ph.D. Thesis, Uni. London, 1960.
144. Toda, M. *et al.*, ''On the Particle Velocities in Solid-Liquid Two-Phase Flow Through Straight Pipes and Bends,'' *J. Chem. Eng.*, Japan, 6, 1973, 140–146.
145. Ohashi, H. *et al.*, ''Average Particle Velocity in Solid-Liquid Two-Phase Flow Through Vertical and Horizontal Tubes,'' *J. Chem. Eng.*, Japan, 13, 1980, 343–349.
146. Govier, G. W. and M. E. Charles, ''The Hydraulics of the Pipeline Flow of Solid-Liquid Mixtures,'' *Engineering J.*, 44, August 1961, 50–57.
147. Wayment, W. R., G. L. Wilhelm, and J. D. Bardill, ''Factors Influending the Design of Hydraulic Backfill Systems, Part I. Friction Head Losses of Sand Slurries During Pipeline Transport,'' *U.S. Bureau of Mines*, Report Investigations 6065, 1962.
148. Ellis, H. S. and G. F. Round, ''Laboratory Studies on the Flow of Nickel-Water Suspensions,'' *Can. Min. Met. Bull.*, 56, 1963, 773–781.
149. Babcock, H. A., ''The State of the Art of Transporting Solids in Pipelines,'' *Chem. Eng. Prog.*, 60, 1964, 36–45.
150. Durand, R., ''Hydraulic Transport of Solid Materials in Pipes,'' (Abstract only), *Chem. Eng. & Min. Rev.*, 60, March 10, 1953, 225–228.
151. Bonnington, S. T., ''Estimation of Pipe Friction Involved in Pumping Solid Material,'' *Brit. Hydromech. Res. Ass.*, TN 708, Dec. 1961.
152. Worster, R. C., ''The Hydraulic Transport of Solids,'' *Part I: Brit. Hydromech. Res. Ass.*, pp. 447 and 482, 1954.
153. Koch, L. W., ''Solids in Pipes,'' *Int. Sci. & Tech.*, No. 26, 68–78, Feb. 1963.
154. Ellis, H. S., P. J. Redberger, and L. H. Bolt, ''Slurries: Basic Principles and Power Requirements,'' *Ind. Eng. Chem.*, 55, Aug. 1963, 18–26.
155. Yen, J. G. and I. Zandi, ''Transport of Slurries in Hetero-

geneous Regime,'' presented at the *Annual Conference of AIM*, Washington, D.C., Feb. 1969.
156. Shih, C. C. S., ''Hydraulic Transport of Solids in a Sloped Pipe,'' *J. Pipeline Div., Proc. ASCE*, 90, 1964, 1–14.
157. Wasp, E. J. *et al.*, ''Cross Country Coal Pipeline Hydraulics,'' *Pipeline News*, pp. 20–28, July 1963.
158. Wasp, E. J. *et al.*, ''Hetero-Homogeneous Solids/Liquid Flow in the Turbulent Regime,'' in *Adv. Solid-Liquid Flow in Pipes and its Applications*, I. Zandi (Ed.), Pergamon, 1971, pp. 199–210.
159. Charles, M. E., ''Transport of Solids by Pipeline,'' *Hydrotransport*, 1, Paper A3, 1971.
160. Turtle, R. B., ''The Hydraulic Conveying of Granular Materials,'' Ph.D. Thesis, Uni. of London, 1952.
161. Abbott, M., ''The Hydraulic Conveying of Solids in Pipelines,'' Ph.D. Thesis, Uni. of London, 1955.
162. Wilson, K. C., M. Streat, and R. A. Bantin, ''Slip Model Correlation of Dense Two-Phase Flow,'' *Hydrotransport*, 2, Paper B1, 1972.
163. Shook, C. A., ''Pipeline Flow of Coarse Particle Slurries,'' in *Handbook of Fluids in Motion* by Cheremisinoff, N. P. and Gupta, R. (Eds.), Ann Arbor, 1983, pp. 929–943.
164. James, J. G. and B. A. Broad, ''Conveyance of Coarse-Particle Solids by Hydraulic Pipeline: Trials with Limestone Aggregates in 102, 156 and 207 mm Diameter Pipes,'' *Transport and Road Research Laboratory*, TRRL supplementary Report #635, 1980.
165. Howard, G. W., ''Transportation of Sand and Gravel in a Four-Inch Pipe,'' *Trans. ASCE*, 104, 1939, 1334–1349.
166. Howard, G. W., ''Effect of Rifling on Four-Inch Pipe Transporting Solids,'' *Trans. ASCE*, 106, 1941, 135–157.
167. Coulson, J. M. and J. F. Richardson, ''Chemical Engineering,'' Vol. 2, Pergamon Press, 3rd edition, 1983.
168. Televantos, Y. *et al.*, ''Flow of Slurries of Coarse Particles at High Solids Concentrations,'' *Can. J. Chem. Eng.*, 57, 1979, 255–262.
169. Boothroyde, J., B. E. A. Jacobs, and P. Jenkins, ''Coarse Particle Hydraulic Transport,'' *Hydrotransport*, 6, Paper E1, 1979.
170. Worster, R. C., ''The Hydraulic Transport of Solids,'' Paper I, *Proc. Colloq. Hydrau. Trans. Coal*, organised by National Coal Board, London, 1953, pp. 5–20.
171. Worster, R. C. and D. F. Denny, ''Hydraulic Transport of Solid Material in Pipes,'' *Proc. Inst. Mech. Engrs.*, 169, 1955, 563–586.
172. Shook, C. A. *et al.*, ''Some Experimental Studies on the Effect of Particle and Fluid Properties Upon the Pressure Drop for Slurry Flow,'' *Hydrotransport*, 2, Paper D2, 1972.
173. Masuyama, T., T. Kawashima, and K. Noda, ''Pressure Loss of Pseudo-Homogeneous Fluid Flow Containing Coarse Particles in a Pipe,'' *Hydrotransport*, 5, Paper D1, 1978.
174. Kenchington, J. M., ''Predictin of Pressure Gradient in Dense Phase Conveyng,'' *Hydrotransport*, 5, Paper D7, 1978.
175. Chhabra, R. P. and J. F. Richardson, ''Hydraulic Transport of Coarse Particles in Viscous Newtonian and Non-Newtonian Media in a Horizontal Pipe,'' *Chem. Eng. Res. Des.*, 63, 1985, 390–397.
176. Farooqi, S. I., ''The Effect of Rheological Properties on the Flow of Gas-Liquid Mixtures,'' Ph.D. Thesis, Univ. College of Swansea, United Kingdom, 1981.
177. Wicks, M., ''Transport of Solids at Low Concentrations in Horizontal Pipes,'' Paper presented at the *Int. Symp. on Solid-Liquid Flow in Pipes*, Uni. of Pennsylvania, Philadelphia, Pa., March 1968.
178. Carstens, M. R., ''A Theory for Heterogeneous Flow of Solids in Pipes,'' *J. Hydraulic Div., Proc. ASCE*, 95, 1969, 275–286.
179. Toda, M. *et al.*, ''Hydraulic Conveying of Solids Through Pipe Bends,'' *J. Chem. Eng.*, Japan, 5, 1972, 4–13.
180. Gregory, W. B., ''Pumping Clay Slurry Through a Four-Inch Pipe,'' *Mech. Eng.*, 49, June 1927, 609–616.
181. Turian, R. M., F. L. Hsu, and M. S. Selim, ''Friction Losses for Flow of Slurries in Pipeline Bends, Fittings, and Valves,'' *Particulate Sci. & Tech.*, 1, 1983, 365–392.
182. Langlinais, J. P. *et al.*, ''Frictional Pressure Losses for Annular Flow of Drilling Mud and Mud-Gas Mixtures,'' *J. Energy, Resources and Tech., Trans. ASME*, 107, 1985, 142.
183. *Journal of Pipelines*, 4 (1984) Number 3 Special issue on theoretical developments in the field of slurry flow.
184. Roco, M. and C. A. Shook, ''Modelling of Slurry Flow: The Effect of Particle Size,'' *Can. J. Chem. Eng.*, 61 (1983) 494–503.
185. Roco, M. and C. A. Shook, ''A Model for Turbulent Slurry Flow,'' *J. Pipelines*, 4 (1984) 3–13.
186. Roco, M. and C. A. Shook, ''Computational Method for Coal Slurry Pipelines with Heterogeneous Size Distribution,'' *Powder Tech.*, 39 (1984) 159–176.

SECTION THREE

Flow with Bed Load

Lateral Bed Load Transport on Side Slopes

SYUNSUKE IKEDA*

Rivers always have lateral bed slopes. Most channel process, therefore, accompanies lateral bed load transport. For example, in alluvial meandering rivers the bed load moves laterally toward the point bars and deposits there, which induces subsequent lateral migration of river courses. Sediment particles on side slope suffer gravitational force laterally beside fluid force, and they are easy to move compared with those placed on level bed. This section will treat the incipient motion of bed materials and the lateral transport rate of bed load on side slopes.

FLUID FORCE EXERTING ON SEDIMENT PARTICLES

Consider a sediment particle moving on the bed of a curved channel, for which a curvilinear coordinate is defined such that s denotes the longitudinal coordinate along the channel centerline, n denotes the lateral normal coordinate, and z denotes the upward normal coordinate (see Figure 1). Let θ denote the lateral angle of bed inclination, which is assumed to be sufficiently small so that $\theta \simeq \tan\theta \simeq \sin\theta$. It is also assumed that a typical value of θ is much larger than the angle of the longitudinal inclination of the bed, ψ.

The existence of lateral gravitational force and secondary flow creates a situation that the sediment particle moves along a course that is skewed an angle β to the s direction, and the secondary flow produces a mean near bed fluid velocity which is skewed an angle δ to the s direction (see Figure 1).

The dynamic fluid force exerting on the particle can be split into two parts, i.e. the drag force, D, and the lift force, L. Assuming sphericity for the bed materials, the drag force can be approximated with [1]

$$D \simeq \frac{1}{2}\rho C_D \frac{\pi}{4} d^2(u_b - v_p)^2 \tag{1}$$

in which ρ = mass density of fluid; C_D = drag coefficient; d = particle size; u_b = near bed fluid velocity; and v_p = particle velocity. The longitudinal and the lateral components of D are given, respectively, by

$$D_s = D\frac{u_{bs} - v_{ps}}{u_b - v_p} \tag{2}$$

$$D_n = D\frac{u_{bn} - v_{pn}}{u_b - v_p} \tag{3}$$

in which u_{bs}, u_{bn} = s and n components of u_b, respectively; and v_{ps}, v_{pn} = s and n components of v_p, respectively. From the definition the following relations hold:

$$\tan\delta = \frac{u_{bn}}{u_{bs}} \tag{4}$$

$$\tan\beta = \frac{v_{pn}}{v_{bs}} \tag{5}$$

The lift force which is assumed to be oriented upward normal to the bed is expressed by

$$L \simeq \alpha D \tag{6}$$

in which α = ratio of lift coefficient to drag coefficient.

*Department of Foundation Engineering, Saitama University
Saitama, 338 Japan

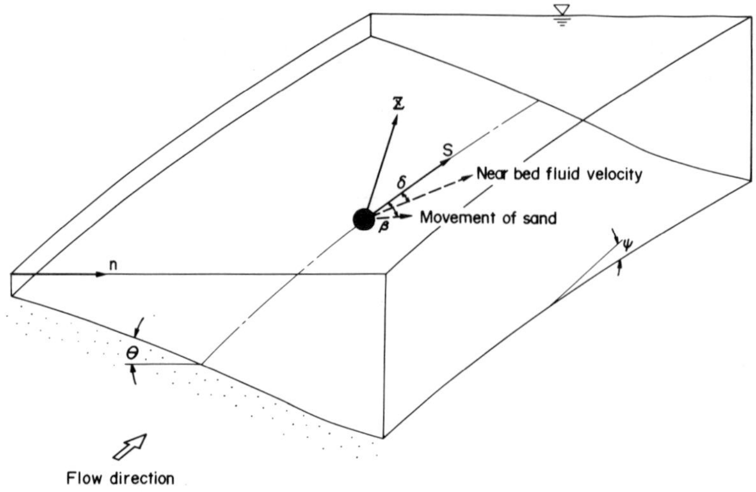

FIGURE 1. Definition sketch for a bed material moving in a curved channel.

INCIPIENT MOTION

Critical Tractive Force on Side Slopes. The equilibrium of forces acting on a static particle placed on a side slope in curved channel is considered. Referring to Figure 2, the force balance gives a relation

$$F^2 = D_s^2 + (D_n^2 + W \sin \theta)^2 \tag{7}$$

in which F = static friction force between the particle and the bed; and W = submerged weight of the particle = $\pi(\rho_s - \rho)gd^3/6$. The static friction is equated to

$$F = (W \cos \theta - L)\tan \phi \tag{8}$$

in which ϕ = angle of repose of the particle. Eliminating F from Equations (7) and (8), it follows that

$$\left(\frac{D_s}{W}\right)^2 + \left(\frac{D_n}{W}\right)^2 + 2\frac{D_n}{W}\sin \theta + \sin^2 \theta - \left(\cos \theta - \frac{L}{W}\right)^2 \tan^2 \phi = 0 \tag{9}$$

The term, D_s/W, is given by

$$\frac{D_s}{W} = \frac{6D_s}{\rho \pi u_*^2 d^2} \frac{\rho u_*^2 d^2}{(\rho_s - \rho)gd^3} = \hat{D}_s \tau_{*c\theta} \tag{10}$$

in which $\tau_{*c\theta} = \rho u_*^2/(\rho_s - \rho)gd$ = critical Shields stress for side slope with inclination of θ; ρ_s = mass density of the particle; and $\hat{D}_s = 6D_s/\rho\pi u_*^2 d^2$. In the same manner, it is obtained that $D_n/W = \hat{D}_s \tau_{*c\theta} \tan \delta$ and $L/W = \alpha \hat{D}_s \tau_{*c\theta}$. Introducing these relations into Equation (9), an equation with respect to $\tau_{*c\theta}$ is derived:

$$\hat{D}_s^2(1 - \alpha^2 \tan^2 \phi)\tau_{*c\theta}^2 + 2\hat{D}_s \times (\tan \delta + \alpha \cos \theta \tan^2 \phi)\tau_{*c\theta} + (\sin^2 \theta - \cos^2 \theta \tan^2 \phi) = 0 \tag{11}$$

in which derivation a smaller term containing $\tan^2 \delta$ is neglected, since it is an order of 0.01. The positive solution

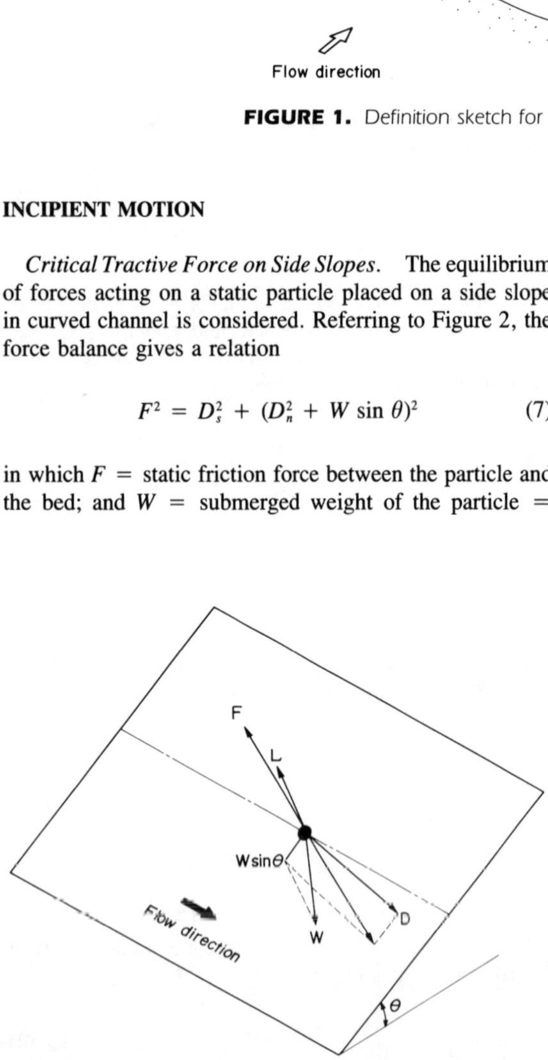

FIGURE 2. Diagram indicating forces acting on a bed material resting on a side slope.

of Equation (11) is

$$\tau_{*c\theta} = \frac{1}{\hat{D}_s} \frac{-\tan\delta - \alpha\cos\theta\tan^2\phi + (2\alpha\tan\delta\cos\theta\tan^2\phi + \tan^2\phi\cos^2\theta + \alpha^2\tan^2\phi\sin^2\theta - \sin^2\theta)^{1/2}}{1 - \alpha^2\tan^2\phi} \quad (12)$$

For level bottom, Equation (12) reduces to

$$\tau_{*c0} = \frac{1}{\hat{D}_s} \frac{-\tan\delta - \alpha\tan^2\phi + (2\alpha\tan\delta\tan^2\phi + \tan^2\phi)^{1/2}}{1 - \alpha^2\tan^2\phi} \quad (13)$$

in which τ_{*c0} = critical Shields stress for level bottom.

To represent the effect of side slope on the critical tractive force, introduction of a slope factor, $K = \tau_{*c\theta}/\tau_{*c0}$, will be useful. From Equations (12) and (13), it follows that

$$K = \frac{-\tan\delta - \alpha\cos\theta\tan^2\phi + (2\alpha\tan\delta\cos\theta\tan^2\phi + \tan^2\phi\cos^2\theta + \alpha^2\tan^2\phi\sin^2\theta - \sin^2\theta)^{1/2}}{-\tan\delta - \alpha\tan^2\phi + (2\alpha\tan\delta\tan^2\phi + \tan^2\phi)^{1/2}} \quad (14)$$

In a special case for which the lift force is negligibly small ($\alpha = 0$) and the secondary flow disappears ($\tan\delta = 0$), Equation (14) reduces to

$$K = \sqrt{1 - \frac{\tan^2\theta}{\tan^2\phi}}\cos\theta \quad (15)$$

which is exactly the same equation as Lane [2] derived.

The validity of Equation (14) was tested with a laterally inclinable wind tunnel [3]. The straight working reach with a length of 4 m and 10 × 30 cm cross-section was designed to rotate laterally from 0° to 40°. The wind tunnel has an advantage that the bed shear stress is nearly uniform both laterally and longitudinally even if the tunnel is laterally tilted. Two kinds of well graded, noncohesive sands were used for the experiment. One has a median diameter of 1.3 mm and the other is a finer one with median size of 0.42 mm, for which the angle of repose was ϕ = 36° and 38°, respectively. In calculating Equation (14), $\tan\delta$ and α should be specified. For straight channels, such as the present case, the secondary flow is essentially absent, and consequently $\tan\delta = 0$. Several laboratory works [4,5,6 and 7] have been made on the measurement of the lift force acting on particles laid on beds. These studies show that α varies from about -0.4 to 0.9, and it is treated as a variable parameter at present. The choice of suitable value for α will be discussed subsequently.

Upon specification of ϕ, δ and α, the slope factor, K, can be determined as a function of θ. The slope factor of the 1.3 mm sand bed are plotted in Figure 3 on which the theoretical curves calculated from Equation (14) for values of α = 0.0 and 0.4 are depicted. It is revealed that the curve of α = 0.0 accurately describes the actual, and the curve of α = 0.4 draws the lower boundary of the incipient motion. The same tendency is found for the 0.42 mm sand (Figure 4).

Critical Tractive Force on Level Bottoms. Test of the validity of the critical tractive force on level bottom expressed by Equation (13) is worth describing. Since $\delta = \theta = 0$ for level bed of straight channels, Equation (13) reduces to

$$\tau_{*c0} = \frac{\tan\phi}{\hat{D}_s(1 + \alpha\tan\phi)} \quad (16)$$

In estimating the dynamic fluid force, $\hat{D}_s$, in Equation (16),

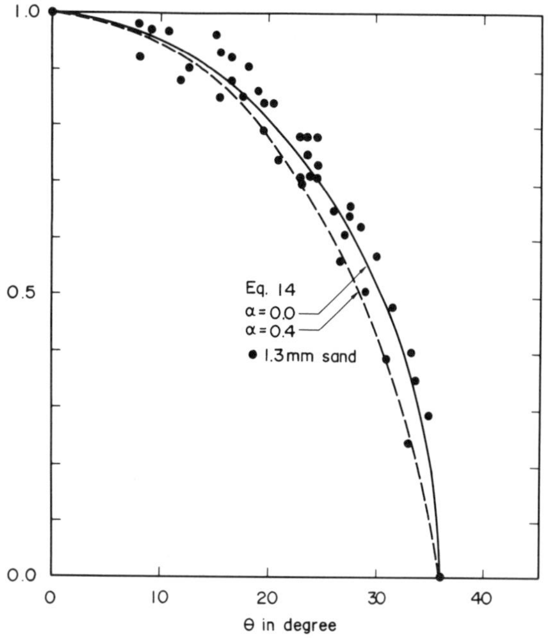

FIGURE 3. Test of slope factor for 1.3 mm sand.

the location of virtual origin of velocity profile should be determined. Coleman [8] reported that the virtual origin of the approach flow to spheres placed on sphere-packing geometry lies $1.016d$ below the top of the spheres. Observation of actual sand beds is supportive of the assumption on the particle configuration. The origin is therefore taken at the bottom of upper particles (Figure 5). The dynamic force, $\hat{D}_s$, exerting of the upper static particle can be calculated using Equations (1) and (2). Assuming $u_b = u_{bs}$, the result is

$$\hat{D}_s = \frac{6D_s}{\rho\pi u_*^2 d^2} \simeq \frac{3}{4} C_D \frac{u_{bs}^2}{u_*^2} = \frac{3}{4} C_D f^2 \tag{17}$$

in which $f = u_{bs}/u_*$. The variables f and C_D included in the right hand side of Equation (17) can be estimated as follows:

1. The dimensionless fluid velocity at the center height of the upper particle, f, can be calculated by using the method developed by Rotta [9] in which the effect of viscous sublayer and the increment of mixing length due to roughness are included. For flows with viscous sublayer, the result is

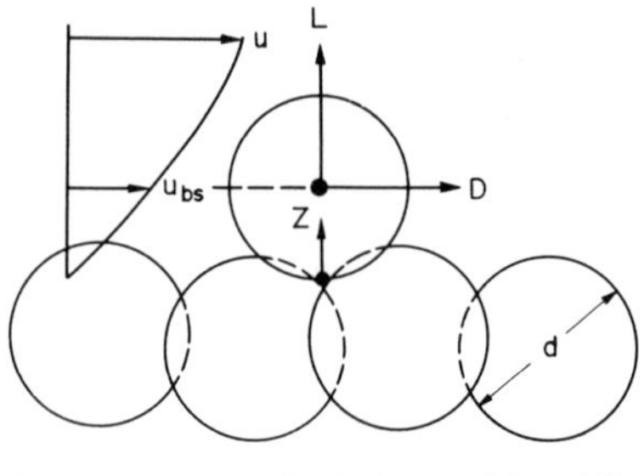

FIGURE 5. Model for bed material and flow.

$$\frac{u}{u_*} = \frac{u_* z}{\nu} \tag{18}$$

$$\frac{u}{u_*} = \frac{1}{\kappa\eta}\left(\frac{1}{2} - \sqrt{\eta^2 + \frac{1}{4}}\right) + \frac{1}{\kappa} \ln 2 \left(\eta + \sqrt{\eta^2 + \frac{1}{4}}\right) + \frac{u_* \Delta}{\nu} \tag{19}$$

in which u = local longitudinal fluid velocity at z; $\eta = lu_*/\nu$; κ = Karman's constant = 0.4; Δ = thickness of viscous sublayer; and l = mixing length = $\kappa(z - \Delta)$ for flows with viscous sublayer. Equation (18) holds in the viscous sublayer, and Equation (19) is applied to the buffer and the inertial zone. For fully rough flows, Rotta's result is

$$\frac{u}{u_*} = \frac{1}{\kappa\eta}\left(\frac{1}{2} - \sqrt{\eta^2 + \frac{1}{4}}\right) - \frac{1}{\kappa\eta_0}\left(\frac{1}{2} - \sqrt{\eta_0^2 + \frac{1}{4}}\right) + \frac{1}{\kappa} \ln \frac{\eta + \sqrt{\eta^2 + \frac{1}{4}}}{\eta_0 + \sqrt{\eta_0^2 + \frac{1}{4}}} \tag{20}$$

in which $\eta_0 = l_0 u_*/\nu$; l_0 = increment of mixing length due to roughness; and thus the mixing length, $l = l_0 + \kappa z$, for fully rough flows. For the inertial zone of tur-

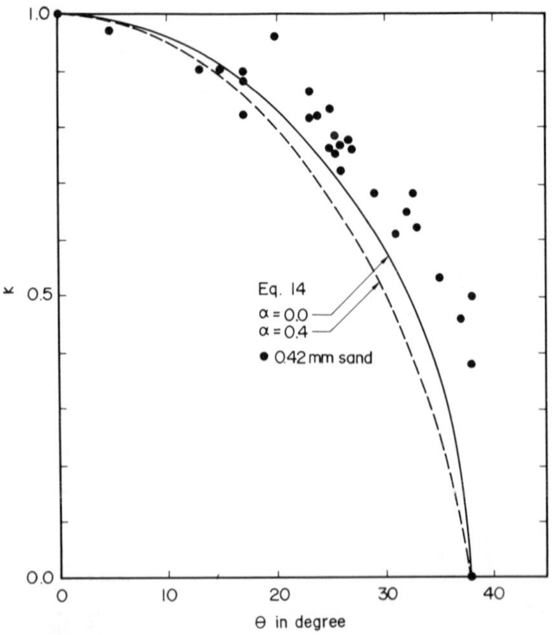

FIGURE 4. Test of slope factor for 0.42 mm sand.

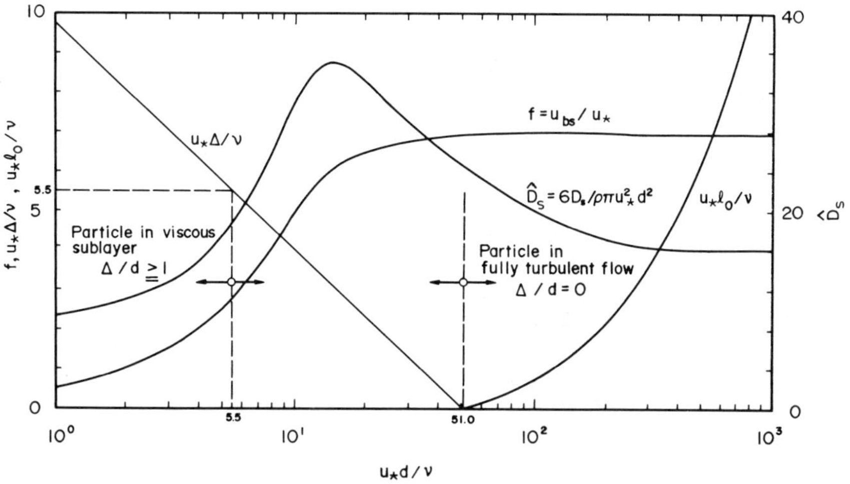

FIGURE 6. Correlation of f, $u_\star\Delta/\nu$, $u_\star \ell_0/\nu$ and $\hat{D}_s$ with $u_\star d/\nu$.

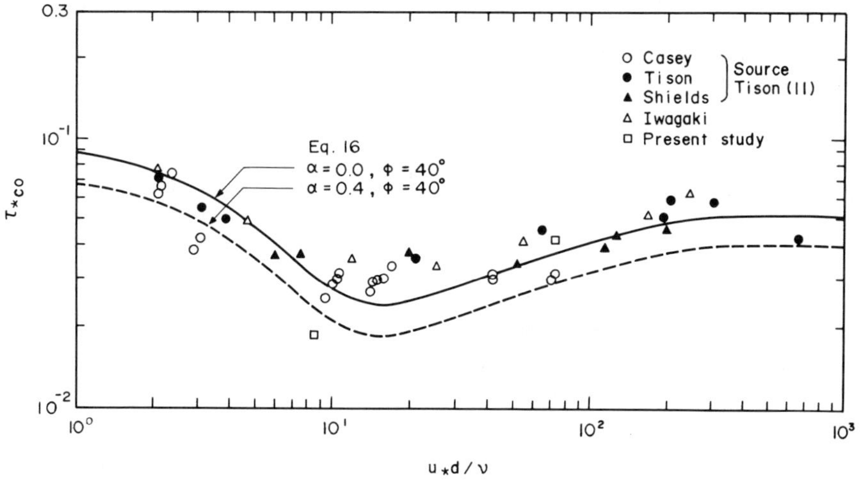

FIGURE 7. Test of critical tractive force on level bottoms.

bulent flow, the velocity profile is generally expressed by

$$\frac{u}{u_*} = A_r + \frac{1}{\kappa}\ln\frac{u_* z}{\nu} - \frac{1}{\kappa}\ln\frac{u_* d}{\nu} \tag{21}$$

in which A_r = constant = 8.5. Comparison of the expanded form of Equation (19) for large η with Equation (21) yields a relation between $u_*\Delta/\nu$ and u_*d/ν, and the same procedure with respect to Equations (20) and (21) produces a relation between u_*l_0/ν and u_*d/ν.

2. The drag coefficient for spheres placed on close-packed spheres, defined against u_{bs}, was found to group around the drag coefficient curve for a sphere in free fall [8], and C_D can be estimated by using the curve for free fall (e.g., Reference 10). The Reynolds number, $u_{bs}d/\nu$, with which C_D is correlated, is equated to fu_*d/ν. C_D is therefore a unique function of u_*d/ν.

   Thus, it is found that the dimensionless drag force, $\hat{D}_s$, is uniquely correlated with u_*d/ν. The dimensionless functions, $\hat{D}_s$, f, $u_*\Delta/\nu$ and u_*l_0/ν, are depicted in Figure 6 as functions of u_*d/ν.

Several data for the critical tractive force on level beds [11 and 12] are plotted in Figure 7 along with theoretical curves of α = 0.0 and 0.4 calculated from Equation (16). In the calculation, the angle of repose, ϕ = 40°, is assumed because it takes the value of about 40° for angular materials. The curve of α = 0.0 accurately describes the actual for a wide range of u_*d/ν, and the curve of α = 0.4 approximately draws the lower boundary of the experimental values on incipient motion for flat beds.

Thus, the effects of lift force seems to be negligible in describing the most probable values of critical tractive force of both level and laterally sloping boundaries as far as the analysis developed here is applied.

LATERAL BED LOAD TRANSPORT

Movement of Sand Particles on Side Slopes. Consider a sand particle moving on a plane inclined by ψ and θ in s and n directions, respectively. Referring to Figure 1, the equations of motion in s and n directions are represented by

$$D_s + (M - m)g\sin\psi - \mu[(M - m)g\cos\theta - L]\frac{v_{ps}}{v_p} = 0 \tag{22}$$

$$D_n + (M - m)g\sin\theta - \mu[(M - m)g\cos\theta - L]\frac{v_{pn}}{v_p} = 0 \tag{23}$$

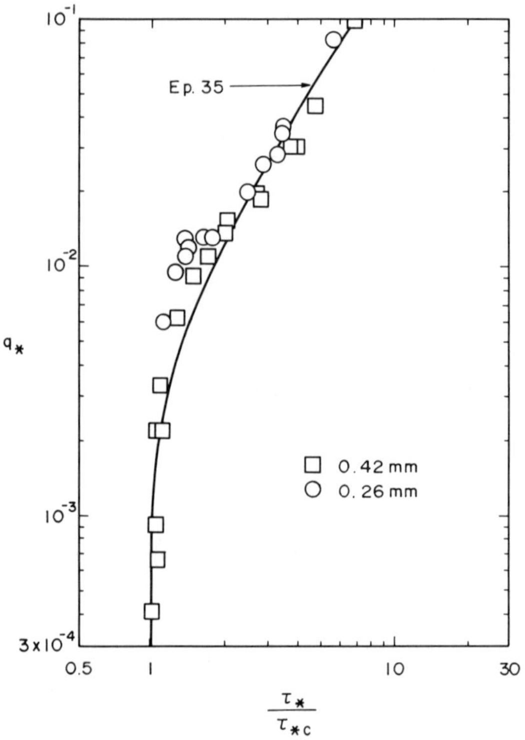

FIGURE 8. Longitudinal bed load, q_*, as a function of τ_*/τ_{*c}.

in which M = mass of spherical sand particle; m = mass of fluid which occupies the same volume as the sand particle; μ = dynamic coefficient of Coulomb friction; and D_s and D_n are given by Equations (2) and (3), respectively. Since $\tan\delta \ll 1$ and $\tan\beta \ll 1$ in Equations (4) and (5), respectively, the following approximations result:

$$\frac{v_{ps}}{v_p} \simeq 1;\ D_s \simeq D \tag{24}$$

Since $\psi \ll \theta \ll 1$, it follows that

$$\sin\psi \simeq 0;\ \cos\theta \simeq 1 \tag{25}$$

Substituting Equations (6), (24) and (25) into Equation (22) and assuming sphericity for the particle, the following relation is derived:

$$u_b - v_p = \sqrt{\frac{4}{3}\frac{\mu R_s g d}{C_D(1 + \alpha\mu)}} \tag{26}$$

in which R_s = submerged specific gravity of sand particles = 1.65 for almost natural sands in water. Substitution of Equations (3) and (26) into (23), with aid of $\sin\theta \cong \tan\theta$ and $\cos\theta \cong 1$, yields an expression for v_{pn}/v_p

$$\frac{v_{pn}}{v_p} = \frac{\sqrt{\frac{3}{4}\frac{\mu C_D}{1+\alpha\mu}}\frac{u_{bn}}{\sqrt{R_s g d}} + \tan\theta}{\sqrt{\frac{3}{4}\frac{\mu C_D}{1+\alpha\mu}}\frac{v_p}{\sqrt{R_s g d}} + \frac{\mu}{1+\alpha\mu}} \tag{27}$$

Eliminating v_p from the right hand side of Equation (27) by using Equation (26), Equation (27) reduces to

$$\frac{v_{pn}}{v_p} = \frac{u_{bn}}{u_b} + \frac{\tan\theta}{\sqrt{\frac{3}{4}\frac{\mu C_D}{1+\alpha\mu}}\frac{u_b}{\sqrt{R_s g d}}} \tag{28}$$

Evaluating u_b at critical conditions for which $v_p = 0$ in Equation (26), it results

$$u_{bc} = \sqrt{\frac{3}{4}\frac{\mu R_s g d}{C_D(1+\alpha\mu)}} \tag{29}$$

in which u_{bc} = near bed fluid velocity at critical conditions. Substitution of Equation (29) into Equation (28) yields

$$\frac{v_{pn}}{v_p} = \frac{u_{bn}}{u_b} + \frac{1+\alpha\mu}{\mu}\frac{u_{bc}}{u_b}\tan\theta \tag{30}$$

Equation (30) correlates the direction of bed load movement with the near bed fluid velocity and the lateral slope.

Lateral Bed Load. A lateral bed load formula is presented herein based on Equation (30). The following rela-

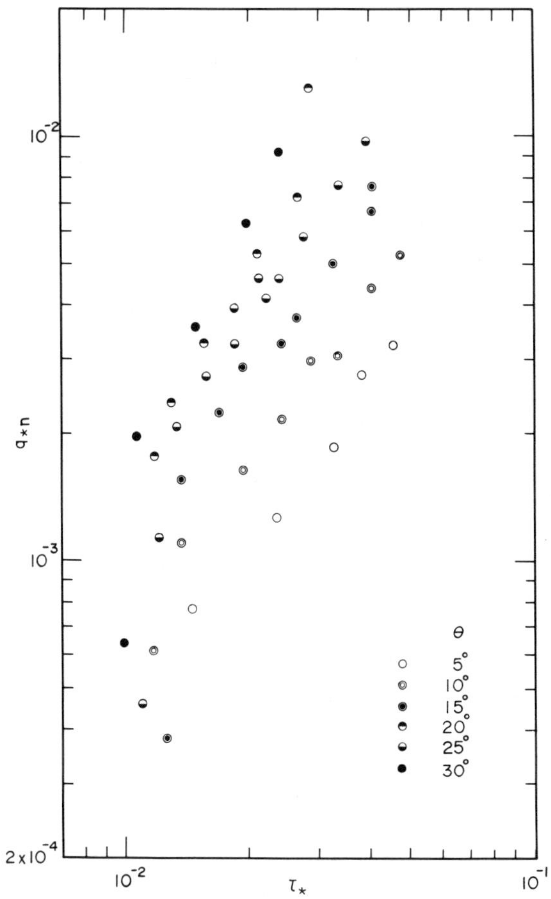

FIGURE 9. Lateral bed load, q_{*n}, plotted against the shields stress, τ_*.

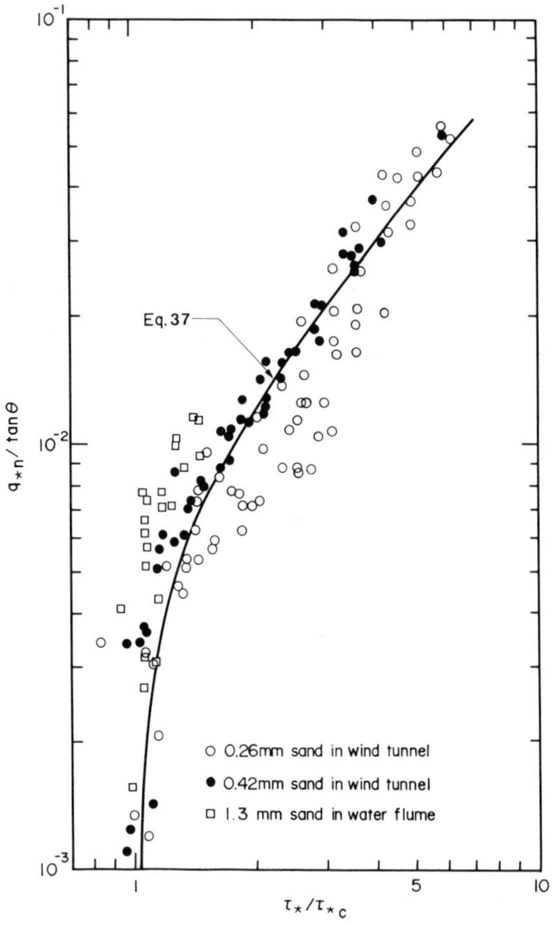

FIGURE 10. Relation between $q_{*n}/\tan\theta$ and τ_*/τ_{*c}.

tions hold in general:

$$\frac{v_{pn}}{v_p} = \frac{q_n}{q} \tag{31}$$

$$\frac{u_{bc}}{u_b} = \sqrt{\frac{\tau_{*c}}{\tau_*}} \tag{32}$$

Then, Equation (30) reduces to

$$\frac{q_n}{q} = \tan\delta + \frac{1 + \alpha\mu}{\mu}\sqrt{\frac{\tau_{*c}}{\tau_*}}\tan\theta \tag{33}$$

in which q = longitudinal volumetric bed load per unit width; q_n = lateral volumetric bed load per unit length; τ_* = Shields stress = $u_*^2/R_s gd$; τ_{*c} = critical Shields stress; and $\tan\delta = u_{bn}/u_{bs} \simeq u_{bn}/u_b$ from Equation (4). The validity Equation (33) is tested in the straight, laterally inclinable wind tunnel, for which it can be reasonably assumed $\tan\delta = 0$. Then, Equation (33) reduces to

$$\frac{q_n}{q} = \frac{q_{*n}}{q_*} = \frac{1 + \alpha\mu}{\mu}\sqrt{\frac{\tau_{*c}}{\tau_*}}\tan\theta \tag{34}$$

in which $q_* = q/\sqrt{R_s gd^3}$; and $q_{*n} = q_n/\sqrt{R_s gd^3}$. Tests were conducted for two kinds of well-graded fine sands [13]. One has a median diameter of 0.42 mm and the other is 0.26 mm sand. In the wind tunnel tests, both q_* and q_{*n} were measured to test the validity of the functional relationship expressed by Equation (34). The dimensionless longitudinal volumetric bed load for level bed, q_*, is plotted against τ_*/τ_{*c} in Figure 8, from which an appropriate expression for q_* is found to take the form [14]

$$q_* = 0.00595\,\frac{\tau_*}{\tau_{*c}}\left(\frac{\tau_*}{\tau_{*c}} - 1\right)^{1/2} \tag{35}$$

Substitution of Equation (35) into Equation (34) provides a relation

$$q_{*n} = 0.00595\,\frac{1 + \alpha\mu}{\mu} \times \left[\frac{\tau_*}{\tau_{*c}}\left(\frac{\tau_*}{\tau_{*c}} - 1\right)\right]^{1/2}\tan\theta \tag{36}$$

As an example, relations between q_{*n} and τ_* measured in the wind tunnel for several values of θ are depicted in Figure 9 (for 0.42 mm sand). It is revealed that q_{*n} is approximately proportional to $\tan\theta$. The same tendency was found for 0.26 mm sand. Using these data, $q_{*n}/\tan\theta$ is plotted against in Figure 10, in which the most probable formula for the lateral bed load is found to take the form

$$\frac{q_{*n}}{\tan\theta} = 0.0085\left[\frac{\tau_*}{\tau_{*c}}\left(\frac{\tau_*}{\tau_{*c}} - 1\right)\right]^{1/2} \tag{37}$$

Equation (37) reduces to Equation (36) if it is assumed $(1 + \alpha\mu)/\mu = 1.43$, which is a reasonable value since $(1 + \alpha\mu)/\mu$ is expected to take a value slightly larger than unity. Thus, the functional relationship proposed in Equation (34) is proved to be correct.

It should be noted that the longitudinal bed load defined in Equation (35) is valid for air flows. Therefore, for water flows another bed load formula such as the Einstein bed load formula should be used for q_*.

REFERENCES

1. Kikkawa, H., Ikeda, S., and Kitagawa, A., "Flow and Bed Topography in Curved Open Channels," *Journal of the Hydraulics Division,* ASCE, Vol. 102, No. HY9, 1976, pp. 1327–1342.
2. Lane, E. W., "Design of Stable Channels," *Transactions,* ASCE, Vol. 120, 1955, pp. 1234–1260.
3. Ikeda, S., "Incipient Motion of Sand Particles on Side Slopes," *Journal of the Hydraulics Division,* ASCE, Vol. 108, No. HY1, 1982, pp. 95–114.
4. Chepil, W. S., "The Use of Evenly Spaced Hemispheres to Evaluate Aerodynamic Forces on a Soil Surface," *Transactions,* American Geophysical Union, Vol. 39, No. 3, 1958, pp. 397–404.
5. Coleman, N. L., "A Theoretical and Experimental Study of Drag and Lift Forces Acting on a Sphere Resting on a Hypothetical Stream Bed," *Proceedings of the Twelfth Congress,* International Association for Hydraulic Research, Fort Collins, Colorado, Vol. 3, 1967, pp. 185–192.
6. Davies, T. R. H. and Samad, M. F. A., "Fluid Dynamic Lift on a Bed Particle," *Journal of the Hydraulics Division,* ASCE, Vol. 104, No. HY8, 1978, pp. 1171–1182.
7. Einstein, H. A. and El-Samni, E. A., "Hydrodynamic Forces on a Rough Wall," *Review of Modern Physics,* Vol. 21, 1949, pp. 520–524.
8. Coleman, N. L., "Bed Reynolds Modelling for Fluid Drag," *Journal of Hydraulic Research,* International Association for Hydraulic Research, Vol. 17, No. 2, 1979, pp. 91–106.
9. Rotta, J., "Das in Wandnähe Gültige Beschwindigkeitsgesetz Turbulenter Strömungen," *Ingenieur Archiv,* 18 Band, 1950, pp. 278–280 (in German).
10. Schlichting, H., *Boundary-Layer Theory,* 6th ed., McGraw-Hill Co., New York, N.Y., 1968, pp. 15–17.
11. Tison, L. J., "Studies of the Critical Tractive Force of En-

trainment of Bed Materials,'' *Proceedings of Minnesota International Hydraulic Convention,* International Association for Hydraulic Research, 1953, pp. 21–35.
12. Iwagaki, Y., ''Fundamental Study on Critical Tractive Force,'' *Proceedings of the Japan Society of Civil Engineers,* No. 41, 1956, pp. 1–21 (in Japanese).
13. Ikeda, S., ''Lateral Bed Load Transport on Side Slopes,'' *Journal of the Hydraulics Division,* ASCE, Vol. 108, No. HY11, 1982, pp. 1369–1373.
14. Parker, G., Discussion of, ''Lateral Bed Load Transport on Side Slopes,'' by S. Ikeda, *Journal of Hydraulic Engineering,* ASCE, Vol. 110, No. 2, 1984, pp. 197–203.

Theoretical Foundations of Form Friction in Sand-Bed Channels

M. I. HAQUE*

INTRODUCTION

In fluvial hydraulics it is assumed that the total resistance to flow consists of two main parts: 1) the skin drag associated with the micro-scale roughness elements and 2) the pressure drag induced by the macro-scale bed features such as ripples and dunes. These two parts are frequently referred to as the *skin friction* and the *form friction*, respectively. The separation of total resistance into its constituent parts is conceptually necessary because each part manifests at totally different scale and plays its own distinctive role in the mechanics of alluvial-channel flows. For example, it is believed that the skin drag is primarily instrumental in the transport of bed material along the surface of the channel bed. The pressure drag on the other hand is considered to play a dominant role in the overall resistance to flow. In an analytical theory, therefore, this division of total resistance into its constituent parts is desirable. But the direct measurement or verification of individual components of total resistance is extremely difficult, if not impossible. It is only the total resistance which has been so far an easily observable quantity. Its decomposition into constituent parts is generally accomplished indirectly. It is assumed that the grain (skin) roughness can be computed on the basis of Prandtl-Karman's resistance equations based on Nikuradse's experimental results, or by their subsequent adaptations due to Keulegan [25]. Since Nikuradse's experimental studies in the early part of this century there has been growing tendency, prompted in part by necessity, to regard the grain roughness as a more easily discernible quantity than the form (pressure) roughness. This perception has been partly responsible for the common practice of subtracting grain roughness from the total resistance in order to arrive indirectly at the form roughness. The efficacy of this procedure is no more certain than our own analytical abilities for predicting the shear stresses acting along the surface of bluff bodies immersed in fluid streams. An adequate determination of skin friction in nonuniform turbulent flows in the presence of large roughness elements still remains as one of the most difficult and unsolved problems of fluid mechanics.

A precise division of total resistance into its various components is further hampered by the fact that the grain and the form roughness are not the only constituent parts of the sum total. Of all the components through which the fluid experiences resistance, the grain and the form roughness are, perhaps, the most important; but never should they be mistaken to represent the whole makeup of resistance to flow in alluvial channels. Surface waves and diffused sediments, to mention a few, further complicate the resistance problem. Their presence is often recognized but out of expediency ignored!

The importance of form drag in fluvial hydraulics was recognized in the early fifties by Einstein and Barbarossa [6]. Since then there have been many attempts to solve this problem in the presence of large-scale roughness elements [1,7,14,26,31,33,37–39,41,43]. In view of the complexities of turbulent flows, most of these studies have been empirical in nature, based largely on dimensionless renderings of the experimental data, guided by intuitive reasoning. In view of the usual scatter in the experimental data in fluvial hydraulics, any empirical inferences drawn from a purely statistical regression analysis of these data is naturally subject to skepticism, unless we provide some theoretical framework for interpreting the experimental results.

It is the purpose of this chapter to present the theoretical basis of an analytical theory of form resistance in fluvial hydraulics, and to compare the theoretical results with the experimental observations. In the following sections we address the problem of form friction in the presence of ripples

*School of Engineering and Applied Science, The George Washington University, Washington, DC

and dunes only. They are by far the most common features in nature. In their presence the form friction factor often achieves such a proportion that the other components of resistance appear insignificant in comparison [4,33,41]. An analytical solution of this problem, therefore, has both academic as well as practical importance.

The subject matter of this chapter is organized into two complementary parts. The first treats the geometry of ripples and dunes as a necessary consequence of fluid mechanics and the incompetency of the sand medium to withstand extreme variations of pressure. The second deals with the determination of form resistance of two-dimensional bluff bodies with well defined points of flow separation. Since the separation points in this case do not shift with the intensity of flow, the coefficient of form drag is independent of the Reynolds number and its value depends on the geometry of the flow region only. Neither the viscosity nor the turbulence of flow has any material effect on the coefficient of form drag in this case. It is one of those fortuitous situations in fluid mechanics when a simple mathematical model of flow yields exceptionally good results.

PART 1. GEOMETRY OF RIPPLES AND DUNES

General Remarks

Although ripples and dunes show some subtle differences, their geometrical appearance bears remarkable similarities. Both are characterized by a gently sloping upstream face which is slightly convex toward the fluid, and a downstream face which is inclined roughly at an angle of repose of the granular material. The abrupt change in the surface gradient at the crest causes the flow to separate and form a leeward eddy (Figure 1). The eddy, or the separation of flow, plays a fundamental role in the initiation and development of ripples and dunes. It provides, in the posterior of the stagnation point, the necessary spatial variation of sediment transport rate for the growth and maturity of succeeding bedforms. Among the most discriminating features of ripples and dunes are their skewed shapes and the presence of the leeward eddies. Any mathematical description of flow over these bedforms must, therefore, take into account adequately the phenomenon of flow separation in the rear of

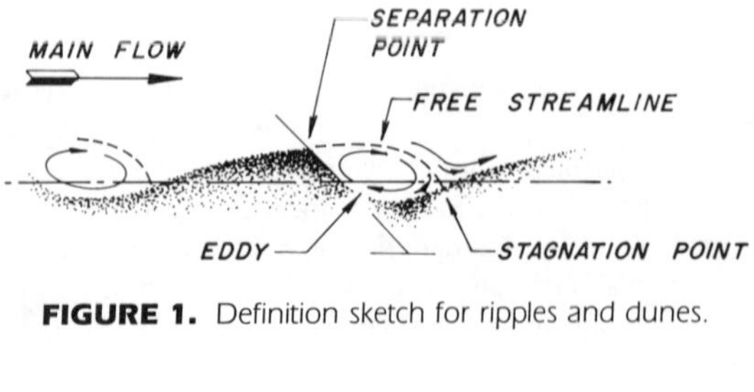

FIGURE 1. Definition sketch for ripples and dunes.

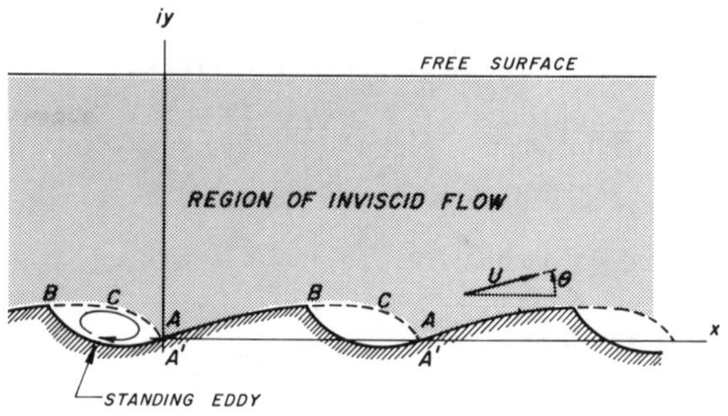

FIGURE 2. Region of inviscid flow.

bedforms. All analytical models of bedform behavior based on the linear theory of instability of flow [9,12,23,35,36,40] fall short of this requirement. Once the separation of flow occurs, the variations of local velocity and pressure in the vicinity of bedforms deviate drastically from those predicted by these analyses.

From the viewpoint of fluid mechanics the flow over a train of ripples or dunes could be regarded as similar to the flow over a train of bluff bodies harboring "standing eddies" in their lee. One of the viable techniques for analysing the steady motion past these bluff bodies is the classical free-streamline method of Helmholtz and Kirchhoff. The Helmholtz-Kirchhoff method is based on the notion of dividing the entire flow region into regions of inviscid motion separated by singular surfaces, consisting partly of solid boundaries and partly of free-streamline surfaces. It is along these singular surfaces that the viscous effects and vorticity are conceived to be concentrated. In the free-streamline surface the pressure and velocity remain constant in the absence of extraneous body forces. The free-streamline method is generally regarded as an adequate procedure for analysing the steady motion of real fluids past bluff bodies at large Reynolds number. The last assertion is made on the premise that as the Reynolds number approaches infinity, the regions of viscous effects shrink into singular surfaces.

Following the Helmholtz-Kirchhoff method, we can analyze meaningfully the flow over a train of bedforms, if proper care is taken to delineate the region of inviscid motion. For a typical flow domain over a sequence of two-dimensional bedforms, the region of inviscid motion and its singular surface are shown in Figure 2. The lower boundary, which represents the singular surface, consists partly of the upstream faces of the bedforms (segments AB in Figure 2) and partly of the free-streamline surfaces detaching from the separation points and terminating at the stagnation points (segments BCA′ in Figure 2). The upper boundary of the inviscid region comprises the free surface which in a mathematical formulation could be at infinity. In this region of inviscid motion, the flow could be described by a complex logarithmic velocity function:

$$W(z) = \ln(Ue^{-i\theta}) = \ln U - i\theta \quad (1)$$

in which

$z = x + iy$, a point in the complex plane of flow
$U = u/u_\infty$, the dimensionless velocity
u_∞ = the undisturbed velocity at infinity
θ = the direction of velocity in radians, measured in counterclockwise direction

The complex velocity function should satisfy the following boundary conditions:

$$\theta = dy/dx \text{ on segments AB} \quad (2)$$

$$U = U_e \text{ on segments BCA}' \quad (3)$$

where

dy/dx = the bedform slope at (x,y)
U_e = the constant velocity acting along the free streamline

In the usual boundary-value problems, the boundary conditions are known a priori and from a mathematical point of view their prescription is arbitrary. However, it is known from previous studies [19,28–30] that an arbitrary assignment of boundary condition on segments AB leads, in general, to a singularity in the flow field at the separation point (point B in Figure 2), where the velocity gradient $\partial U/\partial x$ tends to infinity. Such an abrupt change in velocity is accompanied by a rapidly changing pressure, which cannot be sustained by incompetent granular materials such as sand. Thus from the viewpoint of soil mechanics the presence of this singularity at the crest is unrealistic and its elimination provides a criterion of acceptable bedform shapes. It is this point of view which we intend to exemplify further in the subsequent section.

Flow Over Triangular Wedges

As a point of departure for further discussion, we shall investigate the flow over an infinite sequence of triangular wedges. This problem is of particular importance for at least two reasons: first, it allows us to explore quickly the essential elements of the mathematical analysis without cumbersome details; second, it helps us focus more sharply on the singularity at the separation point, whose existence is otherwise not readily recognized. The material presented in this section is in sufficient detail in order to avoid vagueness which often persists when only the highlights are presented.

Let us consider the flow in a semi-infinite plane $y \geq 0$. The lower boundary of flow comprises an infinite sequence of wedges of height, a, length, b, equispaced at interval L, as shown in Figure 3. Our objective is to find the complex velocity function $W(z)$ which satisfies the appropriate boundary conditions for this problem. Since the problem is periodic with the fundamental period L, the solution need only be investigated in the fundamental strip $0 \leq x \leq L$. To linearize the problem, we shall assume that the wedge slope S $(= a/b)$ is much smaller than unity, so that the boundary conditions can be applied on the real axis, instead of the actual lower boundary of the flow. We are thus seeking a complex function W whose imaginary and real parts satisfy the following conditions:

$$\theta = S \quad \text{on the wedge, } 0 \leq x \leq kL \quad (4)$$

$$\ln U = \ln U_e \quad \text{on the free streamline, } kL \leq x \leq L \quad (5)$$

Here the parameter k denotes the dimensionless length b/L of the wedge, as shown in Figure 3. The boundary conditions 4 and 5 lead to the so-called "mixed" boundary-value problem in which, on some segments of the boundary, the real part of W is specified, while on the remaining parts of the boundary the imaginary part is specified. It is desirable to "unmix" the boundary conditions, for it leads to a more amenable mathematical formulation. This can be accomplished by dividing (or multiplying) the complex velocity function W by another complex function H, which is real on the interval $0 < x < kL$, and purely imaginary on the interval $kL < x < L$. The choice of the function is obviously not unique but the final outcome is independent of H [5]. To

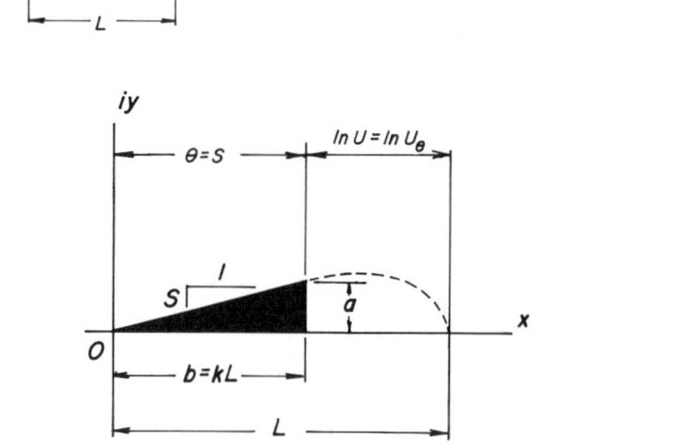

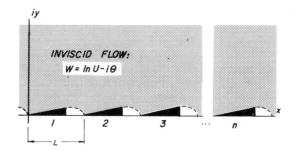

FIGURE 3. Definition sketch for flow over an infinite sequence of triangular wedges.

transform the boundary conditions, we select the following function:

$$H(z) = \left[\cot \frac{\pi}{L} z - \cot k\pi \right]^{1/2} \quad (6)$$

so that the imaginary part of the quotient function

$$\Phi(z) = \phi + i\psi = \frac{W(z)}{H(z)} \quad (7)$$

satisfies the boundary conditions

$$\psi(x,0) = -S \left[\cot \frac{\pi}{L} x - \cot k\pi \right]^{-1/2} \quad \text{on } 0 \leq x \leq kL \quad (8)$$

$$\psi(x,0) = -\ln U_e \left[- \left(\cot \frac{\pi}{L} x - \cot k\pi \right) \right]^{-1/2} \quad (9)$$

$$\text{on } kL \leq x \leq L$$

Next we map the fundamental strip $0 \leq x \leq L$ onto the lower half of a z^*-plane by the following transformation:

$$z^* = \left[\cot \frac{\pi}{L} z - \cot k\pi \right] \quad (10)$$

The purpose of this transformation is to account for the periodicity of the problem. Under this transformation the adjacent strips $nL < x < (n + 1)L$ map onto different branches of an infinitely many n-sheeted Reimann surface, where $n = \pm 1, \pm 2, \pm 3, \ldots \pm \infty$. The salient features of this mapping are shown in Figure 4. Note that under this transformation the separation point B of all wedges plots at the origin $z^* = 0$, the wedges plot along the positive real axis of z^*-plane, and the free streamlines plot along the negative real axis. Now, the function Φ (Equation 7) can also be expressed in terms of the complex variable z^*. Thus in the z^*-plane the imaginary part of

$$\Phi(z^*) = (z^*)^{-1/2} W \quad (11)$$

should satisfy the following boundary conditions:

$$\psi(x^*,0) = -S(x^*)^{-1/2} \qquad x^* > 0 \quad (12)$$

$$\psi(x^*,0) = -\ln U_e(-x^*)^{-1/2} \qquad x^* < 0 \quad (13)$$

This is a typical problem in the theory of functions of a complex variable and its general, piecewise continuous solution is given by the following integral:

$$\Phi(z^*) = \frac{-1}{\pi} \int_{-\infty}^{\infty} \frac{\psi(t,0)}{t - z^*} dt + \phi_\infty \quad (14)$$

where t is a dummy variable of integration and ϕ_∞ is a constant. The integral in the above equation can be evaluated in closed form and the result is

$$\Phi(z^*) = -iS(z^*)^{-1/2} + i \ln U_e(-z^*)^{-1/2} + \phi_\infty \quad (15)$$

which can be transformed back to the z-plane using Equation (10). The result is

$$W(z) = -iS + \ln U_e + \phi_\infty \left[\cot \frac{\pi}{L} z - \cot k\pi \right]^{-1/2} \quad (16)$$

The unknown constants U_e and ϕ_∞ can be evaluated by satisfying the conditions at the free surface, where W vanishes and $\cot \pi z/L = -i$. Making these substitutions we obtain the complex velocity function for the flow field over triangular wedges:

$$W(z) = S \left[\tan \tfrac{1}{2}k\pi - \sqrt{2 \tan \tfrac{1}{2}k\pi \left(\cot \frac{\pi}{L} z - \cot k\pi \right) - i} \right] \quad (17)$$

This concludes the derivation of W. The real part of the above equation gives the dimensionless velocity at any point z in the complex plane of flow, and the imaginary part gives the direction of flow. If we replace z by x in Equation (17), we obtain an expression for the variation of velocity along the lower boundary of flow. The variation of velocity along the surface of a typical wedge with $k = 2/3$ is shown in Figure 5 for three different values of S. It is evident from this figure that $\partial U/\partial x \rightarrow \infty$ as the separation point is approached from the left. Thus, by virtue of the Bernoulli theorem, the pressure gradient $\partial p/\partial x \rightarrow -\infty$. Such a rapid variation in pressure cannot be sustained by sand particles, particularly at the separation point where the granular medium is unconfined from one side.

Some further comments on the existence of singularity are in order. The presence of singularity at the separation point is not unique to triangular wedges. Even in the case of bluff bodies with gradually varying mild surface curvatures, the separation of flow induces the singularity. Only if the body shape is carefully selected, can we ensure the removal of this singularity. Thus the shape of ripples and dunes in a sandy medium is not completely arbitrary. In fact, the kinematics of flow alone imposes severe restrictions on the acceptability of their geometry.

Kinematically Admissible Shapes of Ripples and Dunes

All bedform trains existing in natural watercourses show some degree of randomness in size, shape and celerity of individual bedforms. For theoretical intents and purposes, however, we shall treat them as uniform trains of two-dimensional bedforms of identical shape and size. In such an idealized uniform train the shape of the individual bedform will be considered *kinematically admissible* if it meets the following two criteria:

1. It contains a leeward eddy in its rear with distinct points of separation and re-attachment of flow.
2. The velocity gradient $\partial U/\partial x$ is continuous at the separation point.

The first criterion ensures a more realistic description of the flow field over bedforms such as ripples and dunes. The second criterion is included to preclude the possibility of infinite velocity gradients, which, as we have seen in the previous section, appear generally at the separation points. Two questions of theoretical importance immediately arise. Is it possible to find a family of bedform shapes which meets these criteria? If yes, does it represent a unique set of geometrical shapes? The answer to the second question is not known. We are, however, content with the fact that the existence of at least one family of admissible bedform shapes can be demonstrated. Due to limited space, it is not appropriate to go into the mathematical details of the analysis. We shall, however, discuss at sufficient length the pertinent features of this family of kinematically admissible bedform shape. The reader who is interested in the mathematical analysis should consult the original works [19,28–30].

The family of kinematically admissible bedform shapes is described by the following two equations:

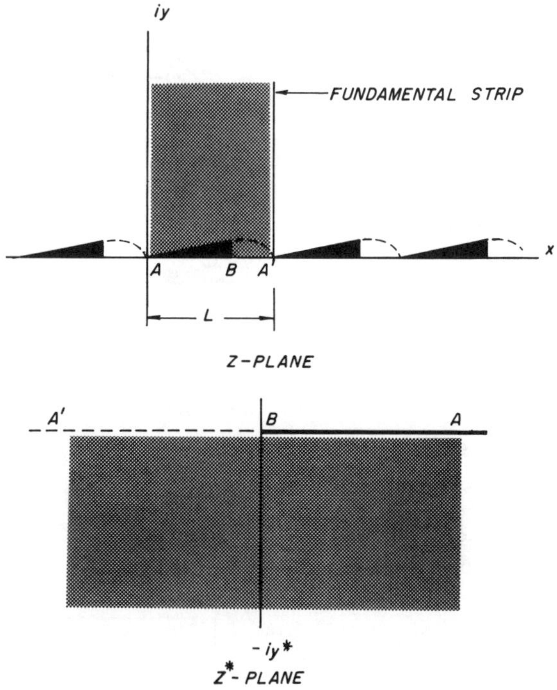

FIGURE 4. Mapping of z-plane onto z*-plane under transformation $z^* = [\cot \pi z/L - \cot k\pi]$.

These equations have been obtained by analyzing the flow over an infinite sequence of bedforms in the semi-infinite plane $y \geq 0$. Equation (18) defines the shape of the upstream face (curves AB in Figure 2) of bedforms. The integration of Equation (19) yields the geometry of the free streamlines. Both equations are valid for arbitrary values of k $(0 < k < 1)$. In these equations the paramater S denotes

$$y = SL\left(\frac{1}{2\pi \sin \frac{k\pi}{2}}\left\{\ln \sin \frac{\pi}{L}\left[x + \frac{L}{2}(1-k)\right] - \ln \sin \frac{\pi}{2}(1-k)\right\} + \frac{x}{L}\right) \qquad (0 \leq x \leq kL) \tag{18}$$

and

$$\frac{dy}{dx} = S\left(1 - \left[2 \tan \frac{k\pi}{2}\left(\cot k\pi - \cot \frac{\pi}{L} x\right)\right]^{1/2} + \frac{1}{2 \sin \frac{k\pi}{2}}\left\{\cot \frac{\pi}{L}\left[x + \frac{L}{2}(1-k)\right] + \frac{\sec^2 \frac{k\pi}{2}\left(\cot k\pi - \cot \frac{\pi}{L} x\right)^{1/2}}{(\csc k\pi)^{1/2}\left(\cot \pi \frac{x}{L} + \tan \frac{k\pi}{2}\right)}\right\}\right) \qquad (kL \leq x \leq L) \tag{19}$$

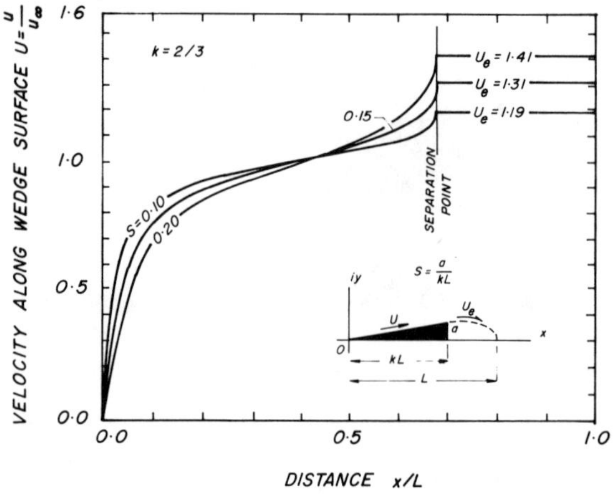

FIGURE 5. Variation of velocity along the surface of wedge (k = 2/3).

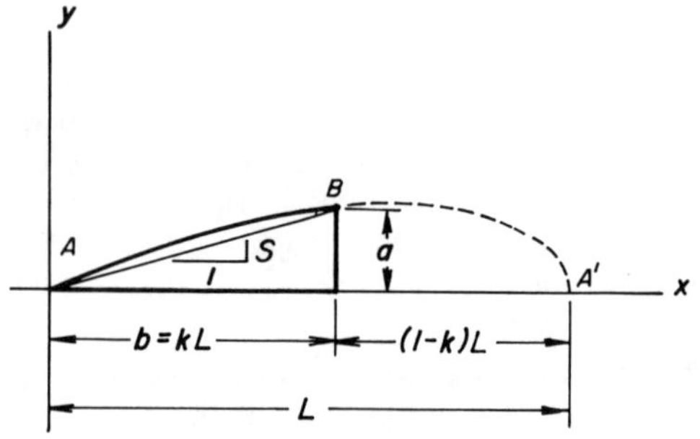

FIGURE 6. Definition sketch.

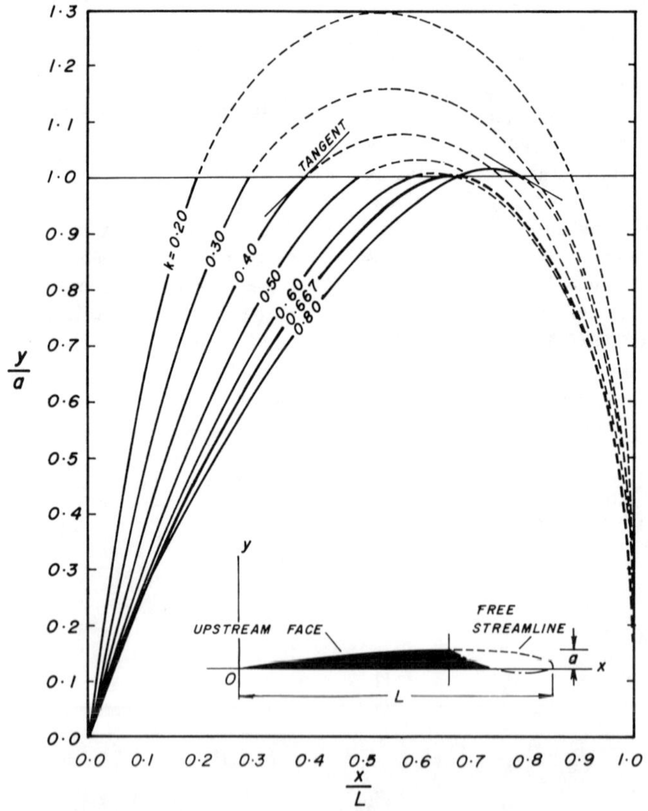

FIGURE 7. A family of kinematically admissible shapes for ripples.

the wedge slope a/b (Figure 6). Its value is arbitrary but small within the context of a linear theory.

The nondimensional shapes of the kinematically admissible bedforms are shown in Figure 7 for various values of k. In this figure the abscissa is normalized by the bedform length L and the ordinate by the wedge height a. The solid curves are the graphs of Equation (18) and the dotted curves, which represent the free streamlines, are obtained by numerical integration of Equation (19).

In order to study the mechanics of ripples and dunes we must examine the variation of velocity along the bedform surface. The velocity u that acts along the upstream face is given by

$$\ln\left(\frac{u}{u_\infty}\right) = S\left\{\tan\frac{k\pi}{2} - \left[2\tan\frac{k\pi}{2}\left(\cot\pi\frac{x}{L} - \cot k\pi\right)\right]^{1/2} + \frac{\left(1-\sec\frac{k\pi}{2}\right)}{2\sin\frac{k\pi}{2}} + \frac{\left(1+\tan^2\frac{k\pi}{2}\right)\left(\cot\pi\frac{x}{L} - \cot k\pi\right)^{1/2}}{2\sin\frac{k\pi}{2}(\csc k\pi)^{1/2}\left(\cot\pi\frac{x}{L} + \tan\frac{k\pi}{2}\right)}\right\} \qquad (20)$$

$$(0 \le x \le kL)$$

The constant velocity U_e acting along the free streamline is given by

$$\ln\left(\frac{u}{u_\infty}\right) = S\left[\tan\frac{k\pi}{2} + \frac{\left(1-\sec\frac{k\pi}{2}\right)}{2\sin\frac{k\pi}{2}}\right] \qquad (21)$$

$$(kL \le x \le L)$$

The effect of bedform geometry on the behavior of local velocity acting along the bedform surface is shown graphically in Figures 8 and 9. As a brief reminder we reiterate that the velocity gradient $\partial U/\partial x$ must be continuous at the separation point for all kinematically admissible bedform shapes. It is clear from Figures 8 and 9 that this condition is satisfied by the cases shown thereon. We can also differentiate Equation (20) with respect to x and evaluate the partial derivative at $x = kL$ to demonstrate in a more general manner that $\partial U/\partial x = 0$ at the separation point for all members of the kinematically admissible family defined by Equations (18) and (19). This fact has some further theoretical implications as explained in the following section.

Sediment Continuity, Crest Slope and Bedform Maturity

Let us consider the longitudinal trace of a two-dimensional train of bedforms as shown in Figure 10. Let 0-x-t be a fixed frame of reference where x denotes the horizontal distance and t the time. The bed elevation from an arbitrary datum is denoted by $y(x,t)$ and the volumetric sediment transport rate (including voids) per unit time per unit width by $q(x,t)$. The conservation of sediment mass immediately leads to the continuity equation

$$\frac{\partial q}{\partial x} = -\frac{\partial y}{\partial t} \qquad (22)$$

which was originally derived by Exner [11]. For further discussion it is convenient to express the above equation with respect to a moving frame of reference $0'-x'-t'$, whose origin $0'$ is translating with velocity c in the positive x-direction. In the moving frame of reference, let t' denote the time. In general both x and t are functions of x' and t'. The variables of the two reference frames are related by

$$x' = x - ct \qquad (23)$$

$$t' = t + t_0 \qquad (24)$$

where t_0 denotes some constant. In the moving frame of reference $y = y'(x',t')$, and $q = q'(x',t')$. It can be shown by chain rule of differentiation that under the given coordinate transformation, the continuity equation becomes

$$\frac{\partial q'}{\partial x'} = c\frac{\partial y'}{\partial x'} - \frac{\partial y'}{\partial t'} \qquad (25)$$

This can also be written as

$$\frac{\partial q}{\partial x} = c\frac{\partial y}{\partial x} - \frac{\partial y'}{\partial t'} \qquad (26)$$

which was previously derived by Mercer [28] in the context of the bedform mechanics.

Now, let us consider the case in which the bed material is transported mainly by local velocity along the bed surface. We can assume in this case that q is a much stronger function of local velocity than other variables such as viscosity, turbulence, bed slope, etc. Thus q can be regarded as a

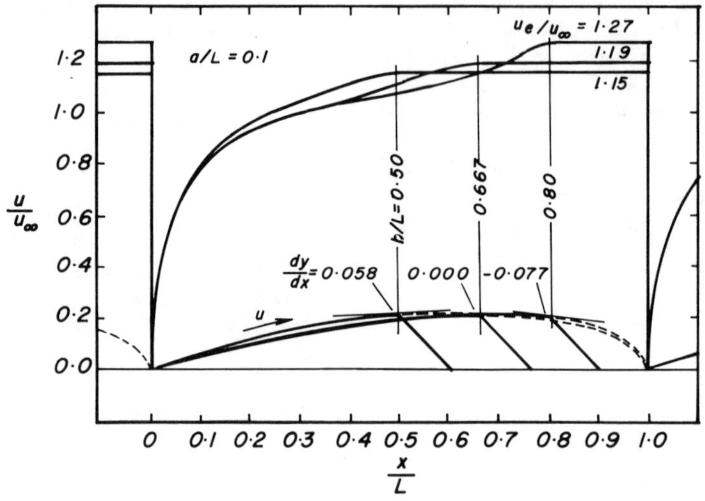

FIGURE 8. The effect of k on the variation of velocity along the bedform surface (a/L = 0.1).

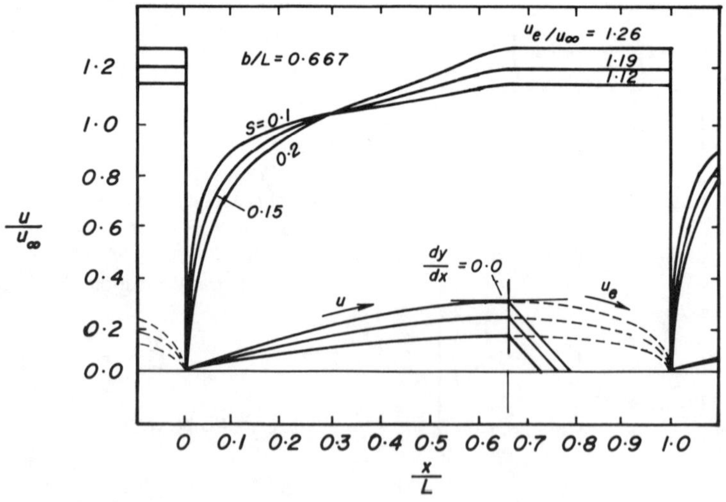

FIGURE 9. The effect of slope on the variation of velocity along the bedform surface (k = 2/3).

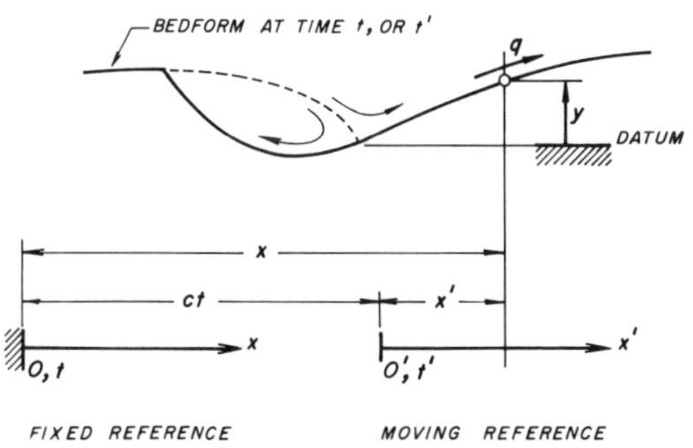

FIGURE 10. Definition sketch for continuity equation.

function of U only,

$$q = q(U) = q[U(x)] \tag{27}$$

and by chain rule we obtain

$$\frac{\partial q}{\partial x} = \frac{\partial q}{\partial U}\frac{\partial U}{\partial x} \tag{28}$$

Since $\partial U/\partial x = 0$ at the separation point of *every* kinematically admissible bedform, it follows from Equation (28) that $\partial q/\partial x = 0$ must be true at the separation point of each and every admissible bedform. Thus, as a consequence of the kinematical admissibility, we conclude from Equation (26) that at the separation point the following equation must hold:

$$\frac{\partial y'}{\partial t'} = c\,\frac{\partial y}{\partial x} \tag{29}$$

This leads to the conclusion that for those bedforms for which the crest slope (i.e., the bedform slope at the separation point) is positive, the height of the separation point must increase with time. And for those bedforms for which the crest slope is negative, the height of the separation point must decrease with time. Only when the crest slope is zero, the bedform height at the separation point will remain steady.

In order to study the variation of the crest slope, we differentiate Equation (18) with respect to x and evaluate the gradient at the separation point $x = kL$. This yields

$$s_k = S\left[1 - \tfrac{1}{2}\sec\frac{k\pi}{2}\right] \tag{30}$$

where s_k denotes the bedform gradient at the separation point. The expression inside the brackets in Equation (30) is positive for $k < 2/3$, negative for $k > 2/3$, and zero for $k = 2/3$. A similar conclusion can also be drawn by a visual examination of tangents at the separation point in Figure 7. Thus three cases of interest can be identified:

1. Growing bedforms, whose amplitude must increase with time, correspond to $k < 2/3$ and $s_k > 0$.
2. Decaying bedforms, whose amplitude must decrease with time, correspond to $k > 2/3$ and $s_k < 0$.
3. Mature bedform, whose amplitude neither increases nor decreases with time, corresponds to $k = 2/3$ and $s_k = 0$.

Finally, in conclusion, we state that out of all the kinematically admissible bedform shapes it is only the mature bedform which has the possibility of maintaining a steady profile. The question whether it will, indeed, maintain a steady profile in a given situation cannot be settled by just examining the continuity of sediment at the separation point only. For this we must study the continuity along the entire upstream face of the mature bedform. We shall revert to this point later in the next section.

The theoretical results are compared with the experimental observations in Figures 11 and 12, which are reproduced from Reference [21]. The nondimensional shape of the mature, kinematically admissible bedform is compared with the bedform shapes observed in large canals in Figure 11. The dark area in this figure represents the range of admissible bedform shapes corresponding to $0.6 \le k \le 0.667$. The dots represent the observed bedform data. Although there appears considerable visual scatter, the data generally fall within $\pm 10\%$ of the bedform amplitude a. The observed statistical distribution of the parameter k is shown in Figure 12. In this figure the piecewise continuous function represents the empirical probability density function obtained from a sample size of 214 bedforms, whose k-values ranged from 0.167 – 0.844 with a mean of 0.59 and variance of 0.024. The smooth curve represents the equivalent beta-distribution. It is evident from this figure that the observed bedform population has the tendency toward maintaining a mature profile.

Limiting Steepness of Ripples and Dunes

The theory presented so far yields only the dimensionless shape of the kinematically admissible profiles. Neither the bedform length nor the bedform height is determined uniquely. To determine these variables we must examine the complete interaction between the bedform and the ensuing flow field. As bedforms develop, their shape continually adjusts to acquire a "limiting-equilibrium" profile, which maintains its steady shape during movement in the downstream direction. When bedforms are in such a steady state, the sediment transported along their surface must meet the continuity condition at every point on their surface. For steady bedforms, which do not change their shape as they

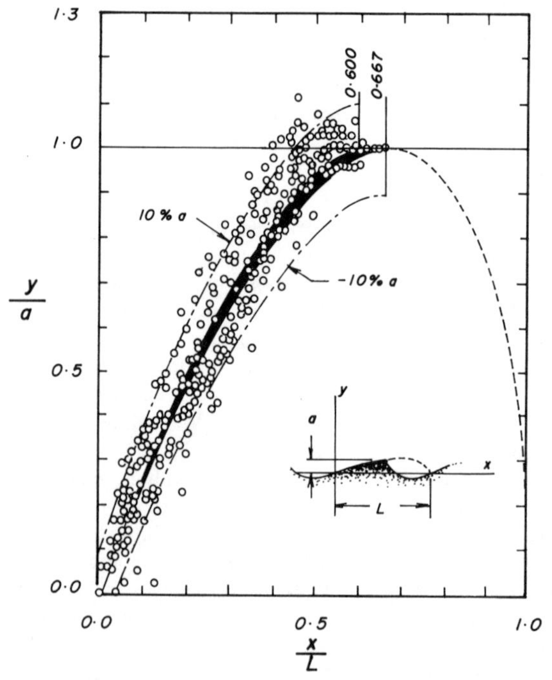

FIGURE 11. Comparison of kinematically admissible shapes with observed dune shapes.

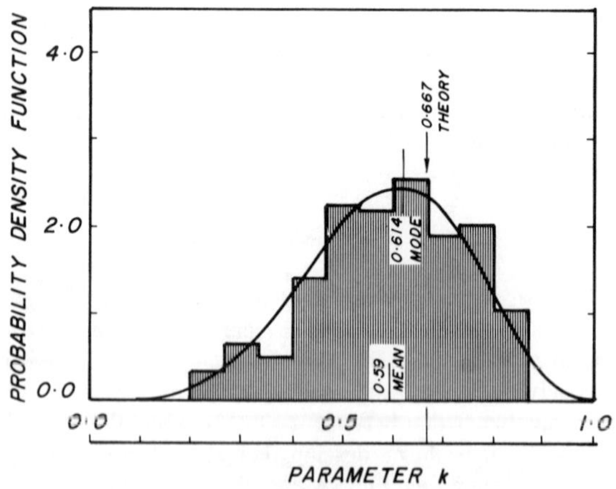

FIGURE 12. Statistical distribution of parameter k.

migrate, the continuity Equation (26) can be integrated to yield

$$q = cy \tag{31}$$

which can be further expressed in a normalized form:

$$\frac{q}{q_c} = \frac{y}{a} \tag{32}$$

where

q_c = the volumetric sediment transport rate per unit time per unit width at the separation point
y = the bedform elevation measured from a datum passing through stagnation points
a = the elevation of the separation point

If we assume that q is proportional to the nth power of the local velocity Equation (32) can be written as

$$\left(\frac{u}{u_c}\right)^n = \frac{y}{a} \tag{33}$$

where

u = the local velocity acting along the bedform surface
u_c = the velocity acting at the separation point

The right-hand side of Equation (33) simply represents the dimensionless shape of the bedform. It will, therefore, be called the *bedform shape function* and will be denoted by N. The left-hand side of Equation (33) is, essentially, a function of local velocity. It will be called the *dimensionless velocity function* and will be denoted by V. For a steady bedform its shape function must, therefore, equal its dimensionless velocity function at every point on its surface. Since, among all the kinematically admissible bedform shapes, only the mature profile can maintain its shape, the shape function N for steady ripples and dunes can be obtained from Equation (18) by substituting $k = 2/3$. Thus in the continuity Equation (33) the right-hand side is a unique function for all steady ripples and dunes, and based on kinematical considerations alone its form is known in advance. It can be easily seen that the shape function is independent of the bedform height, a. The dimensionless velocity function V on the left-hand side of Equation (33) is neither unique nor independent of a. In addition, V depends on the value of the exponent n. For a given value of n, it is possible to find a unique value of the steepness ratio a/L of a mature bedform so that its dimensionless velocity function V gets closest (in some sense) to its shape function N. It is this steepness ratio which the mature bedform must acquire in order to maintain its shape during downstream movement.

The analysis presented so far pertains to the case in which the free surface is at infinity. It is, however, possible to extend the theory to include the case of finite depth of flow, if

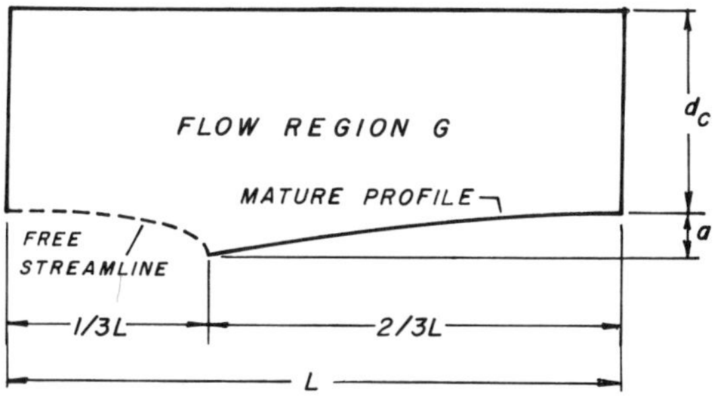

FIGURE 13. A typical flow region over bedform.

some simplifying assumptions are introduced. For instance, it can be assumed that the change in depth of flow has more pronounced effect on V than on N. With this simplification it is possible to analyze the flow in a finite region over a typical mature bedform in order to compute V. Since the effect of depth on N is neglected, the geometry of the flow region (Figure 13) is completely defined by two dimensionless variables d_c/L and a/L, where d_c denotes the depth of flow at the crest. The function V depends now on both the steepness ratio a/L and the relative depth d_c/L. For a given value of d_c/L, we can find a value of a/L, which forces the function V to approach N as closely as possible. This problem has been investigated and the details are given in a forthcoming publication [22]. It is found that for V to remain as close as possible to N, the following relationship must be satisfied:

$$\frac{a}{L} = 0.4\,(n)^{-1.178}\left[1 - \exp\left(-2.5\,\frac{d_c}{L}\right)\right] \quad (34)$$

This shows that the bedforms tend to be steeper in deeper waters. If we assume similarity between velocity profiles in alluvial channels, it can be shown that n equals roughly 6, based on Einstein-Brown's bed material transport equation. Substituting $n = 6$ in the above equation yields $a/L = 0.049$ for bedforms in an infinite depth of flow. This value is within the range of commonly observed values for the maximum steepness of bedforms [45].

PART 2. FORM RESISTANCE

General Remarks on Skin and Pressure Drags

Of the various resistive forces acting on the fluid body, we shall consider only two: the normal and the tangential stresses acting along the wetted surface of the solid boundaries. This restricts us to those circumstances in which there is no spreading of gravity waves, no boundaries that deform, no sediments that diffuse into the fluid body, and no seepage of flow through the permeable boundaries, etc. In this case, there is no fundamental difference between the hydrodynamic behavior of bluff bodies [Figure 14(a)] and that of beforms [Figure 14(b)]. On each elementary area of the wetted surface, there acts a normal force and a tangential force. The normal force is, essentially, due to pressure and it does not depend on the (micro-scale) roughness texture of the wetted surface. The tangential force, on the other hand, depends on the roughness texture.

By examining Figure 14(a), we can see that on the major portion of the surface the tangential force has a strong component in the main flow direction. An integration of this component yields the skin drag (also called the surface or friction drag). Notice that when there is separation of flow, the tangential force on an elementary area in contact with the separation bubble may have a component which acts against the main flow [Figure 14(b)].

Now, consider the normal force acting on the surface. On the anterior part of the body the normal force has a component which acts in the direction of the main flow. On the rear part, with or without flow separation, the component of the pressure force acts always against the main flow. Integrated over the entire surface, this component yields the pressure drag, or the form drag, on the body. When there is no separation of flow, the resultant pressure force on the anterior part almost equals the resultant pressure force on the rear. In this case the pressure drag is practically zero. However, when the flow separates, the pressure at the separation point is generally the lowest, because the velocity in the vicinity of this point is generally the highest. Since it is this pressure which tends to prevail inside the separation zone, there is an imbalance between the resultant pressure forces acting on the anterior and the rear parts of the body. This imbalance causes a net drag on the body. Since there is always separation of flow in the case of ripples and dunes, the pressure drag on these bedforms is usually considerable in magnitude as compared to the skin drag. An idea of pressure variation along the surface of ripples and dunes can be obtained from Figure 15.

Form Resistance of Rigid Ripples and Dunes

In this section we shall discuss the form friction factors of ripples and dunes whose geometry is known in advance, independent of the prevailing flow conditions. We shall again consider an idealized train of two-dimensional bedforms of identical shape and size. Since in this case each roughness element acts independently, the form friction factor can be obtained by integrating the pressure on a single bedform. The possibility of calculating the form friction factor by integrating the pressure on the bedform surface has been suggested previously by Raudkivi [32], Vanoni and Hwang [41] and Mercer [28], among others. However, the idea has been carried out systematically by Haque and Mahmood [20]. We shall review their analysis here.

A schematic representation of pressure variation along a

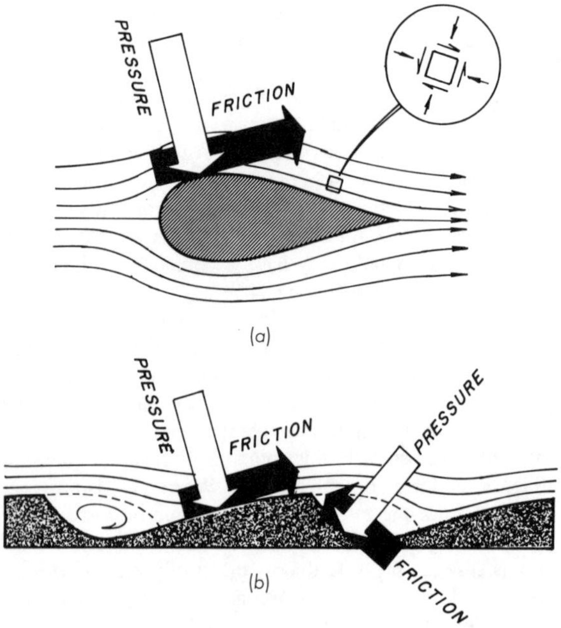

FIGURE 14. Surface tractions on immersed bodies.

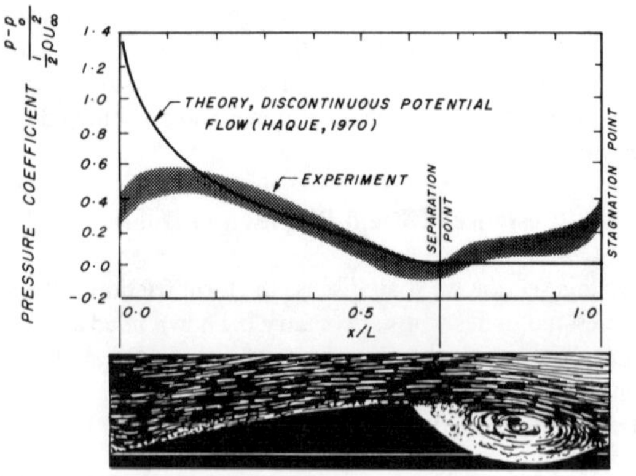

FIGURE 15. Variation of pressure on the surface of a ripple ($a/L = 0.1$; $d_c/L = 0.32, 0.53$).

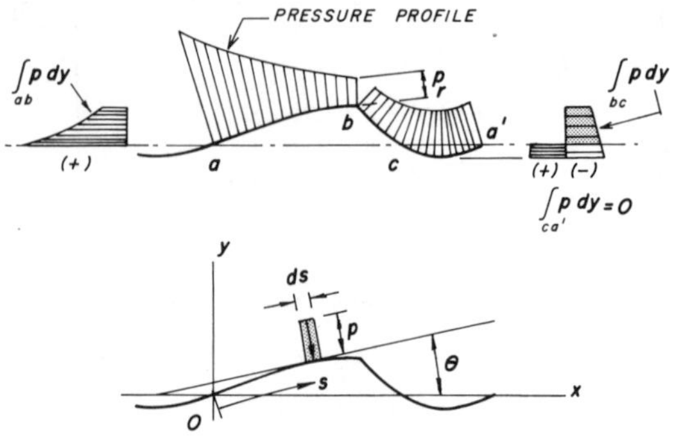

FIGURE 16. Schematic representation of pressure on bedform surface.

two-dimensional ripple or dune surface of unit width normal to the plane of paper is shown in Figure 16. By integrating the downstream component of pressure force acting on an elementary area *ds*, the drag on a single bedform can be obtained:

$$D_p = \int_{abca'} p \sin\theta \, ds \tag{35}$$

where

D_p = the pressure drag per unit width
p = the pressure
θ = the slope
s = the arc length

Since $dy = \sin\theta \, ds$, the above equation can also be written as

$$D_p = \int_{ab} p \, dy + \int_{bc} p \, dy + \int_{ca'} p \, dy \tag{36}$$

where the first, second, and third integrals represent, respectively, the downstream components of pressure forces acting on segmental areas whose traces are identified by *ab*, *bc* and *ca'* in Figure 16. Under the Helmholtz hypothesis of constant pressure inside the separation zone in the absence of extraneous forces—or hydrostatic pressure under gravity—the last integral in Equation (36) can be dropped. The pressure drag can now be written as

$$D_p = \int_{ab} [(p - p_r) - \gamma(y_r - y)] \, dy \tag{37}$$

where γ denotes the specific weight of water, and the subscript r refers to the quantities at an arbitrary reference point (in this study the separation point is taken as the reference point). The above equation indicates that under the assumption of hydrostatic pressure distribution inside the separation zone, the pressure drag on a ripple or a dune depends on the shape of the upstream face only. The configuration of the downstream face does not enter directly into the calculation of the pressure drag. Furthermore, if we assume inviscid flows, the Bernoulli equation can be used to replace the integral in this equation by the kinematic pressure

$$D_p = \int_{ab} \tfrac{1}{2}\varrho(u_r^2 - u^2)\, dy \qquad (38)$$

where u_r denotes the reference velocity at the separation point. The form friction factor can be obtained from

$$f'' = \frac{8\dfrac{D_p}{\varrho L}}{\bar{u}^2} \qquad (39)$$

where

f'' = the form friction factor
ϱ = the mass density of water
$\bar{u}$ = the average velocity based on the average depth of flow

If we know the variation of velocity along the surface of bedform, the determination of f'' becomes a simple exercise in numerically evaluating the integral in Equation (38), and then calculating f'' from Equation (39). In order to determine the velocity field in a typical flow region G, such as shown in Figure 17, the finite element method based on a rotational, inviscid, incompressible flow model can be used quite effectively. The governing equations in this case are

$$\frac{\partial^2\psi}{\partial x^2} + \frac{\partial^2\psi}{\partial y^2} = -\Omega(\psi) \quad \text{in G} \qquad (40)$$

$$\psi = \bar{\psi} \quad \text{On } \partial\text{G} \qquad (41)$$

where

ψ = the stream function
Ω = the vorticity function

and the bar (—) denotes a known function on the boundary. The details of the computational procedure for solving the preceding boundary-value problem can be found elsewhere [20]. We would, however, make some comments on the usefulness of this flow model. The advantages of rotational, inviscid, incompressible flow model are enormous. First and foremost, it is far easier to solve than the Navier-Stokes equations and produces pressure results which agree quite favorably with the experimental observations [30]. In contrast to a potential flow model, it carries the rotational effects, imparted by the oncoming flow, as it accelerates and decelerates over the ripple surface. It naturally does not take into account the shear stresses, nor does it modify the vor-

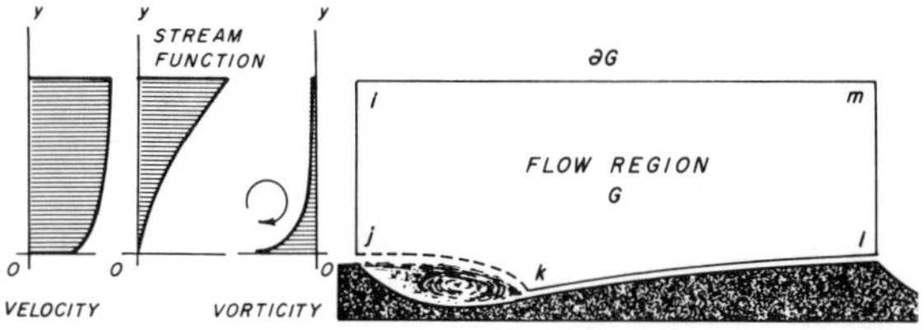

FIGURE 17. Region G and boundary conditions for rotational inviscid incompressible flow model.

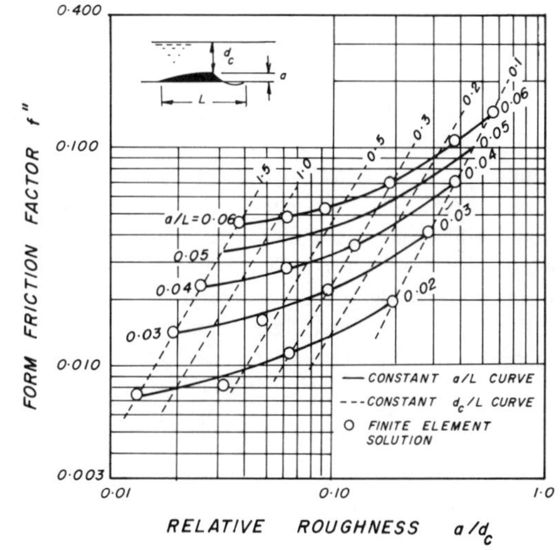

FIGURE 18. Form friction factor f'' as a function of relative roughness a/d_c and steepness ratio a/L.

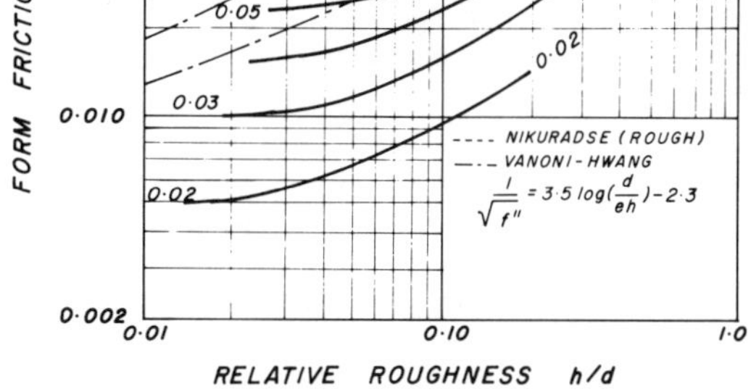

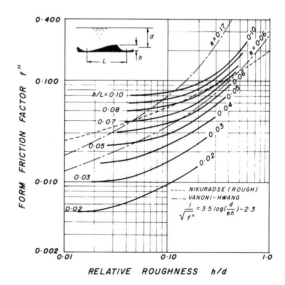

FIGURE 19. Form friction factor as a function of relative roughness h/d and steepness ratio h/L.

ticity vector as it passes over the ripple surface. The cumulative effects of friction of the entire upstream reach on the oncoming flow, is, however, taken into account indirectly by meeting the specified velocity and vorticity conditions at the upstream section ij (Figure 17).

Since f'' in the absence of viscosity does not depend on the intensity of flow, the flow per unit width through the region G can be specified arbitrarily in the preceding boundary-value problem. The form friction factor is also not very sensitive to the shape of the velocity profile assumed at section ij, as shown in Reference [20]. It is, therefore, possible to almost arbitrarily choose some reasonable boundary conditions and solve for velocity and pressure fields. The form friction factor in this case depends on the geometry of the flow region G and not on its absolute size. Since the geometry of flow region depends on two paramaters a/L and d_c/L, the form friction factor is also a function of these parameters; i.e., $f'' = f''(a/L, d_c/L)$. This functional relationship can be established by analyzing the flow in a sequence of flow regions whose geometry is varied parametrically by assigning different values to the parameters a/L and d_c/L. The functional relationship is shown graphically in Figure 18. The abscissa in this figure represents the relative roughness a/d_c and the ordinate f''. Each solid curve corresponds to some constant value of the bedform steepness a/L. By further studying this figure, we can show that in the case of $d_c/L > 0.5$ the form friction factor can be expressed by the following equation:

$$f'' = 4.9\left(\frac{a}{L}\right)^{1.477}\left(\frac{a}{d_c}\right)^{0.176} \cong 5\left(\frac{a}{L}\right)^{3/2}\left(\frac{a}{d_c}\right)^{1/6} \tag{42}$$

Thus for relatively small dunes the form friction factor is a more sensitive function of bedform steepness than of relative roughness.

The analytical curves of Figure 18 are transformed into a set of new curves, which are shown in Figure 19. In this figure the variable h denotes the trough-to-crest height of the bedform, and d the average depth of flow. In transforming the curves from Figure 18 to Figure 19, it has been assumed that $a = 0.8h$. For comparison purpose, Nikuradse's and Vanoni-Hwang's empirical equations are also shown in this figure.

Figure 20 shows the comparison between the theory and the experimental observations in four different plots. The shaded areas in these plots indicate the range of theoretical values. The data are drawn from various sources, covering laboratory flumes to large channels, as indicated on the figure. Each plot represents a certain range of bedform steepness h/L; for instance, the top-left plot is for bedforms with steepness ratio $0.1 \geq h/L > 0.8$, while the bottom-right for $0.04 \geq h/L > 0.02$. Despite some scatter—which is within the range of usual scatter in fluvial hydraulics—the trend of the theoretical curves is remarkably borne by the experimental data.

Form Resistance of Limiting-Equilibrium Ripples and Dunes

In this section we shall discuss the form resistance in the presence of those ripples and dunes whose geometry is affected by the prevailing flow conditions. We shall assume the bedforms are in a state of limiting equilibrium.

In the previous section we have seen that f'' for rigid bedforms depends on two independent, dimensionless parameters d_c/L and a/L. However, in the case of steady bedforms in a limiting-equilibrium state these two parameters are no more independent because a, d_c and L are now related by Equation (34). The constraints imposed by Equation (34) are shown in Figure 21 by dotted curves for various assumed values of n. The solid curves are the same as shown in Figure 19. In plotting the dotted curves we have again assumed that $a = 0.8h$. Each chain-dotted line represents a constant relative depth d/L. The region extending to the left represents smaller bedforms relative to the depth of flow, and the region to the right represents larger bedforms.

Let us examine Figure 21 more closely, because it promises some interesting theoretical explanations for the intriguing bedform behavior and their influence on the resistance problem. To facilitate discussion, we shall assume that f'' is roughly equal to f (the total friction factor) in the presence of ripples and dunes. If we assume that $n = 6$ is a good representative value in the power law of sediment transport rate, then it follows from completely theoretical considerations that the friction factor for channels with relatively small bedforms (length < depth) is about 0.03, as shown in the figure. For most sand-bed channels the observed friction factors fall around this value. For a constant value of n, the friction factor attains a maximum value in the vicinity where the bedform length is roughly equal to the flow depth. In the case of a uniform flow the friction factor is constant and it is represented by a horizontal line in this figure. This line intersects the dotted curve at two points R and D, as shown in the diagram. Thus, in a uniform flow (in an average sense over a sufficiently long channel reach) there are two different lengths of bedforms, one corresponding to the point R and the other to the point D, which can possibly coexist in a channel without disturbing the (average) uniformity of flow. The point R represents the smaller bedform and the point D the larger bedform. For smaller bedforms (length < depth of flow) the dotted curves are almost horizontal, indicating the fact that a variety of small bedforms can possibly coexist in a uniform flow.

When a point A moves along a dotted curve, it represents the variation in friction factor due to changes in the geometry of a sedimentary ripple, or dune, which continually maintains a limiting-equilibrium profile. When such a point

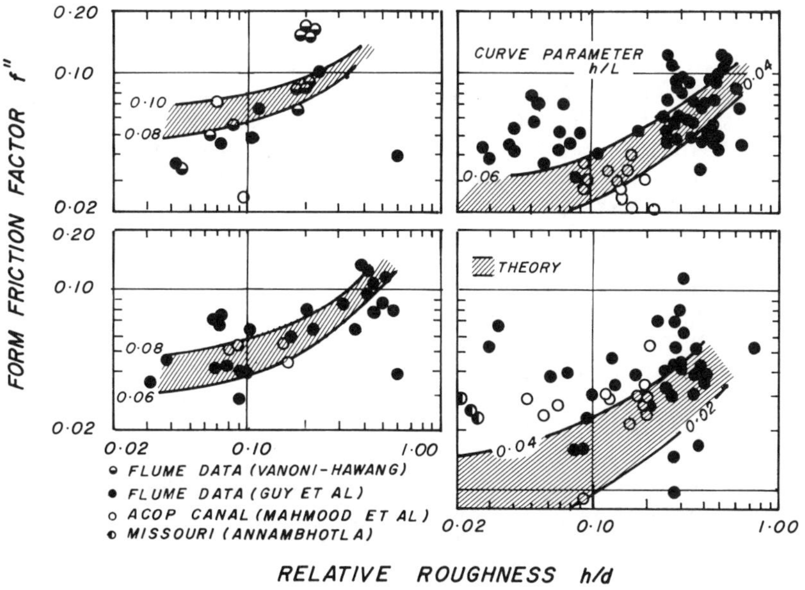

FIGURE 20. Comparison of theory with observed data.

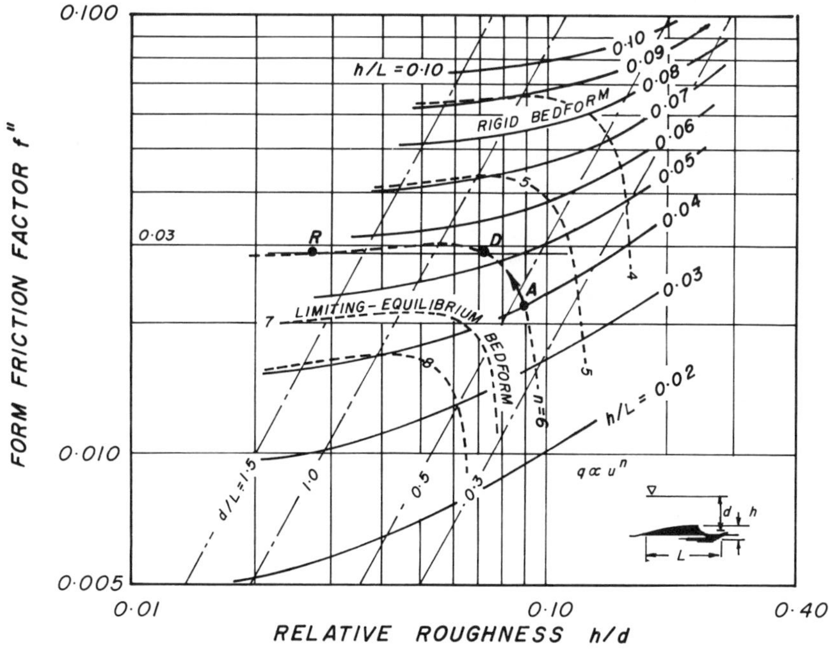

FIGURE 21. Bedform friction factor interaction diagram.

moves along a solid curve, it describes the variation in friction factor of a rigid bedform (with a fixed steepness) due to changes in the flow depth. In a region extending toward left, a dotted curve tends to approach asymptotically a solid curve. This implies that for small bedforms in limiting equilibrium an increase in the flow depth does not cause any change in the bedform steepness. Thus friction factor decreases in this case whenever the flow depth increases. This is in agreement with the experimental observation with regard to ripples [39]. Now, consider the larger bedforms ($L > 2d$). In this case an increase in depth steepens the bedform and thus results in a larger value of f'', as shown by the arrow at point A in Figure 21. This again is in agreement with the experimental observation with regard to dunes [39].

Although this figure explains the physics of the bedform behavior quite remarkably, there are no direct methods available at present to relate the exponent n to other easily measurable flow and sediment variables. In this regard the need for additional research could hardly be overemphasized. Despite these comments, we hope that the analytical development presented in the foregoing will provide a useful theoretical framework for future studies.

NOTATION

a = bedform height measured from stagnation point to separation point
b = length of upstream face of bedform
c = celerity of bedform
D_p = pressure drag on a single bedform
d = average depth of flow
d_c = depth of flow at the crest of bedform
f = total friction factor
f'' = form friction factor
G = inviscid flow region over a typical bedform
h = bedform height measured from trough to crest
k = dimensionless length of the upstream face, b/L
L = bedform length
N = shape function
n = a dimensionless exponent
p = pressure
p_r = reference pressure at the separation point
q = sediment transport rate per unit time per unit width
q_c = sediment transport rate at the separation point
S = wedge slope, a/b
s_k = bedform slope at the separation point
t, t' = time
t_0 = reference time
U = dimensionless velocity, u/u_∞
U_e = velocity along the free streamline
u = local velocity at a point
u_c = velocity at the separation point
u_r = reference velocity at the separation point, same as u_c
u_∞ = undisturbed velocity at infinity
$\bar{u}$ = average velocity of flow
V = dimensionless velocity function for bedform
W = complex velocity function
x,y = coordinates of a generic point
x' = x-coordinate in a moving reference frame
y' = bedform elevation in moving reference frame
z = a point in complex plane
z^* = a point in complex z^*-plane

Greek Symbols

γ = specific weight of water
θ = slope
ϱ = mass density of water
Φ = a complex function
ψ = imaginary part of Φ, or stream function
Ω = vorticity function

REFERENCES

1. Alam, M. Z. and J. F. Kennedy, "Friction Factors for Flow in Sand-Bed Channels," *Journal of Hydraulic Div., ASCE, 95* (HY6), 1973–1992 (Nov. 1969).
2. Anderson, A. G., "The Characteristics of Sediment Waves Formed by Flow in Open Channels," *Proceedings of the Third Mid-Western Conference on Fluid Mechanics,* University of Minnesota, Minneapolis, MN, 379–395 (1953).
3. Annambhotla, V. S., W. W. Sayre, and R. H. Livesey, "Statistical Properties of Missouri River Bed Form," *Journal of the Waterways Harbors and Coastal Engineering Div., ASCE, 98,* 489–510 (Nov. 1972).
4. Brooks, N. H., "Mechanics of Stream With Movable Beds of Fine Sands," *Transactions, ASCE, 123,* 526–594 (1958).
5. Cheng, H. K. and N. Rott, "Generalization of Inversion Formula of Thin Airfoil Theory," *Journal of Rational Mechanics and Analysis,* Indiana University, Bloomington, 357–382 (1954).
6. Einstein, H. A. and N. Barbarossa, "River Channel Roughness," *Transactions, ASCE, 117,* 1121–1146 (1952).
7. Engelund, F., "Hydraulic Resistance of Alluvial Streams," *Journal of Hydraulics Div., 92* (HY2), ASCE, 315–326 (March 1966).
8. Engelund, F., "Hydraulic Resistance of Alluvial Streams," *Closure, Journal of Hydraulics Div., ASCE,* 287–296 (July 1967).
9. Engelund, F., "Instability of Erodible Beds," *J. Fluid Mechanics, 42* (2), 225–244 (1970).
10. Engelund, F. and J. Fredsoe, "Sediment Ripples and Dunes," *Annual Review of Fluid Mechanics, 14,* M. V. Dyke, ed. (1982).
11. Exner, F. M., "Uber die Wechselwirkung Zwischen Wasser

und Geschiebe In Flussen," Sitzenberichte der Academie der Wissenschaften, Wein, Osterreich h, Heft 3-4 (1925).

12. Fredsoe, J., "On the Development of Dunes in Erodible Channels," *J. Fluid Mechanics, 64* (1), 1-16 (1974).
13. Fredsoe, J., "Shape and Dimensions of Stationary Dunes in Rivers," *Journal of the Hydraulics Div., ASCE, 108* (HY8), 932-947 (Aug. 1982).
14. Garde, R. J., and K. G. R. Raju, "Resistance Relationships for Alluvial Channel Flows," *Journal of Hydraulics Div., ASCE, 92* (HY4), 77-100 (July 1966).
15. Gilbert, G. K., "The Transport of Debris by Running Water," U.S. Geological Survey Professional Paper 86 (1914).
16. Gill, M. A., "Height of Sand Dunes in Open Channel Flows," *Journal of Hydraulics Div., ASCE, 97* (HY12), 2067-2074 (Dec. 1971).
17. Graf, W. H., *Hydraulics of Sediment Transport,* McGraw-Hill Inc. (1972).
18. Guy, H. P., D. B. Simons, and E. V. Richardson, "Summary of Alluvial Channel Data From Flume Experiments, 1956-61," Geological Survey Professional Paper No. 462-I, Washington, D.C. (1966).
19. Haque, M. I., "Analytically Determined Ripple Shapes," Thesis presented to Colorado State University, Fort Collins, CO, in partial fulfillment of the requirements for the degree of Master of Science (1970).
20. Haque, M. I. and K. Mahmood, "Analytical Determination of Form Friction Factor," *Journal of Hydraulic Engineering, ASCE, 109* (4), 590-610 (April 1983).
21. Haque, M. I. and K. Mahmood, "Geometry of Ripples and Dunes," *Journal of Hydraulic Engineering, ASCE, 111* (1), 48-63 (Jan. 1985).
22. Haque, M. I. and K. Mahmood, "Analytical Study on Steepness of Ripples and Dunes," *Journal of Hydraulic Engineering, ASCE,* 112 (3) (March 1986).
23. Kennedy, J. F., "The Mechanics of Dunes and Antidunes in Erodible-bed Channels," *J. Fluid Mechanics, 16,* 521-544 (1963).
24. Kennedy, J. F., "The Formation of Sediment Ripples, Dunes and Antidunes," *Annual Review of Fluid Mechanics, 1,* 147-168 (1969).
25. Keulegan, G. H., "Laws of Turbulent Flow in Open Channels," *Journal of Research, National Bureau of Standards, 21,* 707-741 (Dec. 1938).
26. Laursen, E. M., "The Total Sediment Load of Streams," *Journal of Hydraulics Div., ASCE,* HY1, 1530-1536 (Feb. 1958).
27. Liu, H. K., "Mechanics of Sediment Ripple Formation," *Journal of Hydraulics Div., ASCE, 83* (HY2), 1-23 (1957).
28. Mercer, A. G., "Characteristics of Sand Ripples in Low Froude Number Flow," Dissertation presented to University of Minnesota in partial fulfillment of the requirements for the degree of Doctor of Philosophy (June 1964).
29. Mercer, A. G., "Analytically Determined Bed-Form Shape," *Journal of Engineering Mechanics Div., ASCE, 97* (EM1), 175-180 (Feb. 1971).
30. Mercer, A. G. and M. I. Haque, "Ripple Profiles Modeled Mathematically," *Journal of Hydraulics Div., ASCE, 99* (HY3), 441-459 (March 1973).
31. O'Loughlin, E. M., "Resistance to Flow Over Boundaries with Small Roughness Concentrations," Dissertation presented to University of Iowa, Iowa City, in partial fulfillment of the requirements for the degree of Doctor of Philisophy (Aug. 1965).
32. Raudkivi, A. J., "Study of Sediment Ripple Formation," *Journal of Hydraulics Div., ASCE, 89* (HY6), 15-33 (Nov. 1963).
33. Raudkivi, A. J., "Analysis of Resistance in Fluvial Channels," *Journal of Hydraulics Div., ASCE, 93,* 73-84 (Sep. 1967).
34. Raudkivi, A. J., *Loose Boundary Hydraulics,* Pergamon Press Ltd., First edition (1967).
35. Reynolds, A. J., "Waves on the Erodible Bed of an Open Channel," *J. Fluid Mechanics, 22,* 113-133 (1965).
36. Richards, K. J., "The Formation of Ripples and Dunes on an Erodible Bed," *J. Fluid Mechanics, 99* (3), 597-618 (1980).
37. Roberson, J. A., and C. K. Chen, "Flow in Conduits with Low Roughness Concentration," *Journal of Hydraulics Div., ASCE, 96* (HY4), 941-957 (April 1970).
38. Sayre, W. W. and M. L. Albertson, "Roughness Spacing in Rigid Open Channels," *Transactions, ASCE, 128* (1), 343-372 (1963).
39. Simons, D. B. and E. V. Richardson, "Resistance to Flow in Alluvial Channels," U.S. Geological Survey Professional Paper 422J, Washington, D.C. (1966).
40. Sumer, B. M. and M. Bakioglu, "On the Formation of Ripples on an Erodible Bed," *J. Fluid Mechanics, 144,* 177-190 (1984).
41. Vanoni, V. A. and L. Hwang, "Relationship Between Bed Forms and Friction in Streams," *Journal of Hydraulics Div., ASCE, 93* (HY3), 121-144 (May 1967).
42. Vanoni, V. A., ed., *Sedimentation Engineering,* American Society of Civil Engineers, New York (1975).
43. Willis, J. C., "Flow Resistance in Large Test Channel," *Journal of Hydraulic Engineering, ASCE, 109* (12), 1755-1770 (Dec. 1983).
44. Yalin, M. S., *Mechanics of Sediment Transport,* Pergamon Press Ltd., First Edition (1972).
45. Yalin, M. S. and E. Karahan, "Steepness of Sedimentary Dunes," *Journal of Hydraulics Div., ASCE, 105* (HY4), 381-392 (April 1979).
46. Yalin, M. S., "On the Determination of Ripple Geometry," *Journal of Hydraulic Engineering, ASCE, 111* (8), 1148-1155 (Aug. 1985).

Scour At Bridge Sites

B. W. MELVILLE*

INTRODUCTION

The boundaries of alluvial rivers change continually due to erosion by water, the greatest changes occurring during floods. At a bridge site these changes introduce the possibility of undermining of the bridge foundations and the risk of the eventual failure of the bridge (Figure 1). In an extensive study of bridge failures in the United States, the Federal Highway Administration reported in 1978 [13] that damage to bridges and highways from major regional floods in 1964 and 1972 amounted to about $100 million per event.

The theoretical basis for the structural design of bridges is well established. In contrast, the mechanics of flow and erosion in mobile-boundary channels has not been well defined and it is not possible to estimate with confidence the river boundary changes which may occur at a bridge subject to a given flood. This is due not only to the extreme complexity of the problem but also to the fact that stream characteristics, bridge constriction geometry and soil and water interaction are different for each bridge as well as for each flood.

The hydraulic design of a bridge typically involves four main steps:

—selection of crossing site and orientation
—selection of type of foundations (piers and abutments) together with determination of size, spacing and arrangement of foundations
—estimation of maximum scour depth for the design flood
—evaluation of the need for scour protection and river training works

This chapter is a review of information available to the bridge designer faced with carrying out the third step, estimation of maximum scour depth for the design flood.

Types of Bridge Scour

'Scour' is used here to mean the lowering of the level of the river bed by water erosion such that there is a tendency to expose the foundations of a bridge. The amount of this reduction below an assumed natural level (generally the level of the river bed prior to the commencement of the scour) is termed the 'depth of scour' or 'scour depth'.

The types of scour which can occur at a bridge crossing are shown in Figure 2.

General scour occurs irrespective of the existence of the bridge and can occur as either long-term or short-term scour, the two types being distinguished by the time taken for scour development. Short-term general scour develops during a single or several closely spaced floods. Long-term general scour has a considerably longer time scale, normally of the order of several years and includes progressive degradation and lateral bank erosion. Progressive degradation is the quasi-permanent general lowering of the river bed due to hydrometerological changes (e.g. prolonged high flows) or human activities (e.g. dam construction). Lateral erosion of channel banks can lead to the bridge being outflanked by the river or to undermining of the abutments. Lateral bank erosion may result from channel widening, meander migration, a change in the river controls or a sudden change in the river course (e.g. with the formation of a cut-off). The long-term general scour may not be significant during the design life of a bridge if the rate of scour development is relatively slow. The effects of long-term general scour can be felt suddenly, however, e.g. the formation of a cut-off near the bridge site.

General scour can also occur as short-term fluctuations in river bed levels associated with the passage of a single or possibly several closely spaced floods. This type of scour

*Department of Civil Engineering, University of Auckland, Auckland, New Zealand

FIGURE 1. Pier failure at Bulls Bridge (Rangitikei River, N.Z., June 1973)

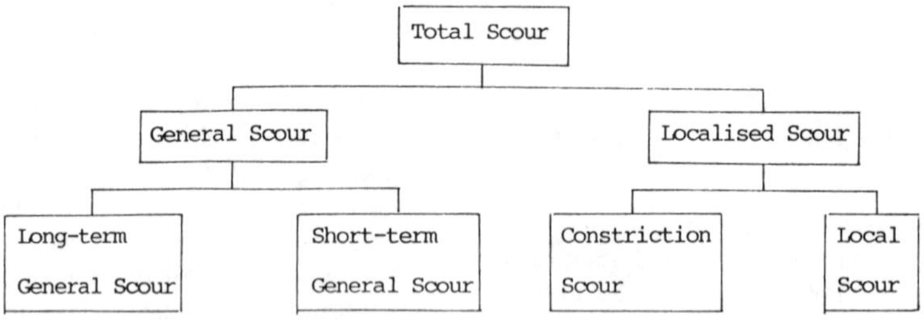

FIGURE 2. Types of bridge scour.

includes that due to convergence of flow, a shift in the channel talweg or braids within the channel and bed form migration.

In contrast to general scour, *localised scour* is directly attributable to the existence of the bridge. Localised scour includes *constriction scour* and *local scour*. Constriction scour may occur if the bridge or its approaches cause a constriction of the flow (e.g. bridge approaches encroaching onto the floodplain of a river). Local scour is caused by the interference of the piers and abutments with the flow and is characterised by the formation of scour holes immediately at the bridge pier or abutment. Scour at spur dykes and other river training works is also local scour.

Localised scour can occur as either 'clear-water scour' or 'live-bed scour'. Clear-water scour occurs when the bed material upstream of the scour area is at rest. The maximum local scour depth is reached when the flow can no longer remove bed material from the scour area. Live-bed scour occurs when there is general sediment transport by the river. The 'equilibrium scour depth' is attained when the time-averaged transport of bed material into the scour hole equals that removed from it.

At a particular bridge crossing, any or all of the different types of scour may occur simultaneously. It is necessary to ensure that the design scour depth includes the scour due to all possible causes. Graham [30] makes the point that it may not be possible to isolate general and localised scour because not only is the scour at a bridge dependent on the fluvial and sediment regimes, but the flow and sediment flux are in turn affected by the bridge. In Graham's words "the structure's design, to some extent, will determine its own seating". For the purpose of design, however, it can generally be assumed that the depths of scour, arising from the different types of scour shown in Figure 2, are cumulative.

Physical Factors Affecting Bridge Scour

Scour at bridge crossings arises from a complex interaction of the river flow, the channel boundary materials and, in the case of localised scour, the bridge structure itself.

Generally the flow in a river could be quantified in terms of mean flow velocity, U and flow depth, y_0. However, in natural river channels, flows are strongly three-dimensional and unsteady, so that mean velocities and depths give at best only a rough indication of the erosive potential of the stream. In wide braided river channels, where low flows occupy only a small channel, the deep water channel can shift substantially and unpredictably during floods. In such a case, whereas the mean velocity and depth may give a reasonable estimate of general scour (assuming the availability of a reliable method of estimation) these parameters may be totally misleading if used to predict local scour depths.

The materials which form the boundaries of river channels may range from boulders to silts and clays and are usually inhomogeneous. The deposits may also have complex grain size distributions and be stratified. It is normal to divide the deposits into cohesive and non-cohesive (or alluvial) sediments. Understanding of the physics of erosion of cohesive materials is very limited and, of necessity, this chapter is primarily concerned with alluvial sediments.

The simplest means of describing alluvial river sediment is the mean particle size (by weight). For uniform materials d_{50} is adequate but is quite inadequate for non-uniform materials. The different size fractions have entirely different resistance to scour and their interactions are complex. Sediment non-uniformity can be described by the standard deviation, σ of the particle size distribution.

A stream bed may also be composed of a series of strata of different resistance to scour. Where a relatively resistant material overlies a more readily erodible material, large scour depths may result if the scour breaks through the resistant material. Conversely, if a highly resistant material is known to exist at a particular level, it may be unnecessary to extend bridge foundations below that level. Variation in material type is not restricted to vertical stratification. Variation of sediment over the cross-section can influence the scouring process.

Localised scour can occur under clear-water or live-bed conditions. Because the maximum depth of scour and the time to attain this differ for the two cases, it is important to identify the condition of movement of bed material. The Shields entrainment function (Figure 3) can be used to determine whether design flow conditions are those of clear-water or live-bed scour. Under live-bed conditions estimation of the size of bed features becomes necessary because these affect the scour depth. The rate at which scour develops is another important parameter. Localised scour and short-term general scour occur with a time scale related to the flood duration. However a given flood may not last long enough for maximum possible scour depths to develop. This is especially likely with erosion-resistant (cohesive) materials and clear-water scour conditions.

Bridge foundations are designed in a large variety of sizes and shapes. Generally the geometry of bridge foundations can be described in terms of the size of the obstruction (e.g. the diameter of a cylindrical bridge pier), together with a shape factor to account for differences in shape, and the angle between the approach flow and the obstruction. Local scour depths are strongly dependent on the size and alignment, and to a lesser extent on the shape of the foundations. During floods, debris rafts may form against bridge piers and abutments, increasing their effective size and changing their shape. Changes in the position or alignment of the main or subsidiary flow channels during flood can adversely affect the alignment of the bridge foundations with respect to the flow.

The width of bridge piers and abutments with respect to the unconstricted channel width determines the degree of flow constriction and consequently the constriction scour at

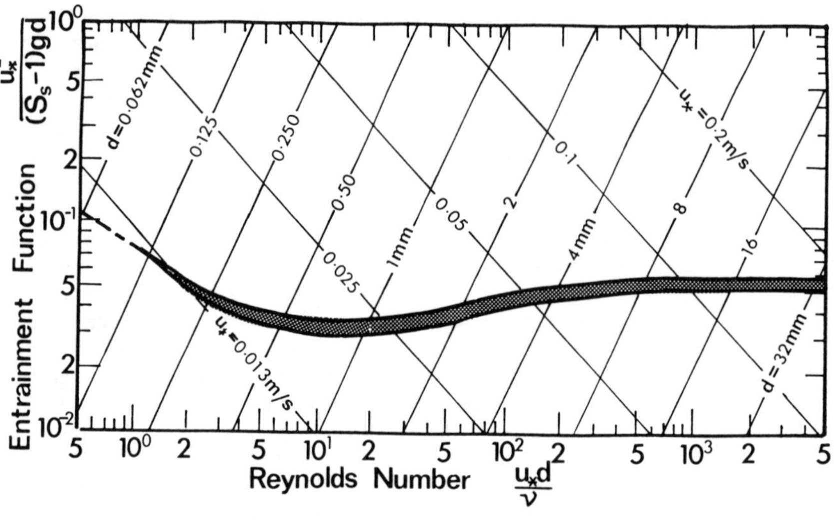

FIGURE 3. Shields entrainment function.

a bridge crossing. Bridge approach embankments extending across wide flood plains can lead to significant constriction of flow. In addition, debris rafting can increase the degree of flow constriction.

GENERAL SCOUR

It has already been stated that the total scour at a bridge crossing comprises general scour and localised scour. It is necessary in design firstly to assess the general scour in the unconstricted channel (i.e. in the absence of the bridge). Once the bed movement in the unconstricted channel is reasonably well understood, the localised scour can be estimated.

Long-term General Scour

River morphology (or geomorphology) is the study of geometries of rivers on time-scales of thousands of years. From a morphologic point of view water and sediment discharges are rarely constant and river channels are constantly trying to adjust their geometry in response to changes in water and sediment load. Recognition of the potential for and existence of such long-term changes is an important aspect of bridge design and it requires an understanding of the processes occurring within the particular catchment. A good review of factors affecting long-term river stability in the context of bridge location and design is given by Trent and Brown [82].

FUNCTIONAL RELATIONSHIPS

During the design life of a bridge, the long-term erosion processes may or may not be significant. If changes in water and sediment load are large enough and if the resulting erosion rates are high enough, general scour may occur at the bridge crossing.

Many authors have studied the functional relationships between water discharge, Q and sediment discharge, Q_s in a channel and the channel geometry (i.e. its width B, depth y_0, slope S, and form in plan view) and bed material size, d_{50}, including Leopold *et al.* [55], Lane [45], Schumm [74], Simons and Senturk [77] and Trent and Brown [82]. Lane [45] summarised these relationships as follows:

$$Q_s d_{50} = QS$$

This equation states that a channel can be maintained in dynamic equilibrium by balancing changes in sediment discharge and sediment size by compensating changes in water discharge and channel slope. The Regime Theory, derived from observations of water and sediment loads in Indian irrigation canals which had no significant scour or deposition, leads to a useful set of such functional relationships. Neill [67] provides a summary of regime formulae which can be used to assess qualitatively the effects on river regime of imposed changes in width, slope, 'dominant' discharge and sediment load (Table 1). Definition of the dominant discharge allows for the fact that rivers usually form their channels to accommodate large but not the maximum flows. The dominant discharge is that discharge assumed to produce the same geometry as the actual fluctuating flows of the river, and is typically taken to have a return period of about 2 years.

CAUSES OF LONG-TERM GENERAL SCOUR

Long-term general scour can occur as the result of natural, hydrogeological changes or artificial "man-induced" changes. Within these categories the causes of long-term general scour include the following:

Natural erosion —extended high-flow periods
Man-induced erosion—agricultural activity, including crop cultivation, animal grazing and deforestation
—urbanisation resulting in increased run-off
—channel straightening giving increased slope
—streambed mining, affecting the sediment movement and supply in a channel
—water resources management such as diversion of flows from one river to another
—dam construction which traps sediment flows
—increased slope associated with lowering of a downstream control (e.g. lowering of lake level or erosion at a waterfall)

It should be noted that natural erosion processes often are difficult to predict because they are so dependent on hydrological events.

It is convenient to divide long-term general scour into that arising from vertical changes in the natural level of the river bed, or 'degradation', and that arising from changes in the plan view form of the river or 'lateral bank erosion'. There is, of course, no clear-cut distinction between these two processes. Channel widening, a lateral erosion process by definition, can occur due to bank slumping as the result of channel degradation.

DEGRADATION

Galay [24] defines degradation as an extensive, often progressive, lowering of the river bed over long distances. Degradation can proceed upstream or downstream.

Downstream progressive degradation generally arises as the result of changes to one or more of water flow (an increase), sediment discharge (a decrease) and bed material size (a decrease). One of the most common examples is that due to sediment trapping behind a new dam which results in the release of clear water downstream, the changed flow having a greater potential for erosion. Construction of the Hoover Dam on the Colorado River resulted in a 7.1 m degradation of the river bed composed of sand and gravels. The degradation extended 111 km downstream of the dam. In some cases, prolonged degradation and coarsening of bed sediment can lead to armouring of the bed and significant reduction in the degradation rate.

Upstream progressive degradation is generally the result of an imposed increase in slope which can be caused by lowering a base level, decreasing river length or removing a control point, e.g. a drop in lake level gives increased slope, increased velocity and this may lead to river bed erosion. Upstream progressive degradation usually proceeds more rapidly than its downstream counterpart. It is important to realise that any type of degradation on a main river generally produces upstream progressive degradation on a tributary and that the degradation of the tributary can interfere with the degradation process in the main river.

It is not yet possible to predict with confidence depths of

TABLE 1. Qualitative Relationship Between Imposed and Consequent Changes in River Regime According to Neill [67].

Imposed change	Principal consequent changes	Rough formula to indicate sensitivity to change*
Reduction in width (as by constriction)	Increase in depth by scour	Depth $\propto 1/B^{3/4}$
Increase in slope (as by short-cut diversion)	Increase in velocity; large increase in sediment tranport from bed and bank erosion; increase in channel shifting upstream and downstream	Velocity $\propto S^{1/2}$ Bed-material load $\propto S^3$
Increase in dominant discharge (as by diversions)	Increase in width, depth, and erosion; reduction in slope by degradation	Width $\propto Q^{1/2}$ Depth $\propto Q^{1/3}$ Slope $\propto 1/Q^{1/4}$
Reduction in sediment load (as by damming)	Downstream reduction in slope by degradation	Highly dependent on nature of bed and load
Increase in sediment load (as by careless land-use changes)	Increase in slope by aggradation; increase in tendency to wander; increase in width	Slope $\propto$ load$^{1/3}$

*The exponents are not to be taken as generally applicable. They are included to give a rough idea of the sensitivity only, and vary according to type of river and numerous other factors. B is the channel width, S the slope, and Q the channel-forming discharge.

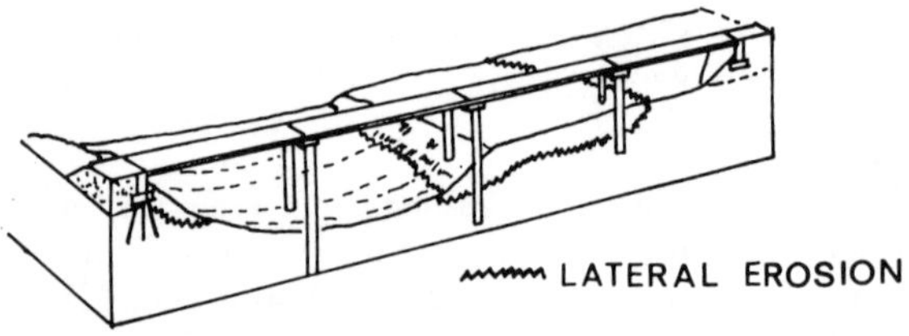

FIGURE 4. Lateral bank erosion at a bridge crossing, after Brice [12].

scour which may occur at a bridge site due to progressive degradation, although assessment of a particular site in terms of the above possible causes should indicate the likelihood of degradation. The existence within a catchment of channel scarps, tributary gullying or high, steep unvegetated channel banks is evidence of degradation. The rate at which the erosion is occurring relative to the design life of the bridge is fundamentally important. If available, time-sequential areal photographs, together with historical measurements of bed levels, are useful in quantifying likely degradation.

Upstream progressive degradation is controlled from downstream. Therefore, the amount of the vertical change at the downstream control point (e.g. the magnitude of a drop in lake level) can be taken as an upper limit to the scour depth further upstream.

In principle, the rate of change of bed elevation, or rate of degradation, is proportional to the rate of change of sediment transport rate, Q_s, in the downstream, x, direction, or

$$\frac{\partial[f(B)]}{\partial t} = \frac{\partial(Q_s)}{\partial x} \tag{1}$$

in which t is time and $f(B)$ is an equation describing the channel boundary shape. This equation has been solved for a few simplified two-dimensional cases. However, any solution is dependent on definition of Q_s and can only be as good as the particular sediment transport relationship used. Raudkivi and Sutherland [72] present a numerical example based on the solution by Komura and Simons [41], who used the Kalinske and Brown sediment transport formula. Other solutions are available including Gessler [28] and Mostafa [65].

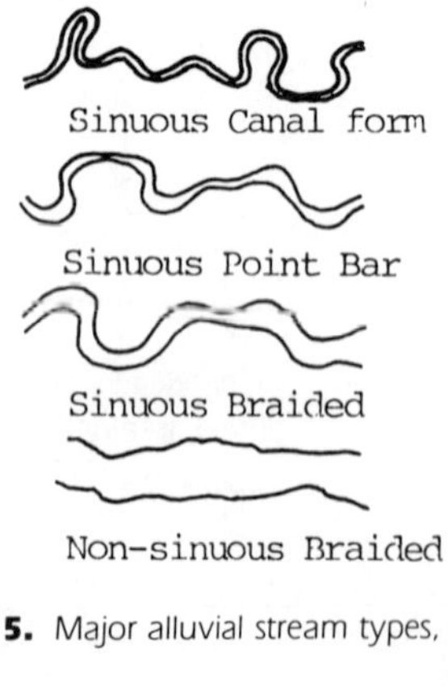

FIGURE 5. Major alluvial stream types, Brice [12].

LATERAL BANK EROSION

Lateral bank erosion may endanger bridge foundations (Figure 4). Both bed and bank scour can be expected to occur at river bends. However, bends can also form at bridge sites which were originally straight. Attempts to limit channel movements at one point can cause changes elsewhere and this may lead to scour at the bridge foundations.

Table 1 shows that either an increase in water discharge or a decrease in sediment discharge may produce a reduction in channel slope. The slope reduction is likely to produce an increased tendency for the channel to wander and consequently lateral bank erosion. In addition, an increase in sediment discharge can result in lateral bank erosion while channel widening (erosion to both banks) may occur with increased flow.

Brice, Blodgett *et al.* [13] analysed 224 bridge sites in U.S. where scour problems had occurred. At 106 sites they attributed the hydraulic problems mainly to lateral bank erosion, concluding that this was the most common cause of scour. Brice [12] compared bank erosion rates with channel width, B and found a tendency for erosion rate to increase with B. In terms of the 4 major alluvial stream types, (Figure 5), he found that sinuous canal form streams had the lowest rates of bank erosion, while braided streams had the highest.

Generally, lateral bank erosion can occur as channel widening, natural meander migration, due to a change in river controls, or as a major river avulsion (e.g. formation of a cut-off). Regarding prediction of erosion rates, some progress has recently been made with quantitative assessment of the mechanics of river meanders [69]. Notwithstanding the best indication of likely problems at a bridge crossing is obtained from comparison of time-sequential photos or maps which enables estimation of historical erosion rates. As an approximation, the maximum existing channel depth at the bridge crossing site can normally be taken as an upper limit to the possible depth of scour which may arise as a result of lateral bank erosion.

Short-Term General Scour

Short-term general scour could be defined as natural short-term fluctuations in the position of the channel boundaries about an equilibrium position and associated with the passage of a flood. The equilibrium position is normally taken as the channel boundary shape at low flow. 'Channel scour and fill' are words often used to describe short-term general scour.

For most streams the magnitude of short-term general scour varies significantly along the channel. Bends and narrows are more susceptible to scour than straight or gently curving reaches. Generally, sediment is distributed both laterally and longitudinally during a flood, but there is little or no overall scour over a long enough reach of a laterally unconstrained channel. This view is supported by many authors including Andrews [2,3], Gerard [27] and Raudkivi [71].

Some data are available concerning the relationship between short-term general scour and flowrate. These have been derived from both field studies [2,3,46,56] and laboratory flume studies [33,86]. In a review paper, Graham [30] states that

> The major channel response to increased discharge appears to be in the width, rather than depth, adjustment in unconstrained sections.

More recently some progress has been achieved with mathematical modelling. However such models are all two-dimensional, and do not adequately describe the complex three-dimensional scour process. A better understanding of the physical processes involved is needed before modelling can be successful.

At an unconstricted bridge crossing, the mechanisms of short-term general scour are:

—convergence of flows
—shift in the talweg or braids
—flow interaction at a confluence
—bed form migration

In principle, sections of a river channel system subject to convergent flows can be expected to have greater velocities than average and be subject to erosion. Conversely, divergent currents are associated with deposition. This is the basis of Leliavsky's convergent-divergent criterion and is borne out by observations of greater scour at pools or deep channel sections than at riffles or shallow sections (Figure 6).

Estimates of short-term scour depths in unconstricted alluvial rivers are possible using calculated flow depths, if water levels at design flood conditions are known. Flow depths can be calculated using any of the stage-discharge formulae for alluvial rivers. Vanoni [84] gives a description of methods of estimation of flow depth, while numerical examples are presented by Raudkivi and Sutherland [72].

Scour design criteria based on the empirical regime formulae of Lacey [44], Blench [7] and others have commonly been used for estimation of scour depths at bridge crossings.

The appropriate equations are:

$$d_s = (z - 1)y_r \tag{2}$$

$$y_r = 0.47\sqrt[3]{Q/F_L} \quad \text{(Lacey)} \tag{3}$$

$$y_r = 1.49\sqrt[3]{q^2/F_B} \quad \text{(Blench)} \tag{4}$$

in which d_s is mean flow depth (or 'regime' depth), Q is flowrate, F_L is Lacey's silt factor equal to $1.59\sqrt{d_{50}}$, F_B is Blench's bed factor equal to $1.9\sqrt{d_{50}}$, d_{50} is mean grain size (mm) and z is the ratio of maximum depth to mean depth. The regime formulae give the mean regime depth or 'stable channel' depth in a straight channel in terms of the discharge. The regime depth typically is multiplied by the z-factor which accounts for departures from the mean; for instance greater depths at bends, local scour at bridge foundations and lateral variations in depth across the channel. Blench [6] provides a comprehensive account of regime theory.

In some situations, past scour depths can be inferred from geotechnical site investigation methods. Sub-bed borehole samples may indicate the lowest level of recent scouring if differences in soil density or type or size of material can be detected. The method is necessarily subject to considerable uncertainty.

Scour due to a shift in the talweg or braids of a channel is a consequence of the three-dimensionality of flows in natural river channels and the possibility that the lateral distribution of flow may vary during the passage of a single flood or succession of floods. Figure 7 shows the damage which occurred in February 1985 to a bridge abutment at a stream near Thames, New Zealand, due to a shift in the

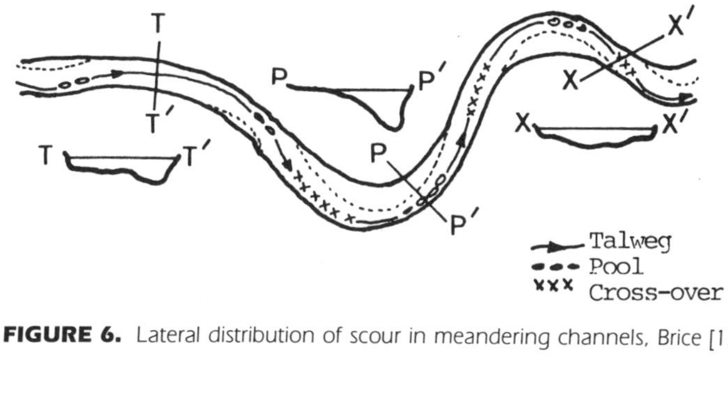

FIGURE 6. Lateral distribution of scour in meandering channels, Brice [12].

FIGURE 7. Scour at a bridge abutment, Thames, N.Z.

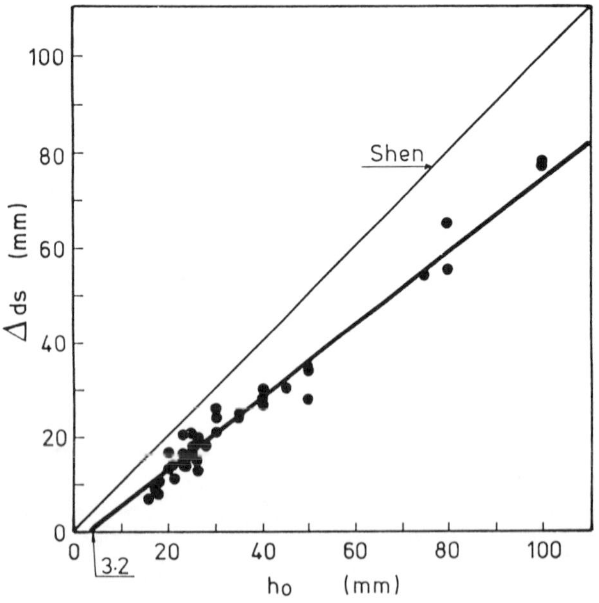

FIGURE 8. Comparison of bed feature height and fluctuation in local scour depth at a pier, Chiew [16].

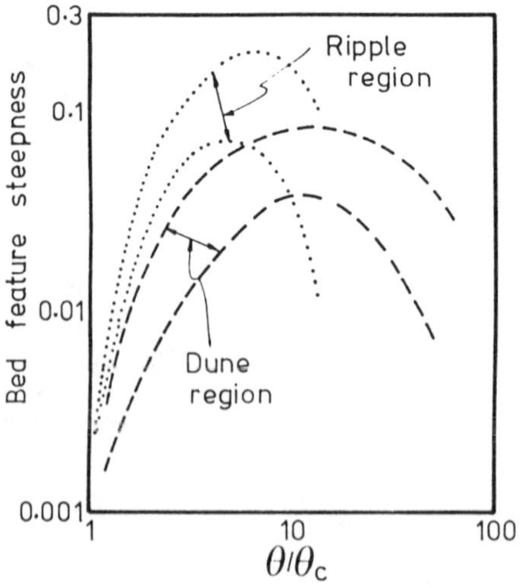

FIGURE 9. Bed feature steepness function after Yalin [88].

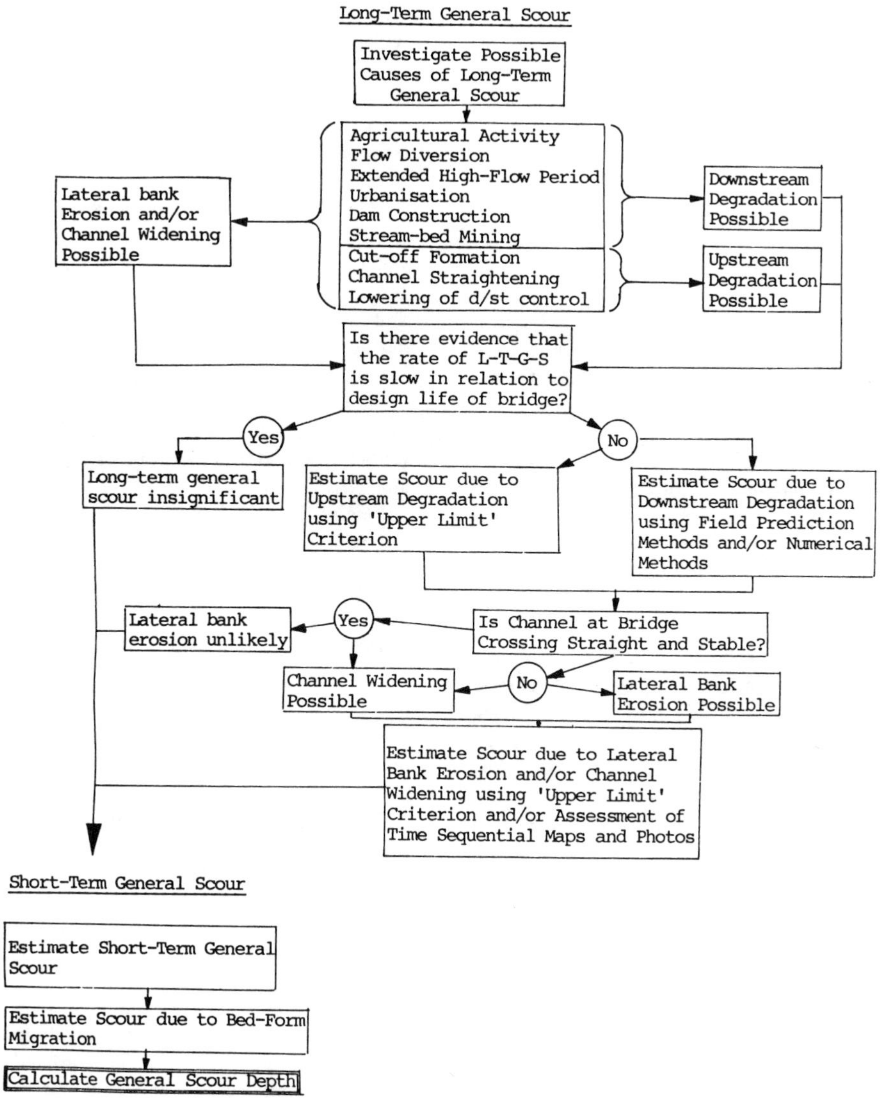

FIGURE 10. Flow chart for general scour depth estimation.

talweg towards the right bank and the corresponding concentration of flow to that side of the channel.

Shift of the talweg with increase in stage is a significant factor in bridge design, not only for estimation of maximum scour depths but also for alignment of piers with flood flow. Brice [12] states that shift of the talweg is greatest with sinuous point-bar streams.

The braids in wide, gravel rivers may shift rapidly during floods giving significant lateral redistribution of bed material. Deep localised stream channels may develop and if these occur at a bridge site, the local scour depth is additive to the minimum bed level of the localised channel. As a general rule, it should be assumed that the lowest point in the channel bed may move laterally to any location within the active channel. \

Where two localised channels meet, particularly deep scour may occur. Raudkivi [71] showed that at the confluence where the two flows meet at an angle to each other, a spiralling motion is generated similar to that at the ends of groynes and larger abutments. Field measurements at Ohau River in New Zealand by Thompson and Davoren [81] showed localised channels of about 3 m depth in a wide gravel river with an average depth of only 0.75 mm.

In sandy rivers, the maximum scour depth occurs as the trough of a bed feature migrates past the bridge. Good discussions of the types of bed forms that may exist in a river channel and their prediction are given by Vanoni [84] and Raudkivi [70].

Shen *et al.* [75] stated that 'based on flume experiments, it is safe to add half the expected dune height to the maximum scour depth as a design criteria'. Chiew [16] carried out an extensive experimental investigation of live-bed local scour at cylindrical bridge piers in which he recorded the height of "well-defined bed features," h_0 and compared this with the corresponding fluctuations in the local scour depth, Δd_s. The results, Figure 8, confirm the Shen *et al.* [75] conclusion. The bed-feature height, h_0 varies with flow conditions. It can be estimated, for example, from experimental data by Yalin [88], Figure 9, which gives the bed feature steepness h_0/λ in terms of the flow function θ/θ_c, where θ is the Shield's Entrainment Function, θ_c is the critical value of θ and λ, the bed feature length, can be approximated by $\lambda = 2\pi y_0$.

Bed form migration in gravel rivers occurs mostly as movement of gravel in bars. At a bridge site, bar migration may deflect and concentrate the flow at the bridge foundations causing scour due to a shift of the talweg or scour at a confluence.

Summary—General Scour

General scour can occur as progressive long term changes (long-term general scour) or as natural short term fluctuations (short-term general scour) in the boundaries of alluvial rivers.

Understanding of long-term general scour rests on knowledge of river geomorphology. Functional relationships for the effects on river regime of imposed changes in width, slope, discharge and bed load enable qualitative prediction of long-term general scour. No reliable quantitative methods of analysis of long-term general scour depths are available and comparison of time-sequential aerial photos and maps remains the most reliable method of assessment.

There is a scarcity of quantitative information concerning short-term general scour. Generally, bed material in a channel is redistributed both laterally and longitudinally during a flood, but there is little or no overall short-term general scour over a long enough reach of a laterally unconstrained channel. In principle, scour depths can be estimated using stage-discharge formulae. This method presupposes that water levels at flood discharge are known. In addition, none of the stage-discharge relationships adequately describes the three-dimensional characteristics of sediment movement during floods, in spite of the fact that three-dimensionality is fundamentally important to the scour process.

Short-term general scour depths due to bed-form migration are normally less than one-half the bed-form height. Reliable methods of estimation of bed-form height are available.

Finally, Figure 10 is a flow chart which gives suggested steps for assessment of general scour depths at a bridge crossing.

CONSTRICTION SCOUR

Constriction scour and local scour are here collectively termed localised scour. The occurrence of localised scour is directly attributable to the existence of the bridge. Bridge construction leads to a disturbance in the flow pattern in the immediate vicinity of the bridge which may lead to constriction scour and local scour. Constriction scour is considered here while local scour is considered in the section which follows.

Introduction

Bridges across waterways are usually located in response to alignment and traffic considerations and the length of the bridge is minimised by extending the approach embankments to reduce costs. As a consequence, the bridge restricts the waterway and this can result in increased scour under flood conditions. Blodgett [8] compared hydraulic data at various sites both with and without bridges and showed that significant scour activity can arise as a direct consequence of flow constriction. Blodgett states:

> During the first 2 or 3 years after construction of a bridge that constricts a channel scour may lower the channel bed several feet. Abutments and piers that create large amounts of constriction may induce changes that require mainte-

nance to prevent erosion. The period of instability may continue for decades.

In principle the constriction due to bridges can be considered in terms of the constriction ratio, β given by

$$\beta = B_1/B_2$$

where B_1 and B_2 are widths of the normal (or unconstricted) and constricted sections respectively. B_2 can also be considered to be the clear opening width at the bridge crossing. Many of the formulae for scour at bridge abutments contain an influence of β or an equivalent parameter.

Rectangular Long Constriction

The presence of abutments and piers reduces the channel width. This reduction is the cause of scouring within the constricted reach due to an increase in the bed shear stress. Most of the published studies have considered the special case of scour in a constriction which is long enough such that it can be assumed that uniform flow exists in both the normal and constricted reaches.

The basic problem of steady uniform flow through a long rectangular constriction is defined in Figure 11. The flow in the normal reach has depth y_1, width B_1 and velocity U_1 while in the constricted reach the corresponding parameters are y_2, B_2 and U_2. The equilibrium scour depth is d_{sc}. The flow rate in both reaches is the same:

$$Q = U_1B_1y_1 = U_2B_2y_2 \qquad (5)$$

Assuming that the head loss through the transition is of the form,

$$H_L = K\frac{U_2^2}{2g} \qquad (6)$$

the one-dimensional energy equation together with Equation (5) gives

$$\frac{y_2}{y_1} = \frac{d_{sc}}{y_1} + 1 - \frac{1}{2}Fr_1^2\left[\frac{(1 + K)\beta^2}{(y_2/y_1)^2} - 1\right] \qquad (7)$$

where Fr_1 is the Froude number in the normal reach given by

$$Fr_1 = U_1/\sqrt{gy_1} \qquad (8)$$

If it is further assumed that the kinetic terms on the right-hand side of Equation (7) can be neglected

$$\frac{y_2}{y_1} = \frac{d_{sc}}{y_1} + 1 \qquad (9)$$

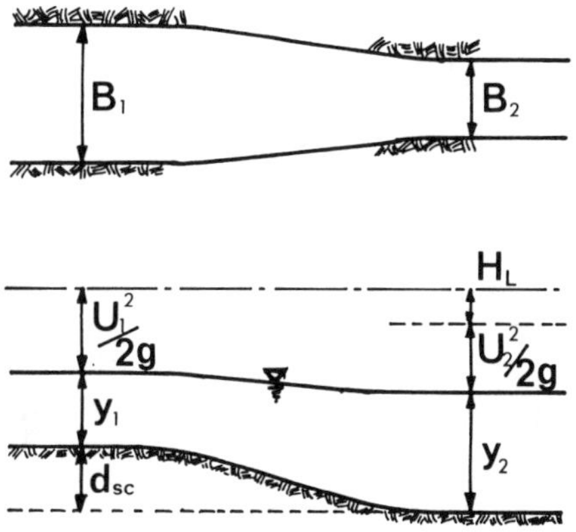

FIGURE 11. Flow at a long rectangular constriction.

or

$$d_{sc} = y_2 - y_1 \qquad (10)$$

Straub [78] presented a simplified one-dimensional analysis of the long constriction. His work was subsequently augmented by Laursen [49,50,51], Komura [40], Ashida [4] and Gill [29]. The basis of this theory is presented here.

It is assumed that the hydraulic resistance, in the normal and constricted reaches, is described by the Manning formula,

$$U = \frac{1}{n}y^{2/3}S^{1/2} \qquad (11)$$

in which n is the Manning roughness coefficient and S is energy gradient. The equation of continuity of sediment discharge is

$$B_1q_{s1} = B_2q_{s2} \qquad (12)$$

in which q_s is the unit sediment discharge and the rate of sediment transport is assumed to be given by

$$q_s = [g(S_s - 1)d_{50}^3]^{1/2}(\tau/\tau_c - 1)^m \qquad (13)$$

in which S_s is specific gravity of the bed material, d_{50} is mean size of bed material, τ is bed shear stress, τ_c is the critical bed shear stress and m is an exponent. At high transport rates, the highest value of m in the known empirical formulae is 3, as in the Einstein–Brown formula, while the lowest value of m is 1.5, as in the Meyer–Peter formula.

For live-bed scour, Equations (10)–(13) can be solved to give

$$\frac{y_2}{y_1} = \beta^{6/7}[\beta^{1/m}(1 - \tau_c/\tau_1) + \tau_c/\tau_1]^{-3/7} \qquad (14)$$

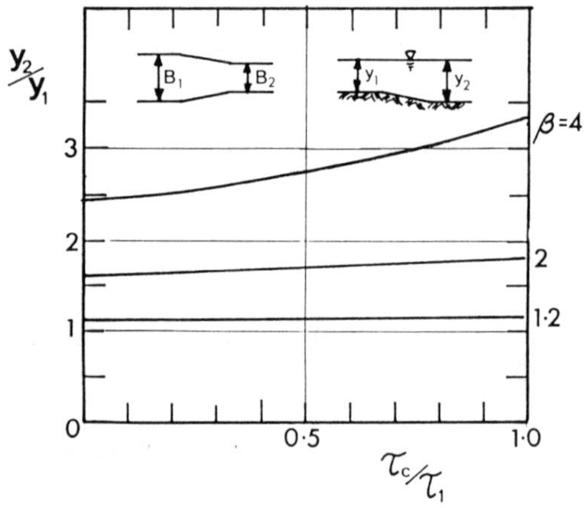

FIGURE 12. Relative depth in a long constriction.

in which τ_1 is the bed shear stress in the normal reach. Equation (14) is plotted in Figure 12 for $m = 2$ which shows that the dependence of scour depth on τ_c/τ_1 (the relative transport rate of bed material in the normal reach) is minimal even at high values of β, the constriction ratio.

For live-bed scour, $\tau_c/\tau_1 < 0$ and two limiting cases can occur. At very high rates of sediment transport, $\tau_c/\tau_1 \rightarrow 0$ and Equation (14) reduces to

$$\frac{y_2}{y_1} = \beta^{(6/7 - 3/7m)} \tag{15}$$

which for $m = 2$, as in the Du Boys formula gives

$$\frac{y_2}{y_1} = \beta^{9/14} \tag{16}$$

Alternatively for $\tau_c/\tau_1 = 1$, the threshold condition in the normal reach, Equation (14) gives

$$\frac{y_2}{y_1} = \beta^{6/7} \tag{17}$$

The exponent in Equations (16) and (17) does not vary much so that for live-bed conditions an approximate solution is given by

$$\frac{y_2}{y_1} = \beta^{m'} \tag{18}$$

where m' varies from 9/14 to 6/7. Intermediate values of m' have been derived empirically [4] and these are shown in Figure 13.

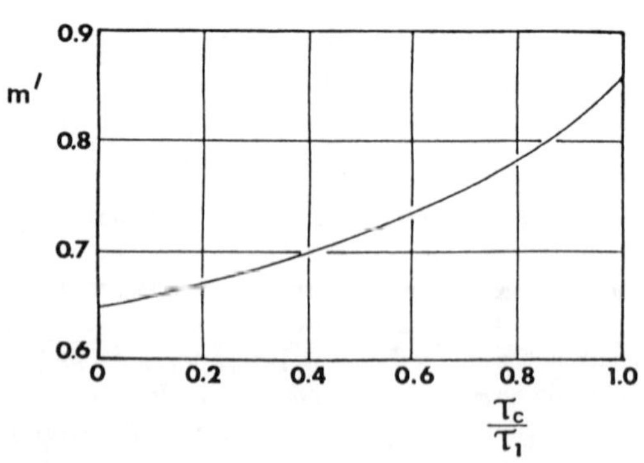

FIGURE 13. Variation of exponent m′ with relative transport rate.

For clear-water scour, $\tau_c/\tau_1 > 0$, and at equilibrium, threshold conditions will exist in the constricted reach. In this case, Equation (14) becomes

$$\frac{y_2}{y_1} = \beta^{6/7}(\tau_c/\tau_1)^{-3/7} \tag{19}$$

Laursen [49,50,51] extended the Straub Method to streams having a main channel and an overbank channel. For live-bed scour, the equation proposed was

$$\frac{y_2}{y_1} = \left(\frac{Q}{Q_m}\right)^{6/7} \beta^{(6/7)(2+z)/(3+z)} \left(\frac{n_2}{n_1}\right)^{(6/7)(z)/(3+z)} \tag{20}$$

where Q is total discharge, Q_m is that part of the total discharge in the main channel, n_1 and n_2 are values of the Manning roughness coefficient in the normal and constricted reaches and z is a coefficient depending on the ratio of shear velocity, u_* to fall velocity, w (Table 2). If the constriction is for overbank flow only, the main channel flow is unaffected and the width ratio is unity. If, in addition the ratio of the Manning n is close to unity, Equation (20) becomes:

$$\frac{y_2}{y_1} = \left(\frac{Q}{Q_m}\right)^{6/7} \tag{21}$$

for all conditions of sediment transport. Alternatively, if the constriction narrows the main channel and $Q = Q_m$, Equation (20) becomes with $n_1 = n_2$;

$$\frac{y_2}{y_1} = \beta^{m''} \tag{22}$$

where $m'' = 0.59, 0.64, 0.69$ for $u_*/\omega < 1/2, 1, > 2$ respectively. Equations (21) and (22) are plotted in Figure 14. Note that Equations (18) and (22) are similar.

An equivalent relationship with an exponent of 0.67 is obtained by manipulation of Blench's regime, Equation (4),

TABLE 2. Values of Coefficient z in Laursen Equation for Live-Bed Constriction Scour.

$\frac{u_*}{\omega}$	z
$<1/2$	1/4
1	1
>2	9/4

if it is assumed that the bed-factor, F_B has the same value in the normal and constricted reaches.

For clear-water scour, Laursen [51] proposed Equation (19) and suggested that the shear stress ratio could be evaluated using;

$$\tau_1/\tau_c = \frac{U_1^2}{36 d_{50}^{2/3} y_1^{1/3}} \tag{23}$$

Komura [40] treated the same problem for both clear-water and live-bed conditions. The equations fitted to experimental data are similar to Equation (22) with the addition of Froude number and sediment terms.

For clear-water scour, the equation proposed is

$$\frac{y_2}{y_1} = 1.6 Fr_1^{0.2} \beta^{0.67} \sigma_g^{-0.5} \tag{24}$$

in which σ_g is standard deviation of the bed material given by,

$$\sigma_g = (d_{84}/d_{16})^{1/2} \tag{25}$$

while for live-bed scour, the equation proposed is

$$\frac{y_2}{y_1} = 1.45 Fr_1^{0.2} \beta^{0.67} \sigma_g^{-0.2} \tag{26}$$

Keller [39] conducted a series of laboratory experiments on clear-water constriction scour and compared the results with the methods of Laursen [49,50,51] and Komura [40]. He concluded that the Laursen equation is in general more reliable than the Komura equation for the range of his experiments. Laursen's method under-predicted the actual results. Similarly Gill [29] found that the depth ratio predicted by Equation (19) needs to be increased by 20% to agree with the results of his experiments.

In summary, the analytical methods for estimating scour in the long rectangular, constriction are all of a similar form with only minor differences in the influence of various parameters. In general, Figure 14 for live-bed scour and Equation (19) for clear-water scour give reasonable results when compared to laboratory data.

Estimation of Constriction Scour Depths

Little is known concerning the applicability of the methods of analysis of the long rectangular constriction to constriction scour at bridge abutments.

Wong [87] measured depths of clear-water scour due to constriction at wing-wall and spill-through abutments in the laboratory and compared the measured depths with depths predicted by the Laursen, Straub and Komura methods. Overall the Laursen method gave reasonable results, while the Straub method predicted significantly deeper scour than was measured.

In addition, comparison of the Komura live-bed constriction scour, Equation (26), with data by Ashida [4] for scour at a short constriction (length to width $\approx$ 1.0) shows that the constant in Equation (26) needs to be reduced from 1.45 to 1.22.

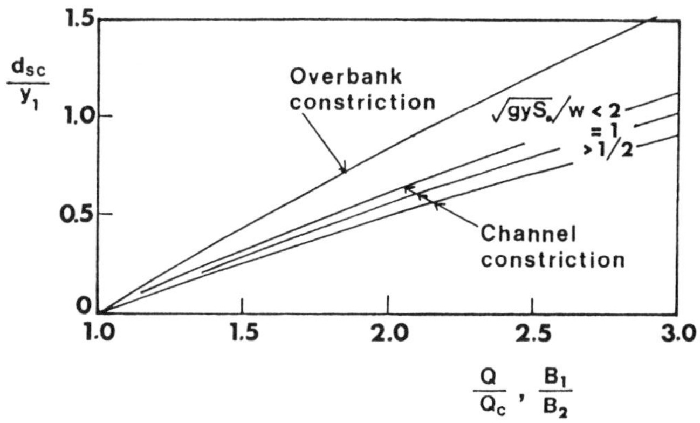

FIGURE 14. Live-bed scour in a long constriction after Laursen [50].

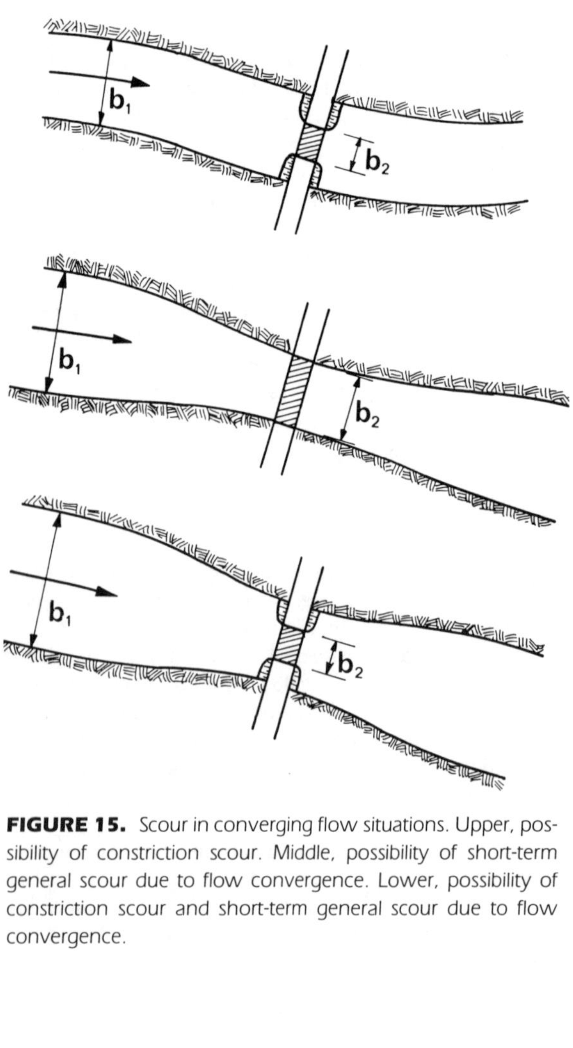

FIGURE 15. Scour in converging flow situations. Upper, possibility of constriction scour. Middle, possibility of short-term general scour due to flow convergence. Lower, possibility of constriction scour and short-term general scour due to flow convergence.

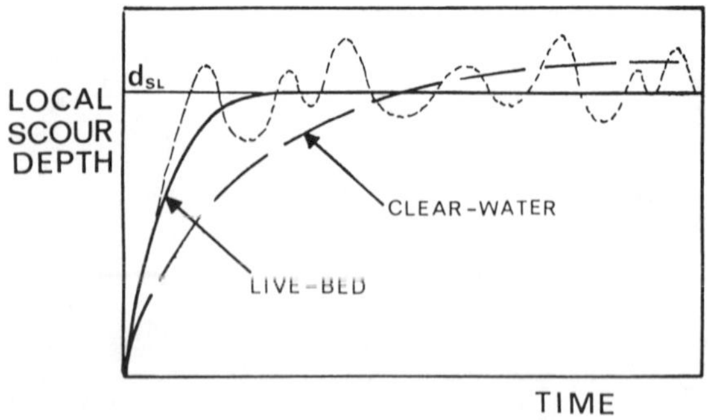

FIGURE 16. Development of local scour depth.

Generally it appears that use of the methods of analysis of the long rectangular constriction is conservative.

Neill [67] outlines 4 methods for estimating constriction scour at bridges. The first method involves direct inference from field measurements of scour at comparable sites. The second method, based on Blench's Equation (4) can be used where limited field measurements are available. The method is virtually equivalent to Equations (18) and (22). The third method is based on the assumption that scour will proceed until the velocity in the constricted waterway opening is equal to the mean velocity in the normal reach. This is equivalent to using an exponent of 1.0 in Equations (18) and (22) and Neill points out that the method can be expected to give greater estimates of scour depth than method 2. The fourth method is based on the assumption that scour will proceed until threshold conditions are attained in the constricted reach. The method is said to be excessively conservative for channels carrying substantial bed load because the threshold condition may not be reached.

Equivalence of Scour at Converging Flow Sections

Short-term general scour due to flow convergence and constriction scour are physically equivalent, the only difference being that the former occurs irrespective of the bridge presence while the latter is a direct consequence of the bridge. Because of this equivalence the analysis of scour in a long rectangular constriction given earlier is also applicable to short-term general scour where the natural river channel converges at the bridge site, so long as the river is laterally constrained within non-erodible banks.

At a particular site if the natural channel has constant width there will be no short-term general scour due to flow convergence, although constriction scour may occur if the bridge foundations and approaches constrict the flow. Alternatively, short-term general scour could occur at a bridge site where constriction scour is unlikely, and it is also possible for both types of scour to occur simultaneously. In the latter case, the net scour due to both cases could be analysed collectively. These situations are illustrated in Figure 15.

LOCAL SCOUR

Local scour occurs at a bridge site when the local flow field near the bridge piers and/or abutments is strong enough to remove bed material. Local scour is a direct consequence of the flow obstruction caused by the bridge, the depth of scour being strongly dependent on the size of obstruction.

Local scour is a time dependent process in which an equilibrium between the erosive capability of the flow and the resistance to motion of the bed materials is progressively attained through erosion of the flow boundary. Generally the equilibrium or final depth of local scour is rapidly attained in live-bed conditions, but rather more slowly in clear-water conditions, Figure 16. Observations of the equilibrium scour depth in live-bed conditions will show fluctuations in depth (the dotted line in Figure 16) due to the effects of bed feature migration.

Flow Patterns at Bridge Foundations

Bridge foundations, when introduced to a flow channel, result in significant changes to the flow pattern. A detailed description of the modified flow is essential to understanding and analysis of the local scour which develops.

CYLINDRICAL PIER

Melville [61,63] conducted a detailed study of the changing flow pattern at a cylindrical pier throughout the development of a local scour hole. The principal features of the flow pattern are shown in Figure 17. The flow decelerates as it approaches the cylinder coming to rest at the face of the pier. The associated stagnation pressures are highest near the surface, where the deceleration is greatest, and decrease downwards. In response to the downwards pressure gradient at the pier face, the flow is directed downwards. The strength of the downflow reaches a maximum just below the bed level.

It is the downflow impinging on the bed which is the main scouring agent. The development of the scour hole around the pier also gives rise to a lee eddy, known as the horseshoe vortex. The horseshoe vortex is effective in transporting dislodged particles away past the pier and the two provide the dominant scour mechanism.

The flow separates at the sides of the pier leading to development of the concentrated 'cast-off' vortices in the interface between the flow and the wake. These vortices are translated downstream by the mean flow and act like vacuum cleaners sucking up sediment from the bed.

Scour hole development commences at the sides of the pier with the two holes rapidly propagating upstream around the perimeter of the cylinder to meet on the centreline. In this way, a shallow hole, concentric with the cylinder, is formed around most of the perimeter of the cylinder, but not in the wake region. The downflow acts like a vertical jet eroding a groove in front of the pier, the eroded material being transported downstream by the flow. During early stages of development of the scour hole the lip of the groove is often very sharp and the face is almost vertical. The groove becomes shallow or disappears completely when scour approaches the equilibrium depth. Excavation of the groove undermines the scour hole slope above, which collapses in local avalanches of bed material such that the slope angle is maintained. The collapsed material is ejected from the groove by the downflow and carried downstream by the flow where a bar develops. The upstream part of the scour hole has the shape of a frustrum of an inverted cone with slope equal to the repose angle of the bed material under erosion conditions.

ABUTMENTS

It has often been suggested that scour at an abutment should be a similar process to scour at a pier of the same dimension as the abutment and its mirror image in the wall of the channel. Shen *et al.* [75] were the first to test this

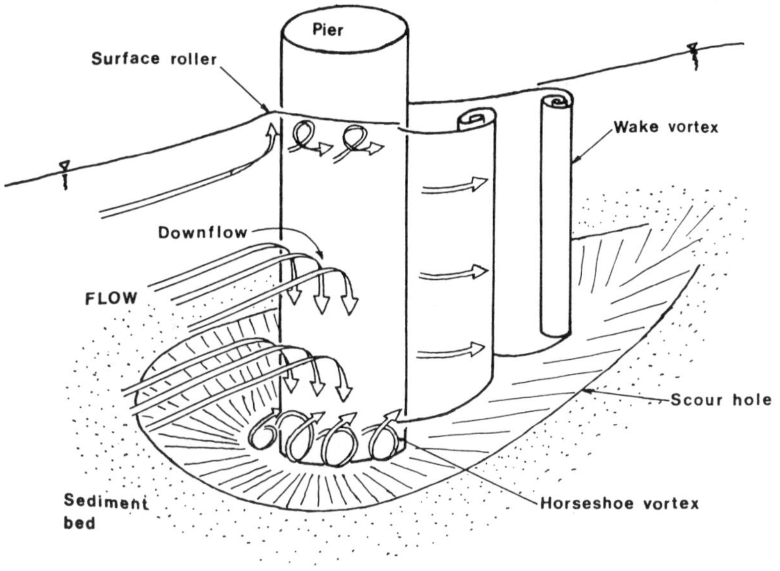

FIGURE 17. The flow pattern adjacent to a cylindrical pier.

hypothesis and from a limited study concluded that it was approximately correct. Lemos [54] drew a similar analogy between the flow structures at an abutment and a pier, although he recognised a basic difference between the two systems. This is the deceleration of the flow by the boundary layer at the (abutment) channel wall.

Liu *et al.* [57] and Wong [87] made detailed observations of scour hole development and flow structures at wing-wall and spill-through abutments. They consider the approach flow to consist of an upper and a lower layer which separate into an upflow and a downflow on hitting the abutment, Figure 18. The upflow forms a surface roller, while the downflow rolls up to form the bottom vortex, called the 'principal vortex', Figure 19.

On approaching the abutment the upper layer tends to divide, Figure 18, part of the flow accelerating around the upstream corner of the abutment, the remaining flow slowly circulating in a near-stagnant pool ahead of the abutment. Within the lower layer, the streamside flow dives diagonally down while the bankside flow dives directly down to the bottom of the scour hole, Figure 18. In this way, the principal vortex, which is analogous to the horseshoe vortex at a pier, is formed. The spiralling motion of the principal vortex induces a weak secondary vortex of opposite sense.

Generally scouring commences at the upstream corner of the abutments, where a shallow depression is excavated due to the flow acceleration at this point. The depression rapidly propagates around the upstream corner of the abutment and a small mound develops downstream of it. As the scour hole develops, bed material is eroded from the entrainment zone adjacent to the abutment and is swept up the slope and around the sides of the abutment by the principal vortex. A ridge forms at the interface between the principal vortex and the secondary vortex, Figure 20. The entrainment zone is analogous to the groove at the cylindrical pier. Above the ridge, the bed material collapses in irregular avalanches of sediment into the entrainment zone. Generally, the scour process exhibits many similarities to that at a pier.

Kwan [43] made detailed observations of the flow past vertical-wall abutments extending well into the flow. Typical results are shown in Figure 21. In principle, the flow structures and scour hole geometry are similar to those recorded by Wong [87] at wing-wall and spill-through abutments, although the stagnation region is more extensive and the scour activity is strongest near the end of the abutment where the spiralling flow is most concentrated. For convenience, the former type of abutment is called a 'long abutment' while the latter type is called a 'short abutment'.

Local Scour Formulae

Many formulae have been proposed for the prediction of local scour depths at bridge crossings and only a summary of the better known ones is included here. Melville [60] and Breusers *et al.* [11] present more comprehensive surveys.

The formulae can be classified into *clear-water* and *live-bed* local scour formulae although there are also formulae which fail to recognise the difference.

Inglis [34] proposed the *Inglis–Lacey* formula for estimating local scour depths at bridge crossings. The formula is a statement that the total scour depth measured from the water surface, $D_{SL} = d_{SL} + y_0$ is proportional to the Lacey regime depth, y_r given by Equation (3). For piers $D_{SL} = 2\ y_r$, while for spurs D_{SL} varies from 1.7 y_r to 3.8 y_r depending on the length, angle and slope. The formula presumably applies to live-bed conditions.

Garde [25] and Garde *et al.* [26] proposed that local scour at bridge crossings should be expressed as

$$\frac{D_{SL}}{y_0} = (4\eta_1\eta_2\eta_3/M)Fr^n \tag{27}$$

where $Fr = U/\sqrt{gy_0}$, U is mean approach flow velocity, y_0 is mean approach flow depth, $\eta_1\eta_2\eta_3$ are functions of the particle drag coefficient C_D, Froude number, F_r and foundation shape respectively and n is a function of C_D as shown in Figure 22. The drag coefficient is expressed as

$$C_D = \frac{4}{3}\frac{\Delta\gamma_s d}{\rho\omega^2} \tag{28}$$

where $\Delta\gamma_s$ is submerged specific weight of the bed material of size d and fall velocity ω and ϱ is water density. The ratio $M = (B - b)/B$ is the opening ratio where B is the channel width at the bridge crossings and b is the width of pier. For abutments of projected length L measured across the channel, $M = (B - L)/B$. For a vertical wall spur or abutment $\eta_3 = 1.0$, while for a semi-circular pier nose $\eta_3 = 0.90$ and for an elliptic or lenticular pier nose $\eta_3 = 0.80$ to 0.70, depending on the length to width ratio of the pier. The method, by implication, applies to live-bed scour.

Laursen and Toch [48] proposed a design curve from model studies for local live-bed scour at a square-nosed pier aligned with the flow. Neill expressed the curve as

$$d_{SL} = 1.5b^{0.7}y_0^{0.3} \tag{29}$$

which is also known as the *Laursen I* formula. For clear-water scour, Laursen and Toch [21] proposed

$$d_{SL} = 1.35b^{0.7}y_0^{0.3} \tag{30}$$

and presented multiplying factors K_θ, for the effect of angle of attack (Figure 23) and K_s, for shape effect (Table 3).

Laursen [49,50,51] extended the solutions for the long rectangular constriction to local scour at abutments and piers using the observation that the depth of local scour does not depend on the constriction ratio until the scour holes from neighboring piers or abutments start to overlap. For sand,

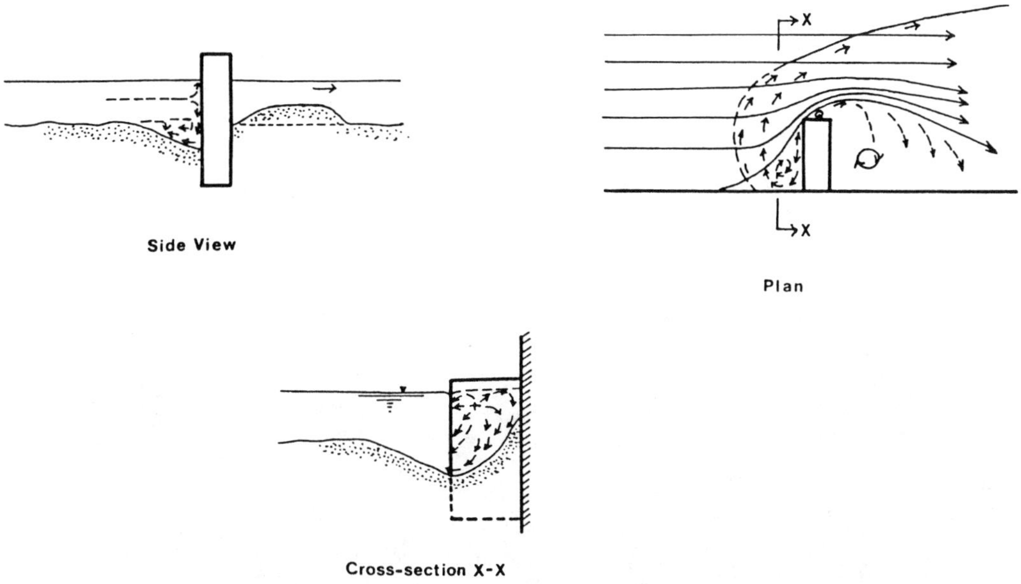

FIGURE 18. Flow patterns in the vicinity of the local scour hole at an abutment, Liu et al. [57].

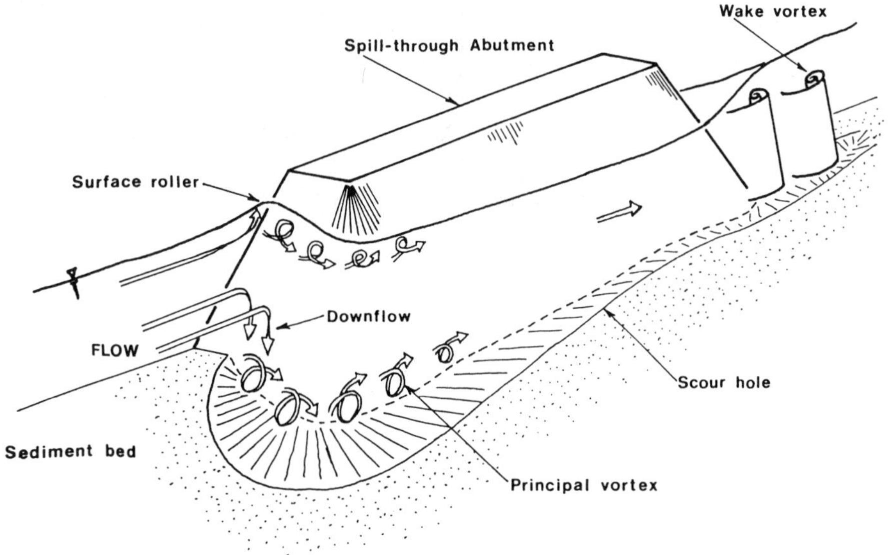

FIGURE 19. Vortex systems at a wing-wall abutment.

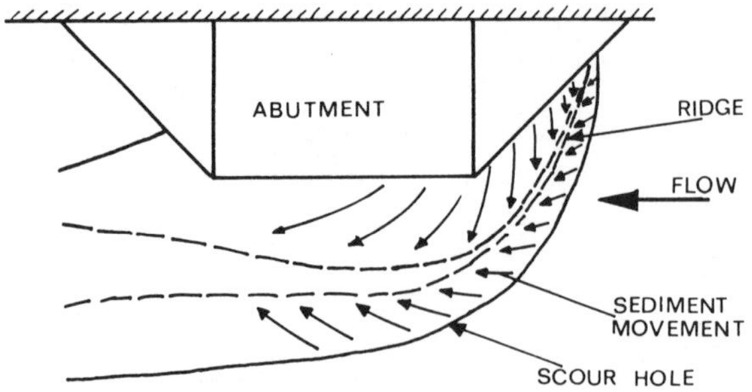

FIGURE 20. Plan-view of local scour features at a wing-wall abutment.

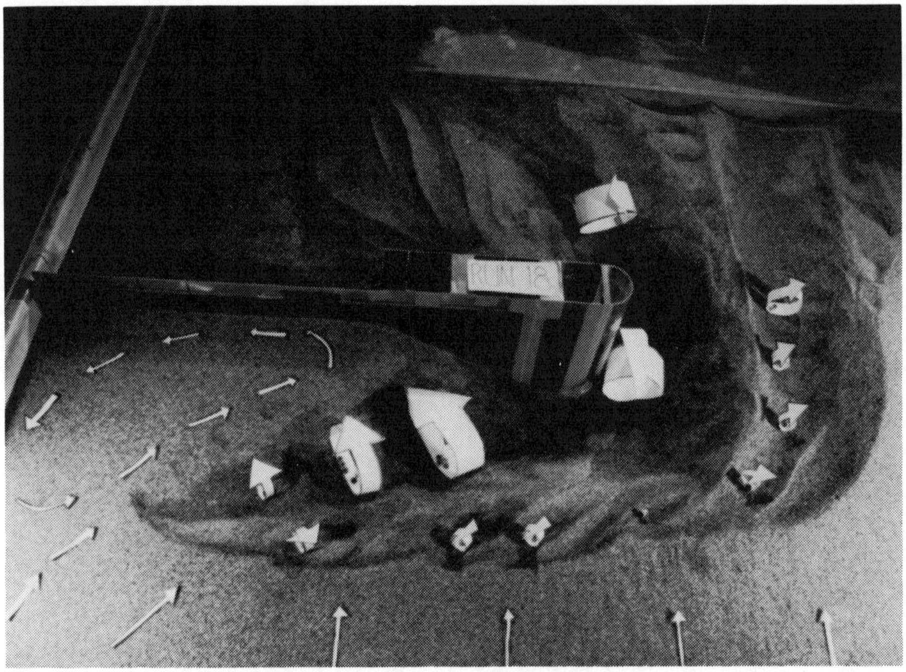

FIGURE 21. Perspective view of vortex systems at a long vertical-wall abutment, Kwan [43].

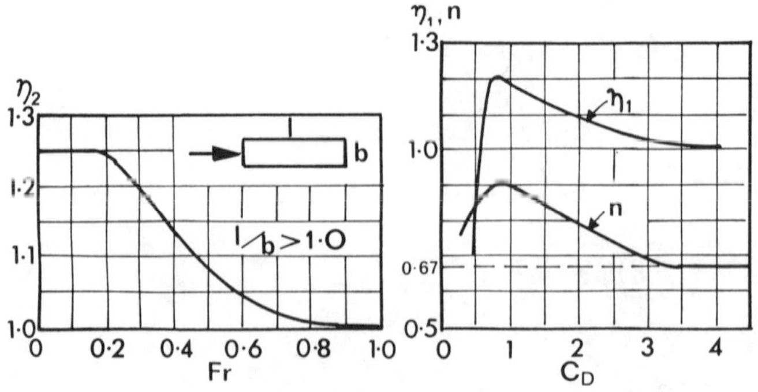

FIGURE 22. Coefficients for use in the Garde [25] equation.

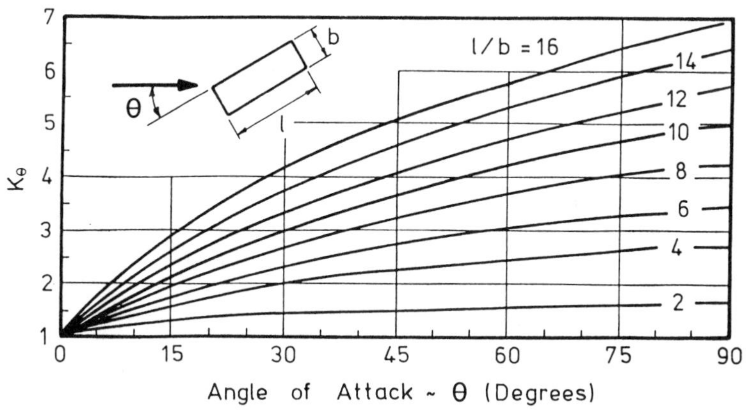

FIGURE 23. Alignment factor K_θ for piers, Laursen [49].

the width of scour hole normal to the flow, B was observed to be about 2.75 d_{SL}. The adaptation of the long constriction solution is shown in Figure 24. The scour in the constriction defined by B is assumed to be a fraction $1/r$ of the scour depth d_{SL} at the abutment, i.e. $d_{SL}/r + y_0$ is the downstream depth. The unknown channel flow Q_m is obtained from

$$\frac{Q_B}{B} = \frac{Q_m}{2.75 d_{SL}} \tag{31}$$

The width B is first estimated and then adjusted by trial to satisfy $B = 2.75\ d_{SL}$. With these substitutions, Equation (21) becomes

$$\frac{Q_0 B}{Q_B y_0} = 2.75 \left[\frac{d_{SL}}{y_0}\left(\frac{d_{SL}}{r y_0} + 1\right)^{7/6} - 1\right] \tag{32}$$

where Q_0 is the flow over the flood plain and Q_B is the flow in the channel. The equation applies to live-bed scour at an overbank bridge constriction. The relationship is shown in Figure 25 as a design curve ($r = 4.1$) and can be simplified to

$$\frac{d_{SL}}{y_0} \simeq 1.13 \left(\frac{Q_0 B}{Q_B y_0}\right)^{1/2} \tag{33}$$

If the overbank flow Q_0 is small and there is no cross flow from the channel to the flood plain immediately upstream of the constriction, the use of the lower limit to the shaded area is recommended. For a skewed abutment, a multiplying factor, K_θ was given, Figure 26.

The same approach is used to deal with an abutment encroaching into the main channel, based on Equation (22). The effective length of the abutment is L and it obstructs a flow Q_L. With $Q_L/(Ly_0) = Q_B/(B/y_0)$, $B_1 = L + 2.75\ d_{SL}$, $B_2 = 2.75\ d_{SL}$, $m'' = 0.59$ and by trial and error adjustment, $B = B_2$, Equation (22) can be written

$$\frac{L}{y_0} = 2.75 \frac{d_{SL}}{y_0}\left[\left(\frac{d_{SL}}{r y_0} + 1\right)^{1.7} - 1\right] \tag{34}$$

which can be simplified to

$$\frac{d_{SL}}{y_0} \simeq 1.57\sqrt{L/y_0} \tag{35}$$

The relationship is shown in Figure 26 for $r = 11.5$. For other values of m'', the d_{SL} from Figure 26 is adjusted by a multiplication factor, K_τ given in Figure 26. Substitution of pier width, $b = 2L$ in Equation (34) leads to the *Laursen*

TABLE 3. Shape Coefficients K_s for Nose Forms (to be Used Only for Piers Aligned with Flow).

Nose Form	Length–Width Ratio	K_s
Rectangular		1.00
Semi-circular		0.90
Eliptical	2:1	0.80
	3:1	0.75
Lenticular	2:1	0.80
	3:1	0.70
Abutments:		
Vertical wall		1.00
Wing wall		0.90
Spill through		0.80

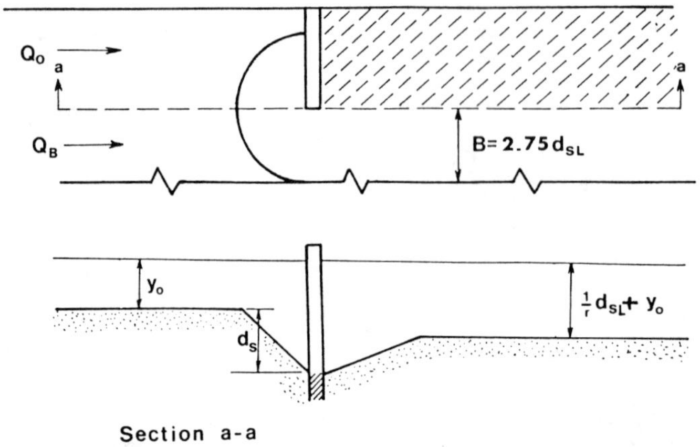

FIGURE 24. Definition sketch of an overbank bridge constriction, Laursen [49].

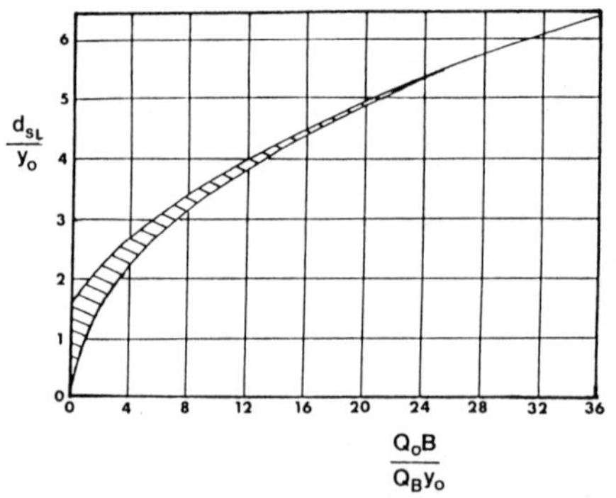

FIGURE 25. Design curve for an overbank bridge constriction, Laursen [49].

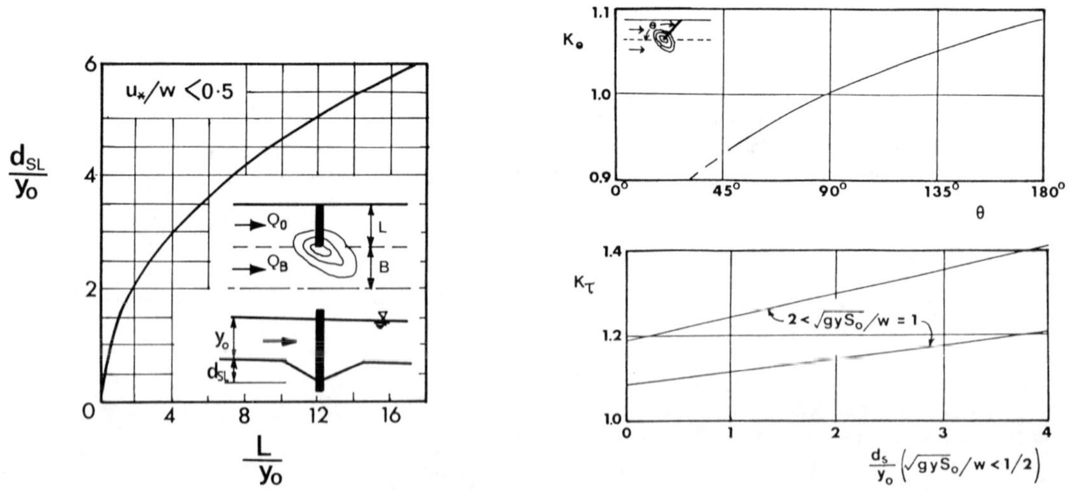

FIGURE 26. Design curve for local scour depth at abutment of length L [Equation (34)] and correction factors K_τ for type of sediment transport and K_θ for angle of attack.

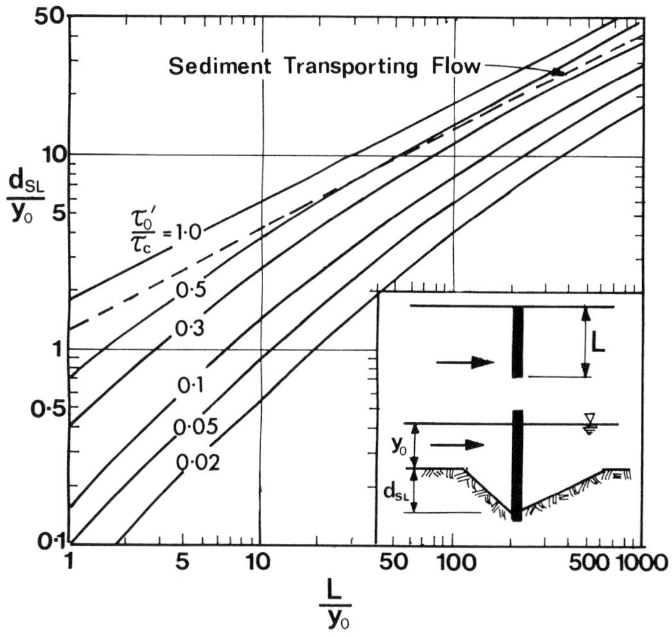

FIGURE 27. Local clear-water scour depths at an abutment [Equations (34) and (37)], Laursen [51].

II formula, for local live-bed scour depth at piers

$$\frac{b}{y_0} = 5.5\,\frac{d_{SL}}{y_0}\left[\left(\frac{d_{SL}}{ry_0} + 1\right)^{1.7} - 1\right] \tag{36}$$

Laursen [51] similarly modified Equation (19) for clear-water scour in a rectangular long constriction. Substituting $B_1 = L + 2.75\,d_{SL}$ and $B_2 = 2.75\,d_{SL}$, as before, gives

$$\frac{L}{y_0} = 2.75\,\frac{d_{SL}}{y_0}\left[\frac{\left(\dfrac{d_{SL}}{ry_0} + 1\right)^{7/6}}{(\tau_0/\tau_c)^{1/2}} - 1\right] \tag{37}$$

for local clear-water scour at an abutment encroaching into the main channel.

Equation (37) for $\tau_0/\tau_c = 1$ can be simplified to

$$\frac{d_{SL}}{y_0} \simeq 1.89\sqrt{L/y_0} \tag{38}$$

Substituting $b = 2L$ in Equation (37) leads to the *Laursen III* formula for local clear-water scour at piers

$$\frac{b}{y_0} = 5.5\,\frac{d_{SL}}{y_0}\left[\frac{\left(\dfrac{d_{SL}}{ry_0} + 1\right)^{7/6}}{(\tau_0/\tau_c)^{1/2}} - 1\right] \tag{39}$$

Equations (34) and (37) are plotted in Figure 27 for $r = 11.5$

Liu *et al.* [57] presented a set of design equations for local scour depth at abutments. For live-bed scour at spill-through abutments, they proposed

$$\frac{d_{SL}}{y_0} = 1.1\left(\frac{L}{y_0}\right)^{0.4} Fr^{0.33} \tag{40}$$

in which L is the projected length of abutment at the undisturbed bed level. For live-bed scour at wing wall or other vertical-wall abutments the equivalent equation is

$$\frac{d_{SL}}{y_0} = 2.15\left(\frac{L}{y_0}\right)^{0.4} Fr^{0.33} \tag{41}$$

while the clear-water scour formula for vertical-wall abutments is

$$\frac{d_{SL}}{y_0} = 12.5 Fr\beta \tag{42}$$

in which β is the constriction ratio, $B/(B - L)$.

Larns [47] proposed for local live-bed scour at piers

$$d_{SL} = 1.05 K_L b^{0.75} \tag{43}$$

TABLE 4. K Values for Equation (43).

Shape of Piers	Elongation of piers*	K_x	K_y 0°	10°	15°	20°	30°	40°
Circular		1.0	1.0	1.0	1.0	1.0	1.0	1.0
Lenticular	2.0	0.91	0.91				1.13	
	3.0	0.76	0.76	0.98	1.02	1.24		
	4.0	0.73	0.76		1.12		1.50	2.02
	7.0	0.41						
Joukowski	4.0	0.86	0.86		1.09		1.40	1.97
	4.1	0.76						
	4.5	0.76					1.36	
Elliptical	2.0	0.91	0.91				1.13	
	3.0	0.83	0.83	0.98	1.06	1.24		
Ogival	4.0	0.92	0.92		1.18		1.51	
Circular Double	4.0	0.95						
Oblong	1.0	1.0						
	1.5	1.0						
	2.0	1.0	1.0				1.17	
	3.0	1.0	1.0	1.02	1.13	1.24		
	4.0	1.03	1.0		1.15		1.52	
	4.5						1.60	
Rectangular chamferred	4.0	1.01						
Rectangular	0.25	1.30						
	2.0		1.11		1.38		1.56	1.65
	4.0	1.40	1.11		1.72		2.17	2.43
	4.5	1.25					2.09	
	5.3	1.40						
	6.0		1.11		2.20		2.69	3.05
	8.0		1.11		2.23		3.03	3.64
	9.3	1.40						
	10.0		1.11		2.48		3.43	4.16

*Pier length in plan to width ratio.

TABLE 5. Coefficient Values for Use in the Maza [50] Formula.

θ	P_θ	Q_1/Q	P_q	Slope S	P_s
30°	0.84	0.10	2.00	0	1.0
60°	0.94	0.20	2.65	0.5	0.91
90°	1.00	0.30	3.22	1.0	0.85
120°	1.07	0.40	3.45	1.5	0.83
150°	1.188	0.50	3.67	2.0	0.61
		0.60	3.87	3.0	0.50
		0.70	4.06		
		0.80	4.20		

TABLE 6. Multiplying Factors for Abutment Shape and Position.

Shape	K_s	Position	K_b
Vertical board	1.00	Straight channel or entry to bend	1.00
Narrow vertical wall	1.00	Concave side of bend	1.10
45° wing wall	0.85	Convex side of bend	0.80
1½:1 full spill through	0.65	Below bend, concave side	
1½:1 spill through with vertical wall below normal bed level	0.80	(i) Sharp bend	1.40
		(ii) Moderate bend	1.10

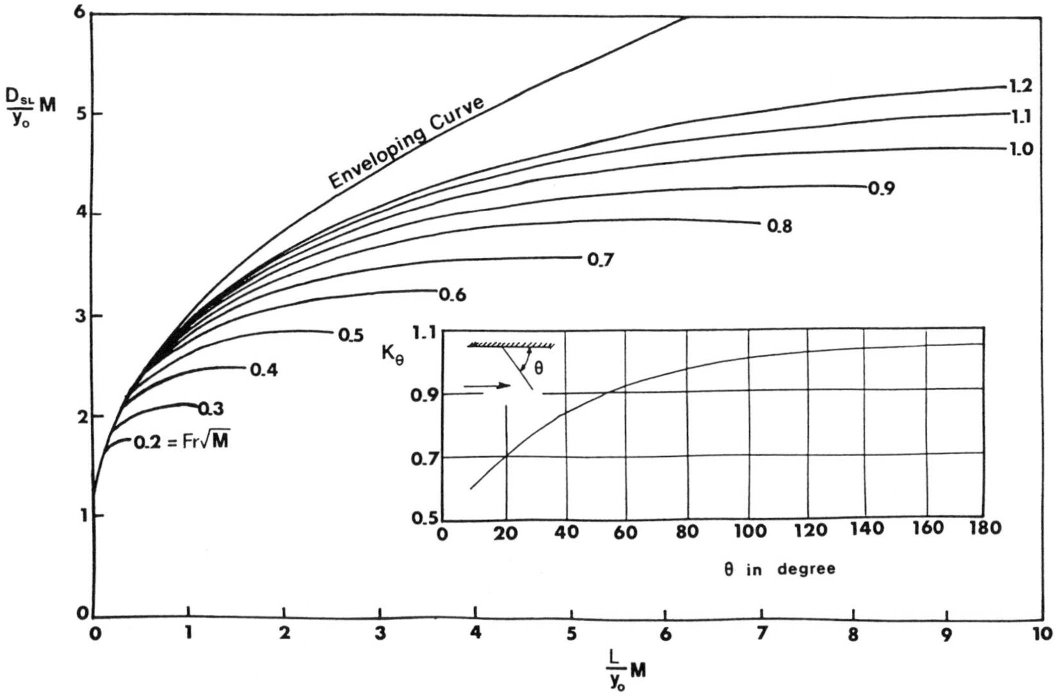

FIGURE 28. Field's design chart for local scour depth at an abutment with correction factor K_θ for angle of attack.

where $K_L = K_x K_y$ is a multiplying factor which accounts for pier shape and angle of attack, Table 4.

For the same conditions, Breusers [10] proposed

$$d_{SL} = 1.4b \tag{44}$$

while the Blench [7] formula also for the same conditions, is

$$\frac{D_{SL}}{y_r} = 1.8\left(\frac{b}{y_r}\right)^{0.25} \tag{45}$$

in which y_r is the regime depth given by Equation (4).

Shen *et al.* [75,76] proposed a clear-water formula in terms of the pier Reynold's number $R_e = Ub/v$, known as the *Shen I* formula

$$d_{SL} = 0.000223R_e^{0.619} \tag{46}$$

The authors also proposed the *Shen IIa* formula for local clear-water scour depth at piers

$$\frac{d_{SL}}{y_0} = 2Fr^{0.43}\left(\frac{b}{y_0}\right)^{0.645} \tag{47}$$

and the *Shen II* formula for live-bed conditions

$$\frac{d_{SL}}{y_0} = 3.4Fr^{2/3}\left(\frac{b}{y_0}\right)^{2/3} \tag{48}$$

For local scour depth at abutments, *Maza* [50] proposed

$$d_{SL} = P_\theta P_q P_s y_0 \tag{49}$$

in which P_θ, P_q and P_s are adjustment factors for angle of attack, overbank flow to total flow ratio, Q_1/Q and slope of abutment, $S(S = 0$ corresponds to vertical wall) given in Table 5.

Field [23] developed a design chart, Figure 28, for local scour depth at piers and abutments based on Liu *et al.* [57] data for vertical wall abutments. Multiplying factors for abutment shape, location and angle of attack are given in Table 6 and Figure 28. For pier scour the multiplying factors by Laursen (Figure 23 and Table 3) were proposed. Clear-water scour conditions were evaluated with y_c and U_c instead of y_0 and U. These are the flow depth and velocity for threshold conditions in the approach channel.

Coleman [18] proposed a formula for local live-bed scour at piers, which can be expressed as

$$\frac{d_{SL}}{y_0} = 1.39Fr^{0.2}\left(\frac{b}{y_0}\right)^{0.9} \tag{50}$$

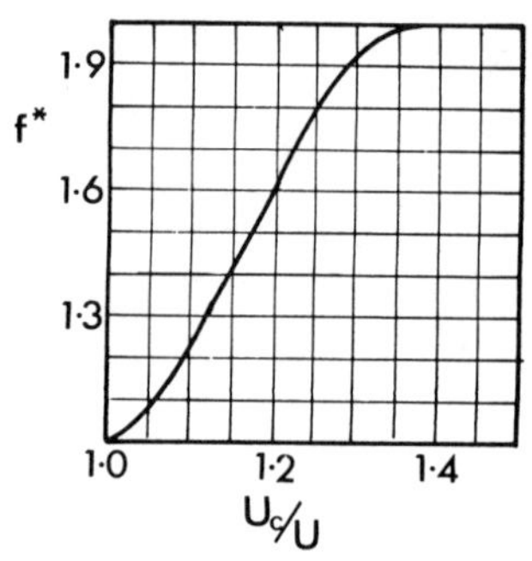

FIGURE 29. Values of f_* for use in the Bonasoundas [9] Equation (53).

Hancu [31] proposed equations for local scour at piers under clear-water conditions, the *Hancu I* formula

$$\frac{d_{SL}}{y_0} = 2.42\left(\frac{2U}{U_c} - 1\right)\left(\frac{U_c^2}{gb}\right)^{1/3} \quad (51)$$

and under live-bed conditions, the *Hancu II* formula

$$\frac{d_{SL}}{y_0} = 2.42\left(\frac{U_c^2}{gb}\right)^{1/3} \quad (52)$$

Bonasoundas [9] proposed for local scour at piers

$$\frac{d_{SL}}{y_0} = a_i\left(\frac{b}{y_0} - 0.60\right)^{1/3} f_* \quad (53)$$

in which

$$a_i = 2.00 - 0.88U_c/U$$

for $U_c/U < 1$ (live bed)

$$a_i = 4.65 - 2.55U_c/U$$

$a_i = 2.55(U_c/U)^{-10/3}$ for $1.6 \leq U_c/U$ (clear-water)

and U_c/U is given by

$$\frac{U_c}{U} = \frac{(S_s - 1)^{1/2}}{Fr}\left(\frac{d}{y_0}\right)^{7/10} \quad (54)$$

in which d is particle size. The function f_* is given in Figure 29.

From data collected at rodk dykes on the Mississippi River, Simons and Senturk [77] proposed

$$\frac{d_{SL}}{y_0} = 4Fr^{0.33} \quad (55)$$

They recommended use of this equation for local scour depth at abutments when $L/y_0 > 25$. For $b < L/y_0 < 25$ they recommended Equation (40).

Breusers *et al.* [11] suggested, for local scour depth at piers

$$\frac{d_{SL}}{b} = f\left(\frac{U}{U_c}\right) 2.0 \tanh\left(\frac{y_0}{b}\right) K_s K_\theta \quad (56)$$

in which

$$f(U/U_c) = 0 \text{ for } U/U_c < 0.5$$

$$= 2U/U_c - 1 \text{ for } 0.5 < U/U_c < 1.0 \quad (57)$$

$$= 1 \text{ for } U/U_c > 1.0$$

They also recommended $K_s = 1.0$ for circular and rounded piers, $K_s = 0.75$ for streamlined shapes and $K_s = 1.3$ for rectangular piers and proposed the Laursen [49] chart for angle of attack (Figure 23).

In Table 7, a summary of the design formulae included here, is presented. Melville [61] and Raudkivi and Sutherland [72] compared the various pier formulae on non-dimensional plots, while Wong [87] compared the abutment formulae in the same way. Melville [61] also compared the results obtained when the pier formulae were used to predict local scour depths under given conditions. Generally, these comparisons indicate substantial differences in the predicted scour depths.

Parameters Influencing Local Scour

The relationshp between the equilibrium depth of local scour d_{SL} and its dependent parameters can be written [79]

$$d_{SL}/b = f(u_*/u_{*c}, y_0/b, b/d_{50}, \sigma, b/B, K_s, \theta) \quad (58)$$

or

$$d_{SL}/y_0 = f(u_*/u_{*c}, b/y_0, b/d_{50}, \sigma, b/B, K_s, \theta) \quad (59)$$

in which $u_* = \sqrt{gy_0S_0}$ is shear velocity, y_0 is mean approach flow depth, S_0 is energy slope, u_{*c} is the threshold value of u_*, b is width of bridge pier measured perpendicular to the flow, d_{50} is mean particle size, B is channel width, K_s is a pier or abutment shape factor and θ is the angle between the pier or abutment and the approach flow. For abutments, the pier width b is replaced by L, the abutment length measured perpendicular to the flow.

Equations (58) and (59) are identical except for the use of b and y_0 respectively to normalise scour depth. It is generally accepted [22] that Equation (58) is appropriate to local scour at piers, i.e. that the local scour depth scales with pier size. The situation at abutments is less clear.

TABLE 7. Summary of Equations for Local Scour Estimation.

	Piers		Abutments	
	Clear-Water	Live-Bed	Clear-Water	Live-Bed
Blench		45		
Bonasoundas	53	53		
Breusers		44		
Breusers et al.	56	56		
Coleman		50		
Garde		27		27
Hancu	51	52		
Inglis-Lacey		3		3
Larras		43		
Laursen	30, 39	29, 36	37	32, 34
Liu et al.			42	40, 41
Maza				49
Shen	46, 47	48		
Simons and Senturk				55

Melville and Raudkivi [64] suggested that the appropriate functional relationship depends on whether the abutment length, L is large or small in relation to the flow depth, y_0. They proposed that Equation (58) may apply to short abutments, i.e. small L/y_0 while Equation (59) may apply to longer abutments, i.e. large L/y_0.

Equation (58) essentially gives the relationship between d_{SL} and y_0 assuming that d_{SL} scales with L. Conversely, Equation (59) gives the relationship between d_{SL} and L, assuming that d_{SL} scales with y_0. If it is accepted that local scour depth at a pier scales with pier size, then it is reasonable to assume, by analogy, that local scour depth at short abutments scales with abutment length. However, for longer abutments, the pier analogy becomes unrealistic and it is unlikely that local scour depth will increase in proportion to abutment length. In this case, it is more likely that local scour depth scales with flow depth and that there is a limiting abutment length, for a given flow depth, beyond which d_{SL} is insensitive to L. Laboratory data, Figure 35, demonstrate this trend.

Effects of Specific Parameters

EFFECT OF FLOW VELOCITY

The depth of local scour at bridge *piers* is closely related to the undisturbed approach flow velocity. It is evident from published literature [22,75,58] that under clear-water conditions, the local scour depth increases almost linearly with velocity to a maximum at the critical velocity.

Early research into the effect of approach flow velocity on local scour depth at piers under *live-bed* conditions produced a number of equations relating d_{SL} to Froude number, Fr. Following this, Chabert and Engeldinger [14], Laursen [51] and others suggested that the local live-bed scour depth is about 10% less than the threshold depth irrespective of the velocity. More recently it has been shown that as the velocity exceeds the threshold velocity, the local scour depth first decreases and then increases again.

Figure 30(a) shows experimental data for local equilibrium scour depths at cylindrical piers in uniform sediments by Chee [15]. A similar plot was produced by Chiew [16], Figure 30(b). These data have been adjusted so that they are independent of flow depth and sediment size effects. Raudkivi [71] summarised the data in a diagram, Figure 31. The band, indicated by the dotted lines, signifies the range of fluctuations in scour depth due to the passage of bed features. Melville [62] demonstrated that data by other researchers [22,31,37,76] when adjusted for flow depth and sediment size effects, compare favourably with Figure 31.

The local scour depth ratio for piers under clear-water conditions increases rapidly with flow velocity (or shear velocity) towards a peak value of about 2.3 at the threshold condition. The data are separated, according to grain size, i.e. ripple-forming ($d < \sim 0.7$ mm) and non-ripple-forming sediments. With the latter, experiments can be run with $u_* \simeq 0.95\, u_{*c}$ without the approach bed deforming. With fine sands, under these conditions, ripples form on the approach bed and a small transport of bed material occurs. Thus, the clear-water conditions are not maintained for sands long enough to reach the peak scour depth as obtained with non-ripple forming sediments.

Under live-bed conditions, there is a flow of sediment into the local scour hole from upstream and the depth of scour is generally less than the threshold peak. The sediment input leads to a reduction in the scour depth which reaches a minimum at about $U/U_c = 1.5 - 2$, when the bed features are relatively short and steep. Thereafter the scour depth increases again and reaches a new peak at the transition flat

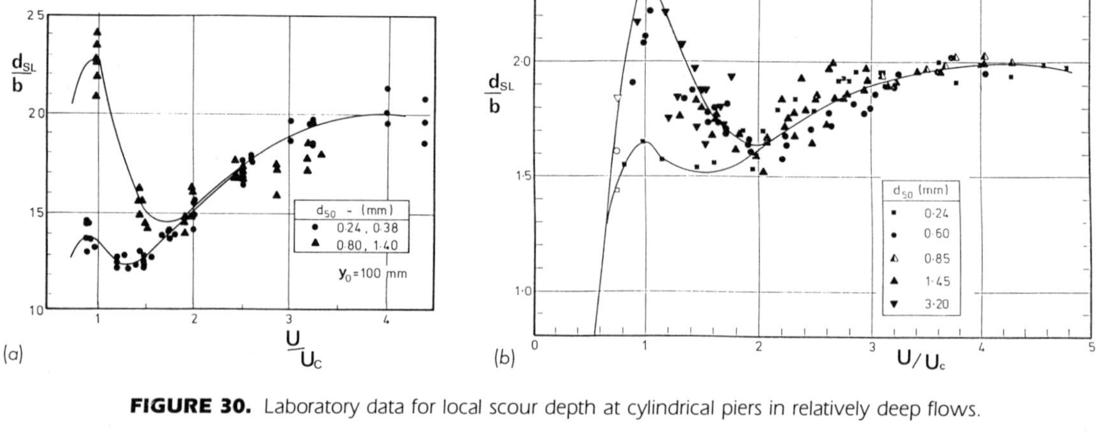

FIGURE 30. Laboratory data for local scour depth at cylindrical piers in relatively deep flows.

bed condition when the greatest fraction of the flow energy goes into sediment transport. At still higher velocities antidunes dissipate some energy and the local scour depth appears to decrease again. The second peak does not exceed the first peak except for fine sands. However, equilibrium is reached much more rapidly under live-bed conditions and the second peak may be the critical condition for design because clear-water conditions may not last long enough for the scour depth, associated with the first peak, to be attained.

The design formula by Breusers *et al.* [11] (Equation (54)) is plotted in Figure 31 for a cylindrical pier at a large value of y_0/b. It is seen that Equation (54) is conservative at all velocities except near the threshold condition for non-ripple-forming sediment when the equation underpredicts by about 15%.

The effect of flow velocity on local scour depth at bridge piers in *non-uniform sediments* is discussed below.

Little is known concerning the relationship between local scour depth at bridge *abutments* and flow velocity ratio.

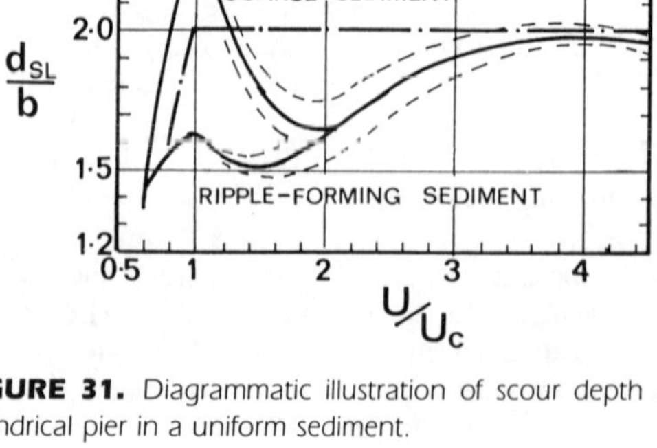

FIGURE 31. Diagrammatic illustration of scour depth at a cylindrical pier in a uniform sediment.

Melville and Raudkivi [64] plotted data by Gill [29] for spur dykes and compared these with Chiew's [16] cylindrical pier data, Figure 32. A similar trend is demonstrated in both sets of results, although no conclusions can be drawn about the relative magnitude of the results because, although the pier data are independent of flow depth effect ($y_0/b \geq 3.5$), the abutment data are assumed to contain an effect of flow depth as measured by the parameter $y_0/2L$.

EFFECT OF FLOW DEPTH

Observations show that at shallow flow depths the local scour at *piers* increases with flow depth but as the water depth increases, the scour depth becomes almost independent of flow depth. This trend is shown in the data of many researchers [5,9,11,16,22,51].

The presence of the pier causes a surface roller around the pier and the horseshow vortex at the base of the pier. The two rollers have opposite direction of rotation. In principle, so long as they do not interfere with each other, the local scour depth is independent of flow depth. With decreasing flow depth, the surface roller becomes more dominant and interferes with the approach flow and downflow which becomes less capable of entraining sediment.

Figure 33 shows laboratory data for local equilibrium scour depth at cylindrical piers in uniform sediments at threshold condition. In Figure 33, $K(y_0/b)$ is given by

$$K(y_0/b) = \frac{\dfrac{d_{SL}}{b}(y_0)}{\dfrac{d_{SL}}{b}(y_0 \longrightarrow \infty)} \tag{60}$$

where the denominator is, for a particular set of data, a constant scour depth ratio measured at a large flow depth for which d_{SL} is insensitive to y_0. $K(y_0/b)$ is thus a measure of the influence of y_0/b on local scour depth for shallow flows. It is seen that the finer the sediment relative to pier size, the smaller is the range of influence of flow depth. For fine sediment, the scour depth may be essentially independent of flow depth for $y_0/b \simeq 1.0$, whereas for coarse sediment, the ratio may be six. Chiew [16] investigated the effect of flow depth on local scour depth at cylindrical piers under live-bed conditions. His data, limited to relatively fine sediments, indicate the same trends for depth dependence as shown in Figure 33.

The flow depth effect, i.e. $\tanh(y_0/b)$, in the Breusers *et al.* [11] design formula, Equation (54), is plotted in Figure 33. It is seen that the function closely approximates to the laboratory data for fine sediments.

Similar results for the influence of flow depth are demonstrated in data for local equilibrium scour depth at abutments [29,43,79,87]. Figure 34 is a dimensionless plot of the influence of y_0/L on local scour depth at *abutments* in *uniform sediment* containing all the known data for $u_*/u_{*c} \simeq 1.0$ and for which there is no influence of relative sediment size. The data for wing-wall and spill-through abutments are reasonably consistent. Scour depths are greater for vertical-wall abutments, presumably reflecting the effects of other parameters including K_s and L/B. Laursen's design formula (Equation 36) for clear-water scour at abutments is plotted in Figure 34 for $\tau_0/\tau_c = 1$ and is seen to be an envelope to all data.

The data in Figure 34 can also be plotted to show the relationship between d_s/y_0 and L/y_0, i.e. the dependence of local scour depth on abutment length, Figure 35. The trend in these data clearly suggest that at large values of L/y_0, scour depth becomes insensitive to abutment length.

EFFECT OF SEDIMENT SIZE

Generally, early laboratory studies [17,42,48,80] of local scour depth at bridge *piers* showed the effect of sediment size to be relatively minor; presumably because of the limited range of data collected in individual projects. More recently, Nicollet and Ramette [68] and Le Clerc [52], demonstrated that sediment size can have a significant effect on local scour. Laboratory data by Ettema [22], Figure 36, show that the maximum value of the clear-water local scour depth in non-ripple-forming sediments is unaffected by sediment size as long as $b/d_{50} \geq 50$. For smaller values, individual grains are large relative to the groove excavated by the downflow and erosion is impeded. The formation of ripples with fine sands ($d_{50} \leq 0.7$) also limits the scour depth as discussed below.

When $b/d_{50} < 8$, individual grains are so large relative to the pier that scour is mainly due to erosion at the sides of the pier. In the coarse range of sediment the grains are still large relative to the groove and the porous bed dissipates some of the energy of the downflow. A peak value occurs at $b/d_{50} \cong 50$ and at higher ratios the scour depth is slightly reduced.

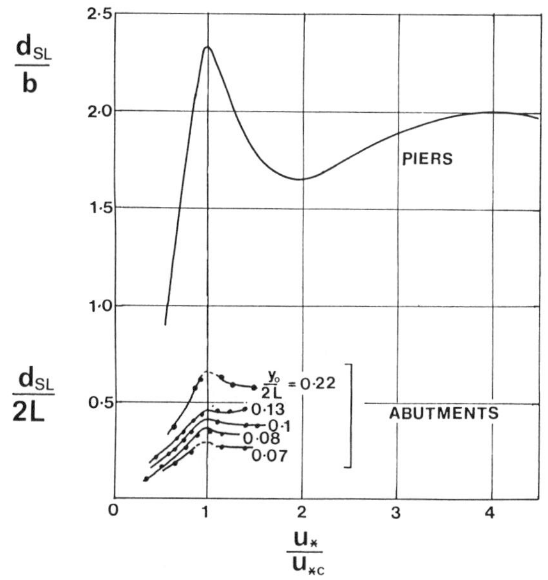

FIGURE 32. Local scour depth variation with shear velocity ratio at piers and abutments.

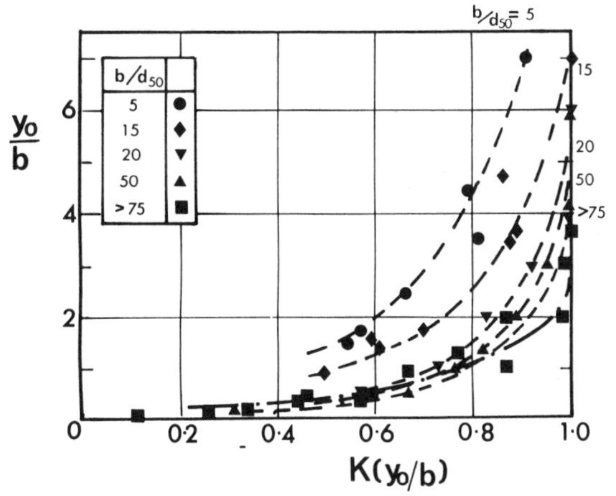

FIGURE 33. Local scour depth at threshold condition versus relative flow depth with relative pier size as parameter.

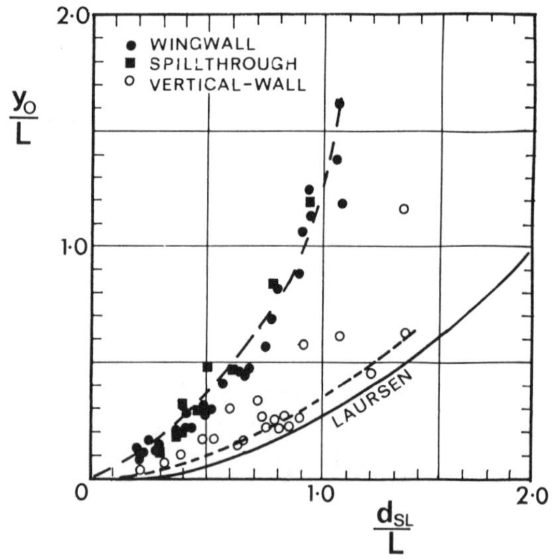

FIGURE 34. Local scour depth at abutments versus relative flow depth.

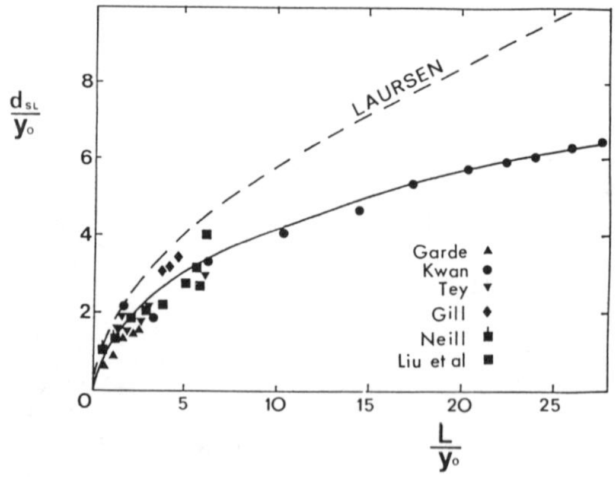

FIGURE 35. Local scour depth at abutments versus relative abutment length.

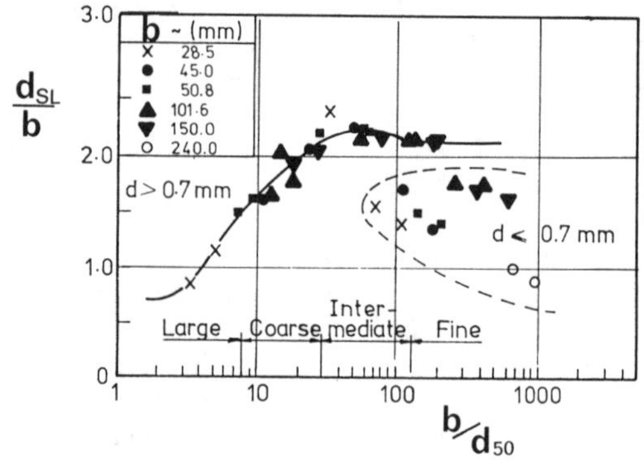

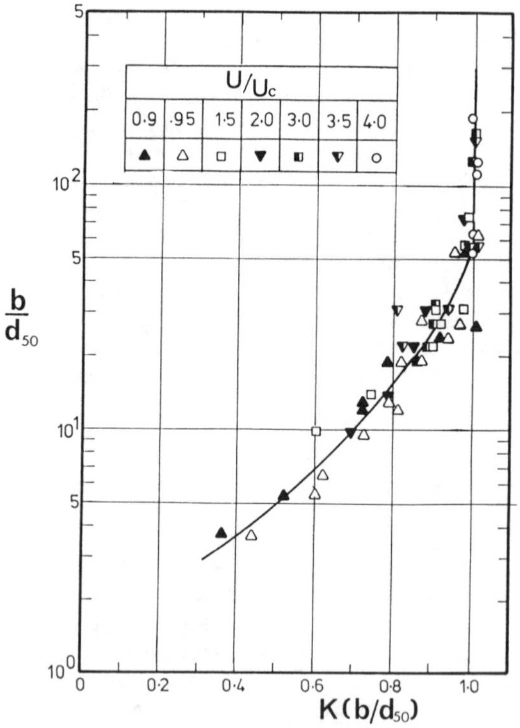

FIGURE 36. Local clear-water scour depth versus relative pier size at threshold condition and the reduction factor $K_i(b/d_{50})$.

Chiew [16] conducted additional experiments under live-bed conditions. His data together with those of Ettema [22] for threshold conditions, when normalised with the peak value at $b/d_{50} = 50$, are also shown in Figure 36. Note that the slight reduction shown in Ettema [22] data is not apparent in Chiew [16] data and has been ignored in this plot. In Figure 36, $K(b/d_{50})$ is given by

$$K(b/d_{50}) = \frac{\frac{d_{SL}}{b}(b/d_{50})}{\frac{d_{SL}}{b}(b/d_{50} = 50)} \qquad (61)$$

and is a measure of the effect on local scour depth of sediment size for coarse sediments.

Limited information is available concerning the influence of sediment size on local scour depth at *abutments*. Early studies of live-bed scour by Laursen and Toch [48] and Ahmad [1] showed that the maximum scour depth is independent of sediment size. However, Lacey [44], Blench [7], Iwasaki [36] and Garde *et al.* [26] all found the contrary to be true, while Laursen determined that sediment size affects clear-water scour but not live-bed scour. More recently, limited data by Gill [28] and Wong [87] suggest that scour depth is slightly deeper for coarser materials, the reverse trend to that shown in pier data. Generally, insufficient data are available to demonstrate systematic trends. The condition $2L/d_{50} \leq 50$ is unlikely in practice, so that in principle (by analogy with piers) local scour depths at abutments should be relatively insensitive to uniform sediment size.

EFFECT OF SEDIMENT GRADING

Ettema [21,22] carried out an extensive study of the effect of sediment gradation on local clear-water scour depth at *cylindrical piers*. The experiments were conducted at the threshold condition for the d_{50} size. He concluded that both the rate of scour and the equilibrium scour depth decrease as the standard deviation of the particle size distribution increases, Figure 37. The reason for the reduced depth of scour is the formation of an armour layer, at the base of the scour hole. In Figure 37, $\sigma_g = (d_{84}/d_{16})^{1/2}$ is geometric standard deviation and K_σ is given by

$$K_\sigma = \frac{\frac{d_{SL}}{b}(\sigma_g)}{\frac{d_{SL}}{b}(\sigma_g \longrightarrow 1.0)} \qquad (62)$$

It is seen that the depth of local scour is dependent on whether the sediment is ripple-forming or not. The reduced value of K_σ for uniform fine sediments is due to the formation of ripples on the approach bed near the threshold condition.

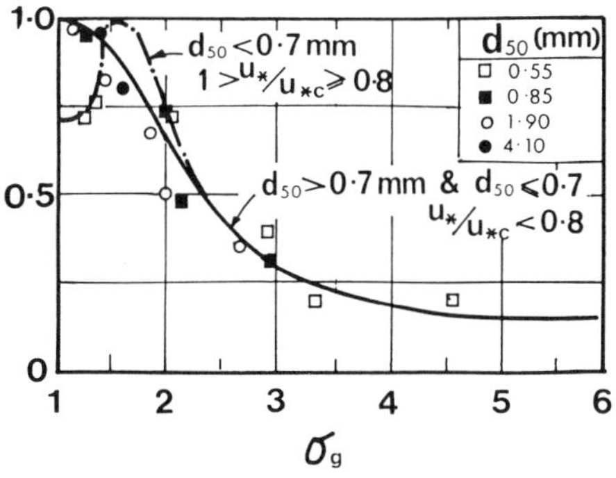

FIGURE 37. Coefficient K_σ as a function of the geometric standard deviation of the particle size distribution.

For $\sigma_g \simeq 1.5$, the approach bed surface for fine material becomes armoured before ripples are formed while no effective armouring occurs at the base of the scour hole. At higher values of σ_g, armouring becomes significant at the base of the scour hole and scour depths are reduced for all sediment sizes. Ettema [21] also determined that under clear-water conditions ($u_*/u_{*c} < 0.8$, based on d_{50} size) the fine sediments behave like coarser materials.

Little is known about the effect of sediment grading on local *live-bed* scour depth at *piers*. Laboratory data [16] indicate that if the applied shear stress is substantially higher than that required to remove the largest particle in the non-uniform sediment, the particle size is only effective in determining the size of the bed features and not the scour depth. If however, the applied shear stress is near to the critical shear stress of the largest particles, armouring by these particles can be expected to occur at the base of the scour hole, hence reducing the maximum scour depth. On the other hand, if the applied shear stress is smaller than the critical shear stress of the coarser fraction, armouring of both the bed surface and the scour hole would occur and conditions similar to those for clear-water scour in non-uniform sediments would be approached. Typical data [16] obtained in a sediment recirculating flume are shown in Figure 38. This condition, where there is a continual supply of transported material from upstream, may occur in natural river systems. Alternatively, sediment supply from upstream could approach zero in which case local scour depths can be expected to be greater as shown by the dotted line in Figure 38 and could approach the clear-water scour depth if the flow conditions at the critical shear stress for the limiting armour layer lasted long enough.

Wong [87] carried out a systematic study of the effects of sediment grading on local scour depth at wing-wall and spill-through *abutments* under clear-water conditions, Figure 39. Generally the results are consistent with Ettema [21] data for cylindrical piers.

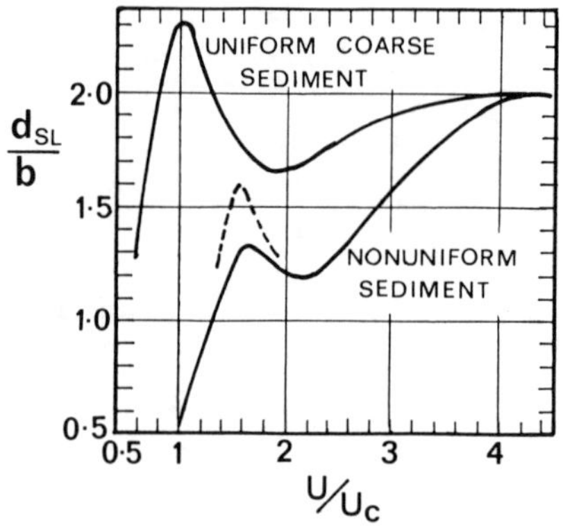

FIGURE 38. Local scour depth at piers in uniform and non-uniform sediments. Full lines in both cases are for recirculating sediment transport. Dashed line shows the armour peak with diminishing sediment transport.

EFFECT OF CONSTRICTION RATIO

The constriction ratio β influences the equilibrium depth of local scour at piers and abutments. Shen *et al.* [76] suggest that, for the purpose of experimental investigation, a flume width B should be at least 8 times the pier size, b. This value is generally accepted for clear-water experiments. However, for live-bed scour, Chiew [16] found that the flume should be at least 10 times the pier size. Otherwise scour depths are reduced due to bed features being modified as they propagate through the constriction.

Concerning abutments, Cunha [19] showed that local scour depth under clear-water conditions is insensitive to constriction ratio as long as $b/B < 0.1$, the same value suggested by Chiew [16] for live-bed scour at piers. For other values of b/B, local clear-water scour increases with constriction ratio as flow conditions approach threshold. Under live-bed conditions, local scour is little affected but a constriction scour may develop.

EFFECT OF SHAPE

The effect of pier or abutment shape is accounted for by a shape factor, K_s. The method is necessarily approximate because of the large variety of shapes, and in particular, because of changes which can occur due to debris trapped during floods or ice jams, Figure 40. In practice shape factors are important only if axial flow can be guaranteed. Even a small angle of attack of the approach flow will eliminate any benefit from shape.

The influence of *pier shape* is well established by a large number of investigations. The general conclusion is; the blunter the pier the deeper the local scour. The shape of the downstream end of pier is shown to be of little significance. Typical pier shapes are shown in Figure 41. Table 8 lists the shape coefficients recommended by different researchers.

The local scour depth also depends on the slope, in elevation, of the leading edge of the pier. For oblong shaped piers, shape factors based on a cylindrical pier of the same diameter as the thickness of the tapered pier are given in Table 9.

The estimation of local scour depth at piers is frequently complicated by the presence of a footing, a caisson or cap-

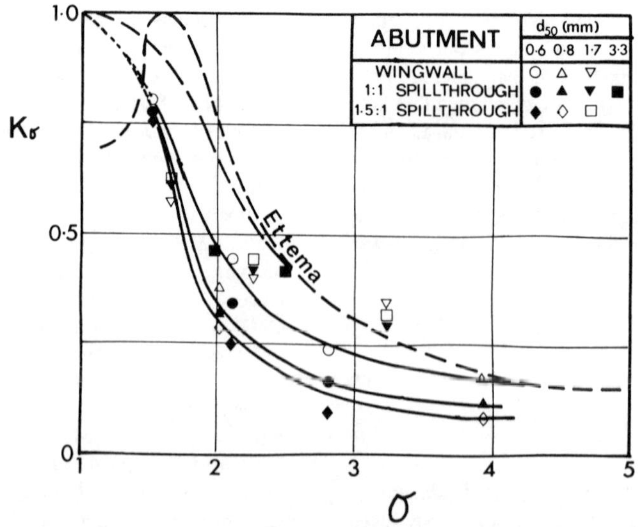

FIGURE 39. Effect of sediment non-uniformity on local scour depth at piers and abutments at threshold condition.

FIGURE 40. Effect of debris trapped at a bridge pier.

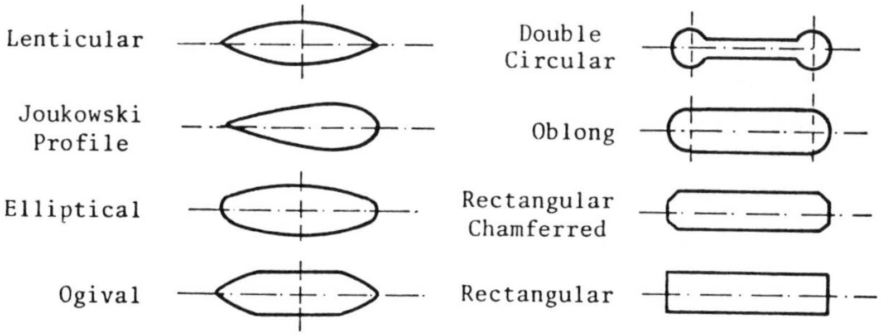

FIGURE 41. Typical pier plan profiles.

TABLE 8. Shape Factors, K_s for Piers.

	length/width	Tison (1940)	Laursen & Toch (1956)	Chabert & Engeldinger (1956)	Garde (1961)	Venkatadri (1965)	Neill (1973)	Dietz (1972)
Circular	1.0	1.0	1.0	1.0		1.0	1.0	1.0
Lenticular	2.0		0.91		0.90			
	3.0		0.76		0.80			
	4.0	0.67		0.73	0.70			
	7.0	0.41					0.80	
Parabolic Nose						0.56	0.80	
Triangular Nose 60°						0.75		0.65
Triangular Nose 90°						1.25		0.76
Elliptical	2.0		0.91		0.80		1.0	0.83
	3.0		0.83		0.75			0.80
	5.0							0.99
Ogival	4.0	0.86		0.92				
Joukowski	4.0			0.86				
	4.1	0.76						
Rectangular	1.0							1.22
	2.0		1.11					
	4.0	1.40		1.11				
	6.0		1.11					
	3.0							1.08
	5.0							0.99
Rectangular chamferred	4.0							1.01
Aerofoil	3.5							0.80

TABLE 9. Shape Factors at Tapered Piers.

	Neill		Chiew	
Shape	α	K_s	α	K_s
Cylinder		1.0		1.0
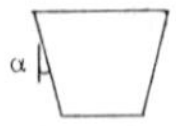	>20°	1.33	22½°	1.20
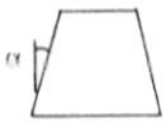	>20°	0.67	22½°	0.76

TABLE 10. Shape Factors for Abutments.

Shape	Field	Maza	Laursen	Tey
Vertical Wall	1.00	1.00	1.00	1.00
45° Wing-wall	0.85			
1½:1 Full spill-through	0.65	0.83		
1½:1 Spill-through with vertical wall below normal bed level	0.80			
Wingwall			0.90	0.70
Spill-through			0.80	0.70
2:1 Full spill-through		0.61		
2:1 Full spill-through		0.51		

ping blocks on piles. Raudkivi [71] explains that the capping or caisson with the top below the general bed level can be effective in reducing the local scour depth by interception of the downflow. However, if the top of the caisson comes to bed level, or even above, the scour depth is increased being governed by the size of the caisson rather than the pier. The practical difficulty is the determination of the lowest bed level and unless definite predictions are possible it is dangerous to rely on limitation of the scour depth due to the caisson.

Hannah [39] made a detailed study of local scour depth at pile groups. Generally he found that the maximum scour depth is closely related to the dimension of the pile group as a whole, as seen from upstream. Hannah [39] also recommended that a single line of piles should be used in preference to piers for angles of attack greater than 8°.

The influence on local scour depth of *abutment shape* is less well understood. In principle, the blunter the face of the abutment as seen from upstream, the deeper the local scour, as shown in laboratory data, Table 10. Overall, spill-through abutments produce lesser scour depths than other shapes, with significant benefits in comparison with vertical-wall types. The shape factors in Table 10 are generally based on data for short abutments. At longer abutments, it is unlikely that similar benefits of shape would be demonstrated.

EFFECT OF ALIGNMENT

The depth of local scour for all shapes of *pier*, except cylindrical, is strongly dependent on the alignment to the flow, θ. As the angle increases, the scour depth increases because the effective frontal width of the pier is increased. Laursen and Toch [48] developed the chart of multiplying factors, K_θ, given in Figure 23, which demonstrates the importance of alignment, i.e. the local scour depth at a rectangular pier, with length to width ratio 16, is tripled at an angle of attack of 15°. At bridge crossings, the angle of attack may change significantly during floods, i.e. in braided channels and also it may change progressively over a period of time, i.e. in meandering channels. Where such changes are possible, the use of cylindrical piers or other shapes with low aspect ratio (length to width ratio) is beneficial.

Field [23] proposed a chart for the effect of *abutment* alignment, Figure 28. Similar results were given by Laursen [50], Figure 26. Generally, K_θ values are much less than those for piers. This is because K_θ for an abutment is defined in terms of a constant projected length; rather than a constant length as for a pier, and also because of the effect of the boundary layer on the channel wall at an abutment. Until recently, available data [1,73,89] confirmed the Field [23] and Laursen [50] design charts, i.e. an increase in local scour depth as an abutment is pointed progressively upstream. Kwan [43] and Kandasamy [38], however, showed that for upstream pointing abutments ($\theta > 90°$) local equilibrium scour depths are less than for an abutment at $\theta =$ 90°, although their data were consistent with the Laursen and Field charts during the early development of the local scour. They showed that other data were drawn from experiments which were stopped before equilibrium was obtained and therefore did not exhibit the reversed trend. Kandasamy [38] explained that for upstream pointing abutments, the curvature of the axis of the principal vortex decreasse in later stages of scour hole development and that therefore the equilibrium local scour hole depths were less than for $\theta = 90°$.

Estimation of Local Scour Depth

BRIDGE PIERS

A generalised design formula for estimation of local scour depth at bridge piers in *uniform* bed material is

$$\frac{d_{SL}}{b} = \frac{d_{SL}}{b}(U/U_c)K\left(\frac{y_0}{b}\right)K\left(\frac{d_{50}}{b}\right)K_\theta K_s \qquad (63)$$

where $d_{SL}/b(U/U_c)$ is the local scour depth ratio at a particular flow stage and can be obtained from Figure 31. Alternatively the Breusers *et al.* [11] flow function, Equation (57), can be used as long as local scour depths for flows near the threshold condition are increased by about 15%. In Equation (63), $K(y_0/b)$ is the flow depth factor, given in Figure 33, for which the Breusers *et al.* [11] function, $\tanh(y_0/b)$, is a good approximation, $K(d_{50}/b)$ is the bed material size factor given in Figure 36, K_θ, for the effect of angle of attack, can be estimated from Laursen's [49] chart, Figure 23 and K_s for shape effects is 1.3 for rectangular piers, 1.0 for circular and rounded piers and 0.75 for streamlined shapes, as recommended by Breusers *et al.* [11].

The maximum value of $d_{SL}/b(U/U_c)$ is 2.3 and, in practice, lesser values should only be used if reliable flow predictions are possible. Similarly $K(y_0/b)$ and $K(d_{50}/b)$ should be assumed to be unity unless they can confidently be shown to have lesser value. The use of K_s values less than 1.3 is only warranted where axial flow can be guaranteed.

The effect of sediment *non-uniformity* on local *clear-water* scour depth is accounted for by the factor K_σ, given in Figure 37. For local *live-bed* scour at piers, insufficient is known about the effect of sediment non-uniformity and no reductions in local scour depths are warranted for design purposes.

In addition, attention should be given to the following special effects, as recommended by Breusers *et al.* [11]:

—*Debris* and *ice* can increase the effective size of the piers
—*Flash floods* can give a greater scour depth because of unsteady transport conditions. Also non-monsoon floods can give relatively large scour depths (Min. of Railways, India, 1968). Laursen and Toch [49] suggest a 50% increase in design scour depth.
—A *cohesive upper layer* can be distributed near the pile

and cause an increase in scour depth because upstream supply is not present (Melville [61]). The same effect is caused by vegetation in a dry period.

Intensive *suspension* of sediment in large fine-bed rivers may invalidate the empirical relations.

—*Bad placement of riprap* can provoke scour.

BRIDGE ABUTMENTS

Design recommendations for estimation of local scour depth at bridge abutments must make allowances for the significant short-comings existing in present knowledge of the subject. Laursen's clear-water formula (with $\tau_0/\tau_c = 1$) is an envelope to the reliable laboratory data for threshold conditions, as demonstrated in Figures (34) and (35). The formula can be simplified to Equation (38) which can be written

$$d_{SL}/L = 1.89(y_0/L)^{1/2} \quad (64)$$

The formula may be excessively conservative for spill-through and wing-wall abutments.

NOTATION

B = width of channel; scour hole width
b = pier width
C_D = particle drag coefficient
D_{SL} = local scour depth below water surface = $d_{SL} + y_0$
d = bed material size
d_{50} = mean particle size by weight
d_s = maximum scour depth below mean bed level
d_{SC} = depth of constriction scour
d_{SL} = depth of local scour
F_B = Blench bed factor = $1.9\sqrt{d_{50}}$
F_L = Lacey silt factor = $1.59\sqrt{d_{50}}$
Fr = Froude number = $U/\sqrt{gy_0}$
g = acceleration of gravity
H_L = head loss
h_0 = bed feature height
L = abutment length measured across the channel
M = opening ratio
m, m', m'' = exponents in constriction scour equations
n = Manning coefficient; exponent
Q = flowrate
Q_B = flowrate in channel for overbank constriction
Q_M = flowrate in main channel
Q_0 = flowrate on the flood plain
Q_s = sediment discharge rate
q_s = unit rate of sediment discharge
S, S_0 = channel slope; energy slope
S_s = specific gravity of bed material
t = time
U = mean velocity
u_* = shear velocity
x = distance downstream
y_0 = mean flow depth
y_r = regime depth of flow
z = coefficient
β = constriction ratio
Δd_s = fluctuation in scour depth due to bed feature migration
$\Delta\gamma_s$ = submerged specific weight of bed material
η_1, η_2, η_3 = parameters in Garde formula
θ = Shields Entrainment Function
θ_c = critical value of θ
ρ = water density
σ = standard deviation of particle size distribution
σ_g = geometric standard deviation of bed material
τ = bed shear stress
τ_c = critical value of τ
ω = fall velocity

REFERENCES

1. Ahmad, M., "Experiments on Design and Behaviour of Spur Dikes," *Proc. Minnesota International Hydraulics Convention*, Minneapolis, 1953.
2. Andrews, E. D., "Hydraulic Adjustment of an Alluvial Stream Channel to the Supply of Sediment," Ph.D. thesis, U.C.B., 1977, p. 152.
3. Andrews, E. D., "Scour and Fill in a Stream Channel, East Fork River, Western Wyoming," *U.S. Geological Survey Prof. Paper 1117*, U.S.G.P.O., 1979.
4. Ashida, K., "On River Bed Variation and Stable Channels in Alluvial Streams," *Bulletin of the Disaster Prevention Research Institute*, Japan, Vol. 14, Part 1, 1964.
5. Basak, V., Basamisli, Y., and Ergun, O., "Maximum Equilibrium Scour Depth around Linear-Axis Square Cross-Section Pier Groups" (in Turkish), *Devlet su isteri genel müdür lügü, Rep. No. 585*, Ankara, 1975.
6. Blench, T., "Regime Behaviour of Canals and Rivers," Butterworths, London, 1957.
7. Blench, T., "Mobile-Bed Fluviology," University of Alberta, Edmonton, Canada, 1966.
8. Blodgett, J. C., "Effect of Bridge Piers on Streamflow and Channel Geometry," *Transportation Research Record 950*, T.R.B., Sept. 1984.
9. Bonasoundas, M., "Strömungsvorgang und Kolkproblem," *Rep. No. 28*, Oscar v. Viller Institut, Techn. Univ. Munich, 1973.
10. Breusers, H. N. C., "Scour around Drilling Platforms," *Hydraulic Research 1964 and 1965, I.A.H.R.*, Vol. 19, p. 276, 1965.
11. Breusers, H. N. C., Nicollet, G., and Shen, H. W., "Local

Scour around Cylindrical Piers,'' *Journal of Hydraulic Research*, Vol. 15, No. 3, pp. 211–252, 1977.

12. Brice, J. C., ''Assessment of Channel Stability at Bridge Sites,'' *Transportation Research Record 950*, T.R.B., Sept. 1984.
13. Brice, J. C., Blodgett, J. C. *et al.*, ''Countermeasures for Hydraulic Problems at Bridges,'' Vols. 1 and 2, Federal Highway Administration, U.S. Dept. of Transportation, 1978.
14. Chabert, J. and Engeldinger, P., ''Etude des Affouillements autour des Piles des Ponts,'' *Laboratoire National d'Hydraulique*, Chatou, France, 1956.
15. Chee, R. K. W., ''Live-Bed Scour at Bridge Piers,'' University of Auckland, Dept. of Civil Engineering, Report No. 290, 79 pp., 1982.
16. Chiew, Y. M., ''Local Scour at Bridge Piers,'' University of Auckland, Dept. of Civil Engineering, Report No. 355, 197 pp., 1984.
17. Chitale, W. S., Discussion of ''Scour at Bridge Crossings,'' by E. M. Laursen, *Trans. A.S.C.E.*, Vol. 127, Pt. 1, pp. 191–196, 1962.
18. Coleman, N. L., ''Analysing Laboratory Measurements of Scour at Cylindrical Piers in Sand Beds,'' *Proc. 14th Congress I.A.H.R.*, Vol. 3, pp. 307–313, 1971.
19. Cunha, L. V., ''Time Evolution of Local Scour,'' *Proc. 16th Congress I.A.H.R.*, Sao Paulo, Vol. 2, pp. 285–299, 1975.
20. DuBoys, P., ''Le Rohne et les Rivieres a Lit Affouillable,'' *Annales des Ponts et Chausses*, Series 5, Vol. 18, pp. 141–195, 1879.
21. Ettema, R., ''Influence of Bed Gradation on Local Scour,'' University of Auckland, School of Engineering, Report No. 124, 1976.
22. Ettema, R., ''Scour at Bridge Piers,'' University of Auckland, School of Engineering, Report No. 216, 1980.
23. Field, W. G., ''Flood Protection at Highway Bridge Openings,'' University of Newcastle, NSW, *Engineering Bulletin CE3*, 1971.
24. Galay, V. J., ''Causes of River Bed Degradation,'' *Water Resources Research*, Vol. 19, pp. 1057–1090, 1983.
25. Garde, R. J., ''Local Bed Variation at Bridge Piers in Alluvial Channels,'' *University of Roorkee Research Journal*, Vol. 4, No. 1, pp. 101–106, 1961.
26. Garde, R. J., Subramanya, K., and Nambudripad, K. D., ''Study of Scour around Spur Dikes,'' *A.S.C.E., Journal of Hydraulics*, Vol. 87, HY6, 1961.
27. Gerard, R., ''Field Observations of Scour on a Small Sand Bed River,'' *Proc. of the Workshop on Bridge Hydraulics*, Banff School of Fine Arts, Alberta, Canada, 1983.
28. Gessler, J., ''Aggradation and Degradation,'' Chapter 8 of ''River Mechanics,'' Edited by H. W. Shen, 1971.
29. Gill, M. A., ''Bed Erosion in Rectangular Long Contraction,'' *Journal of the Hydraulics Division, A.S.C.E.*, Vol. 107, No. HY3, pp. 273–284, 1981.
30. Graham, D. S., ''Review of Status of Knowledge on Scour in Constrained Rivers,'' *Proceedings of the Workshop on Bridge Hydraulics*, Banff School of Fine Arts, Alberta, Canada, 1983.
31. Hancu, S., ''Sur le Calcul des Affouillements Locaux dans la Zone des Piles de Ponts,'' *Proc. 14th I.A.H.R. Congress*, Paris, Vol. 3, pp. 299–313, 1971.
32. Hannah, C. R., ''Scour at Pile Groups,'' University of Canterbury, Civil Engineering Department, Research Report No. 78-3, 92 pp., 1978.
33. Ikeda, S., ''Self-Formed Straight Channels in Sandy Beds,'' *Journal of the Waterways and Harbours Division, A.S.C.E.*, Vol. 107, No. HY4, pp. 389–406, 1981.
34. Inglis, Sir C., ''The Behaviour and Control of Rivers and Canals,'' Central Water-Power Irrigation and Navigation Research Station, Poona, Research Publication No. 13, Part 2, 1949.
35. Irvine, M. H. and Sutherland, A. J., ''A Probabilistic Approach to the Initiation of Movement of Non-Cohesive Sediments,'' *Proc. I.A.H.R. Int. Symposium on River Mechanics*, Bangkok, Vol. 1, pp. 383–394, 1973.
36. Iwagaki, I., Smith, G. L., and Albertson, M. L., ''Analytical Study of the Mechanics of Scour for Three-Dimensional Jet,'' presented at *A.S.C.E. Hydraulics Conference*, Atlanta, Georgia, August 1958.
37. Jain, S. C. and Fischer, E. E., ''Scour around Bridge Piers at High Flow Velocities,'' *A.S.C.E. Journal of Hydraulics*, Vol. 106, HY11, 1980.
38. Kandasamy, J. K., ''Local Scour at Skewed Abutments,'' Auckland University, Department of Civil Engineering, Report No. 375, June 1985.
39. Keller, R. J., ''General Scour in a Long Contraction,'' Report No. 3-77/1, MWD Central Laboratories, 1977.
40. Komura, S., ''Equilibrium Depth of Scour in Long Constrictions,'' *A.S.C.E. Journal of Hydraulics*, Vol. 92, HY5, 1966.
41. Komura, S. and Simons, D. B., ''River-Bed Degradation Below Dams,'' *Proc. A.S.C.E.*, 93, HY4, 1967.
42. Krishnamurthy, M., Discussion on ''Local Scour around Bridge Piers,'' *Proc. A.S.C.E.*, HY7, pp. 1637–1638, 1970.
43. Kwan, T. F., ''Study of Abutment Scour,'' University of Auckland, Department of Civil Engineering, Report No. 328, 225 pp., 1984.
44. Lacey, G., ''Stable Channels in Alluvium,'' *Minutes of Proceedings*, Institution of Civil Engineers, 1930.
45. Lane, E. W., ''Design of Stable Channels,'' *Trans. A.S.C.E.*, Vol. 120, pp. 1234–1279, 1955.
46. Lane, E. W. and Borland, W. M., ''River-Bed Scour During Floods,'' *Trans. A.S.C.E.*, Vol. 119, pp. 1069–1079, 1953.
47. Larras, J., ''Profondeurs Maximales d'erosion des Fonds Mobiles Autor des Piles en Riviere,'' *Annales des Ponts et Chausses*, Vol. 133, No. 4, pp. 411–424, 1963.
48. Laursen, E. M. and Toch, A., ''Scour around Bridge Piers and Abutments,'' Iowa Highway Res. Board, Bulletin No. 4, 60 pp., 1956.
49. Laursen, E. M., ''Scour at Bridge Crossings,'' Iowa Highway Research Board, Bulletin No. 8, 1958.
50. Laursen, E. M., ''Scour at Bridge Crossings,'' *Trans. A.S.C.E.*, Vol. 127, Pt. 1, pp. 166–179, 1962.
51. Laursen, E. M., ''Analysis of Relief Bridge Scour,'' *Proc. A.S.C.E.*, Vol. 89, HY3, pp. 93–118, 1963.

52. Le Clerc, J. P., "Recherche des Lois Regissant Les Phenomenes d'Affouillement au Pied des Piles de Pont, Premiers Resultats," *Proc. 14th Congress I.A.H.R.*, Paris, Vol. 3, pp. 323–330, 1971.
53. Leliavsky, S., "An Introduction to Fluvial Hydraulics," Constable and Co. Ltd., London, 257 pp., 1955.
54. Lemos, F. O., *General Report, Proceedings 16th Congress IAHR*, Sao Paulo, Brazil, 1975.
55. Leopold, L. B., Wolman, M. G., and Miller, J. P., "Fluvial Processes in Geomorphology," W. H. Freeman & Co., San Francisco, 1964.
56. Leopold, L. B. and Maddock, T. Jr., "Relation of Suspended Sediment Concentration to Channel Scour and Fill," *Proc. of the 5th Hydraulic Conference*, Iowa State University, Studies in Engineering Bulletin No. 34, 1953.
57. Liu, H. K., Chang, F. M., and Skinner, M. M., "Effect of Bridge Constriction on Scour and Backwater," Engineering Research Centre, Colorado State University, CER 60 KHL 22, 1961.
58. Maza Alvarez, J. A., "Sovocacion en Cauces Naturales," translated by A. J. Miguel-Rodriguez, University of Auckland, Dept. of Civil Engineering, Report No. 114, 1968.
59. Melville, B. W., "Scour at Bridge Sites," University of Auckland, School of Engineering, Report No. 104, 88 pp., 1974.
60. Melville, B. W., "Scour at Bridge Sites," *Proc. N.R.B. R.R.U. Seminar on Bridge Design and Research*, Wellington, 1974.
61. Melville, B. W., "Local Scour at Bridge Sites," University of Auckland, School of Engineering, Report 117, 227 pp., 1975.
62. Melville, B. W., "Live-Bed Scour at Bridge Piers," *Journal of Hydraulic Engineering, A.S.C.E.*, Vol. 110, No. 9, Sept. 1984.
63. Melville, B. W. and Raudkivi, A. J., "Flow Characteristics in Local Scour at Bridge Piers," *Journal of Hydraulic Research*, Vol. 15, pp. 373–380, 1977.
64. Melville, B. W. and Raudkivi, A. J., "Local Scour at Bridge Abutments," *Road Research Unit Bulletin 73*, National Roads Board, N.Z., 1984.
65. Mostafa, G., "River Bed Degradation below Large Capacity Reservoir," *Transactions A.S.C.E.*, Vol. 112, pp. 688–695, 1957.
66. Neill, C. R., "River-Bed Scour, a Review for Engineers," Canadian Good Roads Assoc., Techn. Pub. No. 23, 1964.
67. Neill, C. R. (Ed.), "Guide to Bridge Hydraulics," Roads and Transportation Association of Canada, University of Toronto Press, 191 pp., 1973.
68. Nicollet, G. and Ramette, M., "Affouillements au Voisinage de Piles des Pont Cylindriques Circulaires," *Proc. 14th Congress IAHR*, Paris, Vol. 3, pp. 315–322, 1971.
69. Parker, G., "Self-Formed Rivers with Stable Banks and Mobile Beds, Parts 1 and 2," *Journal of Fluid Mechanics*, Vol. 89, 1978.
70. Raudkivi, A. J., "Loose Boundary Hydraulics," Pergamon Press, 1976.
71. Raudkivi, A. J. *et al.*, "Scouring," *IAHR Hydraulics Structures Design Manual*, Monograph No. A8, publication pending.
72. Raudkivi, A. J. and Sutherland, A. J., "Scour at Bridge Crossings," Road Research Unit, National Roads Board, N.Z., Bulletin 54, 1981.
73. Sastry, C. L. N., "Effect of Spur-Dike Inclination on Scour Characteristics," M.E. thesis presented to the University of Roorkee, India, 1962.
74. Schumm, S. A., "The Shape of Alluvial Channels in Relation to Sediment Type," Prof. Paper 352-B, U.S. Geological Survey, Washington, DC, 1960.
75. Shen, H. W., Schneider, V. R., and Karaki, S., "Mechanics of Local Scour," Colorado State University Rep. CER66, HWS-VRS-SK22, 1966.
76. Shen, H. W., Schneider, V. R., and Karaki, S., "Local Scour around Bridge Piers," *Proc. A.S.C.E.*, Vol. 95, HY6, pp. 1919–1940, 1969.
77. Simons, D. B. and Senturk, F., "Sediment Transport Technology," *Water Resources Publication*, Fort Collins, Colorado, 1976.
78. Straub, L. G., "Effect of Channel Contraction Works upon Regimen of Movable Bed Streams," Trans. American Geophysics Union, Part II, 1935.
79. Tey, C. B., "Local Scour at Bridge Abutments," University of Auckland, Department of Civil Engineering, Report No. 329, 111 pp., 1984.
80. Thomas, A. R., Discussion on "Scour at Bridge Crossings," *Trans. A.S.C.E.*, Vol. 127, Pt. 1, 1962.
81. Thompson, S. M. and Davoren, A., "Local Scour at a Pier, Ohau River, Otago, N.Z.," *MWD Christchurch Science Centre Report*, 1982.
82. Trent, R. E. and Brown, S. A., "An Overview of Factors Affecting River Stability," *Transportation Research Record 950*, Vol. 2 of Second Bridge Engineering Conf., Transportation Research Board, U.S. Dept. of Transportation, 1984.
83. Vanoni, V. and Nomicos, G. N., "Resistance Properties of Sediment-Laden Streams," *Proc. A.S.C.E.*, Vol. 85, HY5, 1959.
84. Vanoni, V. A. (Ed.), "Sedimentation Engineering," *A.S.C.E. Manual on Engineering Practice*, No. 54, 1975.
85. Webby, M. G., "General Scour at a Contraction," *Road Research Unit Bulletin 73*, National Roads Board, N.Z., 1984.
86. Wolman, M. G. and Brush, L. M., "Factors Controlling the Size and Shape of Stream Channels in Coarse Non-Cohesive Sands," *U.S. Geological Survey Prof. Paper 282-G*, Washington, D.C., pp. 183–209, 1961.
87. Wong, W. H., "Scour at Bridge Abutments," University of Auckland, Department of Civil Engineering, Report No. 275, 109 pp., 1982.
88. Yalin, M. S., "Mechanics of Sediment Transport," Pergamon Press, Oxford, 1972.
89. Zaghoul, N. A., "Local Scour around Spur Dikes," *Journal of Hydrology*, 60, pp. 123–140, 1983.

The Transport of Sand in Equilibrium Conditions: 1. Bed Load Transport; 2. Suspended Load Transport; 3. Bed Forms and Alluvial Roughness

L. C. VAN RIJN*

1. BED LOAD TRANSPORT

INTRODUCTION

The transport of sediment particles by a flow of water can be in the form of bed-load and suspended load, depending on the size of the bed material particles and the flow conditions. The suspended load may also contain some wash load, which is generally defined as that portion of the suspended load which is governed by the upstream supply rate and *not* by the composition and properties of the bed material.

Although in natural conditions there will be *no* sharp division between the bed-load transport and suspended load transport, it is necessary to define a layer with bed-load transport for mathematical representation.

Usually, three modes of particle motion are distinguished: (i) rolling and/or sliding motion; (ii) saltation motion; and (iii) suspended particle motion.

When the value of the bed-shear velocity just exceeds the critical value for initiation of motion, the particles will be rolling and/or sliding in continuous contact with the bed. For increasing values of the bed-shear velocity, the particles will be moving along the bed by more or less regular jumps, which are called saltations. When the value of the bed-shear velocity exceeds the fall velocity of the particles, the sediment particles can be lifted to a level at which the upward turbulent forces will be comparable with or of higher order than the submerged weight of the particles and as a result the particles may go in suspension.

Usually, the transport of particles by rolling, sliding and saltating is called the bed-load transport. For example, Bagnold [3,4,5,6,7] defines the bed-load transport as that in which the successive contacts of the particles with the bed are strictly limited by the effect of gravity, while the suspended load transport is defined as that in which the excess weight of the particles is supported wholly by a random succession of upward impulses imparted by turbulent eddies.

Einstein [11], however, has a somewhat different approach. He defines the bed-load transport as the transport of sediment particles in a thin layer of 2 particle diameters thick just above the bed by sliding, rolling and sometimes by making jumps with a longitudinal distance of a *few* particle diameters. The bed layer is considered as a layer in which the mixing due to the turbulence is so small that it cannot influence the sediment particles, and therefore suspension of particles is impossible in the bed-load layer. Further, Einstein assumes that the average distance travelled by any bed-load particle is a constant distance of 100 particle diameters, independent of the flow conditions, transport rate and the bed composition. In the view of Einstein the saltating particles belong to the suspension mode of transport, because the jump lengths of saltating particles are considerably larger than a few grain diameters. This approach of Einstein is followed by Engelund and Fredsøe [13]. In the present study the approach of Bagnold is followed, which means that the motion of the bed-load particles is assumed to be *dominated* by gravity forces, while the effect of turbulence on the overall trajectory is supposed to be of minor importance. This latter statement means that the trajectory of a saltating particle can be somewhat wavy at certain locations due to the highest (turbulent) fluid velocities of the spectrum. However, the overall dimensions of the trajectory are typically those of a saltating particle.

It is assumed that the (theoretical) maximum saltation height for a given flow condition can be determined from the equations of motion (excluding turbulent fluid forces) for a bed-load particle. As a start of a particle trajectory

*Delft Hydraulics, The Netherlands

will be caused by random fluid forces, the height (and length) of a trajectory are random variables. However, by estimating the most unfavourable initial conditions, a maximum saltation height can be computed.

If for given flow conditions there are sediment particles with a jump height larger than the (computed) *maximum* saltation height present in the flow, then these particles are assumed to be transported as suspended load. All particles with a jump height smaller than the maximum saltation height are transported as bed load. According to Bagnold [6] a particle is suspended when the bed-shear velocity (u_*) exceeds the fall velocity (w_b). Hence, the saltation mode of transport is dominant when the bed-shear velocity is smaller than the fall velocity ($u_*/w_b < 1$).

In the present analysis the bed-load transport is defined as the transport of particles by rolling and saltating along the bed surface. The transport rate (q_b) of the bed load is defined as the product of the particle velocity (u_b), the saltation height (δ_b) and the bed-load concentration (c_b) resulting in $q_b = u_b \delta_b c_b$. To compute the bed-load transport rate, the particle characteristics must be known. In the present analysis new relationships for the particle characteristics are proposed. These relationships are based on the numerical solution of the equations of motion for a saltating particle. Using the proposed relationships and measured bed-load data, the concentration of bed-load particles has been computed and represented by a simple function.

Finally, a verification analysis of predicted and measured bed-load transport rates for particles in the range 200 to 2000 μm is presented.

THE CHARACTERISTIC PARAMETERS

The steady and uniform (two-dimensional) flow of water and sediment particles is defined by seven basic parameters: density of water (ρ), density of sediment (ρ_s), dynamic viscosity coefficient (μ), particle size (D), flow depth (d), channel slope (S) and acceleration of gravity (g). These seven basic parameters can be reduced to a set of four dimensionless parameters being a particle mobility parameter, a particle Reynolds' number, a depth-particle size ratio and a specific density parameter (Yalin, [46]).

In the present analysis, it is assumed that the bed-load transport rate can be described sufficiently accurately by two dimensionless parameters only, being a dimensionless particle parameter (D_*) and a transport stage parameter (T). The D_*-parameter can be derived by eliminating the shear stress from the particle mobility parameter and the particle Reynolds' number, while the T-parameter expresses the mobility of the particles in terms of the stage of movement relative to the critical stage for initiation of motion. The introduction of the D_*- and T-parameters is not new. Similar expressions have been used by Ackers–White [2] and Yalin [46]. The particle and transport stage parameters are defined as follows:

(a) particle parameter,

$$D_* = D_{50} \left[\frac{(s-1)g}{\nu^2} \right]^{1/3} \tag{1}$$

in which: D_{50} = particle size, s = specific density ($\rho_s/$

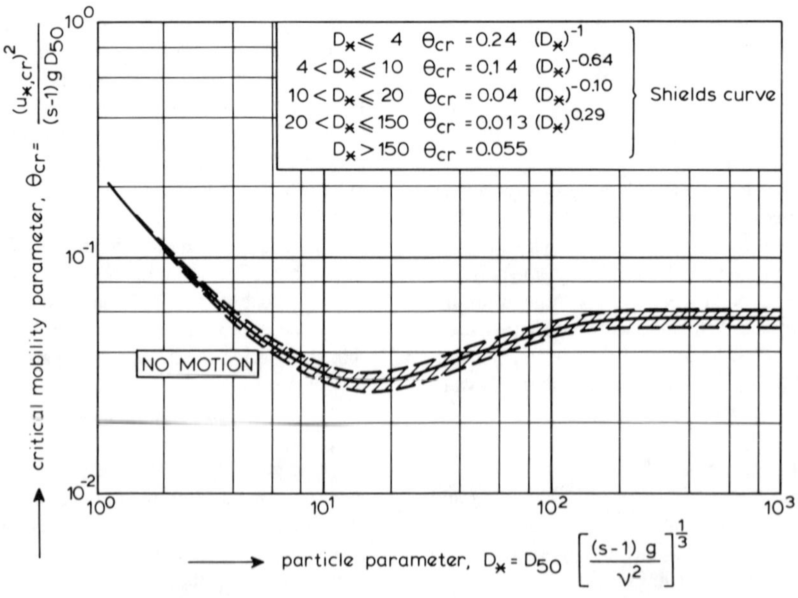

FIGURE 1. Initiation of motion according to Shields.

ρ), g = acceleration of gravity, ν = kinematic viscosity coefficient (μ/ρ).

(b) transport stage parameter,

$$T = \frac{(u'_*)^2 - (u_{*,cr})^2}{(u_{*,cr})^2} \tag{2}$$

in which: $u'_* = (g^{0.5}/C')\,\bar{u}$ = bed-shear velocity related to grains, C' = Chézy-coefficient related to grains, $\bar{u}$ = mean flow velocity, $u_{*,cr}$ = critical bed-shear velocity according to Shields [38] as given in analytical (by the writer) and graphical form in Figure 1.

In the above-given concept the u'_*-parameter is described in terms of the mean flow velocity and a Chézy-coefficient related to the grains of the bed. The u'_*-parameter can also be described in terms of a bed-form factor (μ_b) and the overall bed-shear velocity (u_*), as follows: $u'_* = \mu_b^{0.5} u_*$ with $\mu_b = (C/C')^2$ and C = overall Chézy-coefficient. Later on this method, which is simple and convenient, will be described in more detail and verified using data.

In the present approach it is preferred to use the mean flow velocity instead of the overall bed-shear velocity (or energy gradient) to represent the transport stage because the energy gradient is not an appropriate parameter for morphological computations (mathematical modelling). Other disadvantages of the use of the energy gradient may be: (i) the variations due to non-equilibrium phenomena (rising and falling stages) and (ii) measuring problems in (isolated) field conditions.

THE MOTION OF SOLITARY BED-LOAD PARTICLES FOR A PLANE BED

Characteristics of Particle Saltations

From detailed experimental studies [1,17] the following general characteristics of particle saltations can be inferred. The saltation mode of transport is confined to a layer with a maximum thickness of about 10 particle diameters, in which the particle motion is dominated by gravitational forces, although the particle motion may be initiated by instantaneous turbulent impulses during upward bursts of fluid or just by the effect of shear in the sense that a body in a sheared flow experiences a lift force due to the velocity gradient near the bed. The particles receive their momentum directly from the flow pressure and viscous skin friction. On the rising part of the trajectory, both the vertical component of the fluid drag force and the gravitational force are directed downwards. During the falling part of the trajectory, the vertical component of the fluid drag force opposes the gravitational force. The lift force is always directed upwards as long as the particle lags behind the fluid velocity.

When a particle strikes the bed, it may either impact into the surface or rebound off the surface particles. During the impact of a particle with the bed, most of its momentum is dissipated by the particles of the bed in a sequence of more or less horizontal impulses which may initiate the rolling mode of transport known as surface creep.

Equations of Motion

The forces acting on a saltating particle are a downward force due to its submerged weight (F_G) and hydrodynamic fluid forces, which can be resolved into a lift force (F_L), a drag force (F_D), as shown in Figure 2. The direction of the drag force is opposite to the direction of the particle velocity (v_r) relative to the flow, while the lift component is in the normal direction.

It is assumed that: (i) the particles are spherical and of uniform density, (ii) the forces due to fluid accelerations are of a second order [24], (iii) the lift forces due to rotational motions (Magnus effect) are also of a second order. With these assumptions, the equations of motion can be represented by [45].

$$m\ddot{x} - F_L\left(\frac{\dot{z}}{v_r}\right) - F_D\left(\frac{u - \dot{x}}{v_r}\right) = 0 \tag{3a}$$

$$m\ddot{z} - F_L\left(\frac{u - \dot{x}}{v_r}\right) + F_D\left(\frac{\dot{z}}{v_r}\right) + F_G = 0 \tag{3b}$$

in which: m = particle mass and added fluid mass, $v_r = [(u - \dot{x})^2 + (\dot{z})^2]^{0.5}$ = particle velocity relative to the flow, u = local flow velocity, $\dot{x}$ and $\dot{z}$ = longitudinal and vertical particle velocities, $\ddot{x}$ and $\ddot{z}$ = longitudinal and vertical particle accelerations.

The total mass of the sphere can be represented by

$$m = \frac{1}{6}(\rho_s + \alpha_m \rho)\pi D^3 \tag{4}$$

in which: α_m = added mass coefficient.

Assuming potential flow, the added mass of a perfect sphere is exactly equal to half the mass of the fluid displaced by the sphere. When the flow is separated from the solid sphere, the added mass may be different. In the present analysis an added mass coefficient equal to 0.5 has been used.

The drag force, which is caused by pressure and viscous skin friction forces, can be expressed as:

$$F_D = \frac{1}{2} c_D \rho A v_r^2 \tag{5}$$

in which: c_D = drag coefficient, $A = \frac{1}{4}\pi D^2$ = cross-sectional area of the sphere.

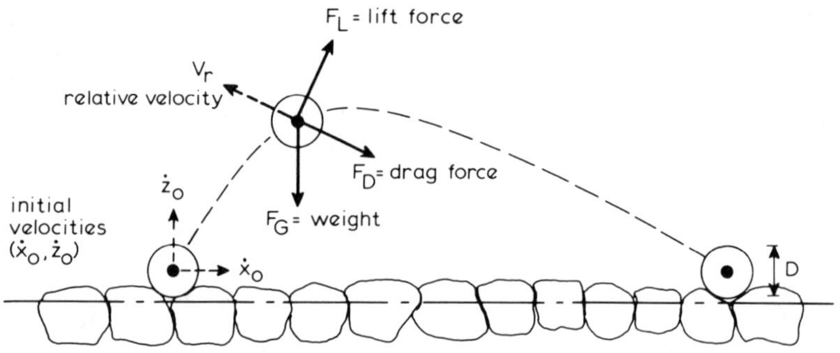

FIGURE 2. Definition sketch of particle saltation.

For the drag coefficient the (empirical) expressions given by Morsi and Alexander [29] were used.

The lift force in a shear flow is caused by the velocity gradient present in the flow (shear effect) and by the spinning motion of the particle (Magnus effect). For a sphere moving in a *viscous* flow, Saffman [37] derived the following expression:

$$F_L \text{ (shear)} = \alpha_L \rho \nu^{0.5} D^2 v_r \left(\frac{\partial u}{\partial z}\right)^{0.5} \tag{6}$$

in which: α_L = lift coefficient (= 1.6 for viscous flow), $\partial u/\partial z$ = velocity gradient.

Equation (6) is only valid for small Reynolds' numbers. The lift force due to the spinning motion in a viscous flow was determined by Rubinow and Keller [34]:

$$F_L \text{ (spin)} = \alpha_L \rho D^3 v_r \omega \tag{7}$$

in which: α_L = lift coefficient (= 0.4 for viscous flow), ω = angular velocity of the particle.

Saffman [37] showed theoretically that for *viscous* flow the lift force due to the particle rotation is less by an order of magnitude than that due to the shear effect. Although the above considerations are only valid for *viscous* flow, it is assumed that also in *turbulent* flow conditions the lift force is mainly caused by the shear effect, which is described by Equation (6) using the lift coefficient α_L as a calibration parameter. This latter approach has been used because in the present state of research an exact expression for the lift force in turbulent flow conditions is not available.

The submerged particle weight can be described by:

$$F_G = \frac{1}{6}\pi D^3(\rho_s - \rho)g \tag{8}$$

The vertical flow velocity distribution is described by:

$$u(z) = \frac{u_*}{\kappa} \ln\left(\frac{z}{z_0}\right) \tag{9}$$

in which: u_* = bed-shear velocity, κ = constant of Von Karman (= 0.4), z_0 = 0.11 (ν/u_*) + 0.03 k_s = zero-velocity level above the bed, k_s = equivalent roughness height of Nikuradse.

Boundary Conditions

The bed level is assumed at a distance of 0.25 D below the top of the particles, as shown in Figure 3. In its initial position a particle is supposed to be resting on a bed surface

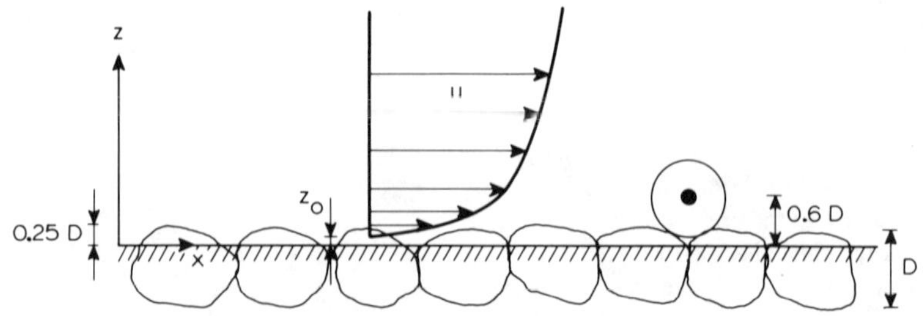

FIGURE 3. Initial position of particle.

of close-packed identical particles. The most stable position will be that of a particle resting above one of the interstices formed by the top layer of the particles of the bed surface, which yields an initial position of about 0.6 D above the bed level (Figure 3). It is evident that this schematization cannot represent the movements of all the bed-load particles. The particles in the compacted bed can only be moved by the highest fluid velocities of the spectrum resulting in somewhat larger saltations. However, the majority of the particles is supposed to be moving over the surface of close-packed particles.

To solve Equation (3), the initial vertical and longitudinal particle velocities must be known. Measurements of particles in a water stream by Francis and Abbott [17,1] indicate an average initial longitudinal and vertical velocity of approximately $2u_*$. White and Schultz [45] analyzed high-speed motion-picture films of saltating particles in air and observed a lift-off velocity varying from u_* to $2.5u_*$ and a lift-off angle varying from 30° to 70°.

Solution Method

Equation (3) has been transformed to a system of ordinary simultaneous differential equations of the first order. This system has been solved numerically by means of an automatic step-change differential equation solver, resulting in a solution accurate to four digits [35].

CALIBRATION AND COMPUTATIONAL RESULTS

As experiments on saltating bed-load particles are very scarce, the mathematical model has not been calibrated extensively; only the experiments of Fernandez Luque were considered for calibration [15,16]. Fernandez Luque carried out flume experiments on bed-load transport. Advanced film techniques were used to measure the average particle velocity, the average saltation length and the average number of particles deposited per unit area and time (in water) as a function of the temporal mean shear stress. He used four different bed materials: sand, gravel, magnetite and walnut grains. The experimental conditions were restricted to relatively low transport stages without bed forms (plane bed). In the present analysis only the experiment with gravel particles ($D = 1800\ \mu$m) and a bed-shear velocity (u_*) of about 0.04 m/s has been used. The ratio of the bed-shear velocity and the particle fall velocity for this experiment is about 0.25. For these conditions the sediment particles can only be transported as bed load (Bagnold [6]). Therefore, the trajectories measured by Fernandez Luque are considered to be trajectories of saltating particles, although the trajectories are somewhat wavy at certain locations (Figure 4).

Two parameters were used to calibrate the model: the lift coefficient (α_L) and the equivalent roughness of Nikuradse (k_s). As input data the following experimental results were used: D = particle diameter = 1800 μm (= 1.8 mm), ρ_s = density of sediment = 2650 kg/m^3, u_* = bed-shear velocity = 0.04 m/s, ν = kinematic viscosity coefficient = 1.10^{-6} m^2/s. The initial longitudinal and vertical particle velocities were assumed to be equal to $2u_* = 0.08$ m/s.

Figure 4 shows measured and computed particle trajectories for various lift coefficients and equivalent roughness heights.

As can be observed, both calibration parameters have a strong influence on the computed trajectories. A reduced lift coefficient results in a reduction of the saltation length. Increasing the roughness height also reduces the saltation length considerably, due to its direct influence on the local flow velocity and thus on the lift and drag forces. As regards the average particle velocity, the "best" agreement between measured and computed values is obtained for $k_s/D = 2$ to 3 (as shown in Table 1), which is a realistic value for plane bed conditions with active sediment transport. In an earlier study [36], the writer has analyzed a large amount of movable bed experiments, which were explicitly indicated as "plane bed" experiments resulting in k_s-values in the range 1 to 10 D_{90} with a mean value of about $3D_{90}$. These values, which are rather large show that a completely plane bed does not exist for conditions with active sediment transport. Probably, the effective roughness is caused by very small irregularities ("bed forms") of the movable bed. Similar k_s-values were reported by Kamphuis: $k_s = 2.5D_{90}$ [26], Hey: $k_s = 3.5D_{84}$ [23], Mahmood: $k_s = 5.1D_{84}$ [27], and Gladki: $k_s = 2.3D_{80}$ [19]. Both the measured saltation height and the measured particle velocity are "best" represented for $\alpha_L = 20$ and $k_s/D = 2$ to 3. For these values, however, a relatively large discrepancy between the measured and computed saltation lengths can be observed (Table 1 and Figure 4). It is supposed that the measured particle trajectories are influenced by turbulent motions resulting in a larger saltation length. Particularly, the wavy pattern of the measured trajectories indicates the influence of the upward fluid forces by turbulent action. Figure 5 presents the computed flow and particle velocity and the computed fluid forces during the trajectory and the influence of the initial particle velocity on the trajectory. The influence of the initial particle velocity is relatively small. The velocity of the particle relative to the flow is greatest after take-off and smallest just before impact. According to Fernandez Luque, the particles must experience a lift force by the shear flow which is approximately equal to their submerged weight, because for the greater part of the trajectory both the observed vertical and longitudinal accelerations of the particles were very small. However, this theory of Fernandez Luque is contradicted by the mathematical model, which shows (Figure 5) that the lift force due to shear is not more than half the submerged weight for the greater part of the trajectory. Only very close to the bed does the lift force exceed the submerged weight considerably up to a value of three times the submerged particle weight for $\alpha_L = 20$. Such high

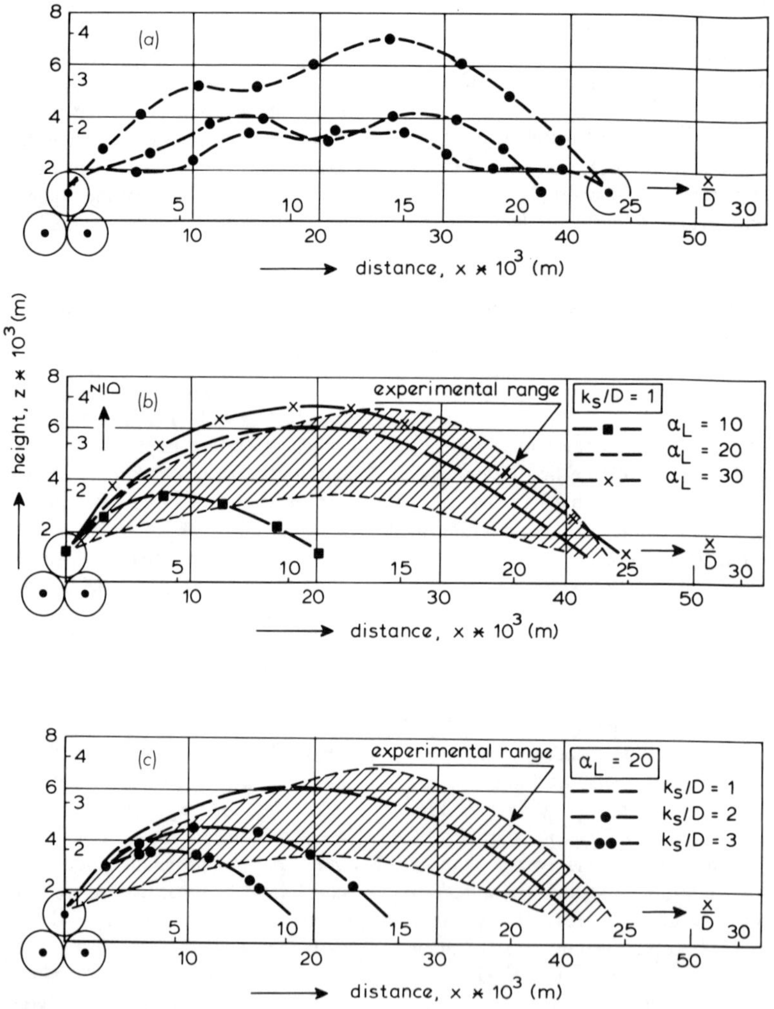

FIGURE 4. Measured (a) and computed (b,c) particle trajectories for experiment of Fernandez Luque.

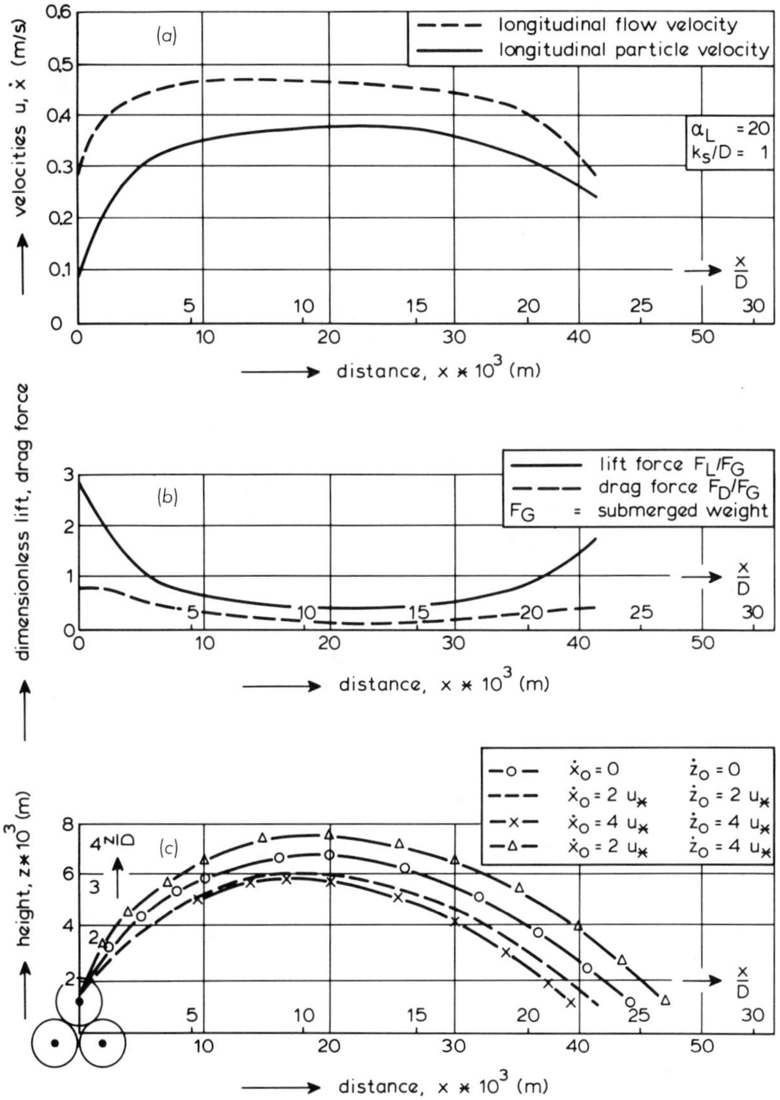

FIGURE 5. Computed particle velocity (a), forces (b) and trajectories (c) for experiment of Fernandez Luque.

TABLE 1. Measured and Computed Saltation Characteristics for Experiments of Fernandez Luque.

Particle Characteristics	Measured	Computed ($\alpha_L = 20$)		
		$\frac{k_s}{D} = 1$	$\frac{k_s}{D} = 2$	$\frac{k_s}{D} = 3$
Particle velocity $\frac{u_b}{u_*}$	5.4	8.1	6.1	4.9
Saltation length $\frac{\lambda_b}{D}$	21–24	23	14.8	10.2
Saltation height $\frac{\delta_b}{D}$	2–3.5	3.5	2.5	1.9

lift forces close to the bed can also be derived from the experiments of Sumer [40] with plastic spheres (diameter = 3000 μm), which were slightly heavier than water (density = 1009 kg/m^3). The particle trajectories were recorded by using a stereo-photogrammetric system. The lift force in each particle direction was computed from the equation of motion for the vertical direction by substituting the particle accelerations determined from the measured time records of the particles. For a Reynolds' number $Re_* = (u_* D/\nu) \simeq 69$, the ratio of the lift force and the submerged particle weight close to the bed was about 5, as shown in Figure 6 (Fernandez Luque experiments: $Re_* \simeq 72$). Figure 6 also expresses the influence of the lift coefficient on the vertical distribution of the lift force. As can be observed, an increase of the lift coefficient (α_L) from 20 to 30 does not significantly influence the lift force in the higher part of the trajectory.

Finally, some remarks must be made with respect to the lift coefficient (α_L). The value $\alpha_L = 20$, which is needed to represent the particle trajectories measured by Fernandez Luque, is rather large compared with the value $\alpha_L = 1.6$ for laminar flow [see Equation (6)]. As the lift coefficient (α_L) is used as a calibration parameter, ir reflects all influences (for example, the fluctuating turbulent motions, additional pressure forces in the proximity of the wall and additional forces due to local fluid accelerations which are not taken into account by the mathematical model.

To get some insight into the saltation mechanism for *small* sand particles, the particle trajectory for a particle of 100 μm was computed. For comparison with a large particle, the flow conditions were taken equal to those in the experiments of Fernandez Luque (D = 1800 μm). Further, it was assumed that the life coefficient (α_L) computed by Saffman [37] for viscous flow ($\alpha_L = 1.6$) also holds for hydraulically smooth flow ($Re_* = u_* D/\nu \leqslant 5$), while for rough flow conditions ($Re_* < 70$) the lift coefficient is supposed to vary linearly from 1.6 to 20.

Figure 7 shows the computed trajectories for various lift coefficients and equivalent roughness heights for a 100 μm-particle while also the longitudinal distribution of the lift and drag forces are presented. As can be observed, the saltation height and length are relatively sensitive to the value of the lift coefficient. The influence of the equivalent roughness height is rather small. Compared with the experiments of Fernandex Luque (equal flow conditions) in which coarse particles were used [Figure 5(c)], the dimensionless saltation length of the 100 μm-particles is much larger, while the saltation height is smaller. These differences can be explained by considering the longitudinal distribution of the lift and drag forces [Figures 5(b) and 7]. For a small particle the saltation process is dominated by the drag force, which is always larger than the lift force. For a large particle this situation is just the opposite with a dominating lift force. Another remarkable phenomenon is the large value ofthe initial lift and drag forces with respect to the submerged particle weight and also the rapid and large decrease of these forces in the higher parts of the trajectory.

COMPUTATION OF SALTATION CHARACTERISTICS FOR VARIOUS FLOW CONDITIONS

To learn more about the relationship between the saltation characteristics and the flow conditions, the saltation characteristics were computed for various particle diameters (D = 100 to 2000 μm) and flow conditions ($u_* = 0.02$ to 0.14 m/s) assuming a flat bed [35].

The following assumptions were used: $k_s = 2D$, $\dot{x}_0 =$

$\dot{z}_0 = 2u_*$, $z(0) = 0.6D$, $\rho_s = 2650$ kg/m³, $\nu = 1.10^{-6}$ m²/s, $\alpha_m = 0.5$, $\alpha_L = 1.6$ for $Re_* \leq 5$, $\alpha_L = 20$ for $Re_* \geq 70$ and $\alpha_L = 1.6$ to 20 (linear) for $Re_* = 5$ to 70, $\kappa = 0.4$.

Saltation Height

For each set of hydraulic conditions, the T- and D_*-parameters were computed and related to the computed (dimensionless) saltation height resulting in a set of curves, as shown in Figure 8. These curves can be approximated with an inaccuracy of about 10% by the following simple expression:

$$\frac{\delta_b}{D} = 0.3D_*^{0.7}T^{0.5} \qquad (10)$$

Figure 8 clearly shows that for small particles the dimensionless saltation height is smaller than for large particles at the same transport stage. Also the dependency on the transport stage is evident.

In this context the experiments of Williams [43] are of importance. Williams carried out flume experiments with bed-load transport ($D_{50} = 1350\ \mu$m) in channels of different widths and depths. According to visual estimation, the bed-load particles moved within a zone of no more than about 8 particle diameters high at the largest transport stage (plane bed). The height of the bed-load layer was independent of the flow depth, but increased as the transport stage decreased. Using the data of Williams ($u_* \simeq 0.09$ m/s, $u_{*,cr} \simeq 0.03$ m/s), Equation (10) predicts a saltation height of about 10 particle diameters, which is remarkably close to the observed value. Francis [17] argues that Williams observations of $8D$ results from an error in defining the top of the bed. According to the measurements of Francis the maximum saltation height is 2 to $4D$. To the opinion of the present writer the experiments of Francis are not representative for natural conditions because of the small flow depths ($d \simeq 0.05$ m) used by Francis. Furthermore, the particle sizes were rather large compared with the height of the

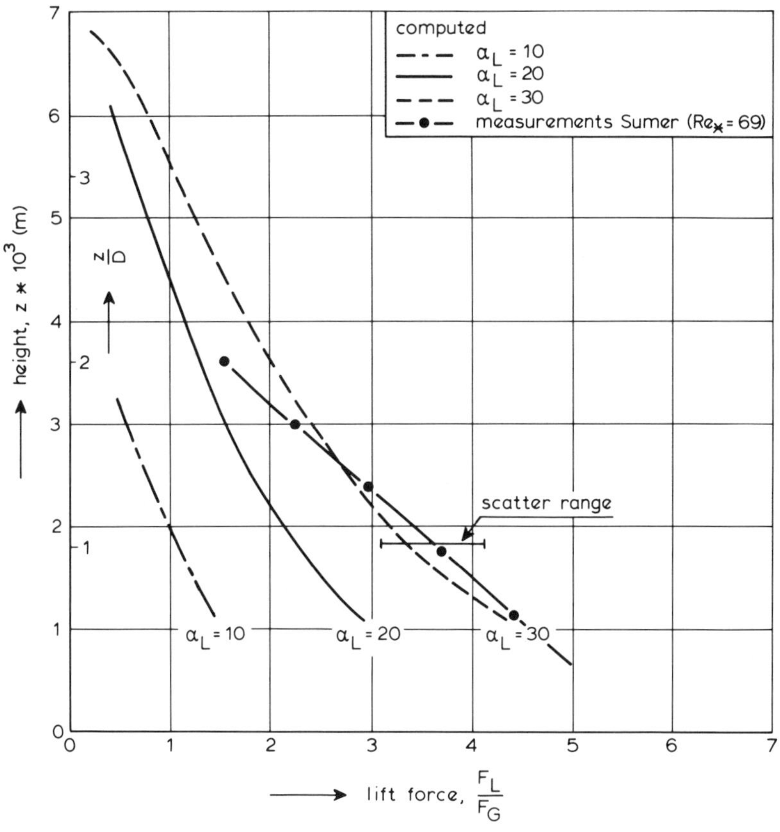

FIGURE 6. Vertical distribution of lift force for experiment of Fernandez Luque.

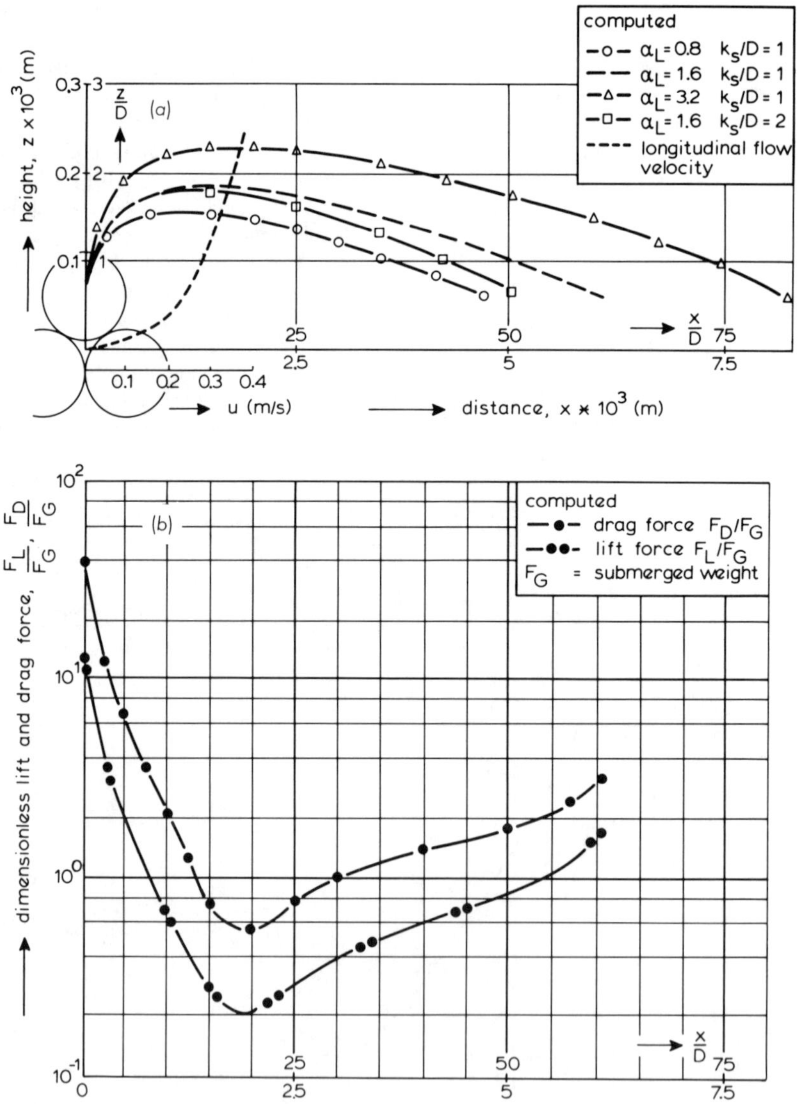

FIGURE 7. Computed particle trajectories (a) and fluid forces (b) for a 100 μm-particle.

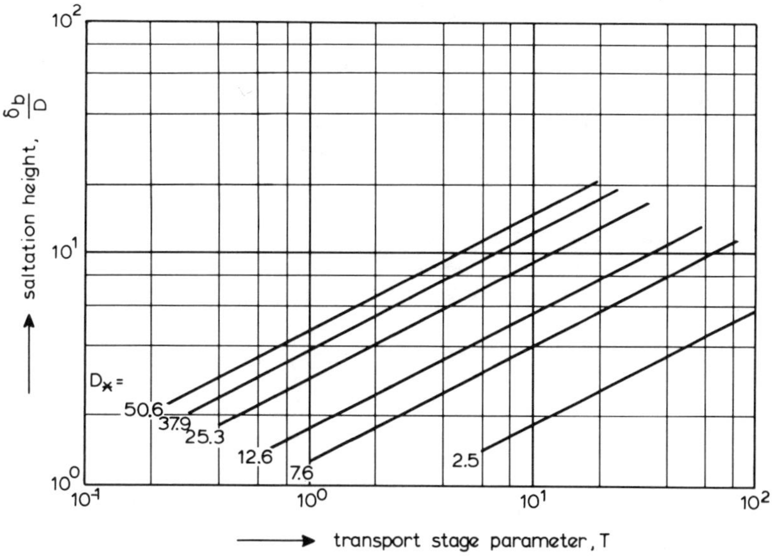

FIGURE 8. Computed saltation height as a function of transport stage and particle parameter.

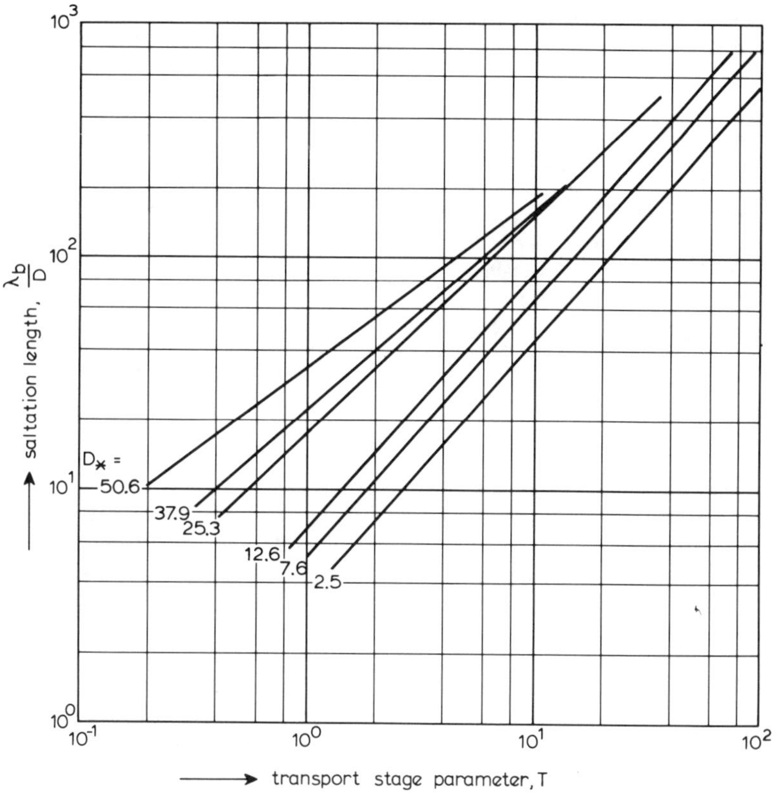

FIGURE 9. Computed saltation length as a function of transport stage and particle parameter.

shearing zone, where large velocity gradients do occur. Therefore, the saltation heights measured by Francis are supposed to be too small to represent natural conditions.

The saltation height according to Equation (10) has also been compared with the results of the models of Reizes [33] and Hayashi–Ozaki [22]. For a gravel particle with $D = 1000\ \mu m$, a bed-shear velocity $u_* = 0.04$ m/s and $\nu = 1.10^{-6}$ m^2/s, Equation (10) predicts $\delta_b/D \simeq 4$, the model of Reizes yields $\delta_b/D \simeq 3$ and that of Hayashi–Ozaki produces $\delta_b/D \simeq 0.7$. The latter result is much smaller than the results of the other two models, which show reasonable agreement. For the experiment of Fernandez Luque ($D = 1800\ \mu m$, $u_* = 0.04$ m/s), the model of Hayashi–Ozaki predicts $\delta_b/D = 0.2$. This value is about 10 times smaller than the observed values (Table 1). It can be concluded that the model of Hayashi–Ozaki yields saltation heights which are much too small. Probably, because the lift forces near the bed surface were not modelled.

Finally, it is remarked that Equation (10) may predict relatively high values for the saltation height, because it is based on computations for individual particles, neglecting the influence of adjacent particles. In the case of collective motion of the particles, the actual saltation dimensions as well as the particle velocity, will be reduced by particle collisions in the bed-load layer.

Saltation Length

Figure 9 presents the saltation length (λ_b) as a function of the transport stage and the particle parameter. At the *same* transport stage a relatively small particle performs a shorter saltation than a large particle, because the bed-shear velocity in the small particle case is much smaller. The curves, as shown in Figure 9, can be approximated (with an inaccuracy of about 50%) by the following simple expression:

$$\frac{\lambda_b}{D} = 3D_*^{0.6}T^{0.9} \tag{11}$$

Experimental support for the computed saltation lengths can be obtained from experiments concerning the sampling efficiency of bed-load samplers carried out by Poreh *et al.* [31]. They observed saltation lengths in the range $5D_{50}$ to $40D_{50}$ ($D_{50} = 1900\ \mu m$, $u_* \simeq 0.04$ to 0.05 m/s). For these conditions the mathematical model predicts a range from $10D_{50}$ to $30D_{50}$ (Figure 9). The model of Hayashi–Ozaki predicts a range from $1D_{50}$ to $3D_{50}$, which is about 10 times smaller than the observed values. Reizes did not present results for the saltation length.

Particle Velocity

Bagnold [7] assumed that in steady continuing saltation the mean longitudinal velocity relative to the flow is that velocity at which the *mean* fluid drag on the particle is in equilibrium with the *mean* longitudinal frictional force exerted by the bed surface on the particles. The relationship of Bagnold can be represented by the following general expression:

$$\frac{u_b}{u_*} = \alpha_1 - \alpha_2 \left[\frac{\theta_{cr}}{\theta}\right]^{0.5} \tag{12}$$

in which: u_* = bed-shear velocity, $\theta = u_*^2/((s-1)gD)$ = particle mobility parameter, θ_{cr} = critical particle mobility parameter according to Shields [38], α_1, α_2 = coefficients.

As the saltation height is a function of the sediment size (D_*), also the mean flow velocity along the particle trajectory, and thus the α_1-coefficient will be a function of the sediment sieze (D_*). Figure 10 represents the computed particle velocity as a function of the flow conditions and sediment size. The curves, as shown in Figure 10, can be approximated (with an inaccuracy of about 10%) by:

$$\frac{u_b}{u_*} = 9 + 2.6 \log D_* - 8 \left[\frac{\theta_{cr}}{\theta}\right]^{0.5} \tag{13}$$

For a particle mobility parameter (θ) approaching the critical value, the particle velocity is supposed to approach zero. Figure 10 shows some data of Fernandez Luque [15] and Francis [17]. As regards the experiments of Fernandez Luque only the sand ($D = 900\ \mu m$) and the gravel ($D = 1800\ \mu m$) experiments for a flat bed-surface slope ($\beta = 0$) were used. From the experiments of Francis only the gravel data were used. As can be observed, the (scarce) data do not confirm the influence of the D_*-parameter as expressed by the mathematical model. More experimental research is necessary to investigate the D_*-influence. The computational results can also be approximated (20% inaccuracy) by the following simple expression:

$$\frac{u_b}{(\Delta g D)^{0.5}} = 1.5T^{0.6} \tag{14}$$

Using an expression similar to Equation (12) and data fitting, Engelund and Fredsøe [13] derived:

$$\frac{u_b}{u_*} = 10 - 7 \left[\frac{\theta_{cr}}{\theta}\right]^{0.5} \tag{15}$$

Equation (15) is shown in Figure 10. The agreement between Equation (15) and the data of Fernandez Luque appears to be less good as presented by Engelund and Fredsøe in their original paper [13].

The influence of particle shape was investigated by Francis [17]. His experiments show that angular particles travel slower than spherical particles. The spherical particles made violent rebounds from the bed and were lifted to higher levels where they experienced higher flow velocities than

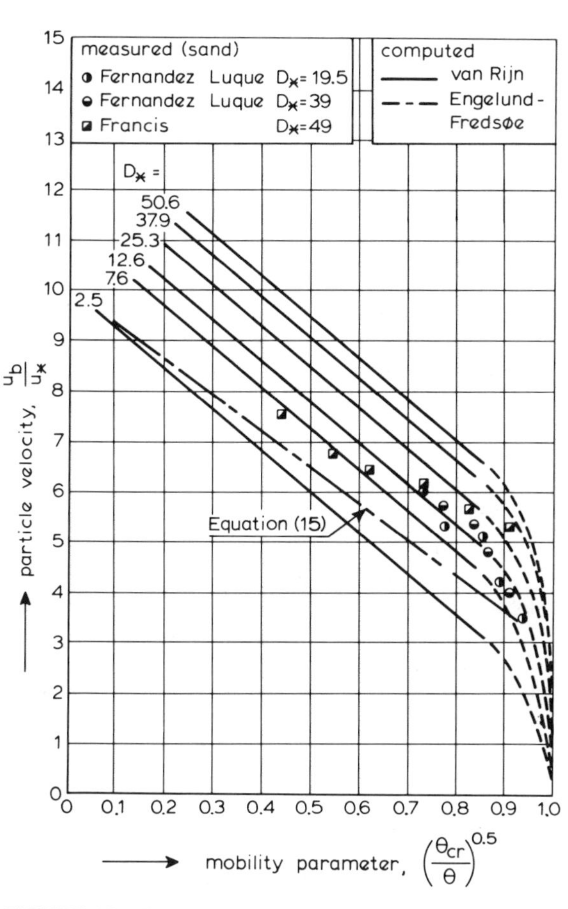

FIGURE 10. Particle velocity as a function of flow conditions and particle parameter.

the angular particles, resulting in a relatively large average particle velocity.

COMPUTATION OF BED-LOAD CONCENTRATION

In the present analysis the bed-load transport is defined as the product of the thickness of the bed-load layer, the particle velocity and the bed-load concentration. Using measured bed-load transport rates, the bed-load concentration can be determined if both other parameters are known. In this Section a general function for the bed-load concentration will be derived both for small and large particles. Firstly, however, an effective grain-shear velocity must be defined to take into account the influence of bed forms.

Effective Grain-Shear Velocity

When the bed forms are present, the influence of the form drag, which does not contribute to the bed-load transport, must be eliminated. As shown in Figure 11, the grain-shear stress along a bed form varies from zero in the bed-form trough to its maximum value at the top. To compute the average bed-load transport, it is necessary to estimate the average grain-shear stress (τ_b') at the upsloping part of the bed form. It is assumed that at or near the location where the average grain-shear stress is acting, the local flow velocity profile is about the same at that for a plane bed flow with the same mean flow velocity ($\bar{u}$) and particle size (D_{50}). Hence, the grain-shear stress is:

$$\tau_b' = \rho g \frac{(\bar{u})^2}{(C')^2} \tag{16}$$

in which: τ_b' = effective grain-shear stress, $\bar{u}$ = cross-section or depth-averaged flow velocity, C' = Chézy-coefficient related to grain roughness.

As the transport stage parameter (T) in Equation (2) is given in terms of the grain-shear velocity (u_*'), Equation (16) is transformed to:

$$u_*' = \frac{g^{0.5}\bar{u}}{C'} \tag{17}$$

The Chézy-coefficient related to the surface (or grain) roughness of the sediment bed is defined as:

$$C' = 18 \log \left(\frac{12R_b}{3D_{90}}\right) \tag{18}$$

in which: R_b = hydraulic radius related to the bed according to the side-wall correction method of Vanoni–Brooks [42].

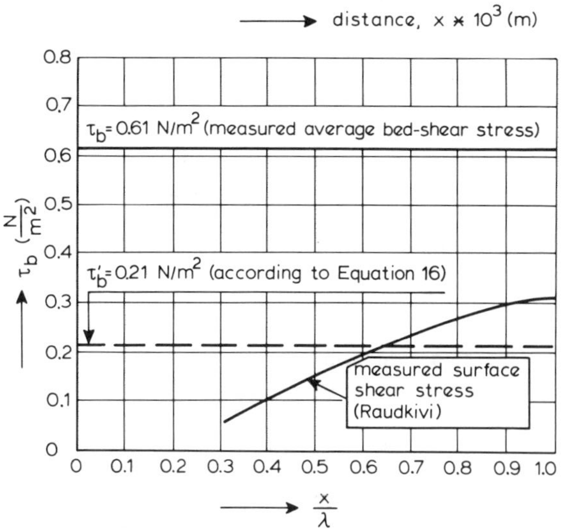

FIGURE 11. Shear stress on bed forms.

As stated before, the effective roughness height of a movable "plane" bed surface may be about $3D_{90}$. This value is based on the analysis of about 100 flume and field data with plane bed conditions [36].

To evaluate the value of the effective bed-shear stress expressed by Equation (16), an experiment of Raudkivi [32] concerning the flow over small ripples is considered. He conducted experiments in a small closed-circuit flume (width = 0.075 m). The bed material consisted of particles with $D_{50} = 400\ \mu m$ and $D_{90} = 580\ \mu m$. The bed was covered with small ripples. To carry out accurate pressure measurements along one individual ripple, the bed forms over the entire length of the flume were replaced by identical bed forms of galvanized metal sheet. Along one ripple the normal pressures (by means of taps) and surface shear stresses (by means of Pitot-tubes) were measured. The average shear stress related to the form drag was determined by averaging the measured pressure distribution over the ripple length (λ), resulting in $\tau_b'' = 0.46\ N/m^2$. In a similar way the average surface shear-stress was found to be $\tau_b' = 0.15\ N/m^2$. The measured shear-stress distribution is shown in Figure 11. Consequently, the total average bed-shear stress is $\tau_b = \tau_b' + \tau_b'' = 0.61\ N/m^2$, which is in good agreement with the average bed-shear stress based on the surface slope (S) and the hydraulic radius (R_b) yielding $\tau_b = 0.65\ N/m^2$ [32].

The effective surface shear-stress was computed according to Equation (16) and

$$C' = 18 \log \left[\frac{11.5}{C'} \frac{\bar{u} R_b}{\nu} \right] \simeq 60\ m^{0.5}/s \qquad (19)$$

resulting in $\tau_b' \simeq 0.21\ N/m^2$. To compute the Chézy-coefficient related to the surface roughness (C'), a resistance Equation (19) for hydraulically smooth flow conditions was used because the ripples consisted of relatively smooth galvanized metal sheet material. In the case of a sediment bed, Equation (18) should be used.

Summarizing, it is concluded that the present method, which is simpler and no more crude than the existing methods, yields an acceptable estimate of the average skin friction at the upsloping part of the bed forms (Figure 11).

Thickness of Bed-Load Layer

According to Einstein [11], the bed-load transport is confined to a layer just above the bed in which the mixing is so small that it cannot affect the particles so that suspension will be impossible. Einstein assumes that this layer has a thickness of two particle diameters for all flow stages. In the present analysis it is assumed that in the case of dominating bed-load transport, the sediment particles are moving along the bed by rolling and/or by saltating. In the case of dominating suspended-load transport, the movement of the (bed-load) particles close to the bed is mainly caused by the *longitudinal* velocity components during the fluid inrush phase (sweeps), while during the turbulent outrush phase (bursts) the bed-load layer is disrupted locally by bursts of fluid with sediment particles moving into suspension.

As a simple approach it is assumed that the *thickness* of the bed-load layer for *all* flow and sediment conditions can be described by Equation (10) with a minimum value of two particle diameters.

Bed-Load Concentration

Using *measured* bed-load transport rates (q_b), the bed-load concentration (c_b) was determined as:

$$c_b = \frac{q_b}{u_b \delta_b} \qquad (20)$$

in which: u_b = particle velocity according to Equation (13), δ_b = thickness of bed-load layer according to Equation (10).

In the study of the motion of solitary particles, the particles were assumed to be of uniform shape, size and density. Here it is assumed that the size distribution of the bed material can be represented by the D_{50} and the geometric standard deviation σ_s.

In all, 130 flume experiments with particle diameters ranging from 200 to 2000 μm were used. Only experiments in which the bed-load transport was measured either directly or indirectly, were selected. The data were those of Guy *et al.* [20], Falkner [14], Gilbert [18], Tsubaki *et al.* [41], Fernandez Luque [15], Willis [44] and Williams [43]. Experiments with a flow depth smaller than 0.1 m and a Froude number larger than 0.9 were not considered. The influence of the side-wall roughness was eliminated by using the method of Vanoni–Brooks [42]. Where the water temperature was not reported, a value of 15°C was assumed.

Extensive analysis of the data showed that the bed-load concentration can be represented by [35]:

$$\frac{c_b}{c_0} = 0.18 \frac{T}{D_*} \qquad (21)$$

in which: c_0 = maximum (bed) concentration = 0.65.

Equation (21), as well as the computed values, are presented in Figure 12. About 80% of the computed values are within the range of half and double the average value according to Equation (21), which is rather good for a sediment transport theory. As only a minority (25%) of the data covers the experiments with particles smaller than about 300 μm, there is a need for further experimental investigation of the bed-load transport in the small particle range to increase the validity of the proposed Equation (21). Finally, it must be stressed that Equation (21) only gives an estimate of the concentration of the bed-load particles in the saltation layer

o Guy et al   (190 μm)   ⊗ Gilbert   (505 μm)
◐ Guy et al   (270 μm)   ⊘ Gilbert   (785 μm)
◑ Guy et al   (280 μm)   ⊖ Gilbert   (1710 μm)
◒ Guy et al   (450 μm)   △ Tsubaki et al   (1260 μm)
◓ Guy et al   (930 μm)   ▲ Tsubaki et al   (1460 μm)
□ Falkner   (180 μm)   ● Fernandez Luque   (900 μm)
▨ Falkner   (280 μm)   ⚏ Fernandez Luque   (1800 μm)
▧ Falkner   (400 μm)   ø Willis   (540 μm)
■ Falkner   (500 μm)   x Williams   (1350 μm)

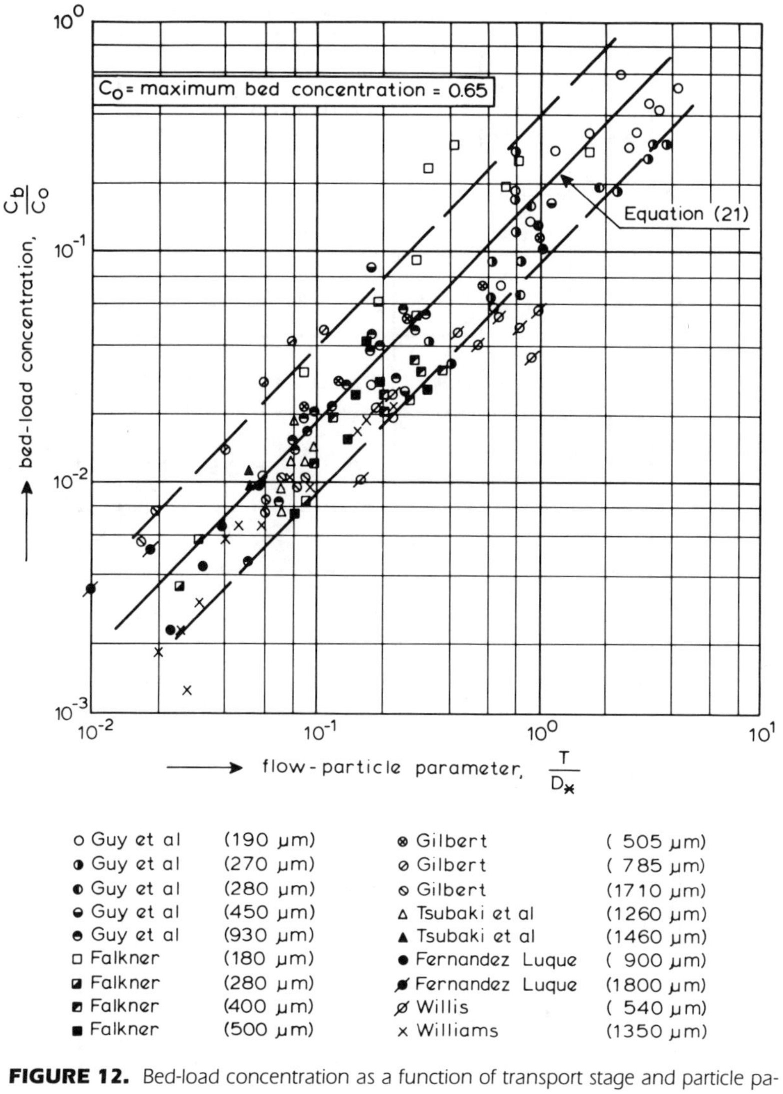

FIGURE 12. Bed-load concentration as a function of transport stage and particle parameter.

TABLE 2. Comparison of Computed and Measured Bed-Load Transport.

							Scores (%) of Predicted Bed Load in Discrepancy Ranges											
							$0.75 \leq r \leq 1.5$				$0.5 \leq r \leq 2$				$0.33 \leq r \leq 3$			
	Source	Number of Tests	Flow Velocity (m/s)	Flow Depth (m)	Particle Diameter ($\times 10^{-6}$ m)	Temperature (°C)	Van Rijn	Engelund/ Hansen	Ackers/ White	Meyer–Peter Müller	Van Rijn	*E–H*	*A–W*	*MPM*	Van Rijn	*E–H*	*A–W*	*MPM*
field data	Japanese Channels (Tsubaki)	12	0.63–0.93	0.20–0.73	1330–1440	—	48%	61%	83%	57%	78%	87%	91%	91%	86%	100%	100%	96%
	Mountain Creek (Einstein)	43	0.49–0.79	0.10–0.43	900	15–25	28	56	21	74	54	81	67	95	84	93	98	100
	Skive-Karup River (Hansen)	1	0.6	1.0	470	10	100	0	0	0	100	0	0	100	100	100	100	100
flume data	Guy *et al.*	22	0.36–1.29	0.15–0.23	320	8–34	64	—	—	51	92	—	—	87	100	—	—	100
	Delft Hydraulics Laboratory	18	0.40–0.87	0.10–0.49	770	12–18	13	17	78	0	75	83	100	71	88	94	100	86
	Stein	38	0.42–1.10	0.10–0.37	400	20–26	27	—	—	4	63	—	—	4	92	—	—	22
	Meyer–Peter	18	0.45–0.88	0.11–0.21	1000–1500	—	6	82	24	6	29	94	59	41	82	94	94	76
	U.S.W.E.S. (sands)	48	0.44–0.58	0.10–0.20	1000	14–18	50	49	80	17	92	100	96	50	100	100	100	60
	U.S.W.E.S. (synth. sands)	183	0.44–0.57	0.15–0.27	500–1100	19–25	50	39	49	30	82	85	86	56	96	99	99	74
	Singh	60	0.31–0.66	0.10–0.20	600	13–20	58	55	80	3	92	97	100	43	100	98	100	78
	Znamenskaya	10	0.53–0.80	0.11–0.20	800	—	30	40	40	50	60	50	70	70	80	90	80	100
	Southampton B	73	0.31–0.70	0.15–0.46	480	22–30	44	11	34	41	87	27	64	73	92	63	84	85
	East Pakistan	21	0.44–0.70	0.15–0.30	470	25–30	10	5	10	14	29	5	10	33	62	10	29	67
	Williams	33	0.46–1.04	0.15–0.22	1350	16–26	20	74	23	37	71	89	60	65	97	100	82	83
	Total	580					42%	43%	48%	23%	77%	76%	77%	58%	93%	90%	92%	76%

at the upsloping part of the bed form. In Part 2 of this study the proposed Equation (21) will be modified so that it can be used to predict the reference concentration for the concentration profile.

COMPUTATION OF BED-LOAD TRANSPORT

Using Equations (10), (14) and (21), the bed-load transport (in m^2/s) for particles in the range 200 to 2000 μm can be computed as:

$$\frac{q_b}{(\Delta g)^{0.5} D_{50}^{1.5}} = 0.053 \frac{T^{2.1}}{D_*^{0.3}} \tag{22}$$

The input data are: mean flow velocity ($\bar{u}$), mean flow depth (d), mean flow width (b), particle diameters (D_{50}, D_{90}), density of water and sediment (ρ, ρ_s), viscosity coefficient (ν) and acceleration of gravity (g).

The computation of the bed-load transport is as follows:

1. Compute particle parameter, D_* using Equation (1).
2. Compute critical bed-shear velocity, $u_{*,cr}$ according to Shields (Figure 1).
3. Compute Chézy-coefficient related to grains, C' using Equation (18).
4. Compute effective bed-shear velocity, u'_* (with a maximum value equal to u_*) using Equation (17).
5. Compute transport stage parameter, T using Equation (2).
6. Compute bed-load transport, q_b using Equation (22).

VERIFICATION

The second phase of the analysis was focused on the verification of Equation (22) *for bed-load data* only. For comparison the formulae of Engelund–Hansen [12] and Ackers–White [2] were also applied (see Appendix). These two formulae gave the best results in a verification study carried out at the H. R. S. Wallingford [25]. The typical bed-load formula of Meyer–Peter and Müller [28] was also used (see Appendix). Most of the flume data used for verification were selected from a compendium of solids transport data compiled by Peterson and Howells [30]. Brownlie [8] has shown that various sets of this data bank contain serious errors. The present writer has eliminated these errors before using the data in the verification analysis. Only experiments with a D_*-value larger than 12 ($\simeq$ 500 μm) were selected, assuming that for these conditions the mode of transport is mainly bed-load transport. For nearly all data the ratio of the (overall) bed-shear velocity and the particle fall velocity was smaller than one ($u_*/w_b < 1$). In addition, data from Guy *et al.* [20], Stein [39] and the Delft Hydraulics Laboratory [9] concerning bed-load transport only were used. In all, *524 flume* tests were available for verification. To account for side-wall roughness, the method of Vanoni–Brooks was used. As regards reliable field data, which are extremely scarce, only three data sets were considered: some small Japanese channels (Tsubaki *et al.* [41], the Mountain Creek in the U.S.A. (Einstein [10]) and the Skive–Karup River in Denmark (Hansen [21]). None of the data were used in the calibration.

To evaluate the accuracy of the computed and measured values, a discrepancy ratio (r) has been used, defined as:

$$r = \frac{q_{b,computed}}{q_{b,measured}} \tag{23}$$

The results are given in Table 2. It is remarked that the formulae of Engelund–Hansen and Ackers–White were not applied to the data of Guy *et al.* and Stein (small particle range) because for these data the formulae will predict the total load and not the bed-load transport.

The proposed method scores an average percentage of 42% in the discrepancy ratio range of 0.75 to 1.5, and an average percentage of 77% in the range 0.5 to 2. The formulae of Engelund–Hansen and Ackers–White have about the same accuracy. In the range 0.5 to 2 the Engelund–Hansen formula scores a percentage of 76%, while the Ackers–White formula has a value of 77%. The formula of Meyer–Peter and Müller yields the least good results with a score of 58% in the range 0.5 to 2. A sensitivity analysis showed that the computed bed-load transport according to the proposed method is rather sensitive to the computed grain-shear velocity (u'_*) in the case of a small flow depth (flumes) and a large particle diameter (D_{50} = 1000 to 2000 μm). For example, the data of Meyer–Peter had a considerably better score for a grain roughness $k_s = D_{90}$ instead of $k_s = 3D_{90}$. An analysis of the computed and measured transport rates showed that the computed values are somewhat too large for a T-parameter in the range 1 to 5. No serious errors in relation to the D_*-parameter are present [35]. Finally, the accuracy of the measured transport rates is discussed. It must be stressed that an investigation of flume experiments performed under similar flow conditions (equal depth, velocity, particle size, temperature) by various research workers showed deviations up to a factor 2 [35]. Thus, even under controlled flume conditions the accuracy of the measured values may be rather low, which may be caused by the influence of the applied width-depth ratio [43], the applied adjustment period to establish uniform flow conditions and the applied experimental method (sand feed or recirculating flume).

Concludingly, it may be stated that it is hardly possible to predict the transport rate with an inaccuracy less than a factor 2.

NOTATION

The following symbols have been used

A = area of cross-section (L^2)
C = Chézy-coefficient ($L^{0.5}T^{-1}$)

C' = Chézy-coefficient related to grain roughness ($L^{0.5}T^{-1}$)
c_b = bed-load concentration (volume = $\rho_s 10^3 c_b$ in ppm)
c_D = drag-coefficient
D = particle diameter (L)
D_* = dimensionless particle parameter
d = flow depth (L)
F_D = fluid drag force (MLT^{-2})
F_G = gravitational force (MLT^{-2})
F_L = fluid lift force (MLT^{-2})
F_w = frictional force (MLT^{-2})
g = acceleration of gravity (LT^{-2})
k_s = equivalent roughness of Nikuradse (L)
m = mass (M)
Q = flow discharge (L^3T^{-1})
q_b = bed-load transport per unit width (L^2T^{-1})
R_b = hydraulic radius (L)
Re = Reynolds number
r = discrepancy ratio
S = slope
s = specific density
T = transport stage parameter
u = longitudinal flow velocity (LT^{-1})
$\bar{u}$ = average flow velocity (LT^{-1})
u_b = particle velocity (LT^{-1})
u_* = bed-shear velocity (LT^{-1})
u'_* = effective bed-shear velocity (LT^{-1})
$u_{*,cr}$ = critical bed-shear velocity according to Shields (LT^{-1})
v_r = relative particle velocity (LT^{-1})
w_b = particle fall velocity of bed material (LT^{-1})
x, z = longitudinal, vertical coordinate (L)
$\dot{x}, \dot{z}$ = longitudinal, vertical particle velocity (LT^{-1})
$\ddot{x}, \ddot{z}$ = longitudinal, vertical particle acceleration (LT^{-2})
z_0 = zero-velocity level (L)
α_L = lift coefficient
α_m = added mass coefficient
δ_b = saltation height (L)
θ = particle mobility parameter
κ = constant of Von Karman
μ = dynamic viscosity coefficient ($ML^{-1}T^{-1}$)
μ_b = bed form factor
λ_b = saltation length (L)
ν = kinematic viscosity coefficient (L^2T^{-1})
ρ = density of fluid (ML^{-3})
ρ_s = density of sediment (ML^{-3})
σ_s = geometric standard deviation of bed material
τ_b = bed-shear stress ($ML^{-1}T^{-2}$)
τ'_b = effective bed shear stress ($ML^{-1}T^{-2}$)
ω = angular velocity (T^{-1})

REFERENCES

1. Abbott, J. E. and Francis, J. R. D., "Saltation and Suspension Trajectories of Solid Grains in a Water Stream," *Proc. Royal Soc.*, London, Vol. 284, A 1321, 1977.
2. Ackers, P. and White, W. R., "Sediment Transport: New Approach and Analysis," *Journal of the Hydraulics Division, ASCE*, HY11, 1973.
3. Bagnold, R. A., "The Physics of Blown Sands and Desert Dunes," *Methuen*, London, 1941.
4. Bagnold, R. A., "Experiments on a Gravity-Free Dispersion of Large Solid Spheres in a Newtonian Fluid Under Shear," *Proc. Royal Soc.*, London, Vol. 225A, 1954.
5. Bagnold, R. A., "The Flow of Cohesionless Grains in Fluids," *Proc. Royal Soc.*, London, Vol. 249, 1956.
6. Bagnold, R. A., "An Approach to the Sediment Transport Problem from General Physics," *Geological Survey Prof. Paper 422-I*, Washington, USA, 1966.
7. Bagnold, R. A., "The Nature of Saltation and of Bed-Load Transport in Water," *Proc. Royal Soc.*, London, A 322, 1973.
8. Brownlie, W. R., "Discussion of Total Load Transport in Alluvial Channels," *Journal of the Hydraulics Division, Proc. ASCE*, No. HY12, 1981.
9. Delft Hydraulics Laboratory, "Verification of Flume Tests and Accuracy of Flow Parameters," Note R 657-VI, The Netherlands, June 1979 (in Dutch).
10. Einstein, H. A., "Bed-Load Transportation in Mountain Creek," U.S. Department of Agriculture, Washington, D.C., USA, 1944.
11. Einstein, H. A., "The Bed-Load Function for Sediment Transportation in Open Channel Flow," United States Department of Agriculture, Washington, D.C., USA, Technical Bulletin No. 1026, page 25, September 1950.
12. Engelund, F. and Hansen, E., "A Monograph on Sediment Transport," *Teknisk Forlag*, Copenhagen, Denmark, 1967.
13. Engelund, F. and Fredsøe, J., "A Sediment Transport Model for Straight Alluvial Channels," *Nordic Hydrology*, 7, pages 294 and 298, 1967.
14. Falkner, H., "Studies of River Bed Materials and their Movement with Special Reference to the Lower Mississippi River," Paper 17, U.S. Waterways Exp. Station, Vicksburg, USA, 1935.
15. Fernandez Luque, R., "Erosion and Transport of Bed-Load Sediment," Dissertation, Krips Repro B.V., Meppel, The Netherlands, 1974.
16. Fernandez Luque, R. and Beek, R. van, "Erosion and Transport of Bed-Load Sediment," *Journal of Hydraulic Research*, Vol. 14, No. 2, 1976.
17. Francis, J. R. D., "Experiments on the Motion of Solitary Grains along the Bed of a Water-Stream," *Proc., Royal Soc.*, London, A332, 1973.
18. Gilbert, G. K., "Transportation of Debris by Running Water," *Prof. Paper 86, U.S. Geol. Survey*, Washington, USA, 1914.
19. Gladki, H., Discussion of "Determination of Sand Roughness for Fixed Beds," *Journal of Hydraulic Research*, Vol. 13, No. 2, 1975.
20. Guy, H. P., Simons, D. B., and Richardson, E. V., "Summary of Alluvial Channel Data from Flume Experiments, 1956–1961," *Geol. Survey Prof. Paper 462-I*, Washington, USA, 1966.

21. Hansen, E., "Bed-Load Investigation in Skive-Karup River," Technical University Denmark, Bulletin No. 12, 1966.
22. Hayashi, T. and Ozaki, S., "On the Unit Step Length of Saltation of Sediment Particles in the Bed-Load Layer," *Third Int. Symp. on Stochastic Hydraulics*, Tokyo, Japan, 1980.
23. Hey, R. D., "Flow Resistance in Gravel-Bed Rivers," *Journal of the Hydraulics Division, Proc. ASCE*, No. HY4, 1979.
24. Hinze, J. O., *Turbulence*, second edition, McGraw-Hill Book Company, 1975.
25. Hydraulics Research Station Wallingford, *Sediment Transport: An Appraisal of Available Methods*, Int. 119, Wallingford, England, November 1973.
26. Kamphuis, J. W., "Determination of Sand Roughness for Fixed Beds," *Journal of Hydraulic Research*, Vol. 12, No. 2, 1974.
27. Mahmood, K., "Flow in Sand Bed Channels," *Water Management Techn. Report*, No. 11, Colorado State University, USA, 1971.
28. Meyer–Peter, E. and Müller, R., "Formulas for Bed-Load Transport," *Proc. Second Congress IAHR*, Stockholm, Sweden, 1948.
29. Morsi, S. A. and Alexander, A. J., "An Investigation of Particle Trajectories in Two-Phase Flow Systems," *Journal of Fluid Mechanics*, Vol. 55, Part 2, page 207, 1972.
30. Peterson, A. W. and Howells, R. F., "A Compendium of Solids Transport Data for Mobile Boundary Channels," Report No. HY-1973-ST3, *Dep. of Civ. Eng.*, Univ. of Alberta, Canada, January 1973.
31. Poreh, M., Sagiv, A., and Seginer, J., "Sediment Sampling Efficiency of Slots," *Journal of the Hydraulics Division ASCE*, HY10, October 1970.
32. Raudkivi, A. J., "Study of Sediment Ripple Formation," *Journal of the Hydraulics Division ASCE*, HY6, November 1963.
33. Reizes, J. A., "Numerical Study of Continuous Saltation," *Journal of the Hydraulics Division, Proc. ASCE*, No. HY9, 1978.
34. Rubinow, S. I. and Keller, J. B., "The Transverse Force on a Spinning Sphere Moving in a Viscous Fluid," *Journal of Fluid Mechanics*, Vol. 11, page 454, 1961.
35. Rijn, L. C. van, "Computation of Bed-Load Concentration and Bed-Load Transport," *Delft Hydraulics Laboratory, Research Report S 487-I*, The Netherlands, March 1981.
36. Rijn, L. C. van, "Equivalent Roughness of Alluvial Bed," *Journal of the Hydraulics Division, ASCE*, No. HY10, 1982.
37. Saffman, P. G., "The Lift on a Small Sphere in a Slow Shear Flow," *Journal of Fluid Mechanics*, Vol. 22, page 393, 1965, Vol. 31, 1968.
38. Shields, A., Anwendung der Ähnlichkeitsmechanik und der Turbulenzforschung auf die Geschiebebewegung, Mitt. der Preuss. Versuchsanst. für Wasserbau und Schiffbau, Heft 26, Berlin, 1936.
39. Stein, R. A., "Laboratory Studies of Total Load and Apparent Bed Load," *Journal of Geophysical Research*, Vol. 70, No. 8, 1965.
40. Sumer, B. M. and Deigaard, R., "Experimental Investigation of Motions of Suspended Heavy Particles and the Bursting Process," Institute of Hydrodynamics and Hydraulic Engineering, Techn. Univ. Denmark, Paper No. 23, page 86, November 1979.
41. Tsubaki, T. and Shinohara, K., "On the Characteristics of Sand Waves Formed upon the Beds of Open Channels and Rivers," *Reports of Research Institute for Applied Mechanics*, Vol. VII, No. 25, 1959.
42. Vanoni, V. A. and Brooks, N. H., "Laboratory Studies of the Roughness and Suspended Load of Alluvial Streams," Sedimentation Laboratory, California Institute of Technology, Pasadena, Report E-68, 1957.
43. Williams, P. G., "Flume Width and Water Depth Effects in Sediment Transport Experiments," Geol. Survey Prof. Paper 562-H, page H5, Washington, D.C., USA, 1970.
44. Willis, J. C., "Suspended Load from Error-Function Models," *Journal of the Hydraulics Division, ASCE*, HY7, July 1979.
45. White, B. R. and Schultz, J. C., "Magnus Effect in Saltation," *Journal of Fluid Mechanics*, Vol. 81, page 507, 1977.
46. Yalin, M. S. *Mechanics of Sediment Transport.* Pergamon Press, pp. 59–61, 1972.

APPENDIX PART 1

The *total load formula of Engelund–Hansen* is applied, as follows:

$$q_t = \frac{0.05\bar{u}^5}{(s-1)^2 g^{0.5} D_{50} C^3}$$

in which

q_t = total load transport per unit width (m²/s)
$\bar{u}$ = mean flow velocity (cross-section) (m/s)
s = 2.65 = relative density (−)
g = acceleration of gravity (m/s²)
D_{50} = particle diameter (m)
C = coefficient of Chézy ($m^{0.5}$/s)

The *total load formula of Ackers–White* is applied, as follows

$$q_t = K\bar{u}D_{35}\left[\frac{\bar{u}}{u_*}\right]^n\left[\frac{Y - Y_{cr}}{Y_{cr}}\right]^m$$

in which

q_t = total load transport per unit width (m²/s)
$K = \exp\{2.86\ \ln(D_*) - 0.434(\ln D_*)^2 - 8.13\}$
$D_* = D_{35}\left[\frac{(s-1)g}{\nu^2}\right]^{1/3}$ (−)
ν = kinematic viscosity coefficient (m²/s)
$\bar{u}$ = mean flow velocity (m/s)
u_* = bed-shear velocity (m/s)
$n = 1 - 0.56 \log(D_*)$ (−)
$m = \frac{9.66}{D_*} + 1.34$ (−)
$Y_{cr} = \frac{0.23}{D_*^{0.5}} + 0.14$ (−)

$$Y = \left[\frac{(u_*)^n}{[(s - 1)gD_{35}]^{0.5}}\right]\left[\frac{\bar{u}}{5.66 \log\left(\frac{10R_b}{D_{35}}\right)}\right]^{1-n} \quad (-)$$

R_b = hydraulic radius related to the bed according to method of Vanoni–Brooks (1967) (m)
D_{35} = particle diameter of bed material (m)

The *bed-load formula of Meyer–Peter and Müller* is applied, as follows:

$$q_b = 8(s - 1)^{0.5}g^{0.5}D_{50}^{1.5}(\theta - 0.047)^{1.5}$$

in which

q_b = bed-load transport per unit width (m^2/s)
$\theta = \frac{\mu R_b S}{(s - 1)D_{50}}$ = mobility parameter (−)
$\mu = \left(\frac{C}{C'}\right)^{1.5}$ = bed-form factor (−)
$C' = 18 \log\left(\frac{12R_b}{D_{90}}\right)$ = Chézy-coefficient related to grains ($m^{0.5}/s$)
C = overall Chézy-coefficient ($m^{0.5}/s$)
S = energy gradient (−)
R_b = hydraulic radius related to the bed according to the method of Vanoni–Brooks (1957) (m)

2. SUSPENDED LOAD TRANSPORT

INTRODUCTION

An essential part of morphological computations in the case of flow conditions with suspended sediment transport is the use of a reference concentration as a bed-boundary condition. Usually, the equilibrium bed concentration is specified [24,25,26,27]. In that approach the equilibrium concentration is computed from the sediment transport capacity as given by a total load transport formula and the relative concentration profile. To improve the bed-boundary condition, a theoretical investigation has been carried out by the author with the aim of determining a relationship which specifies the reference concentration as function of local (near-bed) flow parameters and sediment properties [33,34].

In the present analysis it will be shown that the function for the bed-load concentration as proposed in Part 1, can also be used to compute the reference concentration for the suspended load. Furthermore, the main controlling hydraulic parameters for the suspended load, which are the particle fall velocity and the sediment diffusion coefficient, are studied in detail. Especially investigated and described by new expressions are the diffusion of the sediment particles in relation to the diffusion of fluid particles and the influence of the sediment particles on the turbulence structure (damping effects). Finally, a method to compute the suspended load transport is proposed and verified, using a large amount of flume and field data.

Characteristic Parameters

In the present analysis it is assumed that the bed-load transport and hence the reference concentration are determined by a particle parameter (D_*) and a transport stage parameter (T) as follows:

(a) particle parameter:

$$D_* = D_{50}\left[\frac{(s - 1)g}{\nu^2}\right]^{1/3} \quad (1)$$

in which: D_{50} = particle diameter of bed material, s = specific density, g = acceleration of gravity, ν = kinematic viscosity coefficient.

(b) transport stage parameter:

$$T = \frac{(u'_*)^2 - (u_{*,cr})^2}{(u_{*,cr})^2} \quad (2)$$

in which: $u'_* = (g^{0.5}/C')\bar{u}$ = bed-shear velocity related to grains, $C' = 18 \log(12R_b/3D_{90})$ = Chézy-coefficient related to grains, R_b = hydraulic radius related to the bed according to Vannoni–Brooks [38], $\bar{u}$ = mean flow velocity, $u_{*,cr}$ = critical bed-shear velocity according to Shields [36].

To describe the suspended load transport, a suspension

parameter which expresses the influence of the upward turbulent fluid forces and the downward gravitational forces, is defined as follows:

(c) suspension parameter:

$$Z = \frac{w_s}{\beta \kappa u_*} \tag{3}$$

in which: w_s = particle fall velocity of suspended sediment, β = coefficient related to diffusion of sediment particles, κ = constant of Von Karman, u_* = overall bed-shear velocity.

INITIATION OF SUSPENSION

Before analysing the main hydraulic parameters which influence the suspended load, it is necessary to determine the flow conditions at which initiation of suspension will occur.

Bagnold stated in 1966 [4] that a particle only remains in suspension when the turbulent eddies have dominant vertical velocity components which exceed the particle fall velocity (w_s). Assuming that the vertical velocity component (w') of the eddies are represented by the vertical turbulence intensity ($\hat{w}$), the critical value for initiation of suspension can be expressed as:

$$\hat{w} = [\overline{(w')^2}]^{0.5} \geqslant w_s \tag{4}$$

Detailed studies on turbulence phenomena in boundary layer flow [20] suggest that the maximum value of the vertical turbulence intensity ($\hat{w}$) is of the same order as the bed-shear velocity (u_*). Using these values, the critical bed-shear velocity ($u_{*,crs}$) for initiation of suspension becomes:

$$\frac{u_{*,crs}}{w_s} = 1 \tag{5}$$

which can be expressed as (see Figure 1)

$$\theta_{crs} = \frac{(u_{*,crs})^2}{(s-1)gD_{50}} = \frac{(w_s)^2}{(s-1)gD_{50}} \tag{6}$$

Another criterion for initiation of suspension has been given by Engelund [16]. Based on a rather crude stability analysis, he derived:

$$\frac{u_{*,crs}}{w_s} = 0.25 \tag{7}$$

Finally, some results of experimental research at the Delft Hydraulics Laboratory are discussed. The writer determined the critical flow conditions at which instantaneous upward turbulent motions of the sediment particles (bursts) with jump lengths of the order of 100 particle diameters were observed [11]. The experimental results can be represented by:

$$\frac{u_{*,crs}}{w_s} = \frac{4}{D_*}, \text{ for } 1 < D_* \leqslant 10 \tag{8}$$

$$\frac{u_{*,crs}}{w_s} = 0, 4, \text{ for } D_* > 10 \tag{9}$$

The Equations (6), (7), (8) and (9) are shown in Figure 1. Summarizing, it is suggested that the criterion of Bagnold may define an upper limit at which a concentration profile starts to develop, while the writer's criterion defines an intermediate stage at which locally turbulent bursts of sediment particles are lifted from the bed into suspension.

MATHEMATICAL DESCRIPTION OF CONCENTRATION PROFILES

In a steady and uniform flow the vertical distribution of the sediment concentration profile can be described by:

$$(1-c)cw_{s,m} + \varepsilon_s \frac{dc}{dz} = 0 \tag{10}$$

in which: c = sediment concentration, $w_{s,m}$ = particle fall velocity in a fluid-sediment mixture, ε_s = sediment diffusion coefficient, z = vertical coordinate.

Particle Fall Velocity

In a clear, still fluid the particle fall velocity (w_s) of a *solitary* sand particle smaller than about 100 μm (Stokes-range) can be described by:

$$w_s = \frac{1}{18} \frac{(s-1)gD_s^2}{\nu} \tag{11}$$

For suspended sand particles in the range 100 to 1000 μm, the following type of equation, as proposed by Zanke [42], can be used:

$$w_s = 10 \frac{\nu}{D_s} \left[\left[1 + \frac{0.01(s-1)gD_s^3}{\nu^2} \right]^{0.5} - 1 \right] \tag{12}$$

For particles larger than about 1000 μm the following simple equation can be used [34]:

$$w_s = 1.1[(s-1)gD_s]^{0.5} \tag{13}$$

In Equations (11), (12) and (13) the D_s-parameter ex-

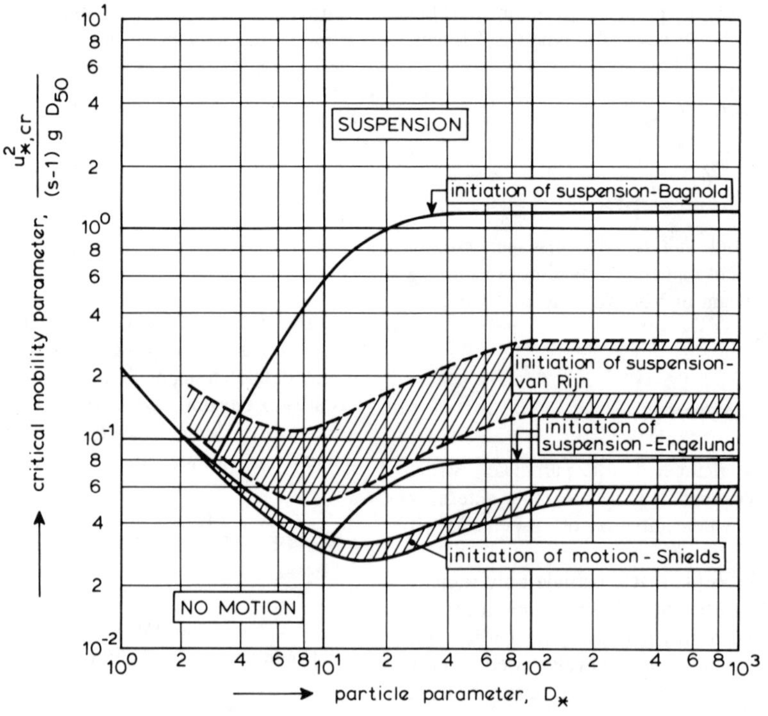

FIGURE 1. Initiation of motion and suspension.

presses the representative particle diameter of the suspended sediment particles, which may be considerably smaller than D_{50} of the bed material, as will be shown later on.

Experiments with high sediment concentrations have shown a substantial reduction of the particle fall velocity due to the presence of the surrounding particles. For normal flow conditions with particles in the range 50 to 500 μm the reduced particle fall velocity can be described by a Richardson–Zaki type equation [34]:

$$w_{s,m} = (1 - c)^4 w_s \qquad (14)$$

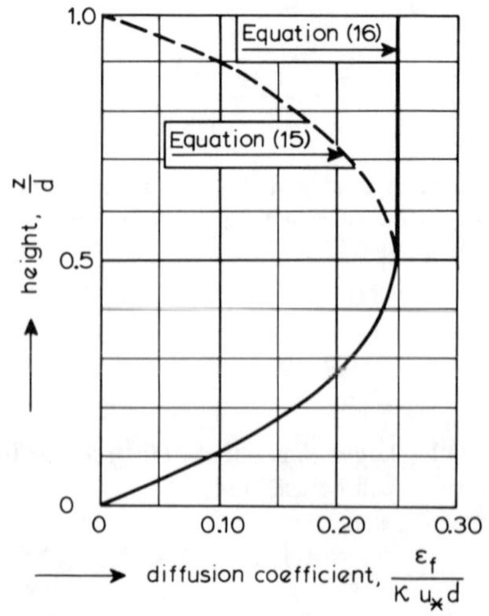

FIGURE 2. Fluid diffusion coefficient.

Diffusion Coefficient

Usually, the diffusion of fluid momentum (ϵ_f) is described by a parabolic distribution over the flow depth (d):

$$\varepsilon_f = \frac{z}{d}\left(1 - \frac{z}{d}\right)\kappa u_* d \qquad (15)$$

In the present analysis a parabolic-constant distribution, which means a parabolic distribution in the lower half of the flow depth and a constant value in the upper half of the flow depth, is used mainly because it may give a better description of the concentration profile. The parabolic-constant distribution reads:

$$\varepsilon_{f,max} = 0.25\kappa u_* d \qquad \text{for } \frac{z}{d} \geq 0.5 \qquad (16a)$$

$$\varepsilon_f = 4\frac{z}{d}\left(1 - \frac{z}{d}\right)\varepsilon_{f,max} \qquad \text{for } \frac{z}{d} < 0.5 \qquad (16b)$$

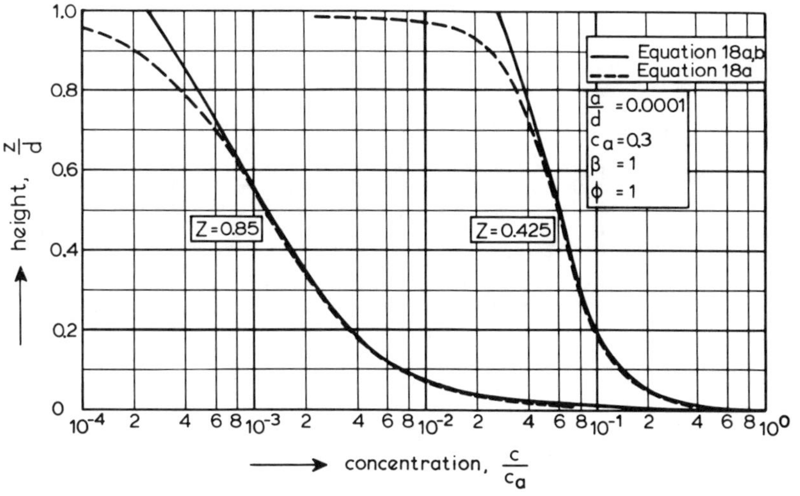

FIGURE 3. Concentration profiles.

Equations (15) and (16) are shown in Figure 2.

The diffusion of sediment particles (ε_s) is related to the diffusion of fluid momentum by:

$$\varepsilon_s = \beta \phi \varepsilon_f \tag{17}$$

The β-factor describes the difference in the diffusion of a discrete sediment particle and the diffusion of a fluid "particle" (or small coherent fluid structure) and is assumed to be constant over the flow depth. The ϕ-factor expresses the damping of the fluid turbulence by the sediment particles and is assumed to be dependent on the local sediment concentration. It will be shown (later on) that the β-factor and the ϕ-factor can be described separately.

Firstly, various expressions for the concentration profile will be given.

Concentration Profiles

(a) Using a parabolic-constant ε_s-distribution according to Equations (16) and (17) with $\phi = 1$ (no damping effect) and a concentration dependent particle fall velocity according to Equation (14), the sediment concentration profile can be obtained by integration of Equation (10) resulting in:

$$\sum_{n=1}^{4}\left[\frac{1}{n(1-c)^n}\right] - \sum_{n=1}^{4}\left[\frac{1}{n(1-c_a)^n}\right] + \ln\left[\frac{(c)(1-c_a)}{(c_a)(1-c)}\right] = \ln\left[\frac{(a)(d-z)}{(z)(d-a)}\right]^Z \quad \text{for } \frac{z}{d} < 0.5 \tag{18a}$$

$$\sum_{n=1}^{4}\left[\frac{1}{n(1-c)^n}\right] - \sum_{n=1}^{4}\left[\frac{1}{n(1-c_a)^n}\right] + \ln\left[\frac{(c)(1-c_a)}{(c_a)(1-c)}\right] = -Z\left[\ln\left(\frac{a}{d-a}\right) + 4\left(\frac{z}{d} - 0.5\right)\right] \quad \text{for } \frac{z}{d} \geq 0.5 \tag{18b}$$

in which: c_a = reference concentration, a = reference level, d = depth, z = vertical coordinate, Z = suspension parameter.

Equation (18a,b) is shown in Figure 3.

For small concentrations ($c < c_a < 0.001$) Equation (18a,b) reduces to:

$$\frac{c}{c_a} = \left[\frac{(a)(d-z)}{(z)(d-a)}\right]^Z, \text{ for } \frac{z}{d} < 0.5 \tag{19a}$$

$$\frac{c}{c_a} = \left[\frac{a}{d-a}\right]^Z [e]^{-4Z(z/d-0.5)}, \text{ for } \frac{z}{d} \geq 0.5 \tag{19b}$$

(b) Using a parabolic ε_s-distribution, the concentration profile for the entire flow depth is described by Equation (18a) or (19a), the latter being the well-known Rouse-expression. Equation (18a) is also shown in Figure 3.

(c) Using a concentration dependent ε_s-distribution ($\phi \neq 1$), the concentration profile can only be computed by numerical integration of Equation (10). In the present

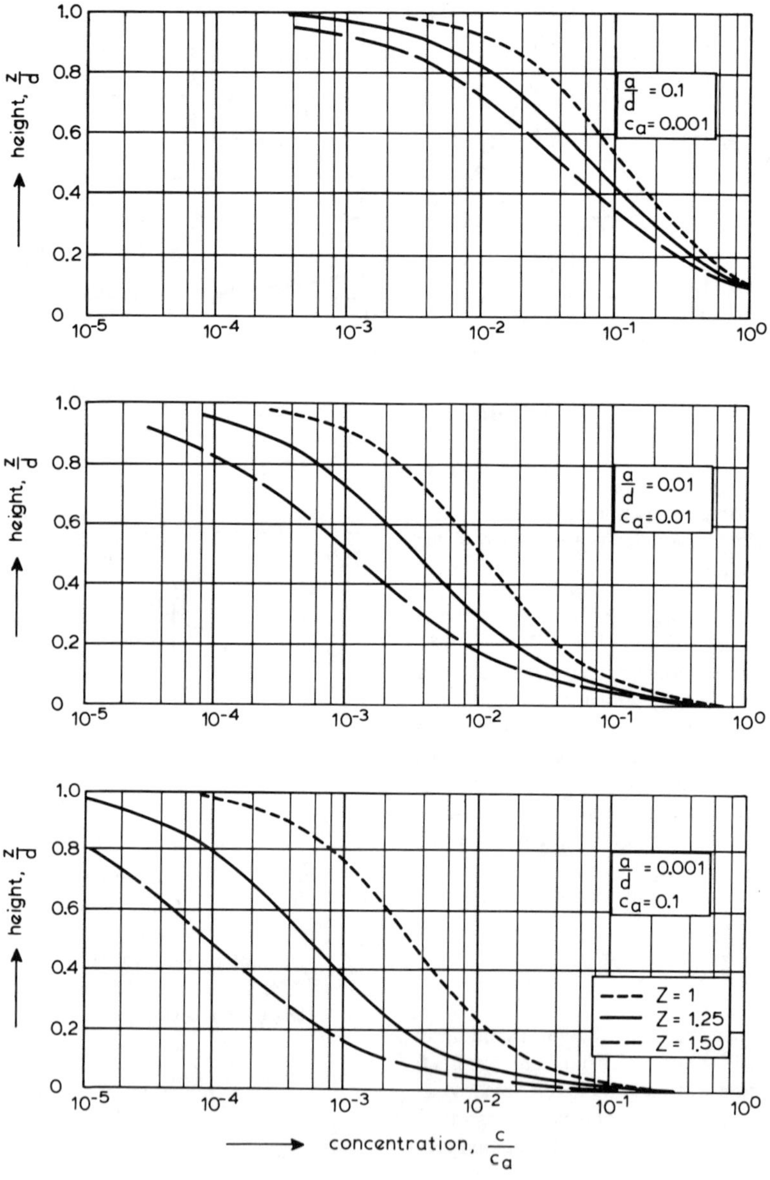

FIGURE 4. Influence of reference level and suspension parameter.

analysis a simple Runga-Kutta method with an automatic step reduction will be used.

Influence of Reference Level and Suspension Parameter

To show the influence of the reference level (a) and the suspension parameter (Z) on the concentration profile, Equation (18a) has been solved for $Z = 1.0$, 1.25 and 1.5 (variation of about 20% with respect to the mean value) and a reference level $a = 0.1d$, $0.01d$ and $0.001d$. The results are shown in Figure 4. To simulate increasing sediment concentrations towards the bed, the reference concentration (c_a) has been increased from $c_a = 0.001$ at $a/d = 0.1$ to $c_a = 0.1$ at $a/d = 0.001$ (see Figure 4).

As can be observed, the concentration profile is relatively sensitive to small variations (about 20%) in the Z-parameter, particularly for the reference level very close to the bed ($a = 0.001d$). It is evident that a reference level smaller than $0.01d$ leads to large errors in the concentration profile, but even for $a = 0.01d$ the prediction of a concentration profile with an error less than a factor 2 requires a Z-parameter with an error of less than 20% which is hardly possible. Although the particle fall velocity and the bed-shear velocity may be estimated with sufficient accuracy, the accuracy of the β-factor is rather poor.

In this context also the approach of Einstein [12], which is followed by Engelund and Fredsøe [17], is discussed because he uses a reference level equal to two particle diameters. In the present analysis it has been shown that the approach of Einstein will lead to large errors in the predicted suspended load. Moreover, in the case of flow conditions with bed forms, Einstein's approach is rather artificial. Therefore, the method of Einstein is not attractive to use as a predictive sediment transport theory.

INVESTIGATION OF SEDIMENT DIFFUSION COEFFICIENT

β-Factor

Some investigators have concluded that $\beta < 1$ because the sediment particles cannot respond fully to the turbulent velocity fluctuations. Others have reasoned that in a turbulent flow the centrifugal forces on the sediment particles (being of higher density) would be greater than those on the fluid particles, thereby causing the sediment particles to be thrown to the outside of the eddies with a consequent increase in the effective mixing length and diffusion rate, resulting in $\beta > 1$. Chien [7] analyzed concentration profiles measured in flume and field conditions. He determined the Z-parameter from the slopes of plotted concentration profiles and compared those values with $Z = w_s/(\kappa u_*)$. In most cases the latter (predicted) Z-values overestimated the Z-values based on the measurements, thereby clearly indicating $\beta > 1$. The results of Chien mainly demonstrate the influence of the β-factor because his results are based on concentrations measured in the upper part of the flow ($z > 0.1d$) where the concentrations are not large enough to cause a significant damping of the turbulence ($\phi \simeq 1$). Information about the β-factor in relation to particle characteristics and flow conditions can be obtained from a study carried out by Coleman [9]. He computed the ε_s-coefficient from the following equation:

$$w_s c + \varepsilon_s \frac{dc}{dz} = 0 \tag{20}$$

His results indicate a sediment diffusion coefficient which is nearly constant in the upper half of the flow for each particular value of the ratio w_s/u_* (see Figure 5).

The writer used the results of Coleman to determine the β-factor, defined as [34]:

$$\beta = \frac{\varepsilon_{s,max}}{\varepsilon_{f,max}} = \frac{\varepsilon_{s,max}}{0.25\kappa u_* d} \tag{21}$$

The maximum value of the ε_f-distribution is the maximum value according to Equation (15) for $z/d = 0.5$. The $\varepsilon_{s,max}$-value was determined as the average value of the ε_s-values in the upper half of the flow (as given by Coleman [9]), where the concentrations and hence the damping of the turbulence (ϕ-factor) are relatively small. In that way only the influence of the β-factor is considered. The computed β-factors can be described by:

$$\beta = 1 + 2\left[\frac{w_s}{u_*}\right]^2, \text{ for } 0.1 < \frac{w_s}{u_*} < 1 \tag{22}$$

as shown in Figure 6. A relationship proposed by Kikkawa and Ishikawa [28], based on a stochastic approach is also shown. According to the present results, the β-factor is always larger than unity, thereby indicating a dominating influence of the centrifugal forces.

ϕ-Factor

The ϕ-factor expresses the influence of the sediment particles on the turbulence structure of the fluid (damping effects). Usually the damping effect is taken into account by reducing the constant of Von Karman (κ). Several investigators have observed that the constant of Von Karman becomes less than the value of 0.4 (clear flow) in the case of a heavy sediment-laden flow over a rigid, flat bed. It has also been observed that the flow velocities in a layer close to the bed are reduced, while in the remaining part of the flow there are larger flow velocities. Apparently, the mixing is reduced by the presence of a large amount of sediment particles.

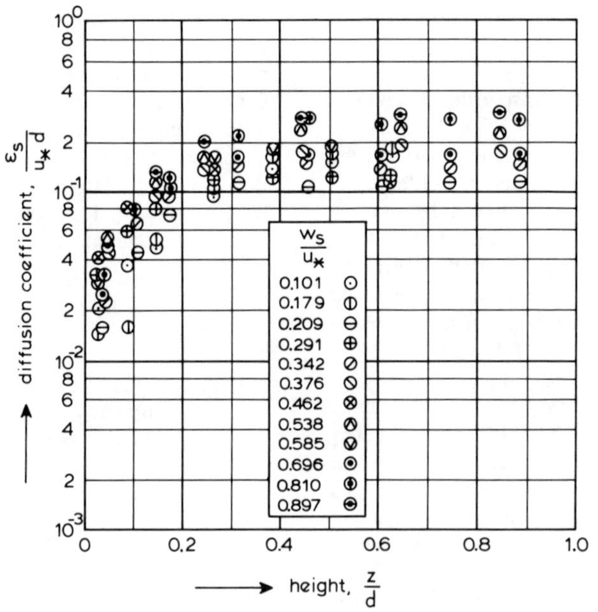

FIGURE 5. Sediment diffusion coefficient according to Coleman.

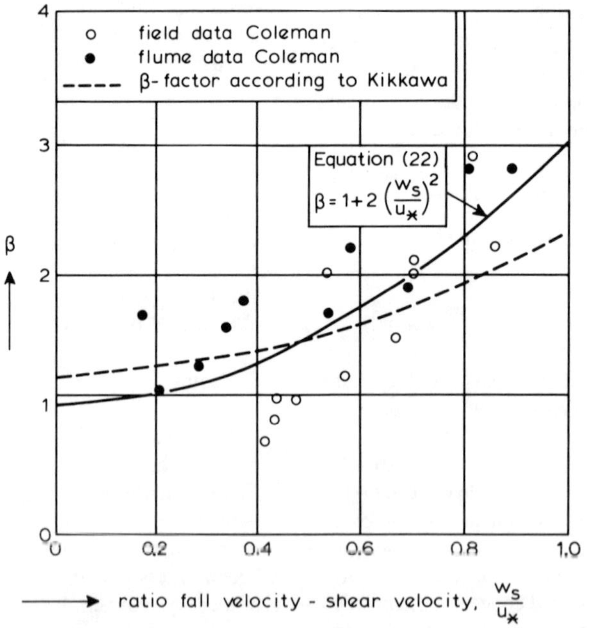

FIGURE 6. *β*-factor.

According to Einstein and Chien [13], who determined the amount of energy needed to keep the particles in suspension, the constant of Von Karman is a function of the depth-averaged concentration, the particle fall velocity and the bed-shear velocity.

Although Ippen [23] supposed that the constant of Von Karman is primarily a function of some concentration near the bed, an investigation of Einstein and Abdel-Aal [14] showed only a weak correlation between the near-bed concentration and the constant of Von Karman.

Coleman [10] questioned the influence of the sediment particles on the constant of Von Karman. He re-analyzed the original data of Einstein–Chien and Vanoni–Brooks and concluded that they used an erroneous method to determine the constant of Von Karman.

In view of these contradictions it may be questioned if the concept of an overall constant of Von Karman for the entire velocity profile is correct for a heavy sediment-laden flow. An alternative approach may be the introduction of a local constant of Von Karman (κ_m) dependent on the local sediment concentration, as has been proposed by Yalin and Finlayson [40]. They analysed measured flow velocity profiles and observed that the local velocity gradient in a sediment-fluid mixture is larger than that in a clear flow. Assuming $\varkappa_m = \phi\varkappa$ (ϕ = damping factor for local concentration, $\varkappa = 0.4$), Yalin and Finlayson finally derived that:

$$\left(\frac{du}{dz}\right)_m = \frac{1}{\phi}\left(\frac{du}{dz}\right) \tag{23}$$

Using Equation (23) and flow velocities measured in a flow with and without sediment particles, Yalin and Finlayson determined some ϕ-values, as shown in Figure 7.

Firstly, it is pointed out that the approach of Yalin and Finlayson is rather simple and based on sometimes rather crude assumptions. Basically, a proper study of the influence of the sediment particles on the velocity and concentration profile requires the solution of the equations of motion and continuity applying a first order closure (mixing length) or a second order (turbulence energy and dissipation) closure. However, as such an approach is far beyond the scope of the present analysis, the writer has modified the approach of Yalin and Finlayson to get a first understanding of the phenomena involved. The modified method is based on the numerical computation of the flow velocity and concentration profile, as follows [34]:

(a) *Velocity profile.* Using the hypothesis of Boussinesq, the flow velocity profile in a fluid-sediment mixture is described by:

$$\tau_m = \rho_m(\varepsilon_m + \nu_m)\left(\frac{du}{dz}\right)_m \tag{24}$$

in which: τ_m = shear stress in a fluid-sediment mixture, ρ_m = local density of mixture, ε_m = diffusion coefficient of mixture, ν_m = viscosity coefficient of mixture.

Taking $\tau_m = (1 - z/d)\bar{\rho}_m u_*^2$, in which $\bar{\rho}_m$ = average density of mixture and assuming $\bar{\rho}_m \simeq \rho_m$, the velocity gradient can be expressed as:

$$\left(\frac{du}{dz}\right)_m = \frac{\left(1 - \frac{z}{d}\right)u_*^2}{\varepsilon_m + \nu_m} \tag{25}$$

The diffusion coefficient for the fluid phase of the

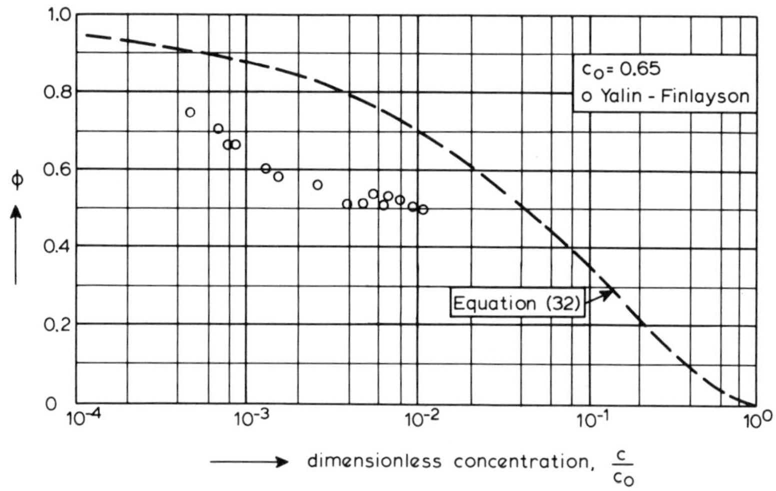

FIGURE 7. ϕ-factor.

mixture (ε_m) is described by:

$$\varepsilon_m = \phi \varepsilon_f \tag{26}$$

$$\varepsilon_f = 4 \frac{z}{d}\left(1 - \frac{z}{d}\right) \varepsilon_{f,max} \tag{27}$$

$$\varepsilon_{f,max} = 0.25 \kappa u_* d \tag{28}$$

The ϕ-factor, which has been used as (free) fit-parameter, is supposed to depend on the local concentration and is described by a simple function:

$$\phi = F\left(\frac{c}{c_0}\right) \tag{29}$$

in which: c = local volumetric concentration, c_0 = 0.65 = maximum volumetric bed concentration.

In a sediment-laden flow, the viscosity coefficient is also modified. Based on experiments with large concentrations, Bagnold [3] derived:

$$\nu_m = \nu(1 + \lambda)(1 + 0.5\lambda) \tag{30}$$

in which: $\lambda = [(0.74/c)^{1/3} - 1]^{-1}$ = dimensionless concentration parameter.

(b) *Sediment concentration profile.* To describe the concentration profile, Equations (10) and (14) are supposed to be valid, resulting in:

$$\frac{dc}{dz} = \frac{w_s c(1 - c)^5}{\varepsilon_s} \tag{31}$$

in which: $\varepsilon_s = \beta \phi \varepsilon_f$ = diffusion coefficient for the sediment phase of the mixture.

The β-factor according to Equation (22) is assumed to be valid, while the ϕ-factor, as stated before, has been used as a fit-parameter to reproduce measured concentration profiles.

(c) *Determination of ϕ-factor.* Three sets of data were used to fit the ϕ-function: (i) the data of Einstein and Chien [13] who measured flow velocity and concentration profiles in a heavy sediment laden flow, (ii) the data of Barton and Lin [5], and (iii) the data of Vanoni and Brooks [38].

By assuming a ϕ-function and then solving Equations (25) and (31) simultaneously by numerical integration and fitting with measured velocity and concentration profiles, the actual ϕ-function was determined. The boundary conditions for the velocity and concentration profiles were: $u = 0$ at $z = 0$ and $c = c_a$ at $z = a$, the latter being the concentration (c_a) measured in the lowest sampling point (a).

Several ϕ-functions were used, but the "best" agreement with measured concentration profiles was obtained by using:

$$\phi = 1 + \left[\frac{c}{c_0}\right]^{0.8} - 2\left[\frac{c}{c_0}\right]^{0.4} \tag{32}$$

Equation (32), shown in Figure 7, gives values which are considerably larger (less damping) than those given by Yalin and Finlayson.

Firstly, the results for the experiments of Einstein and Chien are discussed in more detail. Figure 8 shows measured and computed results for Run *S*-15 with the 275 μm-sediment. It is remarked that the applied reference concentration was assumed to be equal to the concentration measured in the lowest sampling point, resulting in c_a = 625000 ppm (by weight) at a = 0.005 m.

The computed concentration and velocity profiles are based on the numerical integration of Equations (25) and (31) applying a ϕ-factor according to Equation (32).

As can be observed, the applied ϕ-function does not give optimal agreement for the entire profile, probably because Equation (32) is somewhat too simple. Using $\phi = 1$ results in computed concentrations which are an order of magnitude larger than the measured values.

As regards the flow velocity profile, the computed velocities are only of importance in a qualitative sense because the equation of continuity for the fluid has not been taken into account. However, the qualitative trend with reduced flow velocities in the near-bed region is reproduced to some extent. For comparison also the flow velocity profile for a clear flow {Equation (40)} is shown (based on overall flow parameters). Figure 9 shows measured and computed velocity and concentration profiles for an experiment of Barton and Lin [5] with 180 μm-sediment. It may be noted that the concentrations measured by Barton and Lin are considerably smaller than those in the experiments of Einstein and Chien. Finally, an experiment of Vanoni and Brooks [38] is shown in Figure 10. For the latter two experiments the flow velocity profile based on Equations (25), (31) and (32) is not shown because the computed values were close to the values according to the logarithmic profile.

(d) *Simplified method.* The present method is not very suitable for practical use because the concentration profile can only be computed by means of numerical integration of Equation (31). Therefore, a simplified method based on Equation (19) in combination with a modified suspension number (Z'), is introduced. The modified suspension number (Z') is defined as:

$$Z' = Z + \varphi \tag{33}$$

in which: Z = suspension number according to Equation (3) and φ = overall correction factor representing all additional effects (volume occupied by particles, re-

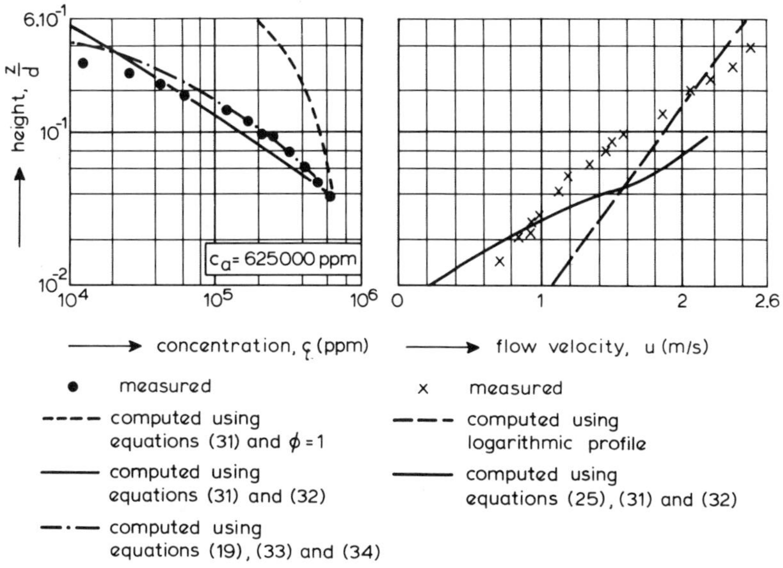

FIGURE 8. Measured and computed concentration and flow velocity profiles for Einstein-Chien experiment (Run S-15).

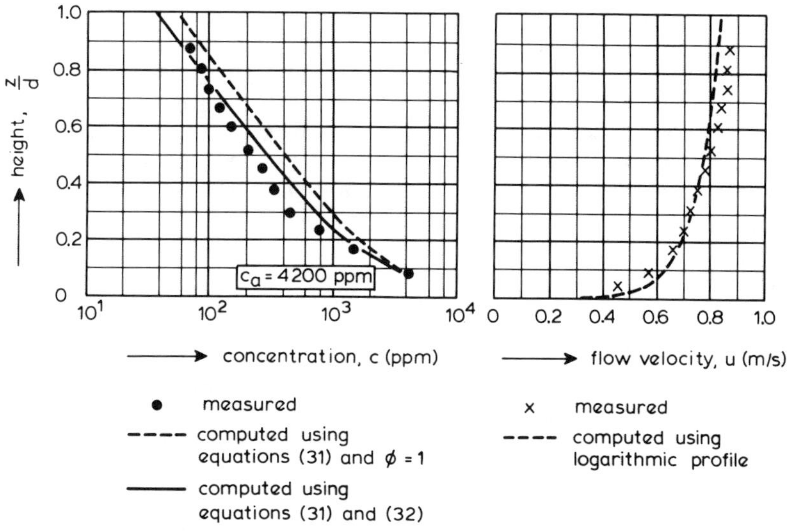

FIGURE 9. Measured and computed concentration and flow velocity profiles for Barton-Lin experiment (Run 31).

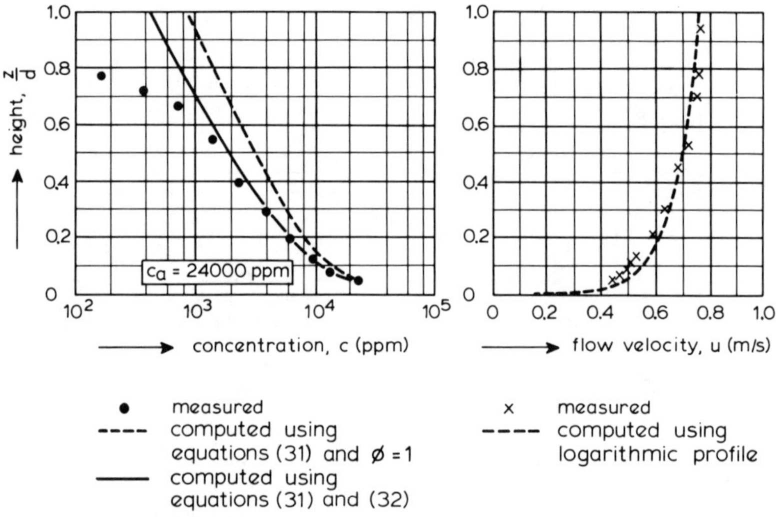

FIGURE 10. Measured and computed concentration and flow velocity profiles for Vanoni-Brooks experiment (Run 3).

duction of particle fall velocity and damping of turbulence). The φ-values have been determined by means of a trial and error method which implies the numerical computation of concentration profiles {Equation (31) and (32)} for various sets of hydraulic conditions and the determination of the φ-value that yields a concentration profile {Equations (19) and (33)} similar to the concentration profile based on the numerical method. Hence for each set of hydraulic conditions (w_s, u_*, c_a) a φ-value is obtained. An analysis of the φ-values showed a simple relationship with the main hydraulic parameters, as follows (inaccuracy of about 25%):

$$\varphi = 2.5\left[\frac{w_s}{u_*}\right]^{0.8}\left[\frac{c_a}{c_0}\right]^{0.4} \quad \text{for } 0.01 \leqslant \frac{w_s}{u_*} \leqslant 1 \qquad (34)$$

Figure 8 shows an example of a concentration profile according to the simplified method.

Summarizing, it is stated that for concentrations larger than about 0.001 (≃ 2500 ppm) the ϕ-factor becomes smaller than about 0.9 (Figure 7) and hence the reduction of the sediment diffusion coefficient must be taken into account. Compared with the concentration profile, the influence of the damping of the turbulence on the velocity profile is relatively small. As stated before, the present analysis only provides a first understanding of the phenomena involved. More research is necessary applying the complete set of equations to compute the concentration and velocity profiles.

Influence of Bed Forms

In a qualitative sense the experiments of Ikeda [22] provide some information about the influence of the bed forms on the concentration profile. He carried out an experiment with a rigid flat bed in which the amount of sediment particles (D_{50} = 180 μm) was controlled so as not to yield deposition and a similar experiment with a movable bed surface. In both experiments the flow conditions were about the same ($w_s/u_* \simeq 0.6$), but the concentration profiles in the movable bed experiment were much more uniform than those in the rigid flat bed experiment, thereby indicating a more intensive mixing process due to the bed forms. The computed β-factor was about 2.4 for the movable bed experiment and about 1.3 to 1.8 for the flat bed experiment. Another remarkable phenomenon observed by Ikeda was the increase of the sediment concentrations by a factor of 10 as the bed forms became three-dimensional, whereas the bed-shear velocity remained nearly constant.

As regards the influence of the bed forms on the concentration profile, and hence on the suspended load, also the approach of Einstein [12] should be mentioned. Einstein assumed that the diffusion coefficient depends on the bed-shear velocity related to grain-roughness (u'_*) instead of the overall bed-shear velocity (u_*), which is stated on pages 9 and 25 of his original report [12]. The approach of Einstein is remarkable because the diffusion of sediment particles in the main part of the flow is merely related to the overall bed-shear velocity in which the turbulence energy generated in the separated flow regions plays an essential role. Therefore the concept of Einstein, which is followed by Engelund and Fredsøe [17], must be rejected.

Finally, it is stated that for most practical situations the present knowledge of the sediment transport is sufficient to do morphological predictions. However, from a scientific point of view further theoretical and experimental research is necessary to extend the knowledge of the sediment diffusivity (β- and ϕ-factor), while also the influence of the bed forms on the vertical distribution of the fluid diffusivity and hence the sediment diffusivity must be studied.

COMPUTATION OF SUSPENDED LOAD

Reference Concentration

In Part 1 (Bed-Load Transport) a function for the bed-load concentration has been proposed. Generally, however, it is not attractive to use the bed-load concentration as the reference concentration for the concentration profile because it prescribes a concentration at a level equal to the saltation height which will result in large errors for the concentration profile as shown in Figure 4. Furthermore, this approach is rather artificial in the case of bed forms because the bed-load concentration is an estimate for the concentration in the bed-load layer at the upstream side of the bed forms. Therefore, another approach [34] is introduced with a reference level (a) related to the bed-form height as shown in Figure 11.

Below the reference level the transport of all sediment particles is considered as bed-load transport (q_b) and an effective reference concentration (c_a) is defined as:

$$q_b = c_b u_b \delta_b = c_a \bar{u}_a a \qquad (35)$$

in which: c_b = bed-load concentration, u_b = velocity of bed-load particles, δ_b = saltation height, $\bar{u}_a$ = effective particle velocity, a = reference level above bed.

Assuming $\bar{u}_a = \alpha_2 u_b$ and using the proposed relationships (Part 1) for the bed-load concentrations (c_b) and the saltation height (δ_b), the reference concentration (c_a) can be expressed as:

$$c_a = \frac{0.035}{\alpha_2} \frac{D_{50}}{a} \frac{T^{1.5}}{D_*^{0.3}} \qquad (36)$$

In the present analysis the reference level is assumed to be equal to half the bed-form height (Δ), or the equivalent roughness height (k_s) if the bed-form dimensions are not known, while a minimum value $a = 0.01d$ is used for reasons of accuracy (Figure 4). Thus,

$$a = 0.5\Delta, \text{ or } a = k_s, \text{ (with } a_{min} = 0.01d) \qquad (37)$$

The actual value of the α_2-factor has been determined by fitting of measured and computed concentration profiles for a range of flow conditions. As only data in the lower flow regime with relatively low concentrations ($\phi \simeq 1$) were selected, the concentration profiles were computed using Equations (19), (22) and (36). The reference level was assumed to be equal to the equivalent roughness height of Nikuradse, because the bed-form heights were not available for all experiments.

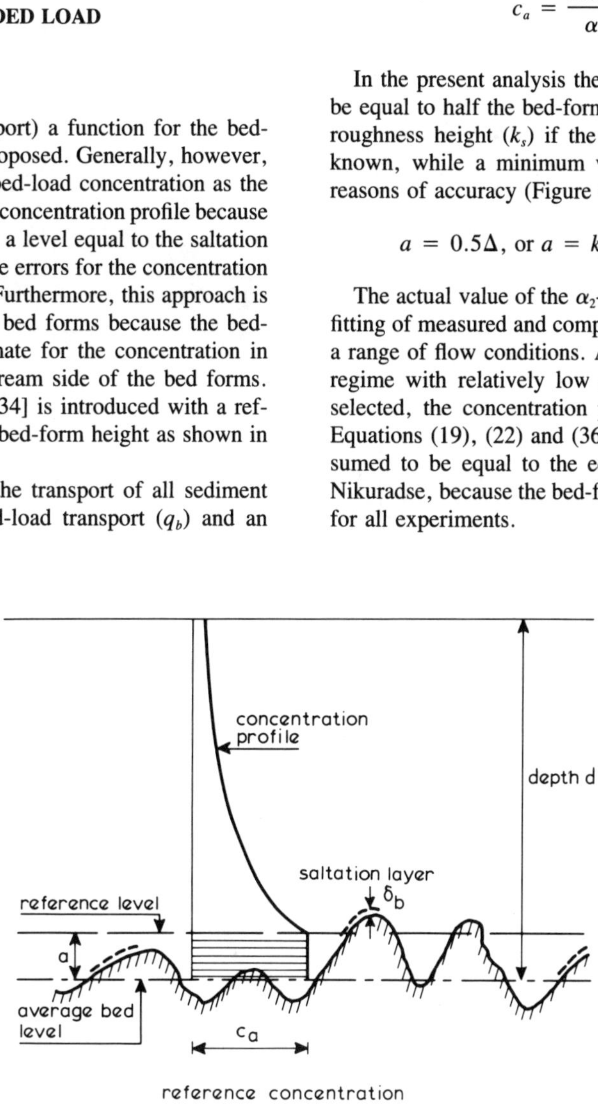

FIGURE 11. Definition sketch for reference-concentration.

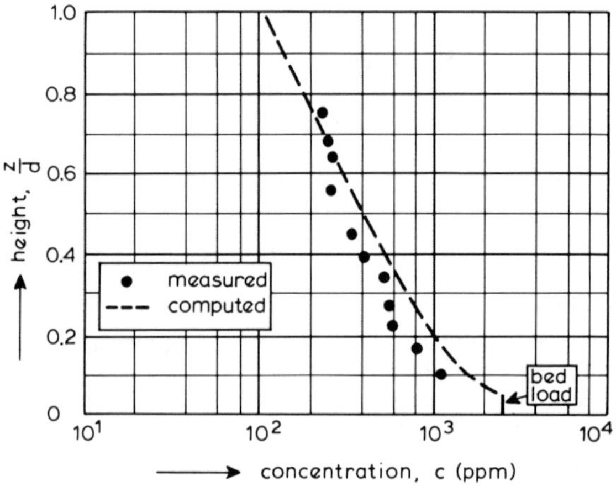

FIGURE 12. Concentration profile for Barton-Lin experiment (Run 7).

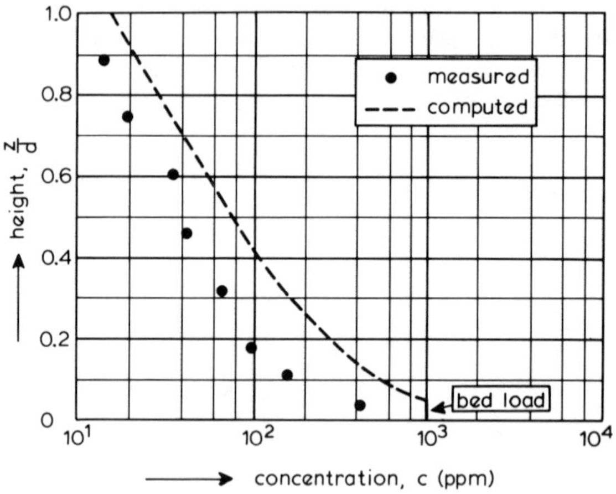

FIGURE 13. Concentration profile for Enoree River (19 February, 1940).

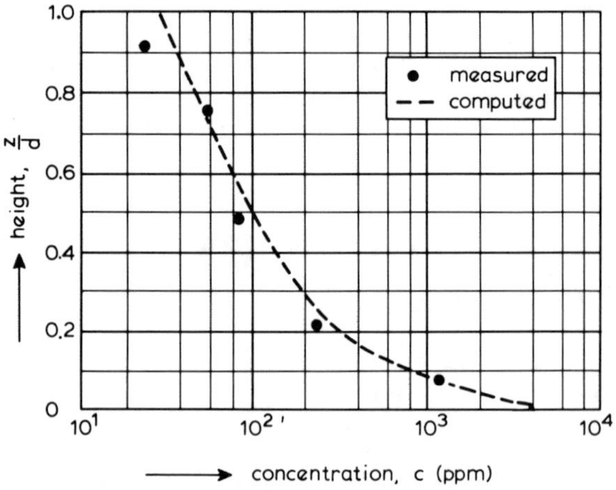

FIGURE 14. Concentration profile for Mississippi River (station 1100, April 1963).

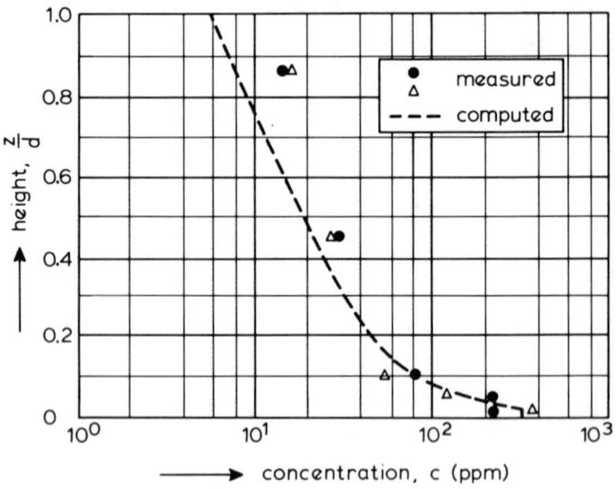

FIGURE 15. Concentration profile for Eastern Scheldt (September 1978).

In all, 20 flume and field data were selected, which were the experiments of Barton–Lin [5] and concentration profiles measured in the Enoree River [2], in the Mississippi River [35] and an estuary in the Netherlands (Eastern Scheldt). The flow depths varied from 0.1 to 25 m, the flow velocity varied from 0.4 to 1.6 m/s and the sediment size from 180 to 700 μm. The "best" agreement between measured and computed concentration profiles for all data was obtained for $\alpha_2 = 2.3$ resulting in:

$$c_a = 0.015 \frac{D_{50}}{a} \frac{T^{1.5}}{D_*^{0.3}} \quad (38)$$

Figures 12, 13, 14 and 15 show some examples of measured and computed concentration profiles for each data set using Equation (38).

In the present state of research the knowledge of the reference concentration is rather limited. Only some graphical results have been presented [18]. However, these curves are not well-defined because the reference level is not specified. Therefore, Equation (38) offers a simple and well-defined expression for the computation of the reference concentration in terms of solids volume per unit fluid volume (or in kg/m^3 after multiplying by the sediment density, ρ_s).

Representative Particle Size of Suspended Sediment

Observations in flume and field conditions have shown that the sediments transported as bed load and as suspended load have different particle size distributions. Usually, the suspended sediment particles are considerably smaller than the bed-load particles. Basically, it is possible to compute the suspended load for any known type of bed material and flow conditions by dividing the bed material into a number of size fractions and assuming that the size fractions do not influence each other. However, a disadvantage of this method, which has been proposed by Einstein [12], is the relatively large computer costs, particularly for time-dependent morphological computations. Therefore, in the present analysis the Einstein-approach is only used to determine a representative particle diameter (D_s) of the suspended sediment [34]. Using the size-fractions method, as proposed by Einstein, the total suspended load has been computed for various conditions, after which by trial and error the representative (suspended) particle diameter was determined that gave the same value for the suspended load as according to the size-fractions method.

In all, six computations were done using two types of bed material with a geometric standard deviation: $\sigma_s = 0.5$ $(D_{84}/D_{50} + D_{16}/D_{50}) = 1.5$ and 2.5. The D_{50} was equal to 250 μm. The mean flow velocities were 0.5, 1.0 and 1.5 m/s. The flow depth was assumed to be 10 m. The concentration profile was computed by means of Equations (19) and (38) with $\beta = 1$, $\phi = 1$ and $\kappa = 0.4$. The reference level was applied at $a = 0.05d$. The flow velocity profile was computed according to the logarithmic law for rough flow conditions. The suspended load transport was computed by means of integration over the flow depth of the product of the local concentration and flow velocity.

The computation results can be approximated by the following expression:

$$\frac{D_s}{D_{50}} = 1 + 0.011(\sigma_s - 1)(T - 25) \quad (39)$$

which is shown in Figure 16. For comparison, some experimental data given by Guy *et al.* [19], are also shown. The scatter of the experimental data is too large to detect any influence of the size gradiation of the bed material. In an average sense the agreement between the measured values and the computed values for $\sigma_s = 2.5$ is reasonably good.

Using the above-given approach, a better representation of the suspended load in the case of a graded bed material

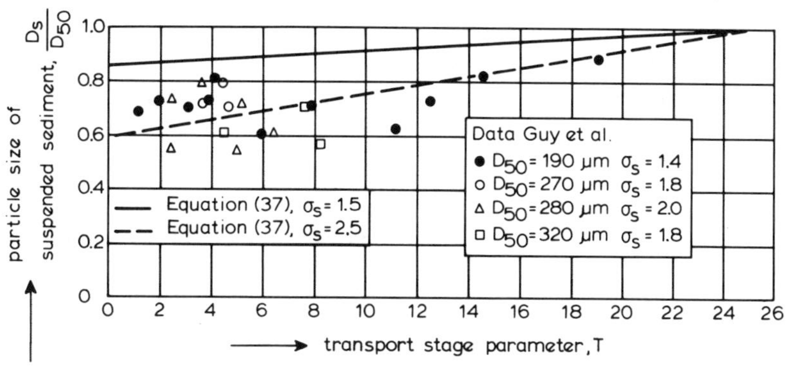

FIGURE 16. Representative particle diameter of suspended sediment.

can be obtained than by taking a fixed particle diameter such as the D_{35}, D_{50} or D_{65} [1,15,12].

Flow Velocity Profile

In a clear fluid with hydraulic rough flow conditions, the flow velocity profile can be described by:

$$\frac{u}{u_*} = \frac{1}{\kappa} \ln\left(\frac{z}{z_0}\right) \tag{40}$$

in which: $z_0 = 0.33k_s$ = zero-velocity level, k_s = equivalent roughness height of Nikuradse.

In the present analysis it has been shown that Equation (40) yields an acceptable representation of the flow velocity profile when the sediment load is not too large (Figures 9 and 10). Therefore, Equation (40) can be applied to compute the suspended load transport. It must be stressed, however, that for very heavy sediment-laden flows the application of Equation (40) may lead to serious errors in the near-bed region (Figure 8). Further research is necessary to determine a simple method for the velocity profile in the case of heavy sediment-laden flows.

Suspended Load Transport

Usually, the suspended load transport per unit width is computed by integration as follows:

$$q_s = \int_a^d cudz \tag{41}$$

Using Equations (19), (33), (34) and (40) to describe the concentration profile and the velocity profile, the suspended load transport follows from Equation (41) resulting in:

$$q_s = \frac{u_* c_a}{\kappa}\left[\frac{a}{d-a}\right]^{Z'}\left[\int_a^{0.5d}\left[\frac{d-z}{z}\right]^{Z'} \ln\left(\frac{z}{z_0}\right) + \int_{0.5d}^{d} [e]^{-4Z'(z/d-0.5)} \ln\left(\frac{z}{z_0}\right) dz\right] \tag{42}$$

The transport of sediment particles below the reference level (a) is considered as bed-load transport (Equation 35).

Equation (42) can be represented with an inaccuracy of about 25% by ($0.3 \le Z' \le 3$ and $0.01 \le a/d \le 0.1$):

$$q_s = F\bar{u}dc_a \tag{43}$$

$$F = \frac{\left[\frac{a}{d}\right]^{Z'} - \left[\frac{a}{d}\right]^{1.2}}{\left[1 - \frac{a}{d}\right]^{Z'}[1.2 - Z']} \tag{44}$$

in which: $\bar{u}$ = mean flow velocity, d = flow depth, c_a = reference concentration. The F-factor is shown for a/d = 0.01, 0.05 and 0.1 in Figure 17.

Summarizing, the complete method to compute the suspended load (volume) per unit width should be applied as follows:

1. Compute particle diameter, D_* by Equation (1)
2. Compute critical bed-shear velocity according to Shields, $u_{*,cr}$ (Part 1)
3. Compute transport stage parameter, T by Equation (2)
4. Compute reference level, a by Equation (37)
5. Compute reference concentration, c_a by Equation (38)
6. Compute particle size of suspended sediment, D_s by Equation (39)
7. Compute fall velocity of suspended sediment, w_s by Equations (11), (12) or (13)
8. Compute β-factor by Equation (22)
9. Compute overall bed-shear velocity, $u_* = (gdS)^{0.5}$
10. Compute φ-factor by Equation (34)
11. Compute suspension parameter Z and Z' by Equation (3) and (33)
12. Compute F-factor by Equation (44)
13. Compute suspended load transport, q_s by Equation (43)

The input data are: $\bar{u}$ = mean flow velocity, d = mean flow depth, b = mean flow width, S = energy gradient, D_{50} and D_{90} = particle sizes of bed material, σ_s = geometric standard coefficient of bed material, ν = kinematic viscosity coefficient, ρ_s = density of sediment, ρ = density of fluid, g = acceleration of gravity, κ = constant of Von Karman.

Ratio of Suspended Load and Total Load

Using Equations (35), (43), (44) and $\varphi = 1$ (low concentrations), the ratio of the suspended and total load transport can be computed as:

$$\frac{q_s}{q_t} = \frac{q_s}{q_s + q_b} = \frac{1}{1 + \left[\frac{q_b}{q_s}\right]} = \frac{1}{1 + \left[\frac{1}{F}\frac{\bar{u}_a}{\bar{u}}\frac{a}{d}\right]} \tag{45}$$

The ratio $\bar{u}_a/\bar{u}$ may be identified as the ratio of the average transport velocity of the bed load and suspended load particles, which varies from about 0.4 for large, steep bed forms in the lower flow regime to about 0.8 for flat bed conditions in the upper flow regime. Figure 18 shows the ratio of the suspended load and the total load as a function of the ratio of the bed-shear velocity and particle fall velocity for different values of $\bar{u}_a/\bar{u}$ and β, with $\kappa = 0.4$ and $a/d = 0.05$. Also an empirical relationship given by Laursen [29] and some data of Guy *et al.* [19] are shown. The β-factor must be known to determine the suspension parameter (Z) and hence the correction factor (F). Two β-functions are applied: β according to Equation (22) and $\beta = 1$. Using Equation

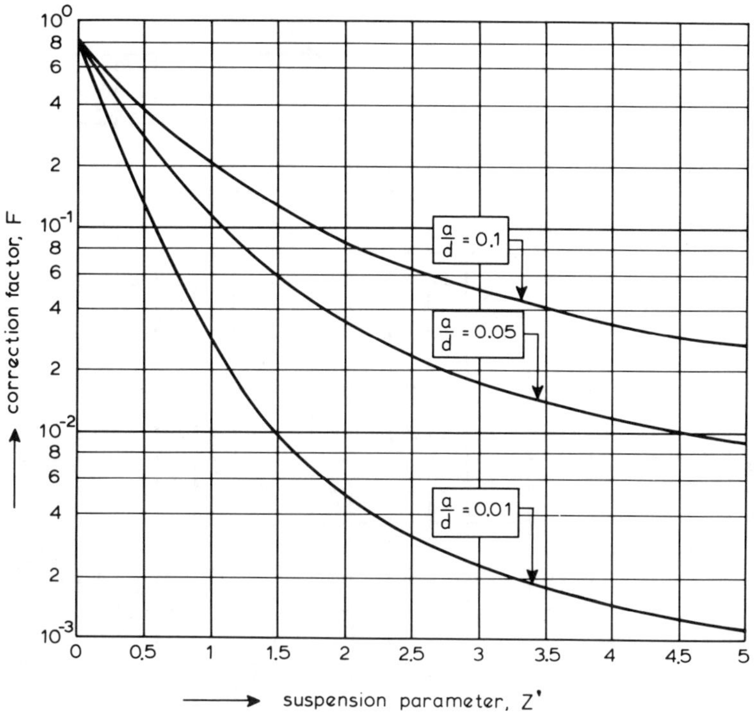

FIGURE 17. F-factor.

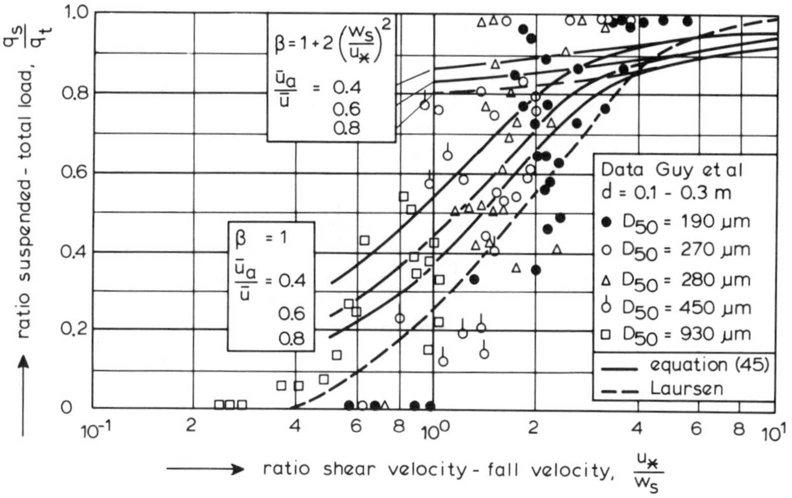

FIGURE 18. Ratio of suspended load and total load transport.

TABLE 1. Comparison of Computed and Measured Total Load Transport.

	Source	Number	Flow Velocity (m/s)	Flow Depth (m)	Particle Diameter (μm)	Temperature (°C)	Scores (%) of Predicted Total Load in Discrepancy Ranges											
							$0.75 \leq r \leq 1.5$				$0.5 \leq r \leq 2$				$0.33 \leq r \leq 3.0$			
							Van Rijn	Engelund–Hansen	Ackers–White	Yang	Van Rijn	*E–H*	*A–W*	Yang	Van Rijn	*E–H*	*A–W*	Yang
Field Data	Various USA-Rivers (corps-eng.)	266	0.4–2.4	0.3–17	120–160	2–35	53%	39%	32%	6%	79%	67%	61%	24%	94%	87%	78%	44%
	Middle Loup River	46	0.65–1.15	0.3–0.65	300–400	0–30	39	13	37	63	78	37	74	94	96	80	98	100
	India-canals	30	0.7–1.6	1.3–3.4	90–310	10–30	30	15	27	3	60	45	48	6	90	73	70	24
	Pakistan canals	87	0.6–1.3	1.4–3.6	110–290	15–35	23	37	34	13	56	71	71	29	91	94	91	48
	Niobrara River	57	0.6–1.3	0.4–0.65	280	0–30	55	13	29	86	95	67	58	98	98	95	98	98
		486					45%	32%	32%	22%	76%	64%	63%	39%	94%	88%	84%	55%
Flume Data	Guy *et al.*	90	0.4–1.2	0.1–0.4	190–470	8–34	40	67	56	68	70	89	85	90	91	98	99	98
	Oxford	84	0.4–1.3	0.1–0.4	100	14–30	37	20	31	45	84	38	59	89	96	70	81	96
	Stein	37	0.4–1.2	0.1–0.4	400	20–30	54	73	81	56	70	95	97	97	97	97	100	100
	Southampton A	33	0.4–0.8	0.15–0.3	150	15–25	64	49	46	49	85	73	79	82	97	91	94	94
	Southampton B	33	0.4–0.55	0.15	480	21	18	12	82	91	81	82	96	97	94	97	100	100
	Barton–Lin	20	0.4–0.95	0.15–0.4	180	15–27	35	60	30	40	65	100	50	65	100	100	100	100
		297					41%	46%	52%	59%	77%	74%	77%	89%	95%	89%	94%	98%
Total		783					43%	37%	40%	36%	76%	68%	68%	58%	94%	88%	88%	71%

(22), the computed transport (q_s/q_t) ratio is much too large compared with the experimental data. Using $\beta = 1$, the agreement between computed and measured values is much better, although the experimental data show that for small u_*/w_s-values the computed transport ratio is still somewhat too large which may be an indication that for small u_*/w_s-values the sediment diffusivity (ε_s) may be relatively small compared with the fluid diffusivity (ε_f) and hence $\beta < 1$. Ultimately, the β-factor may approach zero ($\beta \downarrow 0$) for decreasing u_*/w_s-values. From these results it can be concluded that Equation (22), which predicts an opposite trend with an increasing β-factor (up to $\beta = 3$) for decreasing u_*/w_s-values, is not reliable for small u_*/w_s-values.

Therefore, in the present stage of knowledge it is proposed to use Equation (22) for normal flow conditions ($u_*/w_s > 2$), while for low flow stages just beyond initiation of suspension $\beta = 1$ should be used.

Further research is necessary to investigate the discrepancies between the computed results based on Equation (22) and the experimental data of Guy *et al.* (Figure 18). It is remarked that Equation (22) is based on measured concentration profiles in the u_*/w_s-range from 1 to 10 only (Coleman [9]). Experimental research is necessary to determine a general β-function for all flow conditions from initiation of suspension to the upper flow regime.

VERIFICATION

To verify the proposed method, a comparison of predicted and measured values of the total bed material load has been made. As the present analysis is focussed on the computation of the suspended load transport, only data with particle sizes smaller than about 500 μm ($D_* < 12$) were selected. Other selection criteria were: flow depth larger than 0.1 m, mean flow velocity larger than 0.4 m/s, width–depth ratio larger than 3 and a Froude number smaller than 0.9. Most of the flume and field data (Table 1) were selected from a compendium of Solids Transport compiled by Peterson and Howells [31]. As reported by Brownlie [6], various sets of this data bank contain a number of errors. The data sets used in the present analysis are free of the errors uncovered by Brownlie with exception of the Indian Canal data which contain a 12% error in the sediment concentration. The present writer has eliminated this error before using the data. Therefore, all data used in the present analysis can be considered as reliable data. In addition to the data bank of Peterson and Howells, the writer has used 87 data from the Pakistan Canals [30], 46 data from the Middle Loup River [21] and 57 data from the Niobrara River [8]. In all, 486 field data and 297 flume data were used. Firstly, the field data are described in more detail. The USA-River data collected by the Corps of Engineers consist of 30 data from the Rio Grande River near Bernalillo, 65 data from the Atchafalaya River near Simmesport, 45 data from the Mississippi River near Tarbert Landing, 100 data from the Mississippi River near St. Louis and 26 data from the Red River near Alexandria. The USA-River data include the bed material load in the measured zone and the estimated bed material load in the unmeasured zone. The bed material loads for the Indian and Pakistan Canals as used in the present analysis only include the bed material loads in the measured zone and are, therefore, less than the total bed material loads. The data from the Middle Loup River and the Niobrara River represent the total bed material loads. Finally, it is remarked that the washload is excluded from all field data and that a water temperature of 15°C has been assumed, if not reported. Where the geometric standard deviation of the bed material was not known, a value of 2 was assumed.

A side-wall correction method according to Vanoni–Brooks [38] has been used to eliminate the side-wall roughness.

The suspended load transport (q_s) according to the writer's method is computed from Equation (43), while the bed-load transport (q_b) is computed as given in Part 1. The total load transport is computed as $q_t = q_s + q_b$. For comparison also the total load formulas of Engelund–Hansen [15], Ackers–White [1] and Yang [41] were used (see Appendix). The accuracy of the four methods is given in terms of a discrepancy ratio (r) defined as:

$$r = \frac{q_{t,computed}}{q_{t,measured}} \tag{46}$$

For all data, the score of the predicted values in the ranges $r = 0.75$ to 1.5, 0.5 to 2 and 0.33 to 3 were determined. The results are given in Table 1.

The writer's method yields the best results for the field data and the best results for all data used. This is a remarkably good result particularly when it is remembered that only 20 data were used to calibrate the function for the reference concentration. Analysis of the data showed no systematic errors of the total load in relation to the T- and D_*-parameters [34]. The method of Yang yields excellent results for the flume data and the small-scale river data (Middle Loup and Niobrara River), but very poor results for large-scale rivers (flow depth larger than 1 meter). This cannot be attributed to the quality of the large-scale river data, because the other three methods produce reasonable results for these data sets. Hence, the method of Yang must have serious systematic errors for large flow depths. On the average, the predicted values are much too small.

As regards the experimental data used for calibration and verification, some remarks are made with respect to the accuracy of the measured values. The writer has analyzed some Laboratory experiments performed under similar flow conditions [34]. The results show deviations up to a factor 2. Based on these results, it may be concluded that it seems hardly possible to predict the total load with an inaccuracy of less than a factor 2. The relatively low accuracy of measured transport rates also justifies the use of simple approximation functions with less accuracy to avoid the use of complicated numerical solution methods [Equation (43)]

NOTATION

The following symbols have been used:

a = reference level, (L)
C = overall Chézy coefficient, $(L^{0.5}T^{-1})$
C' = Chézy coefficient related to grains, $(L^{0.5}T^{-1})$
c = concentration (volume = $\rho_s\ 10^3\ c$ in ppm)
c_a = reference concentration
c_b = bed-load concentration
c_0 = maximum (bed) concentration (= 0.65)
D_* = particle parameter
D_{50} = particle diameter of bed material, (L)
D_s = representative particle diameter of suspended sediment, (L)
d = flow depth, (L)
F = correction factor for suspended load
g = acceleration of gravity, (LT^{-2})
k_s = equivalent roughness height of Nikuradse, (L)
q_b = bed-load transport per unit width, (L^2T^{-1})
q_s = suspended load transport per unit width, (L^2T^{-1})
q_t = total load transport per unit width, (L^2T^{-1})
R_b = hydraulic radius of the bed, (L)
Re = Reynolds' number
r = discrepancy ratio
S = slope
s = specific density
T = transport stage parameter
Te = temperature, (°C)
u = local mean longitudinal flow velocity (LT^{-1})
$\bar{u}$ = mean flow velocity, (LT^{-1})
$\bar{u}_a$ = effective velocity of bed-load particles (LT^{-1})
u' = longitudinal flow velocity fluctuation, (LT^{-1})
u_b = transport velocity of bed-load particles (LT^{-1})
u_* = overall bed-shear velocity, (LT^{-1})
u'_* = bed-shear velocity related to grains, (LT^{-1})
$u_{*,cr}$ = critical bed-shear velocity for initiation of motion (LT^{-1})
$u_{*,crs}$ = critical bed-shear velocity for initiation of suspension (LT^{-1})
w = local mean vertical flow velocity, (LT^{-1})
w' = vertical flow velocity fluctuation, (LT^{-1})
w_s = particle fall velocity in clear still fluid, (LT^{-1})
$w_{s,m}$ = particle fall velocity in sediment-fluid mixture, (LT^{-1})
Z, Z' = suspension number
z = vertical coordinate, (L)
z_0 = zero velocity level, (L)
β = ratio of sediment diffusion and fluid diffusion coefficient
δ_b = saltation height, (L)
Δ = bed form height, (L)
ε_f = diffusion coefficient of fluid, (L^2T^{-1})
ε_m = diffusion coefficient of fluid-sediment mixture (L^2T^{-1})
ε_s = diffusion coefficient of sediment (L^2T^{-1})
κ = constant of Von Karman for clear fluid
κ_m = constant for Von Karman for fluid-sediment mixture
λ = bed form length, (L)
μ = dynamic viscosity coefficient $(ML^{-1}T^{-1})$
μ_b = bed form factor
ν = kinematic viscosity coefficient for clear fluid, (L^2T^{-1})
ν_m = kinematic viscosity coefficient for a fluid-sediment mixture (L^2T^{-1})
ρ = density of fluid, (ML^{-3})
ρ_s = density of sediment, (ML^{-3})
σ_s = geometric standard deviation of bed material
τ = shear stress $(ML^{-1}T^{-2})$
ϕ = ratio of diffusion coefficient in fluid-sediment mixture and clear fluid
φ = correction factor for concentration profile

REFERENCES

1. Ackers, P. and White, R., "Sediment Transport: New Approach and Analysis," *Journal of the Hydraulics Division, Proc. ASCE*, No. Hy 11, 1973.
2. Anderson, A. G., "Distribution of Suspended Sediment in a Natural Stream," Transactions, American Geophysical Union, Papers, *Hydrology*, 1942.
3. Bagnold, R. A., "Experiments on a Gravity-free Dispersion of Large Solid Spheres in a Newtonian Fluid Under Shear," Page 60, *Proc. Royal Soc.*, Vol. 225 A, 1954.
4. Bagnold, R. A., "An Approach to the Sediment Transport Problem for General Physics," *Geol. Survey Prof. Paper 422-I*, Washington, USA, 1966.
5. Barton, J. R. and Lin, P. N., "A Study of the Sediment Transport in Alluvial Streams," Civ. Eng. Dep., Colorado College, Report No. 55 JRB2, Fort Collins, USA, 1955.
6. Brownlie, W. R., "Discussion of Total Load Transport in Alluvial Channels," *Journal of the Hydraulics Division, Proc. ASCE*, No. HY12, 1981.
7. Chien, N., "The Present Status of Research on Sediment Transport," *Journal of the Hydraulics Division, Proc. ASCE*, Vol. 80, 1954.
8. Colby, B. R. and Hembree, C. H., "Computations of Total Sediment Discharge Niobrara River Near Cody, Nebraska," *Geol. Survey Water-Supply Paper 1357*, Washington, USA, 1955.
9. Coleman, N. L., "Flume Studies of the Sediment Transfer Coefficient," *Water Resources*, Vol. 6, No. 3, 1970.
10. Coleman, N. L., "Velocity Profiles with Suspended Sediment," *Journal of Hydraulic Research*, Vol. 19, No. 3, 1980.
11. Delft Hydraulics Laboratory, "Initiation of Motion and Suspension, Development of Concentration Profiles in a Steady, Uniform Flow Without Initial Sediment Load," Report M1531-III, The Netherlands, 1982.
12. Einstein, H. A., "The Bed-Load Function for Sediment Transportation in Open Channel Flows," Dept. of Agriculture, Techn. Bulletin No. 1026, Washington, USA, 1950.
13. Einstein, H. A. and Chien, N., "Effects of Heavy Sediment

Concentration Near the Bed on Velocity and Sediment Distribution,'' M.R.D. Sediment Series No. 8, Univ. of California, Berkeley, USA, 1955.

14. Einstein, H. A. and Abdel-Aal, F. M., ''Einstein Bed-Load Function at High Sediment Rate,'' *Journal of the Hydraulics Division, Proc. ASCE*, Vol. 98, No. Hy 1, 1972.
15. Engelund, F. and Hansen, E., ''A Monograph on Sediment Transport in Alluvial Streams,'' *Teknisk Forlag*, Copenhagen, Denmark, 1967.
16. Engelund, F., ''A Criterion for the Occurrence of Suspended Load,'' *La Houille Blanche*, No. 8, page 7, 1965.
17. Engelund, F. and Fredsøe, J., ''A Sediment Transport Model for Straight Alluvial Channels,'' *Nordic Hydrology*, 7, page 294, 1976.
18. Garde, R. J. and Ranga Raju, K. G., ''Mechanics of Sediment Transportation and Alluvial Streams Problems,'' page 171, Wiley Eastern LTD, 1977.
19. Guy, H. P., Simons, D. B., and Richardson, E. V., ''Summary of Alluvial Channel Data from Flume Experiments, 1956–1961,'' *Geol. Survey Prof. Paper, 462-I*, Washington, USA, 1966.
20. Hinze, J. O., ''Turbulence,'' (Second Edition), Page 640–645, McGraw-Hill Book Company, New York, USA, 1975.
21. Hubbell, D. W. and Matejka, D. Q., ''Investigations of Sediment Transportation Middle Loup River at Dunning, Nebraska,'' *Geol. Survey Water-Supply Paper 1476*, Washington, USA, 1959.
22. Ikeda, S., ''Suspended Sediment on Sand Ripples,'' *Third International Symposium on Stochastic Hydraulics*, Tokyo, Japan, 1980.
23. Ippen, A. F., ''A New Look at Sedimentation in Turbulent Streams,'' *Journal of the Boston Soc. of Civ. Engrs.*, Vol. 58, No. 3, 1971.
24. Kerssens, P. J. M., ''Adjustment Length of Suspended Sediment Profiles,'' (in Dutch), River Eng. Dep., Delft Technical University, The Netherlands, 1974.
25. Kerssens, P. J. M. and Rijn, L. C. van, ''Model for Non-Steady Suspended Transport,'' *Seventeenth Congress IAHR*, Baden-Baden, West Germany (Publication No. 91, Delft Hydraulics Laboratory, The Netherlands), 1977.
26. Kerssens, P. J. M., ''Morphological Computations for Suspended Sediment Transport,'' Report S78-VI, Delft Hydraulics Laboratory, The Netherlands, 1978.
27. Kerssens, P. J. M., Prins, A., and Rijn, L. C. van, ''Model for Suspended Sediment Transport,'' *Journal of the Hydraulics Division, Proc. ASCE*, No. HY 5, 1979.
28. Kikkawa, H. and Ishikawa, K., ''Applications of Stochastic Processes in Sediment Transport,'' *Water Resources Publications*, edited by M. Hino, page 8-22, Littleton, Colorado, USA, 1980.
29. Laursen, E. M., ''The Total Sediment Load of Streams,'' *Journal of the Hydraulics Division, Proc. ASCE*, No. HY 1, 1958.
30. Mahmood, K., ''Selected Equilibrium State Data from ACOP Canals,'' Civ. Mech. Envir. Eng. Dept., George Washington University, Report No. EWR-79-2, Washington, USA, 1972.
31. Peterson, A. W. and Howells, R. F., ''A Compendium of Solids Transport Data for Mobile Boundary Channels,'' Report No. HY-1973-ST3, Dept. of Civ. Eng., Univ. of Alberta, Canada, 1973.
32. Rijn, L. C. van, ''Model for Sedimentation Predictions,'' *Nineteenth Congress IAHR*, New Delhi, India (Publication No. 241, Delft Hydraulics Laboratory, The Netherlands), 1981.
33. Rijn, L. C. van, ''Computation of Bed-Load Concentration and Transport,'' Report S487-I, Delft Hydraulics Laboratory, The Netherlands, 1981.
34. Rijn, L. C. van, ''Computation of Bed-Load and Suspended Load,'' Report S487-II, Delft Hydraulics Laboratory, The Netherlands, 1982.
35. Scott, C. H. and Stephens, H. D., ''Special Sediment Investigations Mississippi River at St. Louis, Missouri, 1961–1963,'' *Geol. Survey Water Supply, Paper 1819-J*, Washington, USA, 1966.
36. Shields, A., ''Anwendung der Ähnlichkeitsmechanik und der Turbulenz Forschung auf die Geschiebe Bewegung,'' *Mitt. der Preuss. Versuchsanst. für Wasserbau und Schiffbau*, Heft 26, Berlin, 1936.
37. Subcommittee on Sedimentation, ''Inter-Agency Committee on Water Research,'' Report No. 12, Minneapolis, Minnesota, USA, 1957.
38. Vanoni, V. A. and Brooks, N. H., ''Laboratory Studies of the Roughness and Suspended Load on Alluvial Streams,'' Sedimentation Laboratory, California Inst. of Technology, Report E-68, Pasadena, USA, 1957.
39. Task Committee, ''Sediment Transportation Mechanics: Hydraulic Relatins for Alluvial Streams,'' *Journal of the Hydraulics Division, Proc. ASCE*, No. HY 1, 1971.
40. Yalin, M. S. and Finlayson, G. D., ''On the Velocity Distribution of the Flow-Carrying Sediment in Suspension,'' Symposium to honour Professor H. A. Einstein, page 8–11, Edited by H. W. Shen, USA, 1972.
41. Yang, C. T., ''Incipient Motion and Sediment Transport,'' *Journal of the Hydraulics Division, Proc. ASCE*, Vol. 99, No. HY 10, October 1973.
42. Zanke, U., ''Berechnung der Sinkgeschwindigkeiten von Sedimenten,'' *Mitt. des Franzius-Instituts für Wasserbau*, Heft 46, Seite 243, Techn. Univ. Hannover, West Deutschland, 1977.

APPENDIX PART 2

The *total load formula of Engelund–Hansen* is applied, as follows:

$$q_t = \frac{0.05\bar{u}^5}{(s-1)^2 g^{0.5} D_{50} C^3}$$

in which:

q_t = total load transport per unit width (m^2/s)
$\bar{u}$ = mean flow velocity (cross-section) (m/s)
s = 2.65 = relative density (−)

g = acceleration of gravity (m/s^2)
D_{50} = particle diameter (m)
C = coefficient of Chézy ($m^{0.5}/s$)

The *total load formula of Ackers–White* is applied, as follows

$$q_t = K\bar{u}D_{35}\left[\frac{\bar{u}}{u_*}\right]^n\left[\frac{Y - Y_{cr}}{Y_{cr}}\right]^m$$

in which

q_t = total load transport per unit width (m^2/s)
$K = \exp\{2.86 \ln(D_*) - 0.434(\ln D_*)^2 - 8.13\}$
$D_* = D_{35}\left[\frac{(s-1)g}{\nu^2}\right]^{1/3}$ (−)
ν = kinematic viscosity coefficient (m^2/s)
$\bar{u}$ = mean flow velocity (m/s)
u_* = bed-shear velocity (m/s)
$n = 1 - 0.56 \log(D_*)$ (−)
$m = \frac{9.66}{D_*} + 1.34$ (−)
$Y_{cr} = \frac{0.23}{D_*^{0.5}} + 0.14$ (−)

$$Y = \left[\frac{(u_*)^n}{((s-1)gD_{35})^{0.5}}\right]\left[\frac{\bar{u}}{5.66 \log\left(\frac{10R_b}{D_{35}}\right)}\right]^{1-n} \quad (-)$$

R_b = hydraulic radius related to the bed according to method of Vanoni-Brooks (1967) (m)
D_{35} = particle diameter of bed material (m)

The *total load formula of Yang* is applied, as follows:

$$qt = \frac{10^{-3}}{\varrho_s}\,\bar{u}\,d\,C_t$$

$$\log C_t = A + B \log\left(\frac{\bar{u}S}{w_s} - \frac{\bar{u}_{cr}S}{w_s}\right)$$

$$A = 5.435 - 0.286 \log\left(\frac{w_s D_{50}}{\nu}\right) - 0.457 \log\left(\frac{u_*}{w_s}\right)$$

$$B = 1.799 - 0.409 \log\left(\frac{w_s D_{50}}{\nu}\right) - 0.314 \log\left(\frac{u_*}{w_s}\right)$$

$$\frac{\bar{u}_{cr}}{w_s} = \frac{2.5}{\log\left(\frac{u_* D_{50}}{\nu}\right) - 0.06} + 0.66$$

$$\text{if } 1.2 < \frac{u_* D_{50}}{\nu} < 70$$

$$\frac{\bar{u}_{cr}}{w_s} = 2.05$$

$$\text{if } \frac{u_* D_{50}}{\nu} \geqslant 70$$

in which

C_t = depth-mean sediment concentration (ppm)
$\bar{u}$ = depth-mean flow velocity (m/s)
S = energy gradient (−)
w_s = median fall velocity of bed material (m/s)
D_{50} = median diameter of bed material (m)
u_* = bed-shear velocity (m/s)
ν = kinematic viscosity coefficient (m^2/s)
$\bar{u}_{cr}$ = depth-mean flow velocity at initiation of motion (m/s)

3. *BED FORMS AND ALLUVIAL ROUGHNESS*

INTRODUCTION

In Parts 1 and 2 of the present analysis the bed load and suspended load transport have been investigated, resulting in new relationships which have been verified extensively. However, in those Parts the proposed relationships were not really used in a predictive sense because the applied Chézy-coefficient was based on measured variables (flow velocity, depth and energy gradient). Even in uniform flow conditions the morphological behaviour of an alluvial channel is rather complicated. The fundamental difficulty is that the channel bed characteristics (bed forms), and hence the hydraulic roughness, depend on the flow conditions (flow velocity and depth) and the sediment transport rate, but these flow conditions are in turn strongly dependent on the channel bed configuration and its hydraulic roughness. Therefore, additional information is required for morphological predictions. Firstly, the relationship between the flow conditions and the sediment transport rate should be known, while also the relationship between the hydraulic roughness and the flow conditions should be available. The importance of the application of a reliable roughness prediction method for morphological computations has been pointed out very clearly by de Vries [40].

In the present analysis the hydraulic roughness due to small-scale bed forms has been considered only; other geometrical characteristics such as breading and meandering are not taken into account. Basically, a one-dimensional morphological system can be described by the following 5 equations:

fluid continuity:

$$Q = \bar{u}db \tag{1}$$

fluid motion:

$$\bar{u}\frac{\delta \bar{u}}{\delta x} + g\frac{\delta d}{\delta x} + g\frac{\delta z_b}{\delta x} = -g\frac{\bar{u}|\bar{u}|}{C^2 d} \tag{2}$$

sediment transport:

$$Q_t = F(\mu, \rho, \rho_s, g, D_{50}, \sigma_s, \bar{u}, C, d, b) \tag{3}$$

sediment continuity:

$$\frac{\delta z_b}{\delta t} + \frac{1}{(1-p)b}\frac{\delta Q_t}{\delta x} = 0 \tag{4}$$

alluvial roughness:

$$C = F(\mu, \rho, \rho_s, g, D_{50}, \sigma_s, \bar{u}, d) \tag{5}$$

known variables:

$$\mu, \rho, \rho_s, g, D_{50}, p, \sigma_s, Q, b$$

unknown variables:

$$Q_t, C, \bar{u}, d, z_b$$

in which: Q = discharge, Q_t = sediment transport rate, C = Chézy-coefficient, $\bar{u}$ = mean flow velocity, d = mean flow depth, b = mean flow width, z_b = bed level, D_{50} = particle diameter, σ_s = geometrical standard deviation of bed material, μ = dynamic viscosity coefficient, ρ = density of fluid, ρ_s = density of sediment, g = acceleration of gravity, p = porosity factor and t = time.

The system is completely closed if a relationship to predict the bed-roughness (C) can be specified. In Parts 1 and 2 such a relationship was supposed to be unknown and, therefore, the measured Chézy-coefficient was used to predict the sediment transport rate (Q_t).

In this part of the analysis the attention is focussed on the prediction of the bed-form characteristics and the Chézy-coefficient as a function of the flow variables (mean flow velocity, depth) and sediment properties (size, gradation). The aim of the analysis is to propose simple functions based on reliable flume and field data and that have a good predicting ability for engineering purposes.

Firstly, a classification diagram for determining the type of bed forms in the lower and transitional flow regime is proposed. Thereafter, new relationships for the dune height and length are derived by analyzing flume and field data. Further, new relationships for the grain- and form roughness which have been proposed recently [28] are now verified and compared with the existing methods of Engelund–Hansen [10], White *et al.* [41] and Brownlie [5].

Finally, the proposed method for the hydraulic roughness will be used in predicting the total bed-material load for field data (USA-Rivers).

Characteristic Parameters

It is assumed that the dimensions of the bed forms are mainly controlled by the bed-load transport. In Part 1 it has been shown that the bed-load transport can be described by a dimensionless particle parameter (D_*) and a transport stage parameter (T) as follows:

(a) particle diameter,

$$D_* = D_{50}\left[\frac{(s-1)g}{\nu^2}\right]^{1/3} \tag{6}$$

in which: D_{50} = particle diameter of bed material, s = specific density, ν = kinematic viscosity coefficient.

(b) transport stage parameter,

$$T = \frac{(u'_*)^2 - (u_{*,cr})^2}{(u_{*,cr})^2} \tag{7}$$

in which: $u'_* = (g^{0.5}/C')$, $\bar{u}$ = bed-shear velocity related to grains, $C' = 18\log(12R_b/3D_{90})$ = Chézy-coefficient related to grains, R_b = hydraulic radius related to the bed according to Vanoni–Brooks [38], D_{90} = particle diameter of bed material, $\bar{u}$ = mean flow velocity and $u_{*,cr}$ = critical bed-shear velocity according to Shields (Part 1).

BED-FORM CLASSIFICATION

Usually, the flow conditions in an alluvial channel are classified into [30]:

1. Lower flow regime with plane bed, ripples and dunes
2. Transitional flow regime with washed-out dunes
3. Upper flow regime with plane bed and anti-dunes

In the literature, roughly two groups of classification methods are described. Engelund [10] and Garde–Albertson [12] use the Froude number as a classification parameter, while Liu [20] and Simons–Richardson [30] describe the type of bed forms in terms of a suspension parameter and a particle-related Reynolds' number.

In the present analysis the attention is focussed on the lower and transitional flow regimes only because these regimes are the most important for field conditions. As will be shown, these regimes can be defined quite well without the use of the Froude number. This may also be indicated by the fact that in flume conditions the transitional stage with washed-out dunes is generated for a Froude number of about 0.6, while in field conditions this value may be about 0.2 to 0.3, as observed in the Missouri River [29]. Only in the upper flow regime with anti-dunes is the Froude number of importance because the generation of anti-dunes is mainly governed by free-surface phenomena as indicated by the length of the anti-dunes which is equal to the wave length of the free surface (Kennedy [17], Yalin [44]).

Ripples and dunes have different geometrical characteristics. The ripple height is much smaller than the flow depth and practically independent of the flow depth, while the ripple length may be as large as the flow depth. The generation of ripples seems to depend mainly on the stability of the granular bed surface under the action of turbulent velocity fluctuations. The dune height is strongly dependent on the flow depth with local maximum values up to the flow depth [30], while the dune length is much larger than the flow depth. The formation of dunes may be caused by large-scale eddies as described by Yalin [44]. Due to the presence of large (low frequency) eddies there will be regions at regular intervals with decreased and increased bed-shear stresses, resulting in local deposition and erosion of sediment particles. For increasing stages of flow the particles will go into suspension depending on the ratio of the grain: shear stress and the particle-fall velocity resulting in washed-out dunes. The washing-out process can be described quite well by a transport stage parameter as defined by Equation (7). This parameter expresses the grain–shear velocity (u'_*), which is an estimate for the average shear velocity at the upsloping part of the bed forms, in relation to the critical Shield's value ($u_{*,cr}$). When $u'_* \gg u_{*,cr}$ the sediment particles will go into suspension directly and the bed forms will be washed out. A similar reasoning has been given by Yalin [44]. The applicability of the T-parameter in characterizing the generation of bed forms in the lower and transitional flow regime is also shown by Figure 1, which shows distinct zones for ripples, dunes and washed-out dunes. Figure 1 is based on a large number of reliable flume and field data (Guy *et al.* [14], Ackers [1], Delft Hydraulics Laboratory [8], Stein [32], Williams [42], Znamenskaya [46], Pakistan Canals [22], Dutch Rivers [37], Rio Parana [33], Hii River [34], Missouri River [30]). Dune-type bed forms are present for $T < 15$. However, for particles smaller than about 450 μm ($D_* \simeq 10$) are ripples generated after initiation of motion, but which disappear for $T > 3$. The transitional flow regime with washed-out dunes is present for $15 < T < 25$. For $T > 25$ a flat bed flow will be generated. Finally, it is pointed out that various classification diagrams can be found in literature (Engelund [10], Liu [20], Garde–Albertson [12] and Simons–Richardson [30]). According to Simons and Sentürk [31], the diagrams of Liu and Garde–Albertson do not give acceptable results for field conditions because relatively few field data were used. The diagram of Engelund–Hansen is mainly based on flume data. Simons–Richardson have used some field data, but only of small-scale rivers. Therefore, the writer's diagram is supposed to be more valid for the lower and transitional flow regime in field conditions because a large number of field data with small and large flow depths have been used.

BED-FORM DIMENSIONS

It is assumed that dimensions and migration of the bed-forms are mainly determined by the value of the bed-load transport.

In Part 1 the bed-load transport was described as:

$$q_b = \delta_b u_b c_b \tag{8}$$

in which: q_b = bed-load transport rate per unit width, u_b = velocity of bed-load particles, c_b = bed-load concentration, δ_b = thickness of bed-load layer.

Using simple kinematic considerations and the continuity equation, the bed-load transport rate can also be described as [44]:

$$q_b = (1 - p)\alpha\Delta u_d \tag{9}$$

in which: p = porosity factor, α = shape factor of bed-forms, Δ = bed-form height, u_d = migration velocity of bed-forms.

From Equations (8) and (9) it can be derived that:

$$\frac{\Delta}{d} = \frac{c_b}{(1 - p)\alpha} \frac{u_b}{u_d} \frac{\delta_b}{D_{50}} \frac{D_{50}}{d} \tag{10}$$

Using the functional relationships as given in Part 1, it

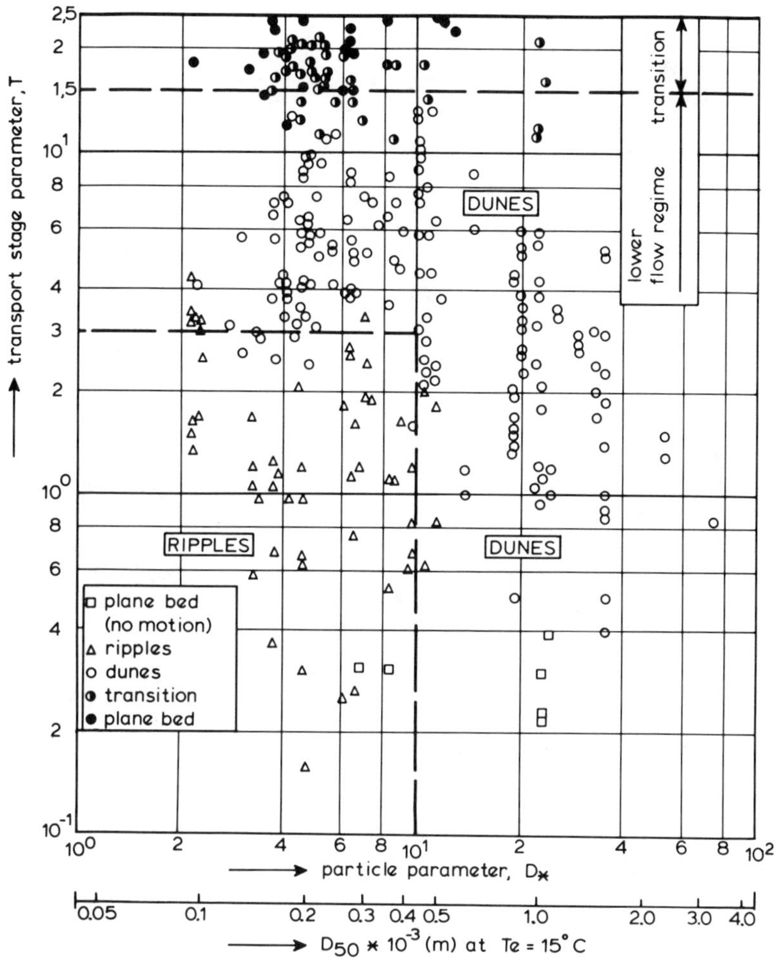

FIGURE 1. Diagram for bed-form classification in the lower and transitional flow regime.

is assumed that:

$$c_b, \frac{\delta_b}{D_{50}}, \frac{u_b}{u_d} = F(D_*, T) \tag{11}$$

resulting in:

$$\frac{\Delta}{d} = F\left(\frac{D_{50}}{d}, D_*, T\right) \tag{12}$$

Likewise, it is assumed that the ratio of the bed-form height (Δ) and length (λ) can be expressed by a similar functional relationship:

$$\frac{\Delta}{\lambda} = F\left(\frac{D_{50}}{d}, D_*, T\right) \tag{13}$$

To determine these functional relationships, a large quantity of experimental data of bed-form dimensions was analyzed. In all, 84 flume experiments with particle diameters in the range 190 to 2300 μm were used [14,42,32,46,8]. Only experiments in the lower and transitional flow regime with dune-type bed forms were considered. Ripple data were not selected bacause ripples are supposed to be independent of the flow depth.

As regards reliable field data, which are extremely scarce, only 22 data with particle diameters in the range 490 to 3600 μm were used from some rivers in the Netherlands: the IJssel, Waal and Rijn [37], the Rio Parana in Argentina [33], the Mississippi River in USA [18] and some Japanese irrigation channels [34,35]. Summarizing, the selection criteria used were as follows: (i) dune-type bed-forms, (ii) width-depth ratio larger than 3, (iii) flow-depth larger than

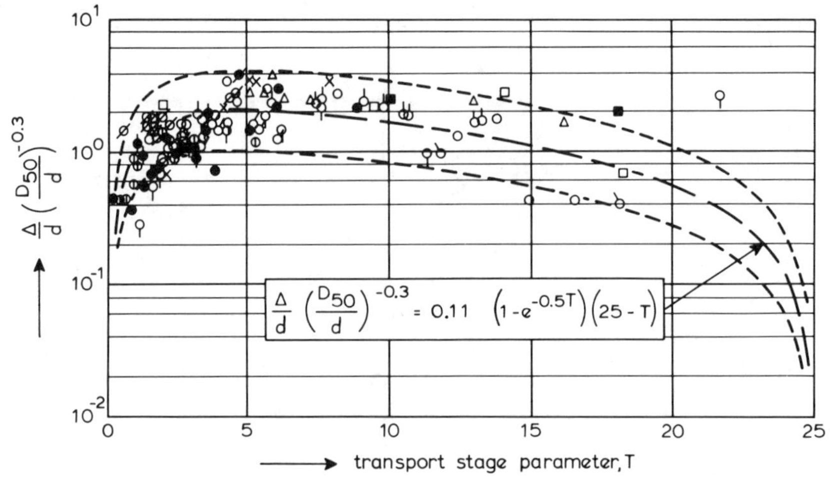

$\frac{\Delta}{d}\left(\frac{D_{50}}{d}\right)^{-0.3} = 0.11 \left(1-e^{-0.5T}\right)(25-T)$
$\frac{\Delta}{d}\left(\frac{D_{50}}{d}\right)^{-0.3}$
transport stage parameter, T

	source	flow velocity $\bar{u}$ (m/s)	flow depth d (m)	particle size D_{50} (μm)	temperature Te (°C)
flume data	○ Guy et al	0.34 - 1.17	0.16 - 0.32	190	8 - 34
	× Guy et al	0.41 - 0.65	0.14 - 0.34	270	8 - 34
	△ Guy et al	0.47 - 1.15	0.16 - 0.32	280	8 - 34
	Guy et al	0.77 - 0.98	0.16	330	8 - 34
	□ Guy et al	0.48 - 1.00	0.10 - 0.25	450	8 - 34
	Guy et al	0.53 - 1.15	0.12 - 0.34	930	8 - 34
	Williams	0.54 - 1.06	0.15 - 0.22	1350	25 - 28
	Delft Hydr. Lab.	0.45 - 0.87	0.26 - 0.49	790	12 - 18
	Stein	0.52 - 0.95	0.24 - 0.31	400	20 - 26
	Znamenskaya	0.53 - 0.80	0.11 - 0.21	800	–
field data	● Dutch Rivers	0.85 - 1.55	4.4 - 9.5	490 - 3600	5 - 20
	Rio Parana	1.0	12.7	400	–
	Japanese Channels	0.53 - 0.89	0.25 - 0.88	1100 - 2300	–
	■ Mississippi River	1.35 - 1.45	6 - 16	350 - 550	–

FIGURE 2. Bed-form height.

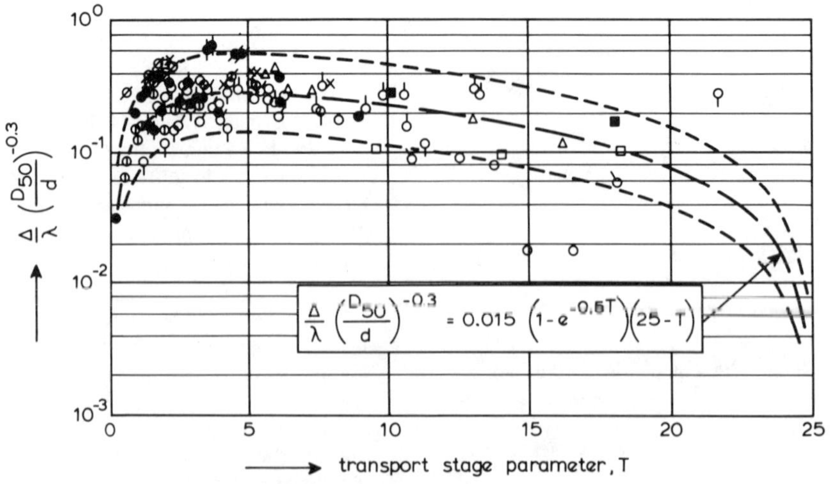

$\frac{\Delta}{\lambda}\left(\frac{D_{50}}{d}\right)^{-0.3} = 0.015 \left(1-e^{-0.5T}\right)(25-T)$
$\frac{\Delta}{\lambda}\left(\frac{D_{50}}{d}\right)^{-0.3}$
transport stage parameter, T

FIGURE 3. Bed-form steepness.

0.1 m and (iv) transport stage parameter (T) smaller than 25.

Where the water temperature was not reported, a value of 15°C was assumed and where the D_{50} of the bed material was not reported, it was computed from the D_{50} and the gradation (σ_s) of the bed material assuming the latter to be equal to 2. The influence of the side-wall roughness was eliminated by using the method of Vanoni–Brooks [38].

In analyzing the data, only for the Missouri River [29] was a clear influence of the particle diameter D_* observed. As reported, a decrease from a temperature of 25°C in early September to 5°C in December (a decrease in D_* of about 30%) caused a remarkable change of the bed configuration from dune-type bed-forms to almost flat bed; although the other hydraulic parameters (discharge, depth, slope and particle size) were essentially unchanged. However, for the large majority of the data such a clear correlationship between the bed-form dimensions and the D_*-parameter could not be detected. Therefore, in the present stage of analysis the D_*-influence was neglected and the bed-form dimensions were related to the particle diameter (D_{50}), the flow depth (d) and the transport stage parameter (T). The best agreement was obtained for [27]:

$$\frac{\Delta}{d} = 0.11 \left[\frac{D_{50}}{d}\right]^{0.3} [1 - e^{-0.5T}][25 - T] \qquad (14)$$

$$\frac{\Delta}{\lambda} = 0.015 \left[\frac{D_{50}}{d}\right]^{0.3} [1 - e^{-0.5T}][25 - T] \qquad (15)$$

It is assumed that for $T \leqslant 0$ and $T \geqslant 25$ the bed surface is almost flat (anti-dunes are not considered). Equations (14) and (15) as well as the error range of a factor 2 are shown in Figures 2 and 3.

Both functions show maximum values for about $T = 5$. From Equations (14) and (15) also an expression for the bed-form length can be derived:

$$\lambda = 7.3d \qquad (16)$$

Equation (16), which is based on a large number of reliable flume and field data, indicates that the dune length is only related to the mean flow depth. This has also been reported by Yalin [44, page 242]. Based on theoretical analysis he derived $\lambda = 2\pi d$, which is close to Equation (16). Yalin also presented experimental data supporting a constant dimensionless dune length. Physically, it means that the bed-form height is reduced for increasing stages of flow, while the bed-form length remains essentially unchanged. Finally, it is pointed out that Equations (14) and (15) are in agreement with the functional relationships for dune characteristics as proposed by Yalin [44, page 237].

To apply Equations (14), (15) and (16) the mean flow velocity, the flow depth and the particle size must be known. The energy gradient need not be known which may be an additional advantage because the energy gradient: (i) may be relatively inaccurate (low flow velocities, rising or falling stages) and (ii) may be difficult to determine in (isolated) field conditions.

Other investigators who have proposed analytical or graphical relationships for the bed-form dimensions, are Tsubaki–Shinohara [35], Yalin [44, 45], Ranga Raju–Soni [25], Allen [3] and Fredsøe [11].

The present writer [27] has compared the available methods by computing the bed-form height for a sand bed with a particle diameter $D_{50} = 600\ \mu m$ ($D_{90} = 1500\ \mu m$), a depth $d = 1$ and 10 m and a mean flow velocity $\bar{u} = 0.5$, 1, 1, 25, and 1.5 m/s. As the methods of Tsubaki–Sinohara, Yalin and Ranga Raju–Soni require the specification of the energy gradient, these methods were used iteratively with an equivalent roughness height equal to half the bed-form height. The results are shown in Figure 4.

As can be observed, the methods of Ranga Raju–Soni, Tsubaki–Shinohara show an increasing trend. The method of Yalin yields values which approach $\Delta/d = 0.167$ for large flow velocities. Only Equation (14) and the method of Fredsøe predict a decreasing dune height for increasing flow velocities (washed-out dunes). For a small depth (d = 1 m) the writer's method and the method of Fredsøe show good agreement, while for a large depth ($d = 10$ m) the method of Fredsøe yields larger values. Finally, it is pointed out that the method of Ranga Raju–Soni produces remarkably small bed-form heights (and also lengths) for large depths as present in field conditions probably because only flume data were used for calibration.

EQUIVALENT ROUGHNESS OF BED FORMS

The hydraulic roughness of a movable bed surface is caused by grain roughness ($k_{s,grain}$) and by form roughness ($k_{s,form}$).

In an earlier study the present writer [28] has shown that the equivalent or effective grain roughness of a flat bed can be related to the D_{90} of the bed material. From about 100 flume and field data it was derived that:

$$k_{s,grain} = 3D_{90} \qquad (17)$$

Similar values have been reported by Kamphuis: $k_s = 2.5D_{90}$ [16], Gladki: $k_s = 2.3D_{80}$ [13], Hey: $k_s = 3.5D_{84}$ [15], and Mahmood: $k_s = 5.1D_{84}$ [21]. As regards the form roughness, the following functional relationship, introduced by Yalin [44, page 235] is assumed to be valid:

$$k_{s,form} = F\left(\Delta, \frac{\Delta}{\lambda}\right) \qquad (18)$$

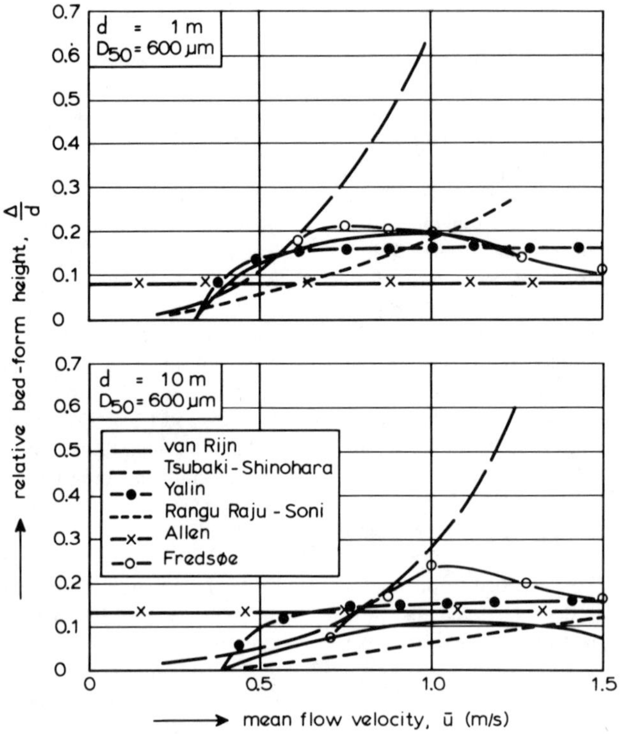

FIGURE 4. *Computed bed-form heights.*

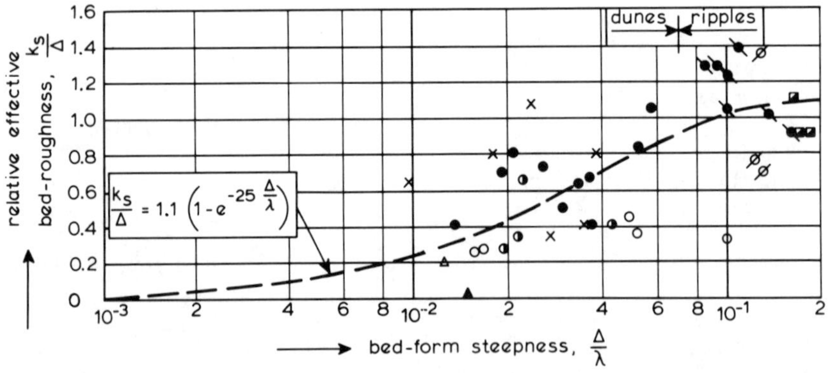

source	d (m)	ū (m/s)	D_{50} (µm)
● Guy et al	0.16-0.32	0.24-0.59	190-930
△ Znamenskaya	0.08	0.49	800
ø Ackers	0.17-0.20	0.35-0.43	180
◑ Stein	0.24-0.30	0.88-1.12	400
◪ Laursen	0.17-0.23	0.42-0.71	100

source	d (m)	ū (m/s)	D_{50} (µm)
● Barton-Lin	0.14-0.26	0.41-0.58	180
x Japanese Channels	0.11-0.43	0.55-0.73	1260-1440
o River Lužnic	0.14-0.75	0.35-0.74	2400
▲ Missouri River	4.5	1.5	240

FIGURE 5. *Equivalent roughness of bed-forms.*

In an earlier study [28], the following expression which is based on flume and field data, was derived:

$$k_{s,form} = 1.1\Delta[1 - e^{-25\psi}] \qquad (19)$$

in which: $\psi = \Delta/\lambda$ = bed-form steepness.

Equation (19), as well as the data, is shown in Figure 5.

Taking into account both the grain roughness and the form roughness, it is proposed to compute the effective roughness (k_s) of a movable bed surface in the lower, transitional and upper flow regime (with exception of anti-dunes) by means of

$$k_s = 3D_{90} + 1.1\Delta[1 - e^{-25\psi}] \qquad (20)$$

It may be noted that for a bed-form steepness (Δ/λ) equal to zero, the flat bed value $k_{s,grain} = 3D_{90}$ is obtained.

Finally, the Chézy-coefficient can be computed by:

$$C = 18 \log\left(\frac{12R_b}{k_s}\right) \qquad (21)$$

in which: R_b = hydraulic radius of the bed according to Vanoni–Brooks [38].

Summarizing, the proposed method is as follows:

1. Compute particle diameter, D_* by Equation (6)
2. Compute critical bed-shear velocity, $u_{*,cr}$ (Part 1)
3. Compute transport stage parameter, T by Equation (7)
4. Compute bed-form height, Δ by Equation (14)
5. Compute bed-form length, λ by Equation (16)
6. Compute equivalent roughness, k_s by Equation (20)
7. Compute Chézy-coefficient, C by Equation (21)

The input data are: mean flow velocity ($\bar{u}$), mean flow depth (d), mean flow width (b), particle diameter of bed material (D_{50}, D_{90}), fluid and sediment density (ρ, ρ_s) and kinematic viscosity coefficient (ν).

As an example, the Chézy-coefficient has been computed for various flow conditions assuming a fluid temperature (Te) of 15°C and a geometric standard deviation of the bed material (σ_s) equal to 2. The results are shown in Figure 6.

Independent support for the validity of the proposed method [Equations (20) and (21)] has been produced by Van Urk [36]. He used an automatic sounding system to determine the dune dimensions in some branches of the Rhine River, a major river in the Netherlands and Europe. Various hydraulic roughness prediction methods based on the bed-form dimensions were selected from the literature. Using the measured dune dimensions, Van Urk computed the Chézy-coefficients for the various methods and compared those values with the Chézy-coefficients derived from the overall hydraulic conditions (mean flow velocity, depth and energy slope). This comparison shows that the prediction method of the present writer and that of Vanoni–Hwang [39] yield the best results for all data.

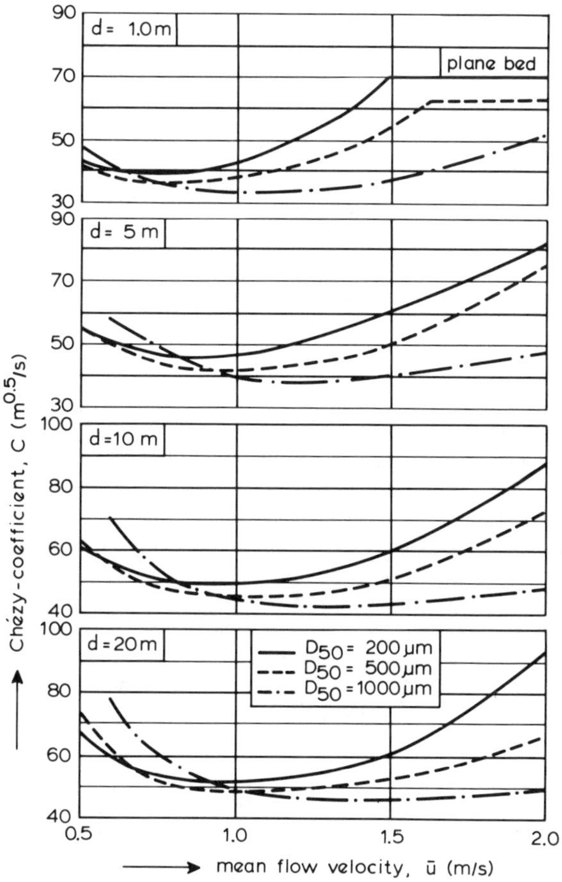

FIGURE 6. Computed Chézy-coefficients according to method of Van Rijn.

Other evidence is presented by Van Urk and Klaassen [37]. They compared various prediction methods for hydraulic roughness using flume data. This study also supports the validity of the present method.

Finally, a discussion of the form drag expression [Equation (19)] by Engel is reported [9], which shows that Equation (19) yields a realistic estimate of the equivalent roughness for dunes.

VERIFICATION

To verify the proposed method, a large amount of flume and field data were selected from a compendium of solids transport data [24]. In addition, about 200 field data reported in literature were used [4,6,7,19,22,23,29]. In all, 786 field data and 758 flume data were selected using the following criteria: (i) flow depth larger than 0.1 m, (ii) width-depth ratio larger than 3, and (iii) particle diameter in the range 100 to 2500 μm.

To eliminate the influence of the side-wall roughness, the method of Vanoni–Brooks [38] was used. Where the water temperature was not reported, a value of 15°C was assumed, and where only the D_{50} of the bed material was reported, the other characteristic particle diameters (D_{35}, D_{65}, D_{90}) were computed assuming a long-normal particle size distribution with a geometric standard deviation $\sigma_s = 2$. None of the data were used by the writer to calibrate his proposed relationships! An investigation into the accuracy of the "measured" Chézy-coefficient has shown that errors up to 20% may be expected [26,27,42,43]. These errors may be caused by: (i) measuring errors, (ii) non-equilibrium phenomena (rising or falling stages) and (iii) three-dimensional effects (alternate bars) in case of a large width-depth ratio.

Therefore, it may hardly be possible to predict the Chézy-coefficient with an inaccuracy less than 20%.

For comparison with the writer's method, the methods of Engelund–Hansen [10] and White *et al.* [41] were also applied to the selected data (see Appendix). These latter two methods gave the best results in an extensive verification study carried out by White *et al.* [41]. It must be stressed that Engelund and Hansen used about 100 flume data (Guy *et al.* [14]) to calibrate their formulae, while White *et al.* used about 1400 flume data and 260 field data for calibration.

For each method the score (in %) of the predicted values in the following error ranges were determined: $C_{measured} \pm 10\%$, $C_{measured} \pm 20\%$, $C_{measured} \pm 30\%$.

The results are shown in Table 1.

For flume conditions the method of Engelund–Hansen produces the best results, while for field conditions the writer's method is superior. Particularly, the method of Engelund–Hansen produces rather bad results for field conditions, probably because only flume data were used for calibration.

The results of the method of White *et al.* are not as good as these investigators reported in their verification study [41]. They gave their results in terms of the ratio of the measured and computed Darcy–Weisbach friction coefficient.

According to White *et al.* their method scored (for about 2500 data) 38% in the 0.8 to 1.25 range (≃10% error range for the Chézy-coefficient) and 89% in the 0.5 to 2.0 range (≃40% error range for the Chézy-coefficient). These results are better than those of the present analysis, probably because in the present analysis also data of the upper flow regime with flat bed conditions were selected, for which the method of White *et al.* is not supposed to be valid. In the verification study of White *et al.* only data in the lower flow

TABLE 1. Comparison of Computed and Measured Chézy-Coefficients.

							Scores (%) of Predicted Chézy-Coefficient in Error Ranges								
							±10% Error			±20% Error			±30% Error		
	Source	Number of Tests	Flow Velocity (m/s)	Flow Depth (m)	Particle Diameter (μm)	Temperature (°C)	van Rijn	Engelund–Hansen	White *et al.*	*R*	*E–H*	*W*	*R*	*E–H*	*W*
FIELD DATA	USA-Rivers (Leopold)	55	0.36–1.26	1.50–4.10	140–420	12–26	29%	25%	32%	65%	34%	56%	80%	45%	78%
	USA-Rivers (Corps of Eng.)	300	0.4–2.4	0.3–16.4	120–560	2–35	40	20	32	68	38	56	82	50	75
	India Canals	31	0.7–1.6	1.3–3.4	90–310	10–30	48	29	45	70	38	64	83	45	70
	Pakistan Canals	142	0.35–1.3	0.7–4.3	110–290	15–35	62	24	52	84	42	80	94	66	92
	Missouri River, USA	21	1.35–1.75	2.8–4.3	190–230	3–22	62	43	0	67	62	0	100	100	24
	East-Fork River, USA	45	0.55–1.4	0.25–2.0	450–1400	15–20	31	22	0	94	67	100	100	100	100
	Elkhorn River, USA	43	0.5–2.0	0.4–2.2	240	2–26	47	23	28	91	42	47	100	72	70
	Rio Grande, USA	79	0.45–2.4	0.3–1.5	180–450	2–29	47	29	18	80	44	40	100	58	70
	Mountain Creek, USA	47	0.5–1.35	0.15–0.45	290–900	15–25	29	27	72	48	91	91	89	91	91
	Japanese Channels	23	0.55–0.95	0.10–0.75	1260–1440	—	34	34	78	73	91	95	82	100	100
		786					43%	25%	33%	74%	47%	58%	89%	62%	79%
FLUME DATA	Guy *et al.*	292	0.2–1.9	0.1–0.4	190–930	8–34	35	30	26	60	61	44	75	70	54
	Stein	54	0.4–1.65	0.1–0.35	400	20–30	38	9	16	62	35	31	83	51	51
	Southampton A	67	0.2–0.8	0.15–0.30	150	15–25	13	52	17	31	82	40	49	86	52
	Southampton B	139	0.2–0.7	0.15–0.30	480	21	48	45	57	73	73	71	79	80	78
	USWES (sand)	128	0.25–0.55	0.1–0.2	180–950	14–18	28	50	46	51	81	79	67	91	86
	Meyer–Peter	22	0.35–0.90	0.1–0.5	400–1500	—	22	22	27	31	36	54	36	63	100
	Barton–Lin	29	0.2–1.1	0.1–0.3	180	15–27	27	27	3	31	51	31	58	62	58
	Laursen	14	0.4–1.0	0.1–0.3	110	20–26	50	35	28	64	50	50	92	64	78
	Williams	13	0.5–1.15	0.15–0.25	1350	16–26	23	46	53	53	61	76	92	69	92
		758					34%	37%	33%	56%	65%	54%	71%	75%	66%
Total		1544					39%	31%	33%	65%	56%	56%	80%	68%	73%

Field Data (small rivers)	
o India Canals	(90 - 310 μm)
x Pakistan Canals	(110 - 290 μm)
ø Missouri River, U.S.A.	(190 - 230 μm)
q East - Fork River, U.S.A.	(450 - 1400 μm)
ъ Elkhorn River, U.S.A.	(240 μm)
♂ Rio Grande, U.S.A.	(180 -450 μm)
● Mountain Creek, U.S.A.	(290 - 900 μm)
□ Japanese Channels	(1260 - 1440 μm)

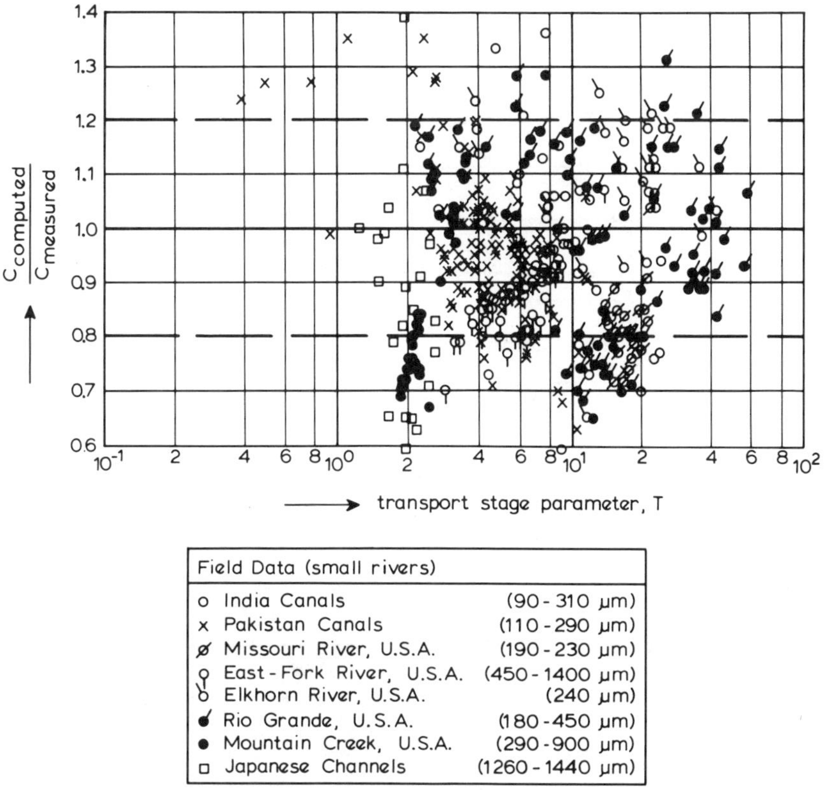

FIGURE 7. Ratio of computed and measured Chézy-coefficients according to method of Van Rijn (small rivers).

Field Data (large rivers)	
● U.S.A. - Rivers (Corps, Eng.),	(120 -560 μm)
o U.S.A. - Rivers (Leopold),	(140 -420 μm)

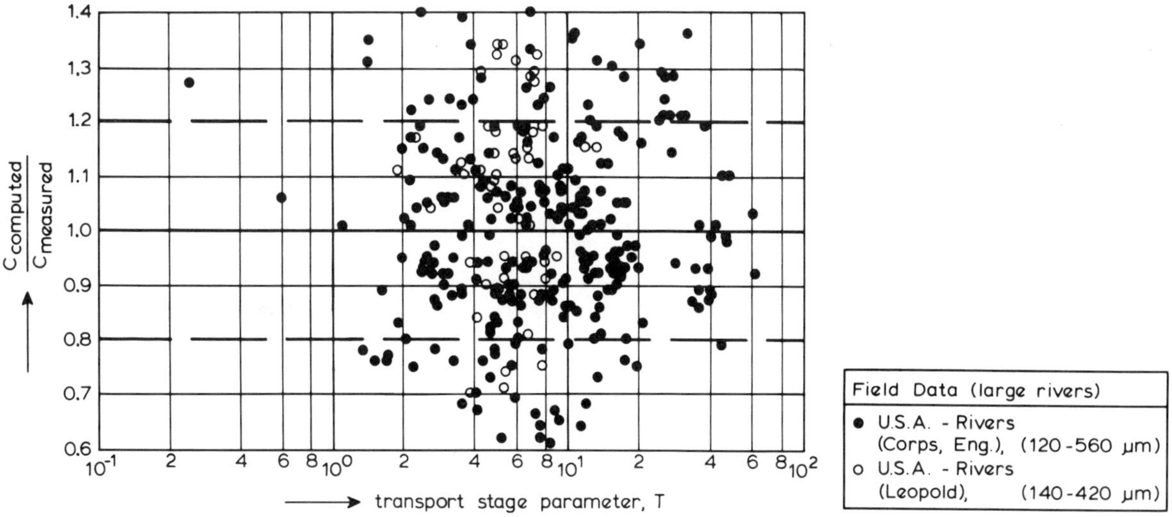

FIGURE 8. Ratio of computed and measured Chézy-coefficients according to method of Van Rijn (large rivers).

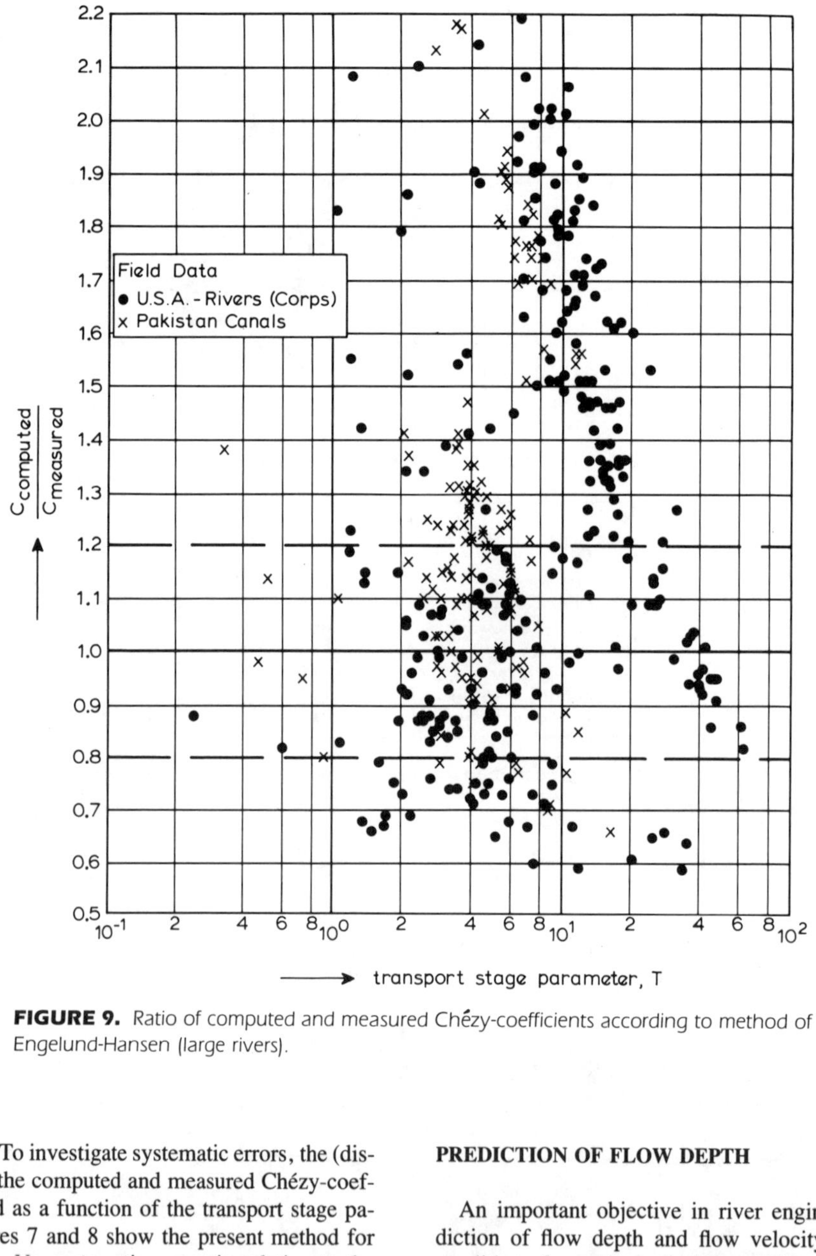

FIGURE 9. Ratio of computed and measured Chézy-coefficients according to method of Engelund-Hansen (large rivers).

regime were used. To investigate systematic errors, the (discrepancy) ratio of the computed and measured Chézy-coefficient were plotted as a function of the transport stage parameter (T). Figures 7 and 8 show the present method for the field data only. No systematic errors in relation to the transport stage can be detected.

Figures 9 and 10 show the discrepancy ratios for the method of Engelund–Hansen and White *et al.* Only the field data from the USA Rivers and the Pakistan Canals are presented. For these data both methods show serious systematic errors. On the average, the predicted values are larger than the measured values. Summarizing, it may be stated that the methods of Engelund–Hansen and White *et al.* are not very good for field conditions.

PREDICTION OF FLOW DEPTH

An important objective in river engineering is the prediction of flow depth and flow velocity in uniform flow conditions for a given discharge (Q), energy gradient (i), width (b), particle diameter (D_{50}) and gradation (σ_s) of bed material, and fluid temperature (Te).

The methods which were used to predict the flow depth, are: Van Rijn, Engelund–Hansen [10], White *et al.* [41] and Brownlie [5], see Appendix. Only the latter method is straightforward; for the other methods an iteration technique is necessary, which is applied as follows:

1. Estimate flow depth, d_j (start value $= 1.5\ d_{measured}$).

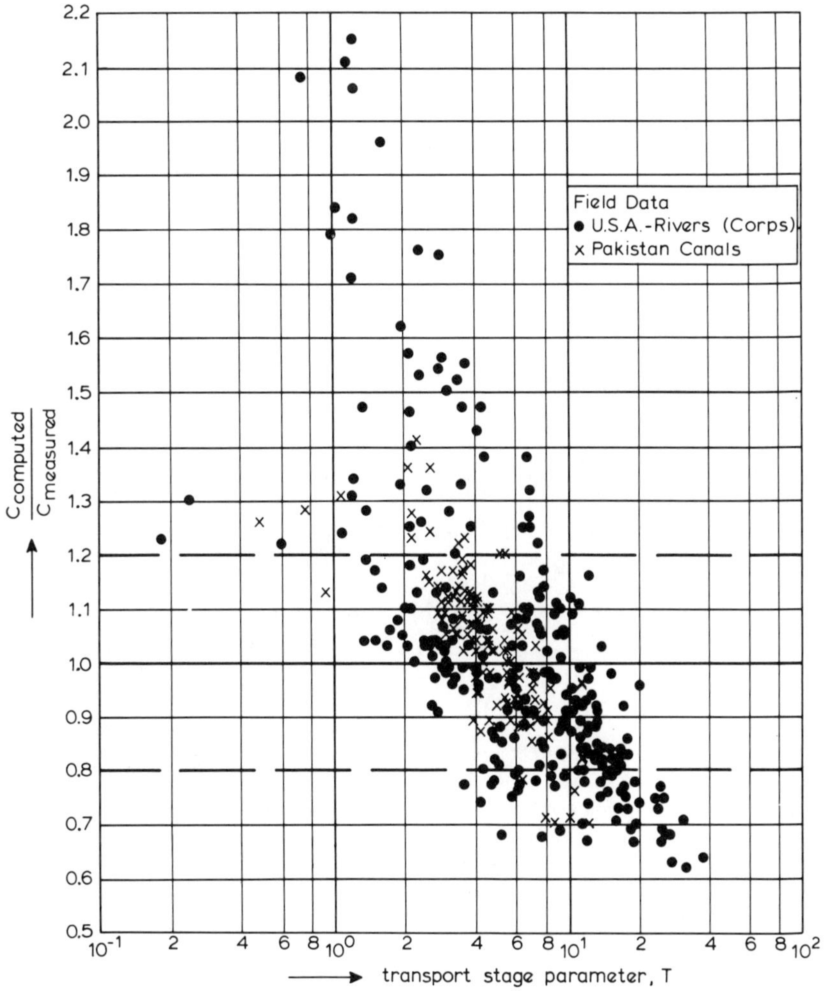

FIGURE 10. Ratio of computed and measured Chézy-coefficients according to method of White et al. (large rivers).

2. Compute flow velocity, $\bar{u}_j$ from the discharge.
3. Compute new flow velocity, $\bar{u}_{j+1}$ according to the prediction methods.
4. Compute new flow depth d_{j+1} from the discharge.
5. Compare new and old flow depths, if

$$|d_{j+1} - d_j| > \frac{d_{measured}}{100},$$

$$\text{then } d_{j+2} = \frac{d_{j+1} + d_j}{2} \text{ and repeat.}$$

As regards the selected data, only those of Guy *et al.* [14,24], Pakistan Canals [22] and the USA rivers [24] were considered, using the following criteria: Guy *et al.*: $d \geqslant 0.1$ m, Pakistan Canals: $d \geqslant 1.0$ m, and USA rivers: $d \geqslant 5.0$ m. None of these data were used by the writer to calibrate the proposed relationships.

Furthermore, only those data were selected which had a width-depth ratio larger than 10 to reduce the influence of the side-wall roughness. No correction method was applied. Where the water temperature (Te) and the geometric standard deviation (σ_s) of the bed material were not given, the values $Te = 15°C$ and $\sigma_s = 2$ were used.

The scores (in %) of the predicted flow depths in the error ranges ±10%, ±20%, and ±30%, are given in Table 2.

As regards the field data, the method of Brownlie is

TABLE 2. Comparison of Computed and Measured Flow Depths.

	Number of Data	Scores (%) of Predicted Flow Depths in Error-Ranges											
		±10% Error				±20% Error				±30% Error			
		Van Rijn	Engelund	White	Brownlie	*R*	*E*	*W*	*B*	*R*	*E*	*W*	*B*
USA-Rivers	240	53%	34%	50%	60%	82%	57%	83%	88%	92%	72%	96%	98%
Pakistan Canals	139	79%	49%	76%	82%	95%	82%	94%	95%	97%	87%	99%	99%
Guy *et al.*	147	57%	47%	29%	60%	79%	71%	47%	89%	88%	78%	60%	97%

slightly better than the writer's method and the method of White *et al.*, while the method of Engelund–Hansen gives the least satisfactory results. Compared with the flume data of Guy *et al.*, the method of White *et al.* is not performing very well.

Finally, it is pointed out that the present verification does not give an independent check of the method of Brownlie because all data used in the present investigation were also used by Brownlie to calibrate his method. Therefore, it merely is a check of Brownlie's calibration. Additional research with data not used by Brownlie is necessary to verify his method.

PREDICTION OF SEDIMENT TRANSPORT

In Parts 1 and 2, a method has been proposed to compute the bed and suspended load transport. As stated in the introduction of this Part 3 the proposed method was not really used in a predictive sense because in addition to the mean flow velocity, the flow depth and width, also the energy-gradient was supposed to be known to compute the overall bed-shear velocity. In this Section the total load transport will be predicted using only the mean flow velocity, depth and width as input data. This is of essential importance for the one-dimensional mathematical modelling of river systems with a movable bed as pointed out by de Vries [40]. Usually, only the flow discharge is known and the local flow depth and flow velocity are predicted by solving Equations (1), (2), (3), (4) and (5). For mathematical modelling the local energy slope is not an appropriate parameter.

The writer's method to predict the total bed-material load is given in Parts 1 and 2, while the Chézy-coefficient is predicted by Equations (14), (16), (20) and (21). For comparison with the writer's method, also the sediment transport theories of Engelund–Hansen [10] and Ackers–White [2] were applied to the selected data. Basically, the sediment transport theory of Engelund–Hansen is defined in terms of a given flow depth, energy gradient and Chézy-coefficient. Therefore, the former two parameters were supposed to be known, while the Chézy-coefficient was predicted as proposed by Engelund–Hansen [10]. The sediment transport theory of Ackers–White is defined in terms of a given flow depth, flow velocity and energy-gradient. In the present analysis the flow depth and the energy-gradient were supposed to be known while the mean flow velocity was predicted by means of the method of White *et al.* [41].

As regards the data, only the 266 field data of the USA rivers collected by the Corps of Engineers [24] were used because this data set represents a wide range of flow conditions for rather large rivers. This data set consist of data from the Rio Grande River, the Mississippi River, the Red River and the Atchafalaya River which can be considered as reliable field data.

The accuracy of the three methods is given in terms of a discrepancy ratio (r), defined as the ratio of the computed and measured total load. The scores (in %) of the predicted values in three error-ranges are given in Table 3. As can be observed the writer's method is superior to both other methods. The relatively poor results of the methods of Engelund–Hansen and Ackers–White are mainly caused by the relatively strong dependence of both methods (in their basic form) on the predicted Chézy-coefficient (or mean flow velocity). Finally, a simplified method is given which can be used to compute the total bed-material load when only the mean flow velocity, flow depth and the particle size are known. This simplified method is based on the computation of the sediment transport as proposed in Parts 1 and 2 in combination with the proposed hydraulic roughness predic-

TABLE 3. Comparison of Computed and Measured Total Load Transport.

	Scores (%) of Predicted Total Load in Discrepancy-Ranges		
	$0.75 \leq r \leq 1.5$	$0.5 \leq r \leq 2$	$0.33 \leq r \leq 3$
Van Rijn	52%	75%	88%
Engelund–Hansen	24%	44%	58%
Ackers–White	22%	44%	62%

tion method (Part 3). Using regression analysis, the computational results for various flow and sediment conditions ($d = 1$–20 m, $\bar{u} = 0.5$–2.5 m/s and $D_{50} = 100$–2000 μm, $\sigma_s = 2$, $Te = 15$°C) were represented by simple power functions.

$$\frac{q_b}{\bar{u}d} = 0.005 \left[\frac{\bar{u} - \bar{u}_{cr}}{((s-1)gD_{50})^{0.5}}\right]^{2.4} \left[\frac{D_{50}}{d}\right]^{1/2} \quad (22)$$

$$\frac{q_s}{\bar{u}d} = 0.012 \left[\frac{\bar{u} - \bar{u}_{cr}}{((s-1)gD_{50})^{0.5}}\right]^{2.4} \left[\frac{D_{50}}{d}\right] [D_*]^{-0.6} \quad (23)$$

in which: q_b = bed-load transport (volume) per unit width, q_s = suspended load transport (volume) per unit width, $\bar{u}_{cr}$ = critical mean flow velocity based on Shield's criterion.

For particles in the range 100 to 2000 μm, the critical mean flow velocity can be computed by:

$$\bar{u}_{cr} = 0.19(D_{50})^{0.1} \log\left(\frac{12R_b}{3D_{90}}\right) \quad (24)$$

for $100 \leq D_{50} \leq 500$ μm

$$\bar{u}_{cr} = 0.5(D_{50})^{0.6} \log\left(\frac{12R_b}{3D_{90}}\right) \quad (25)$$

for $500 \leq D_{50} \leq 2000$ μm

in which: D_{50}, D_{90} = particle diameters of bed material (in meters) and $\bar{u}_{cr}$ in m/s.

The accuracy of Equations (22) and (23) is somewhat less than the original method as given in Parts 1 and 2. For example, the score of the predicted total loads for the 266 data of the USA rivers in the discrepancy ratio 0.5 to 2 is 70% which is somewhat less than according to the original method which scored 79%. However, the score is considerably better than those of Engelund–Hansen and Ackers–White which is 44% for both methods.

NOTATION

The following symbols have been used in this paper:

- a = reference level (L)
- b = width (L)
- C = Chézy-coefficient ($L^{0.5}T^{-1}$)
- C' = Chézy-coefficient related to grains ($L^{0.5}T^{-1}$)
- D = particle diameter (L)
- D_* = particle parameter
- d = depth (L)
- Fr = Froude number
- g = acceleration of gravity (LT^{-2})
- k_s = equivalent roughness of Nikuradse (L)
- p = porosity factor
- Q = discharge (L^3T^{-1})
- Q_t = total load transport (L^3T^{-1})
- q = discharge per unit width (L^2T^{-1})
- q_b = bed-load transport per unit width (L^2T^{-1})
- q_s = suspended load transport per unit width (L^2T^{-1})
- R_b = hydraulic radius related to bed (L)
- S = energy gradient
- S_{cr} = critical energy gradient for initiation of motion
- s = specific density
- T = transport stage parameter
- Te = temperature (°C)
- $\bar{u}$ = mean flow velocity (LT^{-1})
- $\bar{u}_{cr}$ = critical mean flow velocity (LT^{-1})
- u_b = velocity of bed-load particles (LT^{-1})
- u_d = migration velocity of bed forms (LT^{-1})
- u_* = bed-shear velocity (LT^{-1})
- u'_* = bed-shear velocity related to grains (LT^{-1})
- $u_{*,cr}$ = critical bed-shear velocity for initiation of motion (LT^{-1})
- α = coefficient
- β = ratio of sediment and fluid diffusion coefficient
- δ_b = thickness of bed-load layer (L)
- Δ = bed-form height (L)
- λ = bed-form length (L)
- μ = dynamic viscosity coefficient ($ML^{-1}T^{-1}$)
- μ_b = bed-form factor
- ν = kinematic viscosity coefficient (L^2T^{-1})
- ρ = density of fluid (ML^{-3})
- ρ_s = density of sediment (ML^{-3})
- σ_s = geometric standard deviation of bed material
- τ_b = bed-shear stress ($ML^{-1}T^{-2}$)

REFERENCES

1. Ackers, P., "Experiments on Small Streams," *Journal of the Hydraulics Division, ASCE*, HY 4, 1964.
2. Ackers, P. and White, W., "Sediment Transport: New Approach and Analysis," *Journal of the Hydraulics Division, ASCE*, HY 11, 1973.
3. Allen, J. R. L., "Current Ripples," North Holland Publishing Company, Page 139, Amsterdam, The Netherlands, 1968.
4. Beckman, E. W. and Furness, L. W., "Flow Characteristics of Elkhorn River Near Waterloo, Nebraska," *Geol. Survey Water Supply Paper 1498-B*, Washington, USA, 1962.
5. Brownlie, W. R., "Flow Depth in Sand-Bed Channels," W. M. Keck Laboratory of Hydraulics and Water Resources, Techn. Memo 81-3, pages 52–54, California Institute of Pasadena, USA, 1981.
6. Culbertson, J. K. and Dawdy, D. R., "A Study of Fluvial Characteristics and Hydraulic Variables, Middle Rio Grande, New Mexico," *Geol. Survey Water Supply Paper 1498-F*, Washington, USA, 1964.
7. Culbertson, J. K., Scott, C. H., and Bennett, J. P., "Summary of Alluvial Channel Data From Rio Grande Conveyance Channel," *Geol. Survey Prof. Paper 562-J*, Washington, USA, 1972.
8. Delft Hydraulics Laboratory, "Verification of Flume Tests and Accuracy of Flow Parameters," (in Dutch), Note R 657-VI, The Netherlands, 1979.

9. Engel, P., "Discussion of Equivalent Roughness of Alluvial Bed," (in print), *Journal of the Hydraulics Division, Proc. ASCE*, 1983.
10. Engelund, F. and Hansen, E., "A Monograph on Sediment Transport," *Technisk Forlag*, Copenhagen, Denmark, 1967.
11. Fredsøe, J., "Shape and Dimensions of Stationary Dunes in Rivers," *Journal of the Hydraulics Division, Proc. ASCE*, No. HY8, 1982.
12. Garde, R. J. and Albertson, M. L., "Sand Waves and Regimes of Flow in Alluvial Channels," *Proc. IAHR-Congress, Paper 28*, Montreal, Canada, 1959.
13. Gladki, H., "Discussion of Determination of Sand Roughness for Fixed Beds," *Journal of Hydraulic Research*, Vol. 13, No. 2, 1975.
14. Guy, H. P., Simons, D. B., and Richardson, E. V., "Summary of Alluvial Channel Data from Flume Experiments, 1956–1961," *Geological Survey Professional Paper 461-I*, Washington, USA, 1966.
15. Hey, R. D., "Flow Resistance in Gravel-Bed Rivers," *Journal of the Hydraulics Division, Proc. ASCE*, No. HY4, 1979.
16. Kamphuis, J. W., "Determination of Sand Roughness for Fixed Beds," *Journal of Hydraulic Research*, Vol. 12, No. 2, 1974.
17. Kennedy, J. F., "The Mechanics of Dunes and Anti-Dunes in Erodible Bed Channels," *Journal of Fluid Mechanics*, Vol. 16(4), 1963.
18. Lane, E. W. and Eden, E. W., "Sand Waves in the Lower Mississippi River," *Journal Western Soc. Civ. Eng.*, Vol. 45, No. 6, 1940.
19. Leopold, L. B. and Emmett, W. W., "Bed-Load Measurements, East Fork River, Wyoming," *Proc. Nat. Acad. Sci.*, Vol. 73, No. 4, USA, 1976.
20. Liu, H. K., "Mechanics of Sediment Ripple Formation," *Journal of the Hydraulics Division, ASCE*, Vol. 83, HY2, 1957.
21. Mahmood, K., "Flow in Sand Bed Channels," *Water Management Technical Report No. 11*, Colorado State University, USA, 1971.
22. Mahmood, K. *et al.*, "Selected Equilibrium-State Data from Acop Canals," Civil, Mechanical and Environmental Eng. Dept., George Washington Univ., Report No. ERW-79-2, Washington, USA, 1979.
23. Nordin, C. F., "Aspects of Flow Resistance and Sediment Transport, Rio Grande Near Bernalillo, New Mexico," *Geological Survey Water Supply Paper 1498-H*, Washington, USA, 1964.
24. Peterson, A. W. and Howells, R. F., "A Compendium of Solids Transport Data for Mobile Boundary Channels," Report No. HY-1973-ST3, Dept. of Civ. Eng. University of Alberta, Canada, 1973.
25. Ranga Raju, K. G. and Soni, J. P., "Geometry of Ripples and Dunes in Alluvial Channels," *Journal of Hydraulic Research*, Vol. 14, No. 3, The Netherlands, 1976.
26. Rijn, L. C. van and Klaassen, G. J., "Experience with Straight Flumes for Movable Bed Experiments," Delft Hydraulics Laboratory, Publication 255, The Netherlands, 1981.
27. Rijn, L. C. van, "The Prediction of Bed Forms, Alluvial Roughness and Sediment Transport," Delft Hydraulics Laboratory, Report S 487-III, The Netherlands, 1982.
28. Rijn, L. C. van, "Equivalent Roughness of Alluvial Bed," *Journal of the Hydraulics Division, Proc. ASCE*, No. HY10, 1982.
29. Shen, H. W., Mellema, W. J., and Harrison, A. S., "Temperature and Missouri River Stages Near Omaha," *Journal of the Hydraulics Division, ASCE*, HY1, 1978.
30. Simons, D. B. and Richardson, E. V., "Resistance to Flow in Alluvial Channels," *Geological Survey Professional Paper 422-J*, Washington, USA, 1966.
31. Simons, D. B. and Sentürk, F., "Sediment Transport Technology," *Water Resources Publications*, Fort Collins, Colorado, USA, 1977.
32. Stein, R. A., "Laboratory Studies of Total Load and Apparent Bed Load," *Journal of Geophysical Research*, Vol. 70, No. 8, USA, 1965.
33. Stückrath, F., "Die Bewegung von Grossriffeln an der Sohle des Rio Parana," Mitteilungen Franzius Institut, Heft 32, Hannover, West Germany, 1969.
34. Tsubaki, T., Kawasumi, T., and Yasutomi, T., "On the Influences of Sand Ripples upon the Sediment Transport in Open Channels," Research Institute for Applied Mechanics, Kyushu Univ., Vol. II, No. 8, Japan, 1953.
35. Tsubaki, T. and Shinohara, K., "On the Characteristics of Sand Waves Formed upon the Bed of the Open Channels and Rivers," Research Institute for Applied Mechanics, Kyushu Univ., Vol. VII, No. 25, Pages 30–31, Japan 1959.
36. Urk, A. van, "Bed Forms in Relation to Hydraulic Roughness and Unsteady Flow in the Rhine Branches (The Netherlands)," *Euromech Conference*, Mechanics of Sediment Transport, Paper 33, Univ. of Istanbul, Turkey, 1982.
37. Urk, A. van and Klaassen, G. J., "Relationships for Bed Forms and Hydraulic Roughness," (in Dutch), Directorate Upper Rivers, Arnhem, Rijkswaterstaat, Report 61.000.04, The Netherlands, 1982.
38. Vanoni, V. A. and Brooks, N. H., "Laboratory Studies of the Roughness and Suspended Load of Alluvial Streams," Sedimentation Laboratory, California Inst. of Tech., Report E-68, USA, 1957.
39. Vanoni, V. A. and Hwang, L. S., "Relation Between Bed Forms and Friction in Streams," *Journal of the Hydraulics Division, Proc. ASCE*, No. HY3, 1967.
40. Vries, M. de, "A Sensitivity Analysis Applied to Morphological Computations," *Third APD-IAHR Congress*, Bandung, Indonesia, 1982.
41. White, W. *et al.*, "A New General Method for Predicting the Frictional Characteristics of Alluvial Streams," H. R. S. Wallingford, Report No. IT 187, England, 1979.
42. Williams, P. G., "Flume Width and Water Depth Effects in Sediment Transport Experiments," *Geological Survey Professional Paper 562-H*, Washington, USA, 1970.
43. Wijbenga, J. H. A. and Klaassen, G. J., "Changes in Bed Form Dimensions under Unsteady Flow Conditions in a Straight Flume," Delft Hydraulics Laboratory Publication No. 260, The Netherlands, 1981.

44. Yalin, M. S., *Mechanics of Sediment Transport*, Pergamon Press, 1972.
45. Yalin, M. S., "Geometrical Properties of Sand Waves," *Journal of the Hydraulics Division, ASCE*, No. HY5, 1964.
46. Znamenskaya, N. S., "Experimental Study of the Dune Movement of Sediment," *Soviet Hydrology Selected Papers*, No. 3, USSR, 1963.

APPENDIX PART 3

The *roughness predictor of Engelund–Hansen* (1967) is applied, as follows:

1. Compute hydraulic radius of the bed, R_b
2. Compute particle mobility parameters, θ and θ'

$$\theta = \frac{R_b S}{(s-1)D_{50}}$$
$$\theta' = 0.4\theta^2 + 0.06 \text{ for } \theta' \geq 0.55$$
$$\theta' = \theta \text{ for } 0.55 < \theta' < 1$$
$$\theta' = [0.3 + 0.7\theta^{-1.8}]^{-0.56} \text{ for } \theta' > 1$$

3. Compute hydraulic radius related to grains, R'_b

$$R'_b = \frac{(s-1)D_{50}\theta'}{S}$$

4. Compute depth-averaged flow velocity, $\bar{u}$

$$\bar{u} = [gR'_bS]^{0.5}\left[6 + 2.5\ln\left(\frac{R'_b}{2D_{65}}\right)\right]$$

5. Compute bed-shear velocity, u_*

$$u_* = [gR_bS]^{0.5}$$

6. Compute Chézy-coefficient, C

$$C = \frac{g^{0.5}\bar{u}}{u_*}$$

in which

S = energy gradient (−)
s = 2.65 = relative density (−)
R_b = hydraulic radius of the bed according to the method of Vanoni–Brooks (1957) (m)

The *roughness predictor of White et al. (1979)* is applied, as follows:

1. Compute particle parameter D_*, n, Y_{cr}, P

$$D_* = D_{35}\left[\frac{(s-1)g}{\nu^2}\right]^{1/3}$$

$$\left.\begin{aligned} n &= 1 - 0.56\log(D_*) \\ Y_{cr} &= \frac{0.23}{(D_*)^{0.5}} + 0.14 \end{aligned}\right\} \text{for } 1 \leq D_* < 60$$

$$\left.\begin{aligned} n &= 0 \\ Y_{cr} &= 0.17 \\ P &= [\log(D_*)]^{1.7} \end{aligned}\right\} \text{for } D_* \geq 60$$

2. Compute bed-shear velocity, u_*

$$u_* = (gR_bS)^{0.5}$$

3. Compute mobility parameter, Y_{fg}

$$Y_{fg} = \frac{u_*}{[(s-1)\,gD_{35}]^{0.5}}$$

4. Compute mobility parameter, Y_{gr}

$$\frac{Y_{gr} - Y_{cr}}{Y_{fg} - Y_{cr}} = 1 - 0.76[1 - e^{-P}]$$

5. Compute depth-averaged flow velocity, $\bar{u}$ from

$$Y_{gr} = \left[\frac{(u_*)^n}{((s-1)gD_{35})^{0.5}}\right]\left[\frac{\bar{u}}{5.66\log\left(\frac{10R_b}{D_{35}}\right)}\right]^{1-n}$$

6. Compute Chézy-coefficient, C

$$C = \frac{g^{0.5}\bar{u}}{u_*}$$

The *method of Brownlie* [5] is applied, as follows:

Lower flow regime

$$d = 0.3724D_{50}Q_*^{0.6539}S^{-0.2542}\sigma_s^{0.105}$$

Upper flow regime ($S \geq 0.006$ or $Fg \geq 1.74S^{-1/3}$)

$$d = 0.2836D_{50}Q_*^{0.6248}S^{-0.2877}\sigma_s^{0.08013}$$

in which

$$Q_* = \frac{Q}{bg^{0.5}D_{50}^{1.5}}$$

$$Fg = \frac{Q}{b((s-1)gD_{50})^{0.5}}$$

σ_s = geometric standard deviation of bed material (−)
s = 2.65 = relative density (−)
g = acceleration of gravity (m/s^2)
b = flow width (m)
Q = discharge (m^3/s)
d = water depth (m)
S = energy gradient (−)

Lateral Migration Rates of River Bends

EDWARD J. HICKIN*

INTRODUCTION

River channels rarely are straight. Some river reaches exhibit a meandering habit with regularly alternating sinusoidal bends while others deviate from their downvalley courses in an irregular and less periodic fashion. Most single-channeled river planforms, however, display both elements of irregularity and of meandering, as do the subchannels of braided and anastomosed rivers.

Although some general empirical relationships among discharge and various planform elements of meandering channels have been known for some time (see Leopold, et al. [87]), the underlying causes and controls of these regularities are yet to be identified.

Empirical studies of planform geometry apart, research into meandering has been conducted largely independently on three closely related topics: the cause of meandering, the nature of bend flow, and the nature of lateral migration.

The cause of meandering, a fundamental and longstanding question in fluvial studies, remains unanswered to this day. Earlier work, in which helical flow is featured as an important process, is reviewed briefly by Leopold, et al. [87] and more recent developments are discussed by Callander [18]. Currently in favour is the case for dynamic instability of the alluvial channel bed. Stability analyses have been used to argue (for example, see Callander [17], Engelund and Skovaard [34], and Parker [111]) that, because the mobile bed of a straight channel is unstable, any small perturbation will be amplified to produce the pattern of alternating pools and riffles that Friedkin [39] and many others subsequently have noted are the precursers to meandering. Although these rather complex mathematical models have certain basic elements in common (periodic functions for the initial perturbation in which the wavelength of maximum amplification is assumed to be the corresponding meander wavelength) they employ a variety of other assumptions to close the set of flow equations. In spite of the variety of specific solutions, all succeed in predicting wavelengths similar to those found in natural channels; it is not clear whether this is a strength or a weakness of this approach. Certainly the general nature of the approach is appealing; the periodic function does not specify the process of perturbation and therefore can easily accommodate meandering tendencies in a variety of media. Parker [111] argues that, although perturbations in sediment transport are necessary to form meanders in rivers, variations in Coriolis acceleration serve the same purpose in oceanic currents, heat differences in meltwater streams, and surface tensions in water threads on glass surfaces. This generality at the same time leaves stability analyses open to the criticism of being "black box" solutions to complex problems.

Other explanations of meandering proposed in the last couple of decades are less general than stability analyses but are just as lacking in specific process information. For example, Shen and Komura [121] and Quick [116] appeal to a periodic reversal in vorticity, while Yalin [144] argues that meander wavelength scales with the largest macroturbulent eddy that will fit in the channel.

A rather different view of the meandering process is offered by Davies and Tinker [26]. They argue on the basis of a surface-tension meander experiment that the fundamental characteristics of stream meanders (including a non-sinusoidal convex-downstream planform asymmetry; see also Carson and Lapoint [21]) are causally independent of secondary flow. They view the periodic reversal of meanders as the inevitable result of a stream gradually increasing its length by nonavulsive migration; the lateral movement of a particular length of channel then induces migration of the adjacent downstream reach in the opposite direction. This

*Department of Geography and The Institute for Quaternary Research, Simon Fraser University, British Columbia, Canada

notion is very similar to the view that any side bar can initiate meandering by deflecting flow to the opposite side of the channel, thus scouring the bank and bed and producing another bar downstream which in turn will deflect the flow back again across the channel, and so on (see References 44, 68, 69, 117, 138, 139). From this perspective the scale relations among meander and channel parameters do not result from some system-wide perturbation in the flow but rather from local bend effects which are linked by the necessity of continuity. It is a view which is supported by recent work on computer simulation of meander kinematics [61,62]. The debate continues.

But regardless of their form and origin, river bends are of considerable engineering significance. Once flow asymmetry has been initiated, erosion on one side of the channel and concomitant deposition of sediment at the other set in motion a self-enhancing erosional process characterised by strong positive feedback and consequently accelerating bend development through lateral migration of the channel. It is here that the rates of river-bank erosion generally are greatest and where engineering problems caused by lateral shifting of channels are most likely to occur.

The process of lateral migration in open channels, even in the simplest case of single bends, is exceedingly complex. Unfortunately, few theoretical models have been proposed and direct measurement of the process also is far from straightforward. Nevertheless, it is useful to examine lateral migration as an erosional process and to review techniques for its measurement. Furthermore, although no general model of bend migration is available, it is constructive to consider the growing body of empirical data as a basis for design guidelines.

THE PROCESS OF RIVER BEND MIGRATION

The factors controlling lateral migration rate (M) of most river bends probably can be expressed by the qualitative statement [51,103]:

$$M = f(\Omega, b, G, h, T_b) \tag{1}$$

where Ω = stream power (essentially the discharge-slope product), b = a parameter expressing planform geometry, G = sediment supply rate, h = the height of the outer bank (degree of incision), and T_b = the erosional resistance of the outer or concave bank material. Each of these terms in Equation (1) is itself a complex variable.

The driving force of the bend migration process is that proportion of the total stream power available for lateral erosion. The manner and degree to which stream power is partitioned for this purpose depends on the channel planform. Indeed, central to any discussion of bend migration is the nature of bend flow.

The Nature of Bend Flow

The general characteristics of the velocity distribution through a channel bend are well established and shown schematically for a bend of intermediate curvature in Figure 1. As the flow enters the bend (at B) the initially symmetrical isovel pattern (at A) becomes distorted as the near-surface high-velocity filament shifts towards the inner bank. By the time the flow reaches the bend apex at D, however, the high-velocity filament has begun to be displaced in the reverse direction towards the outer bank. Beyond the apex, at E, the high-velocity filament is close to the outer bank and depressed well below the water surface. At this point (E) the bend flow is said to be fully developed [118]. Further downstream (at F and G) the high-velocity filament moves back towards the centre and surface of the flow as it passes out of the bend.

This pattern of bend flow is the outcome of the interaction of two sets of forces: those demanding the conservation of angular momentum and those promoting the lateral transfer of momentum in the flow.

The effect of the first set of forces can be derived from elementary theory (for details, see Reference 42). If the flow is assumed to be irrotational and pressure simply is distributed hydrostatically on any vertical line, balancing the centripetal and pressure forces yields

$$dy/dn = V^2/gr \tag{2}$$

where y is the elevation of the water surface above a horizontal datum, n is a radial distance measured outwards towards the outer bank, V is mean velocity at the vertical section, r is the radius of the local streamline, and g is gravitational acceleration.

The same assumptions permit the energy equation to be cast in the form:

$$y + V^2/2g = H_o \text{ (a constant head)} \tag{3}$$

which differentiated with respect to n yields

$$dy/dn + V/g\ dV/dn = 0 \tag{4}$$

Combining Equations (2) and (4) we obtain

$$dV/dn + V/r = 0 \tag{5}$$

It follows that, since V/r and V^2/r in Equations (5) and (2) are always positive, velocity decreases as y increases (i.e., the water-surface exhibits superelevation from the inner to the outer banks).

If Equations (2) through (5) accurately described bend flow then the high-velocity filament would move close to the inner convex bank in the bend leaving the outer bank enveloped in relatively quieter flow. Indeed, if the streamlines

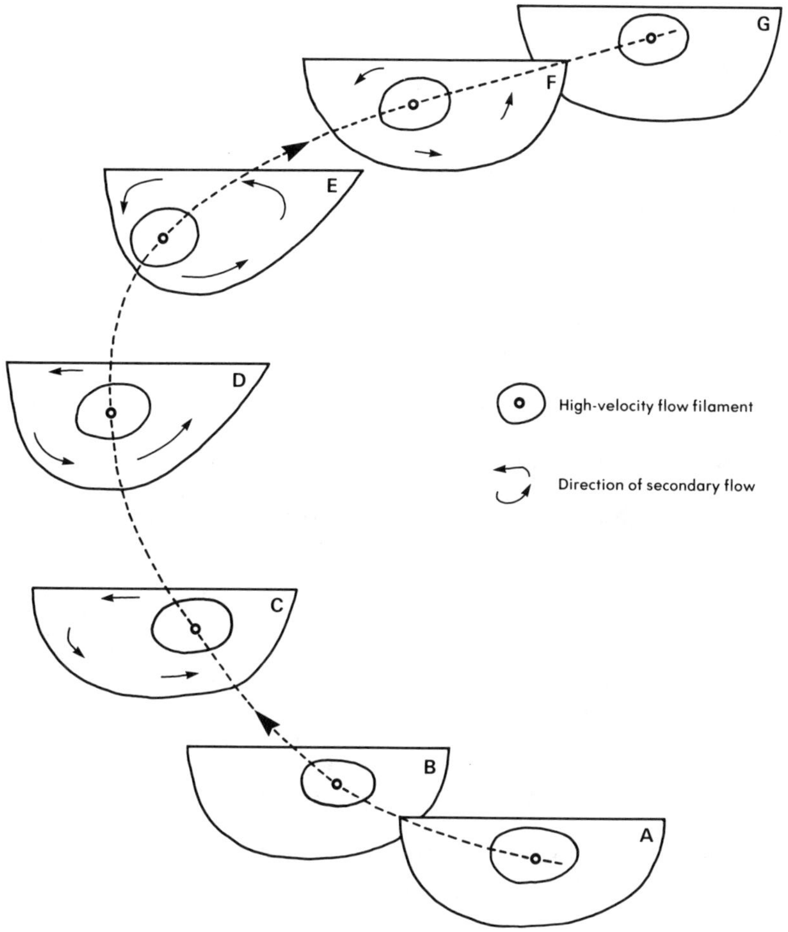

FIGURE 1. Typical streamwise and secondary flow-velocity pattern in an open-channel bend.

and the banklines are concentric to the same center, n and r become identical and the flow conforms exactly to a free vortex.

But as Figure 1 and the practical experience on which it is based dictate, this irrotational flow model must be relaxed to recognise the effects of the secondary flow which always occurs in river bends. These secondary currents essentially are caused by bed friction. Because the pressure distribution sensibly is hydrostatic, the transverse pressure gradient arising from the gradient dy/dn of the inclined water-surface is nearly the same for all points in any vertical section. But the balance of Equation (1) does not exist for all points in a vertical section because velocity and the related centripetal force are not depth invariant. Near the bed where the velocity is retarded, the inward pressure-gradient prevails and induces an inward flow along the bed. This flow is balanced by the outward flow of near-surface water in response to the higher velocity and stronger centripetal force prevailing there. An important consequence of this pattern of helical flow is that high-velocity near-surface flow is moved from the inner to outer banks, then towards the bed. At the same time slower near-boundary flow is moved inwards and upwards to the surface at the outer bank. Obviously these secondary flows can achieve considerable redistribution of momentum in the flow and account for the shift of the high-velocity filament of bend flow from the inner to the outer bank.

These theoretically deduced general properties of bend flow have been confirmed in the many experimental studies of flow in curved conduits that followed the initial observations in the classic works of Mockmore [95] and Rozovskii [118]; for example, see References 2, 32, 37, 38, 60, 65, 122, 145, 146, 147 and the review of Callander [18].

In natural channels, however, this simple picture of bend

flow may be complicated by additional effects related to boundary form, stage variation, bend-flow interaction, and the occurence of minor cells of secondary flow and flow separation.

In an alluvial channel the pattern of boundary shear imposed by the bend flow molds the channel into its typical asymmetrical form with the river talweg skewed to the outer bank. At discharges less than the formative flow, this channel form in turn influences the velocity field in the bend through advective acceleration [47] and relative-roughness controlled flow resistance [29]. The tendency for the high-velocity filament of the flow to shift towards the inner bank as stage increases has been noted by several researchers [7,14,15,29] but the precise cause is difficult to isolate. Indeed, one of the primary difficulties of obtaining useful data on the basic physics of bend flow from real rivers is that the channel boundary almost never is fully equilibrated with the flow being measured. For the same reason, experimental and theoretical studies of equilibrium bend-flow can generate only approximate analogues of the real channels with their complex memories of historical flow conditions.

Bend-flow interaction is a further complication in natural channels that is not well understood. Figure 1 shows the approach flow to the bend as being symmetrical but usually this will not be the case. It is common for the helical flow of a bend to extend into the flow domain of the adjacent downstream bend where it may overlap with the helical flow there. Thus, in the bend entrance the flow consists of two rotating cells of secondary flow. As they move into the bend the inherited cell dissipates as the new cell develops and eventually displaces the former structure. Bridge and Jarvis [14,15] argue that the initially skewed velocity field in the entrance to a bend largely is the result of the pattern inherited from the upstream bend. Certainly it will reinforce the radial acceleration effects and thus delay the establishment of fully developed flow induced by secondary currents [30]. Some of these bend interation effects have been experimentally examined by Khroda [71]. Secondary flow cells related to the primary gyre in bends are discussed by Bathurst, et al. [4].

Analytical studies of bend flow, in which realistic solutions have been sought for the continuity and momentum equations for turbulent flow, have led to successful prediction of its general character. Early research is reviewed by Rozovskii [118] and later contributions are discussed by Callander [18]. Recent advances have improved the theory to the point that it can provide realistic descriptions of the three-dimensional flow field (including the correct sense of helical flow and secondary flow strength of the right order of magnitude), of bed shear-stress distribution through the bend, and of the general bed configuration (for example, see References 27, 28, 29, 33, 67, 72, 88, 146). This theoretical work paves the way for an analytical formulation of bend migration, some notable attempts at which now are beginning to appear and will be discussed later in this chapter.

Bend flow clearly has been one of the more successful areas of theoretical analysis of river behaviour although the theory certainly is not without rather serious limitations. For example, it only applies to bends with a single helical cell and it does not apply to flows which include separation zones.

The latter deficiency represents a severe limitation of theory because flow separation appears to be very important to the process of lateral migration in certain types of channel bend [46]. In a zone of abrupt flow expansion satisfying continuity may not be possible because of the development of adverse pressure gradients and this zone may separate from the main flow field. In other words, flow separates from any boundary which is curved too abruptly for angular momentum to be conserved while simultaneously satisfying continuity. In open channels two likely candidate locations for flow separation are at the inner and outer banks near the apex of the bends. Here the channel commonly flares to its maximum width but the main flow is confined to a narrower region between the zones of separated flow (Figure 2). Although flow separation has been reported in experimental studies of bend flow (for example, see References 65, 118, 145), it likely is suppressed in many studies because of the common but unreal design assumption of constant channel width. In field studies, however, the phenomenon is widely reported (for example, see References 3, 19, 47, 48, 66, 85, 105, 110, 131, 143). Leeder and Bridges [85] examined intertidal meanders on the Solway Firth and observed that, for Froude numbers less than 0.4, there was no flow separation when the bend curvature ratio (radius of centerline bend-curvature/channel width; r/w) exceeded 4.0. It is a common observation that separation zones become more extensive and vigorous as Froude numbers and abruptness of channel transitions increase.

Flow separation is an important influence on channel migration rates because, if formed against the outer bank of a bend, it can provide a protective buffer between the fully developed bend flow and the bank [46]. Indeed, rapid deposition of fines and organic material in these outer bank separation zones can halt or even reverse the migration of a channel bend [46,49]. Separation zones formed at the inner banks similarly may locate zones of rapid deposition and point-bar formation.

Sediment Supply

As sediment is transported into a river bend the bed load is moved to the inner bank by the secondary flow described in the previous section. There it accumulates as a point-bar, gradually aggrading until it becomes an integral part of the floodplain. As a channel bend migrates by eroding the outer bank, this point-bar deposition at the inner bank must keep pace with the lateral erosion if the channel form and the bend flow it carries are to maintain their integrity. Obviously this balance is essential to dynamic equilibrium and

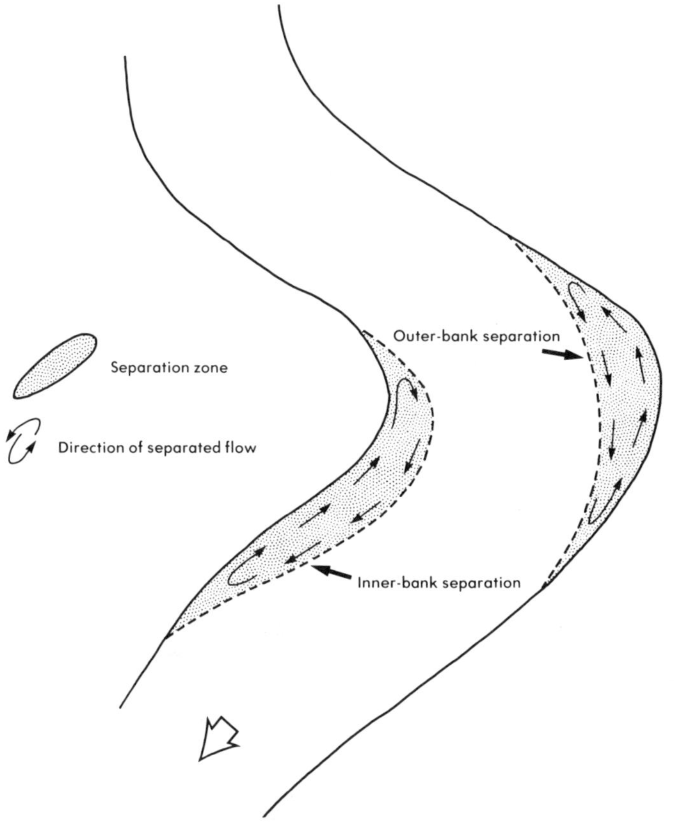

FIGURE 2. Typical locations of flow-separation zones in an open-channel bend of tight curvature.

continued lateral migration. The inclusion of G in Equation (1), however, does not necessarily follow from these relations.

It is a common observation of experimental researchers that the lateral migration rate of model channel bends can be increased by increasing the rate of sediment supply to the flume. On these grounds one might expect that sediment supply rate is an important control on the rate of lateral migration. But it will only be important if, as in the laboratory flume, it is both a limiting condition and a variable independent of the flow system. Of course one can imagine circumstances where these conditions undoubtedly are met; for example, a channel supplied with abundant landslide and debris flow material from steep slopes of weak unconsolidated sediments such as reaches of the Toutle River on Mt. St. Helens in Washington. Here the external sediment supply may dominate the rate of point-bar formation which in turn contracts the channel at bends and conditions and limits the rate of erosion at the outer banks. But on many rivers—perhaps most—sediment supply-rate and the corresponding rate of point-bar formation may be passive factors in the migration process. If the bulk of sediment deposited as point-bars is not externally derived but instead comes from the lateral erosion in channel bends immediately upstream, then one might argue that it is erosion which determines the sediment supply rate and not *vice versa.* Although one might debate this "chicken and egg" question, in such circumstances stream power and sediment supply clearly are not independent variables and one might be justified in omitting G from Equation (1). In this regard the recent work of Neill [106,107] is significant. He analysed data on channel change on the Tanana River in Alaska and found that the rates of bed-load transport measured by bed samplers closely corresponded with the mass transfers of bed-load sized material implied by erosion volumes in upstream bends.

Resistance to Bank Erosion

The remaining variables in Equation (1) represent the resisting forces in the channel migration process. The volume of material to be removed to achieve a given amount of bank

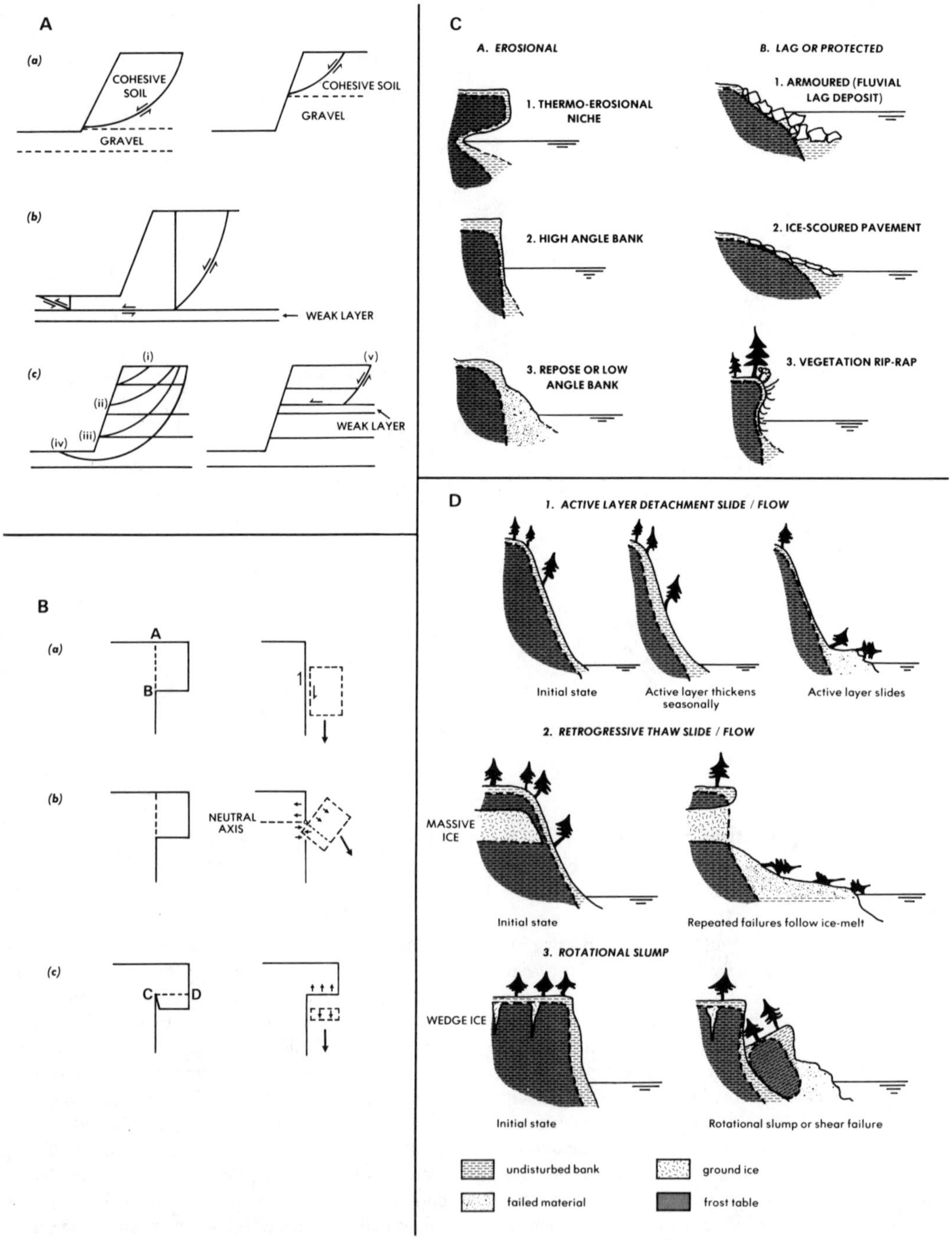

FIGURE 3. River-bank failure mechanisms. A. (a) Rotational slip-failure of the cohesive upper layer of a high composite bank. Location of gravel soil interface determines whether slip is a toe or slope failure. (b) Composite failure surface owing to a weak substratum. (c) Slip failures in a multilayered bank: (i)—(iv) Possible locations of failure surfaces within or between layers depending on soil properties and bank geometry; (v) Critical influence of a weak layer (after Thorne, [135]). B. Mechanisms of failure depending on cantilever geometry: (i) Shear failure along AB (ii) Beam failure about a neutral axis (iii) Tensile failure across CD (after Thorne and Lewin, [136]). C. Erosional and protected river-bank types in regions with permanently frozen ground. D. Riverbank failure mechanisms associated with ground ice (C and D after Church and Miles, [22]).

retreat is proportional to bank height, and the ease with which a unit volume can be eroded depends on its resistance (bank strength).

In the case of an equilibrium channel bend (neither aggrading nor degrading) migrating through its own alluvium, bank height likely is dependent on stream power and could be dropped from Equation (1). In most channels, however, bank height is equal to bankfull flow depth plus a bank-height increment due to incision; the degree of incision is an independent variable.

The resistance term, T_b, in Equation (1) perhaps is the most complex of the independent variables and certainly is not well understood in detail. The general nature of bank erosion has been examined and reviewed by several researchers (for example, see References 20, 135, 136, 137, 140) and the processes generally are regarded as being one of two related types: those caused by fluvial entrainment of particles and those related to slope failure.

Thorne [134,135] introduced from the work of Carson and Kirkby (20) the notion of *basal endpoint control* to characterise the balance of supply and removal of bank failure and basal slope materials. Three states are recognised:

1. Impeded removal: bank failures supply material to the base at a higher rate than it can be removed, resulting in basal accumulation and reduction in bank angle and height. The rate of supply decreases, tending towards the second state.
2. Unimpeded removal: because rates of supply and removal of basal material are balanced there is no change in basal elevation or slope angle and the bank recedes by parallel retreat in response to fluvial activity at the base.
3. Excess basal capacity: bank failure supplies less material to the basal slopes than is removed, resulting in basal lowering and increases in bank angle and height. The rate of supply increases, tending towards the second state.

Bank failure mechanisms vary in type depending on whether the bank sediment is uncohesive, cohesive, or a composite of the two [90,135,136,137]; see Figure 3. In uncohesive sediments failure occurs by the dislodgement of individual grains or by shallow slip along a planar or slightly curved surface [20,130,132,133]. In cohesive sediments failure occurs as rotational slips [8,9,96,97,130], shallow slips [43,94,124], or plane slips [130] not uncommonly associated with tension cracks and consequent slab failure [125,130, 132]. In composite banks combinations of these types occur together with shear, beam, and tensile failures of cantilever structures [90,136]; see Figure 3. Field studies of these types of bank failure are reviewed by Thorne [135]. A special but important case of the composite bank, those which include permafrost and ice lenses, is discussed by Church and Miles [22]; also, see Figure 3.

It is important to recognise, however, that although bank failure obviously is the primary control on short-term adjustments in bank form and position, it is the rate of entrainment of the basal sediments that limits the long-term bend migration rate. Basal sediments may be bed material or they may consist of failed bank material. The former case generally involves the erosion of coarse noncohesive sediments whereas the latter involves the removal of the more sedimentologically complex cohesive material.

Important sediment properties affecting the erodibility of noncohesive material include particle size, intergranular binding by the sediment matrix, and particle fabric (especially imbrication). The way these and other factors influence sediment entrainment is the subject of a vast literature, the most complete review of which can be found in the ASCE Task Committee Reports [127,129].

Important sediment properties affecting the erodibility of cohesive material typically include grain size, type and amount of clay, moisture content, organic content, bulk density, compaction, and a host of other factors which influence adhesion and the formation of particle aggregates or crumbs [40,128,129]. Specifying the erodibility of cohesive material is further complicated by the fact that, typically, these bank sediments are bound together to varying degrees by the roots of riparian vegetation, the strengthening effect of which can be overwhelming [49].

It is hardly surprising, given the complexity of Equation (1), and of T_b in particular, that it has not been possible to derive a general method for successfully predicting the rate of river bend migration. For this reason it is very important that there will be well designed and carefully executed measurement programs to determine empirically the normal range of bend migration rates in rivers. In recent years research has come some distance in achieving this goal.

MEASUREMENT OF LATERAL MIGRATION RATES OF RIVER BENDS

Measurement of channel bend migration rates can be achieved by several methods, some appropriate to short timescales of days to years and others to longer timescales of decades to centuries. They are considered below, ordered from the most direct method to those which provide surrogate measures of migration rates. Some of these techniques recently have been reviewed by Lewin [84,91].

Direct Field Measurement

There have been relatively few direct field surveys of channel bend migration and most of these have been concerned with just a few seasons of record. All have relied on measurement from erosion pins or from survey data along monumented transects (for example, see References 23, 55, 57, 63, 76, 86, 92, 141, 142).

Erosion pins are simple metal rods, usually a few mm in diameter and 10–30 cm in length, and generally are inserted horizontally into the bank. Progressive erosion of the bank is measured by subsequent exposure of the pin. Pins can also be inclined at an angle so that the subsequent burial of a basal sliding washer can yield data on sediment accumulation. These methods are reviewed by Lawler [84].

Clearly, many pins must be employed to yield meaningful patterns of bank erosion but at the same time bank disturbance must be kept to a minimum. Obviously this approach relies on the rate of bank erosion being measureable but not so great that it dislodges the erosion pins. Care must also be exercised when measuring pin exposure at sites where slope failure is evident; the method works best when bank erosion is caused only by particle entrainment.

Direct monitoring of bend migration by successive levelling surveys at monumented transects overcomes many of the problems related to erosion pin measurements. But such surveys demand a very high degree of accuracy with respect to the transect relocation and are not generally as reliable as erosion pin surveys for relatively short time periods involving small amounts of erosion.

Because both types of direct field measurement of channel shifting requires a considerable maintenance effort in the face of the uncertainty of obtaining useful results, they rarely are sustained for more than a few years. It follows that the method has been most useful for describing channel migration processes at relatively short timescales of a few days to just a few years at most. Although this type of direct measurement provides valuable information on the process of lateral migration, the sampling time is far too short to provide a reliable estimate of migration rate for timescales of the order of decades or longer.

Map and Aerial Photograph Records

The great majority of lateral migration measurements, however, have been obtained indirectly from serial cartography [19,98,119] and from aerial photography [25,52,89, 103] of channels. This type of historical record generally extends the period of observation to a few decades in the case of aerial photographs and up to a century or so in the case of cartography. Clearly, its usefulness will vary considerably from place to place. In areas with a long cultural history the cartographic record is likely to be useful (at least five surveys over the last 130 years generally are available in Britain [93]) but is minimally useful in much of the New World. The detailed record of channel pattern for parts of the Mississippi–Missouri system in the United States provides a notable exception to this latter generalisation.

Systematic photogrammetric surveys generally are restricted to the post-war period although specific projects may have been undertaken in the first half of this century in some areas. For example, a typical maximum time interval for the photographic recording of channel migration in most of North America is about 30 years. Except in special cases, most areas are not likely to be aerially photographed any more frequently than once every five to ten years; this practical limit of coverage defines the short-term limit of resolution of channel migration processes by this method.

There is some evidence to suggest that 30 years may not be long enough to establish the long-term migration rate of a channel bend from aerial photography. Nanson and Hickin [103] found that migration rates for a 21-year photographic record of two identical bends of the Chinchaga River in Northern Alberta were, respectively, 0 and 5 m/year whereas the 120-year averages were, respectively, 1.8m/year and 1.4m/year. Clearly, in this case river bend migration rates, at the timescale of the available aerial photography, are highly variable and not representative of long-term averages. These observations suggest that, until the nature of bend migration rate variability at a range of timescales has been clarified, rates determined from sequential aerial photography should be treated with caution. A judicious approach to this problem is to average rates from a number of bends in a single reach, a sampling procedure that will be taken up again later in this chapter.

Geomorphological Indicators

Intuitively one might expect that rapidly migrating channel bends are associated with steep vegetation-free outer banks and that laterally stable channels have lower bank slopes, are well vegetated, and show few or no signs of failures. Field observations on channel bends for which long-term migration rates are known [52,103] confirm that this intuitive view generally is correct. Unfortunately, there appear to be sufficient exceptions to warrant a further note of caution not unrelated to the concern over short-term versus long-term migration rates expressed in the previous section. Although all actively migrating bends have obviously eroding outer banks, the converse appears not always to be true. For example, in the case of the Chinchaga River bends cited earlier, the outer bank of the stable case showed all the usual signs of instability even though the channel had displayed no net migration in the 20 years of record. Similar observations can be made on some of the bends of the Beatton River in British Columbia. The reach of the Murrumbidgee River near the town of Wagga Wagga in Eastern Australia, examined in detail for its concave-bank benches [105,110], has apparently unstable outer banks but is known from photographic and survey records to have a very stable talweg.

These observations suggest that bank position may fluctuate in a minor way in the short term at the same time that the general channel alignment remains sensibly fixed in the long run.

A somewhat more reliable morphological and sedimentological indicator of lateral instability in mixed-load rivers is the condition of the inner bank: the point-bar. Channels which are laterally stable have relatively steep banks and tend to be narrower at the bend apex than their migrating counterparts. Their point-bars typically have grasses and

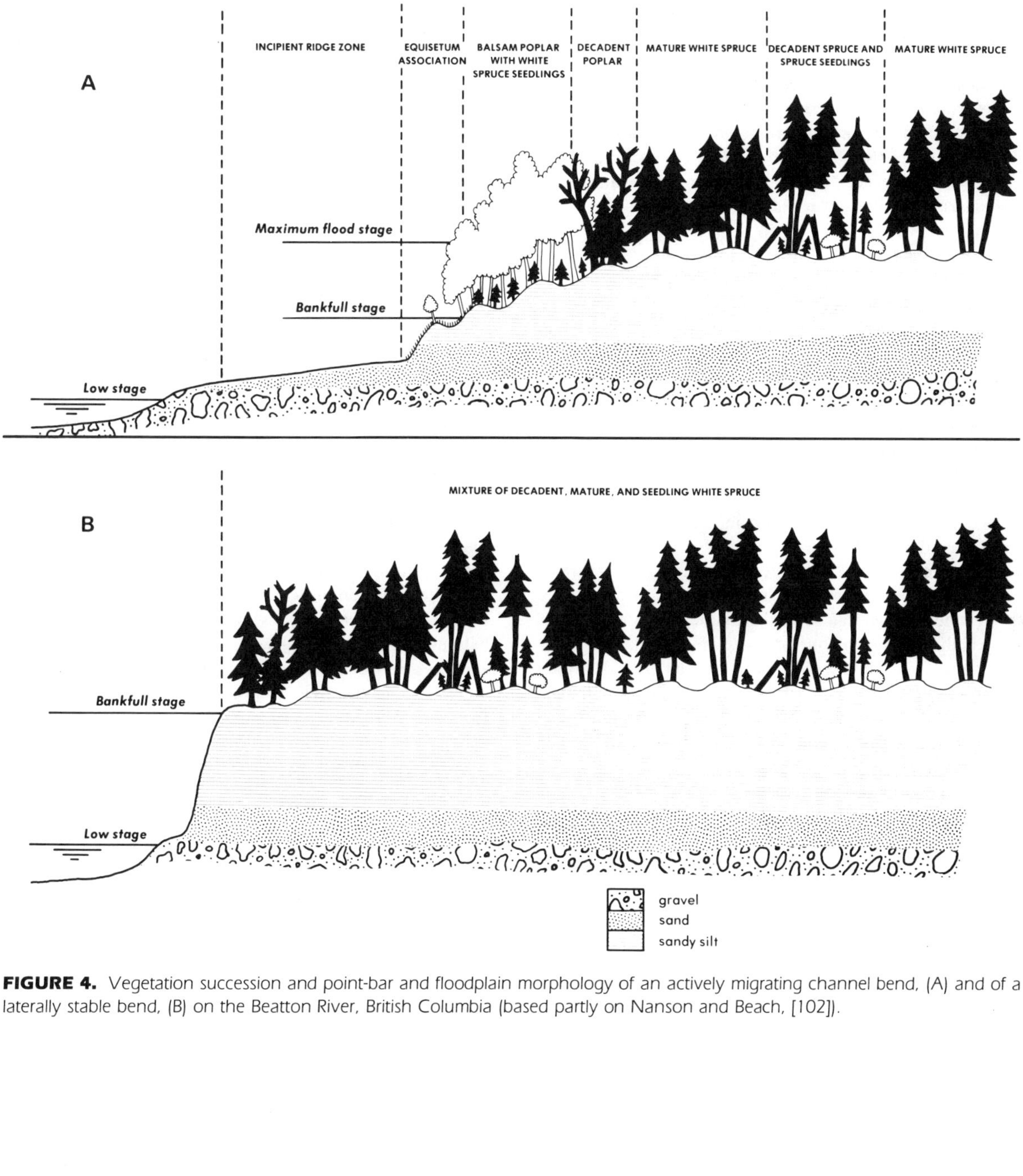

FIGURE 4. Vegetation succession and point-bar and floodplain morphology of an actively migrating channel bend, (A) and of a laterally stable bend, (B) on the Beatton River, British Columbia (based partly on Nanson and Beach, [102]).

other riparian vegetation growing down to the normal high-water level and zones of active sedimentation of the coarse fraction of the sediment load are few and small in extent (see Figure 4). Actively migrating channels, on the other hand, are relatively wide at the bend apex and have point-bars which typically consist of an extensive apron or platform of sand over gravel which rises gradually to the level of the floodplain (Figure 4). These extensive areas of point-bar sediment have little or no vegetation.

Nevertheless, morphological and sedimentological indicators provide only a qualitative guide to the lateral migration activity of a river bend at best. They are useful in identifying unstable bends but less reliable in recognising those characterised by long-term stability.

Riparian Vegetation

In the absence of accurate historical records, the only reliable indicator of long-term migration rates of river bends is the riparian vegetation succession on the point-bar surface (see Figure 4). As a bend migrates and deposits new point-bar sediments, plants colonise this surface very early in the point-bar formation process [102]. Tree species such as willow and cottonwood can survive very high rates of sedimentation [102] and many of them live to maturity. If one assumes that the time between sediment accumulation and plant colonisation is sensibly constant, and that the first colonisers are the oldest trees, then the age gradient in the flood plain trees is a reliable measure of channel migration rate. Obviously the trees have to belong to a species which lays down annual growth rings in order for their ages to be determined; the most common candidates for this task are the cottonwood and spruce. Together they can provide minimum ages of a point-bar surface to about 400 years before present (B.P).

This technique has been applied successfully in several studies of bend migration rates [31,35,51] and remains one of the few to give access to such information at a timescale of centuries.

Dendrochronology obviously is only useful for measuring migration rates if an intact stand of riparian vegetation displaying a successional gradient is present on the point-bar. In cases where migration rates are very slow the forest stand on the floodplain will be even aged and not useable for dating purposes. In other cases forest clearing and fire will eliminate the use of dendrochronology as a measurement tool.

In cases where this type of dendrochronological survey is not possible it still may be feasible to use vegetation response to sedimentation to measure migration rate if certain assumptions about point-bar form can be met. In Figure 5, channel migration is viewed as the lateral displacement of an equilibrium cross-sectional channel form. In this case the lateral displacement, D, and the vertical accumulation of point-bar sediment, Y, during the same period of time, t, is related to the point-bar slope on the line of section ($S = \tan \beta$) as follows:

$$dD/dt = M = S.dY/dt \tag{6}$$

It is possible to estimate sedimentation rate, dY/dt, on some point-bar surfaces by studying wood anatomy changes in trees sensitive to partial burial [99]. Various authors have noted that wood formed in the partially buried trunk of a living tree differs morphologically from wood formed above the level of burial or from that formed prior to burial. The precise nature of this wood anatomy response, however, appears to vary among tree species [77,115,123]. Nanson [99,102] found that sections removed from partially buried balsam poplar trunks growing beside the Beatton River showed abrupt changes in wood anatomy within the space of one or two annual growth rings. This change took the form of a marked increase in vessel size and generally more even and open texture compared to wood formed outside the tree ring marking the point of change (see Figure 6). Using these discontinuities in wood texture as indicators of burial Nanson constructed a vertical sedimentation history which was

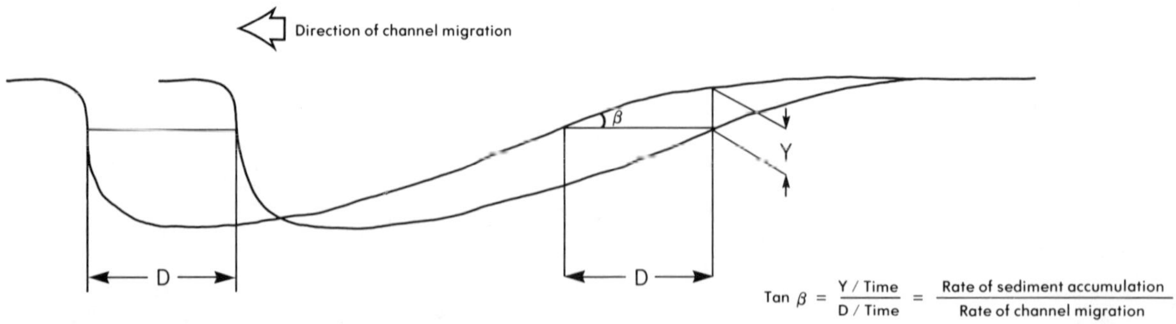

FIGURE 5. The relation between vertical accretion and lateral migration on a river bend of stable cross-sectional form.

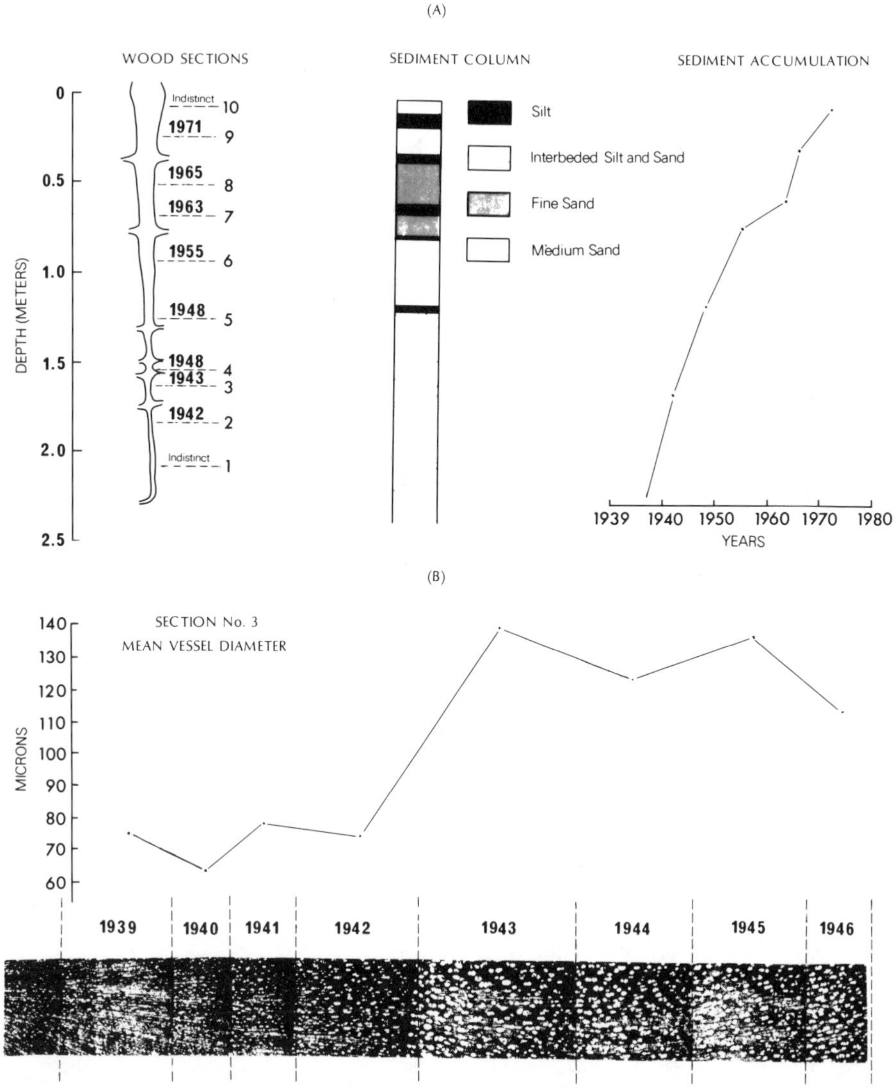

FIGURE 6. An example of the sequential burial of a living balsam poplar trunk on the Beatton River floodplain. (A) Location of sample points on the buried trunk; the associated sediment column and sediment accumulation curve. (B) Change in vessel diameters associated with burial. Burial of this portion of the trunk (Section 3) occurred either in late 1942 or early 1943 (after Nanson and Beach, [102]).

consistent with the independently measured migration rate [99,100,102].

This technique is most useful when applied to an early coloniser which has survived the initial high rates of vertical accretion. It holds much promise for dating long-term lateral movement of channels with only sparse point-bar vegetation; its principal limitation is that it is not well tested.

AN INVENTORY OF RIVER BEND MIGRATION RATES

Although it is likely that considerable bank-erosion data are scattered throughout the unpublished files of government agencies and local engineering and water authorities in most places, the tasks of locating, assessing, and collating such data are simply too costly to pursue and one is obliged to rely on the published data base. Unfortunately, these published data are very few and of limited use for scientific and engineering purposes. Not all relate to river bends and even those appropriate in this respect often are not comparable because there is no measurement standard. Many data sets exclude essential site information and most represent too brief a record to be reliable indicators of stable mean migration rates.

Nevertheless, the published data do provide a sense of the range and variability of river bend migration rates. In Table 1 is a list of such data from readily available sources. It is based on an inventory assembled by Hooke [57], and on additional data obtained by Hooke [54,56], Hickin and Nanson [52], and Nanson and Hickin [104]. As expected these uncontrolled data display great variability although absolute migration rate generally increases as the size of a river increases. Indeed, Hooke [57] plotted rate of bank erosion, M, versus catchment area, A, based on a subset of the data in Table 1 to establish that $M = f(\sqrt{A})$ (Figure 7). The scatter in Figure 7, however, is so large that the relationship disappears among the noise for anything less than about four orders of magnitude in catchment area range. Thus, it has no use as a predictive tool for engineering purposes.

Hooke [56] attempted a multiple regression analysis of ten variables believed to influence the bank erosion rates listed for her in Table 1 but most were not sufficiently independent of each other to obtain meaningful results. From further study [57] she found, as did Daniel [25], that, in addition to catchment area (discharge), erosion rate was related statistically to the percentage of silt and clay in the banks. Her data are 2.5-year erosion-pin measurements, however, and are more variable and average less than longer-term rates estab-

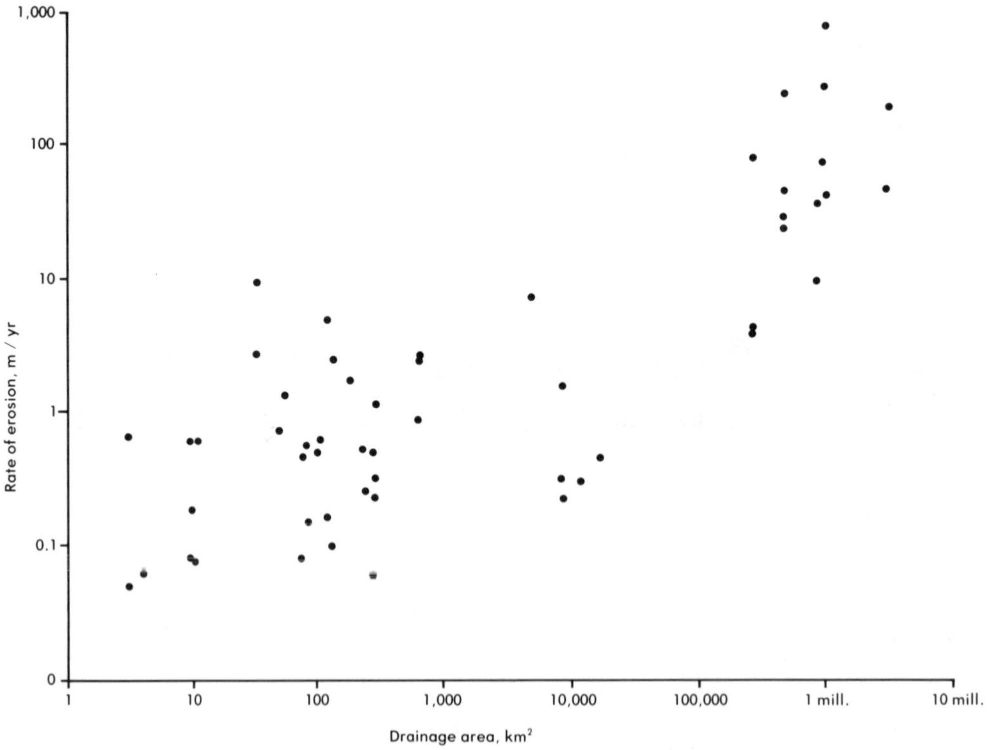

FIGURE 7. Relationship between river-bank erosion rates and catchment area (after Hooke, [57]).

TABLE 1. Channel Migration-Rate Measurements and Related River Data Published in Readily Accessible Sources.

River and Location	Drainage area, km^2	Width, m	Mean discharge, m 3/sec	Rate of movement, m/yr	Period of measurement	Method	Reference Source
Ohio River, Kentucky				0.357	1807-1958	Maps	(1)
R. Endrick, Scotland	97.66	25	6.94	0.5	1896-1957	Maps, mean	(10)
White R., Indiana	6042		66.2	0.67	1937-1968	Maps, mean	(11)
R. Mississippi				23	1722-1971	Maps and historical data, maximum	(16)
R. Mississippi				23	1881-1963	Maps	(19)
R. Brahmaputra	934990	6000-13000	1898	6-275	1952-1963	Maps	(23)
				15-792	1944-1952	Maps	
R. Pembina, Alta, Cda		64	19.2	3.35	1910-1956	Maps	(24)
				0.3		Bedrock channel	
Little Missouri R., Dakota		91.5	16	1.7-7.0	100 + years	Cottonwood trees	(35)
Des Moines R., Iowa				6.6	1880-1970	Maps	(41)
R. Beatton, B.C. Cda	16000	48	225	0.48	250 years	Trees and deposits	(51)
Little Smoky R., Alta, Cda		39	44	0.53	1950-1975		
Milk R., Alta, Cda		47	78	1.68	1945-1978		
Belly R., Alta, Cda		42	101	1.18	1951-1978		
Lesser Slave R., Alta, Cda		42	101	0.86	1950-1970		
Beaver R., Alta, Cda		49	244	1.4	1952-1977		
Waterton R., Alta, Cda		72	247	3.93	1951-1978		
Eagle R. (Upper), B.C., Cda		49	272	1.34	1952-1977		
Eagle R. (Lower), B.C., Cda		48	272	0.71	1951-1977		
Swan R., Alta, Canada		41	272	1.52	1952-1978		
Shuswap R. (Upper), B.C., Cda		63	306	1.73	1951-1975		
Fontas R., B.C., Cda		63	365	2.20	1950-1979	Air photos, average for reach, maximum	(52)
Pembina R., Alta, Cda		79	369	2.60	1949-1974		
Muskwa R., B.C., Cda		49	377	2.65	1948-1971		
Oldman R., Alta, Cda		95	383	7.26	1950-1979		
Notikiwan R., Alta, Cda		54	388	1.21	1950-1970		
Shuswap R. (Lower), B.C. Cda		92	454	1.89	1951-1976		
Clearwater R., Alta, Cda		135	198	1.51	1951-1972		
Chinchaga R., Alta, Cda		91	766	1.03	1950-1978		
Prophet R., B.C., Cda		141	830	2.34	1952-1979		
Sikanni Chief R., B.C., Cda		127	1259	2.92	1950-1979		
Fort Nelson R., B.C., Cda		281	3972	4.44	1952-1979		
West Prairie R., Alta, Cda		30	102	0.86	200 years	Tree dating	
Crawfordsburn R., N. Ireland	3	2-3		0-0.5	1966-1968	Erosion pins	(53)
Clady R., N. Ireland	4	2-2.5		0-0.064	1966-1968	Erosion pins	(53)
R. Exe, Devon, U.K.	620	26-33	140	0.88			
R. Creedy, Devon, U.K.	235	16	55	0.07			
R. Culm, Devon, U.K.	270	13-17	7-18	0.31			
R. Axe, Devon, U.K.	288	16-35	80	0.25	1842-1975	Maps, average	(57)
R. Yarty, Devon, U.K.	51	8.5	2.5	0.24			
R. Colby, Devon, U.K.	74			0.08			
R. Hookmoor, Devon, U.K.	9.6	4.2	6	0.08			
R. Cound, Shropshire	100	17	4	0.64	1972-1974	Pegs, mean	(63)
R. Mississippi				14.9-40.5	1963-1970	Field measurement	(70)
Wisloka R., Poland			22.5	8-11	1970-1972	Field measurement	(74)
Dunajec R., Poland		30-120	40	0.4-1.0		Maps	(75)
R. Bollin-Dean, Cheshire	120	3-12		0.01-0.09	1967-1969	Erosion pins	(76)
R. Mississippi		1200	16800	0.61-305		Maps and deposits	(78)
				100		Maps, maximum	(80)
Russian Rivers				10-15	1897-1958	Maps	(79)
				2.25-3.1		Maps	
R. Ob, USSR			1434	0.15	1897-1958	Maps	(81)
R. Hernad, Czechoslovakia	5400	50-60	10-30	5-10	1937-1972	Maps, mean	(82)
R. Rheidol, Wales	179			1.75	1951-1971	Maps, maximum	(89)
R. Tyfi, Wales	633			2.65	1905-1971	Maps	
R. Bollin-Dean, Cheshire	114	13	39	0.16	1872-1935	Maps	(98)
					1935-1973		
Jaszcze R., Poland	11-74		0.2	Equation	1970-1971	Field measurement	(108)
Jamme R., Poland	9-14						
Chemung R., Pennsylvania				3.05	1938-1955	Air photos, one bend, aver.	(109)
Lower Missouri R.			1120-1400	Over 1/3 floodplain reworked	1879-1954	Maps	(119)
				1.6	1850-1950	Resurvey	
R. Klaralven, Sweden	5420-11820	120	650	0.23	1800-1850	Maps, maximum	(125)
				0.32	1850-1950	ratio	
R. Torrens, S. Australia	78	5-10		0.58	1960-1963	Erosion pins	(141)

lished from serial cartography [57]. The short-term noise leading to this discrepancy obscures the physical relationships of the migration process.

The most internally consistent and refined set of data published to date is that obtained by Hickin and Nanson for gravel-bed rivers in Western Canada [51,52,103]. These were collected in order to specify the nature of Equation (1) for the relatively simple case of single expansion bends. This work is reviewed below.

LATERAL MIGRATION RATES ALONG THE AXIS OF SINGLE BENDS

Scroll-bars on river floodplains form parallel to the inner bank of a migrating channel bend and remain as a geomorphological record of the history of channel alignment (Figure 8). Orthogonals through the scroll-bar field can be visualised as erosion pathlines describing the locus of migration of the various segments of the expanding bend. From an aerial photograph reconnaissance of the scroll-bar pattern of floodplains on northern British Columbian rivers, Hickin [45] observed that relative rates of bend migration (scroll-bar spacing) increased along the erosional axis (the longest erosion pathline) as bend curvature ratio (radius of bend curvature/channel width, r/w) declined and the bend tightened. But as bend curvature attained $r/w < 2.0$, scroll-bar spacing abruptly decreased, implying a rapid reduction in migration rate (see Figure 8).

Hickin and Nanson [51] subsequently tested these interpretations in a field study of channel migration on the Beatton River in Northeastern British Columbia. They selected ten simple expansion bends within a 50 km reach of the river and measured the migration rate along each bend axis from isochrone maps based on a dendrochronological survey of the floodplain vegetation. In other words, the field experiment was designed to isolate the relation $M = f(b) = f(r/w)$ by holding the remaining variables in Equation (1) constant. Additional fieldwork subsequently expanded the data set to 16 bends (Nanson and Hickin, [103]) and the results are shown in Figure 9. These data confirm that, like the driving centripetal force, migration rate is inversely proportional to the radius of curvature in relatively open bends, peaks in the domain $2.0 < r/w < 3.0$, and then rapidly declines in bends for which $r/w < 2.0$.

The reason for the rapid decline in migration rate in tight bends is not known but may well be a complex interaction of several factors. At least four general causes may be involved in addition to local effects such as bedrock control, impingement on high bluffs, etc.: (a) Bagnold's separation-zone collapse mechanism; (b) flow separation at the outer bank; (c) pool deepening at the outer bank; and (d) downstream movement of the point of fully-developed flow.

When the aerial reconnaissance and initial field results from the Beatton River were first reported [45,51] it was suggested that Bagnold's notion [3] of rapid increases in flow resistance related to the collapse of an inner-bank separation zone was the underlying cause of the rapid decline in migration rate at the bend axis. He argued that an inner-bank separation zone forms in developing bends, narrowing the effective width (see Figure 2), increasing the depth of the main-flow zone, and lowering flow resistance through the bend; this separation zone remains stable until the curvature ratio approaches $r/w = 2.0$. As the bend tightens beyond this curvature the separation zone breaks down and the consequent increase in macro-turbulence, flow expansion, and the increase in relative roughness (the mainflow zone now includes the shallow convex side of the channel), causes the flow resistance in the bend to rapidly increase. But as Hickin reported subsequently [46], the main problem with this otherwise appealing idea is that inner-bank flow separation was neither observed to form nor collapse at the inner banks of Beatton River bends with arrested migration rates. On the contrary, flow separation far more commonly forms at the outer bank of tight river bends on the Beatton River (Figure 2).

Flow separation at the outer bank of river bends has been reported from direct measurement [47] and implied from the presence of concave-bank benches [19,48,105,110,131, 143]. There is little doubt that, by isolating the outer bank from the full force of the main flow, separation arrests the migration process in many tight river bends. Indeed, Hickin [46] has noted that, if tightening of the outer-bank curvature does generally induce flow separation and thus protects the bank from further erosion, it suggests a novel approach to bank protection. Current engineering practise deals with laterally unstable channel bends by building the concave bank out into the channel and revetting the artificial surface. This method is a sometimes useful short-term measure but often fails over several years. It might be argued from observations on natural bends that a better solution would be provided by excavating the concave bank further and letting the resulting separation zone do the work of the revetment. It is also the case, however, that not all strongly curved and slowly migrating natural channels display outer bank flow-separation.

The third mechanism, pool deepening at the outer bank, most recently has been proposed by Begin [6]. Essentially this cause recognises that as bend curvature tightens, the channel becomes increasingly asymmetrical to the point where the radial forces in the flow are distributed over such a large (overdeepened) outer-bank area that radial force per unit area and bank erosion-rate rapidly decline. Although the interaction of radial forces and outer-bank enlargement may contribute to declining migration rates in some tight bends, there are several reasons to suspect that it is not a general cause of the penomenon. First, Begin's [6] analysis is based on an expression involving flow depth and radius of curvature (from Rozovscii, [118]) that demands for its conclusion (a relation between force/area at the outer bank and

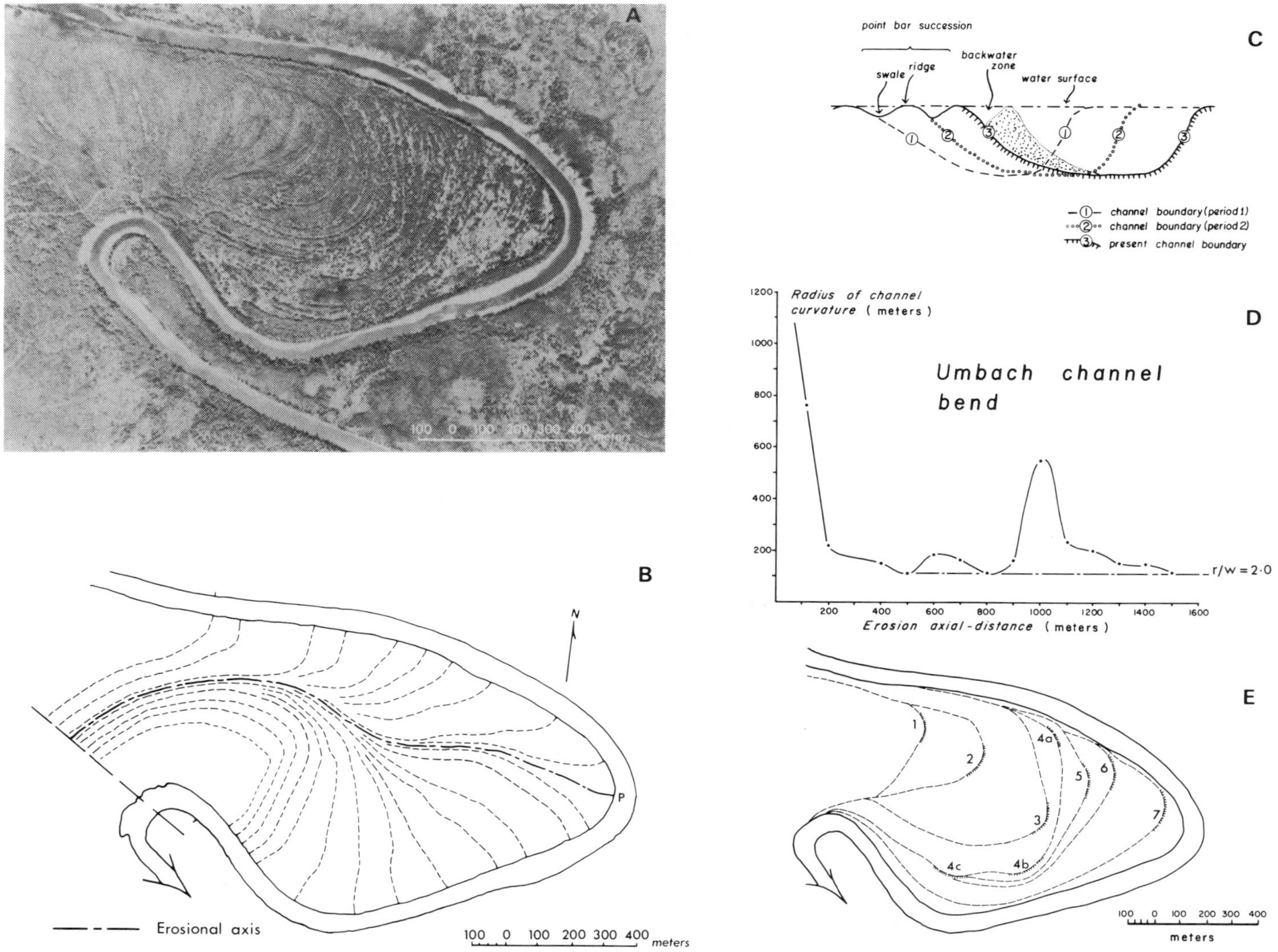

FIGURE 8. (A) Umbach channel bend, Beatton River, British Columbia, with its regular array of scroll-bars marking former channel positions. (B) Erosion pathlines (orthogonals) through the scroll-bar field of Umbach channel bend. (C) Point-bar Formation by the lateral accretion of scroll-bars. (D) The relation of channel curvature to distance along the erosional axis of Umbach channel bend. (E) Scroll-bar pattern and critical-curvature segments (r/w < 2.0) on the floodplain of Umbach channel bend (after Hickin, [45]).

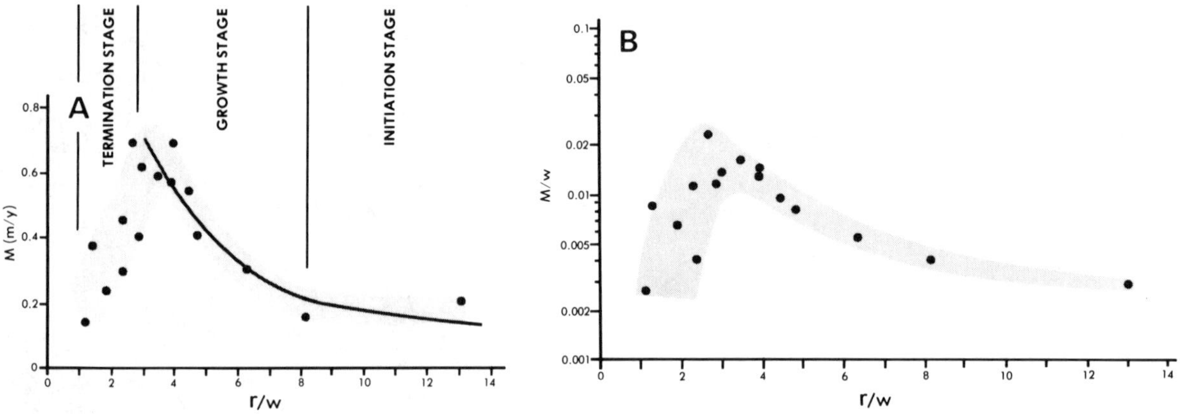

FIGURE 9. Absolute (A) and relative (B) channel migration rate related to curvature ratio of bends on the Beatton River, British Columbia (after Hickin and Nanson, [52]).

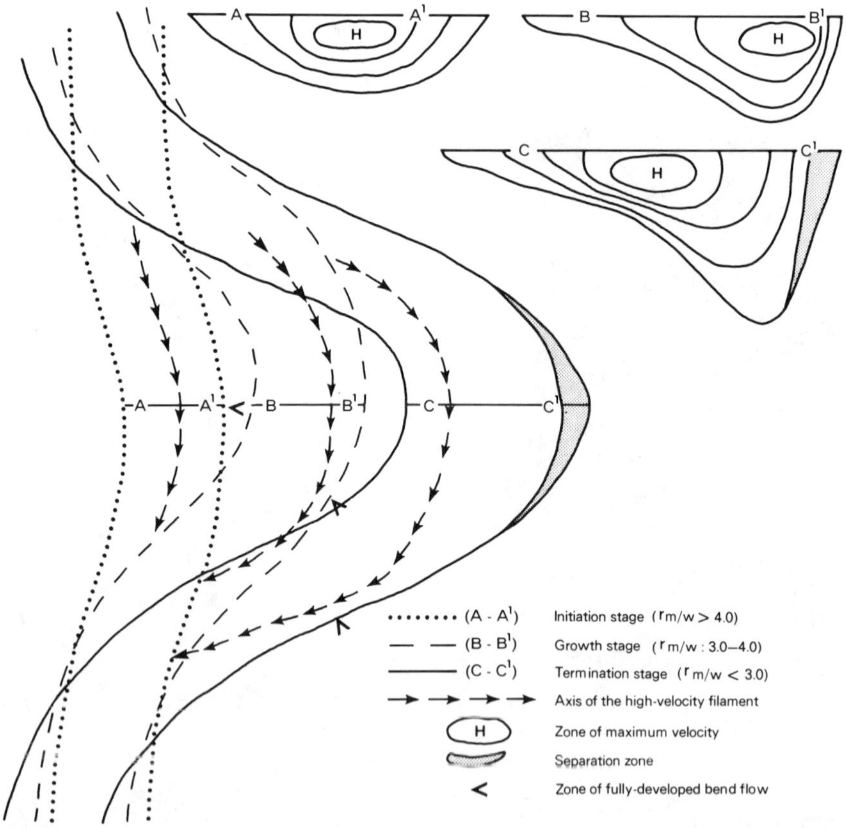

FIGURE 10. Change in the pattern of mean-flow structure as a river bend tightens during channel migration.

bend curvature ratio similar in form to $M = f\ (r/w)$ in Figure 9) channel asymmetries far in excess of those found in most natural channel bends (for example, the Beatton River with which he compares his model).

Second, his analysis considers only the case of radial forces being fully supported by a vertical outer bank. On the one hand it ignores the common case in which lowering of the bed in the concave half of the channel bend is necessary for lateral migration to occur. On the other hand, if bank erosion is the limiting condition for channel migration, it ignores the observations cited earlier that bank failure (and channel migration) is most likely at overdeepened sections of the channel.

The fourth mechanism, downstream movement of the point of fully-developed flow, probably represents the most universal of the four explanations of arrested migration rates in tight bends. As the amplitude of a bend increases, the length of curved channel (arc length) tends to decline as shown in Figure 10. As a consequence, the point of the onset of helical flow moves progressively closer to the bend apex as the bend migrates laterally and tightens. Similarly, the point of fully developed flow and of most vigorous bank erosion, now delayed, moves further downstream from the bend apex, which becomes a relatively slack-water (and possibly separated-flow) zone. This shift in erosional activity downstream of the bend apex during bend growth has long been recognised and is documented in terms of a characteristic erosional pathline form by Hickin [45]. It also is associated with an inward shift of the high-velocity filament at the bend apex [47] which may lead to erosion of the convex bank.

But whatever the cause, the form of $M = f\ (r/w)$ in Figure 9 is taken by Hickin and Nanson [52] to apply generally to simple expansion bends. From observations on migrating rivers in Western Canada they further reasoned that the sediment supply-rate, G, could be dropped from Equation (1) because it is determined by local bank erosion, that planform geometry, b, is best expressed as radius of curvature, r, and that lateral migration rate of river bends essentially is limited by the rate of entrainment and transport of bed and basal bank-material. Hence, they proposed that

$$M = f\,(\omega, T_b, h, r, w) \qquad (7)$$

where ω = stream power per unit bed area, w = channel width, and the other symbols are as previously defined.

Dimensional analysis of Equation (7) yields

$$M = K\ \omega/T_b\ (h/w;\ r/w) \qquad (8)$$

Hickin and Nanson [52] collected migration data for 189 river bends on 23 river reaches in British Columbia and Alberta (see the summary in Table 1). The data were obtained from bends of varying curvature but conform broadly to the relation $M = f\ (r/w)$ in Figure 9. Thus they were standardised to a refence curvature ratio of $r/w = 2.5$ by the transformations in Equations (9) and (10):

$$M_{2.5} = (r/w \cdot M)/2.5 \qquad (r/w > 2.5) \qquad (9)$$

$$M_{2.5} = 1.5\ M/(r/w - 1) \qquad (r/w < 2.5) \qquad (10)$$

Thus, the specific purpose of their study was to seek a correlation $M(\Omega, T_b, h, w)$ for which a dimensionally balanced expression is

$$Mh/w = K\omega/T_b \qquad (11)$$

Here the rate of channel migration, expressed as a sediment volume per channel width and length, is a function of the ratio of driving to resisting forces in the bend.

Migration rate data in part were obtained from a 20–30 year sequential aerial photography record. These were averaged by river reach in order to minimise the problem of sampling and intermittency noted earlier. The results of the Hickin and Nanson [52] study are summarised in Figure 11.

The relation $M_{2.5}h$ versus Ω in Figure 11a discriminates on bank material type. There is an orderly separation of channels with outer banks of gravel from those of sand. This separation suggests that values plotting above the line represent proportionately greater bank resistance than points near or below the line. The 45° separatix implies that $M_{2.5}/\Omega$ = a constant for any given bank-strength condition. This circumstance provides a means for evaluating T_b as a coefficient of proportionality in Equation (11):

$$T_b = \Omega/M_{2.5}h \qquad (12)$$

Although T_b has the dimensions of force/area (Nm^{-2}), it is a coefficient of resistance to lateral migration, presumably dependent largely on bank strength, but also absorbing all the other factors including the statistical variability in stream power and migration rates. In this regard it is analogous to Mannings n, a coefficient of total flow resistance but not directly measurable. Also included in T_b is the constant $K = 3.156 \times 10^7$, the ratio of time units, yr/sec, but for our purposes it is convenient to ignore this constant and retain the use of units in years and seconds where appropriate.

Because T_b is a direct function of $\Omega/M_{2.5}h$, the relation among the variables can be represented by the family of linear graphs shown in Figure 11(b). The results of Figure 11(b) are consistent with the intuitive notion that migration rate should increase as stream power increases, and decrease as bank strength becomes greater, other things remaining constant. Figure 11(b) also indicates that a change of bank strength at high stream power results in a much greater change in absolute migration rate than does the same proportional change in bank strength in a river with low stream power. It further shows that weak bank materials are

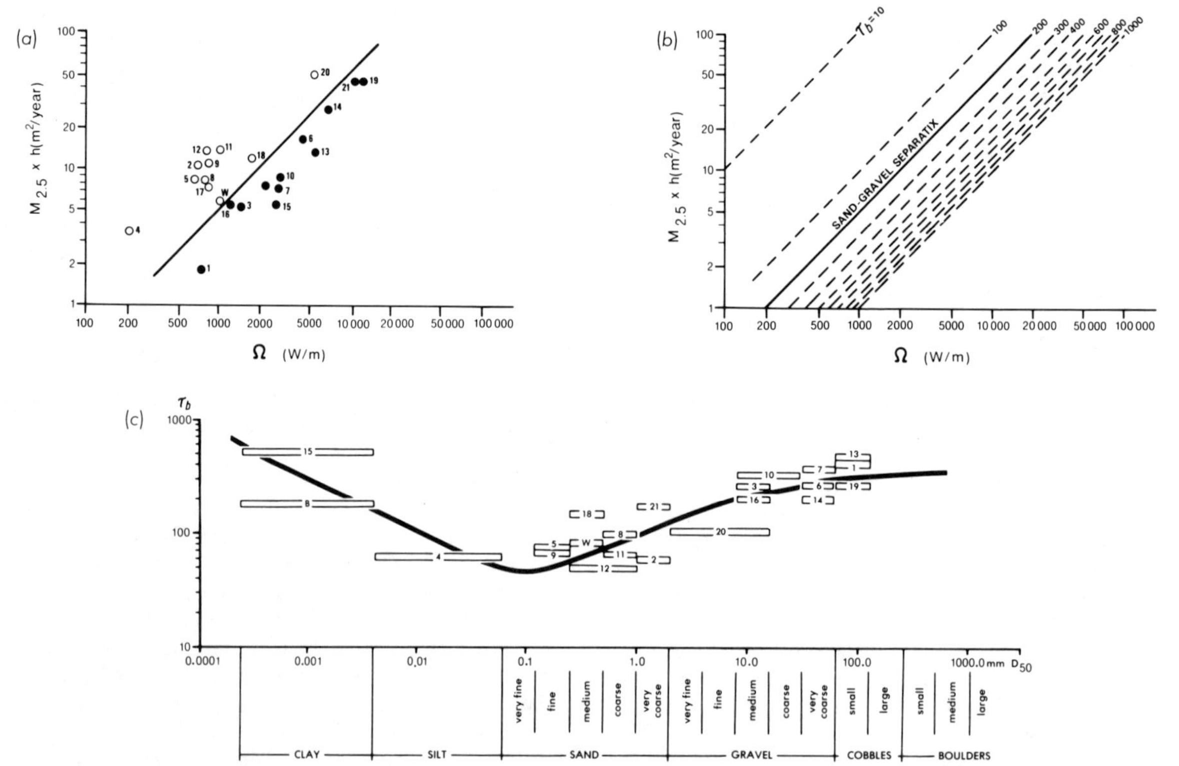

FIGURE 11. Stream power, bank material, and migration rates of Western Canadian rivers. (a) The relations between $M_{2.5}h$, Ω, bank material and stream power. A 45° separatix divides channels with sand banks (open circles) from those with gravel banks (solid circles). (b) The relations between $M_{2.5}h$, Ω, and coefficient of resistance to lateral erosion (T_b). (c) The relation between coefficient of resistance to lateral erosion (T_b) and texture of outer bank sediments (after Hickin and Nanson, [52]).

much more sensitive to increases in stream power than are resistant banks.

Bank strength (T_b) is a composite term for erosion resistance and reflects the range of bank resistance variables described previously. Consequently, an integrated measure of bank strength would be very difficult to derive. Texture (grain size) of the basal sediments was selected as the most important variable affecting bank erosion-resistance, and the results appear to justify this choice.

The relation of bank strength to the textural character of basal sediment in the outer banks of the channel is shown in Figure 11c. Bank texture was assessed in terms of 12 Wentworth textural classes from clay to boulders and are represented as bar lengths. Although there is considerable scatter in Figure 11c, the general form of the relation $T_b = f$(grain size) is clear. As grain size declines from cobbles to fine sand, T_b declines to a minimum. This relationship is conservative in the sense that grain size changes by four orders of magnitude while T_b declines by just one. The left-hand side of the graph is not as well defined but the data indicate that, as grain size declines into the silt-clay range, T_b increases again. Indeed, the curve is remarkably similar to a Hjulstrom or Shields sediment entrainment function, and presumably for the same basic reasons. The relatively high cohesive strength of the fines gives a resistance to lateral erosion equivalent to that of a much coarser material. The combined effects of cohesive and inertial forces of the sediment grains are minimised in the fine-sand range and here resistance to lateral erosion also is at a minimum.

Figure 11c and Equation (12) in its simplified form:

$$M_{2.5}(\text{m/yr}) = pgQs/T_bh \approx 9800\ Qs/T_bh \qquad (13)$$

(where Q = discharge at the 5-year flood and s = water-surface slope) provides a predictive model for maximum migration rates at the bend axis (apex). As with all empirical models, however, it should be noted that it may not apply well to river bends which differ markedly from those on which it is based. In particular, the following limitations should be recognised:

1. The model only applies to fully alluvial mixed-load rivers and to cases of simple expansion bends. It will not

apply well to channels with bank ice (permafrost and lenses) nor to the complex bend migration behaviour discussed in the next section.

2. The model was developed for rivers with forested floodplains (cottonwood and spruce) and it may not apply to channels with markedly different vegetation cover.
3. It predicts only the maximum long-term migration rate at the bend apex, a rate which will need discounting in accordance with Equations (9) and (10) if $r/w \neq 2.5$.
4. Equation (13) utilises the 5-year flood on the annual series as a measure of Q. The rivers in the data set all are slightly incised and $Q_{5.0}$ may represent overbank flow in other settings. Furthermore, in areas hydrologically different from Western Canada (with its strong nival Spring-runoff maximum and secondary Fall pluvial maximum), $Q_{5.0}$ may not provide the same type of summary discharge parameter as that used here.
5. The model has not been tested for regions beyond Western Canada.

THE CHARACTER AND CAUSES OF COMPLEX BEND-MIGRATION

It long has been recognised that meander trains are everywhere similar in general planform regardless of the size of the river. Indeed, almost all modern research assumes that a freely developing river bend will adopt an equilibrium planform scaled to the size of the river. Since the early work of Langbein and Leopold [83] this equilibrium planform generally has been taken to be a sine-generated curve although its assumed universality has not gone unchallenged [21,26]. But regardless of the precise form of a meander bend, the conventional view is that a meander train consists of alternating single bends of regular geometry connected through a series of inflection points. A number of researchers has noted, however, that on many rivers a distinction needs to be drawn between the notion of meander loops and the bends of which they are composed [46,69,101]; Figure 12, based on Beatton River meander development [45], illustrates the point. Here phase 1 of meander loop development is marked by relatively slow bend migration from a slightly curved initial channel. Scroll-bar spacing indicates that, during the growth period, the migration rate reaches a maximum, after which it abruptly declines to zero at the termination stage (compare with Figure 9). The termination stage occurs when any segment of the channel develops a tightly curved reach in which $r/w = 2.0$ (critical curvature ratio).

The arrested migration stage of phase 1 is followed by the initiation stage of phase 2. A new lobe of the floodplain begins to develop on the downstream limb of the phase 1

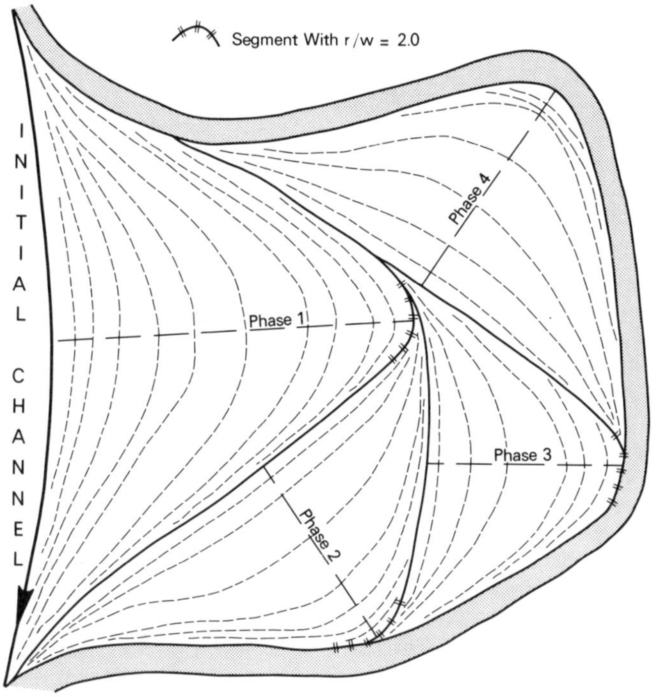

FIGURE 12. A complex meander loop produced by four distinct phases of migration (after Hickin, [46]).

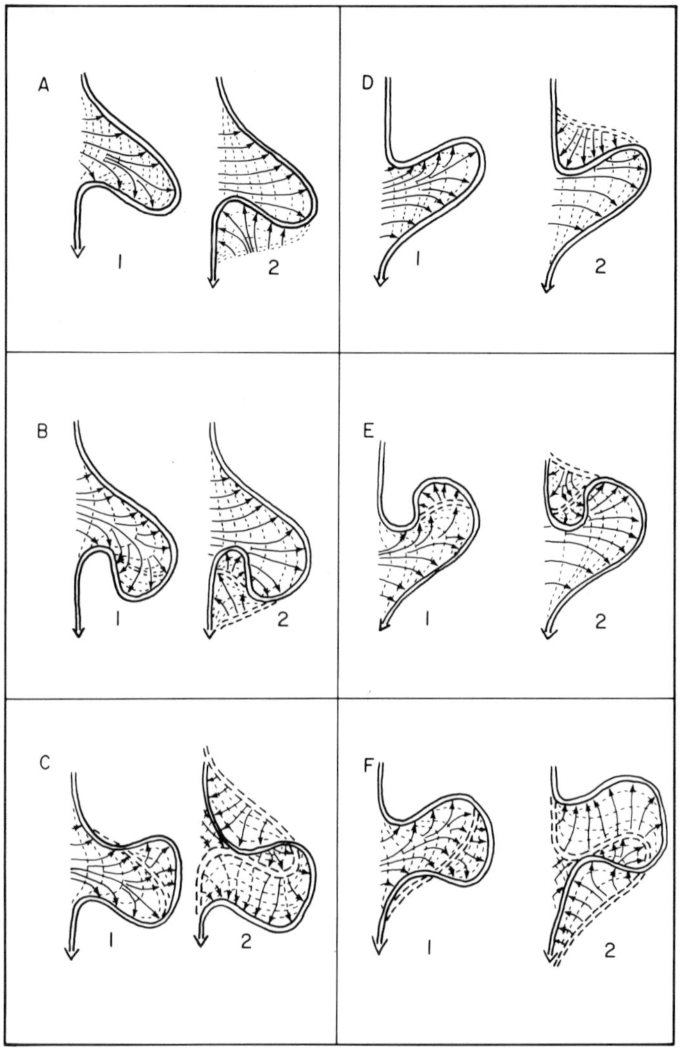

FIGURE 13. The development of typical meander patterns on the Beatton River, British Columbia (after Hickin, [45]).

channel bend. Phase 2 similarly displays a growth period of relatively rapid migration followed by an abrupt cessation of migration when the phase 2 channel achieves the critical curvature ratio. In this particular case the cycle subsequently is repeated through phases 3 and 4. Thus, a single loop may consist of multiple bend phases, a phenomenon common to many unconfined meanders. By this process of channel lengthening the pool and riffle sequence is forced out of phase with the planform oscillations, causing "meanders within meanders" to form [59,69].

Even without these complications of meander loop formation, meander trains can display considerable planform complexity at the level of the individual bend. The migration model for single bends described in the previous section considers the case of a simple expansion bend isolated from the influence of lateral erosion in adjacent meander bends. But in many cases the migration pattern and shape of a given bend are strongly conditioned by flow structure inherited from the upstream bend and by lateral erosion of common channel segments in both upstream and downstream bends. In Figure 13 are shown pairs of morphologically identical channel bends from the Beatton River which have evolved by secondary lobe formation (as in Figure 12) in cases [1], and by interactive bend-migration effects in cases [2]. Each is an example of a complex planform response to accommodate progressive lengthening of the

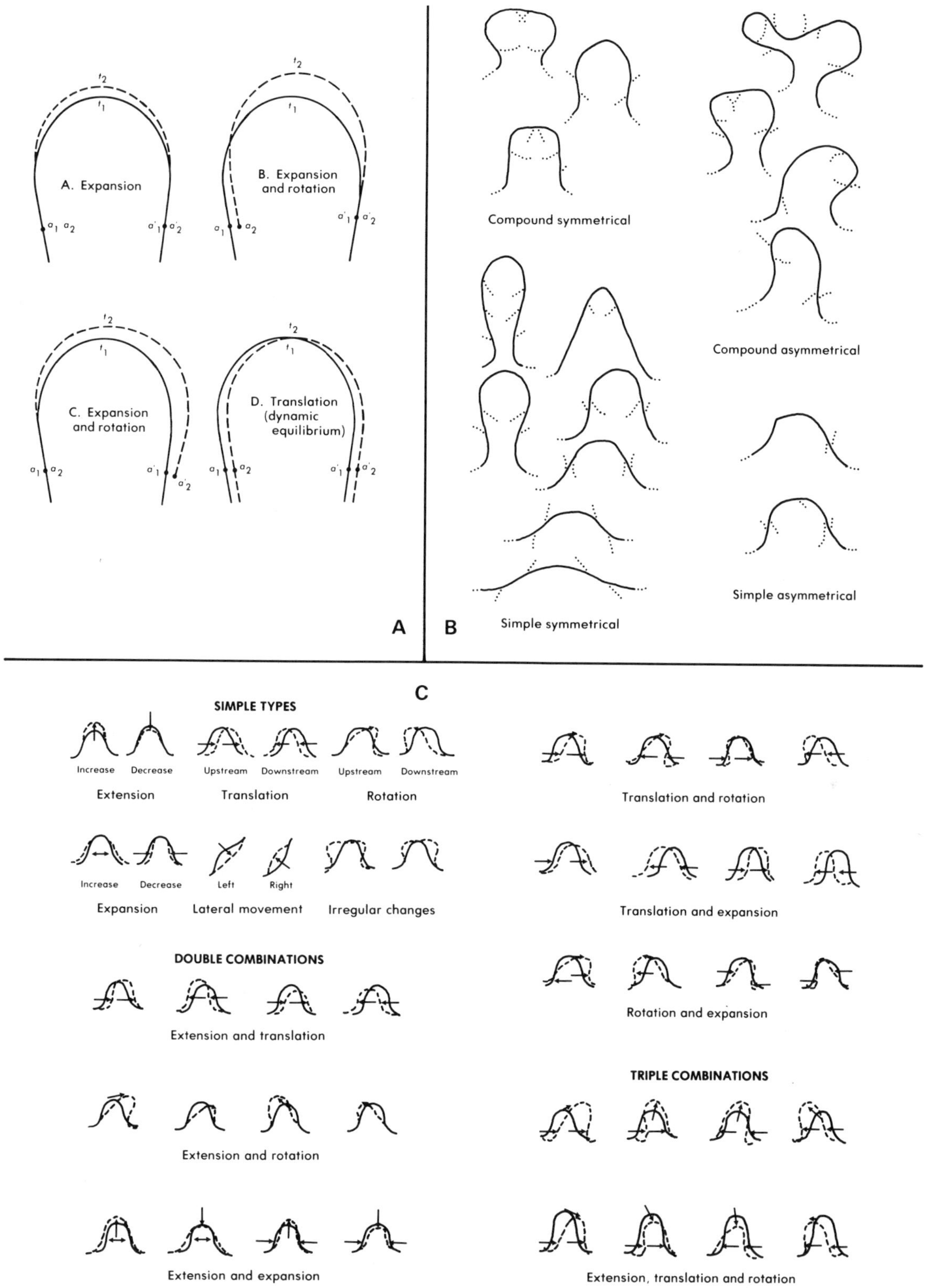

FIGURE 14. Models of complex meander development and morphology. (A) (after Daniel, [25]). (B) (after Brice, [12]). (C) (after Hooke, [54]).

channel. Clearly, if mean water-surface slope is to be maintained in a meander train, channel lengthening by migration, and contraction by cutoff and avulsion, must be balanced over time.

There have been numerous attempts to describe and classify the great variety of bend displacement and deformation styles [11,12,13,25,45,55,59,69,79] and these recently have been reviewed by Hooke [58]. Most styles are combinations of the three basic modes of displacement characterised by Daniel [25] as expansion, translation, and rotation (Figure 14). These basic elements together with others derived from empirical studies, form the basis of Hooke's classification of bend changes [54,58], the most comprehensive scheme yet devised (Figure 14). She determined from a study of 338 river bends in Devon, U.K. that about half were relatively simple cases involving extension/expansion, translation, and rotation, and that the remainder involved more complex changes of the type shown in Figure 14.

Attempts to model complex bend-migration behaviour are recent in origin and are of two types: simulation models of meander kinematics and analytical models based on the flow equations. The first type was noted in the introduction to this chapter and includes the work of Howard [61,62] and Ferguson [36]. These computer graphics models can simulate realistic meander-train evolution and bend migration complexity on the basis of very few assumptions, the most important of which are continuity of the meander trace and that $M = f(r/w)$.

Dynamic modelling of meander migration also in its infancy and remains the concern of just a few practitioners [5,64,112,113,114] Parker and his colleagues [64,113,114], the principal contributors to this approach, have developed a complex but general analytical model of bend deformation incorporating Engelund's bend-flow Equations [43] and orthogonal bend displacement of the type implied by Hickin's concept of erosional pathlines [45]. It indicates a convolutional rather than algebraic relation between migration rate and bend curvature which results in tight bends subsiding while those of long wavelength grow in amplitude as all bends migrate downstream. For bend reaches in which the radius of curvature is nearly constant and bend flow is fully developed, the convolutional relationship reduces to an algebraic form similar to that proposed by Hickin and Nanson [51]. A simplified ideal bend shape is obtained in the form of the Kinoshita equation [73] which, at small amplitude, reduces to a sine-generated curve, but at high amplitude displays pronounced skewing of the type reported by Carson and Lapointe [21] and Davies and Tinker [26].

These models have developed rapidly over the last few years and hold much promise for significantly increasing our understanding of interactive bend-migration effects. At the present time, however, none is capable of predicting rates of bend migration although Parker and Andrews [112] may soon provide the first attempt. Nevertheless, their utility in design and control work will remain limited by our ability to specify the mechanics of lateral erosion and migration processes within the channel.

REFERENCES

1. Alexander, C. S. and N. R. Nunally, "Channel Stability on the Lower Ohio River," *Annals of the Association of American Geographers,* Vol. 62, pp. 411–417 (1972).
2. Apmann, R. P., "Flow Process in Open Channel Bends," *Journal of the Hydraulics Division,* ASCE, Vol. 98, pp. 795–810 (1972).
3. Bagnold, R. A., "Some Aspects of River Meanders," *U.S. Geological Survey Professional Paper,* 282-E, p. 10 (1960).
4. Bathurst, J. C., C. R. Thorne, and R. D. Hey, "Secondary Flow and Shear Stress at River Bends," *Journal of the Hydraulics Division,* ASCE, Vol. 105, pp. 1277–1295 (1979).
5. Beck, S., "Mathematical Modelling of Meander Interaction," *Proceedings, ASCE Conference on River Meandering* (Rivers '83), New Orleans (1983).
6. Begin, Z. B., "Stream Curvature and Bank Erosion; a Model Based on the Momentum Equation," *Journal of Geology,* Vol. 89, pp. 497–504 (1981).
7. Bhowmik, N. G., "Hydraulics of Flow in the Kaskaskia River, Illinois," *Illinois State Water Survey (Urbana), Report on Investigation 91,* p. 116 (1979).
8. Bishop, A. W., "The Use of the Slip Circle in the Stability Analysis of Slopes," *Geotechnique,* Vol. 5, pp. 7–17 (1955).
9. Bishop, A. W. and N. R. Morgenstern, "Stability Coefficients for Earth Slopes," *Geotechnique,* Vol. 10, No. 4, pp. 129–150 (1960).
10. Bluck, B. J., "Sedimentation in the Meandering River Endrick," *Scottish Journal of Geology,* Vol. 7, pp. 73–138 (1971).
11. Brice, J. C., "Meandering Pattern of the White River in Indiana: An Analysis," In *Fluvial Geomorphology,* M. Morisawa (editor), State University of New York, Binghampton, pp. 178–200 (1973).
12. Brice, J. C., "Evolution of Meander Loops," *Geological Society of America Bulletin,* Vol. 85, pp. 581–586 (1974).
13. Brice, J. C., "Planform Properties of Meandering Rivers," *Proceedings, ASCE Conference on River Meandering* (Rivers '83), New Orleans (1983).
14. Bridge, J. S. and J. Jarvis, "Flow and Sedimentary Processes in the Meandering River South Esk, Glen Clova, Scotland," *Earth Surface Processes,* Vol. 1, No. 4, pp. 303–336 (1976).
15. Bridge, J. S. and J. Jarvis, "The Dynamics of a River Bend: A Study of Flow and Sedimentary Processes," *Sedimentology,* Vol. 29, No. 4, pp. 499–541 (1982).
16. Brunsden, D. and R. Kesel, "Slope Development on a Mississippi River Bluff in Historic Time," *Journal of Geology,* Vol. 81, pp. 576–597 (1973).
17. Callander, R. A., "Instability and River Channels," *Journal of Fluid Mechanics,* Vol. 36, pp. 465–480 (1969).
18. Callander, R. A., "River Meandering," *Annual Review of Fluid Mechanics,* Vol. 10, pp. 129–158 (1978).

19. Carey, W. C., "Formation of Floodplain Lands," *Journal of the Hydraulics Division*, ASCE, Vol. 95, pp. 981–994 (1969).
20. Carson, M. A. and M. J. Kirkby, *Hillslope Form and Process*, Cambridge University Press, Cambridge, pp. 475 (1972).
21. Carson, M. A. and M. F. Lapointe, "The Inherent Asymmetry of River Meander Planform," *Journal of Geology*, Vol. 91, pp. 41–45 (1983).
22. Church, M. and M. J. Miles, Discussion of "Processes and Mechanisms of Bank Erosion," by C. R. Thorne, in *Gravel Bed Rivers*, R. D. Hey, J. C. Bathurst, and C. R. Thorne (editors), John Wiley and Sons, New York, pp. 259–268 (1972).
23. Coleman, J. M., "Brahmaputra River: Channel Process and Sedimentation," *Sedimentary Geology*, Vol. 3, pp. 129–239 (1969).
24. Crickmay, C. H., "Lateral Activity in a River of North Western Canada," *Journal of Geology*, Vol. 65, pp. 377–391 (1957).
25. Daniel, J. F., "Channel Movement of Meandering Indiana Streams," *U.S. Geological Survey Professional Paper*, 732-A, p. 18 (1971).
26. Davies, T. R. H. and C. C. Tinker, "Fundamental Characteristics of Stream Meanders," *Geological Society of America Bulletin*, Vol. 95, No. 5, pp. 505–512 (1984).
27. De Vriend, H. J., "A Mathematical Model of Steady Flow in Curved Shallow Channels," *Journal of Hydraulic Research*, Vol. 15, No. 1, pp. 37–54 (1977).
28. De Vriend, H. J., "Velocity Redistribution in Curved Rectangular Channels," *Journal of Fluid Mechanics*, Vol. 107, pp. 423–439 (1981).
29. De Vriend, H. J. and H. J. Geldof, "Main Flow Velocity in Short and Sharply Curved River Bends," *Department of Civil Engineering, Delft University of Technology (The Netherlands), Communications on Hydraulics*, Report No. 83-6, pp. 38 (1983).
30. Dietrich, W. E., J. D. Smith, and T. Dunne, "Flow and Sediment Transport in a Sand Bedded Meander," *Journal of Geology*, Vol. 87, No. 2, pp. 305–315 (1979).
31. Eardley, A. J., "Yukon River Channel Shifting," *Geological Society of America Bulletin*, Vol. 49, pp. 343–358 (1938).
32. Einstein, H. A. and J. A. Harder, Velocity Distribution and the Boundary Layer at Channel Bends," *Transactions, American Geophysical Union*, Vol. 35, pp. 114–120 (1954).
33. Engelund, F., "Flow and Bed Topography in Channel Bends," *Journal of the Hydraulics Division*, ASCE, Vol. 100, pp. 1631–1648 (1974).
34. Engelund, F. and O. Skovaard, "On the Origin of Meandering and Braiding in Alluvial Streams," *Journal of Fluid Mechanics*, Vol. 57, pp. 289–302 (1973).
35. Everitt, B. L., "Use of the Cottonwood in an Investigation of the Recent History of a Floodplain," *American Journal of Science*, Vol. 266, pp. 417–439 (1968).
36. Ferguson, R. I., "Kinematic Model of Meander Migration," *Proceedings, ASCE Conference on River Meandering* (Rivers '83), New Orleans (1983).
37. Fox, J. A. and D. J. Ball, "The Analysis of Secondary Flow in Bends in Open Channels," *Proceedings, Institution of Civil Engineers*, Vol. 39, pp. 467–475 (1968).
38. Francis, J. R. D. and A. F. Asfari, "Velocity Distributions in Wide Curved Open-Channel Flows," *Journal of Hydraulic Research*, Vol. 9, pp. 73–90 (1971).
39. Friedkin, J. F., "A Laboratory Study of the Meandering of Alluvial Rivers," *U.S. Waterways Engineering Experimental Station Report*, pp. 40 (1945).
40. Grissinger, E. H., "Bank Erosion of Cohesive Materials," in *Gravel Bed Rivers*, R. D. Hey, J. C. Bathurst, and C. R. Thorne (editors), John Wiley and Sons, New York, pp. 273–287 (1982).
41. Handy, R. L., "Alluvial Cut-off Dating from Subsequent Growth of a Meander," *Geological Society of America Bulletin*, Vol. 83, pp. 475–480 (1972).
42. Henderson, F. M., *Open Channel Flow*, Macmillan, New York, pp. 250–252 (1966).
43. Henkel, D. J. and A. W. Skempton, "A Landslide at Jackfield, Shropshire, in an Overconsolidated Clay," *Proceedings of the Conference on Stability of Earth Slopes*, Stockholm, Vol. 1, pp. 90–101 (1954).
44. Hickin, E. J., "A Newly Identified Process of Point Bar Formation in Natural Streams," *American Journal of Science*, Vol. 267, pp. 999–1010 (1969).
45. Hickin, E. J., "The Development of Meanders in Natural River Channels," *American Journal of Science*, Vol. 274, pp. 414–442 (1974).
46. Hickin, E. J., "Hydraulic Factors Controlling Channel Migration," *Proceedings of the 5th Guelph Symposium in Geomorphology*, Geomorphology Abstracts, Norwich, England, pp. 59–66 (1977).
47. Hickin, E. J., "Mean Flow Structure in Meanders of the Squamish River, British Columbia," *Canadian Journal of Earth Sciences*, Vol. 15, No. 11, pp. 1833–1849 (1978).
48. Hickin, E. J., "Concave Bank Benches on the Squamish River, British Columbia, Canada," *Canadian Journal of Earth Sciences*, Vol. 16, No. 1, pp. 200–203 (1979).
49. Hickin, E. J., "Vegetation and Channel Dynamics," *The Canadian Geographer*, Vol. 28, No. 2, pp. 111–126 (1984).
50. Hickin, E. J., "Concave-Bank Benches in the Floodplains of Muskwa and Fort Nelson Rivers, British Columbia," *The Canadian Geographer*, (1985).
51. Hickin, E. J. and G. C. Nanson, "The character of Channel Migration on the Beatton River, Northeast British Columbia, Canada," *Geological Society of America Bulletin*, Vol. 86, pp. 487–494 (1975).
52. Hickin, E. J. and Nanson, G. C., "Lateral Migration Rates of River Bends," *Journal of Hydraulic Engineering*, ASCE, Vol. 110, No. 11, pp. 1557–1567 (1984).
53. Hill, A. R., "Erosion of River Banks Composed of Glacial Till Near Belfast, Northern Ireland," *Zeitschrift fur Geomorphologie*, Vol. 17, pp. 428–442 (1973).
54. Hooke, J. M., "An Analysis of Changes in River Channel Matters," *Ph.D. Thesis*, University of Exeter, U.K. (1977).
55. Hooke, J. M., "The Distribution and Nature of Changes in

River Channel Patterns," in *River Channel Changes,* K. J. Gregory (editor), John Wiley and Sons, Chichester, pp. 265–280 (1977).

56. Hooke, J. M., "An Analysis of the Processes of River Bend Erosion," *Journal of Hydrology,* Vol. 42, pp. 39–62 (1979).
57. Hooke, J. M., "Magnitude and Distribution of Rates of River Bank Erosion," *Earth Surface Processes,* Vol. 5, pp. 143–157 (1980).
58. Hooke, J. M., "Changes in River Meanders: A Review of Techniques and Results of Analysis," *Progress in Physical Geography,* Vol. 8, No. 4, pp. 473–508 (1984).
59. Hooke, J. M. and A. M. Harvey, "Meander Changes in Relation to Bend Morphology and Secondary Flows," in *Modern and Ancient Fluvial Systems,* J. D. Collinson and J. Lewin (editors), Oxford, Basil Blackwell, pp. 121–132 (1983).
60. Hooke, P. Le B., "Distribution of Sediment Transport and Shear Stress in a Meander Bend," *Journal of Geology,* Vol. 83, pp. 543–565 (1975).
61. Howard, A. D., "Simulation Model of Stream Meandering," *Proceedings, ASCE Conference on River Meandering,* New Orleans (August, 1983).
62. Howard, A. D. and T. R. Knutson, "Sufficient Conditions for River Meandering: a Simulation Approach," *Water Resources Research,* Vol. 20, No. 11, pp. 1659–1667 (1984).
63. Hughes, D. J., "Rates of Erosion on Meander Arcs," in *River Channel Changes,* K. J. Gregory, (editor), John Wiley and Sons, Chichester, pp. 193–205 (1977).
64. Ikeda, S., G. Parker, and K. Sawai, "Bend Theory of River Meanders. Part I. Linear Development," *Journal of Fluid Mechanics,* Vol. 112, pp. 363–377 (1981).
65. Ippen, A. P. and P. A. Drinker, "Boundary Shear Stress in Curved Trapezoidal Channels," *Journal of the Hydraulics Division,* ASCE, Vol. 88, pp. 143–180 (1962).
66. Jackson, R. G., "Velocity-Bedform-Texture Patterns of Meander Bends in the Lower Wabash River of Illinois and Indiana," *Geological Society of America Bulletin,* Vol. 86, pp. 1511–1522 (1975).
67. Kalkwijk, J. P. T. and H. J. De Vriend, "Computations of the Flow in Shallow River Bends," *Journal of Hydraulic Research,* Vol. 18, No. 4, pp. 327–342 (1980).
68. Keller, E. A., "Pools, Riffles and Meanders: Discussion," *Geological Society of America Bulletin,* Vol. 82, pp. 279–280 (1971).
69. Keller, E. A., "Development of Alluvial Channels," *Geological Society of America Bulletin,* Vol. 83, pp. 1531–1536 (1972).
70. Kesel, R. N., K. C. Dunne, R. C. McDonald, K. R. Allison, and B. E. Spicer, "Lateral Erosion and Overbank Deposition on the Mississippi River in Louisiana caused by 1973 Flooding," *Geology,* Vol. 2, pp. 461–464 (1974).
71. Khroda, G., "Flow Behaviour in Model Open-Channel Bends and Implications for Lateral Migration of Rivers," Ph.D. Thesis, Department of Geography, Simon Fraser University, British Columbia, Canada, pp. 384 (1985).
72. Kikkawa, H., S. Ikedo, and A. Kitagawa, "Flow and Bed Topography in Curved Open Channels," *Journal of the Hydraulics Division,* ASCE, Vol. 102, pp. 1327–1342 (1976).
73. Kinoshita, R., "An Investigation of Channel Deformation of the Ishikari River," *Publication No. 36,* Natural Resources Division, Ministry of Science and Technology of Japan, p. 139 (in Japanese) (1961).
74. Klimek, K., "Retreat of Alluvial Banks," *Geographica Polonica,* Vol. 28, pp. 59–76 (1974).
75. Klimek, K. and K. Trafas, "Young Holocene Changes in the Course of the Dunajec River in the Beskid Sadecki Mountains," *Studia Geomorphologica Carpatho-Balcanica,* Vol. 6, pp. 85–92 (1972).
76. Knighton, A. D., "Riverbank Erosion in Relation to Streamflow Conditions, River Bollin-Dean, Cheshire," *East Midland Geographer,* Vol. 5, pp. 416–426 (1973).
77. Knowlson, H., "A Long Term Experiment on the Radial Growth of the Oak," *Naturalist* (Leeds), pp. 93–99 (April, 1939).
78. Kolb, C., "Sediments Forming the Bed and Banks of the Lower Mississippi and Their Effects on Migration," *Sedimentology,* Vol. 2, pp. 215–226 (1963).
79. Kondtrat'yev, N. E., "Hydromorphic Principles of Computations of Free Meandering and Sign and Indices of Free Meandering," *Soviet Hydrology,* Vol. 4, pp. 309–335 (1968).
80. Kondrat'yev, N. E. and I. V. Popov, "Methodological Prerequisites for Conducting Network Observations on the Channel Process," *Soviet Hydrology,* Vol. 3, pp. 273–297 (1967).
81. Kulemina, N. M., "Some Characteristics of the Process of Incomplete Meandering of the Channel of the Upper Ob River," *Soviet Hydrology,* Vol. 6, pp. 518–534 (1973).
82. Laczay, I. A., "Channel Pattern Changes of Hungarian Rivers," in *River Channel Changes,* K. J. Gregory (editor), John Wiley and Sons, pp. 185–192 (1973).
83. Langbein, W. and L. B. Leopold, "River Meanders: Theory of Minimum Variance," *U.S. Geological Survey Professional Paper* 422-H (1966).
84. Lawler, D. M., "The Use of Erosion Pins in River Banks," *Swansea Geographer,* Vol. 16, pp. 9–17 (1978).
85. Leeder, M. R. and P. H. Bridges, "Flow Separation in Meander Bends," *Nature,* Vol. 253, pp. 338–339 (1975).
86. Leopold, L. B., "River Channel Change with Time: An Example," *Geological Society of America Bulletin,* Vol. 84, pp. 1845–1860 (1973).
87. Leopold, L. B., M. G. Wolman, and J. P. Miller, *Fluvial Processes in Geomorphology,* Freeman, San Francisco, pp. 522.
88. Leschziner, M. and W. Rodi, "Calculation of Strongly Curved Open Channel Flow," *Journal of the Hydraulics Division,* ASCE, Vol. 105, pp. 1297–1314.
89. Lewin, J., "Late Stage Meander Growth," *Nature,* Vol. 240, p. 116 (1972).
90. Lewin, J., "Channel Pattern Changes," in *River Channel Changes,* K. J. Gregory (editor), John Wiley and Sons, Chichester, pp. 167–184 (1977).
91. Lewin, J., "River Channels," in *Geomorphological Tech-*

niques, A. Goudie, (editor), Allen and Unwin, London, pp. 196–212 (1981).
92. Lewin, J. and B. J. Brindle, "Confined Meanders," in *River Channel Changes,* K. J. Gregory (editor), John Wiley and Sons, Chichester, pp. 221–233 (1977).
93. Lewin, J. and D. Hughes, "Assessing Channel Change on Welsh Rivers," *Cambria,* Vol. 3, pp. 1–10 (1976).
94. McGowan, A., A. Saldivar-Sali, and A. M. Radwan, "Fissure Patterns and Slope Failures in Till at Hurlford, Ayrshire," *Quarterly Journal of Engineering Geology,* Vol. 7, pp. 1–26 (1974).
95. Mockmore, C. A., "Flow Around Bends in Stable Channels," *Transactions,* ASCE, Vol. 109, pp. 335–360 (1943).
96. Morgenstern, N. R., "Stability Charts for Earth Slopes During Rapid Drawdown," *Geotechnique,* Vol. 13, No. 2, pp. 121–131 (1963).
97. Morgenstern, N. R. and V. E. Price, "The Analysis of the Stability of General Slip Surfaces," *Geotechnique,* Vol. 15, pp. 79–93 (1965).
98. Mosley, M. P., "Channel Changes on the River Bollin, Cheshire, 1872–1973," *East Midland Geographer,* Vol. 42, pp. 185–199 (1975).
99. Nanson, G. C., "Channel Migration, Floodplain Formation, and Vegetation Succession on a Meandering-River Floodplain in N.E. British Columbia, Canada," *Ph.D. Thesis,* Department of Geography, Simon Fraser University, Burnaby, British Columbia, Canada, pp. 349 (1977).
100. Nanson, G. C., "Point Bar and Floodplain Formation of the Meandering Beatton River, Northeastern British Columbia, Canada," *Sedimentology,* Vol. 27, pp. 3–29 (1980).
101. Nanson, G. C., "A Regional Trend to Meander Migration," *Journal of Geology,* Vol. 88, pp. 100–108 (1980).
102. Nanson, G. C. and H. F. Beach, "Forest Succession and Sedimentation on a Meandering-River Floodplain, Northeast British Columbia, Canada," *Journal of Biogeography,* Vol. 4, pp. 229–251 (1977).
103. Nanson, G. C. and E. J. Hickin, Channel Migration and Incision on the Beatton River," *Journal of Hydraulic Engineering,* ASCE, Vol. 109, No. 3, pp. 327–337 (Mar., 1983).
104. Nanson, G. C. and E. J. Hickin, "A Statistical Analysis of Bank Erosion and Channel Migration in Western Canada," *Internal Report,* Department of Geography, University of Wollongong, N.S.W., Australia, pp. 34 (1983).
105. Nanson, G. C. and K. J. Page, "Lateral Accretion of Fine-Grained Concave Benches on Meandering Rivers," *Special Publications of the International Association of Sedimentologists,* Vol. 6 (1982).
106. Neill, C. R., "Riverbed Transport Related to Meander Migration Rates," *Journal of the Waterways Harbors and Coastal Engineering Division,* ASCE, Vol. 97, pp. 783–786 (1971).
107. Neill, C. R., "Bank Erosion Versus Bedload Transport in a Gravel River," *Proceedings, ASCE Conference on River Meandering* (Rivers '83), New Orleans, p. 8 (1984).
108. Niemrowski, M., "Trends of Action and Intensity of Fluvial Processes Forming the Bottoms of Carpathian Valleys in the Holocene," *Studia Geomorphologica Carpatho-Balcanica,* Vol. 6, pp. 93–103 (1972).
109. Nelson, J. G., "Man and Geomorphic Processes in the Chemung River Valley, New York and Pennsylvania," *Annals of the Association of American Geographers,* Vol. 56, pp. 24–32 (1966).
110. Page, K. J. and G. C. Nanson, "Concave-Bank Benches and Associated Floodplain Formation," *Earth Surface Processes and Landforms,* Vol. 7, pp. 529–543 (1982).
111. Parker, G., "Cause and Characteristic Scales of Meandering and Braiding in Rivers," *Journal of Fluid Mechanics,* Vol. 76, pp. 457–480 (1976).
112. Parker, G. and E. D. Andrews, "On the Time Development of Meander Bends," unpublished manuscript, St. Anthony Falls Hydraulic Laboratory, University of Minnesota, Minnesota.
113. Parker, G., P. Daplas, and J. Akiyama, "Meander Bends of High Amplitude," *Journal of the Hydraulics Division,* ASCE, Vol. 109 (1983).
114. Parker, G., K. Sawai, and S. Ikeda, "Bend Theory of River Meanders. Part 2. Nonlinear Deformation of Finite Amplitude Bends," *Journal of Fluid Mechanics,* Vol. 115, pp. 303–314 (1982).
115. Patel, R. N., "A Comparison of the Anatomy of the Secondary Xylem in Roots and Stems," *Holzforschung,* Vol. 19, pp. 72–79 (1965).
116. Quick, M. C., "Mechanism for Streamflow Meandering," *Journal of the Hydraulics Division,* ASCE, Vol. 100, pp. 741–753 (1974).
117. Quraishy, M. S., "The Origin of Curves in Rivers," *Current Science,* Vol. 13. pp. 36–39 (1944).
118. Rozovskii, I. L., *Flow of Water in Bends of Open Channels,* Academy of Sciences of the Ukrainian SSR, Kiev, (Israel Program for Scientific Translations, 1961), pp. 233 (1957).
119. Ruhe, R. V., *Geomorphology,* Houghton Mifflin, Boston (1975).
120. Schmudde, T. H., "Some Aspects of Landforms of the Lower Mississippi Floodplain," *Annals of the Association of American Geographers,* Vol. 53, pp. 60–73 (1963).
121. Shen, H. W. and S. Komura, "Meandering Tendencies in Straight Alluvial Channels," *Journal of the Hydraulics Division,* ASCE, Vol. 94, pp. 997–1016 (1968).
122. Siebert, W. and W. Gotz, "A Study on the Deformation of Secondary Flow in Models of Rectangular Meandering Channels," *Proceedings, 16th Congress, International Association of Hydraulic Research,* Vol. 2, pp. 141–149 (1975).
123. Sigafoos, R. S., "Botanical Evidence of Floods and Flood-Plain Deposition," *U.S. Geological Survey Professional Paper,* 485-A, pp. 35 (1964).
124. Skempton, A. W., "The Longterm Stability of Clay Slopes," *Geotechnique,* Vol. 14, pp. 90–108 (1964).
125. Spangler, M. G. and R. L. Handy, *Soil Engineering,* 3rd Edition, Intext Educational Publishers, New York and London, pp. 748 (1973).
126. Sundborg, A., "The River Klaralven, a Study of Fluvial Processes," *Geografiska Annaler,* Vol. 38, pp. 127–316 (1956).

127. Task Committee on Sedimentation, "Sediment Transport Mechanics: Initiation of Motion," *Journal of the Hydraulics Division*, ASCE, Vol. 92, pp. 291–314 (1966).
128. Task Committee on Sedimentation, "Erosion of Cohesive Sediments," *Journal of the Hydraulics Division*, ASCE, Vol. 94, pp. 1017–1047 (1968).
129. Task Committee on Sedimentation, *Sedimentation Engineering*, ASCE, V. A. Vanoni (editor), p. 745 (1975).
130. Taylor, D. W., *Fundamentals of Soil Mechanics*, John Wiley and Sons, New York, pp. 700 (1948).
131. Taylor, G., K. A. W. Crook, and K. D. Woodyer, "Upstream Dipping Foreset Cross-Stratification: Origin and Implications for Paleoslope Analysis," *Journal of Sedimentary Petrology*, Vol. 41, pp. 578–581 (1971).
132. Terzaghi, K., *Theoretical Soil Mechanics*, John Wiley and Sons, New York, pp. 510 (1943).
133. Terzaghi, K. and R. B. Peck, *Soil Mechanics and Engineering Practice*, John Wiley and Sons, New York, pp. 727 (1948).
134. Thorne, C. R., "Processes of Bank Erosion in River Channels," Ph.D. Thesis, University of East Anglia, Norwich, U.K., p. 447 (1978).
135. Thorne, C. R., "Processes and Mechanisms of River Bank Erosion," in *Gravel Bed Rivers*, R. D. Hey, J. C. Bathurst, and C. R. Thornes (editors), John Wiley and Sons, New York, pp. 227–259 (1982).
136. Thorne, C. R. and J. Lewin, "Bank Processes, Bed-Material Movement and Planform Development in a Meandering River," in *Adjustments of the Fluvial System*, D. D. Rhodes and G. P. Williams (editors), Kendall/Hunt Publishing Co., Dubuque, Iowa, pp. 117–137 (1979).
137. Thorne, C. R. and N. K. Tovey, "Stability of Composite River Banks," *Earth Surface Processes and Landforms*, Vol. 6, pp. 469–484 (1981).
138. Tinkler, K. J., "Pools, Riffles, and Meanders," *Geological Society of America Bulletin*, Vol. 81, pp. 547–552 (1970).
139. Tinkler, K. J., "Pools, Riffles, and Meanders: a Reply," *Geological Society of America Bulletin*, Vol. 83, pp. 1531–1536 (1972).
140. Turnbull, W. J., M. Krinitsky, and F. J. Werner, "Bank Erosion in Soils of the Lower Mississippi Valley," *Journal of the Soil Mechanics Division*, ASCE, Vol. 92, pp. 121–136 (1966).
141. Twidale, C. R., "Erosion of an Alluvial Bank at Birdwood, South Australia," *Zeitschrift für Geomorphologie*, Vol. 8, pp. 189–211 (1964).
142. Wolman, M. G., "Factors Influencing Erosion of a Cohesive River Bank," *American Journal of Science*, Vol. 257, pp. 204–216 (1959).
143. Woodyer, K. D., "Discussion on Formation of Floodplain Lands" by W. C. Carey, *Journal of the Hydraulics Division*, ASCE, Vol. 96, pp. 849–850 (1970).
144. Yalin, M. S., *Mechanics of Sediment Transport*, 2nd ed., Pergamon Press, Oxford, pp. 290 (1977).
145. Yen, C. L., "Bed Topography Effect on Flow in a Meander," *Journal of the Hydraulics Division*, ASCE, Vol. 96, pp. 57–73 (1970).
146. Yen, C. L., "Spiral Motion of Developed Flow in Wide Curved Open Channels," in *Sedimentation (Einstein)*, H. W. Shen (Editor), Water Resources Publications, Fort Collins, CO, Chapter 22, pp. 1–33 (1972).
147. Yen, C. L. and B. C. Yen, "Water Surface Configuration in Channel Bends," *Journal of the Hydraulics Division*, ASCE, Vol. 97, pp. 303–321 (1971).

Optimal Risk-Based Hydraulic Design of Highway Bridges

YEOU-KOUNG TUNG* AND LARRY W. MAYS**

INTRODUCTION

Important considerations in the hydraulic design of highway bridges include the flood hazard associated with the proposed highway bridge and the impact of the bridge on human lives, property, and stream stability. Hydraulic design of highway bridges specifically involves the selection of bridge's location and the size of its opening, as well as the determination of the embankment height. Conventional design procedures are based on: (1) the selection of flood peak discharge, known as the design flow, for a certain predetermined return period depending on the type of the road; and (2) the size of bridge opening and embankment height which will allow this selected flood flow to pass. Due to many uncertainties involved in the hydraulic design of bridge structures, the conventional design procedures usually assign an additional embankment height as a safety margin to protect highway bridges. Thus, existing conventional design procedures fail to explicitly account for the cost interaction between project components and the expected damages. Neither does it provide a means to systematically account for uncertainties in the design. The optimal design can be defined as the one, based on the least cost principle, which maintains a proper balance between the project cost and potential flood damages or economic losses.

In general, the uncertainties involved in the hydraulic design of structures can be attributed to hydrologic aspects as well as hydraulic aspects. Hydrologic uncertainty can be divided into three categories: (1) natural or inherent uncertainty, due to the stochastic nature of the hydrologic processes involved; (2) the model uncertainty, resulting from the limited amount of data available for assessing the underlying random mechanism of the hydrologic processes involved; and (3) the parameter uncertainty, due to an insufficient amount of data from which the statistical parameters in an assumed probability model are estimated. Hydraulic uncertainties involve the determination of flow capacity through the bridge openings and the calculation of the backwater surface profile. Detailed discussions of uncertainties involved were given by Tung and Mays [1].

Hydraulic engineers have only recently recognized the concept of risk-based hydraulic design for structures. Young et al. [2] developed an optimal design model for highway culverts that considered annual expected flood damage by using appropriate probabilities for each flood. Mays [3] also developed an optimization model for determining culvert sizes by considering the inherent hydrologic uncertainty as well as hydraulic uncertainties. Corry et al. [4] prepared a circular providing procedures for the design of encroachments on floodplains using risk analysis. Schneider and Wilson [5] presented a risk-based procedure for the optimal hydraulic design of bridges which considered an economic risk defined as the additional annual expected flood losses resulting from the presence of a bridge. Tang et al. [6] have applied the optimal risk-based design procedure to other types of hydraulic structures, such as storm sewers. Tung and Mays [7] and Tung [8] presented various types of uncertainties in hydrology as well as in hydraulics and developed a generalized procedure for analyzing an optimal risk-based design of flood levee systems.

Most risk-based design procedures for highway bridges developed consider the inherent hydrologic uncertainty. Other aspects of hydrologic uncertainties are not included. It is the intention of this chapter to present a methodology to integrate various aspects of hydrologic uncertainties in deriving an optimal risk-based design of highway bridges.

*Wyoming Water Research Center & Department of Statistics, University of Wyoming, Laramie, WY

**Department of Civil Engineering, University of Texas, Austin, TX

The optimum herein refers to the design having the minimum total expected annual cost.

HYDROLOGIC UNCERTAINTIES

Hydrologic Parameter Uncertainty

The hydrologic parameters in this context refer to parameters used in probability models to describe the random process of hydrologic events. They are synonymous to the statistical properties of hydrologic processes which, in general, are uniquely related to the parameters in the probability model. In area of hydrology, the true unknown parameters governing the random hydrologic process can generally be estimated by two ways. One is to estimate parameters based on sample data which is referred to as sample statistics, θ_s. Sampling error is usually associated with θ_s which can be large especially when the sample size is small, which is the case typical for most hydrologic data. Secondly, hydrologists have recognized that hydrologic information often can be transferred among watersheds or even river basins. The transferring of information has the function of data augmentation which would increase the reliability of the analysis. This second source of hydrologic information enables us to estimate unknown parameter called the regional estimator, θ_r. To derive a regional parameter estimator, procedures developed in hydrologic regional analysis [9] can be applied. The most popular regionalization technique is the statistical regression analysis. The procedure developed for treating or reducing parameter uncertainty is to develop a generalized parameter estimator through the use of a weighting factor between the sample and the regional parameter estimators as

$$\hat{\theta} = W_\theta \theta_s + (1 - W_\theta)\theta_r \quad (1)$$

where $\hat{\theta}$ is the generalized estimator for unknown parameter θ and W_θ is the weighting function. Since both the sample estimator, θ_s, and the regional estimator, θ_r, are random with associated uncertainties, the generalized parameter estimator $\hat{\theta}$ involves uncertainty. The question is then, how to combine θ_s and θ_r such that the "best $\hat{\theta}$" can be obtained. The "best $\hat{\theta}$" herein represents the one with the least uncertainty. The commonly used measure of uncertainty feature of an estimator is the variance. Assuming that θ_s and θ_r are independent and unbiased estimators of unknown true parameter θ, the optimal weighting factor which minimizes the variance of θ is

$$W_\theta = \frac{\mathrm{Var}(\theta_r)}{\mathrm{Var}(\theta_s) + \mathrm{Var}(\theta_r)} \quad (2)$$

in which $\mathrm{Var}(\theta_r)$ and $\mathrm{Var}(\theta_s)$ are the variances of the regional estimator and the sample estimator, respectively.

The determination of W_θ requires the calculation of $\mathrm{Var}(\theta_r)$ and $\mathrm{Var}(\theta_s)$. The variance of sample parameter estimator can be determined by using two non-parametric statistical methods called the bootstrap method [10,11] and the jackknife method [12]. A detailed application of the two methods can be found elsewhere [13,14]. The appendix briefly describes the algorithm of the two methods. The variance of the regional estimator, $\mathrm{Var}(\theta_r)$, can be derived from a statistical regression analysis in which θ_r is related to various physiographical and meterological variables of the river basins. A detailed description of the methodologies will be given later.

Hydrologic Model Uncertainty

The hydrologic model uncertainty is the uncertainty associated with assessing the hydrologic probability model with only a limited amount of data. In other words, the uncertainty is associated with the choice of model based on an insufficient amount of data. In reality, the probability distribution of hydrologic process involved in risk-based hydraulic design of structure is not known. Engineers have to resort to some statistical inferences based on data collected. The determination of the probability distribution of the annual maximum flood series is an example of this type.

Several conventional statistical goodness-of-fit techniques such as Chi-square, Kolmogorov-Smirnov, or Cramer-von Mises [15,16] can be used to select an appropriate probability model. These methods employ a statistical hypothesis test based on a test statistic for measuring the degrees of fit of an unknown discrete or continuous distribution function to an empirical distribution function. However, many hydrologists discourage the use of above mentioned goodness-of-fit techniques for testing hydrologic frequency distribution because of the importance of the tail behavior of the hydrologic frequency distributions and the insensitivity of these statistical tests on the tails of the distributions. Furthermore, the power of these statistical goodness-of-fit tests, i.e., the probability of rejecting the false hypothesis, is not high. This is especially true for small sample which is the case for most hydrological data.

In regard to model choice, Wood and Iturbe [17] raise an important point: "Most hydrologic processes are so complex that no model yet devised may be true model or that no hydrologic events follow one particular model. Consequently, it could be reasonably expected that a combination of models would better 'explain' the hydrologic process than does a single model." This then lead to the use of Bayes theorem to compute the posterior probability of a model as

$$P''(m_i) = \frac{L(\underline{X}|\underline{\hat{\theta}},m_i)P'(m_i)}{\sum_{m_{i\epsilon M}} L(\underline{X}|\underline{\hat{\theta}},m_i)P'(m_i)} \quad (3)$$

in which $P''(m_i)$ is the posterior probability for the distribution model m_i in the set of candidate models M investigated, $P'(m_i)$ is the prior probability for model m_i, and $L(\underline{X}|\hat{\underline{\theta}},m_i)$ is the likelihood function for model m_i conditioned upon the observations $\underline{X} = \{X_1,X_2, \ldots ,X_n\}$. As noted, the values of generalized parameter estimator, $\hat{\underline{\theta}}$, obtained by Equation (1), are used here in computing the likelihood function. The value of the posterior probability is a measure of the likelihood for a model being the "true" one among all the models considered based on the data collected and the prior knowledge about the model. The use of Bayes theorem enables the incorporation of prior knowledge about the probability distribution, which may be based on experiences or other physical evidences, which is then updated with the data collected.

The Bayesian approach enables the derivation of a composite probability distribution [18] for the flood from different models under investigation and being weighted by their associated posterior probability as

$$f_{m_c}(Q) = \sum_{m_i \epsilon M} P''(m_i) f_{m_i}(Q) \tag{4}$$

in which $f_{m_c}(Q)$ is the composite probability distribution of hydrologic events and $f_{mi}(Q_{m_i})$ is the probability distribution for hydrologic events following the model m_i. This composite probability distribution model is used to account for hydrologic uncertainty. A flowchart for the analysis of hydrologic uncertainties is shown in Figure 1.

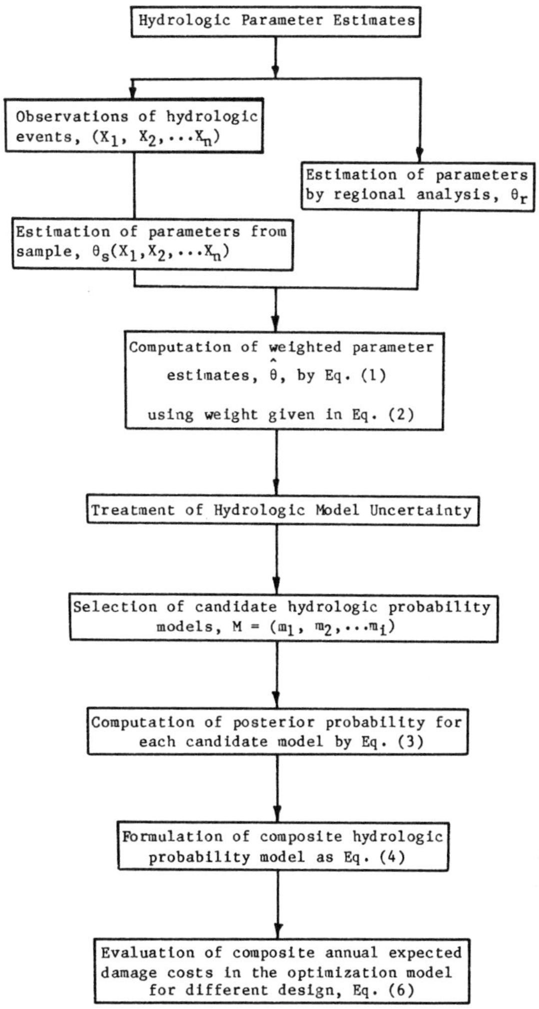

FIGURE 1. Flowchart of hydrologic uncertainty analysis.

COST COMPONENTS IN RISK-BASED DESIGN OF HIGHWAY BRIDGES

To perform risk-based hydraulic design of highway bridges several types of flood related costs, in addition to construction and operation/maintenance costs, must be included. The quantifications and assessments of flood related damage costs requires significantly additional effort in collecting and analyzing data. This section briefly describes some relevant cost components to be considered in risk-based hydraulic design of highway bridges.

Construction Costs (C_1)

This usually is the largest single cost item of the project. Various cost estimation procedures are used by highway agencies to estimate this cost. Construction costs generally include initial embankment, pavement and structures. In general, operation and maintenance costs are included. Any other costs associated with bridge accessaries such as spur dike and riprap for scouring protection should also be included. The construction cost largely depends on the physical layout of structures which usually increase with bridge opening and embankment height. A qualitative representation of construction cost in relation to the bridge opening and embankment height is shown in Figure 2.

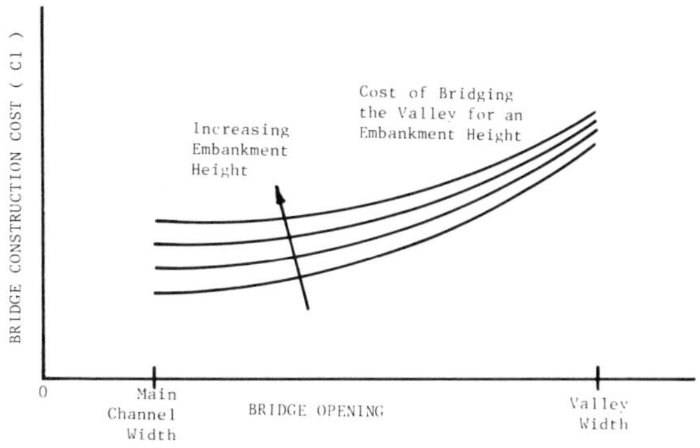

FIGURE 2. Variation of construction costs with bridge opening and embankment height [6].

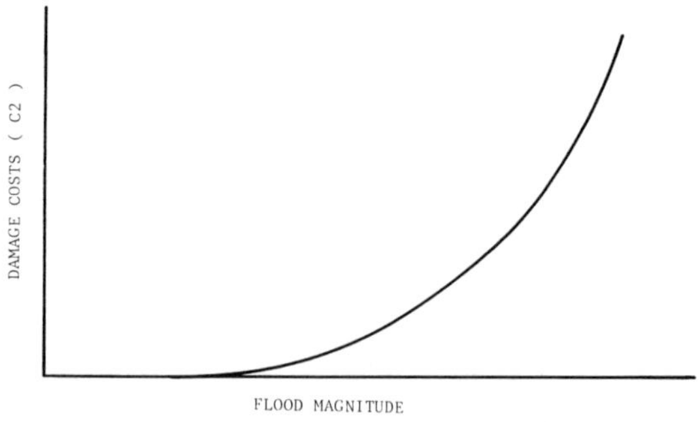

FIGURE 3. Schematic diagram of damage cost corresponds to flood magnitude before project construction [6].

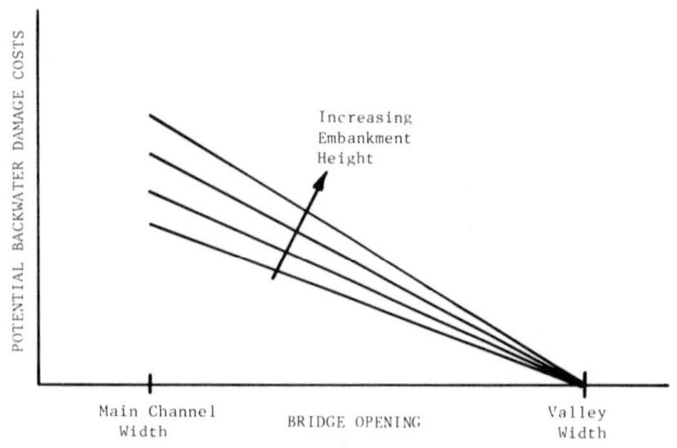

FIGURE 4. Variation of backwater damage with bridge opening and embankment height [6].

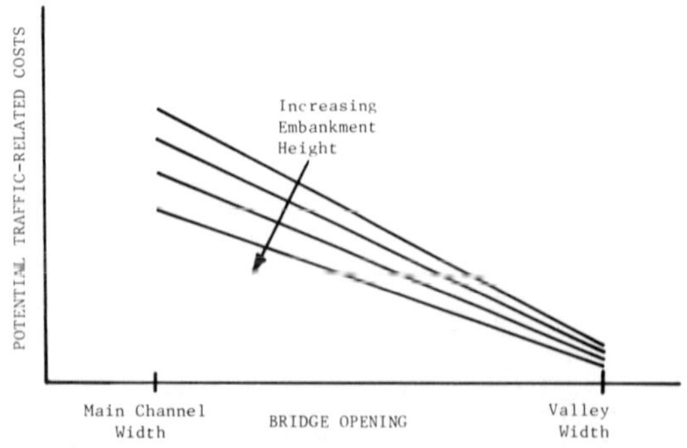

FIGURE 5. Variation of traffic-related costs with bridge opening and embankment height [6].

Flood Damage Cost (C_2)

This is the cost of property damage caused by flood which would not be affected by the project construction. To assess this type of cost, it is necessary to gather the information about the stage-discharge relationships, land use pattern, and property evaluation in the floodplain. Herein, the flood damage cost is evaluated under the natural condition, i.e., no bridge construction. For a given location and with the assumption of steady land use conditions, this flood damage cost solely depends on the magnitude of flood flow as shown qualitatively in Figure 3.

Backwater Damage Cost (C_3)

It is the difference between the flood damage with the bridge of certain layout in place, and the flood damage cost, C_2, under the natural conditions. In other words, it is the added flood damage cost to the floodplain due to the backwater effect caused by the presence of bridge and embankment. Their presence usually raises the surface water elevation upstream and inundates more land under the natural condition. In order to assess this backwater damage cost, backwater computations are needed which, in turn, requires information of hydraulic characteristics of and geometry of floodplain. Backwater damage usually increases with embankment height for a constant bridge opening; however, it decreases with an increase in bridge opening, as shown in Figure 4.

Traffic-Related Cost (C_4)

This is an additional cost of driving a vehicle on a primary detour opposite the usual route when the road is out of service due to inundation of bridge structure. There are basically three types of traffic-related costs [4], namely, (1) increased running cost due to the detour, (2) lost time of vehicle occupants, and (3) increased accidents on the detour. Information required for the assessment of traffic-related costs include, not limited to, average daily traffic, traffic mixture, vehicle running cost, lengths of normal and detour routes, period of inundation and its frequency, fatal rate, unit cost of injuries and property damage, etc. It is not difficult to imagine that some of the cost items involved in calculating traffic-related damage cannot be obtained with accuracy. Since traffic-related damages are mainly caused by traffic interruption due to inundation of bridge deck, this cost usually decreases with an increase in embankment height and bridge opening, as shown in Figure 5. More detailed descriptions are given by Schneider and Wilson [5] and Corry et al. [4].

Embankment and Pavement Repair Cost (C_5)

Floodflow, when overtops bridge and embankment over a sustained period of time, would cause erosion of embank-

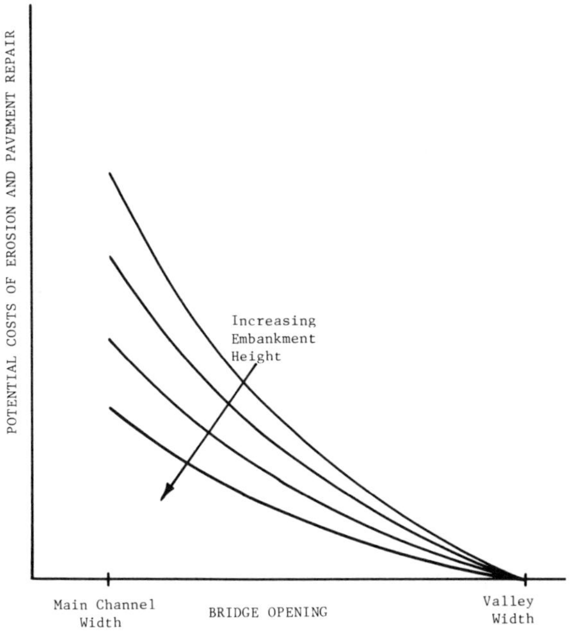

FIGURE 6. Variation of erosion damage potential with bridge opening and embankment height [6].

ment material as well as damaging the pavement. The assessment of embankment repair cost requires an understanding of erosion mechanics to quantify the extent of erosion. The cost of embankment and pavement repair depends on variables such as total volume of embankment and area of pavement damaged, rates of embankment and pavement repair, unit costs of embankment repair and pavement repair, and mobilization cost. In general, the amount of embankment erosion and pavement damage can depend on the duration and depth of overflow. Heuristically, for a given flood hydrograph, C_5 decreases with an increase in embankment height and bridge opening, as shown in Figure 6.

Summary

Considering the above cost components in risk-based hydraulic design of highway bridges, the backwater damage (C_3), the traffic-related cost (C_4), and the embankment and pavement repair cost (C_5) each depends on the flood magnitude, bridge opening, construction method, material, and embankment height. Therefore, for a given flood magnitude, Q, width of bridge opening, W, and embankment height, H, the resulting total damage, $D(Q,W,H)$, can be expressed as

$$D(Q,W,H) = C_2 + C_3 + C_4 + C_5 \qquad (5)$$

The total annual expected damage cost, $E(D)$, for given W and H, then can be expressed as

$$E(D) = \int_{o}^{\infty} D(Q,W,H)f(Q)dQ \qquad (6)$$

in which $f(Q)$ is the probability density function of flood.

OPTIMIZATION OF BRIDGE DESIGN

Objective

The decision variables involved in the hydraulic design of highway bridges are the embankment height, H, and the length of bridge opening, W. The objective of optimization is to seek optimal embankment height and bridge opening which minimize the total expected annual cost ($TEAC$) which is the sum of the annual bridge construction cost and expected annual damage cost. The objective function in

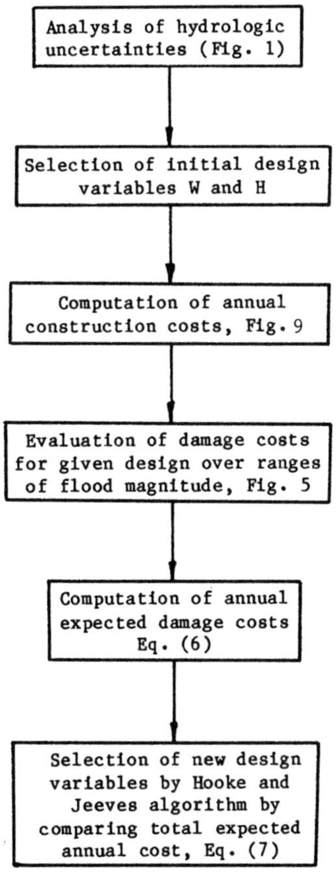

FIGURE 7. Flowchart of optimal risk-based hydraulic design procedure.

terms of expected annual cost is expressed as

$$\underset{W,H}{\text{Minimize}}\ TEAC = C_1(W,H) \cdot CRF + E(D|W,H) \quad (7)$$

where *CRF* is the capital recovery factor which converts a lump sum of construction cost into an annual basis for a specified interest rate and service period of bridge.

Optimization Technique

Because there are only two decision variables, *H* and *W*, an easily programmed accelerated climbing technique called pattern search, developed by Hooke and Jeeves [19], is used. This technique does not require expressions for the gradient of the objective function which is difficult to be obtained in this case. The technique is based on the philosophy that any set of moves (changes in value of the decision variables) which have been successful on early explorations, will be worth trying again. The method starts cautiously with small changes (steps) in the value of decision variables from a starting point. The step size grows with a repeated success. Subsequently failure indicates that a smaller step is in order. If a change in search direction is required, the technique will start over again with a new pattern. In the vicinity of the optimum, the step sizes for all variables become very small to avoid overlooking any promising directions. It should be noted that the search procedure does not guarantee a global optimum. Therefore, several different starting points and step sizes are recommended to be used for the comparison of the final results. A flowchart of Hooke-Jeeves pattern search technique is shown in Figure 8. The procedures for the optimal risk-based hydraulic design of highway bridges is shown in Figure 7.

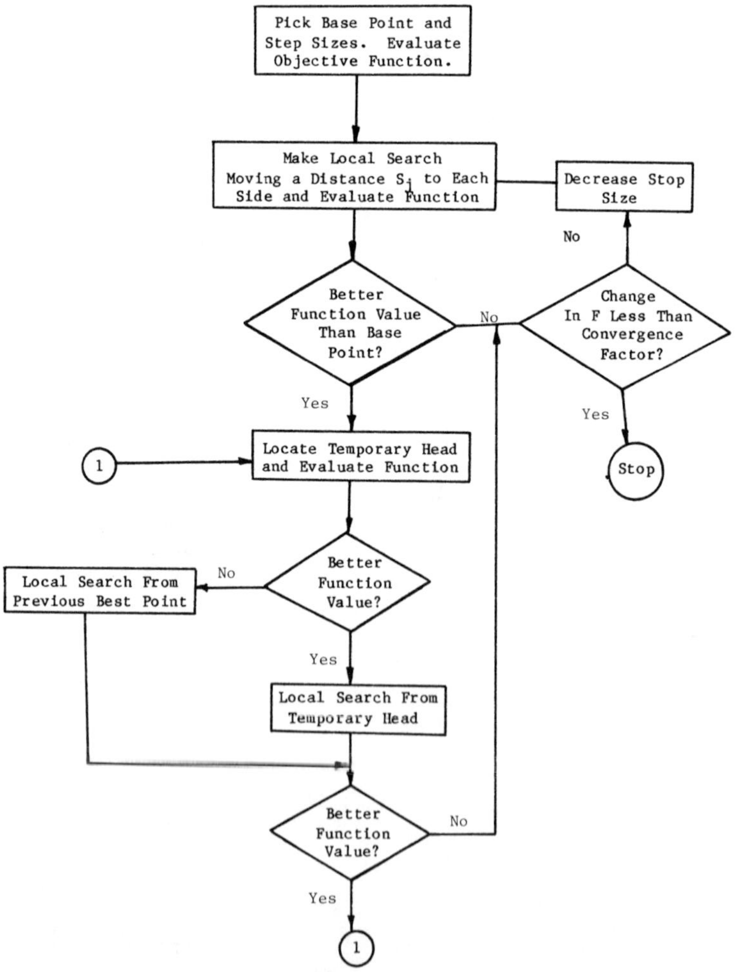

FIGURE 8. Flowchart of Hooke and Jeeves algorithm (from Kuester and Mize [21]).

The constraints in the optimal risk-based design model are the maximum and minimum embankment heights and lengths of bridge opening, which depend on the topography of bridge site, foundation conditions, and technical aspects of construction.

In Schneider and Wilson's work [5], the risk-based hydraulic design of highway bridges considered only the inherent hydrologic uncertainty using a log-Pearson type III distribution. The optimal design was found by comparing the sum of annual capital construction cost and additional annual expected damage cost due to the construction, namely, using $D(Q,W,H) = C_3 + C_4 + C_5$ in Equation (6) for several preselected combinations of bridge openings and embankment heights.

The model considered in this chapter incorporates a hydrologic parameter estimation technique that minimizes the variance of the statistical parameter estimators, and the Bayes theorem to account for the hydrologic model uncertainty into an optimization framework for determining the optimal embankment height and bridge opening such that the total expected annual cost of the project is minimized. An example illustrating the application of the methodology is presented in the following section.

ILLUSTRATION

This application considers a proposed bridge on Leaf River near Hattiesburg, Mississippi. The costs of bridge construction cost (C_1) for different bridge opening and embankment heights, flood damage cost before the construction (C_2), additional damage costs due to the bridge construction such as backwater damage (C_3), traffic-related cost (C_4), and embankment and pavement repair (C_5) for a specific design and flood magnitude are given in detail by Corey et al. [4]. This section will not repeat the processes involved in deriving those cost estimates and will merely summarize them which can be incorporated in the risk-based hydraulic design of highway bridges.

The damage functions for describing the flood damage before the construction (C_2) and that of additional damage due to the construction ($C_3 + C_4 + C_5$) are developed separately as Equations (8)-(9), respectively, based on the data provided by Corry et al. [4] as

$$\ln(C_2) = -2215.48 + 1292.38\sqrt{\ln(Q)} - 187.2\ln(Q) \quad (8)$$

$$\ln(C_3 + C_4 + C_5) = 8194.21 - 0.295\ln(W) - 3487.04\ln(H) + 346.56[\ln(H)]^2 + 99.92\ln(Q) - 4.21[\ln(Q)]^2 \quad (9)$$

with correlation coefficients of 0.993 and 0.940, respectively. Using C_2, C_3, C_4, and C_5, the total damage $D(Q,W,H)$

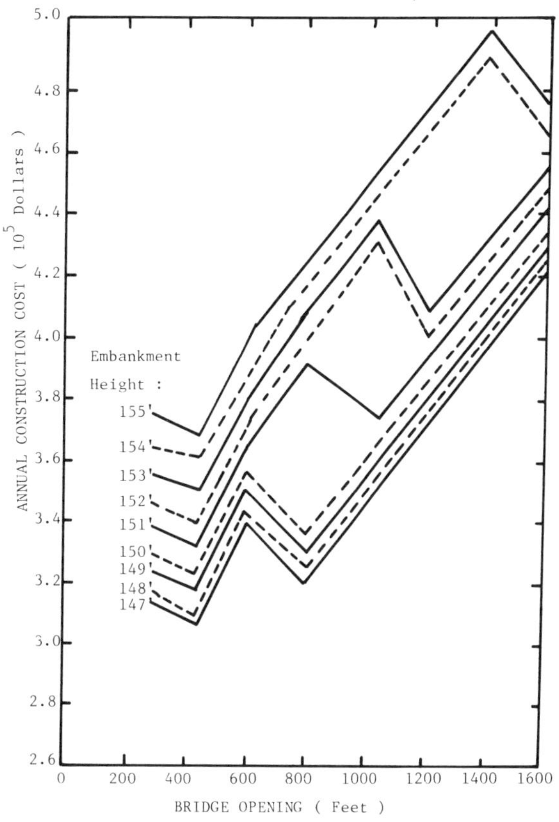

FIGURE 9. Construction costs (Corry et al., 1980).

can be computed by Equation (5) which, in turn, is used in Equation (6) to determine $E(D)$. It should be noted that these two damage cost functions were derived within the range of 147–151 ft. (45–46m) for the embankment height, and 280–1600 ft. (85–488m) for the bridge opening. The cost of bridge construction for different embankment height and bridge opening were also adopted from Corry et al. [4] as shown in Figure 9.

In order to compute the generalized parameter estimates, a regression study was performed to correlate hydrologic statistical parameters of streamflow such as the mean ($\bar{X}_r$), standard deviation ($\bar{S}_r$), and skewness ($\bar{G}_r$) of annual maximum flood series as well as the mean ($\bar{X}_r'$), standard deviation ($\bar{S}_r'$), and skewness ($\bar{G}_r'$) of the log-transformed annual maximum flood series, to physiographic and meterologic characteristics such as the contributing drainage area (A), in square miles, the main channel slope (S), in feet per mile, the main channel length (L), in miles, the percentage of forest cover (F) and mean annual precipitation (P), in inches, for 39 watersheds in the neighborhood of the study area. Stepwise regression procedures [20] were used and the resulting regional equations for the annual maximum flood series of original scale are

$$\bar{X}_r = 16280 + 23.0(A) - 265.6(5) - 53.0(L) - 87.6(F) - 58.6(P) \quad (10)$$

$$S_r = 4934 + 24.6(A) + 91.7(S) - 51.0(L) - 47.4(F) \quad (11)$$

$$G_r = 2.864 + 0.0014(A) - 0.011(L) + 0.011(F) - 0.032(P) \quad (12)$$

and for log-transformed annual maximum flood series

$$\bar{X}_r' = 5.65 - 0.003(A) - 0.16(S) + 0.014(L) + 0.018(F) + 0.054(P) \quad (13)$$

$$S_r' = 2.76 - 0.0008(A) - 0.0058(S) + 0.0019(L) - 0.011(F) - 0.016(P) \quad (14)$$

$$G_r' = -7.34 + 0.0016(A) + 0.088(S) + 0.0048(L) + 0.017(F) + 0.076(P) \quad (15)$$

The regional parameter estimates for the study area and their corresponding variances were determined by the proceding regression equations. At the project site, the U.S. Geological Survey has operated a recording gaging station since 1939. The annual maximum flood series from 1939 to 1980 is obtained from which the sample parameter estimates were computed. The variances of sample statistics computed by the jackknife method and the bootstrap method, and the resulting generalized parameter estimates by Equation (2) are given in Table 1.

For the treatment of hydrologic model uncertainty, Table 2 lists the posterior probabilities of five commonly used hydrologic probability models using equal prior probability for each model and the generalized parameter estimates given in Table 1. The resulting composite hydrologic probability model, with parameters estimated by either the jackknife or the bootstrap method in conjunction with regional regression analysis, is given as

$$f_{m_c}(Q) = 0.38 f_{LN}(Q) + 0.15 f_P(Q) + 0.46 f_{LP}(Q) \quad (16)$$

where f_{LN} is the probability density function for the log-normal distribution, f_p is for the Pearson type III distribution, and f_{LP} is for the log-Pearson type III distribution. This composite probability model for the annual maximum flood series can be used in Equation (6) to evaluate the composite annual expected damage cost.

Tables 3 and 4 list the optimal design parameters, i.e., bridge opening and embankment height, and the total expected annual cost, respectively, for various hydrologic probability models using different methods for parameter estimation. The interest rate used was 5% over a 25-year expected service life of the bridge.

ANALYSIS

Table 3 shows that the resulting optimal bridge openings are of the same length for various probability models and parameter estimation procedures, while the resulting embankment heights vary by approximately 1.8 ft (0.55m). Another observation is that the total expected annual cost is underestimated using sample statistics alone, as compared with using the weighted parameter estimates. This is because the value of the mean and standard deviation (Table

TABLE 1. Results of Hydrologic Parameter Estimation of Mean, Standard Deviation, and Skewness.

(1)		$\bar{X}$ (2)	S (3)	G (4)	$\bar{X}'$ (5)	S′ (6)	G′ (7)
	θ_S	29,600	17,272	1.645	10.154	0.529	0.187
	θ_r	46,544	42,618	2.161	5.178	0.717	0.852
	Var (θ_r)	2.3×10^7	6.2×10^7	1.359	3.848	0.441	2.129
Jackknife	Var (θ_S)	4.5×10^6	5.0×10^6	0.076	0.004	0.002	0.053
	W_θ	0.84	0.92	0.95	1.00	0.99	0.98
	$\hat{\theta}$	32,361	19,360	1.673	10.148	0.530	0.203
Bootstrap	Var (θ_S)	4.4×10^6	5.4×10^6	0.072	0.04	0.004	0.043
	W_θ	0.84	0.92	0.95	0.99	0.99	0.98
	$\hat{\theta}$	32,263	19,297	1.671	10.149	0.530	0.202

Note: 1 cfs = 0.028 m^3/s.

TABLE 2. Distributed Posterior Probabilities for Selected Hydrologic Probability Models Using Bayes Theorem.

(1)	Normal (2)	Log-Normal (3)	Gumbel (4)	Pearson III (5)	Log-Pearson III (6)
Prior probability	0.20	0.20	0.20	0.20	0.20
Sample	0.00	0.37	0.01	0.18	0.44
Jackknife	0.00	0.38	0.00	0.15	0.46
Bootstrap	0.00	0.38	0.01	0.15	0.46

TABLE 3. Optimal Design of Bridge Opening (Width, Height) as Related to Various Hydrologic Probability Models.

Model (1)	Normal (2)	Log-Normal (3)	Gumbel (4)	Pearson III (5)	Log-Pearson III (6)	Composite (7)
Sample	(439.1,147.0)	(439.1,147.2)	(439.1,147.0)	(439.1,147.3)	(439.1,147.7)	(439.1,147.5)
Jackknife	(439.1,147.0)	(439.1,147.1)	(439.1,148.4)	(439.1,148.8)	(439.1,147.8)	(439.1,147.7)
Bootstrap	(439.1,147.0)	(439.1,147.1)	(439.1,148.3)	(439.1,148.8)	(439.1,147.8)	(439.1,147.7)

Note: 1 ft = 0.305 m.

TABLE 4. Total Expected Annual Cost (TEAC) Corresponding to Optimal Design, as Related to Various Hydrologic Probability Models (in Hundreds of Thousands of Dollars).

(1)	Normal (2)	Log-Normal (3)	Gumbel (4)	Pearson III (5)	Log-Pearson III (6)	Composite (7)
Sample	2.434	2.800	2.727	2.848	2.945	2.873
Jackknife	2.723	2.790	3.108	3.251	2.946	2.934
Bootstrap	2.712	2.791	3.095	3.237	2.946	2.933

TABLE 5. Comparison of Annual Expected Damage Costs as Related to Various Hydrologic Probability Models (in Tens of Thousands of Dollars).

E(D) / Model / (W,H) (1)	Normal (2)	Log-Normal (3)	Gumbel (4)	Pearson III (5)	Log-Pearson III (6)	Composite (7)
(439.1,147.0)	5.51	6.19	9.52	11.08	7.80	7.67

1) derived by the weighting procedure is higher than those computed using sample statistics alone. In this example, the jackknife and bootstrap methods result in nearly the same optimal solution.

The effect of using different hydrologic probability models and the composite model in evaluating the annual expected damage costs for a given design can be observed in Table 5. The design variables arbitrarily selected for the comparison were 439.1 ft (133.9 m) for the bridge opening, and 147 ft (44.8 m) for the embankment height. With this specific design under examination, the annual construction cost is $\$2.178 \times 10^5$, and the expected annual damage costs, using different hydrologic probability models, are tabulated in Table 5. The difference in expected damage costs, using normal distribution as an arbitrary basis, among different models ranges from $6,800 annually for log-normal distribution, to $55,700 annually for Pearson III distribution.

SUMMARY AND CONCLUSIONS

A methodology for incorporating the analysis of hydrologic uncertainties into an optimization framework for the hydraulic design of highway bridges is presented. An application is made in which the evaluation of flood damage is simplified through the development of damage functions based on the result of previous study. The model can be expanded to include the backwater computation and a detailed evaluation of various costs involved, leading to an estimate of the total cost of different designs.

From the case study, it was found that the resulting optimal embankment height varies with different hydrologic probability models and with different methods of parameter estimation used. Because of the uncertainty of hydrologic model choices, a composite model is recommended. A composite model is suggested as a better alternative because it is an approximate method to account for hydrologic model uncertainty. As illustrated by the example application, the annual expected damage costs are highly dependent on the hydrologic model considered; therefore, in order to minimize the uncertainty in choosing a hydrologic model, the composite model is recommended. As indicated by this example, the parameter estimation also can effect the computation of annual expected damages; therefore, methods that minimize the parameter uncertainty are important. The weighting procedure introduced herein is such a method. This method uses either the jackknife or bootstrap method to compute variances of sample statistics. This procedure estimates weights using variances of sample and regional parameters for determining generalized weighted parameter estimates. The jackknife method is easier, requires less computer time, and gives approximately the same answers in this case.

REFERENCES

1. Tung, Y. K. and L. W. Mays, *J. of Hydraul. Div.*, ASCE, *106:* HY5 (1980).
2. Young, G. K., M. R. Childrey, and R. E. Trent, *J. of Hydraul. Div.*, ASCE, *100:* HY5 (1974).
3. Mays, L. W., *J. of Hydraul. Div.*, ASCE, *105:* HY5 (1979).
4. Corry, M. L., J. S. Jones, and D. L. Thompson, *Hydraulic Engineering Circular No. 17,* U. S. Department of Transportation, Federal Highway Administration, Washington D.C. (1980).
5. Schneider, V. R. and K. V. Wilson, *Report,* FHWA-TS-80-226, Department of Transportation, Federal Highway Administration, Washington D.C. (1980).
6. Tang, W. H., L. W. Mays, B. C. Yen, *J. of Environ. Engr. Div.*, ASCE, *103:* EE3 (1975).
7. Tung, Y. K. and L. W. Mays, *Wat. Resour. Res.*, AGU, *17:*4 (1981).
8. Tung, Y. K., Ph.D. Thesis, Univ. of Texas at Austin (1980).
9. Kite, G. W., *Frequency and Risk Analysis in Hydrology,* Water Resource Publication, Fort Collins, Colorado (1977).
10. Efron, B., *Technical Report No. 32,* Div. of Biostatistics, Stanford University (1977).
11. Efron, B., *Technical Report No. 39,* Div. of Biostatistics, Stanford University (1978).
12. Miller, R. G., *Biometrika, 6:* 1 (1974).
13. Tung, Y. K. and L. W. Mays, *Wat. Resour. Bull,* AWRA, *17:* 2 (1981).
14. Tung, Y. K. and L. W. Mays, *J. of Wat. Resour. Plan. and Mgmt.*, ASCE, *107:* WR1 (1981).
15. DeGroot, M. H., *Probability and Statistics,* Addison Wesley (1975).
16. Conover, W. J., *Practical Nonparametric Statistics,* John Wiley & Sons (1971).
17. Wood, E. F. and I. Rodriguez-Iturbe, *Wat. Resour. Res.*, AGU, *11:* 6 (1975).
18. Bogardi, I., L. Puckstein, and E. Castano, in *Stochastic Processes in Water Resources Engineering,* (L. Gottschalk, G. Lindh, and L. deMare, ed.) Water Resources Publication, Fort Collins, Colorado (1977).
19. Hooke, R. and T. A. Jeeves, *J. Computer, 8:* 2 (1961).
20. Draper, N. R. and H. Smith, *Applied Regression Analysis,* John Wiley and Sons, New York (1966).
21. Kuester, J. L. and H. H. Mize, *Optimization Techniques With Fortran*, McGraw-Hill, New York (1973).

APPENDIX

Variance of Sample Estimators

Two non-parametric statistical methods, namely the jackknife and bootstrap methods, which make no assumption about the distribution function, are especially useful to determine Var (θ_s). These nonparametric methods do pay a computational price for their freedom from normal distribution theory. However, the use of the computer makes these methods very tractable rather than a nightmare for the user.

JACKKNIFE METHOD

The jackknife, introduced in late 1950, is an intriguing attempt to answer an important statistical problem: "Having computed an estimate of some quantity of interest, what accuracy can be attached to the estimate?" Accuracy here refers to the "$\pm$ something" which often accompanies statistical estimates. The usual $\pm$ quantities are based on normal distribution theory, or occasionally on some other parametric theory, while the jackknife is a nonparametric technique which makes no assumption on the distribution. Miller [12] gave an excellent review of this method. The jackknife method has the function of bias correction which makes this method more favorable. The most commonly used expression for the "accuracy" of an estimate is the standard deviation. The jackknife procedure can be used to compute the standard deviation of parameter estimates as follows:

1. Compute the estimate of interest, θ_s, from the N sample observations.
2. For each observation $i = 1, \ldots, N$, compute $\theta_s^{(i)}$ from $N - 1$ observations with the ith observation deleted from the data set.
3. Compute the accuracy of θ_s using

$$\sigma^{(J)} = \frac{N-1}{N} \sum_{i=1}^{N} [(\theta_s^{(i)} - \theta_s)^2]^{1/2} \tag{16}$$

where $\sigma^{(J)}$ is the jackknifed standard deviation of sample estimate θ_s. For the purpose of bias correction, statisticians usually replace θ_s by

$$\sum_{i=1}^{N} \theta_s^{(i)}/N$$

in Equation (16); otherwise, with no bias correction for θ_s is made.

Hence, for a given set of observations the jackknifed standard deviation of the estimate, θ_s, can be computed by the procedures described above. The variance of θ_s is expressed as

$$\text{Var}^{(J)}(\theta_s) = [\hat{\sigma}^{(J)}]^2 \tag{17}$$

which can be used to calculate the weight in Equation (2). Readers are referred to Efron [11] and Miller [12] for the details of the theory of this method.

BOOTSTRAP METHOD

Efron [10] developed the bootstrap method which can also be used to access the accuracy of any estimate of interest derived from a sample. From the theoretical point of view, the bootstrap is more widely applicable than the jackknife, and also more dependable [10,11]. The procedures of this method are outlined as follows [11]:

1. Let $\hat{F}$ be an empirical distribution of N observed data points, $\{y_i, i = 1, \ldots, N\}$, i.e., $1/N$ is the probability of occurrence assigned to each of the observations.
2. Use a random number generator to draw N new points $\{y_i^*, i = 1, \ldots, N\}$ independently and with replacement from $\hat{F}$, so each new point is an independent random selection of one of the N original data points. This set of N new points is called the bootstrap sample which is a subset of the original data points.
3. Compute the estimate θ_s^* for the bootstrap sample $\{y_i^*, i = 1, \ldots, N\}$.
4. Repeat steps (2) and (3) a large number of times, say $m = 1, \ldots, M_B$ times, with each time using an independent set of new random numbers to generate the new bootstrap sample. The resulting sequence of bootstrap statistics is $\theta_s^{*(m)}$, $m = 1, \ldots, M_B$.
5. The variance of θ_s can be calculated as

$$\text{Var}(\theta_s) = \frac{1}{M_B} \sum_{m=1}^{M_B} [\theta_s^{*(m)} - \bar{\theta}_s^*]^{1/2} \tag{18}$$

in which

$$\theta_s^* = \frac{1}{M_B} \sum_{m=1}^{M_B} \theta_s^{*(m)} \tag{19}$$

Optimal Design of Stilling Basins of Overflow Spillways

YEOU-KOUNG TUNG* AND LARRY W. MAYS**

INTRODUCTION

Many modern dams and their hydraulic structures are of immense size and are required to hold large volumes of water under high pressure. When water is released over the spillway a tremendous amount of potential energy is converted to kinetic energy at the base of the structures. This energy must be dissipated in order to prevent the possibility of severe scouring of the downstream river bed, minimize erosion, and prevent undermining the foundation which endangers the dam safety. Energy dissipators are used at the base of these structures to dissipate excessive energy and to establish safe flow conditions in outlet channels.

Operation of any hydraulic-energy dissipator largely on expending part of the energy in high-velocity flow by some combinations: external friction between the waters and the channel, between water and air, or by internal friction and turbulence. Fundamentally, energy dissipators or stilling basins which perform the energy reduction convert kinetic energy into turbulence and finally into heat. Dissipation of energy in stilling basin is primarily accomplished by means of a hydraulic jump.

A stilling basin reduces the exit supercritical flow from the spillway to a tranquil state by a hydraulic jump which quickly dissipates flow energy through large-scale turbulence such that the flow becomes incapable of scouring the downstream channel. To ensure that a stilling basin serves its function effectively, it is important to design a basin with the objective that, for the range of design discharges, the elevation of the tailwater surface in the downstream channel should be greater than or equal to that of the conjugate depths after the hydraulic jump to prevent sweep out of the jump from the basin. Conversely, if the resulting conjugate depth is too low, the jump will be drowned, losing its purpose as an energy dissipator. The general design objective is to sacrifice jump efficiency in order to avoid downstream river bed scouring.

A perfect hydraulic jump is formed when the elevation of the conjugate depth after the jump corresponds to the elevation of the tailwater depth. Neither the spillway nor the stilling basin will alter the tailwater rating curve for the regime of natural river channel downstream since it is governed by the natural conditions along the channel. Thus the floor elevation and width of stilling basins are selected such that the conjugate depth elevations most nearly agree with the tailwater depth elevations for a design discharge or for a range of discharges.

Stilling basins are usually not designed to confine the entire length of the hydraulic jump because of the expense. Certain accessory devices such as baffle piers, end sills and shute blocks are installed along the basin floor to control and stabilize the jump and increase the turbulence which assists dissipating the energy and to reduce the conjugate depth after jump [1,2]. Also their presence permits shortening of the basin by as much as 60 to 80 percent [3,4] as well as providing a safety factor against the jump being swept out. Laboratory investigations also indicate that the presence of the appurtenances reduces the tailwater depths required. A detailed description of the appurtenances of stilling basins is given by Elevatorski [5]. Rectangular basins are used in preference of trapezoidal basins because of better hydraulic performance even though the latter are more economical to build. In practice, the sidewalls of basins are vertical or as near vertical as practical.

To ensure the cost of the basin is commensurate with its function as an effective energy dissipator, design procedures leading to optimal determination of the floor elevation and

*Wyoming Water Research Center, University of Wyoming, Laramie, WY

**Department of Civil Engineering, University of Texas, Austin, TX basin width are desirable. Various types of procedures have been developed for the generalized design of hydraulic jump stilling basin. These procedures are based upon intensive laboratory experiments, long-term studies of many existing structures and empirical rules. Blaisdell [3] investigated the Saint Antonio Fall (SAF) stilling basin design rules primarily for use on small structures such as those built by the U.S. Soil Conservation Service; Bradley and Peterka [4] summarized their intensive studies and broad investigations on the design rules for the U.S. Bureau of Reclamation Type I, II, III, IV and V basins.

The generalized design practice usually results in safe stilling basins, having the conservative safety factor; however, optimal determination of the floor elevation and basin width for minimum cost designs have seldom been achieved. An optimization model is developed herein to analytically determine the optimal floor elevation and basin width of stilling basins. This model is based on the principle that within a specified range of discharges, the hydraulic jump formed will be both stable and effective in dissipating energy provided that the elevations of tailwater depths are greater than or equal to that of the conjugate depths.

BASIC HYDRAULIC EQUATIONS

The schematic diagram shown in Figure 1 illustrates the system considered which consists of a spillway and a horizontal, hydraulic jump type stilling basin. The principles of conservation of energy and momentum are used to compute the conjugate depth for a given discharge. The discharge over a spillway can either be uncontrolled or controlled by gates. Assuming that the head losses between section 0 (the crest of the spillway) and section 1 (the toe of the spillway) is negligible. By applying Bernoulli's equation, it results in

$$H_{o,i} = Z + y_{1,i} + \frac{V_{1,i}^2}{2g} \tag{1}$$

in which $H_{o,i}$ is the total head at section 0, Z is the elevation of stilling basin, $y_{1,i}$ and $V_{1,i}$ are the depth and velocity of water before the jump, respectively, for a given discharge, Q_i, and g is the gravitational constant.

For an uncontrolled spillway the total head at the spillway crest for a given discharge Q_i is expressed as

$$H_{o,i} = E_o + 1.5y_{c,i} \tag{2}$$

where E_o is the elevation of spillway crest, $y_{c,i}$ is the critical depth at the spillway crest which can be calculated as $y_{c,i} = Q_i^{0.667}L_c^{0.667}g^{-0.333}$, and L_c is the length of the spillway crest. For a controlled spillway, the discharge rating curve which specifies the head-discharge relationship can usually be obtained. The velocity before the jump, $V_{1,i}$, can be computed as

$$V_{1,i} = \frac{Q_i}{Wy_{1,i}} \tag{3}$$

where W is the width of the stilling basin. Combining Equations (1) and (3) results in

$$y_{1,i} + Ay_{1,i}^{-2} = h \tag{4}$$

or can be approximated by

$$y_{1,i} = q_i/\sqrt{2gh} \tag{5}$$

where $A = q_i^2/(2g)$, $h = H_{o,i} - Z$, and $q_i = Q_i/W$.

Because hydraulic jumps involve a high rate of internal energy dissipation by turbulence into heat, only the momentum equation can be applied from sections 1 and 2 of Figure 1. The boundary shear stress and the effect of water weight in the jump are neglected since the jump occurs in a relatively short distance and the basin floor is horizontal. For a flowrate of Q_i, the steady state momentum equation can be expressed as

$$Q_i^2/(gy_{1,i}W) + 0.5y_{1,i}(y_{1,i}W) = Q_i^2/(gy_{2,i}W) + 0.5y_{2,i}(y_{2,i}W) \tag{6}$$

in which $y_{2,i}$ is the conjugate depth of water after jump. Since the steady state continuity equation can be expressed as

$$Q_i = V_{1,i}(y_{1,i}W) = V_{2,i}(y_{2,i}W) \tag{7}$$

it can be combined with Equation (6) to obtain

$$(y_{2,i}/y_{1,i})^2 + (y_{2,i}/y_{1,i}) - 2F_{1,i}^2 = 0 \tag{8}$$

where $F_{1,i}$ is the Froude number $F_{1,i} = V_{1,i}/\sqrt{gy_{1,i}}$ before hydraulic jump. The solution to the above quadratic equation is

$$y_{2,i} = \frac{y_{1,i}}{2}(\sqrt{1 + 8F_{1,i}^2} - 1) \tag{9}$$

from which the conjugate depth, $y_{2,i}$, can be computed.

Performance of a hydraulic jump can be measured by the efficiency, e_i, defined as [6]

$$e_i = \frac{\Delta E_i}{E_{1,i}} = \frac{(\sqrt{8F_{1,i}^2 + 1} - 3)^3}{8(\sqrt{8F_{1,i}^2 + 1} - 1)(2 + F_{1,i}^2)} \tag{10}$$

where $\Delta E_i = E_{1,i} - E_{2,i}$ is the difference between the

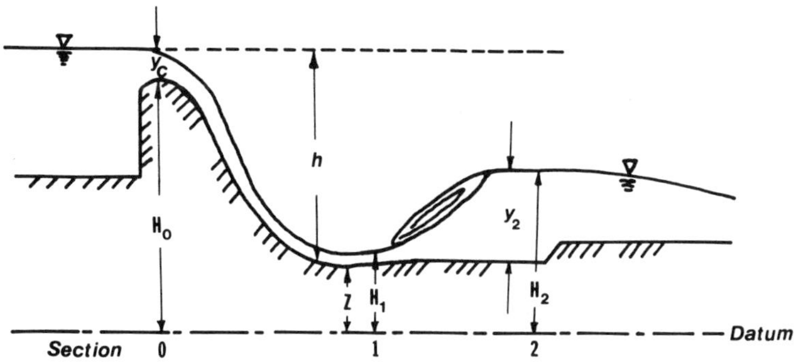

FIGURE 1. Schematic diagram of spillway-stilling basin system.

specific energy before the jump, $E_{1,i} = y_{1,i} + V_{1,i}^2/2g$, and the specific energy after the jump, $E_{2,i}$.

The two physical laws, i.e., the conservation of energy and momentum, presented in this section essentially form the basic constraints to be satisfied in the optimization model for determining the optimal width and elevation of the stilling basin described in the following sections. The performance of hydraulic jump in the stilling basin is also considered so that the scouring of the downstream river bed is prevented.

OPTIMIZATION MODEL

Objective Function

The criterion used in selecting the optimal values of the decision variables, i.e., width and elevation of the stilling basin, is the minimization of the total construction cost of the stilling basin subject to the constraints that the performance of the stilling basin must be satisfied for a range of possible discharges. The cost components for constructing a stilling basin consist of the cost for the excavation and the cost of construction of the apron slab and appurtenances. The recommended dimensions for stilling basin appurtenances, shown in Figure 2, are for high spillways and large outlet works in which $F_1 > 4.5$ and the entering velocity to the basin exceeds 50 ft/sec (15 m/sec). The dimensions shown in Figure 3 are for low and medium high spillways in which $F_1 > 4.5$ and velocity is less than 50 ft/sec (15 m/sec) [5]. The dimensions of the appurtenances are determined based on the design discharge used.

For a given stilling basin width, W, and elevation, Z, the cost of excavation C_e can be expressed as

$$C_e = U_e(E_r - Z + t_s)\, L_b\, W/27 \tag{11}$$

where U_e is the unit cost of excavation \$/yd³; E_r is the river bed elevation in feet; L_b is the length of stilling basin in feet defined in Figures 2 and 3; and t_s is the thickness of the

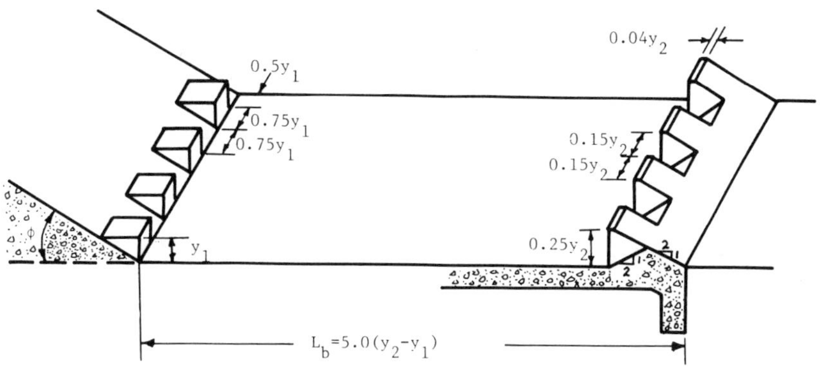

FIGURE 2. Recommended dimensions for stilling basin appurtenances for high spillways and large outlet works (Elevatorski, 1959).

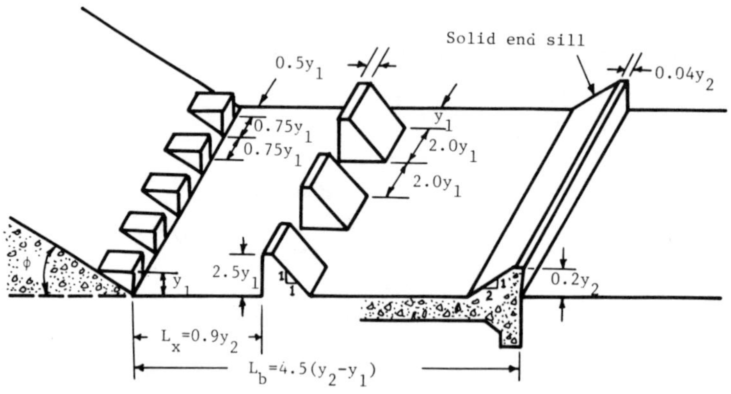

FIGURE 3. Recommended dimensions for stilling basin appurtenances for low and medium-high spillways (Elevatorski, 1959).

basin slab in feet. The cost of material C_c, required for constructing the stilling basin with appurtenances, for high head spillway, shown in Figure 2, is given as

$$C_c = U_c(V_{cb} + V_{ds} + V_{sb})/27 \quad (12)$$

in which U_c is the unit cost of concrete with rebar, in \$/yd^3 of concrete volume; V_{cb} is the volume of concrete for chute blocks, in ft^3; V_{ds} is the volume of concrete for dentrated sills, in ft^3; and V_{sb} is the volume of concrete for basin slab, in ft^3. The cost of construction for the stilling basins of low and medium-head spillways, as shown in Figure 3, is given as

$$C_c = U_c(V_{cb} + V_{bp} + V_{es} + V_{sb})/27 \quad (13)$$

in which V_{bp} is the volume of concrete for baffle piers, in ft^3; and V_{es} is the volume of concrete for the end sill. The volume of concrete for these basin components of a given design can be calculated from the geometry shown in Figures 2 and 3. The cost for the stilling basin wall can be estimated as

$$C_w = 2U_cL_bH_bt_w/27 \quad (14)$$

in which t_w is the wall thickness in feet; and H_b is the height of the basin wall which is the sum of the conjugate depth for the design flow and the freeboard required. The value of freeboard depends on the flow rate [5,7]. For the purpose of demonstrating the methodology, the values given by Elevatorski [5] are used in this chapter. The objective function is to minimize the total construction cost:

$$\min f_2 = C_e + C_c + C_w \quad (15)$$

Model Constraints

Primarily the model presented herein has three sets of constraints. The concepts related to these constraints have been explained in the previous section so only a brief description of the constraints are given here.

The first set of constraints requires that the energy relation (Equation (1)) between the spillway crest and the supercriticial flow before the jump on Section 1 in Figure 1 is satisfied. The value of $y_{1,i}$ for a given Z and W can be computed for various discharges, Q_i, using Equations (4) or (5). This set of constraints is comprised of $i = 1, \ldots, N$ non-linear equations where N is total number of discharges considered.

The second set of constraints, comprising of $i = 1, \ldots, N$ non-linear equations, governs the momentum relation between the supercritical flow before the jump and the subcritical flow after the jump, i.e., Equation (6), from which the computation of $y_{2,i}$ can be made for given Z, W, and Q_i using Equation (9). Referring to Figure 1, the conjugate depth after the jump, with respect to the datum, is given by the expression $(Z + y_{2,i})$ for discharge Q_i.

The third set of constraints considers the performance of the stilling basin over a wide range of flow discharges. Laboratory tests have shown that a satisfactory hydraulic jump will occur if the tailwater depth is made equal to 0.9 $y_{2,i}$ for low and medium high spillways [5]. In the model this criterion is made a little conservative whereas a satisfactory performance of a stilling basin for this case is assumed to be achieved if

$$TW_i - (Z + y_{2,i}) \geq -0.1\, y_{2,i} \quad (16)$$

in which TW_i is the tailwater elevation corresponding to a discharge Q_i. For high spillways and large outlet works, the criterion for satisfactory basin performance is assumed as [5]

$$TW_i - (Z + y_{2,i}) \geq 0.05\, y_{2,i} \tag{17}$$

which is the same as those recommended by the USBR in 1973 for type II basins.

An additional constraint is for the width of the stilling basin specifying that the basin width cannot be narrower than the length of the spillway crest, i.e.,

$$W \geq L_c \tag{18}$$

This constraint is used in the examples described in the later section. However, this constraint could be too restrictive in some applications. Normally, it may be preferable for the stilling basin width of an overpour dam spillway to equal the spillway crest length. However, there may be some conditions where it is expedient to make the basin narrower for economic reasons, such as narrow V-shaped canyon, or economics demand a wide spillway crest length.

Other criteria for basin performance, such as recommended by the USBR [1], could also be incorporated in the model. However, the purpose of this chapter is to develop and demonstrate the analytical optimization model; the performance criteria to be used will be those described above. It should be cautioned that the results derived from this model should not be directly used as the basis for the field construction of a prototype stilling basin. Rather, it serves as a guideline for selecting the prototype size and dimensions of stilling basins to be further tested in the laboratory.

In summary, the model has an objective function to be minimized and is subject to a total of $2N$ equality constraints plus $N + 1$ inequality constraints with only 2 decision variables, Z and W. Since there are only two decision variables in the model and the functional form of the gradients for the objective function and the constraints are not easily obtainable, a search technique which does not require expressions for the gradient is used. An optimization procedure for this problem is described in the next section.

OPTIMIZATION METHODOLOGY

Phase I Optimization

This phase is used to find a solution (W and Z) such that the weighted root mean square difference between the conjugate depth elevations and the tailwater elevations for the N discharges is minimized. In other words, the purpose of this phase is to locate the boundary of the feasible space of the problem. The objective function for this phase can be expressed as

$$\min f_1 = \frac{\sum_{i=1}^{N} (P \cdot \epsilon_i)^2}{N} \tag{19}$$

in which $\epsilon_i = (y_{2,i} + Z) - TW_i$; and P is the penalty assigned when the constraint for describing the performance of the stilling basin, Equations (16) or (17), is not satisfied. The penalty, P, reflects the tolerance of a design to the violation of performance criteria of the basin, and also is used as a means to treat the constraint, Equations (16) or (17), in the adopted search technique. Heuristically, f_1 is minimized if Equations (16) or (17) are satisfied for all Q_i's, i.e., no penalty is imposed.

An easily programmed accelerated climbing technique called pattern search, developed by Hooke and Jeeves [7], is applied. The technique is based on the philosophy that any set of moves (changes in value of the decision variable) which have been successful on early explorations will be worth trying again. The method starts cautiously with short changes or step sizes in the value of the decision variables from a starting point and the step sizes grow with repeated success. Subsequent failure indicates that shorter steps are in order. If a change in direction is required, the technique will start over again with a new pattern. In the vicinity of the optimum the step sizes become very small to avoid overlooking any promising directions. It should be noted that the search procedure of this nature does not guarantee a global optimality being sought. Therefore, several different starting points and step sizes are recommended for a comparison of the final results.

Experience using this optimization phase showed that the final search converges at different solutions when different starting points and step sizes are used. However, it is noted that the feasible space for this problem can be sketched schematically as shown in Figure 4. For a given head (H_o), discharge (Q), and stilling basin width (W), the basin elevation which results in the conjugate depth elevation match-

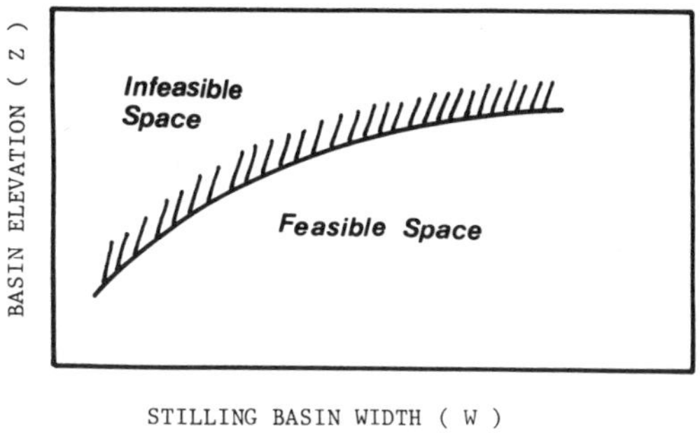

FIGURE 4. Schematic sketch of feasible space in stilling basin optimization problem.

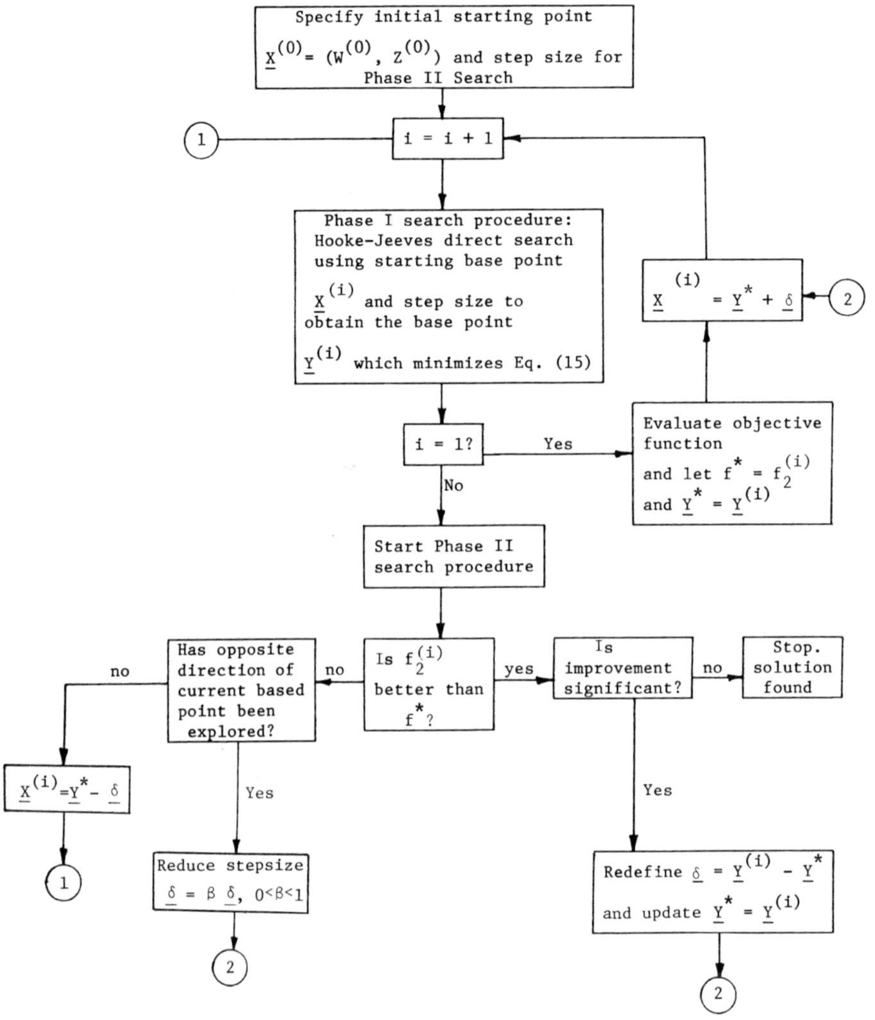

FIGURE 5. Flow chart of optimization procedures.

ing the tailwater elevation is denoted as Z^* (H_o, Q, W). It is shown in the Appendix that the gradient of the conjugate depth elevation, with respect to basin elevation, dH_2/dZ is positive and asymptotic to one for Froude numbers exceeding 1.14, i.e., $dH_2/dZ > 0$, for $F_1 > 1.14$. Thus at Z^* (H_o, Q, W), $dH_2/dZ^* > 0$, then for $dZ > 0$, $dH_2 > 0$ and for $dZ < 0$, $dH_2 < 0$. This means that at $Z = Z^*$ further lowering of the basin elevation Z results in a lowering of the corresponding conjugate depth elevation, H_2. Once H_2 is less than the tailwater elevation, TW (Q), the jump is submerged. Therefore, for a given H_o, Q, and W, the basin elevations which are less than or equal to Z^* are feasible. On the contrary, if the basin elevation is raised above Z^*, i.e., $Z > Z^*$, then the corresponding conjugate depth elevation exceeds the tailwater elevation which could cause downstream scouring such that an unsatisfactory hydraulic jump occurs and the elevation becomes infeasible. This reasoning is also valid when a range of discharges is considered and the optimal basin elevation in this case, minimizes the mean square difference between the conjugate depth elevation and tailwater elevation. Using different values of basin width, it is possible to depict a boundary as shown in Figure 4 under which the solution space is feasible.

Phase II Optimization

From the proceeding analysis a special characteristic of the feasible space boundary can be observed. For a fixed value of basin width the optimal basin elevation falls on the boundary of feasible space both for the best hydraulic performance, i.e., f_1 is minimized, and best economic design, i.e., minimum value of f_2, because further lowering of the basin elevation will cause a submerged jump and also increase the excavation cost. This special characteristic of the boundary of the feasible space is then utilized in this phase to search for the optimal dimension, i.e., width and elevation, of the stilling basin such that the total construction cost is minimized and satisfies the constraints imposed on the problem. The determination of the type of appurtenances to be used in the basin, i.e., either Figures 2 or 3, for computing the costs of material depends on the entering velocity of flow based on the current basin width and elevation, and design discharge.

From the study of the phase I optimization it is found that the stilling basin elevation on the boundary of feasible space for a given width of basin is optimal for both objective functions, i.e., Equation (15) and Equation (19). This phase of the optimization procedure will then utilize this characteristic and search for the optimal basin width and elevation, with objective function Equation (15), along the feasible space boundary shown in Figure 4. The optimization procedure used in this phase has the same basic philosophy as the Hooke-Jeeves technique which starts with an initial base point obtained from the phase I search and initial step size. The step size for phase II search procedure is expanded

TABLE 1. Physical Characteristics of Spillways and Stilling Basins.

Characteristics	Rockwall	Conroe	Twinbutts
Net Length of Spillway Crest, in feet	560.0	200	200
Elevation of Spillway Crest*	409.5	173.0	1969.1
Elevation of Stream Bed*	391.0	131.0	1864.0
Width of Stilling Basin, in feet	664.0	231.0	300.0
Elevation of Stilling Basin	367.0	124.0	1849.0
Thickness of Basin Slab, in feet	3.0	5.0	2.0
Range of Discharge, in 1000 cubic feet per second	40–400	10–110	5–45
Design Discharge, in 1000 cubic feet per second	400	110	45
N (No. of discharges considered)	10	11	9

*Feet above mean sea level.
Note: 1 ft = 0.305 m; 1 cfs = 0.028 m^3/s.

TABLE 2. Results for Rockwall Dam Data.

Characteristics	No Optimization	Minimum f_1 Optimal Design	Minimum f_2 Optimal Design
(W,Z), in ft	Existing design: (664, 367)	(4,238, 388.0)	(568.8, 353)
f_1 (P = 100)	321.266	0.970	24.267
f_1 (P = 1)	6.124	0.970	12.562
f_2 (× 10^6 \$)	8.041	16.66	8.325
$\bar{e}$	0.518	0.726	0.548

Note: 1 ft = 0.305 m.

TABLE 3. Results for Conroe Dam Data.

Characteristics	No Optimization	Minimum f_1 Optimal Design	Minimum f_2 Optimal Design
(W,Z), in ft	Existing design: (231, 124)	(483.7, 139.0)	(230.6, 124)
f_1 (P = 100)	8.326	2.889	7.881
f_1 (P = 1)	8.326	2.889	7.881
f_2 (× 10^6 \$)	2.728	3.213	2.728
$\bar{e}$	0.581	0.643	0.585

Note: 1 ft = 0.305 m.

TABLE 4. Results for Twinbutts Dam Data.

Characteristics	No Optimization	Minimum f_1 Optimal Design	Minimum f_2 Optimal Design
(W,Z), in ft	Existing design: (300, 1,849)	(619.7, 1,860.5)	(231.3, 1846)
f_1 (P = 100)	5.378	0.871	5.703
f_1 (P = 1)	5.378	0.871	5.703
f_2 (× 10 \$)	2.280	3.005	2.113
$\bar{e}$	0.844	0.886	0.827

Note: 1 ft = 0.305 m.

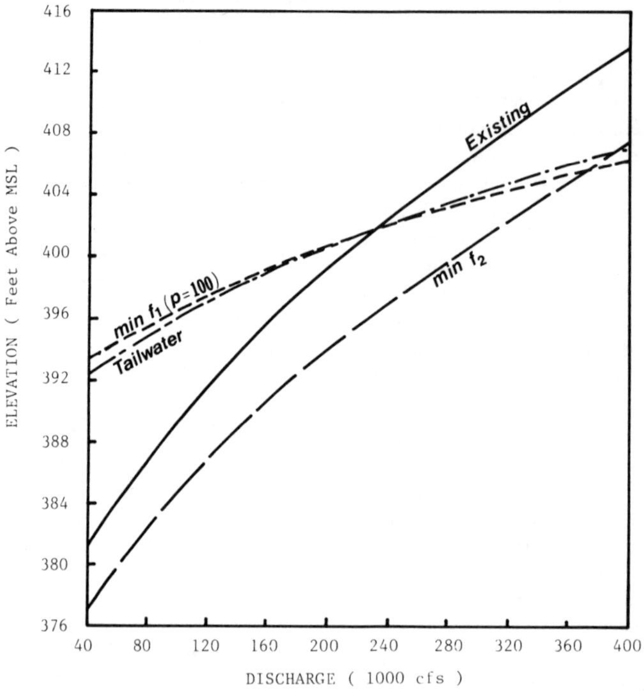

FIGURE 6. Relationship of tailwater rating curves and conjugate depth curves for Rockwall Dam with different designs.

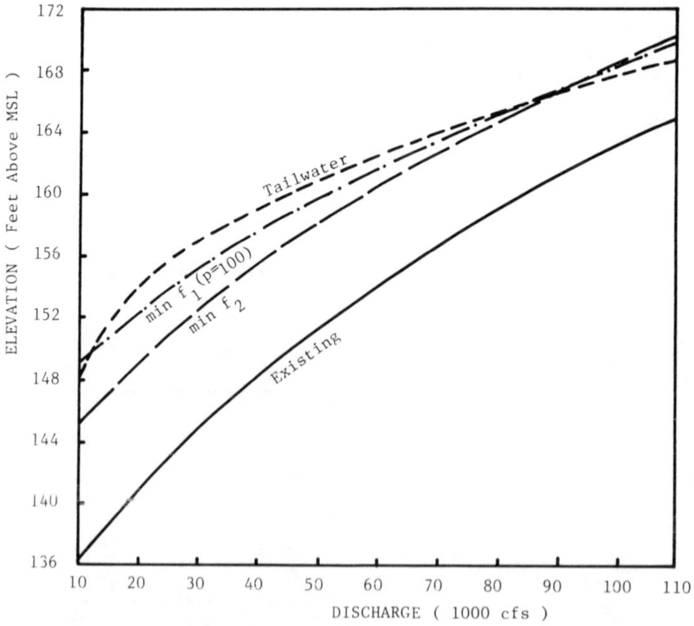

FIGURE 7. Relationship of tailwater rating curve and conjugate depth curves for Conroe Dam with different designs.

when the objective function is improved or is reduced when the objective function value cannot be improved in both directions of the current base point. The procedure stops when either the total number of iterations or the step size reduction has been exceeded, or the improvement of the objective function is insignificant. A flowchart of the optimization procedures is shown in Figure 5.

APPLICATION

Applications of the optimization procedures are made to the stilling basins of three existing dams in Texas to determine the optimal dimensions of the stilling basins and the appurtenances that minimize the cost. Comparisons of hydraulic performance and the construction cost of the stilling basins are made for the existing and optimal designs using the two different objective functions, i.e., Equations (15) and (19). The stilling basins for Rockwall-Forney Dam,

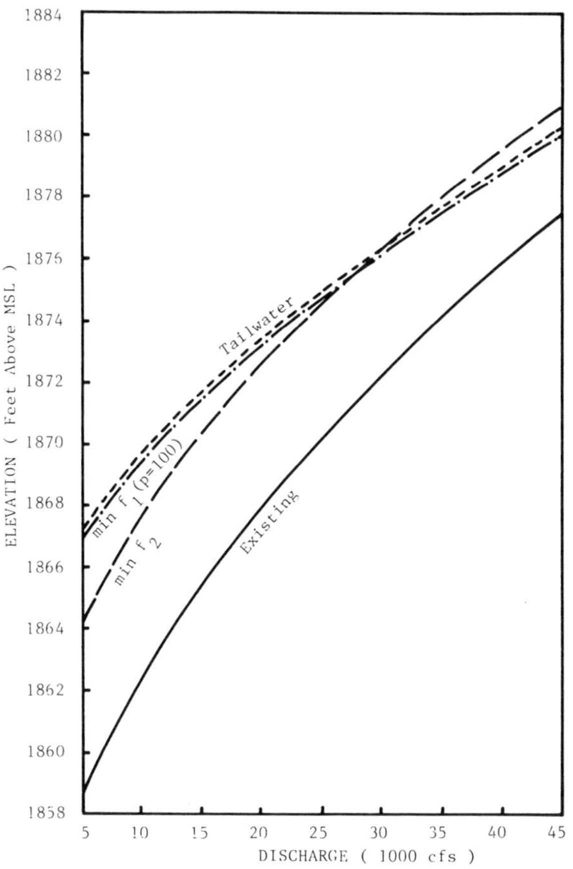

FIGURE 8. Relationship of tailwater rating curve and conjugate depth curves for Twinbutts Dam with different designs.

TABLE 5. Appurtenances for Optimal Designs (Min f_2).

Basin Layout	Name of Dam		
	Rockwall	Conroe	Twinbutts
Spillway classification	High Head	High Head	High Head
Basin length, in ft.	215.1	173.4	151.9
Basin width, in ft.	568.8	230.6	231.3
Height of basin wall, in ft.	69.4	54.9	40.4
Number of chute blocks	32	18	58
Width of chute blocks and spacings, in ft.	8.91	6.38	2.01
Height of chute blocks in ft.	9.03	6.50	2.01
Number of sill dentates	37	19	24
Width of sill dentates and spacings, in ft.	7.84	6.22	4.83
Height of sill dentates, in ft.	13.01	10.29	8.10
Top width of sill dentates, in ft.	2.08	1.65	1.30

Note: 1 ft. = 0.305 m.

Conroe Dam, and Twinbutts Dam are used in the applications. These dams all have concrete ogee spillways with control devices on the crests. The physical characteristics of the spillways and the existing stilling basins are listed in Table 1. Unit prices for concrete (with reinforced steel bars) and excavation of \$400.00/yd^3 and \$2/yd^3, respectively, were based upon contractor bid prices submitted to the Texas Department of Water Resources (TDWR), in 1979, for construction of a stilling basin located in Texas.

The optimization procedures with each objective, f_1 and f_2, were used to compute the optimal dimensions (W, Z) of stilling basins for the three dams. Tables 2, 3 and 4 list results for the existing stilling basin design, the designs using f_1 as the objective function in Equation (19), and the design using f_2 as the objective function in Equation (15). Note that, in Table 2 for the Rockwall Dam, the stilling basin width could be as wide as 4,238 ft. (1,293 m) using the object f_1 which is ridiculously large, thus denoting the inadequacy of the objective function f_1. The values of f_1 with $P = 100$, and $P = 1$, f_2 and average jump efficiency, $\bar{e} = \Sigma_{i=1}^{n} e_i/N$, are listed in Tables 2–4 for the existing designs, and the designs using Min f_1 and Min f_2. The construction costs of stilling basins for the existing designs were computed, and figures were based on the dimensions of basin length and appurtenances recommended in Elevatorski [5] as shown in Figures 2 and 3. The existing layout of all three dams uses baffle piers in the stilling basin. For the existing designs the entering velocity for the design discharge exceeds 50 fps (15 mps) indicating possible cavitation of baffle piers. The relationship of the tailwater rating curve and the conjugate depth curves for different basin designs are shown in Figures 6, 7 and 8 for Rockwall, Conroe and Twinbutts data, respectively.

Obviously, the existing stilling basins for Conroe and Twinbutts Dams have rather conservative designs to prevent potential scouring in the downstream river bed, while for the Rockwall Dam the hydraulic jump will sweep out of the basin when discharge exceeds 230,000 cfs (6,440 cms). From a numerical point of view, it is always possible to obtain a solution to have a very close fit for the tailwater and conjugate depth curves. Column 2 in Tables 2–4 shows the result of minimizing the difference between the tailwater rating curve and the conjugate depth curve without considering any economic feasibility. The resulting widths of the stilling basins range from two to six times as large as the length of the spillway crest which may not be economically and physically justifiable. Because the flow condition at the bottom of the spillway is supercritical, the water cannot possibly spread over the entire width of the stilling basin. This is just to demonstrate the inadequacy of an optimization model considering only the fitting of curves. The recommended dimensions for the basin and the necessary appurtenances for each dam are listed in Table 5. It is also observed that the existing design for Conroe Dam is close to that being obtained using the least cost criterion.

SUMMARY AND CONCLUSION

An optimization model has been developed for determining the least cost stilling basin design. The optimization procedure is divided into two search phases: the phase I optimization considers the hydraulic performance to define various feasible combinations of W, Z, and the phase II optimization determines the basin width and elevation and the corresponding dimension of the basin appurtenances associated with the minimum cost design. Applications of the optimization model were made to existing stilling basin designs of three dams located in Texas. The final design of a stilling basin in many cases requires model studies to examine its hydraulic performance. The optimization model developed in this chapter can be a useful tool in a preliminary study to provide an initial stilling basin design before the physical laboratory model is built and tests are performed.

REFERENCES

1. U.S. Bureau of Reclamation, *Design of Small Dams*, Water Resources Technical Publication (1973).
2. Bhowmik, N. G., *J. of Hydraul. Div.*, ASCE, *101*:HY7 (1975).
3. Blaisdell, F. W., *Transaction*, ASCE, *113* (1948).
4. Bradley, J. N. and A. J. Peterka, *J. of Hydraul. Div.*, ASCE, *83*:HY5 (1957).
5. Elevatorski, E. A., *Hydraulic Energy Dissipators*, McGraw-Hill, New York (1959).
6. Morris, H. M. and J. M. Wiggert, *Applied Hydraulics in Engineering*, John Wiley & Sons (1972).
7. Hooke, R. and Jeeves, T. A., *J. of Computer*, *8*:2 (1961).

APPENDIX

Relationship Between Jump Height Elevation and Apron Elevation for Horizontal Stilling Basins

This appendix is for the purpose of mathematically examining the relationship between jump height elevation and apron elevation for horizontal stilling basins of given width and discharge. The jump height elevation, H_2, is the sum of the apron elevation, Z, and conjugate depth, y_2. From Equation (9) the jump height elevation can be expressed as

$$H_2 = Z + \frac{y_1}{2}(\sqrt{1 + 8F_1^2} - 1) \tag{20}$$

$$\approx Z + \frac{y_1}{2}(2\sqrt{2}F_1 - 1) \tag{21}$$

Equation (21) is used for simplicity with only very small errors for $F_1 > 1$, e.g., with $F_1 = 2$, the error is approximately 1.5 percent. The velocity of flow at the base of the spillway for a given head difference, h, between section 0 and 1 in Figure 1, can be computed as

$$V_1 = \sqrt{2gh} \tag{22}$$

and the depth of flow before jump, y_1, corresponding to a given flow discharge and basin width, can be expressed as

$$y_1 = q/\sqrt{2gh} \tag{23}$$

in which q is the discharge per unit width. The Froude number of flow before the jump can be expressed in terms of q and h as

$$F_1 = \frac{2^{3/4}\, g^{1/4}\, h^{3/4}}{q^{1/2}} \tag{24}$$

Substituting Equations (23) and (24) into Equation (21), the following expression for H_2 is obtained:

$$H_2 = Z + \frac{2^{3/4} q^{1/2} h^{1/4}}{g^{1/4}} - \frac{q h^{-1/2}}{2^{3/2} g^{1/2}} \tag{25}$$

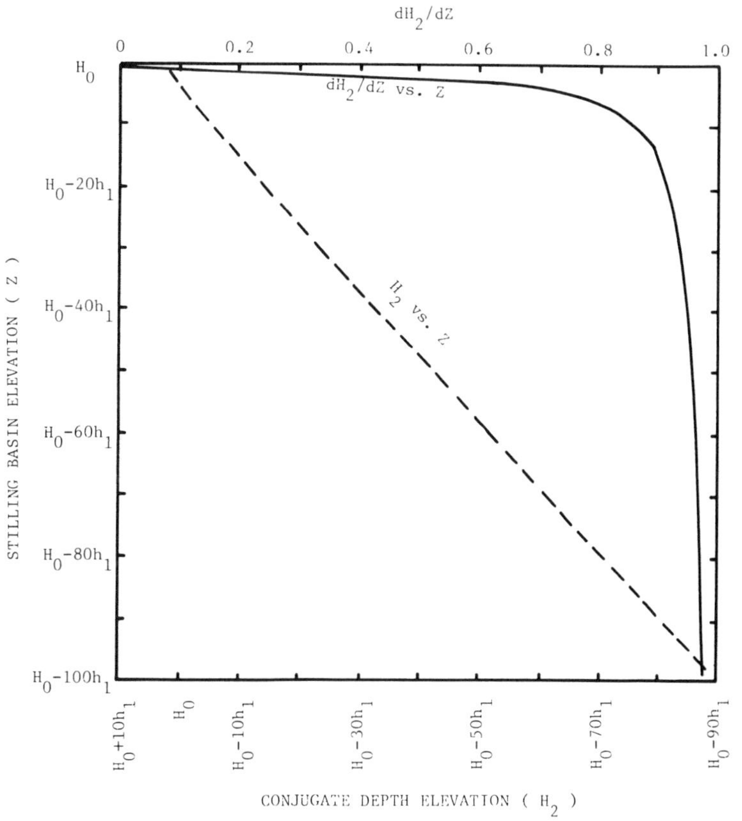

FIGURE 9. Functional relationship between H_2 vs. Z and dH_2/dZ vs. Z.

The change in jump height elevation with respect to apron elevation can be obtained by taking the derivative of H_2 with respect to Z in Equation (25) as

$$\frac{dH_2}{dZ} = 1 + \left(\frac{q^{1/2}\, h^{-3/4}}{2^{5/4}\, g^{1/4}} + \frac{qh^{-3/2}}{2^{5/2}\, g^{1/2}}\right)\frac{dh}{dZ} \qquad (26)$$

where $dh/dZ = -1$ because $Z = H_o - h$ in which H_o is the total head, a constant. Define

$$F_1^2 = \frac{(2g\, h_n)^{3/2}}{g\, q} = n \qquad (27)$$

in which h_n is the head drop corresponding to $F_1^2 = n$ for a given discharge and basin width. From Equation (27) the head drop required to produce a corresponding Froude number ($\sqrt{n}$) at the base of the spillway can be expressed as

$$h = h_n = \frac{(n\, q)^{2/3}}{2g^{1/3}} \qquad (28)$$

Substituting Equation (28) into Equation (26), the following expression for dH_2/dZ is obtained:

$$\frac{dH_2}{dZ} = 1 - \frac{1}{\sqrt{2n}} - \frac{1}{2n} \qquad (29)$$

By further setting Equation (29) equal to zero and solving for n, the value of F_1^2 at which $dH_2/dZ = 0$ is 1.309. From this derivation we can conclude that $dH_2/dZ > 0$ for $F_1 > 1.144$ and approaches to one as the value of F_1 increases. In the design of a stilling basin for spillways, the values of F_1 are usually greater than 1.144.

The expression for H_2 can also be obtained by substituting Equation (28) into Equation (25) as

$$H_2 = H_o - (n^{2/3} - 2^{3/2}\, n^{1/6} + n^{-1/3})h_1 \qquad (30)$$

in which $h_1 = q^{2/3}/2g^{1/3}$. Figure 9 depicts the relationship for H_2 versus Z and dH_2/dZ versus Z.

River Ice Processes

Poojitha D. Yapa* and Hung Tao Shen*

INTRODUCTION

Surface water bodies in a substantial portion of the populated regions of the world freeze either totally or partially in the winter. The presence of ice can seriously interfere with the utilization of lakes and rivers for power generation, navigation, water supply, etc. Ice and its effects are important considerations in the design, operation, and maintenance of hydraulic engineering facilities. In rivers ice can cause serious problems by the damage it does to hydraulic structures, by the flooding that often accompanies ice jams, by interference with hydropower operations, and by impeding inland navigations. A few treatises on ice engineering exist, including those of Barnes [12], Michel [56,57], Pivovarov [74], Shulyakovskii [91], and Ashton [11]. Several review papers and reports on river ice have been published [6,8,9,95,105]. This chapter will review aspects of ice processes in relation to river hydraulics.

HEAT EXCHANGE PROCESSES

A river's thermal-ice condition is affected by heat exchanges through interfaces among the atmosphere, ice cover, river water, and channel bed. A brief summary of the various heat-exchange processes follows.

Energy Exchanges at Air-Ice or Air-Water Interfaces

Heat exchanges with the atmosphere vary with meteorological conditions [65,88]. The surface heat exchange process consists of five major components: (1) Net solar (shortwave) radiation, ϕ_s; (2) longwave radiation, ϕ_b; (3) evapo-condensation, ϕ_e; (4) sensible heat exchange, ϕ_c; and (5) precipitation, ϕ_r.

Shortwave Radiation. The incoming shortwave radiation at the air-ice or air-water interface can be written as

$$\phi_{ri} = [a - b(\Phi - 50)](1 - 0.0065C_c^2) \qquad (1)$$

in which ϕ_{ri} = incoming shortwave radiation, in cal cm^{-2} day^{-1}; Φ = latitude in degrees; C_c = cloud cover in tenths; and a, b = constants which represent variations of solar radiation under a clear sky. Average monthly values for a and b are given in Table 1. Since part of the solar radiation reaching the water surface is reflected back into the atmosphere, the net solar radiation, ϕ_s, in cal cm^{-2} day^{-1}; can be written as

$$\phi_s = (1 - \alpha)\phi_{ri} \qquad (2)$$

in which α = a coefficient approximately equal to 0.1 for the water's surface. Dingman and Assur [38] suggested the following formula for calculating ϕ_s:

$$\phi_s = \phi_{ri} - (0.108\phi_{ri} - 6.766 \times 10^{-5}\phi_{ri}^2) \qquad (3)$$

For the ice surface, the value of the albedo, α, in Equation (2) is dependent on the material behavior of the ice cover. Based on empirical curves developed by Krutskih, *et al.* [52] for snow-free sea ice in the Arctic, Wake and Rumer [106] proposed the following expressions:

$$\begin{aligned} \alpha &= \alpha_i; && \text{for } T_a \leq 0°\text{C} \\ \alpha &= \alpha_a + (\alpha_i - \alpha_a)e^{-\psi T_a}; && \text{for } T_a > 0°\text{C} \end{aligned} \qquad (4)$$

in which α_i, α_a, and ψ = empirical constants; and T_a = air temperature in degrees Celsius. Bolsenga [21] reported

*Department of Civil and Environmental Engineering, Clarkson University, Potsdam, NY

TABLE 1. Values of a and b in Equation (1).

Month	a, in cal − cm^{-2} − day^{-1}	b
Dec.	100	8.2
Jan.	142	11.0
Feb.	228	11.2
Mar.	394	12.7
Apr.	554	8.4

values of albedo for Great Lakes ice covers, and observed that the albedo varies with surface temperatures. For uneven surfaces, the albedo also varies with solar latitude. Table 2 gives albedo values reported by Bolsenga for various ice conditions.

The penetration of shortwave radiation into the ice cover can be considered as an internal heat source. The vertical distribution of the intensity of monochromatic shortwave radiation can be described by the Bouguer–Lambert exponential law [43,90].

$$\phi_p = \phi_s e^{-\tau_i z} \tag{5}$$

in which ϕ_p = intensity of the shortwave radiation at depth z; τ_i = a bulk extinction coefficient which varies between 0.004 cm^{-1} and 0.07 cm^{-1}. The extinction coefficient is strongly wavelength dependent so that the solar heating in ice cannot be represented by a simple exponential law with a constant bulk extinction coefficient [43,10]. However, since ice thickness computations usually do not consider the vertical distribution of solar heating in ice, a simple exponential law with a constant bulk extinction coefficient may be used.

Based on Equation (5), the amount of shortwave radiation that penetrates into the water underneath an ice cover is

$$\phi_{sp} = \beta_i \phi_s e^{-\tau_i \theta} \tag{6}$$

in which β_i = a fraction of absorbed solar radiation that penetrates through the ice-water interface. Due to the small difference in the refractive indices of ice and water, the value of β_i can be taken to be 1.0 [69].

TABLE 2. Albedo of Great Lakes Ice [21].

Ice Type	Albedo, as a Percentage
Clear lake ice (snow free)	10
Bubbly lake ice (snow free)	22
Ball ice (snow free)	24
Refrozen pancake (snow free)	31
Slush curd (snow free)	32
Slush ice (snow free)	41
Brash ice (snow between blocks)	41
Snow ice (snow free)	46

Longwave Radiation. The longwave radiation is the combination of the longwave radiation emitted from the water surface or the ice cover, ϕ_{bs}, and the net atmospheric thermal radiation absorbed by the water body, ϕ_{bn}. Based on the Stefan–Boltzman law of radiation, modified to account for the emissivity of the body's surface, ϕ_{bs} can be represented by

$$\phi_{bs} = \varepsilon \sigma T_{sk}^4 \tag{7}$$

in which σ = the Stefan–Boltzman constant, 1.171×10^{-7} cal cm^{-2} day^{-1} °K^{-4}; T_{sk} = water or ice surface temperature, in degrees Kelvin; and ε = emissivity of the surface, assumed to be 0.97 for both the water and the ice surfaces.

The atmospheric radiation under clear skies, ϕ_{bc}, can be estimated by considering the atmosphere as a gray body:

$$\phi_{bc} = \varepsilon_a \sigma T_{ak}^4 \tag{8}$$

in which T_{ak} = air temperature, in degrees Kelvin; and ε_a = emissivity of the atmosphere. The Brunt formula [65] gives

$$\varepsilon_a = c + d\sqrt{e_a} \tag{9}$$

in which e_a = vapor pressure of air at the temperature T_{ak}, in millibars; and c and d = empirical constants, approximately equal to 0.55 and 0.052, respectively.

Using Bolz's formula the atmospheric radiation under cloudy skies, ϕ_{ba}, can be represented by

$$\phi_{ba} = \sigma T_{ak}^4 (c + d\sqrt{e_a})(1 + k_c C_c^2) \tag{10}$$

in which k_c = empirical constant $\simeq$ 0.0017. Considering that the reflectivity of the ice or water surfaces is 0.03, the net atmospheric radiation is

$$\phi_{bn} = 0.97\phi_{ba} \tag{11}$$

in which ϕ_{bn} = net atmospheric radiation, in cal cm^{-2} day^{-1}. Combining Equations (7) and (11), the effective back radiation becomes

$$\phi_b = \phi_{bs} - \phi_{bn} = 1.1358 \times 10^{-7} \times [T_{sk}^4 - (1 + k_c C_c^2)(c + d\sqrt{e_a}) T_{ak}^4] \tag{12}$$

in which ϕ_b = effective back radiation, in cal cm^{-2} day^{-1}.

Evapo-Condensation. The heat flux from the water surface due to evapo-condensation, ϕ_e, can be estimated by using the Rimsha–Donchenko formula [65]

$$\phi_e = (1.56K_n + 6.08V_a)(e_s - e_a) \tag{13}$$

in which ϕ_e = rate of heat loss due to evaporation, in cal cm^{-2} day^{-1}; V_a = wind velocity at 2 m above the water surface in meters per second; e_s = saturated vapor pressure at temperature T_s; and K_n = a coefficient that accounts for the effect of free convection determined by

$$K_n = 8.0 + 0.35(T_s - T_a) \tag{14}$$

in which T_s = river surface temperature; and T_a = air temperature at 2 m above the water surface, in degrees Celsius. Since the existence of an ice cover on the river surface tends to suppress evaporation [88], the heat flux due to evapo-condensation from the ice surface will be a fraction of that calculated from Equation (13).

Conductive Heat Transfer. The energy conducted from the water surface as sensible heat by air can be determined by the Rimsha–Donchenko formula [65]

$$\phi_c = (K_n + 3.9V_a)(T_s - T_a) \tag{15}$$

in which ϕ_c = rate of conductive heat loss, in cal cm^{-2} day^{-1}. Similar to the evapo-condensation flux, the value of ϕ_c will be reduced when the river surface is covered by ice.

Heat Loss Due to Snow Fall. Heavy snow fall during the ice formation period can increase the amount of ice in the river and affect the ice cover formation significantly. The heat loss due to snow falling on the water surface can be estimated from:

$$\phi_r = A_{sn}[L_i + c_i(T_s - T_a)] \tag{16}$$

in which, A_{sn} = the mass rate of snow fall per unit area of water surface; L_i = the latent heat of fusion of ice, 80 cal g^{-1}; and c_i = specific heat of ice, 1.0 cal g^{-1} °C^{-1}. A_{sn} can be estimated from the visibility by:

$$A_{sn} = 7.85V_p^{-2.375} \tag{17}$$

in which, V_p = the visibility in km; and A_{sn} is in the unit of g cm^{-2} day^{-1}.

Turbulent Heat Transfer from Water to Ice Cover

The turbulent heat transfer from the flowing river water to the ice cover has significant effects on the thickness of an ice cover [7]. The heat flux from the water to the ice cover can be represented by

$$q_{wi} = h_{wi}(T_w - T_f) \tag{18}$$

in which q_{wi} = heat flux from the water to the ice cover; h_{wi} = heat transfer coefficient, in cal cm^{-2} day^{-1} °C^{-1}; T_w = water temperature, in degrees Celsius; and T_f = the freezing point of water, 0°C. By considering the ice-covered river flow as a smooth-walled closed-conduit turbulent flow, the turbulent heat flux can be formulated as [46]:

$$N_u = CR^{0.8}P_r^{0.4} \tag{19}$$

where, N_u = Nusselt Number, $h_{wi}R/k_w$; P_r = Prandtl Number, $\rho v_w C_p/k_w$ = 13.6 for water near 0°C; C_p = specific heat of water, 1.0 cal g^{-1} °C^{-1}; C = empirical coefficient approximately 0.023; R_e = Reynolds Number, UR/v_w; k_w = thermal conductivity of water, 0.552 W m^{-1} °C^{-1}; v_w = viscosity of water, 1.788 × 10^{-6} m^2 s^{-1}; and R = hydraulic radius. The heat transfer coefficient can be evaluated [7,8,9] by the following equation:

$$h_{wi} = C_{wi}\frac{U^{0.8}}{d_w^{0.2}} \tag{20}$$

in which U = flow velocity in meters per second; d_w = flow depth in meters; and C_{wi} = 1622 W sec$^{0.8}$ m$^{-2.6}$ °C^{-1}. Laboratory and field investigations indicate that the coefficient C_{wi} may vary with the resistance of the ice cover [46,24]. A refined form of Equation (20), based on the formulation of Petukhov and Popov [73], is

$$h_{wi} = \frac{2k_w}{d_w}\frac{(f_i/8)R_eP_r}{1.07 + 12.7\sqrt{f_i/8}(P_r^{2/3} - 1)} \tag{21}$$

in which, f_i = friction factor of the ice cover.

Bed Heat Influx

In ice-covered rivers, the heat exchange at the bed, q_{gw}, may become an important component of the river's heat budget [82]. O'Neil and Ashton [63] developed a procedure for analyzing heat transfer at the channel bottom by considering one-dimensional heat conduction in the channel bed. Expressing the normal water temperature during a year as a sinusoidal function, which intercepts 0°C at times t_1 and t_2:

$$T_w(t) = \begin{cases} \bar{T}_w + a\sin(2\pi t/T_y), & t \le t_0 \text{ and } t \ge t_2 \\ 0, & t_1 \le t \le t_2 \end{cases} \tag{22}$$

in which, T_y = the period, 1 year; and $\bar{T}_w$ and a = the mean and amplitude of the sinusoidal function, respectively. The nondimensional river bottom temperature gradient at the bed/water interface can be obtained as shown in Figure 1. The dimensionless variables in Figure 1 are defined as

$$(\theta, \xi, r) \equiv \left(\frac{T_b - \bar{T}_w}{a}, \frac{y_b}{\sqrt{KT_y}}, \frac{\bar{T}_w}{a}\right) \tag{24}$$

in which, y_b = depth of the soil from the bed/water interface, T_b = temperature in the bed; $K = k/\rho_b C_b$ = thermal diffusivity of soil; in which, k, ρ_b, and C_b = thermal conductivity, density and heat capacity of the bed material [50].

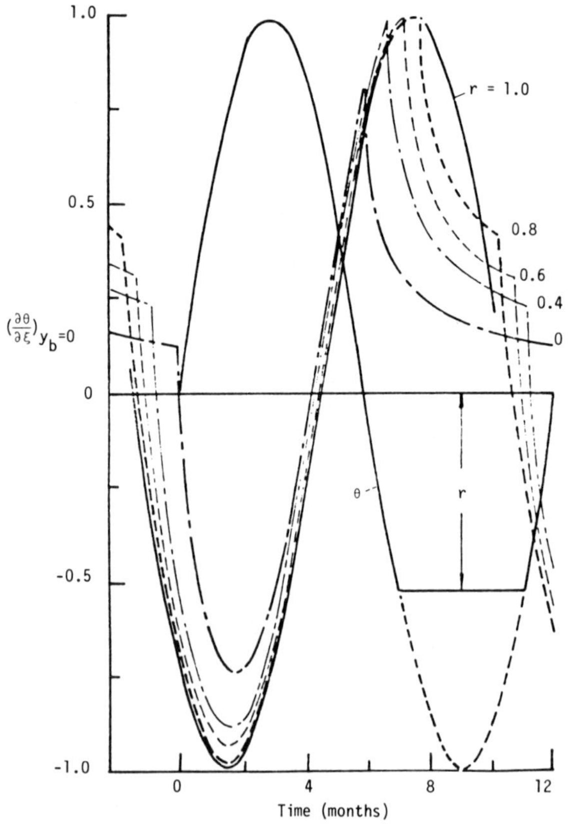

FIGURE 1. Nondimensional river bottom temperature gradient, evaluated at the bed/water interface, as a function of time and r. The nondimensional river bottom temperature is also shown [63].

The actual bed heat flux, $k\partial T_b/\partial y_b|_{y_b=0}$ can be determined from the value of $\partial\theta/\partial\zeta$ obtained from curves in Figure 1.

THERMAL-ICE REGIME

Thermal and ice conditions in rivers are strongly influenced by the ambient weather conditions through heat exchanges. Figure 2 shows a typical example of the normal air and water temperatures in a river. In this section the thermal and ice conditions of rivers will be discussed in relation to water temperature distribution, frazil ice formation, and the growth and decay of ice covers.

River Freeze-Up

Ice begins to form when the water temperature reaches 0°C and continues to lose heat to the atmosphere. Under the consideration of complete mixing over the channel cross section, the conservation of thermal energy in a river reach can be represented by a one-dimensional advection-diffusion equation. For water temperature above freezing, this equation can be written [22] as

$$\frac{\partial}{\partial t}(\rho C_p A T_w) + \frac{\partial}{\partial x}(Q\rho C_p T_w) = \frac{\partial}{\partial x}\left(AE_x\rho C_p\frac{\partial T_w}{\partial x}\right) + B\Sigma\phi \tag{24}$$

in which A = cross-sectional area of river; B = channel width; Q = river discharge; ρ = density of water; C_p = specific heat of water; E_x = longitudinal dispersion coefficient; x = distance along the river; and $\Sigma\phi$ = net heat influx per unit surface area of the river.

A few freeze-up forecast models exist [87] for predicting the time at which the water temperature at a river station drops to the freezing point. By assuming that changes in river discharge are not significant, and neglecting the longitudinal mixing term, Equation (24) can be reduced to a simplified form

$$\frac{\partial T_w}{\partial t} + V\frac{\partial T_w}{\partial x} = \frac{\Sigma\phi}{\rho C_p d_w} \tag{25}$$

in which, V = the average flow velocity, and d_w = the depth of flow. The heat exchange may be approximated by a simple linear function in the form of:

$$\Sigma\phi = h_{wa}(T_a - T_w) \tag{26}$$

where h_{wa} = the energy exchange coefficient. For the St. Lawrence River, h_{wa} is approximately equal to 46.4 cal cm^{-2} day^{-1} °C^{-1}. By representing the air temperature T_a as a combination of normal air temperature approximated by the first harmonic of the Fourier series with a period of one year and the short-term variations, a freeze-up forecasting scheme has been developed based on an analytical solution of Equations (25) and (26) [87]. The short-term variations in air temperature can be obtained from the weather forecast.

Frazil Ice

Frazil ice forms in supercooled turbulent water. Frazil suspension consists of fine spicule, plate or discoid crystals. Present understanding on the formation of frazil ice has been summarized by Osterkamp [64] and Daly [35]. When the water temperature drops to the freezing point, further cooling, usually by a few hundreths of a degree, causes frazil ice to form in the river. At the supercooled condition, the frazil ice crystals are in their active state of growth. Active frazil ice is found to adhere to nearly any object submerged in the water. Frazil ice blockage of water intakes is a common problem in northern rivers. The temporal variation in

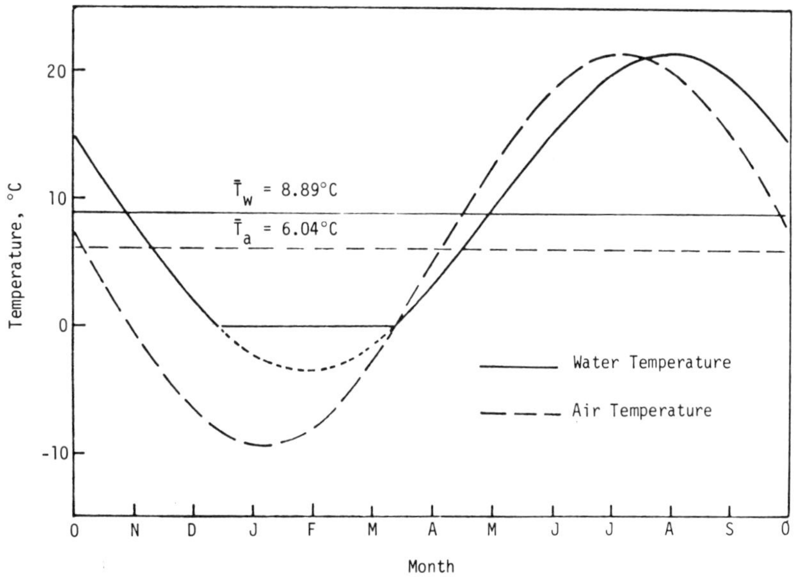

FIGURE 2. Variations of air and water temperature in the St. Lawrence River at Massena, NY.

water temperature for a constant rate of surface heat loss during the supercooling and frazil formation is typically as shown in Figure 3. The initial cooling from time t_e to t_n is governed by the surface heat loss. At t_n initial nucleation begins and the cooling rate decreases as frazil ice is being produced. The temperature reaches its maximum depression ΔT_m before asymptotically approaching a residual supercooling temperature ΔT_r. Flume experiments indicated that the value of ΔT_r approaches zero at a large turbulence level [30,31]. In general the rate of change of the water temperature of a water mass M_w can be determined using an energy budget analysis:

$$\frac{dT_w}{dt} = \frac{-\phi^*}{C_p M_w} \quad \text{when } t < t_n \tag{27}$$

$$\frac{dT_w}{dt} = \left(L_i \frac{dM_i}{dt} - \phi^*\right) \Big/ [C_p(M_w - M_i)] \quad \text{when } t > t_n \tag{28}$$

in which, ϕ^* = net rate of heat loss from the water mass, L_i = latent heat of fusion of ice (80 cal/g), M_i = mass of ice formed. Since frazil ice concentration in rivers is usually on the order of a few percent or less, one can assume $M_i \ll M_w$. Neglecting the effect of dT_w/dt during the short period $t_n < t < t_r$, Equation (28) gives

$$\frac{dM_i}{dt} = \frac{\phi^*}{L_i} \tag{29}$$

For a given river reach at freezing point the rate of ice production can be estimated by

$$\frac{dQ_i}{dt} = \frac{-\Sigma \phi A_s}{\rho_i L_i} \tag{30}$$

where, A_s = surface area of the river reach, Q_i = volume rate of ice production. Figure 4 shows the total ice production per unit open water area versus the total freezing degree-day of the winter for the St. Lawrence River [78]. Recognizing the thermal energy per unit volume in a frazil-laden water as $-\rho_i L_i C_i$, Equation (24) can be used to determine the concentration of frazil ice, C_i, by replacing $\rho C_p T_w$ with $-\rho_i L_i C_i$.

While the general nature of frazil ice formation can be described by the dynamic balance of the heat loss from the water to the ambient environment and the latent heat released by the growing frazil ice crystals, the initial mechanism of frazil nucleation is not completely understood. The available data seems to support that the formation of frazil is started by the introduction of seed crystals into supercooled water. The concept of mass exchange at the air/water interface proposed by Osterkamp [64] is the most probable mechanism by which the seed crystals are introduced. The origin of the seed crystals and the rate at which they are introduced will depend on the local environmental conditions. Water originating in a river and then introduced into the air by bubble bursting, splashing, windspray, evaporation, etc., can freeze and return to the river in the form of seed crystals to initiate the formation of frazil ice in rivers. It is interesting

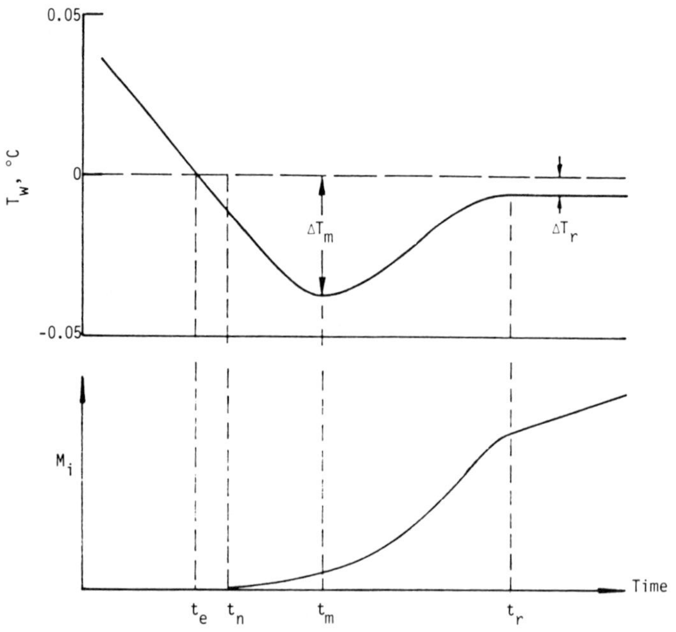

FIGURE 3. Schematic temporal history of water temperature and total ice mass produced in water [64].

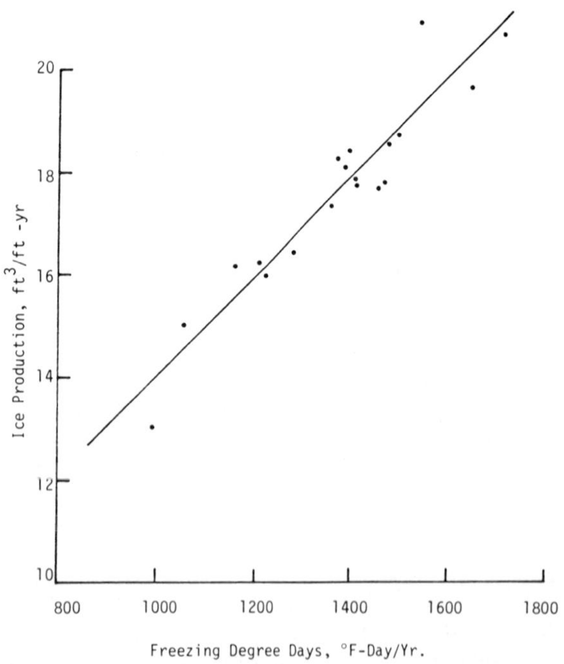

FIGURE 4. Correlation between seasonal values of ice production and air temperature, St. Lawrence River at Massena, NY [78].

to note that minimum air temperatures of −9 to −8°C are often reported as necessary for the production of frazil. These temperatures correspond to the minimum temperatures at which spontaneous heterogeneous nucleation could be expected in water particles suspended in air. Ice particles that originated at some distances above the water body, such as snow, frost, ice particles from trees, shrubs, etc., could be effective seed crystals. Very cold soil particles and cold organic materials at temperatures less than the supercooling level necessary to cause spontaneous nucleation can also be introduced across the air/water interface and may serve to nucleate ice, although their effectiveness is not known.

The mass exchange mechanism provides a reasonable explanation for the observation of ice crystals at the water surface at the start of frazil formation. This mechanism cannot explain the existence of all frazil crystals, however. The number of ice crystals increases rapidly when a crystal is introduced into turbulent supercooled water in which spontaneous nucleation is not possible. This increase in the number of crystals occurs only because of the presence of the original seed crystal and secondary nucleation. Therefore, to determine the rate of increase of frazil ice crystals, the rate of introduction of new crystals and the rate of secondary nucleation must be known. The relative magnitude of these rates will depend on the local environment; however, the rate of secondary nucleation is probably much greater than the rate of introduction.

In regions with very low velocity, frazil ice crystals formed at the surface of the water freeze together and rapidly grow into a thin cover which thickens during the winter. In fast flowing channels, the turbulent mixing will both supercool the water over the entire depth and entrain frazil crystals deep into the flow to prevent the formation of the thin static ice cover [55]. Active frazil ice adheres to river bottom material and leads to the formation of anchor ice [2]. The suspended frazil ice will agglomerate and eventually float to the surface to form a moving surface layer of ice pans, floes and slush ice mixture. This moving surface layer may develop into a continuous ice cover initiated by the formation of an ice bridge or surface ice jam at a river section or from an artificial obstacle.

In rivers with fast flow velocity or steep gradient, a natural stable ice cover will not form. Without the existence of an ice cover the production of frazil ice will continue in order to balance the surface heat loss. Once formed, the frazil ice will be swept underneath a downstream ice cover and eventually accumulate on the underside of the cover. Massive undercover accumulations of ice mass, termed hanging dams, have been found in both large and small rivers [14, 49,60,83,84]. The hanging dams can block the water passage and lead to substantial changes in water levels. The large accumulation of frazil ice will reduce the conveyance capacity of a river and may eventually lead to extensive flooding by forming ice jams in the spring [36].

Growth and Decay of Ice Cover

The thermal growth and melting of the ice cover can take place on both the air-ice and the ice-water interfaces, Figure 5. The growth and decay of the ice cover is governed by the heat conduction across the thickness of the ice cover, which can be described as

$$\rho_i C_i \frac{\partial T}{\partial t} = \frac{\partial}{\partial z}\left(k \frac{\partial T}{\partial t}\right) + \phi_p(z, t) \tag{31}$$

in which ρ_i = density of ice, 0.92 g/cm^3; c_i = specific heat of ice, 1.0 cal g^{-1} °C^{-1}; k = thermal conductivity of ice, 0.0053 cal cm^{-1} S^{-1} °C^{-1}; ϕ_p = rate of internal heating per unit volume due to the adsorption of shortwave penetration, Equation (5); T = temperature in the ice cover, in degrees Celsius; and z, t = space and time variables, respectively.

If the ice surface is assumed to be well drained, the growth of the ice cover will take place only at the ice-water interface. This assumption can be justified by considering the existence of cracks in the ice cover for drainage and furthermore that no flooding of river water can occur over a floating ice cover [106]. At the upper boundary, the boundary condition is

$$k \frac{\partial T}{\partial z} = \Sigma\phi_i - \rho_i L_i \frac{d\theta}{dt}; \text{ at } z = 0 \tag{32}$$

in which $\Sigma\phi_i$ = net heat loss rate at the air-ice interface excluding ϕ_s; θ = thickness of the ice cover, in centimeters. Similarly, the boundary condition at the lower boundary is

$$k \frac{\partial T}{\partial z} = q_{wi} + \rho_i L_i \frac{d\theta}{dt}; \text{ at } z = \theta \tag{33}$$

in which q_{wi} = net flux from water to the ice cover. The

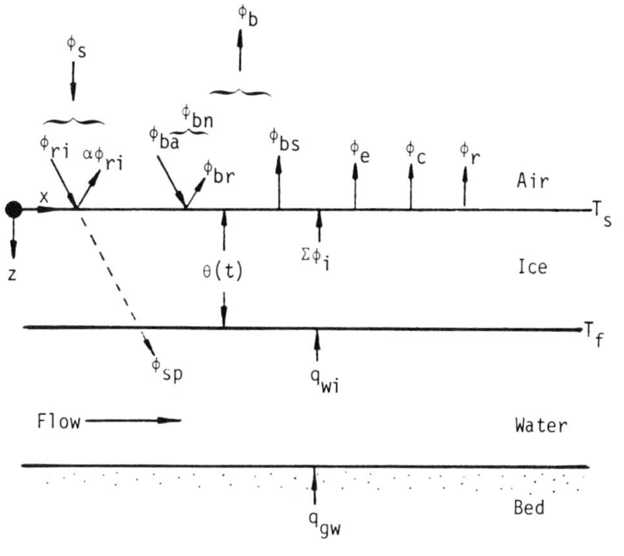

FIGURE 5. Heat exchanges in a river ice cover [88].

heat flux q_{wi} is a function of local water temperature T_w, which can be determined from Equation (25). The boundary value problem defined by Equations (31–33) is a nonlinear problem, which can be solved by finite-differnce techniques [43,44]. However, the time-dependent ice growth and decay at each station along the river can be approximated by a one-dimensional, quasi-steady state linear temperature distribution over the thickness of the ice cover. (The validity of linear temperature disribution for thin ice covers has been examined by Greene [44].) For the quasi-steady state linear temperature distribution, Equation (31) can be approximated at each time step by

$$\frac{\partial T}{\partial z} = \frac{T_f - T_s}{\theta} \tag{34}$$

With this approximation, the shortwave contribution should be included in the heat loss term $\Sigma\phi_i$ in the boundary condition, Equation (32). A detailed numerical model for coupled water temperature and ice thickness distributions exists [88]. For practical applications a simple stepwise integration procedure can be used [7,80]. In this procedure $\Sigma\phi_i$ can be expressed in terms of a heat transfer coefficient, $\Sigma\phi_i = h_{ia}(T_s - T_a)$. The ice surface temperature can then be calculated by

$$T_s = \frac{kT_f + T_a h_{ia}\theta}{h_{ia}\theta + k} \tag{35}$$

If the calculated $T_s < T_f$, then the rate of change of ice cover thickness is

$$\rho_i L_i \frac{d\theta}{dt} = \frac{k(T_f - T_s)}{\theta} - h_{wi}(T_w - T_f) \tag{36}$$

If the calculated $T_s > T_f$, then the top surface of the ice cover is melting. The surface temperature is T_f and

$$\rho_i L_i \frac{d\theta}{dt} = -h_{ia}(T_a - T_f) - h_{wi}(T_w - T_f) \tag{37}$$

in which, T_s = temperature on the top surface of the cover; T_a = air temperature; T_f = melting point, 0°C; k = thermal conductivity of ice; θ = ice cover thickness; h_{ia}, h_{wi} = heat transfer coefficients at the air-ice and ice-water interfaces. Equations (36) and (37) can be used to determine the ice cover thickness through stepwise integrations. The water temperature can be efficiently calculated using a Lagrangian finite-difference solution of Equation (25) with $\Sigma\phi_i$ as the heat exchange term [7]. In the preceding discussion, effects of snow cover above the ice sheet and the accumulation of frazil ice were not considered. Both of these can influence the heat exchange through the ice cover. These heat exchange effects can be accounted for by adding appropriate layers to the ice cover [8,9]. The frazil accumulation tends to accelerate the thickness growth since less water mass is required to be frozen in the frazil layer [23]. The snow cover generally tends to retard the growth in ice thickness through its insulation effect. In regions with heavy snow fall, however, the submerged portion of the thick snow accumulation on the ice cover can lead to the formation of snow ice and compensate for the insulation effect on ice thickness growth.

Further simplifications can lead to an empirical degree-day model [86].

$$\theta = (h_i + \alpha S)^{1/2} - \beta D_t^{\gamma} \tag{38}$$

in which, h_i = initial ice cover thickness; D_t = number of days since the formation of the ice cover, S = cumulative degree-days of freezing since the formation of the ice cover; α = an empirical coefficient which is a constant during the growth period and decreases linearly with the air temperature during the decay period; β and γ = empirical constants which account for the suppression effect of the turbulent heat flux from the river water. The value of γ is approximately 1.0.

Assuming that the turbulent heat transfer q_{wi} is equal to zero, $T_s = T_a$, and the initial thickness h_i is negligible, Equation (38) reduces to the classical degree-day formula for ice thickness growth.

$$\theta = \alpha_0 S^{1/2} \tag{39}$$

Theoretically, the value of α_0 is 1.0 day$^{-1/2}$ °F$^{-1/2}$ for ice covers free of snow. This value is much higher than the actual values due to the assumptions introduced. Empirical values of α_0 are given in Table 3. Equation (39) is applicable only to the growth period, Bilello [19] suggested the use of an accumulated thawing degree-days formula to describe the decay and breakup of ice cover from their maximum thicknesses.

$$\theta = \theta_{max} - \alpha_0' S_T \tag{40}$$

in which, θ_{max} = maximum ice thickness at the beginning

TABLE 3. Typical Values of Coefficient α_0 [56].

Ice Cover Condition	α_0 (in. °F$^{-1/2}$ day$^{-1/2}$)	α_0 (cm °C$^{-1/2}$ day$^{-1/2}$)
Windy lakes with no snow	0.8	2.7
Average lake with snow	0.5 ~ 0.7	1.7 ~ 2.4
Average river with snow	0.4 ~ 0.5	1.3 ~ 1.7
Sheltered small river with rapid flow	0.2 ~ 0.4	0.7 ~ 1.4

of the ice decay; S_T = accumulated thawing degree-days, and α_0' = an empirical constant that is site dependent.

ICE ACCUMULATION AND TRANSPORT

At the beginning of the winter as the ice is being produced and transported along the river, a significant portion of it will rise to the surface to form a moving layer of a mixture of ice floes, pans, and slush. The concentration of this surface layer increases as it moves downstream. The downstream transport of the surface ice will cease when it reaches an artificial obstacle or a river section where an ice bridge across the river is formed by the congestion of the ice. Once an obstacle is reached the incoming surface ice will accumulate at its upstream side and extend the ice cover upstream. Frazil ice that remains in the suspension underneath the surface layer will be transported downstream and deposited on the underside of the ice cover to form a layer of frazil ice accumulation.

The phenomena of ice bridging is not well understood, even though ice bridges usually form at the same location of a river each winter. The formation of an ice bridge at a river section is related to the ice transport capacity of the section and the rate of ice discharge coming from upstream. The maximum rate of ice discharges that can pass through a river while not forming an ice bridge is dependent on the flow velocity, channel top width between banks or border ice boundaries, surface slope, size and concentration of ice in the surface layer [1,26].

The upstream progression of ice cover from an obstacle is dependent on the rate of ice supply from upstream and the river flow conditions. In low velocity reaches, a relatively thin smooth cover can be formed while in other reaches ice jams or hanging dams may form before upstream progression can proceed. The above discussion is presented in the context of ice cover formation in the beginning of the winter. A similar process can occur when a large volume of surface ice floes is released during the breakup of ice cover upstream. In this case, the leading edge of the intact downstream ice cover often acts as the obstacle which initiates the ice accumulation process.

Ice Cover Progression

Once an ice cover is initiated, it will progress upstream through the accumulation of incoming surface ice floes and slush. The rate of progression of the leading edge of an ice cover, V_{cp}, can be calculated from the conservation of mass

$$V_{cp} = \left(1 + \frac{V_{cp}}{V_s}\right)\left(\frac{Q_i^s}{1 - e}\right) \Big/ [h_i(1 - e_p)B_c] \quad (41)$$

in which, Q_i^s = surface ice discharge; h_i = thickness of the ice cover formed by the accumulation; e = porosity of individual ice floes; e_p = porosity in the accumulation between ice floes; B_c = width of the newly accumulated cover; and V_s = mean velocity of the incoming surface ice discharge.

In order for an ice cover to progress, a stability condition for incoming ice floes must be satisfied. The stability of an ice floe, when the leading edge thickness of the ice cover equals that of the ice floe, can be determined by an equilibrium analysis of an arriving ice flow [4,35,103]. Using a "no-spill" condition, the stability criteria was shown to be [4]

$$\frac{V_c}{\left[gt_i\left(1 - \frac{\rho_i}{\rho}\right)\right]^{1/2}} = \frac{2\left(1 - \frac{t_i}{H}\right)}{\left[5 - 3\left(1 - \frac{t_i}{H}\right)^2\right]^{1/2}} \quad (42)$$

in which, V_c = critical velocity upstream of leading edge for underturning and submergence; t_i = thickness of the ice floe; H = upstream flow depth; and g = acceleration of gravity. A similar expression has been obtained by Pariset and Hausser [66,67].

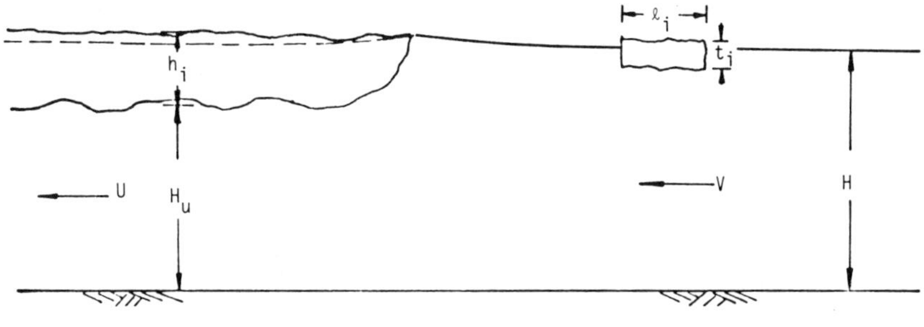

FIGURE 6. Definition sketch for ice cover progression.

$$F_{rc} \equiv \frac{V_c}{\sqrt{gH}} = F\left(\frac{t_i}{l_i}\right) \times \left(1 - \frac{t_i}{H}\right)\sqrt{2\left(\frac{\rho - \rho_i}{\rho}\right)(1 - e)\frac{t_i}{H}} \tag{43}$$

where, $F(t_i/l_i)$ = a form factor which varies between 0.60 and 1.30; l_i = length of the ice flow. Equations (42) and (43) are identical if $F(t_i/l_i) = \sqrt{2}/[5 - 3(1 - t_i/H)^2]^{1/2}$. The effect of porosity, e, of an ice floe is included in Equation (43) [56]. Under the criteria given by Equations (42) or (43), an ice cover of one floe thickness will form by simple juxtaposition.

At a velocity higher than V_c, incoming ice floes will submerge and deposit on the underside of the cover. This leads to a gradual increase of ice cover thickness in the vicinity of the ice front. The velocity that is required for ice floes to submerge below the ice cover increases with the thickness of the ice cover. For a given flow depth and velocity upstream of the ice front, a limiting value exists for the accumulation thickness of an ice cover during its progression. Pariset *et al.* [66,67], Michel [56,57], and Tatinclaux [100] have developed methods for predicting the equilibrium thickness h_i of the ice cover formed by floe accumulation. For a given flow condition, the thickness h_i may be estimated by [57,58]

$$\frac{V}{\sqrt{gH}} = \left(1 - \frac{h_i}{H}\right)\sqrt{2(1 - e_c)\left(1 - \frac{\rho_i}{\rho}\right)\frac{h_i}{H}} \tag{44}$$

From Equation (44), one can show that there is a maximum Froude number for the flow in front of an ice cover at which it becomes unstable and cannot progress further upstream. This critical Froude number, which occurs when h_i/H = 0.33, is

$$F_{rc}^* = \frac{V_c}{\sqrt{gH}} = 0.158\sqrt{1 - e_c} \tag{45}$$

in which, $e_c = e_p + (1 - e_p)e$, is the overall porosity of the ice accumulation. This equation indicates that e_c has an important effect on the progression of an ice cover. Field observations [51] indicated F_{rc}^* can vary from 0.05 to 0.10 for different floe and accumulation characteristics. Newly formed ice floes will have higher porosity and, therefore, smaller F_{rc}^*. If the ice thickness resulting from Equation (44) leads to an undercover velocity U exceeding the critical velocity of entrainment for ice floes on the underside of the cover, the ice cover thickness will then be governed by this undercover entrainment velocity. In this case the ice cover progression will cease. Laboratory flume observations indicate that at F_r close to F_{rc}^* incoming floes could be carried under the accumulated ice cover and swept along the underside of the cover [98]. This leads to a possible physical interpretation of the existence of F_{rc}^*. Detailed discussion of the critical velocity for undercover entrainment will be given later in relation to the formation of hanging dams.

Ice cover thickness formulas derived from the consideration of hydraulic accumulation are often categorized as "narrow river jam" theories [66,67]. This type of formulation can be used to determine the ice cover thickness if the internal resistance of the ice accumulation is able to withstand the streamwise force. For accumulations of loose ice floes or slush ice, local packing of an ice cover at its front is often observed [51]. Michel [59] derived an expression for ice thickness by considering the frontal packing due to hydrodynamic thrust at the ice front.

$$\frac{U}{\sqrt{gh_i}} = \left[\frac{\cos\phi\left(\frac{\rho - \rho_i}{\rho}\right)\frac{\rho_i}{\rho}(1 - e_c)\frac{h_i}{H}}{k_l + \left(1 - \frac{H_u}{H}\right)^2}\right]^{1/2} \tag{46}$$

in which, k_l = minor loss coefficient for the ice cover front; ϕ = angle of friction of granular ice mass.

A graphical summary of formulas for calculating ice cover thickness by accumulation, h_i, is presented in Figure 7. The Froude number upstream of the leading edge, $V/\sqrt{gH}$, is used as the vertical coordinate for the convenience of interpretation.

Mechanical Thickening of Ice Cover

In a wide river the increase in streamwise force may exceed the increase in bank resistance. In this case the internal resistance of the ice accumulation will be unable to resist the increasing streamwise force as the cover progresses upstream. If the stress in the ice accumulation exceeds its internal strength, the cover will collapse and thicken until an equilibrium thickness is reached [58,66,104]. This process of mechanical thickening is commonly known as "shoving," and an accumulation of this kind is often called "wide river jams" [66,67]. When shoving occurs, a relatively long reach of ice cover will suddenly collapse and the leading edge will move a long distance downstream. Based on the analysis of Pariset and Hausser [66,67] and Uzuner and Kennedy [104], the following equation for the equilibrium thickness h_i can be obtained for steady uniform flows.

$$\left[f_i + \frac{\rho_i}{\rho}(f_b + f_i)\frac{h_i}{d_w}\right]\frac{U^2B}{8g} = \frac{2\tau_c h_i}{\rho g} + \mu(1 - e_p)\left(1 - \frac{\rho_i}{\rho}\right)\frac{\rho_i h_i^2}{\rho} \tag{47}$$

in which, f_i and f_b = Darcy–Weisbach friction factors re-

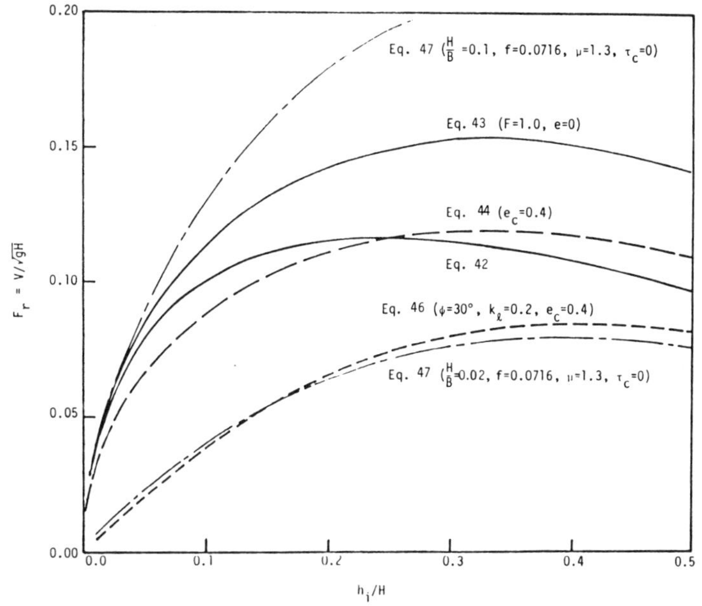

FIGURE 7. Relationship between Froude number and the ice accumulation thickness.

lated to the ice covers and the channel bed, respectively; d_w = depth of flow under the ice cover. Bank resistance per unit length of the ice cover is quantified as $\tau_c h_i + \mu F_s$, where $\tau_c h_i$ is the cohesive contribution; μF_s is the ice-over-ice friction term; and F_s is the streamwise force per unit width of the cover. The coefficient μ has a value of about 1.28, and $\tau_c h_i$ varies between 75 lb/ft and 91 lb/ft [67]. Equation (47) was obtained by considering the ice cover as an accumulation of granular material. A plot for Equation (47) is included in Figure 7 to provide a comparison between ice thickness resulting from different river conditions. Michel [59] has pointed out that it takes only a little freezing to form a solid crust on the top of the ice cover and prevent the occurrence of shoving.

Hanging Ice Dams and Undercover Transport

Large accumulations of ice mass in rivers is commonly known as hanging dams. According to their formation processes, hanging dams may be classified into two categories [83,84]. The first type of hanging dams, which will be referred to as surface ice hanging dams, are accumulations of surface ice floes or frazil ice pans, formed near the leading edge of an ice cover during its upstream progression. The initial thickness of the ice cover formed during the leading edge progression can be estimated by Equations (44) or (46) using the local value of $V/\sqrt{gH}$. Since the depth and velocity varies along (and across) the river, the initial thickness of the ice cover in a river will not be uniform. In fast flowing regions, the initial ice cover will appear as large localized ice accumulations in the form of hanging dams. When the ice cover leading edge reaches a rapid section with an upstream Froude number exceeding F_{rc}^*, incoming surface ice pans and slush will be forced to pass underneath the cover and carried downstream until the velocity of the flow becomes low enough for them to deposit. These depositions will then increase the thickness of the existing ice cover and hanging dams. Mechanical thickening can occur during the above described process when the shoving condition given by Equation (47) is satisfied.

Little is known about the mechanism of the transport and accumulation of ice floes on the underside of the ice cover. Using laboratory data by Filippov [39], Ashton [5] obtained empirical relationships, as shown in Figure 8, for the undercover travel distance of ice floes

$$\frac{L}{l_i} = f\left(\frac{V}{\left[g \cdot \dfrac{\Delta\rho}{\rho} \cdot l_i\right]^{1/2}}, n_i, \frac{t_i}{l_i}\right) \tag{48}$$

in which, L = undercover travel distance of an ice floe before its rest; V = mean velocity upstream of the cover; $\Delta\rho = \rho - \rho_i$; n_i = Manning's roughness coefficient of the underside of the ice cover, and l_i = length of the block. The experimental data indicate that when t_i/l_i decreases or

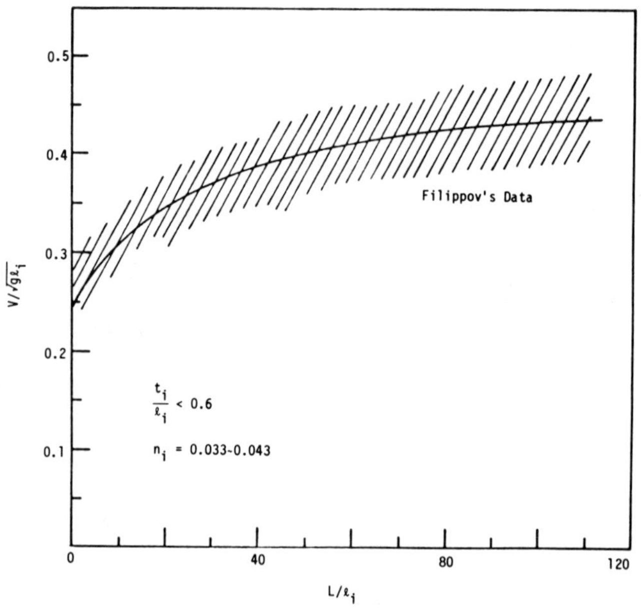

FIGURE 8. Transport distance (L) for ice floes.

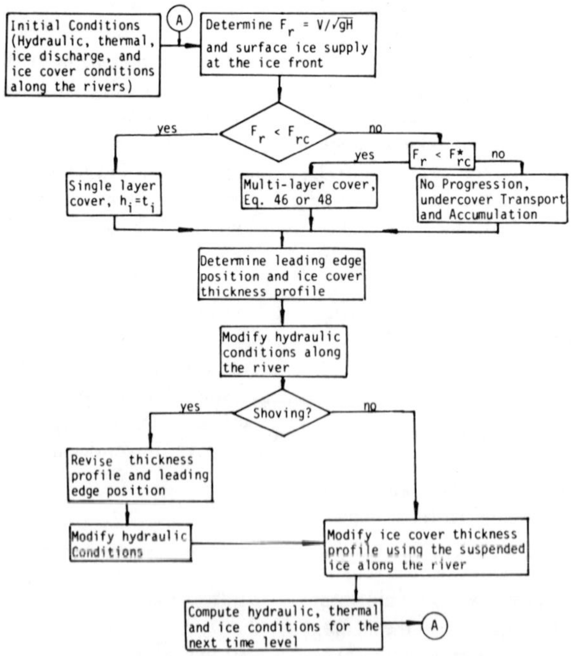

FIGURE 9. Ice cover formation processes.

n_i increases the value of L/l_i decreases. Tatinclaux and Gogus [99] studied the re-entrainment criteria of an ice floe resting under the ice cover behind an artificial obstacle of height δ equal to or less than t_i. Based on laboratory data, they found that the critical Froude number for entrainment can be expressed in terms of the ratio t_i/l_i and size of the obstacle δ

$$F_e \equiv \frac{U}{\sqrt{g \frac{\Delta\rho}{\rho} t_i}} = \left[C_1 \left(\frac{t_i}{l_i}\right)^2 + C_2 \left(\frac{t_i}{l_i}\right) + C_3 \right]^{-1/2} \quad (49)$$

where C_1, C_2, and C_3 = coefficients which are dependent on the size of the obstacle. Suggested values for C_1, C_2 and C_3 are -2.26, 2.14 and 0.015, respectively, for $\delta = 0$. Both Equations (48) and (49) are applicable to a flat ice cover.

The second type of hanging dams, which will be referred to as frazil ice hanging dams, are formed by the deposition of suspended frazil ice particles on the underside of a stable ice cover. These frazil ice particles are produced in open water areas upstream of a stable ice cover during periods of supercooling. In the early stages of their production, frazil particles are active, which means that they very readily stick to any surface with which they come into contact. Therefore, near the leading edge of a river's ice cover, almost all of the frazil particles that reach the ice-water interface are deposited. Frazil particles, which remain entrained in the flow, lose some of their adhesiveness and become inactive. Therefore, further downstream from the leading edge, these frazil particles will deposit on the underside of the ice cover only in regions of relatively low flow velocity. Based on observations in the LaGrande River, Michel and Drouin [61] suggested that the critical velocity for ice deposition ranges between 0.6 and 1.3 m/sec. Shen and VanDeValk [83] found that this critical velocity is about 3 ft/sec in the upper St. Lawrence River. Besides the activeness of the frazil ice, the rate of deposition is dependent upon the buoyant velocity and the concentration of the frazil ice suspension [45]. Neither the buoyant characteristics nor the adhesive properties of frazil ice particles are well understood.

For a surface ice hanging dam, which is located near the leading edge of an ice cover, a relatively soft outer layer of frazil slush could form on the surface due to the accumulation of active frazil ice particles produced in the open water area during the winter. Such hanging dams have been found in the St. Lawrence River [83,84].

Simulation of Ice Cover Formation

A few computer models exist for simulating flow and ice conditions in a river [25,61,70,71,72,89,94]. Ice cover formation is a major component in these models. In applying the existing theories to a natural river, it is important to recognize that the flow conditions in the river are continuously changing in space as well as in time. During the ice formation period, the progression and thickening of the ice cover can cause a rapid change in a river's flow conditions especially in steep shallow rivers. The processes and evolution of the initial ice cover as discussed in this chapter can be summarized as in Figure 9. Figure 9 is developed for one-dimensional models. Two-dimensional effects, which appear to be very important for the distribution of ice cover thickness and the progression of ice fronts are presently being studied [48].

HYDRAULICS OF ICE-COVERED RIVERS

The flow condition in a river not only influences but also interacts with its ice condition. Except in very narrow channels, the ice cover is able to accommodate gradual changes in water level and can be considered as floating in hydrostatic equilibrium. Stage and discharge variations of a gradually varied flow in a river with floating ice covers can be described by the continuity equation

$$\frac{\partial Q}{\partial x} + \frac{\partial A}{\partial t} = 0 \quad (50)$$

and the momentum equation

$$\rho \frac{\partial Q}{\partial t} + \rho \left[\frac{2Q}{A}\frac{\partial Q}{\partial x} - \frac{Q^2}{A^2}\frac{\partial A}{\partial x} \right] + \rho g A \frac{\partial H_e}{\partial x} + (p_i \tau_i + p_b \tau_b) = 0 \quad (51)$$

in which, Q = discharge; A = flow area; $H_e = z_b + d_w + \bar{h}_i$, water level; z_b = bed elevation; d_w = depth of flow; $\bar{h}_i$ = equivalent thickness of the ice cover, $\rho_i\theta/\rho$; p_b, p_i = wetted perimeter formed by the channel bed and the ice cover, respectively; τ_b, τ_i = shear stress at the channel bottom and at the ice-water interface, respectively.

To solve Equations (50) and (51) for Q and H_e, it is necessary to know the geometry of the cover and the additional resistance induced by the ice cover. The estimation of the ice cover geometry including its thickness and areal extent may be obtained according to information discussed in previous chapters. The resistance term in Equation (51) can be expressed in terms of the resistance coefficients of the underside of the ice cover and the channel bed.

Roughness Parameters

In river hydraulics, the friction term in Equation (51) is often expressed in terms of the frictional slope, S_f.

$$S_f = (p_i \tau_i + p_b \tau_b)/(\rho g A) \quad (52)$$

This boundary resistance is usually measured in terms of the Chezy C, the Darcy–Weisbach f, or the Manning n. For uniform flow these three parameters can be related by Equation (53)

$$\frac{C}{\sqrt{g}} = \frac{(1.486)}{\sqrt{g}}\frac{R^{1/6}}{n} = \left(\frac{8}{f}\right)^{1/2} \tag{53}$$

in which R = hydraulic radius. The friction factor f is dimensionless and has the most general range of application. However, it has the drawback of being a function of the Reynolds number and relative roughness both of which are a linear function of R. Chezy's C has the disadvantage of being dimensional and varying significantly with flow. Manning's n, although dimensional and only applicable to fully rough flows, is nearly constant over a wide range of R for a given roughness height. Hence, it has been used for the analysis of most river flows [32,77].

Uzuner [102] and Pratte [75] reviewed different methods of calculating the roughness coefficient for ice-covered channels and showed that the method of Larsen [53] is the most rigorous. By dividing the flow depth into two zones at the maximum velocity plane, assuming zero shear stress at this plane, and using the Karman–Prandtl velocity formula for both zones, Larsen showed that the equivalent Manning coefficient, n_c, of an ice-covered channel can be calculated by Equation (54).

$$\frac{1}{n_c} = \frac{\frac{1}{n_i} d_i^{5/3} + \frac{1}{n_b} d_b^{5/3}}{\left(\frac{1}{2}\right)^{2/3} d_w^{5/3}} \tag{54}$$

in which, n_c = equivalent Manning's coefficient; n_i and n_b = Manning's coefficients of the ice cover and the channel bed; d_w = total flow depth; d_i and d_b = depths of flow above and below the maximum velocity line. The depth ratio d_i/d_b can be expressed as:

$$d_i/d_b = \left[\frac{a(b-1)n_i}{b(a-1)n_b}\right]^{3/2} \tag{55}$$

in which, $a = \ln(30\ d_b/k_b)$; and $b = \ln(30\ d_i/k_i)$. For channels with small relative roughness, Equation (54) can be simplified to the Belokon-Sabaneev formula [62,102].

$$n_c = \left[\frac{1}{2}(n_i^{3/2} + n_b^{3/2})\right]^{2/3} \tag{56}$$

Larsen suggested methods for determining the roughness height from a measured vertical velocity profile. Some of the difficulties involved in determining the roughness height from the field data are discussed by Calkins *et al.* [27]. Tatinclux and Gogus [100] discussed the deficiency of assuming zero shear stress at the maximum velocity plane, and suggested a refined method for analyzing flow in channels with large differences between the ice cover and channel bed roughness.

The Manning equation can be derived based on the "1/6 power law" approximation of the logarithmic velocity profile for free surface flow in a wide rectangular channel with rough walls [93]. For R/k less than 5 the "1/6 power law" is a very poor approximation of this velocity profile. Gerard *et al.* [40,41,42] suggested the use of average roughness height to quantify the boundary resistance. In terms of roughness heights, the Belokon–Sabaneev formula becomes

$$k_t = \left[\frac{1}{2}(k_i^p + k_b^p)\right]^{1/p};\ p = \frac{2m}{2m+1} \tag{57}$$

in which, k_t = the composite roughness; k_i and k_b = the effective ice cover and bed roughness, respectively; and m = exponent of the appropriate power law.

Resistance Due to Ice Cover

The method of analysis described above can be used to evaluate local roughness coefficients from measured velocity profiles. It is not adequate, however, for the purpose of computing backwater profiles or unsteady flow modeling where the gross effect of the ice cover on the flow resistance of a river reach is required. Determination of the resistance coefficient for a river reach from recorded discharge and water level data can be made through the use of flow equations.

For steady-state flow, Equation (51) can be written as

$$\frac{\partial H_e}{\partial x} = \frac{1}{gA}\left[\frac{Q^2}{A^2}\frac{\partial A}{\partial x} - \frac{1}{\rho}(p_i\tau_i + p_b\tau_b)\right] \tag{58}$$

in which, H_e = water level = $z_b + d_w + \bar{h}_i$. For a partially ice-covered channel, shown in Figure 10, the resistance term can be expressed as [79]:

$$p_i\tau_i + p_b\tau_b = \gamma n_b^2\left[\left(\frac{Q^2p^{4/3}}{A^{7/3}}\right)_\alpha + (1 + F')^{4/3}\left(\frac{Q^2p_b^{4/3}}{A^{7/3}}\right)_\beta\right] \tag{59}$$

in which $F' = (n_i/n_b)^{3/2}(p_i/p_b)_\beta$, α and β = subscripts which represent portions of flow areas under free surface and ice cover, respectively. The partial discharges Q_α and Q_β can be obtained from Equation (12) [81].

$$\frac{Q_\alpha}{Q} = \frac{1}{2}\left[\frac{A_\alpha R_\alpha^{2/3}}{AR^{2/3}} + \left(1 - \frac{A_\beta R_\beta^{2/3}}{AR^{2/3}}\right)\right];\ \frac{Q_\beta}{Q} = 1 - \frac{Q_\alpha}{Q} \tag{60}$$

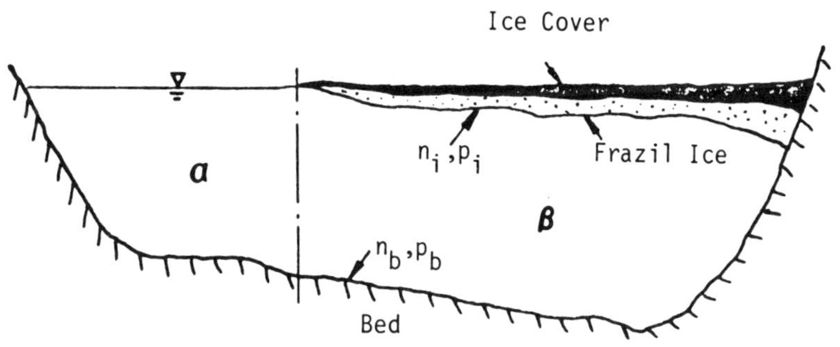

FIGURE 10. Flow cross section of partially covered channel.

Since both the free surface case and the fully covered case can be considered as special cases of the partially covered condition, Equation (59) can be used as a general formula. Terms with the subscript α reduce to zero for the fully covered condition and terms with the subscript β reduce to zero for the free surface condition. Based on Equations (58), (59), and (55), roughness coefficients of river reaches can be determined from recorded discharge and water levels. Computational methods were developed for both single channel or channel networks [68].

It is well recognized that the roughness of the underside of the ice cover varies over a wide range throughout the winter [29,47,101]. Initially, the roughness is generally high as a result of the rough bottom surface of the new ice cover. The roughness gradually diminishes during the winter due to a gradual thickening of the ice cover and smoothing of the ice surface under the influence of the flowing water. The roughness reaches a minimum value and then increases immediately before the spring break-up with the formation of relief features on the ice-water interface [3,29]. The rate of change of ice cover roughness is dependent on the nature of the ice cover, the flow characteristics, and meteorological conditions. The effect of meteorological conditions is particularly important in rivers with large open-water areas. In those rivers, variations of the thermal and ice conditions of the water are predominantly governed by the ambient atmospheric conditions. If the weather is mild, a decrease in ice cover roughness and reduction in frazil accumulation will occur. If the weather is cold, frazil ice will be produced in open-water regions. Accumlations of frazil ice underneath downstream ice covers may then occur causing rapid changes in the roughness of the ice cover and the flow cross sectional area.

Nezhikhovskiy [62] reviewed studies performed by Russian researchers for predicting Manning's roughness coefficient, n_i, on the underside of a river ice cover. In this review, a method for estimating n_i was presented. This method considers the difference between the roughness characteristics of a smooth ice cover formed statically in a slow flowing river and a rough ice cover formed from slush or ice floe accumulations in a fast flowing river. For smooth ice covers, Nezhikhovskiy suggested n_i values of 0.01 ~ 0.012 for the beginning of freeze-up, and 0.008 ~ 0.01 for the middle and end of winter. For rough ice covers, Nezhikhovskiy's proposal takes into consideration meteorological conditions and the nature of the ice cover. Nezhikhovskiy related the initial ice cover roughness coefficient, $n_{i,i}$ to the type of ice which forms the initial ice cover and its thickness, h_i, as shown in Figure 11. The time-dependent variation of n_i during the winter is then described by the following exponential function.

$$n_i = n_{i,e} + (n_{i,i} - n_{i,e})e^{-kt} \tag{61}$$

in which, t = number of days from the beginning of the ice covered period; $n_{i,e}$ = the roughness coefficient at the end of the ice-covered period, which is found to be approximately equal to that of the smooth ice cover, 0.008 ~ 0.012; k = a decay constant which varies from river to river and year to year. Suggested values of k are given in Table 4, which show that the k value is dependent upon the severity of the winter and the relative amount of open water.

Shen and Yapa [85,108,109,110] analyzed data from the upper St. Lawrence River and proposed a conceptual model for the time dependent variation of the ice cover resistance coefficient. By including the effect of hanging dams, the empirical model expresses the resistance coefficient as a combination of three components, i.e. $n_i = n_d + n_t + \tilde{n}$ (Figure 12). The first component, n_d, is a simple function of time which increases monotonically during the freeze-up period and decreases monotonically during the rest of the stable ice-covered period. Expressed in functional form, the component n_d is the smaller value computed from Equations (62) and (63).

$$n_d = \Delta n[t/(\lambda t_F)]^{\gamma} + n_{ed} \tag{62}$$

$$n_d = \Delta n(e^{\psi t_d} - e^{\psi t_L}) + n_{ed} \tag{63}$$

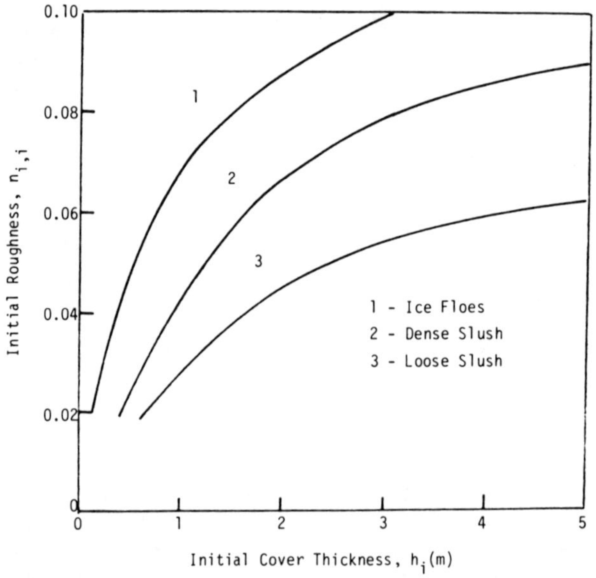

FIGURE 11. Relationships between the initial roughness coefficient of the ice cover and its thickness [62].

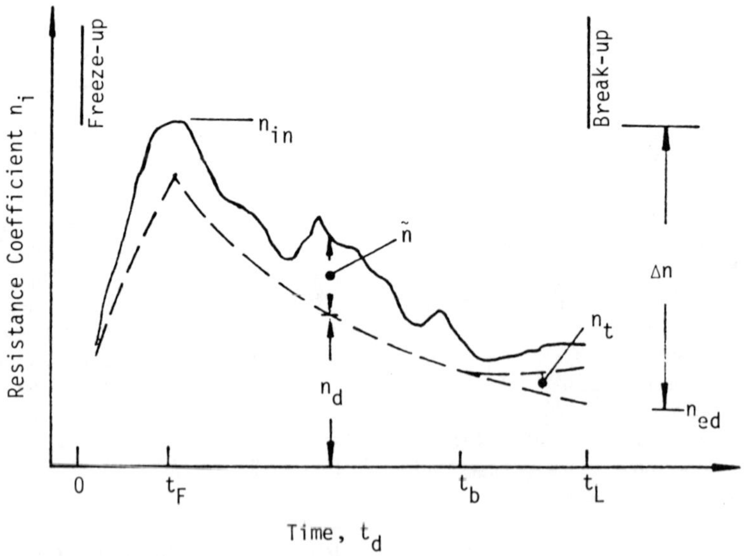

FIGURE 12. Seasonal variation of the resistance coefficient of ice cover.

TABLE 4. Values of Parameter k in Equation (61) [62].

Conditions of the Winter	Characteristics of the Ice Cover		
	Many Open Water Areas	Few Open Water Areas	No Open Water
Severe	0.005	0.010	0.020
Moderate	0.023	0.024	0.025
Mild	0.050	0.040	0.030

where, $\psi = [1/(\delta t_L)]\log_e(\Delta n)$; t_d = time from the beginning of the freeze-up, days; t_F = duration of the freeze-up period; t_L = duration between the beginning of the freeze-up and the beginning of the break-up period; Δn = the difference between the ice roughness coefficients at the end of the freeze-up period, n_{in}, and the beginning of the break-up period, n_{ed}; λ, γ, and δ = empirical constants. Values of n_{in} and n_{ed} in a given river reach can be expressed in empirical relationships of the following forms:

$$n_{in} = f(T_{a0}, T_{w0}, A_s, AI_0, Q_0; t_i); \quad t_i = t_i(T_{w0}, T_{a0}, X) \tag{64}$$

and

$$n_{ed} = an_{in} + b \tag{65}$$

in which, T_{a0} = air temperature during the freeze-up period in freezing-degree-day; T_{w0} = water temperature at the upstream end during the freeze-up period; A_s = the total surface area of the reach; AI_0 = area of the ice cover at the end of the freeze-up period; Q_0 = the discharge during the freeze-up period; t_i = size of the ice floe; X = location of the river reach; and a, b = empirical constants. The parameter t_i is included for the reason that the roughness of a flat ice cover formed by the accumulation of frazil and brash ice will increase with the size and roughness of ice floes.

The increase in n_i during the decay period of the ice season is represented by the component n_t in the simulation model. This component is assumed to be a linear function of time. For the upper St. Lawrence River, the first day to have the effect of n_t was found to be the first day in the late winter to have an average daily temperature rising above 27°F. Let this day be defined as t_b. The functional form n_t is then

$$n_t = 0 \qquad \text{for } t \leq t_b \tag{66}$$

$$n_t = s_n(t - t_b) \qquad \text{for } t \geq t_b \tag{67}$$

where, s_n is a constant which may vary depending on the location and year.

The fluctuating component, $\tilde{n}$, is considered to be governed by the transport and deposition of frazil ice. This component can be obtained from the following empirical relationships.

$$\tilde{n} = \kappa A_0 T_a U^{-\tau} \qquad \text{for } t < t_F \tag{68}$$

$$\tilde{n} = \kappa A_0(T_a + T_r)U^{-\tau} \qquad \text{for } t_F < t \tag{69}$$

in which, T_a = average mean daily air temperature of the three preceding days in terms of the freezing degree-days, °F-day; T_r = a reference temperature; U = average of the flow velocity of the day under consideration and that of the preceding day in *fps*; A_0 = open water area within or immediately upstream of the study reach; κ and τ = empirical constants. Application of the model to reaches of the upper St. Lawrence River [108,110] shows parameters λ, γ, δ, τ, and $\varkappa$ are site independent, although the ice cover condition varies substantially along the river. Values of these constant parameters are presented in Table 5.

Hydraulic Analysis of Ice-Covered Rivers

With the existence of an ice cover, the head loss in a river channel will increase significantly over that of open water conditions at an equivalent discharge. This increase in head loss, which is caused by the increase in wetted perimeter and the displacement effect of the floating cover, often leads to an increase in water level. If the resistance coefficients of the ice cover is known, then conventional backwater or unsteady flow computation techniques can be applied to Equations (50) and (51) to determine the discharge distribution and water surface profiles in rivers and channel networks. Examples of these computed models include those of Calkins *et al.* [28] and Pasquarell [68].

Stage-Discharge Relationships

One important aspect of river engineering is to obtain accurate stream flow data at gaging stations. The stage-discharge relationship at a given gaging station developed for open water conditions is not suitable for ice-covered conditions. Cook and Cerny [33] and Rosenberg and Pent-

TABLE 5. Empirical Constants of the Resistance Model for the Upper St. Lawrence River [108].

Empirical Constant	Estimated Value
λ	2
γ	0.5
δ	2
τ	11 when $U <$ 1.9 *fps*; 9 when $U >$ 1.9 *fps*
κ	0.01

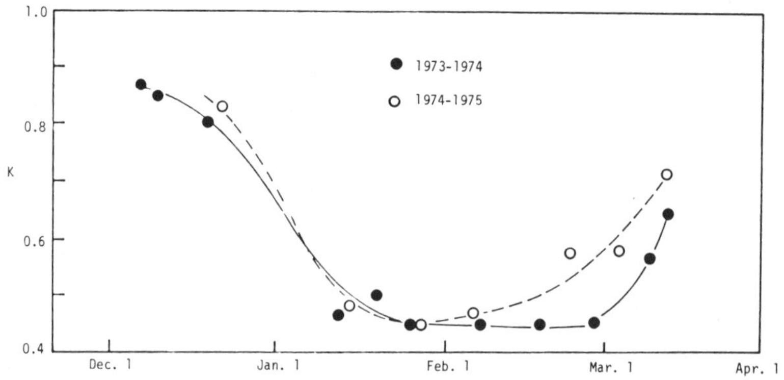

FIGURE 13. Seasonal change of coefficient K at Mukawa Gaging Station, Hokaido, Japan [47].

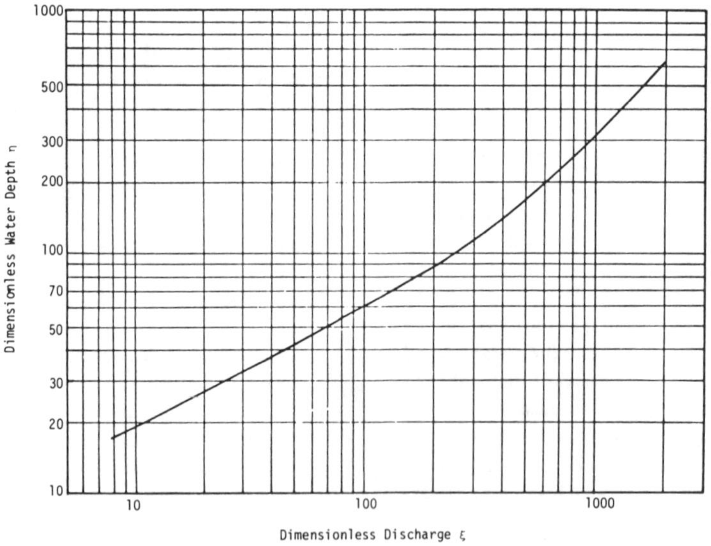

FIGURE 14. Dimensionless water depth due to an equilibrium floating wide channel jam vs. dimensionless discharge [17].

land [76] discussed gaging techniques used in the United States and Canada. With adequate winter data at a gaging station a stage-discharge relationship may be developed. The accuracy is much poorer due to the changing resistance of the ice cover related to the variations of the ice cover roughness and the frazil ice accumulation. Hirayama [47] proposed a procedure for estimating a stage-discharge relationship for ice-covered rivers with no unstable frazil ice accumulation. In this method the winter stage is related to the summer stage of equal discharge by Equation (70)

$$A_i R_i^{2/3} = K A_0 R_0^{2/3} \tag{70}$$

in which, S = channel slope; A = flow cross sectional area; R = hydraulic radius; i and 0 = subscripts representing ice-covered and open water conditions, respectively. The coefficient is defined as

$$K = \frac{n_c}{n_b}\left(\frac{S_0}{S_i}\right)^{1/2} \tag{71}$$

The coefficient K varies as a function of time during a winter. Based on field data, a typical curve for K can be constructed and used in Equation (70) to determine the winter discharge from the winter stage and the summer stage-discharge curve. Figure 13 shows the variation of K for a typical river.

Ice Jam Stage

In cold regions, many flood events are ice-related. Most severe ice-related flood events are those caused by ice jams due to their large thickness and hydraulic resistance. A practical problem related to this is to estimate the stage caused by an ice jam for a given channel geometry and a discharge. Using Equation (47), Beltaos [17] obtained an expression for the stage, H_j, of an equilibrium ice jam as the following:

$$\eta = 0.63 f_c^{1/3}\xi + \frac{5.7}{\mu}\left\{1 + \left[1 + 0.11\mu\frac{f_i}{f_c^{2/3}}\xi\right]^{1/2}\right\} \tag{72}$$

in which, $\eta = H_j/(S_0 B)$; $\xi = (q^2/gS_0)^{1/3}/S_0 B$; $f_c = (f_i + f_b)/2$; f_i, f_b = friction factor of the ice cover and the channel bed, respectively; S_0 = channel slope; B = top width of the flow cross section; $q = Q/B$; Q = flow discharge. Since Equation (72) is weakly dependent on f_i and f_c, a standard curve can be developed with some field data for the relationship between dimensionless stage, η, and dimensionless discharge, ξ, as shown in Figure 14. Gerard and Calkins [42] incorporated a similar method into flood frequency analysis and developed a method for constructing synthetic flood stage probability distribution curves of rivers in cold regions.

DETERIORATION AND BREAKUP OF ICE COVER

Early in the spring or late in the winter, especially after the disappearance of the snow cover, the ice cover melts from both its top and bottom surfaces and disintegrates from inside. This deterioration process causes rapid reduction in the strength of the ice cover and hence its load carrying capacity. The breakup of ice cover is a phenomena contributed by both the deterioration of the ice cover and the increase in discharge [18,37,56]. Ice jams formed during the breakup and the accompanying flooding is a major cause of concern in many northern rivers.

Ice Cover Deterioration

The thermal growth and decay of river ice cover has been discussed in Heat Exchange Processes, above. Methods for determining the reduction in ice cover thickness were presented. In this section, deterioration or the weakening of the integral structural integrity of ice cover will be discussed.

The phenomenon of ice cover deterioration is commonly observed early in the spring. During this period, the internal melting of ice at boundaries of crystal grains by the penetrated solar energy will lead to the loss of strength. Bulatov [20] considered the melting of the ice cover by short-wave penetration and established that the strength of melting ice is dependent on its liquid content. Ashton [10] developed a similar ice deterioration model independent of Bulatov. In Ashton's model the melting of the ice cover is analyzed by considering the combined radiation-conduction heat transfer. The water content is determined by solving the unsteady heat conduction equation, Equation (31), in the intact portions of the ice cover. The strength of the ice cover is then related to the porosity e_m by considering the reduction in effective contact area between grains. The relationship is

$$\frac{\sigma_i}{\sigma_0} = 1 - 2.813 e_m \tag{73}$$

in which, e_m = porosity in the ice cover due to internal melting; σ_i = failure stress; and σ_0 = failure stress when e_m is zero. Ashton also suggested that the elastic modulus will decrease with the increase in porosity. Based on studies in sea ice, the following linear relationship was postulated.

$$E/E_0 = 1 - e_m/e_0 \tag{74}$$

in which, E = Young's modulus; E_0 = values of E when e_m is zero; and e_0 = value of the porosity when E is zero.

Breakup of Ice Cover

The timing of breakup is of importance in the planning of ice-related operations such as the scheduling of a navigation season, the planning of hydropower production, and the assessment of the severity of the potential river flood associated with breakup jams. Breakups can generally be categorized into two types: ''overmature'' breakup and ''premature'' breakup [18,37,56]. The overmature breakup is a slow process due to the deterioration of the ice cover by radiation-conduction heat transfer until it can gradually disintegrate in place by frictional and gravitational forces. This type of breakup often occurs in controlled rivers and is known to cause relatively little negative impact on the river reach. The premature breakup is a faster and more drastic process occurring as a result of a substantial increase in river discharge caused either by intensive snowmelt or rainfall during a warm period. With an increase in discharge, the ice cover first fractures in fast steep sections, breaks into pieces and moves downstream to accumulate at the upstream end of still intact ice sheets. This process forms ice jams, which can be analyzed by theories of ice cover progression as discussed in Ice Accumulation and Transport, above. A continuous warming trend and increase in discharge will lead to successive fractures of ice sheets. This will lead to the shifting of existing ice jams downstream to form even larger jams. The flood stage during a premature breakup increases with the size of jam until the final ice run clears the ice in the channel [56].

The initiation of breakup in a river is influenced by ice cover conditions, weather conditions, flow conditions and the morphology of the river channel [91]. Shulyakovskii [92] formulated an analytical model for the initiation of breakup assuming the ice cover is already separated from the river banks with a water level higher than the maximum winter freezeup level, H_F. The initiation of the breakup is defined as the instant when the strength of the cover is exceeded and transverse cracks form. By considering the force and moment acting on the ice cover, the breakup is shown to initiate when

$$\sigma_i \theta = f_1(H_B - H_F, H_F) \tag{75}$$

Beltaos [16,18], extended this model and shows that

$$H_B/H_F = f_2(\theta/W_F, \sigma_i/\tau_i, \text{dimensionless constants}) \tag{76}$$

in which, θ = ice cover thickness; H_B and H_F = average depth associated with flows breakup initiation and freezup; W_F = maximum freezeup width of the cover during the winter; τ_i = shear stress on the underside of the cover due to flow. Field data shows that H_B/H_F increases with the value of θ/W_F. Beltaos further shows that there exists and ice clearing discharge q_c, which has an upper limit depending on θ, σ_i, H_F, τ_i and channel geomorphology.

$$q_c^* + \frac{1.59}{f_c^{1/3}}\left(\frac{\rho_i}{\rho}\frac{\theta}{H_F}\right) \leq \frac{1.59}{f_c^{1/3}} f \times \left(\frac{\theta}{W_F}, \begin{matrix}\text{other dimensionless river} \\ \text{constants}\end{matrix}\right) \tag{77}$$

in which, q_c^* = dimensionless ice clearing discharge, $q_c^{2/3}/((gs)^{1/3}Y_F)$; s = channel slope; f_c = composite friction factor of the flow under the cover. Equation (75) is confirmed by field data, which shows that q_c increases with θ/W_F. Based on the conceptual model of Beltaos, methods for forecasting the initiation of breakup were developed with moderate success using hydrometric station records [15,97].

ACKNOWLEDGEMENTS

This chapter has been prepared and adopted from Technical Report No. 85-1, entitled "Hydraulics of River Ice" by H. T. Shen, Department of Civil and Environmental Engineering, Clarkson University. Partial support for this work provided by the U.S. Army Cold Regions Research and Engineering Laboratory through a Joint Graduate Research Program in Ice Engineering under Contract No. DACA89-84-K-0008 is greatly appreciated.

REFERENCES

1. Ackermann, N. L. and Shen, H. T., "Mechanics of Surface Ice Jam Formation in River," *CRREL Report 83-31*, U.S. Army Cold Regions Research and Engineering Laboratory, Hanover, N.H., 1983, 14 p.
2. Arden, R. S. and Wigle, T. E., ''Dynamics of Ice Formation in the Upper Niagara River,'' *The Role of Snow and Ice in Hydrology*, Unesco-WMO-IAHS, Sept. 1972, pp. 1296–1313.
3. Ashton, G. D. and Kennedy, J. F., ''Ripples on Underside of River Ice Covers,'' *Journal of the Hydraulics Division*, Vol. 98, No. HY9, Sept. 1972, pp. 1603–1624.
4. Ashton, G. D., ''Froude Criterion for Ice-Block Stability,'' *Journal of Glaciology*, Vol. 13, No. 68, 1974, pp. 307–313.
5. Ashton, G. D., ''Movement of Ice Floes Beneath a Cover,'' *Technical Note*, U.S. Army CRREL, Hanover, N.H., Feb. 1975.
6. Ashton, G. D., ''River Ice,'' *Annual Review of Fluid Mechanics*, Vol. 10, 1978, pp. 369–392.
7. Ashton, G. D., ''Suppression of River Ice by Thermal Effluents,'' *CRREL Report 79-30*, Cold Regions Research and Engineering Lab., U.S. Army, Hanover, N.H., Dec. 1979, 23 p.
8. Ashton, G. D., ''Freshwater Ice Growth, Motion, and De-

cay,'' in *Dynamics of Snow and Ice Masses*, S. C. Colebeck, ed., Academic Press, 1980, pp. 261–304.

9. Ashton, G. D., ''Theory of Thermal Control and Prevention of Ice in Rivers and Lakes,'' *Advances in Hydroscience*, Vol. 13, 1982, pp. 131–185.
10. Ashton, G. D., ''Deterioration of Floating Ice Covers,'' *Third Internatinal Symposium on Offshore Mechanics and Arctic Engineering*, New Orleans, 1984.
11. Ashton, G. D., (ed.), *River and Lake Ice Engineering*, 1986, Water Resources Publications, Littleton, Colorado.
12. Barnes, H. T., *Ice Engineering*, Renouf Publishing Co., Montreal, 1928, 364 p.
13. Bates, R. E. and Bilello, M. A., ''Defining the Cold Regions of the Northern Hemisphere,'' *Technical Report No. 178*, U.S. Army Cold Regions Research and Engineering Lab., Hanover, N.H., 11 p.
14. Beltaos, S. and Dean, A. M., ''Field Investigations of a Hanging Ice Dam,'' Vol. I, *International Symposium on Ice*, Quebec City, Canada, 1981, pp. 475–488.
15. Beltaos, S., ''Study of River Ice Breakup using Hydrometric Stations Records,'' *Workshop on Hydraulics of River Ice*, Fredericton, N.B., June 1984, pp. 41-64.
16. Beltaos, S., "Initiation of River Ice Breakup," *Proc. Fourth Northern Research Basin Symposium Workshop*, Norway, March 1982, pp. 163–177.
17. Beltaos, S., ''River Ice Jams: Theory, Case Studies, and Applications,'' *Journal of the Hydraulics Division*, ASCE, Vol. 109, No. 10, Oct. 1983, pp. 1338–1359.
18. Beltaos, S., ''River Ice Breakup,'' *Canadian Journal of Civil Engineering*, Vol. 11, Sept. 1984, pp. 516–529.
19. Bilello, M. A., ''Maximum Thickness and Subsequent Decay of Lake, River, and Fast Sea Ice in Canada and Alaska,'' *CRREL Report 80-6*, U.S. Army Cold Regions Research and Engineering Laboratory, Hanover, N.H., 1980, 160 p.
20. Bulatov, S. N., ''Calculating the Strength of Thawing Ice Cover and the Beginning of Wind-Activated Ice Drift,'' *Trudy Vypysk 74*, 1970, 120 p.
21. Bolsenga, S. J., ''Total Albedo of Great Lakes Ice,'' *Water Resources Research*, Vol. 5, No. 5, Oct. 1969, pp. 1132–1133.
22. Brocard, D. N. and Harleman, D. R. F., ''One-Dimensional Temperature Predictions in Unsteady Flows,'' *Journal of the Hydraulics Division*, ASCE, Vol. 102, No. HY3, Mar. 1976, pp. 227–240.
23. Calkins, D. J., ''Accelerated Ice Growth in Rivers,'' *CRREL Report 79-14*, Cold Regions Research and Engineering Lab., Hanover, N.H., May 1979, 4 p.
24. Calkins, D. J., ''Ice Cover Melting in a Shallow River,'' *Canadian Journal of Civil Engineering*, Vol. 11, 1984, pp. 255–265.
25. Calkins, D. J., ''Numerical Simulation of Freeze-up on the Ottauquechee River,'' *Workshop on the Hydraulics of River Ice*, Fredericton, N.B., June 1984, pp. 247–277.
26. Calkins, D. J. and Ashton, G. D., ''Arching of Fragmented Ice Covers,'' *Canadian Journal of Civil Engineering*, Vol. 2, No. 4, 1975, pp. 392–399.
27. Calkins, D. J., Deck, D. S., and Martinson, G. R., ''Resistance Coefficients from Velocity Profiles in Ice-Covered Shallow Streams,'' *Canadian Journal of Civil Eng.*, Vol. 9, No. 2, 1982, pp. 236–247.
28. Calkins, D. J., Hayes, R., Daly, S. F., and Montalvo, A., ''Application of HEC-2 for Ice-Covered Waterways,'' *Journal of the Technical Councils of ASCE*, Vol. 108, No. TC2, Nov. 1982, pp. 241–248.
29. Carey, K. L., ''Observed Configuration and Computed Roughness of the Underside of River Ice, St. Croix River, Wisconsin,'' *Prof. Paper 550-B*, U.S. Geological Survey, 1966, pp. B192–B198.
30. Carstens, T., ''Experiments with Supercooling and Ice Formation in Flowing Water,'' *Geofysiske Publikasjoner*, Vol. 26, No. 9, 1966, pp. 3–18.
31. Carstens, T., ''Heat Exchanges and Frazil Formation,'' in *Proceedings of the Symposium on Ice and Its Action on Hydraulic Structures*, Reykjavik, Iceland, International Association for Hydraulic Research, 1970, Paper No. 2.11.
32. Chow, V. T., *Open-Channel Hydraulics*, McGraw-Hill Book Col., 1959, 680 p.
33. Cook, R. E. and Cerny, E. E., ''Patterns of Backwater and Discharge on Small Ice Affected Streams,'' *U.S. Geological Survey Water Supply Paper 1892*, pp. 114–125.
34. Daly, S. F., ''Frazil Ice Dynamics,'' *CRREL Monograph 84-1*, U.S. Army Cold Regions Research and Engineering Laboratory, Hanover, N.H., April 1984, 46 pp.
35. Daly, S. F., ''Ice Block Stability,'' *Water for Resources Development*, ASCE, Coeur d'Alene, Idaho, Aug. 1984, pp. 544–548.
36. Deck, D., ''Controlling River Ice to Alleviate Ice Jam Flooding,'' *Water for Resources Development*, ASCE, Coeur d'Alene, Idaho, Aug. 1984, pp. 524–528.
37. Deslauriers, C. E., ''Ice Break-Up in Rivers, Ice Pressure Against Structures,'' *TM No. 92*, National Research Council, Associate Committee on Geotechnical Research, Ottawa, Mar. 1968, pp. 217–230.
38. Dingman, S. L. and Assur, A., ''The Effect of Thermal Pollution on River Ice Conditions—Part I, A Simplified Method of Calculation,'' *CRREL Report 206*, Cold Regions Research and Engineering Lab., U.S. Army, Hanover, N.H., 1967, 10 p.
39. Filippov, A. M., ''Modeling the Movement of Ice Floes Drawn in Under Ice Cover,'' *Translation TL473*, U.S. Army CRREL, Hanover, N.H., 1974.
40. Gerard, R., ''The Hydraulic Resistance of Ice Covers and Ice Jams: Some Reflections,'' *Proceedings of Workshop on Hydraulic Resistance of River Ice*, Burlington, Ontario, pp. 296–298.
41. Gerard, R. and Andres, D., ''Hydraulic Roughness of Freeze-

up Ice Accumulations: Northern Saskatchewan River Through Edmonton," *Proceedings of the Workshop on Hydraulics of Ice-Covered Rivers*, Edmonton, Alberta, June 1982, pp. 62–87.

42. Gerard, R. and Calkins, D. J., "Ice-Related Flood Frequency Analysis: Application of Analytical Estimates," *Proceedings, Third International Specialty Conference on Cold Regions Engineering*, Edmonton, Alberta, April 1984, pp. 70–85.
43. Goodrich, L. E., "A Numerical Model for Calculating Temperature Profiles in an Ice Cover," *Proceedings, Research Seminar Thermal Regime of River Ice*, Tech. Memo No. 14, National Research Council Canada, Jan. 1975, Ottawa, Canada, pp. 44–59.
44. Greene, G. M., "Simulation of Ice-Cover Growth and Decay in One-Dimensional on the Upper St. Lawrence River," *NOAA TM ERL GLERL-36*, Great Lakes Environmental Laboratory, Ann Arbor, MI, 1981, 82 pp.
45. Halabi, Y. S., Shen, H. T., Papatheodorou, T. S., and Briggs, W. L., "Transport and Accumulation of Frazil Ice Suspensions in Rivers," *Fourth International Conference on Mathematical Modelling*, Zurich, Switzerland, Aug. 1983, pp. 412–417.
46. Haynes, F. D. and Ashton, G. D., "Turbulent Heat Transfer in Large Aspect Channels," *CRREL Report 79-13*, Cold Regions Research and Engineering Lab., U.S. Army, May 1979, 5 pp.
47. Hirayama,, K., "Characteristics of Ice Covered Streams in Connection with Water Discharge Measurements," *IAHR Ice Symposium*, Vol. 2, Lulea, Sweden, Aug. 1978, pp. 195–217.
48. Ho, C.-F., "Two-Dimensional Application of Ice Cover Progression Theories in a Large River," MS Thesis, Clarkson University, Potsdam, N.Y., April 1986.
49. Hopper, H. R. and Raban, R. R., "Hanging Dams in the Manitoba Hydro System," *Proceedings of Workshop on Hydraulic Resistance of River Ice*, Burlington, Sept. 1980, pp. 195–208.
50. Jumikis, A. R., *Thermal Geotechnics*, Rutgers University Press, New Brunswick, N.J., 1977.
51. Kivisild, H. R., "Hydrodynamical Analysis of Ice Floods," *8th IAHR Congress*, Paper 23F, Montreal, Canada, Aug. 1959.
52. Krutskih, B. A., Gudkovic, A. M., and Sokolov, A. L., eds., *Ice Forecasting Techniques for the Arctic Seas*, Amerind Publishing Co. Pvt. Ltd., New Delhi, India, 1976.
53. Larsen, P. A., "Head Losses Caused by an Ice Cover on Open Channels," *Jour. Boston Soc. Civil Engineers*, 1969, pp. 45–67.
54. Larsen, P. A., "Hydraulic Roughness of Ice Covers," *Journal of the Hydraulics Division*, ASCE, Vol. 99, No. HY1, Proc. Paper 9498, Jan. 1973, pp. 111-119.
55. Matousek, V., "Types of Ice Run and Conditions for Their Formation," *Ice Symposium 1984*, IAHR, Hamburg, 1984, pp. 315–327.
56. Michel, B., "Winter Regime of Rivers and Lakes," *Cold Regions Science and Engineering Monograph III-Bla*, Cold Regions Research and Engineering Lab., U.S. Army, Hanover, N.H., Apr. 1971, 131 p.
57. Michel, B., *Ice Mechanics*, Les Presses de L'Universte Laval, Quebec, 1978, 499 p.
58. Michel, B., "Ice Accumulations at Freeze-up or Break-up," *IAHR Symposium on Ice Problems*, Lulea, Vol. 2, 1978, pp. 301–317.
59. Michel, B., "Comparison of Field Data with Theories on Ice Cover Progression in Large Rivers," *Canadian Journal of Civil Engineering*, Vol. 11, 1984, pp. 798–814.
60. Michel, B. and Drouin, N., "Equilibrium of an Underhanging Dam at the LaGrande River," *Report GCS-75-03-01*, Universite d'Laval, 1975, 8 p.
61. Michel, B. and Drouin, M., "Courbes de remous sous les couverts de Glace de la Grande Riviere," *Canadian Journal of Civil Engineering*, Vol. 8, 1981, pp. 351–363.
62. Nezhikhovskiy, R. A., "Coefficients of Roughness of Bottom Surface of Slush Ice Cover," *Soviet Hydrology: Selected Papers*, No. 2, 1964, pp. 127–148.
63. O'Neil, K. and Ashton, G. D., "Bottom Heat Transfer to Water Bodies in Winter," *Special Report 81-18*, Cold Regions Research and Engineering Lab., U.S. Army, Hanover, N.H., 1981.
64. Osterkemp, T. E., "Frazil Ice Formation: A Review," *Journal of the Hydraulics Division*, ASCE, Vol. 104, No. HY9, 1978, pp. 1239–1255.
65. Paily, P. P., Macagno, E. O., and Kennedy, J. F., "Winter-Regime Surface Heat Loss from Heated Streams," *IIHR Report No. 155*, Iowa Institute of Hydraulic Research, Iowa City, Iowa, Mar. 1974, 137 pp.
66. Pariset, E. and Hausser, R., "Formation and Evolution of Ice Covers on Rivers," *Transactions, Engineering Institute of Canada*, Vol. 5, No. 1, 1961, pp. 41–49.
67. Pariset, E., Hausser, R., and Gagnon, A., "Formation of Ice Covers and Ice Jams in Rivers," *Journal of the Hydraulics Division*, ASCE, Vol. 92, Nov. 1966, pp. 1–24.
68. Pasquarell, G. C., "Flow Distribution in Ice Covered Channel Networks," *M.S. Thesis*, Clarkson College, Potsdam, N.Y., 1983.
69. Perovich, D. K. and Grenfell, T. C., "A Theoretical Model of Radiative Transfer in Young Sea Ice," *Journal of Glaciology*, Vol. 28, No. 99, 1982, pp. 341–356.
70. Petryk, S. and Boisvert, R., "Simulation of Ice Conditions in Channels," *Proceedings of the Specialty Conference on Computer Applications in Hydrotechnical and Municipal Engineering*, Toronto, 1978, pp. 239–259.
71. Petryk, S., Panu, U., and Clement, F., "Recent Improvements in Numerical Modelling of River Ice," *International Symposium on Ice*, IAHR, Quebec, Vol. 1, 1981, pp. 426–435.

72. Petryk, S., Panu, U., Kartha, V. C., and Clement, F., ''Numerical Modelling and Predictability of Ice Regime in Rivers,'' *International Symposium on Ice*, IAHR, Quebec, Vol. 1, 1981, pp. 436–449.
73. Petukhov, B. S. and Popov, V. N., ''Theoretical Calculation of Heat Exchange and Frictional Resistance in Turbulent Flow in Tubes of an Incompressible Fluid with Variable Physical Properties,'' *High Temperature*, Vol. 1, No. 1, 1963.
74. Pivovarov, A. A., *Thermal Conditions in Freezing Lakes and Rivers*, John Wiley & Sons, Inc., New York, N.Y., 1973, 135 p.
75. Pratte, B. D., ''Review of Flow Resistance of Consolidated Smooth and Rough Ice Covers,'' *Canadian Hydrological Symposium: 79, Proceedings*, Vancouver, May 1979, pp. 52–92.
76. Rosenberg, H. B. and Pentland, R. L., ''Accuracy of Winter Streamflow Records,'' *Proceeding, Eastern Snow Conference*, Hartford, Connecticut, Feb. 1966, pp. 51–72.
77. Rouse, H., *Elementary Mechanics of Fluids*, John Wiley Sons, Inc., 1946, p. 217.
78. Shen, H. T., ''Surface Heat Loss and Frazil Ice Production in the St. Lawrence River,'' *Water Resources Bulletin*, Vol. 16, No. 6, 1981, pp. 996–1001.
79. Shen, H. T., ''Hydraulic Resistance of River Ice,'' *Frontiers in Hydraulic Engineering*, ASCE, Cambridge, MA, 1983, pp. 224–229.
80. Shen, H. T., ''Mathematical Modeling of River Ice Process,'' *Water for Resource Development*, ASCE, Coeur d'Alene, Idaho, 1984, pp. 554–558.
81. Shen, H. T. and Ackermann, N. L., "Winter Flow Distribution in River Channels," *Journal of the Hydraulics Division*, ASCE, Vol. 106, No. HY5, May 1980, pp. 805–817.
82. Shen, H. T. and Ruggles, R. W., ''Winter Heat Budget and Frazil Ice Production in the Upper St. Lawrence River,'' *Water Resources Bulletin*, Vol. 18, No. 2, Apr. 1982, pp. 251–257.
83. Shen, H. T. and VanDeValk, W. A., ''Field Investigation of St. Lawrence River Hanging Ice Dams,'' *IAHR Ice Symposium*, Hamburg, Aug. 1984, pp. 241–250.
84. Shen, H. T., Ruggles, R. W., and Batson, G. B., ''Field Investigation of St. Lawrence River Hanging Ice Dams, Winter of 1983–84,'' *Report DTSL55-84-C-C0085A*, U.S. Department of Transportation, Washington, D.C., Aug. 1984, 85 p.
85. Shen, H. T. and Yapa, P. D., "Flow Resistance of River Ice Cover," *Journal of Hydraulic Engineering*, ASCE, Vol. 112, No. 2, February 1986, pp. 142–156.
86. Shen, H. T. and Yapa, P. D., ''A Unified Degree-Day Method for River Ice Cover Thickness Simulation,'' *Canadian Journal of Civil Engineering*, Vol. 12, 1985, pp. 54–62.
87. Shen, H. T., Foltyn, E. P., and Daly, S. F., ''Forecasting Water Temperature Decline and Freeze-up in Rivers,'' *CRREL Report 84-19*, U.S. Army Cold Regions Research and Engineering Laboratory, Hanover, N.H., July 1984, 17 p.
88. Shen, H. T. and Chiang, L. A., ''Simulation of Growth and Decay of River Ice Cover,'' *Journal of Hydraulic Engineering*, ASCE, Vol. 110, No. 7, July 1984, pp. 958–971.
89. Shen, H. T. and Yapa, P. D., ''Computer Simulation of Ice Cover Formation in the Upper St. Lawrence River,'' *Workshop on Hydraulics of River Ice*, Fredericton, N.B., June 1984, pp. 227–246.
90. Shishokin, S. A., ''Investigation of the Bouguer–Lambert Formula for the Penetration of Solar Radiation into Ice,'' *Soviet Hydrology: Selected Papers*, Vol. 3, 1969, pp. 287–295.
91. Shulyakovskii, G. L., *Manual of Forecasting Ice-Formation for Rivers and Inland Lakes*, Israel Program for Scientific Translation Ltd., Jerusalem, 1966, 245 p.
92. Shulyakovskii, G. L., ''On a Model of the Ice Breakup,'' *Soviet Hydrology: Selected Papers*, No. 1, 1972, pp. 21–27.
93. Simon, Li and Associates, *Engineering Analysis of Fluvial Systems*, 1982, pp. 6.1–6.35.
94. Simonsen, C. P. S. and Carson, R. W., ''Ice Processes During Construction of Limestone Generating Station,'' *Proceedings of the Third International Hydrotechnical Conference*, Quebec City, 1977.
95. Starosolszky, O. (ed.), ''Multilingual Ice Terminology,'' International Association for Hydraulic Research, Research Center for Water Resources, Budapest, 1977 and 1980.
96. Starosolszky, O., ''Ice in Hydraulic Engineering,'' *Report 70-1*, Norwegian Institute of Technology, Trondheim, 1969, 165 pp.
97. Tang, P. W. and Davar, K. S., ''Forecasting the Initiation of Ice Breakup on the Nashwaak River,'' *Workshop on Hydraulics of River Ice*, Fredericton, N.B., June 1984, pp. 65–94.
98. Tatinclaux, J.-C., ''Equilibrium Thickness of Ice Jams,'' *Journal of the Hydraulics Division*, ASCE, Vol. 103, No. HY9, Sept. 1977, pp. 959–974.
99. Tatinclaux, J.-C. and Gogus, M., ''Stability of Floes Below a Floating Cover,'' Vol. 1, *Internal Symposium on Ice*, Quebec City, Canada, 1981, pp. 298–311.
100. Tatinclaux, J.-C. and Gogus, M., ''Asymmetric Plane Flow with Application to Ice Jams,'' *Journal of Hydraulic Engineering*, ASCE, Vol. 109, No. 11, Nov. 1983, pp. 1540–1554.
101. Tsang, G. and Beltaos, S., *Proceedings of Workshop on Hydraulic Resistance of River Ice*, Environment Canada, Burlington, Canada, 1980, 301 pp.
102. Uzuner, M. S., ''The Composite Roughness of Ice Covered Streams,'' *Journal of Hydraulic Research*, Vol. 13, No. 1, Jan. 1975, pp. 79–101.
103. Uzuner, M. S. and Kennedy, J. F., ''Stability of Floating Ice Blocks," *Journal of the Hydraulics Division*, ASCE, Vol. 98, No. HY12, Dec. 1972, pp. 2117–2133.
104. Uzuner, M. S. and Kennedy, J. F., ''Theoretical Model of

River Ice Jams,'' *Journal of the Hydraulic Division*, ASCE, Vol. 102, Sept. 1976, pp. 1365–1383.

105. U.S. Army Corps of Engineers, ''Ice Engineering,'' *Engineering Manual*, EM1110-2-1612, Washington, D.C., Oct. 1982.

106. Wake, A. and Rumer, R. R., Jr., ''Modeling Ice Regime of Lake Erie,'' *Journal of the Hydraulics Division*, ASCE, Vol. 105, No. HY7, July 1979, pp. 827–844.

107. Wigle, T. E., ''Investigation into Frazil, Bottom Ice and Surface Ice Formation in the Niagara River,'' *IAHR Symposium on Ice and Its Action on Hydraulic Structures*, Reykjavik, Iceland, Sept. 1976, Paper 2.8.

108. Yapa, P. D., ''Unsteady Flow Simulation of Rivers with an Ice Cover," Thesis presented to Clarkson University, at Potsdam, N.Y., in 1983, in partial fulfillment of the requirements for the degree of Doctor of Philosophy.

109. Yapa, P. D. and Shen, H. T., ''Roughness Characteristics of the Upper St. Lawrence River,'' *XXth Congress of the IAHR*, Moscow, U.S.S.R., September 1983.

110. Yapa, P. D. and Shen, H. T., "Unsteady Flow Simulation for an Ice-Covered River," *Journal of Hydraulic Engineering*, ASCE, Vol. 112, No. 11, November 1986, pp. 1036–1049.

SECTION FOUR

Mechanics/Solid Mechanics

Similitude Theory: Plates and Shells Analysis

Boris L. Krayterman*

INTRODUCTION

Analysis of plate-shell type structures very seldom can be done by closed-form solutions. The presence of different irregularities (discrete loads, nonhomogeneous boundary conditions, middle surface discontinuities, material and geometry nonlinearities, etc.) forces the application of numerical methods of analysis and physical modeling.

The disadvantage of physical modeling before the closed-form solutions consists of covering of single cases (particular load, geometry of plates and shells) obtained by the use of expansive computer or experimental techniques. Extrapolation of the single numerical or experimental solution on a certain group of plate (or shell) structures becomes an efficient tool for cost reduction.

The plate-shell structures can be divided into *groups of similar structures.* The stress-strain-load components of these similar structures (or at least part of them) are assumed to be similar, i.e., to have a different scale only. Within one group the structures should satisfy a set of *similitude requirements.* These similitude requirements can be obtained by utilizing either dimensional analysis or similitude theory. In both cases the similitude requirements consist of formulating the minimum nondimensional quantities π (pi) which are equal for one group's structures.

Dimensional analysis [1,2,3] deals with fundamental measures—dimensions of length, force, time, temperature, and electric charge—and requires a good insight into the nature of investigated problems for formulation of pi terms. Buckingham [4] has formulated a general pi theorem which states that *any dimensionally homogeneous equation involving certain physical quantities* $F(X_1, X_2, \ldots, X_n) = 0$ *can be reduced to an equivalent equation* $G(\pi_1, \pi_2, \ldots, \pi_m) = 0$ *which involves a complete set of dimensionless products* π_1. Here: $r = n - m$ is the number of fundamental measures that are involved. By the use of dimensional analysis, variables that are in themselves dimensionless (strain, Poisson's ratio, angles) cannot be chosen in the set of r variables, and obtaining pi terms is open to personal preference. Langhaar [5] summarizes difficulties that arise through the formulation of pi terms with dimensional analysis:

> Frequently the question arises: How do we know that a certain variable affects a phenomenon? To answer this question, one must understand enough about the problem to explain why and how the variable influences the phenomenon. Before one undertakes the dimensional analysis of a problem, he should try to form a theory of the mechanism of the phenomenon. Even a crude theory usually discloses the actions of the more important variables. If the differential equations that govern the phenomenon are available, they show directly which variables are significant.

Similitude theory [6,7,8] is based on the equations related to the physical phenomena and boundary conditions for this phenomenon. In addition to the general pi theorem (first theorem), the following two theorems can be formulated:

1. Second theorem, which proceeds from the equations related to the physical phenomena and establishes *the necessary conditions* of similitude.
2. Third theorem, which proceeds from the boundary conditions related to the physical phenomena and establishes *the sufficient conditions* of similitude.

In general, the use of similitude theory gives less nondimensional quantities π (pi) than those obtained by dimensional analysis; therefore, similitude theory offers more possibilities for modeling and simplification of plates-shells analysis. Formulation of pi terms with similitude theory is much simpler than with dimensional analysis and is rather formal.

*Mechanical Engineering Department, University of Maryland, College Park, MD

Use of differential equations for formulation of dimensionless products π_1 bears the assumptions implemented in the derivation of these equations, i.e., the phenomenon is considered with a special insight into a problem based on excluding some of the second-order importance effects and consideration of first-order importance effects only.

Despite dimensional analysis, which uses the governing differential equations of the phenomenon for formulation of general dimensional equation $F(X_1, X_2, \ldots, X_n) = 0$ as per References 3 and 5, similitude theory uses these governing equations along with prescribed boundary and initial conditions for obtaining similitude requirements, i.e., for obtaining dimensionless products.

If governing differential equations are not available for the phenomenon, dimensional analysis is the only way to formulate both dimensional equation $F(X_1, X_2, \ldots, X_n) = 0$ and nondimensional $G(\pi_1, \pi_2, \ldots, \pi_m) = 0$.

Two numerical or experimental solutions for plates (or shells) that belong to one group of similar structures can be compared by the use of similitude scales. If compared components (stresses, strains, displacements) are *almost* identical, one can have additional information about the precision of utilized numerical or experimental solutions. This control is more effective when similitude theory is used because it gives more opportunities for multiscale similitude.

In some cases, dimensional analysis alone can simplify numerical [8] or experimental [3] solutions of considered problem by specifying the resulting formulas. Similitude theory sometimes allows one to achieve even better results—to discover nonlinear relationships between stress-displacements and acting loads [9].

The nondimensional quantities π obtained with similitude theory can be supplemented by additional π terms obtained with dimensional analysis.

FORMULATION OF THE NECESSARY AND SUFFICIENT CONDITIONS OF SIMILITUDE

The necessary and sufficient conditions of similitude have been expressed in the form of second and third theorems of similitude by M. V. Kirpichev and A. A. Gukhman in the early 1930s for heat transfer problems and later have been generalized in References 6 and 7.

Modeling of large deflected shallow shells with similitude theory has been considered by D. V. Monakhenko and V. B. Proskuryakov in Reference 10. Application of similitude theory to analysis and modeling of thin elastic shells and plates has been considered in References 9 and 11 to 16.

Assume the displacements, stresses, strains, etc., for the plate-shell structures of *one similar group* are T_1 and T_2. Then by determination they relate to each other by the equation:

$$T_2 = C_T \times T_1 \tag{1}$$

where

C_T = the similitude constant or a scale factor
T_1 = components of plate (or shell) 1
T_2 = components of plate (or shell) 2

One of the differential equations that governs the considered similar group can be written in terms of T_1 as follows:

$$\sum_{i=1}^{n} D_i(T_1) = 0 \tag{2}$$

The uniqueness of simultaneous equations [Equation (2)] is determined by boundary condition equations of the following type:

$$\sum_{i=1}^{m} L_i(T_1) = 0 \tag{3}$$

D_i and L_i in Equations (2) and (3) are algebraic, differential, integral, or mixed operators. Identical equations for the plate (or shell) with T_2 components are as follows:

$$\sum_{i=1}^{n} D_i(T_2) = 0 \tag{4}$$

$$\sum_{i=1}^{m} L_i(T_2) = 0 \tag{5}$$

Operators D_i and L_i in Equations (4) and (5) are identical to those in Equations (2) and (3) because both 1 and 2 plates (or shells) belong to one similar group.

With the use of Equation (1), Equations (4) and (5) can be rewritten in terms of T_1 components as follows:

$$\sum_{i=1}^{n} K_i D_i(T_1) = 0 \tag{6}$$

$$\sum_{i=1}^{m} r_i L_i(T_2) = 0 \tag{7}$$

K_i and r_i represent the products of similitude constants C_T raised to suitable powers to match the operators D_i and L_i of the previous equations; these powers are real numbers. Examples of obtaining products K_i (or r_i) are given for different types of operators D_i (or L_i) as follows:

Example 1

Determine K_1 product for operator D_1:

$$D_1 = \frac{Eh^3}{12(1 - \nu^2)} \frac{\partial^t w_1}{\partial x^s \partial y^{t-s}} \tag{8}$$

This operator can be rewritten in terms of components T_1 and T_2:

$$D_1 = \frac{E_1 h_1^3}{12(1 - \nu_1^2)} \frac{\partial^t w_1}{\partial x_1^s \partial y_1^{t-s}};$$
$$D_1 = \frac{E_2 h_2^3}{12(1 - \nu_2^2)} \frac{\partial^t w_2}{\partial x_2^s \partial y_2^{t-s}} \quad (9)$$

After substitution of Equation (1) in Equation (9) (second equation), D_1 operator can be replaced by the following:

$$D_1 = \frac{(C_E E_1)(C_h h_1)^3}{12[C_\nu(1 - \nu_1^2)]} \frac{\partial^t (C_w w_1)}{\partial (C_x x_1)^s \partial (C_y y_1)^{t-s}}$$
$$= \frac{C_E C_h^3 C_w}{C_\nu C_x^s C_y^{t-s}} \frac{E_1 h_1^3}{12(1 - \nu_1^2)} \frac{\partial^t w_1}{\partial x_1^s \partial y_1^{t-s}} \quad (10)$$

In comparing Equation (9) with Equation (10), one concludes that:

$$K_1 = C_E\, C_h^3\, C_w\, C_\nu^{-1}\, C_x^{-s}\, C_y^{s-t} \quad (11)$$

Example 2

Determine K_3 product for operator D_3:

$$D_3 = \sin \omega t \int_0^r q r^s dr \quad (12)$$

Again, this operator can be rewritten in terms of components T_1 and T_2:

$$D_3 = \sin(\omega_1 t_1) \int_0^{r_1} q_1 r_1^s dr_1;$$
$$D_3 = \sin(\omega_2 t_2) \times \int_0^{r_2} q_2 r_2^s dr_2 \quad (13)$$

Substitution of Equation (1) in Equation (13) (second equation) yields:

$$D_3 = \sin[(C_\omega \omega_1)(C_t t_1)] \times \int_0^{C_r r_1} (C_q q_1)(C_r r_1)^s d(C_r r_1)$$
$$= \sin[C_\omega C_t(\omega_1 T_1)] \times C_q C_r^{s+1} \int_0^{r_1} q_1 r_1^s dr_1 \quad (14)$$

Due to the identity of this operator within the considered similar group, the product $C_\omega C_t$ must be equal to unity or:

$$C_\omega = C_t^{-1} \quad (15)$$

Finally:

$$K_3 = C_q C_r^{s+1} \quad (16)$$

If the products K_i and r_i in Equations (6) and (7) are equal to unity, then the plate (or shell) 2 degenerates into plate (or shell) 1. This trivial case is not a subject of our interest for both numerical and experimental solutions of plate-shell structures.

Equations (6) and (7) contain n and m products K_i and r_i of similitude constants C_T. They can be transformed in $(n - 1)$ and $(m - 1)$ products K_i' and r_i' accordingly by dividing all products of the equation by one of them, e.g., K_1 and r_1:

$$D_1(T_1) + \sum_{i=2}^{n} K_i' D_i(T_1) = 0 \quad (17)$$

$$L_1(T_1) + \sum_{i=2}^{m} r_i' L_i(T_1) = 0 \quad (18)$$

In similitude theory, products K_i' and r_i' are known as *indicators of similitude.*

The second theorem of similitude provides the necessary conditions of similitude and states that *indicators of similitude obtained from the basic differential equations that govern the phenomenon must be equal to unity,* that is

$$K_i' = 1 \ (i = 2, \ldots, n) \quad (19)$$

The third theorem of similitude provides the sufficient conditions of similitude and states that *indicators of similitude obtained from the conditions of uniqueness (boundary, initial conditions, etc.) that are formulated for the phenomenon must be equal to unity,* that is:

$$r_i' = 1 \ (i = 2, \ldots, m) \quad (20)$$

As follows from Equation (19), derivation of indicators of similitude requires at least two terms either in the basic equation or in the equation of uniqueness for the phenomenon.

Indicators of similitude can be replaced by ratios of components T_1 and T_2, i.e., expressed by pi terms. Thus, the indicator equation of similitude:

$$C_A^i\, C_B^j\, C_D^k\, C_E^l = 1 \quad (21)$$

in which C_A, C_B, C_D, and C_E = the similitude scale factors expressed as $C_A = A_2/A_1$; $C_B = B_2/B_1$; $C_D = D_2/D_1$; $C_E = E_2/E_1$ can be replaced by the following equation:

$$A_1^i\, B_1^j\, D_1^k\, E_1^l = A_2^i\, B_2^j\, D_2^k\, E_2^l = \text{idem} \quad (22)$$

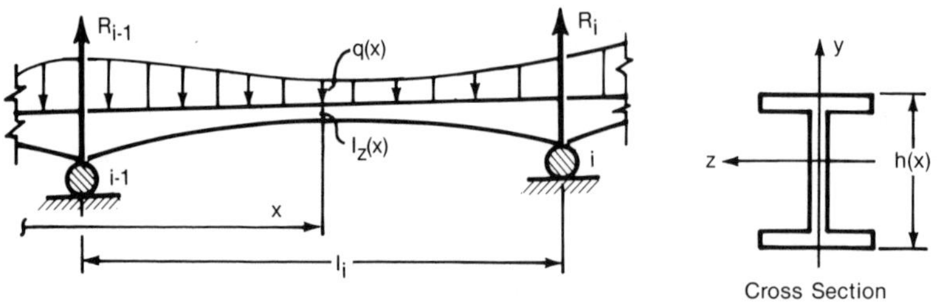

FIGURE 1. Multispan girder example.

Here "idem" [6,7] in Latin means "identical" for all plates (or shells) of the similar group being calculated for identical points of the phenomenon. The corresponding pi term is $A^i\,B^j\,D^k\,E^l$.

The following examples illustrate applications of similitude theory to some of the structures specified in Reference 3.

Example 3

Consider the application of similitude theory to the analysis of a multispan girder subjected to a known unit load $q(x)$ along the spans l_i (Figure 1).

The subjects of investigation are normal stresses σ in the beam cross section and beam deflections w. The governing differential equations for the considered phenomenon can be found in Reference [17]:

$$\frac{d^2}{dx^2}\left[EI_z(x)\frac{d^2w}{dx^2}\right] = -q(x);\ \sigma = Ey\frac{d^2w}{dx^2};$$
$$Q = \frac{d}{dx}\left[EI_z(x)\frac{d^2w}{dx^2}\right] \tag{23}$$

The boundary conditions at supports $(i-1)$ and i reflect equilibrium of vertical forces and compatibility of twist angles and beam curvatures (there are no concentrated moments acting on the girder).

For $x = x_{i-1}$:

$$w = 0;\ \frac{dw_{i-1}}{dx} = \frac{dw_i}{dx};\ \frac{d^2w_{i-1}}{dx^2} = \frac{d^2w_i}{dx^2};$$
$$Q_{i-1} + Q_i + R_{i-1} = 0 \tag{24}$$

For $x = x_i$:

$$w = 0;\ \frac{dw_i}{dx} = \frac{dw_{i+1}}{dx};\ \frac{d^2w_i}{dx^2} = \frac{d^2w_{i+1}}{dx^2};$$
$$Q_i + Q_{i+1} + R_i = 0 \tag{25}$$

Assuming Equation (23) is written for beam 1, using Equation (1) one can rewrite these equations in the form of Equation (17), i.e., with indicators of similitude, as follows:

$$\frac{C_E C_I C_w}{C_x^4 C_q}\frac{d^2}{dx^2}\left[EI_z(x)\frac{d^2w}{dx^2}\right] = -q(x);$$
$$\frac{C_\sigma C_x^2}{C_E C_y C_w}\sigma = Ey\frac{d^2w}{dx^2}; \tag{26}$$
$$\frac{C_Q C_x^3}{C_E C_I C_w}Q = \frac{d}{dx}\left[EI_z(x)\frac{d^2w}{dx^2}\right]$$

Similarly, the boundary conditions in Equations (24) and (25) can be rewritten in the form of Equation (18), as follows:

$$\text{For } x = x_{i-1}:\quad w = 0;\ \frac{C_{w_{i-1}}}{C_{w_i}}\times\frac{dw_{i-1}}{dx} = \frac{dw_i}{dx};$$
$$\frac{C_{w_{i-1}}}{C_{w_i}}\frac{d^2w_{i-1}}{dx^2} = \frac{d^2w_i}{dx^2} \tag{27}$$
$$\frac{C_{Q_{i-1}}}{C_{Q_i}}Q_{i-1} + Q_i + \frac{C_{R_{i-1}}}{C_{Q_i}}R_{i-1} = 0$$

$$\text{For } x = x_i:\quad w = 0;\ \frac{dw_i}{dx} = \frac{C_{w_{i+1}}}{C_{w_i}}\times\frac{dw_{i+1}}{dx};$$
$$\frac{d^2w_i}{dx^2} = \frac{C_{w_{i+1}}}{C_{w_i}}\times\frac{d^2w_{i+1}}{dx^2} \tag{28}$$
$$Q_i + \frac{C_{Q_{i+1}}}{C_{Q_i}}Q_{i+1} + \frac{C_{R_i}}{C_{Q_i}}R_i = 0$$

By equating indicators of similitude from Equation (26) to unity (the second theorem of similitude), the following algebraic equations are obtained:

$$\frac{C_E C_I C_w}{C_x^4 C_q} = 1;\ \frac{C_\sigma C_x^2}{C_E C_y C_w} = 1;\ \frac{C_Q C_x^3}{C_E C_I C_w} = 1 \tag{29}$$

Similarly, from boundary conditions in Equations (27) and (28):

$$C_{w_{i-1}}/C_w = 1;\ C_{Q_{i-1}}/C_Q = 1;\ C_{R_{i-1}}/C_Q = 1$$
$$C_{w_{i+1}}/C_w = 1;\ C_{Q_{i+1}}/C_Q = 1;\ C_{Ri}/C_Q = 1 \qquad (30)$$

Here deflection w_i and shear force Q_i of considered beam span l_i is replaced by w and Q accordingly, as used for basic equations.

Indicators of similitude [Equation (30)] state that all transformations to be done for girder span l_i must be done for all other beam spans. Otherwise, there is no similitude. In addition, C_R scale must be equal to C_Q.

By uniqueness requirements, the third theorem specifies not only boundary conditions but other requirements which justify the use of basic differential equations that govern the phenomenon. For the considered beam example, the following restrictions are to be applied for all beams of the similar group:

1. The beam is slightly curved; the radius of curvature in the plane of bending is at least 10 times the depth [18].
2. The beam is long in proportion to its depth; the span/depth ratio is 8 or more [18].
3. The maximum stress does not exceed the proportional limit.
4. The maximum beam deflection does not exceed ¼ to ½ of its depth (geometrically linear problem).
5. The girders are braced against torsion and their stability and local buckling are provided.

The second and third equations in Equation (29) contain similitude scales of two unknown components ϱ and w, and Q and w, respectively. C_w constant can be excluded from these equations. The beam length scale factor C_l is equal to C_x, and cross section depth scale factor C_h is equal to C_y. Considering this, Equation (29) can be replaced by the following:

$$\frac{C_w C_E C_I}{C_q C_l^4} = 1;\ \frac{C_\sigma C_I}{C_q C_l^2 C_h} = 1;$$
$$\frac{C_Q}{C_q C_l} = 1;\ \frac{C_x}{C_l} = 1 \qquad (31)$$

Equation (31) gives the following pi terms (or generalized variables [14]):

$$\xi = \frac{x}{l};\ \overline{w}(\xi) = \frac{w(x)EI_0}{q_0 l^4};\ \overline{\sigma}(\xi) = \frac{\sigma(x)I_0}{q_0 l^2 h_0};$$
$$\overline{Q}(\xi) = \frac{Q(x)}{q_0 l};\ \overline{R} = \frac{R}{q_0 l};\ \eta = \frac{y}{h_0} \qquad (32)$$

where q_o, l, h_o, I_o = characteristic load intensity, beam span, depth of cross section, and moment of inertia, accordingly.

Pi term $\sigma I_o/q_o l^2 h_o$ in Equation (32) provides more variations than identical pi term $\sigma l/q_o$ [3] because it gives more freedom in the selection of cross-sectional scale factors. Equation (23) can be replaced by an identical equation that is expressed in *generalized variables* [Equation (32)]:

$$\frac{d^2}{d\xi^2}\left[\overline{I}(\xi)\frac{d^2\overline{w}}{d\xi^2}\right] = -P(\xi);$$
$$\overline{\sigma} = \eta\frac{d^2\overline{w}}{d\xi^2}; \qquad (33)$$
$$\overline{Q} = \frac{d}{d\xi}\left[\overline{I}(\xi)\frac{d^2\overline{w}}{d\xi^2}\right]$$

where:

$\overline{I}(\xi) = I_z(x)/I_0$ = nondimensional function of moment of inertia

$I_z(x) = I_0 \times I\,(x/l)$ distribution along the beam length

$P(\xi) = q(x)/q_0$ = nondimensional function of unit load

$q(x) = q_0 \times P(x/l)$ distribution along the beam length

For invariable cross section and uniform unit load:

$$I(\xi) = 1;\ P(\xi) = 1 \qquad (34)$$

The boundary conditions [Equations (24) and (25)] in generalized variables [Equation (32)] are as follows:

For $\xi = \xi_{i-1}\overline{w} = 0$;

$$\frac{d\overline{w}_{i-1}}{d\xi} = \frac{d\overline{w}}{d\xi};$$
$$\frac{d^2\overline{w}_{i-1}}{d\xi^2} = \frac{d^2\overline{w}}{d\xi^2}; \qquad (35)$$
$$\overline{Q}_{i-1} + \overline{Q} + \overline{R}_{i-1} = 0$$

For $\xi = \xi_i\overline{w} = 0$;

$$\frac{d\overline{w}_{i+1}}{d\xi} = \frac{dw}{d\xi};$$
$$\frac{d^2\overline{w}_{i+1}}{d\xi^2} = \frac{d^2\overline{w}}{d\xi^2}; \qquad (36)$$
$$\overline{Q}_{i+1} + \overline{Q} + \overline{R}_i = 0$$

Here similitude theory alone shows that beam deflection w is a linear function of $q_o l^4/EI_o$, normal stress σ is a linear function of $q_o l^2 h_o / I_o$, shear force Q is a linear function of $q_o l$, and just one numerical or experimental solution covers the whole group of similar girders.

Example 4

The natural frequencies of flat elastic plates transverse vibration can be found by solution of the partial differential equation which, with assumption of small deflections w ($w \leq 1/4 - 1/2 h$) can be written as follows:

$$\frac{\partial^4 w}{\partial x^4} + 2\frac{\partial^4 w}{\partial x^2 \partial y^2} + \frac{\partial^4 w}{\partial y^4} + \frac{mh}{Eh^3/12(1-\nu^2)}\frac{\partial^2 w}{\partial t^2} = 0 \tag{37}$$

The plate geometry, coordinates, and boundary conditions are specified in Figure 2.

Initial and boundary conditions for the plate are covered by the following equations:

$$\text{At } t = 0\text{: } w(x, y, o) = 0; \frac{\partial w(x, y, o)}{\partial t} = 0 \tag{38}$$

At $x = 0$ or $x = a$:

$$w = 0;$$

$$\frac{\partial w}{\partial x} = 0 \text{ (for fixed edges)}$$

$$w = 0; M_x = \frac{Eh^3}{12(1-\nu^2)} \times \left(\frac{\partial^2 w}{\partial x^2} + \nu\frac{\partial^2 w}{\partial y^2}\right) = 0 \tag{39}$$

(for hinged edges)

The latter can be replaced by the following:

$$\frac{\partial^2 w}{\partial x^2} = 0 \tag{40}$$

At $y = 0$ or $y = b$:

$$w = 0; \frac{\partial w}{\partial y} = 0 \text{ (for fixed edges)} \tag{41}$$

$$w = 0; \frac{\partial^2 w}{\partial y^2} = 0 \text{ (for hinged edges)} \tag{42}$$

Assuming Equation (37) is written for plate 1 and using Equation (1), one can rewrite this equation in the form of Equation (17), i.e., with indicators of similitude:

$$\frac{\partial^4 w}{\partial x^4} + 2\frac{C_x^2}{C_y^2}\frac{\partial^4 w}{\partial x^2 \partial y^2} + \frac{C_x^4}{C_y^4} \times \frac{\partial^4 w}{\partial y^4} + \frac{C_m C_x^4 C_v}{C_E C_h^2 C_t^2}\frac{mh}{Eh^3/12(1-\nu^2)}\frac{\partial^2 w}{\partial t^2} = 0 \tag{43}$$

Initial and boundary conditions [Equations (38) to (42)] are *one term* equations and therefore do not yield indicators of similitude. As per the second theorem of similitude:

$$Cx/Cy = 1;\ C_m\, C_x^4\, C_\nu / C_E\, C_h^2\, C_t^2 = 1 \tag{44}$$

The third theorem of similitude results in the following for the plates of a similar group:

- identity of initial and boundary conditions
- requirements of plate material elasticity; plate small deflections; the thickness is not more than about one-quarter of the least transverse dimension [18].

The plate plane dimensions scale factors C_a and C_b are equal to C_x and C_y, respectively. As follows from Equation (44) all plates of a similar group must have geometrically similar planes. This means that scale factors C_a and C_b can be replaced by one scale factor, possibly C_a. With these comments, Equation (44) can be replaced by the following:

$$\frac{C_x}{C_a} = 1; \frac{C_y}{C_a} = 1; \frac{C_t C_h C_E^{0.5}}{C_m^{0.5} C_a^2 C_v^{0.5}} = 1 \tag{45}$$

As follows from Equation (44) or (45), the deflection w (x,y,t) cannot be specified for a plate subjected to free transverse vibrations. Note that similitude theory has conformed this statement; there are no C_w scale factors in Equation (45). There is no frequency scale factor C_f in Equation (45) as well as no frequencies $f_1, f_2, \ldots$ in the governing differential equation [Equation (37)]. By knowing the frequency dimension ($\sec^{-1}$), one can express C_f scale factor in terms of C_t scale factor:

$$C_f = C_t^{-1} \tag{46}$$

Equations (45) and (46) provide the structure of the following pi terms or *generalized variables*:

$$\xi - x/a;\ \eta = y/a;$$

$$\theta = \frac{th}{a^2}\sqrt{\frac{E}{m(1-\nu^2)}}; \tag{47}$$

$$\bar{f} = \frac{fa^2}{h}\sqrt{\frac{m(1-\nu^2)}{E}}$$

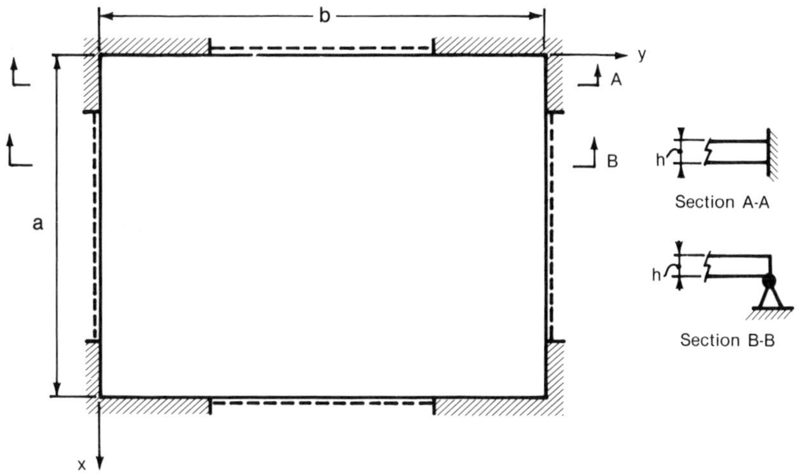

FIGURE 2. Free transverse vibration of a plate.

where:

a = characteristic plate dimension
θ = nondimensional time variable
$\bar{f}$ = nondimensional frequency

Pi term $(fa^2/h)\sqrt{m(1 - \nu^2)/E}$ in Equation (47) provides a wider similar group than pi term $fa\sqrt{m/E}$ [3] because it gives more freedom of selection of plate geometry and material. Equation (37), in terms of generalized variables, is as follows:

$$\frac{\partial^4 w}{\partial \xi^4} + 2\frac{\partial^4 w}{\partial \xi^2 \partial \eta^2} + \frac{\partial^4 w}{\partial \eta^4} + 12\frac{\partial^2 w}{\partial \theta^2} = 0 \qquad (48)$$

The initial and boundary conditions in terms of generalized variables are identical to Equations (38) to (42):

$$\text{At } \theta = 0\text{: } w(\xi,\eta,0) = 0; \ \frac{\partial w(\xi,\eta,o)}{\partial \theta} = 0$$

$$\text{At } \xi = 0 \text{ or } \xi = 1\text{: } w = 0; \ \frac{\partial w}{\partial \xi} = 0 \left(\text{or } \frac{\partial^2 w}{\partial \xi^2} = 0\right)$$

$$(49)$$

$$\text{At } \eta = 0 \text{ or } \eta = b/a\text{: } w = 0; \ \frac{\partial w}{\partial \eta} = 0 \left(\text{or } \frac{\partial^2 w}{\partial \eta^2} = 0\right)$$

A comparison of Equation (48) with Equation (37) shows the effectiveness of similitude theory application for either numerical or experimental analysis of free transverse vibrations of a flat elastic plate. Note that analysis with similitude theory *alone* has shown the following formula for fundamental frequencies:

$$f = \underset{\sim}{K} \times \frac{h}{a^2}\sqrt{\frac{E}{m(1 - \nu^2)}} \qquad (50)$$

Of course, similitude theory could not help to determine the magnitude of vector $\underset{\sim}{K}$ components $K_1, K_2, \ldots$, which yields the frequency spectrum.

Equation (50) can be compared with mathematical solution of differential Equation (37) for different boundary conditions [19]. For the simply supported rectangular plate, natural frequencies can be calculated by the formula:

$$f = \frac{K_n}{2\pi}\sqrt{\frac{Dg}{qa^4}} \qquad (51)$$

where:

$$D = \frac{Eh^3}{12(1 - \nu^2)}$$

$$k_n = \pi^2\left[m_a^2 + \left(\frac{a}{b}\right)^2 m_b^2\right]$$

g = gravitational acceleration
q = uniform load per unit area (including own weight)

For the plate in Figure 2, consider mass density m only, i.e., $m = q/hg$. Finally, after substitution, Equation (51) can be replaced by the following:

$$f = 0.45345[m_a^2 + (a/b)^2 m_b^2] \times \frac{h}{a^2}\sqrt{\frac{E}{m(1 - \nu^2)}} \qquad (52)$$

where:

m_a, m_b = integers characterizing the plate deflected shape
$m_a = 1, 2, \ldots; m_b = 1, 2, \ldots$

Equation (52) is identical to Equation (50) obtained with similitude theory.

The Poisson's ratio influence on frequency f calculation is not a subject of consideration in Example 4. The boundary conditions' influence should be noted. Detailed analysis of the Poisson's ratio influence on plates similitude is considered in the following section.

SIMILITUDE OF THIN, FLAT, ELASTIC PLATES

In most cases, analysis of thin plates can be done either by numerical or by experimental methods. The variety of plate types (isotropic and anisotropic, with small and large deflections, with linearly and nonlinearly elastic material behavior, different plane configurations, under static, dynamic, and temperature loads) allows for a separate consideration of different plates.

Isotropic Linearly Elastic Plates with Small Deflections

The governing differential equation for plates of variable thickness under combined transverse and tangent components of static loads, dynamic loads, and temperature is as follows [19,20,21].

$$D\nabla^4 w + 2\frac{\partial D}{\partial x}\frac{\partial}{\partial x}\nabla^2 w + 2\frac{\partial D}{\partial y} \times \frac{\partial}{\partial y}\nabla^2 w + \nabla^2 D \nabla^2 w - (1-\nu) \times \left(\frac{\partial^2 D}{\partial x^2}\frac{\partial^2 w}{\partial y^2} - 2\frac{\partial^2 D}{\partial x \partial y}\frac{\partial^2 w}{\partial x \partial y} + \frac{\partial^2 D}{\partial y^2}\frac{\partial^2 w}{\partial x^2}\right)$$
$$= q(x, y) + N_x\frac{\partial^2 w}{\partial x^2} + N_y\frac{\partial^2 w}{\partial y^2} + 2N_{xy}\frac{\partial^2 w}{\partial x \partial y} - mh(x, y)\frac{\partial^2 w}{\partial t^2} - kh(x, y)\frac{\partial w}{\partial t} + Z(x, y, t) - \frac{Eh^2(x, y)\alpha}{12(1-\nu)}\nabla^2\lambda \quad (53)$$

where:

$$\nabla^2(\ldots) = \frac{\partial^2(\ldots)}{\partial x^2} + \frac{\partial^2(\ldots)}{\partial y^2};$$

$$\nabla^4(\ldots) = \frac{\partial^4(\ldots)}{\partial x^4} + 2\frac{\partial^4(\ldots)}{\partial x^2 \partial y^2} + \frac{\partial^4(\ldots)}{\partial y^4}$$

$D = \dfrac{Eh^3(x, y)}{12(1-\nu^2)}$ = flexural rigidity of the plate

$q(x, y)$ = intensity of transverse static load per unit area

N_x, N_y, N_{xy} = intensity of tangent statical load per unit length as shown in Figure 3

m = plate mass density; can be replaced by formula $m = q(x, y)/gh(x, y)$ for more general case [19]

$\lambda = -\dfrac{12}{h^2(x, y)}\displaystyle\int_{-h/2}^{h/2} t^{0,z} z dz$ = temperature gradient; for thin plate-shell structures consider linear temperature distribution through the plate thickness $h(x, y)$ shown in Figure 3 and $\lambda(x, y) = [t^0_{top}(x, y) - t^0_{bot}(x, y)]/2$

k = damping coefficient

$Z(x, y, t)$ = perturbation transverse dynamic load per unit area (this load can be harmonic load $Z(x, y, t) = Z(x, y)\sin \omega t$ or step-wise applied constant perturbation load $Z(x, y, t) = Z(x, y)$, etc.).

The plate vibration in tangent to middle plane directions is neglected, and it is assumed the plate has arbitrary plane configuration.

It is assumed that Equation (53) is written for plate 1 and, by substitution of Equation (1), one can equate indicators of similitude to unity and obtain the necessary conditions of similitude.

If plate is subjected to loads q, N_x, N_y, and N_{xy} only, the following equations of similitude can be obtained from Equation (53):

$$C_\nu = 1;\ C_x/C_y = 1;\ C_w C_h^3 C_E/C_q C_l^4 = 1;\ C_N C_l^2/C_E C_h^3 = 1 \quad (54)$$

With consideration of plate characteristic length l ($C_x = C_y = C_l$), Equation (54) can be replaced by the following *generalized variables:*

$$\xi = x/l;\ \eta = y/l;\ \bar{w} = wh_0^3E/q_0l^4;\ \bar{N}_x = N_xl^2/Eh_0^3$$
$$\bar{N}_y = N_yl^2/Eh_0^3;\ \bar{N}_{xy} = N_{xy}l^2/Eh_0^3 \quad (55)$$

where q_0, h_0 = characteristic load intensity (per unit area) and characteristic plate thickness, which are used to describe load and plate thickness by formulas:

$$p(\xi, \eta) = q(x, y)/q_0;$$

$$\bar{h}(\xi, \eta) = h(x, y)/h_0;$$

$$\Omega = \frac{\bar{h}^3(\xi; \eta)}{12(1-\nu^2)}$$

The governing nondimensional differential equation [Equation (53)] in generalized variables from Equation (55) follows:

$$\Omega\nabla^4\overline{w} + 2\frac{\partial\Omega}{\partial\xi}\frac{\partial}{\partial\xi}\nabla^2\overline{w} + 2\frac{\partial\Omega}{\partial\eta} \times \frac{\partial}{\partial\eta}\nabla^2\overline{w} + \nabla^2\Omega\nabla^2\overline{w} - (1-\nu) \times \left(\frac{\partial^2\Omega}{\partial\xi^2}\times\frac{\partial^2\overline{w}}{\partial\eta^2} - 2\frac{\partial^2\Omega}{\partial\xi\partial\eta}\times\frac{\partial^2\overline{w}}{\partial\xi\partial\eta} + \frac{\partial^2\Omega}{\partial\eta^2}\times\frac{\partial^2\overline{w}}{\partial\xi^2}\right) = p(\xi,\eta) + \overline{N}_x\frac{\partial^2\overline{w}}{\partial\xi^2} + \overline{N}_y\frac{\partial^2\overline{w}}{\partial\eta^2} + 2\overline{N}_{xy}\frac{\partial^2\overline{w}}{\partial\xi\partial\eta} \tag{56}$$

For plates loaded with dynamic loads only, the following equations of similitude can be derived from Equation (53):

$$\begin{aligned} &C_\nu = 1;\ C_x/C_y = 1;\\ &C_wC_h^3C_E/C_ZC_l^4 = 1;\\ &C_tC_hC_E^{0.5}/C_m^{0.5}C_l^2 = 1;\\ &C_f = C_t^{-1};\\ &C_kC_l^2/C_E^{0.5}C_m^{0.5}C_h = 1;\\ &C_\omega = C_t^{-1} \text{ (for harmonic load)} \end{aligned} \tag{57}$$

or generalized variables:

$$\begin{aligned} &\xi = x/l;\ \eta = y/l;\\ &w = wh_0^3E/Z_0l^4;\\ &\theta = \frac{th_0}{l^2}\sqrt{\frac{E}{m}};\\ &\overline{k} = \frac{kl^2}{h_0}\cdot\frac{1}{\sqrt{Em}} \end{aligned} \tag{58}$$

where Z_0 = characteristic magnitude of perturbation load used to describe nondimensional loads by the formula:

$$Z(\xi,\eta,\theta) = Z(x,y,t)/Z_0 \tag{59}$$

The governing nondimensional differential equation for plates in values of generalized variables is as follows:

$$\Omega\nabla^4\overline{w} + 2\frac{\partial\Omega}{\partial\xi}\frac{\partial}{\partial\xi}\nabla^2\overline{w} + 2\frac{\partial\Omega}{\partial\eta} \times \frac{\partial}{\partial\eta}\nabla^2\overline{w} + \nabla^2\Omega\nabla^2\overline{w} - (1-\nu) \times \left(\frac{\partial^2\Omega}{\partial\xi^2}\times\frac{\partial^2\overline{w}}{\partial\eta^2} - 2\frac{\partial^2\Omega}{\partial\xi\partial\eta}\times\frac{\partial^2\overline{w}}{\partial\xi\partial\eta} + \frac{\partial^2\Omega}{\partial\eta^2}\times\frac{\partial^2\overline{w}}{\partial\xi^2}\right) = -\overline{h}(\xi,\eta) \times \frac{\partial^2\overline{w}}{\partial\theta^2} - \overline{k}\times\overline{h}(\xi,\eta) \times \frac{\partial\overline{w}}{\partial\theta} + \overline{Z}(\xi,\eta,\theta) \tag{60}$$

If plate is subjected to the thermal loads only, the similitude equations from Equation (53) are as follows:

$$C_\nu = 1;\ C_x/C_y = 1;\ C_wC_h/C_\lambda\, C_\alpha\, C_l^2 = 1 \tag{61}$$

or generalized variables:

$$\xi = x/l;\ \eta = y/l;\ \overline{w} = wh_0/\lambda_0\alpha l^2 \tag{62}$$

where λ_0 = characteristic temperature gradient used to describe nondimensional gradient by the formula:

$$\mu(\xi,\eta) = \lambda(x,y)/\lambda_0 \tag{63}$$

Then Equation (53) can be rewritten in the following nondimensional form:

$$\Omega\nabla^4\overline{w} + 2\frac{\partial\Omega}{\partial\xi}\frac{\partial}{\partial\xi}\nabla^2\overline{w} + 2\frac{\partial\Omega}{\partial\eta} \times \frac{\partial}{\partial\eta}\nabla^2\overline{w} + \nabla^2\Omega\nabla^2\overline{w} - (1-\nu) \times \left(\frac{\partial^2\Omega}{\partial\xi^2}\times\frac{\partial^2\overline{w}}{\partial\eta^2} - 2\frac{\partial^2\Omega}{\partial\xi\partial\eta}\frac{\partial^2\overline{w}}{\partial\xi\partial\eta} + \frac{\partial^2\Omega}{\partial\eta^2}\frac{\partial^2\overline{w}}{\partial\xi^2}\right) = -\frac{\overline{h}^2(\xi,\eta)}{12(1-\nu)}\nabla^2\mu \tag{64}$$

If plates are under the combination of external loads and temperature, consider the following generalized deflections w and temperature gradient $\mu(\xi,\eta)$ in lieu of Equations (62) and (63):

$$\overline{w} = wh_0^3\ E/q_0l^4;\ \mu(\xi,\eta) = \lambda(x,y)\ \alpha h_0^2E/q_0l^2 \tag{65}$$

and add the right-hand term of Equation (64) in Equation (56).

The equations of similitude [Equations (54), (57), and (61)] and pi terms-generalized variables are results of the application of the second similitude theorem to the governing equation [Equation (53)] for isotropic linearly elastic plates with small deflections. Formulas for generalized stresses, moments, transverse shear forces, strains, and curvatures can be obtained from the equations which express the above components in terms of plate deflections w:

$$\sigma_x = -\frac{Eh}{2(1-\nu^2)}\left(\frac{\partial^2 w}{\partial x^2} + \nu\frac{\partial^2 w}{\partial y^2}\right);$$

$$\sigma_y = -\frac{Eh}{2(1-\nu^2)}\left(\frac{\partial^2 w}{\partial y^2} + \nu\frac{\partial^2 w}{\partial x^2}\right);$$

$$\sigma_{xy} = \frac{Eh}{2(1+\nu)}\frac{\partial^2 w}{\partial x \partial y};$$

$$M_x = \sigma_x h^2/6;\ M_y = \sigma_y h^2/6;\ M_{xy} = \sigma_{xy} h^2/6; \tag{66}$$

$$Q_x = -\frac{Eh^3}{12(1-\nu^2)}\frac{\partial}{\partial x}\nabla^2 w;$$

$$Q_y = -\frac{Eh^3}{12(1-\nu^2)}\frac{\partial}{\partial y}\nabla^2 w;$$

$$\varepsilon_x = \frac{h}{2}\frac{\partial^2 w}{\partial x^2};\ \varepsilon_y = \frac{h}{2}\frac{\partial^2 w}{\partial y^2};\ \varepsilon_{xy} = \frac{h}{2}\frac{\partial^2 w}{\partial x \partial y};$$

$$\chi_x = \frac{\partial^2 w}{\partial x^2};\ \chi_y = \frac{\partial^2 w}{\partial y^2};\ \chi_{xy} = \frac{\partial^2 w}{\partial x \partial y}$$

Here the terms with temperature are not included; the proper scale factors are in Equation (61).

The equations of similitude [Equations (54), (57), and (61)] show $C_\nu = 1$, i.e., material of the plates which belong to the similar group must have an identical Poisson's ratio. Therefore, the following *additional equations of similitude* can be written:

$$C_\sigma = C_E C_h C_w / C_l^2;\ C_M = C_E C_h^3 C_w / C_l^2; \tag{67}$$
$$C_Q = C_E C_h^3 C_w / C_l^3;\ C_\varepsilon = C_h C_w / C_l^2;\ C_\chi = C_w / C_l^2$$

where:

$$C_\sigma = C_{\sigma_x} = C_{\sigma_y} = C_{\sigma_{xy}};$$

$$C_M = C_{M_x} = C_{M_y} = C_{M_{xy}};$$

$$C_Q = C_{Q_x} = C_{Q_y}; \tag{68}$$

$$C_\varepsilon = C_{\varepsilon_x} = C_{\varepsilon_y} = C_{\varepsilon_{xy}};$$

$$C_\chi = C_{\chi_x} = C_{\chi_y} = C_{\chi_{xy}}$$

For plates subjected to loads q, N_x, N_y, and N_{xy} only, the following generalized variables can be utilized:

$$\overline{\sigma} = \frac{\sigma h_0^2}{q_0 l^2};\ \overline{M} = \frac{M}{q_0 l^2};\ \overline{Q} = \frac{Q}{q_0 l};$$
$$\overline{\varepsilon} = \frac{\varepsilon E h_0^2}{q_0 l^2};\ \overline{\chi} = \frac{\chi E h_0^3}{q_0 l^2} \tag{69}$$

For dynamic loads, load intensity q_0 should be replaced by load intensity Z_0 in order to obtain generalized variables [Equation (69)].

If plates are subjected to the thermal load only, the generalized variables can be obtained from Equation (67) after replacement of deflection scale factor C_w by using Equation (61):

$$\overline{\sigma} = \frac{\sigma}{E\lambda_0\alpha};\ \overline{M} = \frac{M}{Eh_0^2\lambda_0\alpha};$$

$$\overline{Q} = \frac{Ql}{Eh_0^2\lambda_0\alpha};\ \overline{\varepsilon} = \frac{\varepsilon}{\lambda_0\alpha};\ \overline{x} = \frac{xh_0}{\lambda_0\alpha}$$

If plate is subjected to transverse concentrated loads P and/or linearly distributed loads T, the proper scale factors C_P and C_T can be expressed in terms of C_q (or C_Z for dynamic loads) by using dimensional relationships:

$$C_P = C_q C_l^2;$$
$$C_T = C_q C_l \tag{70}$$
$$(\text{or } C_P = C_Z C_l^2;$$
$$C_T = C_Z C_l \text{ for dynamic loads})$$

For similitude of plate free vibrations, there are no deflection scale factors C_w in the fourth and fifth equations of Equation (57) due to homogeneity of the considered problem.

To apply the third theorem of similitude, consider boundary conditions for plates with curvilinear boundary [20]. If the curvilinear edge of the plate is fixed, the following boundary conditions can be written (refer to Figure 3):

$$w = 0;\ \frac{\partial w}{\partial n} = 0 \tag{71}$$

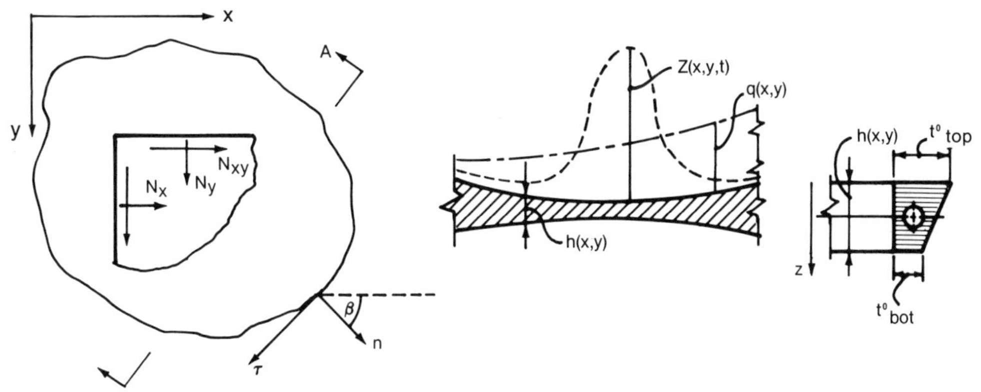

FIGURE 3. General case of a small deflected plate.

For simply supported plate edge, one can write:

$$w = 0$$

$$M_n = M_x \cos^2\beta + M_y \sin^2\beta - 2\, M_{xy} \sin\beta \cos\beta = 0 \qquad (72)$$

If the edge of a plate is free, the boundary conditions are:

$$\nu\nabla^2 w + (1-\nu)\left(\cos^2\beta \frac{\partial^2 w}{\partial x^2} + \sin^2\beta \times \frac{\partial^2 w}{\partial y^2} + \sin 2\beta \frac{\partial^2 w}{\partial x \partial y}\right) = 0$$

$$\cos\beta \frac{\partial}{\partial x}\nabla^2 w + \sin\beta \frac{\partial}{\partial y}\nabla^2 w + (1-\nu) \times \frac{\partial}{\partial s}\left[\cos 2\beta \frac{\partial^2 w}{\partial x \partial y} + \frac{1}{2}\sin 2\beta \times \left(\frac{\partial^2 w}{\partial y^2} - \frac{\partial^2 w}{\partial x^2}\right)\right] = 0 \qquad (73)$$

Initial conditions can be assumed as follows:

$$\text{At } t = 0\text{: } w = w_0;\ \frac{\partial w}{\partial t} = s_0 \qquad (74)$$

The sufficient similitude equations are obtained from Equations (72) to (74) by use of Equation (1):

$$C_\beta = 1;\ C_{M_n}/C_{M_x} = 1;\quad C_{M_n}/C_{M_y} = 1;\ C_{M_n}/C_{M_{xy}} = 1 \qquad (75)$$

$$C_\nu = 1;\ C_\beta = 1;\ C_x/C_y = 1 \qquad (76)$$

$$C_w/C_{w0} = 1;\ C_w/C_s C_t = 1 \qquad (77)$$

Here Equations (75) to (77) correspond to Equations (72) to (74), respectively, and Equation (71) does not "produce" the indicators of similitude. Equations (75) and (76) do not add new information and do not contradict the previously obtained equations of similitude. The initial conditions produce the following generalized initial deflections $\bar{w}_0$ and initial velocity $\bar{s}_0$ of plate vibration:

$$\bar{w}_0 = w_0 h_0^3 E/Z_0 l^4;\ \bar{s}_0 = s_0 \sqrt{Em}\ h_0^2/Z_0 l^2 \qquad (78)$$

Finally, nondimensional presentation of boundary and initial conditions [Equations (71) to (74)] consists of the replacement of W; M_n; M_x; M_y; M_{xy}; $\partial^n w/\partial x^{n-r}\partial y^r$; t; w_0; s_0 by corresponding nondimensional quantities.

Besides boundary and initial conditions, some additional restrictions should be put on the similar group of plates, such as the previously mentioned requirements of plate material elasticity. Deflections should not exceed ¼ to ½ of the plate minimum thickness, and the maximum thickness shall not be more than ¼ of the least plate plane dimension.

Summarizing the application of similitude theory for analysis of isotropic linearly elastic plates with small deflections (general case of the plate geometry), one can notice a variety of "distorted" plates that can be included in the group of similar plates. Regardless of acting loads, all plates should have a material with identical Poisson's ratio and should have geometrically similar plane dimensions. In addition, there is a similitude in all stress-strain-displacement components.

Example 5

In Figure 4, the prestressed reinforced concrete plate is shown to be of complex configuration and variable thickness

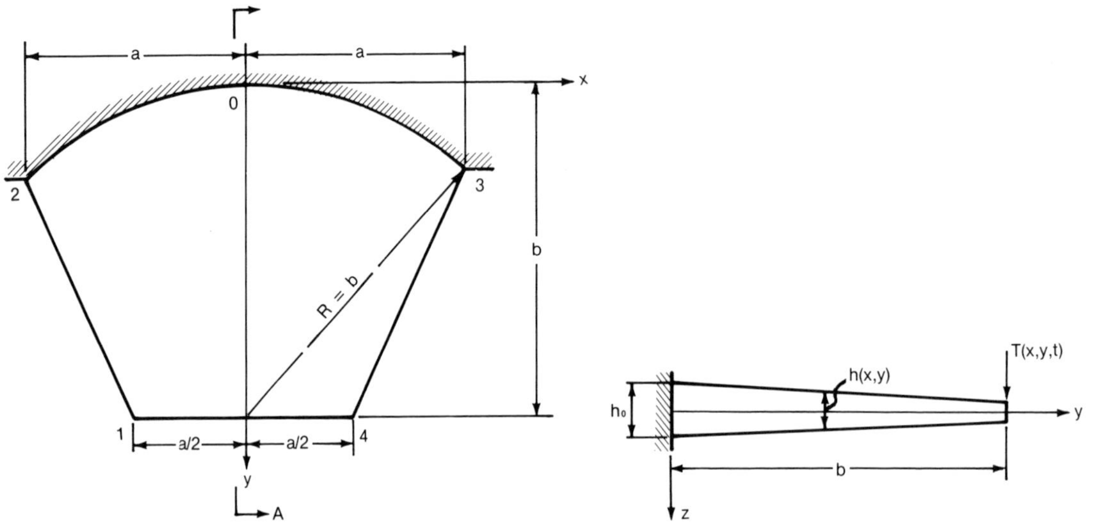

FIGURE 4. Modeling of a concrete plate.

$h(x,y) = h_o\ (1 - y/2b)$. Assume this plate can be considered as isotropic, and linearly elastic plate and crack development is restricted. The plate possesses the following dimensions and material properties:

$$a_1 = 240 \text{ inches (6096 mm)};\ b = 360 \text{ inches (9144 mm)}$$

$$h_{01} = 18 \text{ inches (457 mm)} \tag{79}$$

$$\nu_1 = 0.2;\ E_1 = 3300 \text{ ksi } (2.274 \times 10^7 \text{ kPa})$$

$$\alpha \cong 14 \times 10^{-6} \text{ in/in/}^\circ\text{C}$$

and is subjected to transverse static load $q(x,y)$, temperature gradient $\lambda(x,y)$, and perturbation dynamic harmonic load $T(x,y,t)$, which can be described by the following formulas:

$$q(x,y) = q_0(1 - 2y/3b)$$

$$\lambda(x,y) = \lambda_0 e^{-(x^2/a^2+y^2/b^2)} \tag{80}$$

$$T(x,y,t) = T_0 \sin \omega t$$

where characteristic loads q_o, T_o and temperature gradient are: $q_o = 2$ psi (13.78 kPa); $T_o = 250$ lb/in (43.80 N/mm); $\lambda_o = +50°C$; $\omega = 20\ \text{sec}^{-1}$.

Let an analysis of the above plate be done by experimental means—by modeling. Choose a stiff polyvinyl chloride (viniplast) as a material for the model because of lower modulus of elasticity $E_2 = 570$ ksi (0.393×10^7 kPa) and almost identical Poisson's ratio $\nu_2 = 0.22$ [22]. All other model parameters are as follows:

$$a_2 = 6 \text{ inches (152.4 mm)}$$

$$b_2 = 9 \text{ inches (228.6 mm)}$$

$$\alpha \cong 5 \times 10^{-5} \text{ in/in/}^\circ\text{C}$$

Calculate some of the similitude constants:

$$C_l = a_2/a_1 = 6/240 = 1/40$$

$$C_E = E_2/E_1 = 570/3300 = 0.1727$$

$$C_\nu \cong 1$$

$$C_\alpha = 5 \times 10^{-5}/14 \times 10^{-6} = 3.5714$$

Selection of the thickness scale $C_h = C_l$ leads to $h_{02} = h_{01}/40 = 18/40 = 0.45$ inch (11.4 mm). The maximum thickness to plate plane dimension ratio is equal to: $h_{2_{max}} l_{2_{min}} = 0.45/6 = 1/13.333$; i.e., much higher than ¼ of the least plate dimension. Using a relatively thicker model (for example, twice as thick), i.e., $C_h = 1/20$ and $h_2 = 0.90$ inch, one can easily manufacture more precise models and obtain higher deflections and strains by testing of this thicker model. Even by testing of a geometrically similar plate model ($C_h = C_l = 1/40$), one can take advantage of load increment (up to achievement by the maximum deflection magnitude of one-half of plate thickness). Keeping in mind this additional "tool" of deflections and strains increment [9], obtain C_w scale factor from the deflection

TABLE 1. Comparison of Similitude Constants for Statically Loaded Models.

Model	C_E	C_ν	C_l	C_h	C_w	C_σ	$C_M \times 10^4$	$C_Q \times 10^3$	C_ϵ	C_X	C_q	C_λ
A	.1727	1	1/40	1/40	1/40	.1727	1.0793	4.3175	1	40	.1727	0.280
B	.1727	1	1/40	1/20	1/20	.6908	17.27	69.08	4	80	2.7632	1.120

limiting formula:

$$w \leq (\tfrac{1}{4} - \tfrac{1}{2})\, h \tag{81}$$

which yields:

$$C_w = C_h \tag{82}$$

By substitution of C_l; C_h; C_E; C_w in Equations (54), (61), and (67), one can get the similitude constants for combined statical and thermal loads, stresses, moments, transverse forces, strains, and curvatures. The similitude constants are compared in Table 1 for geometrically similar and distorted models. Comparison shows the effectiveness of model *B*, which is expected to yield more precise deflections, strains, and curvatures.

Boundary conditions for concrete plate (fixed at edge "203," free at other edges) have to be reproduced for the model. No additional equations of similitude are followed from the boundary conditions.

Assume the above plate is subjected to dynamic loads only. The plate 1 and 2 mass densities can be calculated as follows:

$$m_1 = \gamma_1/g = .09044/386.2 = 2.3417 \times 10^{-4}\ \frac{\text{lbs} \times \text{sec}^2}{\text{in}^4}\left(254.86\ \frac{\text{kg} \times \text{sec}^2}{\text{m}^4}\right)$$

$$m_2 = \gamma_2/g = .05065/386.2 = 1.3114 \times 10^{-4}\ \frac{\text{lbs} \times \text{sec}^2}{\text{in}^4}\left(142.72\ \frac{\text{kg} \times \text{sec}^2}{\text{m}^4}\right)$$

that is, C_m scale factor is equal to:

$$C_m = 1.3114 \times 10^{-4}/2.3417 \times 10^{-4} = 0.56$$

The scale factor of perturbation dynamic load C_T, time scale factor C_t, scale factor of frequencies C_ω, and damping resistance of surrounding medium C_k can be calculated from Equations (57), (70) and (82). Some comments can be made above C_m scale factor selection. For the plate-shell structures, the mass density *m* of the model can be increased by the use of additional weight, and no centrifuge is required. Therefore, in addition to model A and B consider model C, which differs from model B by double mass density only, i.e., $C_m = 1.12$.

The plate-shell free vibration is a subject of independent study. It is rational for this particular problem to use a thinner model, e.g., model D. Assume the following scale factors for model D: $C_l = 1/40$; $C_h = 1/160$, or $h_{02} = 18/160 \cong 0.1125$ inch (2.86 mm), $C_m = 0.56$; $C_E = 0.1727$.

The similitude constants for four models subjected to dynamic loads are compared in Table 2.

Based on constants of similitude, the models have to be loaded as follows:

Model A : $q_0 = .1727 \times 2 = .3454$ psi (27.56 kPa); $\lambda_0 = .28 \times 50 = 14°\text{C}$; $T_0 = .00432 \times 250 = 1.08$ lbs/in (.1892 N/mm); $\omega = 22.21 \times 20 = 444.2\ \text{sec}^{-1}$

Model B : $q_0 = 2.7632 \times 2 = 5.5264$ psi (440.96 kPa); $\lambda_0 = 1.12 \times 50 = 56°\text{C}$; $T_0 = 0.06908 \times 250 = 17.27$ lbs/in (3.0255 N/mm); $\omega = 44.43 \times 20 = 888.6\ \text{sec}^{-1}$

Model C : $T_0 = .06908 \times 250 = 17.27$ lbs/in (3.0255 N/mm); $\omega = 31.41 \times 20 = 628.2\ \text{sec}^{-1}$

In the above example, assume that temperature gradients do not change the material properties for the prototype or for the models. The manner of the prototype plate supporting does not contribute to the axial force appearance because of the temperature.

The influence of temperature loads on the plates can take place through the boundary conditions. The reactive temperature moments and transverse shear forces at the plate

TABLE 2. Comparison of Similitude Constants for Dynamically Loaded Models.

Model	C_E	C_ν	C_l	C_h	C_w	C_σ	$C_M \times 10^4$	$C_Q \times 10^3$	C_ϵ	C_X	C_Z	C_T	C_m	C_t	C_ω; C_f	C_k
A	.1727	1	1/40	1/40	1/40	.1727	1.0793	4.3175	1	40	.1727	0.00432	0.56	.04502	22.21	12.44
B	.1727	1	1/40	1/20	1/20	.6908	17.27	69.08	4	80	2.7632	.06908	0.56	.02251	44.43	24.88
C	.1727	1	1/40	1/20	1/20	.6908	17.27	69.08	4	80	2.7632	.06908	1.12	.03183	31.41	35.18
D	.1727	1	1/40	1/160	1/160	—	—	—	—	—	—	—	0.56	.18007	5.55	3.11

boundaries are calculated by scale factors C_M, C_Q, which include scale factor C_W. The axial unit force's scale factor C_N reflects the temperature effects as well as other effects, and can be used for modeling special cases of boundary conditions, such as "spring" type boundary conditions, etc.

RECTANGULAR PLATES OF UNIFORM THICKNESS

This type of small deflected plate is one of the most widespread, and the governing differential equation [Equation (53)] can be simplified as follows:

$$D\nabla^4 w = q(x, y) + N_x \frac{\partial^2 w}{\partial x^2} + N_y \frac{\partial^2 w}{\partial y^2} + 2N_{xy} \frac{\partial^2 w}{\partial x \partial y} - mh \frac{\partial^2 w}{\partial t^2} - kh \frac{\partial w}{\partial t} + Z(x, y, t) - \frac{Eh^2 \alpha}{12(1 - \nu)} \nabla^2 \lambda \tag{83}$$

where $h(x,y) = h =$ constant.

For plates subjected to the statical loads and temperature gradient λ, the similitude equations [Equation (54)] based on application of the *second theorem of similitude* can be modified as follows:

$$C_x/C_y = 1;\ C_w C_h^3 C_E / C_q C_l^4 C_{\nu_0} = 1$$
$$C_N C_l^2 C_{\nu_0} / C_E C_h^3 = 1;\ C\lambda C_\alpha C_h^2 C_E / C_q C_l^2 C_{\nu_1} = 1 \tag{84}$$

or for plates under the thermal loads only:

$$C_w C_h / C_\lambda C_\alpha C_l^2 C_{\nu_2} = 1 \tag{85}$$

where the scale factors for Poisson's ratio C_ν and C_{ν_0}, C_{ν_1}, C_{ν_2} are determined by the formulas:

$$C_\nu = \nu_2/\nu_1;\ C_{\nu_0} = (1 - \nu_2^2)/(1 - \nu_1^2)$$
$$C_{\nu_1} = (1 - \nu_2)/(1 - \nu_1);\ C_{\nu_2} = (1 + \nu_2)/(1 + \nu_1) \tag{86}$$

It is evident that Equations (84) and (85) give free choice of Poisson's ratio selection.

When the dynamic problem is under consideration, the following *necessary* equations of similitude can be released instead of Equation (57):

$$C_x/C_y = 1;\ C_w C_h^3 C_E / C_Z C_l^4 C_{\nu_0} = 1$$
$$C_t C_h C_E^{0.5} / C_m^{0.5} C_l^2\, C_{\nu_0}^{0.5} = 1 \tag{87}$$
$$C_k C_l^2 C_{\nu_0}^{0.5} / C_E^{0.5} C_m^{0.5} C_h = 1;\ C_\omega = C_f = C_t^{-1}$$

Before pi terms or generalized variables formulation, the above equations of similitude must be "approved" or "rejected" by sufficient equations of similitude which follow from boundary conditions.

If the plate edge $x = a$ is fixed [Figure 5(b)], then [20]:

$$w = 0;\ \frac{\partial w}{\partial x} = 0 \tag{88}$$

The simply supported plate edge $x = a$ [Figure 5(c)] can be specified by boundary conditions:

$$w = 0;\ \frac{\partial^2 w}{\partial x^2} + \frac{\partial^2 w}{\partial y^2} = 0 \tag{89}$$

The boundary conditions for the plate free edge $x = a$ [Figure 5(d)] were presented at first by Poisson [20] in the following form:

$$M_x = 0;\ M_{xy} = 0;\ Q_x = 0$$

or after replacements by deflections:

$$\frac{\partial^2 w}{\partial x^2} + \nu \frac{\partial^2 w}{\partial y^2} = 0;$$
$$\frac{\partial^2 w}{\partial x \partial y} = 0; \tag{90}$$
$$\frac{\partial^3 w}{\partial x^3} + \frac{\partial^3 w}{\partial x \partial y^2} = 0$$

But later on, Kirchhoff suggested the following presentation of these boundary conditions [20]:

$$\frac{\partial^2 w}{\partial x^2} + \nu \frac{\partial^2 w}{\partial y^2} = 0;$$
$$\frac{\partial^3 w}{\partial x^3} + (2 - \nu) \frac{\partial^3 w}{\partial x \partial y^2} = 0 \tag{91}$$

Kelvin and Tait [20] have noticed that replacement of Poisson's boundary conditions produces only local changes in the stress distribution at the plate edge; i.e., it does not reflect the actual stress-strain state. The same conclusion can be made for the formulation of boundary conditions for shell structures [23]. As per Saint Venant's principle, replacement of Equation (90) with Equation (91) influences the plate-shell edge only. By application of the third theorem of similitude, one can notice that only the first equation in Equation (90) requires $C_\nu = 1$ and both equations in Equation (91) require identical Poisson's ratio for materials of the similar group. Keeping this in mind and referring to Reference 24, one can expect, in the case of $C_\nu \neq 1$, insignificant

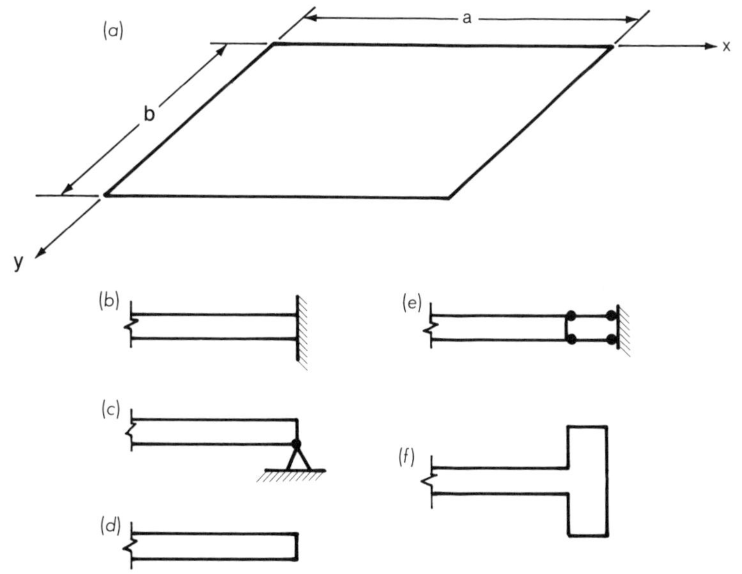

FIGURE 5. Boundary conditions for the edge x = a of a rectangular plate with small deflections.

influence within the boundary and for deflections w rather than for curvatures $\partial^2 w/\partial x^2$; $\partial^2 w/\partial y^2$; $\partial^2 w/\partial x \partial y$.

The boundary conditions for a guided plate edge $x = a$ [Figure 5(e)] are as follows:

$$\frac{\partial w}{\partial x} = 0; \frac{\partial}{\partial x}\left(\frac{\partial^2 w}{\partial x^2} + \frac{\partial^2 w}{\partial y^2}\right) = 0 \quad (92)$$

When the plate is supported by an elastic beam, the boundary conditions for edge $x = a$ [Figure 5(f)] are [20]:

$$B\frac{\partial^4 w}{\partial y^4} = D\frac{\partial}{\partial x}\left[\frac{\partial^2 w}{\partial x^2} + (2 - \nu)\frac{\partial^2 w}{\partial y^2}\right]$$
$$C\frac{\partial}{\partial y}\left(\frac{\partial^2 w}{\partial x \partial y}\right) = -D\left(\frac{\partial^2 w}{\partial x^2} + \nu\frac{\partial^2 w}{\partial y^2}\right) \quad (93)$$

By increment of torsional and decrement of flexural beam rigidity, the boundary conditions [Equation (93)] are transformed in Equation (92). By decrement of torsional and increment of flexural beam rigidity, the boundary conditions [Equation (93)] are transformed in Equation (89). By increment of both torsional and flexural beam rigidity, boundary conditions [Equation (93)] degenerate in Equation (88). On the contrary, by decrement of both torsional and flexural beam rigidity one can expect degeneration of Equation (93) in equations for the plate free edge. The above discussion lets us conclude about the insignificance of the following term in boundary conditions [Equation (93)]:

$$\nu\frac{\partial^2 w}{\partial y^2} \cong 0 \quad (94)$$

Assuming that all the above boundary conditions are written for rectangular plate 1 and using Equation 1, one can conclude the following:

a. For the fixed plate edge there are no additional equations of similitude.
b. For the simple supported plate edge $C_x/C_y = 1$, i.e., there are no changes in Equations (84), (85), and (87).
c. For the free plate edge:

$$C_\nu = 1; C_x/C_y = 1 \quad (95)$$

and by assuming $C_\nu \neq 1$ one includes insignificant error in deflections and curvatures.
d. For the guided plate edge $C_x/C_y = 1$, i.e., there are no contradictions with Equations (84), (85), and (87).
e. For the plate connections with elastic beams oriented parallel to the plate edges:

$$C_x/C_y = 1; C_\nu = 1; C_B/C_D C_l = 1; C_C/C_B = 1 \quad (96)$$

or with assumption [Equation (94)] and after replacement $D = Eh^3/12(1 - \nu^2)$:

$$C_x/C_y = 1; C_B C_{\nu_0}/C_E C_h^3 C_l = 1; C_C/C_B = 1 \quad (97)$$

Strictly speaking, the free plate edge and edge supported by the elastic beam require identical Poisson's ratio for materials of the similar group of plates, and Equations (84), (85), and (87) have to be simplified with consideration of the following:

$$C_\nu = C_{\nu_0} = C_{\nu_1} = C_{\nu_2} = 1 \quad (98)$$

For this case, refer to the section about isotropic linearly elastic plates with small deflections.

When $C_\nu \neq 1$ (and, therefore, $C_{\nu_0} \neq 1$; $C_{\nu_1} \neq 1$; $C_{\nu_2} \neq 1$), there is no similitude of all plate components listed in Equation (66). There is similitude of deflections; shear stresses τ_{xy}; twist moments M_{xy}; shear forces Q_x, Q_y (and stresses caused by them); curvatures X_x, X_y, X_{xy}; and strains ϵ_x, ϵ_y, ϵ_{xy}. The relevant *additional* equations of similitude are as follows:

$$\begin{aligned} C_\tau &= C_E C_h C_w / C_{\nu_2} C_l^2;\ C_{M_{xy}} = C_\tau C_h^2 \\ C_Q &= C_E C_h^3 C_w / C_l^3 C_{\nu_0};\ C_x = C_w / C_l^2 \\ C_{\epsilon_x} &= C_{\epsilon_y} = C_{\epsilon_{xy}} = C_w C_h / C_l^2 \end{aligned} \quad (99)$$

There is no similitude of normal stresses and bending moments. Keeping in mind the case $C_\nu \neq 1$, the following are revised generalized variables [Equations (55), (58), (62), (65), (69), and (70)] for different loading of plates:

a. Plates under statical loads and temperature gradient:

$$\begin{aligned} &\xi = x/l;\ \eta = y/l; \\ &\overline{w} = wh^3E/q_0l^4(1 - \nu^2); \\ &\overline{N}_x = N_xl^2(1 - \nu^2)/Eh^3; \\ &\overline{N}_y = N_yl^2(1 - \nu^2)/Eh^3; \\ &\overline{N}_{xy} = N_{xy}l^2(1 - \nu^2)/Eh^3; \\ &\mu(\xi, \eta) = \lambda(x, y)\alpha h^2E/q_0l^2(1 - \nu); \\ &\overline{\varepsilon} = \varepsilon Eh^2/q_0l^2(1 - \nu^2); \\ &\overline{\tau} = \frac{\tau h^2}{q_0l^2(1 - \nu)}; \\ &\overline{M}_{xy} = \frac{M_{xy}}{q_0l^2(1 - \nu)}; \\ &\overline{Q} = \frac{Q_x}{q_0l};\ \overline{Q}_y = \frac{Q_y}{q_0l}; \\ &\overline{\chi}_x = \frac{\chi_xEh^3}{q_0l^2(1 - \nu^2)}; \\ &\overline{\chi}_y = \frac{\chi_yEh^3}{q_0l^2(1 - \nu^2)}; \\ &\overline{\chi}_{xy} = \frac{\chi_{xy}Eh^3}{q_0l^2(1 - \nu^2)} \end{aligned} \quad (100)$$

b. For plates under dynamic loads, the pi terms $\overline{N}_x$, $\overline{N}_y$, $\overline{N}_{xy}$, and μ should be excluded from Equation (100) and load intensity q_0 replaced by Z_0. Also, time t and damping coefficient k should be presented by the following nondimensional terms:

$$\begin{aligned} \theta &= \frac{th}{l^2}\sqrt{\frac{E}{m(1 - \nu^2)}}; \\ \overline{k} &= \frac{kl^2}{h}\sqrt{\frac{1 - \nu^2}{Em}} \end{aligned} \quad (101)$$

c. Plates under the thermal loads only:

$$\begin{aligned} &\xi = x/l;\ \eta = y/l; \\ &\overline{w} = wh/\lambda_0\alpha l^2(1 + \nu); \\ &\overline{\tau} = \frac{\tau}{E\lambda_0\alpha};\ \overline{M}_{xy} = \frac{M_{xy}}{Eh^2\lambda_0\alpha}; \\ &\overline{Q}_x = \frac{Q_xl(1 - \nu)}{Eh^2\lambda_0\alpha}; \\ &\overline{Q}_y = \frac{Q_yl(1 - \nu)}{Eh^2\lambda_0\alpha}; \\ &\overline{\chi}_x = \frac{\chi_xh}{\lambda_0\alpha(1 + \nu)}; \\ &\overline{\chi}_y = \frac{\chi_yh}{\lambda_0\alpha(1 + \nu)}; \\ &\overline{\chi}_{xy} = \frac{\chi_{xy}h}{\lambda_0\alpha(1 + \nu)}; \\ &\overline{\varepsilon} = \frac{\varepsilon}{\lambda_0\alpha(1 + \nu)}; \\ &\mu(\xi, \eta) = \lambda(x, y)/\lambda_0 \end{aligned} \quad (102)$$

Here all characteristic parameters are identical to those described in the previous section, and generalized strain $\bar{\epsilon}$ can be equally used for $\bar{\epsilon}_x$, $\bar{\epsilon}_y$, $\bar{\epsilon}_{xy}$. The governing nondimensional differential equation [Equation (83)] can be presented for the specified three types of loading and for $C_\nu \neq 1$:

a. For plates under statical loads and thermal loads:

$$\frac{1}{12}\nabla^4\bar{w} = p(\xi, \eta) + \bar{N}_x \frac{\partial^2\bar{w}}{\partial\xi^2} + \bar{N}_y \frac{\partial^2\bar{w}}{\partial\eta^2} + 2\bar{N}_{xy}\frac{\partial^2\bar{w}}{\partial\xi\partial\eta} - \frac{1}{12}\nabla^2\mu \qquad (103)$$

b. For plates under dynamic loads:

$$\frac{1}{12}\nabla^4\bar{w} = -\frac{\partial^2\bar{w}}{\partial\theta^2} - \bar{k}\frac{\partial\bar{w}}{\partial\theta} + \bar{Z}(\xi, \eta, \theta) \qquad (104)$$

c. For plates under the thermal loads only:

$$\frac{1}{12}\nabla^4\bar{w} = -\frac{1}{12}\nabla^2\mu \qquad (105)$$

The initial conditions [Equation (74)] are different for $C_\nu \neq 1$:

$$\bar{w} = \bar{w}_0; \bar{s} = \frac{\partial\bar{w}}{\partial\theta} = \bar{s}_0 \qquad (106)$$

where: $\bar{w} = w_0h^3E/Z_0l^4(1 - \nu^2)$

$$\bar{s}_0 = s_0\sqrt{Em/(1 - \nu^2)h^2/Z_0l^2}$$

The boundary conditions [Equations (88), (89), and (92)] can be replaced by identical but with nondimensional quantities $\bar{w}$, ξ, η. The simplified boundary conditions [Equation (93)] (i.e., without the terms $\nu\ \partial^2w/\partial y^2$) are presented in generalized variables as follows:

$$\bar{B}\frac{\partial^4\bar{w}}{\partial\eta^4} \cong \frac{1}{12}\frac{\partial}{\partial\xi}\left[\frac{\partial^2\bar{w}}{\partial\xi^2} + 2\frac{\partial^2\bar{w}}{\partial\eta^2}\right];$$

$$\bar{C}\frac{\partial}{\partial\eta}\left(\frac{\partial^2 w}{\partial\xi\partial\eta}\right) \cong -\frac{1}{12}\frac{\partial^2\bar{w}}{\partial\xi^2} \qquad (107)$$

where $\bar{B} = B(1 - \nu^2)/Eh^3l$; $\bar{C} = C(1 - \nu^2)/Eh^3l$.

Example 6

Generalize the formulas for calculation of rectangular plates with a particular Poisson's ratio [18] by the use of similitude theory for material with any Poisson's ratio.

Table 26 of Reference 18 contains numerical solutions for rectangular plates with simply supported and fixed edges, which permit the inclusion of Poisson's ratio in generalized deflections and curvatures. Because References 18, 20, and 25 provide bending stresses information, the following formulas can be used for calculation of generalized curvatures $\bar{\chi}_x = \chi_xEh^3/q_0b^2(1 - \nu^2)$ and $\bar{\chi}_y = \chi_yEh^3/q_0b^2(1 - \nu^2)$:

$$\bar{\chi}_x = -\frac{2}{(1 - \nu^2)}(\bar{\sigma}_x - \nu\bar{\sigma}_y);$$

$$\bar{\chi}_y = -\frac{2}{(1 - \nu^2)}(\bar{\sigma}_y - \nu\bar{\sigma}_x) \qquad (108)$$

where:

$\bar{\sigma}_x = \sigma_xh^2/q_0b^2$; $\bar{\sigma}_y = \sigma_yh^2/q_0b^2$
b = plate plane dimension, which replaces l in the above formulas

If a concentrated force P acts, it can be included by replacing q_0b^2 with P.

By knowing generalized curvatures $\bar{\chi}_x$, $\bar{\chi}_y$ one can easily calculate generalized stresses $\bar{\sigma}_x$, $\bar{\sigma}_y$ by the following formulas:

$$\bar{\sigma}_x = -\frac{1}{2}(\bar{\chi}_x + \nu\bar{\chi}_y);$$

$$\bar{\sigma}_y = -\frac{1}{2}(\bar{\chi}_y + \nu\bar{\chi}_x) \qquad (109)$$

The generalized deflections $\bar{w}$ can be expressed by Equation (100), i.e., $\bar{w} = wh^3E/q_0b^4(1 - \nu^2)$, which includes Poisson's ratio.

The nondimensional reaction force $\bar{R}_x$ (or $\bar{R}_y$), normal to the plate surface exerted by the boundary support on the edge of the plate, consists of two parts: $\bar{Q}_x$ (or $\bar{Q}_y$) and $\partial\bar{M}_{xy}/\partial\eta$ (or $\partial\bar{M}_{yx}/\partial\xi$) and can be expressed by the following formulas:

$$\bar{R}_x = \bar{Q}_x - \frac{\partial\bar{M}_{xy}}{\partial\eta}; \bar{R}_y = \bar{Q}_y - \frac{\partial\bar{M}_{yx}}{\partial\xi} \qquad (110)$$

where $\bar{Q}_x$, $\bar{Q}_y$, $\bar{M}_{xy} = -\bar{M}_{yx}$ are determined in Equation (100), and as follows from these equations, they differently depend on Poisson's ratio. Therefore, in order to give informative data for any Poisson's ratio, the values of max. ($\bar{Q}_x$;

$\bar{Q}_y$) and max. ($\bar{M}_{xy}$; $\bar{M}_{yx}$) have to be given separately and then combined for reactions max. ($\bar{R}_x$, $\bar{R}_y$) calculation by Equation (110) after substitution of the particular Poisson's ratio in $\bar{M}_{xy}$, $\bar{M}_{yx}$.

The modified cases 1a, 1b, 1c, 1d, 1e, 3a, 4a, 5a, and 6a of Table 26 in Reference 18 for any Poisson's ratio are shown in Tables 3 to 6. By the use of these tables, one can provide calculations for plates with Poisson's ratios $\nu_1 = 0.15$; $\nu_2 = 0.30$; $\nu_3 = 0.45$ (i.e., all spectrum of ν) as follows. Calculate the maximum deflection, normal stress, and reaction force for the small deflected plate with $a/b = 1.4$ uniformly loaded over the entire surface. As per Table 3, loading 1a:

$$\overline{w} = wh^3E/q_0b^4(1 - \nu^2) = 0.0846;$$

$$\overline{\chi}_x = \chi_x Eh^3/q_0b^2(1 - \nu^2) = -0.7970;$$

$$\overline{\chi}_y = \chi_y Eh^3/q_0b^2(1 - \nu^2) = -0.3633;$$

$$\overline{Q}_x = Q_x/q_0l = 0.411;$$

$$\frac{\partial M_{xy}}{\partial \eta} = \frac{\partial M_{xy}}{\partial y} \Big/ q_0l(1 - \nu) = -0.096$$

For $\nu = 0.15$:

$$w = 0.0846(1 - 0.15^2)\frac{q_0b^4}{Eh^3} = 0.0827\frac{q_0b^4}{Eh^3};$$

$$\overline{\sigma}_x = \sigma_xh^2/q_0b^2 = -\frac{1}{2}(-0.7970 + 0.15 \times (-0.3633)) = 0.4257;$$

$$\overline{\sigma}_y = \sigma_yh^2/q_0b^2 = -\frac{1}{2}(-0.3633 + 0.15 \times (-0.7970)) = 0.2414;$$

$$\text{or } \sigma_{max} = 0.4257\frac{q_0b^2}{h^2}$$

$$\overline{R}_{max} = \frac{R_{max}}{q_0b} = 0.411 - (-0.096 \times (1 - 0.15)) = 0.4926;$$

$$R_{max} = 0.4926q_0b;$$

For $\nu = 0.30$:

$$w = 0.0846(1 - 0.3^2)\frac{q_0b^4}{Eh^3} = \underline{0.0770\frac{q_0b^4}{Eh^3}};$$

$$\overline{\sigma}_x = \sigma_xh^2/q_0b^2 = -\frac{1}{2}(-0.7970 + 0.30 \times (-0.3633)) = 0.4530;$$

$$\overline{\sigma}_y = \sigma_yh^2/q_0b^2 = -\frac{1}{2}(-0.3633 + 0.30 \times (-0.7970)) = 0.3012;$$

$$\text{or } \sigma_{max} = 0.453\frac{q_0b^2}{h^2};$$

$$\overline{R}_{max} = \frac{R_{max}}{q_0b} = 0.411 - (-0.096(1 - 0.30)) = 0.4782;$$

$$\underline{R_{max} = 0.4782q_0b};$$

Here the underlined results coincide with those in Reference 18.

For $\nu = 0.45$:

$$w = 0.0846(1 - 0.45^2)\frac{q_0b^4}{Eh^3} = 0.06747\frac{q_0b^4}{Eh^3};$$

$$\overline{\sigma}_x = \sigma_xh^2/q_0b^2 = -\frac{1}{2}(-0.7970 + 0.45 \times (-0.3633)) = 0.4802;$$

$$\overline{\sigma}_y = \sigma_yh^2/q_0b^2 = -\frac{1}{2}(-0.3633 + 0.45 \times (-0.7970)) = 0.3610;$$

$$\text{or } \sigma_{max} = 0.4802\frac{q_0b^2}{h^2};$$

$$\overline{R}_{max} = \frac{R_{max}}{q_0b} = 0.411 - (-0.096(1 - 0.45)) = 0.4638;$$

$$R_{max} = 0.4638q_0b;$$

TABLE 3. Generalized Maximum Deflections, Curvatures, Shear Forces, and Stresses in Simply Supported Rectangular Plates.

Shape and Supports	Loading	a/b	1.0	1.2	1.4	1.6	1.8	2.0	3.0	4.0	5.0	∞
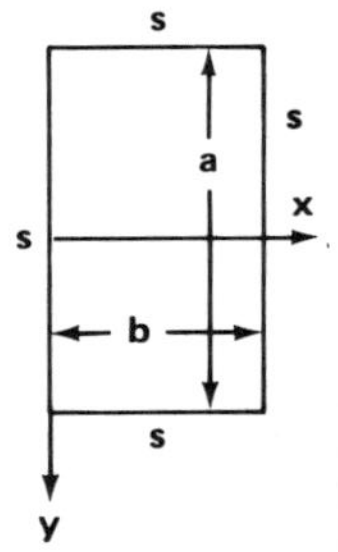	1a Uniform Over Entire Plate	$\bar{w}$	0.0487	0.0677	0.0846	0.0996	.1117	.1216	.1468	.1538	.1556	.1562
		$\bar{\chi}_x$	−0.4422	−0.6286	−0.7970	−0.9421	−1.0606	−1.1575	−1.4073	−1.4767	−1.4947	−1.5000
		$\bar{\chi}_y$	−0.4422	−0.4126	−0.3633	−0.3078	−0.2566	−0.2095	−0.0650	−0.0178	−0.0016	0
		$\bar{Q}_x$	0.338	0.380	0.411	0.435	0.452	0.465	0.493	0.498	0.500	0.500
		$\frac{\partial \bar{M}_{xy}}{\partial \eta}$	−0.117	−0.107	−0.096	−0.080	−0.067	−0.054	−0.017	−0.005	−0.001	0
	1b Uniform Over Small Concentric Circle of Radius r_0 (or $r_0' = \sqrt{1.6r_0^2 + h^2} - 0.675h$ if $r_0 < 0.5h$)	(At center) Max $\sigma = \frac{3P}{2\pi h^2}\left((1+\nu)\ln\frac{2b}{\pi r_0'} + \beta\right)$; $\beta = (\bar{\gamma}_x + \nu\bar{\gamma}_y)/2$										
		a/b	1.0	1.2	1.4	1.6	1.8	2.0	∞			
		$\bar{w}$	0.1392	0.1624	0.1781	0.1884	0.1944	0.1981	0.2034			
		$\bar{\gamma}_x$	0.6692	1.1549	1.4802	1.6877	1.8152	1.8925	2.000			
		$\bar{\gamma}_y$	0.6692	0.4835	0.3259	0.2077	0.1295	0.0782	0			
	1c Uniformly Increasing Along Length 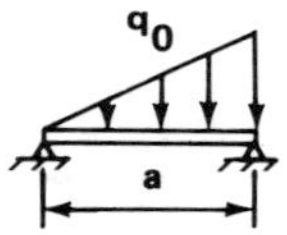	a/b	1.0	1.2	1.5	1.8	2.0	3.0	4.0			
		$\bar{w}$	0.02436	0.3432	0.04788	0.05892	0.06504	0.08484	0.09984			
		$\bar{\chi}_x$	−0.25121	−0.24673	−0.21837	−0.18237	−0.15666	−0.17209	−0.08703			
		$\bar{\chi}_y$	−0.21864	−0.31358	−0.44809	−0.55609	−0.61780	−0.80640	−0.95789			
	1d Uniformly Increasing Along Width 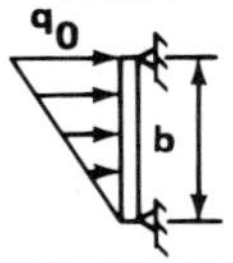	a/b	1.0	1.2	1.5	1.8	2.0	3.0	4.0			
		$\bar{w}$	0.02436	0.03384	0.04632	0.05580	0.06072	0.07344	0.07692			
		$\bar{\chi}_x$	−0.25121	−0.34523	−0.46602	−0.55833	−0.60501	−0.72382	−0.75587			
		$\bar{\chi}_y$	−0.21864	−0.20123	−0.16259	−0.12290	−0.09930	−0.03125	−0.00844			

TABLE 4. Generalized Curvatures in Simply Supported Rectangular Plates Uniformly Loaded Over Central Rectangular Area.

Loading

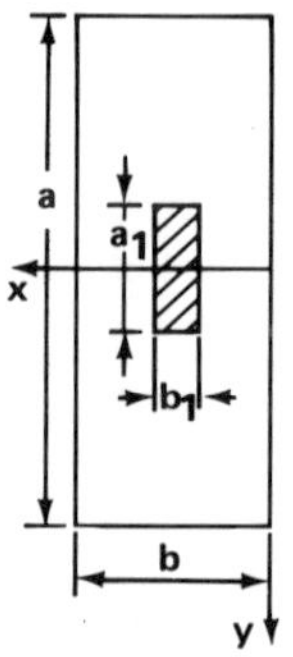

b_1/b \ a_1/b		a = b					
		0	0.2	0.4	0.6	0.8	1
0							
0.2	$\bar{\chi}_x$	−3.06857	−1.97538	———	———	———	———
	$\bar{\chi}_y$	−2.09143	−1.97538				
0.4	$\bar{\chi}_x$	−2.34725	−1.72352	−1.30154	———	———	———
	$\bar{\chi}_y$	−1.45582	−1.41495	−1.30154			
0.6	$\bar{\chi}_x$	−1.92132	−1.47560	−1.14857	−0.91385	———	———
	$\bar{\chi}_y$	−1.11560	−1.08132	−1.01143	−0.91385		
0.8	$\bar{\chi}_x$	−1.60088	−1.25407	−0.99429	−0.80044	−0.64615	———
	$\bar{\chi}_y$	−0.86374	−0.85978	−0.80571	−0.73187	−0.64615	
1	$\bar{\chi}_x$	−1.31077	−1.05231	−0.83341	−0.67121	−0.53934	−0.44308
	$\bar{\chi}_y$	−0.71077	−0.69231	−0.66198	−0.60264	−0.52220	−0.44308

b_1/b \ a_1/b		a = 1.4b					
		0	0.2	0.4	0.8	1.2	1.4
0	$\bar{\chi}_x$		−3.40484	−2.74154	−2.02286	−1.54681	−1.35297
	$\bar{\chi}_y$		−1.93055	−1.30154	−0.73714	−0.48396	−0.41011
0.2	$\bar{\chi}_x$	−2.45670	−2.32879	−2.10857	−1.68000	−1.32396	−1.15253
	$\bar{\chi}_y$	−2.85099	−1.79736	−1.25143	−0.72000	−0.46681	−0.39824
0.4	$\bar{\chi}_x$	−1.83297	−1.76044	−1.66945	−1.40703	−1.11429	−0.97846
	$\bar{\chi}_y$	−2.21011	−1.57187	−1.15516	−0.66989	−0.44571	−0.37846
0.6	$\bar{\chi}_x$	−1.4255	−1.42286	−1.34901	−1.16044	−0.94418	−0.81758
	$\bar{\chi}_y$	−1.7684	−1.33714	−1.02330	−0.61187	−0.41275	−0.35473
0.8	$\bar{\chi}_x$	−1.1697	−1.15780	−1.11692	−0.96659	−0.79121	−0.68703
	$\bar{\chi}_y$	−1.4611	−1.14066	−0.87692	−0.53802	−0.36264	−0.30989
1	$\bar{\chi}_x$	−0.95604	−0.95473	−0.92440	−0.79912	−0.65143	−0.56967
	$\bar{\chi}_y$	−1.21319	−0.93758	−0.71868	−0.45626	−0.30857	−0.26110

(continued)

TABLE 4 (continued).

		a = 2b					
b_1/b \ a_1/b		0	0.4	0.8	1.2	1.6	2.0
0	$\bar{\chi}_x$		−2.94593	−2.25363	−1.83033	−1.49275	−1.23692
	$\bar{\chi}_y$		−1.18022	−0.60791	−0.33890	−0.22418	−0.15692
0.2	$\bar{\chi}_x$	−2.64791	−2.31297	−1.91077	−1.59033	−1.31473	−1.07736
	$\bar{\chi}_y$	−2.73363	−1.13011	−0.59077	−0.33890	−0.21758	−0.16879
0.4	$\bar{\chi}_x$	−2.01099	−1.86066	−1.60747	−1.36747	−1.14989	−0.94813
	$\bar{\chi}_y$	−2.09670	−1.03780	−0.56176	−0.32176	−0.20703	−0.15956
0.6	$\bar{\chi}_x$	−1.59956	−1.52703	−1.36088	−1.17099	−0.98901	−0.82286
	$\bar{\chi}_y$	−1.66813	−0.90989	−0.50374	−0.29670	−0.18330	−0.13714
0.8	$\bar{\chi}_x$	−1.31341	−1.26462	−1.14066	−0.99165	−0.83736	−0.69363
	$\bar{\chi}_y$	−1.38198	−0.78462	−0.43780	−0.25451	−0.16879	−0.12791
1	$\bar{\chi}_x$	−1.07341	−1.03780	−0.94681	−0.82154	−0.70286	−0.58154
	$\bar{\chi}_y$	−1.14198	−0.66066	−0.36396	−0.22154	−0.13714	−0.10154

Comparing calculation results for different Poisson's ratios, one can notice that direct application of formulas in Reference 18 for lower than 0.3 Poisson's ratio causes underestimation of maximum deflections (up to 6.9% for $\nu = 0.15$) and reactions (up to 2.92%) and overestimation of maximum stresses (up to 6.4%). For higher than 0.3 Poisson's ratio, the direct application of formulas 18 causes underestimation of maximum normal stresses (up to 6% for $\nu = 0.45$) and overestimation of maximum deflections (up to 14.1%) and reactions (up to 3.1%).

Calculate the maximum deflection and normal stress for the small deflected plate with $a/b = 1.2$ uniformly loaded by force P over small concentric circle of radius r_0 for $\nu_1 = 0.15$; $\nu_2 = 0.30$; $\nu_3 = 0.45$. As per Table 3, loading lb:

$$\bar{w} = wh^3E/Pb^2(1 - \nu^2) = 0.1624;$$

$$\bar{\gamma}_x = 1.1549; \bar{\gamma}_y = 0.4835$$

For $\nu = 0.15$:

$$w = 0.1624(1 - 0.15^2)\frac{Pb^2}{Eh^3} = 0.1587\frac{Pb^2}{Eh^3};$$

$$\beta = (1.1549 + 0.15 \times 0.4835)/2 = 0.614$$

For $\nu = 0.30$:

$$w = 0.1624(1 - 0.30^2)\frac{Pb^2}{Eh^3} = 0.1478\frac{Pb^2}{Eh^3};$$

$$\beta = (1.1549 + 0.30 \times 0.4835)/2 = 0.65;$$

Here there is a coincidence results versus Reference [18].

For $\nu = 0.45$:

$$w = 0.1624(1 - 0.45^2)\frac{Pb^2}{Eh^3} = 0.1295\frac{Pb^2}{Eh^3};$$

$$\beta = (1.1549 + 0.45 \times 0.4835)/2 = 0.686$$

Calculate the maximum normal stresses in the plate subjected to the concentrated force P uniformly distributed over central rectangular area $a_1 \times b_1$, if $a_1/b = 0.6$; $b_1/b = 0.2$; $a = b$. Because this plate has a square configuration, there is a symmetry about the diagonal in Table 4; i.e., the case of central area $a_1/b = 0.6$; $b_1/b = 0.2$ is equivalent to the case $a_1/b = 0.2$; $b_1/b = 0.6$, which is specified in the above table. For calculation of normal stresses use curvatures from Table 4:

$$\bar{\chi}_x = -1.47560; \bar{\chi}_y = -1.08132$$

and calculate stresses for $\nu = 0.15$; $\nu = 0.30$; and $\nu = 0.45$.

$\nu = 0.15$:

$$\bar{\sigma}_x = -\frac{1}{2}((-1.47560) + 0.15(-1.08132)) = 0.8189$$

TABLE 5. Generalized Deflections and Curvatures in Uniformly Loaded Plates Over Entire Surface Plates with Simply Supported and Fixed Edges.

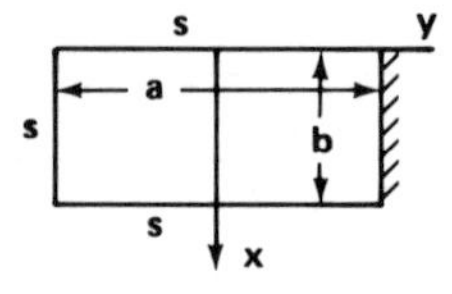

a/b	1	1.1	1.2	1.3	1.5	2.0	∞
$\bar{w}$	0.0336	0.0420	0.0516	0.0600	0.0768	0.1116	0.156
$\bar{\chi}_x$	0	0	0	0	0	0	0
$\bar{\chi}_y$	−1.0080	−1.1040	−1.1760	−1.2480	−1.3440	−1.4640	−1.500

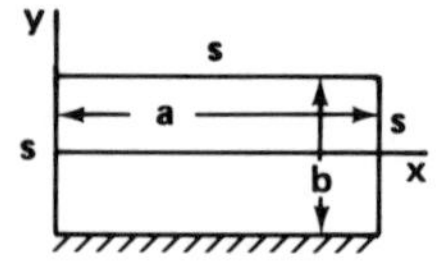

a/b	1	1.1	1.2	1.3	1.5	2.0	∞
$\bar{w}$	0.0336	0.0384	0.0420	0.0456	0.0504	0.0588	0.0624
$\bar{\chi}_x$	0	0	0	0	0	0	0
$\bar{\chi}_y$	−1.0080	−1.1040	−1.1760	−1.2360	−1.3320	−1.4640	−1.500

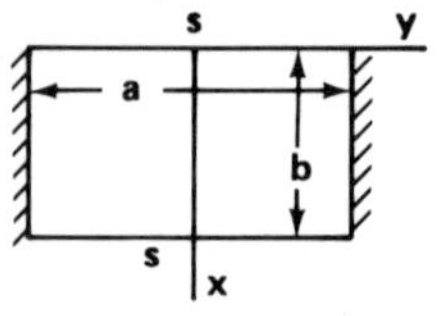

a/b	1	1.2	1.4	1.6	1.8	2.0	∞
$\bar{w}$	0.02308	0.03835	0.05516	0.07231	0.08791	0.10132	0.15626
$\bar{\chi}_x$	0	0	0	0	0	0	0
$\bar{\chi}_y$	−0.8364	−1.0416	−1.1976	−1.3080	−1.3824	−1.4292	−1.500

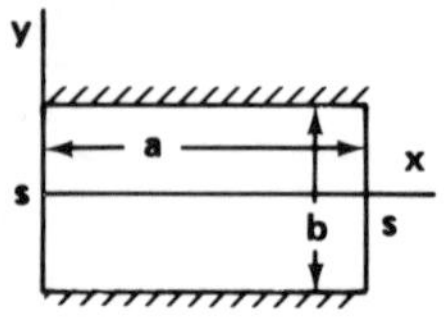

a/b	1	1.2	1.4	1.6	1.8	2.0	∞
$\bar{w}$	0.02308	0.02670	0.02879	0.0300	0.03077	0.03110	0.03132
$\bar{\chi}_x$	0	0	0	0	0	0	0
$\bar{\chi}_y$	−0.8364	−0.9252	−0.9720	−0.9936	−0.9942	−0.9946	−1.00

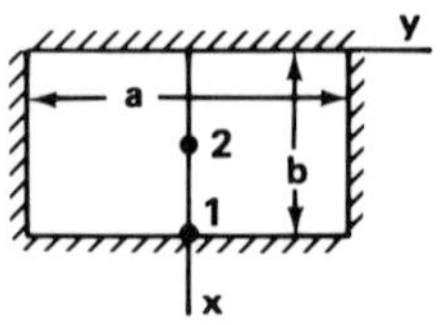

a/b	1	1.2	1.4	1.6	1.8	2.0	∞
$\bar{w}$	0.01512	0.02064	0.02484	0.0276	0.0294	0.03048	0.0312
$\bar{\chi}_{x1}$	−0.6156	−0.7668	−0.8712	−0.9360	−0.9744	−0.9948	−0.9996
$\bar{\chi}_{y1}$	0	0	0	0	0	0	0
$\bar{\chi}_{x2}$	−0.2132	−0.3041	−0.3764	−0.4261	−0.4600	−0.4808	−0.5000
$\bar{\chi}_{y2}$	−0.2132	−0.1824	−0.1415	−0.1038	−0.0708	−0.0454	0

TABLE 6. Generalized Maximum Deflections, Stresses, and Curvatures in Fixed Plates Under Load Uniformly Distributed Over Small Concentric Circle of Radius r_0 (or r'_0).

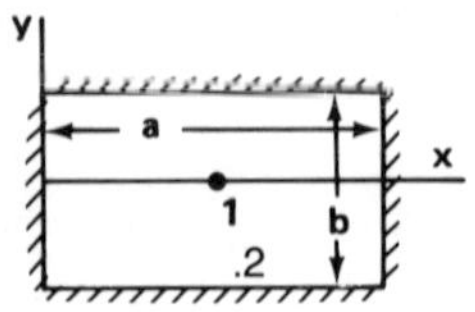

$$\sigma_1 = \frac{3P}{2\pi h^2}\left[(1+\nu)\ln\frac{2b}{\pi r'_0} + \beta_1\right];\ \beta_1 = (\bar{\gamma}_x + \nu\bar{\gamma}_y)/2$$

a/b	1	1.2	1.4	1.6	1.8	2.0	∞
$\bar{\gamma}_{x1}$	−0.36702	−0.00834	0.20176	0.30677	0.34643	0.34590	0.35173
$\bar{\gamma}_{y1}$	−0.36702	−0.48948	−0.58989	−0.65979	−0.69663	−0.71441	−0.72194
$\bar{\chi}_{x2}$	0	0	0	0	0	0	0
$\bar{\chi}_{y2}$	1.5084	1.7880	1.9248	1.9812	2.000	2.008	2.016
$\bar{w}$	0.06714	0.07758	0.08286	0.08538	0.08637	0.08659	0.08692

$$\overline{\sigma}_y = -\frac{1}{2}((-1.08132)$$

$$+ 0.15(-1.4756)) = 0.6513$$

$$\text{or } \sigma_{max} = 0.8189 \frac{P}{h^2}$$

$\nu = 0.30$:

$$\overline{\sigma}_x = -\frac{1}{2}((-1.4756)$$

$$+ 0.30(-1.08132)) = 0.900$$

$$\overline{\sigma}_y = -\frac{1}{2}((-1.08132)$$

$$+ 0.30(-1.47560)) = 0.762$$

$$\text{or } \sigma_{max} = 0.900 \frac{P}{h^2}$$

$\nu = 0.45$:

$$\overline{\sigma}_x = -\frac{1}{2}((-1.4756)$$

$$+ 0.45(-1.08132)) = 0.9811$$

$$\overline{\sigma}_y = -\frac{1}{2}((-1.08132)$$

$$+ 0.45(-1.4756)) = 0.8727$$

$$\text{or } \sigma_{max} = 0.9811 \frac{P}{h^2}$$

As one can see, deviation of maximum stresses for $\nu = 0.15$ and $\nu = 0.45$, as compared with stresses for $\nu = 0.3$, are $+9.9\%$ and -8.3%

The stresses at plate fixed edges are specified in Tables 5 to 9 by generalized curvatures rather than directly to make use of formulas [Equation (109)] unique to all cases covered by Tables 3 to 9.

PLATES OF UNIFORM THICKNESS AND ARBITRARY CONFIGURATIONS

The necessary equations of similitude [Equations (84) to (87)] are independent of plate configuration and are applicable, but the difference is in the boundary conditions. The simply supported plate edge with boundary condition [Equation (72)] does not allow formulation [Equation (89)], and therefore requires material of all plates of the similar group to have an identical Poisson's ratio, i.e.:

$$C_\nu = C_{\nu_0} = C_{\nu_1} = C_{\nu_2} = 1 \tag{111}$$

The fixed plate edge specified by Equation (71), as well as the guided edge with boundary conditions:

$$\frac{\partial w}{\partial n} = 0; \frac{\partial}{\partial x}\left(\frac{\partial^2 w}{\partial x^2} + \frac{\partial^2 w}{\partial y^2}\right)\cos\beta + \frac{\partial}{\partial y}\left(\frac{\partial^2 w}{\partial x^2} + \frac{\partial^2 w}{\partial y^2}\right)\sin\beta \tag{112}$$

does not contradict $C_\nu \neq 1$, and plates of uniform thickness and arbitrary configurations can have a material with a different Poisson's ratio.

In case of an identical Poisson's ratio, the equations of similitude and generalized variables are identical to those originated in the section on isotropic linearly elastic plates with small deflections, and in the case of $C_\nu \neq 1$ the equations of similitude and generalized variables are identical to those originated in the section about rectangular plates of uniform thickness.

Example 7

a. Generalize formulas for calculation of small deflected regular polygonal plates with fixed edges and $\nu = 0.3$ (Reference 18, Table 26, Case 20).

   For considered boundary conditions, $C_\nu \neq 1$ and generalized curvatures $\overline{\chi}_x = \chi_x Eh^3/q_0b^2(1 - \nu^2)$ and $\overline{\chi}_y = \chi_y Eh^3/q_0b^2(1 - \nu^2)$ can be calculated by Equation (108) along with generalized deflections $\overline{w} = wh^3E/q_0b^4(1 - \nu^2)$. They are given in Table 10.

b. Generalize formulas for calculation of maximum resultant stresses (along longer edge toward obtuse corners) and deflections in parallelogram plates with fixed edges and $\nu = 1/3$ (Reference 18, Table 26, Case 16).

   The formula for stresses at the center of the plate is not a subject of generalization because these are the maximum principal stresses and the corresponding minimum principal stresses are not available [26].

   In Table 11 curvatures $\overline{\chi}_x$ and $\overline{\chi}_y$ along with deflection $\overline{w}$ are calculated for skew angles $\theta = 15$ degrees, 30 degrees, 45 degrees, and 60 degrees.

c. Generalize formulas for calculation of maximum deflection $\overline{w}$ and curvatures $\overline{\chi}$ (and stresses) in solid circular plates subjected to the uniformly distributed load over the shaded segment. The plate is rigidly clamped and $\nu = 1/3$ (Reference 18, Table 24, Case 25).

   The stress information in Reference 18 is not enough for calculation of generalized curvatures. Therefore, refer to the origin [27] and use its information for calculation of deflections and curvatures given in Table 12. Here the maximum shearing force $\overline{Q}_r = Q_r/q_0a$ is added.

TABLE 7. Generalized Maximum Deflections and Curvatures in Fixed Plates Under Hydrostatic Pressure.

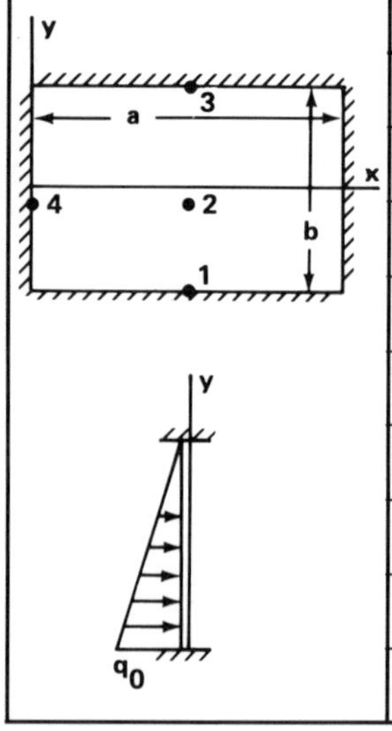

a/b	0.6	0.8	1.0	1.2	1.4	1.6	1.8	2.0
$\bar{\chi}_{x1}$	0	0	0	0	0	0	0	0
$\bar{\chi}_{y1}$	0.22644	0.35568	0.47304	0.55548	0.60072	0.61848	0.61992	0.61356
$\bar{\chi}_{x2}$	−0.11445	−0.10952	−0.11002	−0.08875	−0.06013	−0.02906	−0.01498	−0.00265
$\bar{\chi}_{y2}$	−0.04805	−0.09375	−0.14087	−0.18098	−0.20756	−0.24232	−0.22687	−0.22876
$\bar{\chi}_{x3}$	0	0	0	0	0	0	0	0
$\bar{\chi}_{y3}$	0.04128	0.09936	0.17952	0.24972	0.29640	0.32304	0.33588	0.34188
$\bar{\chi}_{x4}$	0.26088	0.28716	0.33732	0.36012	0.36900	0.37476	0.38028	0.38160
$\bar{\chi}_{y4}$	0	0	0	0	0	0	0	0
$\bar{w}$	0.001758	0.005165	0.008132	0.010659	0.012418	0.013846	0.014615	0.014945

TABLE 8. Generalized Curvatures in Plates with Three Fixed Edges and One Simply Supported Edge Under the Uniformly Distributed and Uniformly Decreasing Loads.

1a Uniformly Loaded Over Entire Plate

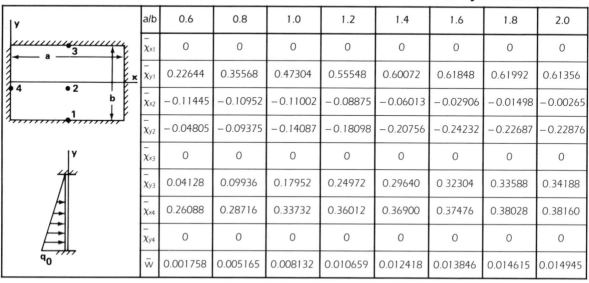

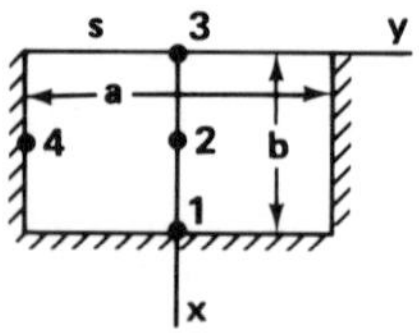

a/b	0.25	0.50	0.75	1.0	1.5	2.0	3.0
$\bar{\chi}_{x1}$	−0.04	−0.162	−0.346	−0.614	−1.078	−1.314	−1.436
$\bar{\chi}_{y1}$	0	0	0	0	0	0	0
$\bar{\chi}_{x2}$	0.00176	0.00066	−0.05846	−0.19033	−0.51604	−0.72418	−0.86352
$\bar{\chi}_{y2}$	−0.03253	−0.12220	−0.21846	−0.25890	−0.17319	−0.05275	+0.06505
$\bar{\chi}_{x4}$	0	0	0	0	0	0	0
$\bar{\chi}_{y4}$	−0.062	−0.242	−0.484	−0.686	−0.834	−0.796	−0.636

(continued)

TABLE 8 (continued).

	a/b	0.25	0.50	0.75	1.0	1.5	2.0	3.0
1b Uniformly Loaded Over ⅔ of Plate From Fixed Edge 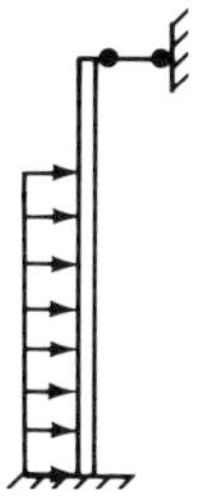	$\bar{\chi}_{x1}$	−0.04	−0.160	−0.328	−0.548	−0.890	−1.050	−1.132
	$\bar{\chi}_{y1}$	0	0	0	0	0	0	0
	$\bar{\chi}_{x2}$	0.00132	−0.00681	−0.04330	−0.13319	−0.35033	−0.49385	−0.58505
	$\bar{\chi}_{y2}$	−0.02440	−0.08396	−0.14901	−0.17604	−0.11890	−0.03385	−0.04352
	$\bar{\chi}_{x4}$	0	0	0	0	0	0	0
	$\bar{\chi}_{y4}$	−0.062	−0.222	−0.394	−0.510	−0.568	−0.526	−0.408

	a/b	0.25	0.50	0.75	1.0	1.5	2.0	3.0
1c Uniformly Loaded Over ⅓ of Plate From Fixed Edge 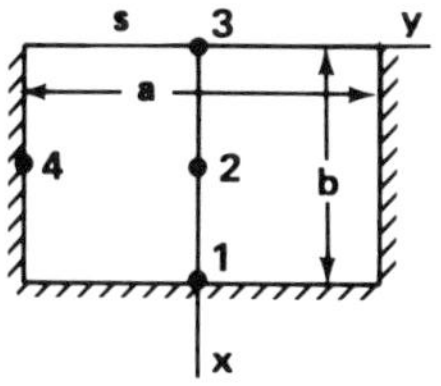 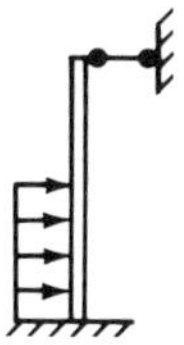	$\bar{\chi}_{x1}$	−0.02	−0.136	−0.216	−0.296	−0.388	−0.426	−0.444
	$\bar{\chi}_{y1}$	0	0	0	0	0	0	0
	$\bar{\chi}_{x2}$	−0.00242	−0.03868	−0.07626	−0.09275	−0.09275	−0.08286	−0.07604
	$\bar{\chi}_{y2}$	−0.02527	−0.04440	−0.03912	−0.02418	−0.00418	+0.00286	+0.00681
	$\bar{\chi}_{x4}$	0	0	0	0	0	0	0
	$\bar{\chi}_{y4}$	−0.052	−0.126	−0.158	−0.158	−0.136	−0.112	−0.074

	a/b	0.25	0.50	0.75	1.0	1.5	2.0	3.0
1d Uniformly Decreasing Load From Fixed to Simply Supported Edge 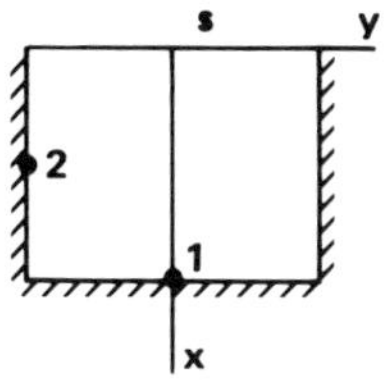	$\bar{\chi}_{x1}$	−0.036	−0.128	−0.240	−0.388	−0.606	−0.712	−0.764
	$\bar{\chi}_{y1}$	0	0	0	0	0	0	0
	$\bar{\chi}_{x2}$	0	0	0	0	0	0	0
	$\bar{\chi}_{y2}$	−0.038	−0.136	−0.248	−0.322	−0.362	−0.336	−0.264

TABLE 9. Generalized Curvatures in Plates with Three Fixed Edges and One Simply Supported Edge Under the Uniformly Decreasing Load on the Part of Plate Surface.

Uniformly Decreasing From Fixed Edge to Zero at ⅔ b

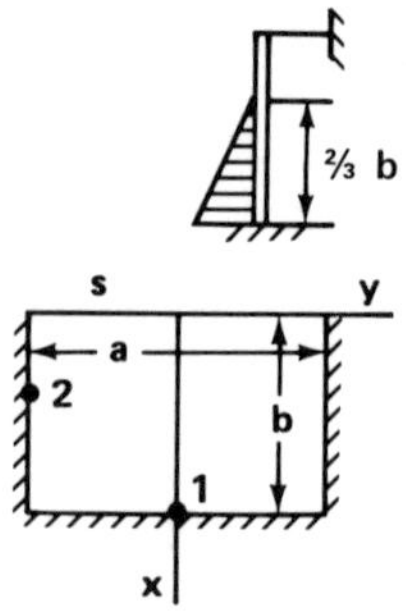

a/b	0.25	0.50	0.75	1.0	1.5	2.0	3.0
$\bar{\chi}_{x1}$	−0.034	−0.112	−0.190	−0.280	−0.402	−0.456	−0.482
$\bar{\chi}_{y1}$	0	0	0	0	0	0	0
$\bar{\chi}_{x2}$	0	0	0	0	0	0	0
$\bar{\chi}_{y2}$	−0.038	−0.100	−0.136	−0.196	−0.212	−0.194	−0.148

Uniformly Decreasing From Fixed Edge to Zero at ⅓ b

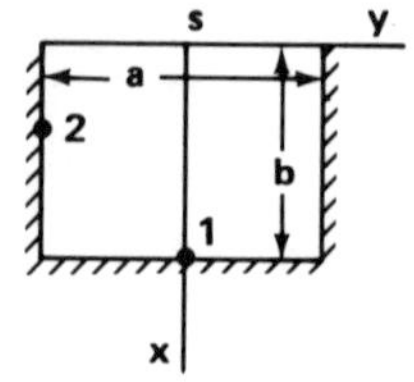

a/b	0.25	0.50	0.75	1.0	1.5	2.0	3.0
$\bar{\chi}_{x1}$	−0.028	−0.070	−0.094	−0.122	−0.150	−0.160	−0.164
$\bar{\chi}_{y1}$	0	0	0	0	0	0	0
$\bar{\chi}_{x2}$	0	0	0	0	0	0	0
$\bar{\chi}_{y2}$	−0.020	−0.048	−0.062	−0.060	−0.050	−0.040	−0.026

d. Generalize formulas for calculation of stresses given for solid semicircular plates with all fixed edges and subjected to uniformly distributed loads over the entire surface. These formulas are given in Reference 18, Table 24, Case 30 for $\nu = 0.2$. By using the information in Reference 28, one can calculate generalized curvatures $\bar{\chi}_r$, $\bar{\chi}_t$ and maximum shearing forces $\bar{Q}_r = Q_r/q_0a$ shown in Table 13.

It should be noted that generalized curvatures $\bar{\chi}_r$ and $\bar{\chi}_t$ for circular (item c) and semicircular (item d) plates are calculated by formulas:

$$\bar{\chi}_r = \frac{\chi_r Eh^3}{q_0a^2(1 - \nu^2)};\ \bar{\chi}_t = \frac{\chi_t Eh^3}{q_0a^2(1 - \nu^2)} \tag{113}$$

BUCKLING OF FLAT PLATES

It is assumed that plates remain flat (no geometrical irregularities) up to critical load (stress) development. The membrane stresses that appear through the buckling process are negligible as compared with bending stresses. Also, the deflections are assumed to be less than one-quarter to one-half of the plate thickness. With these assumptions consider simplified Equations (53) and (83) as follows:

$$D\nabla^4 w + 2\frac{\partial D}{\partial x}\frac{\partial}{\partial x}\nabla^2 w + 2\frac{\partial D}{\partial y}\frac{\partial}{\partial y}\nabla^2 w + \nabla^2 D \nabla^2 w - (1 - \nu) \times \left(\frac{\partial^2 D}{\partial x^2} \times \frac{\partial^2 w}{\partial y^2} - 2\frac{\partial^2 D}{\partial x \partial y} \times \frac{\partial^2 w}{\partial x \partial y} + \frac{\partial^2 D}{\partial y^2}\frac{\partial^2 w}{\partial x^2}\right) + N_x\frac{\partial^2 w}{\partial x^2} + N_y\frac{\partial^2 w}{\partial y^2} + 2N_{xy}\frac{\partial^2 w}{\partial x \partial y} = 0 \tag{114}$$

$$D\nabla^4 w + N_x\frac{\partial^2 w}{\partial x^2} + N_y\frac{\partial^2 w}{\partial y^2} + 2N_{xy}\frac{\partial^2 w}{\partial x \partial y} = 0 \tag{115}$$

TABLE 10. Generalized Curvatures and Maximum Deflections in Regular Polygonal Plates with All Fixed Edges, and Uniformly Loaded Over Entire Plate (e = at Center of Straight Edge; c = at Center of Plate).

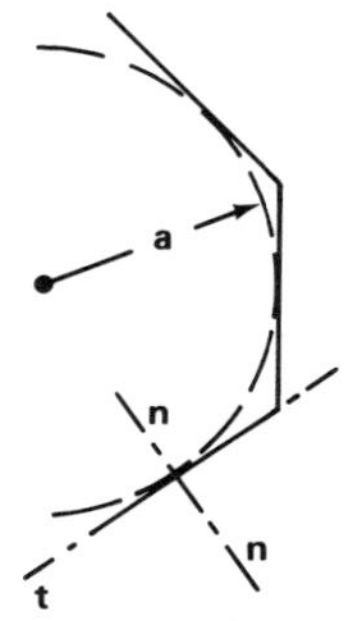

Number of Sides	3	4	5	6	7	8	9	10	∞
$\bar{\chi}_{xc}$	−0.90615	−0.84615	−0.81538	−0.79692	−0.78615	−0.77846	−0.77385	−0.76923	−0.7500
$\bar{\chi}_{yc}$	−0.90615	−0.84615	−0.81538	−0.79692	−0.78615	−0.77846	−0.77385	−0.76923	−0.7500
$\bar{\chi}_{ne}$	−2.846	−2.164	−2.264	−2.136	−2.046	−1.980	−1.928	−1.888	−1.500
$\bar{\chi}_{te}$	0	0	0	0	0	0	0	0	0
$\bar{w}$	0.29011	0.24286	0.22308	0.21319	0.20659	0.20220	0.2000	0.19780	0.18791

TABLE 11. Generalized Curvatures and Maximum Deflections in Parallelogram Plate With All Fixed Edges, and Uniformly Loaded Over Entire Plate.

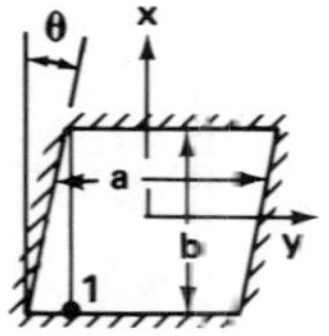

θ	a/b	1.00	1.25	1.50	1.75	2.00	2.25	2.50	3.00
15°	$\bar{\chi}_{x1}$	−0.640	−0.824	−0.966	−1.062	−1.106	—	—	—
	$\bar{\chi}_{y1}$	0	0	0	0	0	—	—	—
	$\bar{w}$	0.01429	+0.02126	0.02610	0.02891	0.03071	—	—	—
30°	$\bar{\chi}_{x1}$	—	−0.800	−0.990	−1.094	−1.136	−1.160	—	—
	$\bar{\chi}_{y1}$	—	0	0	0	0	0	—	—
	$\bar{w}$	—	0.01890	0.02453	0.02801	0.03015	0.03161	—	—
45°	$\bar{\chi}_{x1}$	—	—	−0.788	−0.940	−1.062	−1.150	−1.202	—
	$\bar{\chi}_{y1}$	—	—	0	0	0	0	0	—
	$\bar{w}$	—	—	0.01856	0.02340	0.02723	0.02981	0.03195	—
60°	$\bar{\chi}_{x1}$	—	—	—	—	−0.620	−0.900	−1.076	−1.226
	$\bar{\chi}_{y1}$	—	—	—	—	0	0	0	0
	$\bar{w}$	—	—	—	—	0.01530	0.01924	0.02228	0.02756

TABLE 12. Generalized Maximum Deflections, Curvatures, and Shearing Forces in Solid Circular Plate With Fixed Edge Subjected to the Uniformly Distributed Load Over the Segment (Shaded).

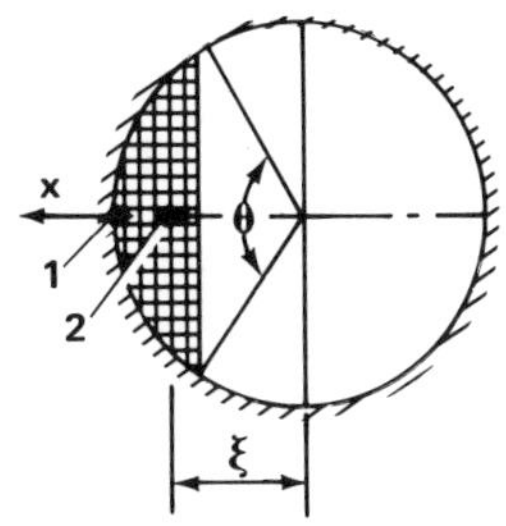

Component Type	θ					
	90°		120°		180°	
	$\xi = r/a$	Component	$\xi = r/a$	Component	$\xi = r/a$	Component
$\bar{w}$	0.5	0.0041301	0.4	0.019385	0.2	0.101437
$\bar{\chi}_{r1}$	—	—	1.0	0.570	—	—
$\bar{\chi}_{t1}$	—	—	1.0	0	—	—
$\bar{\chi}_{r2}$	—	—	0.6	−0.19845	—	—
$\bar{\chi}_{t2}$	—	—	0.6	−0.11025	—	—
$\bar{Q}_r$	—	—	1.0	0.3333	—	—

In Equations (114) and (115), the compressive membrane forces N_x, N_y, and N_{xy} are assumed to be positive.

By application of the second theorem of similitude one can obtain the similitude equations:

From Equation (114):

$$C_\nu = 1;\ C_N C_l^2 / C_E C_h^3 = 1 \qquad (116)$$

From Equation (115):

$$C_N C_l^2 C_{\nu_0} / C_E C_h^3 = 1 \qquad (117)$$

which are identical to Equations (54) and (84) with the only difference that N_x, N_y, and N_{xy} are critical loads. The critical stresses are equal to:

$$\sigma_x = N_x/h;\ \sigma_y = N_y/h;\ \sigma_{xy} = N_{xy}/h \qquad (118)$$

and:

$$C_\sigma = C_N/C_h$$

The decision about Poisson's ratio influence is based on boundary conditions and has already been considered.

The generalized critical stresses and loads are as follows:

$$\text{For } C_\nu = 1:\ \bar{\sigma}_{crit} = \frac{\sigma l^2}{Eh^2};\ \bar{N}_{crit} = \frac{Nl^2}{Eh^3} \qquad (119)$$

$$\text{For } C_\nu \neq 1:\ \bar{\sigma}_{crit} = \frac{\sigma l^2(1 - \nu^2)}{Eh^2};$$

$$\bar{N}_{crit} = \frac{Nl^2(1 - \nu^2)}{Eh^3} \qquad (120)$$

By substitution of Equation (118) in Equations (116) and (117), one can obtain the following similitude constants for critical stresses:

$$C_\nu = 1;\ C_\sigma = \frac{C_E C_h^2}{C_l^2} \qquad (121)$$

$$C_\nu \neq 1;\ C_\sigma = \frac{C_E C_h^2}{C_l^2 C_{\nu 0}} \qquad (122)$$

These similitude equations alone [without solution of differential Equations (114) and (115)] have shown the follow-

TABLE 13. Generalized Curvatures and Maximum Shearing Forces in Solid Semicircular Plate With All Edges Fixed, Under Uniformly Distributed Load Over the Entire Surface.

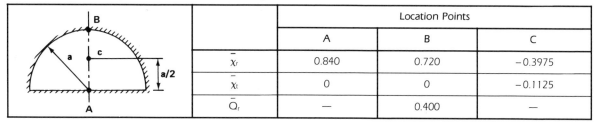

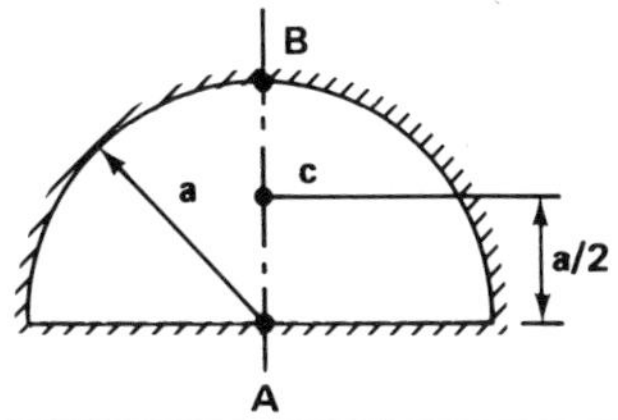

	Location Points		
	A	B	C
$\bar{\chi}_r$	0.840	0.720	−0.3975
$\bar{\chi}_t$	0	0	−0.1125
$\bar{Q}_r$	—	0.400	—

ing formulas for calculation of critical stresses:

$$C_\nu = 1;\ \sigma_{crit} = A\,\frac{Eh^2}{l^2} \tag{123}$$

$$C_\nu \neq 1;\ \sigma_{crit} = A\,\frac{Eh^2}{l^2(1 - \nu^2)} \tag{124}$$

where A = a constant obtained either by numerical or experimental analysis.

Equation (124) coincides with formulas for critical stresses given in Reference 18, Table 35, Cases 1 to 4, 6, and 8 to 12.

If concentrated force P is applied in the plate plane, the suitable similitude constant C_p can be expressed through C_N by formula:

$$C_p = C_N C_l \tag{125}$$

or:

$$C_\nu = 1;\ C_p = \frac{C_E C_h^3}{C_l}$$

$$C_\nu \neq 1;\ C_p = \frac{C_E C_h^3}{C_p C_{\nu 0}}$$

and finally:

$$C_\nu = 1;\ P_{crit} = A\,\frac{Eh^3}{l} \tag{126}$$

$$C_\nu \neq 1;\ P_{crit} = A\,\frac{Eh^3}{l(1 - \nu^2)} \tag{127}$$

Equation (127) coincides with the formula in Table 35, Case 7 of Reference 18.

If the flat plate is subjected to temperature $T(x,y)$, which is invariable through plate thickness h (the plane stress problem of thermoelasticity), the following equation [21] gives additional information about the relationships between membrane forces (stresses) in Equations (114) and (115) and temperature $T(x,y)$:

$$\frac{1}{E}\,\nabla^4\phi = -\alpha\nabla^2 T(x, y) \tag{128}$$

where:

$$N_x = h\,\frac{\partial^2\phi}{\partial y^2};\ N_y = h\,\frac{\partial^2\phi}{\partial x^2};\ N_{xy} = -h\,\frac{\partial^2\phi}{\partial x\partial y} \tag{129}$$

Equations (118), (128), and (129) lead to the following:

$$C_\sigma = C_E C_\alpha C_T \tag{130}$$

where $C_\sigma = C_{\sigma x} = C_{\sigma y} = C_{\sigma xy}$.

Equation (128) covers cases when $T(x,y)$ is at least the second order polynom. If $T(x,y)$ is a constant or linear function of x,y, then $\nabla^2 T(x,y) \equiv 0$ and $\phi(x,y) \equiv 0$. But the temperature influence can occur through the boundary conditions if they restrict the plate thermal expansion. It could be shown that Equation (130) works for any function $T(x,y)$.

By substituting Equation (130) in Equations (121) and (122), one can derive similitude constants for critical temperature:

$$C_\nu = 1;\ C_T = \frac{C_h^2}{C_l^2 C_\alpha} \tag{131}$$

$$C_\nu \neq 1;\ C_T = \frac{C_h^2}{C_l^2 C_{\nu 0} C_\alpha} \tag{132}$$

These similitude equations lead to the following formulas for critical temperatures:

$$C_\nu = 1;\ T_{crit} = A\,\frac{h^2}{l^2\alpha} \tag{133}$$

$$C_\nu \neq 1;\ T_{crit} = A\,\frac{h^2}{l^2\alpha(1 - \nu^2)} \tag{134}$$

The formula [Equation (134)] for critical temperature coincides with the formula:

$$T_{1,crit} = \frac{K_T}{1 - \nu^2}\,\frac{1}{\alpha}\left(\frac{h}{b}\right)^2 \tag{135}$$

which is a result of the solution of differential Equations (115) and (128) given in Reference 21.

Anisotropic Linearly Elastic Plates with Small Deflections

Experimental investigation of some materials shows different elastic behavior in different directions; an anisotropic material must be assumed [29]. It is assumed that the material of the plate has three planes of symmetry with respect to its elastic properties [20,30]. If these planes are the coordinate planes, the relationships between the stress and strain components for the case of plane stress in the xy plane can be written as follows:

$$\sigma_x = E_x'\epsilon_x + E''\epsilon_y;\ \sigma_y = E'_y\epsilon_y + E''\epsilon_x;\ \sigma_{xy} = G\epsilon_{xy} \tag{136}$$

Here four constants, E_x', E_y', E'', and G are needed to characterize the elastic properties of a material.

These four constants can be replaced by another set of four constants [21], namely, moduli of elasticity E_1 and E_2 along x and y axes, shear modulus G, and Poisson's ratio ν_1, which corresponds to ϵ_y. The second Poisson's ratio ν_2, which corresponds to ϵ_x, can be expressed through ν_1 by the formula:

$$\nu_2 = \frac{E_2}{E_1}\nu_1 \tag{137}$$

Both groups of constants are related to each other as follows:

$$E_x' = \frac{E_1}{(1 - \nu_1\nu_2)}; \quad E_y' = \frac{E_2}{(1 - \nu_1\nu_2)}; \quad E'' = \frac{E_1\nu_2}{(1 - \nu_1\nu_2)} \tag{138}$$

Keeping in mind that the subjects of our consideration are the plates of uniform thickness h, the governing differential equation for plates under combined transverse and tangent static load components, transverse dynamic load, and temperature can be written as follows [20,21]:

$$\begin{aligned} D_x\frac{\partial^4 w}{\partial x^4} &+ 2H\frac{\partial^4 w}{\partial x^2\partial y^2} + D_y\frac{\partial^4 w}{\partial y^4} \\ &= q(x, y) + N_x\frac{\partial^2 w}{\partial x^2} + N_y\frac{\partial^2 w}{\partial y^2} \\ &+ 2N_{xy}\frac{\partial^2 w}{\partial x\partial y} - mh\frac{\partial^2 w}{\partial t^2} - kh\frac{\partial w}{\partial t} \\ &+ Z(x, y, t) - \frac{h^2\alpha}{12(1 - \nu_1\nu_2)} \\ &\times \left[E_1(1 + \nu_2)\frac{\partial^2\lambda}{\partial x^2} + E_2(1 + \nu_1)\frac{\partial^2\lambda}{\partial y^2}\right] \end{aligned} \tag{139}$$

where:

$$D_x = \frac{E_x'h^3}{12}; \quad D_y = \frac{E_y'h^3}{12}; \quad D_1 = \frac{E''h^3}{12}; \quad D_{xy} = \frac{Gh^3}{12}; \quad H = D_1 + 2D_{xy} \tag{140}$$

Plate material is assumed to have uniform mass density m and thermally invariable material properties E_1, E_2, ν_1, ν_2, α. The addition to Equation (139) plate stress-strain components can be expressed in terms of deflection w:

$$\begin{aligned} \sigma_x &= \frac{E_1}{1 - \nu_1\nu_2}(\varepsilon_x + \nu_2\varepsilon_y); \\ \sigma_y &= \frac{E_2}{1 - \nu_1\nu_2}(\varepsilon_y + \nu_1\varepsilon_x); \\ \sigma_{xy} &= G\varepsilon_{xy}; \\ M_x &= -D_x(\chi_x + \nu_2\chi_y) + \frac{E_1h^2(1 + \nu_2)}{12(1 - \nu_1\nu_2)}\alpha\lambda; \\ M_y &= -D_y(\chi_y + \nu_1\chi_x) + \frac{E_2h^2(1 + \nu_1)}{12(1 - \nu_1\nu_2)}\alpha\lambda; \\ M_{xy} &= -2D_{xy}\chi_{xy}; \\ \chi_x &= \frac{\partial^2 w}{\partial x^2}; \quad \chi_y = \frac{\partial^2 w}{\partial y^2}; \\ \chi_{xy} &= \frac{\partial^2 w}{\partial x\partial y}; \quad \varepsilon_x = \frac{h}{2}\chi_x; \\ \varepsilon_y &= \frac{h}{2}\chi_y; \quad \varepsilon_{xy} = \frac{h}{2}\chi_{xy}; \\ Q_x &= -\frac{\partial}{\partial x}(D_x\chi_x + H\chi_y); \\ Q_y &= -\frac{\partial}{\partial y}(D_y\chi_y + H\chi_x) \end{aligned} \tag{141}$$

Assuming that Equations (138) to (141) are written for plate 1, by substitution of Equation (1) one can equate indicators of similitude to unity and obtain the necessary conditions of similitude.

Thus, if plate is subjected to loads q, N_x, N_y, N_{xy} only, the similitude equations from Equation (139) are as follows:

$$\begin{aligned} &C_{\nu1}C_{\nu2} = 1 \text{ (or } C_{\nu12} = 1); \\ &C_{E_2}/C_{E_1}C_{\nu_2}^2 = 1; \quad C_G/C_{E_1}C_{\nu_2} = 1; \\ &C_xC_{\nu_2}^{0.5}/C_y = 1; \quad C_wC_h^3C_{E_1}/C_qC_l^4 = 1; \end{aligned} \tag{142}$$

$$C_{N_x}C_l^2/C_{E_1}C_h^3 = 1; C_{N_y}C_l^2/C_{E_1}C_h^3C_{v_2} = 1;$$

$$C_x/C_l = 1; C_{N_{xy}}C_l^2/C_{E_1}C_h^3C_{v_2}^{0.5} = 1$$

where:

$$C_{v_1} = v_{1,2}/v_{1,1}; C_{v_2} = v_{2,2}/v_{2,1};$$

$$C_{v_{12}} = (1 - v_1v_2)_2/(1 - v_1v_2)_1$$

They can be replaced by generalized variables:

$$\begin{aligned}
&\bar{v} = \sqrt{v_1v_2}; \bar{E} = E_2/E_1v_2^2;\\
&\bar{G} = G/E_1v_2; \xi = x/l;\\
&\eta = y/lv_2^{0.5}; \bar{w} = wh^3E_1/q_0l^4;\\
&\bar{N}_x = N_xl^2/E_1h^3; \bar{N}_y = N_yl^2/E_1h^3v_2;\\
&\bar{N}_{xy} = N_{xy}l^2/E_1h^3v_2^{0.5}
\end{aligned} \tag{143}$$

and nondimensional Equation (139) is as follows:

$$\begin{aligned}
&\frac{\partial^4\bar{w}}{\partial\xi^4} + 2[1 + 2(1 - \bar{v}^2)\bar{G}]\frac{\partial^4\bar{w}}{\partial\xi^2\partial\eta^2}\\
&+ \bar{E}\frac{\partial^4\bar{w}}{\partial\eta^4} = 12(1 - \bar{v}^2)\\
&\times\left[P(\xi, \eta) + \bar{N}_x\frac{\partial^2\bar{w}}{\partial\xi^2} + \bar{N}_y\frac{\partial^2\bar{w}}{\partial\eta^2}\right.\\
&\left.+ 2\bar{N}_{xy}\frac{\partial^2\bar{w}}{\partial\xi\partial\eta}\right]
\end{aligned} \tag{144}$$

If plate is additionally subjected to thermal load λ, one more indicator of similitude follows from Equation (139):

$$C_{\nu_1} = C_{\nu_2} = 1 \tag{145}$$

and this indicator simplifies Equations (142) to (144) as follows:

$$\begin{aligned}
&C_{E_2}/C_{E_1} = 1;\\
&C_G/C_{E_1} = 1;\\
&C_x/C_y = 1;\\
&C_wC_h^3C_{E_1}/C_qC_l^4 = 1;\\
&C_{N_x}C_l^2/C_{E_1}C_h^3 = 1;\\
&C_{N_y}C_l^2/C_{E_1}C_h^3 = 1;\\
&C_x/C_l = 1;\\
&C_{N_{xy}}C_l^2/C_{E_1}C_h^3 = 1;\\
&C_\lambda C_\alpha C_h^2C_{E_1}/C_qC_l^2 = 1
\end{aligned} \tag{146}$$

$$\begin{aligned}
&\bar{E} = E_2/E_1;\\
&\bar{G} = G/E_1;\\
&\xi = x/l;\\
&\eta = y/l;\\
&\bar{w} = wh^3E_1/q_0l^4;\\
&p(\xi, \eta) = \frac{q(x, y)}{q_0}\\
&\bar{N}_x = N_xl^2/E_1h^3;\\
&\bar{N}_y = N_yl^2/E_1h^3;\\
&\bar{N}_{xy} = N_{xy}l^2/E_1h^3;\\
&\mu(\xi, \eta) = \lambda(x, y)\alpha h^2E_1/q_0l^2
\end{aligned} \tag{147}$$

$$\begin{aligned}
&\frac{\partial^4\bar{w}}{\partial\xi^4} + 2[v_2 + 2(1 - v_1v_2)\bar{G}]\frac{\partial^4\bar{w}}{\partial\xi^2\partial\eta^2}\\
&+ \bar{E}\frac{\partial^4\bar{w}}{\partial\eta^4} = \underline{12(1 - v_1v_2)}\\
&\underline{\times\left[p(\xi, \eta) + \bar{N}_x\frac{\partial^2\bar{w}}{\partial\xi^2} + \bar{N}_y\frac{\partial^2\bar{w}}{\partial\eta^2} + 2\bar{N}_{xy}\frac{\partial^2\bar{w}}{\partial\xi\partial\eta}\right]}\\
&- \left[(1 + v_2)\frac{\partial^2\mu}{\partial\xi^2} + \bar{E}(1 + v_1)\frac{\partial^2\mu}{\partial\eta^2}\right]
\end{aligned} \tag{148}$$

If plate is subjected to thermal load λ only, the similitude indicators [Equation (61)] and generalized variables [Equation (62)] can be utilized and Equation (148) has to be simplified by exclusion of the underlined terms.

For plates subjected to dynamic loads only, the following

equations of similitude can be derived from Equation (139):

$$
\begin{aligned}
C_{\nu_1}C_{\nu_2} &= 1 \text{ (or } C_{\nu_{12}} = 1);\\
C_{E_2}/C_{E_1}C_{\nu_2}^2 &= 1;\\
C_G/C_{E_1}C_{\nu_2} &= 1;\\
C_xC_{\nu_2}^{0.5}/C_y &= 1;\\
C_x/C_l &= 1;\\
C_wC_h^3C_{E_1}/C_zC_l^4 &= 1;\\
C_tC_hC_{E_1}^{0.5}/C_m^{0.5}C_l^2 &= 1;\\
C_f &= C_t^{-1};\\
C_\omega &= C_t^{-1} \text{ (for harmonic load)};\\
C_kC_l^2/C_{E_1}^{0.5}C_m^{0.5}C_h &= 1
\end{aligned}
\tag{149}
$$

which shows the following structure of generalized variables:

$$
\begin{aligned}
&\bar{\nu} = \sqrt{\nu_1\nu_2};\ \bar{E} = E_2/E_1\nu_2^2;\\
&\bar{G} = G/E_1\nu_2;\ \xi = x/l;\\
&\eta = y/l\nu_2^{0.5};\ \bar{w} = wh^3E_1/Z_0l^4;\\
&\theta = \frac{th}{l^2}\sqrt{\frac{E_1}{m}};\ \bar{k} = \frac{kl^2}{h}\frac{1}{\sqrt{E_1m}}
\end{aligned}
\tag{150}
$$

Following is the governing nondimensional differential equation:

$$
\frac{\partial^4\bar{w}}{\partial\xi^4} + 2[1 + 2(1 - \bar{\nu}^2)\bar{G}]\frac{\partial^4\bar{w}}{\partial\xi^2\partial\eta^2} + \bar{E}\frac{\partial^4\bar{w}}{\partial\eta^4} = 12(1 - \bar{\nu}^2) \times \left[\bar{Z}(\xi, \eta, \theta) - \frac{\partial^2\bar{w}}{\partial\theta^2} - \bar{k}\frac{\partial\bar{w}}{\partial\theta}\right]
\tag{151}
$$

Consider boundary conditions in which the plate has an arbitrary plane configuration and the plate edge is fixed. This boundary condition [Equation (71)] does not give additional indicators and therefore does not have any influence.

The guided edge can be characterized by the following equations:

$$
\frac{\partial w}{\partial n} = 0;\ Q_x\cos\beta + Q_y\sin\beta = \frac{\partial}{\partial x}(D_x\chi_x + H\chi_y)\cos\beta + \frac{\partial}{\partial y}(D_y\chi_y + H\chi_x)\sin\beta = 0
\tag{152}
$$

These equations yield the similitude indicators:

$$
C_x/C_y = 1;\ C_{\nu_1} = C_{\nu_2} = 1;
$$

$$
C_{E_2}/C_{E_1} = 1;\ C_G/C_{E_1} = 1
$$

i.e., they are identical to Equations (145) and (146). These similitude indicators are peculiar to the simply supported, free, and elastically supported edge of a plate with arbitrary configuration.

In the case of rectangular anisotropic plates and guided plate edges, the following equations can be written:

For $x = a$:

$$
\frac{\partial w}{\partial x} = 0;\ Q_x = \frac{\partial}{\partial x}(D_x\chi_x + H\chi_y) = 0
$$

For $y = b$:

$$
\frac{\partial w}{\partial y} = 0;\ Q_y = \frac{\partial}{\partial y}(D_y\chi_y + H\chi_x) = 0
\tag{153}
$$

which yield the similitude indicators:

$$
C_xC_{\nu^2}^{0.5}/C_y = 1;\ C_{E_2}/C_{E_1}C_{\nu_2}^2 = 1
$$

The latter can be found in Equation (142).

For simply supported edges of rectangular plates, boundary conditions $M_x = 0$ and $M_y = 0$ can be replaced by $\partial^2w/\partial x^2 = 0$ and $\partial^2w/\partial y^2 = 0$ accordingly due to $w = 0$ along the edge. Therefore, there is no additional indicator of similitude for these edges.

When the buckling of flat anisotropic plates is under consideration, the basic equation (139) can be used [identical transformation of Equation (83) in Equation (115)].

Applying the second theorem of similitude, one can obtain the similitude equations:

$$
C_{\nu_1}C_{\nu_2} = 1 \text{ (or } C_{\nu_{12}} = 1);
$$

$$C_{E_2}/C_{E_1}C_{\nu_2}^2 = 1;\ C_G/C_{E_1}C_{\nu_2} = 1;$$

$$C_xC_{\nu_2}^{0.5}/C_y = 1;\ C_{N_x}C_l^2/C_{E_1}C_h^3 = 1; \qquad (154)$$

$$C_{N_y}C_l^2/C_{E_1}C_h^3C_{\nu_2} = 1;$$

$$C_{N_{xy}}C_l^2/C_{E_1}C_h^3C_{\nu_2}^{0.5} = 1;\ C_x/C_l = 1$$

As one can see, they are identical to Equation (142). Therefore, generalized variables $\bar{\nu}$, $\bar{E}$, $\bar{G}$, ξ, η, $\bar{N}_x$, $\bar{N}_y$, and $\bar{N}_{xy}$ in Equation (143) can be utilized as well as nondimensional Equation (144) [without $P(\xi,\eta)$].

Isotropic Elastic Plates with Large Deflections

The governing differential equations for large deflected plates of uniform thickness under combined static and dynamic loads and temperature can be written in Von Kármán's approximation as follows [20,21,31]:

$$\frac{Eh^2}{12(1-\nu^2)}\Delta^4 w = q(x,y) + \frac{\partial^2\phi}{\partial y^2} \times \frac{\partial^2 w}{\partial x^2} + \frac{\partial^2\phi}{\partial x^2} \times \frac{\partial^2 w}{\partial y^2} - 2\frac{\partial^2\phi}{\partial x\partial y}\frac{\partial^2 w}{\partial x\partial y} - mh\frac{\partial^2 w}{\partial t^2} - kh\frac{\partial w}{\partial t} + Z(x,y,t) - \frac{Eh^2\alpha}{12(1-\nu)}\nabla^2\lambda \qquad (155)$$

$$\nabla^4\phi = Eh\left[\left(\frac{\partial^2 w}{\partial x\partial y}\right)^2 - \frac{\partial^2 w}{\partial x^2}\frac{\partial^2 w}{\partial y^2}\right] - Eh\alpha\nabla^2 T \qquad (156)$$

where, in addition to Equation (53) values:

$$T = [t^0_{top}(x,y) + t^0_{bot}(x,y)]/2;$$

$$N_x = \frac{\partial^2\phi}{\partial y^2};\ N_y = \frac{\partial^2\phi}{\partial x^2};$$

$$N_{xy} = -\frac{\partial^2\phi}{\partial x\partial y};$$

N_x, N_y, N_{xy} = membrane forces due to large deflections. Assuming that Equations (155) and (156) are written for plate 1, by substitution of Equation (1) one can obtain necessary conditions of similitude by equating similitude indicators to unity in case of statical loads [16,32] and thermal loads:

$$C_x/C_y = 1;\ C_\phi C_{\nu_0}/C_E C_h^3 = 1;$$

$$C_wC_{\nu_0}^{0.5}/C_h = 1;$$

$$C_xC_{\nu_0}^{0.375}C_q^{0.25}/C_E^{0.25}C_h = 1; \qquad (157)$$

$$C_\lambda C_\alpha C_{\nu_2}C_E^{0.5}/C_{\nu_0}^{0.25}C_q^{0.5} = 1;$$

$$C_NC_{\nu_0}^{0.25}/C_E^{0.5}C_hC_q^{0.5} = 1;$$

$$C_TC_\alpha C_E^{0.5}C_{\nu_0}^{0.25}/C_q^{0.5} = 1$$

where $C_N = C_{N_x} = C_{N_y} = C_{N_{xy}}$; C_{ν_0}, C_{ν_2} = value specified by Equation (86).

Equation (157) can be replaced by generalized variables:

$$\bar{w} = w(1-\nu^2)^{0.5}/h;$$

$$\psi = \phi(1-\nu^2)/Eh^3;$$

$$\xi = x(1-\nu^2)^{0.375}q_0^{0.25}/E^{0.25}h;$$

$$\eta = y(1-\nu^2)^{0.375}q_0^{0.25}/E^{0.25}h;$$

$$\mu(\xi,\eta) = \lambda(x,y)\alpha(1+\nu)E^{0.5}/q_0^{0.5}(1-\nu^2)^{0.25}; \qquad (158)$$

$$\bar{N}_x = N_x(1-\nu^2)^{0.25}/E^{0.5}hq_0^{0.5};$$

$$\bar{N}_y = N_y(1-\nu^2)^{0.25}/E^{0.5}hq_0^{0.5};$$

$$p(\xi,\eta) = q(x,y)/q_0;$$

$$\bar{N}_{xy} = N_{xy}(1-\nu^2)^{0.25}/E^{0.5}hq_0^{0.5};$$

$$\bar{T}(\xi,\eta) = T(x,y)\alpha E^{0.5}(1-\nu^2)^{0.25}/q_0^{0.5}$$

Nondimensional Equations (155) and (156), in terms of Equation (158), are as follows:

$$\frac{1}{12}\nabla^4\bar{w} = p(\xi,\eta) + \frac{\partial^2\psi}{\partial\eta^2}\frac{\partial^2\bar{w}}{\partial\xi^2} + \frac{\partial^2\psi}{\partial\xi^2}\frac{\partial^2\bar{w}}{\partial\eta^2} - 2\frac{\partial^2\psi}{\partial\xi\partial\eta}\frac{\partial^2\bar{w}}{\partial\xi\partial\eta} - \frac{1}{12}\nabla^2\mu \qquad (159)$$

$$\nabla^4\psi = \left(\frac{\partial^2\bar{w}}{\partial\xi\partial\eta}\right)^2 - \frac{\partial^2\bar{w}}{\partial\xi^2}\frac{\partial^2\bar{w}}{\partial\eta^2} - \nabla^2\bar{T} \qquad (160)$$

It is evident that Equations (159) and (160) give free choice of Poisson's ratio selection, but a final conclusion will be made after analysis of boundary and initial conditions, i.e., formulation of sufficient conditions of similitude.

In case of the only thermal loads λ, T equations [Equation (157)] have to be replaced by the following:

$$C_x/C_y = 1;\ C_\phi C_{\nu_0}/C_E C_h^3 = 1;$$
$$C_w C_{\nu_0}^{0.5}/C_h = 1;$$
$$C_x C_\alpha^{0.5} C_\lambda^{0.5} C_{\nu_0}^{0.75}/C_h C_{\nu_1}^{0.5} = 1; \qquad (161)$$
$$C_T C_{\nu_1}/C_\lambda C_{\nu_0}^{0.5} = 1;$$
$$C_N C_{\nu_1}/C_E C_h C_\alpha C_\lambda C_{\nu_0}^{0.5} = 1$$

where $C_N = C_{N_x} = C_{N_y} = C_{N_{xy}}$; C_{ν_1} = value specified by Equation (86).

From Equation (161) follow the generalized variables:

$$\bar{w} = \frac{w(1 - \nu^2)^{0.5}}{h};$$
$$\psi = \frac{\phi(1 - \nu^2)}{Eh^3};$$
$$\xi = \frac{x\alpha^{0.5}\lambda_0^{0.5}(1 - \nu^2)^{0.75}}{h(1 - \nu)^{0.5}};$$
$$\eta = \frac{y\alpha^{0.5}\lambda_0^{0.5}(1 - \nu^2)^{0.75}}{h(1 - \nu)^{0.5}};$$
$$\mu(\xi, \eta) = \lambda(x, y)/\lambda_0; \qquad (162)$$
$$\bar{N}_x = \frac{N_x(1 - \nu)}{Eh\alpha\lambda_0(1 - \nu^2)^{0.5}};$$
$$\bar{N}_y = \frac{N_y(1 - \nu)}{Eh\alpha\lambda_0(1 - \nu^2)^{0.5}};$$
$$\bar{N}_{xy} = \frac{N_{xy}(1 - \nu)}{Eh\alpha\lambda_0(1 - \nu^2)^{0.5}};$$
$$\bar{T}(\xi,\eta) = \frac{T(x, y) \times (1 - \nu)}{\lambda_0(1 - \nu^2)^{0.5}}$$

Nondimensional Equations (159) and (160) can be utilized for calculations ignoring the term $p(\xi,\eta)$.

When the dynamic problem is under consideration, the following necessary equations of similitude can be released:

$$C_x/C_y = 1;$$
$$C_\phi C_{\nu_0}/C_E C_h^3 = 1;$$
$$C_w C_{\nu_0}^{0.5}/C_h = 1;$$
$$C_x C_{\nu_0}^{0.375} C_Z^{0.25}/C_E^{0.25} C_h = 1; \qquad (163)$$
$$C_N C_{\nu_0}^{0.25}/C_E^{0.5} C_h C_Z^{0.5} = 1;$$
$$C_t C_{\nu_0}^{0.25} C_Z^{0.5}/C_m^{0.5} C_h = 1;$$
$$C_k C_h/C_m^{0.5} C_{\nu_0}^{0.25} C_Z^{0.5} = 1$$

or generalized variables:

$$\bar{w} = w(1 - \nu^2)^{0.5}/h;$$
$$\psi = \phi(1 - \nu^2)/Eh^3;$$
$$\xi = x(1 - \nu^2)^{0.375}Z_0^{0.25}/E^{0.25}h;$$
$$\eta = y(1 - \nu^2)^{0.375}Z_0^{0.25}/E^{0.25}h;$$
$$\bar{N}_x = N_x(1 - \nu^2)^{0.25}/E^{0.5}hZ_0^{0.5};$$
$$\bar{N}_{xy} = N_{xy}(1 - \nu^2)^{0.25}/E^{0.5}hZ_0^{0.5}; \qquad (164)$$
$$\bar{N}_y = N_y(1 - \nu^2)^{0.25}/E^{0.5}hZ_0^{0.5};$$
$$\bar{Z}(\xi, \eta, \theta) = Z(x, y, t)/Z_0;$$
$$\theta = \frac{t(1 - \nu^2)^{0.25}Z_0^{0.5}}{m^{0.5}h};$$
$$\bar{k} = kh/m^{0.5}(1 - \nu^2)^{0.25}Z_0^{0.5}$$

and term "$-(1/12)\nabla^2\mu$" in Equation (159) should be replaced by the following terms:

$$-\frac{\partial^2\bar{w}}{\partial\theta^2} - \bar{k}\frac{\partial\bar{w}}{\partial\theta} + \bar{Z}(\xi, \lambda, \theta) \qquad (165)$$

If plate with large deflections subjected to free vibrations and both cases (with and without damping) should be covered, the following necessary similitude equations can be written:

$$C_x/C_y = 1;$$
$$C_x/C_l = 1;$$

$$C_\phi C_{v_0}/C_E C_h^3 = 1;$$
$$C_w C_{v_0}^{0.5}/C_h = 1; \quad (166)$$
$$C_t C_h C_E^{0.5}/C_m^{0.5} C_l^2 C_{v_0}^{0.5} = 1;$$
$$C_k C_{v_0}^{0.5} C_l^2/C_m^{0.5} C_E^{0.5} C_h = 1$$

which give generalized variables:

$$\xi = x/l; \ \eta = y/l;$$
$$\psi = \phi(1 - \nu^2)/Eh^3;$$
$$\overline{w} = w(1 - \nu^2)^{0.5}/h; \quad (167)$$
$$\theta = thE^{0.5}/m^{0.5}l^2(1 - \nu^2)^{0.5};$$
$$\overline{k} = k(1 - \nu^2)^{0.5}l^2/m^{0.5}E^{0.5}h$$

and differential equations similiar to Equations (159), (160), and (165).

Generalized variables [Equations (158), (162), (164)] assume invariable nondimensional loads $p(\xi,\eta)$; $\overline{Z}(\xi,\eta,\theta)$ regardless of loads $q(x,y)$, $Z(x,y,t)$ intensity, but consider variable generalized plate dimensions ξ_0 and η_0, such as:

$$\xi_0 = a(1 - \nu^2)^{0.375}q_0^{0.25}/E^{0.25}h;$$
$$\eta_0 = b(1 - \nu^2)^{0.375}q_0^{0.25}/E^{0.25}h \quad (168)$$

These variables can be replaced by new ones if we assume

$$C_x = C_l$$

as in Equations (166) and (167) and earlier in Equation (54). Thus, instead of Equation (157) one can write:

$$C_x/C_y = 1;$$
$$C_x/C_l = 1;$$
$$C_\phi C_{v_0}/C_E C_h^3 = 1;$$
$$C_w C_{v_0}^{0.5}/C_h = 1;$$
$$C_q C_l^4 C_{v_0}^{1.5}/C_E C_h^4 = 1; \quad (169)$$
$$C_N C_l^2 C_{v_0}/C_E C_h^3 = 1;$$
$$C_\lambda C_\alpha C_l^2 C_{v_0}^{0.5} C_{v_2}/C_h^2 = 1;$$
$$C_T C_\alpha C_l^2 C_{v_0}/C_h^2 = 1$$

or generalized variables

$$\overline{w} = w(1 - \nu^2)^{0.5}/h$$
$$\psi = \phi(1 - \nu^2)/Eh^3$$
$$\xi = x/l$$
$$\eta = y/l$$
$$p(\xi,\eta) = q(x,y)l^4(1 - \nu^2)^{1.5}/Eh^4 \quad (170)$$
$$\overline{N}_x = N_x l^2(1 - \nu^2)/Eh^3$$
$$\overline{N}_{xy} = N_{xy} l^2(1 - \nu^2)/Eh^3$$
$$\overline{N}_y = N_y l^2(1 - \nu^2)/Eh^3$$
$$\mu(\xi,\eta) = \lambda(x,y)\alpha(1 + \nu)(1 - \nu^2)^{0.5}l^2/h^2$$
$$\overline{T}(\xi,\eta) = T(x,y)\alpha(1 - \nu^2)l^2/h^2$$

and nondimensional Equations (159) and (160).

Instead of Equation (161) the following can be written:

$$C_x/C_y = 1;$$
$$C_x/C_l = 1;$$
$$C_\phi C_{v_0}/C_E C_h^3 = 1;$$
$$C_w C_{v_0}^{0.5}/C_h = 1; \quad (171)$$
$$C_\lambda C_\alpha C_l^2 C_{v_0}^{1.5}/C_h^2 C_{v_1} = 1;$$
$$C_T C_\alpha C_l^2 C_{v_0}/C_h^2 = 1;$$
$$C_N C_l^2 C_{v_0}/C_E C_h^3 = 1$$

or generalized variables:

$$\xi = x/l;$$
$$\eta = y/l;$$
$$\overline{w} = w(1 - \nu^2)^{0.5}/h;$$
$$\psi = \phi(1 - \nu^2)/Eh^3;$$
$$\mu(\xi, \eta) = \lambda(x, y)\alpha l^2(1 - \nu^2)^{1.5}/h^2(1 - \nu); \quad (172)$$
$$\overline{N}_x = N_x l^2(1 - \nu^2)/Eh^3;$$
$$\overline{N}_y = N_y l^2(1 - \nu^2)/Eh^3;$$

$$\overline{N}_{xy} = N_{xy}l^2(1 - \nu^2)/Eh^3;$$

$$\overline{T}(\xi, \eta) = T(x, y)\alpha l^2(1 - \nu^2)/h^2$$

Equations (159) and (160) can be utilized after ignoring $p(\xi,\eta)$. For dynamic loads, Equation (163) can be replaced by the following:

$$\begin{aligned} C_x/C_y &= 1; \\ C_x/C_l &= 1; \\ C_\phi C_{\nu_0}/C_E C_h^3 &= 1; \\ C_w C_{\nu_0}^{0.5}/C_h &= 1; \\ C_Z C_l^4 C_{\nu_0}^{1.5}/C_E C_h^4 &= 1; \\ C_N C_l^2 C_{\nu_0}/C_E C_h^3 &= 1; \\ C_t C_h C_E^{0.5}/C_m^{0.5} C_l^2 C_{\nu_0}^{0.5} &= 1; \\ C_k C_l^2 C_{\nu_0}^{0.5}/C_E^{0.5} C_m^{0.5} C_h &= 1 \end{aligned} \tag{173}$$

with generalized variables:

$$\begin{aligned} \xi &= x/l; \\ \eta &= y/l; \\ \overline{w} &= w(1 - \nu^2)^{0.5}/h; \\ \psi &= \phi(1 - \nu^2)/Eh^3; \\ \overline{Z}(\xi, \eta, \theta) &= Z(x, y, t)l^4(1 - \nu^2)^{1.5}/Eh^4; \\ \overline{N}_x &= N_x l^2(1 - \nu^2)/Eh^3; \\ \overline{N}_y &= N_y l^2(1 - \nu^2)/Eh^3; \\ \overline{N}_{xy} &= N_{xy} l^2(1 - \nu^2)/Eh^3; \\ \theta &= thE^{0.5}/m^{0.5}l^2(1 - \nu^2)^{0.5}; \\ \overline{k} &= kl^2(1 - \nu^2)^{0.5}/E^{0.5}m^{0.5}h \end{aligned} \tag{174}$$

and Equations (159), (160), and (165).

For Equations (170), (172), and (174), nondimensional loads $p(\xi,\eta)$; $\overline{Z}(\xi,\eta,\theta)$ are variable but the generalized plate dimensions ξ_0, η_0 are invariable:

$$\xi_0 = a/l; \ \eta_0 = b/l \tag{175}$$

Von Kármán's equations [Equations (155) and (156)] do not consider all stress-strain state components. Equation (66) characterizes only bending components. The membrane component relationships are as follows:

$$\begin{aligned} N_x &= \frac{Eh}{1 - \nu^2}(\varepsilon_x^{mem} + \nu\varepsilon_y^{mem}); \\ N_y &= \frac{Eh}{1 - \nu^2}(\varepsilon_y^{mem} + \nu\varepsilon_x^{mem}); \\ N_{xy} &= \frac{Eh}{2(1 + \nu)}\varepsilon_{xy}^{mem}; \\ \varepsilon_x^{mem} &= \frac{\partial u}{\partial x} + \frac{1}{2}\left(\frac{\partial w}{\partial x}\right)^2; \\ \varepsilon_y^{mem} &= \frac{\partial v}{\partial y} + \frac{1}{2}\left(\frac{\partial w}{\partial y}\right)^2; \\ \varepsilon_{xy}^{mem} &= \frac{\partial u}{\partial y} + \frac{\partial v}{\partial x} + \frac{\partial w}{\partial x}\frac{\partial w}{\partial y}; \\ \sigma_x^{mem} &= N_x/h; \\ \sigma_y^{mem} &= N_y/h; \\ \sigma_{xy}^{mem} &= N_{xy}/h \end{aligned} \tag{176}$$

As Equations (66) and (176) show, if all plate components are required to be similar, the Poisson's ratio for material of all plates which belong to the group of similar structures must be the same, i.e., $C_\nu = 1$. In the case of $C_\nu \neq 1$, there is a similitude of deflections w; curvatures χ_x, χ_y, χ_{xy}; bending strains ϵ_x, ϵ_y, ϵ_{xy}; transverse shear forces Q_x, Q_y; torsional moments M_{xy} and stresses σ_{xy}; membrane forces N_x, N_y, N_{xy} and stresses σ_x^{mem}, σ_y^{mem}, σ_{xy}^{mem}; and membrane shear strains ϵ_{xy}^{mem}.

Therefore, the type of boundary condition predetermines if condition $C_\nu \neq 1$ is acceptable. If the boundary conditions are specified in components that are similar in case of $C_\nu \neq 1$, the sufficient conditions of similitude are satisfied.

Examples of boundary conditions are listed below for the rectangular plate edge $x = a$ and are shown in Figure 6.:

a. Fixed and held edge [Figure 6(b)]:

$$w = 0; \frac{\partial w}{\partial x} = 0; u = 0; v = 0 \tag{177}$$

b. Fixed but not held edge [Figure 6(c)]:

$$w = 0; \frac{\partial w}{\partial x} = 0; N_x = 0; N_{xy} = 0 \tag{178}$$

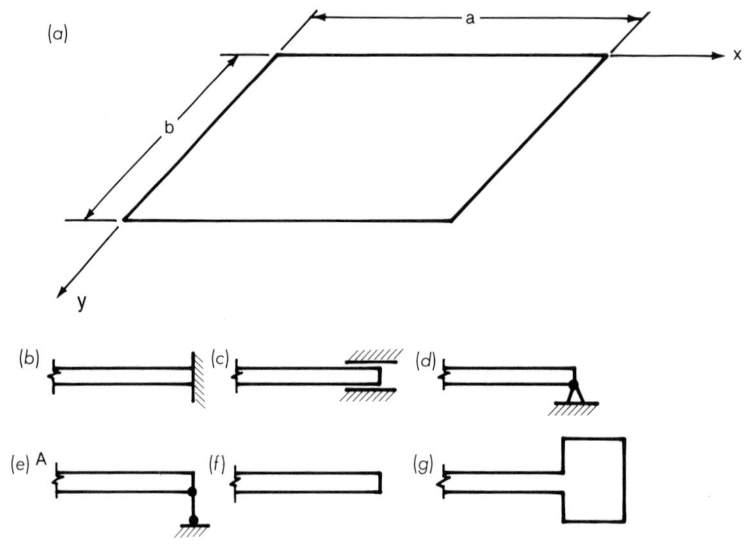

FIGURE 6. Boundary conditions for the edge x = a of a rectangular plate with large deflections.

c. Hinged edge [Figure 6(d)]:

$$w = 0; \frac{\partial^2 w}{\partial x^2} = 0; u = 0; v = 0 \tag{179}$$

d. Simply supported (neither fixed nor held) edge [Figure 6(e)]:

$$w = 0; \frac{\partial^2 w}{\partial x^2} = 0; N_x = 0; N_{xy} = 0 \tag{180}$$

e. Free edge [Figure 6(f)]:

$$M_x = 0; Q_x - \frac{\partial M_{xy}}{\partial y} = 0; \quad N_x = 0; N_{xy} = 0 \tag{181}$$

f. Guided edge [Figure 6(g)]:

$$\frac{\partial w}{\partial x} = 0; Q_x - \frac{\partial M_{xy}}{\partial y} = 0; \quad N_x = 0; N_{xy} = 0 \tag{182}$$

The boundary conditions [Equations (178) and (180)] are formulated in terms of deflections and membrane forces and therefore do not contradict $C_\nu \neq 1$. To find out the influence of boundary conditions $u = 0$ and $v = 0$ in Equations (177) and (179), one proceeds from Equation (176) as follows:

$$N_{xy} = \frac{Eh}{2(1 + \nu)} \times \left(\frac{\partial u}{\partial y} + \frac{\partial v}{\partial x} + \frac{\partial w}{\partial x} \times \frac{\partial w}{\partial y}\right)$$

and obtains the following indicators of similitude:

$$C_u/C_v = 1;$$

$$C_u C_l/C_w^2 = 1 \text{ (or } C_u C_l C_{\nu_0}/C_h^2 = 1);$$

$$C_N C_l^2 C_{\nu_0} C_{\nu_2}/C_E C_h^3 = 1.$$

The last indicator of similitude contradicts Equation (169):

$$C_N C_l^2 C_{\nu_0}/C_E C_h^3 = 1$$

and requires $C_\nu = 1$.

The boundary condition $Q_x - \partial M_{xy}/\partial y = 0$ in Equations (181) and (182) consists of two components, Q_x and M_{xy}, with similitude scales calculated by the following formulas [Equation (99)]:

$$C_{M_{xy}} = C_E C_h^3 C_w/C_{\nu_2} C_l^2;$$

$$C_{Q_x} = C_E C_h^3 C_w/C_l^3 C_{\nu_0}$$

and yields the following indicator of similitude:

$$C_{Q_x} C_l / C_{M_{xy}} = 1 \text{ or}$$

$$C_{v_2}/C_{v_0} = 1 \text{ or}$$

$$C_\nu = 1$$

That is, for the above boundary condition the Poisson's ratios for similar plates have to be identical. But if the influence of torsional moments is insignificant it can be assumed that the boundary condition $Q_x - \partial M_{xy}/\partial y = 0$ does not require $C_\nu = 1$.

The boundary condition $M_x = 0$ in Equation (181) can be replaced by thc following:

$$\frac{\partial^2 w}{\partial x^2} + \nu \frac{\partial^2 w}{\partial y^2} = 0$$

which requires $C_\nu = 1$.

If $C_\nu = 1$ the generalized variables obtained earlier can be simplified by exclusion of Poisson's ratio functions $(1 - \nu^2)$, $(1 - \nu)$, and $(1 + \nu)$. These variables have to be supplemented by values obtained from additional equations [Equations (66) and (176)] as well as for the case of $C_\nu \neq 1$.

The particular case of axisymmetrical bending of large deflected circular plates is a subject of practical importance, and corresponding Von Kármán differential equations can be written as follows [20,21]:

$$\frac{Eh^3}{12(1 - \nu^2)} \frac{\partial}{\partial r} \times \left[\frac{1}{r}\frac{\partial}{\partial r}\left(r\frac{\partial w}{\partial r}\right)\right] = \frac{1}{r} \times \left(\left(\int_b^r q(r) r dr + V \times b\right) + N_r \frac{\partial w}{\partial r} - \frac{1}{r}\left(mh \int_b^r \frac{\partial^2 w}{\partial t^2} r dr + kh \int_b^r \frac{\partial w}{\partial t} r dr - \int_b^r Z(r, t) r dr\right) - \frac{Eh^2 \alpha}{12(1 - \nu)} \times \frac{d\lambda(r)}{dr}\right) \quad (183)$$

$$r\frac{\partial^2 N_r}{\partial r^2} + 3\frac{\partial N_r}{\partial r} = -\frac{Eh}{2r}\left(\frac{\partial w}{\partial r}\right)^2 - Eh\alpha \frac{dT(r)}{dr} \quad (184)$$

where:

$$w = w(r, t); N_r = N_r(r, t)$$

These equations can be supplemented by the following:

$$N_t = N_r + r\frac{\partial N_r}{\partial r};$$

$$\sigma_r = -\frac{Eh}{2(1 - \nu^2)}\left(\frac{\partial^2 w}{\partial r^2} + \frac{\nu}{r}\frac{\partial w}{\partial r}\right);$$

$$\sigma_t = -\frac{Eh}{2(1 - \nu^2)}\left(\frac{1}{r}\frac{\partial w}{\partial r} + \nu\frac{\partial^2 w}{\partial r^2}\right);$$

$$M_r = \sigma_r h^2/6;$$

$$M_t = \sigma_t h^2/6;$$

$$Q_r = -\frac{Eh^3}{12(1 - \nu^2)}\frac{\partial}{\partial r}\left(\frac{\partial^2 w}{\partial r^2} + \frac{1}{r}\frac{\partial w}{\partial r}\right);$$

$$\varepsilon_r = \frac{h}{2} \times \frac{\partial^2 w}{\partial r^2}; \quad (185)$$

$$\varepsilon_t = \frac{h}{2} \times \frac{1}{r}\frac{\partial w}{\partial r};$$

$$\chi_r = \frac{\partial^2 w}{\partial r^2};$$

$$\chi_t = \frac{1}{r}\frac{\partial w}{\partial r};$$

$$u = \frac{r}{Eh}\left[(1 - \nu)N_r + r\frac{\partial N_r}{\partial r}\right];$$

$$\sigma_r^{mem} = N_r/h;$$

$$\sigma_t^{mem} = N_t/h$$

The indicators of similitude from Equations (183), (184) and (185) are identical to those obtained earlier, and some of the changes should be done in the generalized variables. In the variables from Equations (170), (172), and (174), characteristic length l has to be replaced by the outer radius a; membrane forces $\bar{N}_x$, $\bar{N}_y$, $\bar{N}_{xy}$ by $\bar{N}_r$, $\bar{N}_t$; there are no membrane functions ϕ, ψ; no variable η; and $\xi = r/a$. The nondimensional equations are as follows:

$$\frac{1}{12}\frac{\partial}{\partial\xi}\left[\frac{1}{\xi}\frac{\partial}{\partial\xi}\left(\xi\frac{\partial\overline{w}}{\partial\xi}\right)\right]$$
$$= \frac{1}{\xi}\left(\int_{\xi_0}^{\xi} p(\xi)\xi d\xi + \overline{V}\times\xi_0\right) + \overline{N}_r\frac{\partial\overline{w}}{\partial\xi}$$
$$- \frac{1}{\xi}\left(\int_{\xi_0}^{\xi}\frac{\partial^2\overline{w}}{\partial\theta^2}\xi d\xi + \overline{k}\int_{\xi_0}^{\xi}\frac{\partial\overline{w}}{\partial\theta}\xi d\xi\right.$$
$$\left. - \int_{\xi_0}^{\xi}\overline{Z}(\xi,\theta)\xi d\xi\right) - \frac{1}{12}\frac{d\mu(\xi)}{d\xi} \tag{186}$$

$$\xi\frac{\partial^2\overline{N}_r}{\partial\xi^2} + 3\frac{\partial\overline{N}_r}{\partial\xi} = -\frac{1}{\xi}\left(\frac{\partial\overline{w}}{\partial\xi}\right)^2 - \frac{d\overline{T}(\xi)}{d\xi} \tag{187}$$

For $C_\nu = 1$ the Poisson's ratio functions $(1 - \nu^2)$, $(1 - \nu)$ appear in nondimensional Equations (186) and (187) and disappear in generalized variables.

Keeping in mind the case of $C_\nu \neq 1$, the additional generalized variables can be written from Equations (66) and (176):

$$\begin{aligned}\overline{\varepsilon} &= \varepsilon l^2(1 - \nu^2)^{0.5}/h^2;\\ \overline{\chi} &= \chi l^2(1 - \nu^2)^{0.5}/h;\\ \overline{Q} &= Q(1 - \nu^2)^{1.5}l^3/Eh^4;\\ \overline{\sigma}^{mem} &= \sigma^{mem}l^2(1 - \nu^2)/Eh^2\end{aligned} \tag{188}$$

For $C_\nu = 1$ the generalized variables for plates with large deflections are as follows:

$$\begin{aligned}\overline{w} &= w/h;\\ \psi &= \phi/Eh^3;\\ \xi &= x/l;\\ \eta &= y/l;\\ p(\xi,\eta) &= q(x,y)l^4/Eh^4;\\ \overline{N} &= Nl^2/Eh^3;\\ \mu(\xi,\eta) &= \lambda(x,y)\alpha l^2/h^2;\\ \overline{T}(\xi,\eta) &= T(x,y)\alpha l^2/h^2;\\ \overline{Z}(\xi,\eta,\theta) &= Z(x,y,t)l^4/Eh^4;\\ \theta &= thE^{0.5}/m^{0.5}l^2;\\ \overline{k} &= kl^2/E^{0.5}m^{0.5}h;\\ \overline{\sigma} &= \frac{\sigma l^2}{Eh^2};\\ \overline{\sigma}^{mem} &= \frac{\sigma^{mem}l^2}{Eh^2};\\ \overline{M} &= Ml^2/Eh^4;\\ \overline{Q} &= Ql^3/Eh^4;\\ \overline{\varepsilon} &= \varepsilon l^2/h^2;\\ \overline{\chi} &= \chi l^2/h;\\ \overline{\varepsilon}^{mem} &= \varepsilon^{mem}l^2/h^2;\\ \overline{u} &= ul/h^2;\\ \overline{v} &= vl/h^2;\\ \overline{\sigma}^{mem} &= \sigma^{mem}l^2/Eh^2\end{aligned} \tag{189}$$

Example 8

Generalize the formulas for calculation of large deflected uniformly loaded circular plates with clamped edges and $\nu = 0.3$ (refer to Reference 33).

Consider fixed but not held outer contour ($r = a$ or $\xi = 1$):

$$w(a) = \frac{dw(a)}{dr} = N_r(a) = 0$$

which permits $c_\nu \neq 1$, and consider the following formula for all components (regardless of Poisson's ratio):

$$Z = \overline{A}_0 + \overline{A}_1(w/h) + \overline{A}_2(w/h)^2 + \overline{A}_3(w/h)^3 \tag{190}$$

where $w = w_{max}$.

Keeping in mind the generalized variables:

$$\begin{aligned}\overline{\sigma}^{bend} &= \sigma^{bend}a^2/Eh^2;\\ \overline{\sigma}^{mem} &= \sigma^{mem}a^2(1 - \nu^2)/Eh^2;\\ p &= qa^4(1 - \nu^2)^{1.5}/Eh^4;\\ \overline{\chi} &= \chi a^2(1 - \nu^2)^{0.5}/h;\\ \overline{w} &= w(1 - \nu^2)^{0.5}/h;\\ \overline{w}' &= w'a(1 - \nu_0^2)^{0.5}/h\end{aligned} \tag{191}$$

TABLE 14. $\bar{A}_i$ Coefficients for Loads, Twist Angles, Curvatures, and Membrane Stress Calculations by Equation (190) in Uniformly Loaded Circular Plate with Fixed but Not Held Outer Edge.

Z	$\bar{A}_0$	$\bar{A}_1$	$\bar{A}_2$	$\bar{A}_3$
P	0.026503	4.785664	0.525452	0.674068
$\bar{W}'_{max.}$	0.002330	−1.495967	0.042775	−0.030497
$\bar{\chi}_{r(t)}(0)$	0.024467	−4.158126	0.672565	−0.019739
$\bar{\sigma}^{mem}_{r(t)}(0)$	−0.002015	0.012631	0.462553	−0.043498
$\bar{\chi}_r(1)$	0.032345	7.235659	0.737351	0.428313
$\bar{\chi}_t(1)$	0	0	0	0
$\bar{\sigma}^{mem}_r(1)$	0	0	0	0
$\bar{\sigma}^{mem}_t(1)$	0.000695	−0.014687	−0.265447	−0.033925

the following formulas can be used for calculation of generalized loads p, twist angles $\bar{w}^1$, curvatures $\bar{\chi}_r$, $\bar{\chi}_t$, and membrane stresses $\bar{\sigma}^{mem}$ for $\nu = 0.3$:

$$\bar{p} = (1 - 0.3^2)^{1.5}[A_0 + A_1 w/h + A_2(w/h)^2 + A_3(w/h)^3];$$

$$\bar{w}' = (1 - 0.3^2)^{0.5}[A_0 + A_1 w/h + A_2(w/h)^2 + A_3(w/h)^3];$$

$$\bar{\chi}_r = -2(1 - 0.3^2)^{0.5} \times (\bar{\sigma}_r^{bend} - 0.3\bar{\sigma}_t^{bend}); \tag{192}$$

$$\bar{\chi}_t = -2(1 - 0.3^2)^{0.5}(\bar{\sigma}_t^{bend} - 0.3\bar{\sigma}_r^{bend});$$

$$\bar{\sigma}_r^{mem} = (1 - 0.3^2)\sigma_r^{mem}a^2/Eh^2;$$

$$\bar{\sigma}_t^{mem} = (1 - 0.3^2)\sigma_t^{mem}a^2/Eh^2$$

where A_0, A_1, A_2, A_3 = coefficients of Reference 33, Table 3.

The generalized table with $\bar{A}_0$, $\bar{A}_1$, $\bar{A}_2$, $\bar{A}_3$ coefficients calculated with the formulas in Equation (192) allows us to calculate deflections, twist angles, and stresses for any Poisson's ratio. The bending stress $\bar{\sigma}_r^{bend}$, $\bar{\sigma}_t^{bend}$ have to be calculated with the use of $\bar{\chi}_r$, $\bar{\chi}_t$ by the following formulas:

$$\bar{\sigma}_r^{bend} = -\frac{1}{2(1 - \nu^2)^{1.5}}(\bar{\chi}_r + \nu\bar{\chi}_t);$$

$$\bar{\sigma}_t^{bend} = -\frac{1}{2(1 - \nu^2)^{1.5}}(\bar{\chi}_t + \nu\bar{\chi}_r) \tag{193}$$

Calculate load q, bending and membrane stresses for the metal circular plate with radius $a = 100$ inches, thickness $h = 0.5$ inches, modulus of elasticity $E = 29{,}000$ ksi, and Poisson's ratio $\nu = 0.3$, if the maximum deflection w equals 1.5 inches (or $w/h = 3$). This plate is calculated in Reference 33, Table 5.

From Table 14, Equations (190) and (193):

$$p = 0.026503 + 4.785664 \times 3 + 0.525452 \times 3^2 + 0.674068 \times 3^3 = 37.3124 \text{ or}$$

$$q = pEh^4/a^4(1 - \nu^2)^{1.5}$$

$$= 37.3124 \times 29{,}000 \times 0.5^4/100^4 \times (1 - 0.3^2)^{1.5}$$

$$= 7.79057 \times 10^{-4} \text{ ksi} = 0.779057 \text{ psi}$$

$$\bar{\chi}_r(0) = \bar{\chi}_t(0) = 0.024467 - 4.158126 \times 3 + 0.672565 \times 3^2 - 0.019739 \times 3^3$$

$$= -6.929779$$

$$\bar{\sigma}_r^{bend}(0) = \bar{\sigma}_t^{bend}(0) = -\frac{1 + \nu}{2(1 - \nu^2)^{1.5}}\bar{\chi}_r(0)$$

$$= 5.188844 \text{ or}$$

$$\sigma_r^{bend}(0) = \sigma_t^{bend}(0) = \bar{\sigma}_r^{bend}(0)Eh^2/a^2$$

$$= 5.188844 \times 29{,}000 \times 0.5^2/100^2$$

$$= 3.762 \text{ ksi } \bar{\sigma}_r^{mem}(0) = \bar{\sigma}_t^{mem}(0)$$

$$= -0.002015 + 0.012631 \times 3$$

$$+ 0.462553 \times 3^2 - 0.043498 \times 3^3$$

$$= 3.024409 \text{ or}$$

$$\sigma_r^{mem}(0) = \sigma_t^{mem}(0) = \overline{\sigma}_r^{mem}(0)Eh^2/a^2(1 - \nu^2)$$

$$= 3.024409 \times 29{,}000 \times 0.5^2/100^2$$

$$\times (1 - 0.3^2)$$

$$= 2.410 \text{ ksi}$$

$$\overline{\chi}_r(1) = 0.032345 + 7.235659 \times 3$$

$$+ 0.737351 \times 3^2 + 0.428313 \times 3^3$$

$$= 39.939932$$

$$\overline{\chi}_t(0) = 0;$$

$$\overline{\sigma}_r^{bend}(1) = -\frac{1}{2(1 - 0.3^2)^{1.5}} \times 39.939932$$

$$= -23.004629 \text{ or}$$

$$\sigma_r^{bend}(1) = \overline{\sigma}_r^{bend} \times Eh^2/a^2$$

$$= 23.004629 \times 29{,}000 \times 0.5^2/100^2$$

$$= 16.678 \text{ ksi}$$

$$\overline{\sigma}_r^{mem}(1) = 0$$

The calculated components coincide with results in Reference 33.

As shown in Reference 32, if the outer edge of a circular plate is fixed and held there is an approximate similitude. Generalized variables [Equations (170) and (188)] should be used rather than the variables in Equation (189).

ISOTROPIC ELASTIC MEMBRANES

If plate material bending stiffness is very low or its deflections (w/h) are very large, the bending components in Von Kármán's equations [Equations (155) and (183)] can be ignored, specifically the terms:

$$\nabla^4 w \cong 0;$$

$$\nabla^2 \lambda \cong 0;$$

$$\frac{\partial}{\partial r}\left[\frac{1}{r}\frac{\partial}{\partial r}\left(r\frac{\partial w}{\partial r}\right)\right] \cong 0; \tag{194}$$

$$\frac{d\lambda(r)}{dr} \cong 0$$

The generalized variables obtained earlier can be used for membranes, but it is better to replace them with the following variables:

$$\begin{aligned}
\xi &= x/l;\\
\eta &= y/l;\\
\psi &= \phi/\sqrt[3]{Ehl^8q_0^2};\\
\overline{w} &= w\sqrt[3]{Eh/l^4q_0};\\
\overline{N}_x &= N_x/\sqrt[3]{Ehl^2q_0^2};\\
\overline{N}_y &= N_y/\sqrt[3]{Ehl^2q_0^2};\\
\overline{N}_{xy} &= N_{xy}/\sqrt[3]{Ehl^2q_0^2};\\
\theta &= tE^{1/6}Z_0^{1/3}/m^{0.5}h^{1/3}l^{2/3};\\
\overline{k} &= kh^{1/3}l^{2/3}/m^{0.5}E^{1/6}Z_0^{1/3};\\
\overline{T}(\xi, \eta) &= T(x, y)\alpha[Eh/q_0l]^{2/3};\\
\overline{u} &= \frac{u}{l^{5/3}}\left(\frac{Eh}{q}\right)^{2/3};\\
\overline{v} &= \frac{v}{l^{5/3}}\left(\frac{Eh}{q}\right)^{2/3}
\end{aligned} \tag{195}$$

For circular membranes replace l with a, replace ξ and η with $\xi = r/a$, and for dynamic loads replace q_0 with Z_0. The nondimensional equations in variables [Equation (195)] are simplified equations [Equations (159), (160), (186), and (187)]. As follows from generalized variables [Equation (195)], application of similitude theory alone (without solution of differential equations) has shown formulas for deflections w; tangent displacements u, v; membrane forces N_x, N_y, N_{xy} in membranes:

$$\begin{aligned}
w &= K_1\sqrt[3]{q};\\
u &= K_2\sqrt[3]{q^2};\\
v &= K_3\sqrt[3]{q^2};\\
N_x &= K_4\sqrt[3]{q^2};\\
N_y &= K_5\sqrt[3]{q^2};\\
N_{xy} &= K_6\sqrt[3]{q^2};
\end{aligned} \tag{196}$$

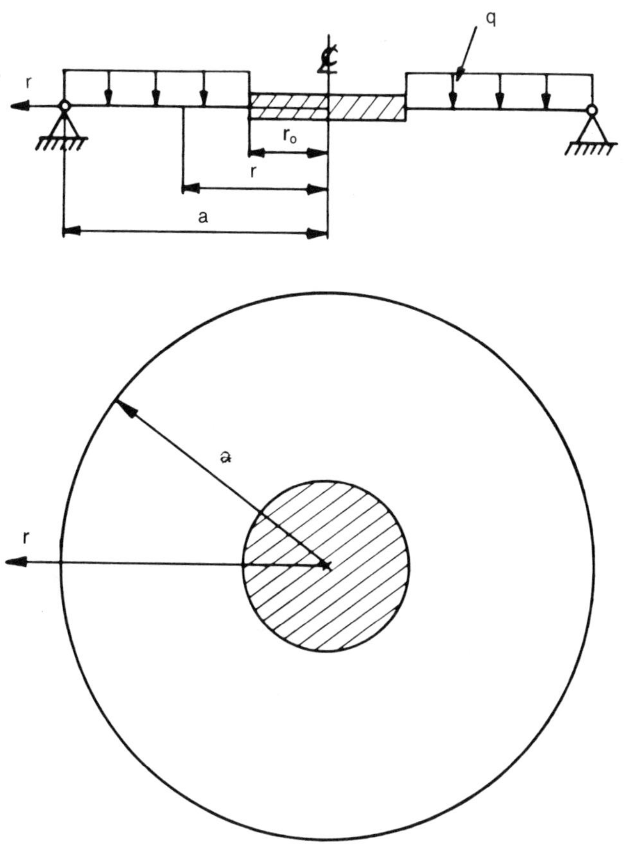

FIGURE 7. Axisymmetrically loaded circular membrane.

THE ACCURACY OF CALCULATIONS PERFORMED WITH NUMERICAL METHODS

Numerical methods are used widely for plates and shells problems. Use of variational methods does not indicate the degree of approximate function sufficient for the solution. Finite difference or finite element methods do not indicate the order of the algebraic equations as sufficient for an acceptable solution. The theorems of similitude presented here can be used to control the accuracy of such solutions. Similar plates and shells can be considered with both small and large loads and then can be compared by proper scale factors using these theorems. Thus, the results from such analysis help establish divergence limits and permissible accuracies.

Example 9

As an example, we consider an elastic membrane with a central circular hole ($0.1 \le \xi_0 = r_0/a \le 0.5$), subjected to the uniform load q_q as shown in Figure 7. The Runge-Kutta method is used with the control on boundary conditions at the hole edge ($r = r_0$). The selected calculation step was equal to 0.001 in the boundary condition ($\nu = 0.3$):

$$0.7N_r + r_0 \frac{dN_r}{dr} = 0$$

in which N_r = membrane force. The boundary condition was satisfied with error 0.001. The results of the elastic membrane calculation are given in Table 15.

In this table, the dimensionless load p, deflection $\overline{w}_0$ (near the edge of the hole with radius $\xi_0 = r_0/a$), rotation angle $\overline{w}'(1)$ (near the edge of the plate $r = a$), and membrane stress $\overline{\sigma}_r^0$ (near the edge $r = r_0$) are the similitude criteria obtained from indicators of similitude as follows:

$$p = \frac{qa}{Eh};$$

$$\overline{w}_0 = \frac{w_0}{a};$$

$$\overline{w}' = w';$$

$$\overline{\sigma}_r^0 = \sigma_r^0/E$$

TABLE 15. Application of Similitude Theory for Precision Control.

ξ_0	p		$\bar{w}_0$	Divergence	$\bar{w}'(1)$	Divergence	σ_r^0	Divergence
0.1	0.003		0.08912	—	−0.2154	—	0.01322	—
	0.192	a	0.3571	0.17%	−0.8637	0.24%	0.2112	0.15%
		b	0.3565		−0.8616		0.2115	
0.5	0.003		0.04641	—	−0.1915	—	0.00851	—
	0.192	a	0.1872	0.83%	−0.7734	0.96%	0.1351	0.78%
		b	0.1856		−0.7660		0.1362	

The direct calculations are given in line a of Table 15, while the calculations obtained from the "small" loaded plates ($p = 0.003$) are given in line b using the similitude scales:

$$C_p = 0.192/0.003 = 64;$$

$$C_w = C_{w'} = (C_p)^{1/3} = 4;$$

$$C_{\sigma_r} = C_p^{2/3} = 16$$

These calculations were made for a range of loads ($0.003 \le p \le 0.192$) and for dimensions of hole radii ($0.1 \le \xi_0 \le 0.5$). If the divergence of 1% is acceptable, we can calculate the stresses by the Runge-Kutta method with chosen precision.

Flat Plates with Nonlinear Elastic Material

In the above analysis it was assumed that plate material is linearly elastic. If we assume the material of plates to be nonlinearly elastic, the relationship between stress intensity σ_i and strain intensity ϵ_i can be expressed as follows [21,34]:

$$\sigma_i = \phi(\epsilon_i) \times \epsilon_i \tag{197}$$

in which $\phi(\epsilon_i)$ = the strain function, which is determined experimentally:

$$\sigma_i = \sqrt{\sigma_x^2 - \sigma_x\sigma_y + \sigma_y^2 + 3\sigma_{xy}^2};$$

$$\varepsilon_i = \frac{2}{3(1-\nu)}\sqrt{(1-\nu+\nu^2)(\varepsilon_x^2+\varepsilon_y^2) - (1-4\nu+\nu^2)\varepsilon_x\varepsilon_y + \frac{3(1-\nu)^2}{4}\varepsilon_{xy}^2};$$

$$\sigma_x - \bar{\sigma} = 2/3\phi(\varepsilon_i) \times \varepsilon_x;$$

$$\sigma_y - \bar{\sigma} = 2/3\phi(\varepsilon_i) \times \varepsilon_y; \tag{198}$$

$$\sigma_{xy} = 1/3\phi(\varepsilon_i) \times \varepsilon_{xy};$$

$$\bar{\sigma} = (\sigma_x + \sigma_y)/3$$

If Poisson's ratio can be assumed to be zero, as is often considered in plate-shell theory [23,35], the strain intensity can be described by the following formula [9]:

$$\epsilon_i = \tfrac{2}{3}\sqrt{\epsilon_x^2 - \epsilon_x\epsilon_y + \epsilon_y^2 + \tfrac{3}{4}\,\epsilon_{xy}^2} \tag{199}$$

Applying similitude theory one can obtain the following indicators of similitude for plates:

$$C_\nu = 1;\ C_\sigma/C_\phi C_\epsilon = 1;\ C_h/C_l = 1$$

$$C_N/C_\sigma C_l = 1;\ C_w/C_l = 1 \tag{200}$$

Here distortion of the strain function ϕ can be used, i.e., $C_\phi \neq 1$, but all plates must be geometrically similar with identical Poisson's ratio.

SIMILITUDE OF THIN ELASTIC SHELLS

Similitude of Membrane Shells

Basic conditions of membrane shells are: no shear forces and moments at the shell edges; no restraint against rotation, and displacements to be normal on shell edges; and the shell shape, thickness, and loads (all) must be continuous. In spite of the fact that these conditions cannot be fulfilled, the state of stress in shells is often close to a membrane state; therefore, the membrane theory can be applied in many cases for modeling shells.

We begin with the basic equations of equilibrium of shells, namely:

$$\frac{1}{A_1A_2}\left(\frac{\partial A_2N_x}{\partial \alpha_1} + \frac{\partial A_1N_{xy}}{\partial \alpha_2} + \frac{\partial A_1}{\partial \alpha_2}N_{xy} - \frac{\partial A_2}{\partial \alpha_1}N_y\right) + q_x = 0$$

$$\frac{1}{A_1A_2}\left(\frac{\partial A_2N_{xy}}{\partial \alpha_1} + \frac{\partial A_1N_y}{\partial \alpha_2} + \frac{\partial A_2}{\partial \alpha_1}N_{xy} - \frac{\partial A_1}{\partial \alpha}N_x\right) + q_y = 0 \qquad (201)$$

$$N_x/R_1 + N_y/R_2 = q_n,\ \sigma_x = N_x/h;$$

$$\sigma_y = N_y/h;$$

$$T = N_{xy}/h$$

Similarly, the displacement equations for shells can be written as:

$$\varepsilon_x = \frac{1}{A_1}\frac{\partial u_1}{\partial \alpha_1} + \frac{1}{A_1A_2}\frac{\partial A_1}{\partial \alpha_2}u_2 + \frac{w}{R_1};$$

$$\varepsilon_y = \frac{1}{A_2}\frac{\partial u_2}{\partial \alpha_2} + \frac{1}{A_1A_2}\frac{\partial A_2}{\partial \alpha_1}u_1 + \frac{w}{R_2}; \qquad (202)$$

$$\varepsilon_{xy} = \frac{A_2}{A_1}\frac{\partial}{\partial \alpha_1}\left(\frac{u_2}{A_2}\right) + \frac{A_1}{A_2}\frac{\partial}{\partial \alpha_2}\left(\frac{u_1}{A_1}\right)$$

If we assume the material of shells to be nonlinearly elastic, the relationship between stress intensity, σ_i, and strain intensity, ϵ_i, can be expressed as:

$$\sigma_i = \phi(\epsilon_i) \times \epsilon_i \qquad (203)$$

in which

$$\sigma_i = \sqrt{\sigma_x^2 - \sigma_x\sigma_y + \sigma_y^2 + 3\sigma_{xy}^2};$$

$$\varepsilon_i = \frac{2}{3(1-\nu)}\sqrt{(1 - \nu + \nu^2)(\varepsilon_x^2 + \varepsilon_y^2) - (1 - 4\nu + \nu^2)\varepsilon_x\varepsilon_y + \frac{3(1-\nu)^2}{4}\varepsilon_{xy}^2};$$

$$\sigma_x - \overline{\sigma} = 2/3\phi(\varepsilon_i) \times \varepsilon_x; \qquad \sigma_{xy} = 1/3\phi(\varepsilon_i) \times \varepsilon_{xy};$$

$$\sigma_y - \overline{\sigma} = 2/3\phi(\varepsilon_i) \times \varepsilon_y; \qquad \overline{\sigma} = (\sigma_x + \sigma_y)/3$$

and $\phi(\epsilon_i)$ is the strain function, which is determined experimentally. In a special case of a linearly elastic material, relations for stresses are:

$$\sigma_x = \frac{E}{1-\nu^2}(\varepsilon_x + \nu\varepsilon_y);$$

$$\sigma_y = \frac{E}{1-\nu^2}(\varepsilon_y + \nu\varepsilon_x); \qquad (204)$$

$$\sigma_{xy} = \frac{E}{2(1+\nu)}\varepsilon_{xy}$$

In Equations (201) and (202), A_i and A_2 are Lame's parameters, which relate to the radii of principal curvatures R_i and R_2 as expressed by the following equations:

$$\frac{\partial}{\partial \alpha_1}\left(\frac{A_2}{R_2}\right) = \frac{1}{R_1}\frac{\partial A_2}{\partial \alpha_1};$$

$$\frac{\partial}{\partial \alpha_2}\left(\frac{A_1}{R_1}\right) = \frac{1}{R_2}\frac{\partial A_1}{\partial \alpha_2} \qquad (205)$$

$$\frac{\partial}{\partial \alpha_1}\left(\frac{1}{A_1}\frac{\partial A_2}{\partial \alpha_1}\right) + \frac{\partial}{\partial \alpha_2}\left(\frac{1}{A_2}\frac{\partial A_1}{\partial \alpha_2}\right) = \frac{-A_1A_2}{R_1R_2}$$

N_x, (σ_x), N_y, (σ_y) = membrane normal forces (stresses) along α_1, α_2

N_{xy}, σ_{xy} = membrane shear force and stress

ϵ_x, ϵ_y, ϵ_{xy} = deformations corresponding to σ_x, σ_y, σ_{xy}

u_1, u_2, w = displacements along α_1, α_2 and normals n to the shell mean surface

q_x, q_y, q_n = components of a surface load along α_1, α_2, n

In accordance with the second theorem of similitude, we can obtain the necessary conditions of similitude as follows:

$$C_A\ C_\alpha/C_R = 1;\ C_qC_R/C_N = 1;\ C_\sigma C_h/C_N = 1 \qquad (206)$$

in which

$C_R = C_{R_1} = C_{R_2}$; $C_A = C_{A_1} = C_{A_2}$; $C_\alpha = C_{\alpha_1} = C_{\alpha_2}$; $C_q = C_{q_1} = C_{q_2} = C_{q_n}$; $C_N = C_{N_x} = C_{N_y} = C_{N_{xy}}$; $C_\sigma = C_{\sigma_x} = C_{\sigma_y} = C_{\sigma_{xy}}$

To determine all components of stresses and strains in the shell, one must solve the systems of Equations (201) and (202). Equations (203) and (204) give stress-strain relationships for linearly elastic and nonlinearly elastic shells.

Using the third theorem of similitude and Equations (202), (203), and (204), one can obtain the similitude conditions related to boundary conditions:

a. For linearly elastic shells:

$$C_\nu = 1;\ C_\sigma/C_\epsilon\, C_E = 1;\ C_u/C_\epsilon\, C_R = 1 \qquad (207)$$

b. For nonlinearly elastic shells:

$$C_\sigma/C_\phi\, C_\epsilon = 1;\ C_u/C_\epsilon\, C_R = 1 \qquad (208)$$

in which

$$C_\epsilon = C_{\epsilon_x} = C_{\epsilon_y} = C_{\epsilon_{xy}};\ C_u = C_{u_1} = C_{u_2} = C_w$$

It follows from Equation (207) that the linearly elastic material of similar shells should have an identical Poisson ratio. On the other hand, Equation (208) states that the strain functions $\phi(\epsilon_i)$ for shells with nonlinearly elastic material should be scaled by a factor of C_ϕ. Two cases are considered: a) if the similar shells have identical deformations ($\epsilon_i^m = \epsilon_i^p$), and b) if they have different deformations ($\epsilon_i^m = C_\epsilon\, \epsilon_i^p$).

Example 10

Let us consider an example of a long cylindrical shell as a roof made of reinforced concrete. This shell is experimentally discussed by Harris [36], who concludes that for $L/S \geq 4$, a shell corresponds to a membrane and works as a beam with a span L. For the shell shown in Figure 8:

$$\alpha_1 = x;\ \alpha_2 = s;\ A_1 = A_2 = 1;\ R_1 = \infty;\ R_2 = R$$

Considering the similitude conditions in Equation (206) to (208), one can conclude that shells are geometrically similar if:

$$C_A = 1;\ C_x = C_s;\ C_s = C_R \qquad (209)$$

Assume that the considered deformations (ϵ_1 and ϵ_2) and moduli of elasticity (E_1 and E_2) are identical for both shells; then:

$$C_\epsilon = 1;\ C_E = 1 \qquad (210)$$

In this case, both assumptions about the shell material (linearly and nonlinearly elastic) give identical similitude indicators as follows:

$$C_\sigma = 1;\ C_u/C_R = 1;\ C_q\, C_R/C_h = 1;\ C_T/C_h = 1 \qquad (211)$$

The similitude indicators [Equation (209) to (211)] can be

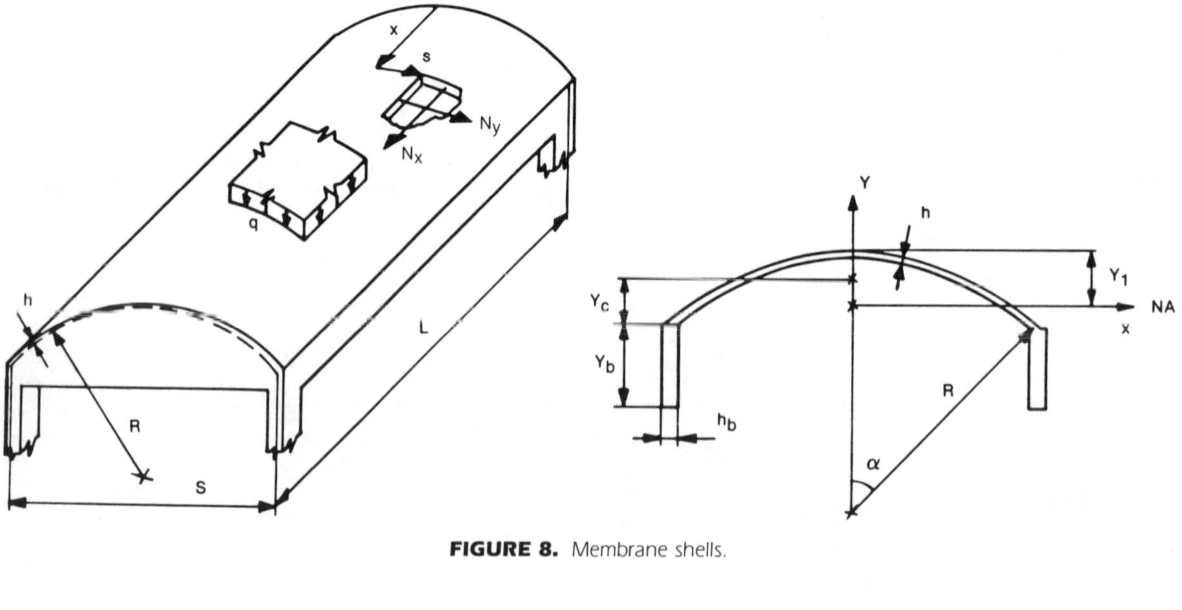

FIGURE 8. Membrane shells.

written in a form of criteria:

$$\varepsilon_2 = \varepsilon_1;$$
$$E_2 = E_1;$$
$$\left(\frac{S}{L}\right)_2 = \left(\frac{S}{L}\right)_1;$$
$$\sigma_2 = \sigma_1;$$
$$\left(\frac{R}{S}\right)_2 = \left(\frac{R}{S}\right)_1 \tag{212}$$
$$\left(\frac{u}{R}\right)_2 = \left(\frac{u}{R}\right)_1;$$
$$\left(\frac{qR}{h}\right)_2 = \left(\frac{qR}{h}\right)_1;$$
$$\left(\frac{T}{h}\right)_2 = \left(\frac{T}{h}\right)_1;$$
$$\nu_2 = \nu_1$$

Comparing these conditions with the requirements given in literature [36], it can be concluded that the relations [Equation (212)] allow different scales for mean surface and shell thickness. The model can be relatively thicker than that required by the true similitude phenomenon.

To obtain the edge beam configuration for a relatively thicker model, we can consider the deflection, w, and the deformation, ϵ_x, for the shell as for a beam:

$$w = K_1 \frac{ql^4}{EI}; \; \varepsilon_x = \frac{M}{ES} = K_2 \frac{ql^2}{ES} \tag{213}$$

But using Equation (202):

$$\varepsilon_x = \frac{\partial u_1}{\partial x} \tag{214}$$

and using Equations (213) and (214), $C_w = C_{u_1} = C_u$.

The similitude condition then follows:

$$C_I / C_S = C_l \tag{215}$$

Equation (215) indicates that the position of neutral axis y_1 will not be changed for cylindrical shells with different thicknesses and identical plane dimensions since $h << R$. For the curved portion of the shell, cross-sectional properties can be obtained using:

$$A = 2\alpha hR;$$
$$Y_c \cong R(\sin \alpha/\alpha - \cos \alpha); \tag{216}$$
$$I_x \cong \frac{hR^3}{2}\left(2\alpha + \sin 2\alpha - \frac{4 \sin^2 \alpha}{\alpha}\right)$$

As follows from Equation (216), to model long cylindrical shells by thicker models, use scale C_h for h_b and scale C_l for y_b dimensions of edge beams. Using this distorted model, the results can be transferred to the prototype shell.

In the particular case $C_h = C_R$, the similitude conditions [Equation (212)] coincide with those given in Reference 36.

Similitude of Shallow Shells

H. Aron [37] has made the suggestion to neglect the terms with tangential deflections u_1 and u_2 in formulas for curvatures χ_x, χ_y, χ_{xy}. A valid application of this assumption has been done by L. Donnell [38]. K. Marquerre [39] has presented his calculation results for so-called shallow shells and conformed H. Aron's approach. He has also derived the shell equations of the mixed method. V. Z. Vlasov [35] has developed a general theory of shallow shells in the form of the mixed method [38] and has suggested the following criterion:

$$d/l_{min} \leq 1/5 \tag{217}$$

where l_{min} = minimum plane dimension of the shallow shell. This criterion [Equation (217)] has to be considered in combination with other criteria and assumptions for shallow shells:

a. The squares of slope angles $\partial z/\partial x$ and $\partial z/\partial y$ can be neglected as compared with unity, i.e.:

$$\left(\frac{\partial z_1}{\partial x}\right)^2 \ll 1 : \left(\frac{\partial z_1}{\partial y}\right)^2 \ll 1 \tag{218}$$

where $z_1 = z_1(x,y)$ = the equation of shell middle surface.

b. Because of Equation (218), the geometry of a shallow shell middle surface can be replaced by the Euclidian geometry in the base plane, i.e.:

$$ds^2 \cong dx^2 + dy^2$$

and the curvatures K_x, K_y, K_{xy} can be calculated by formulas:

$$K_x = \frac{\partial^2 z_1}{\partial x^2}; \; K_y = \frac{\partial^2 z_1}{\partial y^2}; \; K_{xy} = \frac{\partial^2 z_1}{\partial x \partial y} \tag{219}$$

The governing differential equations for shallow shells of uniform thickness and large deflections can be written in Von Kármán's approach as follows [21,31]:

$$\frac{Eh^3}{12(1-\nu^2)}\nabla^4 w = q(x, y) + \left(K_x + \frac{\partial^2 w}{\partial x^2}\right)\frac{\partial^2 \phi}{\partial y^2} + \left(K_y + \frac{\partial^2 w}{\partial y^2}\right) \times \frac{\partial^2 \phi}{\partial x^2} - 2\left(K_{xy} + \frac{\partial^2 w}{\partial x \partial y}\right) \times \frac{\partial^2 \phi}{\partial x \partial y} - mh\frac{\partial^2 w}{\partial t^2} - kh\frac{\partial w}{\partial t} + Z(x, y, t) - \frac{Eh^2\alpha}{12(1-\nu)}\nabla^2\lambda \tag{220}$$

$$\nabla^4\phi = Eh\left[\left(\frac{\partial^2 w}{\partial x \partial y}\right)^2 - \frac{\partial^2 w}{\partial x^2}\frac{\partial^2 w}{\partial y^2} - K_x\frac{\partial^2 w}{\partial y^2} - K_y\frac{\partial^2 w}{\partial x^2} + 2K_{xy}\frac{\partial^2 w}{\partial x \partial y}\right] - Eh\alpha\nabla^2 T \tag{221}$$

Here the shell middle surface curvatures K_x, K_y, K_{xy} can be either variables or constants. Some references [23,40] consider K_x, K_y, K_{xy} to be constant regardless of shell middle surface equation z_1 (x,y). But existing algorithms for shallow shells calculations [41] and comparative calculations for very shallow shells of revolution [42] show that actual curvatures K_x, K_y, K_{xy} should be considered.

The most popular and practical shallow shell middle surfaces are:

a. Surface of translation of circular curve:

$$K_x = \frac{8d_1}{a^2[1 + 4(d_1/a)^2]};$$

$$K_y = \frac{8d_2}{b^2[1 + 4(d_2/b)^2]};$$

$$K_{xy} = 0$$

b. Elliptic paraboloid:

$$z_1 = -d\left[1 - \left(d_1/d \cdot \frac{4x^2}{a^2} + d_2/d \cdot \frac{4y^2}{b^2}\right)\right]$$

$$K_x = \frac{8d_1}{a^2};$$

$$K_y = \frac{8d_2}{b^2};$$

$$K_{xy} = 0$$

c. Spherical surface:

$$K_x = K_y = \frac{8d}{a^2[1 + (b/a)^2 + 4(d/a)^2]};$$

$$K_{xy} = 0$$

For all above surfaces curvatures are constant. They are considered for modeling with Equations (226) and (227).

Additional relationships for components other than those in Equations (220) and (221) are identical to Equation (66) for bending and to Equation (176) for membrane components, but formulas for membrane strains should consider curvatures K_x, K_y, K_{xy}:

$$\varepsilon_x^{mem} = \frac{\partial u}{\partial x} - K_x w + \frac{1}{2}\left(\frac{\partial w}{\partial x}\right)^2;$$

$$\varepsilon_y^{mem} = \frac{\partial v}{\partial y} - K_y w + \frac{1}{2}\left(\frac{\partial w}{\partial y}\right)^2; \tag{222}$$

$$\varepsilon_{xy}^{mem} = \frac{\partial u}{\partial y} + \frac{\partial v}{\partial x} - 2K_{xy}w + \frac{\partial w}{\partial x} \times \frac{\partial w}{\partial y}$$

In addition:

$$N_x = \frac{\partial^2\phi}{\partial y^2};\; N_y = \frac{\partial^2\phi}{\partial x^2};\; N_{xy} = -\frac{\partial^2\phi}{\partial x \partial y} \tag{223}$$

By application of similitude theory one can obtain the identical to Equations (169), (171), and (173) necessary equations of similitude from Equations (220) and (221), and an additional relationship for the middle surface curvatures:

$$C_k C_l^2 C_{\nu_0}^{0.5}/C_h = 1 \tag{224}$$

and equations identical to Equations (170), (172), and (174) generalized variables and generalized middle surface curvatures:

$$\bar{K}_x = K_x l^2(1 - \nu^2)^{0.5}/h$$

$$\bar{K}_y = K_y l^2(1 - \nu^2)^{0.5}/h \tag{225}$$

$$\bar{K}_{xy} = K_{xy} l^2(1 - \nu^2)^{0.5}/h$$

The nondimensional form of Equations (220) and (221) is identical to Equations (159), (160), and (165), but the latter must be supplemented by $\overline{K}_x$, $\overline{K}_y$, $\overline{K}_{xy}$:

$$\frac{1}{12}\nabla^4\overline{w} = p(\xi, \eta) + \left(\overline{K}_x + \frac{\partial^2\overline{w}}{\partial\xi^2}\right) \times \frac{\partial^2\psi}{\partial\eta^2} + \left(\overline{K}_y + \frac{\partial^2\overline{w}}{\partial\eta^2}\right) \times \frac{\partial^2\psi}{\partial\xi^2} - 2\left(\overline{K}_{xy} + \frac{\partial^2\overline{w}}{\partial\xi\partial\eta}\right) \times \frac{\partial^2\psi}{\partial\xi\partial\eta} - \frac{\partial^2\overline{w}}{\partial\theta^2} - \overline{k}\frac{\partial\overline{w}}{\partial\theta} + \overline{Z}(\xi, \eta, \theta) - \frac{1}{12}\nabla^2\mu \tag{226}$$

$$\nabla^4\psi = \left(\frac{\partial^2\overline{w}}{\partial\xi\partial\eta}\right)^2 - \frac{\partial^2\overline{w}}{\partial\xi^2}\frac{\partial^2\overline{w}}{\partial\eta^2} - \overline{K}_x\frac{\partial^2\overline{w}}{\partial\eta^2} - \overline{K}_y\frac{\partial^2\overline{w}}{\partial\xi^2} + 2\overline{K}_{xy}\frac{\partial^2\overline{w}}{\partial\xi\partial\eta} - \nabla^2\overline{T} \tag{227}$$

Identical to elastic plates with large deflections, if material of similar shells differs in Poisson's ratio ($C_\nu \neq 1$), there is a partial similitude of shell components and the thrid similitude theorem requires boundary conditions to be specified in similar components. Otherwise, $C_\nu = 1$ for exact similitude with all shell components similar.

Equation (224) predetermines the geometry range of similar shallow shells and shows that they can be distorted. Therefore, Vlasov's conditions [Equation (217)] has to be checked for the shells of a similar group. If the shell middle surface curvatures K_x, K_y, K_{xy} are variables, the terms in parentheses of Equation (220):

$$\left(\frac{\partial^2 z_1}{\partial x^2} + \frac{\partial^2 w}{\partial x^2}\right);$$

$$\left(\frac{\partial^2 z_1}{\partial y^2} + \frac{\partial^2 w}{\partial y^2}\right);$$

$$\left(\frac{\partial^2 z_1}{\partial x\partial y} + \frac{\partial^2 w}{\partial x\partial y}\right)$$

give the indicator of similitude: $C_z/C_w = 1$, i.e., similitude requires geometrical similarity of shallow shells. In case of $C_\nu \neq 1$, there is approximate geometrical similarity because of $C_z = C_h/C_{\nu_0}^{0.5}$ and $(C_{\nu_0}^{0.5})$ max. $= 0.866$.

For shallow shells with small deflections (w_{max}/h not higher than 0.25 to 0.50), the equations of similitude can be obtained from linearized governing differential Equation (220) and (221):

$$C_x/C_y = 1;\ C_x/C_l = 1;\ C_wC_h^2C_E/C_qC_l^4C_{\nu_0} = 1$$
$$C_kC_l^2C_{\nu_0}^{0.5}/C_h = 1;\ C_\phi C_h/C_qC_l^4C_{\nu_0}^{0.5} = 1$$
$$C_NC_h/C_qC_l^2C_{\nu_0}^{0.5} = 1;\ C_tC_hC_E^{0.5}/C_m^{0.5}C_l^2C_{\nu_0}^{0.5} = 1 \tag{228}$$
$$C_kC_l^2C_{\nu_0}^{0.5}/C_E^{0.5}C_m^{0.5}C_h = 1;\ C_\lambda C_\alpha C_h^2C_E/C_qC_l^2C_{\nu_1} = 1$$
$$C_TC_\alpha C_EC_h^2/C_qC_l^2C_{\nu_0}^{0.5} = 1$$

where static, dynamic, and thermal problems are combined; therefore, for dynamic problems similitude scale C_q should be replaced by C_Z. For shells under thermal loads only, the similitude equations with C_w, C_ϕ, C_N, and C_T scales should be replaced by the following:

$$C_wC_h/C_\lambda C_\alpha C_l^2C_{\nu_2} = 1;\ C_\phi C_{\nu_0}^{0.5}/C_\lambda C_EC_\alpha C_hC_{\nu_2}C_l^2 = 1$$
$$C_N\ C_{\nu_0}^{0.5}/C_\lambda C_EC_\alpha C_hC_{\nu_2} = 1;\ C_TC_{\nu_1}/C_\lambda C_{\nu_0}^{0.5} = 1 \tag{229}$$

The generalized variables corresponding to Equation (228) are:

$$\xi = x/l;$$
$$\eta = y/l;$$
$$\overline{w} = wh^2E/q_0l^4(1 - \nu^2)\ (\text{or } Z_0);$$
$$\overline{K}_x = K_xl^2(1 - \nu^2)^{0.5}/h;$$
$$\overline{K}_y = K_yl^2(1 - \nu^2)^{0.5}/h;$$
$$\overline{K}_{xy} = K_{xy}l^2(1 - \nu^2)^{0.5}/h;$$
$$\psi = \phi h/q_0l^4(1 - \nu^2)^{0.5}\ (\text{or } Z_0);$$
$$\overline{N}_x = N_xh/q_0l^2(1 - \nu^2)^{0.5};$$
$$\overline{N}_y = N_yh/q_0l^2(1 - \nu^2)^{0.5}; \tag{230}$$
$$\overline{N}_{xy} = N_{xy}h/q_0l^2(1 - \nu^2)^{0.5}\ (\text{or } Z_0);$$
$$\theta = thE^{0.5}/m^{0.5}l^2(1 - \nu^2)^{0.5};$$
$$\overline{k} = kl^2(1 - \nu^2)^{0.5}/E^{0.5}m^{0.5}h;$$

$$p(\xi, \eta) = q(x, y)/q_0;$$

$$\overline{Z}(\xi, \eta) = Z(x, y)/Z_0;$$

$$\mu(\xi, \eta) = \lambda(x, y)\alpha h^2 E/q_0 l^2(1 - \nu);$$

$$\overline{T}(\xi, \eta) = T(x, y)\alpha E h^2/q_0 l^2(1 - \nu^2)^{0.5}$$

For the only thermal loads, generalized deflections $\overline{w}$; membrane function ψ and forces $\overline{N}_x$, $\overline{N}_y$, $\overline{N}_{xy}$; and mean temperature $\overline{T}$ are as follows:

$$\overline{w} = wh/\lambda_0 \alpha l^2(1 + \nu);$$

$$\psi = \phi(1 - \nu^2)^{0.5}/\lambda_0 E\alpha h l^2(1 + \nu);$$

$$\overline{N}_x = N_x(1 - \nu^2)^{0.5}/\lambda_0 E\alpha h(1 + \nu);$$

$$\overline{N}_y = N_y(1 - \nu^2)^{0.5}/\lambda_0 E\alpha h(1 + \nu); \qquad (231)$$

$$\overline{N}_{xy} = N_x(1 - \nu^2)^{0.5}/\lambda_0 E\alpha h(1 + \nu);$$

$$\overline{T}(\xi, \eta) = T(x, y)(1 - \nu)/\lambda_0(1 - \nu^2)^{0.5};$$

$$\mu(\xi, \eta) = \lambda(x, y)/\lambda_0$$

The nondimensional governing differential equations [Equations (226) and (227)] can be used for calculations after linearization.

If boundary conditions do not allow $C_\nu \neq 1$, all the above generalized variables for shallow shells do not contain functions $(1 - \nu^2)$, $(1 - \nu)$, and $(1 + \nu)$. They appear in the governing nondimensional differential equations [Equation (226) and (227)] identically to dimensional equations [Equations (220) and (221)].

One can expect approximate partial similitude of shallow shells if they have different Poisson's ratios, but this is a matter of special investigation.

Example 11

As an example, two shallow shells which belong to the group of similar structures are considered (refer to Figure 9) [16]. Assuming the case of large deflections of these shells, referring to the indicators of similitude [Equations (169) and (224)] and assuming their middle surface can be described by equation for an elliptic paraboloid, one can calculate the following constants of similitude using the information provided in Table 16:

Shell A:

$$K_x = \frac{8d_1}{a^2} = \frac{8 \times 1.569}{50^2}$$

$$= 5.0208 \times 10^{-3} \text{ (in.}^{-1});$$

$$K_y = \frac{8d_2}{b^2} = \frac{8 \times 2.098}{50^2}$$

$$= 6.7136 \times 10^{-3} \text{ (in.}^{-1});$$

$$K_{xy} = 0; \; h = 1 \text{ inch};$$

$$a = b = 50 \text{ inches}; \; E = 29{,}000 \text{ ksi}; \; \nu = 0.3$$

Acting surface load $q(x,y) = q_o \times p(\xi,\eta)$ causes maximum displacements w_o, u_0; fiber stresses σ_0; and maximum fiber strains ϵ_0. Here the bending and fiber components are not split because of the case of $C_\nu = 1$:

$$C_\sigma^{mem} = C_\sigma^{bend}; \; C_\epsilon^{mem} = C_\epsilon^{bend}$$

Shell B:

$$h = 2.15 \text{ inches}, \; a = b = 50 \text{ inches};$$

$$E = 29{,}000 \text{ ksi};$$

$$\nu = 0.3, \text{ that is, } C_h = 2.15/1.0 = 2.15;$$

$$C_l = C_a = C_b = 50/50 = 1;$$

$$C_E = 29{,}000/29{,}000 = 1;$$

$$C_\nu = 0.3/0.3 = 1$$

The shell B elevation selection must satisfy the indicator of similitude [Equation (224)]:

$$C_K = C_h/C_l^2 C_{\nu_0}^{0.5} = 2.15/1 \times 1 = 2.15$$

Calculate curvatures K_x, K_y, K_{xy}:

$$K_x = C_k \times 5.0208 \times 10^{-3} = 0.0107947 \text{ (in.}^{-1})$$

$$K_y = C_k \times 6.7136 \times 10^{-3} = 0.0144342 \text{ (in.}^{-1})$$

$$K_{xy} = 0$$

that is, elevations d_1 and d_2 should be:

$$d_1 = \frac{K_x a^2}{8} = \frac{0.0107947 \times 50^2}{8} = 3.373 \text{ inches}$$

$$d_2 = \frac{K_y a^2}{8} = \frac{0.0144342 \times 50^2}{8} = 4.511 \text{ inches}$$

Calculate other similitude scales:

$$C_q = C_E C_h^4/C_l^4 C_{\nu_0}^{1.5} = 1 \times 2.15^4/1 \times 1^{1.5} = 21.3675$$
$$C_w = C_h/C_{\nu_0}^{0.5} = 2.15/1^{0.5} = 2.15$$

$$C_u = C_w^2/C_l = 2.15^2/1 = 4.6225$$
$$C_\sigma^{mem} = C_E C_h^2/C_l^2 C_{\nu_0} = 1 \times 2.15^2/1 \times 1 = 4.6225$$
$$C_\epsilon^{mem} = C_\sigma^{mem}/C_E = 4.6225/1 = 4.6225$$
$$C_\epsilon^{bend} = C_w C_h/C_l^2 = 2.15 \times 2.15/1^2 = 4.6225$$
$$C_\sigma^{bend} = C_\epsilon^{bend} \times C_E = 4.6225 \times 1 = 4.6225$$

Check Vlasov's condition [Equation (217)] for both shells:

Shell A: $d = d_1 + d_2 = 1.569 + 2.098 = 3.667$ inches; $l_{min} = 50$ inches; $d/l_{min} = 3.667/50 = 1/13.64 < 1/5$

Shell B: $d = d_1 + d_2 = 3.373 + 4.511 = 7.884$ inches; $l_{min} = 50$ inches; $d/l_{min} = 7.884/50 = 1/6.342 < 1/5$

Comparing displacements, strains, and stresses for both shells A and B, it can be concluded that:

a. For physical modeling the second model (shell B) is preferable because it can be manufactured more precisely due to its greater thickness. This model gives larger displacements and strains, i.e., higher accuracy for their calculations.
b. If analysis by any numerical method for shell A should be done, the additional analysis of shell B, loaded as shown in Figure 9, provides calculation accuracy control by comparison of two calculations with scales C_w, C_u, C_ϵ, C_σ. In case of closed-form solutions, scales C_w, C_u, C_ϵ, C_σ would be exactly equal to those calculated by use of similitude theory.

Similitude of Cylindrical Shells

These shells are used widely due to the ease of fabrication. If $L/S < 4$, then these cannot be termed as membrane shells [23]. The equations for the behavior of cylindrical shells are written as:

$$\Delta\Delta\tilde{T} + \frac{\partial^2\tilde{T}}{\partial\alpha_2^2} + i2b^2\frac{\partial^2\tilde{T}}{\partial\alpha_1^2} = i2b^2R\left(\Delta q_n + \frac{\partial q_2}{\partial\alpha_2} - \frac{\partial q_1}{\partial\alpha_1}\right) \quad (232)$$

In addition to Equation (232), other relationships for the behavior are given as follows:

$$\tilde{T} = \tilde{T}_1 + \tilde{T}_2;$$

$$\tilde{T}_1 = N_x - i\frac{2b^2}{R}\frac{M_y - \nu M_x}{1 - \nu^2};$$

$$\tilde{T}_2 = N_y - i\frac{2b^2}{R}\frac{M_x - \nu M_y}{1 - \nu^2}$$

$$\tilde{S} = N_{xy} + i\frac{2b^2}{R}\frac{M_{xy}}{1 - \nu};$$

$$2b^2 = \frac{R}{h}\sqrt{12(1 - \nu^2)};$$

$$\Delta(\ldots) = \frac{\partial^2(\ldots)}{\partial\alpha_1^2} + \frac{\partial^2(\ldots)}{\partial\alpha_2^2}$$

$$(1 + \nu)Q_x = \frac{1}{R}\frac{\partial(M_x + M_y)}{\partial\alpha_1};$$

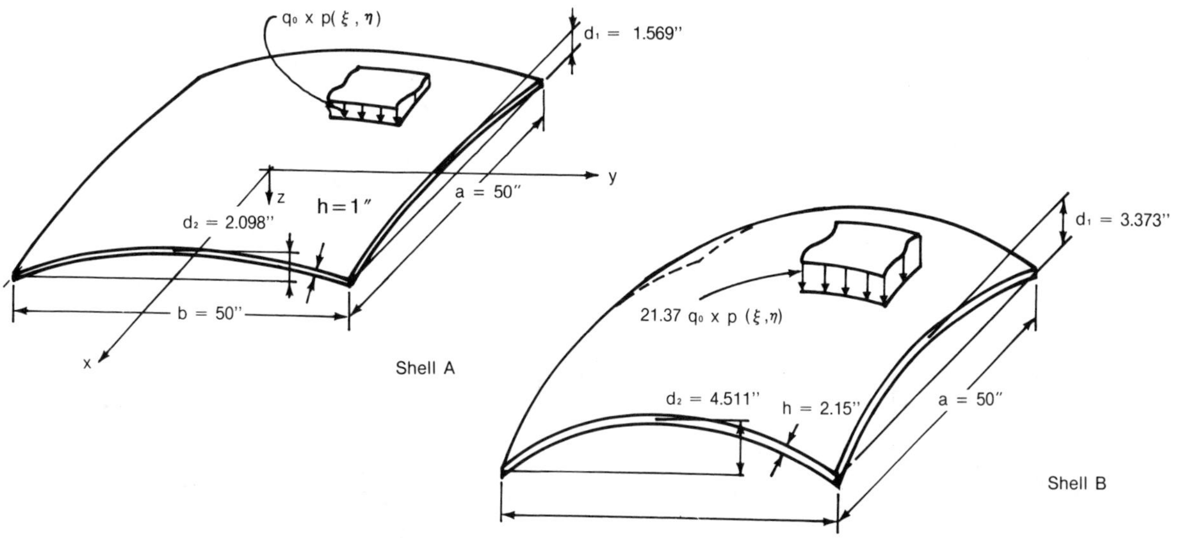

FIGURE 9. Compared geometrically dissimilar shallow shells which belong to the group of similar structures.

$$(1 + \nu)Q_y = \frac{1}{R}\frac{\partial(M_x + M_y)}{\partial \alpha_2};$$

$$\alpha_1 = \frac{x}{R};$$

$$\alpha_2 = \frac{s}{R};$$

$$\varepsilon_x^m = \frac{1}{R}\frac{\partial u_1}{\partial \alpha_1};$$

$$\varepsilon_y^m = \frac{1}{R}\frac{\partial u_2}{\partial \alpha_2} + \frac{w}{R};$$

$$\varepsilon_{xy} = \frac{1}{R}\left(\frac{\partial u_1}{\partial \alpha_2} + \frac{\partial u_2}{\partial \alpha_1}\right)$$

$$\chi_x = -\frac{1}{R^2}\frac{\partial^2 w}{\partial \alpha_1^2};$$

$$\chi_y = -\frac{1}{R^2}\frac{\partial^2 w}{\partial \alpha_2^2} + \frac{1}{R^2}\frac{\partial u_2}{\partial \alpha_2};$$

$$\chi_{xy} = -\frac{1}{R^2}\frac{\partial^2 w}{\partial \alpha_1 \partial \alpha_2} + \frac{1}{R^2}\frac{\partial u_2}{\partial \alpha_1} \qquad (233)$$

For nonlinearly elastic shell material, one can refer to Equation (203); on the other hand, for linearly elastic material, one must add the following conditions to Equation (204):

$$M_x = \frac{Eh^3}{12(1 - \nu^2)}(\chi_x + \nu\chi_y);$$

$$M_y = \frac{Eh^3}{12(1 - \nu^2)}(\chi_y + \nu\chi_x); \qquad (234)$$

$$M_{xy} = \frac{Eh^3}{12(1 + \nu)}\chi_{xy}$$

Equating the similitude indicators obtained from Equations (232) and (233) to unity we obtain the similitude conditions. Thus,

$$c_\nu = 1;$$

$$c_\alpha = 1;$$

$$\frac{C_N}{C_q C_l} = 1;$$

$$\frac{C_M}{C_T C_l} = 1;$$

$$C_\varepsilon = \frac{C_u}{C_l}; \qquad (235)$$

$$C_\chi = \frac{C_u}{C_l^2};$$

$$\frac{C_\sigma C_l}{C_T} = 1$$

Additional similitude conditions for stress-strain are:

a. For linearly elastic shells:

$$C_\sigma = C_E\, C_\epsilon \qquad (236)$$

b. For nonlinearly elastic shells:

$$C_\sigma = C_\phi\, C_\epsilon \qquad (237)$$

In Equations (235) to (237), quantities are defined as follows:

$C_l = C_R = C_x = C_s = C_h$ — geometrical scale factors of similitude

$C_\chi = C_{\chi_x} = C_{\chi_y} = C_{\chi_{xy}}$ — scale factors for changes in bending and twisting curvatures

TABLE 16. Comparison of Components for Shallow Shells A and B.

Shell	a (in.)	b (in.)	E (ksi)	ν	h (in.)	K_x (l/in.)	K_y (l/in.)	K_{xy}	q(x,y) ksi	w (in.)	u (in.)	σ ksi	ϵ
A	50	50	29,000	0.3	1.0	5.0208 $\times 10^{-3}$	6.7136 $\times 10^{-3}$	0	q_0	w_0	u_0	σ_0	ϵ_0
B	50	50	29,000	0.3	2.15	0.0107947	0.0144342	0	21.3675 $\times q_0$	2.15 w_0	4.6225 $\times u_0$	4.6225 $\times \sigma_0$	4.6225 $\times \epsilon_0$

$C_N = C_{N_x} = C_{N_y} = C_{N_{xy}} = C_{Q_x} = C_{Q_y}$ scale factors for membrane, bending, and shear forces

$C_M = C_{M_x} = C_{M_y} = C_{M_{xy}}$ moment scale factors of similitude

These indicate that model and prototype behaviors should be geometrically similar. For elastic model materials, we can assume $C_\epsilon \neq 1$.

For an intermediate cylindrical shell with an L/S ratio of less than 4, these relations can be written as:

$$\frac{12R}{h^2}\frac{\partial^2 M_y}{\partial \alpha_1^2} + \frac{\partial^4 N_x}{\partial \alpha_2^4} + \frac{\partial^2 N_x}{\partial \alpha_2^2} = 0;$$

$$R\frac{\partial^2 N_x}{\partial \alpha_1^2} - \frac{\partial^4 M_y}{\partial \alpha_2^4} - \frac{\partial^2 M_y}{\partial \alpha_2^2}$$

$$= R^2\left(\frac{\partial^2 q_n}{\partial \alpha_2^2} + \frac{\partial q_2}{\partial \alpha_2} - \frac{\partial q_1}{\partial \alpha_1}\right);$$

$$N_x = Eh\varepsilon_x^{mem};$$

$$M_y = \frac{Eh^3}{12}\chi_y;$$

$$\varepsilon_1^{mem} = \frac{1}{R}\frac{\partial u_1}{\partial \alpha_1};$$

$$\chi_y = -\frac{1}{R^2}\left(\frac{\partial^2 w}{\partial \alpha_2^2} - \frac{\partial u_2}{\partial \alpha_2}\right); \tag{238}$$

$$\varepsilon_{xy} = \frac{1}{R}\left(\frac{\partial u_1}{\partial \alpha_2} + \frac{\partial u_2}{\partial \alpha_1}\right) = 0$$

$$\varepsilon_y^{mem} = \frac{1}{R}\left(\frac{\partial u_2}{\partial \alpha_2} + w\right) = 0;$$

$$Q_y = \frac{1}{R}\frac{\partial M_y}{\partial \alpha_2};$$

$$\frac{\partial N_x}{\partial \alpha_1} + \frac{\partial N_{xy}}{\partial \alpha_2} + q_1 R = 0;$$

$$N_y = \frac{\partial Q_y}{\partial \alpha_2} + q_n R;$$

$$\varepsilon_y^{bend} = \chi_y h/2$$

Using the last two theorems of similitude, we get the indicators of similitude from Equation (238) as follows:

$$C_{\alpha_2} = 1;$$

$$\frac{C_R}{C_h C_{\alpha_1}^2} = 1;$$

$$\frac{C_{M_y}}{C_{N_x} C_h} = 1;$$

$$\frac{C_{M_y}}{C_{Q_y} C_R} = 1;$$

$$\frac{C_{N_x}}{C_{N_{xy}} C_{\alpha_1}} = 1;$$

$$\frac{C_{N_y} C_{\alpha_1}^2}{C_{N_x}} = 1;$$

$$\frac{C_{q_n} C_R^2}{C_{N_x} C_h} = 1;$$

$$\frac{C_{q_2}}{C_{q_n}} = 1;$$

$$\frac{C_{q_1}}{C_{q_n} C_{\alpha_1}} = 1; \tag{239}$$

$$\frac{C_{N_x}}{C_{\sigma_x^m} C_h} = 1;$$

$$\frac{C_{\sigma_y^b}}{C_{\sigma_x^m}} = 1;$$

$$\frac{C_{\sigma_x^m}}{C_{\alpha_1} C_{\sigma_{xy}^m}} = 1;$$

$$\frac{C_{\sigma_x^m}}{C_{\alpha_1}^2 C_{\sigma_y^m}} = 1;$$

$$\frac{C_w}{C_{u_2}} = 1;$$

$$\frac{C_{u_1} C_{\alpha_1}}{C_{u_2}} = 1;$$

$$C_{\chi_y} = \frac{C_w}{C_R^2};$$

$$C_{\varepsilon_x^m} = \frac{C_{u_1}}{C_R C_{\alpha_1}};$$

$$\frac{C_{\varepsilon_y^b}}{C_{\varepsilon_x^m}} = 1$$

in which:

σ_x^m, σ_y^m, σ_{xy}^m, ϵ_x^m = membrane stresses and deformations
σ_y^b, ϵ_y^b = bending stresses and deformations
Q_y = shear force, which is perpendicular to the mean shell surface

In the similitude conditions [Equation (239)], the geometrical scale is $C_{\alpha_2} = 1$; i.e., the shells have similar cross sections ($C_s = C_R$ in Figure 8). But scales C_{α_1} can differ from unity, which means that the model and the prototype can be geometrically dissimilar.

The geometrical indicator of similitude $C_R/C_h\ C_{\alpha_1}^2$ yields the thickness scale.

If $C_h \neq C_R$, i.e., $C_{\alpha_1} \neq 1$, we have different scales for stress $C_{\sigma_x^m}$, $C_{\sigma_y^m}$, and $C_{\sigma_{xy}}$ (but $C_{\sigma_x^m} = C_{\sigma_y^b}$). In this case indirect elastic modeling can be used [43] to get increased thickness and accuracy. For direct elastic modeling ($C_\epsilon = 1$) and for nonlinear material approximation [Equation (237)], we should have all stress scales equal to each other; thus,

$$C_{\sigma_x^m} = C_{\sigma_y^m} = C_{\sigma_{xy}} = C_\sigma$$

which results in $C_{\alpha_1} = 1$ and $C_R = C_h$. The model and the phenomenon should therefore be geometrically similar.

Consider now the short cylindrical shell with *L/S* less than or equal to 1.0. The corresponding equations are written from Reference 23:

$$\Delta\Delta\tilde{T} + i2b^2 \frac{\partial^2 \tilde{T}}{\partial \alpha_1^2} = i2b^2 R \times \left(\Delta q_n + \frac{\partial q_2}{\partial \alpha_2} - \frac{\partial q_1}{\partial \alpha_1}\right) \tag{240}$$

The only equations for changes in curvature which differ from Equation (233) are:

$$\chi_y = -\frac{1}{R^2}\frac{\partial^2 w}{\partial \alpha_2^2}; \qquad \chi_{xy} = -\frac{1}{R^2}\frac{\partial^2 w}{\partial \alpha_1 \partial \alpha_2} \tag{241}$$

The similitude conditions differing from Equations (235) and (239) are:

$$C_\nu = 1; \quad \frac{C_R C_\alpha^2}{C_h} = 1; \quad \frac{C_N}{C_{q_n} C_R} = 1; \quad \frac{C_M}{C_N C_h} = 1; \quad \frac{C_{q_n}}{C_{q_2} C_\alpha} = 1; \quad \frac{C_{q_1}}{C_{q_2}} = 1; \quad \frac{C_x}{C_s} = 1; \quad \frac{C_u}{C_w C_\alpha} = 1; \quad C_\varepsilon = \frac{C_w}{C_R}; \quad C_\chi = \frac{C_w}{C_R C_h}; \quad C_\sigma = \frac{C_N}{C_h} \tag{242}$$

in which

$$C_\alpha = C_{\alpha_1} = C_{\alpha_2};$$

$$C_N = C_{N_x} = C_{N_y} = C_{N_{xy}};$$

$$C_M = C_{M_x} = C_{M_y} = C_{M_{xy}};$$

$$C_u = C_{u_1} = C_{u_2};$$

$$C_\varepsilon = C_{\varepsilon_x^m} = C_{\varepsilon_y^m} = C_{\varepsilon_{xy}} = C_{\varepsilon_x^b} = C_{\varepsilon_y^b};$$

$$C_\chi = C_{\chi_x} = C_{\chi_y} = C_{\chi_{xy}}.$$

Here the model and the phenomenon can be geometrically dissimilar. The geometrical scales can be found from the geometrical indicators of similitude $C_R\ C_\alpha^2/C_h$. Because of identical scales C_ϵ, the difference in geometrical scales C_h, C_R, and C_α does not affect the assumption of linear elastic behavior of shell material (direct or indirect modeling) and nonlinear elastic behavior of shell material.

Example 12—(Refer to Reference 9)

In Figure 10, the reinforced concrete vessel for liquid gas is shown. It consists of a metal cylindrical mouthpiece (a) and pre-stressed concrete vessel (b), with steel strands (d) and reinforcement in the circular direction. The dimensions are as shown in the figure. The material constants are: $E_c = 4000$ ksi (2.8×10^7 kPa), $\nu_p = 0.2$, $q_1^p = q_2^p = 0$, $q_n^p = 156$ psig (1100 kPa g). Both ends of the cylinder (b) are closed by metal diaphragms (c).

A plexiglas model was chosen to investigate the behavior of this structure. The properties of the model are $E_m = 0.354$ ksi (2500 kPa) and $\nu_m = 0.45$, and are tested up to a temperature of 212°F (100°C). Because of different Poisson ratios, similitude scale $C_\nu \neq 1.0$ and similitude conditions of Equation (235) were not satisfied. It also follows from Equations (232) and (233) that there is only *a partial similitude* between the model and the prototype, but the membrane forces N_x, N_y, and N_{xy}, displacements u_2 and w, and deformations ϵ_x^b, ϵ_y^b, χ_x, χ_{xy}, and χ_y are similar. The similitude conditions in Equation (235) should also be changed accordingly:

$$
\begin{aligned}
C_\alpha &= 1; \\
C_\nu^{1/2} C_l / C_h &= 1; \\
\frac{C_N}{C_q C_l} &= 1; \\
\frac{C_{\sigma^m} C_h}{C_T} &= 1; \\
C_\chi &= \frac{C_w}{C_l^2} \\
C_w &= C_{u_2}; \\
\frac{C_q}{C_\varepsilon C_\chi C_h} &= 1; \\
C_{\varepsilon^b} &= C_w C_h / C_l^2 .
\end{aligned}
\qquad (243)
$$

in which:

$$
\begin{aligned}
C_l &= C_x = C_R; \\
C_\chi &= C_{\chi_x} = C_{\chi_y} = C_{\chi_{xy}}; \\
C_\nu &= (1 - \nu_m^2)/(1 - \nu_p^2); \\
C_N &= C_{N_x} = C_{N_y} = C_{N_{xy}}; \\
C_q &= C_{q_1} = C_{q_2} = C_{q_n}; \\
C_{\sigma^m} &= C_{\sigma_x^m} = C_{\sigma_y^m} = C_{\sigma_{xy}}; \\
C_{\varepsilon^b} &= C_{\varepsilon_x^b} = C_{\varepsilon_y^b}; \\
C_\alpha &= C_{\alpha_1} = C_{\alpha_2}
\end{aligned}
$$

To increase the experimental accuracy, it is desirable to use indirect modeling [43]. It was found experimentally that model load was 0.99 psi (6.82 kPa), which could cause the maximum deflection:

$$
w_{max}^{model} < 0.5 \times h_{model} \qquad (244)
$$

which corresponds to Equations (232) and (233) for the linear shell theory.

As a variation, let us use the geometrical scale factor $C_l = 0.082$. We can also calculate the following:

$$
C_E = 0.354/4000 = 0.88 \times 10^{-4};
$$

$$
C_q = \frac{0.99 \times 10^{-3}}{0.156} = 6.4 \times 10^{-3}
$$

$$
C_\nu = (1 - 0.45^2)/(1 - 0.2^2) = 0.83
$$

Substituting C_l, C_E, C_q, and C_ν we obtain other similitude scales:

$$
C_h = 0.075,\ C_N = 5.2 \times 10^{-4},
$$

$$
C_{\sigma^m} = 7 \times 10^{-3},\ C_\chi = 970
$$

$$
C_w = C_{u_2} = 6.5,\ C_{\varepsilon^b} = 73
$$

The model thus designed is shown in Figure 11. The similitude scales were obtained without considering the boundary conditions, i.e., without using the last theorem of similitude.

There are two kinds of boundary conditions: connection to the diaphragms at the end, and connection of the mouthpiece to the vessel. The first kind of boundary condition is shown in Figure 12, in which the bending stiffness of

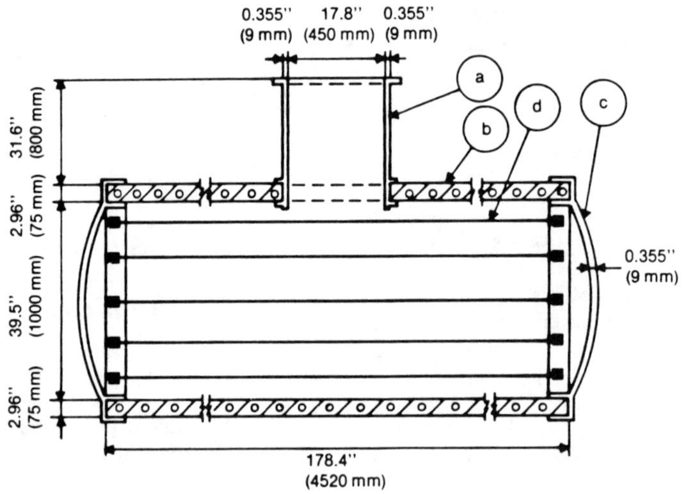

FIGURE 10. Reinforced concrete vessel for liquid gas.

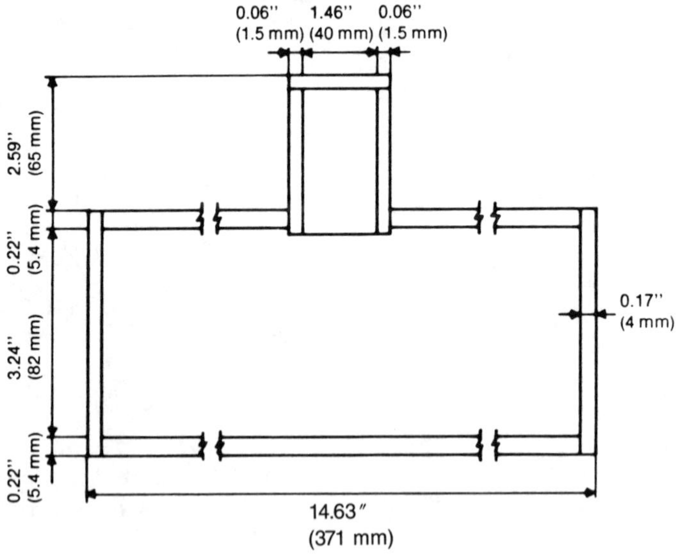

FIGURE 11. Model of the vessel for liquid gas.

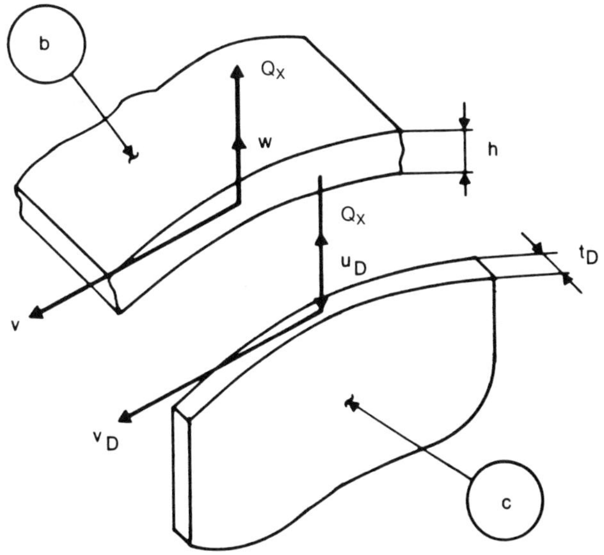

FIGURE 12. The vessel connection to the diaphragms at the end.

a diaphragm is neglected and only tension caused by shear Q_x from the shell is considered. The formulas for shears Q_x and Q_y in Equation (233) can be written in terms of changes in curvatures χ_x and χ_y:

$$Q_x = \frac{Eh^3}{12(1 - \nu^2)R} \frac{\partial(\chi_x + \chi_y)}{\partial\alpha_1};$$
$$Q_y = \frac{Eh^3}{12(1 - \nu^2)R} \frac{\partial(\chi_x + \chi_y)}{\partial\alpha_2} \tag{245}$$

Due to the continuity between the shell and diaphragm:

$$v_D = v;\ u_D = w \tag{246}$$

Relationship between acting forces Q_x and the diaphragm displacements u_D and v_D are expressed as follows:

$$Q_x \cong \frac{E_D t_D}{1 - \nu_D^2}\left(\frac{du_D}{dr} + \nu_D \frac{u_D}{r}\right) \tag{247}$$

which can be simplified to the following form using symmetry of axis for load:

$$Q_x \cong \frac{E_D t_D}{1 - \nu_D^2} \frac{du_D}{dr} \tag{248}$$

From Equations (246) and (248), we obtain additional similitude conditions:

$$c_{t_D} = \frac{C_E C_h^3 C_{\nu_D}}{C_{E_D} C_l^2 C_\nu} \tag{249}$$

in which the similitude scales C_{E_D} and C_{ν_D} for metal diaphragms are:

$$C_{E_D} = 0.354/29000 = 1.22 \times 10^{-5};$$

$$C_{\nu_D} = (1 - .45^2)/(1 - .3^2) = 0.88$$

and from Equation (249), $C_{t_D} = 0.48$ or $t_D = 0.17$ in. (4 mm).

Next, consider the modeling of the connection of the metal cylindrical mouthpiece to the concrete portion. It deforms due to moments M_x and M_y and membrane forces N_x and N_y acting in the shell (Figure 13). Because of the continuity between the two parts, the following conditions are obtained:

a. Along the generator of shell:

$$\tilde{w} = u_1 \tag{250}$$

b. Along the circumference:

$$\tilde{w} \cong u_2 \tag{251}$$

In Equations (250) through (253), the components that are marked with the symbol ~ relate to the cylindrical

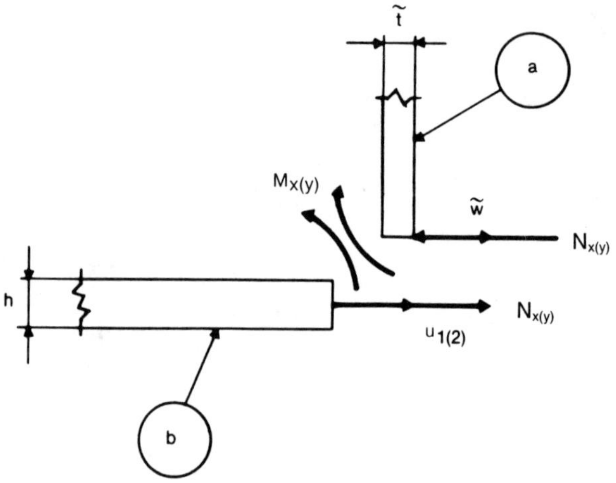

FIGURE 13. Connection of the mouthpiece to the vessel.

mouthpiece. Similarly, equations for bending using these conditions are:

a. Along the generator of the shell:

$$M_x = \frac{\tilde{E}\tilde{t}^3}{12(1 - \tilde{\nu}^2)} \times \left[-\frac{1}{R^2}\frac{\partial^2 u_1}{\partial \bar{\alpha}_1^2} + \tilde{\nu}\frac{1}{R^2}\left(\frac{\partial u_2}{\partial \bar{\alpha}_2} - \frac{\partial^2 u_1}{\partial \bar{\alpha}_2^2}\right)\right];$$

$$N_x = \frac{\tilde{E}\tilde{t}^3}{12(1 - \tilde{\nu}^2)R}\frac{\partial}{\partial \bar{\alpha}_1} \times \left[-\frac{1}{R^2}\frac{\partial^2 u_1}{\partial \bar{\alpha}_1^2} + \frac{1}{R^2}\frac{\partial u_2}{\partial \bar{\alpha}_2} - \frac{1}{R^2}\frac{\partial^2 u_1}{\partial \bar{\alpha}_2^2}\right] \quad (252)$$

b. Along the circumference of the shell:

$$M_y = \frac{\tilde{E}\tilde{t}^3}{12(1 - \tilde{\nu}^2)} \times \left[-\frac{1}{R^2}\frac{\partial^2 u_2}{\partial \bar{\alpha}_1^2} + \tilde{\nu}\frac{1}{R^2}\left(\frac{\partial u_1}{\partial \bar{\alpha}_2} - \frac{\partial^2 u_2}{\partial \bar{\alpha}_2^2}\right)\right];$$

$$N_y = \frac{\tilde{E}\tilde{t}^3}{12(1 - \tilde{\nu}^2)R}\frac{\partial}{\partial \bar{\alpha}_2} \times \left[-\frac{1}{R^2}\frac{\partial^2 u_2}{\partial \bar{\alpha}_1^2} + \frac{1}{R^2}\frac{\partial u_2}{\partial \bar{\alpha}_2} - \frac{1}{R^2}\frac{\partial^2 u_2}{\partial \bar{\alpha}_2^2}\right] \quad (253)$$

Equations (252) and (253) require the prototype and model to have the same Poisson's ratio. If we assume approximate similitude at the connection and also $C_{u_1} = C_{u_2}$, then from Equations (252) and (253) we obtain two similitude indicators:

$$c_t = \sqrt[3]{\frac{C_N C_{\tilde{\nu}} C_l^3}{C_{u_2} C_{\tilde{E}}}};$$

$$c_t = \sqrt[3]{\frac{C_M C_{\tilde{\nu}} C_l^2}{C_{u_2} C_{\tilde{E}}}} \quad (254)$$

Let the similitude scale for moments be determined from:

$$C_M \cong C_N\, C_l\, C_{\tilde{\nu}}$$

in which $C_N = 5.2 \times 10^{-4}$; $C_l = 0.082$; $C_{\tilde{E}} = C_{E_D} = 1.22 \times 10^{-5}$; $C_{\tilde{\nu}} = C_{\nu_D} = 0.88$; and $C_{u_2} = 6.5$.

Using these quantities, we obtain similitude scales for mouthpiece thickness $\tilde{t}$:

$$C'_{\tilde{t}} = 0.15 \text{ and } C''_{\tilde{t}} = 0.14$$

The scale of 0.15 was used in the actual model.

The model shown in Figure 11 was tested by the double tensometric net method [44]. The tensometer segments were 0.12 to 0.16 inches (3 to 4 mm) in length, i.e., $(0.6 \div 0.8)h_{mod}$ to investigate the stress concentration at the mouth. The measurements were performed both along the generator and the circumference of the shell. The model was loaded with air. To investigate the influence of the length of the mouthpiece, two tests were conducted using $l_1 = 2.59$ inches (65 mm) and $l_2 = 1.30$ inches (33 mm). Results are

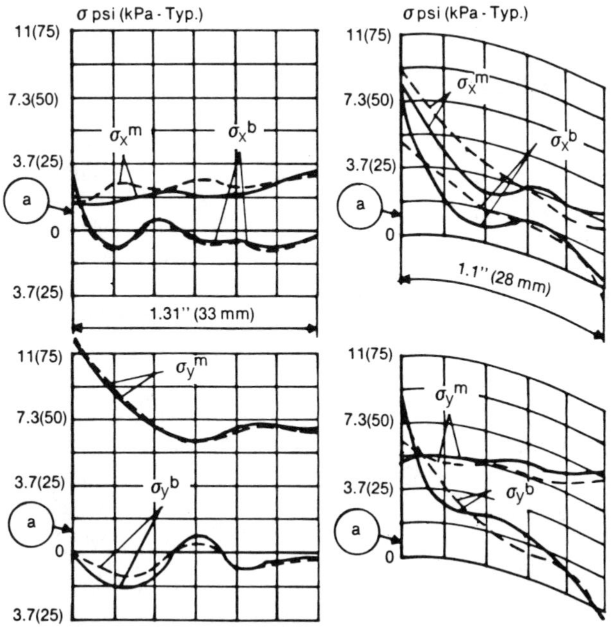

FIGURE 14. The model stresses at the mouthpiece.

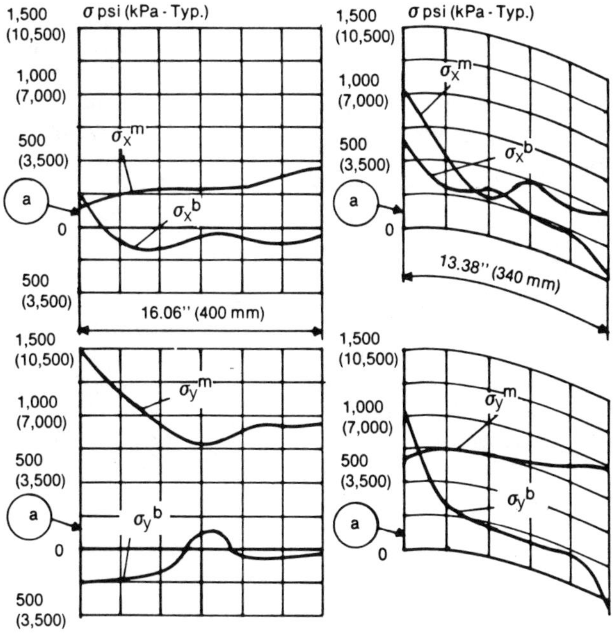

FIGURE 15. The prototype stresses at the mouthpiece.

shown for membrane stresses σ_x^m and σ_y^m and bending stresses σ_x^b and σ_y^b in Figure 14. The model stresses and curvature changes were converted using the similitude scales:

$$C_{\sigma_M} = (7.0 \times 10^{-3})^{-1} = 143; \; C_x = (970)^{-1} = 0.00103$$

The prototype stresses were obtained using the equations developed earlier and are shown in Figure 15.

The prototype was tested under pressure of 156 psig (1080 kPa). The measurements of outside fiber strains were made using gages near the mouthpiece. Comparison of stresses obtained both from the similitude scales and actual test is as follows:

Gage Position	Type of Stress	Model Tests*	Prototype Tests*	Discrepancy (%)
Shell generator	σ_x	0.412(2840)	0.270(1860)	52
	σ_y	1.312(9040)	1.180(8130)	11
Shell Circumference	σ_x	1.764(12200)	1.605(11100)	9.9
	σ_y	1.738(12000)	1.531(10600)	13.5

*All values are in ksi (kPa).

The above comparison shows that partial similitude gives fairly good results using the model. Based on these model studies, the following recommendations were made:

1. The stress concentration factor was less than the existing recommendation, although the dimensions of the mouthpiece were comparable to that of the vessel. The values along the generator and circumference were 1.66 and 3.3. The bending stresses, therefore, can be neglected along the generator near the mouth, but not in the circumferential direction.
2. The stress concentration zone near the mouth extends to about 0.6 to 0.8 of its diameter. Beyond this range the stresses are as in the membrane with values of $\sigma_x^m = qR/2h = 0.535$ ksi (3700 kPa) and $\sigma_y^m = qR/h = 1.069$ ksi (7370 kPa) along the generator and circumference, respectively. The values from tests were 0.492 ksi (3400 kPa) and 0.985 ksi (6800 kPa), respectively.
3. Large stress concentration factor near the mouth caused the mouthpiece to be moved from the side surface of the vessel.

ACKNOWLEDGEMENTS

Special thanks to Mr. Don P. Armstrong, Mrs. Sharon Jordan, and Mrs. Irene Cooperman, who contributed essentially to this work.

NOTATION

The following symbols are used in this chapter.

T_1, T_2 = general components
C_T = similitude constant or scale factor
K_i, r_i = similitude products
K_i', r'_i = indicators of similitude
E, ν = modulus of elasticity, Poisson's ratio
Q_x, Q_y = transverse shear force
w = deflection
I_z = moment of inertia, in.^{4}
x,y = coordinates
$q(x)$ = surface load per unit area
R = reaction
$\sigma_x; \sigma_y; \sigma_{xy}$ = dimensional fiber stresses
ξ, η = nondimensional coordinates
$I(\xi)$ = nondimensional moment of inertia
q_0 = characteristic load intenstity per unit area
l = characteristic length
h_0 = characteristic thickness
I_0 = characteristic moment of inertia
$\bar{R}$ = nondimensional reaction
$\bar{\sigma}$ = nondimensional fiber stress
$\bar{w}$ = nondimensional deflection
p = nondimensional load intensity per unit area
a,b = plane dimensions for rectangular plates or shallow shells on a rectangular plane
h = plate-shell thickness or beam depth
$\bar{h}$ = nondimensional plate-shell thickness
m = mass density (kip $\times$ sec^2/in^4)
t = time or tangential direction
$\underset{\sim}{K}$ = $\{K_1, K_2, \ldots\}$ = vector coefficient for frequency calculations
f = frequency (sec^{-1})
θ = nondimensional time
$\bar{f}$ = nondimensional frequency
D = $Eh^3/12(1 - \nu^2)$ = flexural rigidity of plate-shell
N_x, N_y, N_{xy} = intensity of tangent statical load per unit length (for plates with small deflections only) or membrane forces in plates with large deflections and shells
λ = temperature gradient, degrees
g = gravitational acceleration
k = damping coefficient (kip $\times$ sec/in^4)
$Z(x,y,t)$ = perturbation dynamic load (per unit area)
Z_0 = characteristic magnitude of the above load
Ω = $\bar{h}^3/12(1 - \nu^2)$ = nondimensional flexural rigidity of plate or shell
α = coefficient of thermal expansion
λ_0 = characteristic temperature gradient
μ = nondimensional temperature gradient
$\bar{k}$ = nondimensional damping coefficient

$\bar{N}_x, \bar{N}_y, \bar{N}_{xy}$ = nondimensional intensity of tangent statical load per unit length (for plates with small deflections only) or membrane forces in plates with large deflections and shells;
w_0 = initial deflections
s_0 = initial velocity of vibration
ω = frequency of a perturbation harmonic load
t^0_{top} = function of temperature distribution at the top plate or shell surface
t^0_{bot} = function of temperature distribution at the bottom plate or shell surface
M_x, M_y, M_{xy} = bending and twist moments
Q_x, Q_y = nondimensional transverse forces
$\epsilon_x, \epsilon_y, \epsilon_{xy}$ = fiber strains
$\chi_x, \chi_y, \chi_{xy}$ = bending and twist curvatures
γ = specific weight
B,C = flexural and torsional rigidity of the beam
β = angle between x axis and the normal n to the plate edge
a = outer radius of circular (or annular plate), or shallow shell of revolution projection
b = inner radius of annular plate or shallow shell of revolution projection
ν = tangential displacement
r = radial location of quantity being evaluated or radial direction
T = temperature at plate (shell) middle surface
T_0 = characteristic temperature at plate (shell) middle surface
ϕ = dimensional membrane stress function
G = shear modulus
ψ = nondimensional membrane stress function
V = unit load (force per unit length)
d = elevation of shallow shells
K_x, K_y, K_{xy} = shallow shell middle surface curvatures
α_1, α_2 = coordinates for shells (other than shallow)

REFERENCES

1. Bridgman, P. W., *Dimensional Analysis,* Yale University Press, New Haven, Conn. (1922).
2. Ipsen, D. C., *Units, Dimensions, and Dimensionless Numbers,* McGraw-Hill Book Company, New York (1960).
3. Sabnis, G. M., H. G. Harris, R. N. White, and M. S. Mirza, *Structural Modeling and Experimental Techniques,* Prentice-Hall Civil Engineering and Engineering Mechanics Series (1983).
4. Buckingham, E., "On Physically Similar Systems," *Physics Review,* London, Vol. 4, No. 345 (1914).
5. Langhaar, H. L., *Dimensional Analysis and Theory of Models,* John Wiley & Sons, Inc., New York (1951).
6. Kirpichev, M. V., *Similitude Theory,* IZD-VO AN SSSR (Academy of Science of the USSR), Moscow (1953).
7. Gukhman, A. A., *Introduction to Similitude Theory,* Vysshaya Shkola, Moscow (in Russian) (1963).
8. Sedov, L. I., *Methods of Similitude and Dimension in Mechanics,* Nauka, Moscow (in Russian) (1972).
9. Krayterman, B. and G. M. Sabnis, "Similitude Theory: Plates and Shells Analysis," *Journal of Engineering Mechanics,* Vol. 110, No. 9 (September 1984).
10. Monakhenko, D. V. and V. B. Proskuryakov, "Modeling of the Stress State of Thin Shallow Shells," *Izv. AN SSSR OTN,* No. 6, (in Russian) (1960).
11. Krayterman, B. L. and A. M. Trukhlov, "Application of Similitude Theory for Analysis of a Circular Cylindrical Shell," *Izv. VUZOV, MV SSO SSSR Stroitelstvo i Arkhitektura,* No. 5, (in Russian) (1965).
12. Krayterman, B. L., "Determination of Critical Loads by Modeling for Buckling of Cylindrical Shells," *Stroitelnaya Mekhanika i Raschot Sooruzhenii,* N2, Moscow, (in Russian) (1967).
13. Krayterman, B. L., "Investigation of the Disk Valve Strain State for Main Gas Pipelines with Modeling," *Experimental Investigations of Engineering Structures,* No. 17, Kuibishev, SSSR (in Russian) (1969).
14. Krayterman, B. L., "Application of the π Similitude Theorem to the Investigation of Shallow Shells with Large Deflections," translated from *Prikladnaya Mekhanika,* Vol. 5, No. 8, pp. 114–116 (August 1969).
15. Krayterman, B. L., "Application of Similitude Theory for the Optimum Nondimensional Transformation of Shallow Shell Equations," *Izv. VUZOV, MV SSO SSSR Stroitelstvo i Arkhitektura,* No. 7 (in Russian) (1973).
16. Krayterman, B. L. and G. M. Sabnis, "Application of the Similitude Theory for Numerical Solution of Plates and Shells with Large Deflections," Engineering Mechanics in Civil Engineering, Proceedings of the Fifth Engineering Mechanics Division Specialty Conference, University of Wyoming (August 1–3, 1984).
17. Timoshenko, S., *Strength of Materials,* 3rd ed., Van Nostrand Company, Inc., Princeton, N. J. (1956).
18. Roark, R. J. and W. C. Young, *Formulas for Stress and Strain,* 5th ed., McGraw-Hill Book Company, New York (1975).
19. Leissa, A. W., *Vibration of Plates,* NASA SP-160, National Aeronautics and Space Administration (1969).
20. Timoshenko, S., and S. Woinowsky-Krieger, *Theory of Plates and Shells,* 2nd ed., McGraw-Hill Book Company, New York (1959).
21. Volmir, A., *Stability of Elastic Systems,* Fizmatgiz, Moscow (in Russian) (1963).
22. Belyaev, N. M., *Strength of Materials,* Nauka, Moscow (in Russian) (1976).
23. Novozhilov, V. V., *The Theory of Thin Shells,* Groningen, The Netherlands, Erwin P. Noorhaff, Ltd. (1959).
24. Bares, R., *Tables for the Analysis of Plates, Slabs, and Diaphragms Based on the Elastic Theory,* Bauverlag GmbH. (1969).

25. Odley, E. G., "Deflections and Moments of a Rectangular Plate Clamped on all Edges and Under Hydrostatic Pressure," *Journal of Applied Mechanics,* Vol. 14, No. 4 (December 1947).
26. Kennedy, J. and S. Ng, "Linear and Nonlinear Analyses of Skewed Plates," *Journal of Applied Mechanics,* Vol. 34, No. 2 (June 1967).
27. Bassali, W. A. and M. Nassif, "Stresses and Deflections in an Elastically Restrained Circular Plate Under Uniform Normal Loading Over a Segment," *Journal of Applied Mechanics,* Vol. 26, No. 1 (March 1959).
28. Jurney, W. H., "Displacements and Stresses of a Laterally Loaded Semicircular Plate with Clamped Edges," *Journal of Applied Mechanics,* Vol. 26, No. 2 (June 1959).
29. Boussinesq, J., *Journal of Mathematics,* Ser. 3, Vol. 5 (1879).
30. Lekhnitskii, S. G., *Theory of Elasticity of an Anisotropic Elastic Body,* Holden Day, San Francisco (1963).
31. Von Kármán, *Encyklopädie der Mathematischen Wissenschaften,* Vol. IV, p. 349 (1910).
32. Krayterman, B. L., "On Modeling the Stress State of Flexible Plates with Different Poisson Rates," translated from *Prikladnaya Mekhanika,* Vol. 10, pp. 122–125 (June 1974).
33. Krayterman, B. L. and C. C. Fu, "Nonlinear Analysis of Clamped Circular Plates," *ASCE Journal of the Structural Division*, Vol. 111, No. 11 (November 1985).
34. Lubahn, J. D. and R. P. Felgar, *Plasticity and Creep of Metals,* John Wiley & Sons, Inc. (1961).
35. Vlasov, V. Z., *General Theory of Shells and Its Applications in Engineering,* National Aeronautics and Space Administration, Washington, D.C., NASA TTF-99 (April 1964).
36. Harris, H. G. and R. N. White, "The Inelastic Analysis of Concrete Cylindrical Shell and its Verification Using Small Scale Models," Structural Research Report, Cornell University (September 1967).
37. Aron, H., "Das Gleichgewicht und Die Bewegung Einer Unendlich Dünnen Beliebig Gekrümmten Elastischen Schale," *Journal für Reine und Angew. Math.*, Bd. 78, p. 196 (1874).
38. Donnell, L., *Stability of Thin Walled Tubes Under Torsion,* National Aeronautics and Space Administration, Washington, D.C., NASA Report No. 479 (1933).
39. Marguerre, K., "Zur Theorie der Gekrümmten Platte Grosser Formänderung," Proceedings of the Fifth International Congress for Applied Mechanics, pp. 93–101 (1938).
40. V. G. Rekach, *Static Theory of Thin-Walled Space Structures,* Mir Publishers, Moscow (1978).
41. Zienkiewicz, O. C., *The Finite Element Method,* 3rd ed., McGraw-Hill Book Co., New York (1972).
42. Krayterman, B. L., and G. M. Sabnis, "Large Deflected Plates and Shells with Loading History," *Journal of Engineering Mechanics,* Vol. III, No. 5, pp. 653-663 (May 1985).
43. Sabnis, G. M., and N. FitzSimons, (eds.), "Experimental Methods in Concrete Structures for Practitioners," Proceedings of Symposium held in ACI Fall Meeting, Washington, D.C. (October 1978).
44. Krayterman, B. L., "Design of Thin Shells by Tensometric Net Method with Application of the Similitude Theory," *Stroitelnaya Mekhanika i Raschot Sooruzhenii,* N5, pp. 7–12 (in Russian) (1963).

Multiple Mode Nonlinear Analysis of Plates

M. SATHYAMOORTHY*

The material covered in this chapter is limited to elastic behavior of plates undergoing moderately large deflection nonlinear vibration. A nonlinear plate theory in the sense of von Kármán is presented for moderately thick plates. The plate equations of motion are reduced to a system of two nonlinear equations for the lateral displacement and stress function. The effects of transverse shear deformation and rotatory inertia are included in the plate equations of motion. Nonlinear equations in terms of the three displacement components of the plate are also presented. Several illustrative examples are given, using multiple mode nonlinear analysis procedure, to demonstrate the effects of transverse shear deformation and rotatory inertia on the nonlinear flexural vibrations of plates of various geometries. For the sake of generality, the dynamic plate equations are derived for a single layered moderately thick anisotropic plate using oblique coordinates. For a background study on classical linear plate theory and geometrically nonlinear theory of plates the reader is referred to References [1–8].

EQUATIONS OF MOTION

Consider a skew plate of constant thickness, h, composed of homogeneous, anisotropic and elastic material whose coordinate system is shown in Figure 1. The origin of the coordinate system is located at the center of the midplane of the undeformed plate. The stress-strain relations for the plate may be written as [9]:

$$\begin{Bmatrix} \sigma_\xi \\ \sigma_\eta \\ \sigma_z \\ \sigma_{\eta z} \\ \sigma_{z\xi} \\ \sigma_{\xi\eta} \end{Bmatrix} = \begin{bmatrix} a_{11} & a_{12} & a_{13} & 0 & 0 & a_{16} \\ a_{12} & a_{22} & a_{23} & 0 & 0 & a_{26} \\ a_{13} & a_{23} & a_{33} & 0 & 0 & a_{36} \\ 0 & 0 & 0 & a_{44} & a_{45} & 0 \\ 0 & 0 & 0 & a_{45} & a_{55} & 0 \\ a_{16} & a_{26} & a_{36} & 0 & 0 & a_{66} \end{bmatrix} \begin{Bmatrix} \epsilon_\xi \\ \epsilon_\eta \\ \epsilon_z \\ \epsilon_{\eta z} \\ \epsilon_{z\xi} \\ \epsilon_{\xi\eta} \end{Bmatrix} \quad (1)$$

or written in an alternate form as

$$\{\epsilon_{ij}\} = [b_{ij}]\{\sigma_i\} \quad (2)$$

where the coefficients a_{ij} and b_{ij} are such that $a_{ij} = a_{ji}$ and $b_{ij} = b_{ji}$. Since only free vibration is considered here, the normal stress, σ_z, is assumed to be zero. Also, the normal strain is assumed to be zero because of the assumption that w is independent of z. The elastic stiffnesses in Equation (1) are defined as

$$\begin{aligned} a_{11} &= c_{11}/c^3,\ a_{12} = c_{11}s^2/c^3 + c_{12}/c - 2sc_{16}/c^2 \\ a_{13} &= 0,\ a_{16} = c_{16}/c^2 - c_{11}s/c^3 \\ a_{22} &= cc_{22} + c_{12}s^2/c - 2sc_{26} - s^2a_{12} - 2sa_{26},\ a_{23} = 0 \\ a_{26} &= c_{26} - sc_{12}/c - s^2a_{16} - 2sa_{66},\ a_{33} = a_{36} = 0 \\ a_{44} &= cc_{66} - sc_{56}/c - sa_{45},\ a_{45} = cc_{56} - sc_{55}/c \\ a_{55} &= c_{55}/c,\ a_{66} = c_{66}/c - c_{16}s/c^2 \\ &\quad - sa_{16},\ c = \cos\theta,\ s = \sin\theta \end{aligned} \quad (3)$$

The coefficients b_{ij} in Equation (2) are defined in Reference [9]. c_{ij} in Equation (3) are the material constants of the anisotropic skew plate with reference to the x, y coordinate system and θ is the skew angle. These constants may be expressed in terms of the orthotropic properties along arbitrary principal directions (L, T) by a coordinate transforma-

*Department of Mechanical and Industrial Engineering, Clarkson University, Potsdam, NY tion as

$$\begin{Bmatrix} c_{11} \\ c_{12} \\ c_{22} \\ c_{16} \\ c_{26} \\ c_{55} \\ c_{56} \\ c_{66} \end{Bmatrix} = \begin{bmatrix} m^4 & 2m^2n^2 & n^4 & 0 & 0 & 4m^2n^2 \\ m^2n^2 & m^4 + n^4 & m^2n^2 & 0 & 0 & -4m^2n^2 \\ n^4 & 2m^2n^2 & m^4 & 0 & 0 & 4m^2n^2 \\ nm^3 & mn(n^2 - m^2) & -mn^3 & 0 & 0 & 2mn(n^2 - m^2) \\ mn^3 & mn(m^2 - n^2) & -nm^3 & 0 & 0 & 2mn(m^2 - n^2) \\ 0 & 0 & 0 & m^2 & n^2 & 0 \\ 0 & 0 & 0 & -mn & mn & 0 \\ m^2n^2 & -2m^2n^2 & m^2n^2 & 0 & 0 & (m^2 - n^2) \end{bmatrix} \begin{Bmatrix} E_L/\mu \\ \nu_{LT}/\mu \\ E_T/\mu \\ G_{LZ} \\ G_{TZ} \\ G_{LT} \end{Bmatrix} \tag{4}$$

in which E_L and E_T are the major and minor Young's moduli, ν_{LT} and ν_{TL} are the Poisson's ratios and G_{LT}, G_{LZ}, G_{TZ} are the shear moduli and in which ϕ is the orientation angle of filaments such that

$$\begin{aligned} m &= \cos\phi \\ n &= \sin\phi \\ \mu &= 1 - \nu_{LT}\nu_{TL} \\ \nu_{LT}E_T &= \nu_{TL}E_L \end{aligned} \tag{5}$$

In order to take into account the effects of the transverse shear deformation and rotatory inertia the displacement components in the oblique coordinates (ζ,η) at a distance z away from the midsurface may be taken in the following form:

$$\begin{aligned} u(\zeta,\eta,z,t) &= cu^0(\zeta,\eta,t) + sv^0(\zeta,\eta,t) + z\alpha(\zeta,\eta,t) \\ v(\zeta,\eta,z,t) &= v^0(\zeta,\eta,t) + z\beta(\zeta,\eta,t) \\ w(\zeta,\eta,z,t) &= w^0(\zeta,\eta,t) = w(\zeta,\eta,t) \end{aligned} \tag{6}$$

The midsurface strains in terms of the displacement components u^0, v^0 and w may be written as

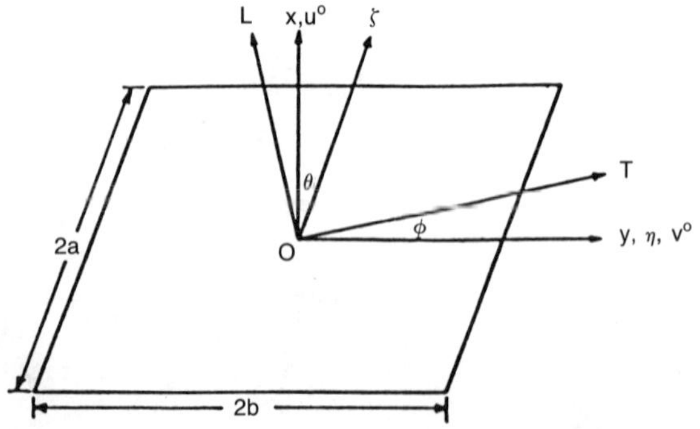

FIGURE 1. Geometry and coordinate system of skew plate.

$$\begin{aligned} \epsilon_\zeta^0 &= cu_{,\zeta}^0 + sv_{,\zeta}^0 + (w_{,\zeta})^{2/2} \\ \epsilon_\eta^0 &= v_{,\eta}^0 + (w_{,\eta})^{2/2} \\ \epsilon_{\zeta\eta}^0 &= cu_{,\eta}^0 + sv_{,\eta}^0 + v_{,\zeta}^0 + w_{,\zeta}w_{,\eta} \end{aligned} \tag{7}$$

The strains at a distance z measured from the midsurface are assumed to be

$$\begin{aligned} \epsilon_\zeta &= \epsilon_\zeta^0 + z\alpha_{,\zeta} \\ \epsilon_\eta &= \epsilon_\eta^0 + z\beta_{,\eta} \\ \epsilon_\zeta &= 0,\ \epsilon_{\zeta\eta} = \epsilon_{\zeta\eta}^0 + z(\alpha_{,\eta} + \beta_{,\zeta}) \\ \epsilon_{\zeta z} &= \alpha + w_{,\zeta},\ \epsilon_{\eta z} = \beta + w_{,\eta} \end{aligned} \tag{8}$$

The stress resultants N_{ij} and moment resultants M_{ij} by definition are

$$N_{ij} = \int_{-h/2}^{h/2} \sigma_{ij}dz, \quad M_{ij} = \int_{-h/2}^{h/2} \sigma_{ij}zdz \tag{9}$$

Equations (1) and (8) are substituted into Equation (9) and integrations are performed to obtain

$$\begin{Bmatrix} N_\zeta \\ N_\eta \\ N_{\zeta\eta} \end{Bmatrix} = h \begin{bmatrix} a_{11} & a_{12} & a_{16} \\ a_{12} & a_{22} & a_{26} \\ a_{16} & a_{26} & a_{66} \end{bmatrix} \begin{Bmatrix} \epsilon_\zeta^0 \\ \epsilon_\eta^0 \\ \epsilon_{\zeta\eta}^0 \end{Bmatrix} \tag{10}$$

$$\begin{Bmatrix} M_\zeta \\ M_\eta \\ M_{\zeta\eta} \end{Bmatrix} = \frac{h^3}{12} \begin{bmatrix} a_{11} & a_{12} & a_{16} \\ a_{12} & a_{22} & a_{26} \\ a_{16} & a_{26} & a_{66} \end{bmatrix} \begin{Bmatrix} \alpha_{,\zeta} \\ \beta_{,\eta} \\ \alpha_{,\eta} + \beta_{,\zeta} \end{Bmatrix} \tag{11}$$

Using the expressions for the strain energies due to stretching and bending of the plate and the kinetic energy of the plate, the equations of motion are derived from Hamilton's principle. The corresponding Euler's equations constitute a system of thirteen equations in terms of membrane forces N_{ij}, moment resultants M_{ij}, transverse shear forces Q, displacement components u^0, v^0, w and slope functions α and

β. In these equations the so-called tracing constants T_s and R_i are introduced to identify the terms which characterize the effects of the transverse shear deformation and rotatory inertia, respectively. Assuming that the in-plane inertia effects are negligibly small, and introducing a stress function F such that

$$N_\zeta = hF_{,\eta\eta}, \quad N_\eta = hF_{,\zeta\zeta}, \quad N_{\zeta\eta} = -hF_{,\zeta\eta} \tag{12}$$

the equation of motion in the lateral direction of the vibrating plate is obtained by solving the system of thirteen equations and eliminating α and β as

$$L(J_1 + J_2) + M(w) = 0 \tag{13}$$

where the differential operators L and M are defined as

$$\begin{aligned} L = {} & r_1 \frac{\partial^4}{\partial\zeta^4} + r_2 \frac{\partial^4}{\partial\zeta^2\partial\eta^2} + r_3 \frac{\partial^4}{\partial\eta^4} \\ & + r_4 \frac{\partial^4}{\partial\zeta^3\partial\eta} + r_5 \frac{\partial^4}{\partial\zeta\partial\eta^3} + r_6 \frac{\partial^4}{\partial t^4} \\ & + r_7 \frac{\partial^4}{\partial\zeta^2\partial t^2} + r_8 \frac{\partial^4}{\partial\zeta\partial\eta\partial t^2} + r_9 \frac{\partial^4}{\partial\eta^2\partial t^2} \\ & + r_{10} \frac{\partial^2}{\partial\zeta^2} + r_{11} \frac{\partial^2}{\partial\eta^2} + r_{12} \frac{\partial^2}{\partial\zeta\partial\eta} - r_{13} \frac{\partial^2}{\partial t^2} - 1 \end{aligned} \tag{14}$$

$$\begin{aligned} M = {} & N + e_9 R_i \left(\frac{\partial^4}{\partial\zeta^2\partial t^2} + \frac{\partial^4}{\partial\eta^2\partial t^2} \right) + e_{13} \frac{\partial^6}{\partial\zeta^2\partial\eta^2\partial t^2} \\ & + e_{14} \frac{\partial^6}{\partial\zeta^4\partial t^2} + e_{15} \frac{\partial^6}{\partial\zeta^3\partial\eta\partial t^2} + e_9 R_i d_{15} \frac{\partial^6}{\partial\zeta^2\partial t^4} \\ & + e_{16} \frac{\partial^6}{\partial\zeta\partial\eta\partial t^4} + e_{17} \frac{\partial^6}{\partial\zeta\partial\eta^3\partial t^2} + e_{18} \frac{\partial^6}{\partial\eta^4\partial t^2} \\ & + e_9 R_i d_7 \frac{\partial^6}{\partial\eta^2\partial t^4} \end{aligned} \tag{15}$$

with N given by

$$\begin{aligned} N = {} & e_1 \frac{\partial^4}{\partial\zeta^4} + e_{10} \frac{\partial^4}{\partial\zeta^2\partial\eta^2} + e_{11} \frac{\partial^4}{\partial\zeta^3\partial\eta} + e_8 \frac{\partial^4}{\partial\eta^4} \\ & + e_{12} \frac{\partial^4}{\partial\zeta\partial\eta^3} + e_{19} \frac{\partial^6}{\partial\zeta^6} + e_{20} \frac{\partial^6}{\partial\zeta^4\partial\eta^2} \\ & + e_{21} \frac{\partial^6}{\partial\zeta^5\partial\eta} + e_{22} \frac{\partial^6}{\partial\zeta^3\partial\eta^3} + e_{23} \frac{\partial^6}{\partial\zeta^2\partial\eta^4} \\ & + e_{24} \frac{\partial^6}{\partial\zeta\partial\eta^5} + e_{25} \frac{\partial^6}{\partial\eta^6} \end{aligned} \tag{16}$$

In Equation (13) J_1 and J_2 are given by

$$J_1 = q(\zeta,\eta) - \varrho h w_{,tt} \tag{17}$$

$$J_2 = h(F_{,\eta\eta}w_{,\zeta\zeta} + F_{,\zeta\zeta}w_{,\eta\eta} - 2F_{,\zeta\eta}w_{,\zeta\eta}) \tag{18}$$

The compatibility condition in terms of F is obtained by virtue of Equation (12), as

$$\begin{aligned} & b_{22}F_{,\zeta\zeta\zeta\zeta} - 2b_{26}F_{,\zeta\zeta\zeta\eta} + (2b_{12} + b_{66})F_{,\zeta\zeta\eta\eta} \\ & \quad - 2b_{16}F_{,\zeta\eta\eta\eta} + b_{11}F_{,\eta\eta\eta\eta} \\ & \quad = (w_{,\zeta\eta})^2 - w_{,\zeta\zeta}w_{,\eta\eta} \end{aligned} \tag{19}$$

In Equations (17) and (18) q is the lateral applied load per unit area of the plate, h is the thickness and ϱ is the mass density of the material of the plate. Equations (13) and (19) constitute a system of two coupled nonlinear governing differential equations applicable for the large amplitude flexural vibrations of anisotropic skew plates. These equations are valid for moderately thick plates where the effects of transverse shear deformation and rotatory inertia are important.

The system of nonlinear differential Equations (13) and (19) is of the tenth order. Therefore, five boundary conditions, as against four in the classical von Kármán plate theory are required in the present theory. The five boundary conditions along each edge consist of three corresponding to the out-of-plane conditions of the linear theory (with the thickness-shear flexibility taken into account) and two corresponding to the in-plane conditions of the nonlinear theory. These conditions which are obtained from the variational technique may be taken as a combination of the out-of-plane and in-plane conditions.

In the case of orthotropic plates, $\phi = 0$ and hence $m = 1$, $n = 0$. The corresponding a_{ij} are obtained from Equations (3)–(5). For isotropic plates $m = 1$, $n = 0$, and $E_L = E_T = E$, $\nu_{LT} = \nu_{TL} = \nu$, $G_{LT} = G_{LZ} = G_{TZ} = G = E/2(1 + \nu)$, $\mu = 1 - \nu^2$.

When the effects of the transverse shear deformation and rotatory inertia are both neglected in the analysis, the corresponding tracing constants T_s and R_i are equated to zero. With appropriately simplified coefficients for a_{ij}, the governing Equations (13) and (19) will reduce to those applicable for isotropic and orthotropic thin plates.

THIN RECTANGULAR AND SQUARE PLATES

For a rectangular orthotropic plate of length a in the x direction, width b in the y direction and uniform thickness h with material axes of symmetry assumed to be parallel to the edges of the plate, the differential equations governing

TABLE 1. Numerical Values of Elastic Constants.

	E_L/E_T	G_{LT}/E_T	ν_{LT}
Glass-epoxy	3.0	0.50	0.25
Boron-epoxy	10.0	1/3	0.22
Graphite-epoxy	40.0	0.50	0.25

the nonlinear free flexural vibration of the plate may be readily derived from Equations (13) and (19) as

$$c_1F_{,\zeta\zeta\zeta\zeta} + 2r^2c_2F_{,\zeta\zeta\eta\eta} + r^4F_{,\eta\eta\eta\eta} = c_1r^2\{(W_{,\zeta\eta})^2 - W_{,\zeta\zeta}W_{,\eta\eta}\} \tag{20}$$

$$c_1W_{,\zeta\zeta\zeta\zeta} + 2r^2c_3W_{,\zeta\zeta\eta\eta} + r^4W_{,\eta\eta\eta\eta} + \mu r^4W_{,\tau\tau} = \mu r^2(W_{,\zeta\zeta}F_{,\eta\eta} + W_{,\eta\eta}F_{,\zeta\zeta} - 2W_{,\zeta\eta}F_{,\zeta\eta}), \tag{21}$$

where

$$\begin{aligned} \zeta &= x/a \\ \eta &= y/b \\ r &= a/b \\ W &= w/h \\ F &= \Phi/E_Th^3 \\ \tau &= t\sqrt{E_Th^2/\varrho b^4} \\ \mu &= 12(1 - \nu_{LT}\nu_{TL}) \\ c_1 &= E_L/E_T \\ c_2 &= (E_T/G_{LT} - 2\nu_{LT})/2 \\ c_3 &= \nu_{LT} + \mu G_{LT}/6E_T \end{aligned} \tag{22}$$

The stress resultants are defined in terms of stress function as

$$(N_x,N_y,N_{xy}) = \{E_Th^3/b^2\}(F_{,\eta\eta}, F_{,\zeta\zeta}/r^2, - F_{,\zeta\eta}/r) \tag{23}$$

If the edges of the plate are assumed to be free from applied membrane forces, the boundary conditions for a clamped plate are

$$\begin{aligned} W = W_{,\zeta} = F_{,\eta\eta} = F_{,\zeta\eta} = 0 \quad &\text{at} \quad \zeta = 0,1 \\ W = W_{,\eta} = F_{,\zeta\zeta} = F_{,\zeta\eta} = 0 \quad &\text{at} \quad \eta = 0,1 \end{aligned} \tag{24}$$

In the case of a simply supported plate the corresponding boundary conditions are

$$\begin{aligned} W = W_{,\zeta\zeta} = F_{,\eta\eta} = F_{,\zeta\eta} = 0 \quad &\text{at} \quad \zeta = 0,1 \\ W = W_{,\eta\eta} = F_{,\zeta\zeta} = F_{,\zeta\eta} = 0 \quad &\text{at} \quad \eta = 0,1 \end{aligned} \tag{25}$$

Solutions to Equations (20) and (21) are taken in the following form of a double series [10]

$$F = \sum_m \sum_n F_{mn}(\tau)X_m(\zeta)Y_n(\eta) \tag{26}$$

$$W = \sum_p \sum_q W_{pq}(\tau)\phi_p(\zeta)\psi_q(\eta) \tag{27}$$

in which X_m, Y_n, ϕ_p and ψ_q are

$$\begin{aligned} X_m(\zeta) &= \cosh\alpha_m\zeta - \cos\alpha_m\zeta - \gamma_m(\sinh\alpha_m\zeta - \sin\alpha_m\zeta) \\ Y_n(\eta) &= \cosh\alpha_n\eta - \cos\alpha_n\eta - \gamma_n(\sinh\alpha_n\eta - \sin\alpha_n\eta) \\ \phi_p(\zeta) &= A_p(\cosh\beta_p\zeta - \cos\beta_p\zeta) + B_p(\sinh\beta_p\zeta + \sin\beta_p\zeta) \\ \psi_q(\eta) &= A_q(\cosh\beta_a\eta - \cos\beta_a\eta) + B_q(\sinh\beta_a\eta + \sin\beta_a\eta) \end{aligned} \tag{28}$$

In Equation (28), the coefficients α, β, γ, A and B depend upon the boundary conditions [10].

Substituting Equations (26) and (27) into Equations (20) and (21), multiplying the first of the resulting Equations by $X_i(\zeta)Y_j(\eta)$ and the second by $\phi_i(\zeta)\psi_j(\eta)$, integrating and using the orthogonal properties of the beam functions it can be shown that [10]

$$F_{ij}(\tau) = H_{ij}^{pqrs}W_{pq}(\tau)W_{rs}(\tau) \tag{29}$$

in which

TABLE 2. Frequency Ratio of Fundamental Vibration of Square Graphite-Epoxy Plate.

Simply Supported Plate			Clamped Plate		
	w_o/h			w_o/h	
$\omega_{11}/\omega_{11}^{(o)}$	Single-Mode	Four-Mode	$\omega_{11}/\omega_{11}^{(o)}$	Single-Mode	Four-Mode
1.010	0.9808	0.9907	1.011	1.4446	1.4832
1.022	1.4334	1.4728	1.020	1.8535	1.9418
1.037	1.8773	1.9462	1.030	2.2637	2.3891
1.055	2.3012	2.4136	1.035	2.4271	2.6103
1.076	2.7037	2.8776	1.041	2.6039	2.8305

$$[H_{ij}^{pqrs}] = [A_{mn}^{ij}]^{-1}[B_{mn}^{pqrs}] \tag{30}$$

$$A_{ij}^{mn} = \delta_i^m\delta_j^n(c_1\alpha_m^4 + r^4\alpha_n^4) + 2r^2c_2\alpha_m^2\alpha_n^2K_1^{im}L_1^{jn}$$
$$B_{ij}^{pqrs} = r^2c_1\{\beta_p\beta_q\beta_r\beta_sK_2^{ipr}L_2^{jqs} - \beta_p^2\beta_s^2K_3^{ipr}L_3^{jsq}\} \tag{31}$$

Substituting Equation (29) into the integrated result of Equation (21) leads to the following system of nonlinear differential equations in the time-dependent deflection coefficients:

$$\mu r^4d^2W_{ij}/d\tau^2 + C_{ij}^{pq}W_{pq} + M_{ij}^{pqrskl}W_{pq}W_{rs}W_{kl} = 0 \tag{32}$$

where

$$[M_{ij}^{pqrskl}] = [G_{ij}^{mnpq}][H_{mn}^{rskl}] \tag{33}$$

$$G_{ij}^{pqrs} = -\mu r^2(\alpha_n^2\beta_r^2K_4^{imr}L_5^{jns} + \alpha_m^2\beta_s^2K_5^{imr}L_4^{jns} - 2\alpha_m\alpha_n\beta_r\beta_sK_6^{imr}L_6^{jns}) \tag{34}$$

Using the method of harmonic balance, an approximate solution to the system of Equations (32) is presented by expanding the deflection coefficients into Fourier Cosine Series in τ as

$$W_{ij}(\tau) = \bar{W}_{ij}^{(1)}\cos\omega\tau + \bar{W}_{ij}^{(3)}\cos 3\omega\tau + \ldots, \tag{35}$$

Substituting Equation (35) in Equation (32) and equating the coefficients of each term of cos $n\omega t$ ($n = 1,3, \ldots$) to zero, a system of simultaneous nonlinear algebraic equations are obtained in terms of $\bar{W}_{ij}^{(n)}$. Solutions to this system of equations will determine the deflection coefficients for a given set of nonlinear frequency and plate parameters.

Numerical results are presented for glass-epoxy (GE), boron-epoxy (BE) and graphite-epoxy (GrE) square and rectangular plates in terms of variations of nonlinear frequency with the nondimensional amplitude of vibration of the plate. The material constants for orthotropic plates are given in Table 1. The ratios of the nonlinear frequency ω_{mn} to the corresponding linear frequency $\omega_{mn}^{(0)}$ are presented in Table 2 for simply supported (*SS*) and clamped (*CC*) square graphite-epoxy plates at various amplitudes of vibration. Similar results are also graphically presented in Figures 2 and 3. Two terms in the time series (Equation 35) and first four symmetric deflections corresponding to $m,n = 1,3$ have been used in getting the numerical results in Table 2 and Figure 2. It can be seen that the effect of coupling of vibrating modes is significant for orthotropic plates and the frequency–amplitude relationship clearly indicates hardening type of nonlinearity.

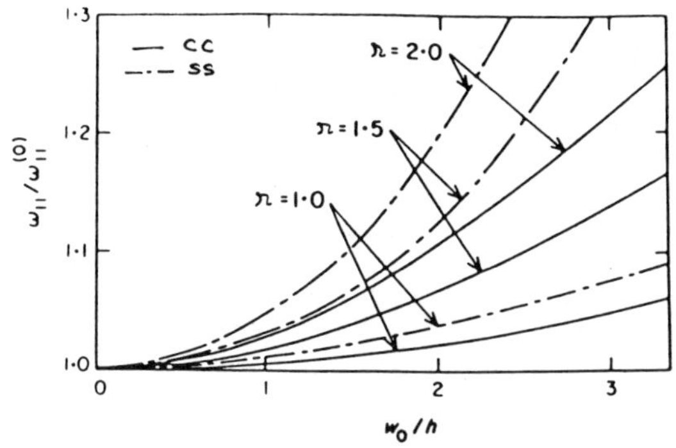

FIGURE 2. Effect of aspect ratio r on non-linear frequency of rectangular graphite-epoxy plate.

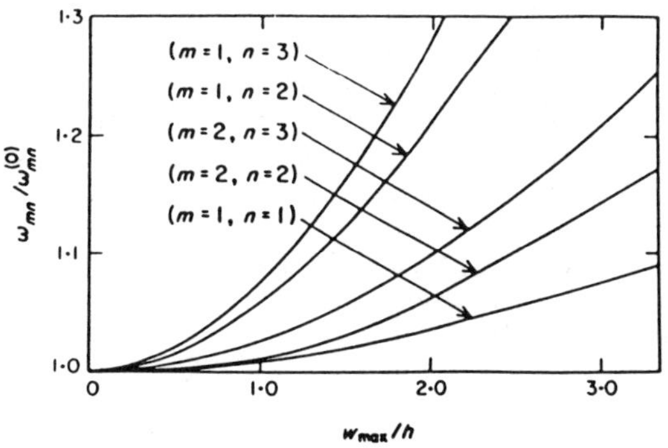

FIGURE 3. Non-linear frequencies of various vibrating modes of simply supported square boron-epoxy plate.

THIN ELLIPTICAL AND CIRCULAR PLATES

For an elliptical plate of semi-major axis a and semi-minor axis b and constant thickness h made of rectilinearly orthotropic material, the governing nonlinear equations in terms of F and w are obtained from Equations (13) and (19) as

$$F_{,xxxx} + k^2F_{,yyyy} + m^2F_{,xxyy} = E_y(w_{,xy}^2 - w_{,xx}w_{,yy}) \tag{36}$$

$$D_1[w_{,xxxx} + k^2w_{,yyyy} + 2(p^2 + s^2)w_{,xxyy}]$$
$$= q(x,y) - \varrho hw_{,tt} + h(F_{,yy}w_{,xx} + F_{,xx}w_{,yy} - 2F_{,xy}w_{,xy}) \tag{37}$$

where

$$k^2 = \frac{E_y}{E_x}, \; q^2 = \nu_{yx}, \; p^2 = \frac{\mu G_{xy}}{E_x},$$

$$m^2 = (k^2 - q^4 - 2p^2q^2)/p^2, \; s^2 = p^2 + q^2, \tag{38}$$

$$D_1 = \frac{E_xh^3}{12\mu}, \; \mu = (1 - \nu_{xy}\nu_{yx})$$

Equations (36) and (37) can be readily specialized for isotropic plates by substituting $k^2 = 1$, $q^2 = \nu$ and $p^2 = (1 - \nu)/2$. In Equation (38) E_x and E_y are the elastic moduli along the x (semi-major axis) and y (semi-minor axis) directions, respectively. ν_{xy} and ν_{yx} are the Poisson's ratios and G_{xy} is the shear modulus of the orthotropic plate material.

The lateral displacement w is assumed in the following form to satisfy the clamped immovable boundary conditions given below [11]:

$$w(x,y,\tau) = \frac{h}{a^4}(x^2 + r^2y^2 - a^2)^2 \left[B_1(\tau) + \frac{B_2(\tau)}{a^2}(x^2 + r^2y^2 - a^2) \right] \tag{39}$$

$$u^o = v^o = w = w_{,x} = w_{,y} = 0 \quad \text{along} \quad x^2 + r^2y^2 - a^2 = 0 \tag{40}$$

In Equation (39), $B_1(\tau)$ and $B_2(\tau)$ are unknown functions of the nondimensional time τ defined as $\tau^2 = t^2(E_x/\varrho a^2)$. t is the real time, ϱ is the mass density of the plate material and r is the aspect ratio a/b. In the case of static problems, B_1 and B_2 are constants and hence independent of τ.

Equation (39) is substituted into Equation (36) and an expression for the stress function F is assumed in the following polynomial form.

$$\begin{aligned} F(x,y,\tau) &= a_1x^{12} + a_2x^{10}y^2 + a_3x^8y^4 + a_4x^6y^6 + a_5x^4y^8 \\ &+ a_6x^2y^{10} + a_7y^{12} + a_8x^{10} + a_9x^8y^2 \\ &+ a_{10}x^6y^4 + a_{11}x^4y^6 + a_{12}x^2y^8 + a_{13}y^{10} \\ &+ a_{14}x^8 + a_{15}x^6y^2 + a_{16}x^4y^4 + a_{17}x^2y^6 \\ &+ a_{18}y^8 + a_{19}x^6 + a_{20}x^4y^2 + a_{21}x^2y^4 \\ &+ a_{22}y^6 + a_{23}x^4 + a_{24}x^2y^2 + a_{25}y^4 \\ &+ a_{26}x^2 + a_{27}y^2 \end{aligned} \tag{41}$$

The coefficients $a_1 - a_{27}$ in Equation (41) are obtained by solving twenty-seven algebraic equations which are obtained by substituting F from Equation (41) and w from Equation (39), in Equation (36) and equating the coefficients of the like terms on both sides. Equating the like term will give fifteen algebraic equations, whereas the fulfillment of the in-plane boundary conditions given by Equation (40) will result in the remaining twelve equations. These systems of equations are solved uniquely to determine the coefficients a_i in Equation (41). These coefficients will be functions of orthotropic and geometric parameters of the plate. It should be noted that the polynomial expression for F in Equation (41) is an exact solution of equation (36). The solution for F also satisfies all the boundary conditions exactly.

The next step in solving the problem is to satisfy the equation of motion in the lateral direction of the vibrating plate, namely Equation (37). Equations (39) and (41) are substituted in Equation (37) to determine the error function of space and time. Integrating this error function over the area of the plate with appropriate normalizing functions $\partial w/\partial B_1$, $\partial w/\partial B_2$ the following set of time-differential equations in B_1 and B_2 are obtained.

$$c_1B_{1,\tau\tau} + c_2B_{2,\tau\tau} + c_3B_1 + c_4B_2 + c_5B_1^2B_2 + c_6B_1B_2^2 + c_7B_1^3 + c_8B_2^3 = q_o^* \tag{42}$$

$$d_1B_{1,\tau\tau} + d_2B_{2,\tau\tau} + d_3B_1 + d_4B_2 + d_5B_1^2B_2 + d_6B_1B_2^2 + d_7B_1^3 + d_8B_2^3 = q_o^* \tag{43}$$

q_o^* is the non-dimensional lateral load q_oa^4/E_xh^4. For static problems, B_1 and B_2 are time-independent and therefore Equations (42) and (43) reduce to a system of nonlinear algebraic equations as given below.

$$c_3B_1 + c_4B_2 + c_5B_1^2B_2 + c_6B_1B_2^2 + c_7B_1^3 + c_8B_2^3 = q_o^* \tag{44}$$

$$d_3B_1 + d_4B_2 + d_5B_1^2B_2 + d_6B_1B_2^2 + d_7B_1^3 + d_8B_2^3 = q_o^* \tag{45}$$

The coefficients c_i and d_i in Equations (42) and (43) are tabulated in Table 3 for certain selected plate parameters. The coupled nonlinear differential Equations (42) and (43) are

TABLE 3. Modal Equation Coefficients for Elliptical Plates with Axes Ratio 2.

	Isotropic		Boron-Epoxy	
i	c_i	d_i	c_i	d_i
1	59.99	66.66	599.99	666.66
2	−49.99	−57.14	−499.99	−571.43
3	43.22	43.22	56.95	56.95
4	−32.42	−51.87	−42.71	−68.34
5	−57.06	−74.38	−75.54	−98.74
6	55.74	75.22	73.93	100.14
7	20.21	25.36	26.69	33.57
8	−18.81	−26.27	−25.04	−35.10

solved on a digital computer, by means of an IMSL subroutine DVERK, which uses the Runge-Kutta-Verner fifth and sixth order integration method. Numerical integration has been carried out with time step size $\Delta\tau = 0.0001$. Period versus amplitude curves are plotted from these values. In the case of static problems the coupled nonlinear algebraic Equations (44) and (45) are solved numerically by means of Newton's iteration method to obtain the load-deflection values.

Numerical results are presented for certain orthotropic and isotropic elliptical plates in Figures 4, 5 and 6. The material constants for glass-epoxy (GE), boron-epoxy (BE) and graphite-epoxy (GrE) plates are listed in Table 4. In the case of static problems, the variation of non-dimensional central deflection with non-dimensional load is presented. For dynamic problems the variation of the ratio of the nonlinear period T to the corresponding linear period T_o with the non-dimensional amplitude is given. Solutions obtained by a two-term deflection function predict higher loads at any given non-dimensional central deflection as can be seen from Figure 4. From Figures 5 and 6 it is clear that the effect of coupling of modes is to reduce the period ratio or to increase the frequency ratio at any given amplitude of vibration and this trend increases with the amplitude. It is also interesting to note that the multiple-mode solutions exhibit the same type of general nonlinear behavior (hardening type nonlinearity) as predicted by single mode solutions.

In the case of circular plates with immovable boundary conditions [12], the value of the aspect ratio r is equal to unity. The corresponding numerical results are presented in Figure 7 and Table 5.

For circular plates with stress free (SF) boundaries [12], the motion of the plate is such that there are no in-plane forces along the boundary of the circular plate. Therefore,

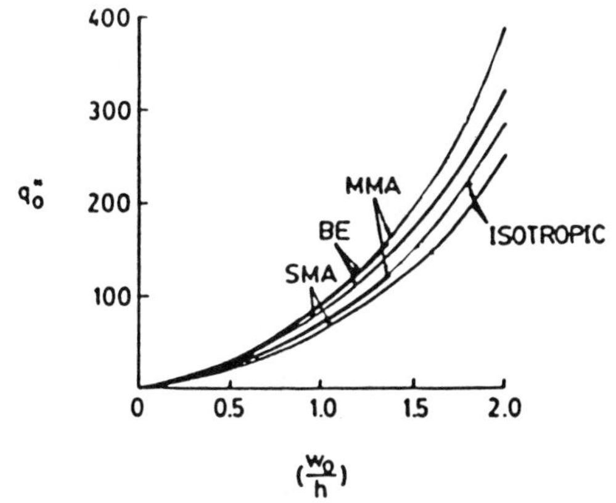

FIGURE 4. Load-deflection curves for isotropic and boron-epoxy elliptical plates with axes ratio 2. SMA: Single-mode approach, MMA: Multiple-mode approach.

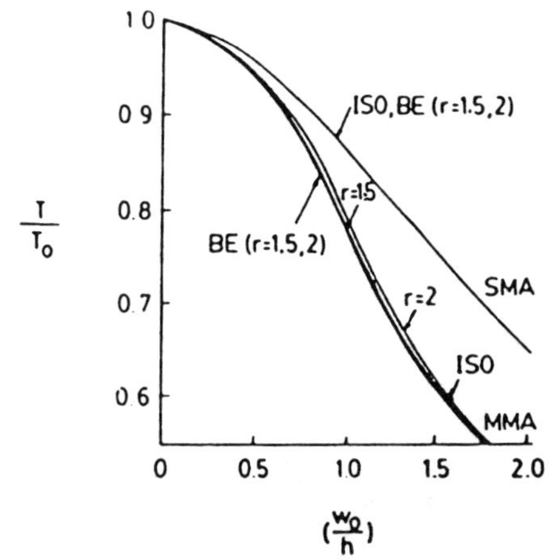

FIGURE 5. Variation of period ratio with amplitude for isotropic and boron-epoxy elliptical plates with various axes ratio, r. SMA: Single-mode approach, MMA: Multiple-mode approach.

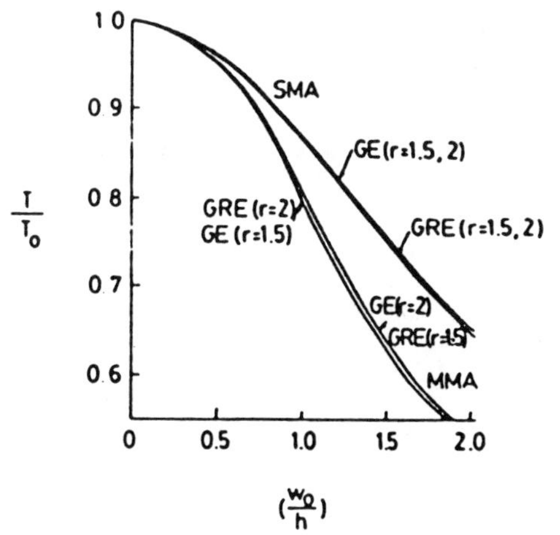

FIGURE 6. Period-amplitude curves for glass-epoxy and graphite-epoxy elliptical plates with various axes ratio, r. SMA: Single-mode approach, MMA: Multiple-mode approach.

TABLE 4. Numerical Values for Elastic Constants.

Material	E_x/E_y	ν_{xy}	G_{xy}/E_x	G_{yz}/E_x
Glass-Epoxy	3	0.25	0.2	0.1
Boron-Epoxy	10	0.22	0.033	0.0165
Graphite-Epoxy	40	0.25	0.015	0.0075
Isotropic	1	0.3	0.385	0.385

TABLE 5. Values of Non-Dimensional Frequency, $\omega/\omega_o\ (10)^4$, for Immovable Circular Plates.

	Isotropic		Glass-Epoxy		Boron-Epoxy		Graphite-Epoxy	
w_{max}/h	SMS*	MMS**	SMS*	MMS**	SMS*	MMS**	SMS*	MMS**
0	10,000	10,000	10,000	10,000	10,000	10,000	10,000	10,000
0.5	10,431	10,495	10,416	10,461	10,428	10,441	10,426	10,427
1.0	11,617	12,723	11,564	12,640	11,605	12,614	11,600	12,577
1.5	13,343	16,461	13,239	16,368	13,319	16,423	13,309	16,428

*SMS—Single-mode solution.
**MMS—Multiple-mode solution.

TABLE 6. Values of Non-Dimensional Frequency, $\omega/\omega_o\ (10)^4$, for Stress-Free Circular Plates.

	Isotropic		Glass-Epoxy		Boron-Epoxy		Graphite-Epoxy	
w_{max}/h	SMS*	MMS**	SMS*	MMS**	SMS*	MMS**	SMS*	MMS**
0	10,000	10,000	10,000	10,000	10,000	10,000	10,000	10,000
0.5	10,136	10,184	10,117	10,183	10,046	10,090	10,018	10,046
1.0	10,532	10,856	10,460	10,779	10,181	10,346	10,070	10,178
1.5	11,157	12,859	11,005	12,363	10,401	10,769	10,158	10,381

*SMS—Single-mode solution.
**MMS—Multiple-mode solution.

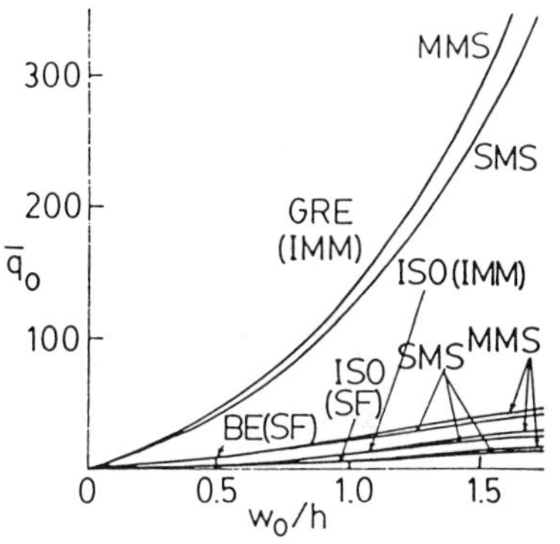

FIGURE 7. Load-deflection curves for isotropic and orthotropic circular plates. SMS: Single-mode solution, MMS: Multiple-mode solution, SF: Stress-free, IMM: Immovable.

the boundary conditions are

$$\left.\begin{aligned} x\ F_{,yy} - y\ F_{,xy} &= 0 \\ y\ F_{,xx} - x\ F_{,xy} &= 0 \end{aligned}\right\} \quad \text{along } x^2 + y^2 = a^2 \tag{46}$$

The solution procedure that was used for immovable elliptical plates is repeated for stress-free circular plates replacing the boundary conditions in terms of u^o and c^o given by Equation (40) by Equation (46). Numerical results for static nonlinear problems are given in Figure 7 and are designated by *SF*. For nonlinear dynamic problems the non-dimensional frequency ratios are presented in Table 6 for various values of the non-dimensional amplitude. It is clear that the nonlinearity is less pronounced in the case of stress-free plates and the effects of nonlinearity and modal coupling are substantial for immovably clamped circular plates.

LAMINATED RECTANGULAR PLATES

A rectangular plate of length a in the x-direction, width b in the y-direction and uniform thickness h in the z-direction is considered. It is assumed that the plate consists of an even number of layers of thin orthotropic sheets of the same thickness and elastic properties. Both cross-ply and angle-ply configurations are considered. In the case of unsymmetric cross-ply plates the orthotropic axes of symmetry of each layer are oriented alternately at 0 and 90 degrees to the plate axes, whereas for unsymmetric angle-ply plates the orthotropic axes in each layer are oriented alternately at angles of $+\theta$ and $-\theta$ to the plate axes. The equations of motion for generally laminated plates expressed in terms of stress function and lateral displacement are [8,13]:

$$\bar{A}_{22}^*F_{,\zeta\zeta\zeta\zeta} + r^2(2\bar{A}_{12}^* + \bar{A}_{66}^*)F_{,\zeta\zeta\eta\eta} + r^4\bar{A}_{11}^*F_{,\eta\eta\eta\eta} + L(W) = r^2\{W_{,\zeta\eta}^2 - W_{,\zeta\zeta}W_{,\eta\eta}\} \tag{47}$$

$$\bar{D}_{11}^*W_{\zeta\zeta\zeta\zeta} + 2r^2(\bar{D}_{12}^* + 2\bar{D}_{66}^*)W_{,\zeta\zeta\eta\eta} + r^4\bar{D}_{22}^*W_{,\eta\eta\eta\eta} + r^4W_{,\tau\tau} - L(F) = r^2\{W_{,\zeta\zeta}F_{,\eta\eta} + W_{,\eta\eta}F_{,\zeta\zeta} - 2W_{,\zeta\eta}F_{,\zeta\eta}\} \tag{48}$$

$$\zeta = x/a, \quad \eta = y/b, \quad r = a/b, \quad W = w/h$$

$$F = \phi/A_{22}h^2, \quad \tau = (t/b^2)\sqrt{A_{22}h/\varrho}$$

$$(A_{ij},B_{ij},D_{ij}) = \int_{-h/2}^{h/2} C_{ij}^{(k)}(1,z,z^2)dz \quad (i,j = 1,2,6)$$

$$[\bar{A}^*] = A_{22}[A]^{-1}, \quad [\bar{B}^*] = -[A]^{-1}[B]/h$$
$$[\bar{D}^*] = ([D] - [B][A]^{-1}[B])/A_{22}h \tag{49}$$

for cross-ply $L(\ ) = \bar{B}_{12}^*(\ )_{,\zeta\zeta\zeta\zeta} - r^4\bar{B}_{12}^*(\ )_{,\eta\eta\eta\eta}$ and for angle-ply plate $L(\ ) = r(\bar{B}_{61}^* - 2\bar{B}_{26}^*)(\ )_{,\zeta\zeta\zeta\eta} + r^3(\bar{B}_{62}^* - 2\bar{B}_{16}^*)(\ )_{,\zeta\eta\eta\eta}$

In Equation (49), w is the lateral displacement, t is the time, ϕ is the stress function, ϱ is the mass density, and $C_{ij}^{(k)}$ are the anisotropic elastic constants for the kth layer of the plate. The nondimensional stress function is defined in terms of the stress resultants as

$$[F_{,\eta\eta},F_{,\zeta\zeta},F_{,\zeta\eta}] = (A_{22}h^2/b^2)[N_x,r^2N_y, -rN_{xy}] \tag{50}$$

The following boundary conditions are considered:

Clamped Cross-Ply and Angle-Ply Plates:

$$\begin{aligned} W = W_{,\zeta} = F_{,\eta\eta} = F_{,\zeta\eta} = 0 \quad &\text{at} \quad \zeta = 0,1 \\ W = W_{,\eta} = F_{,\zeta\zeta} = F_{,\zeta\eta} = 0 \quad &\text{at} \quad \eta = 0,1 \end{aligned} \tag{51}$$

Simply Supported Angle-Ply Plate:

$$\begin{aligned} W = W_{,\zeta\zeta} = F_{,\eta\eta} = F_{,\zeta\eta} = 0 \quad &\text{at} \quad \zeta = 0,1 \\ W = W_{,\eta\eta} = F_{,\zeta\zeta} = F_{,\zeta\eta} = 0 \quad &\text{at} \quad \eta = 0,1 \end{aligned} \tag{52}$$

Simply Supported Cross-Ply Plate:

$$\begin{aligned} W &= F_{,\eta\eta} = F_{,\zeta\eta} = 0, \\ W_{,\zeta\zeta} &= (\bar{B}_{12}^*/\bar{D}_{11}^*)F_{,\zeta\zeta} \quad \text{at} \quad \zeta = 0,1 \\ W &= F_{,\zeta\zeta} = F_{,\zeta\eta} = 0, \\ W_{,\eta\eta} &= -(\bar{B}_{12}^*/D_{11}^*)F_{,\eta\eta} \quad \text{at} \quad \eta = 0,1 \end{aligned} \tag{53}$$

Solutions to the governing nonlinear Equations (47) and (48) are obtained using a double series for lateral displacement w and the stress function F. Following a similar procedure that was outlined earlier for single layered rectangular plates, numerical results are obtained for laminated rectangular and square plates. The values of the elastic constants for different orthotropic materials are given in Table 1.

The nonlinear fundamental frequencies for square angle-ply graphite-epoxy plates are presented in Table 7 for various values of the non-dimensional amplitude. Both simply supported (SS) and clamped (CC) plates are considered. Similar numerical results are graphically presented in Figures 8, 9, 10 and 11. In Figure 8, N stands for the number of orthotropic layers in the laminated plate. It is clear that the nonlinear frequency of a laminated plate (cross-ply or angle-ply) increases with the amplitude of vibration and hence a hardening type of nonlinearity is noticed. For a given amplitude the nonlinear frequency decreases with the aspect ratio or the orientation angle of the filaments but increases with the total number of layers or the tensile moduli of the plate. Since the accuracy of a single-mode solution decreases at higher amplitudes, a multiple-mode solution is necessary for laminated plates.

A multiple-mode nonlinear analysis of cross-ply and angle-ply rectangular plates is also reported in [14]. It was found that the nonlinear frequency ratio increased at all amplitudes of vibration due to coupling of higher modes.

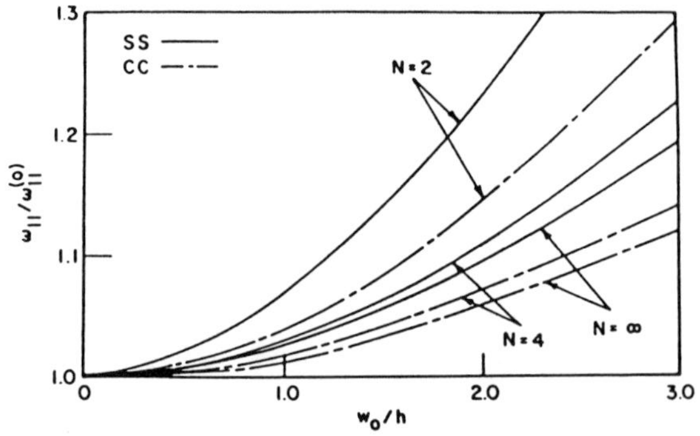

FIGURE 8. Effect of number of layers on fundamental frequency of square graphite-epoxy cross-ply plate.

TABLE 7. Comparison of Nonlinear Fundamental Frequencies of Square Two-Layer ±45° Angle-Ply Graphite-Epoxy Plate.

	Simply Supported	
	w_o/h	
$\omega_{11}/\omega_{11}^{(0)}$	Single Mode	Four Modes
1.034	0.9836	0.9839
1.071	1.4274	1.4506
1.139	2.0491	2.1075
1.163	2.2305	2.3159
1.213	2.5801	2.7187
	Clamped	
1.037	0.9303	0.9610
1.075	1.3449	1.3971
1.143	1.9031	1.9803
1.168	2.0370	2.1574
1.217	2.2519	2.4903

MODERATELY THICK ELLIPTICAL AND CIRCULAR PLATES

For a rectilinearly orthotropic elliptical plate of semi-major axis a and semi-minor axis b and uniform thickness h, the two coupled nonlinear differential equations that take into account the effects of transverse shear deformation and rotatory inertia can be derived from Equations (13) and (19) as [15]:

$$\begin{aligned} &F_{,\zeta\zeta\zeta\zeta} + k^2r^4F_{,\eta\eta\eta\eta} + 2m^2r^2F_{,\zeta\zeta\eta\eta} \\ &\quad = E_yr^2(w^2_{,\zeta\eta} - w_{,\zeta\zeta}W_{,\eta\eta}) \end{aligned} \tag{54}$$

$$\begin{aligned} &a_1(\bar{I}_{,\zeta\zeta\zeta\zeta} + r^4\bar{I}_{,\eta\eta\eta\eta}) + a_2r^2\bar{I}_{,\zeta\zeta\eta\eta} + a_3a^4\bar{I}_{,tttt} \\ &+ a_4a^2(\bar{I}_{,\zeta\zeta tt} + r^2\bar{I}_{,\eta\eta tt}) + a_5a^2(\bar{I}_{,\zeta\zeta} + r^2I_{,\eta\eta}) \\ &+ a_6a^4\bar{I}_{,tt} - a^4\bar{I} + a_7a^6(W_{,\zeta\zeta tt} + r^2w_{,\eta\eta tt}) \\ &+ a_8a^4(w_{,\zeta\zeta\zeta\zeta} + r^4w_{,\eta\eta\eta\eta}) + a_9a^4r^2w_{,\zeta\zeta\eta\eta} \\ &+ a_{10}a^4r^2w_{,\zeta\zeta\eta\eta tt} + a_{11}a^4(w_{,\zeta\zeta\zeta\zeta tt} + r^4w_{,\eta\eta\eta\eta tt}) \\ &+ a_{12}a^6(w_{,\zeta\zeta tttt} + r^2w_{,\eta\eta tttt}) + a_{13}a^2(w_{,\zeta\zeta\zeta\zeta\zeta\zeta} + r^6w_{,\eta\eta\eta\eta\eta\eta}) \\ &+ a_{14}a^2(r^2w_{,\zeta\zeta\zeta\zeta\eta\eta} + r^4w_{,\eta\eta\eta\eta\zeta\zeta}) = 0 \end{aligned} \tag{55}$$

where

$$\bar{I} = a^4[q(\zeta,\eta) - \varrho h w_{,tt}] + r^2(F_{,\eta\eta}w_{,\zeta\zeta} + F_{,\zeta\zeta}w_{,\eta\eta} - 2F_{,\zeta\eta}w_{,\zeta\eta})$$

$$\zeta = \frac{x}{a}, \quad \eta = \frac{y}{b}, \quad r = \frac{a}{b}$$

$$k^2 = E_y/E_x, \quad q^2 = \nu_{xy}, \quad m^2 = (k^2 - q^4 - 2p^2q^2)/p^2$$
$$p^2 = G_{xy}\nu'/E_x, \quad \nu' = 1 - \nu_{xy}\nu_{yx} \qquad (56)$$

The coefficients a_1 to a_{14} are functions of aspect ratio r, orthotropic material constants k^2, p^2, q^2, and tracing constants T_s and R_i.

The clamped immovable boundary conditions of an elliptical plate are represented by means of the following equations:

$$w = w_{,\zeta} = w_{,\eta} = 0$$
$$\text{along } 1 - \zeta^2 - \eta^2 = 0 \qquad (57)$$
$$u^o = v^o = 0$$

The in-plane boundary conditions in terms of u^o and v^o can be transformed in terms of the lateral displacement w and the stress function F as

$$u_o = \int u^o_{,\zeta} d\zeta = \int \left(c_{11}r^2F_{,\eta\eta} - c_{12}F_{,\zeta\zeta} - \frac{1}{2}w^2_{,\zeta}\right) d\zeta = 0 \qquad (58)$$

$$v^o = \int v^o_{,\eta} d\eta = \int \left(\frac{c_{22}}{r^2}F_{,\zeta\zeta} - c_{12}F_{,\eta\eta} - \frac{1}{2}w^2_{,\eta}\right) d\eta = 0 \qquad (59)$$

along the boundary and where

$$c_{11} = k^2\nu'/E'_x, \quad c_{12} = q^2\nu'/E'_x, \quad c_{22} = \nu'/E'_x$$
$$E'_x = E_x(k^2 - q^4)$$

Solutions to Equations (54) and (55) are sought in terms of polynomial functions which would satisfy all the in-plane and out-of-plane boundary conditions. An expression for the lateral mode w is assumed in the following separable form

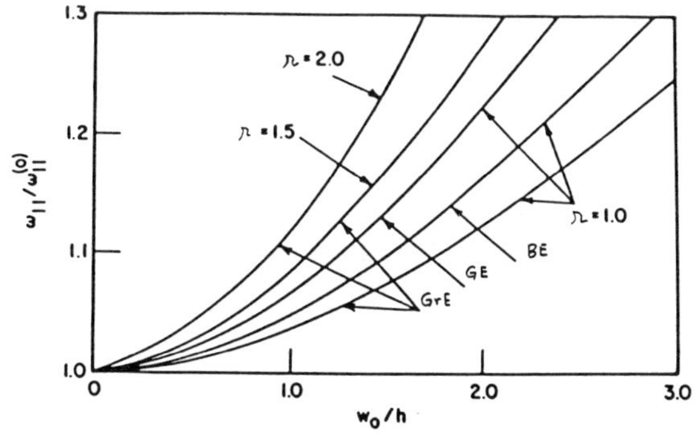

FIGURE 9. Effect of aspect ratio and elastic properties on fundamental frequency of simply supported two-layer ±45° angle-ply plate.

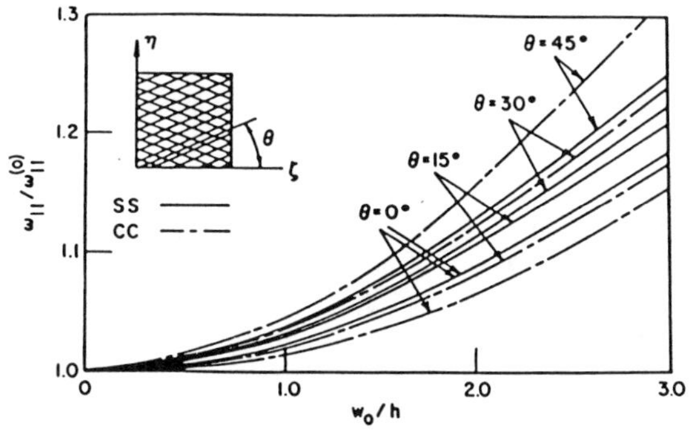

FIGURE 10. Effect of orientation angle on fundamental frequency of square two-layer graphite-epoxy angle-ply plate.

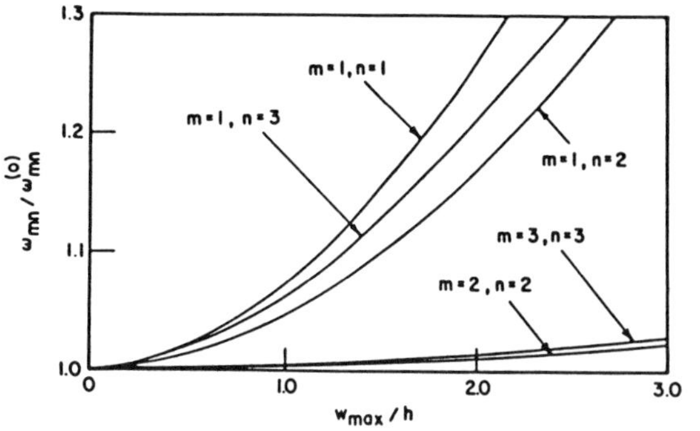

FIGURE 11. Frequencies of square simply supported two-layer boron-epoxy cross-ply plate for various vibrating modes.

consisting of unknown time functions $A(\tau)$ and $B(\tau)$.

$$w(\zeta,\eta,\tau) = h(\zeta^2 + \eta^2 - 1)^2[A(\tau) + B(\tau)(\zeta^2 + \eta^2 - 1)] \quad (60)$$

In Equation (60) τ is the nondimensional time as defined under Equation (40). As was done in the case of thin elliptical plates, Equation (60) is substituted in Equation (54) and an expression for the stress function is taken in the following polynomial form.

$$\begin{aligned} F(\zeta,\eta,\tau) = {} & b_1\zeta^{12} + b_2\zeta^{10}\eta^2 + b_3\zeta^8\eta^4 + b_4\zeta^6\eta^6 \\ & + b_5\zeta^4\eta^8 + b_6\zeta^2\eta^{10} + b_7\eta^{12} + b_8\zeta^{10} \\ & + b_9\zeta^8\eta^2 + b_{10}\zeta^6\eta^4 + b_{11}\zeta^4\eta^6 + b_{12}\zeta^2\eta^8 \\ & + b_{13}\eta^{10} + b_{14}\zeta^8 + b_{15}\zeta^6\eta^2 + b_{16}\zeta^4\eta^4 \\ & + b_{17}\zeta^2\eta^6 + b_{18}\eta^8 + b_{19}\zeta^6 + b_{20}\zeta^4\eta^2 \\ & + b_{21}\zeta^2\eta^4 + b_{22}\eta^6 + b_{23}\zeta^4 + b_{24}\zeta^2\eta^2 \\ & + b_{25}\eta^4 + b_{26}\zeta^2 + b_{27}\eta^2 \end{aligned} \quad (61)$$

The coefficients b_i in Equation (61) are uniquely determined following the same procedure that was described earlier. It must be noted that the semi-inverse procedure that has been used here to evaluate the coefficients b_i enables one to solve Equation (54) exactly for the lateral mode given in Equation (60). It is possible to obtain such exact solutions for any number of appropriate additional modes included in Equation (60). The complexity of the solution procedure, however, increases enormously for each additional mode included.

Having the stress function completely defined, it is now required to satisfy the equation of motion in the lateral direction of the vibrating plate, namely Equation (55). As done before, the procedure involves multiplying Equation (55) by the two spatial functions $(\zeta^2 + \eta^2 - 1)^2$ and $(\zeta^2 + \eta^2 - 1)^3$ and integrating the resulting expressions over the area of the plate. Performing these integrations, a set of two coupled ordinary nonlinear time-differential equations are obtained in terms of the time functions $A(\tau)$ and $B(\tau)$. These equations are

$$\begin{aligned} & A_1(A^3)_{,\tau\tau\tau\tau} + A_2(A^2B)_{,\tau\tau\tau\tau} + A_3(AB^2)_{,\tau\tau\tau\tau} + A_4(B^3)_{,\tau\tau\tau\tau} \\ & + A_5(A)_{,\tau\tau\tau\tau\tau\tau} + A_6(B)_{,\tau\tau\tau\tau\tau\tau} + A_7(A^3)_{,\tau\tau} \\ & + A_8(A^2B)_{,\tau\tau} + A_9(AB^2)_{,\tau\tau} + A_{10}(B^3)_{,\tau\tau} \\ & + A_{11}(A)_{,\tau\tau\tau\tau} + A_{12}(B)_{,\tau\tau\tau\tau} + A_{13}A^3 + A_{14}A^2B \\ & + A_{15}AB^2 + A_{16}B^3 + A_{17}(A)_{,\tau\tau} + A_{18}(B)_{,\tau\tau} \\ & + A_{19}A + A_{20}B = q_o^* \end{aligned} \quad (62)$$

$$\begin{aligned} & B_1(A^3)_{,\tau\tau\tau\tau} + B_2(A^2B)_{,\tau\tau\tau\tau} + B_3(AB^2)_{,\tau\tau\tau\tau} + B_4(B^3)_{,\tau\tau\tau\tau} \\ & + B_5(A)_{,\tau\tau\tau\tau\tau\tau} + B_6(B)_{,\tau\tau\tau\tau\tau\tau} + B_7(A^3)_{,\tau\tau} \\ & + B_8(A^2B)_{,\tau\tau} + B_9(AB^2)_{,\tau\tau} + B_{10}(B^3)_{,\tau\tau} \\ & + B_{11}(A)_{,\tau\tau\tau\tau} + B_{12}(B)_{,\tau\tau\tau\tau} + B_{13}A^3 + B_{14}A^2B \\ & + B_{15}AB^2 + B_{16}B^3 + B_{17}(A)_{,\tau\tau} + B_{18}(B)_{,\tau\tau} \\ & + B_{19}A + B_{20}B = q_o^* \end{aligned} \quad (63)$$

Some typical numerical values for coefficients A_i and B_i in Equations (62) and (63) are tabulated in Table 8 considering certain plate and material parameters as well as tracing constants. The nondimensional load q_o^* in Equations (62) and (63) is given by $q_o^* = q_o\beta^4/E_x$ where β is the thickness parameter, a/h, and q_o is the intensity of the uniformly distributed load. In the case of a thin elliptical plate, due to the fact that T_s and R_i are zero, Equations (62) and (63) will simplify considerably resulting in the following two equations

$$A_{13}A^3 + A_{14}A^2B + A_{15}AB^2 + A_{16}B^3 + A_{17}(A)_{,\tau\tau} + A_{18}(B)_{,\tau\tau} + A_{19}A + A_{20}B = q_o^* \quad (64)$$

$$B_{13}A^3 + B_{14}A^2B + B_{15}AB^2 + B_{16}B^3 + B_{17}(A)_{,\tau\tau} + B_{18}(B)_{,\tau\tau} + B_{19}A + B_{20}B = q_o^* \quad (65)$$

Equations (64) and (65) are the same as those given by Equations (42) and (43) for thin elliptical plates. It is, therefore, possible to obtain the time-differential equations applicable for thin plates from those derived for thick plates by simply ignoring the effects of transverse shear deformation ($T_s = 0$) and rotatory inertia ($R_i = 0$). In a similar manner, by equating T_s and R_i to zero and eliminating the time-derivative terms in Equations (62) and (63) a set of nonlinear algebraic equations applicable for thin elliptical plates can be readily obtained. Those equations will be exactly the same as those given by Equations (44) and (45). As before, numerical integration of Equations (62) and (63) is performed by means of the Runge-Kutta-Verner integration technique.

Numerical results are presented for moderately thick orthotropic as well as isotropic elliptical plates. Orthotropic plates are made of glass-epoxy (GE), boron-epoxy (BE) and

TABLE 8. Coefficients of Modal Equations for Boron-Epoxy Elliptical Plate with $r = 2(T_s = R_i = 1)$.

i	A_i	B_i	A_i	B_i
	$\beta = 10$		$\beta = 30$	
1	0.049009	0.061646	0.000605	0.000761
2	−0.138734	−0.181347	−0.001713	−0.002239
3	0.135773	0.183904	0.001676	0.002270
4	−0.045995	−0.064458	−0.000568	−0.000796
5	1.101922	1.224363	0.122436	0.136040
6	−0.918265	−1.049455	−0.102029	−0.116606
7	2.982468	3.731926	0.276422	0.347454
8	−8.351297	−11.085650	−0.781359	−1.023445
9	8.123972	11.343490	0.764071	1.039129
10	−2.713010	−4.000415	−0.258359	−0.364513
11	60.621450	67.897590	55.220320	61.416190
12	−50.922770	−59.239590	−46.061750	−52.758170
13	39.402700	49.156690	28.032820	35.200790
14	−110.401200	−148.429400	−79.144620	−103.830600
15	107.272500	153.705500	77.360970	105.563100
16	−35.733100	−54.811900	−26.109260	−37.062220
17	743.175000	835.837400	5531.398000	6157.414000
18	−626.872500	−743.245800	−4618.023000	−5300.570000
19	56.951900	56.951900	56.951900	56.951900
20	−57.288200	−82.917290	−44.332670	−69.961540

TABLE 9. Values of Period Ratio T/T_o for Isotropic and Orthotropic Elliptical Plates with $r = 2$ (Multiple-Mode Solution).

	$\beta = 20$			$T_s = R_i = 0$		
w_o^*	Isotropic	GE	GrE	Isotropic	GE	GrE
0	1.0296	1.0359	1.1040	1.0000	1.0000	1.0000
0.5	0.9810	0.9854	1.0252	0.9588	0.9581	0.9586
1.0	0.7825	0.7790	0.7479	0.7912	0.7887	0.7908
1.5	0.6162	0.6157	0.6211	0.6101	0.6071	0.6084
2.0	0.5405	0.5450	0.5545	0.5358	0.5331	0.5340

TABLE 10. Load-Deflection Values for Elliptical Plates with $r = 2$.

	Values of q_o^*	
w_o^*	Single-Mode	Multiple-Mode
0	0	0
0.5	24.14	24.52
1.0	63.43	67.05
1.5	133.04	148.91
2.0	248.12	298.47

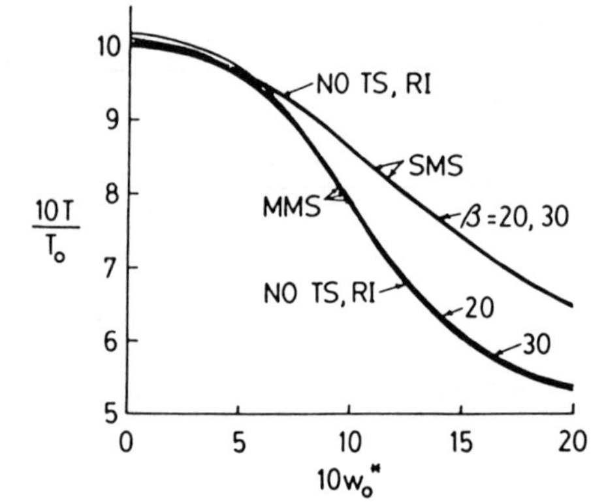

FIGURE 12. Variation of period ratio with amplitude for isotropic plates with r = 1.5; MMS: Multiple-mode solution, SMS: Single-mode solution.

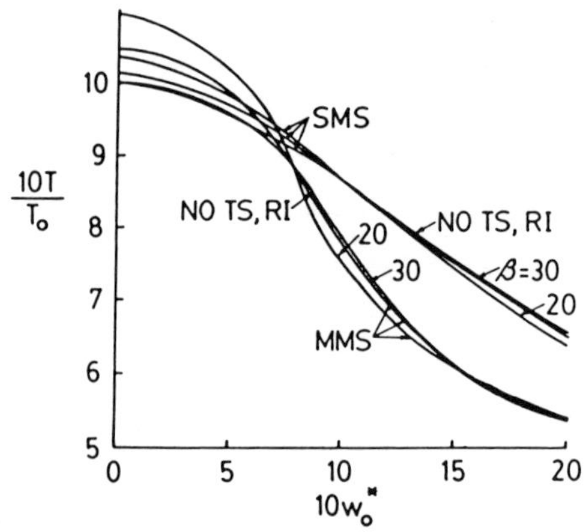

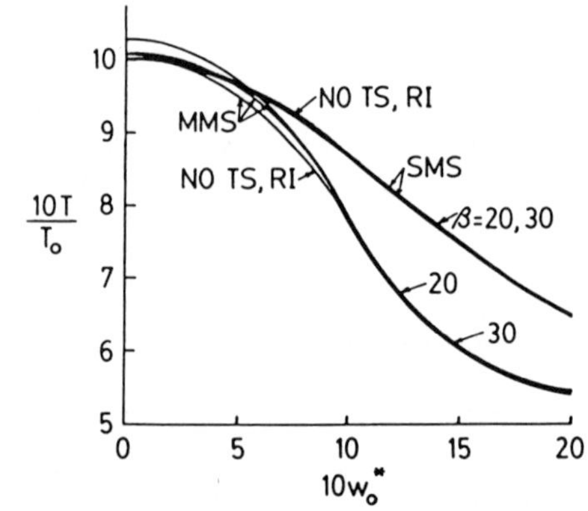

FIGURE 13. Relation between period ratio and amplitude for glass-epoxy plates with r = 1.5; MMS: Multiple-mode solution, SMS: Single-mode solution.

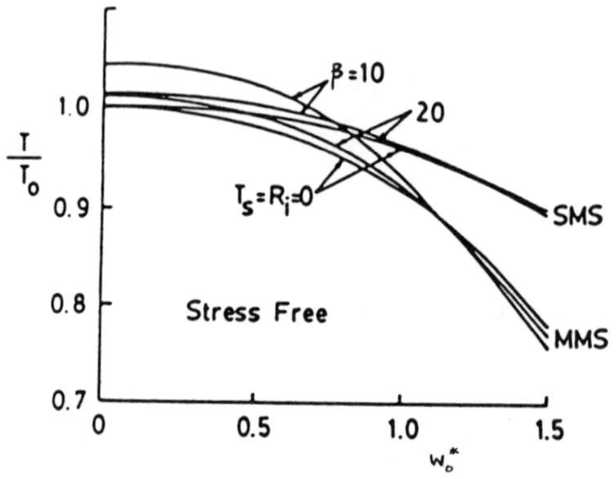

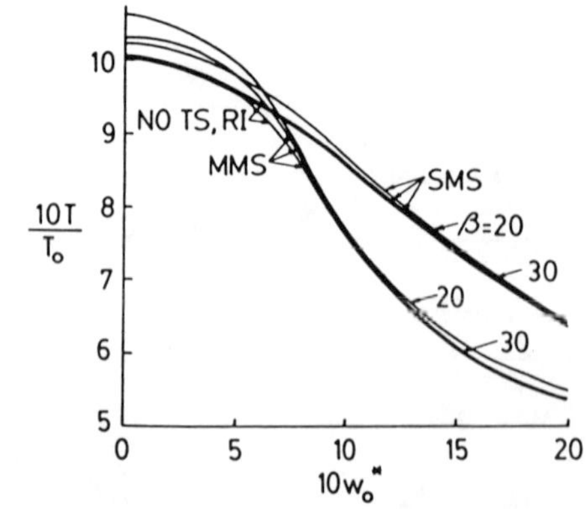

FIGURE 14. Variation of period ratio with amplitude for boron-epoxy plates with r = 1.5; MMS: Multiple-mode solution, SMS: Single-mode solution.

FIGURE 15. Relation between period ratio and amplitude for graphite-epoxy plates with r = 1.5; MMS: Multiple-mode solution, SMS: Single-mode solution.

FIGURE 16. Variation of period ratio with amplitude for isotropic stress-free circular plates with $T_s = R_i = 1$, MMS: Multiple-mode solution, SMS: Single-mode solution.

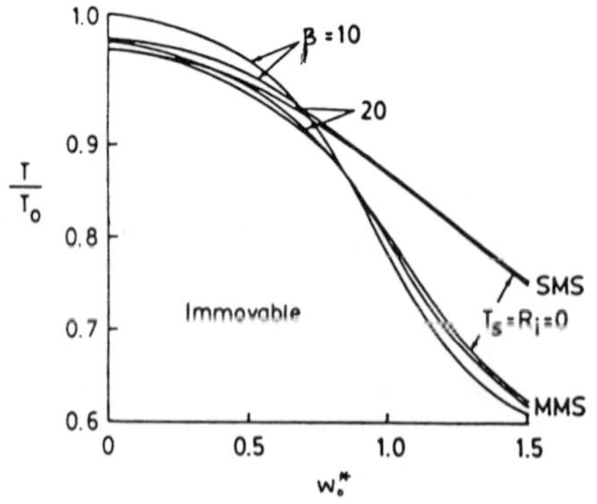

FIGURE 17. Amplitude–period response curves for isotropic immovable circular plates with $T_s = R_i = 1$, MMS: Multiple-mode solution, SMS: Single-mode solution.

TABLE 11. Load Deflection Values for Isotropic Circular Plates, $T_s = 0$.

	Nondimensional Load q_0^*			
	Stress-Free		Immovable	
w_{max}/h	Two-Term	Single-Term	Two-Term	Single-Term
0	0	0	0	0
0.5	3.061	3.038	3.328	3.276
1.0	6.885	6.718	9.055	8.623
1.5	12.147	11.684	19.358	18.112
2.0	19.319	18.578	35.913	33.816

TABLE 12. Values of $(T/T_0)10^4$ for Various Values of Amplitude for Stress-Free Circular Plates.

	$T_s = R_i = 0$		$T_s = R_i = 1, \beta = 10$	
w_{max}/h	Two-Mode	Single-Mode	Two-Mode	Single-Mode
0	10,000	10,000	10,468	10,159
0.5	9,819	9,866	10,320	10,017
1.0	9,211	9,495	9,337	9,623
1.5	7,776	8,963	7,602	9,062
2.0	6,589	8,353	6,635	8,424

TABLE 13. Coefficients of Modal Equations (62) and (63) for Immovable Circular Plates, $T_s = R_i = 1, \beta = 10$.

i	A_i	B_i
1	0.000019	0.000024
2	−0.000053	−0.000070
3	0.000035	0.000064
4	−0.000018	−0.000025
5	0.000406	0.000451
6	−0.000338	−0.000386
7	0.014774	0.018630
8	−0.041879	−0.055089
9	0.033255	0.053296
10	−0.013954	−0.019803
11	0.316889	0.352534
12	−0.264398	−0.303008
13	2.921170	3.681003
14	−8.269774	−10.903200
15	7.425312	10.932950
16	−2.748966	−3.929678
17	61.914580	68.956740
18	−51.717050	−59.440710
19	5.860757	5.860753
20	−4.606479	−7.243891

graphite-epoxy (GrE) materials whose material constants are presented in Table 4. The ratios of G_{xz}/E_x are assumed to be the same as those of G_{xy}/E_x and the ratios of G_{yz}/E_x are taken to be one-half of the ratios of G_{xz}/E_x. The ratios of the nonlinear period T of vibration, including the effects of the transverse shear and rotatory inertia, to the corresponding linear period T_o of the classical plate (for which $T_s = R_i = 0$) are computed for some plate aspect ratios, thickness parameters, and material constants at different non-dimensional amplitudes of vibration. These results are graphically presented in Figures 12 and 15 and tabulated in Table 9. w_o^* is the non-dimensional maximum deflection. Nonlinear load-deflection values are also presented in Table 10 for static problems. It should be noted that for moderately thick plates, the nonlinear periods are dependent on the thickness parameter β whereas they are independent of β for classical thin plates. It is seen that there is a significant difference in numerical values between single-mode and multiple-mode solutions suggesting that the coupling effect between modes is significant. This is found to be true for both moderately thick as well as thin plates. Period ratios are higher at any given amplitude when the effects of transverse shear and rotatory inertia are accounted for. It is also clear that the effect of transverse shear is much more important than the effect of rotatory inertia, while both of them contribute to an increase in the numerical value of the period ratio at any given amplitude.

Results for immovable and stress free isotropic circular plates [16] are presented in Tables 11, 12 and Figures 16 and 17. Coefficients of the time-differential Equations (62) and (63) for a unit aspect ratio are also tabulated in Table 13. As pointed out earlier under thin plates, it is seen that the nonlinearity is much less pronounced in the case of circular plates with stress-free in-plane boundary conditions. The transverse shear effect will substantially change the period ratios at all amplitudes of vibration for stress-free plates whereas its effect is predominant only at small amplitudes of vibration for immovable circular plates.

DISPLACEMENT FIELD EQUATIONS

In all previous sections the governing nonlinear equations for plates have been presented in terms of the lateral displacement w and the stress function F. Von Kármán-type field equations which are in terms of the three displacement components u°, v° and w are given in this section. Included in these field equations are the effects of transverse shear deformation and rotatory inertia such that they can readily be used for moderately thick plates of any plate geometry.

For a rectilinearly orthotropic elliptical plate of uniform thickness h, semi-major axis a, and semi-minor axis b, the governing nonlinear dynamic equations in terms of u°, v° and w are [17]:

$$\begin{aligned} &au^\circ_{,\zeta\zeta} + p^2r^2au^\circ_{,\eta\eta} + s^2rav^\circ_{,\zeta\eta} \\ &= w_{,\zeta}(w_{,\zeta\zeta} + p^2r^2w_{,\eta\eta}) - s^2r^2w_{,\eta}w_{,\zeta\eta} \end{aligned} \tag{66}$$

$$\begin{aligned} &k^2r^2av^\circ_{,\eta\eta} + p^2av^\circ_{,\zeta\zeta} + s^2rau^\circ_{,\zeta\eta} \\ &= -rw_{,\eta}(k^2r^2w_{,\eta\eta} + p^2w_{,\zeta\zeta}) - s^2rw_{,\zeta}w_{,\zeta\eta} \end{aligned} \tag{67}$$

$$\begin{aligned} &a_1I_{,\zeta\zeta\zeta\zeta} + a_2r^2I_{,\zeta\zeta\eta\eta} + a_3r^4I_{,\eta\eta\eta\eta} + a_4a^4I_{,tttt} + a_5a^2I_{,\zeta\zeta tt} \\ &+ a_6r^2a^2I_{,\eta\eta tt} + a_7a^2I_{,\zeta\zeta} + a_8r^2a^2I_{,\eta\eta} + a_9a^4I_{,tt} \\ &- a^4I + a_{10}a^2(w_{,\zeta\zeta tt} + r^2w_{,\eta\eta tt}) + a_{11}w_{,\zeta\zeta\zeta\zeta} \\ &+ a_{12}r^2w_{,\zeta\zeta\eta\eta} + a_{13}r^4w_{,\eta\eta\eta\eta} + a_{14}r^2w_{,\zeta\zeta\eta\eta tt} \\ &+ a_{15}w_{,\zeta\zeta\zeta\zeta tt} + a_{16}a^2w_{,\zeta\zeta tttt} + a_{17}r^4w_{,\eta\eta\eta\eta tt} \\ &+ a_{18}a^2r^2w_{,\eta\eta tttt} + \frac{a_{19}}{a^2}w_{,\zeta\zeta\zeta\zeta\zeta\zeta} \\ &+ \frac{a_{20}}{b^2}w_{,\zeta\zeta\zeta\zeta\eta\eta} + a_{21}\frac{r^4}{a^2}w_{,\eta\eta\eta\eta\zeta\zeta} \\ &+ a_{22}\frac{r^4}{b^2}w_{,\eta\eta\eta\eta\eta\eta} = 0 \end{aligned} \tag{68}$$

where

$$k^2 = \frac{E_\eta}{E_\zeta}, \quad q^2 = \nu_{\zeta\eta}, \quad p^2 = \frac{\mu G_{\zeta\eta}}{E_\zeta}$$

$$s^2 = p^2 + q^2, \quad r = a/b, \quad \mu = 1 - \nu_{\zeta\eta}\nu_{\eta\zeta}$$

$$\begin{aligned} I &= q(\zeta,\eta) - \varrho hw_{,tt} + \frac{N_{\zeta\zeta}}{a^2}w_{,\zeta\zeta} \\ &+ \frac{N_{\eta\eta}}{b^2}w_{,\eta\eta} + 2\frac{N_{\zeta\eta}}{ab}w_{,\zeta\eta} \end{aligned} \tag{69}$$

$$\zeta = \frac{x}{a}, \quad \eta = \frac{y}{b}$$

The stress resultants, N_{ij}, in Equation (69) are defined in terms of the median surface strains of the plate as

$$\begin{aligned} N_{\zeta\zeta} &= \frac{E_\zeta h}{\mu}(\epsilon_1^\circ + q^2\epsilon_2^\circ) \\ N_{\eta\eta} &= \frac{E_\zeta h}{\mu}(q^2\epsilon_1^\circ + k^2\epsilon_1^\circ) \\ N_{\zeta\eta} &= hG_{\zeta\eta}\gamma \end{aligned} \tag{70}$$

where

$$\epsilon_1^o = \frac{1}{a} u^o{}_{,\zeta} + \frac{1}{2a^2} w_{,\zeta}^2$$

$$\epsilon^o{}_2 = \frac{1}{b} v^o_{,\eta} + \frac{1}{2b^2} w_{,\eta}^2 \tag{71}$$

$$\gamma = \frac{1}{b} u^o_{,\eta} + \frac{1}{a} v^o_{,\zeta} + \frac{1}{ab} w_{,\zeta} w_{,\eta}$$

In Equation (68), the coefficients a_i are functions of orthotropic parameters as well as tracing constants T_s and R_i which are used to represent the effect of transverse shear and rotatory inertia, respectively.

For a moderately thick clamped elliptical plate with an immovable edge, the conditions to be satisfied along the boundary are

$$w = w_{,\zeta} = w_{,\eta} = u^o = v^o = 0$$
$$\text{along } 1 - \zeta^2 - \eta^2 = 0 \tag{72}$$

The boundary conditions in terms of w will be satisfied by the following multiple-mode polynomial function.

$$w(\zeta,\eta,\tau) = h(1 - \zeta^2 - \eta^2)^2 [A(\tau) + B(\tau)(1 - \zeta^2 - \eta^2)] \tag{73}$$

In Equation (73) $A(\tau)$ and $B(\tau)$ are unknown functions of nondimensional time τ defined as $\tau^2 = t^2(E_\zeta/\varrho a^2)$. Substituting Equation (73) into Equations (66) and (67) and solving for the in-plane displacements u^o and v^o by means of a semi-inverse procedure it can be shown that

$$\begin{aligned} u^o = {} & c_1\zeta^{11} + c_2\zeta^9 + c_3\zeta^7 + c_4\zeta^5 + c_5\zeta^3 + c_6\zeta \\ & + c_7\zeta^9\eta^2 + c_8\zeta^7\eta^4 + c_9\zeta^5\eta^6 + c_{10}\zeta^3\eta^8 + c_{11}\zeta\eta^{10} \\ & + c_{12}\zeta^7\eta^2 + c_{13}\zeta^5\eta^4 + c_{14}\zeta^3\eta^6 + c_{15}\zeta\eta^8 + c_{16}\zeta^5\eta^2 \\ & + c_{17}\zeta^3\eta^4 + c_{18}\zeta\eta^6 + c_{19}\zeta^3\eta^2 + c_{20}\zeta\eta^4 + c_{21}\zeta\eta^2 \end{aligned} \tag{74}$$

$$\begin{aligned} v^o = {} & d_1\eta^{11} + d_2\eta^9 + d_3\eta^7 + d_4\eta^5 + d_5\eta^3 + d_6\eta \\ & + d_7\eta^9\zeta^2 + d_8\eta^7\zeta^4 + d_9\eta^5\zeta^6 + d_{10}\eta^3\zeta^8 \\ & + d_{11}\eta\zeta^{10} + d_{12}\eta^7\zeta^2 + d_{13}\eta^5\zeta^4 + d_{14}\eta^3\zeta^6 \\ & + d_{15}\eta\zeta^8 + d_{16}\eta^5\zeta^2 + d_{17}\eta^3\zeta^4 + d_{18}\eta\zeta^6 + d_{19}\eta^3\zeta^2 \\ & + d_{20}\eta\zeta^4 + d_{21}\eta\zeta^2 \end{aligned} \tag{75}$$

In Equations (74) and (75) the coefficients c_i, d_i are functions of orthotropic plate parameters and time. To determine these forty-two coefficients explicitly, Equations (74) and (75) are substituted in Equations (66) and (67) along with Equation (73) for w. After expanding each term, the coefficients of thiry like terms are compared in these two equations to come up with thirty linear algebraic equations in terms of c_i and d_i. The remaining twelve algebraic equations are generated by imposing the in-plane boundary conditions which are written as

$$u^o = \int u^o_{,\zeta}\, d\zeta = \int \left[e_1(k^2 N_{\zeta\zeta} - q^2 N_{\eta\eta}) - \frac{1}{2} w_{,\zeta}^2 \right] d\zeta = 0$$
$$v^o = \int v^o_{,\eta}\, d\eta = \int \left[\frac{e_1}{r} (N_{\eta\eta} - q^2 N_{\zeta\zeta}) - \frac{1}{2} w^2{}_{,\eta} \right] d\eta = 0 \tag{76}$$

where

$$e_1 = \beta\mu[E_\zeta(k^2 - q^4)], \quad \beta = \frac{a}{h} \tag{77}$$

Thus, for the immovable boundary condition considered here, a total of forty-two simultaneous equations are solved to obtain the unknown coefficients c_i and d_i in Equations (74) and (75). It must be pointed out here that u^0 and v^0 thus obtained are exact solutions of Equations (66) and (67) satisfying the given immovable boundary conditions completely.

Since the in-plane displacements u^o and v^o are known, these are substituted along with w from Equation (73) into Equation (68) which represents the motion of the plate in the lateral direction. The resulting equation is multiplied by two functions, $(1 - \zeta^2 - \eta^2)^2$ and $(1 - \zeta^2 - \eta^2)^3$, and integrated over the area of the plate. This procedure will result in a set of two ordinary nonlinear time-differential

TABLE 14. Values of Nondimensional Frequency Ratio $(\omega/\omega_o)10^4$ for Glass-Epoxy Elliptical Plates with $r = 1.5$, $T_s = R_i = 0$.

w_o^*	Single-Mode	Multiple-Mode
0	10,000	10,000
0.5	10,416	10,435
1.0	11,562	12,694
1.5	13,236	16,412

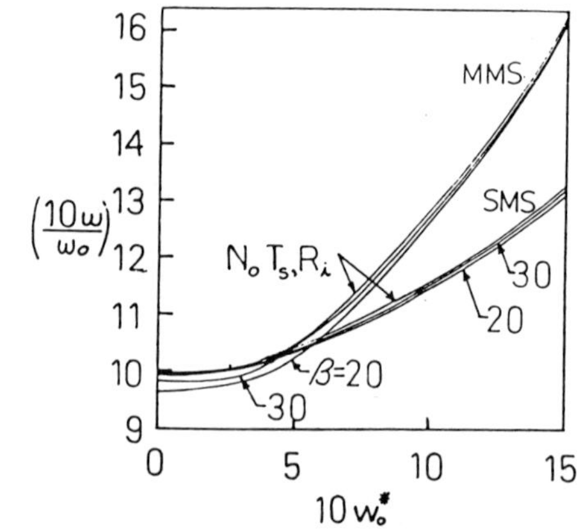

FIGURE 18. Variation of frequency ratio with amplitude for isotropic elliptical plates with r = 2, MMS: Multiple-mode solution, SMS: Single-mode solution.

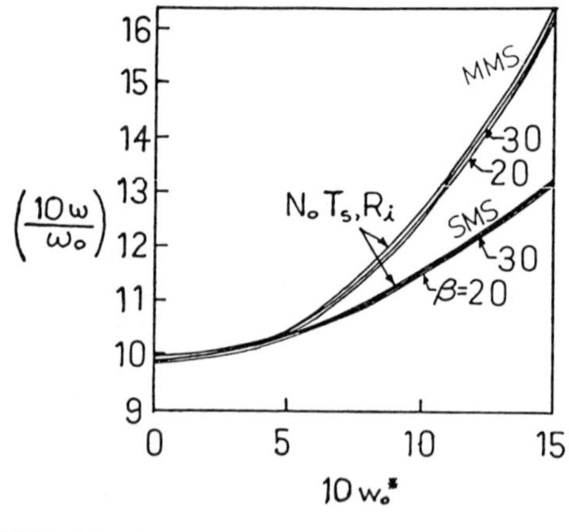

FIGURE 19. Frequency ratio versus amplitude for glass-epoxy elliptical plates with r = 2, MMS: Multiple-mode solution, SMS: Single-mode solution.

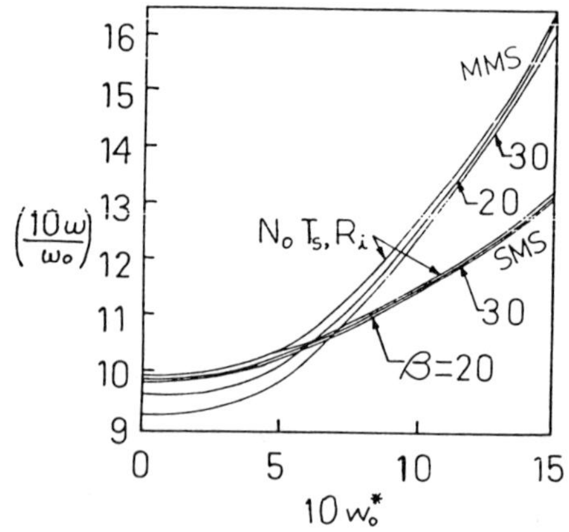

FIGURE 20. Variation of frequency ratio with amplitude for boron-epoxy elliptical plates with r = 2, MMS: Multiple-mode solution, SMS: Single-mode solution.

equations in $A(\tau)$ and $B(\tau)$ which will be coupled and are as given below:

$$\begin{aligned} & e_1(A^3)_{,\tau\tau\tau\tau} + e_2(A^2B)_{,\tau\tau\tau\tau} + e_3(AB^2)_{,\tau\tau\tau\tau} + e_4(B^3)_{,\tau\tau\tau\tau} \\ & + e_5(A)_{,\tau\tau\tau\tau\tau\tau} + e_6(B)_{,\tau\tau\tau\tau\tau\tau} + e_7(A^3)_{,\tau\tau} \\ & + e_8(A^2B)_{,\tau\tau} + e_9(AB^2)_{,\tau\tau} + e_{10}(B^3)_{,\tau\tau} \\ & + e_{11}(A)_{,\tau\tau\tau\tau} + e_{12}(B)_{,\tau\tau\tau\tau} + e_{13}A^3 + e_{14}A^2B \\ & + e_{15}AB^2 + e_{16}B^3 + e_{17}(A)_{,\tau\tau} + e_{18}(B)_{,\tau\tau} \\ & + e_{19}A + e_{20}B = q_o^* \end{aligned} \tag{78}$$

$$\begin{aligned} & g_1(A^3)_{,\tau\tau\tau\tau} + g_2(A^2B)_{,\tau\tau\tau\tau} + g_3(AB^2)_{,\tau\tau\tau\tau} + g_4(B^3)_{,\tau\tau\tau\tau} \\ & + g_5(A)_{,\tau\tau\tau\tau\tau\tau} + g_6(B)_{,\tau\tau\tau\tau\tau\tau} + g_7(A^3)_{,\tau\tau} \\ & + g_8(A^2B)_{,\tau\tau} + g_9(AB^2)_{,\tau\tau} + g_{10}(B_3)_{,\tau\tau} \\ & + g_{11}(A)_{,\tau\tau\tau\tau} + g_{12}(B)_{,\tau\tau\tau\tau} + g_{13}A^3 + g_{14}A^2B \\ & + g_{15}AB^2 + g_{16}B^3 + g_{17}(A)_{,\tau\tau} + g_{18}(B)_{,\tau\tau} \\ & + g_{19}A + g_{20}B = q_o^* \end{aligned} \tag{79}$$

Equations (78) and (79) are the same as Equations (62) and (63) obtained by means of the stress function approach. For any chosen geometric and material plate parameters, coefficients e_i and g_i of Equations (78) and (79) will be identical to the corresponding coefficients A_i and B_i of Equations (62) and (63).

In the case of dynamic problems, the ratio of the

nonlinear frequency ω, including the effects of transverse shear and rotatory inertia, to the corresponding linear frequency ω_o of the classical thin plate is computed for various non-dimensional amplitudes and thickness parameters. The results are presented in Table 14, Figures 18, 19 and 20. Similar results for immovable clamped circular plates can be readily generated by equating the aspect ratio to unity. By following a similar analysis procedure, results for stress-free circular plates can be obtained [18]. The results provided by the multiple-mode approach predict higher frequency ratios at any given amplitude of vibration as compared with the results of the single-mode approach. As can be expected, a hardening type of nonlinearity is observed for moderately thick as well as thin plates.

STRESSES DURING VIBRATION

It is important to be able to evaluate the state of stress in the plate during large amplitude vibration. For a plate undergoing large amplitude vibrations, the membrane and extreme-fibre bending stresses are

$$\sigma_x^m = \frac{E_x}{\mu}(\epsilon_1^o + q^2\epsilon_2^o), \quad \sigma_y^m = \frac{E_x}{\mu}(q^2\epsilon_1^o + k^2\epsilon_2^o)$$

$$\sigma_{xy}^m = G_{xy}\epsilon_{12}^o, \quad \sigma_x^b = \frac{E_x h}{2\mu}(w_{,xx} + q^2 w_{,yy}) \tag{80}$$

$$\sigma_y^b = \frac{E_x h}{2\mu}(q^2 w_{,xx} + k^2 w_{,yy}), \quad \sigma_{xy}^b = G_{xy} h w_{,xy}$$

in which the subscripts b and m refer to the bending effect and membrane action, respectively. The median surface strains ϵ_1^o, ϵ_2^o and $\epsilon_{12}^o(\gamma)$ are given by Equation (71) in terms of the three displacement components of the plate.

For any chosen amplitude of vibration [such as B_1 and B_2 in Equation (39)] the values of bending and membrane stresses can be obtained from Equation (80). The median surface strains in Equation (80) are calculated with the aid of the expressions for the lateral displacement w and the stress function F or the in-plane displacements u^o and v^o. Finally, these stresses may be written in the nondimensional form as

$$\bar{\sigma}_{ij} = \frac{b^2}{E_y h^2}\sigma_{ij} \tag{81}$$

BEAMS

The large amplitude vibrations of a beam whose ends are restrained from axial displacement have been treated extensively in the literature [19,20]. The most common approach in solving these problems is to assume some form of the spatial solution, usually a single term mode shape, and then solve the nonlinear time-differential equation that results for the time variable. However, some limited amount of research in this area has been directed towards multiple-mode approach to solving nonlinear beam vibration problems. McDonald [21] has treated the vibration of a uniform beam with hinged ends which are restrained, and which has arbitrary initial conditions using an expansion of elliptic functions. Dynamic coupling of the modes was found to be important in this study. Srinivasan [22] has applied a general modal approach in treating a simply supported beam and has obtained response solutions including several modes. Bennett and Eisley [23] have investigated the response and stability of multiple-mode, large amplitude, transverse vibrations of a beam using analytical and experimental techniques. All these investigations seem to indicate the need for a multiple-mode analysis if accurate results are desired.

REFERENCES

1. Timoshenko, S. and S. Woinowsky-Krieger, *Theory of Plates and Shells,* McGraw-Hill Book Co., New York (1959).
2. Ashton, J. E. and J. M. Whitney, *Theory of Laminated Plates,* Technomic Publishing Co., Inc. (1970).
3. Szilard, R., *Theory and Analysis of Plates,* Prentice Hall Inc., New Jersey (1974).
4. Vinson, J. R., *Structural Mechanics: The Behavior of Plates and Shells,* John Wiley and Sons, New York (1974).
5. Leissa, A. W., "Vibration of Plates," NASA Special Publication 160, Washington, D.C. (1969).
6. Donnell, L. H., *Beams, Plates and Shells,* McGraw-Hill Book Co., New York (1976).
7. Ugural, A. C., *Stresses in Plates and Shells,* McGraw-Hill, New York (1981).
8. Chia, C. Y., *Nonlinear Analysis of Plates,* McGraw-Hill Book Co., New York (1980).
9. Sathyamoorthy, M. and C. Y. Chia, "Effects of Transverse Shear and Rotatory Inertia on Large Amplitude Vibration of Anisotropic Skew Plates—Part I: Theory," *ASME Journal of Applied Mechanics,* Vol. 47, pp. 128–132 (1980).
10. Prabhakara, M. K. and C. Y. Chia, "Nonlinear Flexural Vibrations of Orthotropic Rectangular Plates," *Journal of Sound and Vibration,* Vol. 52, pp. 511–518 (1977).
11. Sathyamoorthy, M., "Nonlinear Dynamic Analysis of Orthotropic Elliptical Plates," *Fibre Science and Technology,* Vol. 20, pp. 135–142 (1984).
12. Sathyamoorthy, M., "Nonlinear Analysis of Composite Circular Plates," *Proceedings of the AIAA/ASME/ASCE/AHS 24th Structures, Structural Dynamics and Materials Conference,* Lake Tahoe, Nevada, pp. 693–696 (1983).
13. Chia, C. Y. and M. K. Prabhakara, "A General Mode Approach to Nonlinear Flexural Vibrations of Laminated Rectangular Plates," *ASME Journal of Applied Mechanics,* Vol. 45, pp. 623–628 (1978).

14. Chandra, R. and B. Basava Raju, "Large Deflection Vibration of Angle Ply Laminated Plates," *Journal of Sound and Vibration*, Vol. 40, pp. 393–408 (1975).
15. Sathyamoorthy, M. and M. E. Prasad, "A Multiple Mode Approach to Large Amplitude Vibration of Moderately Thick Elliptical Plates," *ASME Journal of Applied Mechanics*, Vol. 51, pp. 153–158 (1984).
16. Sathyamoorthy, M. and M. E. Prasad, "Multiple Mode Nonlinear Analysis of Circular Plates," *ASCE Journal of Engineering Mechanics*, Vol. 109, pp. 1114–1123 (1983).
17. Sathyamoorthy, M., "Multiple Mode Nonlinear Dynamic Analysis of Composite Moderately Thick Elliptical Plates," *Proceedings of the AIAA/ASME/ASCE/AHS 26th Structures*, Structural Dynamics and Materials Conference, Orlando, Florida, pp. 201–207 (1985).
18. Sathyamoorthy, M., "Multiple Mode Large Amplitude Vibration of Thick Orthotropic Circular Plates," *International Journal of Nonlinear Mechanics*, Vol. 19, pp. 341–348 (1984).
19. Sathyamoorthy, M., "Nonlinear Analysis of Beams Part I: A Survey of Recent Advances," *The Shock and Vibration Digest*, Vol. 14, No. 8, pp. 19–35 (1982).
20. Nayfeh, A. H. and D. T. Mook, *Nonlinear Oscillations*, John Wiley, New York (1979).
21. McDonald, P. H., "Nonlinear Dynamic Coupling in a Beam Vibration," *ASME Journal of Applied Mechanics*, Vol. 22, pp. 573–578 (1955).
22. Srinivasan, A. V., "Nonlinear Vibrations of Beams and Plates," *International Journal of Nonlinear Mechanics*, Vol. 1, pp. 179–191 (1966).
23. Bennett, J. A. and J. G. Eisley, "A Multiple Degree-of-Freedom Approach to Nonlinear Beam Vibrations," *AIAA Journal*, Vol. 8, pp. 734–739 (1970).

Timoshenko Beams and Mindlin Plates on Linear Viscoelastic Foundations

K. Sonoda* and H. Kobayashi*

INTRODUCTION

The Winkler's foundation model consists of a system of closely spaced spring element which behaves independently of the adjacent elements and deforms proportionally to a pressure undergone from superstructures. Therefore, this model cannot express the continuum characteristics of foundation. Though a real soil-foundation does not behave in such a discontinuous way, the Winkler's model has been one of the most widely used foundation model for the analysis of engineering problems, because of the simplest mathematical model leading to a reasonable estimate of the response of beam and plate resting on a soil-foundation [1].

On the other hand, many researchers in the field of soil mechanics have taught us a significance of time-dependent behavior due to consolidation or creep in a real soil-foundation. Considerations of these time-dependent characteristics taken into the foundation models have therefore been tried by many structural researchers. At present, one of successful attempts seems to be an adoption of linear viscoelastic model, because the theory of viscoelasticity has been well established in the field of continuum mechanics. The simplest type among linear viscoelastic models applied to the foundation models may be an analogy to the Winkler's model. This type of model consists of a system of closely spaced linear viscoelastic spring elements [2,3].

Since Freudenthal and Lorsch's pioneer work [4] for beams on the linear viscoelastic foundations of Maxwell, Kelvin (Voigt), and Standard linear solid types, many studies has been done for beams and plates [3,5–9] within the scope of classical bending theories such as Bernoulli–Euler's beam theory and Kirchhoff–Love's thin plate theory, which disregard the effect of shear deformation on the flexure. On the other hand, a number of attempts have been made to improve those theories in order to include the effect of shear deformation. The most refined theories widely used are theories of Timoshenko [10] for beams and Reissner [11] and Mindlin [12] for plates. In this chapter, we devote to the presentations of general solutions for Timoshenko beams of finite length and rectangular and circular Mindlin plates on linear viscoelastic foundations [13–16].

It is well known that the application of the correspondence principle is successfully convenient for solving linear viscoelastic problems [3,17]. For effective use of this principle, however, an elastic solution easily applicable to the inverse Laplace transforms must be obtained. However, the solution forms derived from the governing differential equations for the elastic foundation problem contain the transcendental functions including fourth root of the elastic modulus of foundation in their arguments, i.e., (1) hyperbolic and circular functions in both Timoshenko beams [18,19] and rectangular Mindlin plates [16,20], and (2) Bessel functions in circular Mindlin plates [16,21]. When the correspondence principle is applied, therefore, the inverse Laplace transforms of these functions cannot be performed analytically and then must be relied upon a numerical method, because of lacks of the effective inverse formulae available [22].

In this chapter, to avoid the difficulty of the inverse Laplace transforms, the solutions for the elastic foundation problem are given in series forms of the eigenfunctions, which are derived from the free vibration analysis of the associated Timoshenko beam and Mindlin plate. This technique has been successfully developed by the authors in the studies on Euler beam and thin plate problems [5,8,9,16]. Such a solution-form has the following advantages: (1) There is no need to solve simultaneous equations to determine the unknown constants depending on the boundary conditions and, consequently, the solution-form can be given by a single

*Department of Civil Engineering, Osaka City University, Osaka, Japan

expression valid for the entire length of beam (or the entire surface of plate) irrespective of the kinds of boundary conditions and also loading conditions; and (2) the parameters indicating the properties of beam (or plate) and foundation explicitly appear in the final solution-form not included in the arguments of the eigenfunctions.

TIMOSHENKO BEAM

Basic Equations for Elastic Foundation Problem

Consider a beam of length $2l$ whose coordinate system is shown in Figure 1. The governing equation for flexure of a Timoshenko beam on the Winkler type foundation is given in the following coupled forms:

$$EI\frac{d^2\psi}{dx^2} + \kappa FG\left(\frac{dw}{dx} - \psi\right) = 0 \qquad (1a)$$

$$\kappa FG\left(\frac{d\psi}{dx} - \frac{d^2w}{dx^2}\right) + kw = q \qquad (1b)$$

in which $w(x)$ is total deflection due to both bending and shear; $\psi(x)$ is bending slope; x is distance along the beam; E is Young's modulus; G is shear modulus; I is moment of inertia of cross section; F is cross-sectional area; κ is shear coefficient; k is modulus of foundation; and $q(x)$ is lateral load. The shear coefficient is introduced to take into account the fact that the distribution of shear stress is not constant through the beam cross section. The value of κ depends on both the shape of the cross section and Poisson's ratio [23].

Bending moment M and shearing force Q are expressed as

$$M(x) = -EI\frac{d\psi}{dx} \qquad (2a)$$

$$Q(x) = \kappa FG\left(\frac{dw}{dx} - \psi\right) \qquad (2b)$$

The three standard types of boundary conditions at a beam end are written as follows:

(1) Simply supported end: $w = M = 0$ (3a)

(2) Clamped end: $w = \psi = 0$ (3b)

(3) Free end: $M = Q = 0$ (3c)

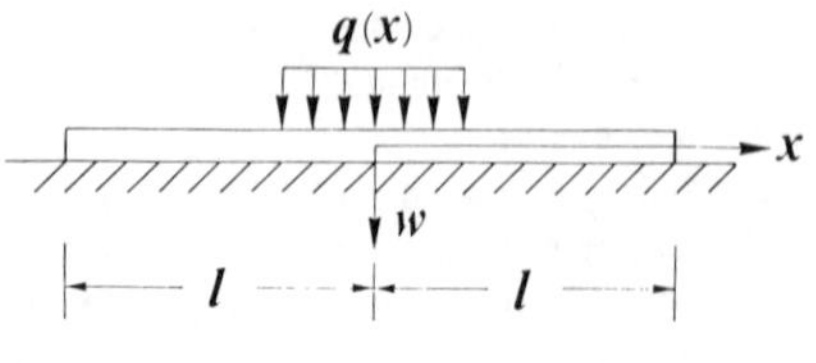

FIGURE 1. Coordinate system of Timoshenko beam.

Eigenfunction

Original dynamic Timoshenko beam equation includes both the effects of shear deformation and rotatory inertia [10]. In this quasistatic analysis, the latter effect can be deleted. Thus, the differential equation governing a free vibration of Timoshenko beam without the effect of rotatory inertia (after separation of the time variable) is reduced to

$$EI\frac{d^2\Psi}{dx^2} + \kappa FG\left(\frac{dW}{dx} - \Psi\right) = 0 \qquad (4a)$$

$$\kappa FG\left(\frac{d\Psi}{dx} - \frac{d^2W}{dx^2}\right) - CW = 0 \qquad (4b)$$

in which $C = EI(\lambda_m/l)^4$ and λ_m is the eigenvalue.

Solving Equations (4), one obtains the general expressions for the eigenfunctions as [24]:

$$W_m(x) = A_1\cosh a_m\frac{x}{l} + A_2\sinh a_m\frac{x}{l} + A_3\cos b_m\frac{x}{l} + A_4\sin b_m\frac{x}{l} \qquad (5a)$$

$$\Psi_m(x) = \frac{b_m^2}{la_m}\left(A_1\sinh a_m\frac{x}{l} + A_2\cosh a_m\frac{x}{l}\right) - \frac{a_m^2}{lb_m}\left(A_3\sin b_m\frac{x}{l} - A_4\cos b_m\frac{x}{l}\right) \qquad (5b)$$

in which

$$\begin{pmatrix} a_m \\ b_m \end{pmatrix} = \frac{1}{\sqrt{2}}\lambda_m^2[\mp S^2 + (S^4 + 4\lambda_m^{-4})^{1/2}]^{1/2}$$

$$S = \sqrt{\frac{EI}{\kappa FGl^2}} \qquad (6)$$

Substitution of Equations (5) into the boundary conditions at beam ends gives a set of four homogeneous equations with four unknowns A_i. The characteristic equation determining the eigenvalues is obtained by setting the determinant of coefficient matrix in these homogeneous equations equal to zero. To each root (eigenvalue) λ_m, only relative ratios among four unknowns are determined and then the

TABLE 1. Eigenfunctions and Characteristic Equations of Timoshenko Beam ($\xi = x/l$).

Boundary Conditions at $\xi = \pm 1$	Type of Mode About Origin	Eigenfunctions $W_m(\xi)$ and $\Psi_m(\xi)$	Characteristic Equation
Simply supported	Symmetric	$W_m(\xi) = \cos b_m\xi$ $\Psi_m(\xi) = -\dfrac{a_m^2}{lb_m}\sin b_m\xi$	$\cos b_m = 0$
	Antisymmetric	$W_m(\xi) = \sin b_m\xi$ $\Psi_m(\xi) = \dfrac{a_m^2}{lb_m}\cos b_m\xi$	$\sin b_m = 0$
Clamped	Symmetric	$W_m(\xi) = \dfrac{\cosh a_m\xi}{\cosh a_m} - \dfrac{\cos b_m\xi}{\cos b_m}$ $\Psi_m(\xi) = \dfrac{b_m^2}{la_m}\left(\dfrac{\sinh a_m\xi}{\cosh a_m} + \dfrac{a_m^3}{b_m^3}\dfrac{\sin b_m\xi}{\cos b_m}\right)$	$\dfrac{b_m^3}{a_m^3}\dfrac{\cot b_m}{\coth a_m} + 1 = 0$
	Antisymmetric	$W_m(\xi) = \dfrac{\sinh a_m\xi}{\sinh a_m} - \dfrac{\sin b_m\xi}{\sin b_m}$ $\Psi_m(\xi) = \dfrac{b_m^2}{la_m}\left(\dfrac{\cosh a_m\xi}{\sinh a_m} - \dfrac{a_m^3}{b_m^3}\dfrac{\cos b_m\xi}{\sin b_m}\right)$	$\dfrac{b_m^3}{a_m^3}\dfrac{\tan b_m}{\tanh a_m} - 1 = 0$
Free	Symmetric	$W_m(\xi) = \dfrac{\cosh a_m\xi}{\cosh a_m} + \dfrac{b_m^2}{a_m^2}\dfrac{\cos b_m\xi}{\cos b_m}$ $\Psi_m(\xi) = \dfrac{b_m^2}{la_m}\left(\dfrac{\sinh a_m\xi}{\cosh a_m} - \dfrac{a_m}{b_m}\dfrac{\sin b_m\xi}{\cos b_m}\right)$	$\dfrac{b_m}{a_m}\dfrac{\tan b_m}{\tanh a_m} + 1 = 0$
	Antisymmetric	$W_m(\xi) = \dfrac{\sinh a_m\xi}{\sinh a_m} + \dfrac{b_m^2}{a_m^2}\dfrac{\sin b_m\xi}{\sin b_m}$ $\Psi_m(\xi) = \dfrac{b_m^2}{la_m}\left(\dfrac{\cosh a_m\xi}{\sinh a_m} + \dfrac{a_m}{b_m}\dfrac{\cos b_m\xi}{\sin b_m}\right)$	$\dfrac{b_m}{a_m}\dfrac{\cot b_m}{\coth a_m} - 1 = 0$

eigenfunctions (mode functions) are obtained with an unknown multiplicative factor. Their explicit forms are tabulated in Table 1, in which the symmetric and antisymmetric eigenfunctions about the origin are separately listed for the convenience of reference.

The orthogonality relation of the eigenfunctions of Timoshenko beam without the effect of rotatory inertia is given by

$$\int_{-l}^{l} W_m(x)W_n(x)dx = 0 \quad \text{for } m \neq n \qquad (7)$$
$$= N_m \quad \text{for } m = n$$

Symbol N_m represents the norm of the eigenfunctions.

Series Solution for Elastic Foundation Problem

The governing equations (1) can be separated into two ordinary differential equations of fourth order with respect to w and ψ, respectively, and therefore, the solutions are given in the closed form [18,19]. As mentioned in Introduction, however, the solutions obtained are not appropriate for the application of correspondence principle. Therefore, an alternative solution-form of Equations (1) is given in the form of an infinite series of the eigenfunctions $W_m(x)$ and $\Psi_m(x)$. A general form of solution may be written as

$$w(x) = \frac{A + Bx}{k} + \sum_{m=1}^{\infty} w_m W_m(x) \qquad (8a)$$

$$\psi(x) = \frac{B}{k} + \sum_{m=1}^{\infty} w_m \Psi_m(x) \qquad (8b)$$

The first term on the right-hand sides of Equations (8a–b), respectively, represents a rigid-body displacement and rotation for a beam with both ends free. The constants A and B are determined from the conditions of equilibrium of a rigid beam as follows:

For symmetric loading about the origin,

$$A = \frac{1}{l}\int_0^l q(x)dx, \quad B = 0 \tag{9a}$$

For antisymmetric loading about the origin,

$$A = 0, \quad B = \frac{2}{l^2}\int_0^l q(x)xdx \tag{9b}$$

The eigenfunctions are orthogonal, and thus the lateral load $q(x)$ can be expanded into a series of the eigenfunctions as

$$q(x) = \sum_{m=1}^{\infty} q_m W_m(x) \tag{10}$$

In order to find the coefficient q_m, both sides of Equation (10) are first multiplied by $W_n(x)$ and then are integrated over the length of the beam. Making use of the orthogonality relation of the eigenfunctions, q_m is given by

$$q_m = \frac{1}{N_m}\int_{-l}^{l} q(x)W_m(x)dx \tag{11}$$

Substituting Equations (8) and (10) into Equations (1) and using Equations (4), the unknown coefficient w_m is determined from Equation (1b) as follows:

$$w_m = \frac{l^4}{EI}\frac{q_m}{\lambda_m^4 + K^4} \tag{12}$$

in which

$$K = (kl^4/EI)^{1/4} \tag{13}$$

Thus, the general solutions of Equations (1) are obtained in the forms

$$w(x) = \frac{A + Bx}{k} + \frac{l^4}{EI}\sum_{m=1}^{\infty}\frac{q_m}{\lambda_m^4 + K^4} W_m(x) \tag{14a}$$

$$\psi(x) = \frac{B}{k} + \frac{l^4}{EI}\sum_{m=1}^{\infty}\frac{q_m}{\lambda_m^4 + K^4}\Psi_m(x) \tag{14b}$$

The foundation reaction $p(x)$ is determined from the relation, $p = kw$, as

$$p(x) = A + Bx + \sum_{m=1}^{\infty}\frac{q_m K^4}{\lambda_m^4 + K^4} W_m(x) \tag{15}$$

Once the displacement components are obtained, the bending moment and shearing force are determined from Equations (2) as follows:

$$M(x) = -l^4 \sum_{m=1}^{\infty}\frac{q_m}{\lambda_m^4 + K^4} M_m(x) \tag{16a}$$

$$Q(x) = \left(\frac{l}{S}\right)^2 \sum_{m=1}^{\infty}\frac{q_m}{\lambda_m^4 + K^4} Q_m(x) \tag{16b}$$

in which

$$M_m(x) = d\Psi_m/dx \text{ and } Q_m(x) = dW_m/dx - \Psi_m \tag{17}$$

Solution for Linear Viscoelastic Foundation Problem

Figure 2 shows three types of linear viscoelastic models to be used here. By virtue of linearity of the problem, the application of the correspondence principle [3,17] to the elastic solutions yields the Laplace transforms of the viscoelastic solutions with respect to time t. For brevity only the Laplace transforms on the deflection, bending slope and foundation reaction are described as follows:

$$\overline{w}(x, s) = \frac{\overline{A}(s) + \overline{B}(s)x}{\overline{k}(s)} + \sum_{m=1}^{\infty}\frac{\overline{q}_m(s)}{EI\left(\frac{\lambda_m}{l}\right)^4 + \overline{k}(s)} W_m(x) \tag{18a}$$

$$\overline{\psi}(x, s) = \frac{\overline{B}(s)}{\overline{k}(s)} + \sum_{m=1}^{\infty}\frac{\overline{q}_m(s)}{EI\left(\frac{\lambda_m}{l}\right)^4 + \overline{k}(s)} \Psi_m(x) \tag{18b}$$

$$\overline{p}(x, s) = \overline{A}(s) + \overline{B}(s)x + \sum_{m=1}^{\infty}\frac{\overline{q}_m(s)\overline{k}(s)}{EI\left(\frac{\lambda_m}{l}\right)^4 + \overline{k}(s)} W_m(x) \tag{18c}$$

in which a bar over the symbols denotes the Laplace transform defined as [22]

$$\overline{F}(s) = \int_0^{\infty} F(t)\exp(-st)dt$$

$$F(t) = \frac{1}{2\pi i}\int_{\gamma - i\infty}^{\gamma + i\infty}\overline{F}(s)\exp(st)ds \tag{19}$$

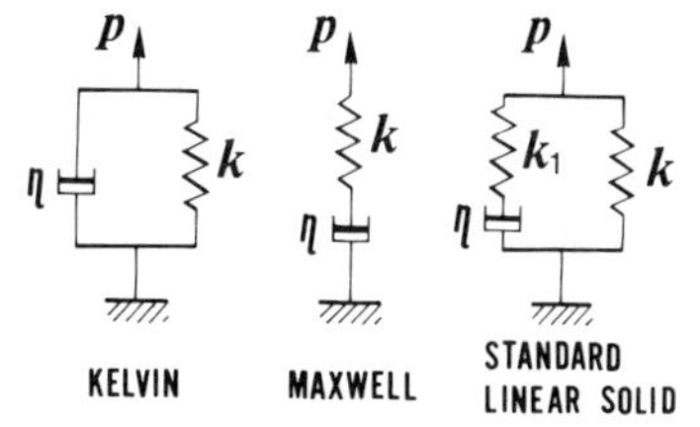

FIGURE 2. Three types of viscoelastic models.

The symbol $\bar{k}(s)$ is referred to as a viscoelastic operator [3] and is generally expressed by a rational function of the transformed variable, s. The explicit forms of $\bar{k}(s)$ are given in Table 2.

In the case of a time-independent load expressed by the unit step function, the following results are obtained:

$$\bar{A}(s) = A/s,\ \bar{B}(s) = B/s, \text{ and } \bar{q}_m(s) = q_m/s \quad (20)$$

Substituting $\bar{k}(s)$ in Table 2 and Equation (20) into Equations (18), and taking the inverse Laplace transforms of the resulting equations, the viscoelastic solutions are obtained as

$$w(x, t) = w(x) - \frac{A + Bx}{k} T_1(t) - \frac{l^4}{EI} \sum_{m=1}^{\infty} \frac{q_m T_2(t)}{\lambda_m^4 + K^4} W_m(x) \quad (21a)$$

$$\psi(x, t) = \psi(x) - \frac{B}{k} T_1(t) - \frac{l^4}{EI} \sum_{m=1}^{\infty} \frac{q_m T_2(t)}{\lambda_m^4 + K^4} \Psi_m(x) \quad (21b)$$

$$p(x, t) = p(x) + K^4 \sum_{m=1}^{\infty} \frac{q_m T_3(t)}{\lambda_m^4 + K^4} W_m(x) \quad (21c)$$

In the above expressions, $w(x)$ and $\psi(x)$ are the static part of solution given by Equations (14) and $p(x)$ is the foundation reaction given by Equation (15); and $T_1(t)$, $T_2(t)$ and $T_3(t)$ are functions of time t only, which are explicitly presented in Table 2 for each viscoelastic foundation model. Furthermore, the bending moment and shearing force can be determined by the same manner described above or by substituting Equations (21a) and (21b) into Equations (2).

Numerical Consideration

The method developed in the previous sections is applied to the case of a beam with both ends free and subjected to a uniformly distributed load q symmetrically placed over the finite region $-c \le x \le c$.

EIGENVALUE

The characteristic equations, which are highly transcendental (except for a beam with both ends simply supported), yield an infinite number of roots (eigenvalues) for a given parameter S. Three sets of a_m, b_m and λ_m in Equation (6) are not independent from each other but are connected by the relations

$$a_m = \frac{b_m}{\sqrt{1 + (b_m S)^2}}, \text{ and } \lambda_m = \frac{b_m}{\sqrt[4]{1 + (b_m S)^2}} \quad (22)$$

Therefore, the problem of obtaining the eigenvalues is reduced to the determination of b_m. The characteristic equation is easily solved numerically by use of a trial-and-error method such as the Regula–Falsi method.

The first twenty eigenvalues of a beam with both ends free are given in Table 3 for a parameter S varying from 0 to 0.20 with an interval 0.05. In $S = 0$, the eigenvalues coincide with those from the elementary beam (Euler beam) theory. It may be noticed from the table that the eigenvalues do not differ appreciably between the Timoshenko theory and Euler theory for the first few modes. The difference appears in higher modes as S becomes larger. In other words, the effect of shear deformation is considerable in higher modes.

CONVERGENCE OF SERIES

The quasi-static solutions consist of two parts: (1) the static part, corresponding to the solution for an elastic foundation problem, and (2) the time-dependent part. The series for the deflection of the static part converges by taking only the first several terms of the series. While the series for the bending moment and shearing force of the static part are not so rapidly convergent. Therefore, they should be transformed to a rapidly convergent form as follows:

$$M(x) = M^R(x) + (lK)^4 \sum_{m=1}^{\infty} \frac{q_m}{\lambda_m^4(\lambda_m^4 + K^4)} M_m(x) \quad (23a)$$

$$Q(x) = Q^R(x) - K^4 \left(\frac{l}{S}\right)^2 \sum_{m=1}^{\infty} \frac{q_m}{\lambda_m^4(\lambda_m^4 + K^4)} Q_m(x) \quad (23b)$$

in which $M^R(x)$ and $Q^R(x)$, respectively, are the bending moment and the shearing force of a rigid beam on the elastic foundation with the same modulus and subjected to the given lateral load. They can be exactly evaluated from the elementary statics.

Table 4 shows the results of a convergency study for the

TABLE 2. Viscoelastic Operator $\bar{k}(s)$ and Time Functions $T_1(t)$, $T_2(t)$ and $T_3(t)$.

	Kelvin	Maxwell	Standard Linear Solid
$\bar{k}(s)$	$k(1 + \tau_c s)$	$k\left(\frac{\tau_r s}{1 + \tau_r s}\right)$	$k\left(\frac{1 + \tau_c s}{1 + \tau_r s}\right)$
$T_1(t)$	$\exp\left(-\frac{t}{\tau_c}\right)$	$-\frac{t}{\tau_r}$	$\left(\frac{K_1}{K_2}\right)^4 \exp\left(-\frac{t}{\tau_c}\right)$
$T_2(t)$	$T(t)$	$\left(\frac{K}{\lambda_m}\right)^4 [T(t) - 1]$	$\frac{(K_1/\lambda_m)^4}{1 + (K_2/\lambda_m)^4} T(t)$
$T_3(t)$	$\left(\frac{\lambda_m}{K}\right)^4 T(t)$	$T(t) - 1$	$\frac{(K_1/K)^4}{1 + (K_2/\lambda_m)^4} T(t)$
$T(t)$	$\exp\left\{-\left[1 + \left(\frac{\lambda_m}{K}\right)^4\right]\frac{t}{\tau_c}\right\}$	$\exp\left\{-\left[\frac{1}{1 + (K/\lambda_m)^4}\right]\frac{t}{\tau_r}\right\}$	$\exp\left\{-\left[\frac{(K_2/K)^4 + (K_2/\lambda_m)^4}{1 + (K_2/\lambda_m)^4}\right]\frac{t}{\tau_c}\right\}$
Remarks	$\tau_c = \frac{\eta}{k}, K^4 = \frac{kl^4}{EI}$	$\tau_r = \frac{\eta}{k}, K^4 = \frac{kl^4}{EI}$	$\tau_c = \left(1 + \frac{k}{k_1}\right)\frac{\eta}{k}, \tau_r = \frac{\eta}{k_1}, K^4 = \frac{kl^4}{EI}, K_1^4 = \frac{k_1 l^4}{EI}, K_2^4 = K_1^4 + K^4$

TABLE 3. Eigenvalues λ_m for Symmetric Mode of Timoshenko Beam with Both Ends Free.

	S				
m	0*	0.05	0.1	0.15	0.2
1	2.365020	2.360495	2.347164	2.325725	2.297228
2	5.497804	5.415953	5.202106	4.921734	4.629158
3	8.639380	8.315682	7.611333	6.882850	6.259801
4	11.780972	10.996147	9.599610	8.412874	7.514326
5	14.922565	13.441050	11.271356	9.680416	8.560466
6	18.064158	15.660632	12.716581	10.779170	9.475684
7	21.205750	17.678378	13.997914	11.761220	10.300138
8	24.347343	19.521634	15.157071	12.657454	11.057136
9	27.488936	21.216576	16.222122	13.487482	11.761462
10	30.630528	22.786153	17.212470	14.264512	12.423149
11	33.772121	24.249639	18.141949	14.997882	13.049362
12	36.913714	25.622895	19.020725	15.694463	13.645435
13	40.055306	26.918866	19.856476	16.359476	14.215465
14	43.196899	28.148113	20.655140	16.997001	14.762682
15	46.338492	29.319278	21.421411	17.610295	15.289691
16	49.480084	30.439472	22.159070	18.202008	15.798625
17	52.621677	31.514591	22.871210	18.774333	16.291264
18	55.763270	32.549557	23.560400	19.329109	16.769107
19	58.904862	33.548512	24.228802	19.867896	17.233436
20	62.046455	34.514969	24.878251	20.392034	17.685354

*Elementary beam.

TABLE 4. Convergence of Series of Equation (14a) for Maximum Deflection $(w)_{x=0}$ and Equations (23) for Maximum Bending Moment $(M)_{x=0}$ and Shearing Force $(Q)_{x=0.21}$.

	S = 0.1					
	K = 2			K = 4		
m	w^a	M^b	Q^c	w^a	M^b	Q^c
1	0.29787	0.57221	−0.14138	0.45346	0.21005	−0.11178
2	0.30506	0.56994	−0.14022	0.54062	0.18249	−0.09767
3	0.30614	0.56980	−0.14009	0.55681	0.18038	−0.09586
4	0.03636	0.56978	−0.14008	0.56019	0.18014	−0.09567
5	0.30638	—	—	0.56039	0.18014	−0.09567
6	0.30634	—	—	0.55984	0.18015	−0.09566
7	0.30630	—	—	0.55927	0.18016	−0.09563
8	0.30628	—	—	0.55889	0.18017	−0.09562
9	0.30627	—	—	0.55872	—	−0.09561
10	0.30627	—	—	0.55871	—	−0.09561
Exact [19]	0.30628	0.56978	−0.14008	0.55893	0.18017	−0.09560
Elementary beam	0.30158	0.57575	−0.14008	0.53405	0.18956	−0.09934

	S = 0.2					
	K = 2			K = 4		
m	w^a	M^b	Q^c	w^a	M^b	Q^c
1	0.30569	0.55635	−0.13991	0.46122	0.19780	−0.11035
2	0.31738	0.55295	−0.13804	0.58560	0.16161	−0.09041
3	0.31967	0.55266	−0.13779	0.61720	0.15757	−0.08699
4	0.32021	0.55262	−0.13776	0.62535	0.15704	−0.08656
5	0.32023	0.55263	−0.13776	0.62568	0.15702	−0.08656
6	0.32012	—	−0.13776	0.62386	0.15707	−0.08650
7	0.31999	—	−0.13775	0.62192	0.15712	−0.08642
8	0.31991	—	—	0.62062	0.15714	−0.08637
9	0.31987	—	—	0.62003	—	−0.08636
10	0.31987	—	—	0.62001	—	−0.08636
Exact [19]	0.31993	0.55263	−0.13775	0.62085	0.15714	−0.08632
Elementary beam	0.30158	0.57575	−0.14008	0.53405	0.18956	−0.09934

[a]Multiplier = q/k.
[b]Multiplier = $10^{-1}\, ql^2$.
[c]Multiplier = ql.

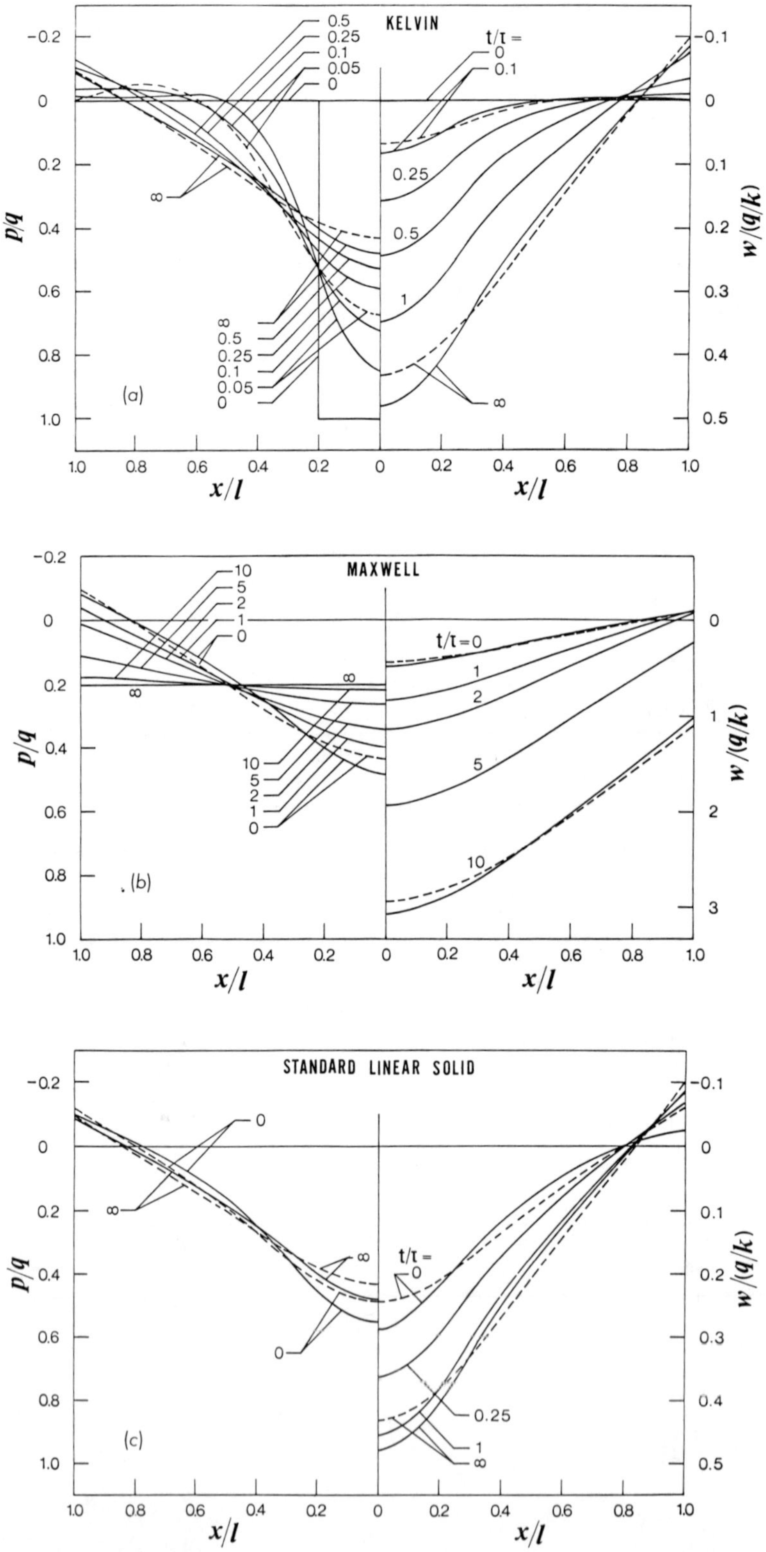

FIGURE 3. Deflection w and foundation reaction p for beam on: (a) Kelvin-type foundation; (b) Maxwell-type foundation; (c) Standard linear solid-type foundation. (——), Timoshenko beam; (---), elementary beam; $k_1 = k$; and $\tau = \eta/k$.

series (21a) and (23) under the parameters: $c/l = 0.2$; $S = 0.1, 0.2$: and $K = 2, 4$. From the table, it can be understood that the numerical values obtained for any combinations of S and K converge very rapidly and the differences from the exact values are within 1% when m is larger than 3. Also, it is noted that the convergence of the bending moment and shearing force are faster than that of the deflection because the rapidly convergent forms of Equations (23) are used.

On the other hand, as the series of the time-dependent part in Equations (21) contains the terms of $\exp[-F(\lambda_m)t]$, as shown in Table 2, this part converges rapidly.

EFFECT OF SHEAR DEFORMATION

It can be seen from Table 4 that the effect of shear deformation on the flexure of beam becomes more significant as S and K increase. For example, the maximum deflectin given by Timoshenko beam theory is greater by appro-proximately 16%, in the case of $S = 0.2$ and $K = 4$, than that given by the elementary beam theory [25]. While the maximum bending moment and the maximum shearing force are reduced by approximately 17% and 15%, respectively.

QUASISTATIC RESPONSE

Figures 3(a–c) show the variations in space and time of the deflection and the foundation reaction for Timoshenko beams on the Kelvin, Maxwell and Standard linear solid-types viscoelastic foundations, respectively. The characteristic constants are taken as $c/l = 0.2$, $S = 0.2$, $K = 3$, and $k_1 = k$, and the terms of the series of eigenfunctions are truncated by $m = 10$. The results from the elementary beam theory are also plotted by the dotted lines for comparison.

It is observed that the initial and final states for the Kelvin-type foundation become a beam on a rigid foundation and a beam on an elastic foundation, respectively. For the Maxwell-type foundation, the initial state becomes a beam on an elastic foundation and the final state becomes a floating beam. Finally, the initial and final states for the Standard linear solid-type foundation becomes beams on elastic foundations with moduli of the foundation being $2k(=k_1 + k)$ and k, respectively.

RECTANGULAR MINDLIN PLATE

Basic Equations for Elastic Foundation Problem

The governing equation of a rectangular Mindlin plate on the Winkler-type elastic foundation becomes

$$\frac{D}{2}\left[(1 - \nu)\nabla^2\psi_x + (1 + \nu)\frac{\partial\phi}{\partial x}\right] + \kappa Gh\left(\frac{\partial w}{\partial x} - \psi_x\right) = 0 \tag{24a}$$

$$\frac{D}{2}\left[(1 - \nu)\nabla^2\psi_y + (1 + \nu)\frac{\partial\phi}{\partial y}\right] + \kappa Gh\left(\frac{\partial w}{\partial y} - \psi_y\right) = 0 \tag{24b}$$

$$\kappa Gh(\nabla^2 w - \phi) = kw - q \tag{24c}$$

in which

$$\phi = \frac{\partial\psi_x}{\partial x} + \frac{\partial\psi_y}{\partial y} \tag{24d}$$

In the above equations, $w(x, y)$ is the deflection at the plate middle surface; $\psi_x(x, y)$ and $\psi_y(x, y)$ are the angular rotations of the normal to the middle surface in the x- and y-coordinates directions, respectively; D is the flexural rigidity given by $D = Eh^3/12(1 - \nu^2)$; E is Young's modulus; G is shear modulus; ν is Poisson's ratio; h is plate thickness; κ is shear coefficient similar to the one introduced in Timoshenko beam and is taken equal to 5/6; k is elastic modulus of the foundation; $q(x, y)$ is surface load; and $\nabla^2(=\partial^2/\partial x^2 + \partial^2/\partial y^2)$ is Laplace operator.

Stress couples and resultants are expressed in terms of the displacement components as follows:

$$M_x = -D\left(\frac{\partial\psi_x}{\partial x} + \nu\frac{\partial\psi_y}{\partial y}\right) \tag{25a}$$

$$M_y = -D\left(\frac{\partial\psi_y}{\partial y} + \nu\frac{\partial\psi_x}{\partial x}\right) \tag{25b}$$

$$M_{xy} = -\frac{1 - \nu}{2}D\left(\frac{\partial\psi_x}{\partial y} + \frac{\partial\psi_y}{\partial x}\right) \tag{25c}$$

and

$$Q_x = \kappa Gh\left(\frac{\partial w}{\partial x} - \psi_x\right) \tag{25d}$$

$$Q_y = \kappa Gh\left(\frac{\partial w}{\partial y} - \psi_y\right) \tag{25e}$$

Figure 4 shows the plate geometry and the coordinate system used here. We now restrict our problem to the plate with two opposite edges ($x = 0, a$) simply supported and other remaining two edges ($y = \pm b/2$) restrained in any manner such as simply supported, clamped, or free. In Mindlin's plate theory, three boundary conditions along the edge of plate should be specified, because the governing Equations (24) may be reduced to the sixth order differential equation. For example on an edge y = constant, these

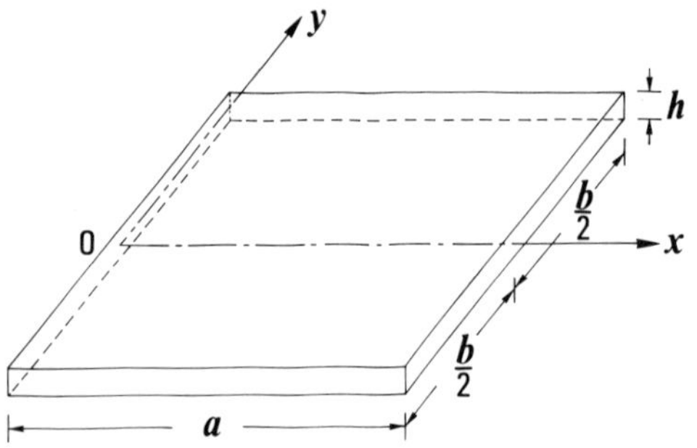

FIGURE 4. Coordinate system of rectangular Mindlin plate.

boundary conditions are written as follows:

(1) Simply supported edge:

$$M_y = \psi_x = w = 0 \tag{26a}$$

(2) Clamped edge:

$$\psi_x = \psi_y = w = 0 \tag{26b}$$

(3) Free edge:

$$M_y = M_{xy} = Q_y = 0 \tag{26c}$$

Similar expressions can be written down for an edge $x =$ constant.

Eigenfunction

For a free vibration of rectangular Mindlin plate neglecting the effect of rotatory inertia, the eigenfunctions are governed by the following set of equations [12]:

$$\frac{D}{2}\left[(1 - \nu)\nabla^2\Psi_x + (1 + \nu)\frac{\partial\Phi}{\partial x}\right] + \kappa Gh\left(\frac{\partial W}{\partial x} - \Psi_x\right) = 0 \tag{27a}$$

$$\frac{D}{2}\left[(1 - \nu)\nabla^2\Psi_y + (1 + \nu)\frac{\partial\Phi}{\partial y}\right] + \kappa Gh\left(\frac{\partial W}{\partial y} - \Psi_y\right) = 0 \tag{27b}$$

$$\kappa Gh(\nabla^2 W - \Phi) + CW = 0 \tag{27c}$$

in which

$$\Phi = \frac{\partial\Psi_x}{\partial x} + \frac{\partial\Psi_y}{\partial y}, \quad C = D(\lambda_{mn}/a)^4 \tag{27d}$$

where λ_{mn} are eigenvalues to be determined from the boundary conditions. The eigenfunction $W_{mn}(x, y)$ corresponds to the deflection and $\Psi_{x,mn}(x, y)$ and $\Psi_{y,mn}(x, y)$ are angular rotations. According to the procedure proposed by Mindlin [12], Equations (27) can be reduced to the uncoupled forms:

$$\begin{aligned} &[\nabla^2 + (\delta_1/a)^2]w_1 = 0 \\ &[\nabla^2 - (\delta_i/a)^2]w_i = 0, \; i = 2, 3 \end{aligned} \tag{28}$$

provided that three functions $w_i (i = 1, 2, 3)$ are related to $W_{mn}(x, y)$, $\Psi_{x,mn}(x, y)$ and $\Psi_{y,mn}(x, y)$ as follows:

$$W_{mn}(x, y) = w_1 + w_2 \tag{29a}$$

$$\Psi_{x,mn}(x, y) = (1 - \sigma_1)\frac{\partial w_1}{\partial x} + (1 - \sigma_2)\frac{\partial w_2}{\partial x} + \frac{\partial w_3}{\partial y} \tag{29b}$$

$$\Psi_{y,mn}(x, y) = (1 - \sigma_1)\frac{\partial w_1}{\partial y} + (1 - \sigma_2)\frac{\partial w_2}{\partial y} - \frac{\partial w_3}{\partial x} \tag{29c}$$

The symbols appearing in Equations (28) and (29) mean

$$\begin{aligned} &\delta_1^2, \delta_2^2 = \frac{1}{2}\lambda_{mn}^4[\pm S + (S^2 + 4\lambda_{mn}^{-4})^{1/2}] \\ &\delta_3^2 = 2/[S(1 - \nu)] \\ &\sigma_1, \sigma_2 = S(\delta_2^2, -\delta_1^2) \end{aligned} \tag{30}$$

where $S = D/\kappa Gha^2$.

Considering the simply supported condition at the edges $x = 0$ and $x = a$, the solutions of Equations (28) can be taken in the forms:

(1) Symmetric mode about x-axis

$$\begin{aligned} w_1 &= A_1 \cosh(\eta_1 y/b)\sin(m\pi x/a) \quad \text{for } \delta_1 < m\pi \\ &= A_1 \cos(\eta_1 y/b)\sin(m\pi x/a) \quad \text{for } \delta_1 > m\pi \\ w_2 &= A_2 \cosh(\eta_2 y/b)\sin(m\pi x/a) \\ w_3 &= A_3 \sinh(\eta_3 y/b)\cos(m\pi x/a) \end{aligned} \tag{31}$$

TABLE 5. Constants A_1, A_2, A_3 and Characteristic Equation for Symmetric Mode About x-axis of Rectangular Mindlin Plate with Two Opposite Edges Simply Supported.

Boundary Conditions at $y = \pm b/2$		Constants A_1, A_2, and A_3	Characteristic Equation
Simply supported	$\delta_1 < m\pi$		There exist no eigenvalue problem.
	$\delta_1 > m\pi$	$A_1 = 1, A_2 = A_3 = 0$	$\cos(\eta_1/2) = 0, \eta_1 = (2n - 1)\pi$
Clamped	$\delta_1 < m\pi$		There exist no eigenvalue problem.
	$\delta_1 > m\pi$	$A_1 = 1/\cos(\eta_1/2)$ $A_2 = -1/\cosh(\eta_2/2)$ $A_3 = (\sigma_1 - \sigma_2)m\pi\xi/[\eta_3 \cosh(\eta_3/2)]$	$(1 - \sigma_1)\eta_1\eta_3 \tan(\eta_1/2) + (1 - \sigma_2)\eta_2\eta_3 \tanh(\eta_2/2)$ $-(\sigma_1 - \sigma_2)(m\pi\xi)^2 \tanh(\eta_3/2) = 0$
Free	$\delta_1 < m\pi$	$A_1 = 1/\sinh(\eta_1/2)$ $A_2 = -\dfrac{\eta_1[\eta_2^2 - \nu(m\pi\xi)^2]}{\eta_2[\eta_1^2 - \nu(m\pi\xi)^2]\sinh(\eta_2/2)}$ $A_3 = \dfrac{(1 - \nu)(\sigma_1 - \sigma_2)\eta_1 m\pi\xi}{[\eta_1^2 - \nu(m\pi\xi)^2]\sinh(\eta_3/2)}$	$(1 - \sigma_1)\eta_2[\eta_1^2 - \nu(m\pi\xi)^2]^2 \coth(\eta_1/2)$ $-(1 - \sigma_2)\eta_1[\eta_2^2 - \nu(m\pi\xi)^2]^2 \coth(\eta_2/2)$ $+(1 - \nu)^2(\sigma_1 - \sigma_2)\eta_1\eta_2\eta_3(m\pi\xi)^2 \coth(\eta_3/2) = 0$*
	$\delta_1 > m\pi$	$A_1 = 1/\sin(\eta_1/2)$ $A_2 = -\dfrac{\eta_1[\eta_2^2 - \nu(m\pi\xi)^2]}{\eta_2[\eta_1^2 + \nu(m\pi\xi)^2]\sinh(\eta_2/2)}$ $A_3 = \dfrac{(1 - \nu)(\sigma_1 - \sigma_2)\eta_1 m\pi\xi}{[\eta_1^2 + \nu(m\pi\xi)^2]\sinh(\eta_3/2)}$	$(1 - \sigma_1)\eta_2[\eta_1^2 + \nu(m\pi\xi)^2]^2 \cot(\eta_1/2)$ $+(1 - \sigma_2)\eta_1[\eta_2^2 - \nu(m\pi\xi)^2]^2 \coth(\eta_2/2)$ $-(1 - \nu)^2(\sigma_1 - \sigma_2)\eta_1\eta_2\eta_3(m\pi\xi)^2 \coth(\eta_3/2) = 0$

*Equation has only one root corresponding to $n = 1$ for each 'm'.

(2) Antisymmetric mode about x-axis

$$
\begin{aligned}
w_1 &= A_1 \sinh(\eta_1 y/b)\sin(m\pi x/a) \quad \text{for } \delta_1 < m\pi \\
&= A_1 \sin(\eta_1 y/b)\sin(m\pi x/a) \quad \text{for } \delta_1 > m\pi \\
w_2 &= A_2 \sinh(\eta_2 y/b)\sin(m\pi x/a) \\
w_3 &= A_3 \cosh(\eta_3 y/b)\cos(m\pi x/a)
\end{aligned}
\tag{32}
$$

In the above expressions,

$$
\begin{aligned}
\eta_1 &= \xi(m^2\pi^2 - \delta_1^2)^{1/2} \quad \text{for } \delta_1 < m\pi \\
&= \xi(\delta_1^2 - m^2\pi^2)^{1/2} \quad \text{for } \delta_1 > m\pi \\
\eta_i &= \xi(\delta_i^2 + m^2\pi^2)^{1/2}, \; i = 2, 3
\end{aligned}
\tag{33}
$$

where $\xi(=b/a)$ is the plate aspect ratio.

Performing the eigenvalue analysis by an usual way as described in Timoshenko beam problem, the characteristic equation and the corresponding unknowns A_i can be determined as tabulated in Table 5.

The eigenfunctions possess the orthogonality properties,

$$
\int_{-b/2}^{b/2} \int_0^a W_{mn}(x, y)W_{ij}(x, y)dxdy = 0 \quad \text{for } i \neq m \text{ or } j \neq n
$$
$$
= N_{mn} \quad \text{for } i = m \text{ and } j = n \tag{34}
$$

Series Solution for Elastic Foundation Problem

The governing Equations (24) can be separated into two partial differential equations on the deflection w and the stress function $\psi(=\partial\psi_x/\partial y - \partial\psi_y/\partial x)$, respectively, one of which is of fourth order, and the other of second order. Therefore, the general solution of Equations (24) for a rectangular Mindlin plate with two opposite edges simply supported can be derived in the form of Levy-type single series [16]. As mentioned in Introduction, however, this solution form cannot be utilized as an elastic solution when the correspondence principle is applied. Then we write the solution form alternately in the form of an infinite series of the eigenfunctions:

$$\begin{bmatrix} w(x, y) \\ \psi_x(x, y) \\ \psi_y(x, y) \end{bmatrix} = \sum_{m=1}^{\infty} \sum_{n=1}^{\infty} Z_{mn} \begin{bmatrix} W_{mn}(x, y) \\ \Psi_{x,mn}(x, y) \\ \Psi_{y,mn}(x, y) \end{bmatrix} \tag{35}$$

where Z_{mn} is unknown coefficient. In this case, a rigid body displacement does not occur.

The surface load $q(x, y)$ is expanded into a double series of the eigenfunctions:

$$q(x, y) = \sum_{m=1}^{\infty} \sum_{n=1}^{\infty} q_{mn} W_{mn}(x, y) \tag{36}$$

where

$$q_{mn} = \frac{1}{N_{mn}} \int_{-b/2}^{b/2} \int_0^a q(x, y) W_{mn}(x, y) dx dy \tag{37}$$

Substituting Equations (35) and (36) into the governing Equations (24) and taking into account Equations (27), we obtain

$$Z_{mn} = \frac{q_{mn}}{D(\lambda_{mn}/a)^4 + k} \tag{38}$$

and thus the expressions for the displacement components and bending moments are determined as follows:

$$\begin{bmatrix} w(x, y) \\ \psi_x(x, y) \\ \psi_y(x, y) \end{bmatrix} = \frac{a^4}{D} \sum_{m=1}^{\infty} \sum_{n=1}^{\infty} \frac{q_{mn}}{\lambda_{mn}^4 + K^4} \begin{bmatrix} W_{mn}(x, y) \\ \Psi_{x,mn}(x, y) \\ \Psi_{y,mn}(x, y) \end{bmatrix} \tag{39a}$$

$$\begin{bmatrix} M_x(x, y) \\ M_y(x, y) \end{bmatrix} = -a^4 \sum_{m=1}^{\infty} \sum_{n=1}^{\infty} \frac{q_{mn}}{\lambda_{mn}^4 + K^4} \begin{bmatrix} M_{x,mn}(x, y) \\ M_{y,mn}(x, y) \end{bmatrix} \tag{39b}$$

where the nondimensional parameter K is defined by

$$K = (ka^4/D)^{1/4} \tag{40}$$

and the symbols $M_{x,mn}(x, y)$ and $M_{y,mn}(x, y)$ mean

$$M_{x,mn}(x, y) = \partial \Psi_{x,mn}/\partial x + \nu \partial \Psi_{y,mn}/\partial y$$
$$M_{y,mn}(x, y) = \partial \Psi_{y,mn}/\partial y + \nu \partial \Psi_{x,mn}/\partial x \tag{41}$$

The foundation reaction $p(x, y)$ is obtained from the relation, $p = kw$, as

$$p(x, y) = \sum_{m=1}^{\infty} \sum_{n=1}^{\infty} \frac{q_{mn} K^4}{\lambda_{mn}^4 + K^4} W_{mn}(x, y) \tag{42}$$

Solution for Linear Viscoelastic Foundation Problem

By the same procedure described in Timoshenko beam problem, the deflection and foundation reaction under a unit step function load are determined as follows:

$$w(x, y, t) = w(x, y) - \frac{a^4}{D} \sum_{m=1}^{\infty} \sum_{n=1}^{\infty} \frac{q_{mn} T_2(t)}{\lambda_{mn}^4 + K^4} W_{mn}(x, y) \tag{43a}$$

$$p(x, y, t) = p(x, y) + K^4 \sum_{m=1}^{\infty} \sum_{n=1}^{\infty} \frac{q_{mn} T_3(t)}{\lambda_{mn}^4 + K^4} W_{mn}(x, y) \tag{43b}$$

where $w(x, y)$ and $p(x, y)$ are the elastic solutions given by Equations (39a) and (42). The time functions $T_2(t)$ and $T_3(t)$ have the same expressions with those of Timoshenko beam problem presented in Table 2, provided that the associated parameters λ_{mn}, $K^4 = ka^4/D$ and $K_1^4 = k_1 a^4/D$ are used. Similarly, the bending moments and shearing forces are also obtained through the same procedure.

Numerical Consideration

For an illustration of the application of the method developed, a square plate with two opposite edges simply supported and the other two free is selected as a typical example. The plate is subjected to a uniform load q over the square area $(0.2a \times 0.2a)$ in the middle portion of plate. Poisson's ratio and nondimensional foundation modulus are fixed as $\nu = 1/6$ and $K = 3$, respectively, for numerical computations.

EIGENVALUE

The eigenvalues which are roots of the characteristic equation are calculated by the Regula–Falsi method. The first thirty eigenvalues are given in Table 6 for a thickness-side ratio h/a varying from 0 to 0.2 with an interval 0.05. The eigenvalues for $h/a = 0$ correspond to those of thin plate. For the effect of shear deformation on the eigenvalues, a similar tendency to the case of Timoshenko beam can be obtained.

CONVERGENCE OF SERIES

For the same reason mentioned in the section of Timoshenko beam, the series expressions of the elastic solution may be transformed to a rapidly convergent form as follows:

$$w(x, y) = w^*(x, y) - \frac{a^4}{D} \times \sum_{m=1}^{\infty} \sum_{n=1}^{\infty} \frac{q_{mn} K^4}{\lambda_{mn}^4 (\lambda_{mn}^4 + K^4)} W_{mn}(x, y) \tag{44a}$$

$$\begin{bmatrix} M_x(x, y) \\ M_y(x, y) \end{bmatrix} = \begin{bmatrix} M_x^*(x, y) \\ M_y^*(x, y) \end{bmatrix} + a^4 \times \sum_{m=1}^{\infty} \sum_{n=1}^{\infty} \frac{q_{mn}K^4}{\lambda_{mn}^4(\lambda_{mn}^4 + K^4)} \begin{bmatrix} M_{x,mn}(x, y) \\ M_{y,mn}(x, y) \end{bmatrix} \tag{44b}$$

In the above, $w^*(x, y)$ is deflection of the plate with the same boundary condition but without foundation and carrying a given surface load. Similarly, $M_x^*(x, y)$ and $M_y^*(x, y)$ also are the bending moments in the same plate. Following the approach proposed by Marguerre and Woernle [26], they can be obtained in the forms of Levy-type single series.

In order to compare the convergence of the series (44) with that of the series (39), the numerical results for maximum deflection and maximum bending moments are shown in Table 7 for the thickness side ratio $h/a = 0.15$. From consideration of symmetry with respect to the axis $x = a/2$, only odd numbers 1, 3, 5, . . . are taken in the term 'm' of series.

It is found from this table that all the values calculated from the series (44) are very rapidly convergent and a sufficient accuracy within three significant figures can be obtained when both m and n are taken up to three terms, while in the series (39) 39 terms for m and 21 terms for n are required to obtain the same degree of accuracy.

EFFECT OF SHEAR DEFORMATION

Table 8 shows values of the deflections and the bending moments at the center and at the middle point of free edge for the plates with different values of h/a. It is observed from the table that w and M_y at the center and M_x at the middle point of free edge increase as h/a increases, while w at the middle point of free edge and M_x at the center decrease as h/a increases. However, the effect of shear deformation on the bending moments is small in comparison with the deflection.

TABLE 6. Eigenvalues λ_{mn} for Symmetric Mode About x-axis of Square Mindlin Plate with Two Opposite Edges Simply Supported and Other Edges Free (ν = 1/6 and κ = 5/6).

Mode	$h/a = 0$*		$h/a = 0.05$		$h/a = 0.1$		$h/a = 0.15$		$h/a = 0.2$	
No.	Eigenvalue	m	Eigenvalue	m	Eigenvalue	m	Eigenvalue	m	Eigenvalue	m
1	3.130950	1	3.125720	1	3.111603	1	3.089194	1	3.059383	1
2	6.156034	1	6.091196	1	5.982891	1	5.842448	1	5.681569	1
3	6.271825	2	6.233964	2	6.129626	2	5.974241	2	5.786442	2
4	8.531693	2	8.407578	2	8.149881	2	7.813958	2	7.449219	2
5	9.413196	3	9.290069	3	8.967349	3	8.533944	3	8.068789	3
6	11.156372	3	10.932209	3	10.420983	3	9.790218	3	9.158064	3
7	11.599612	1	11.379641	1	10.877922	1	10.248562	1	9.608827	1
8	12.554640	4	12.271696	4	11.584094	4	10.761396	4	9.970605	4
9	13.114967	2	12.757668	2	12.027866	2	11.183753	2	10.378935	2
10	13.948232	4	13.549568	4	12.655725	4	11.647285	4	10.719440	4
11	15.115532	3	14.584268	3	13.525494	3	12.386298	3	11.366290	3
12	15.696113	5	15.161486	5	13.972600	5	12.702877	5	11.589231	5
13	16.847422	5	16.182848	5	14.784158	5	13.352428	5	12.131610	5
14	17.394003	4	16.633964	4	15.151891	4	13.668603	4	12.415218	4
15	17.634846	1	16.951400	1	15.504795	1	14.016982	1	12.746801	1
16	18.632975	2	17.786111	2	16.125089	2	14.415075	6	13.002760	6
17	18.837600	6	17.947086	6	16.146676	6	14.488806	2	13.127683	2
18	19.814393	6	18.781301	6	16.777584	6	14.909715	6	13.414164	6
19	19.864187	5	18.795957	5	16.801236	5	14.947680	5	13.460769	5
20	20.118157	3	19.031716	3	17.042088	3	15.185430	3	13.691356	3
21	21.944905	4	20.554372	4	18.129889	7	15.947391	7	14.265436	7
22	21.979095	7	20.620963	7	18.145080	4	16.020870	4	14.368849	4
23	22.476514	6	21.005147	6	18.421758	6	16.189970	6	14.478558	6
24	22.825386	7	21.314216	7	18.632392	7	16.336855	7	14.589531	7
25	23.810716	1	22.235865	1	19.348108	5	16.929413	5	15.108250	5
26	24.015837	5	22.250670	5	19.437721	1	17.048800	1	15.234665	1
27	24.532778	2	22.788286	2	19.814033	2	17.329658	2	15.413928	8
28	25.120594	8	23.179825	8	19.948099	8	17.338291	8	15.459823	7
29	25.195599	7	23.217880	7	19.988081	7	17.382961	7	15.463510	2
30	25.665638	3	23.654086	3	20.357175	8	17.653726	8	15.677105	8

*Thin plate.

TABLE 7. Comparison of Convergence of the Series (39) and (44) for Deflection and Bending Moments at the Center Plate (ν = 1/6, b/a = 1, c/a = 1/5, h/a = 0.15 and K = 3).

	w		M_x		M_y	
m, n	Using Equation (39a)	Using Equation (44a)	Using Equation (39b)	Using Equation (44b)	Using Equation (39b)	Using Equation (44b)
1, 2	0.53268	0.58703	0.52633	0.81519	0.31554	0.56360
3, 3	0.57390	0.58663	0.71561	0.81320	0.48235	0.56235
5, 4	0.58454	0.58661	0.79389	0.81302	0.55095	0.56221
7, 5	0.58765	0.58660	0.82230	0.81299	0.57386	0.56219
9, 6	0.58806	—	0.82742	—	0.57616	0.56219
11,7	0.58759	—	0.82349	—	0.57123	0.56220
19, 11	0.58638	—	0.81060	—	0.55989	—
39, 21	0.58657	—	0.81264	—	0.56186	—
59, 31	0.58659	—	0.81288	—	0.56209	—
79, 41	0.58660	—	0.81295	—	0.56215	—
99, 51	—	—	0.81297	—	0.56217	—
Multiplier	$10^{-3}\ qa^4/D$		$10^{-2}\ qa^2$		$10^{-2}\ qa^2$	

TABLE 8. Deflection and Bending Moments at the Center and at the Middle Point of Free Edge of Plate for Different Values of Thickness to Side Ratio h/a (ν = 1/6, b/a = 1, c/a = 1/5 and K = 3).

	at $x = a/2,\ y = 0$			at $x = a/2,\ y = \pm b/2$	
h/a	w	M_x	M_y	w	M_x
0*	0.5345	0.8244	0.5603	0.3487	0.3488
0.05	0.5407	0.8236	0.5615	0.3464	0.3552
0.10	0.5581	0.8198	0.5621	0.3443	0.3593
0.15	0.5866	0.8130	0.5622	0.3425	0.3613
0.20	0.6258	0.8037	0.5618	0.3409	0.3610
Multiplier	$10^{-3}\ qa^4/D$	$10^{-2}\ qa^2$		$10^{-3}\ qa^4/D$	$10^{-2}\ qa^2$

*Thin plate.

QUASISTATIC RESPONSE

Numerical calculations for viscoelastic foundation problems are carried out using the thickness side ratio $h/a = 0.15$ and $k_1 = 2k$. In Figure 5, the time histories of deflection w, bending moment M_x and foundation reaction p at the center of the plate are shown in terms of the nondimensional time $t/\tau(\tau = \eta/k)$. The results from a thin plate theory [27] are also plotted by the dotted lines for comparison.

The initial states of the plates on the Kelvin, Maxwell, and Standard linear solid-type foundation are the states of plates on rigid foundation and on elastic foundation with the moduli of the foundation being k and $3k$ $(=k_1 + k)$, respectively. The deflections, bending moments, and foundation reactions for the Kelvin and Standard linear solid-type foundations asymptotically approach to those for an elastic foundation with modulus of foundation being k. A state attained after about 10τ elapses becomes almost constant and therefore it may be regarded as the final state from a practical point of view. On the other hand, the final state of the plate on the Maxwell type foundation is the state of plate without an elastic foundation, because the foundation reaction vanishes (Figure 5(c)). The time for reaching to a state almost regarded as the final state is at 20τ.

CIRCULAR MINDLIN PLATE

Basic Equations for Elastic Foundation Problem

Transforming Equations (24) into a form with polar coordinates (r, θ) as shown in Figure 6, one obtains

$$\frac{D}{2}\left\{(1-\nu)\left[\left(\nabla^2 - \frac{1}{r^2}\right)\psi_r - \frac{2}{r^2}\frac{\partial\psi_\theta}{\partial\theta}\right] + (1+\nu)\frac{\partial\phi}{\partial r}\right\} + \kappa Gh\left(\frac{\partial w}{\partial r} - \psi_r\right) = 0 \quad (45a)$$

$$\frac{D}{2}\left\{(1-\nu)\left[\left(\nabla^2 - \frac{1}{r^2}\right)\psi_\theta + \frac{2}{r^2}\frac{\partial\psi_r}{\partial\theta}\right] + (1+\nu)\frac{1}{r}\frac{\partial\phi}{\partial\theta}\right\} + \kappa Gh\left(\frac{1}{r}\frac{\partial w}{\partial\theta} - \psi_\theta\right) = 0 \quad (45b)$$

$$\kappa Gh(\nabla^2 w - \phi) = kw - q \quad (45c)$$

in which

$$\phi = \frac{\partial\psi_r}{\partial r} + \frac{1}{r}\frac{\partial\psi_\theta}{\partial\theta} + \frac{1}{r}\psi_r \quad (45d)$$

and $w(r, \theta)$ is the deflection; $\psi_r(r, \theta)$ and $\psi_\theta(r, \theta)$ are the angular rotations of the normal to the middle surface in

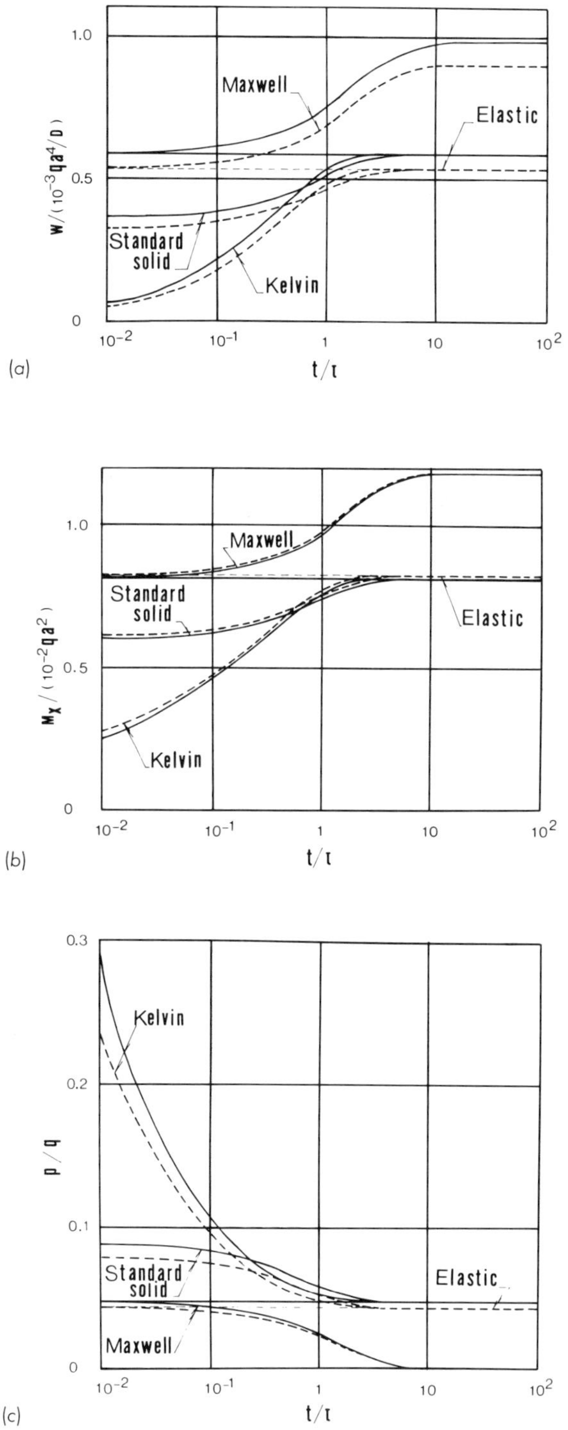

FIGURE 5. (a) Center deflection w; (b) center bending moment M_x; (c) center foundation reaction p versus time. (———), Mindlin plate; (---), thin plate; $k_1 = 2k$; and $\tau = \eta/k$.

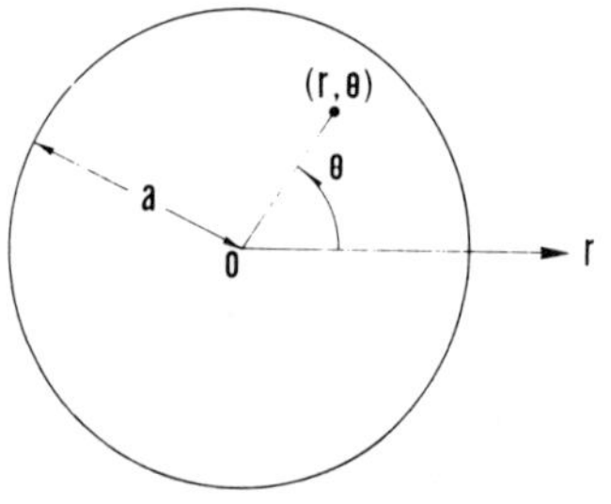

FIGURE 6. Coordinate system of circular Mindlin plate.

radial and circumferential directions, respectively; and ∇^2 is Laplace operator in polar coordinates (r, θ) given by $\nabla^2 = \partial^2/\partial r^2 + \partial/r\partial r + \partial^2/r^2\partial\theta^2$.

Expressions for moments M_r, M_θ, $M_{r\theta}$ and shear forces Q_r, Q_θ are written as

$$M_r = -D\left[\frac{\partial\psi_r}{\partial r} + \frac{\nu}{r}\left(\psi_r + \frac{\partial\psi_\theta}{\partial\theta}\right)\right] \tag{46a}$$

$$M_\theta = -D\left[\frac{1}{r}\left(\psi_r + \frac{\partial\psi_\theta}{\partial\theta}\right) + \nu\frac{\partial\psi_r}{\partial r}\right] \tag{46b}$$

$$M_{r\theta} = -\frac{1-\nu}{2}D\left[\frac{1}{r}\left(\frac{\partial\psi_r}{\partial\theta} - \psi_\theta\right) + \frac{\partial\psi_\theta}{\partial r}\right] \tag{46c}$$

and

$$Q_r = \kappa Gh\left(\frac{\partial w}{\partial r} - \psi_r\right) \tag{46d}$$

$$Q_\theta = \kappa Gh\left(\frac{1}{r}\frac{\partial w}{\partial\theta} - \psi_\theta\right) \tag{46e}$$

Three boundary conditions at the edge $r = a$ are

(1) Simply supported edge:

$$M_r = M_{r\theta} = w = 0 \tag{47a}$$

(2) Clamped edge:

$$\psi_r = \psi_\theta = w = 0 \tag{47b}$$

(3) Free edge:

$$M_r = M_{r\theta} = Q_r = 0 \tag{47c}$$

Eigenfunction

The transformed governing equations for a free vibration problem are expressed as

$$\frac{D}{2}\left\{(1-\nu)\left[\left(\nabla^2 - \frac{1}{r^2}\right)\Psi_r - \frac{2}{r^2}\frac{\partial\Psi_\theta}{\partial\theta}\right] + (1+\nu)\frac{\partial\Phi}{\partial r}\right\} + \kappa Gh\left(\frac{\partial W}{\partial r} - \Psi_r\right) = 0 \tag{48a}$$

$$\frac{D}{2}\left\{(1-\nu)\left[\left(\nabla^2 - \frac{1}{r^2}\right)\Psi_\theta + \frac{2}{r^2}\frac{\partial\Psi_r}{\partial\theta}\right] + (1+\nu)\frac{1}{r}\frac{\partial\Phi}{\partial\theta}\right\} + \kappa Gh\left(\frac{1}{r}\frac{\partial W}{\partial\theta} - \Psi_\theta\right) = 0 \tag{48b}$$

$$\kappa Gh(\nabla^2 W - \Phi) + CW = 0 \tag{48c}$$

in which

$$\Phi = \frac{\partial\Psi_r}{\partial r} + \frac{1}{r}\frac{\partial\Psi_\theta}{\partial\theta} + \frac{1}{r}\Psi_r,\ C = D(\lambda_{mn}/a)^4 \tag{48d}$$

Since the solution of Equations (27) is given by Equations (29) together with Equations (28), the solution of Equations (48) in polar coordinates may be written as:

$$W_{mn}(r,\theta) = w_1 + w_2 \tag{49a}$$

$$\Psi_{r,mn}(r,\theta) = (1-\sigma_1)\frac{\partial w_1}{\partial r} + (1-\sigma_2)\frac{\partial w_2}{\partial r} + \frac{1}{r}\frac{\partial w_3}{\partial\theta} \tag{49b}$$

$$\Psi_{\theta,mn}(r,\theta) = (1-\sigma_1)\frac{1}{r}\frac{\partial w_1}{\partial\theta} + (1-\sigma_2)\frac{1}{r}\frac{\partial w_2}{\partial\theta} - \frac{\partial w_3}{\partial r} \tag{49c}$$

For a solid circular plate under consideration, $w_i (i = 1, 2, 3)$ are given in the forms:

$$\begin{aligned} w_1 &= A_1 J_n(\delta_1 r/a)\cos n\theta \\ w_2 &= A_2 I_n(\delta_2 r/a)\cos n\theta \\ w_3 &= A_3 I_n(\delta_3 r/a)\sin n\theta \end{aligned} \tag{50}$$

in which A_i are arbitrary constants; $J_n(x)$ and $I_n(x)$ denote, respectively, the ordinary and modified Bessel functions of the first kind of order n with real argument x. The symbols σ_i and δ_j appearing in Equations (49) and (50) have the similar forms to those given in Equation (30).

Substitution of Equations (49), under the consideration of Equation (50), into the boundary conditions in Equations

(47) leads to the following set of homogeneous equations for A_i:

$$\begin{bmatrix} C_{11}C_{12}C_{13} \\ C_{21}C_{22}C_{23} \\ C_{31}C_{32}C_{33} \end{bmatrix} \cdot \begin{bmatrix} A_1 \\ A_2 \\ A_3 \end{bmatrix} = 0 \qquad (51)$$

Elements C_{ij} in the coefficient matrix are given for the following cases:

(1) Simply supported edge:

$$C_{11} = (1 - \sigma_1) \times [\delta_1^2 J_n''(\delta_1) + v\delta_1 J_n'(\delta_1) - vn^2 J_n(\delta_1)]$$

$$C_{12} = (1 - \sigma_2) \times [\delta_2^2 I_n''(\delta_2) + v\delta_2 I_n'(\delta_2) - vn^2 I_n(\delta_2)]$$

$$C_{13} = (1 - v)n[\delta_3 I_n'(\delta_3) - I_n(\delta_3)] \qquad (52a)$$

$$C_{21} = -2(1 - \sigma_1)n[\delta_1 J_n'(\delta_1) - J_n(\delta_1)]$$

$$C_{22} = -2(1 - \sigma_2)n[\delta_2 I_n'(\delta_2) - I_n(\delta_2)]$$

$$C_{23} = -[\delta_3^2 I_n''(\delta_3) - \delta_3 I_n'(\delta_3) + n^2 I_n(\delta_3)]$$

$$C_{31} = J_n(\delta_1),\ C_{32} = I_n(\delta_2),\ \text{and}\ C_{33} = 0$$

(2) Clamped edge:

$$C_{11} = (1 - \sigma_1)\delta_1 J_n'(\delta_1)$$

$$C_{12} = (1 - \sigma_2)\delta_2 I_n'(\delta_2)$$

$$C_{13} = nI_n(\delta_3)$$

$$C_{21} = -(1 - \sigma_1)nJ_n(\delta_1) \qquad (52b)$$

$$C_{22} = -(1 - \sigma_2)nJ_n(\delta_2)$$

$$C_{23} = -\delta_3 I_n'(\delta_3)$$

$C_{3j}(j = 1, 2, 3)$ are the same as for a simply supported edge.

(3) Free edge:

$C_{ij}(i = 1, 2$ and $j = 1, 2, 3)$ are the same as those for a simply supported edge.

$$C_{31} = \sigma_1\delta_1 J_n'(\delta_1)$$

$$C_{32} = \sigma_2\delta_2 I_n'(\delta_2) \qquad (52c)$$

$$C_{33} = -nI_n(\delta_3)$$

In the above expressions (52a–c), the prime denotes differentiation of Bessel functions $J_n(x)$ and $I_n(x)$ with respect to x.

For a non-trivial solution, the characteristic equation will be derived from Equation (51) by usual manner. In the case of axially symmetric bending, displacements and resultant forces depend on r only, and therefore all partial derivatives with respect to θ vanish. For this case, the eigenfunctions take the following simple forms:

$$W_m(r) = J_0(\delta_1 r/a) + \mu_m I_0(\delta_2 r/a) \qquad (53a)$$

$$\Psi_{r,m}(r) = -(1 - \sigma_1)(\delta_1/a)J_1(\delta_1 r/a) + \mu_m(1 - \sigma_2)(\delta_2/a)I_1(\delta_2 r/a) \qquad (53b)$$

The corresponding characteristic equations and the constant μ_m are listed in Table 9 for three types of boundary conditions.

The orthogonality relation for the eigenfunctions of a circular Mindlin plate may be expressed as

$$\int_{-\pi}^{\pi}\int_0^a W_{mn}(r, \theta)W_{ij}(r, \theta)rdrd\theta = 0 \quad \text{for } i \neq m \text{ or } j \neq n$$
$$= N_{mn} \quad \text{for } i = m \text{ and } j = n \qquad (54)$$

With the aid of the integral formulae for the product of Bessel functions [28], the above integral is easily evaluated.

Series Solutions for Elastic Foundation Problem

Here, we restrict our problem to a case where the surface load is symmetrically distributed about the lines $\theta = 0$ and π.

For the same reason described in the foregoing two sections, the solution to Equations (45) should be expressed by using the eigenfunctions. In addition to the displacements due to elastic flexure, the rigid body displacement is considered in the plate with free edge. Thus the complete form of solution can be taken in the form:

$$w(r, \theta) = \frac{A + Br\cos\theta}{k} + \sum_{m=1}^{\infty}\sum_{n=0}^{\infty} Z_{mn}W_{mn}(r, \theta) \tag{55a}$$

$$\psi_r(r, \theta) = \frac{B\cos\theta}{k} + \sum_{m=1}^{\infty}\sum_{n=0}^{\infty} Z_{mn}\Psi_{r,mn}(r, \theta) \tag{55b}$$

$$\psi_\theta(r, \theta) = -\frac{B\sin\theta}{k} + \sum_{m=1}^{\infty}\sum_{n=0}^{\infty} Z_{mn}\Psi_{\theta,mn}(r, \theta) \tag{55c}$$

From the condition of equilibrium of a rigid circular plate, the constants A and B are determined as

$$A = \frac{1}{\pi a^2}\int_{-\pi}^{\pi}\int_0^a q(r, \theta) r dr d\theta$$
$$B = \frac{4}{\pi a^4}\int_{-\pi}^{\pi}\int_0^a q(r, \theta) r^2 \cos\theta dr d\theta \tag{56}$$

Next, the surface load $q(r, \theta)$ is also expressed by a double series of the eigenfunctions:

$$q(r, \theta) = \sum_{m=1}^{\infty}\sum_{n=0}^{\infty} q_{mn}W_{mn}(r, \theta) \tag{57}$$

where

$$q_{mn} = \frac{1}{N_{mn}}\int_{-\pi}^{\pi}\int_0^a q(r, \theta)W_{mn}(r, \theta) r dr d\theta \tag{58}$$

Substituting Equations (55) and (57) into Equations (45), the first two equations of Equation (45) are identically satisfied and then the unknown coefficient Z_{mn} is determined from the last equation as follows:

$$Z_{mn} = \frac{a^4}{D}\frac{q_{mn}}{\lambda_{mn}^4 + K^4} \tag{59}$$

in which $K^4 = ka^4/D$.

Then, the general solutions of Equations (45) are obtained in the forms

$$w(r, \theta) = \frac{A + Br\cos\theta}{k} + \frac{a^4}{D} \times \sum_{m=1}^{\infty}\sum_{n=0}^{\infty}\frac{q_{mn}}{\lambda_{mn}^4 + K^4}W_{mn}(r, \theta) \tag{60a}$$

$$\psi_r(r, \theta) = \frac{B\cos\theta}{k} + \frac{a^4}{D} \times \sum_{m=1}^{\infty}\sum_{n=0}^{\infty}\frac{q_{mn}}{\lambda_{mn}^4 + K^4}\Psi_{r,mn}(r, \theta) \tag{60b}$$

$$\psi_\theta(r, \theta) = -\frac{B\sin\theta}{k} + \frac{a^4}{D} \times \sum_{m=1}^{\infty}\sum_{n=0}^{\infty}\frac{q_{mn}}{\lambda_{mn}^4 + K^4}\Psi_{\theta,mn}(r, \theta) \tag{60c}$$

The foundation reaction $p(r, \theta)$ is determined from the relation, $p = kw$, as

$$p(r, \theta) = A + Br\cos\theta + \sum_{m=1}^{\infty}\sum_{n=0}^{\infty}\frac{q_{mn}K^4}{\lambda_{mn}^4 + K^4}W_{mn}(r, \theta) \tag{61}$$

Solution for Linear Viscoelastic Foundation Problem

Applying the correspondence principle to the elastic solutions, the deflection and foundation reaction under a unit step function load are obtained as follows:

$$w(r, \theta, t) = w(r, \theta) - \frac{A + Br\cos\theta}{k}T_1(t) - \frac{a^4}{D}\sum_{m=1}^{\infty}\sum_{n=0}^{\infty}\frac{q_{mn}T_2(t)}{\lambda_{mn}^4 + K^4}W_{mn}(r, \theta) \tag{62a}$$

$$p(r, \theta, t) = p(r, \theta) + K^4 \times \sum_{m=1}^{\infty}\sum_{n=0}^{\infty}\frac{q_{mn}T_3(t)}{\lambda_{mn}^4 + K^4}W_{mn}(r, \theta) \tag{62b}$$

where the first term on the right hand side of each expression means the elastic solution. The time functions $T_1(t)$, $T_2(t)$ and $T_3(t)$ have the same expressions given in Table 2, provided that the associated parameters λ_{mn}, $K^4 = ka^4/D$ and $K_1^4 = k_1a^4/D$ are used.

TABLE 9. Constant μ_m and Characteristic Equation for Axially Symmetric Mode (n = 0) of Circular Mindlin Plate.

Boundary Condition at $r = a$	μ_m	Characteristic Equation
Simply supported	$-\dfrac{J_0(\delta_1)}{I_0(\delta_2)}$	$(1-\sigma_1)\delta_1 \dfrac{J_1(\delta_1)}{J_0(\delta_1)} + (1-\sigma_2)\delta_2 \dfrac{I_1(\delta_2)}{I_0(\delta_2)} = \dfrac{\delta_1^2+\delta_2^2}{(1-\nu)}$
Clamped	$-\dfrac{J_0(\delta_1)}{I_0(\delta_2)}$	$(1-\sigma_1)\delta_1 \dfrac{J_1(\delta_1)}{J_0(\delta_1)} + (1-\sigma_2)\delta_2 \dfrac{I_1(\delta_2)}{I_0(\delta_2)} = 0$
Free	$\dfrac{\delta_2 J_1(\delta_1)}{\delta_1 I_1(\delta_2)}$	$(1-\sigma_1)\delta_1^3 \dfrac{J_0(\delta_1)}{J_1(\delta_1)} + (1-\sigma_2)\delta_2^3 \dfrac{I_0(\delta_2)}{I_1(\delta_2)} = (1-\nu)(\delta_1^2+\delta_2^2)$

TABLE 10. Eigenvalues λ_m for Axially Symmetric Mode of Circular Mindlin Plate with Free Edge (ν = 1/6 and $\varkappa$ = 5/6).

	h/a				
m	0*	0.1	0.2	0.3	0.4
1	2.947926	2.939242	2.913899	2.873867	2.821961
2	6.176678	6.064437	5.779567	5.421696	5.063369
3	9.352393	8.959816	8.133748	7.310182	6.623506
4	12.511580	11.617182	10.074048	8.796202	7.842041
5	15.663888	14.036327	11.713196	10.038500	8.868557
6	18.812690	16.233541	13.137587	11.122310	9.772247
7	21.959465	18.234028	14.405821	12.095250	10.589623
8	25.104962	20.064998	15.556859	12.985887	11.342228
9	28.249602	21.751847	16.617047	13.812566	12.043898
10	31.393640	23.316643	17.604721	14.587757	12.704115
11	34.537238	24.777883	18.533041	15.320345	13.329704
12	37.680504	26.150802	19.411737	16.016910	13.925775
13	40.823515	27.447866	20.248191	16.682479	14.496272
14	43.966326	28.679272	21.048143	17.320993	15.044319
15	47.108975	29.853394	21.816146	17.935611	15.572439
16	50.251493	30.977143	22.555875	18.528912	16.082710
17	53.393903	32.056265	23.270352	19.103035	16.576865
18	56.536223	33.095570	23.962087	19.659776	17.056371
19	59.678466	34.099115	24.633198	20.200666	17.522481
20	62.820645	35.070350	25.285488	20.727015	17.976276

*Thin plate.

TABLE 11. Convergence of Maximum Deflection, Bending Moment and Shear Force (ν = 1/6, h/a = 0.481, K = 2.5, and b/a = 0.4).

m	$(w)_{r=0}$	$(M_r)_{r=0}$	$(Q_r)_{r=0.4a}$
1	0.37661	0.20479	−0.33470
2	0.42270	0.20102	−0.32601
3	0.42126	0.20108	−0.32600
4	0.41520	0.20121	−0.32561
5	0.41337	0.20123	−0.32556
10	0.41502	0.20121	−0.32547
20	0.41538	—	−0.32545
30	0.41543	—	—
40	0.41545	—	—
50	—	—	—
Panc [21]	0.4155	0.2012	−0.3254
Multiplier	q/k	qb^2	qb

Numerical Consideration

In this section, we shall consider axially symmetric bending of a circular plate with a free edge, subjected to a uniformly distributed load q over a circular area of radius b. Numerical calculations are made for the following parameters: $\nu = 1/6$, $b/a = 0.4$ and $K = 2.5$.

EIGENVALUE

The first twenty eigenvalues calculated by Regula–Falsi method are presented in Table 10 for a thickness-radius ratio h/a varying from 0 to 0.4 with an interval 0.1. The eigenvalues for $h/a = 0$ correspond to those of thin circular plate. The effect of shear deformation on the eigenvalues has a similar tendency to the case of Timoshenko beam.

CONVERGENCE OF SERIES

The rapidly convergent series for resultants forces takes the forms

$$M_r(r) = M_r^R(r) + (aK)^4 \times \sum_{m=1}^{\infty} \frac{q_m}{\lambda_m^4(\lambda_m^4 + K^4)} M_m(r) \tag{63a}$$

$$Q_r(r) = Q_r^R(r) - K^4 \frac{a^2}{S} \times \sum_{m=1}^{\infty} \frac{q_m}{\lambda_m^4(\lambda_m^4 + K^4)} Q_m(r) \tag{63b}$$

in which

$$M_m(r) = d\Psi_{r,m}/dr + \Psi_{r,m}/r$$
$$Q_m(r) = dW_m/dr - \Psi_{r,m} \tag{64}$$

and M_r^R and Q_r^R are, respectively, the radial bending moment and the transverse shear force in the plate subjected to both the given surface load and the foundation reaction as the rigid circular plate. They can be accurately calculated from Marguerre and Woernle's formulation for static Mindlin plates [26].

First, the values of the maximum deflection, radial bending moment, and transverse shear force for the elastic foundation problem are compared with those of the solution of the differential equation of Panc's component theory [21] which is quite similar to the static Mindlin theory when the shear coefficient κ is taken as 5/6. Table 11 shows the results of convergency study for the thickness to radius ratio h/a = 0.481 used by Panc. From the table it follows that all

TABLE 12. Deflection, Bending Moment, and Shear Force for Different Values of Thickness to Radius Ratio h/a (ν = 1/6, K = 2.5, and b/a = 0.4).

	w		M_r		Q_r
h/a	r/a: 0	0.4	0	0.4	0.4
0*	0.3494	0.2647	0.2217	0.0881	−0.3474
0.1	0.3525	0.2656	0.2207	0.0876	−0.3463
0.2	0.3616	0.2685	0.2179	0.0862	−0.3433
0.3	0.3765	0.2730	0.2133	0.0840	−0.3384
0.4	0.3963	0.2791	0.2071	0.0811	−0.3318
Multiplier	q/k		qb^2		qb

*Thin plate.

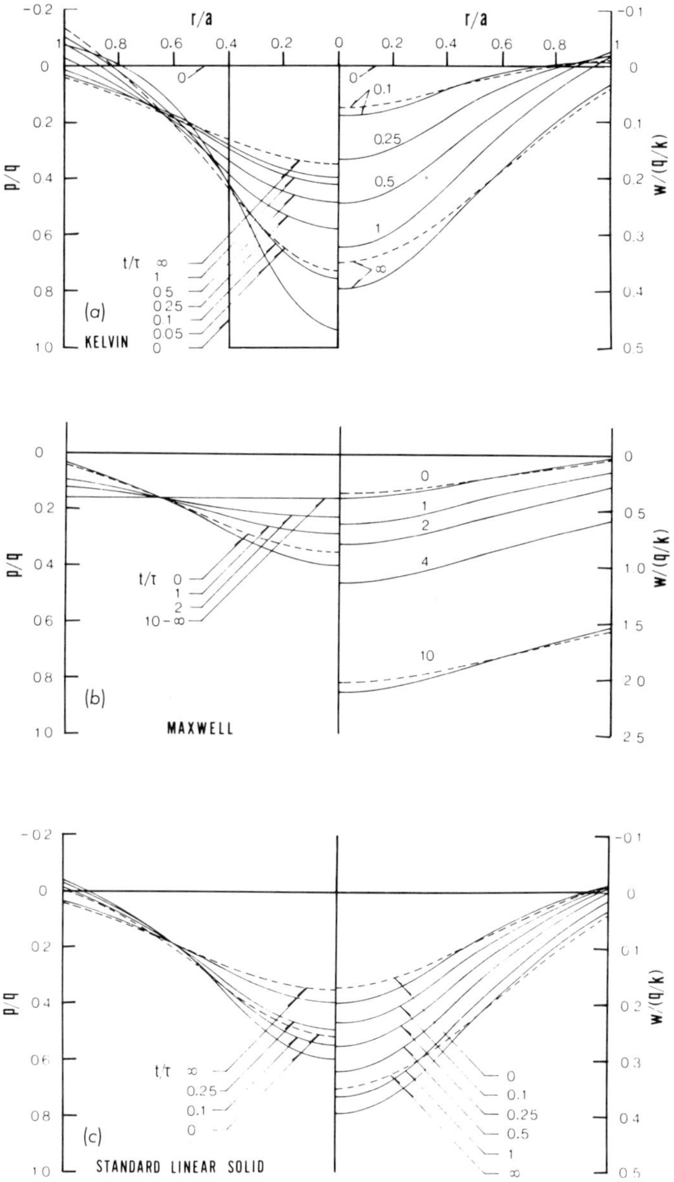

FIGURE 7. Deflection w and foundation reaction p for circular plate on: (a) Kelvin-type foundation; (b) Maxwell-type foundation; (c) Standard linear solid-type foundation. (——), Mindlin plate; (---), thin plate; $k_1 = 2k$; and $\tau = \eta/k$.

the values obtained by the present method are very rapidly convergent and have good agreement with Panc's results when up to ten terms of a series are taken in the numerical calculation. Also, it is noted that the convergence of the bending moment and the transverse shear force is faster than that of the deflection, because the rapidly convergent forms of Equations (63) are used.

EFFECT OF SHEAR DEFORMATION

Table 12 shows the deflection, radial bending moment, and transverse shear force at the center of the plate and at the edge of the uniformly loaded portion for the different thickness to radius ratios. It is observed from the table that the deflection increases as the thickness increases, while the bending moment and transverse shear force decrease as the thickness increases.

QUASISTATIC RESPONSE

Figures 7(a–c) show the variations in space and time of the deflection and the foundation reaction for a thick circular plate on the Kelvin, Maxwell, and Standard linear solid-types of viscoelastic foundations, respectively, where the characteristic constants are taken as $h/a = 0.4$ and $k_1 = 2k$. The results from thin plate theory [29] are also presented by the dotted lines for comparison.

It is observed that the initial and final states for the Kelvin-type foundation become the initial state for the perfectly rigid foundation and the final state for the elastic foundation, respectively. For the Maxwell-type foundation, the initial state is the state of a circular plate on an elastic foundation and the final state becomes a floating circular plate. Finally, the circular plate on an elastic foundation with the moduli of the foundation being $3k$ $(=k_1 + k)$ and k, respectively.

CONCLUDING REMARKS

General solutions of the quasi-static bending problems of finite-length Timoshenko beams and rectangular and circular Mindlin plates on linear viscoelastic foundations are given in the series forms of the eigenfunctions, which are derived from the free vibration problems of the associated Timoshenko beams and Mindlin plates without the presence of foundation.

Although only three elementary models of viscoelastic foundation are treated in this chapter, the method of solution developed herein has applicability to a wide range of models of linear viscoelastic foundation.

REFERENCES

1. Selvadurai, A. P. S., *Elastic Analysis of Soil-Foundation Interaction*, (Development in Geotechnical Engineering, Vol. 17), Elsevier Scientific Publishing, Amsterdom-Oxford-New York, 1979.
2. Keer, A. D., ''Elastic and Viscoelastic Foundation Models,'' *Journal of Applied Mechanics*, Vol. 31, No. 3, 1964, pp. 491–498.
3. Flügge, W., *Viscoelasticity*, 2nd ed., Springer-Verlag, Berlin/Heidelberg, 1975.
4. Freudenthal, A. M. and Lorsch, H. G., ''The Infinite Elastic Beam on a Linear Viscoelastic Foundation,'' *Journal of the Engineering Mechanics Division*, ASCE, Vol. 83, No. EM1, 1957, pp. 1–22.
5. Sonoda, K. and Kobayashi, H., ''Multispan Beams on Linear Viscoelastic Foundations,'' *Journal of Structural Mechanics*, Vol. 9, No. 4, 1981, pp. 451–464.
6. Hoskin, B. C. and Lee, E. H., ''Flexible Surface on Viscoelastic Subgrades,'' *Journal of the Engineering Mechanics Division*, ASCE, Vol. 85, No. EM4, 1959, pp. 11–30.
7. Pister, K. S. and Williams, M. L., ''Bending of Plates on a Viscoelastic Foundation,'' *Journal of the Engineering Mechanics Division*, ASCE, Vol. 86, No. EM5, 1960, pp. 31–44.
8. Sonoda, K., Kobayashi, H., and Ishio, T., ''Circular Plates on Linear Viscoelastic Foundations,'' *Journal of the Engineering Mechanics Division*, ASCE, Vol. 104, No. EM4, 1978, pp. 819–827.
9. Sonoda, K. and Kobayashi, H., ''Rectangular Plates on Linear Viscoelastic Foundations,'' *Journal of the Engineering Mechanics Division*, ASCE, Vol. 106, No. EM2, 1980, pp. 323–338.
10. Timoshenko, S. P., ''On the Correction for Shear of the Differential Equation for Transverse Vibrations of Prismatic Bars,'' *Philosophical Magazine*, Series 6, Vol. 41, 1921, pp. 744–746.
11. Reissner, E., ''The Effect of Transverse Shear Deformation on the Bending of Elastic Plates,'' *Journal of Applied Mechanics*, Vol. 12, No. 2, 1945, pp. 67–77.
12. Mindlin, R. D., ''Influence of Rotatory Inertia and Shear on Flexural Motions of Isotropic, Elastic Plates,'' *Journal of Applied Mechanics*, Vol. 18, No. 1, 1951, pp. 31–38.
13. Kobayashi, H. and Sonoda, K., ''Timoshenko Beams on Linear Viscoelastic Foundations,'' *Journal of Geotechnical Engineering*, ASCE, Vol. 109, No. 6, 1983, pp. 832–844.
14. Kobayashi, H. and Sonoda, K., ''Thick Circular Plates on Linear Viscoelastic Foundations,'' *Theoretical and Applied Mechanics*, Vol. 31, University of Tokyo Press, 1982, pp. 153–164.
15. Kobayashi, H. and Sonoda, K., ''Rectangular Thick Plates on Linear Viscoelastic Foundations,'' *Proceedings of the Japan Society of Civil Engineers*, No. 341, 1984, pp. 33–39.
16. Kobayashi, H., ''Applications of Eigenfunctions to Bending Analysis of Beams and Plates on Linear Viscoelastic Foundations,'' Thesis of Doctor of Engineering presented to Osaka City University, Osaka, Japan, Dec., 1985.
17. Bland, D. R., *The Theory of Linear Viscoelasticity*, Pergamon Press, Oxford, London, New York, Paris, 1960.
18. Essenburg, F., ''Shear Deformation in Beams on Elastic Foundations,'' *Journal of Applied Mechanics*, Vol. 29, No. 2, 1962, pp. 313–317.

19. Mikkola, M. and Ylinen, A., ''Effect of Shearing Force on the Deflection of a Beam of Finite Length on an Elastic Foundation,'' *Acta Polytechinca Scandinavica, Civil Engineering and Building Construction Series No. 23*, Finland Institute of Technology, Helsinki, 1964.
20. Frederick, D., ''Thick Rectangular Plates on an Elastic Foundation,'' *Transaction of ASCE*, Vol. 122, 1957, pp. 1069–1085.
21. Panc, V. C., *Theories of Elastic Plates*, Noordhoff International Publishing, Leyden, 1975.
22. Churchill, R. V., *Operational Mathematics*, 3rd ed., McGraw-Hill, New York, 1972.
23. Cowper, G. R., ''The Shear Coefficient in Timoshenko's Beam Theory,'' *Journal of Applied Mechanics*, Vol. 33, No. 2, 1966, pp. 335–340.
24. Huang, T. C., ''The Effect of Rotatory Inertia and of Shear Deformation on the Frequency and Normal Mode Equations of Uniform Beams With Simple End Conditions,'' *Journal of Applied Mechanics*, Vol. 28, No. 4, 1961, pp. 579–584.
25. Hetényi, M., *Beams on Elastic Foundation*, The University of Michigan Press, Ann Arbor, Michigan, 1946.
26. Marguerre, K. and Woernle, H. T., *Elastic Plates*, Blaisdell Publishing, Waltham, Mass., 1969.
27. Timoshenko, S. P. and Woinowsky–Krieger, S., *Theory of Plates and Shells*, 2nd ed., McGraw-Hill, New York, 1955.
28. McLachlan, N. W., *Bessel Functions for Engineers*, 2nd ed., Oxford University Press, London, 1955.
29. Schleicher, F., *Kreisplatten auf elastischer Unterlage*, Springer-Verlag, Berlin, 1926.

Solution Techniques for the Dynamic Response of Linear Systems

J. L. HUMAR*

ABSTRACT

A survey of the recent developments in computational techniques for the dynamic analysis of linear systems is presented. The principle techniques used in the solution of equations of motion are described. The most direct solution procedure is the numerical integration of the equations. A number of methods of such integration are presented and their stability and accuracy are briefly discussed. For the analysis of large systems, the concept of Ritz vectors can be utilized with advantage to reduce the size of the problem. The eigenvectors of a system are known to be very effective as Ritz shapes. A large volume of research has therefore been carried out in recent years on developing efficient eigensolution techniques. These developments are discussed in brief and several important procedures for solving the eigenvalue problems are described. Finally, it is pointed out that with the development of Fast Fourier Transform algorithms, frequency domain analysis has proved itself to be a highly efficient tool in the dynamic analysis of linear systems. The recent developments in this field are briefly discussed.

INTRODUCTION

Modern civil engineering structures are often subject to severe dynamic forces imposed by the environment. Such forces result from the effects of wind, earthquake, and water waves. The severity and complexity of the forces and the possible disastrous consequences of a failure under their effect pose a serious challenge to the designer.

All structures located in zones of seismic activity must be designed for the forces of earthquake. However, particular attention must be given to the design of nuclear power plants, dams, tall buildings and important bridges. In a similar manner, long span flexible bridge structures, tall buildings and towers, and off-shore structures must be carefully designed for the forces of wind. The forces caused by water waves are of particular importance in the design of off-shore structures, bridge piers and intake towers.

The complexity inherent in the evaluation of dynamic response of structures makes it very difficult to obtain a reasonably precise estimate of the response. However, with the advent of powerful high speed digital computers and the development of numerical techniques suitable for computer solution, it is possible to obtain the dynamic response of large and complicated systems subjected to the forces of earthquakes, wind and waves. Thus, systems involving thousands of degrees of freedoms can be solved for their dynamic response history defined at hundreds of time points.

Although the computation methods are now fairly well developed, uncertainties still persist in defining the dynamic forces of nature and in modelling the material properties. The representation of material behaviour is particularly difficult in the inelastic range. Furthermore, nonlinear material behaviour complicates the analysis method and makes response computations very expensive. For these reasons a majority of dynamic response computations are based on the assumption that the system being analyzed is linear. This is quite reasonable if the structure would be expected to remain elastic under the design forces. Such would be the case, for example, in the design of nuclear power plants subjected to the forces of earthquake. On the other hand, if a structure will be strained into the inelastic range under the action of dynamic forces, the assumption of linearity is no longer valid. However, even in such a case, a linear analysis may be used to provide an assessment of the non-linear response, or a non-linear analysis may be accomplished by a step by step integration scheme in which the structure is assumed to remain linear over a sufficiently short interval of time. It is evident that the response analysis

*Department of Civil Engineering, Carleton University, Ottawa, Canada

of linear systems plays an important role in the design of structures for dynamic loads.

This chapter is meant to serve as a state-of-the-art review of the present practice in the numerical analysis of the dynamic response of linear systems. A very large volume of literature now exists in this general area and it is impossible to survey all the work that has been done. Emphasis is therefore placed on the comparatively more recent developments in the solution techniques. Furthermore, because the formulation of the equations of motion through a finite element discretization has become a standard part of the literature, it is treated only cursorily.

The chapter discusses the solution of the equations of motion in both the time domain and the frequency domain. A time domain analysis can be carried out by a direct integration of the equations of motion in the physical coordinates of the system. Alternatively, the equations can be reformulated in terms of generalized coordinates before a solution is attempted. The generalized coordinate or Ritz vector approach may lead to considerable reduction in the size of the problem. Further, when the natural modes of the system are used as the Ritz vectors, the resulting equations are uncoupled and are thus much easier to evaluate. The determination of the mode shapes requires an eigenvalue solution, which is by no means a trivial task. Considerable attention is therefore given in this chapter to efficient eigensolution techniques.

Recent developments have established the frequency domain analysis as an extremely effective method in evaluating the dynamic response of linear systems. The application of the Fast Fourier Transform algorithms has made frequency domain analysis particularly efficient. The chapter therefore presents a review of the basic techniques involved in the frequency domain analysis procedure.

EQUATIONS OF MOTION

A finite element discretization is currently the most effective means of formulating the equations of motion of complex systems. The procedure involves the modelling of the structural systems by a series of elements of finite size interconnected at nodes.

Frames and other skeletal structures may be represented by beam elements; but for structures that consist of two or three dimensional continuum, plate, shell and solid elements must be used. Examples of such discretization are shown in Figure 1. In each case, the displacement field within the structure is represented in terms of the nodal values of the displacements which then serve as the unknowns in the problem. When expressed in terms of the nodal parameters, the equations of motion for a linear system take the form

$$M\ddot{r} + C\dot{r} + Kr = p(t) \tag{1}$$

in which M is the mass matrix, C the damping matrix, K the stiffness matrix, p the vector of externally applied nodal forces, r the vector of nodal displacements and overdots represent differentiation with respect to time. Equation (1) is, in fact, an expression of dynamic equilibrium between the internal forces given by the product of mass and acceleration, the velocity proportional damping forces, $C\dot{r}$, the internal forces arising from the stiffness of the structure, and the applied forces. For linear systems, matrices M, C and K are constant.

The stiffness matrix K is a positive definite symmetric matrix; it is also usually banded, with the non-zero terms lying in a narrow band around the diagonal. The mass matrix M is also banded but its shape depends on the type of formulation used. If the mass of the system is assumed to be lumped at the finite element nodes, M is diagonal, where some of the elements on the diagonal may be zero. When a consistent mass formulation is used, M is symmetric, positive definite and banded, and the band width is equal to that of the stiffness matrix.

In the finite element method, the stiffness and mass matrices are first derived for the individual elements on the basis of the known physical properties of the latter and then assembled to give the global matrices. A similar approach cannot be used for the damping matrix because damping characteristics of the individual elements are not known. As will be pointed out later, the damping matrix need not be specified explicitly when a mode superposition method is used. The damping forces are in that case obtained by selecting an appropriate damping ratio for each mode of vibration. When a damping matrix must be specified explicitly, suitable procedures that lead to reasonable values of energy dissipating forces are employed in the construction of the matrix.

The details of the methods used for constructing the physical property matrices are well described in the literature and will not be given here [Clough, 1971, Clough and Bathe, 1972, Clough and Penzien, 1975, Zeinkiwicz, 1977].

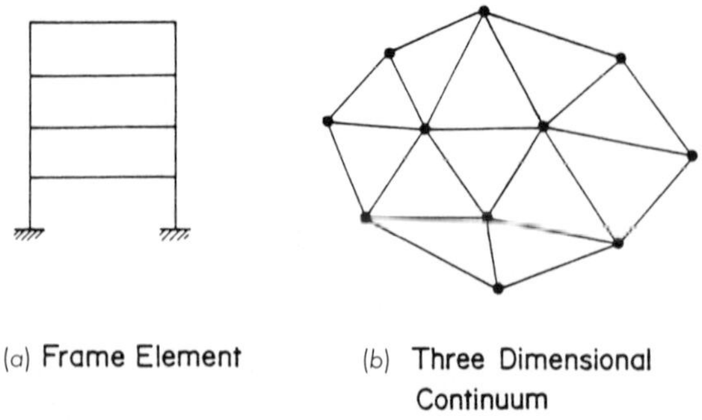

FIGURE 1. Finite element discretization of structural systems: (a) frame element; (b) three dimensional continuum.

SOLUTION OF THE EQUATIONS OF MOTION

The techniques used in the solution of Equation (1) can be divided into two broad categories: analysis in the physical coordinates of the system, and analysis in a set of generalized coordinates. Within each category, the equations can be solved either in the time domain or in the frequency domain.

Analysis in the Physical Coordinates

The time domain solution in the physical coordinates of the system is carried out by one of the several techniques of numerical integration described in a later section. Alternatively, the equations can be transformed into the frequency domain and solved by the application of Fourier transform. A description of the Fourier transform analysis is also presented later in this chapter.

Analysis in Generalized Coordinates

For large systems with thousands of degrees of freedom, the computations involved in direct solutions in the physcial coordinates of the system are very large. Fortunately, in many situations, it is possible to reduce the size of the problem considerably by expressing the displacements in the physical coordinates as a superposition of a set of appropriately selected displacement shapes each multiplied by a weighting factor. Mathematically this can be expressed as

$$r = \sum_{i=1}^{n} y_i q_i \tag{2}$$

in which q_i represents a displacement shape and y_i the corresponding weighting factor. If n linearly independent vectors q are selected, where n is the dimension of r, then the set of vectors q are said to constitute a complete basis spanning an n dimensional space. It is easily shown that for such a set of vectors the weighting factors y_i can be uniquely determined. Thus if Q is a matrix of size n by n whose ith column is the vector q_i and y is the vector of weighting factors, then Equation (2) can be expressed as

$$r = Qy \tag{3}$$

Because vectors q_i are linearly independent, Q is nonsingular and Equation (3) can be solved to give y

$$y = Q^{-1}r \tag{4}$$

Vectors q_i are called the Ritz shapes and the weighting factors y_i the Ritz coordinates or generalized coordinates. The concept of Ritz shapes is best illustrated by an example.

Consider the displaced shape shown in Figure 2a for the four storey building of Figure 1a. The displaced shape is completely defined by the given values of the four unknown coordinates r_1, r_2, r_3, r_4. Also shown in Figure 2 are four possible displacement patterns, which are linearly independent of each other, so that none of them can be derived by a linear combination of the others. It can be easily verified that the Ritz coordinate vector is given by $y^T = <1, 0.2, 0.04, 0.01>$.

Equation (3) represents a transformation from generalized coordinates to physical coordinates. Substitution of Equation (3) in Equation (1) and premultiplication by Q^T gives the following transformed equation.

$$\tilde{M}\ddot{y} + \tilde{C}\dot{y} + \tilde{K}y = \tilde{p} \tag{5}$$

in which

$$\tilde{M} = Q^TMQ$$

$$\tilde{C} = Q^TCQ$$

$$\tilde{K} = Q^TKQ$$

$$\tilde{p} = Q^Tp$$

Equation (5) is an alternate expression of the equations of motion. The matrices involved in this equation are still of order n by n. The solution of the set of Equation (5) is, therefore, in general, no easier than that of Equation (1). However, when the Ritz shapes are so chosen that the response of the structure can be expressed as a superposition of only m of the n vectors, $m << n$, the contributions from the remaining vectors being negligible, the problem can be considerably simplified. In such a situation, matrix Q will be of size $n \times m$ and the number of generalized coordinates is reduced to m. All of the matrices $\tilde{M}$, $\tilde{C}$, and $\tilde{K}$ are now of size $m \times m$ and the resulting Equation (5) are of a significantly lower order.

Referring to the Example of Figure 2, it is readily seen that it is sufficiently accurate to represent the displaced shape as a superposition of only the first two displacement patterns, the contribution from the remaining two being small.

Selection of Ritz Shapes

It is apparent from the discussion presented in the foregoing paragraphs that the success of analysis through a set of generalized coordinates depends on the selection of appropriate Ritz shapes. Because the mode shapes of an associated undamped system undergoing free vibrations are known to provide a very good choice for Ritz vectors, it is of interest to discuss the free vibration problem in some

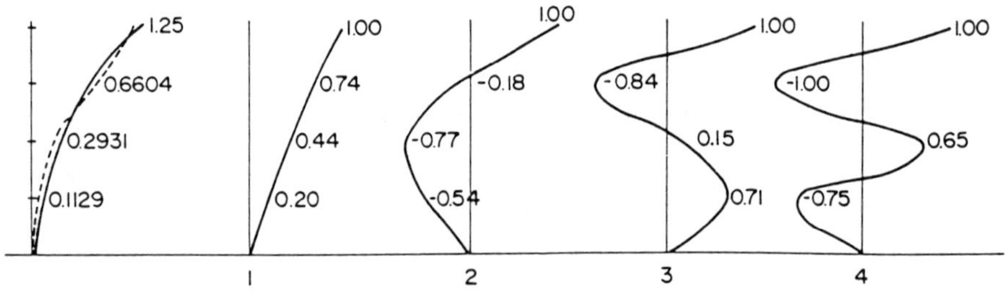

FIGURE 2. Ritz shape representation.

detail. The free vibration equations are obtained from Equation (1) by setting C and p as zero.

$$M\ddot{r} + Kr = 0 \tag{6}$$

A possible solution of Equation (6) is $r = q\sin(\omega t + \phi)$. Substitution into Equation (6) gives

$$(K - \omega^2 M)q = 0 \tag{7}$$

It is easily shown by applying Cramer's rule for the solution of simultaneous equations that Equation (7) will give non-zero values for q only when the determinant of matrix $K - \omega^2 M$ is zero. The determinant can be expressed as an n order polynomial in λ where $\lambda = \omega^2$. If this polynomial is equated to zero and solved for λ, n real and positive values of λ will be obtained. These values of λ are called eigenvalues, and the corresponding values of ω the frequencies of the system. Equation (7) can now be solved for q for each value of λ. Since $|K - \lambda M|$ is zero, infinitely many solutions are possible. However, the elements of the vector q bear a constant relationship to each other, and therefore if one of them is fixed all the rest are also determined. The n values of q are called the eigenvectors or the mode shapes of the system. The mode shapes have the following interesting orthogonality properties [Clough and Penzien, 1975].

$$\left.\begin{aligned} q_i^T K q_j &= 0 \\ q_i^T M q_j &= 0 \end{aligned}\right\} \qquad i \neq j \tag{8}$$

The eigenvectors of the system serve as very effective Ritz shapes for the solution of the equations of motion. In many physical problems, such as, for example, vibration due to earthquake forces, the response is adequately represented by only the first few mode shapes, and the size of the problem, Equation (5), can be considerably reduced. Further, because of the orthogonality relationships, Equation (8), matrices $\tilde{M}$ and $\tilde{K}$ in Equation (5) are both diagonal. Then if the damping matrix is so chosen as to satisfy the orthogonality relationship

$$q_i^T C q_j = 0 \qquad i \neq j \tag{9}$$

Equation (5) becomes completely uncoupled reducing to n single degree-of-freedom equations, each relating to a distinct Ritz coordinate. The solution of these uncoupled equations is much simpler than that of the original Equation (1).

The analysis through generalized coordinates, in which the mode shapes of the system are chosen as the Ritz shape, is called the mode superposition method of analysis. It relies on the determination of some or all of the frequencies and mode shapes of the system. The determination of the eigenvalues and eigenvectors is by no means a trivial exercise, and considerable effort has been directed towards developing computational procedures for solving the problem. Some of these procedures are discussed in a subsequent section of this chapter.

Recent research [Bathe and Wilson, 1983, Wilson, et al., 1982] has shown that the mode shapes of a system are not necessarily the most appropriate Ritz vectors. Besides, the solution of an eigenvalue problem to obtain the mode shapes is often very time consuming and expensive. In the new approach developed recently, a sequence of Ritz vectors is generated by a simple approach which is less costly. Further, these Ritz vectors are shown to give more accurate results when compared to those obtained by using an equal number of mode shapes. A brief description of this new approach is presented later.

EIGENVALUE SOLUTION

The eigenvector of a matrix has the property that when multiplied by the matrix, it yields a vector which is proportional to itself, the constant of proportionality being called an eigenvalue. Mathematically, this is expressed as

$$Aq = \lambda q \tag{10}$$

in which A is a square matrix of order n, q is a right eigenvector and λ is the corresponding eigenvalue.

Equation (10) is referred to as the standard eigenvalue problem; its solution will yield n eigenvalues, some of which may be repeated. Except in special cases, each eigenvalue is associated with a separate eigenvector. When A is real, the eigenvalues are either real or occur in complex conjugate pairs.

The eigenvalues of a transposed real matrix are the same as those of the original matrix; the eigenvectors are, however, in general, different. Thus,

$$A^T p = \lambda p \tag{11}$$

Equation (11) can be expressed in the alternate form

$$p^T A = \lambda p^T \tag{12}$$

Vectors p can therefore be called the left eigenvectors of A. The left and the right eigenvectors satisfy the following orthogonality relationships.

$$\left.\begin{aligned} p_j^T q_i &= 0 \\ p_j^T A q_i &= 0 \end{aligned}\right\} \qquad \lambda_i \neq \lambda_j \tag{13}$$

The case when A is symmetric as well as real is of special interest. The eigenvalues of a real symmetric matrix are all real. Also the left and right eigenvectors are the same, so that the orthogonality relationships of Equation (13) become

$$\left.\begin{aligned} q_j^T q_i &= 0 \\ q_j^T A q_i &= 0 \end{aligned}\right\} \qquad \lambda_i \neq \lambda_j \tag{14}$$

As stated earlier, the elements of an eigenvector are indeterminate in an absolute sense but their relative values are uniquely defined. It is therefore possible to scale the elements of an eigenvector in any arbitrary manner as long as the same scale factor is applied to all the elements. The scaled vector still remains an eigenvector. The process of scaling is called normalization. One possible way of normalization is to divide all the elements by the element which has the largest absolute value. Alternatively, an eigenvector can be scaled so that the product $q_i^T q_i$ called the Euclidean norm of the vector is equal to 1. For vectors scaled in this manner

$$Q^T Q = I \tag{15}$$

in which Q is a square matrix whose ith column is equal to q_i. A matrix that satisfies the relationship Equation (15) is called an orthogonal matrix; its transpose is equal to its inverse.

Linearized Eigenvalue Problem

The free vibration problem can now be identified as a special form of the eigenvalue problem:

$$Kq = \lambda M q \tag{16}$$

in which $\lambda = \omega^2$.

The special form, Equation (16), is known as a linearized eigenvalue problem. It can be readily converted into the standard form of Equation (10) using one of the following two methods:

$$M^{-1} K q = \lambda q \tag{17a}$$

or

$$Eq = \lambda q \tag{17b}$$

Alternatively,

$$K^{-1} M q = \frac{1}{\lambda} q \tag{18a}$$

$$Dq = \gamma q \tag{18b}$$

in which $E = M^{-1}K$, $D = K^{-1}M$, $\lambda = \omega^2$ and $\gamma = 1/\omega^2$. It should be noted that, although K and M are symmetric, D and E are both unsymmetric. It is, however, possible to convert the linearized eigenvalue into a standard form in which the matrix involved is symmetric. This can be achieved by factorizing the K matrix into a lower triangular matrix L_K and its transpose, so that

$$K = L_K L_K^T \tag{19}$$

Equation (18) can now be expressed as

$$L_K^{-1} M (L_K^T)^{-1} L_K^T q = \gamma L_K^T q \tag{20a}$$

$$\tilde{D}\tilde{q} = \gamma \tilde{q} \tag{20b}$$

in which $\tilde{D} = L_K^{-1} M (L_K^T)^{-1}$ is a symmetric matrix and the modified eigenvectors $\tilde{q}$ are related to q such that

$$L_K^T q = \tilde{q} \tag{21}$$

If the mass matrix is positive definite, the following alternative procedure can be used.

$$M = L_M L_M^T$$

$$L_M^{-1} K (L_M^T)^{-1} L_M^T q = \lambda L_M^T q \quad (22)$$

$$\tilde{E}\tilde{q} = \lambda \tilde{q}$$

in which $\tilde{E} = L_M^{-1} K (L_M^T)^{-1}$ and $L_M^T q = \tilde{q}$.

It should be noted that the mass matrix will be positive definite when it is diagonal and there are no zeroes on the diagonal, or whenever a consistent mass formulation is used.

Solution Methods for the Eigenvalue Problem

Prior to 1954, Jacobi diagonalization method was the only practical method for the evaluation of the eigenvalues of a matrix. With the advent of digital computer, and the increasing emphasis on numerical computations, very significant development took place in the numerical evaluation of the eigenvalues and a number of new methods were devised. These methods can be classified into the following broad categories:

1. Transformation methods
2. Determinant search methods
3. Iterative methods

The transformation methods are useful when the matrices are of comparatively small order and are more or less fully populated or have a large band width. They lead to the simultaneous evaluation of all eigenvalues. The eigenvectors can be evaluated either by a process of inverse transformation or by using an appropriate iterative process.

Transformation methods are not particularly appropriate when the matrices involved are of a large order and only a few eigenvalues need to be determined. Also, because transformations destroy the sparsity of a matrix, transformation methods are not efficient in dealing with sparse matrices. In such situations, the determinant search method or the iterative methods prove to be very effective.

A large body of literature now exists on eigenvalue solutions. In this paper only the methods that are of particular interest to structural vibration problems are reviewed. Because only those eigenvalue problems that lead to real and positive eigenvalues are of interest in structural vibrations, the presentation in the following sections implicitly assumes that the eigenvalues being sought are all real and positive. A list of some of the relevant references is included [Bathe and Wilson, 1972, 1973a, 1973b, 1976, Jennings, 1977, Wilkinson, 1965]. They may be consulted for a more detailed coverage of the subject matter.

Transformation Methods of Eigenproblem Solution

These methods rely on the fact that under certain types of transformation the eigenvalues of a matrix remain unchanged. A series of such transformations are carried out on the matrix being analyzed, till it is reduced to such a form that the eigenvalues can be obtained by inspection or are easy to compute. A majority of transformation methods operate on symmetric matrices, but variants that apply to the linearized eigenvalue problem are available in some cases.

The most general type of transformation that leaves the eigenvalues unaltered is called a similiarity transformation. Thus if N is an arbitrary non-singular transformation matrix of the same order as matrix A, then from Equation (10)

$$N^{-1} A N N^{-1} q = \lambda N^{-1} q$$

or

$$\tilde{A}\tilde{q} = \lambda \tilde{q} \quad (23)$$

in which the transformed matrix $\tilde{A} = N^{-1} A N$ and the modified eigenvectors $\tilde{q} = N^{-1} q$. The eigenvalues of $\tilde{A}$ are the same as those of A.

When A is symmetric, an orthogonal transformation is used so that $N^{-1} = N^T$ and Equation (23) reduces to

$$N^T A N N^T q = \lambda N^T q \quad (24)$$

It will be noted that matrix $\tilde{A} = N^T A N$ is also symmetric. After a series of transformations involving matrices N_1, N_2, . . . N_m,

$$\tilde{A} = N_m^T \ldots N_2^T N_1 A N_1 N_2 \ldots N_m \quad (25)$$

and

$$\tilde{q} = N_m^T \ldots N_2^T N_1^T q \quad (26a)$$

so that

$$q = N_1 N_2 \ldots N_m \tilde{q} \quad (26b)$$

The transformation methods available for the solution of an eigenvalue problem are listed below:

1. Jacobi Diagonalization
2. Givens' Tridiagonalization
3. Householders Transformation
4. Hessenberg Transformation
5. The LR Transformation
6. The QR Transformation

Some of the transformations are described in the following paragraphs. A more complete description can be found in Jennings (1975) and Wilkinson (1965).

Jacobi Diagonalization

Although it is the oldest transformation method available, its conceptual simplicity and the resulting ease with which it can be programmed on a computer makes it still attractive for the solution of eigenvalue problems of small size. It is often used for obtaining the eigenvalues of the secondary matrices encountered in the subspace iteration technique described later.

In its standard form, the method depends on the application of repeated orthogonal transformations on a symmetric matrix till the latter becomes diagonal. Each transformation is designed to eliminate two symmetric off-diagonal elements. The form of the transformation matrix for reducing the elements a_{ij} and a_{ji} to zero is

$$[N] = \begin{array}{c} \\ \\ \\ i \\ \\ j \\ \\ \\ \end{array} \begin{bmatrix} & & & i & & j & & \\ 1 & 0 & \cdot & 0 & \cdot & 0 & \cdot & 0 \\ 0 & 1 & \cdot & 0 & \cdot & 0 & \cdot & 0 \\ \cdot & \cdot & \cdot & \cdot & \cdot & \cdot & \cdot & \cdot \\ 0 & 0 & \cdot & \cos\alpha & \cdot & -\sin\alpha & \cdot & 0 \\ \cdot & \cdot & \cdot & \cdot & \cdot & \cdot & \cdot & \cdot \\ 0 & 0 & \cdot & \sin\alpha & \cdot & \cos\alpha & \cdot & 0 \\ \cdot & \cdot & \cdot & \cdot & \cdot & \cdot & \cdot & \cdot \\ 0 & 0 & 0 & 0 & 0 & 0 & \cdot & 1 \end{bmatrix} \tag{27}$$

It is easily proved that angle α should be chosen so that

$$\tan 2\alpha = \frac{2a_{ij}}{a_{ii} - a_{jj}} \quad 0 \leq \alpha \leq \frac{\pi}{2} \tag{28}$$

Each transformation affects all elements of rows and columns i and j. Apparently, therefore, an off-diagonal element reduced to zero during a transformation will become nonzero during subsequent transformations. The process is therefore iterative, and with repeated iterations, the off-diagonal elements will be reduced in magnitude till they are negligible in comparison to the diagonal values.

Since the transformed A matrix is diagonal, its eigenvalues are equal to the elements on the diagonal. These are also the eigenvalues of the original matrix. The required eigenvectors are obtained by using Equation (26b) which gives

$$Q = N_1 N_2 \ldots N_m \tilde{Q} \tag{29}$$

In Equation (29), because the eigenvectors of a diagonal matrix are unit vectors, $\tilde{Q} = I$, the identity matrix.

In the standard Jacobi diagonalization procedure, the off-diagonal element with the largest modulus is chosen for elimination at each step. In a computer implementation, the process of finding the largest element may take considerable time. A variation of the process in which the algorithm does not involve such a search is therefore generally preferred.

One possible alternative is to eliminate off-diagonal elements in a systematic order, for example $(i,j) = (1,2), (1,3) \ldots (1,n)$ followed by $(2,3), (2,4) \ldots (2,n)$ and so on. This process called serial Jacobi still converges to the right solution, although the convergence may be slower.

In another variation, a threshold value is established, and although the elements are eliminated in a serial order, elimination is skipped if the modulus of the off-diagonal elements under consideration lies below the threshold. After a complete pass through the matrix, the threshold value is lowered and the process repeated till the transformed matrix is nearly diagonal.

For its solution by the standard Jacobi diagonalization process, the eigenvalue problem of Equation (16) must first be reduced to the standard symmetric form by using either Equation (20) or Equation (22). However, a variation of the method called the Generalized Jacobi method, can be applied directly to Equation (16) [Bathe and Wilson, 1972]. The procedure consists of repeated simultaneous transformations of the K and M matrices till they are reduced to a diagonal form. Mathematically, the transformation can be expressed as

$$N^T K N N^{-1} q = \lambda N^T M N N^{-1} q$$

or

$$\tilde{K}\tilde{q} = \lambda \tilde{M}\tilde{q} \tag{30}$$

in which $\tilde{K} = N^T K N$, $\tilde{M} = N^T M N$ and $q = N\tilde{q}$.

The transformation matrix N is of the form

$$\begin{array}{c} \\ \\ i \\ \\ j \\ \\ \end{array} \begin{bmatrix} & i & & j & \\ 1 & 0 & \cdot & 0 & 0 \\ 0 & 1 & \cdot & \gamma & 0 \\ \cdot & \cdot & \cdot & \cdot & \cdot \\ 0 & \alpha & \cdot & 1 & 0 \\ 0 & 0 & \cdot & 0 & 1 \end{bmatrix} \tag{31}$$

It can be easily shown that γ and α in Equation (31) are given by

$$a = k_{jj} m_{ij} - k_{ij} m_{jj}$$

$$b = k_{ii} m_{jj} - k_{jj} m_{ii}$$

$$c = k_{ii} m_{ij} - k_{ij} m_{ii}$$

$$\alpha = \frac{1}{2a}\left[b + \sqrt{b^2 + 4ac} \right], \; a \neq 0 \tag{32}$$

$$\alpha = \frac{-c}{b}, \; a = 0$$

$$\gamma = -\alpha \frac{a}{c}$$

where k_{ij} and m_{ij} represent elements of K and M, respectively, and $i < j$.

Householder's Transformation

Householder's method reduces a symmetric matrix to tridiagonal form by the application of a series of orthogonal transformations. The eigenvalues and eigenvectors of the tridiagonal matrix are then computed by either the LR or the QR method. The eigenvalues so computed are the same as those of the original matrix, while the eigenvectors of the tridiagonal matrix are related to those of the original by a relationship of the form of Equation (26b).

Figure 3 shows the transformed matrix A after reduction has been completed to row No. $k - 1$. The kth transformation expressed by

$$A^{(k+1)} = [N^{(k)}]^T A^k N^{(k)} \tag{33}$$

reduces to zero all of the elements enclosed in boxes in Figure 3a. Let N be partitioned as shown in Figure 3b and the submatrix R be of the form

$$R = I - 2ww^T \tag{34}$$

where w is a column vector of size $n - k$ with Euclidean norm equal to unity so that

$$w^T w = 1 \tag{35}$$

By taking the transpose of Equation (34), it is readily shown that $R^T = R$ and therefore $N^T = N$. Thus both R and N are symmetric. Further,

$$\begin{aligned} RR^T &= (I - 2ww^T)(I - 2ww^T) \\ &= I - 4ww^T + 4ww^T ww^T \end{aligned} \tag{36}$$

Substitution of Equation (35) into (36) gives

$$RR^T = I \tag{37}$$

By using the relationship Equation (37), it can easily be proved that $NN^T = I$. This implies that the transformation specified by Equation (33) is orthogonal.

Expansion of the basic transformation Equation (33) using the partitioned forms of A and N shown in Figure 3 gives

$$A^{(k+1)} = \left[\begin{array}{c|c} b & 0 \\ & c^T R \\ \hline 0 \; Rc & RdR \end{array}\right] \tag{38}$$

It is noted from Equation (38) that the transformation leaves the submatrix b unaltered, that is the computations in the kth step do not change that portion of A which has already been tridiagonalized. Also if R has been appropriately chosen, the following relationship should be satisfied.

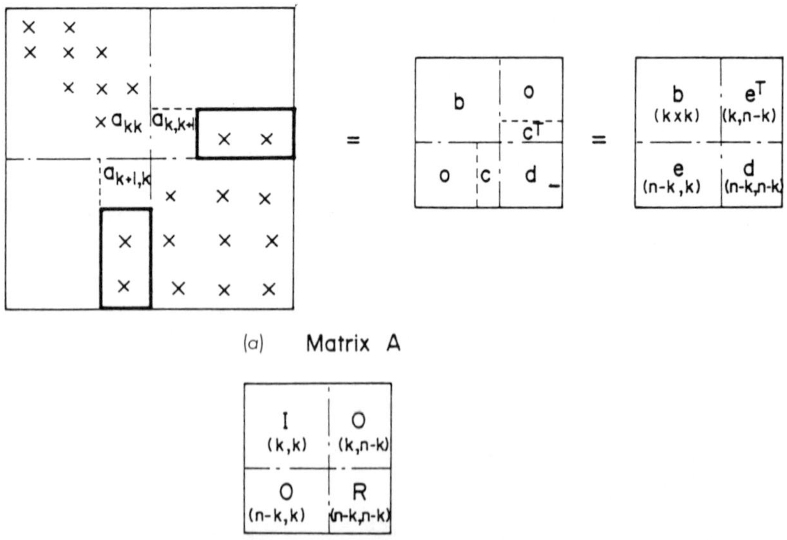

FIGURE 3. Householder's transformation.

$$Rc = [I - 2ww^T]\, c = g \tag{39}$$

in which $g^T = [r\, 0\, 0\, 0 \ldots 0]$, r and w are as yet undetermined and

$$c^T = [a_{k,k+1}\; a_{k,k+2} \ldots a_{k,n}] \tag{40}$$

It follows from Equation [39] that

$$g^T g = c^T R^T R c$$

or

$$r^2 = c^T c \tag{41}$$

Now premultiplying both sides of Equation (39) by c^T and denoting the scalar quantity $w^T c = c^T w$ by h,

$$c^T c - 2h^2 = c^T g \tag{42}$$

Substitution of Equations (40) and (41) and the value of g in Equation (42) gives

$$2h^2 = r^2 - a_{k,k+1} r \tag{43}$$

Equations (41) and (43) provide the values of r and h. The sign of r is chosen so that the product $a_{k,k+1} r$ in Equation (43) is negative. Now using Equation (39)

$$2hw = c - g = v \tag{44}$$

Finally, Equation (34) leads to the transformation matrix R:

$$R = I - \frac{1}{2h^2} vv^T \tag{45}$$

in which v is given by Equation (44) and $2h^2$ by Equation (43).

QR Transformation

This transformation method is applied to a tridiagonal matrix to reduce it to a diagonal form whose eigenvalues and eigenvectors are easily obtained by inspection. The kth step in the iterative process can be expressed as

$$A^{(k)} = Q_k R_k \tag{46}$$

$$A^{(k+1)} = R_k Q_k \tag{47}$$

in which Q_k is an orthogonal transformation matrix ($Q^T = Q^{-1}$), R_k is an upper triangular matrix and A is the tridiagonal matrix being reduced. Substitution of R_k from Equation (46) into Equation (47) gives

$$A^{(k+1)} = Q_k^T A^{(k)} Q_k \tag{48}$$

Equation 48 represents a transformation of the form of Equation (24) under which the eigenvalues are preserved. If Q_k is appropriately chosen, then under repeated transformations described by Equations (46) and (47), matrix A will be reduced to a diagonal form.

To obtain Q_k, it is represented as the product of a series of orthogonal transformation matrices N_i of the form

$$[N_i] = \begin{array}{c} \\ \\ \\ \\ i \\ i+1 \\ \\ \\ \end{array} \begin{array}{c} \begin{array}{cccccc} & & i & i+1 & & \end{array} \\ \begin{bmatrix} 1 & 0 & \cdot & 0 & 0 & \cdot & 0 \\ 0 & 1 & \cdot & 0 & 0 & \cdot & 0 \\ \cdot & \cdot & \cdot & \cdot & \cdot & \cdot & \cdot \\ 0 & 0 & \cdot & \cos\alpha & -\sin\alpha & \cdot & 0 \\ 0 & 0 & \cdot & \sin\alpha & \cos\alpha & \cdot & 0 \\ \cdot & \cdot & \cdot & \cdot & \cdot & \cdot & \cdot \\ 0 & 0 & 0 & 0 & 0 & \cdot & 1 \end{bmatrix} \end{array} \tag{49}$$

$$Q_k = N_1 N_2 \ldots N_{n-1} \tag{50}$$

The value of α in matrix N_i is selected so that the product $N_i^T A$ eliminates the sub-diagonal element $a_{i+1,i}$ in the current version of matrix A. Substitution of Equations (49) and (50) into Equations (46) and (47) gives

$$N_{n-1}^T \ldots N_2^T N_1^T A^{(k)} = R_k \tag{51}$$

and

$$A^{(k+1)} = R_k N_1 N_2 \ldots N_{n-1} \tag{52}$$

Representing $N_{i-1}^T \ldots N_2^T N_1^T A^{(k)}$ by $A_i^{(k)}$ it can be easily shown that the value of α in N_i is given by

$$\tan\alpha = \frac{a_{i+1,i}}{a_{i,i}} \tag{53}$$

and that the transformation $N_i^T A_i^{(k)}$ does not affect the sub-diagonal elements already reduced to zero in $A_i^{(k)}$. When the kth iteration is completed by carrying out the matrix operations indicated in Equation (52), the subdiagonal elements will in general be restored to non-zero values, but their magnitude would have reduced. Eventually after sufficient number of iterations, matrix A would become almost diagonal, the elements on the diagonal being equal to the eigenvalues of the original matrix A. If required, the eigenvectors of the tridiagonal matrix can be obtained by using an expression similar to Equation (29).

In the QR iteration procedure described above, eigenvalues are obtained in increasing order of magnitude, the least dominant (λ_1) being obtained first. After the ith eigenvalue has been obtained, further iterations can be carried out on a deflated matrix obtained by omitting row and column $n - i + 1$ from the current version of A.

Assuming that eigenvalues $\lambda_1, \lambda_2, \ldots, \lambda_{i-1}$ have already been obtained, convergence to λ_i depends on the ratio $|\lambda_i/\lambda_{i+1}|$. This suggests that convergence can be improved by using a modified procedure called *QR* iteration with a shifted origin. The procedure is based on the fact that the eigenvalues $\tilde{\lambda}$ of the matrix $\tilde{A} = A - \mu I$, where μ is any scalar value, are related to the eigenvalues of A by the expression

$$\tilde{\lambda}_j = \lambda_j - \mu \tag{54}$$

Equation (54) can be easily proved by noting that

$$\begin{aligned} \tilde{A}q &= Aq - \mu q \\ &= (\lambda - \mu)q \end{aligned} \tag{55}$$

In the *k*th iteration of *QR* method with shifted origin for obtaining λ_i, an estimate is obtained for the eigenvalue λ_i. The estimated value denoted by μ_k may be taken as equal to the element a_{ii} of the current version of A. The transformation is then carried out on matrix $(A^{(k)} - \mu_k I)$ so that

$$Q_k^T(A^{(k)} - \mu_k I) = R_k \tag{56}$$

$$\tilde{A}^{(k+1)} = R_k Q_k \tag{57}$$

The transformed matrix is then restored by removing the shift. Thus,

$$A^{(k+1)} = \tilde{A}^{(k+1)} + \mu_k I \tag{58}$$

A new estimate is now made for λ_i and the process (Equations 56, 57, 58) repeated. It is evident that if the shift is close to the value of λ_i, the ratio $|\tilde{\lambda}_i/\tilde{\lambda}_{i+1}|$ will be small and the convergence will be significantly speeded up.

Determinant Search Method

When the matrices involved in the eigenvalue problem have a comparatively small band width and only a few of the eigenvalues are desired, the determinant search method can be very effective in solving the eigenvalue problem.

Consider the linearized eigenvalue problem:

$$(K - \lambda M)q = 0 \tag{59}$$

Equating the determinant of $K - \lambda M$ to zero gives an n order polynomial in λ whose roots represent the n eigenvalues. The determinant search method uses an iterative procedure to locate as many zeroes of the polynomial as are desired. Suppose that the least dominant eigenvalue λ_1 is desired and two rough estimates of the eigenvalue are available. Let these estimates be μ_{k-2} and μ_{k-1} where $\mu_{k-2} < \mu_{k-1} < \lambda_1$. The procedure requires that the determinant $|K - \lambda M|$ be evaluated for these two values of λ. Let the two values of the polynomial be denoted by p_{k-2} and p_{k-1}. Referring to Figure 4a, a new estimate, μ_k, for the zero of the polynomial can be obtained by fitting a straight line to the two points (μ_{k-2}, p_{k-2}) and (μ_{k-1}, p_{k-1}). The eigenvalue estimate becomes

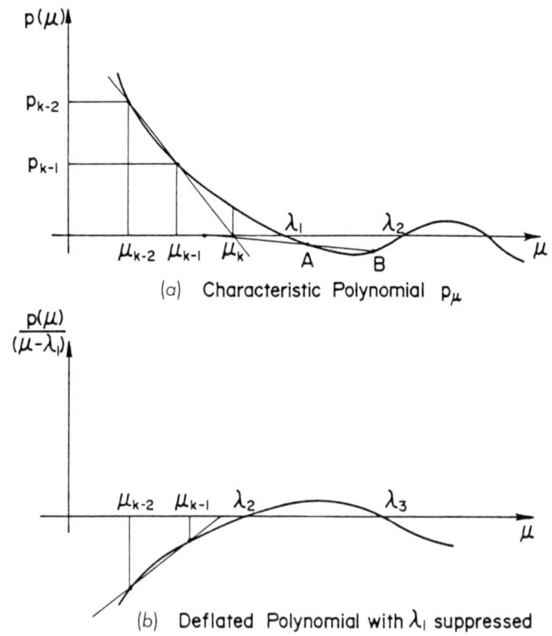

FIGURE 4. Determinant search method: (a) characteristic polynomial p_μ; (b) deflated polynomial with λ_1 suppressed.

$$\mu_k = \mu_{k-1} + \eta \frac{\mu_{k-1} - \mu_{k-2}}{p_{k-1} - p_{k-2}} p_{k-1} \tag{60}$$

in which $\eta = 1$.

The polynomial is evaluated for the new estimate μ_k. If p_k is sufficiently close to zero, the iteration process has converged and $\mu_k \approx \lambda_1$. If convergence has not been achieved, the entire process is repeated with the two previous values p_{k-1} and p_k.

A major computation effort is required in evaluating the determinant for a given value $\lambda = \mu$. The evaluation uses a modified Gaussian elimination process in which $K - \mu M$ is factored so that

$$K - \mu M = LDL^T \tag{61}$$

where L is a unit lower triangular matrix and D is a diagonal matrix. The determinant of the matrix $K - \mu M$ is then given by taking the product of the diagonal elements of D.

With the linear interpolation or the secant iteration scheme of Equation (60), the root of the polynomial is approached only from one side, p does not change sign and there is no chance of jumping the root. However, the convergence may at times be slow. To accelerate the convergence, a value of 2 or more may be used for η. When $\eta > 2$, there is the possibility of jumping over one or two roots. Fortunately, this can be detected by using the Sturm sequence property, so that if necessary the iteration can be retracted.

The Sturm sequence property implies that when the determinant is evaluated for a particular value μ by using the triangular factorization Equation (61), then the number of negative elements in D is equal to the number of eigenvalues less than μ. It is therefore possible to know whether one or more roots have been jumped by counting the negative elements in D.

The determinant search method can also be employed to converge to higher roots. An additional complication arises this time. For the first root if the iteration began with two values of μ both less than λ_1 (no negative elements in D), it should converge towards λ_1. However, for the higher roots this is not guaranteed. For example, iteration with points A and B in Figure 4a will converge back to λ_1. To overcome this difficulty the deflated polynomial $p_j(\mu)$ is used instead of $p(\mu)$ where

$$p_j(\mu) = \frac{p(\mu)}{\prod_{i=1}^{j} (\mu - \lambda_i)} \tag{62}$$

and $\lambda_i (i = 1 \text{ to } j)$ are the eigenvalues already determined. The deflated polynomial for the second root is shown in Figure 4b.

In practice, the determinant search method is used to obtain only a close estimate of the desired eigenvalue. A more accurate estimate of the eigenvalue as well as the corresponding eigenvector is then determined by using the inverse power method with shifts described later in this chapter.

Iterative Methods

Iterative methods are ideally suited for large eigenvalue problems with sparse or banded matrices, particularly when only the lowest or the highest few eigenvalues are required. The following iterative methods can be effectively applied to practical solutions of eigenvalue problems in the vibration of engineering structures.

1. Power method
2. Power method with shifts
3. Sub-space iteration
4. Lanczos method

A brief description of each of the methods is provided below.

Power Method

The method allows successive evaluation of as many eigenvalues and eigenvectors as required, starting from the most dominant one.

Consider the standard eigenvalue value problem Equation (10) and let the eigenvalues be arranged so that $|\lambda_1| < |\lambda_2| < |\lambda_3| \ldots < |\lambda_n|$. It can be shown that any arbitrary vector u of dimension n can be expressed as a linear combination of the n eigenvectors of A. Thus,

$$u = c_i q_1 + c_2 q_2 + \cdots + c_n q_n \tag{63}$$

Coefficient c_i in Equation (63) can be obtained on premultiplying both sides by p_i^T and using the orthogonality relationship Equation (13). This leads to the expression

$$c_i = \frac{p_i^T u}{p_i^T q_i} \tag{64}$$

When A is symmetric, Equation (64) reduces to

$$c_i = \frac{q_i^T u}{q_i^T q_i} \tag{65}$$

For the linearized eigenvalue problem of Equation (16) the corresponding expression is easily shown to be

$$c_i = \frac{q_i^T M u}{q_i^T M q_i} \tag{66}$$

Premultiplication of Equation (63) by A gives

$$\begin{aligned} Au &= c_1 A q_1 + c_2 A q_2 + \cdots + c_n A q_n \\ &= c_1 \lambda_1 q_1 + c_2 \lambda_2 q_2 + \cdots + c_n \lambda_n q_n \end{aligned} \tag{67}$$

It is evident that k such multiplications will give

$$\begin{aligned} A^k u &= c_1 \lambda_1^k q_1 + c_2 \lambda_2^k q_2 + \cdots + c_n \lambda_n^k q_n \\ &= \lambda_n^k \left\{ c_1 \left(\frac{\lambda_1}{\lambda_n}\right)^k q_1 + c_2 \left(\frac{\lambda_2}{\lambda_n}\right)^k q_2 + \cdots + c_n q_n \right\} \end{aligned} \tag{68}$$

Now because $|\lambda_i/\lambda_n| < 1$ for $i = 1,2, \ldots n - 1$, in the limit as k becomes large

$$A^k u \approx c_n \lambda_n^k q_n \tag{69}$$

and

$$A^{k+1} u \approx c_n \lambda_n^{k+1} q_n \tag{70}$$

The resulting vector is thus proportional to the most dominant mode shape and the ratio of the value of an element in say the $(k + 1)$th iteration to the value of the same element in the kth iteration approaches the most dominant eigenvalue.

In practice, because an eigenvector may be arbitrarily scaled, it is convenient to normalize the product vector at the end of each iteration. Normalization is usually carried out by dividing the vector by the value of the element which has the largest modulus.

The method can be applied to the linearized eigenvalue problem of Equation (16) by reducing the latter to either of the two forms given in Equations (17b) and (18b). The form (17b) will converge to the most dominant value of λ; that is to the highest frequency ω. The form (18b) will converge to the highest value of γ; that is the lowest ω or λ. In practical vibration problems, the lower frequencies are of greater interest and therefore form (18b) is preferred. Iteration with form (18b) is usually referred to as inverse iteration.

It is evident that when applied as above, the inverse power method will always converge to the least dominant eigenvalue and the corresponding eigenvector. Convergence to the second least dominant eigenvalue can be achieved if the trial vector is suitably modified so that in the expansion of the modified vector, coefficient c_1 is zero. The process of such modification is normally referred to as purification. Starting from an arbitrary trial vector u, the purified vector u_2 can be obtained as follows.

$$u_2 = u - c_1 q_1 \tag{71}$$

in which it is assumed that q_1 has already been determined and c_1 is obtained from one of Equations (64), (65) and (66). In particular, for the linearized eigenvalue problem, Equation (71) can be expressed as

$$u_2 = u - \frac{q_1 q_1^T M u}{q_1^T M q_1} \tag{72a}$$

$$= S_1 u \tag{72b}$$

where S_1, called the first mode sweeping matrix, is given by

$$S_1 = I - \frac{q_1 q_1^T M}{q_1^T M q_1} \tag{73}$$

Theoretically, once the iterations have been commenced with a purified vector u_2, the process should converge to the next most dominant eigenvalue. In practice, however, small errors in numeric computation during iterations will result in the trial vector being contaminated by the first eigenvector, and purification must therefore be repeated at the end of each iteration. This is conveniently achieved by modifying D as follows

$$D_2 = D S_1 \tag{74}$$

A procedure similar to the one described above can be used to obtain successive eigenvectors. Thus to obtain the third eigenvector, the trial vector should be purified so as to remove the components of the first and second eigenvectors. The required sweeping matrix S_2 is given by

$$S_2 = I - \frac{q_1 q_1^T M}{q_1^T M q_1} - \frac{q_2 q_2^T M}{q_2^T M q_2} \tag{75}$$

The sweeping matrix technique can also be used with the direct iteration method in which case the eigenvalues and eigenvectors will be obtained in order from the most dominant to the least dominant.

Power Method with Shifts

The inverse iteration procedure described in the foregoing paragraphs can be used to obtain the least dominant eigenvalue. In addition, it can be made to converge to the higher eigenvalues by sweeping out all lower eigenvectors from the trial vector selected for iteration. Because the process of sweeping depends on an accurate determination of the lower eigenvectors, numerical accuracies become progressively larger as the eigenvector dominance increases. In addition, it is seen that the number of iterations required to converge to the higher eigenvectors also increases, making the procedure inefficient and expensive. An inverse power method with shifts overcomes these difficulties and provides an efficient means of determining the higher eigenvalues and eigenvectors.

The shift method is described here with reference to the linearized eigenvalue problem of free vibration. Introduction of a shift μ in the origin of the eigenvalues axis gives (Figure 5):

$$\lambda = \mu + \delta \tag{76}$$

where δ is the modified eigenvalue from the shifted origin.

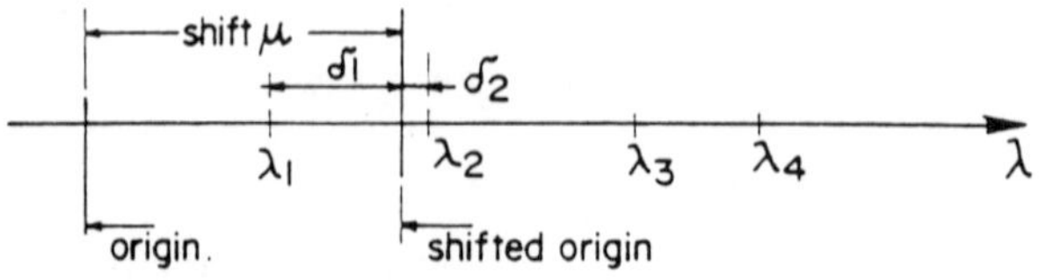

FIGURE 5. Eigenvalues measured from a shifted origin.

Substitution of Equation (76) into Equation (16) gives

$$(K - \mu M)q = \delta Mq \tag{77a}$$

$$(K - \mu M)^{-1}Mq = \frac{1}{\delta} q \tag{77b}$$

$$\hat{D}q = \hat{\gamma} q \tag{77c}$$

Iteration with Equation (77c) will lead to the eigenvalue corresponding to the largest γ; that is to the lowest δ. It is evident that by choosing a shift point close to the desired eigenvalue, a more accurate estimate of such eigenvalue as well as of the corresponding eigenvector can be obtained after a relatively small number of iterations.

The shift point can be located provided that a rough estimate is available for the desired eigenvalue. This can be achieved by a few iterations of the inverse power method in conjunction with the sweeping technique using the previous shift point as the origin.

A more detailed description of the method has been provided by Humar (1978), and Clough and Penzien (1975).

Sub-Space Iteration

The sub-space or simultaneous iteration procedure is one of the most efficient methods for the eigensolution of large systems when only the lower modes are of interest. Conceptually, it is similar to the inverse power method, except that the iteration is performed simultaneously on a number of trial vectors. This number m is much less than n, the order of the matrices involved, but is larger than p, the number of modes to be determined. In practice m is chosen to be the smaller of $2p$ and $p + 8$.

The linearized eigenvalue problem can be expressed as

$$KQ = MQ\Lambda \tag{78}$$

in which Q is an n by m matrix of the first m eigenvectors, and Λ is a diagonal matrix, the elements on the diagonal representing the first m eigenvalues. As in the case of inverse power method, the first step in the iteration is

$$\tilde{V}^{(1)} = K^{-1}MV^{(0)} \tag{79}$$

where $V^{(0)}$ is matrix of m trial shapes.

Before proceeding with next iteration step, the vectors in $\tilde{V}^{(1)}$ must be orthogonalized so that they will converge to different eigenvectors and not all to the least dominant one. Further, as in the case of power method, the eigenvectors should be normalized in some way so that the numbers remain within reasonable bounds. This is achieved by forming a new eigenvalue problem as follows

$$K_1^* = [\tilde{V}^{(1)}]^T K\tilde{V}^{(1)} \tag{80a}$$

$$M_1^* = [\tilde{V}^{(1)}]^T M\tilde{V}^{(1)} \tag{80b}$$

$$K_1^*\tilde{Z}^{(1)} = M^*\tilde{Z}^{(1)}\Gamma \tag{81}$$

The eigenvalue problem of Equation (81) is of a comparatively small size being of the order of m by m and may be conveniently solved by any one of the transformation methods. For example, the Jacobi diagonalization procedure may be used quite effectively. The modal vector $\tilde{Z}$ is then normalized so that

$$Z^{(1)T}M^*Z^{(1)} = I \tag{82}$$

The improved trial vectors for use in the next iteration are now given by

$$V^{(1)} = \tilde{V}^{(1)}Z^{(1)} \tag{83}$$

It is easily shown that these improved trial vectors are orthogonal with respect to the stiffness and mass matrices of the original system and will therefore converge to different mode shapes. Thus,

$$\begin{aligned} V^{(1)T}KV^{(1)} &= Z^{(1)T}\tilde{V}^{(1)T}K\tilde{V}^{(1)}Z^{(1)} \\ &= Z^{(1)T}K^*Z^{(1)} \\ &= \Gamma \end{aligned} \tag{84}$$

where the last relationship follows from Equations (81) and (82).

If the iteration procedure described above is repeated a sufficient number of times, V will converge to the first m eigenvectors and Γ to the first m frequencies. Iteration may, in fact, be terminated when the first p of the m eigenvectors do not change between iteration by more than a specified tolerance.

Lanczos Iteration

When first developed in 1950, Lanczos method was looked upon as an orthogonal transformation that reduced a symmetric matrix into a tridiagonal form. With the development of more efficient methods of tridiagonalization, interest in the Lanczos method waned. Research in the recent years has, however, shown that the method has great potential in the partial eigensolution of large systems when the matrices involved are sparse. Interest in the method has therefore revived and currently, it is considered to be as effective a tool as the sub-space iteration method for the solution of large vibration problems.

The Lanczos transformation matrix Y consists of a series of mutually orthogonal vectors $y_1, y_2, \ldots, y_n$ such that

$$Y^TY = I \tag{85}$$

Denoting the tridiagonal matrix by T, the transformation can be expressed as

$$Y^T A Y = T \tag{86a}$$

or

$$AY = YT \tag{86b}$$

when written in its expanded form Equation (86) becomes

$$[A][y_1 y_2 \ldots y_n] =$$

$$[y_1 y_2 \ldots y_n] \begin{bmatrix} \alpha_1 & \beta_1 & & & \\ \beta_1 & \alpha_2 & \beta_2 & & \\ & \cdot & \cdot & \cdot & \\ & & \beta_{n-2} & \alpha_{n-1} & \beta_{n-1} \\ & & & \beta_{n-1} & \alpha_n \end{bmatrix} \tag{87}$$

or

$$\begin{aligned} Ay_1 &= \alpha_1 y_1 + \beta_1 y_2 \\ Ay_2 &= \beta_1 y_1 + \alpha_2 y_2 + \beta_2 y_3 \\ &\cdots\cdots\cdots\cdots\cdots\cdots \\ Ay_j &= \beta_{j-1} y_{j-i} + \alpha_j y_j + \beta_j y_{j+1} \\ &\cdots\cdots\cdots\cdots\cdots\cdots \\ Ay_n &= \beta_{n-1} y_{n-1} + \alpha_n y_n \end{aligned} \tag{88}$$

The process of finding the Lanczos vectors begins with an arbitrary selection for y_1 but with the condition that y_1 satisfy the relationship

$$y_1^T y_1 = 1 \tag{89}$$

If α_1 is selected so that

$$\alpha_1 = y_1^T A y_1 \tag{90}$$

then premultiplication of the first of Equation (88) by y_1^T and substitution of Equations (89) and (90) gives the relationship

$$y_1^T y_2 = 0 \tag{91}$$

implying that vectors y_1 and y_2 are mutually orthogonal. The first of Equation (88) now gives

$$\hat{y}_2 = \beta_1 y_2 = Ay_1 - \alpha_1 y_1 \tag{92}$$

From Equation (92)

$$\beta_1^2 y_2^T y_2 = \hat{y}_2^T \hat{y}_2 \tag{93}$$

so that if y_2 is selected to satisfy the relationship $y_2^T y_2 = 1$, β_1 is obtained from Equation (93) and Equation (92) gives

$$y_2 = \frac{\hat{y}_2}{\beta_1} \tag{94}$$

In the second of Equation (88), if α_2 is chosen so that

$$\alpha_2 = y_2^T A y_2 \tag{95}$$

then $y_2^T y_3$ will be zero and β_2 and y_3 can be obtained by expressions similar to Equations (93) and (94). The complete iteration process can be summarized as

$$\begin{aligned} \beta_{j-1}^2 &= \hat{y}_j^T \hat{y}_j \\ y_j &= \frac{\hat{y}_j}{\beta_{j-1}} \\ \alpha_j &= y_j^T A y_j \\ \hat{y}_{j+1} &= Ay_j - \alpha_j y_j - \beta_{j-1} y_{j-1} \end{aligned} \tag{96}$$

The iteration begins with $\beta_0 = 1$ and an arbitrary value of y_1 that satisfies the relationship $y_1^T y_1 = 1$.

The eigenvalues and eigenvectors of the tridiagonal matrix can be evaluated by any of the standard methods such as the LR or the QR method. The eigenvalues of matrix A are the same as those of matrix T, while the eigenvectors of the two matrices are related as follows:

$$q = Ys \tag{97}$$

in which s is an eigenvector of T.

Theoretically, it can be proved that the Lanczos vectors $y_1, y_2, \ldots, y_n$ are mutually orthogonal to each other. However, because of round off errors during computation, orthogonality relationship breaks down when vectors are sufficiently separated from each other. For large systems this source of error makes Lanczos method unstable. It is therefore necessary to reorthogonalize a new Lanczos vector by sweeping off any contribution from vectors previously determined. The suggested procedure to achieve this is

$$\bar{y}_j = y_j - \sum_{k=1}^{j-1} (\bar{y}_k^T y_j) \bar{y}_k \tag{98}$$

where $\bar{y}_j$ represents an orthogonalized vector.

The computations involved in the reorthogonalization procedure are very substantial. This makes Lanczos procedure much less efficient as compared to say the Householder's method for the reduction of symmetric matrices. The advantages of the Lanczos method become apparent when it is used to obtain the first few eigenvalues of large symmetric matrices. Thus if only p eigenvalues are required where $p << n$, the tridiagonalization may be terminated after the first m Lanczos vectors have been found, m being sufficiently larger than p. The n by m transformation matrix Y_m can now be used to give a tridiagonal matrix T_m.

$$T_m = Y_m^T A Y_m \tag{99}$$

An eigensolution of T_m will give a good estimate of the p most dominant eigenvalues.

To verify whether convergence has been achieved for the required eigenvalues, it may be necessary to solve the eigenvalue problem for T_{m-1} as well, and to compare the magnitude of the desired eigenvalues for the two successive iterations.

It should be noted that if the starting vector does not contain contribution from a specific mode, that mode will be completely missed. A procedure in which the individual elements of the starting vector are selected by a random number generator will avoid the likelihood of missed eigenvectors. However, when the system has multiple eigenvalues, the Lanczos method will give only one of the family of corresponding eigenvectors.

In the application of Lanczos method to vibration problem, the linearized eigenvalue problem is converted to the form of Equation (20b). Matrix $\tilde{D}$ in that equation plays the same role as matrix A in the foregoing discussion.

Considerable amount of research has been directed towards the Lanczos method in the recent years [Jennings, 1977, 1981, Parlett, 1980, Scott, 1981]. Many of the problems associated with the initial version of the method have been solved so that currently the method provides a very efficient tool for the partial eigensolution of large banded systems.

SOLUTION BY MODE SUPERPOSITION

The mode superposition method is an effective means of solving the equations of motion of a linear system. When the applied forces are such that only the first few modes are excited, the response can be fairly accurately represented by the superposition of these mode shapes. The transformation between the physical coordinates r and the modal coordinates y are then given by the transformation Equation (3) where Q is a matrix of order n by m, formed by the first m significant modes starting from the lowest. The transformed equations of motion are given by Equation (5).

$$\tilde{M}\ddot{y} + \tilde{C}\dot{y} + \tilde{K}y = \tilde{p} \tag{5}$$

in which $\tilde{M}$ and $\tilde{K}$ are diagonal. If the damping matrix $\tilde{C}$ is chosen to satisfy the orthogonality relationship Equation (9), Equation (5) represents m uncoupled single degree of freedom equations, the jth equation being

$$m_j\ddot{y}_j + c_j\dot{y}_j + k_jy_j = p_j \tag{100}$$

in which $m_j = q_j^TMq_j$, $k_j = q_j^TKq_j$, $c_j = q_j^TCq_j$ and $p_j = q_j^Tp$.

Because the damping characteristics of a system are difficult to define, c_j is not evaluated explicitly; rather an appropriate damping coefficient ζ is selected. Since $c_j = 2\zeta_j \omega_j m_j$, Equation (100) becomes

$$\ddot{y}_j + 2\zeta_j\omega_j\dot{y}_j + \omega_j^2 y_j = \frac{p_j}{m_j} \tag{101}$$

where ω_j and ζ_j are, respectively, the frequency and the damping ratio in the jth mode. The single degree-of-freedom Equation (101) can be solved by using the Duhamel's convolution integral

$$y_j(t) = \int_0^t p_j(\tau)h_j(t - \tau)d\tau \tag{102}$$

in which h_j called the unit impulse response function is given by

$$h_j(t - \tau) = \frac{1}{m_j\omega_{dj}} e^{-\zeta_j\omega_j(t-\tau)} \sin\omega_{dj}(t - \tau) \tag{103}$$

and $\omega_{dj} = \omega_j\sqrt{1 - \zeta_j^2}$ is the damped natural frequency.

The Duhamel's integral can be evaluated by any standard method of calculus or by using the numerical methods for finding the area under a curve. These latter methods include the Trapezoidal and the Simpson's rules.

An alternative method of solving Equation (101) is through a direct numerical integration by any one of the several available procedures, some of which are discussed in a later section. Lastly, the equation can be equally effective solved in the frequency domain.

Once y_j have been evaluated for $j = 1$ to m, the response in the physical coordinates can be obtained by using the transformation Equation (3).

RITZ VECTOR SOLUTION

Although the mode superposition method is quite effective in solving for the dynamic response of linear systems, the computational efforts involved in the eigen solution of large systems is very substantial. Further, it will be seen from Equation (100) that the magnitude of modal coordinate

y_j depends on the product $q_j^T p$. As an extreme case, if p is orthogonal to q_j, the jth mode will not be excited and y_j will be zero. In many cases, the applied forces are such that they excite the higher modes of vibration. In such a situation, a large number of modes must be included in the analysis and the analysis procedure becomes expensive.

Recent research [Bathe and Wilson, 1983, Wilson, Yuan and Dickens, 1982] has shown that an alternative approach in which a sequence of Ritz vectors is generated by a simple procedure is less expensive and at the same time more accurate than the mode superposition method.

To find the desired sequence of Ritz vectors, let it be assumed that the forcing function p can be represented as the product of a shape function f and a function of time $g(t)$ so that

$$p = fg(t) \tag{104}$$

The first vector in the sequence is derived by solving the equation

$$K\hat{x}_1 = f \tag{105}$$

The vector $\hat{x}_1$ is now normalized so that

$$x_1 = \frac{\hat{x}_1}{\hat{x}_1^T M \hat{x}_1} \tag{106}$$

Subsequent vectors in the series are derived by using the following recurrence relations:

$$K\hat{x}_j = Mx_{j-1} \tag{107}$$

The vectors are then orthogonalized and normalized at each step so that

$$\bar{x}_j = \hat{x}_j - \sum_{k=1}^{j-1} (x_k^T M \hat{x}_j) x_k \tag{108}$$

and

$$x_j = \frac{\bar{x}_j}{\bar{x}_j^T M \bar{x}_j} \tag{109}$$

If the sequence of vectors is grouped to form an n by m matrix X then using the transformation $r = Xy$, the equations of motion take the form of Equation (5) with Q replaced by X. Although these equations are not uncoupled, they are only of order m where $m << n$. The reduced set of equations can be solved by a direct numerical integration.

As an alternative to direct integration, the following eigenvalue problem can be solved to obtain the frequencies ω_i and the mode shapes Z of the reduced system.

$$(\tilde{K} - \omega_i^2 \tilde{M}) z_i = 0 \tag{110}$$

The frequencies ω_i are approximations to m frequencies of the complete system, while estimates of m eigenvectors of the complete system can be obtained by using the relationship

$$Q = XZ \tag{111}$$

The approximate eigenvectors are orthogonal to K and M matrices of the complete system. Therefore if they are used as the new Ritz vectors, the transformed equations will be uncoupled and can be solved by any standard method.

NUMERICAL INTEGRATION OF THE EQUATIONS OF MOTION

Both the modal coordinate and the Ritz vector approach allow the uncoupling of the equations of motion. The resulting single degree-of-freedom equations can be solved by evaluating the Duhamel's integral either by using the standard methods of calculus, or when that is not possible, by numerical methods of calculating the area under a curve.

In many problems, the determination of mode shapes or appropriate Ritz vectors may involve a very substantial amount of calculation, particularly when the representation of displaced shapes requires a large number of vectors. In such situations, direct numerical integration of the equations of motion may prove to be more useful.

It should be noted that when a direct integration of the equations is desired, the damping matrix should be explicitly defined. Methods are, however, available for constructing damping matrices which will provide suitable energy dissipation mechanism in the system [Clough and Bathe, 1972, Clough and Penzien, 1975].

Finally, although not directly relevant to the topic of this chapter, it may be mentioned that integration of the equations is the only effective means of analyzing nonlinear systems in which the physical property matrices are not constant but change with time.

A number of different methods of numerical integration have been devised and are described in the literature [Humar and Wright, 1974, Newmark, 1959, Norris, et al., 1959]. Most of the methods start by dividing the total time interval of interest into a number of small steps. By making an appropriate assumption regarding the manner in which the accelerations vary over a time step, the differential equations of motion are converted into algebraic equations which relate the response parameters at the end of the time step to their values at the begining of the step.

Timoshenko's Method

Let the time step be denoted by h and the displacements at the beginning of a time step by r_n and at the end by r_{n+1}.

If the accelerations are assumed to remain constant during a time step at a value which is average of their values at the beginning and the end of the step, the following relationships hold between the velocities and the displacements at the end of the time step and those at its begining.

$$\dot{r}_{n+1} = \dot{r}_n + \frac{h}{2}(\ddot{r}_n + \ddot{r}_{n+1}) \tag{112a}$$

$$r_{n+1} = r_n + h\dot{r}_n + \frac{h^2}{4}\ddot{r}_n + \frac{h^2}{4}\ddot{r}_{n+1} \tag{112b}$$

In addition, the response parameters at the end of the step are related by the equations of motion:

$$M\ddot{r}_{n+1} + C\dot{r}_{n+1} + Kr_{n+1} = p_{n+1} \tag{113}$$

Provided r_n, $\dot{r}_n$ and $\ddot{r}_n$ are known, Equations (112a), (112b) and (113) can be solved simultaneously for the three unknown vectors: r_{n+1}, $\dot{r}_{n+1}$, $\ddot{r}_{n+1}$. Thus the substitution of $\ddot{r}_{n+1}$ and $\dot{r}_{n+1}$ from Equation (112) into Equation (113) gives

$$K^* r_{n+1} = p^*_{n+1} \tag{114}$$

where

$$K^* = \frac{4}{h^2}M + \frac{2}{h}C + K$$

$$p^* = p_{n+1} + M\left(\frac{4}{h^2}r_n + \frac{4}{h}\dot{r}_n + \ddot{r}_n\right) + C\left(\dot{r}_n + \frac{2}{h}r_n\right) \tag{115}$$

Equation (114) is solved for r_{n+1} and the resulting value is substituted in Equation (112) to obtain acceleration and velocity vectors at point $n + 1$.

The numerical integration procedure defined by Equations (112) and (113) is usually referred to as the Timoshenko's method or the Newmark's β method with $\beta = 1/4$.

Linear Acceleration Method

Another method which is commonly used for the direct integration of the equations of motion is the linear acceleration or the Newmark's β method with $\beta = 1/6$. As its name implies, the method is based on the assumption that the accelerations vary linearly over the time step of integration. This assumption will lead to the following relationships between the displacement and the velocities at the end of the step to those at the beginning.

$$\dot{r}_{n+1} = \dot{r}_n + \frac{h}{2}(\ddot{r}_n + \ddot{r}_{n+1}) \tag{116a}$$

$$r_{n+1} = r_n + h\dot{r}_n + \frac{h^2}{3}\ddot{r}_n + \frac{h^2}{6}\ddot{r}_{n+1} \tag{116b}$$

Combination of Equations (116a) and (116b) with Equation (113) will lead to a relationship similar to Equation (114) in which

$$K^* = \frac{6}{h^2}M + \frac{3}{h}C + K$$

and

$$p^* = p_{n+1} + M\left[\frac{6}{h^2}r_n + \frac{6}{h}\dot{r}_n + 2\ddot{r}_n\right] + C\left[\frac{3}{h}r_n + 2\dot{r}_n + \frac{h}{2}\ddot{r}_n\right] \tag{117}$$

Wilson θ-Method

As a final example of the methods of numerical integration, the Wilson θ-method is worthy of consideration [Clough and Bathe, 1972]. The method is based on the assumption that the accelerations vary linearly over an extended duration of time θh, where $\theta = 1.37$. Equation (114) can now be written for the displacements $\hat{r}$ at the end of time θh.

$$K^*\hat{r} = p^* \tag{118}$$

where K^* and p^* are similar to those in Equation (117) except that h is now replaced by θh, and p_{n+1} by $p_{n+\theta}$.

Once $\hat{r}$ has been obtained by solving Equation (118), the accelerations at the end of the extended time step can be calculated from an equation similar to (116b) but with h replaced by θh. The accelerations at the end of the normal

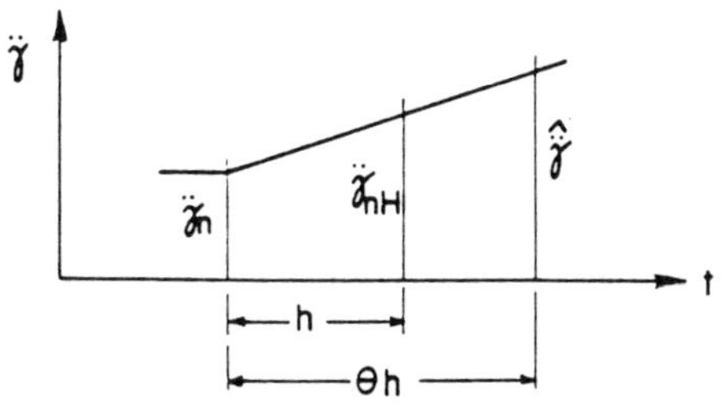

FIGURE 6. Wilson θ-method.

time step are now obtained by linear interpolation:

$$\ddot{r}_{n+1} = \frac{\theta - 1}{\theta}\ddot{r}_n + \frac{1}{\theta}\hat{\ddot{r}} \tag{119}$$

Substitution of r_{n+1} into Equation (116) then gives r_{n+1} and $\dot{r}_{n+1}$.

Selection of the Numerical Method of Integration

The selection of a numerical method of integration is based on an assessment of the errors involved. These errors are of three types: (a) errors due to round-off, (b) errors due to truncation, and (c) propagated errors [Humar and Wright, 1974].

Round-off errors are common to all methods and depend mainly on the precision used in the computation. On the other hand, errors due to truncation as well as the propagated errors depend on the integration method selected.

The expressions for displacement and velocity, such as for example Equations (112) and (116), can be looked upon as Taylor series expansions in which only a finite number of terms have been used. Errors introduced because of the truncation of the series are known as truncation errors. For the same value of h, the linear acceleration method and the Wilson θ-method have a lower truncation error than do several other methods including the Timoshenko method.

Propagated errors are introduced because the numerical integration techniques, in fact, rely on the representation of a differential equation by its finite difference equivalent. Propagated errors introduced during a certain time step can get magnified during a subsequent step, if the magnitude of time step used in the integration is too large. When this happens, the results get unbounded and the method is said to have become unstable. Obviously, instability must be avoided on all accounts.

For a single degree-of-freedom system, the truncation errors can be minimized if the product $h\omega$ is significantly small. Also, when instability is possible it can be avoided by keeping the value of $h\omega$ below a certain limit. For the linear acceleration method this limit is $\sqrt{12}$. On the other hand, the Timoshenko's method and the Wilson θ-method are both unconditionally stable.

The response of a multi degree-of-freedom is the sum total of the contributions from the individual modes of the system. Therefore, truncation errors should be minimized for all those modes that make significant contribution to the response. Since the higher frequency modes, in general, make only a small contribution to the total response, truncation errors in such modes are not too important and the restriction on the value of $h\omega$ is controlled only by the lower modes for which ω is small. The restriction placed on h is therefore within practicable limits. On the other hand, stability must be ensured even for the highest mode for which ω is large, and therefore h has to be kept small to keep the product $h\omega$ within stability limits. For complex systems with a large number of degrees of freedom, the highest mode frequency may be so large that the restriction imposed on h may be too severe for practical implementation of a method that is likely to become unstable. In such situations, an unconditionally stable method, such as the Timoshenko method or the Wilson θ-method, must be used in the integration of the equations of motion.

ANALYSIS IN THE FREQUENCY DOMAIN

The method of time domain analysis described in the earlier sections of this paper can be effectively used to obtain the response of any single or multi degree-of-freedom system. However, when the system being analyzed is linear, the efficiency of the analytical procedure may in many cases be greatly improved by solving the problem in the frequency domain. The following paragraphs describe the frequency domain analysis procedure with reference to a single degree of freedom system [Meek and Veletsos, 1972, Brigham, 1974, Hall, 1982]. The procedures are easily extended to systems with multiple degrees of freedom.

Briefly, a frequency domain analysis requires the resolution of the exciting force into its harmonic components, and the solution of the equations of motion once for each such harmonic component followed by a Fourier synthesis of the resulting responses.

The representation of a general nonperiodic load by the sum of a series of harmonic components requires the use of a Fourier integral given by

$$P(\Omega) = \int_{-\infty}^{\infty} p(t)e^{-i\Omega t}\,dt \tag{120}$$

in which $P(\Omega)$ is a continuous function of the harmonic amplitudes and $p(t)$ is the forcing function.

If $H(\Omega)$ denotes the response of the system to a unit harmonic load ($e^{i\Omega t}$), the total response is obtained by the following integral which represents the sum of component responses.

$$r(t) = \frac{1}{2\pi}\int_{-\infty}^{\infty} P(\Omega)H(\Omega)e^{i\Omega t}d\Omega \tag{121}$$

The integral in Equation (120) is called the Fourier transform of $p(t)$; Equation (121) is the inverse Fourier transform of the product of two transforms: $P(\Omega)$ and $H(\Omega)$. Together, they give the response as a function of time, and provide an alternative to the direct solution of the convolution integral

$$r(t) = \int_0^t p(\tau)h(t - \tau)d\tau \qquad (102)$$

It can be proved that Equations (121) and (102) are entirely equivalent, and that $H(\Omega)$, called the complex frequency response function, is the Fourier transform of the unit impulse response $h(t)$, so that

$$H(\Omega) = \int_{-\infty}^{\infty} h(t)e^{-i\Omega t}\, dt \qquad (122)$$

Discrete Convolution

For numeric computations, the convolution integral must be converted into its discrete form. This is achieved by first constructing periodic versions of the functions $p(t)$ and $h(t)$ with an appropriately chosen period T_0 as shown in Figure 7. The period is then divided into N equal intervals of time each equal to Δt. The convolution integral can now be expressed in its discrete form:

$$r(k\Delta t) = \sum_{j=0}^{N-1} p(j\Delta t)h[(k - j)\Delta t]\Delta t \qquad (123)$$

Equation (123) is different from Equation (102) in two respects. First the integral has been replaced by a rectangular summation. The errors introduced in the results due to this will be negligible if the time interval Δt is small enough. The second difference arises from the fact that Equation (123) represents a convolution of the periodic versions of the original functions and not of the original functions themselves.

To appreciate the difference between a continuous convolution of two functions and the discrete convolution of their periodic version, consider again the functions shown in Figure 7. Figure 7(a) represents an exciting force while Figure 7(b) represents the unit impulse response function of a damped system, truncated after a certain period of time equal to T_h. Figure 7(c) and (d) represent the periodic versions of $p(t)$ and $h(t)$. The period T_0 has been chosen to be equal to $T_p + T_h$ and sufficient number of zero values have been added in both $p(j\Delta t)$ and $h(j\Delta t)$ to make up the period T_0.

Figure 7(e) through (f) provide a graphical illustration of the discrete convolution. The results of the discrete convolution for t from 0 to T_h are similar to those of a continuous convolution. This is true because of the additional zeroes in the periodic versions of p and h. If there were no added zeroes, ordinates from a previous period of h would overlap the exciting force function p and the discrete convolution will not give the same results as a continuous convolution. Even with the addition of zeroes, convolution values beyond T_h are still incorrect because of the truncation applied to h. However, if damping has caused the unit impulse response to decay to a small magnitude after the elapse of time T_h, the errors caused by the truncation of the h would be small and the discrete convolution would closely approximate the continuous convolution over the entire period T_0.

The method described above, of using discrete convolution results to approximate a continuous convolution, is sometimes referred to as the Fast convolution technique. Its main disadvantage is that, in general, convolution needs to be carried out over an extended time interval of $T_p + T_h$ to obtain useful results over a time T_h.

When the duration of the excitation function is so long that convolution over $T_p + T_h$ is not possible within the memory capacity of the computer, the excitation function must be sectioned. Let each section of p be of a length t_p and let T_h be sufficiently large so that the unit impulse response decays to a small value. Also let t_p be larger than T_h. Then the convolution results over the first t_p seconds are correct. However, the results over the remaining $T_0 - t_p$ seconds must be overlapped on the first $T_0 - t_p$ seconds of convolution with the next segment of p. This procedure called overlap add sectioning [Brigham, 1974] is illustrated in Figure 8.

Discrete Fourier Transforms

An alternative to the direct numerical evaluation of convolution integral Equation (102) is to use discrete versions of the Fourier transform integrals Equations (120), (122) and (121). Like discrete convolution, they apply to the periodic versions of function p and h and can be expressed as

$$P(n\Delta\Omega) = \Delta t \sum_{k=0}^{N-1} p(k\Delta t)e^{-2\pi ikn/N} \qquad (124)$$

$$H(n\Delta\Omega) = \Delta t \sum_{k=0}^{N-1} h(k\Delta t)e^{-2\pi ikn/N} \qquad (125)$$

$$r(k\Delta t) = \frac{\Delta\Omega}{2\pi} \sum_{n=0}^{N-1} P(n\Delta\Omega)H(n\Delta\Omega)e^{2\Omega ikn/N} \qquad (126)$$

where $\Delta\Omega = 2\pi/T_0$.

Functions p and h should be sampled at an interval that is sufficiently small so that the sampled functions provide a

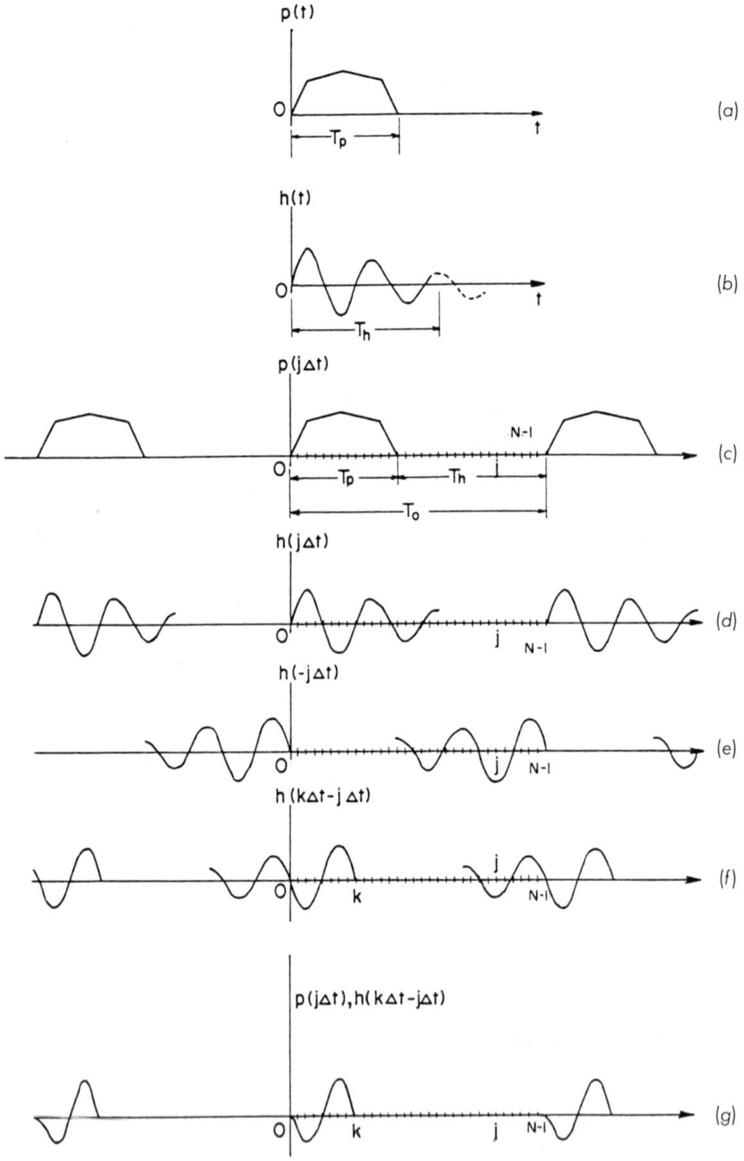

FIGURE 7. Discrete convolution.

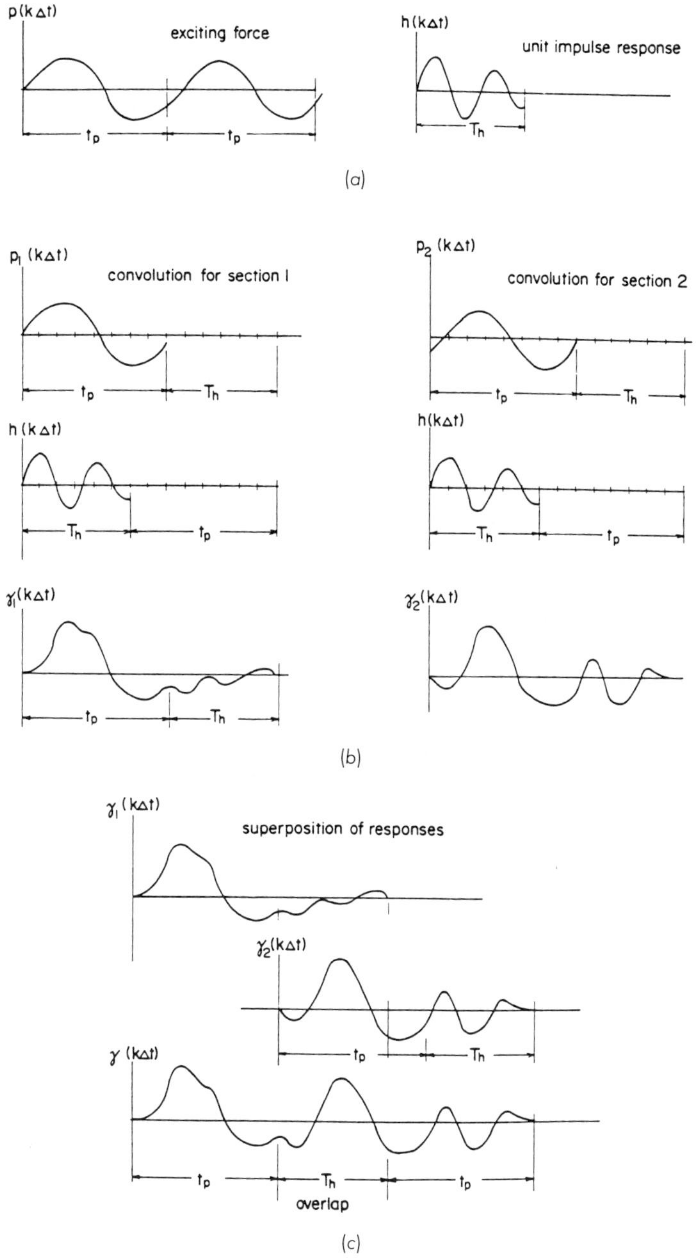

FIGURE 8. Overlap add sectioning.

TABLE 1. Computing Time in Seconds.

N	Direct Method	FFT Method	Speed Factor
16	0.0008	0.0015	0.54
32	0.003	0.0035	0.86
64	0.012	0.0075	1.6
128	0.048	0.0165	2.8
256	0.19	0.0365	5.2
512	0.76	0.08	9.4
1024	2.70	0.18	15.0
2048	11.0	0.39	28.2
4096	43.7	0.84	52.0

good representation of the real functions. This criterion will determine the value of Δt. If the original functions have many oscillations, Δt would have to be smaller. It is of interest to note that the maximum frequency present in the Fourier transforms occurs at $n = N/2$ and is given by $\Omega_{max} = \pi N/T_0 = \pi/\Delta t$. If a time function being convolved has many oscillations, it should have significant high frequency component and a larger value of Ω_{max} must be used in the Fourier transform. This is automatically ensured by choosing a small Δt.

In many cases the complex frequency response $H(n\Delta\Omega)$ is obtained directly rather than through a Fourier transform of the impulse function. In such situations, $\Delta\Omega$ should be chosen small enough and Ω_{max} large enough to accurately represent $H(n\Delta\Omega)$. It should be noted that the selection of two of the quantities T_0, Δt, $\Delta\Omega$ and N determines the other two as well as Ω_{max}, and some trial and error studies may be needed to determine the best values for the parameters involved.

Ordinarily, the computations involved in obtaining the discrete Fourier transforms of the functions being convolved, taking the product of there transforms and then evaluating the discrete inverse Fourier transform, are no less than those in a direct evaluation of the discrete convolution. However, the development of special algorithms called Fast Fourier Transforms (FFT) has completely changed this picture. The FFT algorithms which derive their efficiency by exploiting the harmonic property of the discrete transforms usually cut down the computations by several orders of magnitude and make frequency domain analysis highly efficient.

The exact amount of computer time saved in using FFT approach rather than the conventional approach of evaluating Equation (102) directly depends on the number of sample points N as well as on the details of the FFT convolution program being employed. The results quoted in Table 1 from Brigham (1974) give an indication of the comparative speed of FFT. The table shows as a function of N the time required to convolve two real functions both by the direct and the FFT approach. The data given in the table is based on computations performed on a GE 630 computer. The savings in computer time made possible by the use of FFT are quite evident from the figures given in the table.

The Fast Fourier Transform algorithm was first published by Cooley and Turkey (1965). The dramatic reduction in the speed made possible by this algorithm revolutionized the computing methods in many scientific and engineering applications. The FFT has become one of the most powerful methods for analyzing the dynamic response of linear structural systems, and FFT algorithms are now available at virtually every computer installation.

Meek Veletsos Method

As stated earlier, a limitation on the use of discrete Fourier transforms is that they apply to periodic versions of the functions involved. The convolution obtained through the use of discrete Fourier transforms therefore is not representative of the corresponding continuous convolution, unless special computation procedures, such as for example the Fast convolution method and the overlap-add method are used. Although quite effective, these methods decrease computation efficiency and result in increased cost of computation. More recently, Meek and Veletsos (1972) have suggested a method that does not rely on the addition of a long grace band of zeros to the functions being convolved and is therefore more efficient than the existing methods.

The first step in the method proposed by Meek and Veletsos is to evaluate the exact response of the system to a periodic extension of the prescribed excitation. Let $\bar{h}_k = \bar{h}(k\Delta t)$ be the periodic impulse response of the system defined as the response at time $t_k = k\Delta t$ due to a sequence of unit impulses applied at $t = 0, \pm T_0, \pm 2T_0$, etc. Then it can be shown that

$$\bar{h}_k = \sum_{m=0}^{-\infty} h(k\Delta t - mT_0) \tag{127a}$$

$$= \sum_{m=0}^{\infty} h(k + mN)\Delta t \tag{127b}$$

The periodic response of the system $\bar{r}(k\Delta t)$ to a periodic extension of the exciting force is then given by

$$\bar{r}(k\Delta t) = \sum_{j=0}^{N-1} p(j\Delta t)\bar{h}[(k - j)\Delta t]\Delta t \tag{128}$$

Equation (128) can be evaluated in terms of discrete Fourier transform as follows

$$\bar{r}(k\Delta t) = \frac{\Delta\Omega}{2\pi} \sum_{n=0}^{N-1} P(n)\bar{H}(n)e^{2\Omega ikn/N} \tag{129}$$

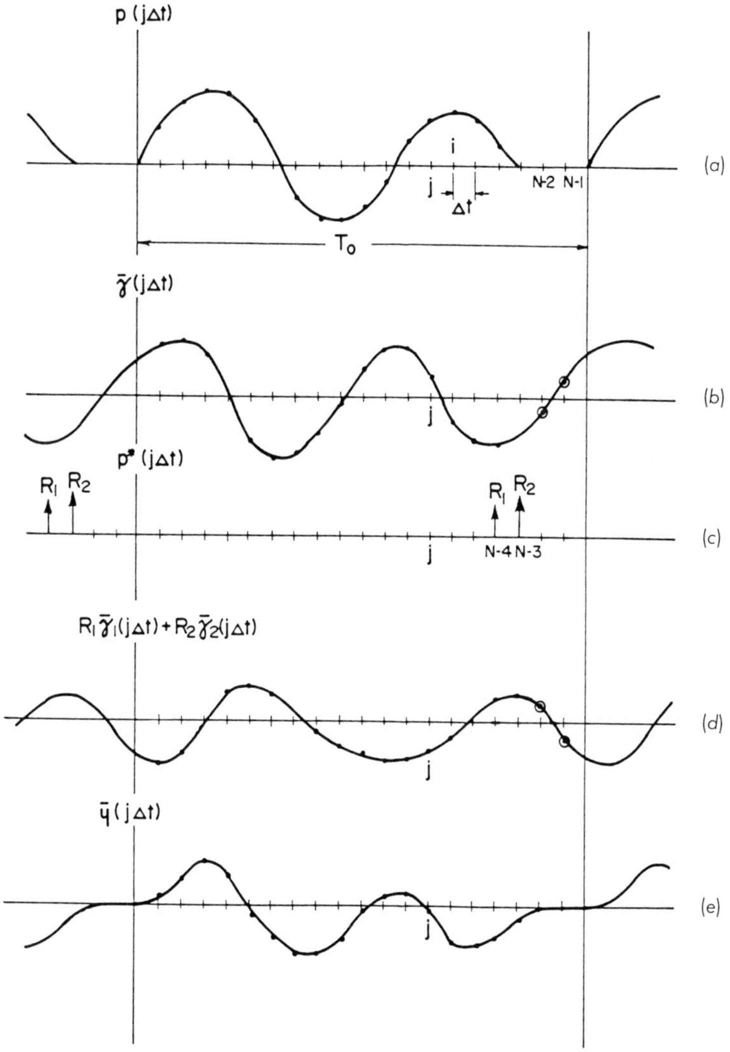

FIGURE 9. Discrete convolution correction by Meek and Veletsos Method.

where $P(n)$ is the discrete Fourier transform of $p(k\Delta t)$, and $\bar{H}(n)$ is the discrete Fourier transform of $\bar{h}(k\Delta t)$.

In general $\bar{r}(k\Delta t)$ is not the same as $r(k\Delta t)$ unless the periodic response is such that the system essentially comes to rest at the end of each period, so that the initial conditions at the start of the next period are zero. A corrective periodic response is therefore superposed on $\bar{r}(k\Delta t)$ so as to bring the system to rest at the end of each excitation period. This ensures that a single period of the sum of the two periodic responses is identical to the response induced by one-time application of the sum of the prescribed excitation and corrective excitation. By confining the corrective excitation to the last few time steps of the original excitation, the true response at all other time steps is obtained.

In the case of a single degree of freedom system, the exciting force is first augmented by adding two zeroes at time points $N - 2$ and $N - 1$. A periodic unit impulse is now applied at $N - 4$ and the response $\bar{r}_1$ of the system to this impulse is determined. In a similar manner, a unit periodic impulse is applied at $N - 3$ and the resulting response $\bar{r}_2$ determined. The total response due to $\bar{p}$ and impulses of magnitudes R_1 and R_2 is now given by

$$\bar{y} = \bar{r} + R_1\bar{r} + R_2\bar{r}_2 \qquad (130)$$

The combined response $\bar{y}$ is then set as equal to zero at time points $N - 2$ and $N - 1$ giving two simultaneous equations from which R_1 and R_2 can be determined. Substitution of these values into Equation (130) gives $\bar{y}$. Because the system response $\bar{y}$ has two zeroes at the end of the period, implying that the system has returned to rest, $\bar{y}$ is the true transient response under the combined action of p and the corrective impulses. Since no corrective impulses are applied between 0 and $N - 4$, this part of $\bar{y}$ also represents the true response under p. The principle involved in the method is illustrated in Figure 9.

SUMMARY AND CONCLUSIONS

The advent of powerful digital computers has had a significant impact on the methods used for the dynamic analysis of linear systems. A large volume of work has been done on the development of efficient procedures of computation which can effectively employ the tremendous power of the modern computers. The development work has resulted in significant advances in computational techniques, so that it is now possible to routinely analyze highly complex systems involving thousands of degrees of freedom.

A major progress in the dynamic analysis of structures has been the development of finite element methods for the efficient modelling of the systems and the formulation of the equations of motion.

The number of equations involved in the dynamic analysis of a complex system can be very large. Hence much research has been carried out on devising efficient techniques for (1) reducing the size of the problem and (2) for the numerical solution of the equations of motion. The first of these two requirements has led to the development of Ritz vector concept, along with efficient Ritz vectors for representing the displaced shapes of the vibrating system. The mode shapes or eigenvectors of a system have been shown to be very good candidates for Ritz vectors. Much research effort has therefore been directed towards the development of efficient procedures for the eigensolution of large systems involving symmetric positive-definite matrices that are often very sparse.

The development of reliable techniques for the numerical integration of the equations of motion has progressed side by side and many such methods are now well recognized and widely used.

More recently, the frequency domain analysis techniques employing Fast Fourier Transform algorithms have been demonstrated to be highly efficient in the analysis of linear systems, and the use of these techniues is progressively increasing.

REFERENCES

1. Newmark, N. M., "A Method of Computation for Structural Dynamics," *Journal of Engineering Mechanics Division, ASCE*, 85, 67–94 (1959).
2. Norris, C., R. J. Hansen, M. J. Holley, J. M. Biggs, S. Namyet, and J. K. Minami. *Structural Design for Dynamic Loads*. McGraw Hill Book Co. Inc., New York (1959).
3. Cooley, J. W. and J. W. Tukey, "An Algorithm for the Machine Calculation of Complex Fourier Series," *Math. Comput.*, 19, 297–301 (1965).
4. Wilkinson, J. H. *The Algebraic Eigenvalue Problem*. Clarendon Press, Oxford, U.K. (1965).
5. Clough, R. W., "Analysis of Structural Vibrations and Dynamic Response," *1st US-Japan Symposium, Recent Advances in Matrix Methods of Structural Analysis and Design*, University of Alabama Press, University of Alabama, Huntsville, AL (1971).
6. Bathe, K. J. and E. L. Wilson, "Large Eigenvalue Problems in Dynamic Analysis," *Journal of Engineering Mechanics Division, ASCE*, 98, 1471–1485 (1972).
7. Clough, R. W. and K. J. Bathe, "Finite Element Analysis of Dynamic Response," *2nd US-Japan Seminar on Matrix Methods of Structural Analysis and Design, Advances in Computational Methods in Structural Mechanics and Design*, University of Alabama Press, University of Alabama, Huntsville, AL (1972).
8. Meek, J. W. and A. S. Veletsos, "Dynamic Analysis by Extra Fast Fourier Transform," *Journal of Engineering Mechanics Division*, ASCE, 18, 367–384 (1972).
9. Bathe, K. J. and E. L. Wilson, "Solution Methods for Eigenvalue Problems in Structural Mechanics," *International Journal of Numerical Methods in Engineering*, 6, 213–226 (1973a).

10. Bathe, K. J. and E. L. Wilson, "Eigensolutions of Large Structural Systems with Small Bandwidth," *Journal of Engineering Mechanics Division, ASCE,* 99, 467–479 (1973b).
11. Brigham, E. O. *The Fast Fourier Transform*. Prentice-Hall Inc., Englewood Cliffs, New Jersey (1974).
12. Humar, J. L. and E. W. Wright, "Numerical Methods in Structural Dynamics," *Canadian Journal of Civil Engineering*, 1, 179–193 (1974).
13. Clough, R. W. and J. Penzien. *Dynamics of Structures*. McGraw Hill, New York (1975).
14 .Bathe, K. J. and E. L. Wilson. *Numerical Methods in Finite Element Analysis*. Prentice Hall (1976).
15. Jennings, A. *Matrix Computations for Engineers and Scientists*. John Wiley and Sons, Chichester, U.K. (1977).
16. Zienkiewicz, O. C. *The Finite Element Method*. McGraw Hill, London, England (1977).
17. Humar, J. L., "Eigenvalue Programs for Building Structures," *Computers and Structures*, 8, 75–91 (1978).
18. Parlett, B. N. *The Symmetric Eigenvalue Problem*. Prentice Hall (1980).
19. Jennings, A., "Eigenvalue Method and the Analysis of Structural Vibration," in *Sparce Matrices and Their Uses,* I. S. Duff, ed., Academic Press, London, pp. 109–138 (1981).
20. Scott, D. S., "The Lanczos Algorithm," in *Sparce Matrices and Their Uses*, I. S. Duff, ed., Academic Press, London, pp. 139–159 (1981).
21. Hall, J. F., "An FFT Algorithm for Structural Dynamics," *Earthquake Engineering and Structural Dynamics*, 10, 797–811 (1982).
22. Wilson, E. L., M. W. Yuan, and J. M. Dickens, "Dynamic Analysis by Direct Superposition of Ritz Vectors," *Earthquake Engineering and Structural Dynamics*, 10 (6), 813–821 (1982).
23. Bayo, E. P. and E. L. Wilson, "Numerical Techniques for the Solution of Soil-Structure Interaction Problems in the Time Domain," Report UCB/EERC-P3/04, Earthquake Engineering Research Center, University of California, Berkeley (1983).

SECTION FIVE

Fluid Mechanics

Turbulent Surface Jets in Stagnant and Moving Ambients

N. RAJARATNAM*

INTRODUCTION

The considerable increase in the generation of electric power in the last two decades has resulted in the release of large quantities of heated-water or thermal discharges into lakes and rivers. One economical mode of discharge is to discharge the heated-water as a surface discharge from either an open-channel or a pipe. To design and operate these outfalls efficiently without damaging the aquatic life and the other uses of the water bodies, it is necessary to understand the mixing of these surface discharges with the water in the recipient water body. Other examples of surface discharges are rivers flowing into lakes and reservoirs, rivers flowing into the oceans, storm water and certain industrial discharges into rivers.

In the case of thermal discharges released as surface jets, the surface discharge is buoyant and the mixing and growth of these surface jets would be affected by the Richardson number Ri_0 at the outfall in addition to other parameters. Hence in the first part of this work, we will consider non-buoyant surface discharges or jets so that we can study their behaviour under rather simpler conditions and bring in the effects of buoyancy in the second part. We will also consider first plane surface jets and then consider bluff surface jets.

After studying these surface discharges into stagnant ambients, we will present a section on bluff surface jets discharged into coflowing streams. This will be followed by the discharge of surface jets into crossflows as in river outfalls and in lakes where there are noticeable ambient currents.

For all our discharges, it is assumed that the Reynolds number of the jets is such that the mixing is turbulent and the viscous stresses are negligible in the equations of motion. It should be pointed out that in the case of buoyant surface jets when the source Richardson number is rather large turbulence is suppressed by the dominant buoyancy forces and the surface jet degenerates to stratified flow. For this stratified flow, the motion in the vicinity of the interface between the diluted surface discharge and underlying body of water could be laminar.

This review is limited to the near-field where the mixing is dominated by the jet momentum and buoyancy with the turbulence in the receiving water body and heat loss to the atmosphere being of very little importance. The longitudinal extent of this near field is of the order of 100 $\sqrt{A_0}$ where A_0 is the area of cross section of the flow at the outfall. It is not very complete and the reader would benefit by also studying the reviews by Harleman and Stolzenbach [8], Koh and Brooks [13], Edinger, Brady and Geyer [3] and Miller and Brighouse [16] for obtaining a comprehensive picture on the behaviour of thermal discharges including submerged discharges and the far-field where the mixing is controlled by the currents and turbulence in the water body as well as by the heat loss to the atmosphere.

NON-BUOYANT PLANE SURFACE JETS

Let us consider non-buoyant plane turbulent surface jets. With reference to the definition sketch (Figure la), b_0 is the thickness and U_0 is the (almost) uniform velocity of the jet whose length (perpendicular to the plane of the paper$-z$ direction) is much larger than its thickness. Let us assume that the ambient fluid (which is identical to the jet fluid) has a depth h which is much larger than b_0. On the underside of the surface jet, soon after it leaves the outfall, a shear layer develops and in most of the practical situations, this shear layer will be turbulent essentially from the beginning. The surface jet would have a potential core of length of approx-

*Department of Civil Engineering, University of Alberta, Edmonton, Alberta, Canada

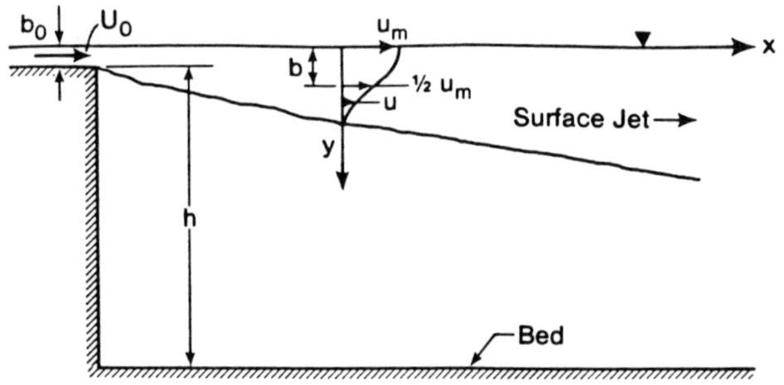

(a) Plane Surface Jet

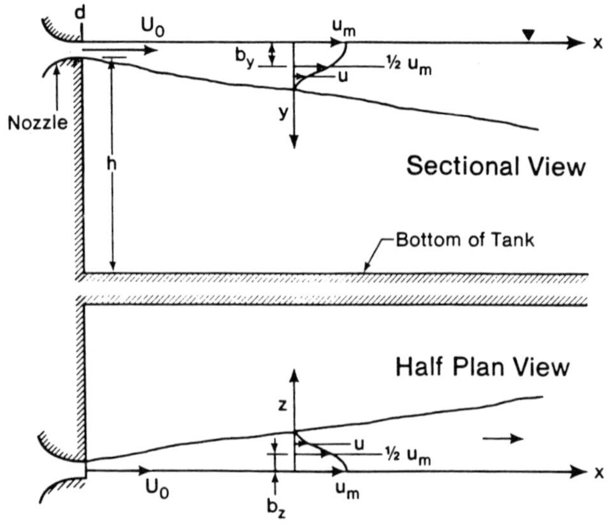

(b) Circular Surface Jet

FIGURE 1. Definition sketches for non-buoyant surface jets.

imately 12 b_0 and beyond the end of this potential core, the mixing would have penetrated to the surface.

Experimental observations on turbulent surface jets [30] have shown that in this region of fully developed flow beyond the end of the potential core, the profile of the time-averaged longitudinal velocity $u(y)$ is "similar" and that the surface jet belongs to the class of slender turbulent flows. If we want to predict time-averaged velocity field of the surface jet, we could accomplish this by working with the continuity equation and the Reynolds equations along with the slender flow approximations as discussed in the following sections.

With reference to the co-ordinate system shown in Figure 1(a), using the slender flow approximations, we can reduce the continuity and Reynolds equations to [26]

$$\frac{\partial u}{\partial x} + \frac{\partial v}{\partial y} = 0 \tag{1}$$

and

$$u\frac{\partial u}{\partial x} + v\frac{\partial u}{\partial y} = \frac{1}{\varrho}\frac{\partial \tau}{\partial y} \tag{2}$$

wherein v is the mean velocity at any point in the y direction, ϱ is the mass density of the fluid and τ is the turbulent shear stress. The corresponding laminar stress has been assumed to be negligible.

Experimental observations have shown that the $u(y)$ profiles at different x-stations are similar. Writing this similarity profile as

$$\frac{u}{u_m} = f(\eta) \tag{3}$$

wherein u_m is the velocity scale, equal to the maximum value of u at any x-station [see Figure 1(a)] and $\eta = y/b$, b

being the length scale defined as the value of y where $u = u_m/2$. Experimental observations have also shown that the exponential equation

$$f(\eta) = \exp(-0.693\eta^2) \tag{4}$$

describes the velocity profiles satisfactorily. Then to predict the velocity field of the surface jet, all we have to do is to predict the variation of u_m and b with the longitudinal distance x measured from a suitable virtual origin, which for practical purposes is assumed to be located at the outfall itself.

Integrating Equation (1) with respect to y, from the surface where $y = 0$ to a large value of y outside the jet, denoted simply as infinity, we get

$$\frac{d}{dx}\int_0^\infty u\,dy = -v_e \tag{5}$$

wherein v_e is the so-called entrainment velocity. Using the Taylor entrainment hypothesis [26]

$$v_e = -\alpha_e u_m \tag{6}$$

wherein α_e is the Taylor entrainment coefficient.

Let us assume that

$$u_m = k_1 x^p \tag{7}$$

$$b = k_2 x^q \tag{8}$$

wherein p and q are unknown exponents and k_1 and k_2 are x-independent coefficients. We can now rewrite Equation (5) as

$$\frac{d}{dx} u_m b = \frac{\alpha_e}{I_1} u_m \tag{9}$$

wherein $I_1 = \int_0^\infty f d\eta$. Equation (9) can be simplified to

$$(p + q)\, k_2 x^{q-1} = \frac{\alpha_e}{I_1} \tag{10}$$

from which one can write that $q = 1$ and $\alpha_e = k_2 (p + q) I_1$.

Integrating the momentum equation, one can obtain:

$$\frac{d}{dx}\int_0^\infty \varrho u^2 dy = 0 \tag{11}$$

which says that the x-momentum flux in the surface jet is invariant in the x direction.

Integrating Equation (11) further, we obtain

$$\int_0^\infty \varrho u^2 dy = M_o \text{ (constant)} \tag{12}$$

wherein M_o is the momentum flux per unit length from the source which is equal to $b_0 \varrho U_0^2$. Equation (12) can be rewritten as

$$b u_m^2 = \frac{b_0\, U_0^2}{I_2} \tag{13}$$

wherein $I_2 = \int_0^\infty f^2 d\eta$. Equation (13) can be further rewritten as

$$k_2 x\, k_1^2\, x^{2p} = \frac{b_0\, U_0^2}{I_2} \tag{14}$$

from which we can deduce that $p = -\frac{1}{2}$ and $k_1^2 = (b_0 U_0^2)/(k_2 I_2)$. Hence, we see that for a plane turbulent surface jet

$$u_m = \frac{k_1}{\sqrt{x}} \tag{15a}$$

$$b = k_2 x \tag{15b}$$

which are very similar to the corresponding equations for the plane turbulent (submerged) jet [26].

Experiments on plane turbulent surface jets have been performed by Chu and Vanvari [2] and Rajaratnam and Humphries [30]. Figure 2(b) reproduced from Reference 30 shows that the length scale b grows linearly with the longitudinal distance x and $db/dx = 0.07$ which happens to be essentially the same as the growth rate of plane turbulent wall jets on smooth walls [26]. The virtual origin appears to be located a distance of about 10 b_0 behind the outfall.

Considering the decay of the velocity scale, Equation 15a can be recast as

$$\frac{u_m}{U_0} = \frac{1}{\sqrt{k_2 I_2}} \frac{1}{\sqrt{x/b_0}} \tag{16}$$

Using the experimental results on the velocity scale, Rajaratnam and Humphries found that Equation (16) takes the form

$$\frac{u_m}{U_0} = \frac{3.1}{\sqrt{x/b_0}} \tag{17}$$

even though substitution of $k_2 = 0.07$ and $I_2 = 0.753$ (assuming the exponential expression for the velocity profile) in Equation (16) gives a larger value for the coefficient

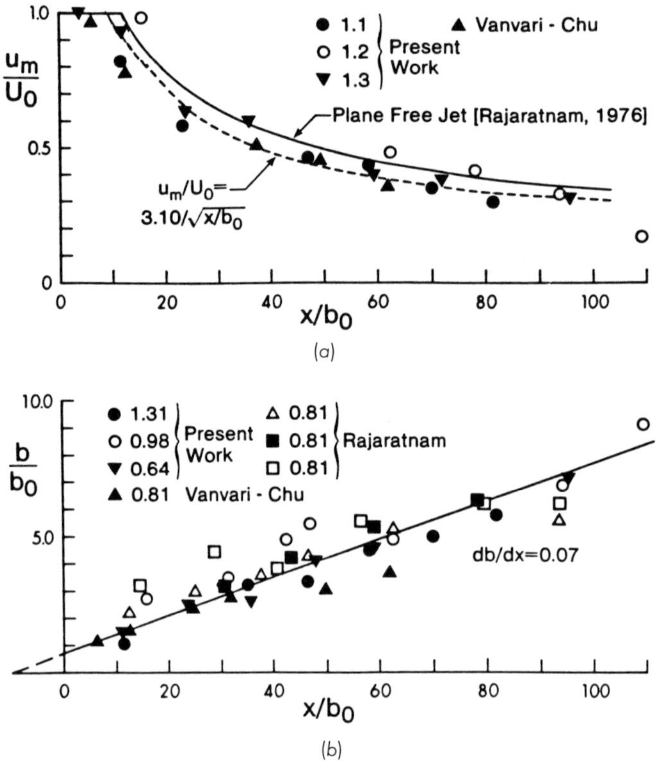

FIGURE 2. Velocity and length scales for plane non-buoyant surface jets [30].

in Equation (17). For rather small values of x/b_0, u_m/U_0 can be read directly from a mean line drawn through the experimental results in Figure 2(a). Concerning the entrainment coefficient α_e, using Equation (10), $k_2 = 0.07$ and $I_1 = 1.065$, $\alpha_e = 0.037$ compared to 0.053 for a plane turbulent (submerged) jet [26].

If the surface jet carries a passive pollutant, based on our understanding of pollutant spreading in submerged jets, we can write down the structure of the main equation describing the pollutant distribution in surface jets. If c is the time averaged concentration of the pollutant at any point, we can assume that

$$\frac{c}{c_m} = h(\eta) = \exp\left[-0.693\left(\frac{y}{kb}\right)^2\right] \tag{18}$$

wherein c_m is the concentration scale, equal to the maximum concentration at any x section and k is a coefficient approximately equal to 1.15. Further we can also write

$$\frac{c_m}{c_0} = \frac{C_3}{\sqrt{x/b_0}} \tag{19}$$

where c_0 is the pollutant concentration at the outfall and c_3 is a constant which is equal to 3.47 for plane (submerged) jets.

NON-BUOYANT BLUFF SURFACE JETS

Having considered plane surface jets in the previous section, let us now consider a circular surface jet of diameter d issuing with a uniform velocity of U_0 into a deep and large stagnant body of the identical fluid as shown in Figure 1(b). We could refer to such a circular jet as a bluff surface jet so that we could include in this class, rectangular jets with the length to thickness (or width) (aspect) ratio not very different from unity, semi-circular jets and a few others. Experimental observations by Rajaratnam and Humphries [30] have shown that bluff surface jets belong to the class of slender flows and show similarities to bluff wall jets, which have been studied in considerable detail [14,26].

Experimental observations by Rajaratnam and Humphries [30] on circular and rectangular surface jets of length to width (or thickness) ratio of 1.6 have shown that in the region of fully-developed flow (downstream of the end of the potential core), the u velocity profiles in the central plane

($z = 0$) are similar; that is $u/u_m = f(\eta_y)$ where $\eta_y = y/b_y$ with the length scale b_y being equal to y where $u = u_m/2$. In the transverse direction, in the level at which u_m occurs, the velocity profiles are again similar; that is $u/u_m = f(\eta_z)$ with $\eta_z = z/b_z$ where b_z is the corresponding length scale. Both the $f(\eta_y)$ and $f(\eta_z)$ functions are reasonably well described by the exponential expression, unless there is excess wave generation at the water surface, in which case the maximum velocity in the centerplane occurs somewhat below the water surface.

For turbulent non-buoyant bluff surface jets, with the coordinate system shown in Figure 1(b), the continuity equation and the Reynolds equations, simplified using the slender flow approximations and the assumptions of hydrostatic pressure distribution in the vertical direction, reduce to

$$\frac{\partial u}{\partial x} + \frac{\partial v}{\partial y} + \frac{\partial w}{\partial z} = 0 \tag{20}$$

$$u\frac{\partial u}{\partial x} + v\frac{\partial u}{\partial y} + w\frac{\partial u}{\partial z} = -\left\{\frac{\partial}{\partial y}\overline{u'v'} + \frac{\partial}{\partial z}\overline{u'w'}\right\} \tag{21}$$

Integrating Equation (21) with respect to y and z from $y = 0$ to $y \to \infty$ and $z = 0$ to $z \to \infty$, we can reduce Equation (21) to the form:

$$\frac{d}{dx}\int_0^\infty \varrho u^2 dy\, dz = 0 \tag{22}$$

and

$$\int_0^\infty \varrho u^2\, dy\, dz = \text{constant } M_0 \tag{23}$$

where M_0 is the momentum flux from the outfall.

Following the work of Rajaratnam and Pani [25,26] on bluff wall jets, let us assume

$$\frac{u}{u_s} = f_1\left(\frac{y}{b_y}\right) = f_1(\eta_y)$$
$$\frac{u_s}{u_m} = f_2\left(\frac{z}{b_z}\right) = f_2(\eta_z) \tag{24}$$

where u_s is a surface velocity and u_m is the surface velocity in the centerplane. Then

$$u = u_m f_1 f_2 \tag{25}$$

Let us assume

$$u_m = k_1 x^p \tag{26a}$$

$$b_y = k_{2y} x^{q_y} \tag{26b}$$

$$b_z = k_{2z} x^{q_z} \tag{26c}$$

With these assumptions, the integral momentum equation can be reduced to:

$$\frac{d}{dx}[\varrho u_m^2\, b_y\, b_z] \int\int_0^\infty f_1^2 f_2^2\, d\eta_y d\eta_z = 0 \tag{27}$$

which gives the relation

$$2p + q_y + q_z = 0 \tag{28}$$

Integrating now the continuity equation in a similar manner,

$$\frac{d}{dx}\int_0^\infty \int u\, dy\, dz = \int_0^\infty (-v_{ez})dz + \int_0^\infty (-w_{ey})dy \tag{29}$$

where v_{ez} and w_{ey} are the entrainment velocities, respectively, at z and y locations. Further,

$$v_{ez} = -\alpha_{e1}\, u_m\, f_2(\eta_z)$$
$$w_{ey} = -\alpha_{ez}\, u_m\, f_1(\eta_y) \tag{30}$$

Substituting Equation (30) into Equation (29), it can be shown that $q_y = q_z = 1$. Then from Equation (28), $p = -1$. Thus for bluff surface jets

$$u_m \propto \frac{1}{x}$$
$$b_z \text{ and } b_y \propto x \tag{31}$$

Experimental observations by Rajaratnam and Humphries [30] on circular surface jets have shown that both the length scales b_y (in the central plane) and b_z (in the plane of u_m) grow linearly with the longitudinal distance as shown in Figure 3(b) and (c). For the transverse length scale, its growth rate $db_z/dx = 0.09$ (which is less than half the corresponding rate for bluff wall jets) and the virtual origin (where b_z is zero) is located at the outfall itself. For the vertical length scale b_y, its rate of growth $db_y/dx \cong 0.044$ (which is essentially the same as that for bluff wall jets on smooth walls). Further, the velocity scale was found to vary

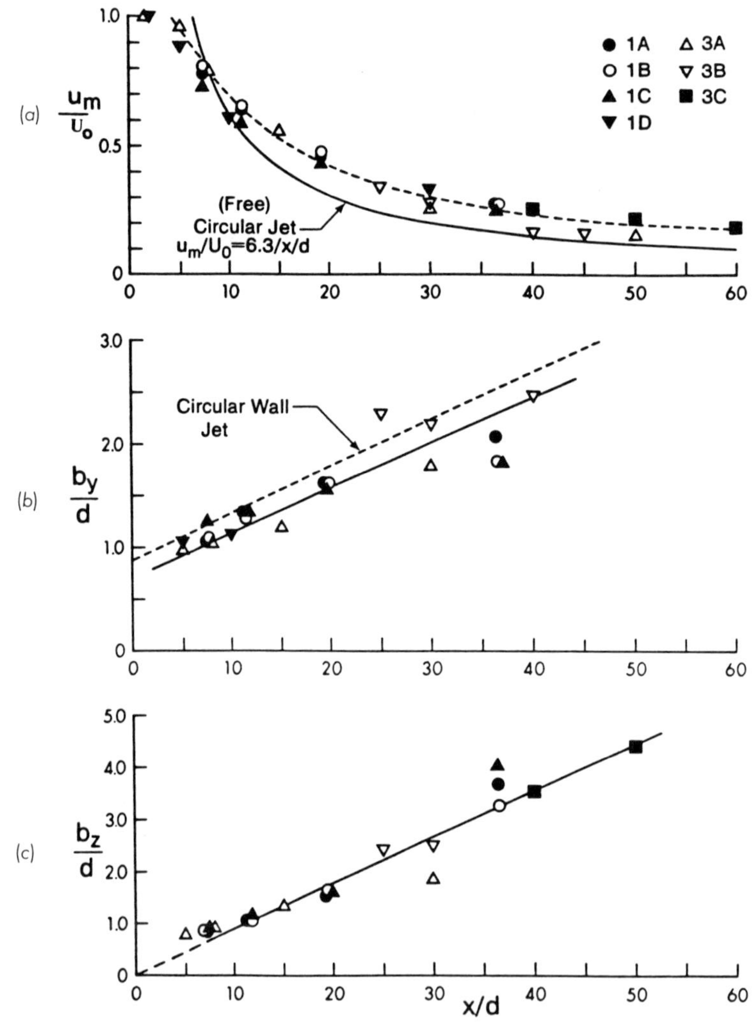

FIGURE 3. Scales for non-buoyant circular surface jets [30].

inversely with x and the variation of u_m/U_0 with x/d is reproduced in Figure 3(c).

Experimental observations on rectangular surface jets with length to thickness of 1.6 agreed very well with the results for circular jets, if d is replaced by $\sqrt{4A_0/\pi}$, A_0 being the area of the rectangular outfall. Apparently very little is known regarding the turbulence characteristics of these surface jets.

NON-BUOYANT SURFACE JETS IN COFLOWING STREAMS

Consider a circular surface jet of diameter d issuing with a uniform velocity of U_0 parallel to a surrounding flow with a mean velocity of U_1 as shown in Figure 4 with a depth of h and width of B such that h and B are much larger than d. As the jet mixes with the surrounding stream, in the region of developed flow, the experimental observations of Rajaratnam [31] have shown that the distribution of the time-averaged longitudinal velocity u with respect to the surrounding freestream velocity U_1 is similar in the centerplane; that is $(u - U_1)/(u_m - U_1) = f(y/b_y)$ wherein u_m is the maximum value of u at any x section and b_y is the vertical length scale equal to y where $(u - U_1) = (u_m - U_1)/2$.

Further, this similarity profile is well described by the exponential equation. Secondly the transverse distribution of the u velocity, in the horizontal plane on which u_m occurs (which generally is the water surface unless there is significant surface wave generation), with respect to the freestream velocity was also found to be similar. That is, $(u - U_1)/(u_m - U_1) = f_2\ (z/b_z)$ with the exponential expression being a good approximation to the f_2 function.

In the light of these findings on the velocity field relative to the freestream, in order to describe the velocity field completely, one has to be able to predict the behaviour of the velocity scale $(u_m - U_1)$ and the length scales b_y and b_z.

Considering the decay of the velocity scale $(u_m - U_1)$, using the concept of excess momentum radius θ, from studies on jets in coflowing streams, defined by the equation

$$M_0 = \frac{\pi d^2}{4}\ \varrho U_0\ (U_0 - U_1) = \pi\ \theta^2 \varrho U_1^2 \qquad (32)$$

values of $(u_m - U_1)/U_1$ were plotted against x/θ as shown in Figure 5(a). In Figure 5(a), the data points covering the values of α from about 5 to about 18 where α is the ratio of the jet to free stream velocity are well described by the corresponding curve of circular (submerged) jet in a coflowing stream [22].

Regarding the length scale, Figure 5(b) shows that the growth of the dimensionless vertical length scale b_y/θ with x/θ is well described by the corresponding curve of the circular jet in a coflowing stream [22]. Figure 5(c) shows that the growth of the dimensionless transverse length b_z/θ is well described by the corresponding curve of the circular wall jet in a coflowing stream [29]. The range of x/θ was increased by photographic observations and these results are shown in Figure 6 wherein $\bar{b}_z$ is the half-width and $\bar{b}_y$ is the thickness of the surface jet in the centerplane (as measured from photographs of the dyed jet). Based on experimental measurements, $\bar{b}_y \cong 2.2\ b_y$ and $\bar{b}_z \cong 2.5\ b_z$.

Rajaratnam [31] also found that the correlations for the length scales described above were satisfactory for $h/d = 10.6$ but for $h/d = 2.8$, the shallow depth affected the growth rates significantly. Hence, at this stage, it can be said that the correlations presented herein can be used for h/d greater than about 10 in the x/θ range studied.

It appears that plane surface jets in coflowing streams have not been studied but since plane surface jets in stagnant ambients behaved very much like plane wall jets and circular surface jets had many similarities to circular wall jets in coflowing streams, it appears, at least as a first approximation, one can suggest that plane surface jets in coflowing streams might behave like plane wall jets in coflowing streams. A number of methods have been developed [26] to handle the plane wall jet in a coflowing stream; but because of the preliminary nature of our attack, it is suggested that Figures 12–14 of Reference 26 be used first to predict (U_m/U_1) at any x/θ where $U_m = u_m - U_1$ and then Equa-

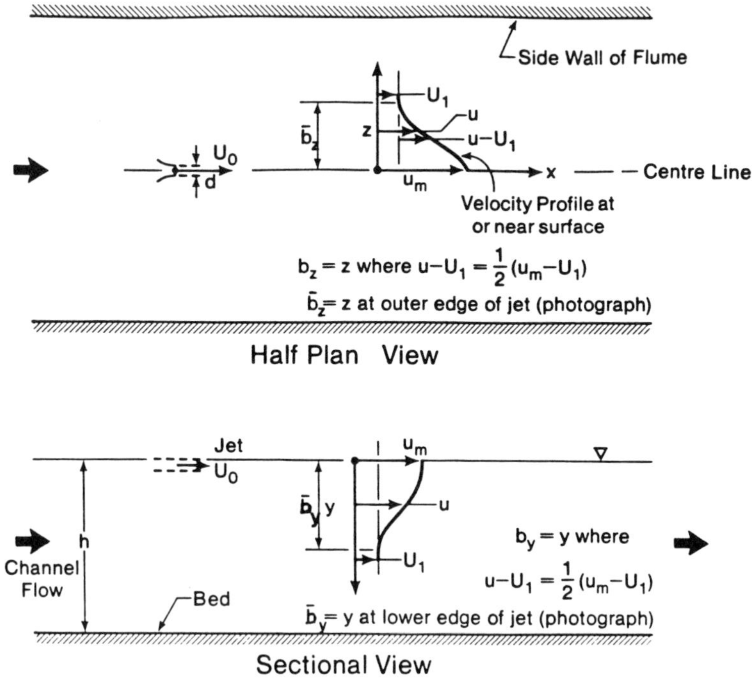

FIGURE 4. Definition sketch of non-buoyant circular surface jet in a coflowing stream.

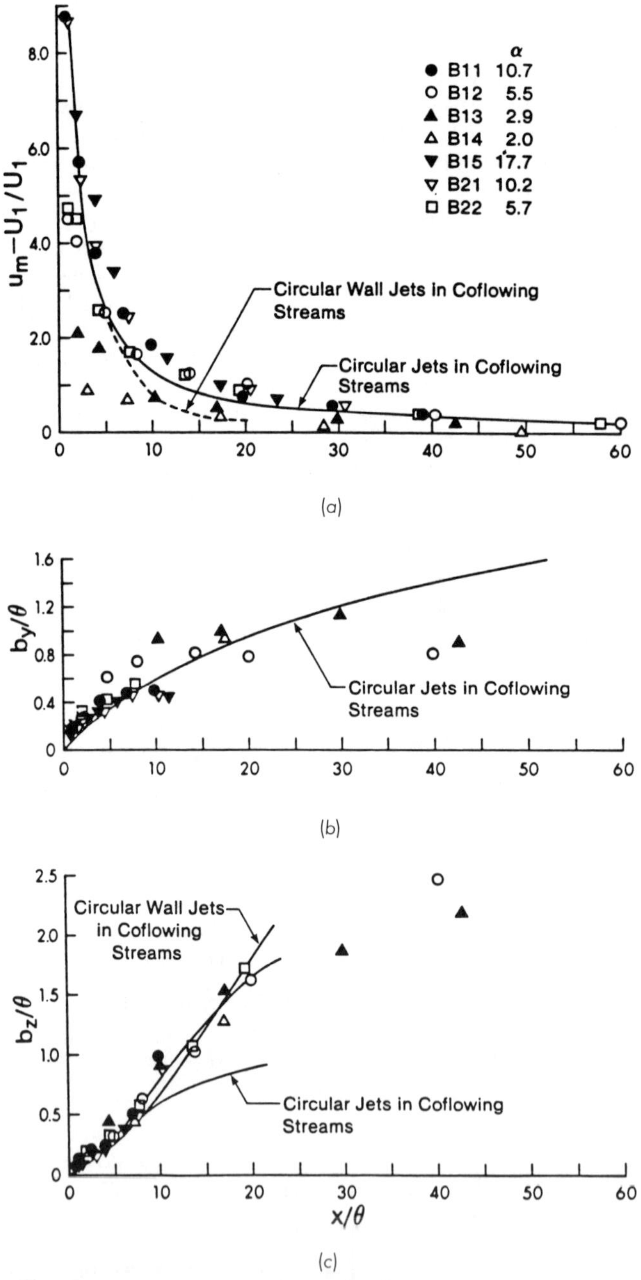

FIGURE 5. Correlation of velocity and length scales for non-buoyant circular surface jets in coflowing streams [31].

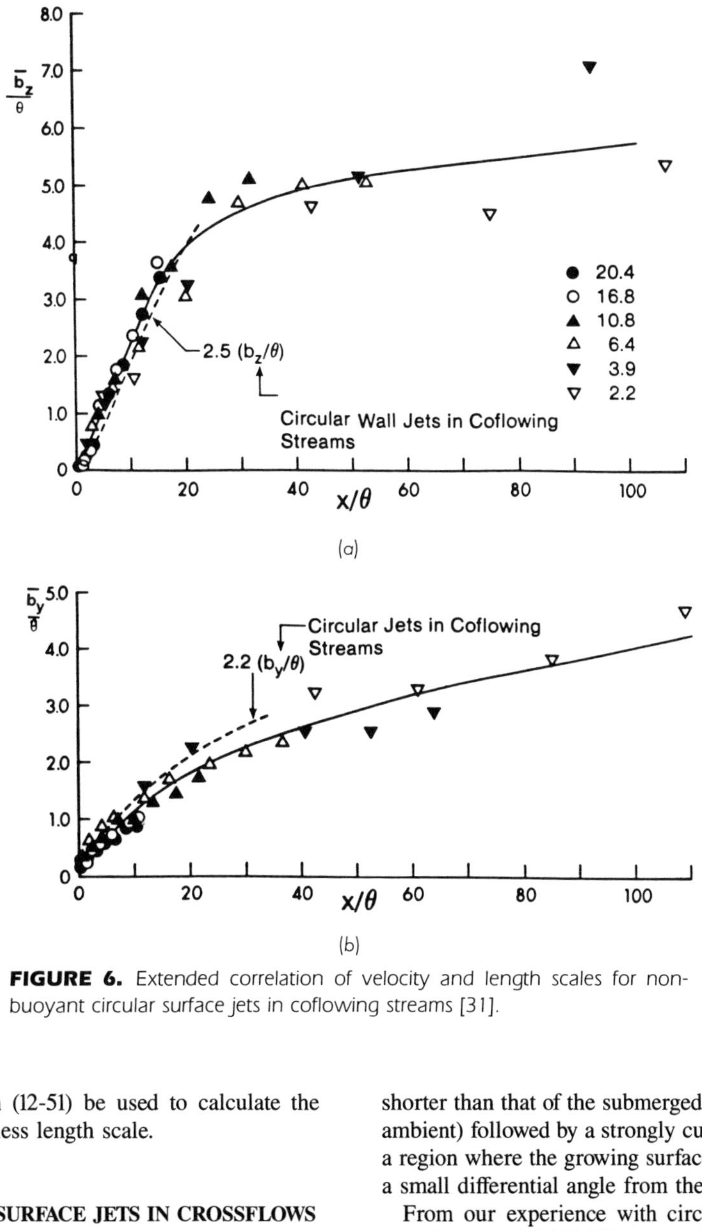

FIGURE 6. Extended correlation of velocity and length scales for non-buoyant circular surface jets in coflowing streams [31].

tion (12-50) or Equation (12-51) be used to calculate the growth of the dimensionless length scale.

NON-BUOYANT BLUFF SURFACE JETS IN CROSSFLOWS

Let us consider the non-buoyant surface discharge from a circular outfall perpendicular to a channel flow with a mean velocity of U_1 (see Figure 7). Herein U_0 is the jet velocity, d is its diameter and the depth of flow h is much larger than the jet diameter d. Based on our experience with circular (submerged) jets in crossflow [26], we know that the jet will be deflected by the cross-flow with the deflection and effect of the crossflow depending upon the magnitude of α, the ratio of U_0 to U_1. For a value of α say equal to about 5, the surface jet will have a flow-development region (somewhat shorter than that of the submerged circular jet in a stagnant ambient) followed by a strongly curvilinear region and then a region where the growing surface jet is flowing with only a small differential angle from the surrounding flow.

From our experience with circular (submerged) jets in crossflows, [1,6,26,27,28] we would expect enhanced mixing, compared to the surface jet in a stagnant ambient. The entrainment coefficient α_e would vary along the length of the jet [28]. The momentum flux in the jet would change because of entrainment of the surrounding fluid as well as due to the drag exerted by the surrounding flow on the diffusing jet. We would present the outline of a computational procedure of the integral type and point out the difficulties in choosing the different physical parameters like α_e and the coefficient of drag in the drag exerted by the ambient flow. Further, it is unrealistic to assume that the jet

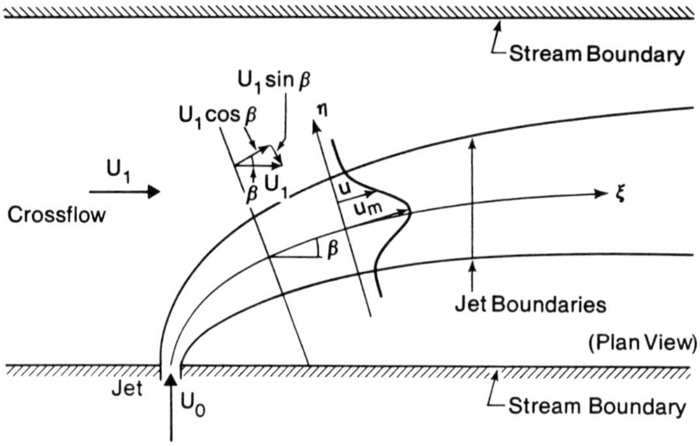

FIGURE 7. Definition sketch of circular surface jet in a crossflow.

grows linearly, even though this convenient assumption has been made by a number of investigators.

Following the method of Motz and Benedict [17] with reference to Figure 7, we write the integral continuity of equation as

$$\frac{d}{d\xi}\int_A u\,dA = Pv_e \tag{33}$$

wherein A is the cross-sectional area of the deflected jet, P is its perimeter for entrainment and v_e is the entrainment velocity. In many computational models, v_e has been written as

$$v_e = -\alpha_e\,(u_m - U_1\cos\beta) \tag{34}$$

where u_m is the maximum velocity in the jet (assumed to be at its center), β is the angle of the deflected jet with the ambient stream and α_e is the entrainment coefficient. As shown for circular (submerged) jets in crossflow [27], α_e is likely to vary along ξ, even though it has been assumed constant in many mathematical models.

Next, we write the equation for the rate of change of the jet momentum in the x or stream direction as

$$\frac{d}{d\xi}\int_A \varrho u^2 dA\cos\beta = (Pv_e\varrho U_1) + C_D\frac{\varrho U_1^2\sin^3\beta}{2}\bar{b}_y \tag{35}$$

where C_D is a drag coefficient and $\bar{b}_y$ is an average thickness of the jet. The drag force component is likely to be important only in the initial region where the bending of the jet is significant and here again C_D will have to be chosen carefully. For the other (z) component

$$\frac{d}{d\xi}\int_A \varrho u^2 dA\sin\beta = -\frac{C_D\varrho U_1^2\sin^2\beta}{2}\bar{b}_y\cos\beta \tag{36}$$

From the geometry of the deflected jet

$$\frac{dx}{d\xi} = \cos\beta \tag{37}$$

$$\frac{dz}{d\xi} = \sin\beta \tag{38}$$

To solve the above set of equations, one will have to know $\alpha_e(\xi)$, $C_D(\xi)$, the cross-sectional geometry of the deflected surface jet and the velocity field in the surface jet which is unlikely to be symmetrical about its (geometrical) axis (except perhaps for very large values of α).

In the absence of any (reliable) information on these aspects for bluff surface jets in crossflows, it is unproductive to proceed any further with the numerical solution of the above set of differential equations and instead we present herein an approximate empirical procedure to predict the characteristics of non-buoyant circular surface jets in crossflows based on some (unpublished) experimental results (of the author).

In these experiments, photographic observations were made of the growth of the half-width $\bar{b}_z$ and thickness $\bar{b}_y$ of non-buoyant circular surface jets issuing from circular nozzles of diameter d equal to 0.63, 1.25 and 2.58 cm into an open-channel type crossflow with an approximate depth of 25 cm with α varying from about 2 to 25.

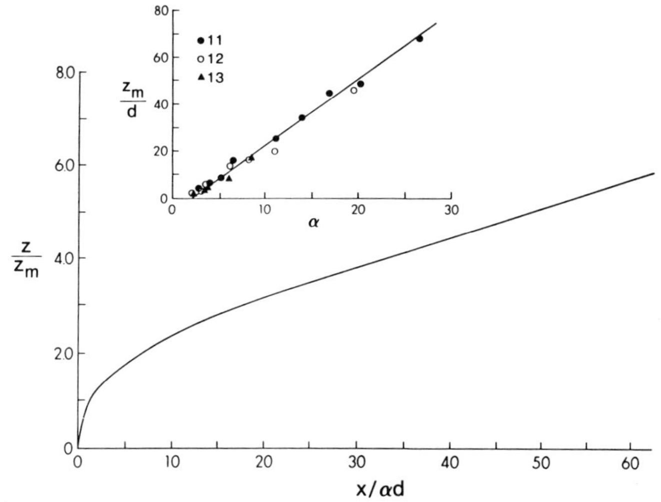

FIGURE 8. Trajectory of the centerline of non-buoyant circular surface jets in crossflow.

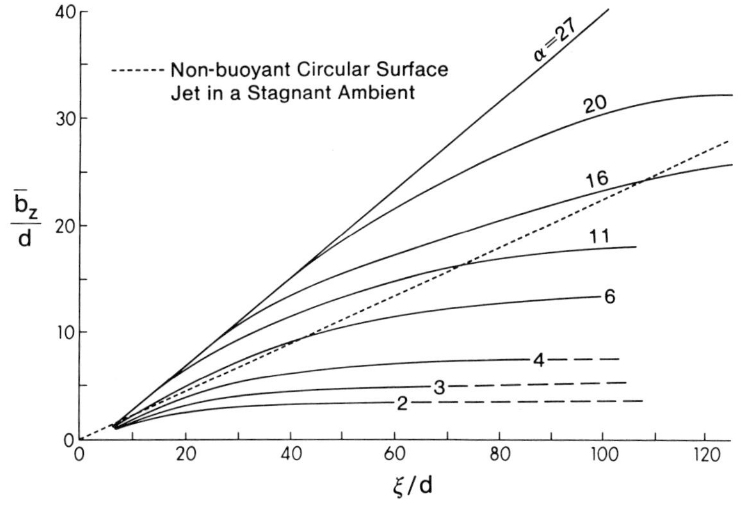

FIGURE 9. Growth of half-width of non-buoyant circular surface jets in crossflow.

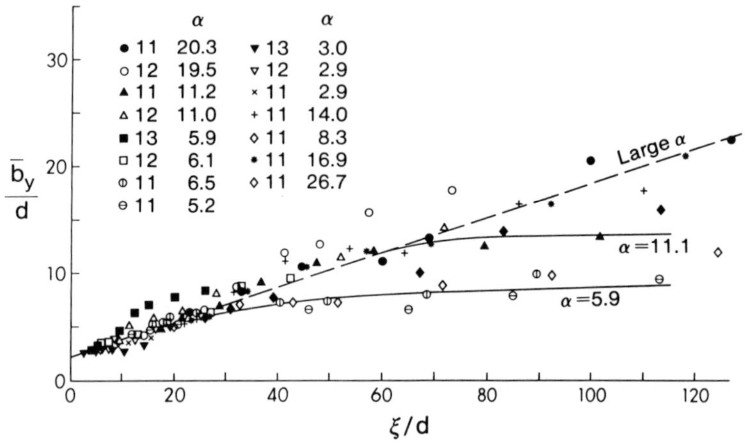

FIGURE 10. Growth of thickness of non-buoyant circular surface jets in crossflow.

Based on these results, Figure 8a presents the trajectory of the centerline of the deflected surface jet at the water surface. In Figure 8a, z/z_m is plotted against $x/\alpha d$ wherein z_m is the value of z where $x = \alpha d$. This scheme correlates the results for α in the range of about 2 to 25 reasonably well with z_m/d increasing linearly with α as shown in Figure 8b. The relation between z_m/d and α for $\alpha \geq 3.5$ can be described by the equation

$$z_m/d = 2.75\ (\alpha - 2.0) \tag{39}$$

If $\bar{b}_z$ is the half-width of the deflected surface jet at any distance ξ along its (deflected) centerline, the growth of $\bar{b}_z/d$ with ξ/d for different values of α is shown in Figure 9 wherein mean curves through experimental results (not shown herein) are shown. In Figure 9, it is seen that for large values of $\alpha(\cong 27)$, $\bar{b}_z$ grows linearly with ξ/d (for ξ/d up to about 100) whereas for smaller values of α, the growth is definitely curvilinear.

Similar results on the growth of the thickness $\bar{b}_y/d$ along ξ are shown in Figure 10 wherein average curves are drawn in for $\alpha = 5.9$, 11.1 and the growth appears to be linear for larger values of α.

It is suggested that these empirical curves be used for practical purposes (with some caution) until more extensive results are available. For these results, the depth and width of the crossflow are believed to be large enough to neglect the effects of the channel bed and sides on the surface jet.

PLANE TURBULENT BUOYANT SURFACE JETS

Let us consider a plane buoyant surface jet discharged from an outfall (as shown in Figure 11) with a depth (or thickness) of b_0, velocity of U_0 and density of ϱ_0. Let the density of the deep stagnant ambient fluid be ϱ_a. Let us assume that the Reynolds number of the surface jet $U_0 b_0/\nu$ (wherein ν is the kinematic viscosity of the fluid) is large enough for the shear layers on the underside of the jet to become turbulent at a relatively small distance from the outfall and in our discussion, we are only concerned with turbulent surface jets. In our discussion we are mainly in-

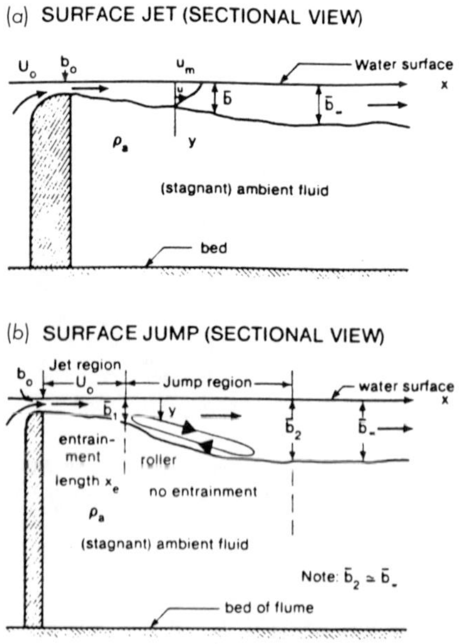

FIGURE 11. Definition sketch of plane buoyant surface jet and jump.

terested in the region of fully-developed flow which starts from the end of the potential core.

Experimental observations have indicated that the behaviour of the buoyant turbulent surface jet is determined jointly by the (bulk) Richardson number Ri_0 at the outfall and the depth of the surface stratified layer $\bar{b}_\infty$ maintained further downstream (at the end of our zone, referred to as the near-field) by (some) downstream control or by surface evaporation (in the case of heated surface discharges) with or without downstream control. This Richardson number is defined as

$$Ri_0 = \frac{g\, b_0 \dfrac{\Delta\varrho_0}{\varrho_a}}{U_0^2} \tag{40}$$

wherein $\Delta\varrho_0 = (\varrho_a - \varrho_0)$ and g is the acceleration due to gravity. Another parameter, used frequently, is the densimetric Froude number at the source F_0 defined as

$$F_0 = \frac{U_0}{\sqrt{g\, b_0 \dfrac{\Delta\varrho_0}{\varrho_a}}} \tag{41}$$

It is easy to see that $F_0 = 1/\sqrt{Ri_0}$. For the non-buoyant surface jet, $Ri_0 = 0$ and $F_0 = \infty$. For the buoyant surface jet, Ri_0 could vary from about zero to reasonably large values like 5 to 10.

Experimental observations indicate that for a buoyant surface jet with given Ri_0 depending upon the value of $\bar{b}_\infty/b_0$ we can have a surface jet, eventually degenerating to a stratified surface layer far downstream or a surface jump at the outfall or a surface jet followed by a surface jump (see Figure 11) eventually ending in a stratified surface layer. For larger values of $\bar{b}_\infty/b_0$ we could even have a drowned jump. All these possibilities make the study of the buoyant surface jet rather complicated.

Let us consider the case where the surface discharge behaves like a surface jet and eventually degenerates to a stratified surface layer. We already know that if $Ri_0 = 0$, the surface jet grows linearly. The results of a (photographic) study [33] on the growth rate of buoyant surface jets for a range of Ri_0 are shown in Figure 12. Figure 12 shows that for any given Ri_0 (or F_0), the experimental results follow the line of the non-buoyant surface jet up to some characteristic distance and then the growth rate slows down eventually approaching a constant thickness. For large values of Ri_0 one can see a hump soon after the departure from the non-buoyant surface jet line.

The pioneering study of Ellison and Turner [4] and the work of Chu and Vanvari [2] have shown clearly that the entrainment of the ambient fluid into the expanding surface jet gets continuously reduced as the local (bulk) Richardson number $\overline{Ri}$ defined as

$$\overline{Ri} = \frac{\text{kinematic buoyancy flux in surface jet}}{u_m^3} \tag{42}$$

(u_m being the maximum velocity in the surface jet at any x station) increases from zero. The observations of Chu and Vanvari [2] indicate that the entrainment coefficient α_e falls to zero when $\overline{Ri} \cong 0.2$.

Let us present a theoretical framework for analysing plane turbulent buoyant surface jets. With reference to the coordinate system shown in Figure 11, for the time-averaged flow, the continuity equation is

$$\frac{\partial u}{\partial x} + \frac{\partial v}{\partial y} = 0 \tag{43}$$

Assuming that the pressure distribution in the vertical (y) direction is hydrostatic, using slender-flow approximations and neglecting the viscous stresses, the Reynolds equation in the longitudinal (x) direction reduces to

$$u \frac{\partial u}{\partial x} + v \frac{\partial u}{\partial y} = \frac{g}{\varrho} \frac{\partial}{\partial x} \int_0^y \Delta\varrho \, dy - \frac{\partial}{\partial y} \overline{u'v'} \tag{44}$$

where $\Delta\varrho = \varrho_a - \varrho$, ϱ being the time-averaged density at any point in the surface jet.

An order of magnitude analysis on the "normalized" forms of Equation (43) and Equation (44) shows that in Equation (44) the first term on the right side is of the order of a bulk Richardson number Ri. If Ri is of the order of one, Equation (44) will reduce to

$$u \frac{\partial u}{\partial x} + v \frac{\partial u}{\partial y} = \frac{g}{\varrho} \frac{\partial}{\partial x} \int_0^y \Delta\varrho \, dy \tag{45}$$

If Ri is equal to zero, Equation (44) will reduce to the corresponding equation of the non-buoyant surface jet. When Ri is much smaller than one, in the early part of the near-field Equation (44) will be

$$u \frac{\partial u}{\partial x} + v \frac{\partial u}{\partial y} = - \frac{\partial}{\partial y} \overline{u'v'} \tag{46}$$

but as the jet moves further away from the outfall, the bulk Richardson number will increase and hence over a certain length, the full equation [i.e., Equation (44)] will have to be used. In the final stages of the near-field, if the Richardson number approaches one, then the equation of motion will be Equation (45). Thus as the jet moves away from the outfall, there is a progressive increase in the local (bulk) Richardson number which affects the characteristics of the jet. Further,

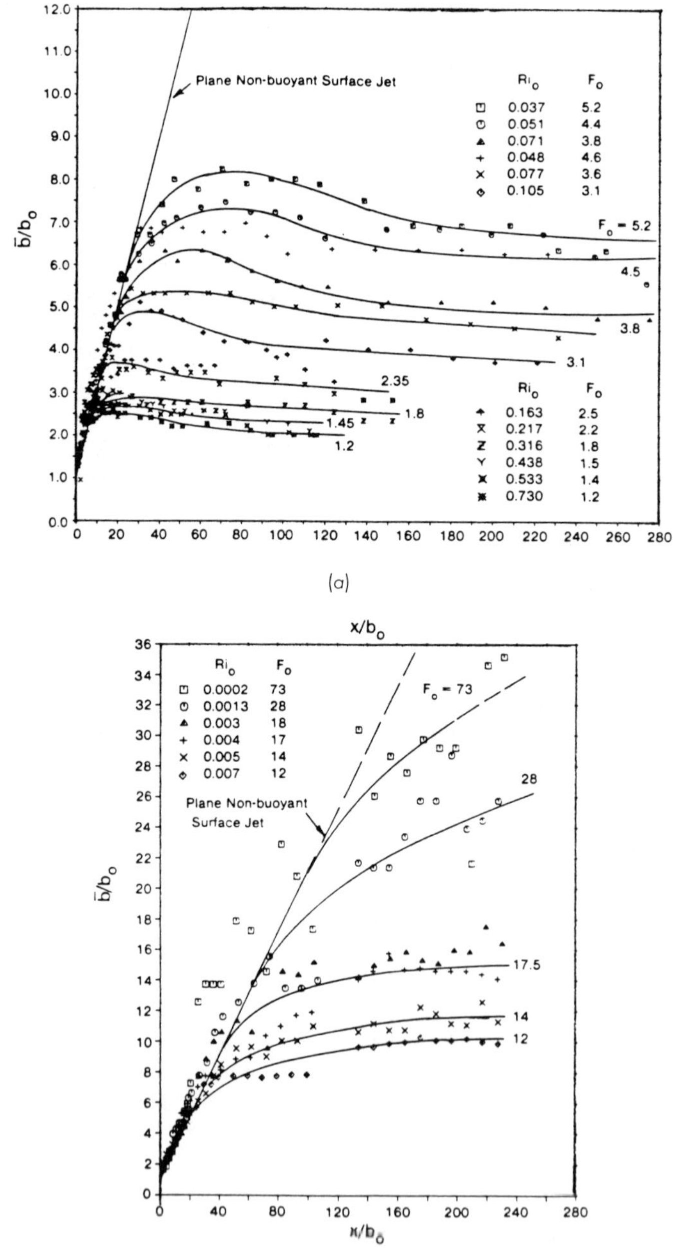

FIGURE 12. Growth of thickness of plane buoyant surface jets [33].

even though the profiles of $u(y)$ at different x stations appear to be similar, the similarity curve or function appears to vary noticeably depending upon the source Richardson number.

All these changing features make integral analysis [12] of buoyant surface jet very difficult. The same comments can also be directed towards the turbulence models like the $\kappa - \epsilon$ models [15]. We will present herein a more practical approach based mainly on some careful experimental results, dimensional arguments and intuitive extrapolations from somewhat related situations.

For the surface jet, we presented the experimental results of Rajaratnam and Subramanyan [33] in Figure 12 on the growth of the thickness of the surface jet $\bar{b}$ with the distance x from the outfall. In Figure 12, we see that for any Ri_0 (or F_0), the growth of $\bar{b}$ with x follows the growth line of the non-buoyant jet described by the equation

$$\frac{\bar{b}}{b_0} = 1.0 + 0.2\,(x/b_0) \tag{47}$$

up to some value of x/b_0 and then curves to the right indicating a continuous reduction in the rate of growth to eventually approach an approximately constant thickness $\bar{b}_\infty$. For a buoyant surface jet, the relation between $\bar{b}_\infty/b_0$ and Ri_0 is given by the equation

$$\frac{\bar{b}_\infty}{b_0} = 1.82\,Ri_0^{-0.36} \tag{48}$$

If x_∞ is distance where $\bar{b} \cong \overline{b_\infty}$, x_∞/b_0 is given by the equation

$$\frac{x_\infty}{b_0} = 80 - 60 \log Ri_0 \tag{49}$$

The u velocity profiles in the surface jet at different x stations were similar but while the similarity curve followed the exponential equation for very small values of Ri_0 like 0.0039, for values of $Ri_0 = 0.4$ and 0.63, the velocity profiles were considerably different as shown in Figure 13. Figure 14(a) shows the growth rate of the length scale b in terms of b_0 (where $b = y$ where $u = (1/2)u_m$) with x/b_0. Here again the experimental results for b follow the non-buoyant jet line up to some section and then depart away to the right.

The decay of the dimensionless maximum velocity u_m/U_0 with x/b_0 is shown in Figure 14(b) where for large values of x/b_0, u_m/U_0 tends to approach an approximately constant value of 0.45.

Considering the entrainment coefficient for the buoyant surface jet, we present the results of Chu and Vanvari [2] in Figure 15 where α_e decreases continuously as the local bulk Richardson number $\overline{Ri}$ increases and it appears to fall to zero for $\overline{Ri} \cong 0.2$.

Experimental investigations have shown that for a buoyant surface with a given Ri_0, if the depth of the stratified layer $\bar{b}_\infty$ is greater than the asymptotic thickness of the surface jet, as given by Equation (48), then a surface jump is formed at some distance x_e from the outfall. For a certain value of $\bar{b}_\infty$, a surface jump would occur right at the outfall. Further, if $\bar{b}_\infty$ is greater than this value for the jump to

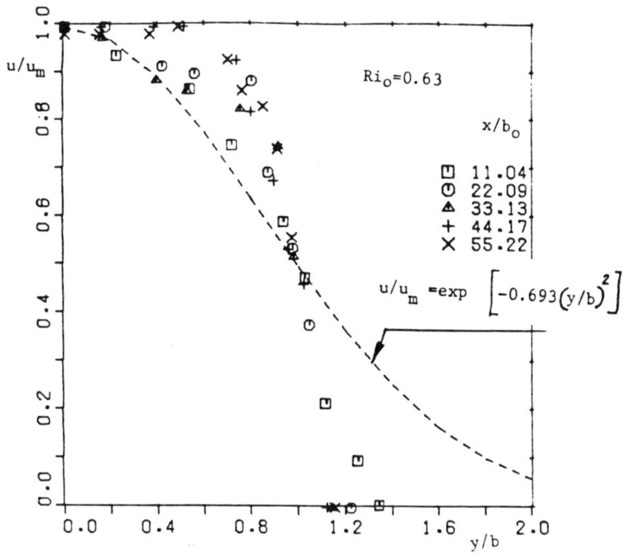

FIGURE 13. Similarity of velocity profiles for Ri_0 = 0.63 [33].

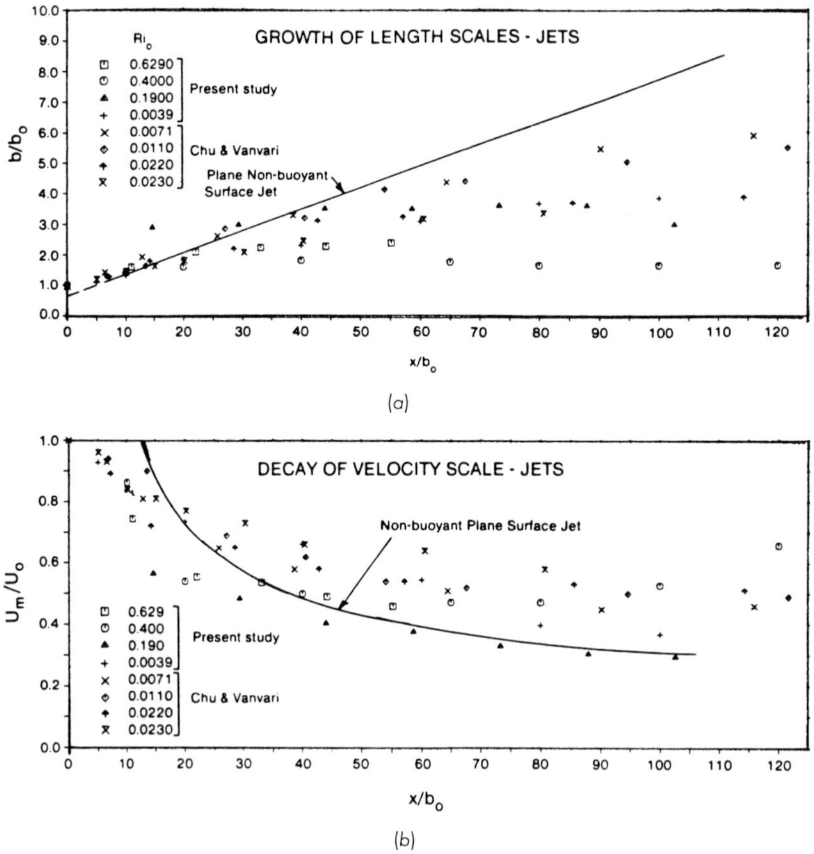

FIGURE 14. Length and velocity scales for plane buoyant surface jet [33].

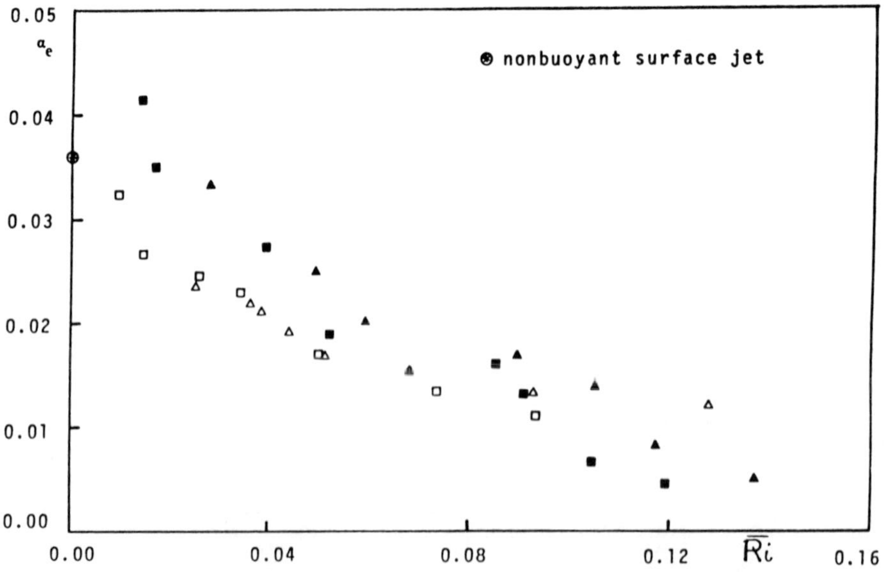

FIGURE 15. Reduction of the entrainment coefficient with Richardson Number [2].

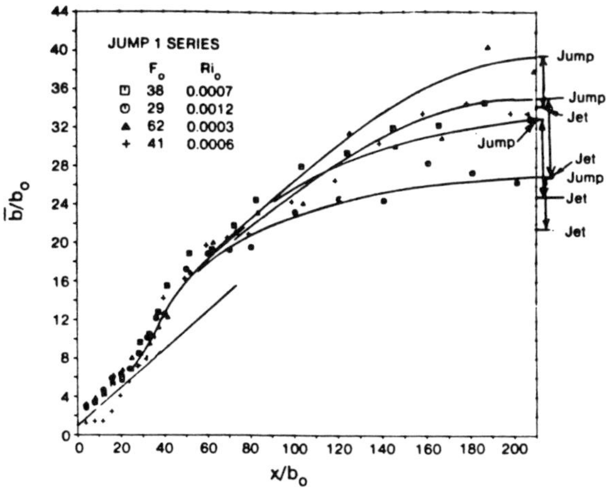

FIGURE 16. Growth of thickness for plane buoyant surface jumps [33].

occur right at the outfall, then the jump would be drowned. Surface jumps have been studied by Wilkinson and Wood [39] and Rajaratnam and Subramanyan [33]. Using dye injection techniques, Rajaratnam and Subramanyan found that there was hardly any entrainment in the jumps they studied (F_1, the densimetric Froude number at the start of jump was varied from about 3 to 9).

To bring out the rather subtle difference between buoyant surface jets and surface jumps, we have reproduced from Reference 33, Figure 16. This figure shows that firstly for a given value of b_0 and Ri_0, the thickness of the stratified flow downstream $\bar{b}_\infty$ for a jump is noticeably larger than the asymptotic thickness of the surface jet. Secondly, it can be noticed in Figure 16 that the results of the growth of the thickness of the jumps are located (characteristically) above the corresponding non-buoyant surface jet line.

If the thickness of the surface jet at the section where a surface jump starts is $\bar{b}_1$ and F_1 is the densimetric Froude number at that section, equal to $U_1/\sqrt{g\,\bar{b}_1\,(\overline{\Delta\varrho}/\varrho_a)}$ (wherein U_1 is the mean velocity, $\overline{\Delta\varrho}$ is the mean density difference at that section) and $\bar{b}_2$ is the thickness at the downstream end of the jump, essentially equal to $\bar{b}_\infty$, these two depths are related approximately by the Belanger-type Equation (24):

$$\frac{\bar{b}_2}{\bar{b}_1} = \frac{1}{2}\left[\sqrt{1 + 8F_1^2} - 1\right] \tag{50}$$

which was derived by Yih and Guha [42].

Rajaratnam and Subramanyan noticed that the profile of the surface jump was essentially the same as that of the corresponding open-channel hydraulic jump [24]. For surface jumps for two cases with $Ri_0 = .0016$ and 0.007, they found that the u velocity profiles at different sections were similar and followed closely the exponential profile in the forward flow region. They also found that the decay of the dimensionless velocity scale u_m/U_0 with x/b_0 was essentially the same as that of plane non-buoyant surface jets dropping to about 0.25 at $x/b \cong 100$. They observed that the presence of the low velocity recirculating roller on the underside of the

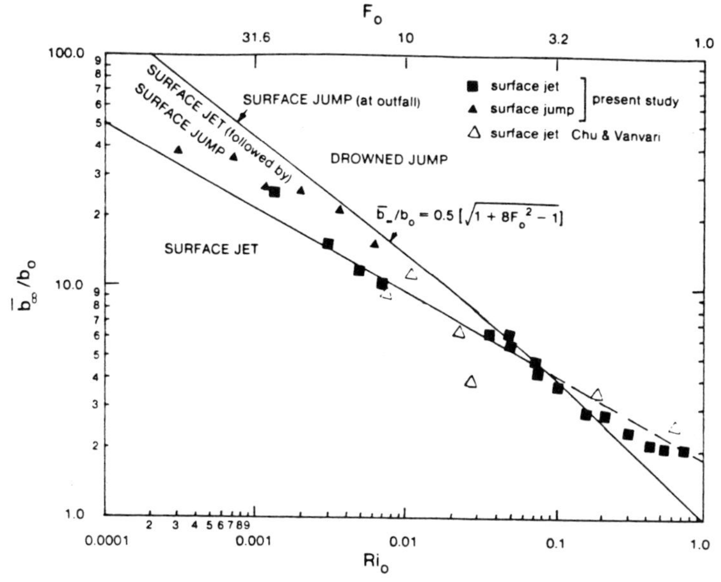

FIGURE 17. Characteristic diagram for plane buoyant surface jets and jumps [33].

surface jump was responsible for the inability of the jump to entrain the ambient fluid.

To present a comprehensive picture of the flow possibilities for the buoyant surface jet, we present in Figure 17 a rough attempt to predict the nature of the flow at an outfall for given values of b_0, Ri_0 (or F_0) and $\bar{b}_\infty$. In Figure 17, if for a given case, the point set of $\bar{b}_\infty/b_0$ (wherein $\bar{b}_\infty$ is the depth of the stratified surface layer at the end of the near-field) and Ri_0 is located in the surface jet area, we will have a surface jet. If it is located between the two lines, we will have a surface jet followed by a surface jump. If the point set is located above the surface jump (at outfall) line, we will have a drowned jump. In Figure 17, there is some uncertainty for Ri_0 in the range of 0.1 to 1.0 and based on some preliminary observations (by the author) it appears that the surface jet line will merge with the jump line.

BLUFF TURBULENT BUOYANT SURFACE JETS

Having discussed the characteristics of plane turbulent buoyant surface jets in the previous section, let us consider a circular surface jet of diameter of d with a velocity of U_0 and density of ϱ_0 enter a deep and wide stagnant fluid of density of ϱ_a as shown in Figure 18. Let us assume that the Reynolds number of the jet is large enough for the viscous effects to be negligible. Experimental observations [9,10,18,21,32,34,35,36,37,41] have shown that the behaviour of the jet is strongly affected by the Richardson number Ri_0 at the outfall defined as

$$Ri_0 = \frac{g\, d\, \dfrac{\Delta\varrho_0}{\varrho_a}}{U_0^2} \tag{51}$$

wherein $\Delta\varrho_0 = (\varrho_a - \varrho_0)$. For very small values of Ri_0, less than about 0.1, the surface jet mixes with the ambient fluid on the bottom as well as the sides; however, even in this narrow range for Ri_0 from 0 to 0.1, as the Ri_0 approaches 0.1, the rate of growth of its thickness reduces considerably. For Ri_0 in the large Richardson number class, which could be taken to extend upwards approximately from 0.1, there is hardly any mixing on the underside of the jet with all the mixing and entrainment of the ambient fluid taking place on the sides of the jet. Similar observations have been made on the behaviour of rectangular jets with length-to-thickness ratio of about three. Hence, it appears to be a good idea to treat the bluff buoyant surface jets in (at least) two classes namely the low Ri_0 class and large Ri_0 class [20].

Let us construct a theoretical framework for analysing bluff buoyant surface discharges. With reference to the coordinate system shown in Figure 18, for the time-averaged flow, the continuity equation is

$$\frac{\partial u}{\partial x} + \frac{\partial v}{\partial y} + \frac{\partial w}{\partial z} = 0 \tag{52}$$

For small Richardson number flows, using the slender-flow

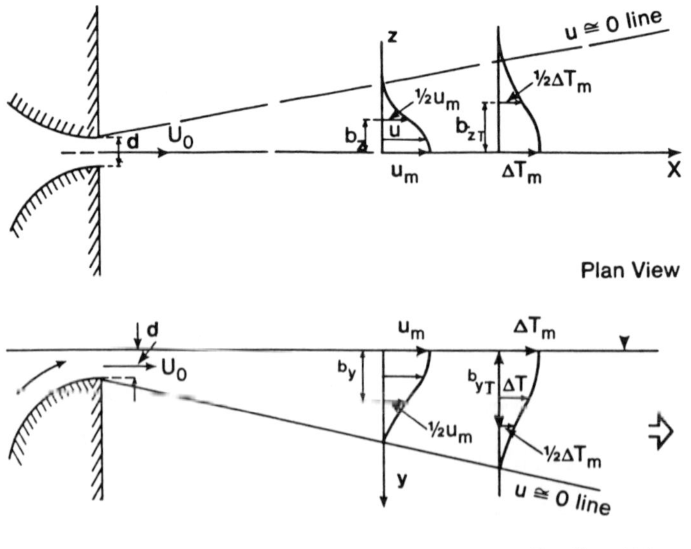

FIGURE 18. Definition sketch of buoyant circular surface jet in a stagnant ambient.

approximation and the assumption of hydrostatic pressure distribution in the vertical or y direction, the x direction Reynolds equation reduces to

$$u\frac{\partial u}{\partial x} + v\frac{\partial u}{\partial y} + w\frac{\partial u}{\partial z} = \frac{g}{\varrho}\frac{\partial}{\partial x}\int_0^y \Delta\varrho\, dy - \left\{\frac{\partial}{\partial y}\overline{u'v'} + \frac{\partial}{\partial z}\overline{u'w'}\right\} \tag{53}$$

In Equation (53), for a flow with a small source Richardson number, for a portion of the near-field, the buoyancy related term will be negligible whereas as the jet moves away from the outfall, the Richardson number increases and the buoyancy term will begin to become important and eventually dominate over the turbulence stress terms.

For bluff surface jets with large Ri_0, the x and z direction Reynolds equations can be reduced to

$$u\frac{\partial u}{\partial x} + v\frac{\partial u}{\partial y} + w\frac{\partial u}{\partial z} = \frac{g}{\varrho}\frac{\partial}{\partial x}\int_0^y \Delta\varrho\, dy - \left\{\frac{\partial}{\partial y}\overline{u'v'} + \frac{\partial}{\partial z}\overline{u'w'}\right\} \tag{54}$$

$$u\frac{\partial w}{\partial x} + v\frac{\partial w}{\partial y} + w\frac{\partial w}{\partial z} = \frac{g}{\varrho}\frac{\partial}{\partial z}\int_0^y \Delta\varrho\, dy - \left\{\frac{\partial}{\partial y}\overline{v'w'} + \frac{\partial}{\partial z}\overline{w'^2}\right\} \tag{55}$$

Pande and Rajaratnam [20], based on a similarity analysis of the simplified equations of motion including the integral energy equation predicted that for small Richardson number flows, the velocity scale u_m (in the centerplane—see Figure 18) will vary inversely with the longitudinal distance x; the length scales in the vertical direction b_y as well as in the transverse direction b_z will grow linearly with x. If ΔT_m is the scale for temperature excess in the centerplane, ΔT_m was predicted to vary inversely with x. For flows with large Richardson number, assuming that b_y is constant, it was predicted that $u_m \propto x^{-1/3}$, $b_z \propto x$, and $\Delta T_m \propto x^{-2/3}$.

A number of other analytical models have been developed including that of Engelund and Pedersen [5], the turbulence model of McGuirk and Rodi [15] and others [8,36]. In the following section, we present a practical treatment of bluff surface jets using selected experimental results, dimensional arguments and some predictions from similarity and integral analyses.

Considering bluff surface discharges with small Richardson number, Rajaratnam [32] has performed experiments on rectangular outfalls with length-to-thickness ratio varying from 1.06 to 1.55 and Ri_0 varying from 0.002 to 0.091. In this study, the time averaged velocity field was measured with the hydrogen bubble method along with some measurements of the excess temperature field. Other investigations are those of Jen et al. [10], Stefan [34], Stolzenbach and Harleman [36], Wiuff [41], Notsopoulos et al. [18], and others.

Based on these studies, we can say that the velocity profiles $u(y)$ in the centerplane are similar and are well described by the exponential expression with the maximum velocity u_m (generally) occurring at the water surface. If b_y is the length scale defined as the value of y where $u = u_m/2$, b_y grows linearly with x, the longitudinal distance from the outfall. But the rate of increase of b_y with x, i.e., db_y/dx decreases with Ri_0. For $Ri_0 = 0$, we know, from the section on non-buoyant bluff jets, that this rate is 0.044. For $Ri_0 = 0.038$ it is about 0.02 and falls off to zero at $Ri_0 \cong 0.09$. Hence, it appears that when $Ri_0 \cong 0.09$ (or 0.1 for convenience), the vertical growth stops thereby implying the absence of vertical mixing and vertical entrainment of the ambient fluid.

The velocity profiles $u(z)$ just below the water surface were also found to be similar and well described by the exponential expression. If b_z is the corresponding length scale where u goes to half the maximum value occurring on the centerline, it has been found that b_z grows linearly with x. However, as shown in Figure 19, db_z/dx increases continuously and almost linearly with Ri_0 for Ri_0 up to the end of the small Ri_0 class.

Considering the velocity scale u_m, the variation of u_m/U_0 with $x/\sqrt{A_0}$ is given by the expression

$$\frac{u_m}{U_0} = \frac{15.3\, Ri_0^{1/10}}{\sqrt{x/A_0}} \tag{56}$$

where A_0 is the area of cross section at the outfall. The excess temperature (ΔT) profiles in the central vertical plane as well as in the transverse (z) direction near the surface have been found to be similar and well described by the exponential expression. If b_{yT} is the temperature length scale where $\Delta T = (1/2)\Delta T_m$, ΔT_m being the maximum value of ΔT, $b_{yT} \cong b_y$ whereas for the transverse profiles near the surface $b_{zT} \cong 1.15\, b_z$.

For practical purposes, it appears that we can call flows with Ri_0 greater than 0.1 as the large Richardson number flows. In this large Richardson number class, the main point to note is that there is hardly any vertical mixing and hence no vertical entrainment of the ambient fluid by the surface

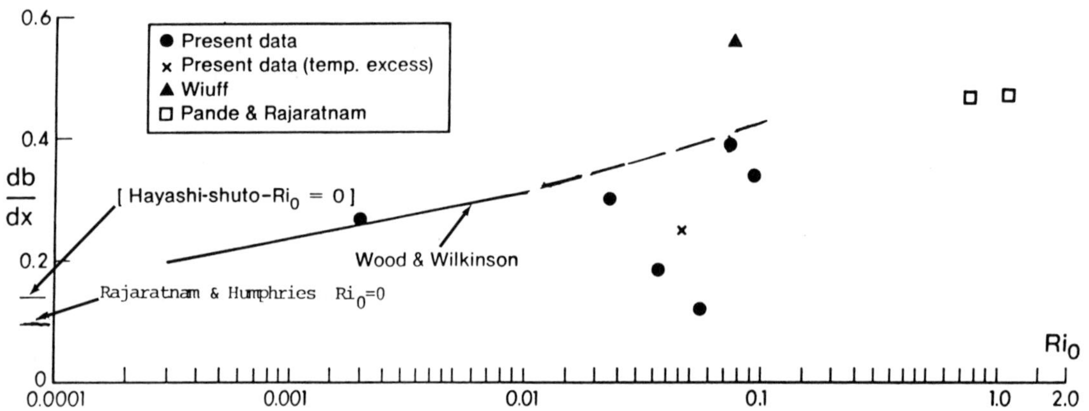

FIGURE 19. Growth of the transverse length scale for buoyant circular surface jet—small Ri_0 class [32].

jet. The mixing and entrainment occur on the sides of the jet.

A number of experimental investigations have been made on the behaviour of large Ri_0 bluff surface jets. These include the work of Tamai, Wiegel and Tornberg [37] (circular outfalls flowing full and partly full with Ri_0 varying from 0.0078 to 0.174; Hayashi and Shuto [9] (square outfall with Ri_0 varying from 0 to 0.54); Stefan and Schiebe [35] (rectangular outfall with Ri_0 varying from 0.019 to 2.60); Stefan [34] (rectangular outfall with an aspect ratio of 3.03 and Ri_0 = 0.072, 0.53 and 2.60); Stolzenbach and Harleman [36] (rectangular outlet with aspect ratio varying from 0.54 to 5.8 and Ri_0 varying from 0.007 to 1.0) and Pande and Rajaratnam [21] (rectangular outfall, aspect ratio varying from 1.06 to 3.22 and Ri_0 = 0.15, 0.35, 0.56, 0.79 and 1.14). Amongst these, the work of Pande and Rajaratnam [21] is generally regarded as definitive because of the approximate steady state maintained by proper cooling water circulation and because of the extensive measurements of the velocity and temperature fields.

Based on the work of Pande and Rajaratnam [21] we can say that the u velocity profiles in all vertical lines in the surface jet are similar and well described by the exponential expression. Further, the u velocity profiles across the jet, near the water surface are also similar and well described by the exponential expression. To express these statements analytically, if u_s is the value of u (at or) near the surface,

$$\frac{u}{u_s} = f_1\left(\frac{y}{b_y}\right) \tag{57}$$

and

$$\frac{u_s}{u_m} = f_2\left(\frac{z}{b_z}\right) \tag{58}$$

Combining these two expressions, in any transverse plane at any x station, the velocity u at any point is expressed by the equation

$$\frac{u}{u_m} = f_1\left(\frac{y}{b_y}\right) f_2\left(\frac{z}{b_z}\right) \tag{59}$$

In these expressions, u_s is the value of u near the surface on any vertical line.

At any x station, the variation of b_y with z was minimal and hence we can take an average value and denote it as b_y*. If $\bar{b}_y$ is the vertical thickness of the surface jet with the bottom boundary defined as the location where $u \cong 0$, it appears that $\bar{b}_y \cong 2.2\, b_y$. In the large Richardson number class, the length scale b_y* or the thickness of the jet remained essentially constant over most of the near-field and b_y* is given by the equation

$$\frac{b_y*}{b_0} = \frac{0.26}{Ri_0^{1/8}} \tag{60}$$

The velocity length scale b_z at the surface also grows (approximately) linearly with x and can be described by the equation

$$\frac{b_z}{\sqrt{A_o}} = 0.54\left(\frac{x}{\sqrt{A_o}} + 2.0\right) \tag{61}$$

If u_m is the maximum velocity along the centerplane of the plume at any distance, for the large Richardson number case, u_m has been predicted to vary as $1/x^{1/3}$. Even though rather detailed equations have been recommended in Reference 21 for different sub-ranges of Ri_0, for practical pur-

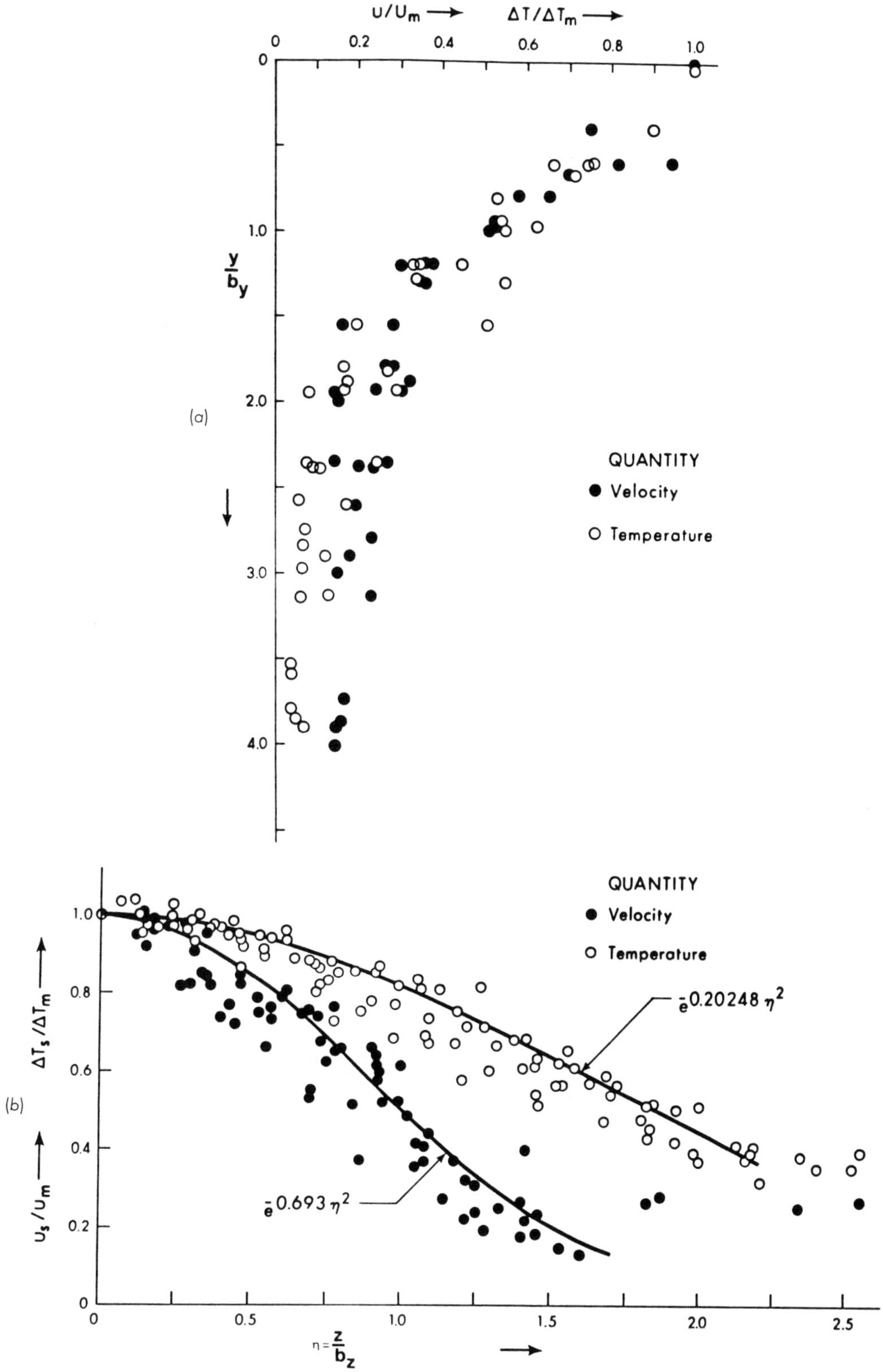

FIGURE 20. Difference between velocity and excess temperature profiles in the vertical and transverse directions for bluff buoyant surface jet for Ri_0 = 0.79 [21].

poses, we suggest the following approximate equation

$$\frac{u_m}{U_0} = \frac{1.25}{\left(\frac{x}{\sqrt{A_o}} - 5.0\right)^{1/3}} \tag{62}$$

Concerning the scales for the excess temperature field in the warm surface jet, if ΔT_m is the maximum temperature excess for any x, occurring at or near the surface, a simple relation for practical purposes is

$$\frac{\Delta T_m}{\Delta T_0} = \frac{2.83}{[x/\sqrt{A_0}]^{2/3}} \tag{63}$$

The length scale for temperature excess in the vertical direction, averaged across the width of the jet is given by the relation

$$\frac{b_{yT}*}{b_0} = \frac{0.29}{Ri_0^{1/8}} \tag{64}$$

This means, $b_{yT}* \cong 1.12\ b_y*$. If b_{zT} is the length scale for the temperature excess profiles in the z direction at the water surface,

$$\frac{b_{zT}}{\sqrt{A_0}} = 0.87\left(\frac{x}{\sqrt{A_0}} + 1.3\right) \tag{65}$$

From Equations (61) and (65), it appears that $b_{zT} \cong 1.6\ b_z$ and this indicates that the temperature field spreads much more rapidly than the velocity field in the horizontal direction. Figure 20(a) and (b) illustrate this difference in the spreading rates between the velocity and temperature fields in the vertical and horizontal directions in a very direct manner.

It appears, in the light of our earlier more detailed presentation of plane buoyant surface jet that our treatment of bluff buoyant surface jet is limited to only the case when the depth of stratified surface layer maintained by downstream control and/or evaporation is such that the surface discharge behaves like a surface jet. If the downstream stratified surface is thicker than a certain value for given thickness b_0 and Ri_0, it is possible to have a jump (similar to the corresponding open-channel case [24]) or even a submerged jump. At the present time, our understanding and knowledge of these cases is not detailed enough to attempt a presentation herein.

Another aspect that should be pointed out here is that we have seen how to analyse a plane surface jet or a bluff surface jet. But if one has a surface jet with a length to thickness of say 10 or 20, then for a certain length after the end of the potential core in the near-field the maximum velocity in the centerplane will decay as in a plane surface jet and after a certain characteristic length, it will decay as in a circular (or bluff) surface jet [26]. Such a surface jet can for convenience be termed the slender surface jet. A preliminary analysis for such a slender surface jet can be made following the analysis of the non-buoyant slender jet [26]. But the analysis will have to be done in two parts, one for small Ri_0 and the other for the large Ri_0. The effect of the downstream stratified surface layer, if it is larger than a certain value, will also have to be considered.

BLUFF BUOYANT SURFACE JETS IN COFLOWING STREAMS

Consider a circular buoyant surface jet of diameter d (or a bluff jet of area of A_0) entering a parallel stream of large depth and width with a velocity of U_0, and let the stream velocity be U_1 with ϱ_0 and ϱ_a being the densities of the jet fluid and surrounding fluid, respectively. Rajaratnam [31] performed an experimental study of such a jet system with α the ratio of the jet to stream velocity ratio varying from about 1.4 to about 20 and the source Richardson number Ri_0 of the surface jet varying from about 0.00015 to 0.03. As will be seen later, even with such small Richardson numbers, vertical mixing becomes negligible even at α being as high as 5.

Let us consider the centerplane of the growing surface jet. Based on the results of Reference 31, it appears that u velocity profiles, at different distances from the nozzle, with respect to the freestream velocity U_1 are similar. That is,

$$\frac{u - U_1}{u_m - U_1} = f\left(\frac{y}{b_y}\right) \tag{66}$$

wherein u_m is the maximum value of u occurring either at and a little below the surface and b_y is the length scale equal to y where $u - U_1 = (u_m - U_1)/2$. It was also found that the function f is well described by the exponential equation. The u velocity profiles in the transverse (z) direction in the (horizontal) plane of u_m were also found to be similar if expressed in the form of $(u - U_1)/(u_m - U_1)$ against z/b_z wherein b_z is the value of z where $(u - U_1) = 1/2\ (u_m - U_1)$. Again the exponential equation was found to describe this similarity profile satisfactorily.

To analyse the variation of the velocity and length scales, following the procedure used earlier in the section on non-buoyant jets in coflowing streams and buoyant jets flowing into stagnant ambients, $(u_m - U_1)/U_1$ was plotted against x/θ in Figure 21(a) wherein θ is the excess momentum thickness (defined earlier in the section about non-buoyant surface jets in coflowing streams). As can be seen in Figure 21a the results are well correlated by the corresponding curve of circular (submerged) jets in coflowing streams [22] and that the effects of buoyancy are negligible.

Figure 21(b) shows the growth of the dimensionless transverse length scale b_z/θ with x/θ along with the curve for

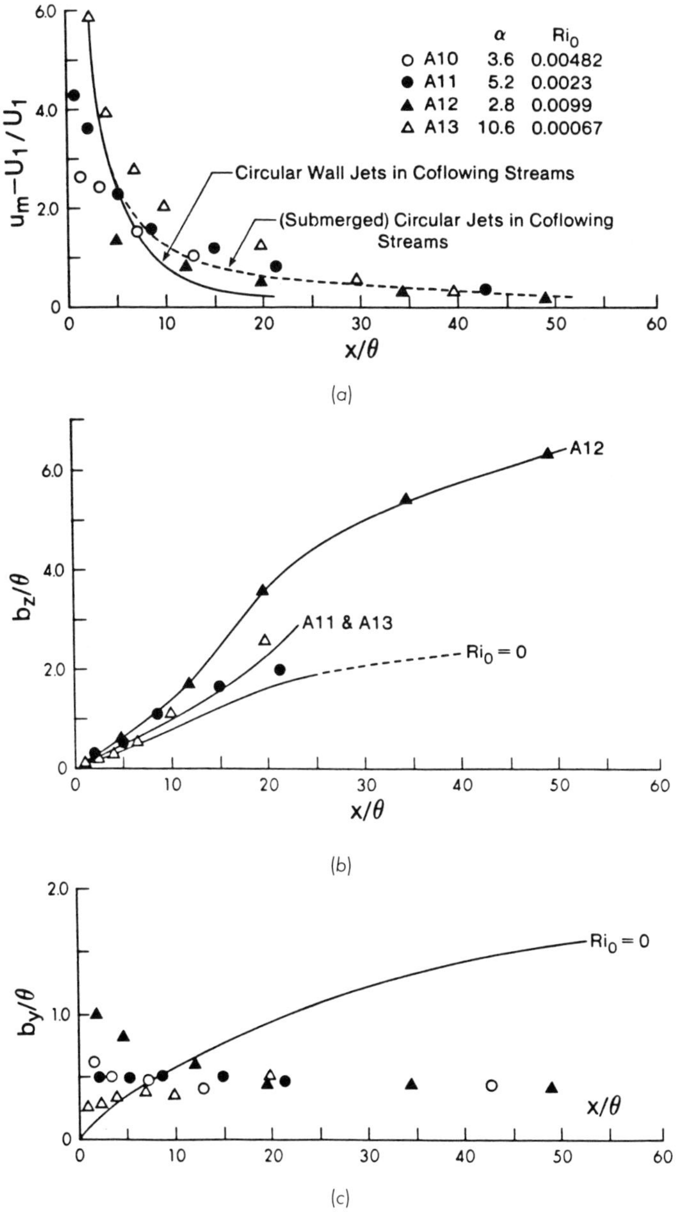

FIGURE 21. Correlation for velocity and length scales for circular buoyant surface jets in coflowing stream [31].

$Ri_0 = 0$ from the above-mentioned section. It appears possible to draw separate curves for different values of Ri_0. It is important to note the significant increase in transverse spreading even for $Ri_0 \cong 0.01$. Figure 21(c) shows the growth of b_y/θ with x/θ along with the curve for $Ri_0 = 0$. It is significant to note that even for the small values of Ri_0 studied, there is hardly any growth in the vertical length scale for x/θ greater than about 10.

Rajaratnam [31] increased the range of his study on buoyant circular surface jets in coflowing streams by presenting the results of a photographic study wherein α was varied from 1.4 to 20 and Ri_0 was increased from 0.00015 to 0.033. The results for the transverse half-width at the surface $\bar{b}_z$ and vertical thickness of the jet $\bar{b}_y$ in terms of θ are shown in Figure 22(a) and (b). In Figure 22(a) and (b) are also shown the corresponding curves for $Ri_0 = 0$. In Figure 22(a) we see three mean curves: one for $Ri_0 = 0$, another for $Ri_0 = 0.0018$ to 0.013 with an intermediate one. In Figure 22(a) again one notices the significant increase in transverse spreading even for the small values of Ri_0 studied. Figure 22(b) shows that $\bar{b}_y/\theta$ grows up to a certain value and then remains constant. (For the experiment with $\alpha = 2.1$, the momentum thickness does not appear to be able to correlate the results even for non-buoyant jets and hence these points are not considered.)

Another interesting observation from Reference 31 is that

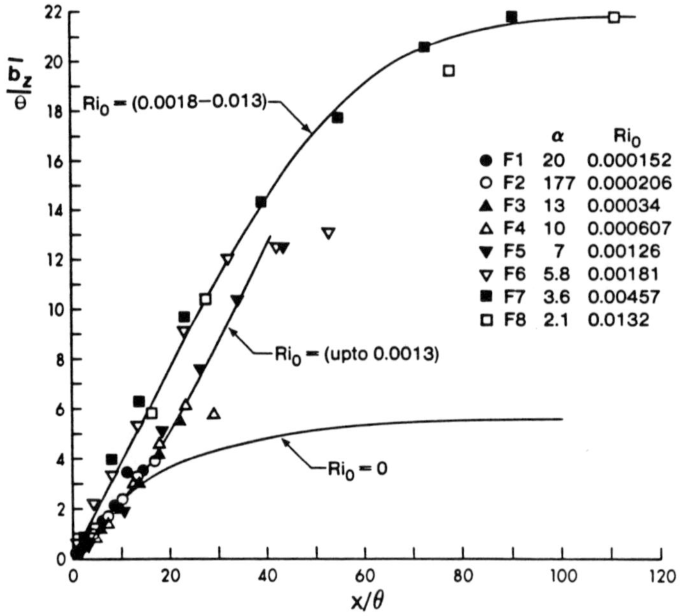

(a)

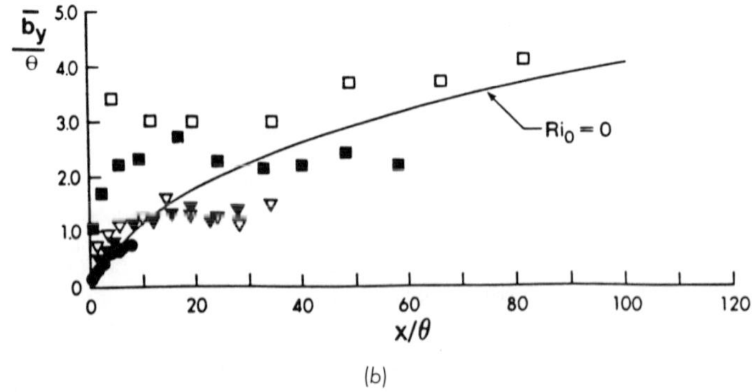

(b)

FIGURE 22. Extended correlation for length scales of circular buoyant surface jets in coflowing streams [31].

even when the depth of the coflowing stream was only 2.8 times the jet diameter, the transverse spreading was not affected by the shallowness of the stream and even in such a shallow stream the jet did not grow thick enough to occupy the full depth. Before closing this section, it should be pointed out that for practical problems, the results of circular surface jets can be extended to bluff surface jets (of area A_0 at the outfall) by defining an equivalent diameter d equal to $\sqrt{4A_0/\pi}$. It also appears that very little has been published on the behaviour of plane turbulent buoyant surface jets in coflowing streams.

BLUFF BUOYANT CIRCULAR SURFACE JET IN A CROSSFLOW

Let us consider a circular buoyant surface jet of diameter d issuing with a velocity of U_0 into a deep crossflow with a velocity of U_1 as shown in Figure 7. Let ϱ_0 be the density of the surface jet with ϱ_a being the density of the crossflow. We would assume that the motion is turbulent. The behaviour of the buoyant surface jet will be somewhat similar to that of the non-buoyant surface jet discussed in the section about non-buoyant bluff surface jets in crossflows with the source Richardson number $Ri_0 = (gd\ \Delta\varrho_0/\varrho_a)U_0^2$ (where $\Delta\varrho_0 = \varrho_a - \varrho_0$) becoming another important parameter. Based on our discussions on the effect of motion of the ambient and buoyancy effects, the growth of the half-width will be enhanced and the growth of the jet thickness will be reduced in comparison to the corresponding non-buoyant surface jet in a crossflow.

A number of predictive models of the integral type have been proposed for predicting the behaviour of warm water discharges into lakes and rivers and a critical assessment of these various methods has been made by Policastro and Tokar [19,23]. We will present herein the outline of the method of Motz and Benedict [17] (with some minor modifications) to give an idea of the structure of these prediction methods and point out their inadequacies.

Motz and Benedict [17] neglect mixing in the vertical direction and treat the flow as a two-dimensional problem; this means it would be applicable only above a certain value of the parameter set of Ri_0 and α where α is the ratio of the jet velocity U_0 to the crossflow velocity U_1. (More will be said about this later.) They assume a constant value for the entrainment coefficient α_e which, from the behaviour of circular (submerged) jets in crossflow, would be expected to vary along the jet length with ξ. The velocity distribution across the jet at any ξ is assumed to be symmetrical and here again, one would notice disparities with experimental observations.

In the Motz-Benedict method, we write the integral continuity equation as

$$\frac{d}{d\xi}\int_A u\,dA = Pv_e \tag{33}$$

where P is the entraining perimeter and

$$v_e = -\alpha_e\,(u_m - U_1\cos\beta) \tag{34}$$

where u_m is the maximum velocity on the ξ axis and β is the angle of the deflected surface jet with the crossflow. Considering the momentum flux of the deflected jet in the x direction,

$$\frac{d}{d\xi}\int_A \varrho u^2 dA\cos\beta = (Pv_e\varrho U_1) + \frac{C_D\varrho U_1^2}{2}\sin^3\beta\bar{b}_y \tag{35}$$

where C_D is the (conventional) drag coefficient and $\bar{b}_y$ is the average thickness of the surface jet.

For the momentum flux in the perpendicular z direction, we write

$$\frac{d}{d\xi}\int_A \varrho u^2 dA\sin\beta = -\frac{C_D\varrho U_1^2\sin^2\beta}{2}\bar{b}_y\cos\beta \tag{36}$$

If ΔT is the (time-averaged) temperature excess at any point in the surface jet with respect to the ambient crossflow, neglecting surface heat loss,

$$\frac{d}{d\xi}\int_A u\,dA\,\Delta T = 0 \tag{67}$$

From geometrical considerations,

$$\frac{dx}{d\xi} = \cos\beta \tag{37}$$

$$\frac{dz}{d\xi} = \sin\beta \tag{38}$$

Motz and Benedict non-dimensionalized the above equations and then empirically allowed for the flow-development region. They picked the values of α_e and C_D by matching the jet trajectory and excess temperature decay along the jet centerline.

Policastro et al. [19,23] found that most of the computational methods are still very unsatisfactory. Further the crossflow would be very different from a channel flow if it is a wind-generated current in a deep lake. It appears that a

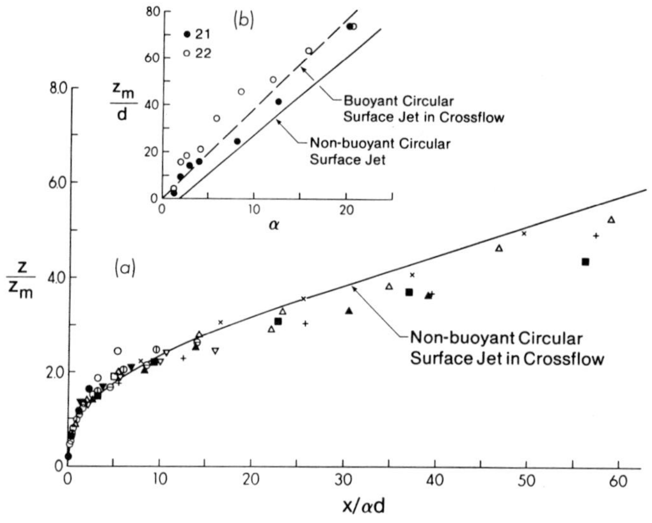

FIGURE 23. Trajectory of circular buoyant surface jet in a crossflow.

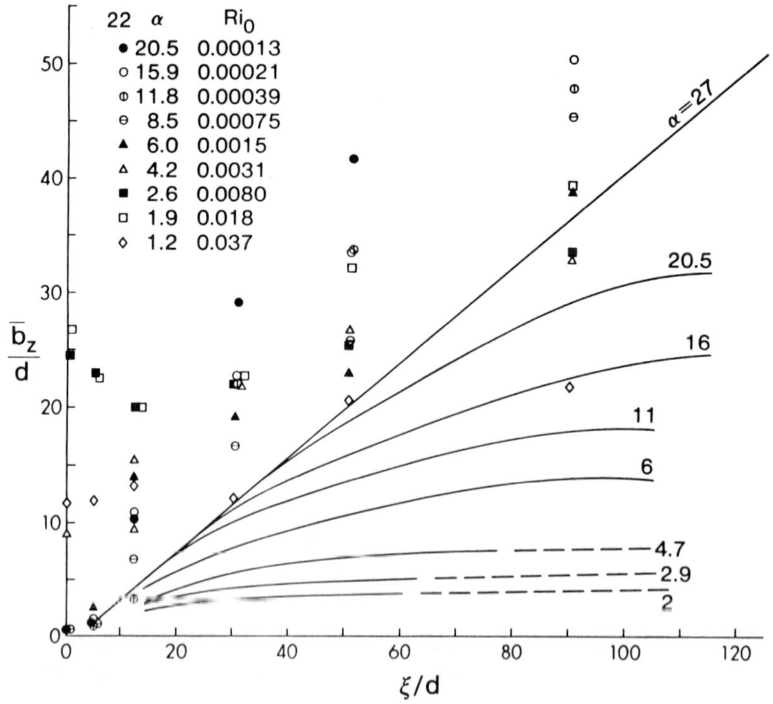

FIGURE 24. Growth of transverse length scale for circular buoyant surface jets in crossflows.

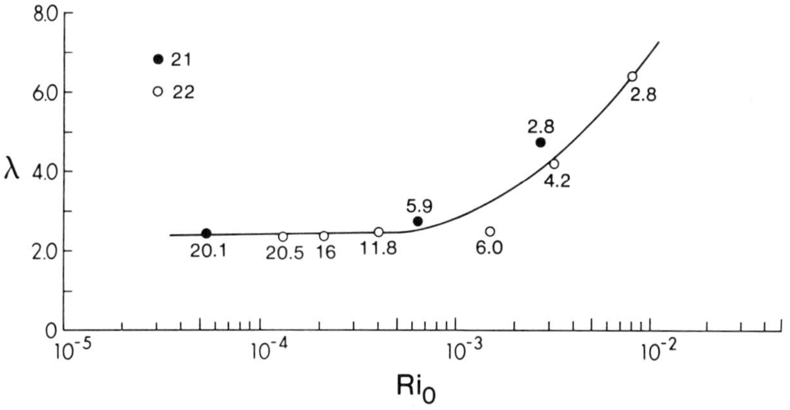

FIGURE 25. Variation of λ with Ri_0 and α for buoyant circular surface jets in crossflows.

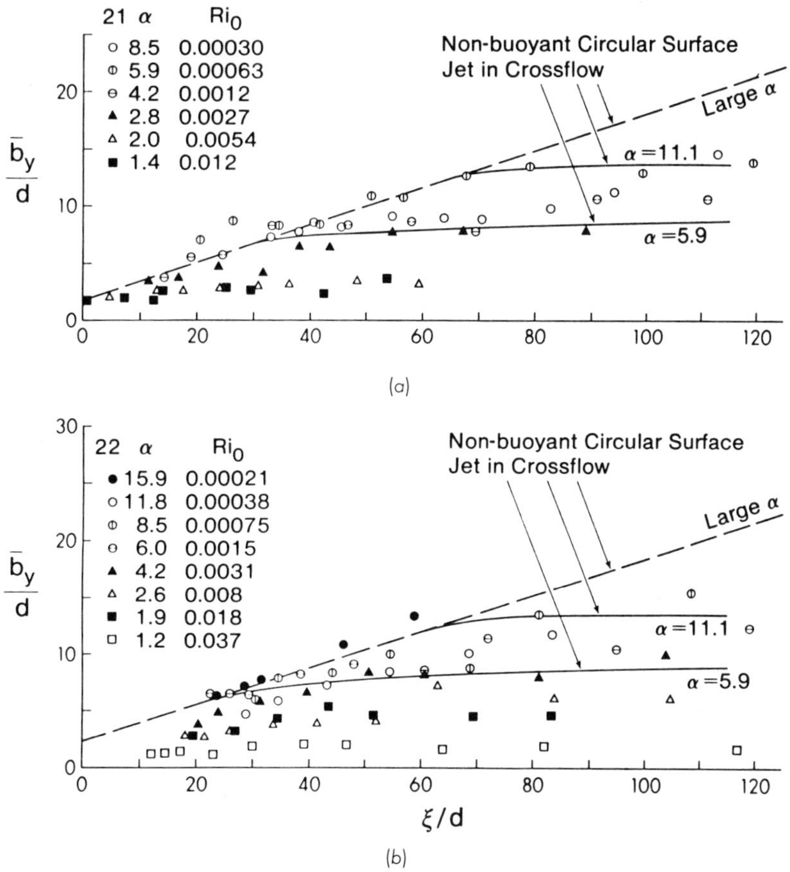

FIGURE 26. Growth of thickness of circular buoyant surface jets in a crossflow.

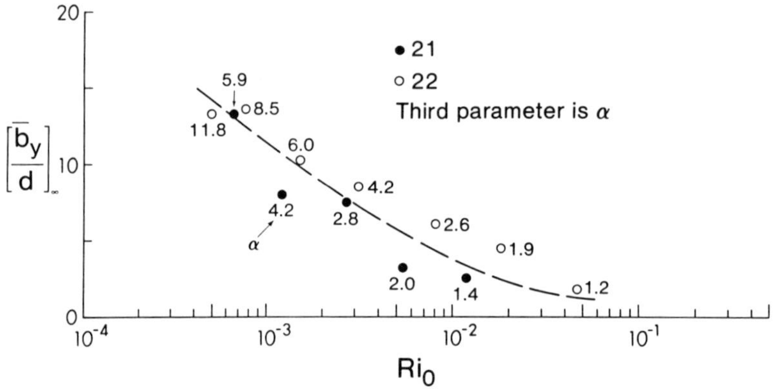

FIGURE 27. Variation of the asymptotic thickness with Ri_0 for buoyant circular surface jets in crossflows.

method like that of Motz and Benedict could be improved by improving upon the assumptions made based on results of suitably designed experiments in the laboratory and field observations. We will present now a practical method for handling buoyant bluff discharges into a crossflow, which is simulated by a channel flow, based on some (unpublished) experimental results of the author.

These experimental results were obtained by discharging circular surface jets from a nozzle of diameter d equal to 0.63 cm into a channel with a width of 1.22 m and flow depth of 24.4 cm with $\alpha = U_0/U_1$ varying from 1.4 to 20 and the source Richardson number Ri_0 for the surface jet varying from 0.00015 to 0.033.

We present first the results for trajectory of the centerline of the deflected jet in Figure 23a. We find that the corresponding curve for the non-buoyant surface jet in crossflow describes the present results well for $x/\alpha d$ up to about 15, beyond which there is some disparity. Figure 23b shows the growth of the length scale z_m/d (defined as before in the section about non-buoyant bluff surface jets in crossflows) which is described by the equation

$$\frac{z_m}{d} = 3.8\,\alpha \tag{68}$$

which is also somewhat different from the corresponding equation for non-buoyant jets.

As shown (partially) in Figure 24, the growth of the half-width of the buoyant jets is considerably more than that of the corresponding non-buoyant jets. If λ is the ratio between the $(\bar{b}_z/d)$ of the buoyant jet and that of corresponding non-buoyant jet (that is, same α), we see the variation of λ with the two parameters α (written beside the symbol) and Ri_0 in Figure 25. It appears possible, at least for practical purposes, to draw one curve in Figure 25 (covering the experimental range of α) which shows that $\lambda \cong 2.4$ for Ri_0 from 0.00005 to 0.0005 and then increases with Ri_0 to reach a value of about 7 for $Ri_0 = 0.01$. This plot shows the dominant effect of Ri_0 on the behaviour of the circular buoyant jet in a crossflow.

Similar results on the growth of the jet thickness along its centerline are shown in Figure 26. In Figure 26, we find that in addition to α, Ri_0 also slows down the growth of the jet thickness. If $(\bar{b}_y/d)_\infty$ is the constant value of the surface jet, reached after some distance from the source, the variation of $(\bar{b}_y/d)_\infty$ with α and Ri_0 is shown in Figure 27. In Figure 27, it appears possible to draw one curve, to cover the experimental range of α, which indicates that $(\bar{b}_y/d)_\infty$ decreases continuously with Ri_0, falling from about 14 for $Ri_0 = 0.005$ to about 1.5 for $Ri_0 = 0.02$. The results presented herein can be used to predict, at least as a first approximation, the behaviour of circular or bluff buoyant surface jets in crossflows which are similar to channel flows.

If we have a rectangular surface jet with a reasonably large aspect ratio of say 20, if Ri_0 is such that some turbulent mixing occurs on the underside of the jet to destroy the potential core, after the end of the potential core, we would have a so-called characteristic decay region similar to that in a plane surface jet modified by the crossflow if the vertical mixing is negligible. In this region, the method of attack outlined earlier in this section can be used so long as proper functions are used for describing velocity field, and it is also important to note that the mixing and entrainment will occur only on the (essentially) vertical sides of the deflected jet. After the mixing from the sides has penetrated to the centerline of the jet, the velocity profile functions will again have to be changed to recognize this penetration. If mixing on the underside exists, then this aspect will have to be considered using perhaps the kind of velocity field formulation used by Pande and Rajaratnam for buoyant surface jets [21].

CONCLUDING REMARKS

In this review, we have presented a reasonably coherent treatment on the near-field behaviour of turbulent surface jets discharged into deep and large water bodies. We have considered plane surface jets as well as bluff surface jets. We have included the effects of currents in the receiving water body. For buoyant surface discharges, we have developed a feel for the size of the source Richardson number beyond which vertical mixing across the underside of the surface discharge ceases. We have relied mainly on an integral treatment combined with dimensional arguments and carefully selected experimental results. Until we obtain enough field measurements to check the validity of the results presented in this review, it is suggested that the methods and results presented in this review be used with some judgement and caution. This review also indicates the areas where further research work is needed.

NOTATION

A = area of cross section
A_0 = area of cross section of jet at outfall
B = width of channel
b = length scale (with different suffixes)
b_0 = thickness of surface jet at outfall
$\bar{b}$ = half-width, or thickness of jet (with suffixes)
d = diameter of circular jet
F = Froude number
f = function
g = acceleration due to gravity
h = depth of flow
$\varkappa$ = coefficient (with suffixes)
M_0 = momentum (or excess momentum) flux
m = suffix used to denote maximum value
P = perimeter
p = exponent
q = exponent
R = Reynolds number
Ri = Richardson number (Bulk)
Ri_0 = Ri at source
T = temperature
ΔT = excess temperature
U_0 = jet velocity at outfall
U_1 = freestream velocity or velocity of crossflow
u = time-averaged velocity at a point in x or ξ direction
u_m = maximum value of u
v = time averaged point velocity in y direction
v_e, w_e = entrainment velocities
x = longitudinal distance
y = co-ordinate distance perpendicular to x
z = transverse coordinate distance
α = ratio of jet to free-stream velocity
α_e = entrainment coefficient
β = angle of deflected jet
λ = ratio of buoyant to non-buoyant jet width
ϱ = mass density of fiuid
ϱ_0 = ϱ at outfall
ϱ_a = ϱ of ambient fluid
$\Delta\varrho_0$ = $(\varrho_a - \varrho_0)$
θ = momentum thickness

REFERENCES

1. Anwar, H. O., "Behaviour of Buoyant Jet in Calm Fluid," *Proc. ASCE., J. Hyd. Div.*, Vol. 95, pp. 1289–1303 (1969).
2. Chu, V. H. and M. R. Vanvari, "Experimental Study of Turbulent Stratified Shearing Flow," *Proc. ASCE, J. Hyd. Div.*, Vol. 102, HY6, pp. 691–706 (1976); (also see their report, McGill Univ., Dept. of Civil Engrg., 74-2, 1974).
3. Edinger, J. K., D. K. Brady, and J. C. Geyer, Heat Exchange and Transport in the Environment, Rept. 14, Dept. of Geography and Env. Engrg., Johns Hopkins Univ., Baltimore, Maryland (1974).
4. Ellison, T. H. and J. S. Turner, "Turbulent Entrainment in Stratified Flow," *Journal of Fluid Mechanics*, Vol. 6, Part 3, pp. 423–448 (Oct. 1959).
5. Engelund, F. and F. B. Pedersen, "Surface Jet at Small Richardson Number," *Proc. ASCE, J. of Hyd. Div.* (March 1973).
6. Fan, L. N., Turbulent Buoyant Jets into Stratified or Flowing Ambients, W. M. Kech Lab. Rept., Caltech, Pasadena, California, 196 p. (June 1967).
7. Harleman, D. R. F., "Stratified Flow," in *Handbook of Fluid Dynamics*, Ed. by V. L. Streeter, McGraw-Hill, N.Y. (1961).
8. Harleman, D. R. F. and K. D. Stolzenbach, "Fluid Mechanics of Heat Disposal from Power Generation," *Ann. Rev. of Fluid Mech.*, Vol. 4, pp. 7–32 (1972).
9. Hayashi, T. and N. Shuto, "Diffusion of Warm Water Jets Discharged Horizontally at the Water Surface," *International Assoc. for Hydraulic Research, 12th Congress Proc.*, Vol. 4, pp. 47–59, Fort Collins, Colo. (1967).
10. Jen, Y., R. L. Wiegel, and I. Mobarek, "Surface Discharge of Horizontal Warm-water Jet," *ASCE, Journal of the Power Div.*, Vol. 92, pp. 1–30 (Apr. 1966).
11. Jirka, G. H., "Turbulent Buoyant Jets in Shallow Fluid Layers," in *Turbulent Buoyant Jets and Plumes*, Ed. by W. Rodi, Pergamon Press, pp. 69–119 (1982).
12. Koh, R. C. Y., "Two Dimensional Surface Warm Jet," *Proc. ASCE, J. of Hyd. Div.*, Vol. 97, HY6, pp. 819–836 (June 1971).
13. Koh, R. C. Y. and N. H. Brooks, "Fluid Mechanics of Waste Disposal in the Ocean," *Ann. Rev. of Fluid Mech.*, Vol. 17, pp. 187–211 (1975).
14. Launder, B. E. and W. Rodi, "The Turbulent Wall Jet—Measurement and Modelling," *Ann. Rev. of Fluid Mech.*, 13, pp. 429–459 (1983).
15. McGuirk, J. J. and W. Rodi, "Mathematical Modelling of

Three Dimensional Heated Surface Jets," *J. of Fluid Mech.*, Vol. 95, pp. 609–633 (1979).
16. Miller, D. S. and B. A. Brighouse, Thermal Discharges, British Hydromech. Res. Assoc., 221 p. (1984).
17. Motz, L. H. and B. A. Benedict, "Surface Jet Model for Heated Discharges," *Proc. ASCE, J. of Hyd. Div.*, pp. 181–199 (Jan. 1972).
18. Noutsopoulos, G., J. Demeteriou, D. Adractas and G. Sakellarkis, "Experiments on Three Dimensional Surface Buoyant Jets," *Euromech 130* Colloquim, Belgrade, Yugoslavia (June 30–July 3, 1980).
19. Paddock, R. A., A. J. Policastro, A. A. Frigo, D. E. Frye, and J. V. Tokar, "Temperature and Velocity Measurements and Predictive Model Comparisons in the Near-Field Region of Surface Thermal Discharges," *ANL/Es-25.*
20. Pande, B. B. L. and N. Rajaratnam, "Similarity Analysis of Buoyant Surface Jets into Quiscent Ambients," *International Assoc. for Hydraulic Research,* 16th Congress, Brazil, Vol. 3, C8, pp. 62–70 (1975).
21. Pande, B. B. L. and N. Rajaratnam, "An Experimental Study of Bluff Turbulent Buoyant Surface Jets," *Journal of Hydraulic Research International Assoc. for Hydraulic Research,* Vol. 15, No. 3, pp. 261–275 (1977).
22. Pande, B. B. L. and N. Rajaratnam, "Turbulent Jets in Coflowing Streams," *Proc. ASCE, J. Engrg. Mech. Div.*, pp. 1025–1038 (Dec. 1979).
23. Policastro, A. J. and J. V. Tokar, "Heated Effluent Dispersion in Large Lakes: State of the Art of Analytical Modeling. Critique of Model Formulations." Argonne National Lab, ANL-ESS-H (1972).
24. Rajaratnam, N., "Hydraulic Jumps," in *Advances in Hydroscience,* Vol. 4, Ed. by V. T. Chow, Academic Press, N.Y., pp. 197–280 (1967).
25. Rajaratnam, N. and B. S. Pani, "Three Dimensional Turbulent Wall Jets," *Journal of Hydraulic Division, ASCE,* Vol. 100, pp. 69–83 (Jan. 1974).
26. Rajaratnam, N., *Turbulent Jets,* Elsevier Publishing Co., Amsterdam, 304 p. (1976).
27. Rajaratnam, N. and T. Gangadharaiah, "Scales for Circular Jets in Crossflow," *Proc. ASCE, J. Hyd. Div.*, Vol. 107, No. HY4, pp. 497–500 (Apr 1981).
28. Rajaratnam, N. and T. Gangadharaiah, "Entrainment by Circular Jets in Crossflow," *J. of Wind Engrg.*, Vol. 9, pp. 251–255 (1982).
29. Rajaratnam, N. and M. J. Stalker, "Circular Wall Jets in Coflowing Streams," *Proc. ASCE, J. Hyd. Div.*, Vol. 108, pp. 187–198 (Feb. 1982).
30. Rajaratnam, N. and J. A. Humphries, "Turbulent Non-Buoyant Surface Jets," *J. of Hyd. Research, IAHR,* Vol. 22, No. 2, pp. 103–114 (1984).
31. Rajaratnam, N., "Non-Buoyant and Buoyant Circular Surface Jets in Co-Flowing Streams," *J. of Hyd. Research, IAHR,* Vol. 22, No. 2, pp. 117–144 (1984).
32. Rajaratnam, N., An Experimental Study of Bluff Surface Discharges with Small Richardson Number," *J. of Hydraulic Research, IAHR,* Vol. 23, No. 1, pp. 47–55 (1985).
33. Rajaratnam, N. and S. Subramanyan, "Plane Turbulent Buoyant Surface Jets and Jumps," *J. of Hyd. Research, IAHR,* Vol. 23, No. 2, pp. 131–146 (1985).
34. Stefan, H., "Spread and Dilution of Three Dimensional Rectilinear Heated Water Surface Jets," International Symposium on Stratified Flows, Novosibirsk (1972).
35. Stefan, H. and F. R. Schiebe, "Heated Discharge from Flume into Tank," *Journal of Sanitary Engineering Division, ASCE,* Vol. 96, SA6, Proc. Paper 7762, pp. 1415–1453 (Dec. 1970).
36. Stolzenbach, K. D. and D. R. F. Harleman, "An Analytical and Experimental Investigation of Surface Discharges of Heated Water," Ralph M. Parson's Laboratory, M.I.T., Cambridge (1971).
37. Tamai, N., R. L. Wiegel, and G. F. Tornberg, "Horizontal Surface Discharge of Warm Water Jets," *ASCE, Journal of the Power Div.*, Vol. 95, pp. 253–276 (Oct. 1969).
38. Turner, J. S., *Buoyancy Effects in Fluids,* Cambridge Univ. Press (1972).
39. Wilkinson, D. L. and I. R. Wood, A rapidly varied phenomenon in a two-layer flow, *J. of Fluid Mechanics,* Vol. 47, pp. 241–256 (1971).
40. Wood, I. R. and D. L. Wilkinson, Disc. of "Surface Discharge of Horizontal Warm-Water Jet," *ASCE, Journal of the Power Div.*, pp. 149–151 (Mar. 1967).
41. Wiuff, R., "Experiments on Surface Buoyant Jet," *ASCE, Journal of the Hydraulics Div.*, Vol. 104, pp. 667–679 (May 1978).
42. Yih, C. S. and G. R. Guha, *Hydraulic Jump in a Fluid System of Two Layers,* Tellus, Vol. 7, pp. 358–366 (1955).

Stokes and Rayleigh Layers in Presence of Naturally Permeable Boundaries

N. C. Sacheti* and B. S. Bhatt**

ABSTRACT

The present theoretical investigation aims to discuss unsteady flow of incompressible Newtonian fluid, caused by linear oscillations of a naturally permeable wall. The presence of a naturally permeable boundary gives rise to a coupled fluid flow problem and we have carried out a detailed review of such problems, including a critical assessment of present state of (1) equations of motion governing flow inside a porous medium and (2) boundary conditions, particularly at a free fluid–porous medium interface. In the process we have also derived slip conditions applicable at a lubricated porous surface.

Solutions of the governing Navier-Stokes equations, subject to Beavers-Joseph modified slip condition at a porous interface have been discussed by considering oscillatory solutions and an initial-value problem. The effect of the permeability of the porous medium enters through a slip grouping parameter, β, whose influence on the Stokes layer and transient Stokes flow has been shown graphically. As a particular case of the initial-value problem we have obtained Rayleigh layer over a naturally permeable wall which has been shown graphically for a range of a nondimensional parameter, δ. Two useful extensions to the basic problems referred to above have been sought; these refer to cases, namely, (a) when the fluid is bounded above by an impermeable wall, and (b) when the fluid above the permeable wall is of finite height. In particular, we have discussed in detail the effect of β on the modified Stokes layers.

BACKGROUND INFORMATION ON COUPLED FLUID FLOWS INVOLVING POROUS BODIES

The flow of fluids through porous media has many technical and engineering applications. Contributions to this subject have been noted in:

1. The simultaneous flow of water and gas in porous rock which is important in connection with the production of oil from oil-fields
2. Fixed-bed reactors in laboratories, refineries or chemical plants
3. The production and conservation of water, in the movement of subsurface water (treatment of water and wastewater [22])
4. The sedimentation of floculated clay minerals which are porous aggregations of particles
5. Geothermal operations—the study is useful to understand the mechanism of transfer of heat of the earth to a shallow depth in the geothermal regions which is of vital importance in the present day grave power crisis
6. Nuclear industries—study of natural convections in porous media in the evaluation of capacity of heat removal from a hypothetical accident in a nuclear reactor
7. Transpiration cooling and building thermal insulations
8. Biofluidmechanics—blood flows in a human body, transportation of ovum in the oviducts, human joint lubrication, etc.

One can divide the study of flow through porous media in broad areas as:

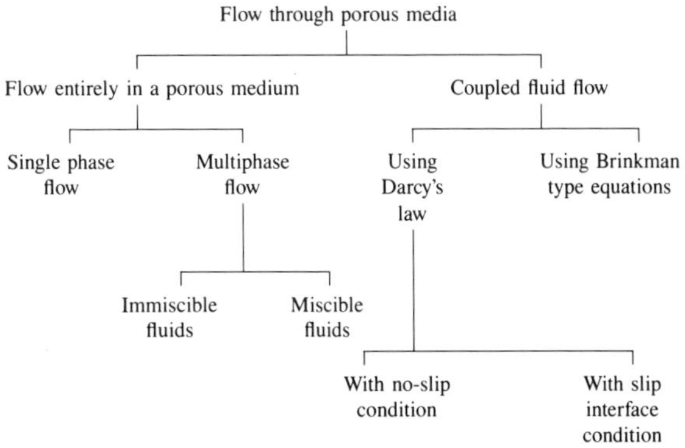

*Department of Mechanical Engineering, University of Jodhpur, Jodhpur-342001, India

**Department of Mathematics, University of the West Indies, St. Augustine, Trinidad (W.I.)

In the present chapter we shall be concerned with coupled fluid flows in the presence of naturally permeable boundaries. There seems to be a number of books dealing with various aspects of flow in a porous medium (e.g., [2,17,21, 23,25,49,84,85,87]), but the literature on the coupled fluid flows in presence of permeable bodies, though fairly exhaustive, seems to be scattered. The coupled fluid flow is encountered in many practical situations and readers are referred to works of Bhatt [10], Blake, Vann and Winet [16], Fitzgerald [27], Fung [28], Fung and Tang [29], Huh and Mason [35], Lin [42], Mow [45], PUri [65,67], and Tang and Fung [92] for details.

Coupled fluid flows can be visualized as simultaneous flow in an open space, called free fluid region (also frequently referred to as "clear region" in literature), and in a porous medium (or porous region). For a viscous Newtonian fluid Navier-Stokes equations govern the flow in the free fluid region, whilst the Darcy's law/Brinkman type equation generally governs the flow in the porous region. In such flows the interface conditions become extremely important and we shall discuss them in a later section.

The theoretical study of coupled fluid flow through porous bodies seems to have gained momentum in last 2 decades or so, particularly after Beavers and Joseph [3] proposed a physically plausible but empirical slip condition for the tangential component of the velocity, at a porous interface. This slip criterion has had a great bearing on the subsequent investigations which covered wide ranging aspects of fluid flow, namely, (a) flow through porous channels/bodies [4,5,37,51,55,71,73,81,89,94,96–100], (b) lubrication [15,30,46–48,63,64,68–70,75,78,90,103,105], (c) natural convection/heat-transfer [76,77,102], (d) stability [53,,54], (e) waves over porous bodies [65–67], (f) biomechanics [10,86,93], (g) MHD effects [62] (Sacheti, 1983, unpublished) and (h) non-Newtonian effects [86,101]. In these aforementioned investigations Darcy's law [24] was generally employed, in conjunction with slip condition [3]. In recent years there has been increasing use of Brinkman's equation to describe flow through a porous medium (particularly one with high porosity). In such coupled fluid/flow problems where Brinkman model [18] has been employed, a different set of boundary conditions at the porous interface has been used to find solution for velocity, etc. (e.g., [31,52,72,79,83,95,106]. It may finally be remarked that an exhaustive review of coupled fluid flows in the presence of permeable bodies/boundaries will be beyond the scope of this chapter.

Although the principal objective in this chapter is to look at the structure of Stokes and Rayleigh layers in presence of naturally permeable walls—based on one of our papers [80]—we feel that any exhaustive treatment of a coupled fluid flow problem involving a porous material is rather incomplete unless we critically review (i) equations governing flow through a porous medium and (ii) boundary conditions at the interface, since there is a lot of controversy regarding them. With this latter objective, too, in mind, we have given a complete update of flow equations in porous media in the third section of this chapter, Equations of Motion Governing Flow Through a Porous Medium. The following section (i.e., Boundary Conditions) deals with boundary condition, particularly at the interface. The title problem, along with a couple of useful extensions, has been dealt with, in detail, in Some Unsteady Problems Involving Naturally Permeable Boundaries. Now we move to the second section, where we briefly mention some definitions to describe a porous medium and give some examples.

SOME DEFINITIONS

Aquifer: An aquifer is a geological formation or a stratum that contains water and permits significant amounts of water to move through it under ordinary field conditions.

Aquiclude: An aquiclude is a formation that may contain water (even in good amount) but is incapable of transmitting a significant quantity under ordinary field conditions. Clay is an example. For practical purposes an aquiclude is taken to be an impervious formation.

Aquitard: An aquitard is a semipervious geological formation transmitting water at a very slow rate as compared to aquifer. However, over a large area it may allow the passage of large amounts of water between adjacent aquifers. It is usually known as a leaky formation.

Aquifuge: An aquifuge is an impervious formation that neither contains water nor transmits it.

Groundwater: This is a term used to denote all water found beneath the ground surface. However, a groundwater hydrologist who is primarily concerned with the water contained in the zone of saturation uses the term groundwater to describe water in this zone.

Pore-space: This is that portion of a rock which is not occupied by solid matter. Extremely small pores in a solid are known as molecular interstices; very large ones are called caverns. Pores are taken to be intermediate of interstices and caverns. The pores in a porous medium may be interconnected or non-interconnected. Flow in interstices is possible only if at least part of the pore-space is interconnected. The interconnected part of the pore system is called the effective pore-space of the porous medium.

Porosity: The ratio of void to the total volume of a porous medium is defined as porosity. If the calculation of porosity is based upon the interconnected pore space instead of total pore-space, the resulting value is defined as effective porosity.

We cannot give a proper definition for a porous medium; however, one may attempt to understand it by noting the following characteristics:

1. It is a portion of space occupied by heterogeneous or multiphase matter. At least one of the phases comprising

the matter should be other than solid; they may be gas and/or liquid phases. The solid phase is known as solid matrix. The portion of the space which is not solid is referred to as void space.

2. The solid phase should be distributed throughout the porous medium within the domain occupied by a porous medium. Solid must be present inside each representative elementary volume. An essential feature of a porous medium is that the specific surface (see Reference 2) of the solid matrix is relatively high. Another characteristic of a porous medium is that the various openings comprising the void space are relatively narrow.

Examples of porous medium: Towers packed with pebbles, beryl saddles; rasching rings; beds formed of sand, granules, lead shot, limestone, pumice, dolomite; fibrous aggregates such as cloth, felt, filter paper, etc., catalytic particles containing extremely fine micro-pores; large geological formations of karstic limestones where the open passages may be of substantial size and far apart.

For further details of this section readers are referred to Bear [2], Muskat [49] and Scheidegger [85].

EQUATIONS OF MOTION GOVERNING FLOW THROUGH A POROUS MEDIUM

In this section we propose to discuss Darcy's law which governs very slow flow of a viscous Newtonian fluid in a porous medium and examine various theories/models justifying Darcy's law and its generalizations. We shall also refer to more rigorous treatment of flow equations and, furthermore, look at some recent developments. Finally we shall make a brief review of the present state of the equations of motion.

Darcy's Law

Darcy [24] considered the steady flow of a viscous incompressible fluid through a porous bed of height h, bounded by horizontal plane areas of equal size A. If h_1 and h_2 be the pressure heads of lower and upper ends then the total volume Q of the fluid passing through a cross section in unit time satisfies the relation

$$Q = -\frac{KA(h_2 - h_1)}{h} = -\frac{K'A(\bar{p}_2 - \bar{p}_1)}{h} \tag{1}$$

where K'—a constant depending upon the properties of the fluid and the porous medium—is known as permeability. Later on Nutting [57] found that

$$K' = \frac{k}{\mu} \tag{2}$$

where k is known as the specific permeability (but now commonly referred to as permeability), and μ is the viscosity of the fluid.

Using relation (2), the Equation (1) gives

$$q = \frac{Q}{A} = -\frac{k}{\mu}\frac{(\bar{p}_2 - \bar{p}_1)}{h} \tag{3}$$

or in general

$$\vec{q} = -\frac{k}{\mu}(\text{grad}\,\bar{p} - \varrho\vec{g}) \tag{4}$$

which is known as Darcy's law. Here $\vec{q}$ is known as filter velocity vector, $\vec{g} = g\underline{i}$ (g being acceleration due to gravity and $\underline{i}$ is unit vector in the downward vertical direction).

Most of the textbooks dealing with flow through porous media (cf. the first section) have given a good account of Darcy's law and its derivation using various models/theories for the porous medium. Here we shall only briefly mention them, but for details of derivation, etc., readers are again referred to Bear [2] or Scheidegger [85].

1. *Capillaric model:* Here a porous medium is represented by a bundle of straight parallel capillaries of uniform diameter $\bar{\delta}$. In this model the relation between permeability (k) and porosity (P) can be shown to be [85]

$$k = \frac{P\bar{\delta}^2}{32} \tag{5}$$

For actual porous medium $\bar{\delta}$ can be taken as average pore diameter.

2. *Parallel type model:* In this model we put one-third of the capillaries in each of the three perpendicular directions. Then

$$k = \frac{P\bar{\delta}^2}{96} \tag{6}$$

3. *Serial type model:* The porous medium is taken to be composed of straight channels shown in Figure 1.
4. *Branching type model:* In this model instead of taking a straight channel, we may have branches like the pore doublet [6] shown in Figure 2.
5. *Hydraulic radius theory:* The theory is based on the simple fact that the permeability of the porous medium has dimension of area or squared length. Thus definition of length plays an important role in the porous medium. A possible measure is taken to be a hydraulic radius which is defined as the ratio of volume to the surface area of the pore-space.

This theory is based upon the assumptions that (a) no

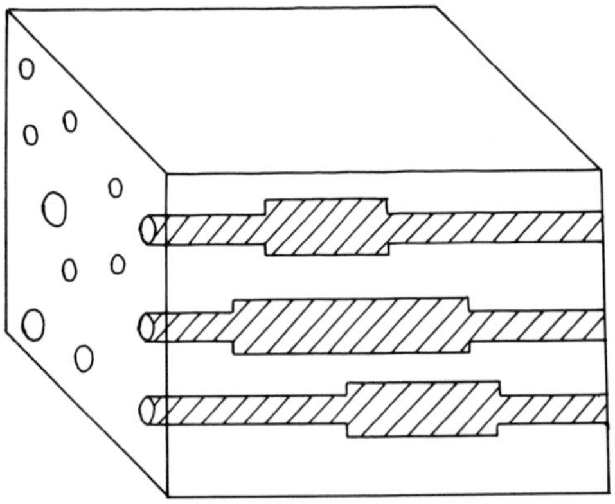

FIGURE 1. Serial type model.

pores are sealed off; (b) pores are distributed at random; (c) pores are reasonably uniform in size; (d) porosity is not too high; (e) diffusion phenomenon is absent; (f) fluid motion is occurring like motion through a batch of capillaries.

There are two theories accounting for hydraulic radius, namely, Kozeny theory and modified Kozeny theory, which have both been adequately discussed in some standard texts (e.g., [2,85]).

Modifications of Darcy's Law

In the models/theories referred to above, one main aim is usually to derive Darcy's law by representing the porous medium with a suitable model. But generally speaking, such a modelling is far from being satisfactory. Therefore, researchers came up with more appropriate theories which led to generalizations of Darcy's law. In the following we briefly outline two drag theories of permeability [85] which take account of viscous forces.

IBERALL'S THEORY

In this theory the walls of the pores are treated as obstacles to an otherwise straight flow. The drag force on each portion of the walls is estimated from the Navier-Stokes equations and then sum of all such drags is equated to the term representing resistance to the porous medium. Iberall [36] considered the model of random distribution of circular cylindrical fibres of the same diameter and found the permeability on the basis of the drag on individual elements. It was assumed that the flow resistivity of all random distributions of the fibres per unit volume will not differ and that it was the same as that obtained by equi-partition of fibres in three perpendicular directions. It was further assumed that the separation between fibres and the length of individual fibres were large compared to the fibre diameter, and also that the disturbance due to adjacent fibres on the flow around any particular fibres was negligible.

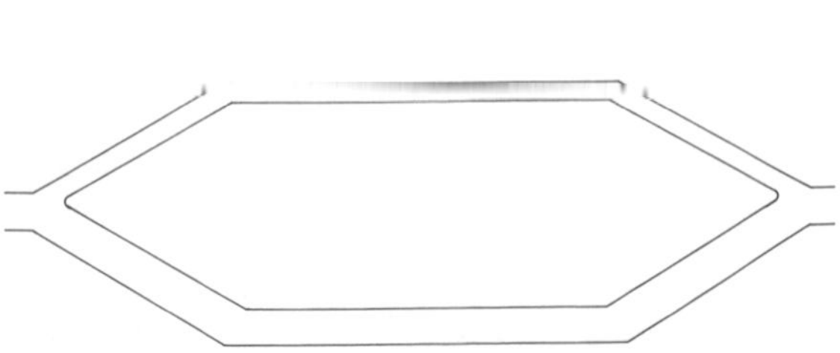

FIGURE 2. Pore doublet [after Benner et al. (6)].

One important point emerging from this theory is that permeability turns out to be a variable quantity. Besides this we get a modification of Darcy's law, as stated earlier.

BRINKMAN'S THEORY

Brinkman [18] on the other hand assumed that the particles in the fluid can be treated as spheres of radius R, kept in position by external forces as in a bed of closely packed particles which support each other by contact. He (refer to [85] for related papers of Brinkman) considered the flow of an incompressible viscous fluid through such a swarm and arrived at the following equation:

$$- \text{grad}\, \bar{p} + \mu \nabla^2 \vec{q} - \frac{\mu}{k} \vec{q} = 0 \qquad (7)$$

This equation may, in fact, be further modified to take account of inertial terms and body forces. But one may notice that the Equation (7) is a modification of Darcy's law through the inclusion of the term $\mu\nabla^2 \vec{q}$ (∇^2 being Laplacian operator). For high particle densities the term $\mu\nabla^2 \vec{q}$ can easily be shown to be negligible to the last term in the left-hand side of Equation (7). Thus Darcy's law, i.e., Equation (4) with body forces absent, is the limiting form of the Brinkman's Equation (7) for low permeabilities.

Averaging Technique

Bear [2] has derived the equations of motion for flow through porous media by defining an average of some fluid property at a point. He also considered inertia effects, in addition to viscous forces, in deriving the flow equation. Slattery [87], on the other hand, analysed the equations of motion for flow through porous media by taking local volume average of Navier-Stokes equations and the equation of continuity, neglecting inertial forces. For details of averaging techniques readers are referred to References 2 and 87. However, it may be noted that the equations of motion derived by Bear [2] and Slattery [87] are similar to Brinkman's equation described in the previous section. Hinch [33], Howells [34],

Lundgren [44] and Tam [91] have also derived similar equations, thus giving a rigorous theoretical justification for the proposed Brinkman's [18] generalization to Darcy's law.

Mixture Theory

Williams [104] has derived equations describing flow of an incompressible viscous fluid through a rigid porous medium on the basis of mixture theory, by accounting for capillary forces, drag forces and viscous forces—by far the most general consideration. There are three unspecified coefficients in the linearized equations he has obtained. These coefficients have been obtained by considering three different situations; in the first the equations reduce to Darcy's law, in the second to diffusion equation and in the third to the problem of Beavers and Joseph [3].

Statistical Theories

There are various statistical theories, namely, Gibb's model, velocity dispersion model, random medium model, etc., to describe flow of fluids through porous media. These models have been dealt with by Scheidegger [85].

A review of coupled fluid flows reveals that there are generally two groups of researchers who use either Darcy's law or Brinkman's equation for describing flow through a porous medium. Darcy's law has lower order terms as compared to Navier-Stokes equations, and hence it cannot account for shear effects either. On the other hand, Brinkman's model does not seem to fit well into some physical situations, e.g., flow caused in a porous medium by motion of bounding surfaces of the porous medium. Nield [54] has critically analysed the behaviour of Brinkman's equation and pointed out that the use of Brinkman's equation within the bulk of porous medium whose porosity is not close to unity, is not always justifiable. He [54] rather regards the Brinkman model as useful for the treatment of flow past a very sparse collection of obstacles, and for flows in porous media where velocity is constant except in regions near boundaries. He also expressed doubts over use of the Brinkman equation in regions close to the boundaries. Similar doubts have also been expressed by Haber and Mauri [31]. Levy [41] has claimed that the Brinkman model really holds only where the particles are of order η_1^3, where η_1 $(1 \leq 1)$ is the distance between two neighbouring particles; for large particles the fluid velocity should be governed by Darcy's law (the smaller particles do not influence the flow). Beavers and Joseph [3], on the other hand, expressed their doubts for the validity of Darcy's law near the boundaries, particularly at the fluid–porous interface, as it cannot account for the boundary layer formed immediately inside the porous medium.

As concluding remarks we would like to emphasize that one cannot ignore the importance of Darcy's law/Brinkman model, for these have been successfully used in, of course, different physical situations. To frame any fixed ideas about applicability of either model will have to await further rigorous investigations.

BOUNDARY CONDITIONS

We pointed out earlier, in the first section, that boundary conditions, particularly those at the free fluid–porous interface assume great significance in any theoretical study involving coupled fluid flows in the presence of porous material. Joseph and Tao [39] have given a detailed account of coupled fluid flow problem investigated until middle sixties or so. In these problems continuity of normal velocity and pressure, together with adherence condition for the free fluid tangential velocity component at the interface, have been assumed. Of course, it is difficult to have a true boundary for the permeable surface, but we can always define a nominal boundary (see Reference 3) by joining outermost perimeters of corrugated porous surface. The adherence condition can be regarded, at best, as only an approximation for fluid particles just above the nominal boundary, for one can always argue that there will be some net tangential drag due to the transfer of forward momentum across the permeable interface. In fact, if we deal with the true velocity in the porous medium, there will be no slip between the free fluid and the fluid immediately within the porous boundary. But in this case the discontinuity of tangential velocity component could not be allowed.

Slip Phenomenon

Beavers and Joseph [3] performed experiments for two-dimensional Poiseuille flow over a fluid saturated porous block, with an aim to estimate interfacial velocity. The results of their experiments indicated that the effects of viscous shear penetrated into the porous medium through a boundary layer region, giving a velocity distribution as shown in Figure 3.

They postulated that the slip velocity (u_B) at the permeable interface differs from the mean filter velocity (q) within the porous medium, and two are related by the empirical formula

$$\left.\frac{du}{dy}\right|_{y=0_+} = \frac{\alpha}{\sqrt{k}}\,(u_B - q) \qquad (8)$$

where 0_+ is the boundary limit point from the exterior fluid, α is a dimensionless quantity independent of the viscosity of the fluid but depending on the material parameters that characterize the structure of the permeable material within

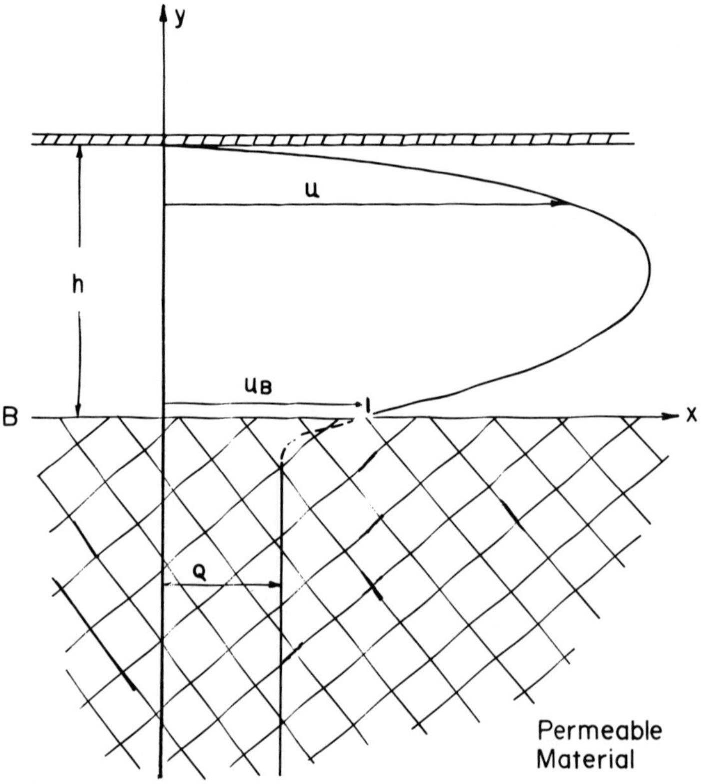

FIGURE 3. Velocity profile for the rectilinear flow in a horizontal channel formed by a permeable wall (y = 0) and an impermeable upper wall (y = h) [after Beavers and Joseph (3)].

the boundary region, u is the free fluid axial velocity, and k is the permeability of the porous medium. They solved the problem theoretically as well, and compared the results with the experiments. Their [3] experiments have shown consistently that the increase in mass flow rate through the gap (i.e., channel cross section) is accompanied by an increase in mass flow rate through the permeable block. This increase, expressed as a fraction of the flow rate predicted by the Darcy's law, has been noticed to be a relatively much smaller effect than the corresponding effect in the free fluid; however, this fractional increment in mass flow does increase with gap size. They [3] regarded these results to be consistent with the hypothesis that a boundary layer in the porous block may strongly influence the rectilinear flow in the gap above the block.

Beavers and Joseph [3] also attempted to correlate α with the average pore diameter. They measured pore diameters over a large part of the upper surface of permeable block and averaged. It was found that for each sample the range of pore sizes about the mean was not great, so that the concept of an "average pore diameter" for this type of material was not too unrealistic. An estimate of effective pore size for each block has been given in Table 1.

Some theoretical support for the Beavers-Joseph slip condition, i.e., Equation (8), has been lent by the results of Taylor [94] and Richardson [73] based on an analogue model of a porous medium.

Saffman [81] applied statistical approach and derived a

TABLE 1. Values of Permeability and Average Pore Size [After Beavers and Joseph (3)].

Block	k (in.^{2})	α	Average pore size (inch)
Foametal A	1.5×10^{-5}	0.78	0.016
Foametal B	6.1×10^{-5}	1.45	0.034
Foametal C	12.7×10^{-5}	4.0	0.045
Aloxite	1.0×10^{-6}	0.1	0.013
Aloxite	2.48×10^{-6}	0.1	0.027

boundary condition for a permeable surface of arbitrary geometry:

$$u = \frac{\sqrt{k}}{\alpha}\frac{\partial u}{\partial n} + 0(k) \qquad (9)$$

where n refers to the direction normal to the boundary and u is the tangential velocity. For Beavers and Joseph model [3], Equation (9) assumes the form:

$$u = \frac{\sqrt{k}}{\alpha}\frac{du}{dy} + 0(k) \qquad (10)$$

Thus it lends further support to the Beavers and Joseph slip condition (also referred to as BJ condition in the literature). Beavers, Sparrow and Magnuson [4] and Sparrow, Beavers and Hwang [90] verified the BJ condition for Poiseuille flows of oil and water through large aspect ratio rectangular ducts having one porous wall.

Jones [37] proposed a modification of BJ condition for curved surfaces, realising that the slip condition of Beavers and Joseph is essentially a relationship involving shear stress rather than just velocity shear. Therefore, he conjectured that the shear stress at a curved porous surface is proportional to $\alpha/\sqrt{k}\,(u_t - q_t)$, where α and k are the same as that in BJ condition, and u_t and q_t represent the tangential components of velocities in the free fluid and the porous regions, respectively. For a spherical boundary, the proposed modification of Jones [37] gives

$$e_{r\theta} = \frac{\alpha}{\sqrt{k}}(u_\theta - q_\theta) \qquad (11)$$

where $e_{r\theta}$ is a rate of strain tensor. For $k \to 0$, Equation (11) reduces to

$$u_\theta = \frac{\sqrt{k}}{\alpha}\, r\frac{\partial}{\partial r}\left(\frac{\partial u_\theta}{\partial r}\right) + 0(k) \qquad (12)$$

on substituting for $e_{r\theta}$. The Equation (12) so obtained is the Saffman's [81] simplification to the Equation (11) up to the $0(k)$.

Boundary Layer Concept

The work of Beavers and Joseph [3] revealed that the viscous shear effects penetrate into the porous medium through a boundary layer region where rapid changes in velocity occur. Neale and Nader [52] pointed out that Darcy's law is not compatible with the existence of boundary layer because no macroscopic shear term is associated with Darcy's equation. These authors [52] considered also the problem of Beavers and Joseph [3] by employing the Brinkman model for the porous region, subject to the following boundary conditions (which they considered to be physically realistic and consistent):

$$\frac{dp}{dx} = \frac{d\bar{p}}{dx} = \text{constant} \qquad (13)$$

$$u = 0 \quad \text{at } y = h \qquad (14)$$

$$\left.\begin{aligned} u &= q \\ \mu\frac{du}{dy} &= \tilde{\mu}\frac{dq}{dy} \end{aligned}\right| \quad \text{at } y = 0 \qquad (15)$$

$$\lim_{y\to -\infty} q = -\frac{k}{\mu}\frac{d\bar{p}}{dx} = \bar{U} \qquad (16)$$

where u and q are fluid velocities in the free fluid and porous regions, respectively; p and $\bar{p}$ are corresponding pressures; and μ and μ are the coefficients of viscosity and effective viscosity, respectively. The most noteworthy point to make here is that the conditions expressed in Equation (15) do acknowledge the fact that the effects of viscous shear within the channel penetrate into the porous medium to form a "boundary layer region," and together these conditions replace the BJ condition [i.e., Equation (8)]. The results obtained here are identical with the predictions of Beavers, Sparrow and Magnuson [4] who employed Darcy's law together with BJ condition, provided that $\alpha = \sqrt{\tilde{\mu}/\mu}$. The authors [52] regard their results to be restrictive for porous media in general, in the sense that the effective viscosity $\tilde{\mu}$ introduced in their analysis has received insufficient theoretical attention; Lundgren [44] has, however, tried to determine the ratio $(\tilde{\mu}/\mu)$ for a dilute swarm of stationary monosized spheres.

Neale, Epstein and Nader [51] have discussed creeping flow of a viscous incompressible fluid relative to an isolated permeable sphere, using both Darcy's law and the Brinkman model for flow in the porous medium. They opine that the most satisfactory results are obtained by employing the Brinkman model together with a set of appropriate boundary conditions. Using Happel's [32] free-surface model they [51] also generalized their solution to describe flow relative to a swarm of permeable spheres.

Liu [43] also took up the Beavers and Joseph [3] problem to support the boundary layer concept, using an equation similar to the Brinkman equation. He probably did not come across Neale and Nader's work [52]. He used the following equation for the boundary layer region:

$$\frac{d\bar{p}}{dx} + \frac{\mu}{k}\tilde{u} - \frac{\mu\epsilon}{P}\frac{d^2\tilde{u}}{dy^2} = 0 \qquad (17)$$

where $\mu\epsilon$ is the effective viscosity and $\epsilon \to 1$ as $P \to 1$. The free-fluid region has been taken to be governed, as usual, by Navier-Stokes equations:

$$\frac{d^2u}{dy^2} = \frac{1}{\mu}\frac{dp}{dx} \tag{18}$$

The solution for the boundary layer region has been worked out by taking

$$\tilde{u} = \bar{u}_1 + q \tag{19}$$

where

$$q = -\frac{k}{\mu}\frac{d\bar{p}}{dx} \tag{20}$$

Therefore, one will obtain

$$\frac{d^2\bar{u}_1}{dy^2} - \frac{P}{\epsilon k}\bar{u}_1 = 0 \tag{21}$$

The boundary conditions used are:

$$u = 0 \quad \text{at } y = h \tag{22}$$

$$\left.\begin{aligned} u &= \bar{u}_1 \\ \mu\frac{du}{dy} &= \epsilon\,\mu\frac{d\bar{u}_1}{dy} \end{aligned}\right| \quad \text{at } y = 0 \tag{23}$$

$$\tilde{u} = 0 \quad \text{at } y = -H \tag{24}$$

One may like to compare the set of Equations (14)–(16) with the set of Equations (22)–(24) to get some insight into the treatment of boundary layer region by Neale and Nader [52] and Liu [43]. Liu also compared his theoretical predictions with the experimental results of Beavers and Joseph [3] and found a fine agreement. He further concluded that the boundary layer structure recognized in his study depends strongly on the properties of the material in the boundary region.

O'Brien [58] has shown that the parallel flows with permeable walls depend on the interface porosity rather than porosity of the whole matrix as suggested by Liu's [43] results. In a later paper he [59] argued that although most porous materials are too complicated to attempt a detailed internal viscous flow solution, a slow steady flow solution is guaranteed to exist under a steady pressure gradient whatever be the geometry.

The results of Haber and Mauri [31] should be of great significance. They have proposed novel boundary conditions applicable at a porous surface, by considering interfaces between porous and free fluid regions (i.e., clear media) and porous and solid media. A boundary condition equivalent to BJ condition and applicable at the interface between porous and impermeable media, namely,

$$\underset{\sim}{V} \cdot \underset{\sim}{n} = \sqrt{k}\,\nabla_t \cdot \underset{\sim}{V}_t \tag{25}$$

where $\underset{\sim}{V}$ is the velocity field inside the porous medium, $\underset{\sim}{n}$ denotes a unit vector normal to the interface pointing towards the porous medium, and subscript t refers to components tangential to the interface, has been derived. The boundary condition so proposed has been applied to the problem of flow fields exterior to a porous spherical particle and interior to it, assuming that the particle has a rigid concentric spherical core and that the submerging flow field is Newtonian, Stokesian and uniform at infinity. A major highlight of this paper lies in the simultaneous use of the Brinkman equation and Darcy's law—possibly giving impetus to the fact that both these models are quite effective and important in describing a variety of flows through porous media. In a very recent paper Nield [56] has analysed boundary layer effects in the case of a porous medium bounded by a "rigid" boundary. He has discussed extensions of Darcy's law to account for this.

From the discussion of the two previous sections, it appears to emerge that one can work with either Darcy's law together with BJ condition or the Brinkman model [i.e., Equation (7)] in conjunction with the continuity of velocities and stresses at the interface. The results would, in general, be similar up to the $0(\sqrt{k})$.

Bhatt and Sacheti [12,13] have brought out some careful observations in the use of Saffman's condition, namely Equation (10). They have pointed out that one cannot apply suction/injection at the "naturally" permeable wall(s) together with the use of Equation (10). One needs to apply a small permeability approximation and display various results correctly up to the $0(\sqrt{k})$ while using the Saffman's condition. Bhatt [7] has noted similar observations in the case of squeeze film between rotating porous annular discs.

Analogy in Slip Flows

Bhatt and Sacheti [14] have noticed that the Saffman's condition [cf. Equation (10)] is analogous to the slip boundary condition which is applied to the solid smooth boundaries (see Lance and Rogers [40]), namely,

$$\frac{du}{dy} = L'u \tag{26}$$

where

$$L' = \frac{f_1}{(2 - f_1)L} \tag{27}$$

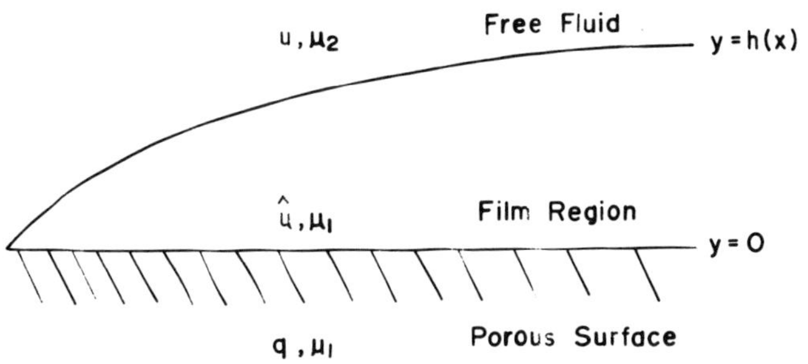

FIGURE 4. Flow over a lubricated porous surface.

L is the mean free path and f_1 is the Maxwell's relaxation coefficient.

From Equations (10) and (26) it may be easily noted that if L' is replaced by $\alpha/\sqrt{k}$ (in both the cases the flow is governed by the Navier-Stokes equations), all the results of viscous incompressible fluid flow through and around solid smooth boundaries can directly be carried over to the corresponding flows involving permeable boundaries. However, in the latter case one has to apply a simplification due to small permeability (k). Similarly, if one wants to derive the results for the first case from those of the second $\alpha/\sqrt{k}$ should be replaced by L' in the results before applying the simplification due to small permeability.

The analogy pointed above will generally hold (1) for curved surfaces, (2)when the bounding walls are in motion, (3) for unsteady flows, (4) when MHD effects are present. The analogy has been displayed for a few cases by Bhatt and Sacheti [14] and Bhatt [8].

Slip Condition for Lubricated Porous Surface

In this section we propose to extend the analysis of Joseph [38] for a naturally permeable surface; i.e., we consider the effect of lubricating film between a flowing Newtonian liquid and a porous surface (see Figure 4).

STATIONARY POROUS SURFACE

Following Joseph's notation, the boundary conditions in the present case are:

$$u\,[x,\, h(x)] = \hat{u}[x, h(x)] = V(x) \tag{28}$$

$$\mu_2 \frac{\partial u}{\partial y} = \mu_1 \frac{\partial \hat{u}}{\partial y} \quad \text{at } y = h(x) \tag{29}$$

$$\frac{\partial \hat{u}}{\partial y} = \frac{\alpha}{\sqrt{k}} (\hat{u} - q) \quad y = 0 \tag{30}$$

The last condition, i.e., Equation (30), is the consequence of there being a porous surface at $y = 0$. The flow in the thin layer ($0 < y < h$) is governed by

$$p'(x) \equiv \frac{dp(x)}{dx} = \mu_1 \frac{\partial^2 \hat{u}(x,y)}{\partial y^2} \tag{31}$$

whose solution using the conditions (28) and (30) can be obtained as

$$\hat{u} = \frac{p'}{2\mu_1} y^2 + \left(\frac{V - q - \dfrac{p'h^2}{2\mu_1}}{h + \dfrac{\sqrt{k}}{\alpha}} \right) \left(y + \frac{\sqrt{k}}{\alpha} \right) + q \tag{32}$$

Equation (29) then gives

$$u^2 \frac{\partial u}{\partial y} = \mu_1 \left[\frac{p'}{\mu_1} h + \frac{V - q - \left(\dfrac{p'}{2\mu_1}\right) h^2}{h + \dfrac{\sqrt{k}}{\alpha}} \right] \tag{33}$$

and the flow rate in the film region

$$Q = \int_0^{h(x)} \hat{u}\, dy = \frac{p'}{6\mu_1} h^3 + \left[\frac{V - q - \left(\dfrac{p'}{2\mu_1}\right) h^2}{h + \dfrac{\sqrt{k}}{\alpha}} \right] \left(\frac{h^2}{2} + \frac{\sqrt{k}\, h}{\alpha} \right) + qh \tag{34}$$

For $h(x)$ to be small, Equations (32)–(34) give

$$\hat{u} = \left(\frac{V - q}{h + \dfrac{\sqrt{k}}{\alpha}} \right) \left(y + \frac{\sqrt{k}}{\alpha} \right) \tag{35}$$

$$\mu_2 \frac{\partial u}{\partial y} = \mu_1 \left(\frac{V - q}{h + \frac{\sqrt{k}}{\alpha}} \right) \tag{36}$$

$$Q = (V + q) \frac{h}{2} \tag{37}$$

Eliminating $h(x)$ in Equation (36) with the help of (37), and writing $V(x)$ as $u(x,h)$ [cf. Equation (28)], we get

$$\mu_2 \frac{\partial u(x,h)}{\partial y} = \mu_1 \left\{ \frac{u^2(x,h) - q^2}{2Q + \left(\frac{\sqrt{k}}{\alpha}\right)[u(x,h) + q]} \right\} \tag{38}$$

and since $h(x) \rightarrow 0$ the condition at $y = 0$ can be taken as

$$\frac{\partial u}{\partial y} = \frac{\mu_1(u^2 - q^2)}{2\mu_2 Q \left[1 + \left(\frac{\sqrt{k}}{2\alpha Q}\right)(u + q) \right]} \tag{39}$$

For small k, Equation (39) reduces to

$$\frac{\partial u}{\partial y} = \lambda_1 u^2(1 - \lambda_2 u) + 0(k) \tag{40}$$

where

$$\lambda_1 = \frac{\mu_1}{2\mu_2 Q}$$

$$\lambda_2 = \frac{\sqrt{k}}{2\alpha Q}$$

MOVING POROUS SURFACE

When the porous surface is moving with velocity U parallel to itself, then the condition given by Equation (30) is replaced by

$$\frac{\partial \hat{u}}{\partial y} = \frac{\alpha}{\sqrt{k}} (u - U - q) \quad \text{at } y = 0 \tag{41}$$

whilst the other conditions given by Equations (28) and (29), and also the flow equation given by Equation (31) remain the same. In this case we get the condition corresponding to the Equation (38) as

$$\mu_1 \frac{\partial u(x,h)}{\partial y} = \frac{\mu_1[u^2(x,h) - (q - U)^2]}{\left\{ 2Q + \frac{\sqrt{k}}{\alpha} [u(x,h) + q + U] \right\}} \tag{42}$$

and since $h(x) \rightarrow 0$, we take the condition at $y = 0$ as

$$\frac{\partial u}{\partial y} = \frac{\lambda_1[u^2 - (q + U)^2]}{[1 + \lambda_2(u + q + U)} \tag{43}$$

where λ_1 and λ_2 are same as before.

For small k, Equation (43) simplifies to

$$\frac{\partial u}{\partial y} = \lambda_1(u^2 - U^2)[1 - \lambda_2(u + U)] + 0(k) \tag{44}$$

and for $k = 0$ Equation (43) or (44) reduces to

$$\frac{\partial u}{\partial y} = \lambda_1(u^2 - U^2) \tag{45}$$

At this juncture we would like to emphasize that when the bounding surface (porous/impermeable) is in "motion," one needs to exercise care in using the slip condition given by either of the Equations (43) and (45). May it be remarked that these conditions cannot be derived directly from the stationary case by replacing u with $(u - U)$. As an illustration, if we consider a solid surface and replace u with $(u - U)$ in the Joseph's [38] condition, namely,

$$\frac{\partial u}{\partial y} = \lambda_1 u^2 \quad \text{at } y = 0 \tag{46}$$

we shall end up with

$$\frac{\partial u}{\partial y} = \lambda_1(u - U)^2 \text{ at } y = 0 \tag{47}$$

which is, quite clearly, different to the valid condition derived earlier, i.e., the Equation (45).

SOME UNSTEADY PROBLEMS INVOLVING NATURALLY PERMEABLE BOUNDARIES

Introduction

The theoretical consideration of flow of viscous incompressible fluids in the presence of naturally porous bodies seem to have generated a good deal of interest among researchers, apparently because of numerous scientific and industrial applications. While most of the problems hitherto considered belong to steady class, the literature on the unsteady flow problems is rather scanty. Thus the scope of carrying out investigations concerning unsteady flow problems is large; the importance of such investigations stems particularly from the fact that, in many practical situations, one needs to consider unsteadiness that may be caused, for example, by oscillations of the permeable boundary or by a time-dependent pressure gradient [82].

In order to have some insight into the features of unsteady flow problems involving permeable boundaries, it is natural to consider, in the first place, well known Stokes first and second problems. One important aim of such a study will obviously be to determine as to how Stokes and Rayleigh layers are modified because of permeable boundaries. And this exactly forms the basis of the problems that we shall consider in this section. For the case of flow which takes place entirely within a porous medium, the two problems referred to above have been dealt with by Narasimha Murthy [50]. On the other hand, Verma [95] has studied Stokes layer in the presence of a bounding porous medium (of finite thickness) whose porosity is closed to unity. It may, however, be remarked that giving oscillations to the lower boundary of the porous medium in this investigation [95] may need justification on physical grounds. In fact, the whole of bounding porous medium needs to oscillate parallel to the bounding interface. And we have considered this aspect in the unsteady flow problems to follow.

In the following section we have dealt with basic problems, namely, flow generated in a semi-infinite incompressible viscous fluid by oscillating permeable wall or by impulsively started permeable wall with constant velocity. An initial-value problem has also been considered to examine the transient flow. In the following sub-sections we have attempted a couple of useful extensions to the basic problem and discussed modifications to the Stokes layer in detail. It is worthwhile to remark that the permeable material considered in the problems investigated here is one of low permeability and, further, that the amplitude and frequency of the oscillations of the permeable wall are of moderate values. The permeable material is assumed to be performing linear oscillations of the type

$$u = U \cos wt \tag{48}$$

in which U = the amplitude, and w = the frequency of the oscillations.

Stokes and Rayleigh Layers over Naturally Permeable Boundaries

In this sub-section we analyze the Stokes and Rayleigh layers over a naturally permeable wall (of finite thickness) bounding semi-infinite extent of an incompressible Newtonian fluid.

GOVERNING EQUATIONS OF MOTION

In the unsteady flow problems to be considered in the present section the only component of velocity is that parallel to the bounding porous interface. A naturally permeable wall at $y = 0$ is oscillating parallel to its bounding interface. We take the axis of x to be along the interface, and thus the axis of y is at right angle to it and its positive direction is towards the fluid outside the porous medium.

The equations governing the free fluid region ($y > 0$) are familiar Navier-Stokes equations, whereas the flow inside the permeable material is governed by extended Darcy's law which takes account of unsteadiness.

In the absence of any external pressure gradient (i.e., $-\partial p/\partial x = 0$), the momentum equation in the free fluid region is

$$\frac{\partial u}{\partial t} = \nu \frac{\partial^2 u}{\partial y^2} \tag{49}$$

in which $u(y,t)$ = the velocity of fluid particle, parallel to the surface of the permeable material; and ν = the kinematic coefficient of viscosity.

For a permeable material (of low porosity) saturated with the incompressible Newtonian fluid, the flow (q) inside it does not make a significant contribution to the exterior flow in the absence of any external pressure gradient; therefore, we shall ignore the former. The effect of the permeability, k, of the porous medium will, thus, enter through the slip boundary condition [cf. Equation (53)]. But for a porous medium whose porosity is very high we shall need to consider the effect of flow inside the medium (e.g., see Reference 95) in determining the exterior flow. Thus we note that the "governing" equation in all unsteady flow problem in this section will be the Equation (49), but with different boundary and other flow conditions.

OSCILLATORY SOLUTION (STOKES LAYER)

Suppose that a semi-infinite Newtonian fluid is at rest initially and bounded by a naturally permeable wall. Our problem here is to find the fluid motion after the boundary has been oscillating [cf. Equation (48)] parallel to its bounding interface, for a considerably "large time." In this case the solution of the problem will be "repetitive" oscillations found by ignoring the initial state of the fluid.

We therefore seek a periodic solution of the form

$$u(y,t) = \varnothing(y)e^{iwt} \tag{50}$$

in which we shall consider only real part.

The boundary conditions in this case are:

$$\frac{\partial u}{\partial y} = \frac{\alpha}{\sqrt{k}}(u - U\cos wt - Q) \quad \text{at } y = 0 \tag{51}$$

and

$$u \to 0 \text{ as } y \to \infty \tag{52}$$

Since the flow inside the porous medium is negligible, we

have $q = 0$. Thus the boundary condition now simplifies to

$$\frac{\partial u}{\partial y} = \frac{\alpha}{\sqrt{k}}(u - U\cos wt) \quad \text{at } y = 0 \tag{53}$$

It may be remarked that Equation (51) [or Equation (52)] is a generalized form of the BJ condition [i.e., Equation (8)] accounting for the boundary oscillations. On plugging Equation (50) into Equation (49), we shall obtain an ordinary linear differential equation for $\varnothing(y)$ whose general solution may be written as:

$$\varnothing(y) = c_1 e^{\lambda y} + c_2 e^{-\lambda} y \tag{54}$$

Here c_1 and c_2 are constants and

$$\lambda^2 = \frac{iw}{\nu} \tag{55}$$

In view of Equation (50), the transformed boundary conditions are:

$$\frac{\partial \varnothing}{dy} = \frac{\alpha}{\sqrt{k}}(\varnothing - U) \tag{56}$$

$$\varnothing \rightarrow 0 \text{ as } y = \infty \tag{57}$$

From Equations (54) through (57), we have the desired particular solution:

$$\varnothing(y) = \frac{U\alpha}{\alpha + \left(\frac{ikw}{\nu}\right)^{1/2}} \exp[-\sqrt{(iw/\nu)}y] \tag{58}$$

Now using Equations (50) and (58) and collecting the real part from the resulting expression (a bit of algebra is involved here), one can express the velocity as

$$u(y,t) = \frac{U\alpha e^{-\eta y}}{(\alpha^2 + 2k\eta^2 + 2\alpha\sqrt{k}\,\eta)}\{a\cos(wt - \eta y) + \sqrt{k}\,\eta\,[\cos(wt - \eta y) + \sin(wt - \eta y)]\} \tag{59}$$

where

$$\eta = \sqrt{\frac{w}{2\nu}} \tag{60}$$

By introducing a non-dimensional prarameter B_S, given by

$$B_S = \frac{wk}{\nu} \tag{61}$$

we can rewrite Equation (59) in the following form:

$$\frac{u}{U} = \frac{e^{-\eta y}\left[\cos(wt - \eta y) + \frac{1}{\sqrt{2}}\frac{\sqrt{B_S}}{\alpha}\{\cos(wt - \eta y) + \sin(wt - \eta y)\}\right]}{1 + \left(\frac{\sqrt{B_S}}{\alpha}\right)^2 + \sqrt{2}\left(\frac{\sqrt{B_S}}{\alpha}\right)} \tag{62}$$

In Equation (62) we may note the grouping of $\sqrt{B_S}/\alpha$ which may well be regarded as a slip grouping parameter (cf. Sparrow, Beavers and Hung [89]). This parameter is likely to arise in unsteady problems relating to incompressible Newtonian/non-Newtonian flows past bodies with naturally permeable boundaries and may be regarded as a characteristic of oscillating porous medium.

By rearranging various terms in the numerator of Equation (62) and with some manipulation, we finally obtain

$$\bar{u} = \frac{e^{-Y}\cos(T - Y - \delta_1)}{s_1} \tag{63}$$

in which we have now introduced non-dimensional physical variables

$$\bar{u} = \frac{u}{U}, \quad T = wt, \quad Y = \eta y \tag{64}$$

and, further,

$$s_1 = (1 + \beta^2 + \sqrt{2}\,\beta)^{1/2} \tag{65}$$

$$\delta_1 = \tan^{-1}\left[\frac{\beta}{(\sqrt{2} + \beta)}\right] \tag{66}$$

For $\beta \rightarrow 0$ (i.e., $k \rightarrow 0$), we easily arrive at

$$\bar{u} = e^{-Y}\cos(T - Y) \tag{67}$$

the classical Stokes layer (cf. Pai [60]).

It is easily seen from Equation (63) that the structure of the Stokes layer over a permeable boundary is similar to that in the case of a rigid boundary [cf. Equation (67)]; in the former case the amplitude of the nondimensional velocity is decreased by a factor $1/(1 + \beta^2 + \sqrt{2}\,\beta)^{1/2}$ as compared to the latter, and further there is a phase-lag of $\tan^{-1}[\beta/(\sqrt{2} +$

β)] with respect to impermeable (i.e., rigid boundary) case. Table 1 gives values of the factor [$= 1/(1 + \beta^2 + \sqrt{2}\beta)^{1/2}$] and associated phase-lag for a range of relatively small values of the slip grouping parameter, β. As in the impermeable case, two fluid layers, a distance $2\pi(2\nu/w)^{1/2}$ apart, oscillate in phase. Further, if $y = \varrho$ is the distance from the permeable boundary where the amplitude [i.e., $e^{-Y}/(1 + \beta^2 + \sqrt{2}\beta)^{1/2}$] has decreased to 1% of the wall value, then it is easily seen that the thickness of such a region will be of order $Y \approx 4.5$, which, on converting to dimensional variables, gives

$$\varrho \approx 4.5\left(\frac{2\nu}{w}\right)^{1/2} = 6.364\left(\frac{\nu}{w}\right)^{1/2} \qquad (68)$$

Clearly, this thickness (the depth to which viscosity makes itself felt, i.e., viscous length) is independent of the parameter characterizing our porous medium.

The term $\cos(T - Y - \delta_1)$ in Equation (63) gives a wave-like behaviour. A specific point on the wave can be given by assigning a fixed value to $(T - Y - \delta_1)$. In fact, this point will travel through the flow region ($Y > 0$) according to (in terms of original variables)

$$y = C + \sqrt{(2\,w)}t - (2\nu/w)^{1/2}\tan^{-1}\left[\frac{\sqrt{wk}}{(\alpha\sqrt{2\nu} + \sqrt{wk})}\right] \qquad (69)$$

in which C is a constant. One may notice additional term in Equation (69) (i.e., the last term on the R.H.S.), because of permeable boundary. The wave velocity is easily calculated to be $(2\nu w)^{1/2}$, the value in the impermeable case.

The velocity distribution $\bar{u}$ (vs. Y) for $T = 0, \pi/4, \pi/2, 3\pi/4$ and π (i.e., during half-cycle) has been shown in Figures 5 through 9, respectively, for a range of values of parameter β. One may notice that the effect of increase in β is to enhance the magnitude of the wall-slip velocity. As one would expect it is the intermediate region where behaviour is a complex one, apparently due to combined effect of phase-lag and factor (cf. Table 2). Roughly around $Y > 3.0$ the oscillating amplitude of velocity begins to die out, this feature being more conspicuous for increasing values of β. Figure 10 shows velocity distribution $\bar{u}$ (vs. T) at different heights, indicating similar type of behaviour as observed earlier.

Skin-Friction

One of the parameters of practical interest in any fluid flow problem is skin-friction or wall shearing stress, τ. It is given by the expression

$$\tau = \mu\left(\frac{\partial u}{\partial y}\right)_w \qquad (70)$$

where the subscript w stands for the wall under consideration. We can now define a non-dimensional quantity, called the coefficient of skin-friction and denoted by τ^* which in the present case is given by

$$\tau^* = \frac{\tau}{uU\eta} \qquad (71)$$

For the permeable wall (at $y = 0$) it has been obtained as

$$\tau^* = \frac{-\sqrt{2}}{s_1}\cos\left(T + \frac{\pi}{4} - \delta_1\right) \qquad (72)$$

on using the expressions (63), (64), and (70). The variation of the coefficient of skin-friction at the surface of the bounding permeable wall [i.e., given by Equation (72)], with slip grouping parameter β, is depicted in Figure 11 for a range of values of T in half-cycle. The behaviour seems to be consistent with wall-slip velocity observations (cf. Figures 5–9).

FLOW OVER A NATURALLY PERMEABLE WALL WHEN THE FREE STREAM AT INFINITY OSCILLATES

The velocity solution obtained in the previous section has an interesting interpretation. The fact that the equations governing incompressible flow are invariant under an "unsteady" Galilean transformation (see, for example, Reference 61), enables us to rephrase our problem and we now say that the permeable wall (or boundary) is "stationary," whilst the fluid at far distance from the permeable boundary is "oscillating." Of course, an oscillating pressure gradient from external sources is required to cause the free stream to oscillate.

TABLE 2. Variation of Factor and Phase-Lag with β.

β	Factor	δ_1 (degrees)
0	1.000	0
0.1	0.932	3.78
0.2	0.869	7.06
0.3	0.813	9.92
0.4	0.761	12.43
0.5	0.715	14.64
0.6	0.673	16.59
0.7	0.635	18.32
0.8	0.601	19.86
0.9	0.570	21.25
1.0	0.541	22.50

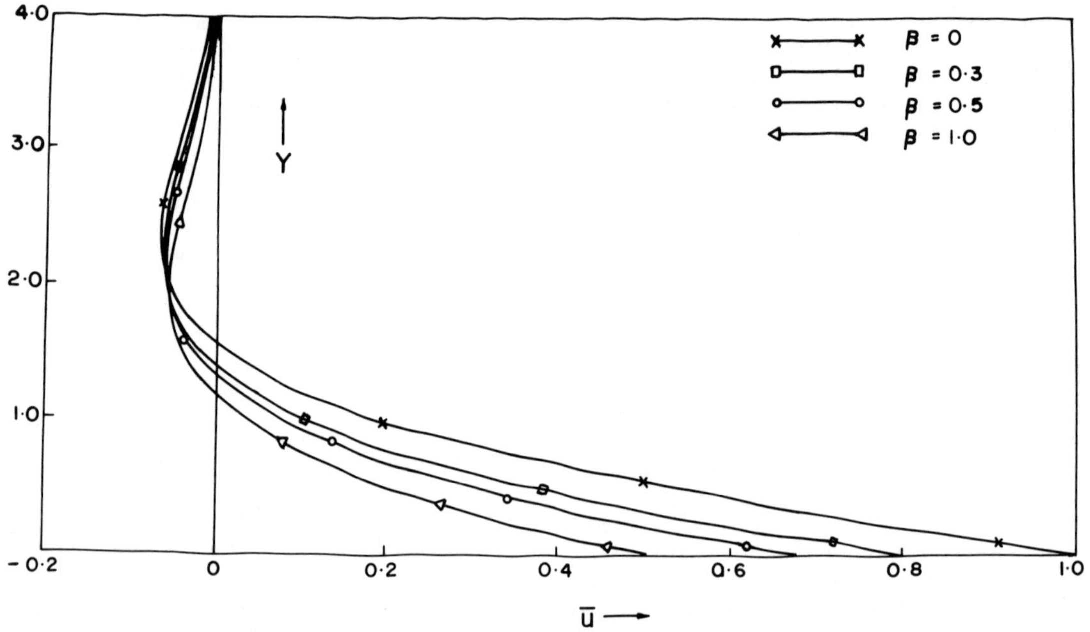

FIGURE 5. Velocity distribution above oscillating permeable wall for T = 0.

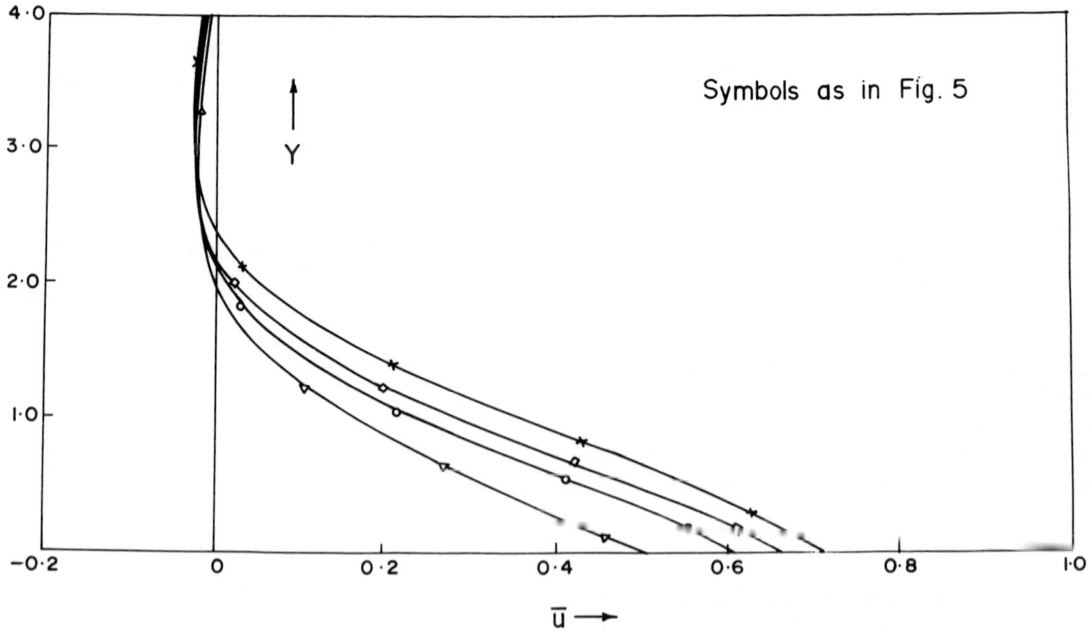

FIGURE 6. Velocity distribution above oscillating permeable wall for T = $\pi/4$.

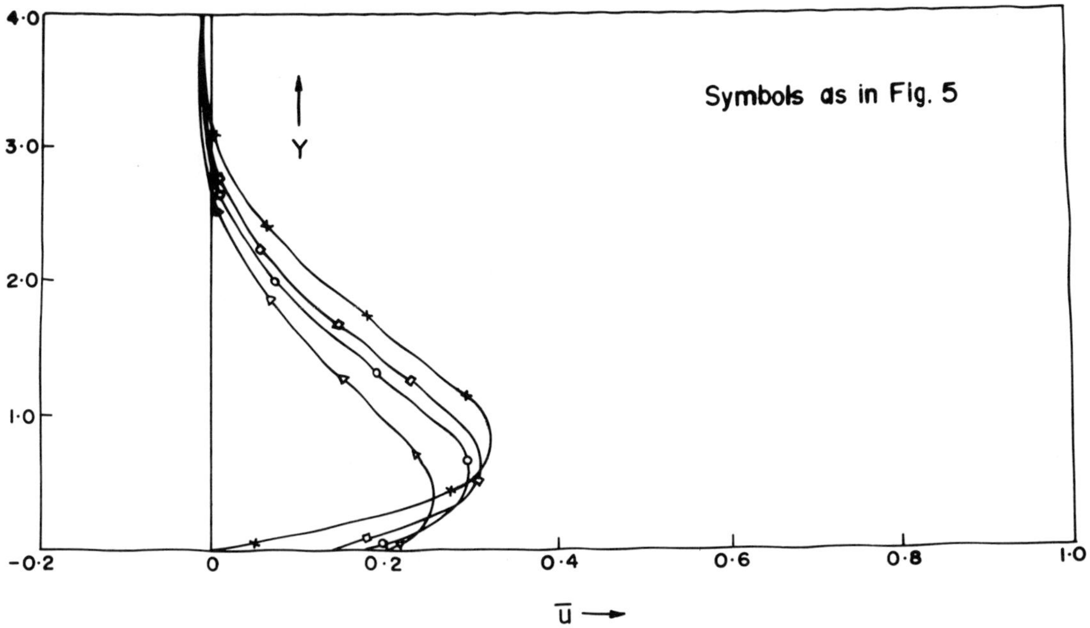

FIGURE 7. Velocity distribution above oscillating permeable wall for T = $\pi/2$.

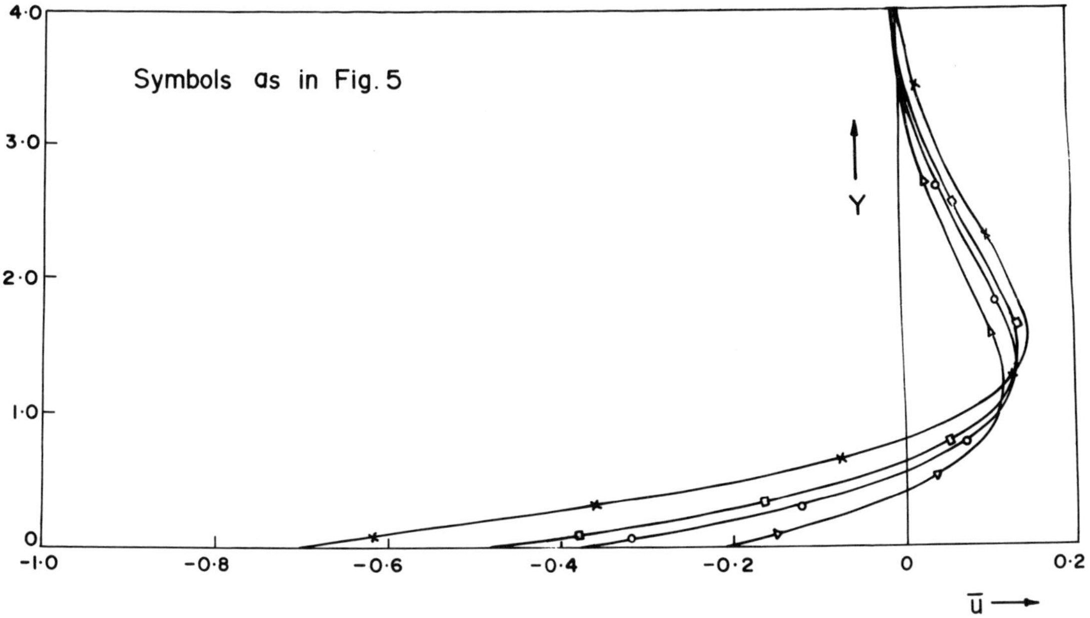

FIGURE 8. Velocity distribution above oscillating permeable wall for T = $3\pi/4$.

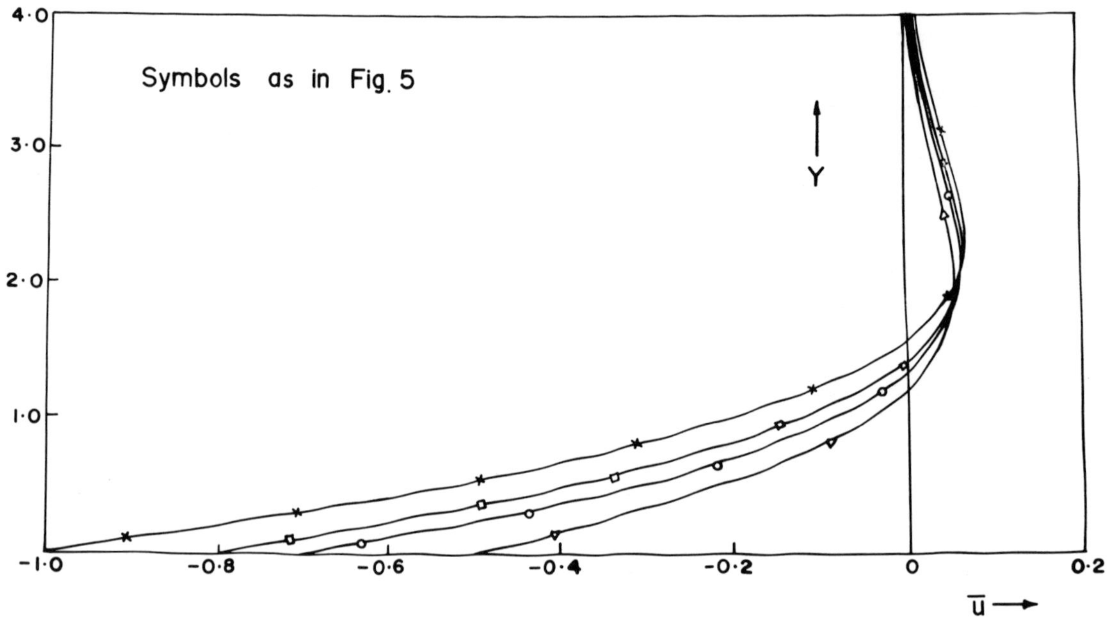

FIGURE 9. Velocity distribution above oscillating permeable wall for $T = \pi$.

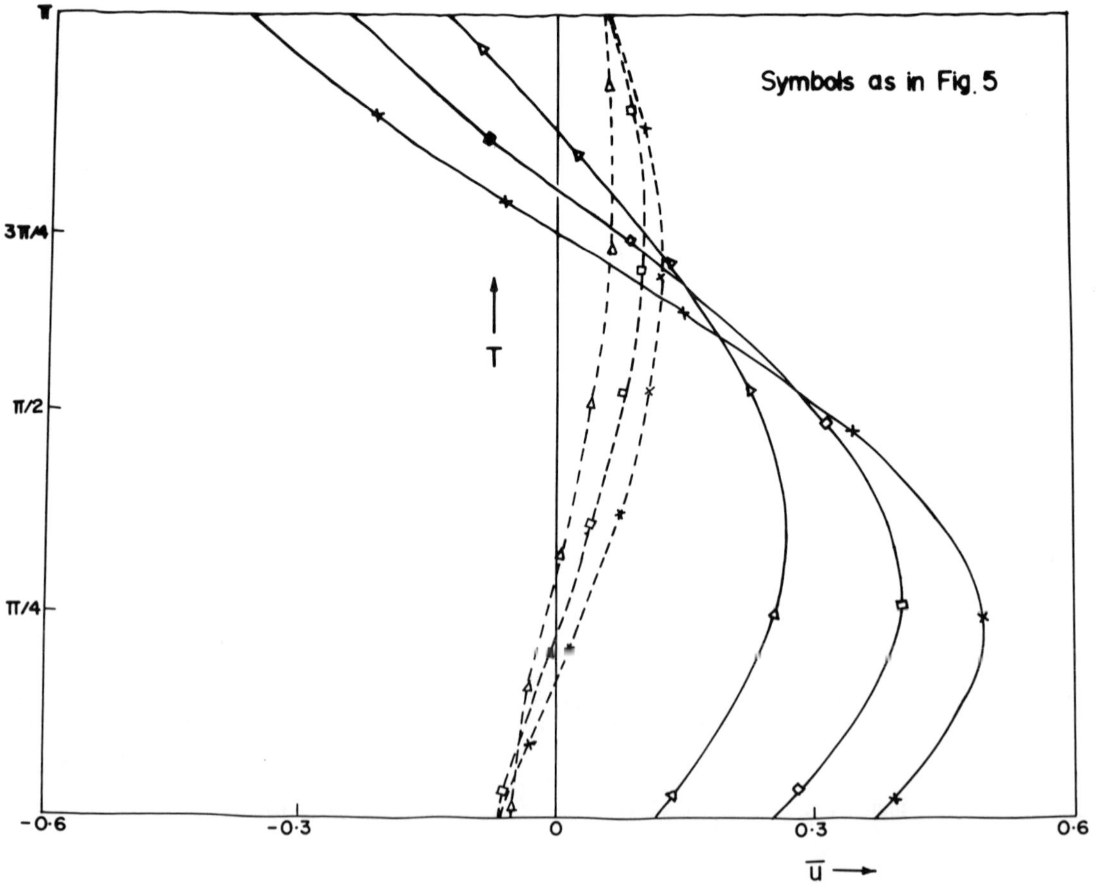

FIGURE 10. Velocity distribution above oscillating permeable wall (—$Y = 1/\sqrt{2}$, --- $Y = 3/\sqrt{2}$).

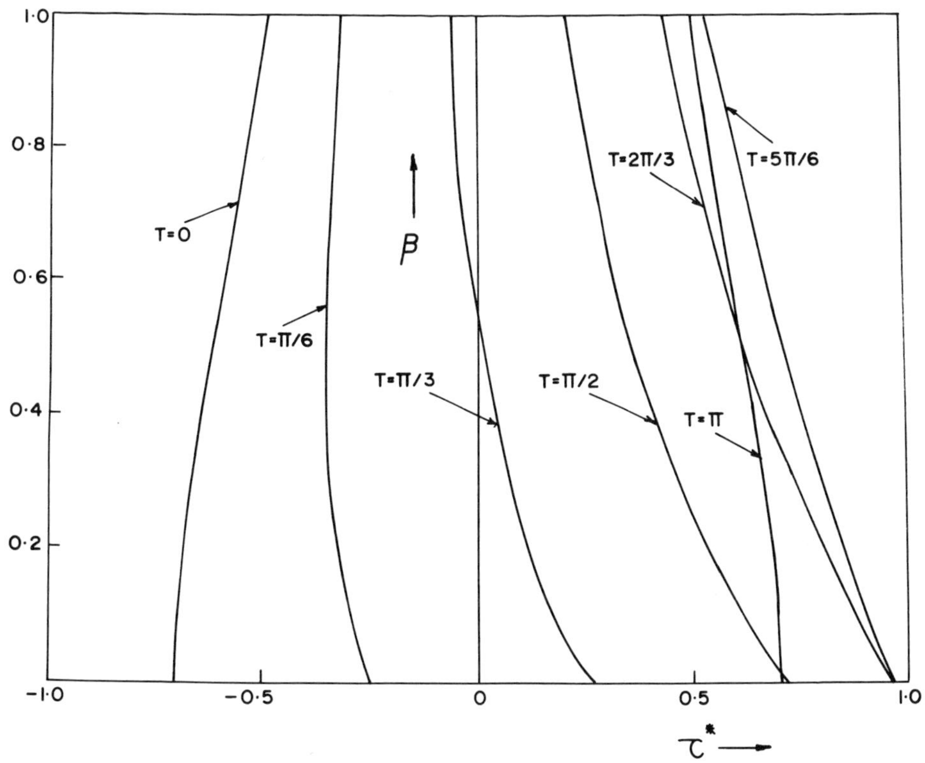

FIGURE 11. Coefficient of skin-friction at permeable wall.

Following Panton [61] closely, we analyse the oscillating wall problem of the previous sub-section as follows. Let the fluid velocity be denoted by $\bar{v}$ so that the fluid at infinity (i.e., $Y \to \infty$) is oscillating with a velocity

$$\bar{v}(Y \to \infty, T) = \cos T \tag{73}$$

Now since the wall ($Y = 0$) is stationary,

$$\frac{\partial \bar{v}}{\partial Y} = \frac{1}{\beta}\bar{v} \quad \text{at } Y = 0 \tag{74}$$

Note that Equation (74) is nothing but slip condition [i.e., Equation (8)] translated into non-dimensional variables [cf. Equation (64)]. We can now choose a new coordinate system that is fixed on a particle at infinity. The velocity v_∞ of the new system will be the same as the fluid velocity given by Equation (73). Thus,

$$v_\infty = \cos T \tag{75}$$

Now fluid velocity measured in the new system—denoted by $\hat{v}$—is related to the fluid velocity and coordinate-system velocity by

$$\hat{v} = \bar{v} - v_\infty \tag{76}$$

Therefore, the boundary conditions given by Equations (73) and (74) will assume the following form:

$$\hat{v}(\hat{Y} \to \infty, T) = \cos T - \cos T = 0 \tag{77}$$

$$\frac{\partial \hat{v}}{\partial \hat{Y}} = \frac{1}{\beta}(\hat{v} + \cos T) \quad \text{at } \hat{Y} = 0 \tag{78}$$

where $\hat{Y}$ is the non-dimensional vertical distance in the new coordinate system.

It may be noted that the boundary condition (78) in the new coordinate system is the boundary condition at the permeable wall (at $\hat{Y} = 0$) when the wall has been subject to $-\cos T$ type of notion. And to our aid is now the fact stated earlier that the governing equations in this new coordinate system will be unchanged in form. This immediately enables us to write the solution ($\hat{v}$) of the new problem as the "negative" of the solution given by Equation (63). Therefore, in view of Equations (75) and (76) solution to the origi-

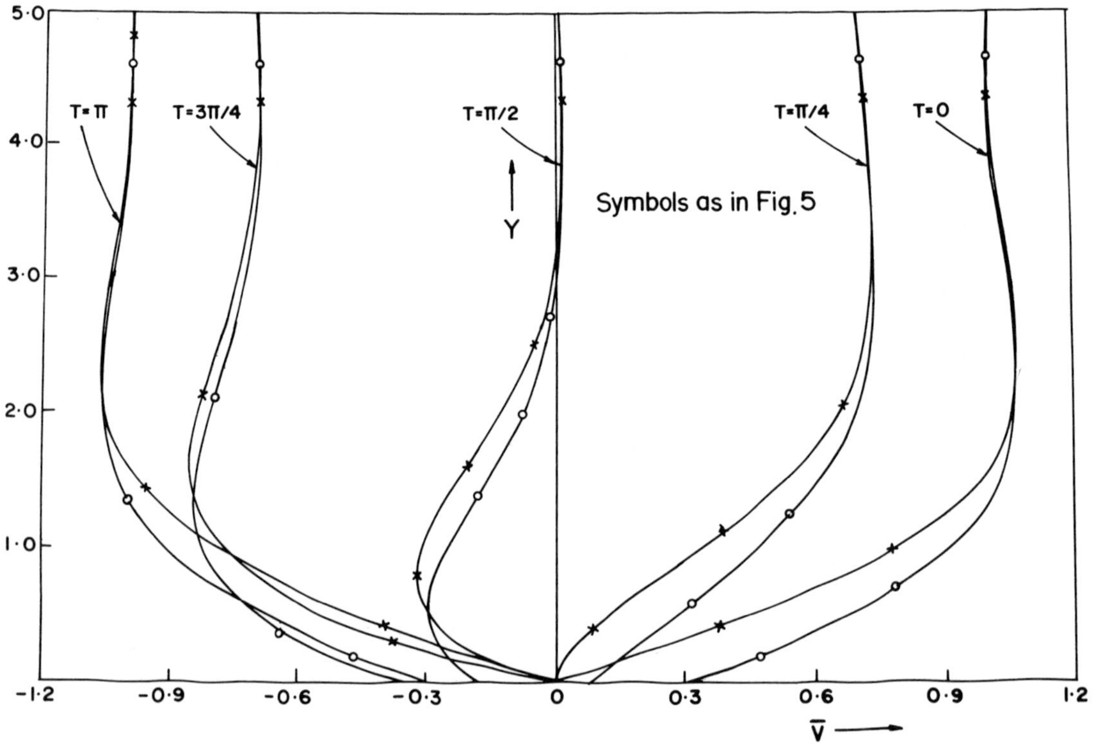

FIGURE 12. Velocity distribution above permeable wall when free stream at infinity oscillates.

nal problem (i.e., uniform stream oscillating above a stationary permeable wall) may be written as

$$\bar{v} = \cos T - \frac{e^{-Y}\cos(T - Y - \delta_1)}{s_1} \tag{79}$$

In Figure 12 are plotted velocity profiles [cf. Equation (79)] for various values of time. The most interesting observation here is the occurrence of the phenomenon of overshoot in the case of permeable boundary as well. It occurs at about $Y = 1.3$ (for $\beta = 0.5$). Thus the effect of bounding permeable wall is to displace the point of overshoot towards the wall as compared to the impermeable case where such an overshoot occurs at $Y \approx 1.6$ — about one-quarter a viscous length away from the wall.

Finally, the expression (79) above may be rearranged, using properties from trigonometry, to give a more convenient form:

$$\bar{v} = A'\cos(T + \Psi) \tag{80}$$

where

$$A' = \left[\frac{s_1^2 - 2s_1e^{-Y}\cos(Y + \delta_1) + e^{-2Y}}{s_1^2}\right]^{1/2} \tag{81}$$

and

$$\Psi = \tan^{-1}\left[\frac{e^{-Y}\sin(Y + \delta_1)}{s_1 - e^{-Y}\cos(Y + \delta_1)}\right] \tag{82}$$

TRANSIENT OSCILLATORY FLOW

Suppose that the fluid and the porous medium are initially at rest and then, for $t > 0$, oscillations of the type given by Equation (48) are imposed on the bounding porous medium. During the first cycle of the oscillations the velocity profile in the free fluid region will differ significantly from the steady-state solution (i.e., repetitive oscillations) given by Equation (63). We thus solve the governing Equation (49) subject to the following conditions:

$$u = 0 \quad \text{at } t = 0 \tag{83}$$

$$\frac{\partial u}{\partial y} = \frac{\alpha}{\sqrt{k}}(u - U\cos wt) \quad \text{at } y = 0,\ t > 0 \tag{84}$$

$$u \to 0 \quad \text{as } y \to \infty,\ t > 0 \tag{85}$$

The resulting solution in this case will be a sum of a steady-state solution and a transient solution. Note that the only additional condition introduced here is that given by Equation

(83); the others, i.e., Equations (83) and (84), are identical to those considered in the oscillatory solution case. We thus have an initial-value problem (I.V.P.) to solve. One of the most effective methods to solve such problems is by integral transform technique. We define the Laplace transform of velocity $u(y, t)$ by

$$\bar{u}(y,s) = \int_o^\infty e^{-st}u(y,t)dt \tag{86}$$

where s is a parameter.

The Laplace transforms of Equations (49), (84) and (85) are:

$$\frac{d^2\bar{u}}{dy^2} - \left(\frac{s}{\nu}\right)\bar{u} = 0 \tag{87}$$

$$\frac{d\bar{u}}{dy} = \frac{\alpha}{\sqrt{k}}\left(\bar{u} - \frac{Us}{w^2 + s^2}\right) \quad \text{at } y = 0 \tag{88}$$

$$\bar{u} \to 0 \quad \text{as } y \to \infty \tag{89}$$

The solution of Equation (87), subject to (88) and (89), is

$$\bar{u}(y,s) = \frac{U\alpha}{2\sqrt{k}}\left[\frac{e^{-\sqrt{(s/\nu)}y}}{s - iw} + \frac{e^{-\sqrt{(s/\nu)}y}}{s - iw}\right]\left(\frac{\alpha}{\sqrt{k}} + \sqrt{\frac{s}{\nu}}\right)^{-1} \tag{90}$$

Inverse Laplace transform of Equation (90) can be obtained by using standard tables of inverse transforms (e.g., Reference 19). The inversion of (90) gives velocity $u(y,t)$ as

$$\begin{aligned} u(y,t) = U\Bigg[&\frac{1}{2\sqrt{2}}e^{iT}\Bigg\{[(\sqrt{2} + \beta) \\ &+ i\beta]^{-1}e^{-Y(1+i)}\text{erfc}\left(\frac{Y}{\sqrt{2T}} - \sqrt{iT}\right) + [(\sqrt{2} - \beta) \\ &- i\beta]^{-1}e^{-Y(1+i)}\text{erfc}\ \frac{Y}{\sqrt{2T}} + \sqrt{iT}\Bigg\} \\ &+ \frac{1}{2\sqrt{2}}e^{-iT}\Bigg\{[(\sqrt{2} + \beta) \\ &- i\beta]^{-1}e^{-Y(1-i)}\text{erfc}\left(\frac{Y}{\sqrt{2T}} - \sqrt{-iT}\right) + [(\sqrt{2} \\ &- \beta) + i\beta]^{-1}e^{Y(1-i)}\text{erfc}\left(\frac{Y}{\sqrt{2T}} + \sqrt{-iT}\right)\Bigg\} \\ &- (1 + \beta^4)^{-1}e\left(\frac{\sqrt{2}\,Y}{\beta} + \frac{T}{\beta^2}\right)\text{erfc}\left(\frac{Y}{\sqrt{2T}} + \frac{\sqrt{T}}{\beta}\right)\Bigg] \end{aligned} \tag{91}$$

in which we have introduced non-dimensional variables given by Equation (64), and only real part has any meaning. Using the properties of complementary error function, we can rearrange various terms in Equation (91) to express $u(y,t)$ as:

$$\frac{u(y,t)}{U} = u_{st} + u_{tr} \tag{92}$$

where u_{st} is the steady-state solution given by Equation (63), and u_{tr}, the transient solution, is given by

$$\begin{aligned} u_{tr} = Re\Bigg[&\frac{1}{\sqrt{2}}e^{-iT}\Bigg\{[(\sqrt{2} - \beta) \\ &+ i\beta]^{-1}e^{Y(1-i)}\text{erfc}\left(\sqrt{-iT} + \frac{Y}{\sqrt{2T}}\right) - [(\sqrt{2} \\ &+ \beta) - i\beta]^{-1}e^{-Y(1-i)}\text{erfc}\left(\sqrt{-iT} - \frac{Y}{\sqrt{2T}}\right)\Bigg\}\Bigg] \\ &- (1 + \beta^4)^{-1}e\left(\frac{\sqrt{2}\,Y}{\beta} + \frac{T}{\beta^2}\right)\text{erfc}\left(\frac{Y}{\sqrt{2T}} + \frac{\sqrt{T}}{\beta}\right) \end{aligned} \tag{93}$$

In (93), Re means real part of the quantity within parentheses. Note that $u_{tr} \to 0$ as $T \to \infty$.

Particular Case

On taking the limit $k \to 0$ (i.e., $\beta \to 0$), the solution (92) gives corresponding transient Stokes flow for impermeable case as

$$\begin{aligned} u(y,t) = U\Bigg[&e^{-Y}\cos(T - Y) \\ &+ Re\Bigg\{\frac{1}{2}e^{-iT}\Bigg[e^{Y(1-i)}\ \text{erfc}\left(\sqrt{-iT} + \frac{Y}{\sqrt{2T}}\right) \\ &- e^{-Y(1-i)}\ \text{erfc}\left(\sqrt{iT} - \frac{Y}{\sqrt{2T}}\right)\Bigg]\Bigg\}\Bigg] \end{aligned} \tag{94}$$

the last term of the right-hand side of (93) tending to zero on employing asymptotic expansion of complementary error function.

Computation of u_{tr} from Equation (93) may be carried out by using tables of error function for complex arguments

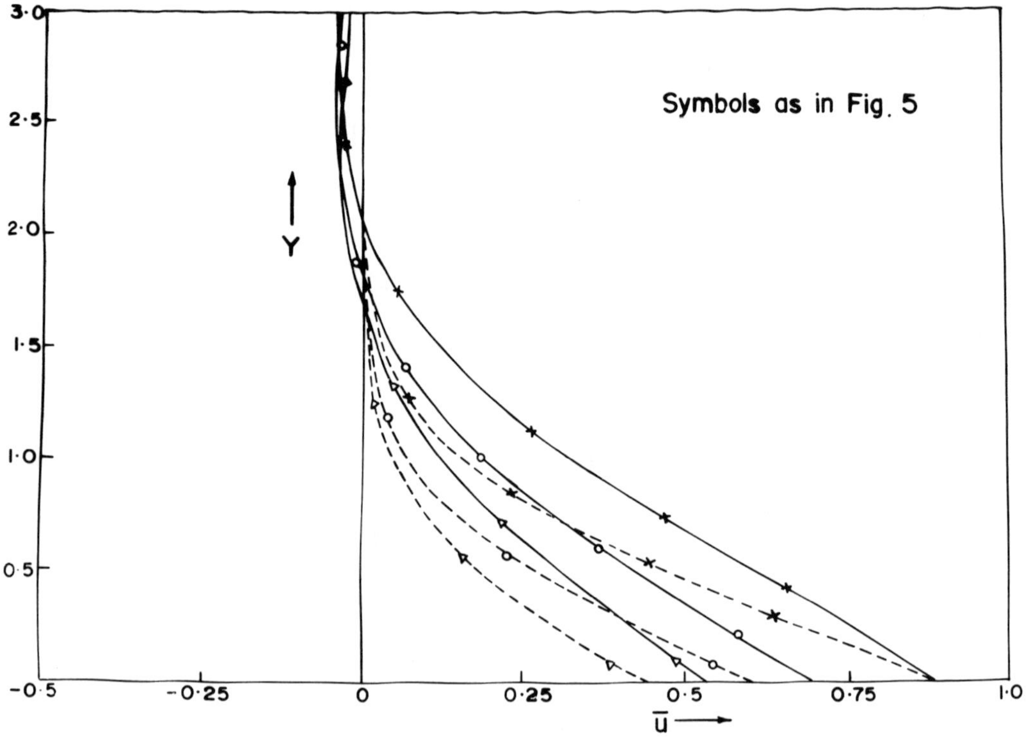

FIGURE 13. Velocity distribution above an oscillating permeable wall: Transient Stokes flow at T = 0.5 (—steady state flow, --- transient flow).

(e.g., Abramowitz and Stegun [1]) and resorting to linear interpolation for reasonable accuracy. The effect of parameter β on the combined solution given by Equation (92) has been depicted for $T = 0.5$ (Figure 12) and $T = 2.0$ (Figure 13). The corresponding steady state velocity profile [cf. Equation (63)] has also been plotted for a direct comparison. A comparison of Figure 12 and Figure 13 reveals that the combined solution approaches steady state value fairly quickly. In fact, the solution given by (93) can be shown to be significant only during the first cycle ($0 < T < 2\pi$) of the wall oscillations.

FLOW CAUSED BY AN IMPULSIVELY STARTED PERMEABLE WALL (RAYLEIGH LAYER)

For $w \to 0$, the Laplace inverse of Equation (90) [26] gives the velocity distribution associated with a porous medium moving with an impulsive velocity U, as

$$\frac{u}{U} = \left[\mathrm{erfc}\left(\frac{y}{2\sqrt{\nu t}}\right) - \exp\left(\frac{\alpha}{\sqrt{k}}\, y + \frac{\alpha^2 t\nu}{k}\right) \cdot \mathrm{erfc}\left(\frac{y}{2\sqrt{\nu t}} + \frac{\alpha}{\sqrt{k}}\sqrt{\nu t}\right)\right]$$

Introducing non-dimensional variables

$$\xi = \frac{y}{2\sqrt{\nu t}}$$

$$\delta = \frac{(k/\nu t)^{1/2}}{\alpha}$$

Equation (95) becomes

$$\frac{u}{U} = \left[\mathrm{erfc}(\xi) - \exp\left(\frac{2\xi}{\delta} + \frac{1}{\delta^2}\right) \cdot \mathrm{erfc}\left(\xi + \frac{1}{\delta}\right)\right] \tag{96}$$

The velocity distribution given by Equation (96) corresponds to Rayleigh layer associated with a porous medium bounding an infinite extent of viscous fluid. We are now interested in the case when δ is very small, i.e., when the permeability of the medium is very small. Using the asymptotic expansion for *erfc*(x) for large x, we obtain from Equation (96)

$$\frac{u}{U} \approx \left[\operatorname{erfc}(\xi) - \exp\left(\frac{2\xi}{\delta} + \frac{1}{\delta^2}\right) \times \frac{\exp\left(-\left\{\xi + \frac{1}{\delta}\right\}^2\right)}{\sqrt{\pi}\left(\xi + \frac{1}{\delta}\right)} \right] \tag{97}$$

$$= \operatorname{erfc}(\xi) - \frac{\exp(-\xi^2)}{\sqrt{\pi}\left(\xi + \frac{1}{\delta}\right)}$$

Note that we have neglected cubic and higher powers of $1/(\xi + 1/\delta)$ as they would have contributed very little for the range of values of δ we are interested in. From Equation (97) we can now easily arrive at the classical Rayleigh layer associated with a rigid wall, namely (Rosenhead [74]),

$$\frac{u}{U} = \operatorname{erfc}(\xi) \tag{98}$$

as

$$\frac{\exp(-\xi^2)\delta}{\sqrt{\pi}\,(\xi\delta + 1)} \to 0$$

when $\delta \to 0$ in Equation (97).

The velocity distribution as given by Equation (97) has been shown graphically for small values of δ (Figure 14), and one can easily notice that the effect of the permeability is to cause a significant reduction in the magnitude of velocity at all points in the neighbourhood of the permeable wall. But the overall pattern of the velocity profiles in permeable and impermeable cases is quite similar as one would expect for a moderately permeable medium.

Flow Between an Oscillating Permeable Wall and a Stationary Impermeable Upper Wall

Here we consider an extension of the problem investigated in the previous sub-section by assuming that the fluid is

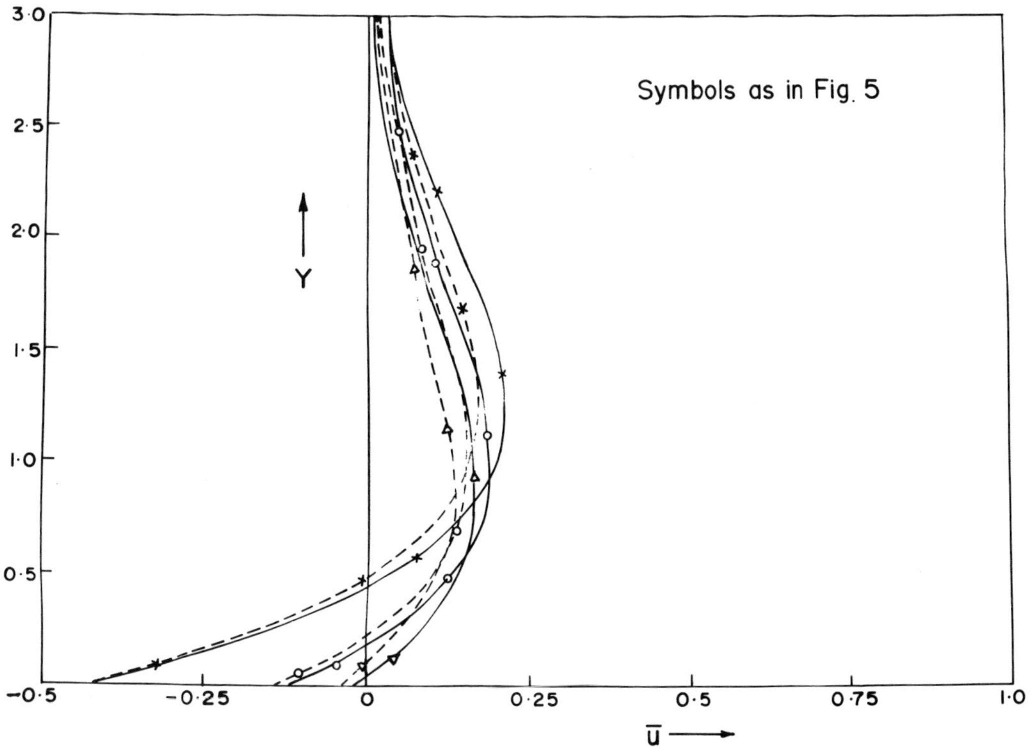

FIGURE 14. Velocity distribution above an oscillating permeable wall: Transient Stokes flow at T = 2.0 (—steady state flow, ---transient flow).

bounded above by an impermeable (i.e., rigid) wall. We shall give more emphasis to the oscillatory solution.

OSCILLATORY SOLUTION (MODIFIED STOKES LAYER)

Our analysis here is essentially similar to that given in the section Oscillatory Solution (Stokes Layer), and, therefore, we shall omit details. Because of an upper boundary, the boundary conditions given by Equations (52) and (53) now assume the form:

$$u = 0 \quad \text{at } y = d_1 \tag{99}$$

$$\frac{\partial u}{\partial y} = \frac{\alpha}{\sqrt{k}}(u - U\cos wt) \quad \text{at } y = 0 \tag{100}$$

where d_1 is the distance between two walls. Note that the condition stipulated in (99) refers to "no-slip" condition. The governing equation is still given by Equation (49). The final expression for the velocity $\bar{u}$ $(=u/U)$ can be written as

$$\bar{u} = \frac{1}{Dr_1}[e^{-Y}\{s_1\cos(T - Y - \delta_1) - s_2e^{-\theta}\cos(T - Y + \theta + \delta_2)\} + e^{Y}\{s_2e^{-2\theta}\cos(T + Y + \delta_2) - s_1 e^{-\theta}\cos(T + Y - \theta - \delta_1)\}] \tag{101}$$

where

$$Dr_1 = (1 + e^{-2\theta} - 2e^{-\theta}\cos\theta) + \beta^2(1 + e^{-2\theta} + 2e^{-\theta}\cos\theta)\sqrt{2}\,\beta(1 - e^{-2\theta} + 2e^{-\theta}\sin\theta) \tag{102}$$

$$s_2 = (1 + \beta^2 - \sqrt{2}\,\beta)^{1/2} \tag{103}$$

$$\delta_2 = \tan^{-1}\left[\frac{\beta}{(\sqrt{2} - \beta)}\right] \tag{104}$$

and

$$\theta = 2\eta d_1 \tag{105}$$

All other non-dimensional variables appearing in Equation (101) have been defined earlier. For $k \to 0$ (i.e., $\beta \to 0$) we can obtain from (101) the velocity distribution between two rigid plates, the lower one performing linear oscillations and the upper one at rest:

$$\bar{u} = \left\{ e^{-Y}[\cos(T - Y) - e^{-\theta}\cos(T - Y + \theta)] + \frac{e^{Y}[e^{-2\theta}\cos(T + Y) - e^{-\theta}\cos(T + Y + \theta)]}{(1 + e^{-2\theta} - 2e^{-\theta}\cos\theta)} \right\} \tag{106}$$

In Equation (101) one may notice substantial modification to the Stokes layer solution [i.e., Equation (63)] because of the presence of an upper boundary. Figures 16 and 17 represent velocity variations (vs. Y) at different times for cases $\theta = 0.5$ and $\theta = 1.0$, respectively. It is to be noted that as a consequence of increase in gap width (2 times here), the velocity at each point of the flow region increases. However, the increase in gap width tends to lower wall-slip values. The velocity profiles in both cases seem to be broadly linear, although the effect of both θ and β is to make them slightly non-linear, generally in the central region. It may be remarked that for $T = \pi/2$, the velocity profiles showed insignificant variation with parameter β, and hence have not been shown graphically.

Figures 18 and 19 show plots of velocity variation with T (during half-a-cycle) at $Y = 0.05$ and $Y = 0.15$, respectively, for several values of β and different gap-widths. We observe that keeping Y fixed, any increase in β causes magnitude of velocity to decrease at all times. Similar effect is observed with regard to increase in Y (Figure 19) for all T. At about $T = \pi/2$ velocities tend to become zero for all values of β—a finding which seems to be consistent with the observation made above.

Skin-Friction

The coefficient of skin-friction, τ^* [cf. Equation (71)] at the permeable boundary may be calculated using the velocity distribution given by Equation (101). Its expression in the present case is

$$\tau^* = \frac{\sqrt{2}}{Dr_1}\left\{ -s_1\left[e^{-\theta}\cos\left(T - \theta - \delta_1 + \frac{\pi}{4}\right) + \cos\left(T - \delta_1 + \frac{\pi}{4}\right)\right] + s_2\left[e^{-2\theta}\cos\left(T + \delta_2 + \frac{\pi}{4}\right) + e^{-\theta}\cos\left(T + \theta + \delta_2 + \frac{\pi}{4}\right)\right]\right\} \tag{107}$$

τ^* given by (107) has been plotted against β for $\theta = 0.5$ and $\theta = 1.0$ in Figures 20 and 21, respectively, at several times. It is noted that both gap-width and β have a tendency to decrease the magnitude of τ^*. Wall-slip effects shown in Figures 16 and 17 seem to agree well with τ^* curves for relevant values of β.

INITIAL-VALUE PROBLEM

One can easily extend the analysis of the section Transient Oscillatory Flow to the present case. But our aim here is to derive an expression for the modified Rayleigh layer, rather than discussing transient Stokes solution in detail which is considered beyond the scope of present investigation.

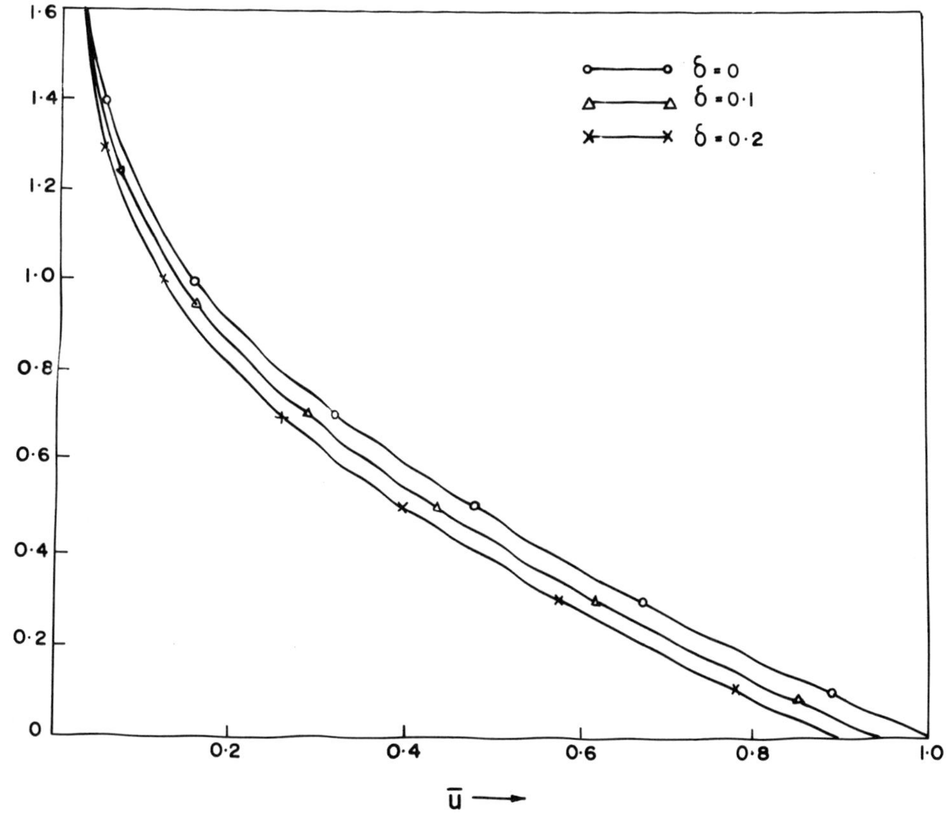

FIGURE 15. Velocity distribution above a suddenly accelerated permeable wall.

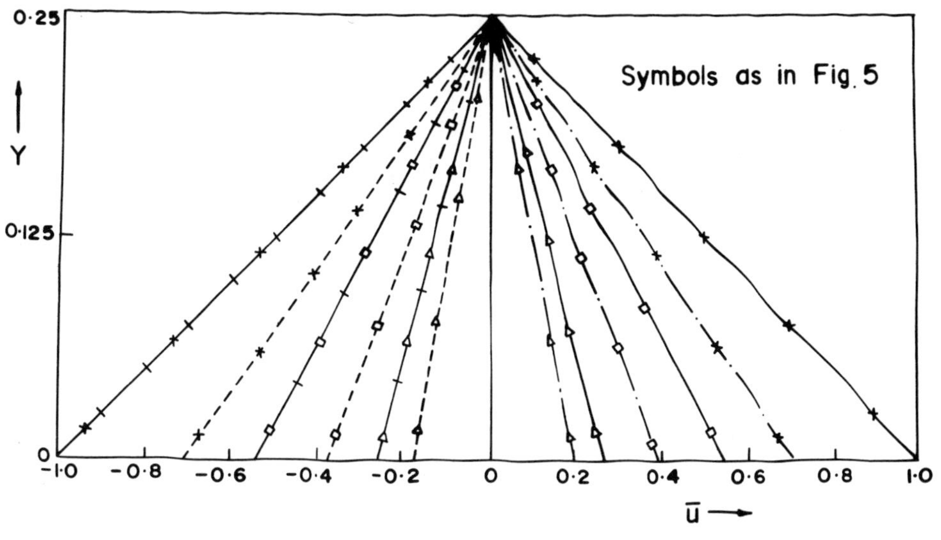

FIGURE 16. Velocity distribution between a permeable oscillating wall and a rigid wall ($\theta = 0.5$) (—T = 0, —·—·— T = $\pi/4$, --- T = $3\pi/4$, — T = π).

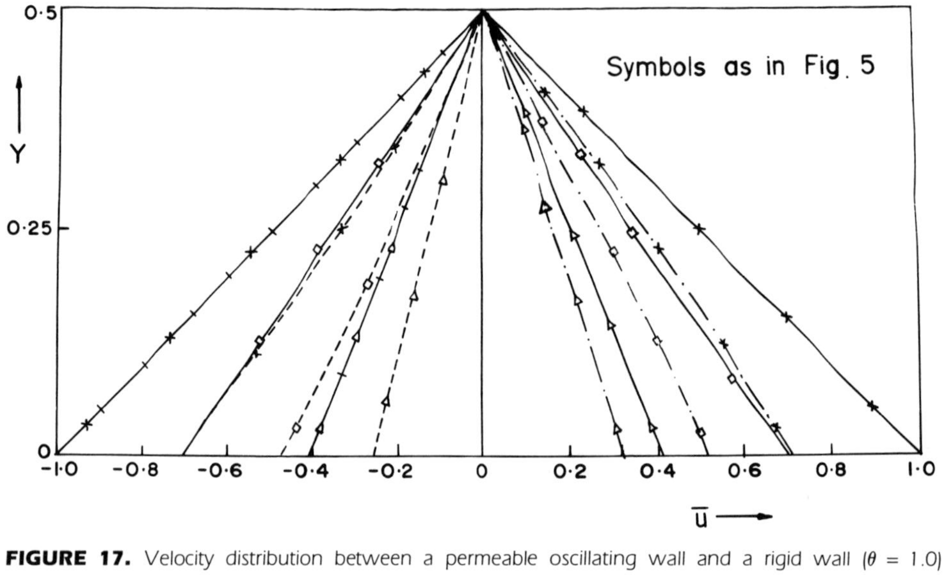

FIGURE 17. Velocity distribution between a permeable oscillating wall and a rigid wall (θ = 1.0) (—T = 0, —·—·— T = π/4, --- T = 3π/4, — T = π).

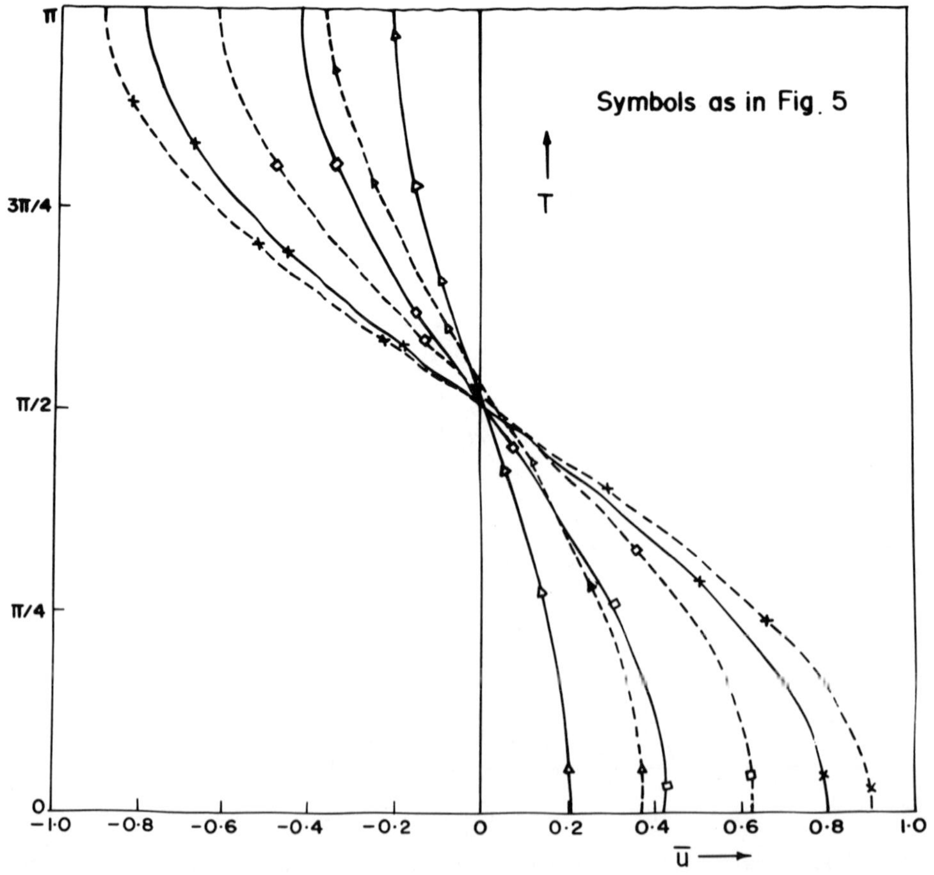

FIGURE 18. Velocity distribution between a permeable oscillating wall and a rigid wall (Y = 0.05) (--- θ = 1.0, — θ = 0.5).

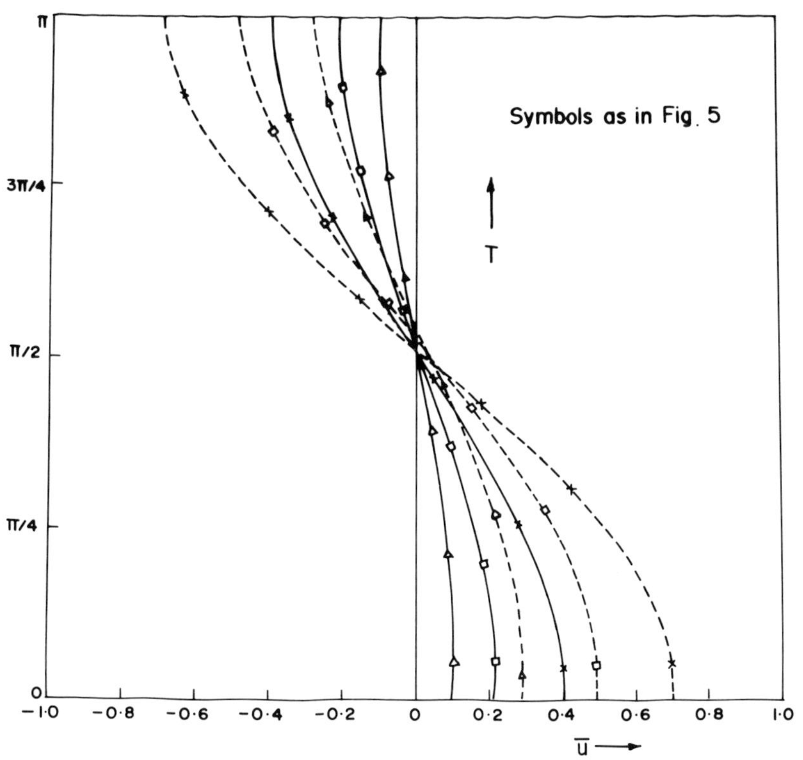

FIGURE 19. Velocity distribution between a permeable oscillating wall and a rigid wall (Y = 0.15) (--- θ = 1.0, — θ = 0.5).

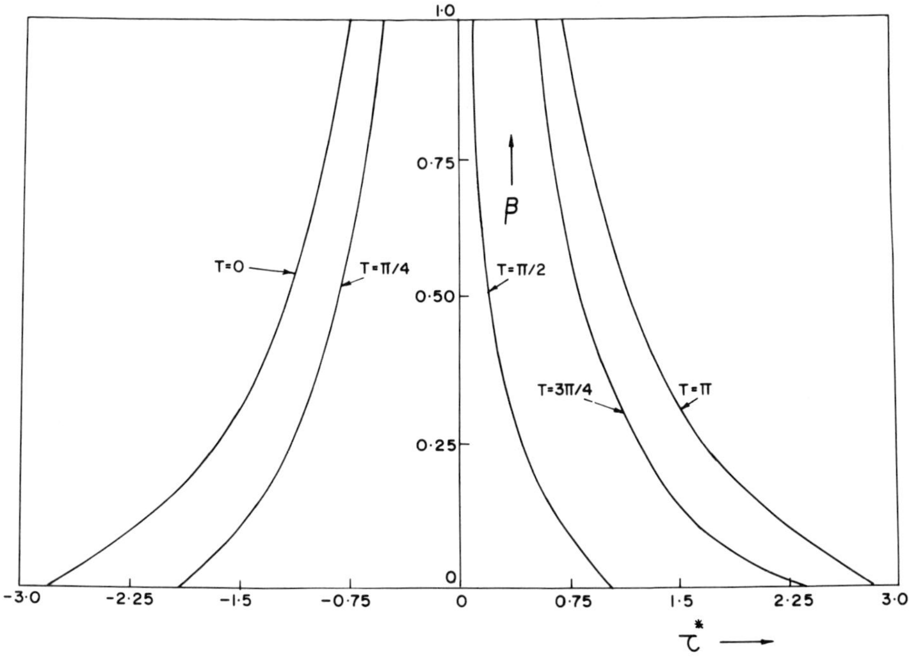

FIGURE 20. Coefficient of skin-friction at lower (permeable) wall (θ = 0.5).

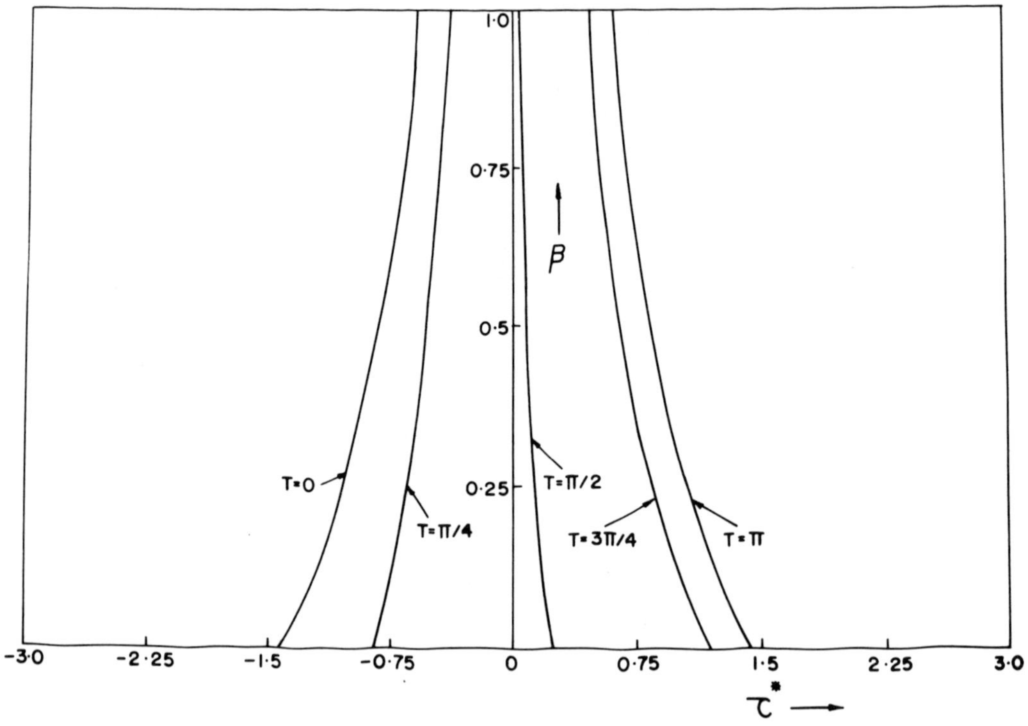

FIGURE 21. Coefficient of skin-friction at lower (permeable) wall ($\theta = 1.0$).

Governing equation still being the same [i.e., Equation (49)], the conditions corresponding to (83) to (85) will now be

$$u = 0 \quad \text{at } t = 0 \tag{108}$$

$$\frac{\partial u}{\partial y} = \frac{\alpha}{\sqrt{k}}(u - U\cos wt) \quad \text{at } y = 0,\ t > 0 \tag{109}$$

$$u = 0 \quad \text{at } y = d_1,\ t > 0 \tag{110}$$

The only change to be noticed is in Equation (110) which replaces condition at far distance from the plate. Following the various steps outlined in Transient Oscillatory Flow, we shall get an expression corresponding to Equation (90), i.e., transform $\bar{u}(y,s)$ as

$$\bar{u}(y,s) = \frac{\alpha U s}{\sqrt{k}} \frac{e^{\sqrt{(s/\nu)}(y-d_1)} - e^{-\sqrt{(s/\nu)}(y-d_1)}}{(s^2+w^2)\left\{\left(\frac{\alpha}{\sqrt{k}} - \sqrt{\frac{s}{\nu}}\right)e^{-\sqrt{(s/\nu)}d_1} - \left(\frac{\alpha}{\sqrt{k}} + \sqrt{\frac{s}{\nu}}\right)e^{-\sqrt{(s/\nu)}d_1}\right\}} \tag{111}$$

We shall not make any attempt to find inverse Laplace transform of the right-hand side of (111). We shall rather find its inverse transform in a particular case.

IMPULSIVE MOTION OF PERMEABLE WALL

On letting $w \to 0$ in Equation (111), we shall get

$$\bar{u}(y,s) = \frac{\alpha}{\sqrt{k}} U \frac{e^{\sqrt{(s/\nu)}(d_1-y)} - e^{-\sqrt{(s/\nu)}(d_1-y)}}{2s\frac{\alpha}{\sqrt{k}}\sinh\sqrt{\frac{s}{\nu}}d_1 + \sqrt{\frac{s}{\nu}}\cosh\sqrt{\frac{s}{\nu}}d_1} \tag{112}$$

We now attempt to invert Equation (112) for small values of time (i.e., for large values of s), the case in which we are more interested.

For large s approximation we can show that Equation (112) takes the form:

$$\bar{u}(y,s) = \frac{\alpha U}{\sqrt{k}\, s\left(\frac{\alpha}{\sqrt{k}} + \sqrt{\frac{s}{\nu}}\right)}\Big[e^{-\sqrt{(s/\nu)}y} - e^{-\sqrt{(s/\nu)}(2d_1-y)} + e^{-\sqrt{(s/\nu)}(2d_1+y)} - e^{-\sqrt{(s/\nu)}(4d_1-y)} \tag{113}$$

$$- \frac{2\sqrt{(s/\nu)}}{\left(\frac{\alpha}{\sqrt{k}} + \sqrt{\frac{s}{\nu}}\right)} \{ e^{-\sqrt{(s/\nu)}(2d_1 + y)} - e^{-\sqrt{(s/\nu)}(4d_1 - y)} \}$$

$$+ \ldots\ldots\Big]$$

The inverse Laplace transform (e.g., Reference 26) of the first four terms of the right-hand side of (113) then gives solution as:

$$\bar{u} = \Bigg\{ \mathrm{erfc}\left(\frac{Z}{2\sqrt{T'}}\right) - \exp(\sigma Z + \sigma^2 T')\mathrm{erfc}\left(\frac{Z}{2\sqrt{T'}} + \sigma\sqrt{T'}\right) - \mathrm{erfc}\left(\frac{2 - Z}{2\sqrt{T'}}\right) + \exp[\sigma(2 - Z) + \sigma^2 T']\mathrm{erfc}\left[\frac{(2 - Z)}{2\sqrt{T'}} + \sigma\sqrt{T'}\right] + \mathrm{erfc}\left(\frac{2 + Z}{2\sqrt{T'}}\right) + \exp[\sigma(2 + Z) + \sigma^2 T']\mathrm{erfc}\left[\frac{(2 + Z)}{2\sqrt{T'}} + \sigma\sqrt{T'}\right] - \mathrm{erfc}\left(\frac{4 - Z}{2\sqrt{T'}}\right) + \exp[\sigma(4 - Z) + \sigma^2 T']\mathrm{erfc}\left[\frac{(4 - Z)}{2\sqrt{T'}} + \sigma\sqrt{T'}\right]\Bigg\} \quad (114)$$

where we have introduced new dimensionless variables given by

$$Z = \frac{y}{d_1}$$

$$T' = \frac{\nu t}{d_1^2} \quad (115)$$

$$\sigma = \frac{\alpha d_1}{\sqrt{k}}$$

The expression (114) can be modified using the case of very small k. This entails use of asymptotic expansions of various complementary functions appearing in the Equation (114) for large σ. We shall finally get the following expression:

$$\bar{u} = \Bigg\{ \mathrm{erfc}\left(\frac{Z}{2\sqrt{T'}}\right) + \mathrm{erfc}\left(\frac{2 + Z}{2\sqrt{T'}}\right) - \mathrm{erfc}\left(\frac{2 - Z}{2\sqrt{T'}}\right) - \mathrm{erfc}\left(\frac{4 - Z}{2\sqrt{T'}}\right) - \frac{2\sqrt{T'}}{\sqrt{\pi}}\Bigg[(Z + 2\sigma T')^{-1} \exp\left(\frac{-Z^2}{(4T')}\right) + (2 + Z + 2\sigma T')^{-1} \exp\left(-\frac{(2 + Z)^2}{(4T')}\right) - (2 - Z + 2\sigma T')^{-1} \exp\left(-\frac{(2 - Z)^2}{(4T')}\right) - (4 - Z + 2\sigma T')^{-1} \exp\left(-\frac{(4 - Z)^2}{(4T')}\right)\Bigg]\Bigg\} \quad (116)$$

The Equation (114) or (116) gives modified Rayleigh layer for the flow between two walls, caused by the impulsive motion of the lower permeable wall. This flow may also be termed as "transient Couette flow" (cf. Reference 61). We have not shown graphically the velocity distribution given by Equation (114) or (116), but it can be easily plotted using tables of error functions (e.g., [1]).

Finally we can obtain Rayleigh layer between two impermeable plates by letting $\sigma \to \infty$ (i.e., $k \to 0$) in Equation (116). We shall get

$$\bar{u} = \left[\mathrm{erfc}\left(\frac{Z}{2\sqrt{T'}}\right) + \mathrm{erfc}\left(\frac{2 + Z}{2\sqrt{T'}}\right) - \mathrm{erfc}\left(\frac{2 + Z}{2\sqrt{T'}}\right) - \mathrm{erfc}\left(\frac{4 - Z}{2\sqrt{T'}}\right)\right] \quad (117)$$

Flow Caused by an Oscillating Permeable Wall Bounding a Fluid Layer of Constant Height

This is another extension of the problem investigated in sub-section Stokes and Rayleigh Layers over Naturally Permeable Boundaries. It is assumed here that the height of the fluid layer above the permeable wall is finite.

OSCILLATORY SOLUTION (MODIFIED STOKES LAYER)

We need to solve the governing Equation (49) subject to the following boundary conditions:

$$\frac{\partial u}{\partial y} = \frac{\alpha}{\sqrt{k}}(u - U \cos wt) \quad \text{at } y = 0 \quad (118)$$

$$\frac{\partial u}{\partial y} = 0 \quad \text{at } y = d_1 \quad (119)$$

The second of the above conditions, i.e., Equation (119), expresses the fact that shearing stress at the free-surface boundary must vanish.

The solution method here exactly followed that in Oscillatory Solution (Stokes Layer). We can express the final form of velocity as:

$$\bar{u} = \frac{1}{Dr_2}[e^{-Y}\{s_2e^{-\theta}\cos(T - Y + \theta + \delta_2) + s_1\cos(T - Y - \delta_1)\} + e^{-Y}\{s_1e^{-\theta}\cos(T + Y - \theta - \delta_1) + s_2e^{-2\theta}\cos(T + Y + \delta_2)\}] \quad (120)$$

where

$$Dr = (1 + 2e^{-\theta}\cos\theta + e^{-2\theta}) + \beta^2(1 - 2e^{-\theta}\cos\theta + e^{-2\theta}) + \sqrt{2}\,\beta(1 - 2e^{-\theta}\sin\theta - e^{-2\theta}) \quad (121)$$

and all non-dimensional quantities appearing in the Equation (120) have been defined earlier. For $k \to 0$ (or $\beta \to 0$) in the Equations (120) and (121) we can obtain velocity distribution in a fluid layer of finite thickness (or height) bounded by an oscillating "impermeable" boundary:

$$\bar{u} = \{e^{-Y}[e^{-\theta}\cos(T - Y + \theta) + \cos(T - Y)] + \frac{e^{Y}[e^{-\theta}\cos(T + Y - \theta) + e^{-2\theta}\cos(T + Y)]\}}{(1 + 2e^{-\theta}\cos\theta + e^{-2\theta})} \quad (122)$$

The velocity distribution given by Equation (120) may be thought of as a modified Stokes layer. Figures 22 and 23 show velocity variation with Y for $\theta = 0.5$ and $\theta = 1.0$, respectively, at different times. The most conspicuous feature here is that there is only little variation in velocity in the first case, i.e., $\theta = 0.5$. On the other hand the doubling of the fluid layer thickness seems to cause significant variation in velocity profiles. Between $T = 0$ to $T = \pi/4$ there seems to be a transition in the behaviour of velocity with regard to β—this feature is similarly indicated in both Figures 22 and 23. In the following figures (24 and 25) is shown variation of velocity with time, T, at different heights. Of particular interest is the region between $T = 0$ to $T = \pi/4$ in these plots, apparently accounting for the transition pointed above. In the post transition period the velocity magnitude increases with increase in parameter β, at both heights, i.e., $Y = 0.05$ and $Y = 0.15$.

Skin-Friction

The coefficient of skin-friction, τ^*, at the bounding lower permeable boundary in the present case is given by [using Equation (120)]:

$$\tau^* = \frac{\sqrt{2}}{Dr_2}\left\{s^1\left[e^{-\theta}\cos\left(T - \theta + \frac{\pi}{4} - \delta_1\right) - \cos\left(T + \frac{\pi}{4} - \delta_1\right)\right] + s_2\left[e^{-2\theta}\cos\left(T + \frac{\pi}{4} + \delta_2\right) - e^{-\theta}\cos\left(T + \theta + \frac{\pi}{4} + \delta_2\right)\right]\right\} \quad (123)$$

One may easily derive the corresponding expression for τ^* in the impermeable case by letting $\beta \to 0$. It is given by

$$\tau^*_{imp} = \sqrt{2}\left[e^{-\theta}\cos\left(T - \theta + \frac{\pi}{4}\right) - \cos\left(T + \frac{\pi}{4}\right) + \frac{e^{-2\theta}\cos\left(T + \frac{\pi}{4}\right) - e^{-\theta}\cos\left(T + \theta + \frac{\pi}{4}\right)}{(1 + 2e^{-\theta}\cos\theta + e^{-2\theta})}\right] \quad (124)$$

noting that for $\beta \to 0$, s_1 and $s_2 \to 1$ whilst $\delta_1, \delta_2 \to 0$. The plots of τ^* giving latter's variation with β at several values of time during half-a-cycle have been shown in Figures 26 and 27 for $\theta = 0.5$ and $\theta = 1.0$, respectively. One may notice that the increase in the thickness of the fluid layer above the permeable wall causes τ^* to vary rather non-linearly with β (cf. Figure 27).

INITIAL-VALUE PROBLEM

Here we solve Equation (49) subject to the initial and boundary conditions

$$u = 0 \quad \text{at } t = 0 \quad (125)$$

$$\frac{\partial u}{\partial y} = \frac{\alpha}{\sqrt{k}}(u - U\cos wt) \quad \text{at } y = 0, \quad t > 0 \quad (126)$$

$$\frac{\partial u}{\partial y} = 0, \quad \text{at } y = d_1,\ t > 0 \quad (127)$$

Once again we can follow the steps of Transient Oscillatory Flow, and obtain the following transformed equation for velocity:

$$\bar{u}(y,s) = \frac{\alpha U s}{\sqrt{k}} \frac{\cosh\sqrt{\frac{s}{\nu}}(d_1 - y)}{\left(\frac{\alpha}{\sqrt{k}}\cosh\sqrt{\frac{s}{\nu}}\,d_1 + \sqrt{\frac{s}{\nu}} \times \sinh\sqrt{\frac{s}{\nu}}\,d_1\right)(w^2 + s^2)} \quad (128)$$

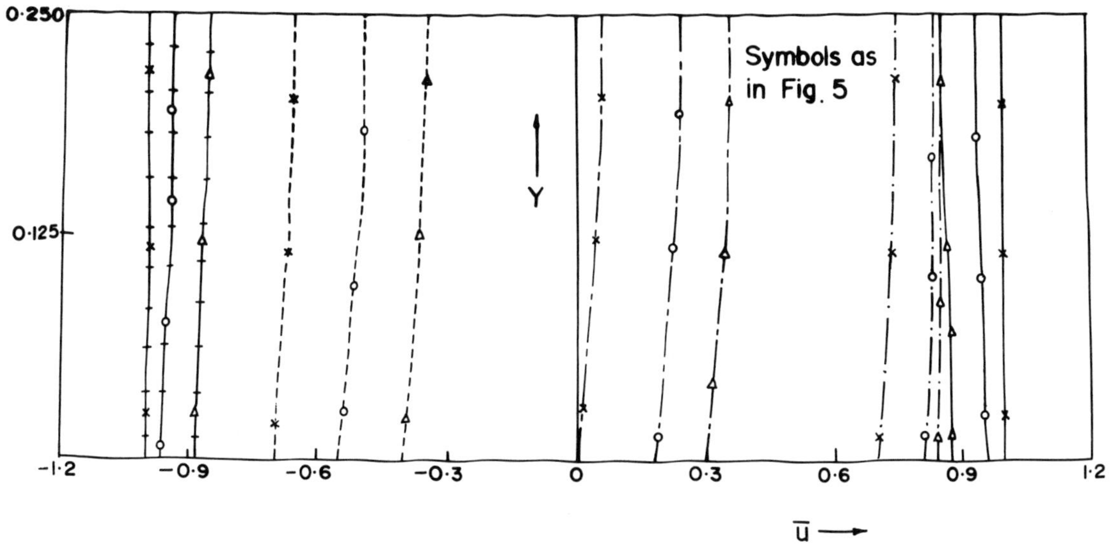

FIGURE 22. Velocity distribution in a fluid layer of finite thickness, bounded by a permeable wall (θ = 0.5) (—T = 0, —·—·— T = π/4, - - - T = π/2, --- T = 3π/4, —T = π).

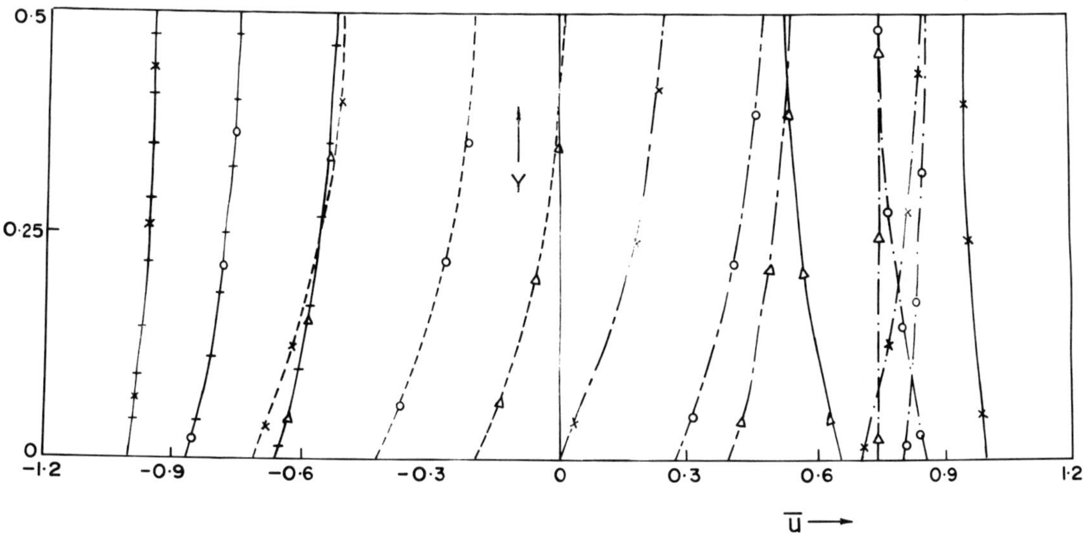

FIGURE 23. Velocity distribution in a fluid layer of finite thickness, bounded by a permeable wall (θ = 1.0) (—T = 0, —·—·— T = π/4, - - - T = π/2, --- T = 3π/4, —T = π). Note: Other symbols follow Figure 5.

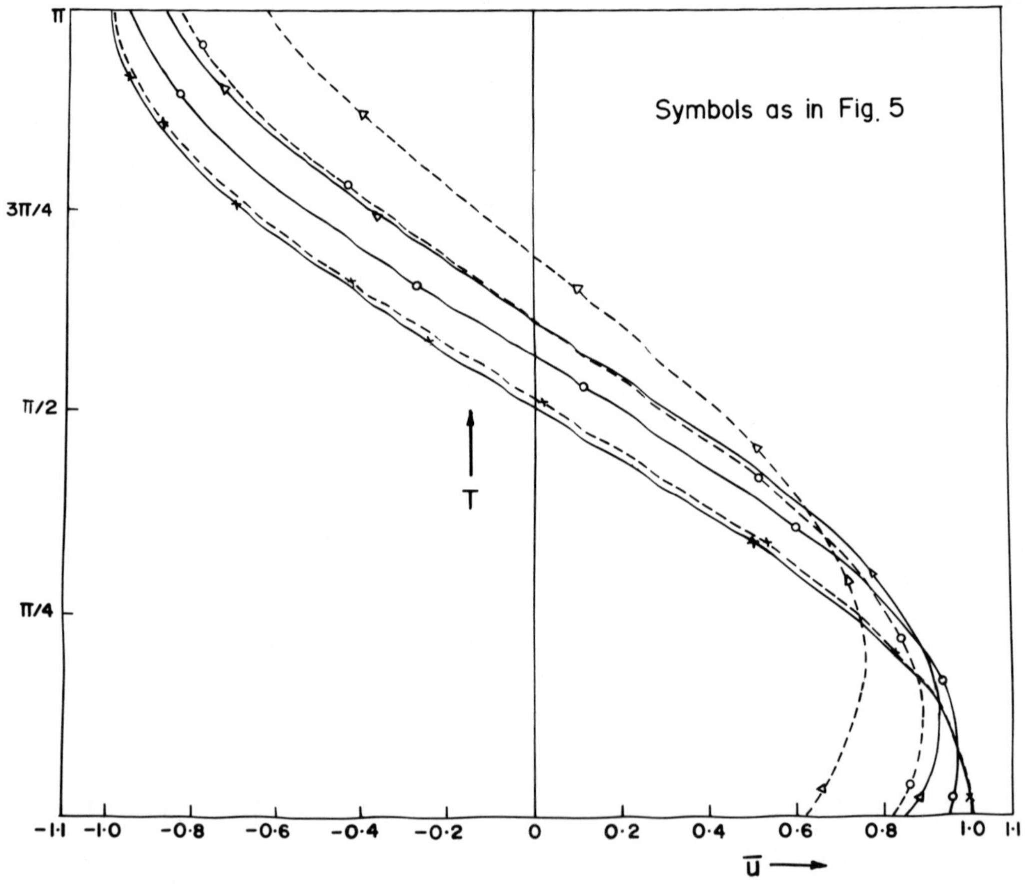

FIGURE 24. Velocity distribution in a fluid layer of finite thickness at fixed height (Y = 0.05). (--- θ = 1.0, — θ = 0.5).

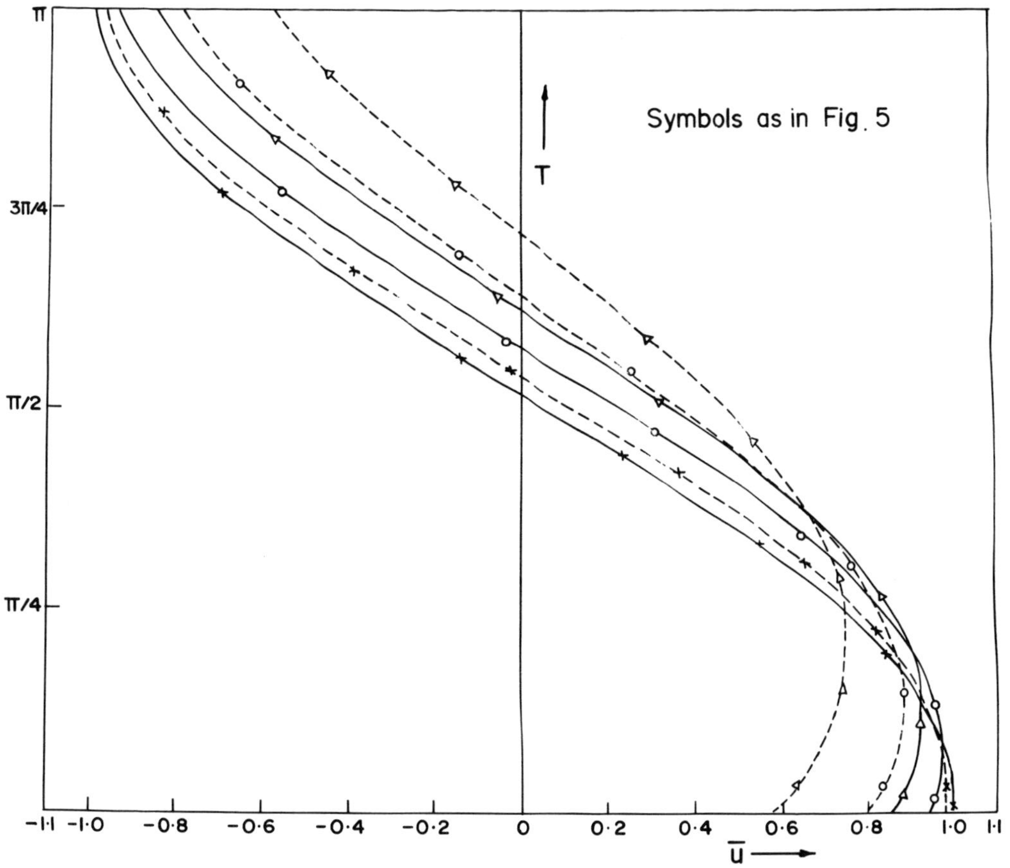

FIGURE 25. Velocity distribution in a fluid layer of finite thickness at a fixed height ($Y = 0.15$). (--- $\theta = 1.0$, — $\theta = 0.5$).

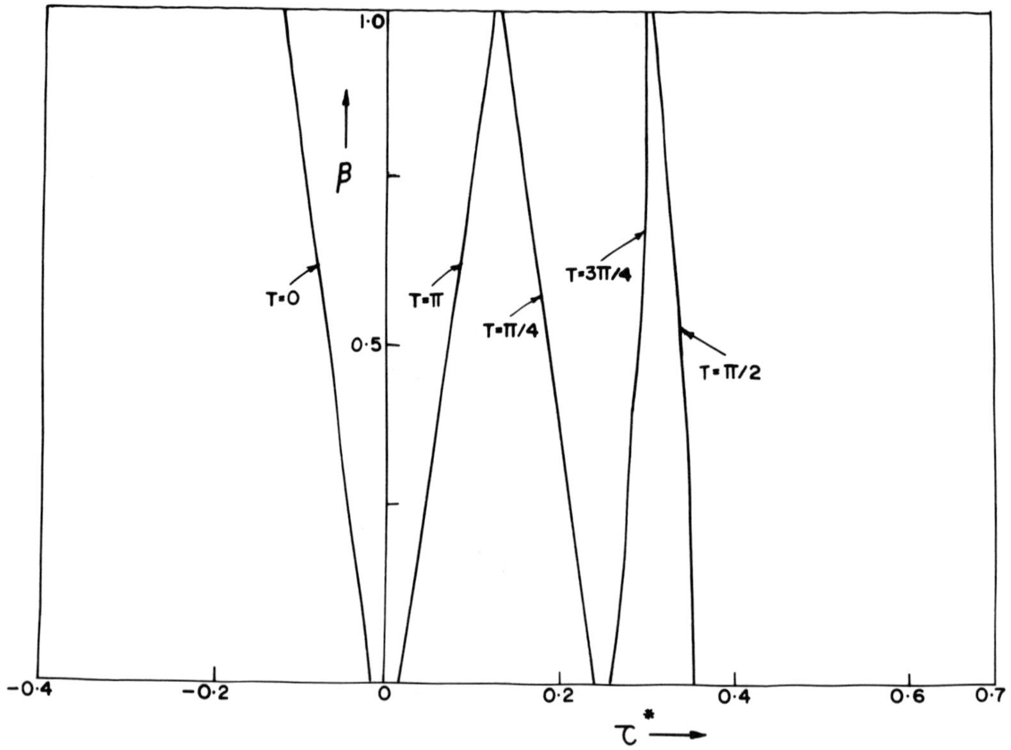

FIGURE 26. Coefficient of skin-friction at bounding permeable wall (θ = 0.5).

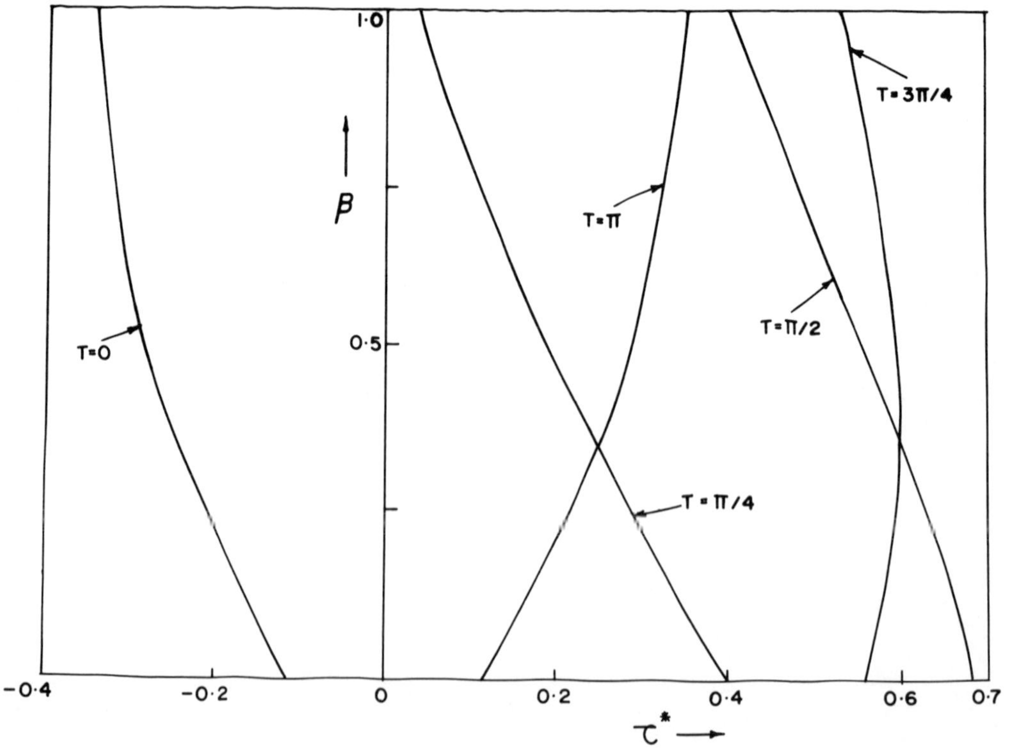

FIGURE 27. Coefificient of skin-friction at bounding permeable wall (θ = 1.0).

As indicated before, we shall be interested in the inverse transform of Equation (128) for a particular case. We may, however, remark that the inverse transform of (128) can be attempted for small values of time by using large s approximation in the right-hand side of Equation (128). We give below the resulting expression $\bar{u}(y,s)$ for such an approximation:

$$\bar{u}(y,s) = \frac{\alpha U}{\sqrt{k}}\left[\frac{1}{s - iw} + \frac{1}{s + iw}\right] \times \left\{ \frac{[e^{-\sqrt{(s/\nu)}\,(2d_1 - y)} + e^{-\sqrt{(s/\nu)}\,y}]}{\left(\frac{\alpha}{\sqrt{k}} + \sqrt{\frac{s}{\nu}}\right)} - \frac{[e^{-\sqrt{(s/\nu)}\,(4d_1 - y)} + e^{-\sqrt{(s/\nu)}\,(2d_1 + y)}]}{\left(\frac{\alpha}{\sqrt{k}} + \sqrt{\frac{s}{\nu}}\right)} + \ldots\ldots \right\} \tag{129}$$

Inversion of Equation (129) can now be affected by using tables of inverse Laplace transform (e.g., Reference 19). The resulting velocity distribution then can be analysed for steady-state and transient parts, following Transient Oscillatory Flow. But we shall leave it to the interested readers.

IMPULSIVE MOTION OF PERMEABLE WALL

We let $w \to 0$ in Equation (128) and obtain

$$\bar{u}(y,s) = \frac{\alpha U \cosh \sqrt{\frac{s}{\nu}}\,(d_1 - y)}{\sqrt{k}s\left[\frac{\alpha}{\sqrt{k}} \cosh \sqrt{\frac{s}{\nu}}\, d_1 + \sqrt{\frac{s}{\nu}} \sinh \sqrt{\frac{s}{\nu}}\, d_1\right]} \tag{130}$$

The inversion of Equation (130) for large s (i.e., small values of time) can be affected in a manner similar to that followed in the sub-section Impulsive Motion of Permeable Wall. Alternatively, one can readily obtain approximated expression $\bar{u}(y,s)$ by putting $w = 0$ in Equation (129) and the resulting expression may then be inverted. In either case we shall obtain the velocity distribution (after inversion) as

$$\begin{aligned}\bar{u} = \Bigg[&\operatorname{erfc}\left(\frac{Z}{2\sqrt{T'}}\right) - \exp(\sigma Z + \sigma^2 T')\operatorname{erfc}\left(\frac{Z}{2\sqrt{T'}} + \sigma\sqrt{T'}\right) + \operatorname{erfc}\left(\frac{2-Z}{2\sqrt{T'}}\right) - \exp\{\sigma(2-Z) \\ &+ \sigma^2 T'\}\operatorname{erfc}\left\{\frac{(2-Z)}{2\sqrt{T'}} + \sigma\sqrt{T'}\right\} - \operatorname{erfc}\left(\frac{2+Z}{2\sqrt{T'}}\right) \\ &+ \exp\{\sigma(2+Z) + \sigma^2 T'\}\operatorname{erfc}\left\{\frac{(2+Z)}{2\sqrt{T'}} + \sigma\sqrt{T'}\right\} \\ &- \operatorname{erfc}\left(\frac{4-Z}{2\sqrt{T'}}\right) + \exp\{\sigma(4-Z) \\ &+ \sigma^2 T'\}\operatorname{erfc}\left\{\frac{(4-Z)}{2\sqrt{T'}} + \sigma\sqrt{T'}\right\}\Bigg]\end{aligned} \tag{131}$$

The velocity distribution given by Equation (131) gives Rayleigh layer over a permeable wall which bounds a fiuid layer of finite thickness. The problem considered above has indirect analogy with conduction of heat in a bar of finite length allowing for radiation at either ends (cf. Carslaw and Jaeger [20]). Equation (131) can also be subjected to small k (i.e., large σ) approximation, and expressions corresponding to Equations (116) and (117) can be obtained.

ACKNOWLEDGEMENTS

Authors wish to dedicate this chapter to their late fathers who were a perennial source of inspiration. One of us (NCS) thankfully acknowledges financial assistance received from CSIR (India). He is grateful to Prof. M. L. Mathur, Head of the Mechanical Engineering Department, University of Jodhpur, Jodhpur, for encouragement and for providing facilities; to Mr. S. K. Bhargava, Assistant Professor, Department of Mechanical Engineering, University of Jodhpur, Jodhpur, for his help during several stages of this work. The other author (BSB) wishes to acknowledge kind assistance received from the Department of Mathematics, University of the West Indies, St. Augustine, Trinidad, West Indies; in particular he would like to thank Dr. David Owen.

NOTATION

- A = area of cross section
- A' = amplitude [Equation (81)]
- B_s = nondimensional parameter
- d_1 = distance in vertical direction
- Dr_1 = nondimensional parameter [Equation (102)]
- Dr_2 = nondimensional parameter [Equation (121)]
- $e_{r\theta}$ = rate of strain tensor
- f_1 = Maxwell's relaxation coefficient

g = acceleration due to gravity
h = height of film thickness
h_1,h_2 = pressure heads
H = bed thickness
k = permeability
K = a constant
K' = a constant
L = mean free path
L' = given by Equation (27) (constant)
m = number of capillaries
n = direction normal to surface
$\underset{\sim}{n}$ = unit normal vector
O_+ = boundary limit point
p = pressure in free fluid
p_1,p_2 = pressures at two specific points
$\bar{p}$ = pressure in porous medium
p' = pressure gradient ($= dp/dx$)
P = porosity
$\underset{\sim}{q}$ = filter velocity
$\vec{q}$ = filter velocity vector
$\underset{\sim}{Q}$ = flux
$\bar{Q}$ = average filter velocity
s = Laplace transform parameter
t = time
T = nondimensional time ($= wt$)
T' = nondimensional time ($= \nu t/d_1^2$)
u = velocity in free fluid region
u_B = free fluid velocity at boundary
$\bar{u}(y,s)$ = Laplace transform of u
$\bar{u}$ = nondimensional velocity ($= u/U$)
$\hat{u}$ = velocity in boundary layer (free fluid region)
$\tilde{u}$ = velocity in boundary layer (porous region)
$\bar{u}_1$ = velocity in boundary layer [Equation (19)]
$\underset{\sim}{U}$ = constant velocity
$\bar{U}$ = interstitial-averaged fluid velocity within porous region
v_∞ = velocity of the coordinate system fixed on a particle at infinity
$\bar{v}$ = nondimensional fluid velocity
$\hat{v}$ = nondimensional fluid velocity in a coordinate system fixed on a particle at infinity
V = velocity vector
$\underset{\sim}{w}$ = frequency of oscillations
x = axial coordinate
y = ordinate
Y = nondimensional ordinate ($= \eta y$)
$\hat{Y}$ = nondimensional ordinate in a coordinate system fixed on a particle at infinity
Z = nondimensional ordinate ($= y/d_1$)
ϵ = a parameter [Equation (17)]
α = slip parameter
β = slip grouping parameter ($= \sqrt{B_s}/\alpha$)
δ = nondimensional parameter [$= (k/\nu t)^{1/2}/\alpha$]
δ_1 = phase lag ($= \tan^{-1}[\beta/(\beta + \sqrt{2})]$
δ_2 = phase lag ($= \tan^{-1}[\beta/(-\beta + \sqrt{2})]$
$\bar{\delta}$ = average pore diameter
η = parameter ($= \sqrt{w/2\nu}$)
η_1 = distance between two neighbouring particles
λ = parameter [$= (iw/\nu)^{1/2}$]
λ_1 = parameter
λ_2 = parameter
μ = coefficient of viscosity
$\tilde{\mu}$ = effective viscosity
μ_1,μ_2 = viscosities of two different fluids
ν = kinematic coefficient of viscosity
$\varnothing$ = unknown function of y
σ = nondimensional parameter ($= \alpha d_1/\sqrt{k}$)
Ψ = phase angle [defined by Equation (82)]
ξ = parameter [$= y/2\sqrt{\nu t}$)]
τ = skin-friction
τ^* = coefficient of skin-friction
θ = nondimensional parameter ($= 2\eta d_1$)

Subscript

imp = impermeable
∞ = infinity
r,θ = spherical polar coordinate system
st = steady-state
tr = transient
t = tangential
w = wall

REFERENCES

1. Abramowitz, M. and I. A. Stegun, eds. *Handbook of Mathematical Functions.* New York:Dover Publications (1965).
2. Bear, J. *Dynamics of Fluids in Porous Media.* New York: American Elsevier Pub. Co. (1972).
3. Beavers, G. S. and D. D. Joseph, "Boundary Conditions at a Naturally Permeable Wall," *Journal of Fluid Mechanics,* Vol. 30, 197–207 (1967).
4. Beavers, G. S., E. M. Sparrow and R. A. Magnuson, "Experiments on Coupled Parallel Flows in a Channel and Bounding Porous Medium," *Journal of Basic Engineering, Transactions ASME,* Vol. 92, 843–848 (1970).
5. Beavers, G. S., E. M. Sparrow and B. A. Masha, "Boundary Conditions at a Porous Surface which Bounds a Fluid Flow," *AIChE Journal,* Vol. 20, 596–597 (1974).
6. Benner, F. C., W. W. Riches and F. E. Bartell, in *Fundamental Research on Occurrence and Recovery in Petroleum,* American Petroleum Institute, p. 74 (1943).
7. Bhatt, B. S., "Comments on the Effect of Velocity Slip on the Squeeze Film Between Rotating Porous Annular Discs," *Wear,* Vol. 56, 421–422 (1979).

8. Bhatt, B. S., "On the Analogy in Slip Flows—II," *Applied Scientific Research,* Vol. 35, 37–41 (1979).
9. Bhatt, B. S., "Relative Motion of Liquid and Porous Spheres," presented at the August 8–12, 1983, American Mathematical Society 805 meeting, held at Albany, New York.
10. Bhatt, B. S., "The Movement of Large Liquid Bubbles in Vertical Porous Tubes," *Rheological Acta,* Vol. 22, 588-591 (1983).
11. Bhatt, B. S., "Slow Rotation of a Porous Sphere," presented at the November 11–12, 1983, American Mathematical Society 807 Meeting, held at San Luis Obispo, California.
12. Bhatt, B. S. and N. C. Sacheti, "Comments on Channel and Tube Flows With Surface Mass Transfer and Velocity Slip," *Physics of Fluids*, Vol. 20, 1380–1381 (1977).
13. Bhatt, B. S. and N. C. Sacheti, "Comments on the Limitations of the Reynolds Equation for Porous Thrust Bearing," *Wear*, Vol. 53, 377–379 (1979).
14. Bhatt, B. S. and N. C. Sacheti, "On the Analogy in Slip Flows," *Indian Journal of Pure and Applied Mathematics*, Vol. 10, 303–306 (1979).
15. Bhatt, B. S. and R. L. Verma, "The Generalized Reynolds Equation for Porous Bearings," *Wear*, Vol. 59, 345–354 (1980).
16. Blake, J. R., P. G. Vann and H. Winet, "A Model of Ovum Transport," *Journal of Theoretical Biology*, Vol. 102, 145–166 (1983).
17. Brenner, H. and P. M. Adler, *Transport Process in Porous Media*, New York:McGraw-Hill (1983).
18. Brinkman, H. C., "On the Permeability of Media Consisting of Closely Packed Porous Particles," *Applied Scientific Research*, Vol. A1, 81–86 (1947).
19. Campbell, G. A. and R. M. Foster, *Fourier Integrals for Practical Applications*, New York:Bell Telephone Laboratories (1931).
20. Carslaw, H. S and J. C. Jaeger, *Operational Methods in Applied Mathematics*, New York:Dover Publications, Inc., p. 276 (1948).
21. Collins, R. E., *Flow of Fluids Through Porous Materials*, New York:Reinhold Publishing Co. (1961).
22. Cooper, P. F. and B. Atkinson, *Biological Fluidised Bed Treatment of Water and Wastewater*, Ellis Horwood Limited, West Sussex, England (1984).
23. Cunningham, R. E. and R. J. Williams, *Diffusion in Gases and Porous Media*, New York:Plenum Press (1980).
24. Darcy, H., "Les Fontaines Publiques de La Ville de Dijon," Dalmont, Paris (1856).
25. De Wiest, R. J. M., *Flow Through Porous Media*, New York:Academic Press (1969).
26. Erdelyi, A., W. Magnus, F. Oberhettinger and F. G. Tricomi, *Tables of Integral Transforms*, Vol. 1, New York:McGraw-Hill (1954).
27. Fitzgerald, J. M., "Mechanics of Red Cell Motion Through Very Narrow Capillaries," *Proceedings of the Royal Society*, Vol. B 174, 193–227 (1969).
28. Fung, Y. C., "On the Foundation of Biomechanics," *Journal of Applied Mechanics, Transactions ASME*, Vol. 50, 1003–1009 (1983).
29. Fung, Y. C. and H. T. Tang, "Solute Distribution in the Flow in a Channel Bounded by Porous Layes—A Model of the Lung," *Journal of Applied Mechanics, Transactions ASME*, Vol. 42, 531–535 (1975).
30. Goldstein, M. E. and W. H. Braun, "Effect of Velocity Slip at a Porous Boundary on the Performance of Incompressible Porous Bearings," NASA Technial Note, D 6181, 1–36 (1971).
31. Haber, S. and R. Mauri, "Boundary Conditions for Darcy's Flow Through Porous Media," *International Journal of Multiphase Flow*, Vol. 9, 561–574 (1983).
32. Happel, J., *AIChE Journal*, Vol. 4, p. 197 (1958).
33. Hinch, E. J., "An Averaged Equation Approach to Particle Interactions in a Fluid Suspension," *Journal of Fluid Mechanics,* Vol. 83, 695–720 (1977).
34. Howells, I. D., "Drag Due to the Motion of a Newtonian Fluid Through a Sparse Random Array of Small Fixed Rigid Objects," *Journal of Fluid Mechanics,* Vol. 64, 449–475 (1974).
35. Huh, C. and S. G. Mason, "The Steady Movement of Liquid Meniscus in a Capillary Tube," *Journal of Fluid Mechanics*, Vol. 81, 401–419 (1977).
36. Iberall, A. S., "Permeability of Glass Wool and Other Highly Porous Media," *Journal Res. Nat. Bur. Stand.*, Vol. 45, 398–406 (1950).
37. Jones, I. P., "Low Reynolds Number Flow Past a Porous Spherical Shell," *Proceedings of the Cambridge Philosophical Society,* Vol. 73, 231–238 (1973).
38. Joseph, D. D., "Boundary Conditions for Thin Lubrication Layers," *Physics of Fluids,* Vol. 23, 2356–2358 (1980).
39. Joseph, D. D. and L. N. Tao, "Lubrication of a Porous Bearings—Stokes' Solution," *Journal of Applied Mechanics, Transactions ASME,* Vol. 33, 753–760 (1966).
40. Lance, G. N. and M. H. Rogers, "Unsteady Slip Flow Over a Flat Plate," *Proceedings of the Royal Society*, Vol. A 266, 109–121 (1962).
41. Levy, T., "Loi de Darcy ou loi de Brinkman?," C. R. Acad. Sci., Paris, Série II, Vol. 292, 872–874 (1981).
42. Lin, F. C., "Lubrication of Animal Joints," *Journal of Lubrication Technology, Transactions ASME*, Vol. 91, 329–341 (1969).
43. Liu, P. I. F., "Permeable Wall Effects on Poiseuille Flow," *Journal of Engineering Mechanics, ASCE*, Vol. 105, 470–476 (1979).
44. Lundgren, T. S., "Slow Flow Through Stationary Random Beds and Suspensions of Spheres," *Journal of Fluid Mechanics*, Vol. 51, 273–299 (1972).
45. Mow, V. C., "The Role of Lubrication in Biomechanical Joints," *Journal of Lubrication Technology, Transactions ASME*, Vol. 91, 320–328 (1969).
46. Murti, P. R. K., "Some Aspects of Slip Flow in Porous Bearings," *Wear*, Vol. 19, 123–129 (1972).

47. Murti, P. R. K., "Effect of Slip in Narrow Porous Bearings," *Journal of Lubrication Technology, Transactions ASME,* Vol. 95, 518–523 (1973).
48. Murti, P. R. K., "Effect of Slip Flow in Narrow Porous Bearings with Arbitrary Wall Thickness," *Journal of Applied Mechanics, Transactions ASME*, Vol. 42, 305–310 (1975).
49. Muskat, M., *The Flow of Homogeneous Fluids Through Porous Media*, New York:McGraw-Hill (1937).
50. Narasimha Murthy, S., "Stokes First and Second Problems in Porous Medium," *Indian Journal of Technology,* Vol. 12, 515–516 (1974).
51. Neale, G., N. Epstein and W. Nader, "Creeping Flow Relative to Permeable Spheres," *Chemical Engineering Science,* Vol. 28, 1865–1874 (1973).
52. Neale, G. and W. Nader, "Practical Significance of Brinkman's Extension of Darcy's Law: Coupled Parallel Flows Within a Channel and a Bounding Porous Medium," *The Canadian Journal of Chemical Engineering,* Vol. 52, 475–478 (1974).
53. Nield, D. A., "Onset of Convection in a Fluid Layer Overlying a Layer of a Porous Medium," *Journal of Fluid Mechanics,* Vol. 81, 513–522 (1977).
54. Nield, D. A., "The Boundary Correction for the Rayleigh-Darcy Problem: Limitations of the Brinkman Equation," *Journal of Fluid Mechanics,* Vol. 128, 37–46 (1983).
55. Nield, D. A., "An Alternative Model for the Wall Effect in Laminar Flow of a Fluid Through Packed Column," *AIChE Journal*, Vol. 29, 688–689 (1983).
56. Nield, D. A., "Non-Darcy Effects in Convection in a Saturated Porous Medium," *Proceedings of the CSIRO/DSIR Seminar on Convective Flows in Porous Media*, DSIR, Wellington, New Zealand, 129–139 (1985).
57. Nutting, P. G.,"Physical Analysis of Oil Sands," *Bulletin of American Association of Petroleum Geologists,* Vol. 14, 1337–1349 (1930).
58. O'Brien, V., "Viscous Flow at the Boundary of a Porous Material," *Applied Physics Laboratory Report TG 1285,* John Hopkins University, Laurel, MD (Sept. 1975).
59. O'Brien, V., "Permeable Wall Effects on Poisuille Flows," *Journal of Engineering Mechanics, ASCE*, Vol. 107, 719–721 (1981).
60. Pai, S. I., *Viscous Flow Theory I—Laminar Flow,* Princeton, NJ:D. Van Nostrand Co. (1956).
61. Panton, R. L., *Incompressible Flow,* John Wiley and Sons (1984).
62. Patel, K. C., "Hydromagnetic Squeeze Film with Slip Velocity Between Two Porous Annular Disks," *Journal of Lubrication Technology, Transactions ASME,* Vol. 97, 644–647 (1975).
63. Patel, K. C. and J. L. Gupta, "Hydrodynamic Lubrication of a Porous Slider Bearing with Velocity Slip," *Wear*, Vol. 85, 309–317 (1983).
64. Prakash, J. and S. K. Vij, "Analysis of Narrow Porous Journal Bearing Using Beavers-Joseph Criteron of Velocity Slip," *Journal of Applied Mechanics, Transactions ASME*, Vol. 41, 348–354 (1974).
65. Puri, K. K., "Damping of Gravity Waves over Porous Bed," *Journal of Hydrology, ASCE*, Vol. 106, 303–312 (1980).
66. Puri, K. K., "Damping of Waves Due to a Soluble Film," *International Journal of Mathematical Sciences*, Vol. 3, 302–312 (1980).
67. Puri, K. K., "Gravity Waves Over a Permeable Bed," *International Journal of Engineering Sciences,* Vol. 12, 527–535 (1983).
68. Puri, V. K. and C. M. Patel, "Analysis of an Anisotropic Porous Slider Bearing Considering Slip Velocity," *Wear*, Vol. 78, 279–283 (1982).
69. Puri, V. K. and C. M. Patel, "Analysis of a Porous Composite Slider Bearing with Slip Velocity," *Wear*, Vol. 78, 337–341 (1982).
70. Puri, V. K. and C. M. Patel, "Analysis of a Composite Porous Slider Bearing with Anisotropic Permeability and Slip Velocity," *Wear*, Vol. 84, 33–38 (1983).
71. Rajasekhara, B. M., N. Rudraiah and B. K. Ramaiah, "Couette Flow Over a Naturally Permeable Bed," *Journal of Mathematical and Physical Sciences,* Vol. 9, 49–56 (1975).
72. Ramacharyulu, N. Ch. P., "A Non-Darcian Approach to Flow Through Porous Media," Internal Report to IAEA and UNESCO IC/76,122, International Centre for Theoretical Physics (1976).
73. Richardson, S., "A Model for the Boundary Condition of a Porous Material. Part 2," *Journal of Fluid Mechanics*, Vol. 49, 327–336 (1971).
74. Rosenhead, L., *Laminar Boundary Layers*, Oxford:Clarendon Press (1963).
75. Rouleau, W. T. and L. I. Steiner, "Hydrodynamic Porous Journal Bearings. Part I. Finite Full Bearings," *Journal of Lubrication Technology, Transactions ASME*, Vol. 96, 346–353 (1974).
76. Rudraiah, N. and S. T. Nagraj, "Natural Convection Through a Vertical Porous Stratum," *International Journal of Engineering Sciences,* Vol. 15, 589–600 (1977).
77. Rudraiah, N. and R. Veerabhadraiah, "Temperature Distribution in Couette Flow Past a Permeable Bed," *Proceedings of the Indian Academy of Sciences*, Vol. 86, 537–547 (1977).
78. Sacheti, N. C., "Comments on 'Externally Pressurized Porous Thrust Bearings'," *Wear*, Vol. 58, 393–395 (1980).
79. Sacheti, N. C., "Application of Brinkman Model in Viscous Incompressible Flow Through a Porous Channel," *Journal of Mathematical and Physical Sciences,* Vol. 17, 567–578 (1983).
80. Sacheti, N. C. and B. S. Bhatt, "Stokes and Rayleigh Layers in Presence of Naturally Permeable Boundaries," *Journal of Engineering Mechanics*, ASCE, Vol. 110, 713–722 (1984).
81. Saffman, P. G., "On the Boundary Condition at the Surface of a Porous Medium," *Studies in Applied Mathematics*, Vol. 50, 93–101 (1971).
82. Sai, K. S., "On the Unsteady Flow of Incompressible Fluid

Over a Naturally Permeable Bed," *Journal of Mathematical and Physical Sciences*, Vol. 14, 599–609 (1980).

83. Sarma, B. K. D., "Flow in a Horizontal Circular Cylinder Bounded by a Porous Medium," *Acta Mechanica*, Vol. 34, 251–255 (1979).
84. Scheidegger, A. E., "Hydrodynamics in Porous Media," *Encyclopedia of Physics*, Vol. 9, Springer, Germany (1960).
85. Scheidegger, A. E., *The Physics of Flow Through Porous Media*, Toronto, Canada:University of Toronto Press (1960).
86. Sinha, P., C. Singh and K. R. Prasad, "Lubrication of Human Joints–A Microcontinuum Approach," *Wear*, Vol. 80, 159–181 (1982).
87. Slattery, J. C., *Momentum, Energy and Mass Transfer in Continua*, New York:McGraw-Hill (1972).
88. Sparrow, E. M., G. S. Beavers, T. S. Chen and J. R. Lloyd, "Breakdown of the Laminar Flow Regime in Permeable Walled Ducts," *Journal of Applied Mechanics, Transactions ASME*, Vol. 40, 337–342 (1973).
89. Sparrow, E. M., G. S. Beavers and L. Y. Hung, "Flow About a Porous-Surfaced Rotating Disk," *International Journal of Heat and Mass Transfer*, Vol. 14, 993–996 (1971).
90. Sparrow, E. M., G. S. Beavers, and L. T. Hwang, "Effect of Velocity Slip on Porous Walled Squeeze Films," *Journal of Lubrication Technology, Transactions ASME*, Vol. 95, 518–523 (1973).
91. Tam, C. K. W., "The Drag on a Cloud of Spherical Particles in Low Reynolds Number Flow," *Journal of Fluid Mecahnics*, Vol. 38, 537–546 (1969).
92. Tang, H. T. and Y. C. Fung, "Fluid Movement in a Channel with Permeable Walls Covered by Porous Media–A Model of Lung Alveolar," *Journal of Applied Mechanics, Transactions ASME*, Vol. 42, 45–50 (1975).
93. Tandon, P. N. and R. S. Gupta, "Effect of Rotation on the Lubrication Characteristic of Knee Joint During Normal Articulation," *Wear*, Vol. 80, 183–195 (1982).
94. Taylor, G. I., "A Model for the Boundary Condition of a Porous Material. Part 1," *Journal of Fluid Mechanics*, Vol. 49, 319–326 (1971).
95. Verma, A. K., "A Note on Oscillatory Horizontal Layer Over a Permeable Bed," *International Journal of Engineering Sciences*, Vol. 21, 289–295 (1983).
96. Verma, P. D. and B. S. Bhatt, "On the Flow of Immiscible Incompressible Fluids Between Naturally Permeable Walls," *Zeitschrift für Angewandte Mathematik und Mechanik*, Vol. 54, 817–819 (1974).
97. Verma, P. D. and B. S. Bhatt, "Flow Past a Porous Spherical Shell Using Matched Asymptotic Technique," *Journal de Mecánique*, Vol. 14, 421–434 (1975).
98. Verma, P.D. and B. S. Bhatt, "On The Steady Flow Between Rotating and a Stationary Naturally Permeable Disc," *International Journal of Engineering Sciences*, Vol. 13, 869–879 (1975).
99. Verma, P. D. and B. S. Bhatt, "Low Reynolds Number Flow Past a Heterogeneous Porous Sphere Using Matched Asymptotic Technique," *Applied Scientific Research*, Vol. 32, 61–72 (1976).
100. Verma, P.D. and B. S. Bhatt, "Flow Past a Circular Cylinder at Small Reynolds Number," *Indian Journal of Pure and Applied Mathematics*, Vol. 8, 908–914 (1977).
101. Verma, P. D. and N. C. Sacheti, "On Two-Dimensional Flow of Power Law Fluids Through Ducts with Naturally Permeable Walls," *Zeitschrift für Angewandte Mathematik und Mechanik*, Vol. 55, 475–478 (1975).
102. Vidyanidhi, V., A. Sithapati and P. C. L. Narayana, "Heat Transfer for Flow Past Permeable Bed," *Proceedings of the Indian Academy of Sciences*, Vol. 86, 557–561 (1977).
103. Wang, C. Y., "Limitation of the Reynolds Equation for Porous Thrust Bearings," *Journal of Lubrication Technology, Transactions ASME*, Vol. 97, 642–643 (1975).
104. Williams, W. O., "Constitutive Equations for Flow of an Incompressible Viscous Fluid Through a Porous Medium," *Quarterly of Applied Mathematics*, Vol. 36, 255–267 (1978).
105. Wu, H., "Effect of Velocity Slip on the Squeeze Film Between Porous Rectangular Plates," *Wear*, Vol. 20, 67–71 (1972)
106. Yamamoto, K. and N. Iwamura, "Flow with Convective Acceleration Through a Porous Medium," *Journal of Engineering Mathematics*, Vol. 10, 41–54 (1976).

SECTION SIX

Solid–Fluid Interaction

Fluid–Structure Interaction

Enzo Levi*, Neftalí Rodríguez-Cuevas, Gabriel Echavez[†]**

INDUCED MOTIONS GENERATED BY WATER (ENZO LEVI)

INTRODUCTION

Structural elements impinged by fluid currents may be compelled to vibrate either with their own frequency (*free vibration*) or with a frequency imposed by the flow (*forced vibration*). Only the second case will be considered in the following.

Forced vibrations are frequent, but usually they are of little consequence. They can be dangerous when the induced frequency happens to be close to the proper frequency of the structural element. In fact, then the oscillations of the latter can be amplified more and more by resonant effect, and eventually lead to a failure.

Therefore, the designer needs to take into account all the possible sources of flow-induced vibrations and to predict with a good approximation the corresponding frequencies. On the other side, when such vibrations appear during the operation, the main problem faced by the engineer is to discover where the cause of the vibrations lies, in order to suppress them, or at least to reduce their intensity.

A tentative classification for fluid-instability induced excitations appears, as suggested by E. Naudascher [14], in Figure 1. This scheme considers four *base flow modules*, arranged according to three *basic excitation mechanisms*. The *flow modules* are

(a) *wake-type flows*, i.e., those periodical motions provoked by the presence of a solid obstacle obstructing the main flow;
(b) *impinging shear layers*, i.e., oscillations provoked by shear effect in lateral cavities;
(c) *bistable flows*, i.e., jetlike flows that dispose themselves alternately in either of two equally stable positions;
(d) *two-phase systems*, usually water jets falling through the air, when the latter participates in the oscillation of the jet.

The *excitation mechanisms* are of the following kinds:

(a) *fluid dynamic*, resulting from a periodical oscillation produced inside the main current ad drawn along by it;
(b) *fluid-resonant*, resulting from oscillations induced and sustained by the main flow in fluid bodies placed at its sides;
(c) *body-resonant*, induced by the vibration of a solid body submerged or touching the water.

Figure 1 shows graphically examples of events that interest hydraulic engineering, for each one of the twelve combinations of flow modules and excitation mechanisms. Although—as Naudascher acknowledges—much can and must be improved in this classification scheme and matrix representation, it offers a sound basis for identification and preliminary assessment of sources of flow-induced vibrations.

The topic of vibrations induced by water flow in hydraulic structures will be treated in Part A of this Article. For more information and bibliography, consult [13,15,21,30].

Another source of water-structure interaction are water waves, when they actuate on offshore structures. These structures are mainly related to oil and gas recovery and harbor engineering. In this case, wave frequency is not usually considered a risk factor, but wave force is; essentially, the additional load that the action of waves can induce

*Instituto Mexicano de Tecnología del Agua
**Instituto de Ingeriería, Universidad Nacional Autónoma de México
[†] Facultad de Ingeníeria, Universidad Nacional Autónoma de México

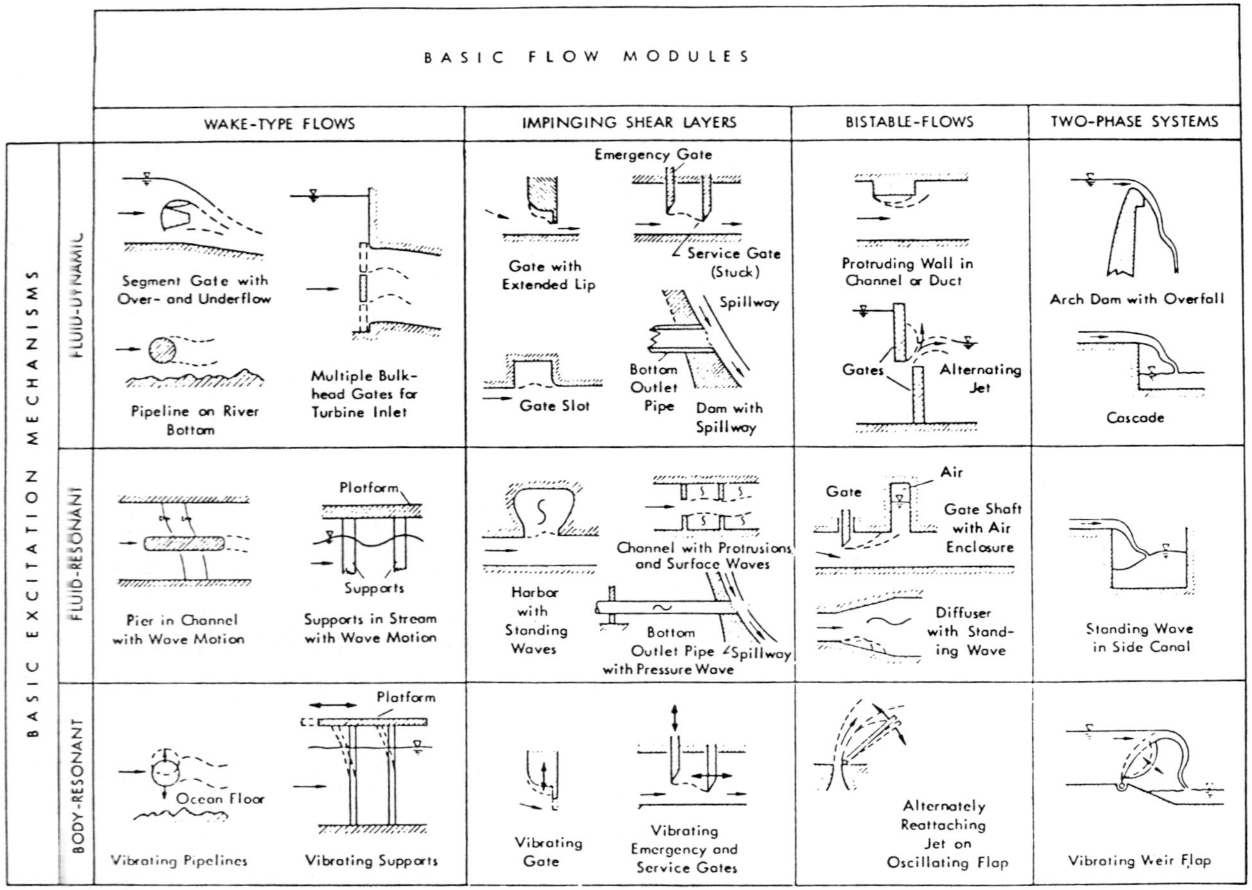

FIGURE 1. Basic excitation mechanism and flow modules leading to instability-induced excitation, from Naudascher [14].

on the structure. A brief introduction to this complex problem will be given in Part B. An up-to-date treatment can be found in [28].

Part A. Water Currents as an Exciting Force

CYLINDER WAKE

V. Strouhal [31] investigated more than a century ago *aeolian tones,* i.e., the vibrations produced when the wind strikes a stretched string. He found that, contrary to the common opinion that those vibrations occur—as those produced by violin bow—in the plane containing the direction of the wind, they are transverse to it. The pitch of the aeolian tone was found to be independent of the length and tension of the string, but to vary with its diameter D and with the approach velocity V of the wind. Strouhal found that, within certain limits, the relation between the frequency of vibration f, D and V is expressible as

$$f = 0.185V/D \tag{1}$$

The nondimensional parameter fD/V was subsequently called *Strouhal number.*

It has to be stressed also that Strouhal found out that any wind velocity gives rise to a vibration, independently that its frequency should or not coincide with one of the proper tones of the string; although, in the first case, the sound produced is greatly reinforced.

An interesting result of Strouhal's investigation was that it is not essential to the production of sound that the string take part in the vibration. The effect does not change if the string is replaced by a rigid rod of circular cross-section. The reason for this became clear when H. Bénard pointed out in 1908 [1] that, within the wake of a cylindrical body submerged in a fluid flow of a not too low speed, two rows of vortices appear, detaching alternately from the two sides of the cylinder, and progressing in downstream direction (Figure 2). The frequency measured by Strouhal, Equation (1), was in fact the one associated with the mentioned alternate vortex shedding.

Examining from this new point of view the conditions under which Strouhal's experiments had been performed, and trying to explain the poor agreement of some of the results with Equation (1), Lord Rayleigh [32] suggested that also the Reynolds number VD/ν (ν being the kinematic viscosity of the fluid) should be taken into account. Actually, he recommended the new equation

$$\frac{fD}{\nu} = 0.195\left(1 - \frac{20.1\nu}{VD}\right) \tag{2}$$

FIGURE 2. Initiation of vortex shedding past a circular cylinder, from Sarpkaya and Shoaff [28].

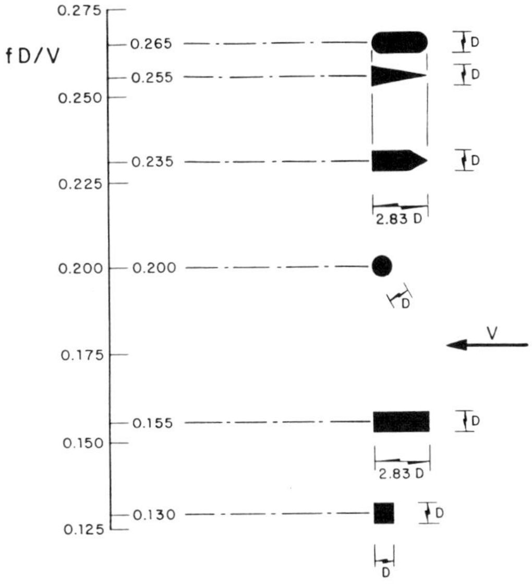

FIGURE 3. Values of the Strouhal number fD/V, for cylindrical obstacles of different cross-sections, at great values of the Reynolds number VD/ν, after Teissie-Solier [11].

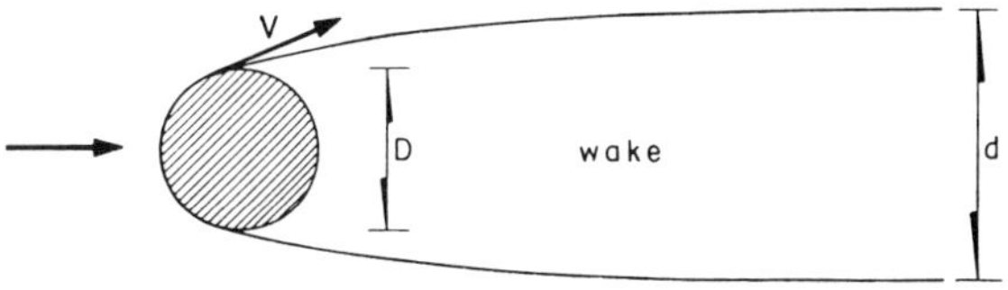

FIGURE 4. How to measure obstacle width D and wake breadth d.

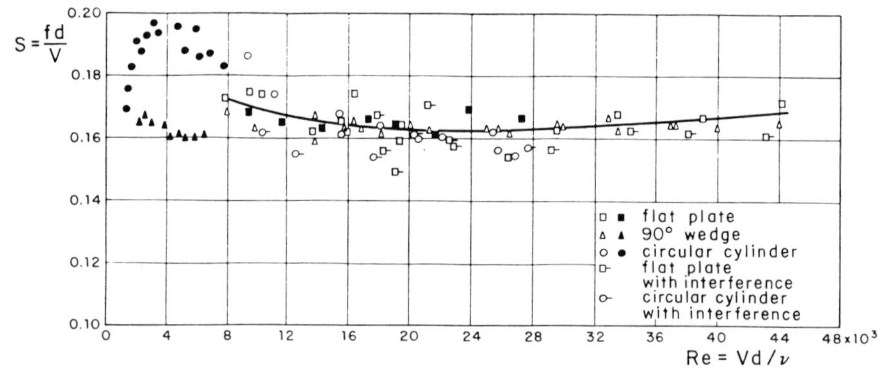

FIGURE 5. Strouhal number S = fd/V for three different kinds of obstacles, both without and with a plate splitting the wake, from Roshko [23].

that was found to fit better Strouhal's experimental results. For great Reynolds numbers, Equation (2) gives the *asymptotic value*

$$fD/V = 0.195 \tag{3}$$

This value corresponds to a cylindrical obstacle of circular cross-section. For other cross-sections, the asymptotic value varies with their geometry. M. Teissie–Solier found out that this value rises from the value of 0.130 for a square bar to 0.265 for one rectangular, rounded at both extremities. Values for some shapes of interest for hydraulic applications appear in Figure 3 [11].

In order to get a Strouhal number S independent of the shape of the cross-section of the obstacle, G. Birkhoff [3] recommended to take

$$\boxed{S = fd/V} \tag{4}$$

where d is the breadth of the wake, measured far enough from the obstacle (Figure 4). A. Roshko experimented with a flat plate, a 90° wedge and a circular cylinder, their wake being both free and splitted by a plate set along its center line [23]. In accordance with Birkhoff foresight, he found that, for values of $Re = Vd/\nu$ between 8×10^3 and 48×10^3 (V being now, as shown by Figure 3, the free velocity at the separation generant), all the corresponding S-values fall approximately on the same curve [nearly parallel to Re axis (Figure 5)]. Their mean value results to be

$$fd/V = 0.16 \tag{5}$$

UNIVERSAL STROUHAL LAW

Having observed that, if a weir is set across a laboratory channel, a pair of counter-rotating erect vortices is formed with a nearly periodical intermittency behind the weir, E. Levi tried to determine their frequency, f. He found that practically the same Equation (5) is able to relate f with d, taken as the flow depth upstream from the weir, and V, taken as the velocity of approach to it. Subsequently it was realized that approximately the same values $S = 0.16$ associates with a wide assortment of evidently unrelated phenomena, such as wing autorotation, jet flapping and puffing, shock-wave and vortex-breakdown oscillation, and periodical deformation of cavitation bubbles. In all the cases, f is representing the frequency of the oscillation of a fluid layer of width d, restrained in its motion, in the presence of an external free flow of velocity V [9].

The law expressed by Equation (5) can be justified by supposing that the system composed by the restrained fluid layer and the external flow behaves as a simple undamped oscillator. In fact, if d represents the displacement of the oscillator and f its frequency,

$$E = \frac{1}{2}(2\pi d f)^2 \tag{6}$$

results to be its specific mechanical energy, while the available specific kinetic energy is $V^2/2$. By equating both energies, one gets that

$$\boxed{\frac{fd}{V} = \frac{1}{2\pi}} \tag{7}$$

i.e., Equation (5), since $1/2\pi = 0.159$. This law was proposed in 1980 by Levi, who called it *universal Strouhal law* [9].

From Equation (7) it follows also a simple formula for the wavelength of the local oscillation, if convected by the flow. In fact, the traveling oscillation will look to a stationary observer as a progressive wave, of length

$$\boxed{\lambda = V/f = 2\pi d} \quad (8)$$

The universal Strouhal law offers the engineer the means to solve the following two problems of practical interest:

(a) *To find the frequency* of a flow-induced vibration, when the width d of the restrained fluid body (considered as *the oscillator*) and the velocity V of the external current that sets and maintains it in motion are known.
(b) *To find the oscillator,* when the velocity V and the frequency of vibration f induced in a structural element are known, but one does not know *which is* the water body that oscillates. In fact, in this case, Equation (7) allows to ascertain the approximate width d of this water body, making it usually easy to recognize it [10].

In what follows, many examples of hydraulic problems concerning flow-induced vibrations, which have proved to be susceptible of a quantitative approach, will be given. They will be ordered following the basic-flow-module classification of Naudascher (Figure 1).

WAKE TYPE FLOW

Trashracks

Trashrack bars can break up when forced to vibrate with their own frequency. Damages for this reason are common [11]. To avoid resonance conditions, it is essential that vortex shedding frequencies of the trashrack elements be outright different from their natural frequencies of vibration. Therefore these last frequencies have to be calculated at first, taking into account that natural frequency of such elements, when submerged into water, can be about 2/3 of their natural frequency in the air [4].

As for the calculation of the flow-induced frequency, two main points have to be reminded [11]:

(a) That water usually crosses trashracks with high speed. Therefore asymptotic values may be assumed for Strouhal numbers, such as the ones recorded in Figure 3.
(b) That Strouhal number increases if bar spacing L is close to bar width D. Figure 6 shows how the asymptotic Strouhal number value (0.265 in this case) is modified with decreasing L/D, for a quite usual kind of trashrack bars.

Moreover, it is indispensable to remind that trash accumulation causes velocities through the rack, and therefore also vibration frequencies, to increase. A trashrack designed so that vibration is avoided when it is clean, can reach resonance conditions because of trash accumulation.

Example [16]. The average velocity through the net flow area of a trashrack, whose bars had a thickness of 19 mm, was calculated as $V = 1.57$ m/s. Taking for the bar Strouhal number the value of 0.265 (Figure 3), one gets for the vortex-shedding frequency the value

$$f = 0.265 \times 1.57/0.019 = 21.9 \text{ Hz}$$

This frequency was considered safe, since the natural frequency of the trashrack bars had been calculated to be about 36 Hz. However, with trash accumulation, the flow velocity was increased to 2.51 m/s. This gives

$$f = 0.265 \times 2.51/0.019 = 35 \text{ Hz}$$

that is very close to the bar natural frequency. Actually, fragments of steel bar downstream of a service gate evidenced the trashrack failure.

As a remedy, a tie bar was welded to the midpoint of the trashrack bars, so increasing their natural frequency to about 147 Hz. Anyway, as a general rule, it has to be recommended to minimize trash accumulation, maintaining always the rack as clean as possible, in order to avoid unsafe vibration frequencies to be reached.

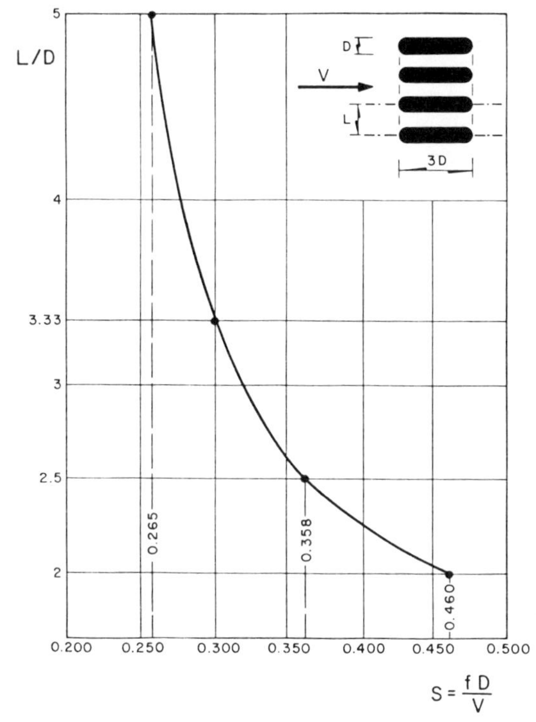

FIGURE 6. Variation of Strouhal number according to the spacing of trashrack bars, according to Crausse [11].

Butterfly Valves

Butterfly valves, although fully open, interfere permanently with the flow, producing vortex shedding whose oscillations are transmitted to the current.

Example [16]. A major failure of vanes of a Howell–Bunger valve was the object of a detailed analysis. A spectrum of the frequencies of pressure fluctuations upstream from the valve revealed a clear peak at a frequency of about 3 Hz. The flowrate at failure conditions was 166 m^3/s. Upstream from the Howell–Bunger valve there was a butterfly valve of 0.91 m width, set there for emergency closure. The diameter of the pipe leading to the butterfly valve being of 3.34 m, the approach velocity of water was 19 m/s. Now, if we put, in Equation (7), $V = 19$ m/s, $f = 3$ Hz, we get

$$d = 0.16 \times 19/3 = 1.0 \text{ m}$$

that is, a value very close to the butterfly-valve width. This result points to the butterfly valve as being responsible for the Howell–Bunger valve failure.

Bridge Piers

A bridge pier, placed centrally in a steady, subcritical open-channel flow, can give rise to violent self-excited oscillating surface waves. They result from a combination of a standing wave in the transverse direction and a progressive wave in the longitudinal direction, travelling upstream. The considerable wave height at the walls of the channel near the obstacle can cause flow to top the wall lining.

Example [22]. Experiments performed in a laboratory rectangular flume showed that oscillating waves were present for

$$0.068 \leq D/B \leq 0.18,$$

$$1.00 \leq L/D \leq 6.00,$$

$$0.200 \leq h/B \leq 1.00$$

the dimensions B, L, D, h being shown in Figure 7, provided that

$$0.08 \leq Fr \leq 0.23$$

where Fr is the Froude number $Fr = V/\sqrt{gh}$ and V the approach velocity. Oscillation frequency measurements showed that in those conditions the Strouhal number fD/V remained always inside the range

$$0.20 \leq fD/V \leq 0.25$$

According to Figure 3, this points clearly to the vortex

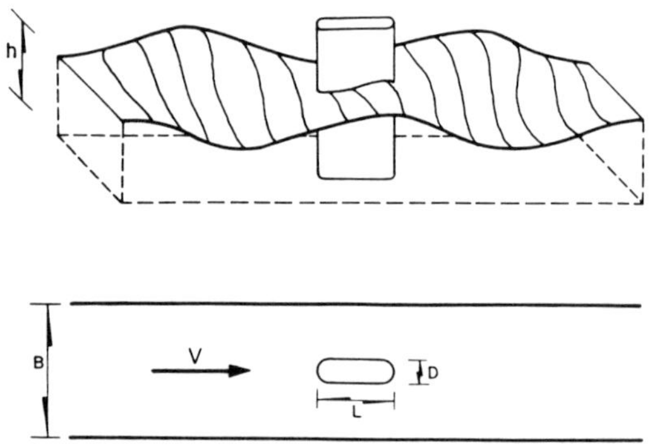

FIGURE 7. Self-excited surface waves around a bridge pier.

shedding from the pile as being the primary cause of oscillations.

To avoid oscillations, it is recommended to modify at least one of the influential parameters L, D, B, h, Fr so that critical conditions be avoided.

IMPINGING SHEAR LAYERS

Underflow Gates With Free Discharge

Freely discharging underflow gates are exposed to oscillate with frequencies close to the value $V/2\pi d$ [Equation (7)], d being the water head acting over the gate.

Example 1 [6]. A Tainter gate of lighter design collapsed and was carried suddenly downstream during a test run. The gate had been operated with the reduced opening of 30 cm during 3 hours, just before the accident. Then the gate had been closed down at a speed of about 30 cm/min. The accident happened at the time that the gate was almost close. Since the collapse occurred suddenly and the situation could not be reconstructed clearly, model tests were carried out in order to understand the failure mechanism. A sketch of the model is given in Figure 8. With an opening $b = 5$ mm and a head $d = 12.6$ cm, intense self-excited vibrations were recorded, with the frequency $f = 2.9$ Hz. Since the weir was broad-crested, the flow velocity can be computed as

$$V = 0.7\sqrt{2gd} = 1.10 \text{ m/s}$$

Therefore, one gets that

$$fd/V = (2.9 \times 0.126)/1.10 = 0.332 = 2 \times 0.166$$

A result like this, giving for the Strouhal number a value of about 0.32 instead of 0.16, according to Equation (7), occurs sometimes, suggesting the presence of the first har-

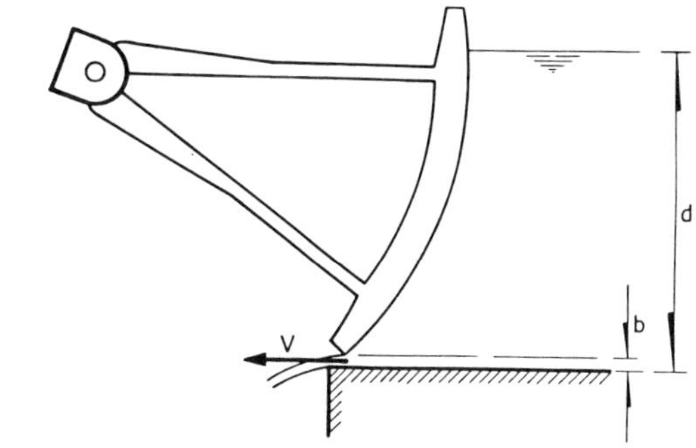

FIGURE 8. Partially opened Tainter gate with free underflow.

monic of the frequency [9]. Taking this into account, one may conclude that the critical flow-induced vibration frequency could have been predicted through Equation (7), taking as d the static water head over the crest.

Example 2 [24]. A Tainter gate, acting as emergency gate in the flood tunnel of a reservoir, was tested through a laboratory model. Special care was given to the fact that flow enters through a morning-glory spillway, thus being composed of an air-water mixture. Severe periodic pressure fluctuations were observed during the test, upstream and downstream from the gate, when the gate openings were between 20 and 40%, at two-phase flow condition. No pressure fluctuations occurred if only water was flowing. The model worked under the static head $d = 2$ m. The tunnel geometry suggests to compute the flow velocity as

$$V = 0.6\sqrt{2gd} = 3.76 \text{ m/s}$$

Then Equation (7) gives

$$f = V/2\pi d = 0.3 \text{ Hz}$$

Actually, the model showed a frequency of pressure fluctuations between 0.2 and 0.4 Hz.

BISTABLE FLOWS

Underflow Gates With Submerged Discharge

When the gate is drowned, its vibration can be excited by the submerged jet that takes rise under the gate. In fact, submerged jets have a tendency to oscillate with a frequency close to the one given by Equation (7), taking as d the jet thickness [9]. In the case of drowned gates, one has to take as d the gate opening.

Example 1 [7]. A drowned gate of rectangular edge was tested in a hydraulic model. At a $d = 13.5$ mm lift, the strongest vibrations occurred with a $H = 40$ cm difference of head acting upon the gate. Frequencies of vibration were not measured; but personal impressions pointed to 20 Hz. Now, taking, as suggested by gate geometry and disposition,

$$V = 0.6\sqrt{2gH} = 1.68 \text{ m/s}$$

one gets by Equation (7)

$$f = V/2\pi d = 19.8 \text{ Hz}$$

Example 2 [20]. A drowned sluice (Figure 9) was found to vibrate with a 10 Hz frequency, when it worked with a $d = 12$ cm opening and $H = 5.2$ m head difference. Sluice geometry and arrangement suggest to take

$$V = 0.75\sqrt{2gH} = 7.58 \text{ m/s}$$

Therefore, from Equation (7)

$$f = V/2\pi d = 10 \text{ Hz}$$

as measured.

The vibrations were imputed to the fact that the high flood had been released through the sluice over periods of time so long that the wooden bottom beam had become rounded at the front due to floating debris. The vibrations were eliminated by replacing the wooden beam by a rubber seal.

Protruding Duct Walls

Protrusions in duct walls can be the source of bistable flows. The frequency of the resulting oscillations, which can be transmitted to structural elements both downstream and upstream from the protrusion, can be predicted through the universal Strouhal law.

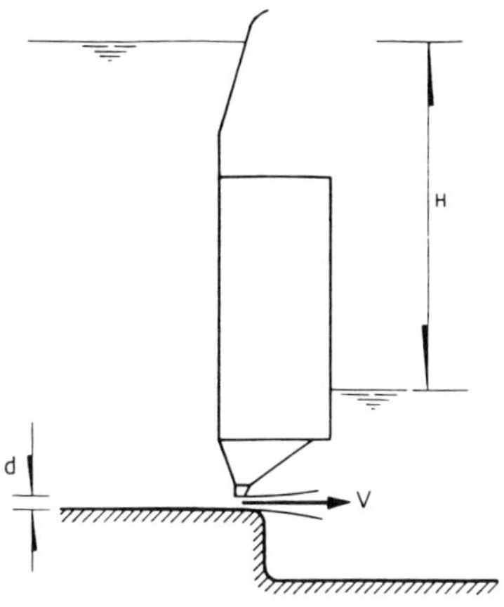

FIGURE 9. Sluice excited by the underlying water jet.

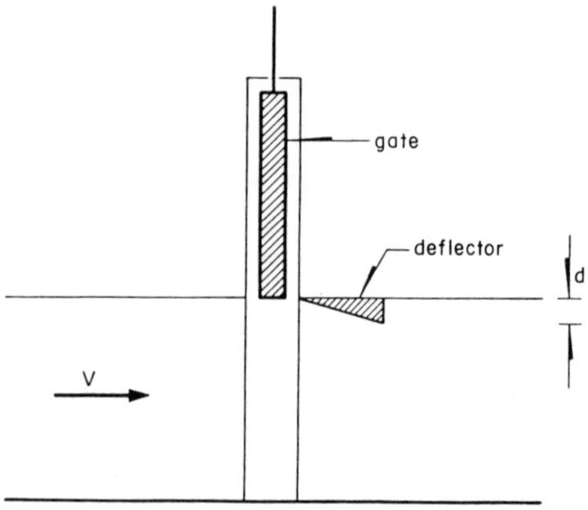

FIGURE 10. Fully opened gate excited by a bistable flow produced by the presence of a deflecting wedge at the tunnel roof.

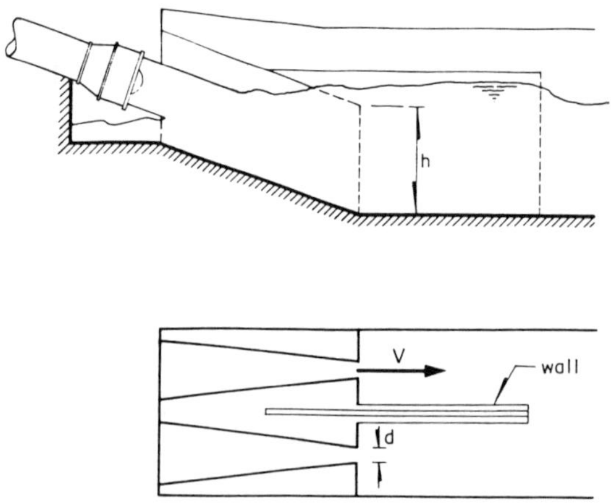

FIGURE 11. Twin stilling basin with central wall.

Example [29]. Horizontal girders supporting the hoist cylinder of a wheel gate were found to experience severe vibrations, in the frequency range of 18 to 20 Hz, when the gate was in its fully raised position, and therefore did not interfere with flow. A deflecting wedge, whose maximum thickness was 30.5 cm, had been provided on the roof of the tunnel (Figure 10), just downstream from the gate grooves, to avoid negative pressures in the transition from circular to rectangular tunnel cross-section. It becomes, therefore, natural to suspect this deflector to be the cause of the vibrations.

Girder vibrations occurred under a water head $H = 65$ m. Taking

$$V = \sqrt{2gH} = 35.7 \text{ m/s}$$

and $d = 0.305$ m the wedge thickness, one gets from Equation (7)

$$f = V/2\pi d = 18.6 \text{ Hz}$$

which falls inside the measured vibration frequency range. This proves that the deflector was in fact the cause of trouble. The deflector was 3.05 m long. A reduction of its length to 1.80 m, and therefore of its thickness to 18 cm, proved to be good for eliminating vibrations.

Stilling Basins

A sudden expansion at the entrance of stilling basins causes a submerged jet to be formed, able to oscillate within the body of water contained in the basin. This oscillation can induce vibrations and consequent damages to the stilling basin structure. As before, the vibration frequency can be predicted through Equation (7), taking the inlet width as the characteristic length d.

Example [5]. A stilling basin was fed by two hollow-jet valves discharging into a common structure. Since the valves might operate either singly or together, their respective flows were kept separate in the basin by a central wall. The annular flow ejected by the valves was transformed into a solid jet before entering the stilling basin, by wedges located on the sides of both sections of the basin (Figure 11). After a prolonged operation, fatigue cracks appeared at the bottom of the central wall.

Vibrations were detected when a single valve discharged 50 m^3/s.

Since the wedge height was $h = 6.25$ m and the inlet width was $d = 0.63$ m, we can assume

$$V = 50/(6.25 \times 0.63) = 12.7 \text{ m/s}$$

Therefore, Equation (7) gives

$$f = V/2\pi d = 3.2 \text{ Hz}$$

Actually, measured frequencies fell between 3.0 and 4.6 Hz.

Revetment Failures

High-speed currents making contact with concrete or steel linings may produce their failure by vibration. Typical was the collapse of the concrete floor of the stilling basin of a spillway able to discharge up to 6000 m^3/s, after discharging only 3000 m^3/s during two weeks. The stilling basin was 26 m deep and its floor was about 96 m lower than spillway crest. The floor was protected with concrete slabs of 12 ×

12 m horizontal cross-section and 2 m height, weighing each one about 720 ton, and anchored to the underlying rock with twelve 1¼ ϕ steel bars. These slabs had been cast in site. The joints between them had been carefully packed with bituminous refill.

The failure, shown in Figure 12, can be reconstructed as follows. The central slabs at the chute heel were overturned and removed. These slabs, falling downstream, had disorganized the flow, provoking in turn the uplift of the other slabs. The anchorage of removed slabs was of course broken out, and showed in places signs of tension failure [25].

Laboratory experiments permitted to understand the basic mechanism of failure. For the uplift of a slab to take place, the presence of an upper high-velocity free or submerged jet and of a water layer under the slab, both interconnected through gaps between adjacent slabs are necessary. With jet velocities great enough, oscillations of the underlying water layer arouse, whose frequency can be determined with a good approximation by means of Equation (7), applied taking as V the jet velocity and as d the length of the layer, i.e., the length of the slab. The oscillations can cause the slab to vibrate with the same frequency.

Example [8]. Tests performed with a concrete slab of 20 × 20 cm horizontal dimensions and 4 cm height, free to move inside a slightly larger cavity of a laboratory channel floor, proved that it vibrates with frequencies between about 0.8 and 1.8 Hz, independently of the water jet submergence. The frequencies showed a tendency, in average, to increase with discharge. Jet velocities varied from 1.1 to 2.2 m/s. Taking $d = 0.20$ m, that is the slab length, Equation (7) gives for $V = 1.1$ m/s, $f = 0.87$ Hz; for $V = 2.2$ m/s, $f = 1.75$ Hz.

Due to the oscillations of the slab, zones of higher underpressure are created within the underlying water layer, close to the upstream and downstream slab edges. Since the amplitude of the oscillations increases with discharge, a discharge can be reached that forces the upstream face of the slab to protrude enough to produce a flow separation covering nearly all the length of the slab, and, as a consequence, to spread the higher upstream underpressures to all of the lower face of the slab. This is enough for the slab to be lifted up [8].

The author opines that this kind of failure can be avoided if the upstream front of the oscillating slab is impeded to protrude over the floor level. This can be achieved by superimposing the downstream extremity of each one of the slabs that support high-velocity flow on the upstream extremity of the following one. Also, a small scaling of the slabs, so that each one of them be slightly lower than the upstream slab, could produce a similar effect. When selecting the second solution, one has to make sure that cavitation risks are avoided.

TWO-PHASE SYSTEMS

Aerated Falling Nappes

The oscillations of a sheet of water that overflows the crest of a dam (Figure 13) can create a hazardous condition when they approach the natural frequency of vibration of the structure. In the following, it will be shown how Equations (7) and (8) can help to predict the frequency at which the nappe oscillates.

Example [2]. Careful laboratory experiments were performed by letting a sheet of water fall vertically from a

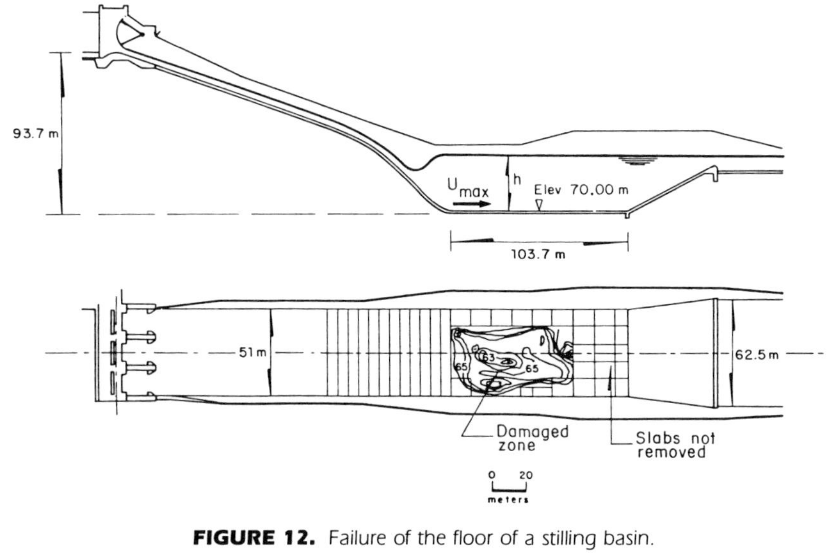

FIGURE 12. Failure of the floor of a stilling basin.

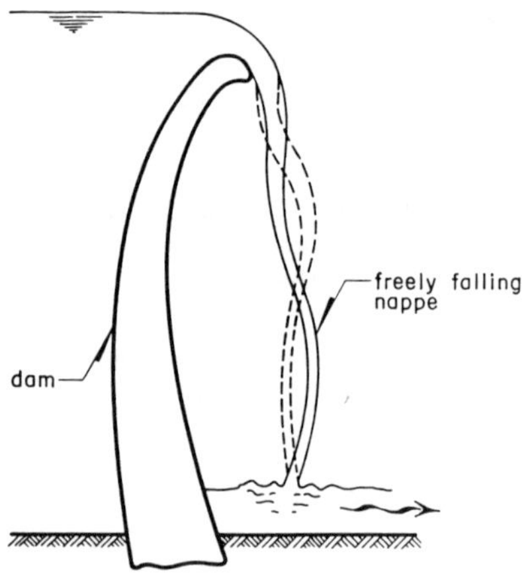

FIGURE 13. Oscillating water nappe overflowing an arch dam.

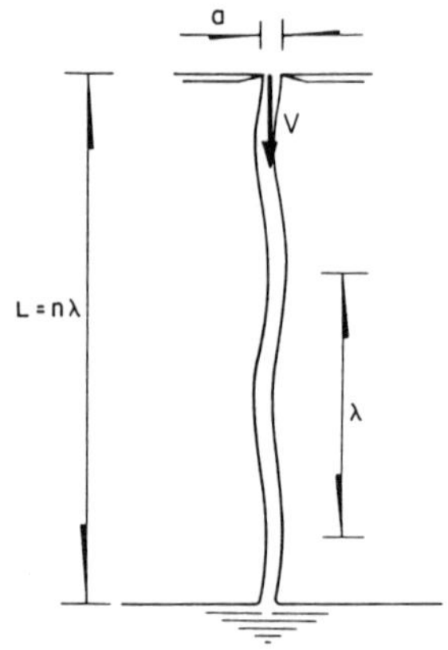

FIGURE 14. Oscillating water nappe falling from a slit.

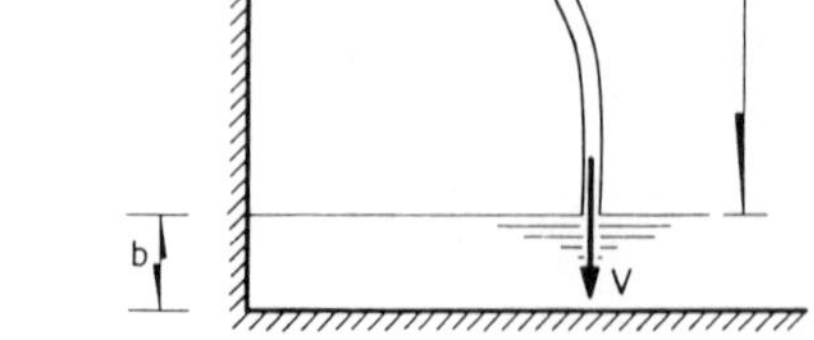

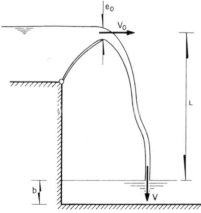

FIGURE 15. Non-aerated water nappe overflowing a fixed flap gate.

horizontal slit, whose width a could be varied. Other related parameters were (Figure 14) the drop height L, the unit discharge q and the nappe oscillation mode n, i.e. the number of waves that were present in the nappe profile.

The wavelength λ being equal to L/n, one gets from Equation (8) that

$$d = L/2\pi n \tag{9}$$

Moreover $V = q/a$. Substituting into Equation (7), one gets that

$$f = \frac{nq}{La} \tag{10}$$

Now, laboratory results agree with Equation (10). For a slit width $a = 0.38$ mm, it was found, for instance, that with $L = 42.4$ cm, $q = 10.9$ cm²/s, $n = 2$, one had $f = 13$ Hz [Equation (10) gives $f = 13.5$ Hz]; with $L = 45$ cm, $q = 14.8$ cm²/s, $n = 4$, one had $f = 33.5$ Hz [Equation (10) gives $f = 34.6$ Hz]; with $L = 50$ cm, $q = 23.1$ cm²/s, $n = 5$, one had $f = 60.5$ Hz [Equation (10) gives $f = 60.8$ Hz]; and so on. It is interesting to recall that part of those measurements were taken with atmospheric ambient pressure, part by placing the apparatus within a compression chamber, in order to experiment under reduced ambient pressures. Equation (10) is equally valid for all of them.

Non-Aerated Falling Nappes

One says that the nappe is non-aerated when the water sheet confines completely the air body that lies between it and the structure, hindering its communication with outer atmosphere. It is believed that this condition favors the transfer of vibrations from the falling nappe to the structure. Therefore, it is usual to recommend in this case to set aerodynamically shaped piers over the dam crest, in order to split the nappe into separate jets. Anyway, we have at our disposal also useful data concerning the case of non-aeration.

Example [18]. The oscillations of the water nappe overflowing a fixed flap gate were studied in a laboratory channel. Water discharge, drop height L and pool depth b (Figure 15) could be varied. In all the tests, the water curtain touched the channel walls, confining the hinder air. Twenty-six modes of oscillation of the nappe were observed, each one with various discharges, and the corresponding Strouhal numbers $S_e = fL/V$ were calculated, V representing the nappe velocity as it enters the pool. For all the modes n between 4 and 27, S_e/n resulted to be nearly constant, its mean value being $S_e/n = 0.532$. As a consequence,

$$\frac{S_e}{\pi n} = \frac{fL}{\pi n V} = 0.169 \tag{11}$$

Moreover, it was found that the nappe is likely to vibrate if

$$e_0/L < 0.1335 - 0.102Fr_0 \qquad (12)$$

e_0 being the head over the crest, V_0 the corresponding velocity and $Fr_0 = V_0/\sqrt{ge_0}$.

Flap Gates

Overflowed flap gates free to oscillate, usually do not vibrate with a frequency induced by the flow, but with their own natural frequency. See for example [17]. A broad analysis of the problem, taking into account the mutual effects of a movable flap having a given structural elasticity, the oscillating water nappe and the volume of air enclosed between the gate and the water curtain, can be found in [19].

Part B. Water Waves as an Exciting Force

SMALL-AMPLITUDE WAVE THEORY

General Equations

Let us suppose that a cylindrical wave train, whose *wavelength* is λ, advances over a horizontal seabed, with *celerity* c. Let coordinate axes x, y be set as Figure 16 shows, h being the mean water depth.

Let us admit that a two-dimensional velocity potential ϕ (x, y, t) exists, t being the time, such that

$$\frac{\partial^2\phi}{\partial x^2} + \frac{\partial^2\phi}{\partial y^2} = 0 \qquad (13)$$

Let us call η (x, t) the instantaneous elevation of the free water surface over the level $y = 0$. The Bernoulli equation at the surface can be written

$$\frac{\partial\phi}{\partial t} + \frac{p}{\rho} + g\eta = 0 \qquad \text{at } y = \eta \qquad (14)$$

By assuming for atmospheric pressure the value $p = 0$ and supposing the wave amplitude to be very small, Equation (14) can be simplified as

$$\frac{\partial\phi}{\partial t} + g\eta = 0 \qquad \text{at } y = 0 \qquad (15)$$

The vertical component of the velocity of a surface particle can be represented either by $\partial\eta/\partial t$ or by $\partial\phi/\partial y$. Therefore, we can write

$$\frac{\partial\phi}{\partial y} = \frac{\partial\eta}{\partial t} \qquad \text{at } y = 0 \qquad (16)$$

Equations (15) and (16) combine as

$$\frac{\partial^2\phi}{\partial t^2} + g\frac{\partial\phi}{\partial y} = 0 \qquad \text{at } y = 0 \qquad (17)$$

In order to envisage wave train as steady, the observer needs to follow it at the speed c. This implies to substitute variables x, t by their combination $x - ct$. If moreover one accepts that the potential ϕ can be represented by separating variables, one will write

$$\phi = Y(y)X(x - ct) \qquad (18)$$

Substituting this expression into Equation (13), one gets the following ordinary differential equations

$$\frac{\partial^2 Y}{\partial y^2} - k^2Y = 0 \qquad (19)$$

$$\frac{\partial^2 X}{\partial x^2} + k^2X = 0 \qquad (20)$$

The general solutions of Equations (19) and (20) are re-

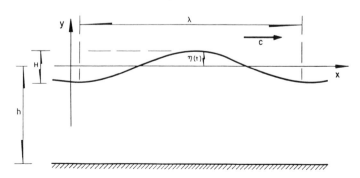

FIGURE 16. Definition sketch for a progressive wave train.

spectively linear combinations of hyperbolic and natural sinuses and cosinuses. Considering that at seabed no vertical motion is possible and assuming that $\phi = 0$ when $x = 0$, $t = 0$, one obtains that

$$Y = m \cosh k(y + h), \quad X = n \sin k(x - ct) \tag{21}$$

m, n being constant coefficients. Substituting expressions (21) into Equation (18) and putting $mn = a$, it results

$$\phi = a \cosh k(y + h) \sin k(x - ct) \tag{22}$$

Introducing the expression (22) into Equation (15) one gets that

$$\eta = \frac{H}{2} \cos k(x - ct) \tag{23}$$

where

$$H = (2akc \cosh kh)/g \tag{24}$$

Therefore,

$$a = \frac{gH}{2kc \cosh kh} \tag{25}$$

Finally, by substituting ϕ derivatives into Equation (17), one gets that

$$c^2 = \frac{g}{k} \tanh kh \tag{26}$$

which is the *Airy formula* for wave celerity.

Main Parameters

Making in Equation (23) $\cos k(x - ct) = 1$, one deduces that H represents the height of the wave, as shown by Figure 16. Also, Equation (23) shows that the *wave number* k has to be

$$k = 2\pi/\lambda \tag{27}$$

since an abscissa increase $\Delta x = \lambda$ has to leave unchanged the elevation η. Moreover,

$$c = \lambda/T \tag{28}$$

T being the *wave period;* and, from Equations (27), (28)

$$T = 2\pi/kc \tag{29}$$

Finally, substituting Equations (26) and (29) into Equation (25), one gets that

$$a = \frac{\pi H}{kT \sinh kh} \tag{30}$$

Velocities and Accelerations

Taking into account Equation (30), Equation (22) can be written

$$\phi = \frac{\pi H}{kT} \frac{\cosh k(y + h)}{\sinh kh} \sin \theta \tag{31}$$

where

$$\theta = k(x - ct) \tag{32}$$

is the *wave phase-angle*. The x-component of the velocity, $u = \partial\phi/\partial x$, is therefore

$$u = \frac{\pi H}{T} \frac{\cosh k(y + h)}{\sinh kh} \cos \theta \tag{33}$$

This formula shows that, at a given vertical section, velocities not only change with time, but also become smaller and smaller with increasing depth. Deriving Equation (33) and making use of Equation (29), one gets

$$\frac{du}{dt} = \frac{2\pi^2 H}{T^2} \frac{\cosh k(y + h)}{\sinh kh} \sin \theta \tag{34}$$

formula that expresses the particle acceleration at any point (x, y).

Deep-Water Conditions

Previous equations simplify when $h > \lambda/2$. Then one says to have deep-water conditions. This is usually the case for offshore structures.

With deep-water conditions, one may assume that y has great values. Therefore, one writes, with a good approximation,

$$\sinh ky \cong \cosh ky \cong e^{ky} \tag{35}$$

So Equation (26) becomes

$$c = \sqrt{g/k} \tag{36}$$

and Equations (33), (34) become

$$u = \frac{\pi H}{T} e^{ky} \cos \theta \tag{37}$$

$$\frac{du}{dt} = \frac{2\pi^2 H}{T^2} e^{ky} \sin \theta \tag{38}$$

Finally, we will obtain a useful expression for wavelength λ. From Equation (36), we have that $kc = \sqrt{gk}$. This, substituted into Equation (29), gives $k = 4\pi^2/gT^2$, which, compared with Equation (27), gives

$$\lambda = gT^2/2\pi \tag{39}$$

Morison Equation

When a fluid body flows against a solid obstacle, it applies to it a force which depends on the density ρ of the fluid, its mean velocity V and the area A that the obstacle opposes to flow. Dimensional considerations show that this *drag force* F_d has to be expressed through the *Newton formula*

$$F_d = \frac{1}{2} \rho C_d A V^2 \tag{40}$$

where C_d is the *drag coefficient*. If moreover the fluid motion is not permanent, as in the case of waves, an additional *inertia force* F_i appears, associated with flow aceleration dV/dt:

$$F_i = \rho C_m \mathcal{V} dV/dt \tag{41}$$

C_m being the *inertia coefficient* and $\mathcal{V}$ a volume representative of the actuating fluid body.

Morison and co-workers [12] suggested in 1950 to express accordingly the *wave force* F on a vertical structural element of width D and height h as

$$F = \left[\frac{1}{2} \rho C_d D V|V| + \rho C_m \frac{\pi D^2}{4} \frac{dV}{dt}\right] h \tag{42}$$

Equation (42) is called *Morison equation*. This equation is considered to be valid if the structural element is relatively narrow in comparison with wave length; usually one limits its validity to $D < \lambda/5$. $V|V|$ has been written in Equation (42) instead of V^2 of Equation (40) because, due to the swaying motion of the wave, velocity periodically changes its direction, and its wave force along with it. The product $V|V|$ has at the same time the magnitude of V^2 and the sign of V.

When deep-water conditions for small-amplitude waves are fulfilled, one has, making use of Equation (37),

$$\begin{aligned} V|V| &= \pm V^2 = \pm \frac{1}{h} \int_{-h}^{0} u^2 dy \\ &= \pm \frac{\pi^2 H^2}{2khT^2} (1 - e^{-2kh}) \cos^2 \theta \end{aligned} \tag{43}$$

and making use of Equation (38),

$$\frac{dV}{dt} = \frac{1}{h} \int_{-h}^{0} \frac{du}{dt} dy = \frac{2\pi^2 H}{khT^2} (1 - e^{-kh}) \sin \theta \tag{44}$$

Therefore, Morison Equation (42) can be written

$$F = \pm m \cos^2 \theta + n \sin \theta \tag{45}$$

with

$$m = \rho C_d \frac{\pi^2 H^2 D}{4kT^2} (1 - e^{-2kh}) \tag{46}$$

$$n = \rho C_m \frac{\pi^3 H D^2}{2kT^2} (1 - e^{-kh}) \tag{47}$$

the sign $\pm$ being the same as the sign of $\cos \theta = \cos k(x - ct)$.

Equation (45) shows that wave force F varies with phase angle θ. Normally, the designer is simply interested in determining the *maximum value* of F. To get this value, one has to put $\partial F/\partial \theta = 0$. Now, from Equation (45) it results that

$$\partial F/\partial \theta = (\pm 2m \sin \theta + n) \cos \theta \tag{48}$$

Therefore, the peak corresponds to one of the values of θ for which

$$\sin \theta = \pm \frac{n}{2m} \tag{49}$$

Drag and Inertia Coefficients

Coefficients C_d and C_m depend on both wave and flow characteristics. Wave motion can be represented by its period T, or better, in non-dimensional form, by the *Keulegan–Carpenter number*

$$K = u_{max} T/D \tag{50}$$

u_{max} representing the maximum value reached by velocity u. Now, if u is given by Equation (33), its maximum value is attained when $y = 0$, $\theta = 0$. Therefore,

$$u_{max} = \frac{\pi H}{T \tanh kh} \tag{51}$$

Substituting into Equation (50), we have that

$$K = \frac{\pi H}{D \tanh kh} \tag{52}$$

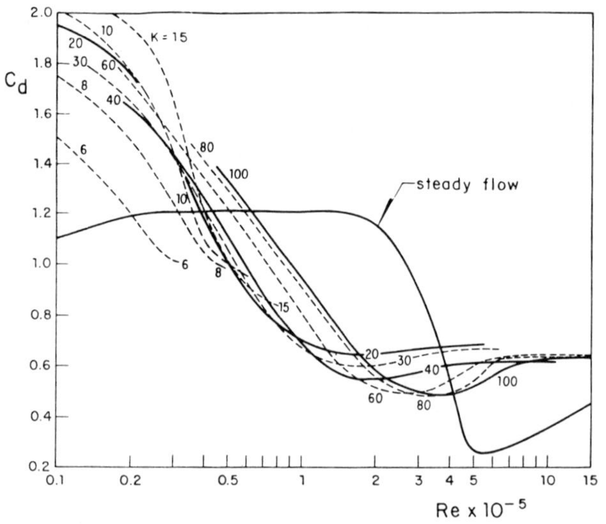

FIGURE 17. Drag coefficient C_d as a function of Keulegan-Carpenter number K and Reynolds number R_e, for circular cylindrical obstacles, from Sarpkaya [27].

Flow characteristics are represented by Reynolds number

$$Re = u_{max}D/\nu \tag{53}$$

ν being the kinematic viscosity coefficient. By Equation (51), one can write

$$Re = \frac{\pi HD}{\nu T \tanh kh} \tag{54}$$

Under deep-water conditions, Equations (52), (54) are simplified as

$$K = \pi H/D, \qquad Re = \pi HD/\nu T \tag{55}$$

Figures 17 and 18, taken from [27], gives values of C_d and C_m as functions of K and Re.

Example of Application

Calculate the peak load imposed upon a cylindrical pile of 20 cm diameter by a 1.80 m high wave, with 3 s period.

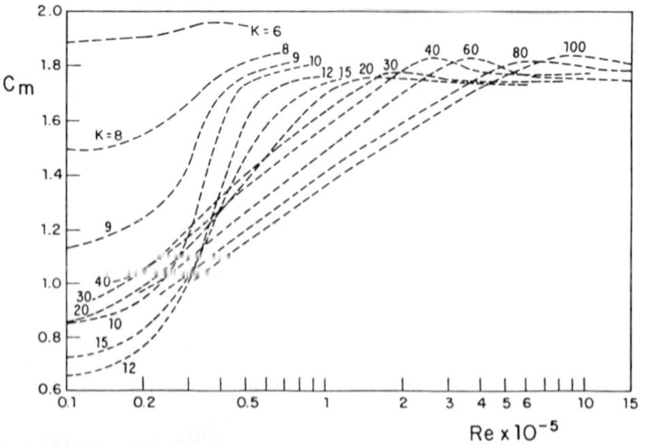

FIGURE 18. Inertia coefficient C_m as function of Keulegan-Carpenter number K and Reynolds number R_e, for circular cylindrical obstacles, from Sarpkaya [27].

Mean water depth is 9 m. Assume the pile dynamic response to be negligible. We have

$$D = 0.2 \text{ m}, \quad H = 1.8 \text{ m}, \quad T = 3 \text{ s}, \quad h = 9 \text{ m}$$

From Equation (39) it results

$$\lambda = gT^2/2\pi = 14.0 \text{ m}$$

Deep-water assumption requires that $h > \lambda/2 = 7$ m. This condition is satisfied. Also satisfied is the condition for applying Morison equation, $D < \lambda/5 = 2.80$ m.

Now we have to obtain coefficients C_d, C_m. Taking, for sea water at 20°C, $\nu = 1.06 \times 10^{-6}$ m²/s, we have from Equations (55) that

$$K = \pi H/D = 28.2$$

$$Re = \pi DH/\nu T = 3.5 \times 10^5$$

With those values, Figures 17, 18 give

$$C_d = 0.65, \quad C_m = 1.74$$

Since by Equation (27)

$$k = 2\pi/\lambda = 0.45 \text{ m}^{-1}$$

taking for sea water the density $\rho = 1025\ Ns^2/m^4$ and substituting the corresponding values for the other variables, from Equations (46), (47) we get $m = 263.0\ N$, $n = 483.0\ N$. Therefore Equation (45) gives the wave force

$$F = \pm 263 \cos^2\theta + 483 \sin\theta \qquad (56)$$

The peak value of wave force corresponds, according to Equation (49), to

$$\sin\theta = n/2m = 0.9182$$

$$\cos^2\theta = 1 - \sin^2\theta = 0.1569$$

and, substituting into Equation (56), one gets finally that the maximum load on the pile is

$$F_{max} = 263 \times 0.1569 + 483 \times 0.9182 = 484.7N$$

NOTATION

a = width; constant coefficient
B = breadth
c = wave celerity
C_d = drag coefficient
C_m = inertia coefficient
d = width of restrained fluid layer
D = width of cylindrical obstacle
f = frequency
F = force
F_d = drag force
F_i = inertia force
Fr = Froude number
g = acceleration due to gravity
h = water depth
H = difference in level; wave height
k = wave number
K = Keulegan–Carpenter number
L = length
m = constant coefficient
n = constant coefficient; oscillation mode
q = discharge per unit width
Re = Reynolds number
S = Strouhal number
t = time
T = period
u = x-component of the velocity
V = velocity
x, y = Cartesian coordinates
η = oscillating-surface elevation
θ = phase angle
λ = wave length
ν = kinematic viscosity
ρ = water density
ϕ = velocity potential

REFERENCES

1. Bénard, H., ''Formation des centres de giration â l'arrière d'un obstacle en mouvement,'' *Comptes Rendus de l'Académie des Sciences de Paris*, Vol. 174, 1908, pp. 839–842.
2. Binnie, A. M., ''Resonating Waterfalls,'' *Proceedings Royal Society*, London, Vol. A339, 1974, pp. 435–449.
3. Birkhoff, G., ''Formation of Vortex Streets,'' *Journal of Applied Physics*, Vol. 24, 1953, No. 1, pp. 98–103.
4. Castex, L., ''Etude de la vibration des barreaux de grille sous l'influence des tourbillons alternés,'' *Proceedings VII Congress, International Association for Hydraulic Research*, Lisbon, 1957, Vol. 1, paper C21.
5. Falvey, H. T., ''Bureau of Reclamation Experience with Flow-induced Vibrations,'' *Practical Experiences with Flow-induced Vibrations*, Springer, Berlin, 1980, pp. 386–398.
6. Ishii, N., Imaichi, K., and Hirose, A., ''Dynamic Instability of Tainter Gates,'' *Practical Experiences with Flow-induced Vibrations*, Springer, Berlin, 1980, pp. 452–460.
7. Kolkman, P. A., ''Development of Vibration-free Gate Design. Learning from Experience and Theory,'' *Practical Experiences with Flow-induced Vibrations*, Springer, Berlin, 1980, pp. 351–385.

8. Levi, E. and Del Risco, E., "A Search for the Cause of High-Speed-Channel Revetment Failures," to be published in the *Journal of Performance of Constructed Facilities, ASCE.*
9. Levi, E., ''A Universal Strouhal Law,'' *Journal of Engineering Mechanics, ASCE*, Vol. 109, 1983, No. 3, pp. 718–727.
10. Levi, E., ''A Universal Strouhal Law-Closure,'' *Journal of Engineering Mechanics, ASCE*, Vol. 110, 1984, No. 5, pp. 841–845.
11. Levin, L., ''Etude hydraulique des grilles de prise d'eau,'' *Proceedings VII Congress, International Association for Hydraulic Research*, Lisbon, 1957, Vol. 1, Paper C11.
12. Morison, J. R. *et al.*, ''The Force Exerted by Surface Waves on Piles,'' *Petrol Transactions, AIME*, Vol. 189, 1950, pp. 149–154.
13. Naudascher, E. (ed.), *Flow-induced Structural Vibrations*, Springer, Berlin, 1974.
14. Naudascher, E., ''On Identification and Preliminary Assessment of Sources of Flow-induced Vibrations,'' *Practical Experiences with Flow-induced Vibrations*, Springer, Berlin, 1980, pp. 520–522.
15. Naudascher, E. and Rockwell, D. (Ed.), *Practical Experiences with Flow-induced Vibrations*, Springer, Berlin, 1980.
16. Neilson, F. M. and Pickett, E. B., ''Corps of Engineers Experiences with Flow-induced Vibrations,'' *Practical Experiences with Flow-induced Vibrations*, Springer, Berlin, 1980, pp. 399–413.
17. Ogikara, K. and Ueda, S., ''Flap Gate Oscillation,'' *Practical Experiences with Flow-induced Vibrations*, Springer, Berlin, 1980, pp. 466–470.
18. Partenscky, H. W. and Sar Khloeung, I., ''Discussion of Paper No. S6,'' *Proceedings XII Congress, International Association for Hydraulic Research*, Fort Collins, 1967, Vol. 5, pp. 593–600.
19. Partenscky, H. W. and Swain, A., ''Theoretical Study of Flap Gate Oscillation,'' *Proceedings XIV Congress, International Association for Hydraulic Research*, Paris, 1971, Vol. 2, pp. 213–220.
20. Petrikat, K., "Seal Vibration," *Practical Experiences with Flow-induced Vibrations*, Springer, Berlin, 1980, pp. 476–497.
21. *Proceedings XIV Congress, International Association for Hydraulic Research*, Paris, 1971, Vol. 2.
22. Rohde, F. G., Rouvé, G., and Pasche, E., ''Self-excited Oscillatory Surface Waves around Cylinders,'' *Practical Experiences with Flow-induced Vibrations*, Springer, Berlin, 1980, pp. 514–519.
23. Roshko, A., ''On the Drag and Shedding Frequency of Two-dimensional Bluff Bodies,'' *NACA Technical Note 3169*, July 1954.
24. Rouvé, G. and Traut, F. J., ''Vibrations due to Two-phase Flow below a Tainter Gate,'' *Practical Experiences with Flow-induced Vibrations*, Springer, Berlin, 1980, pp. 461–465.
25. Sánchez–Bribiesca, J. L. and Capella, A., ''Turbulence Effects on the Lining of Stilling Basins,'' *Proceedings, XI Congrès de Grands Barrages*, Madrid, 1973, Paper Q.41, R.83.
26. Sarpkaya, T., ''Vortex Shedding and Resistance in Harmonic Flow about Smooth and Rough Circular Cylinders at High Reynolds Numbers," Report No. NPS-59 SL76021, Naval Postgraduate School, Monterey, CA 1976.
27. Sarpkaya, T. and Isaacson, M., *Mechanics of Wave Forces on Offshore Structures*, Van Nostrand, New York, 1981.
28. Sarpkaya, T. and Shoaff, R. L., ''An Inviscid Model of Two-dimensional Vortex Shedding for Transient and Asymptotically Steady Separated Flow over a Cylinder,'' *Proceedings XVII Aerospace Sciences Meeting, AIAA*, New Orleans, 1979, Paper 79-0281.
29. Saxena, P. C. *et al.*, ''Causes and Remedy of Vibrations of a High-head Gate,'' *Practical Experiences with Flow-induced Vibrations*, Springer, Berlin, 1980, pp. 435–438.
30. Stephens, H. S. and Warren, G. B. (Ed.), *Flow-induced Vibrations in Fluid Engineering*, BHRA Fluid Engineering, Cranfield, UK, 1982.
31. Strouhal, V., ''Über eine besondere Art der Tonerregung,'' *Annalen der Physik und Chemie*, Vol. 5, 1878, pp. 216–251.
32. Strutt, J. W. (Lord Rayleigh), ''Aeolian Tones,'' *Scientific Papers*, Dover, New York, 1964, Vol. 6, pp. 315–325.

MOTIONS GENERATED BY WIND–STRUCTURE INTERACTION (NEFTALÍ RODRÍGUEZ-CUEVAS)

INTRODUCTION

Wind is the result of the motion of masses of air with different velocities, with kinetic energy produced through thermodynamic changes of the energy radiated from the sun.

The unequal distribution of energy radiated through the atmosphere, produces changes in temperature and humidity, as well as in pressure distribution, causing movement of air masses due to unbalance in thermal and mechanical actions.

The periodical change of the position of the Earth with

respect to the sun, induces unstable motions of moist air. When the motion generates circulation in local areas, where warm air flows in vertical direction, cyclones appear and move over the oceans. In Figure 19 are shown the common paths of hurricanes and isotherms at the oceans, and two barographs obtained during tropical hurricanes, to show pressure modification generated by strong winds.

A tropical cyclone resembles a heat machine, that is constantly receiving energy from water steam. Energy is released in vast quantities, when water condenses, falling as strong rain and producing strong winds. In middle latitudes, extratropical cyclones and anticyclones move, causing damages when their winds interact with structures.

Often energy is released in small areas, generating tornadoes, where the highest wind velocities appear, producing disasters.

CHARACTERISTICS OF ATMOSPHERE DURING STRONG WINDS

As air moves along Earth surface, friction with the ground generates a boundary layer, where energy is dissipated through eddy viscosity, creating turbulent winds, whose characteristics vary with height and weather conditions.

Temperature and pressure gradients, changes in mass density of the air, effects of angular velocity of Earth rotation, the latitude of the observation point, as well as the viscosity of the air, cause changes in wind velocity, both in its mean value as in its variation with time.

In Figure 20 are shown isolines of equal velocity during a strong wind, on which eddies define gusts of different sizes [19], giving a picture of air motion at a given place of observation.

Analyses of records of wind velocities have shown an increase of average wind speed as height above ground increases, and a decrease in wind index of turbulence, I, defined by the coefficient of variation of wind velocity. Different authors have developed theories to model turbulent winds [11,20,22].

As far as average speed is concerned, two main trends have been used to describe average wind velocity variation with height, along a line normal to the ground:

(a) Mixing length theory [22]
(b) Energy balance theory [11]

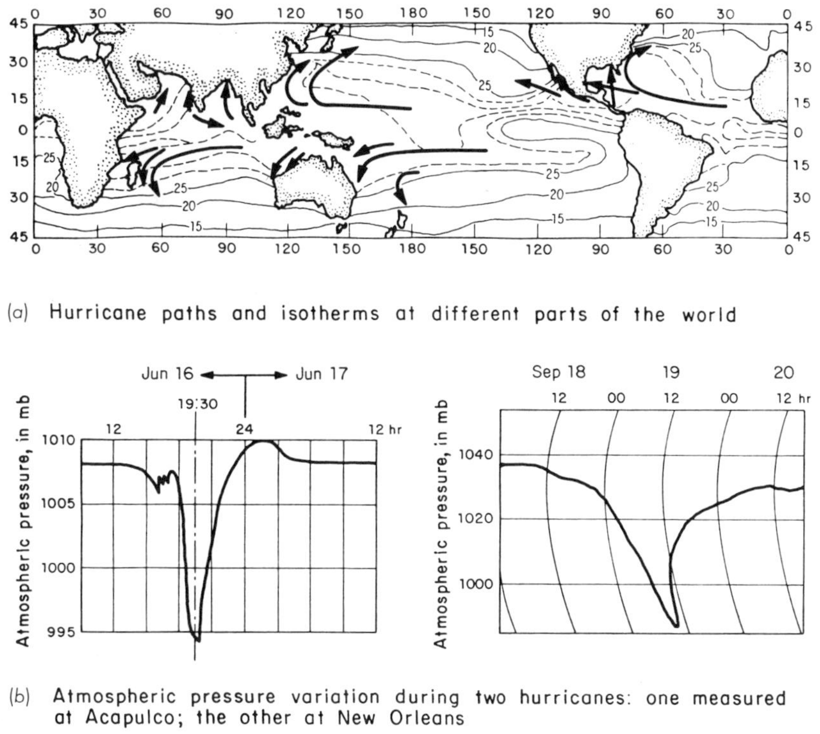

FIGURE 19. Cyclone paths over the oceans; isothermes at sea surface during cyclones and barograph records obtained during tropical cyclones.

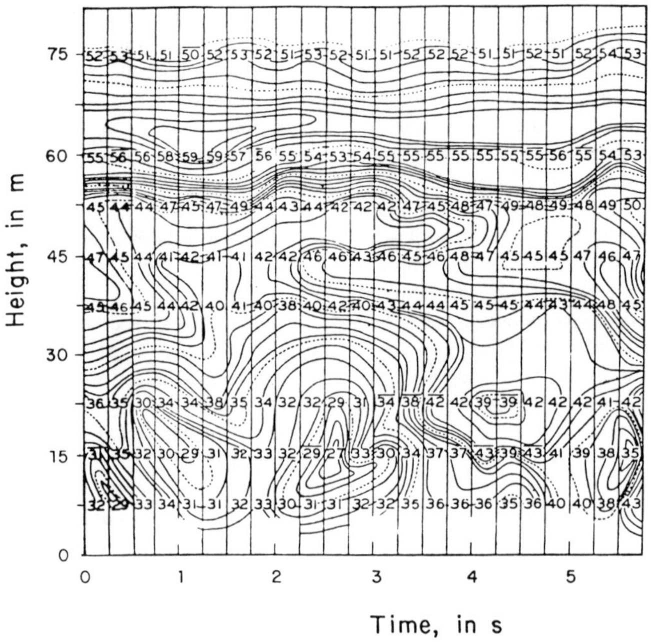

FIGURE 20. Equal wind velocity lines observed at a measuring station during strong winds, from Sherlock [19].

The mixing length theory provides a power law variation of average velocity with height, as follows:

$$\overline{V}_z = V_{10}\left(\frac{h}{10}\right)^{\alpha}$$

where

$\overline{V}_z$, $\overline{V}_{10}$ = average speed at heights z and 10 m above ground
h = height where $\overline{V}_z$ is computed
α = dimensionless number

Common values for α recommended in wind engineering practice are shown in Table 1. Its value depends on topographic features at the ground and on the stability of the atmosphere, measured by the vertical thermal gradient. Wind flow, under neutral stratification, reduces turbulent energy budget to a condition of zero heat flux, and a logarithmic wind profile is obtained from the energy balance theory, to compute average wind speed at different heights, given by:

$$\overline{V}_z = \frac{V^*}{k_*}\ln\left(\frac{z - z_d}{z_0}\right) \tag{2}$$

where

k_* = Von Kármán's constant, equal to 0.4
V^* = friction velocity = $\sqrt{\tau/\varrho}$
z_d = displacement length
z_0 = roughness length

Values for parameters z_d and z_0 are given in Table 1.

Careful statistical studies are done to define the average speed, or the characteristic velocity, from records obtained at climatological stations.

Extreme-value statistics are useful to define regional wind velocities with the same return period.

Using Fisher–Tippett type II extreme distribution, the probability of occurrence of a given wind speed level is given by:

$$F(V_i) = e^{-(V_i/\beta)^{-\gamma}} \tag{3}$$

where

$F(V_i)$ = probability of occurrence of V_i
V_i = speed wind level
β, γ = statistical parameters

Through the knowledge of β and γ at a site, it is possible to define average wind speeds for different return periods, R. It is a common engineering practice to select, for structures without human occupants or negligible risk for human life, R = 5 years; for permanent structures, R = 30 years; and for structure on which life should be preserved after a disaster, R = 100 years.

From statistical studies at several meteorological stations, Figure 21 shows the result of a numerical evaluation of the regional wind velocity V_R, for different return periods.

Once V_R is selected for a given structure, it should be multiplied by correcting factors to define the design velocity. They depend on exposure of the structure and on local characteristics of the topography surrounding the structure [3].

For a structural design, wind velocity is averaged over an interval of time T. The averaging time T should be:

(a) Long enough to allow wind vibrations to decrease to a minimum
(b) Long enough to allow for a structural analysis that develops the maximum response
(c) Short enough for a clear understanding of the action of short wind gusts
(d) Long enough for a gust to cover the entire building volume under wind action

Different building codes recommend values for averaging period T; from the structural point of view, it is necessary to use average times related to the dynamic characteristics of the structure under analysis, which usually vary between 3 to 30 s; some authors [8,10,13] have given expressions for relations between averaging period T, and the corresponding average speed, taking into consideration the turbulence index at the site.

For the description of turbulence, average spectra have been found for different roughnesses and heights above ground; the following expression has been suggested:

$$nS^v(n) = 4K\overline{V}_{10}^2 \left(\frac{h}{10}\right)^{\alpha} \phi \left(\frac{nL}{\overline{V}_{10}}\right) \tag{4}$$

where

$S^v(n)$ = power spectral density, at frequency n
K = roughness coefficient (Table 1)
ϕ = a dimensional function of frequency, scale length L and design velocity at 10 m above ground

Harris [7] suggests that

$$\phi = \frac{\dfrac{nL}{V_{10}}}{\left[2 + \left(\dfrac{nL}{\overline{V}_{10}}\right)^2\right]^{5/6}} \tag{5}$$

whereas Davenport [5] recommends

$$\phi = \frac{\left(\dfrac{nL}{\overline{V}_{10}}\right)^2}{\left[1 + \left(\dfrac{nL}{\overline{V}_{10}}\right)^2\right]^{4/3}} \tag{6}$$

Wind turbulence spectra are shown in Figure 22 for comparison. Careful studies of wind power spectral densities show increased densities as the roughness of ground in-

TABLE 1. Parameters Relating Wind Structure Near the Ground with Surface Roughness, on Neutral Atmosphere.

	Type of Exposure	Power Law Exponent, α	Surface Roughness Coefficient, K
Power law	1. Open terrain with very few obstacles, i.e.: tundra, open farmland, desert, seashores	0.16	0.005
	2. Surface with obstacles 10–15 m in height, uniformly distributed, i.e.: small towns, suburbs, etc.	0.28	0.015
	3. Center of cities, with large and irregular objects, as buildings or tall trees and towers	0.40	0.050
	Type of Exposure	**z_0 (meters)**	**z_d**
Logarithmic law	1. Open waters or seashores	0.005–0.01	0
	2. Open terrain without obstacles	0.03–0.10	0
	3. Small towns, suburbs	0.20–0.30	0
	4. Center of towns	0.35–0.45	0
	5. Center of large cities ($\overline{H}$, average height of surrounding buildings)	0.60–0.80	20 m or 0.75 $\overline{H}$

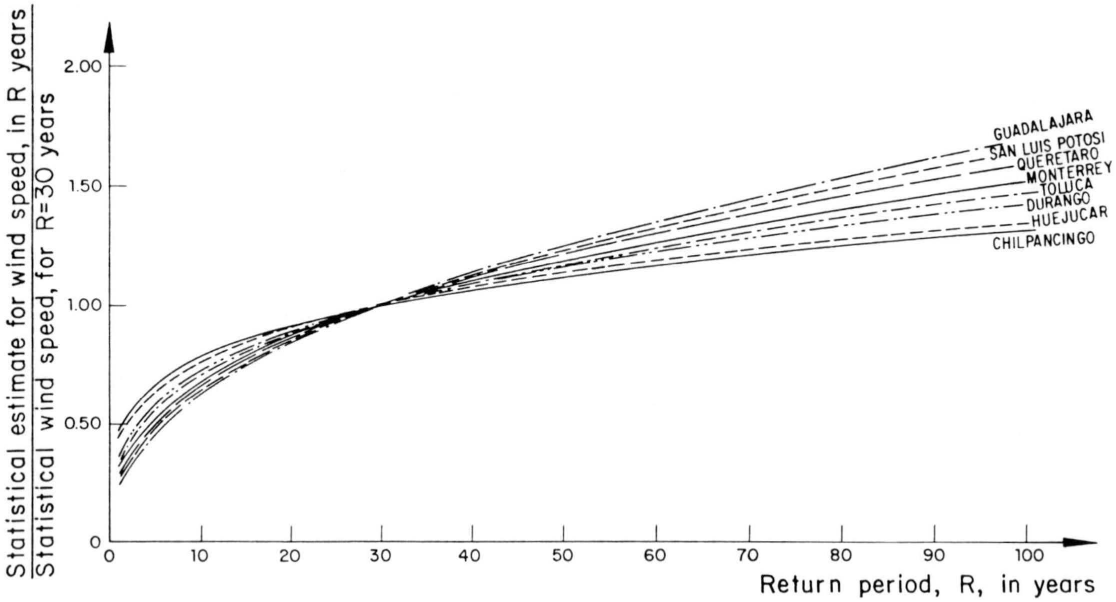

FIGURE 21. Numerical evaluation of regional velocity V_R at different cities of Mexico.

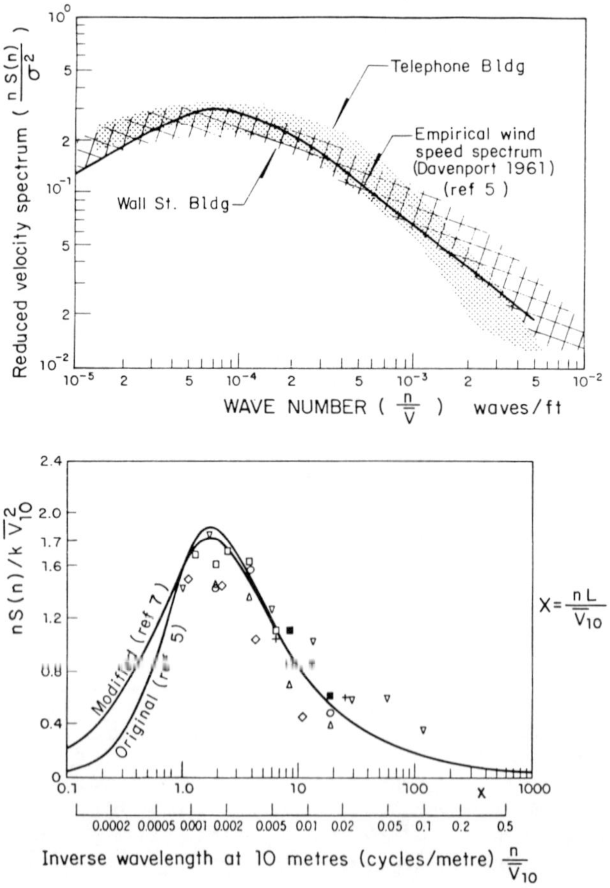

FIGURE 22. Turbulence spectra suggested by two authors, compared with measured values at different observation sites [5,7].

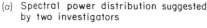

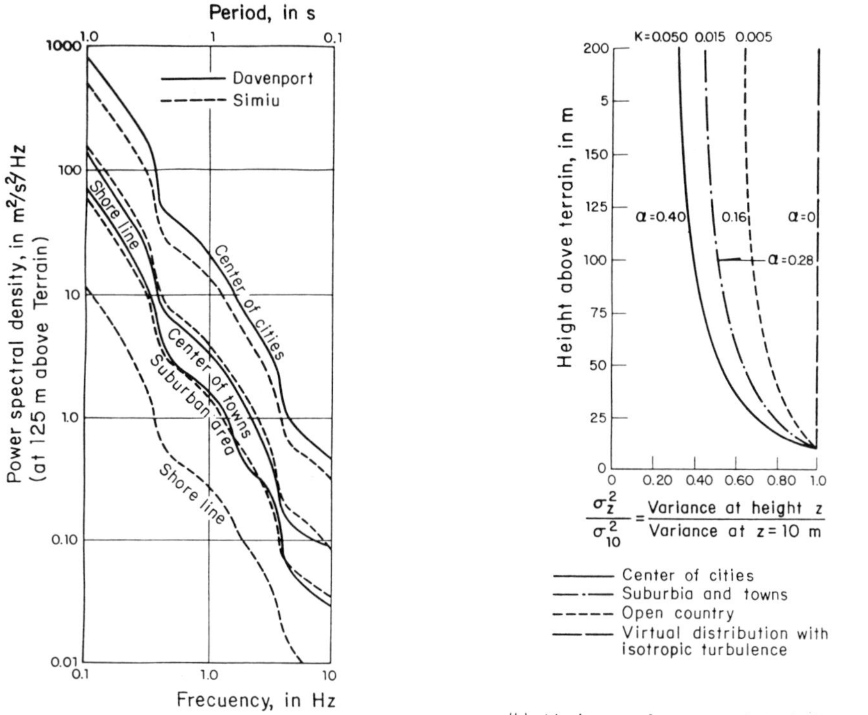

(a) Spectral power distribution suggested by two investigators

(b) Variance of power of turbulence as function of height for different roughness

FIGURE 23. Power spectral densities for different surface roughnesses.

creases and the frequency of a Fourier component of wind velocity has a lower value (as shown in Figure 23). Therefore, the effect of wind turbulence shall produce strong motions on high buildings in center of cities. For small buildings with small natural period at farmlands, wind turbulence effects are negligible [16].

WIND PROPERTIES FOR MOTION ANALYSIS

Atmospheric conditions, topographic features around a structure, and dynamic and geometric characteristics of the structure should be selected for the analysis of the motion that wind-structure interaction produces in a given structure.

To carry out a good estimate of the action of wind it is necessary to know:

(a) State of stability of the atmosphere, for a good selection of wind velocity variation with height above ground
(b) Topographic features around a building, to define a roughness coefficient
(c) Turbulence spectra at different heights
(d) Correlation matrix of the spatial distribution of wind turbulence
(e) Statistical records of wind velocities
(f) Structural characteristics of the building, related to mass distribution, stiffness matrix, damping values and exposed areas.

Based in the above mentioned information, it is possible to identify four basic types of interaction effects induced by wind action.

Figure 24 shows a schematic description of the motions generated by the basic action of wind on structures, and the types of structures sensitive to wind effects.

Laminar flow, vortex generation, turbulence of the air and separation of boundary layers around bodies, are the basic motions generated by strong winds. Their interaction with flexible structures generates their dynamic response, that should be analysed following procedures described in this chapter.

ANALYSIS OF LAMINAR FLOW INTERACTION WITH RIGID STRUCTURES

Rigid structures have low natural periods of vibration that make them insensitive to turbulence effects. Thus, when wind interacts with these structures in a known direction, in the general and local stability analysis of the structure,

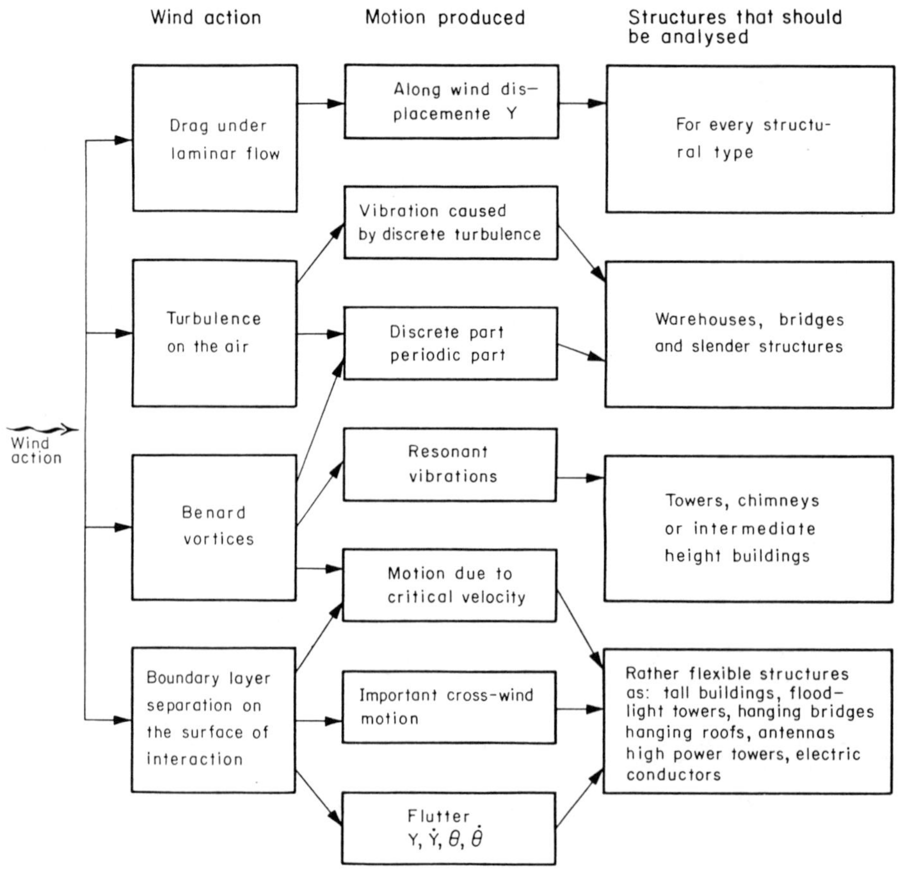

FIGURE 24. Different motions produced by wind-structure interaction.

it is common to consider only the effects generated by laminar flow.

For general stability analysis, wind is considered to produce stable forces, that may be computed by the application of Bernoulli's principle. Two types of forces may be developed at the faces of a structure:

(a) longitudinal force

$$F_L = \frac{1}{2}\rho(V_d)^2 C_{DL} A_L \quad (7)$$

(b) transverse force

$$F_T = \frac{1}{2}\rho(V_d)^2 C_{DT} A_T \quad (8)$$

where

F_L, F_T = longitudinal and transverse wind forces
ρ = density of air
C_{DL}, C_{DT} = longitudinal and transverse drag coefficients, obtained from wind tunnel measurements
A_L, A_T = exposed area for windwise and transverse direction computations

These forces are applied at the pressure center of the structure exposed area, that is not coincident with the centroid of the area, to define the overturning moment to be used for analysis of the stability of the foundation of the structure where wind is acting, to avoid lateral or vertical movements.

Drag coefficients C_{DL} and C_{DT} are highly dependent on Reynolds number [$Re = (Vd/v_d)$]. For laminar flow analyses, it is common to use values obtained from measurements on the range $10^4 \leq Re \leq 10^5$ on which the drag coefficients are rather stable. C_D values are given on building codes and technical literature [1,2,3,6,12,15,21], for structures of different dimensions and shapes. Their values are defined by integration of pressures developed along the surface of bodies under wind-structure interaction.

Pressure distribution around a body is described by

$$p = \frac{1}{2}\rho V_D^2 C_p \tag{9}$$

where

p = pressure at a point at the surface
ρ = density of air
V_D = design velocity, at free flow locations
C_p = pressure coefficient at a point

Pressure coefficients are obtained from model testing at wind tunnels with laminar flow, and they are shown by isobaric lines drawn at the surface of the body, where measuring points connected to piezometric pipes define the pressures induced by wind.

Often, for shell analysis of chimneys, silos or towers in power stations, it is useful to express C_p values as Fourier series, that allow simple mathematical analysis of shell equilibrium under wind effects.

Figure 25 shows pressure distribution around a metallic silo, caused by laminar wind flow.

It is also important to know the distribution of internal pressures generated in permeable bodies, in order to make a proper analysis of local effects produced by wind action, on structural parts of a building.

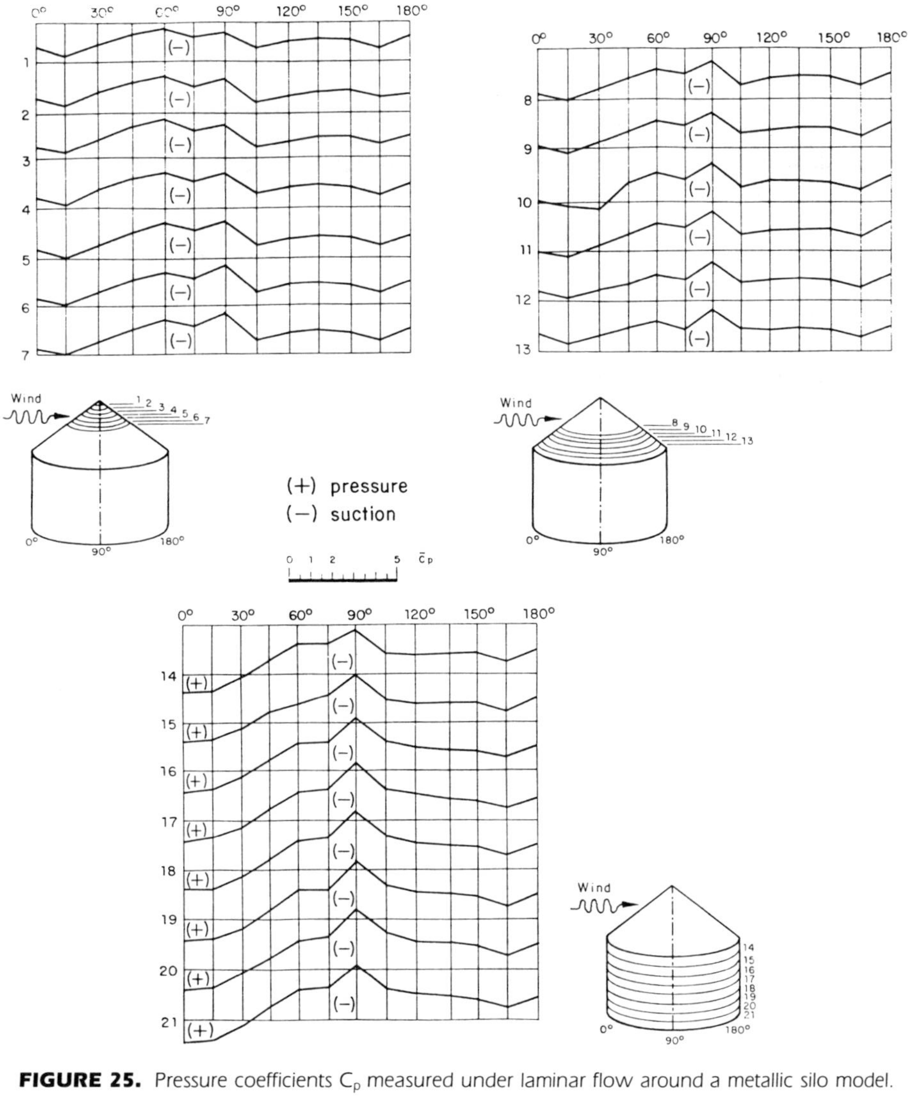

FIGURE 25. Pressure coefficients C_p measured under laminar flow around a metallic silo model.

Building codes usually define local wind effects created along the edges of rectangular buildings, or at eaves, parapets, chimney tops, skylights, roof sheetings, with increased C_p values in specific areas, in order to design fastenings able to withstand wind forces.

Wind action on window panels can break them. Research work in this subject has been carried out [4] in order to avoid damage, by appropriate design rules based on experimental evidence.

Once the pressure distribution is known, structural analysis is carried out to define displacements generated by wind action; from them, internal forces and moments can be found, and wind-resistant design rules can be applied to define adequate dimensions for the structure.

VORTEX EXCITATION ON FLEXIBLE STRUCTURES

Physical evidence has shown that vortex generation produces a cross-wind motion.

Smooth circular cylinders under wind action develop vortices of different type, depending on Reynolds number, *Re*. Three ranges of vortex formation have been described in the literature:

(a) Subcritical range, with periodic vortex shedding, when $Re < 3 \times 10^5$

(b) Critical range, with random vortex shedding, when $3 \times 10^5 < Re < 3.5 \times 10^6$

(c) Transcritical range, with periodic vortex shedding, for $Re > 3.5 \times 10^6$

On ranges where periodic vortex shedding appears, pairs of vortices of equal strength but with reverse rotation, are alternatively shed within the wake, from opposite sides of the cylinder.

As a result, the cylinder undergoes cross-wind oscillations, generated by periodic changes in pressure distribution around the perimeter.

When the vortex frequency coincides with a natural frequency of the structure, harmonic excitation results.

If the vortex frequency is some multiple of a natural frequency, the resulting excitation is subharmonic.

In order to compute vortex frequency, the use of Strouhal number ($S = nd/V$, being n the frequency of vortex; V, the average velocity) has been useful when $Re < 3.5 \times 10^5$.

The shape of the bodies under wind action defines a particular numerical value of the Strouhal number at which vortex shedding appears; S values for different shapes are shown in Figure 26.

When $Re \doteq 3 \times 10^5$, the Strouhal number increases sharply; the drag coefficient C_D shows a sharp drop (Figure 27) and the wake narrows, without a visible vortex formation.

The structure vibrates in an erratic or spasmodic manner, always at their natural frequency.

The equation of motion of a single-degree-of-freedom model, without external excitation can be described by:

$$m\ddot{x}(t) + C\dot{x}(t) + kx(t) = 0 \qquad (10)$$

where

$x(t), \dot{x}(t), \ddot{x}(t)$ = displacement from equilibrium position, velocity and acceleration, respectively
C = coefficient of viscous damping
k = spring constant
m = mass of the structure

when $C < 2\sqrt{mk}$, it is possible to define:

$$\omega_0 = \sqrt{k/m} \quad \text{and} \quad \beta_0 = C/(2\sqrt{mk}) \qquad (11)$$

where ω_0 is the undamped natural circular frequency of the structure, and β_0, the damping ratio between C and the critical value $C_{cr} = 2\sqrt{mk}$.

Vortex excitation induces an aerodynamic exciting force $F(t)$. Its value per unit length, for a simple harmonic motion, with semi-amplitude x_0 and frequency ω, can be expressed by

$$F(t) = (A \text{ sen } \omega t + B \cos \omega t)\eta_0 \rho V^2 d \qquad (12)$$

where

η_0 = x_0/d dimensionless amplitude
d = cross-wind dimension
ρ = air density
A, B = constants

Through algebra, Equation (12) can be rewritten as

$$F(t) = H_a x(t) + K_a \dot{x}(t) \qquad (13)$$

where

$H_a x(t)$ = is an in-phase component with vortex excitation
$K_a \dot{x}(t)$ = is an out-phase component

Introducing these forces in the equation of motion governing cross-wind oscillations:

$$m\ddot{x}(t) + (C - K_a)\dot{x}(t) + (k - H_a)x(t) = 0 \qquad (14)$$

It resembles an equation of motion without external excitation, in which K_a, expresses the contribution to total damping of the system, of the aerodynamic excitation. Using the logarithmic decrement of oscillation δ, defined as the ratio between amplitudes at two successive cycles, i.e. $\delta = 2\pi\beta_0$, K_a can be expressed as:

$$K_a = 2mn\delta_a \qquad (15)$$

H_a expresses the effect of the aerodynamic excitation upon

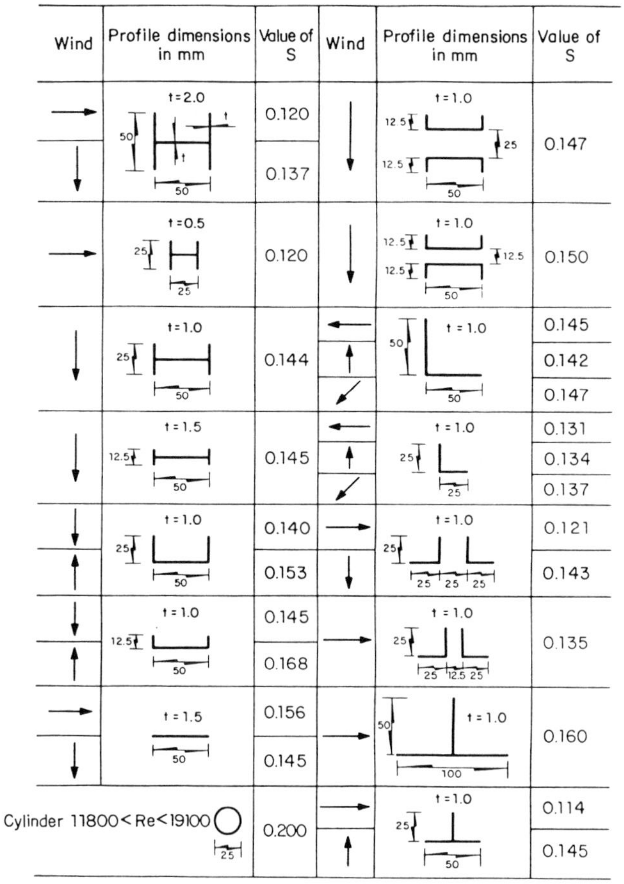
Wind
Profile dimensions in mm
Value of S
Wind
Profile dimensions in mm
Value of S
t=2.0
50
50
0.120
0.137
t=1.0
12.5
25
12.5
50
0.147
t=0.5
25
25
0.120
t=1.0
12.5
12.5
12.5
50
0.150
t=1.0
25
50
0.144
t = 1.0
50
50
0.145
0.142
0.147
t=1.5
12.5
50
0.145
t=1.0
25
25
0.131
0.134
0.137
t=1.0
25
50
0.140
0.153
t=1.0
25
25 25 25
0.121
0.143
t=1.0
12.5
50
0.145
0.168
t = 1.0
25
25 12.5 25
0.135
t=1.5
50
0.156
0.145
t=1.0
50
100
0.160
Cylinder 11800<Re<19100
25
0.200
t=1.0
25
50
0.114
0.145

FIGURE 26. Strouhal numbers for different cross-sections of prismatic bodies under wind action [21].

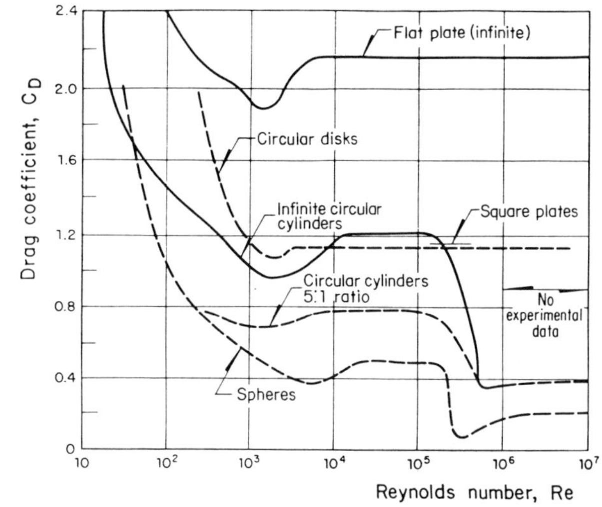

FIGURE 27. Drag coefficient C_D variation with Reynolds number [21].

the natural frequency of the structure, n_0; it is defined by

$$H_a = 4\pi^2 mn^2 \left(1 - \frac{n_0^2}{n^2}\right) \quad (16)$$

where

m = equivalent mass per unit length of the structure
n_0 = natural frequency, in Hz
n = frequency of oscillation, in Hz

When stable vibration generates in a flexible structure, the aerodynamic force can be computed by:

$$F(t) = 4\pi n^2 dm\delta\eta_0\rho \cos(2\pi nt) \quad (17)$$

It causes displacements described by

$$x = x_0 \text{ sen } \omega t$$

These forces and displacements can be decreased using three different devices:

(a) Alteration of the structure, when its natural frequency is similar to that obtained from the Strouhal number, by the use of wind bracing
(b) Increase of the effective damping, through the installation of mechanical damping devices to reduce the amplitude of vibrations
(c) Use of helical strakes spiraled around a circular cylinder, or of a perforated circular shroud, surrounding a circular cylindrical member [21,25]

Strakes or shrouds have proven to be effective for reduction of the problem of vortex-induced vibrations of circular cylinders, by alteration of flow pattern around their surface. Increases in longitudinal drag coefficient C_{DL} should be considered to obtain along-wind forces, when strakes or shrouds are used.

WIND LOADS ON STRUCTURES UNDER TURBULENT FLOW

Turbulence in the atmospheric boundary layer induces unsteady velocity fields on it, modifying wind velocity profiles at a given location, as shown in Figure 28. These changing velocities produce variable response of the structures under wind action.

Two methods have been suggested for assessment of wind turbulence response of structures:

(a) Statistical method, using a stochastic approach
(b) Deterministic method, using random walk processes to generate wind-structure interaction.

Mathematical treatment in both methods shows that wind velocity fluctuations are dependent on dynamic characteristics of the structure and the environment surrounding the structure. This result has been substantiated through experimental work [17].

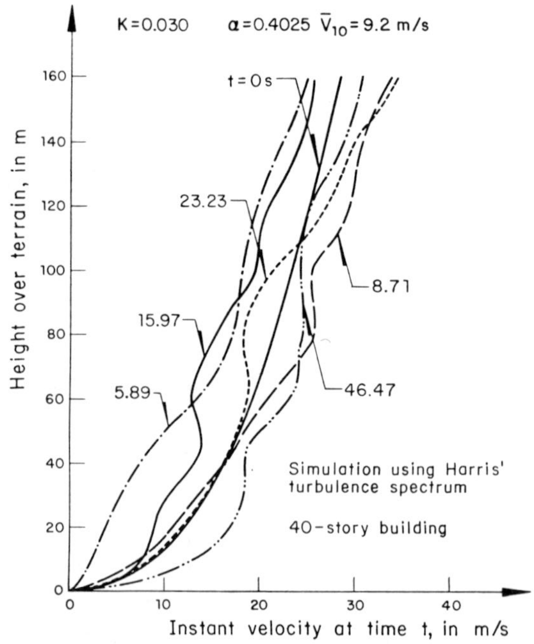

FIGURE 28. Wind velocity profiles generated by turbulent wind as compared with its average value [16].

Stochastic Approach for the Statistical Method

Davenport [5] developed the statistical method. He considered energy transfer from wind turbulence to define a generalized force spectrum. Through the concept of mechanical admittance and a lineal relationship between variable pressure and Reynolds velocity, he develops a response spectrum for the displacement of a single degree of freedom model, in order to obtain an estimate of the maximum response.

Figure 29 shows graphically the basis of the statistical method for a particular structure, from a power density spectrum of time-varying wind velocity. From a cross-correlation function obtained for the structures it is possible to obtain a power density spectrum of time varying wind force that gives a stochastic description of the structure excitation.

Multiplying the ordinates of the power density spectrum of wind force, by those of the mechanical admittance of the structure, the power density spectrum of the time-varying structure deflection gives a stochastic description of the structural response.

The method has been applied in [12] for simplified computations, using the concept of overall gust response factor

G, defined by

$$G = 1 + g\frac{\sigma_{Y'}}{\overline{Y}}$$

$$= 1 + g\left(\frac{2\alpha + 2}{\alpha + 2}\right)[4K^{1/2}]\left(\frac{10}{H}\right)^{\alpha} \tag{18}$$

$$\times \sqrt{\frac{\pi C^2(n_0, A)\left(\frac{1200n_0}{\overline{V}_{10}}\right)^2}{4\beta_0\left[1 + \left(\frac{1200n_0}{\overline{V}_{10}}\right)^2\right]^{4/3}} + \int_0^{\infty}\frac{C^2(n, A)\left(\frac{1200n}{\overline{V}_{10}}\right)^2 dn}{n\left[1 + \left(\frac{1200n}{\overline{V}_{10}}\right)^2\right]^{4/3}}}$$

where

G = gust response factor

$g = \sqrt{2 \ln \nu T} + \dfrac{0.577}{\sqrt{2 \ln \nu T}}$

ν = mean frequency of Y'

$\overline{Y}, Y'$ = average and time-varyng displacement along wind

α = exponent of power law describing wind speed variation with height

T = averaging time

K = roughness coefficient

H = total height of the structure

$$C^2(n, A) = \frac{1}{(1 + \alpha)^2}\left[\frac{1}{1 + \frac{8}{3}\frac{nH}{\overline{V}_H}}\right]\left[\frac{1}{1 + 10\frac{nb}{\overline{V}_H}}\right]$$

= cross-correlation function

n_0 = first mode frequency of the structure

$\overline{V}_{10}$ = average speed at 10 m height

$\sigma_{Y'}$ = standard deviation of Y'

Once the gust response factor G is known [1] for a given structure, and the response $\overline{Y}$ under laminar wind flow has been computed, it is possible to compute the dynamic response under turbulent wind, by the product $\overline{Y}G$.

Random Walk Simulation of Turbulent Wind for Use in a Deterministic Method

By numerical simulation it is possible to generate a wind velocity field whose statistical properties are similar to those obtained from field measurements [16].

For random walk simulation, seeds at different heights above ground are selected to begin random number sequences. By numerical filters, new events are generated with statistical properties that resemble those described by a turbulence spectrum that corresponds to large gust action, whose frequency is related to that of the first fundamental mode of the structure under analysis.

Because the spectrum obtained from large gusts does not correspond to any of those defined by Equations (5) or (6), a corrective random process with frequency content limited by small gust size is generated, and added to the events corresponding to large gusts; the mixture can give any of the spectra used to define wind turbulence.

The method gives wind velocities with spatial correlation attached to natural turbulent winds and turbulence spectrum of known characteristics, in the range of frequencies described by the dynamic properties of the structure.

Figure 30 schematically describes the flow diagram for a computer program that generates turbulent winds associated to a building. Figure 31 shows an ensemble of wind velocities at different times, for the deterministic analysis of the dynamic response of a 9-mass-system idealization of a structure.

The deterministic analysis provides information about the displacement, velocity and acceleration of the masses of the model representing a structure, for an evaluation of wind-structure interaction.

WIND ACTIONS UNDER UNSTABLE FLOWS

Light-weight slender structures with low structural damping, may be damaged when flow instability is generated by wind-structure interaction.

Instabilities may occur when the wind flow pattern is changed by the motion of the structure. Modification of flow pattern changes the aerodynamic forces around a body and unstable fluctuations may induce resonant vibrations or rapid changes on aerodynamic interaction.

Flow instability may appear in smooth and turbulent flow, although the importance of vibrations is dependent on the turbulence of the air.

Four different types of instability have been found:

(a) Resonance due to vortex shedding
(b) Galloping
(c) Interference and proximity effects (buffeting)
(d) Stalling and classical flutter

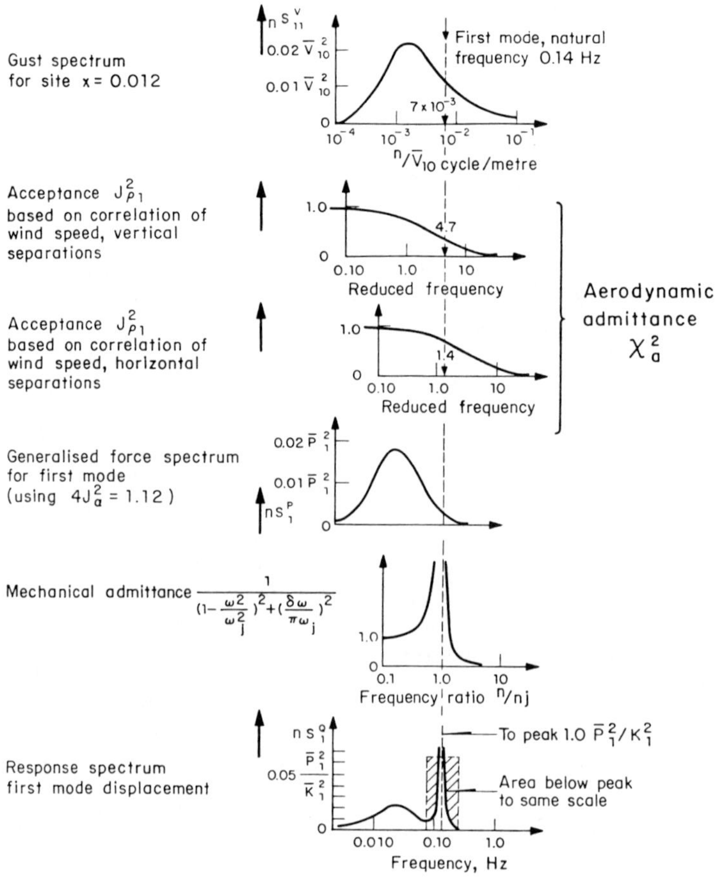

FIGURE 29. Statistical procedure developed by Davenport for evaluation of turbulent response of flexible structures under wind action [26].

In so far only quantitative predictions are available for the description of these phenomena, based on the following parameters:

Structural Parameters

1. Structural logarithmic decrement δ_s

$$\delta_s = \ln \frac{a_N}{a_{N+1}} \tag{20}$$

2. Non-dimensionalized mass

$$M_r = \frac{1}{\rho d^2} \frac{\int_0^1 mf(\xi)d\xi}{\int_0^1 f^2(\xi)d\xi} \tag{21}$$

3. Normalized amplitude of cross-wind oscillation

$$\eta = \frac{x}{d} \tag{22}$$

In expressions (20) to (22)

ρ = density of the air
d = cross dimension of the structure
x = maximum transversed semiamplitude
$mf(\xi)$ = mass function variation per unit length
ξ = dimensionless parameter h/H

Aerodynamic Parameters

1. Reynolds number

$$Re = \frac{Vd}{\nu_d} \tag{23}$$

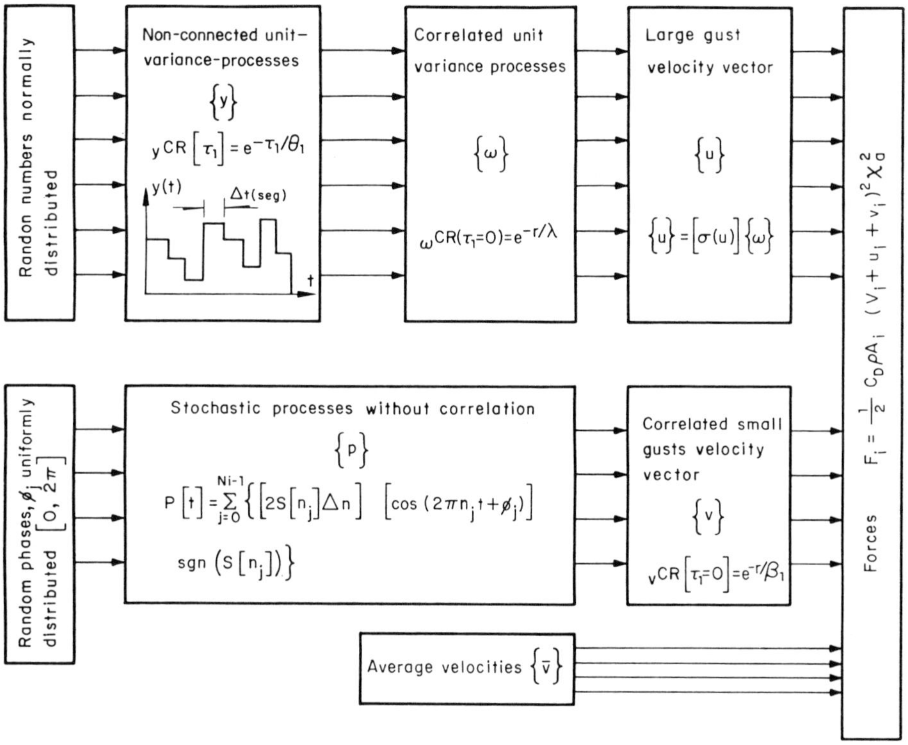

FIGURE 30. Flow diagram for a computer program developed for assessment of structural dynamic response under turbulent wind, using random walk processes.

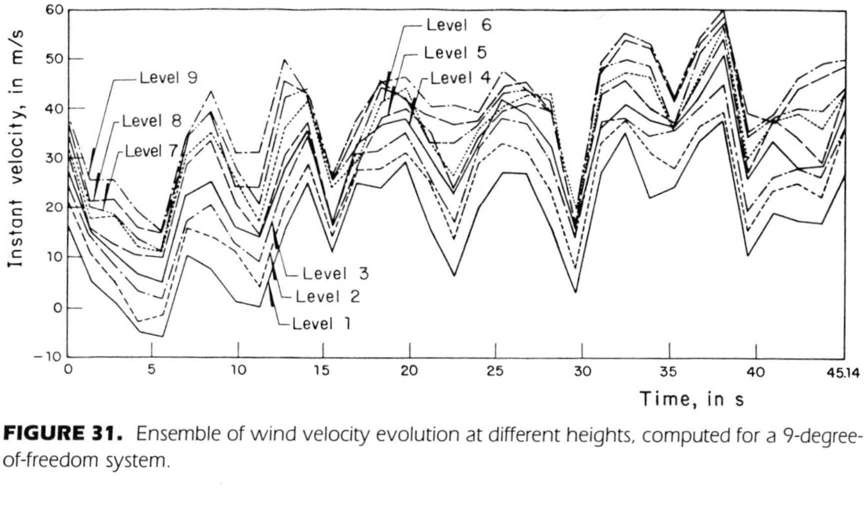

FIGURE 31. Ensemble of wind velocity evolution at different heights, computed for a 9-degree-of-freedom system.

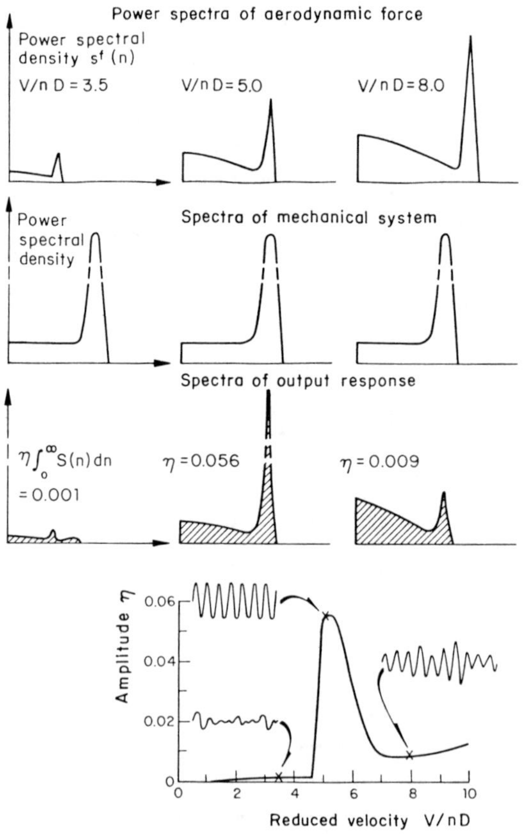

FIGURE 32. Lateral displacement of a flexible structure under different reduced velocities [25].

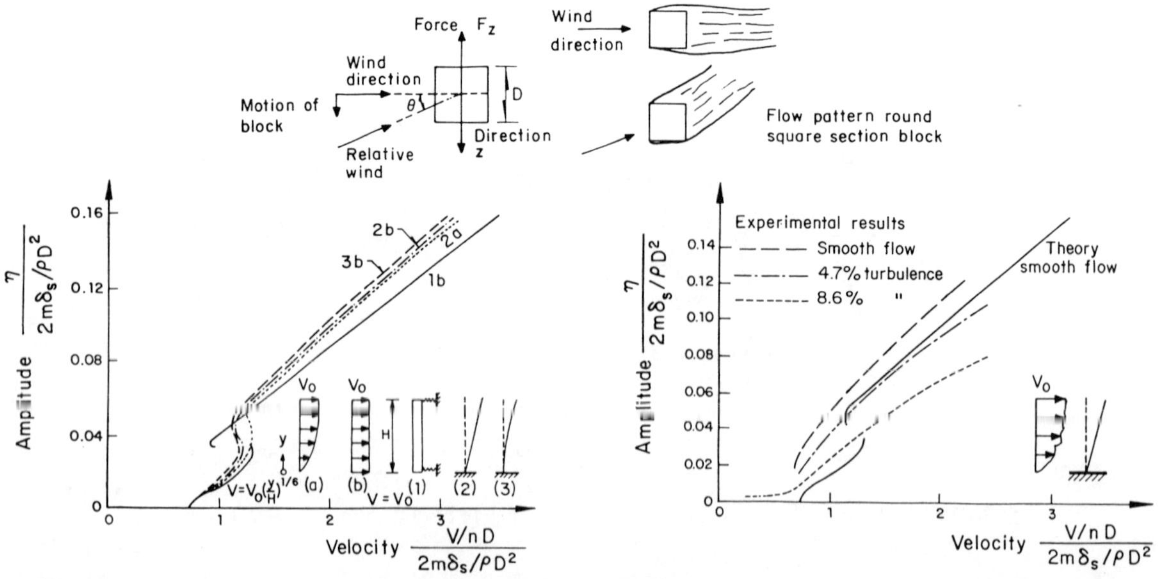

(a) Theoretical response curves, with three different modes of oscilation (ref 14)

(b) Influence of turbulence on experimental results (ref 14)

FIGURE 33. Universal curves for evaluation of structural galloping, under different atmospheric conditions, from Novak [14].

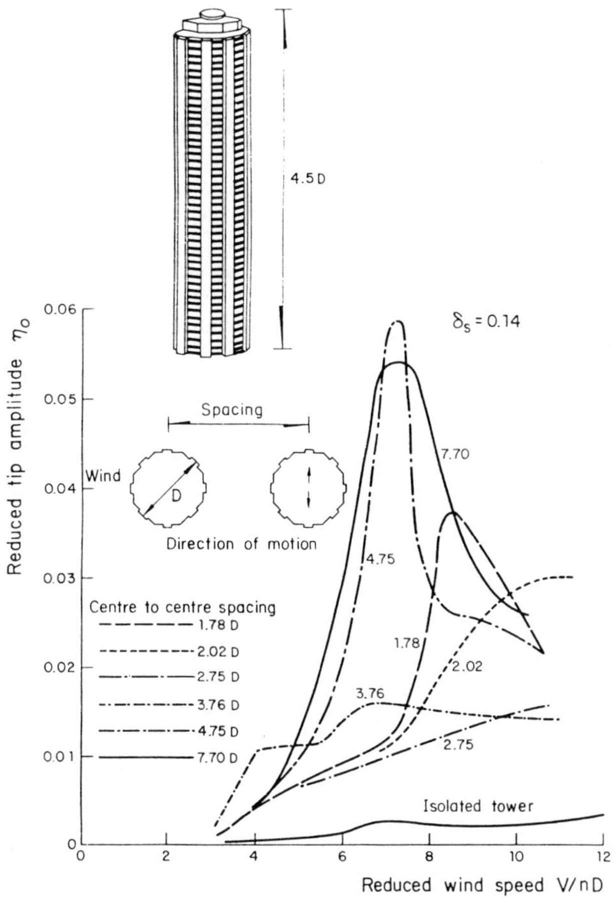

FIGURE 34. Proximity effect on structures of same shape, at different separation and reduced velocities [24].

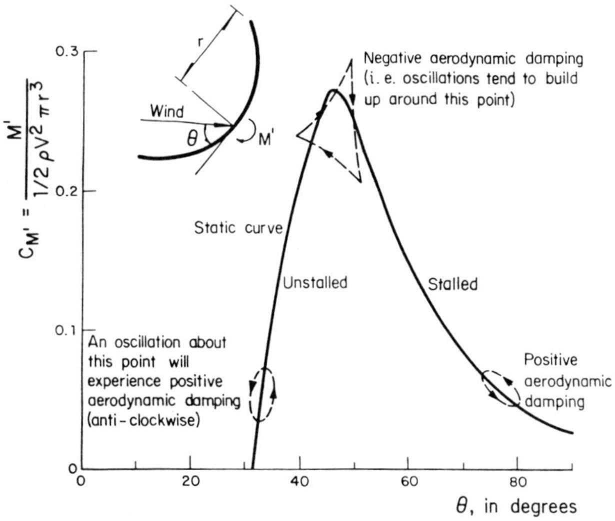

FIGURE 35. Stalling of a radio antenna under varying reduced velocity, from Scruton [18].

V = air velocity
d = cross wind dimension
ν_d = kinematic viscosity coefficient

2. Reduced velocity, V_r

$$V_r = \frac{V}{nd} = \frac{1}{S} \tag{24}$$

n = frequency of vortex shedding
S = Strouhal number

When wind flows past a bluff body, it was previously mentioned that cross-wind vibrations may appear; Figure 32 shows power spectra of the aerodynamic force, and the output response for different reduced velocities. When the reduced velocity reaches $1/S$, corresponding to vortex shedding initiation, a large resonant motion may be induced, and the amplitude η may have large values with periodic vortex shedding.

For smaller or larger reduced velocities, the amplitude η shows different histories, as shown in Figure 32.

If the reduced velocity is further increased and reaches a critical value, galloping may be produced at the structure, with continuous increase in oscillation amplitude as wind velocity increases.

This motion is caused by a negative increase of the aerodynamic side force, with increasing incidence of wind action.

A universal galloping response curve for square section bars is shown in Figure 33, for uniform and shear flow with three different modes of oscillation. Turbulence index can change the response curve, but it is evident that galloping appears when the reduced velocity reaches the critical value defined by twice the value of M_R, multiplied by the structural logarithmic decrement. After this critical velocity, cross-wind vibrations increase in amplitude, and their magnitude may be computed by a procedure described in [14].

The presence of other structures upstream from a body may affect the behavior of a structure under wind action. For instance, the effect of spacing two similar structure on the oscillation of the downwind structure, is shown in Figure 34.

Motion of the structures shows cross-wind movement, whose amplitude depends on spacing of the structures and the structural damping. For comparison, the results obtained for a single tower are also shown.

When wind acts in normal direction to a line of tubes, convergent and divergent flows through alternate gaps, and flow pattern can reverse as alternate tubes are displaced in the fore and aft direction, with a hysteresis lag in the switching from one regimen to the other. These motions may only appear when the tubes are closer than 2.5 diameters.

Stalling and classical flutter may arise in microwave reflectors on radar antennae due to an hysteretic process in the attachment and subsequent reattachment of wind flow, as the reflector oscillates near the stalling angle. Figure 35 shows the change on moment around the center of the reflector, M', produced by wind action as a function of the angle θ formed by the reflector axis of symmetry and wind direction, taken from [18]. Moment M' may reach a maximum value when the structure has stalled.

Classical flutter is associated with long slender structures, as suspension bridge decks, and it is produced by flow instability created by rotational movement of the deck; the enegy to promote the instability is extracted from the wind, by coupling between torsional and flexural natural modes of oscillation, and flow pattern changes.

NOTATION

a_N, a_{N+1} = maximum semiamplitude of vibration at cycles N and $N + 1$, respectively, m
A = constant, dimensionless; exposed area, m^2
b = cross-wind dimension of the exposed area, m
B = constant, dimensionless
C = coefficient of viscous damping, Kgs/m
C_{cr} = critical coefficient of viscous damping, Kgs/m
C_p = pressure coefficient, dimensionless
C_{DL} = drag coefficient for along-wind force estimation, dimensionless
C_{DT} = drag coefficient for cross-wind force estimation, dimensionless
$C^2(n, A)$ = cross correlation function, dimensionless
d, D = cross-wind dimension of a body, m
$f(\xi)$ = dimensionless function for description of mass variation in a structure with height, dimensionless
F_L = longitudinal force due to wind action, Kg
F_T = transverse force generated by wind action, Kg
$F(t)$ = function describing wind force variation with time, Kg
$F(v_i)$ = probability of occurrence of velocity level v_i, dimensionless
g = maximum deviation factor for response computation, dimensionless
G = gust response factor, dimensionless
I = index of turbulence, dimensionless
h = height above ground, m
$\overline{H}$ = average height of buildings surrounding a structure, m
H = total height of a structure, m
H_a = spring stiffness due to aerodynamic excitation, Kg/m
k = spring stiffness constant, Kg/m
k_* = Von Karman's constant, dimensionless
K = surface roughness coefficient evaluated at 10 m, dimensionless
K_a = contribution to total damping of the aerodynamic excitation, Kgs/m
L = scale length, m

m = mass of a single-degree-of-freedom system, Kg/ms^{-2}
M' = moment produced by wind, Kg m
M_R = mass parameter for an equivalent one-degree of freedom system, dimensionless
n, n_0 = frequency, Hz
N = number to identify a cycle, dimensionless
p = pressure, Kg/m^2
R = return period for occurrence of V_R, years
Re = Reynolds number, dimensionless
r = length, m
$S^V(n)$ = velocity power spectral density at frequency n, $m^2\ Hz^{-1}/s^2$
t = time, s
T = averaging time, s
u, v = wind velocities, m/s
w = velocity from a correlated unit variance process, m/s
z = height above ground, m
z_d = displacement length, m
z_0 = roughness length, m
α = exponent related to topography and stability of the atmosphere, dimensionless
β = parameter on Fisher–Tippett distribution, m/s
β_0 = damping ratio, dimensionless
β_1 = small gust scale length, m
γ = parameter on Fisher–Tippett distribution, dimensionless
δ = dynamic logarithmic decrement, dimensionless
η, η_0 = dimensionless amplitude
θ = angle of rotation, degree
θ_1 = time scale, s
λ = large gust scale length, m
ν = effective frequency, Hz
ν_d = kinematic viscosity, m^2/s
ξ = ratio between height z and total height H, dimensionless
ρ = air density, Kgs^2/m^4
σ = standard deviation
τ = horizontal shearing stress, Kg/m^2
τ_1 = time lag, s
ϕ = a dimensional function for spectral power density description, dimensionless
ϕ_j = phase angle, radians
χ_a^2 = aerodynamic admittance, dimensionless
ω, ω_0 = frequency, s^{-1}

REFERENCES

1. ANSI A 58.1—1972, "American National Standard Building Code Requirements for Minimum Design Loads in Buildings and Other Structures." New York, 1972.
2. AS CA 34: Part II, 1971, "Minimum Design Loads on Structures, Part II, Wind Forces," *Standards Association of Australia,* North Sidney, 1971.
3. BSI CP3: Ch V: Part 2, "Code of Basic Data for the Design of Buildings, Wind Loads," *British Standards Institution*, London, 1972.
4. "Canadian Wind Engineering Association and National Research Council of Canada," *Fourth Canadian Workshop on Wind Engineering*, Toronto, Canada, Nov. 1984.
5. Davenport, A. G., "The Dependence of Wind Loads on Meteorological Parameters," Paper 2, *Proceedings International Seminar on Wind Effects on Buildings and Structures*, University of Toronto Press, 1968, pp. 19–62.
6. Ghiocel, D. and Lungu, D., "Wind, Snow and Temperature Effects on Structures Based on Probability," Abacus Press, Tunbrige Wells, Kent, 1975.
7. Harris, R. I., "The Nature of Wind," Paper 3 on *The Modern Design of Wind-Sensitive Structures*, Ciria Publication, London, 1971.
8. Kolousek, V., Pirner, M., Fisher, O., and Náprsteck, J., "Wind Effects on Civil Engineering Structures," Elsevier, Amsterdam, 1984.
9. Lumley, J. L. and Panofsky, H. A., "The Structure of Atmospheric Turbulence," Interscience Publ., New York, 1964.
10. Mackey, S., Pius, Y., and Ko, K. L., "Spatial Configuration of Gusts," *Proceedings 4th International Conference on Wind Effects on Buildings and Structures*, Hearthrow, 1975, Cambridge University Press, 1977.
11. Monin, A. S. and Obuchov, A. M., "Fundamentale Gesetzmässigkeiten der Turbulent Vermischung in der bondennahen Schicht der Atmosphäre," *Trudy Geophysics Institute*, ANSSSR No. 24, 1954, p. 163.
12. National Research Council of Canada, "National Building Code of Canada, 1980," NRCC No. 17303, Canada, 1981.
13. Newbury, C. W., Eaton, K. J., and Mayne, J. R., "Wind Loading of a Tall Building in an Urban Environment," *Building Research Station Current Paper 59/68*, Garston, 1968.
14. Novak, M., "Aeroelastic Instability of Prismatic Bodies in Smooth and Turbulent Flow," University of Western Ontario, *Research Report BLWI 3-68 and BLWI 7-68*, 1968.
15. Règles NV65, Rèvisèes 1967, "Règles Définissant les Effects de la Neige et du Vent sur les Constructions," *Société de Diffusion des Techniques du Bâtiment et des Travaux Publics*, Paris, Jan. 1968.
16. Rodríguez Cuevas, N., "La Simulación de la Acción Turbulenta del Viento en las Vibraciones Longitudinales de Edificios Altos," *Proceedings VI National Congress in Earthquake Engineering*, Puebla, Nov. 1983, pp. 227–236.
17. Rodríguez Cuevas, N., "Admitancia Aerodinámica en Régimen Turbulento," *Internal Report 4737*, Instituto de Ingeniería, UNAM, Mar. 1985.
18. Scruton, C., "Some Considerations of Wind Effects on Large Structures," *Structures Technology for Large Radio and Radar Telescope Systems*, (eds.) J. W. Mar and H. Liebowitz, MIT Press, 1969.
19. Sherlock, R. H., "Variation of Wind Velocity and Gusts with

Height,'' *Transactions ASCE*, Vol. 78, No. 126, 1952, pp. 463–500.
20. Singer, I. A., Busch, N. E., and Fizzola, J. A., ''The Micrometeorology of the Turbulent Flow Field in the Atmospheric Boundary Layer,'' Paper 21, *Proceedings International Seminar on Wind Effects on Buildings and Structures*, University of Toronto Press, 1968, pp. 557–594.
21. Task Committee on Wind Forces, ''Wind Forces on Structures,'' Paper 3269, *Transactions ASCE*, Vol. 126, Part II, 1961, pp. 1124–1198.
22. Taylor, G. I., ''Diffusion by Continuous Movements,'' *Proceedings of the Royal Society of London*, Vol. 148, 1938.
23. Vellozzi, I. and Cohen, E., ''Gust Response Factors,'' *Journal of the Structural Division, ASCE*, June 1968, pp. 1295–1313.
24. Whitbread, R. E. and Wootton, L. R., ''An Aerodynamic Investigation for the Five 50-Story Tower Blocks Proposed for the Ping Shek Estate, Hong Kong,'' NPL Aero Special Report 002, Sept. 1967.
25. Wootton, L. R. and Scruton, C., ''Aerodynamic Stability,'' Paper 5 on *The Modern Design of Wind-Sensitive Structures*, Ciria Publication, London, 1971, pp. 65–81.
26. Wyatt, T. A., ''The Calculation of Structural Response,'' Paper 6 on *The Modern Design of Wind-Sensitive Structures*, Ciria Publication, London, 1971, pp. 83–93.

CAVITATION (GABRIEL ECHÁVEZ)

INTRODUCTION

The word ''cavitation'' implies the presence of gaseous cavities in a flow. However, usually it is only applied when the cavities are mainly filled with the vapor of the same liquid in which they are formed. In water, low pressures, frequently associated to high velocities, can lead to the formation of cavities that collapse violently inducing high pressure shocks applied in very small areas. This phenomenon produces undesirable noise and vibrations and, if it occurs near walls, damages even the hardest surfaces.

HISTORY

Based on theoretical considerations, L. Euler (1707–1783) advanced the idea of cavity formation in liquids. Three centuries later, the failure of the more and more bigger and faster ship propellers to produce the design thrust and the appearance on them of a characteristic pitting, focused again the attention of the engineers in this phenomenon. Thus, at the end of the last century, O. Reynolds [1] and Ch. Parsons [2] made the first experimental work in cavitation.

In 1917 Rayleigh [3], calculating the energy of collapse of the bubbles, showed the possibility of high intensity pressures, and in 1919 a report by a propeller sub-committee was published.

In the thirties, Ackeret, Haller and Boetcher proposed theories of damage in metals and, in 1941, Beeching improved the Rayleigh equations and established that the metal surfaces are disrupted by local stresses due to the collapse of vapor pockets [1].

In 1948, Knapp and Hollander [4], with experiments followed through high-velocity photographs, and Plesset [5], by theoretical means, studied the bubble evolution and found agreement with Rayleigh calculations.

Knapp, in 1955, suggested the formation of ''water hammer'' pressures, and in the same year Plesset and Ellis [1] detected the plastic deformation of the metallic surfaces close to the collapsing bubbles, and Guth photographed the shock wave radiated from the bubble.

During the sixties, the study of cavitation provoked by ultrasound and of the one arising in two-phase flow commenced to be studied, and numerical techniques were incorporated. In those years, works by Naude and Ellis [6], Plesset and Chapman [7] and others, can be mentioned [1].

It is not until 1970 that a book, gathering the material thus far scattered in papers and reports, is published by Knapp *et al.* [2]. From that year on, a growing interest in cavitation on high-head spillways and on ways to protect concrete surfaces against it arises. This interest leads to ways for predicting risks in waterworks [8,9], to the development of aerators aimed at protecting the surfaces [10,11,12], and to the study of special concretes, e.g., reinforced with polymers, steel fibers or with special treatments, that can be used in repairs of cavitation damages [13,14].

BASIC CONCEPTS

From a practical point of view, cavitation can be defined as the formation and collapse of vapor bubbles in a liquid [15].

Although cavitation is a very complex phenomenon whose

detailed description can be found elsewhere [2], there are practical criteria to predict its appearance and to avoid the damage caused by it.

A parameter that shows the possibility, as well as the successive stages, of cavitation, is the Sigma, or Thoma, number, given as:

$$\sigma = \frac{P_0 - P_v}{\rho V_0^2/2} \qquad (1)$$

where

P_0 = absolute static pressure
V_0 = reference velocity
P_v = vapor pressure
ρ = density of liquid

Both P_0 and V_0 must be measured at a point close or directly related to the zone in which cavitation occurs. P_v depends on the temperature and purity of the water, and can be estimated for common water, from Figure 36 with $P_v = \gamma h_v$.

For problems in which the gravity intervenes, i.e., for ship propellers or bodies changing in elevation as they move, the numerator of Equation (1) is modified as:

$$P_0 - P_v + \gamma(h_0 - h_{min})$$

where

h_0 = elevation of the reference point
h_{min} = minimum elevation
γ = specific weight of liquid

The smaller σ is, the worse the flow characteristics with respect to cavitation are.

When dealing with cavitation, several operating limits have to be considered:

1. Onset of cavitation
2. Tolerable noise or vibration level
3. Onset of damage
4. Fully choked flow

and different cavitation limits, such as: incipient, critical, choking etc., have been proposed [16].

Once found the cavitation number for a flow condition through experimentation, physical models, or in the literature, the corresponding stage can be approximately predicted.

For a specific problem it is necessary to identify and select the adequate limit. In problems with rigid boundaries, the above mentioned classifications can be used. In free-surface flows, the interest is focused on the damages found in the walls, and the noise and vibrations are of no consequence, so only an incipient-damage cavitation index is used.

In the next part, the cavitation performance for different rigid-boundary problems is shown; and in the last part the cavitation on high head spillways will be treated.

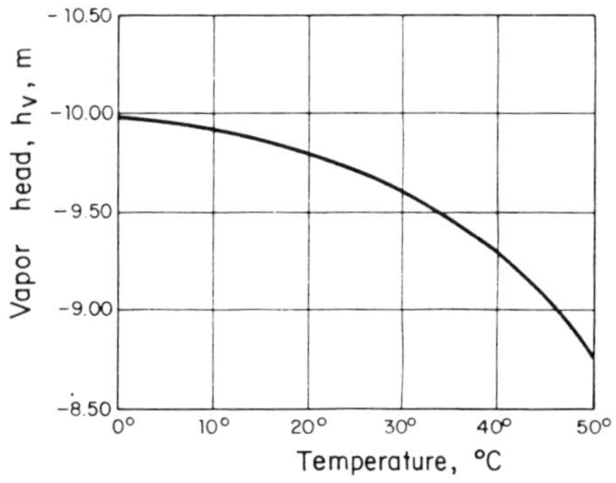

FIGURE 36. Vapor head against temperature for common water [8].

CAVITATION IN RIGID-BOUNDARY PROBLEMS

The cavitation limits considered in this type of problems are [16]:

1. Incipient cavitation, in which the cavitation consists of occasional light pops, does not give rise to objectionable noise or damage.
2. Critical cavitation. There is continuous light cavitation; noise and vibrations are minor, and only small damages can be expected. Usually this limit is used as a design criterion.
3. Incipient damage. The cavitation is intense; noise and vibration are objectionable and surface damage, even for short periods of operation, appears.
4. Choking cavitation. The discharge does not increase with further decrease in the downstream pressure. Noise and vibration reach a maximum and then reduce.

CAVITATION ON DIFFUSERS

For smooth diffusers with no irregularities, Equation (1) applies, with P_0 and V_0 at the throat of the diffuser. For design purposes, a cavitation limit between 0.2 and 0.3 can be used.

CAVITATION ON BENDS

For bend angles greater than 60°, Equation (1) and Figure 37 can be used.

Example: Calculate the maximum velocity of water, at 20°C, allowed in a 90° bend for a 25.4 cm (10 inches) pipe, with a radius of curvature of 40 cm, if the pressure head is 5 m of water.

From Figure 36, for a temperature of 20°C, $hv = -9.8$ m. From Figure 37, with a ratio $r/d = 1.5$, the incipient cavitation number is 1.2.

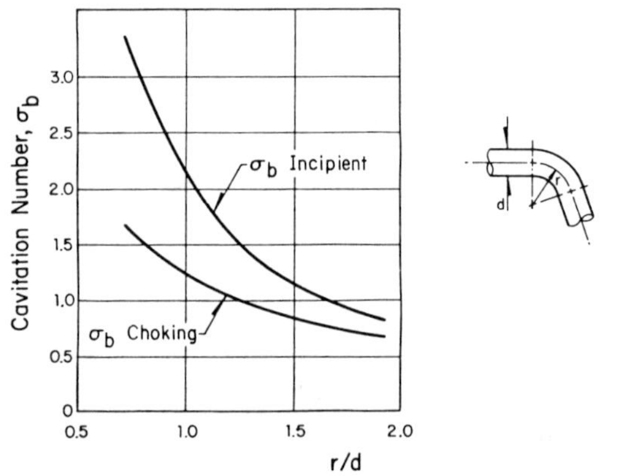

FIGURE 37. Choking and incipient cavitation number for bends [17].

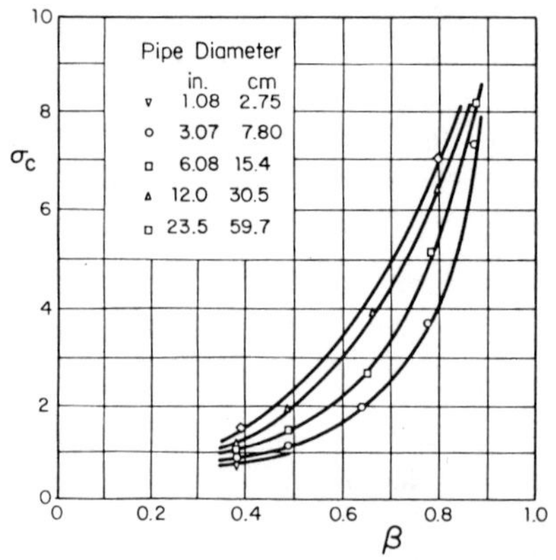

FIGURE 39. Critical cavitation number against $\beta = D_o/D_p$, for orifices [16].

Rearranging Equation (1)

$$V = \left(2g\,\frac{hp - hv}{\sigma i_c}\right)^{1/2} = 15.6 \text{ m/s}$$

assuming negligible scale effects.

CAVITATION IN ORIFICES

Orifices are used as pressure regulators or as flow measurement devices. The cavitation index proposed by Tullis and Govindarajan [16] is

$$\sigma = \frac{P_d - P_v}{P_u - P_d}$$

where

P_d = downstream pressure
P_u = upstream pressure
P_v = vapor pressure

and both P_u and P_d are estimated near the orifice, from the values measured at some distance from it, adjusted for line losses.

In Figures 38 and 39, the incipient and the critical cavitation numbers for several pipe diameters, ranging from 2.75 to 59.7 cm, are shown.

Example: A pressure-head loss of 5 m is suitable in a 12.7 cm (5 in) pipe, where water at 30°C flows at 6 m/s. If the pressure head upstream is 20 m, how many head-reducing orifices would you recommend?

From Figure 36, it results that $h_v = p_v/\gamma = -9.6$ m.

(a) Assuming $\beta = D/D_0 = 0.7$ and an orifice head loss $k = 0.3$, we have

$$\frac{P_d}{\gamma} = \frac{P_u}{\gamma} + \frac{V_u^2}{2g} - \frac{V_d^2}{2g}(1 + k)$$

$$\frac{P_d}{\gamma} = 20 + \frac{6^2}{19.6} - \frac{12.24^2}{19.6}(1 + 0.3) = 11.9 \text{ m}$$

Then

$$\sigma = \frac{P_d - P_v}{P_u - P_d} = \frac{11.9 + 9.6}{20 - 11.9} = 2.65$$

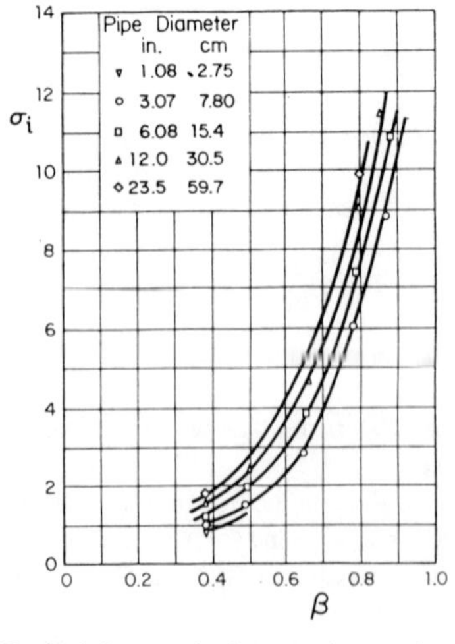

FIGURE 38. Incipient cavitation number against $\beta = D_o/D_p$, for orifices [16].

From Figure 38, with $\beta = 0.7$ and $D = 12.7$ cm, it results

$$\sigma_i = 4.8 < \sigma = 2.65$$

then cavitation occurs.

(b) Assuming $\beta = 0.8$ and an orifice head loss $k = 0.24$, we have

$$\frac{P_d}{\gamma} = 20 + \frac{6^2}{19.6} - \frac{9.38^2}{19.6}(1 + 0.24) = 16.27 \text{ m}$$

Then

$$\sigma = \frac{16.27 + 9.6}{20 - 16.27} = 6.94$$

From Figure 38, with $\beta = 0.8$ and $D = 12.7$ cm, we get

$$\sigma_i = 6.8 < \sigma = 6.94$$

In this case, there is no risk of cavitation. Since the head loss produced by a single orifice is $kV_d^2/2g = 1.08$ m, and the head loss suitable is 5 m, five orifices with $\beta = 0.8$ will be necessary.

CAVITATION ON VALVES

Globe Valves

A critical velocity, as shown in Figure 40, can be used. If this velocity is exceeded, cavitation will occur.

To correct this velocity to any other head condition, use the formula

$$V_c = V_{cr}\left(\frac{h_c - h_v}{51}\right) 0.46$$

where

h_c is the new head, in m

Ball Valves

In Figure 41, critical velocities for ball valves are presented. To correct these velocities to any other head condition, use the formula

$$V_c = V_{cr}\left(\frac{h_c - h_v}{50}\right)^{0.4}$$

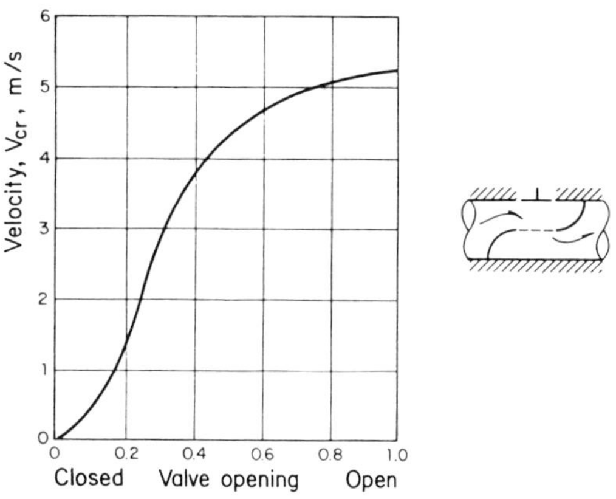

FIGURE 40. Critical velocity against valve opening, for globe valves.

CAVITATION IN PUMPS

For cavitation in pumps, the Thoma number, given by

$$T = \frac{h_a - h_v - h_s}{H}$$

is used, where

h_a = atmospheric pressure
h_v = vapor pressure
h_s = suction head
H = pump head

The test consists in keeping the rotational speed and the head constant, and at the same time to increase the suction

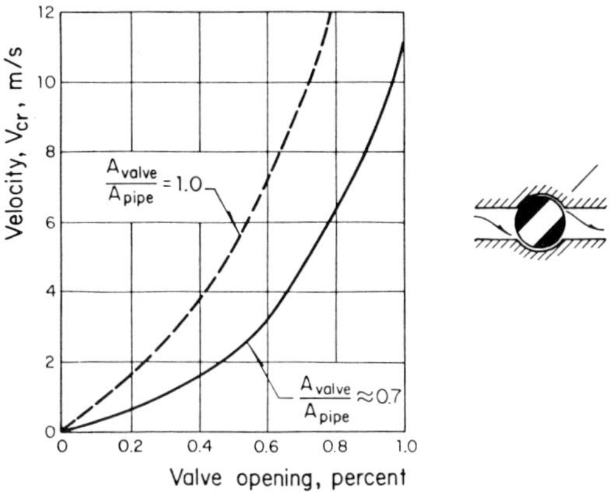

FIGURE 41. Critical velocity against valve opening, for ball valves.

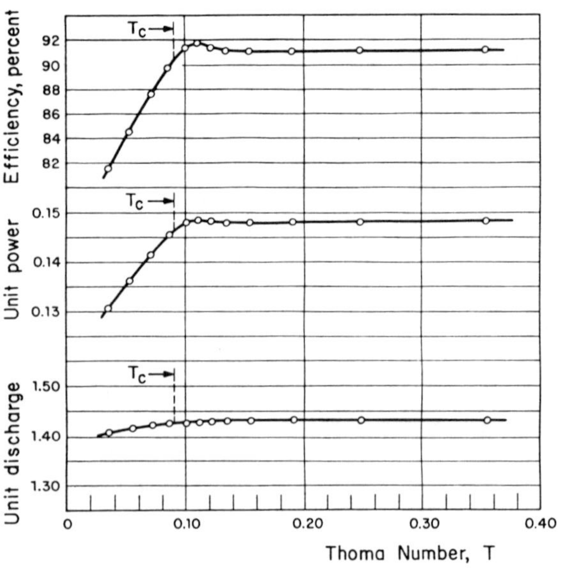

FIGURE 42. Typical test curves of efficiency, unit power, and unit discharge against Thoma number, showing evidence of cavitation at the critical value of T [17].

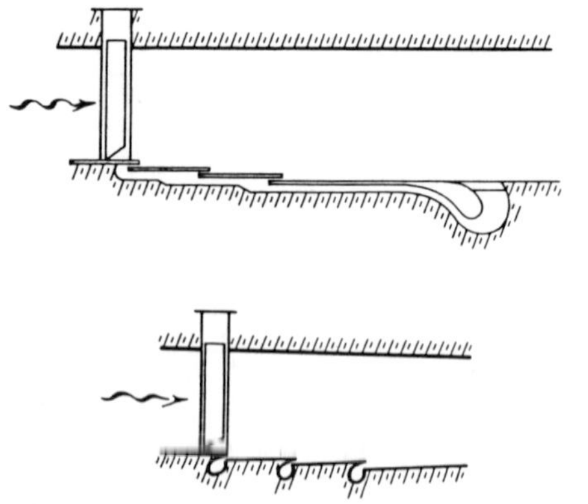

FIGURE 43. Installation for development of low velocity wall-flow [20].

head (that is, decreasing T) by means of a valve in the suction pipe, until the onset of cavitation, recognizable by a characteristic noise and a decrease in the discharge, is reached. The corresponding value of T is called the "critical" coefficient of cavitation.

CAVITATION IN TURBINES

In reaction turbines, the procedure is similar to that in pumps. The head, speed and the opening of the wicket gates is kept constant, while h_s is increased and T plotted against discharge, efficiency and power. A significant drop in all or any of these quantities is usually considered the critical Thoma number for that particular vane setting of the turbine, Figure 42 [17].

CAVITATION IN FREE-SURFACE FLOWS

In high head dams, a limiting factor in design, and a serious problem during operation, is the erosion of the surfaces due to cavitation. Several major damages in outlet works, have been reported [9,10,18,19]. They can be classified in two groups:

1. Local damages associated with special structures, such as nozzles, gates, and, in general, changes in the geometry of the conduit
2. Distributed cavitation that can appear at any point of the surface, due to the natural roughness of the concrete finish, or to protuberances, such as those produced by steps and slots of the casting molds

In the case of free-surface flows, the different limits of cavitation are irrelevant, and it is more realistic to use the terms

1. Destructive cavitation
2. Non-destructive cavitation

The expression "non-destructive cavitation" means that the exposed surfaces have been protected by aerating the flow, or by using designs that shift cavitation far away from the surface (supercavitation).

LOCAL CAVITATION

Obstructions, such as slots, steps, or changes in the geometry of the conduit, can produce zones of low pressure, with risk of cavitation damage if the water velocity exceeds 12 m/s. Though each case needs to be studied separately, if possible through a hydraulic model, three ways of protecting these zones appear to be the more workable: to aerate the flow, to induce a low velocity wall-flow, or to create supercavitation.

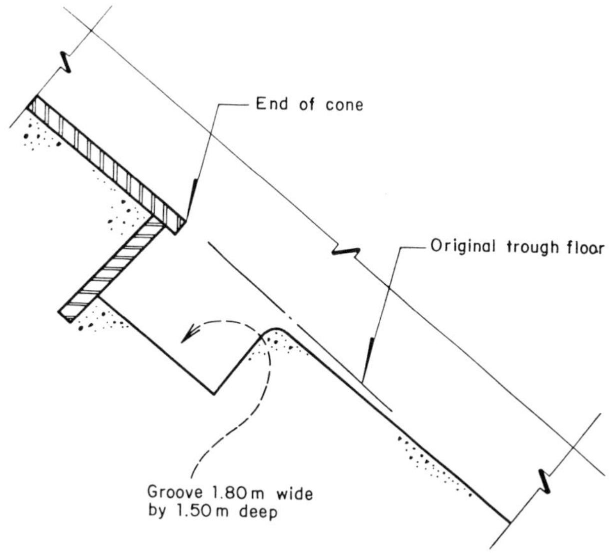

FIGURE 44. Aeration groove in outlet tube at Grand Coolee Dam [25].

If there are ways to supply enough air, the design may include aerating slots, ramps, or steps that incorporate air to the flow. For heads over 120 m or when it is not possible to supply air, a low velocity wall-flow can be created using special designs, as shown in Figure 43 [20].

Outlet nozzles that discharge into atmosphere should be designed so that no separation zones should arise; and, if necessary, an aerating step at the end, to avoid damages in the surfaces downstream has to be provided, see Figure 44.

For gates that operate partially open, a supercavitation design can be used, in which the cavity is shifted away from the wall surface. If the cavitation is present only when the gate is raised or lowered, the use of high-resistance concretes may be enough [21,22].

DISTRIBUTED CAVITATION

Irregularities common in concrete can lead to cavitation damage. To adopt stringent construction specifications to avoid the problem, seems to be both impractical and uneconomical.

Since even small amounts of air (2 to 8% of the water volume) present at the walls inhibit cavitation damages [23], a viable solution is to use small specific flows (i.e., discharges for unit width), e.g., less than 8–9 m^3/s-m [24] allowing the natural aeration to reach the bottom; or to entrain air from below, by means of aerating jumps or slots, as shown in Figure 45.

To attack the problem it is necessary, first, to detect the

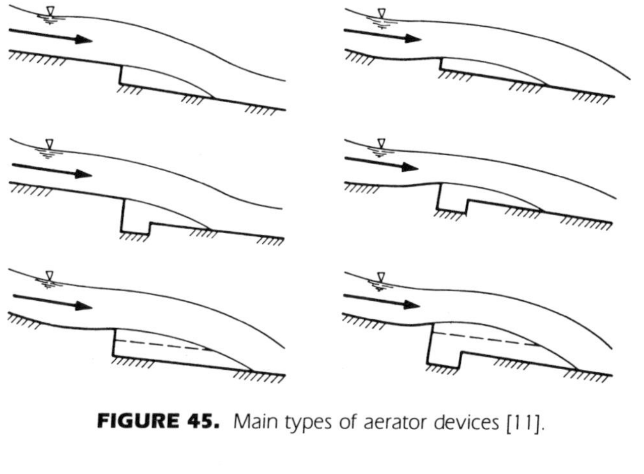

FIGURE 45. Main types of aerator devices [11].

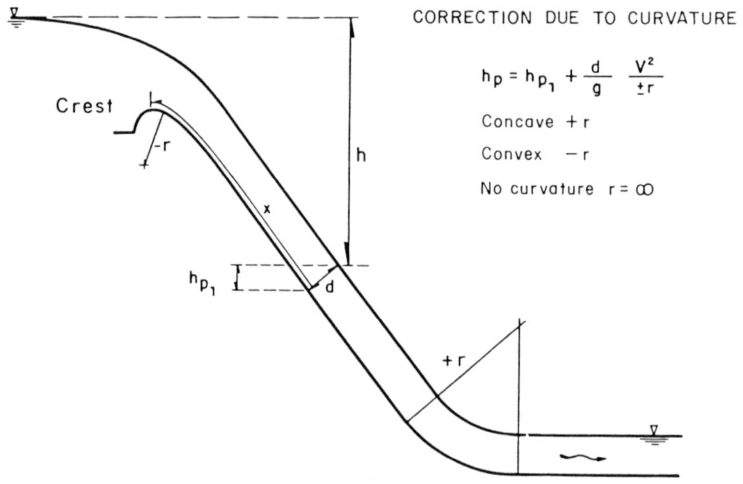

FIGURE 46. Spillway profile, to show the variables d, h, r, x, and h_{p1} [8].

TABLE 2. Values of k, for Different Materials.

Material	*k* (mm)
Glass	0.05 to 0.90
Cement very smooth	0.2 to 1.2
Concrete very smooth	0.3 to 1.5
Unfinished	1.5 to 12
Steel forms	0.6 to 1.5
Smooth wood forms	0.6 to 3.2
Rough wood forms	2.2 to 14
Gunite	3.2 to 15
Very rough	5 to 20

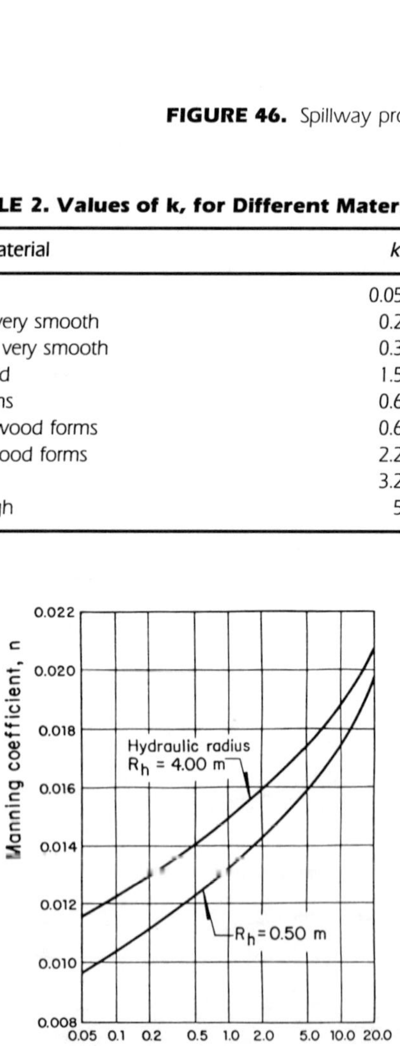

FIGURE 47. Nikuradse roughness against Manning coefficients, for hydraulic radius between 0.5 and 4.0 m [8].

zones exposed to damages and, second, to protect them with suitably placed aerators, whose number has to be kept to a minimum, in order to avoid an excessive increase in water depths and reduce construction costs.

A procedure aimed to locate the zones exposed to cavitation damages in high-head spillways, when the geometry and surface roughness of the spillway, and the discharge and water depth along the spillway surface are known, is the following. Divide the spillway into short longitudinal reaches, e.g., one each 20 m, and compute, for each of them:

(a) The distance, x, of its end section from the spillway crest; the height difference, h, between the reservoir level and the water surface at the section; and the depth, d, of flow, taken perpendicularly to the floor (see Figure 46).

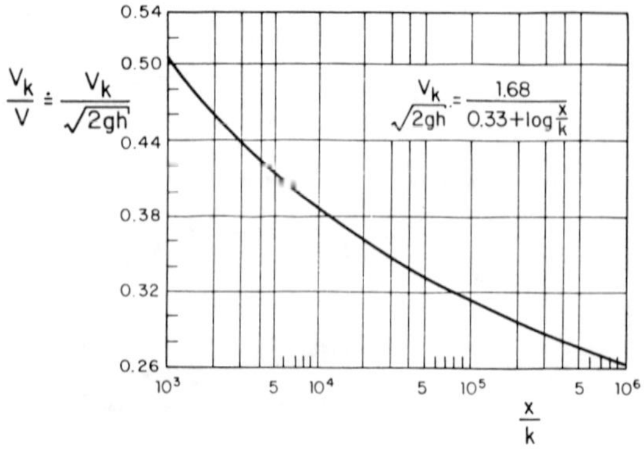

FIGURE 48. Dimensionless local velocity against x/k [8].

Irregularity	Characteristic velocity	Incipient cavitation index
	V_b	2.4
	V_k	1.1
	V_k	1.5
	V_k	1.4
	V_b	1.4
	V_b	1.6

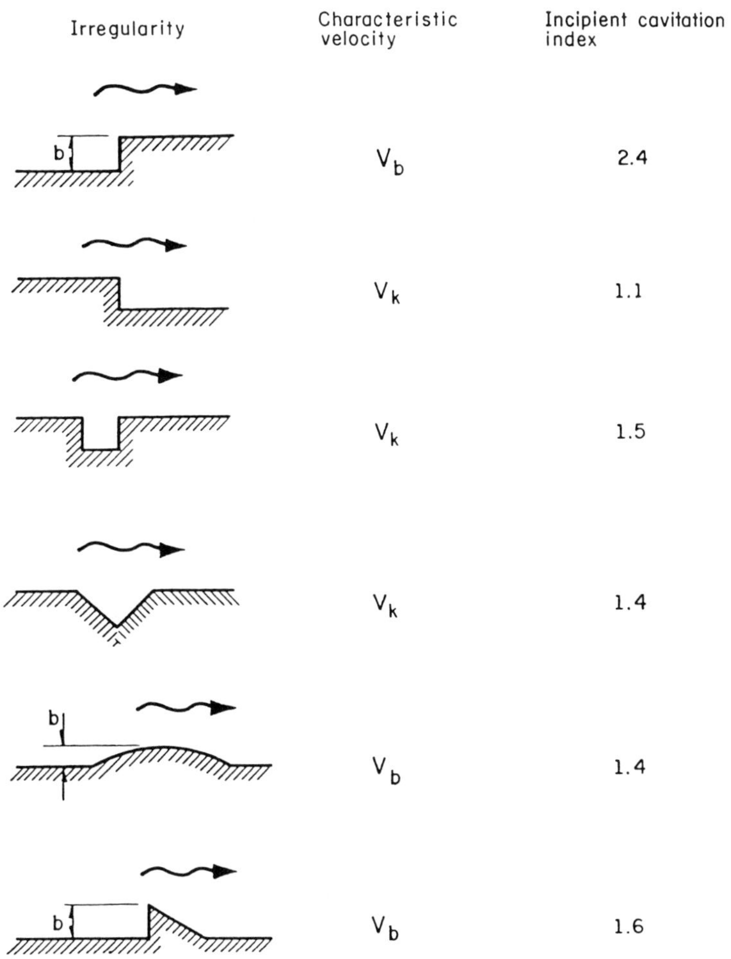

FIGURE 49. Characteristic velocity and incipient cavitation index for different irregularities [8].

$$\frac{V_b}{V_k} = 0.68 \log \frac{b}{k} + 1$$

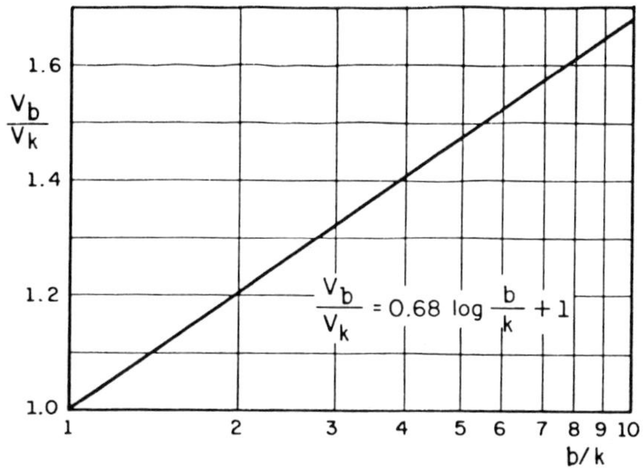

FIGURE 50. Velocity at a height b, as a function of V_k, b, and k [8].

(b) Pressure head, h_p, in each section, using the equation

$$h_p = h_{p_1} + \frac{d}{g}\frac{V^2}{r}$$

where

d = water depth, perpendicular to the flow
g = gravity acceleration
h_{p_1} = vertical projection of the water depth
V = mean velocity at the section
r = radius of curvature, being

$r > 0$ for concave curvature

$r < 0$ for convex curvature

$r = \infty$ for rectilinear boundary

(c) Vapor head, from Figure 36
(d) Roughness k, from Table 2; or assign an n, and through Figure 47 estimate k.
(e) For each section, compute x/k and, from Figure 48, evaluate $V_k/\sqrt{2gh}$, where

V_k = local velocity, close to the surface.

From this expression and the values of h obtained in step b, compute the local velocity head, $V_k^2/2g$.

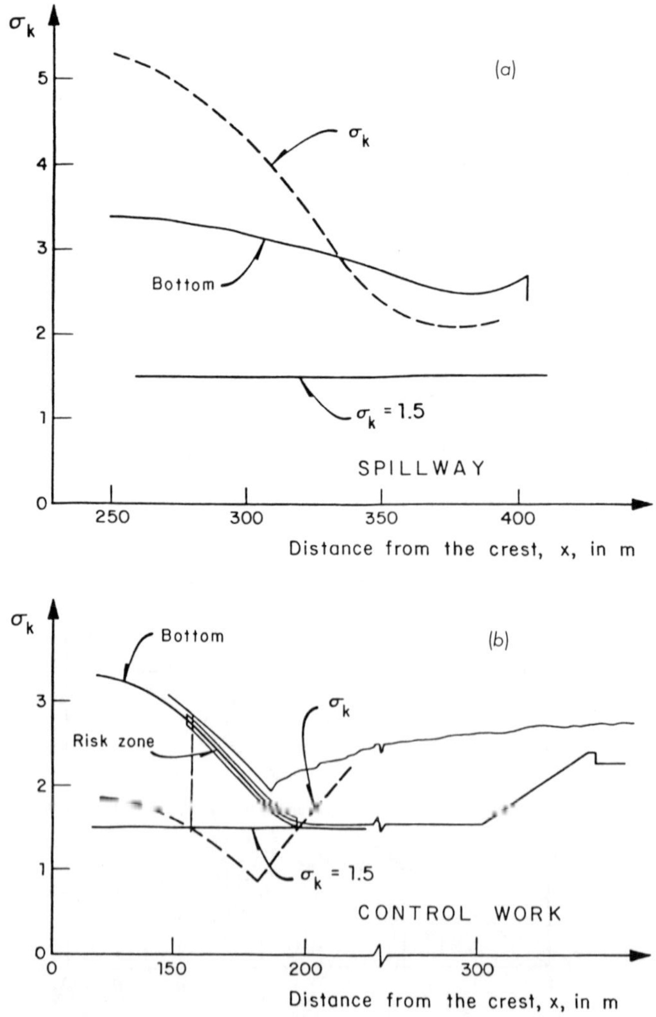

FIGURE 51. Graph of σ_{kp} along: a) the spillway, and b) control work for Netzahualcoyotl Dam [8].

TABLE 3. Cavitation Indexes for Netzahualcoyotl Dam (Mexico)
[Q = 2500 m³/s, k = 2 mm (n ≐ 0.015), h_v = −9.75 m, $r_{flip\ bucket}$ = 60 m].

Spillway										
x (m)	h (m)	d	h_{pl}	$\frac{dV^2}{gr}$	h_p	$x/k(10^{-3})$	$V_k/\sqrt{2gh}$	$V_k^2/2g$	σ_k	σ_{kp}
251.6	19.15	2.34	2.30	—	2.30	50.2	0.345	2.27	5.29	5.29
280.0	21.00	2.23	2.20	—	2.20	56.0	0.345	2.46	4.85	4.85
311.6	26.10	1.99	1.90	—	1.90	63.2	0.340	3.03	3.85	3.85
340.0	37.60	1.67	1.60	—	1.60	68.0	0.338	4.30	2.64	2.64
370.8	44.40	1.54	1.50	—	1.50	74.2	0.335	5.00	2.25	2.25
385.0	45.70	1.52	1.50	3.00	4.50	77.0	0.333	5.10	2.83	2.10

Control Spillway									
x (m)	h (m)	d	h_{pl}	$\frac{dV^2}{gr}$	h_p	$x/k(10^{-3})$	$V_k/\sqrt{2gh}$	$V_k^2/2g$	σ_k
125.0	38.5	2.00	2.05	−2.05	0.0	25.0	0.370	5.25	1.86
150.0	43.0	1.89	1.80	−1.80	0.0	30.0	0.365	5.81	1.68
175.0	74.0	1.44	1.35	−1.35	0.0	35.0	0.359	9.45	1.03
183.8	87.3	24.10	24.10	−1.10	0.0	36.7	0.355	11.0	0.89
200.0	93.3	35.00	35.00	0	35.0	40.0	0.351	11.50	3.90

(f) With the values obtained in (b), (c), and (d), compute

$$\sigma_k = \frac{h_p - h_v}{V_k^2/2g}$$

along the spillway.

(g) Where this cavitation index results to be less than 1.5, damages can be expected. In this case, it is recommended to protect the corresponding zone with aerators. For higher values, no serious damages will occur.

(h) In concave curves, though the mean velocity close to the bottom decreases, the local velocity increases, due to the reduction of the boundary-layer thickness. In such regions it is recommended to multiply the cavitation index found in (f) by 0.8, and to compare the result with the cavitation index given in (g).

ISOLATED IRREGULARITIES

To estimate the possibility of cavitation at isolated irregularities projecting at right angle into the flow, that are common in concrete finishes, compute the cavitation index either with V_k or V_b where

V_k = velocity close to the surface, from Figure 48
V_b = characteristic velocity for steps against the flow, see Figure 49

and compare this index with the incipient cavitation index shown in Figure 49. If the flow index is less than the incipient one, there will be cavitation.

To find V_b, given V_k, b and k, use Figure 50.

Example: Cavitation analysis of the spillway and control works of the Netzahualcóyotl Dam (Mexico).

In Table 3 are the computations to determine both, the local cavitation index, σ_k, and the same index, corrected by curvature, σ_{kp}, as suggested in (f), for a discharge of 2500 m³/s.

In Figure 51(a) is plotted the σ_{kp} along the spillway. Since this index is always greater than 1.5, there is no risk of serious cavitation damages. In Figure 51(b) there is the same graph for the control work. From it, one infers that it exists a zone with possibility of cavitation damages at the foot of the spillway.

REFERENCES

1. Govinda Rao, N. S. and Thiruvengadam, A., "Prediction of Cavitation Damage," *Journal of the Hydraulic Division, ASCE*, HY5, Proc. Paper 2925, Sept. 1961, pp. 37–62.
2. Knapp, R. T., Daily, J. W., and Hammitt, F. G., *Cavitation*, McGraw Hill, Inc., New York, 1970.
3. Gimenez, G., "Essai de Synthèse des Recherches sur la Cavitation," *La Houille Blanche*, No. 5, 1984, p. 326.

4. Knapp, R. T. and Hollander, A., "Laboratory Investigations of the Mechanism of Cavitation," *Trans. ASME*, 70, 1948, pp. 419–435.
5. Plesset, M. S., "The Dynamics of Cavitation Bubbles," *Trans. ASME, Journal of Applied Mechanics*, 16, 1948, pp. 228–231.
6. Naudé, C. F. and Ellis, A. T., "On the Mechanism of Cavitation Damage by Nonhemispherical Cavities Collapsing in Contact with a Solid Boundary," *Trans. ASME*, 83, Ser. D, *Journal Basic Engineering*, 1961, pp. 648–656.
7. Plesset, M. S. and Chapman, R. B., "Collapse of a Vapor Cavity in the Neighborhood of a Solid Wall," California Institute of Technology, Div. of Engrg. and Appl. Sci., Rep. 85-48, December 1969.
8. Echávez, G., "Cavitación en Vertedores," Instituto de Ingeniería, Universidad Nacional de México, Reporte 415, February 1979.
9. Falvey, H. T., "Predicting Cavitation in Tunnel Spillways," *Water Power and Dam Construction*, August 1982.
10. Galperin, R. S., Oskolkov, A. G., Semenkov, V. M., and Tsedrov, G. N., "Cavitation in Hydraulic Structures," *Energiya*, Moscow, 1977.
11. Pinto, N. L. de S., Neidert, S. H., and Ota, J. J., "Aeration at High Velocity Flows," *Water Power and Dam Construction*, February 1982.
12. Eccher, L. and Siegenthaler, A., "Spillway Aeration of the San Roque Project," *Water Power and Dam Construction*, Sept. 1982.
13. Inozemtsev, Y. P., "Cavitational Erosion Resistance of Hydrotechnical Concretes of Cement and Polymer Binders," *Proc. XI Congress, International Association for Hydraulic Research*, Leningrad, 1965.
14. USBR, "Concrete-Polymer Materials, First Topical Report," BNL Report 50134 (T-509), Brookhaven National Laboratory, New York, 1968.
15. Hamilton, W. S., "Preventing Cavitation Damages to Hydraulic Structures," *Water Power and Dam Construction*, Nov. 1983.
16. Tullis, J. P. and Govindarajan, R., "Cavitation and Size Scale Effects for Orifices," *Journal of the Hydraulic Division, ASCE*, HY3, March 1973.
17. Davis, C. V. and Sorensen, K. E., (editors), *Handbook of Applied Hydraulics*, 3rd. ed., McGraw-Hill Book Co., New York, 1969.
18. Colgate, D., "Hydraulic Model Studies of Aeration Devices for Yellowtail Dam Spillway Tunnel," U.S. Department of the Interior, Bureau of Reclamation, REC-ERC-71-47, December 1971.
19. Pugh, C. A., "Modeling Aeration Devices for Glen Canyon Dam," *ASCE Hydraulic Division*, Coeur D'Alene, Idaho, 1984.
20. Khlopentov, P. R. and Chepaikin, G. A., "Anticavitation Protection of Pressure Outlets Through Regulation of Velocities in Wall Layer of the Flow," *Hydrotechnical Construction*, UDC 627.83:620.193.16,1976.
21. Kallas, D. H. and Lichtman, J. Z., "Cavitation Erosion," *Environmental Effects on Polymeric Materials*, Vol. 1, Chapter 2, Interscience, New York, 1968.
22. Kukacka, L. E., Steinberg, M., and Manowitz, B., "Preliminary Cost Estimate for the Radiation-Induced Plastic Impregnation of Concrete," BNL Report 11263, Brookhaven Natl. Lab., New York, 1967.
23. Rasmussen, R. E. H., "Some Experiments on Cavitation Erosion in Water Mixed with Air," *Proc. 1955 NPL Symp. on Cavitation in Hydrodynamics*, Paper 20, H. M. Stationery Office, London, 1956.
24. Semenkov, V. M. and Lentyaev, L. D., "Spillway with Nappe Aeration," *Hydrotechnical Construction*, UDC 627.83, Transl. from Gidrotekhnicheskoe Stroitel'stvo, No. 5, May 1973.
25. Dominy, F. E., "Applied Research in Cavitation in Hydraulic Structures," *Proc. XI Congress, International Association for Hydraulic Research*, Leningrad, 1965.

Stress Optimized Suspended Liquid Containers

N. JNANASEKARAN* AND D. STEPHEN SANDEGREN**

INTRODUCTION

Liquid containers are classified according to the shape of the side wall and base, materials used and the type of supports provided. R.C. containers which are rectangular in plan and having flat bottom are subjected to heavy bending moments. In cylindrical tanks with flat bottom, the sidewall resists liquid pressure by direct tension and bending, their values depending on the degree of fixity with the roof and bottom slabs. The bottom slab also suffers heavy bending moment. By making the bottom conical, as in Intze tank, the bending moments are reduced in the bottom of the tank. In steel tanks with hemispherical bottom, the bending effects are virtually a minimum. Drop tanks [2,5] shown in Figure la and spheroidal oil tanks [1] are the outcome of efforts to eliminate bending moment in the walls of the tank. While tanks supported on staging or on ground are prevalent, suspended tanks [3] are currently not so common.

Regarding the choice of material for water tanks, the type of environment, availability of suitable material overall economy and ease of construction are some of the major factors for consideration. Small capacity plastics tanks, cylindrical in shape, and made by rotational moulding, are finding increasing applications in storing acid and water.

In this chapter, the shape of a tension tank which is stress optimized is developed both for fully suspended and partly suspended cases (Figure lb and lc) using the membrane theory through a simple computer program. For the membrane solution to be valid everywhere, the supports must have freedom of radial movement so that the assumed pattern of stress distribution exists. Shapes that are economical, aesthetic and are suitable to meet other practical considerations are also suggested in this chapter.

*Department of Civil Engineering, Government College of Engineering, Salem-636 011, India

**Department of Civil Engineering, College of Engineering, Anna University, Madras-600 025, India

LIQUID CONTAINERS MADE OF FLEXIBLE MEMBRANE

In order to generate the surface of liquid containers made up of thin membranes such as flexible plastic sheets or thin gauge stainless steel sheets, the simple membrane theory of shells is applied in this section to derive the lead equations.

Geometry

A surface of rotation is chosen for the liquid container and this surface is obtained by rotating a plane curve about an axis lying in the plane of the curve. This curve is called the meridian of the surface and the intersection of this surface with any plane to which the axis of rotation is normal is called a parallel. For such a surface, the lines of principal curvature are along the meridian and the parallel, respectively [refer to Figure 2(a)].

Any point A on the surfac shown in Figure 2(b) is fixed by the z and r system. The outward normal to the surface is called the n direction. The notations used in the derivation are explained below with reference to Figure 2.

- z, r: Coordinates of any point A on the meridian with reference to the (z, r) system.
- φ: Angle between z-axis and normal to the surface at any point A on the surface.
- R_φ: Radius of curvature of the meridian at point A.
- R_θ: Length of normal from A to the axis of rotation, which is the other principal radius of curvature.

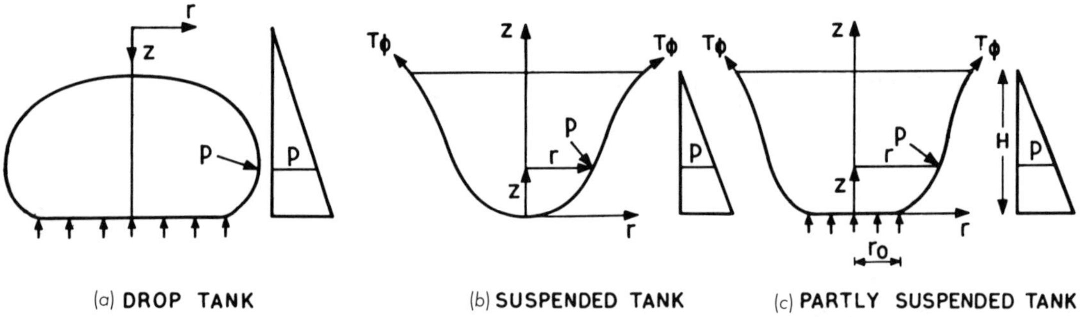

FIGURE 1. Tension tank.

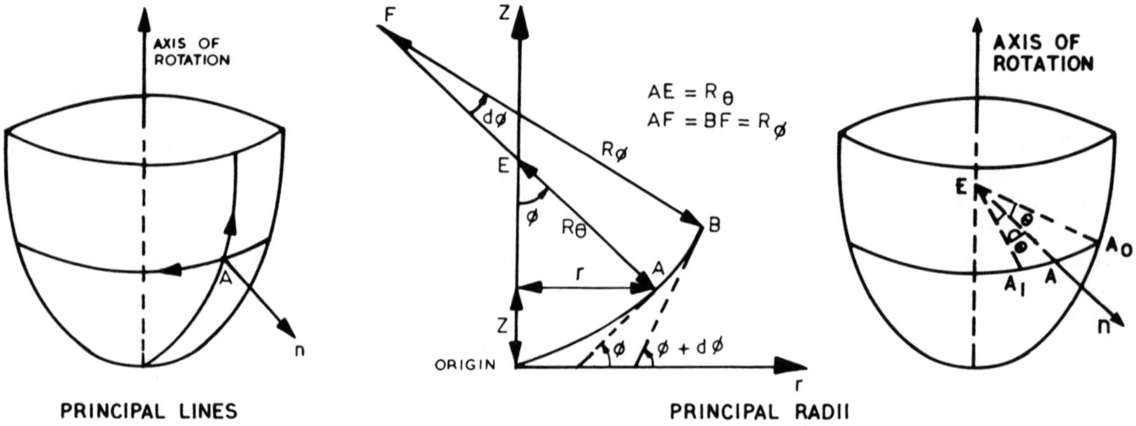

FIGURE 2. Principal curvatures.

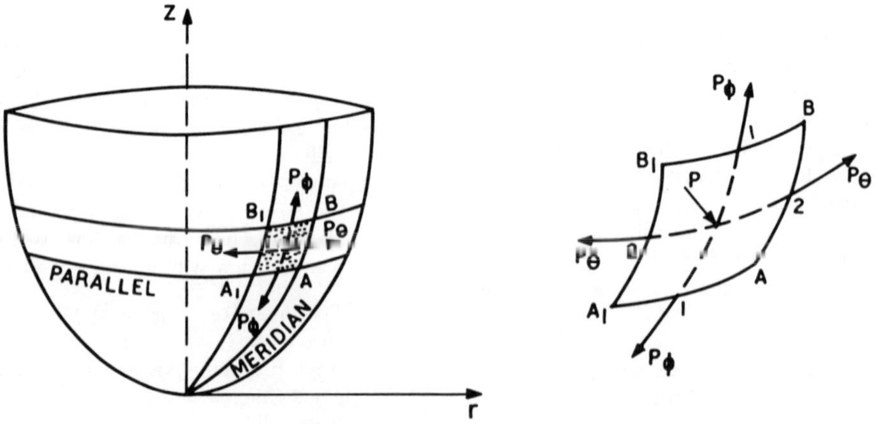

FIGURE 3. Elemental area on the surface.

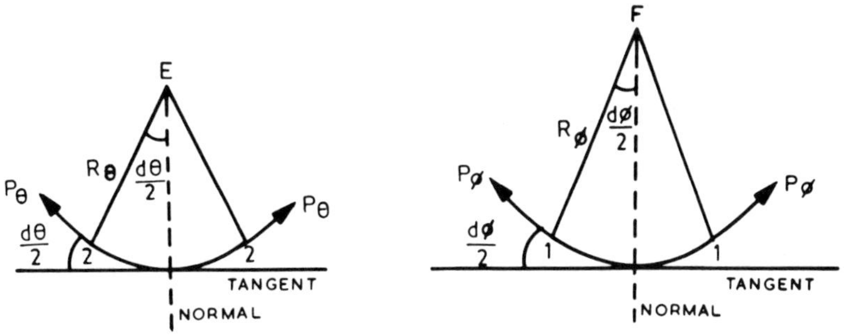

FIGURE 4. Equilibrium of elemental area.

External Load

The self weight of flexible membranes is usually small and the membrane forces introduced by virtue of this are therefore ignored. Liquid pressure which is the predominant force on the membrane is acting normal to its surface. No other external load is assumed to act on the membrane.

Internal Forces

Internal forces are analysed by considering the equilibrium of an element bounded by two meridians and two parallels as shown in Figure 3.

Since the container wall is a flexible membrane, its bending resistance is negligible and it cannot resist compressive forces. Since the container surface and the liquid loads are symmetric with reference to z-axis, the shear forces along the meridian and parallel are zero. Liquid load is resisted by developing only tension in the membrane.

Equations of Equilibrium

The free body diagram for the elemental surface ABB_1A_1 is shown in Figure 3. From Figure 2, AB is the arc length along meridian equal to $R_\varphi\, d\varphi$. AA_1 is the arc length along parallel equal to $R_\Theta\, d\Theta$. Area of element is $(R_\varphi\, d\varphi) \times (R_\Theta\, d\Theta)$. p is the liquid pressure acting on the element. Let P_Θ be the total resistant force acting on AB and A_1B_1, P_φ be the total resistant force acting on AA_1 and BB_1, and P be total force due to liquid pressure acting on the elemental area ABB_1A_1. Resolving the force acting on the element parallel to the normal to the surface (Refer to Figure 4).

$$2\, P_\varphi \sin \frac{d\varphi}{2} + 2\, P_\Theta \sin \frac{d\Theta}{2} = P \tag{1}$$

For small angles, Equation (1) is written as

$$2\, P_\varphi \frac{d\varphi}{2} + 2\, P_\Theta \frac{d\Theta}{2} = P \tag{2}$$

which is

$$P_\varphi\, d\varphi + P_\Theta\, d\Theta = P \tag{3}$$

Let T_φ be the meridional tension per unit length acting on faces AA_1 and BB_1 and T_Θ be the hoop tension per unit length acting on AB and A_1B_1. Therefore

$$P_\varphi = T_\varphi \times AA_1 = T_\varphi \times R_\Theta\, d\Theta \tag{4}$$

$$P_\Theta = T_\Theta \times AB = T_\Theta \times R_\varphi\, d\phi \tag{5}$$

$$P = p \times AB \times AA_1 = p \times R_\Theta\, d\Theta \times R_\varphi\, d\varphi \tag{6}$$

Substituting in Equation (3),

$$T_\varphi\, R_\Theta\, (d\Theta\, d\varphi) + T_\Theta\, R_\varphi\, (d\Theta\, d\varphi) = p\, (R_\Theta\, R_\varphi)\, (d\Theta\, d\varphi) \tag{7}$$

Simplifying, the familiar equation available in literature [2,5] is obtained as below:

$$\frac{T_\varphi}{R_\varphi} + \frac{T_\Theta}{R_\Theta} = p \tag{8}$$

Another equation for T_φ can be obtained by considering the equilibrium of the portion of the container below the parallel through the point A under consideration. Let p be the pressure intensity at this level and W be the weight of liquid below this level, as shown in Figure 5. In the case of the fully suspended tank of Figure 5(a), equilibrium requires

$$2\pi r\, T_\varphi \sin \varphi = W + \pi r^2\, p \tag{9}$$

Therefore

$$T_\varphi = \frac{W + \pi r^2 p}{2\pi r \sin\varphi} \tag{10}$$

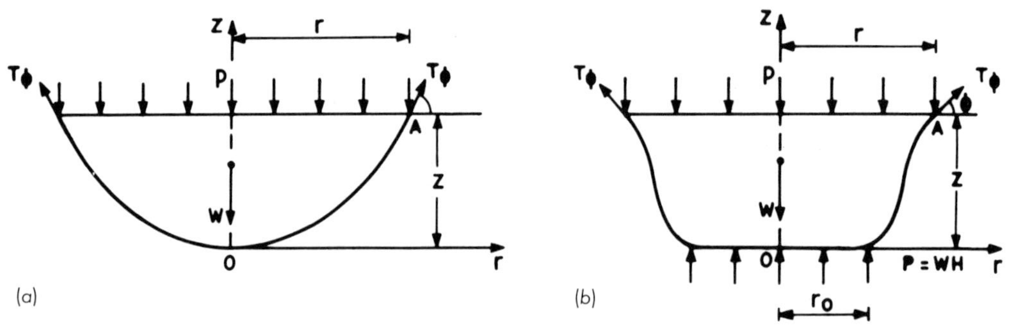

FIGURE 5. Equilibrium of container below a given parallel.

In the case of the partly suspended tank of Figure 5(b),

$$2\pi r\, T_\varphi \sin\varphi = W + \pi r^2 p - \pi r_0^2\, wH \tag{11}$$

Therefore

$$T_\varphi = \frac{W + \pi r^2 p - \pi r_0^2\, wH}{2\pi r \sin\varphi} \tag{12}$$

Equations (8) and (10) or (12) are sufficient to solve the unknown internal tensions T_φ or T_Θ for a known profile of the tank. It is to be emphasized that the above equations are operative for that part of the tank which is filled with liquid.

SHAPE OF A TENSION TANK

If the tank shape is fully defined and the mode of support or suspension specified, it is simple to determine the unit forces T_Θ and T_φ using the master equations derived in the previous section. However, the converse process of deriving the tank shape for a given distribution of membrane forces is by no means straightforward. This section deals with a solution for this problem using a simple computer programme that generates a variety of tank shapes suitable for different imposed conditions.

General

The meridional and hoop forces in a tension tank depend on the radii of curvature of the given shape as seen in Equation (8). It should be noted from Equations (10) and (12) that in a tension tank of any shape T_φ is always positive. The radius of curvature R_Θ is also always positive, as it is the length of normal from the point to the axis of rotation. However, the radius of curvature R_φ can take positive or negative values. If a tank shape with only positive values for R_φ is chosen for the meridian, the hoop force varies from tension at the bottom to compression at the top. This is so because $p = 0$ at the water surface and Equation (8) now becomes

$$\frac{T_\varphi}{R_\varphi} + \frac{T_\Theta}{R_\Theta} = 0 \tag{13}$$

Since T_φ, R_φ and R_Θ are positive, T_Θ has to be negative, which denotes a compressive force that the membrane is incapable of resisting. Therefore, the sign for R_φ has to be negative at the top. In the investigation that follows, the lower region for the tank shape is always started with a positive value for R_φ. Thus the shapes generated have a positive R_φ at the bottom reaches that changes to negative values in the top portion.

Stress Optimized Shape

As discussed above, with the sign for R_φ varying from positive to negative from bottom to top of the tank, it is possible to maintain T_Θ positive throughout which is a necessity. With no other condition imposed on the radii, it will be observed that the magnitude of T_Θ and T_φ can be varying widely for points on the membrane at different levels. Therefore, if the maximum stress is made to correspond to the limiting permissible stress, the stresses elsewhere can be much less, signalling lack of economy.

In order to economise on the material, it is necessary to make T_Θ and T_φ equal to each other at any point and the same value made to prevail over the entire tank wall. This is a case of stress optimization and the equation for the shape of such a tank will have to be evolved on this basis.

Shape Generation for a Stress Optimization Tank

In this subsection, a procedure for developing the shape of a stress optimized tank using an iterative process is outlined.

EQUATIONS GOVERNING THE SHAPE

In a stress optimizing tank, $T_\Theta = T_\phi = T$, a constant that depends on the strength and thickness of material used. Equation (8) now becomes

$$\frac{1}{R_\phi} + \frac{1}{R_\Theta} = \frac{p}{T} \tag{14}$$

For an analytical solution of this equation, it is modified as follows using the geometrical properties of the short arc length AB ($=ds_A$) of a meridian shown in Figure 6. Using the relations $R_\phi\, d\phi_A = ds_A$, $r_A = R_\Theta \sin\phi_A$ and $p = w(H - z_A)$, Equation (14) is written as

$$\frac{d\phi_A}{ds_A} + \frac{\sin\phi_A}{r_A} = \frac{w(H - z_A)}{T} \tag{15}$$

This is a nonlinear equation and a numerical solution is presented below. The immediate object of the exercise is to find the coordinates of point B (z_B, r_B) and the slope of the meridian at B (ϕ_B) if these three values for the adjacent point A are known.

For the determination of the coordinates of the point B, the triangle ABD in Figure 6 is considered [4]. $AD = dz_A$, $DB = dr_A$ and the slope of the chord $AB = \phi_{AB}$. From triangle ABD,

$$\text{Chord length } AB = \frac{DB}{\cos\phi_{AB}} = \frac{dr_A}{\cos\phi_{AB}} \tag{16}$$

Since the arc length AB can be approximated to chord length AB,

$$ds_A = \frac{dr_A}{\cos\phi_{AB}} \tag{17}$$

Substituting this value of ds_A in Equation (15),

$$\frac{\cos\phi_{AB}\, d\phi_A}{dr_A} + \frac{\sin\phi_A}{r_A} = \frac{w(H - z_A)}{T} \tag{18}$$

is obtained. Also, from triangle ABD,

$$\frac{AD}{BD} = \tan\phi_{AB} \tag{19}$$

Substituting the values of AD and DB,

$$\frac{dz_A}{dr_A} = \tan\phi_{AB} \tag{20}$$

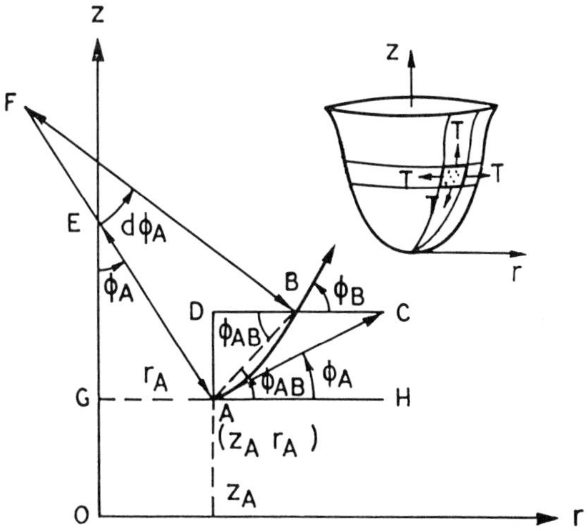

FIGURE 6. Geometry of arc AB on meridian.

Moreover, it will be seen from the same Figure 6 that

$$\phi_{AB} = \phi_A + \frac{d\phi_A}{2} \tag{21}$$

Equations (18), (20), (21) are sufficient to solve for the coordinates of the point B and the slope B as explained below.

SOLUTION OF EQUATIONS

Taking z as the independent variable throughout, and assuming equal intervals of dz_A, the corresponding values of dr_A and $d\phi_A$ are now determined as follows: Since ϕ_{AB} is as yet unknown quantity, its initial value is assumed to be equal to ϕ_A and dr_A is calculated from Equation (20). Substituting in Equation (18), $d\phi_A$ is obtained. The second value for ϕ_{AB} is obtained from Equation (21). Achieving higher degrees of accuracy for φ_{AB} involves a process of iteration which is easily performed in a computer, the iteration being terminated when the prescribed degree of accuracy is reached. The coordinates of the point B are calculated as follows:

$$\begin{aligned} z_B &= z_A + dz_A \\ r_B &= r_A + dr_A \\ \phi_B &= \phi_A + d\phi_A \end{aligned} \tag{22}$$

The solution is started in the computer by assuming the initial conditions discussed earlier. But the next subsection is devoted to explain a nondimensional approach to develop the stress-optimized shape.

NONDIMENSIONALISATION OF EQUATIONS

In order to facilitate designing a tank for a given capacity, certain nondimensional parameters are worked out as follows:

$$\text{Taking } z_A = \alpha_A H \quad \text{and} \quad r_A = \beta_A H$$

$$dz_A \text{ becomes } = d\alpha_A H \quad \text{and} \quad dr_A = d\beta_A H$$

Substituting these values in Equation 18,

$$\frac{\cos \phi_{AB} d\phi_A}{d\beta_A} + \frac{\sin \phi_A}{\beta_A} = \frac{wH^2}{T}(1 - \alpha_A) \tag{23}$$

Calling the nondimensional parameter wH^2/T as the shape constant C,

$$\frac{\cos \phi_{AB}\, d\phi_A}{d\beta_A} + \frac{\sin \phi_A}{\beta_A} = C(1 - \alpha_A) \tag{24}$$

Also,

$$\tan \phi_{AB} \text{ which is } \frac{dz_A}{dr_A} \text{ becomes } \frac{d\alpha_A}{d\beta_A} \tag{25}$$

The volume v of such a nondimensional tank shape is obtained as $\Sigma\, \pi\beta^2\, (\Delta\, \alpha)$. The relationship between the volume of a tank (V) and that of the nondimensional shape (v) is as follows:

$$\begin{aligned} V &= \Sigma\, \pi(\beta H)^2\, (\Delta\, \alpha H) \\ &= H^3\, \Sigma\, \pi\beta^2(\Delta\, \alpha) = H^3 v \end{aligned} \tag{26}$$

It is to be noted that if a particular nondimensional curve, designated by its volume v, is chosen, then, the actual tank shape obtained from this will be geometrically similar to it. Using Equation (26) and assuming any two parameters, the third is evaluated. To find the tank wall thickness t, the C value corresponding to the chosen nondimensional shape is used as below: $C = wH^2/T$ in which $T = \sigma \times t$, σ being the allowable tensile stress in the material.

Therefore

$$t = \frac{wH^2}{C\sigma} \tag{27}$$

INITIAL CONDITIONS

The type of tank generated depends upon the initial values of β and ϕ at $\alpha = 0$. In a fully suspended tank, $\beta = 0$ and ϕ is taken to be equal to zero at $\alpha = 0$. Because of symmetry at the origin, $R_\phi = R_\theta$ and Equation (14) becomes

$$\frac{2}{R_\phi} = \frac{wH}{T} \tag{28}$$

Taking $R_\phi = r_\phi H$ where r_ϕ is the radius of curvature of the nondimensional tank shape at $\alpha = 0$.

$$\frac{2}{r_\phi} = \frac{wH^2}{T} = C \tag{29}$$

Therefore

$$r_\phi = \frac{2}{C} \tag{30}$$

Knowing r_0 the very first incremental value for β, namely $d\beta$ can be determined using the geometry of a circle with intersecting chords. Thus $(d\beta)^2 = (2r_0 - d\alpha)\, d\alpha$. Ignoring terms of trivial value, $d\beta = \sqrt{2r_0 d\alpha}$. The other parameters $d\varphi$ needed for proceeding further is obtained from $r_0 d\varphi = d\beta$ from which $d\varphi = d\beta/r_0$.

In a partly suspended and partly supported tank, r_0 is chosen and ϕ is taken to be equal to zero at the origin. From Equation (18),

$$\frac{\cos \phi_{AB}\, d\phi_A}{R_\phi} = \frac{wH}{T} \tag{31}$$

Since the left hand side of equation is equal to $1/R_\phi$,

$$\frac{1}{R_\phi} = \frac{wH}{T} \tag{32}$$

As before, taking $R_\phi = r_\phi H$,

$$r_\phi = \frac{1}{C} \tag{33}$$

The coordinates of the second point are now obtained along with the corresponding $d\varphi$ as discussed above for the fully suspended case.

DESIGN CONSIDERATIONS

This section deals with some of the major design considerations that govern the shape and economy of the tank. The treatment is made separately for fully suspended and partly suspended tanks and the design procedure is explained in each case by an illustrative example.

Fully Suspended Tank

In order to economize the use of material for the tank construction, the ratio ϱ of the tank volume, V, to the volume of material V_m is now considered. Since V_m = surface area of tank $S \times$ thickness t,

$$\varrho = \frac{V}{V_m} = \frac{V}{S \times t} \tag{34}$$

But $S = 2\pi(\beta H)(\Delta \ell H) = H^2 2\pi\beta(\Delta \ell) = H^2 \times$ surface area of nondimensional tank shape, s.

$$S = H^2 s \tag{35}$$

Substituting for V, t and S from Equations (26), (27) and (35), respectively,

$$\varrho = \frac{vH^3 C\sigma}{sH^2 wH^2} = \frac{vC\sigma}{swH} \tag{36}$$

Substituting for H from Equation (26)

$$\varrho = \left(\frac{vC\sigma}{sw}\right)\frac{v_{1/3}}{V^{1/3}}$$
$$= \frac{v^{4/3}C}{s}\frac{\sigma}{wV^{1/3}} = \eta\frac{\sigma}{wV^{1/3}} \tag{37}$$

where η is another nondimensional parameter equal to $v^{4/3}C/s$. For a given material defined by σ, liquid w, and a tank volume V, ϱ is maximum when η is maximum. Therefore, a plot connecting η and v is drawn as shown in Figure 7. It is seen from this graph that the value of ϱ is a maximum when v is nearly 1.5. Also, the variation of ϱ is seen on the graph to be negligible from v = 1.0 to 1.5. Since higher values for v result in flatter tank shapes, from the point of view of economy alone, a value of v = 1.0 is recommended.

The nondimensional parameter C introduced earlier and representing the quantity wH^2/T is of importance in that every value of C defines a unique tank shape. Thus, various shapes are generated corresponding to different C values. For every shape, the volume of the tank v is also computed which enables the tank shape to be defined either by C or v. The coordinates of points defining the tank shape namely, α and β are calculated through the computer programme and tabulated for each set of C and v values in Table 1. The corresponding curves are shown plotted in Figure 8 in which the curve representing the recommended value of v = 1 is also included. Other curves shown there will enable the designer to have his choice varied to suit conditions of aesthetics or any other practical consideration.

By way of numerical example, the wall thickness of a suspended water tank of 300 cu ft (8.5m³) capacity with a permissible tensile stress of 1500 psi (10.3 N/mm²) can be worked out as follows: Taking v = 1.0 for material economy, from $V = vH^3$, H works out to 6.7 ft (2.01 m). From Table 1, values of β giving r as βH are now obtained for different α values. The width of tank at top is 10.5 ft (3.2 m). Since $t = wH^2/C\sigma$, thickness t works out to 0.00314 ft (0.96 mm) using C = 4.12 corresponding to v = 1.0. Working on the above lines the results of the design of suspended tanks for two different materials and using two different shapes are summarized in Table 2 for two diverging capacities.

Partly Suspended Tank

For the nondimensional tank shape, let v_1 be the volume of liquid supported by the suspension and v_2 be the volume transferred to the support.

$$v = v_1 + v_2 \tag{38}$$

It is suggested that in partly suspended tank, a share of 50% to 70% of the total volume may be apportioned to v_1 and the rest directly borne by the bottom support. Calling the ratio of v_1 to v as ϵ, the above statement means that ϵ may vary from 0.5 to 0.7.

For purposes of economy, the factor ϱ is considered here also. From Figure 1,

$$S = \Sigma\, 2\pi r\,(dS_A) + \pi r_0^2 \tag{39}$$

Expressing r, r_0 and dS_A in terms of H and simplifying

$$S = H^2[s_0 + s] \tag{40}$$

where s_0 and s in the nondimensional tank shape are the surface area of the bottom support and the surface area of the suspension, respectively. The expression for ϱ can be written as:

$$\varrho = \frac{v^{4/3}C}{(s + s_0)}\left[\frac{\sigma}{w\,V^{1/3}}\right] \tag{41}$$

Calling $\dfrac{v^{4/3}C}{(s + s_0)}$ as η_0, $\varrho = \eta_0\left(\dfrac{\sigma}{w\,V^{1/3}}\right)$

For maximum economy, ϱ must be maximised. For a given material defining σ, volume V and density of liquid w, ϱ is maximum when η_0 is a maximum. In order to maximise η_0 an indirect procedure is now adopted. Accordingly, a certain β value for β_0 is chosen and different curves are generated for a range of C values. Now, for each C value, a unique tank shape is obtained since β_0 is held constant. For each shape, the value of ϵ is computed and those curves which have ϵ falling outside the range specified earlier namely 0.5 to 0.7 are discarded. For the remaining curves, the η_0 value is worked out from v, s, s_0 computed and C for the curve. A plot connecting ϵ and η_0 is now drawn for the chosen β_0.

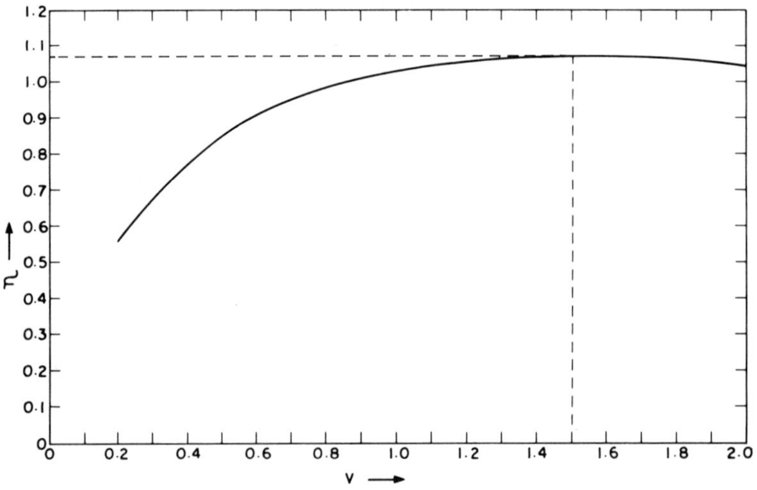

FIGURE 7. Variation of η with v.

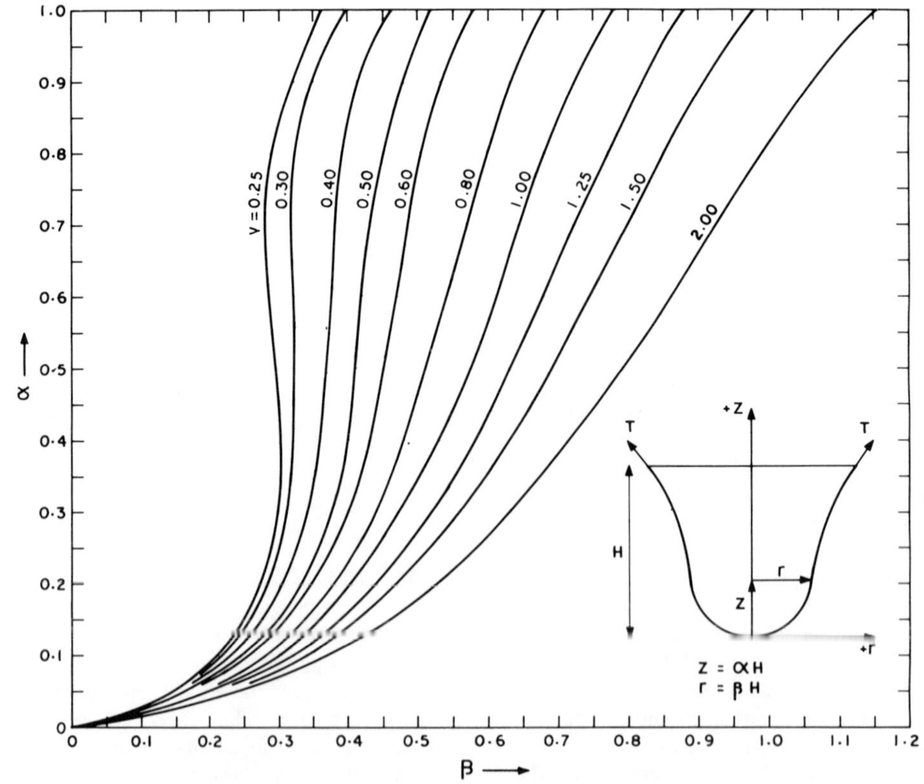

FIGURE 8. Nondimensional tank parameters (fully suspended).

TABLE 1. Parameters α and β for Fully Suspended Tanks.

v	β Values 0.25	0.30	0.40	0.50	0.60	0.80	1.00	1.25	1.50	2.00
C	7.45	7.00	6.30	5.76	5.32	4.64	4.12	3.63	3.24	2.68
(1)	(2)	(3)	(4)	(5)	(6)	(7)	(8)	(9)	(10)	(11)
α										
0.00	0.000	0.000	0.000	0.000	0.000	0.000	0.000	0.000	0.000	0.000
0.10	0.213	0.221	0.235	0.248	0.259	0.280	0.299	0.321	0.341	0.378
0.20	0.272	0.285	0.308	0.327	0.345	0.377	0.406	0.439	0.470	0.524
0.30	0.298	0.316	0.347	0.373	0.397	0.440	0.478	0.521	0.560	0.631
0.40	0.304	0.327	0.367	0.401	0.431	0.484	0.532	0.584	0.632	0.717
0.50	0.297	0.326	0.376	0.417	0.454	0.518	0.574	0.636	0.693	0.793
0.60	0.286	0.321	0.379	0.428	0.471	0.546	0.611	0.682	0.748	0.862
0.70	0.280	0.318	0.383	0.439	0.488	0.572	0.647	0.727	0.801	0.929
0.80	0.287	0.326	0.395	0.455	0.509	0.602	0.685	0.774	0.856	0.997
0.90	0.314	0.350	0.418	0.481	0.538	0.639	0.729	0.826	0.916	1.071
1.00	0.368	0.397	0.461	0.524	0.583	0.689	0.785	0.889	0.986	1.154

TABLE 2. Summary of Design of Suspended Tanks.

	V = 300 cft (8.5m³) σ = 1500 psi (10.3 N/mm²)		V = 3000 cft (85.0m³) σ = 15000 psi (103.0 N/mm²)	
Salient Tank Dimensions	C = 7.45 v = 0.25	C = 4.12 v = 1.00	C = 7.45 v = 0.25	C = 4.12 v = 1.00
Height of tank	10.63 ft. (3.24 m)	6.70 ft. (2.01 m)	22.89 ft. (6.98 m)	14.42 ft. (4.40 m)
Top width of tank	7.82 ft. (2.38 m)	10.50 ft. (3.20 m)	16.85 ft. (5.14 m)	22.64 ft. (6.80 m)
Thickness of tank wall	0.00438 ft. (1.34 mm)	0.00314 ft. (0.96 mm)	0.00203 ft. (0.62 mm)	0.00146 ft. (0.45 mm)

TABLE 3. Parameters α and β for Economic Shapes (Partly Suspended).

v	β Values 0.40	0.60	0.90	1.30	1.60	2.10	2.40	2.60	2.80	3.00	3.20
C	6.45	5.80	5.12	4.48	4.21	3.73	3.61	3.64	3.68	3.74	3.82
(1)	(2)	(3)	(4)	(5)	(6)	(7)	(8)	(9)	(10)	(11)	(12)
α											
0.00	0.200	0.250	0.300	0.350	0.400	0.450	0.500	0.550	0.600	0.650	0.700
0.10	0.366	0.425	0.488	0.554	0.610	0.675	0.729	0.777	0.824	0.871	0.918
0.20	0.409	0.475	0.545	0.620	0.680	0.753	0.808	0.854	0.899	0.943	0.987
0.30	0.421	0.492	0.571	0.655	0.718	0.800	0.856	0.899	0.942	0.984	1.025
0.40	0.410	0.488	0.577	0.671	0.738	0.829	0.886	0.926	0.966	1.005	1.042
0.50	0.384	0.469	0.569	0.675	0.745	0.846	0.904	0.941	0.978	1.013	1.046
0.60	0.348	0.441	0.553	0.671	0.744	0.856	0.914	0.948	0.981	1.012	1.040
0.70	0.314	0.411	0.534	0.664	0.740	0.863	0.921	0.950	0.979	1.005	1.028
0.80	0.292	0.387	0.517	0.658	0.736	0.869	0.927	0.952	0.977	0.997	1.014
0.90	0.290	0.377	0.509	0.659	0.738	0.880	0.937	0.958	0.977	0.991	1.002
1.00	0.317	0.388	0.516	0.669	0.748	0.898	0.954	0.969	0.983	0.992	0.996

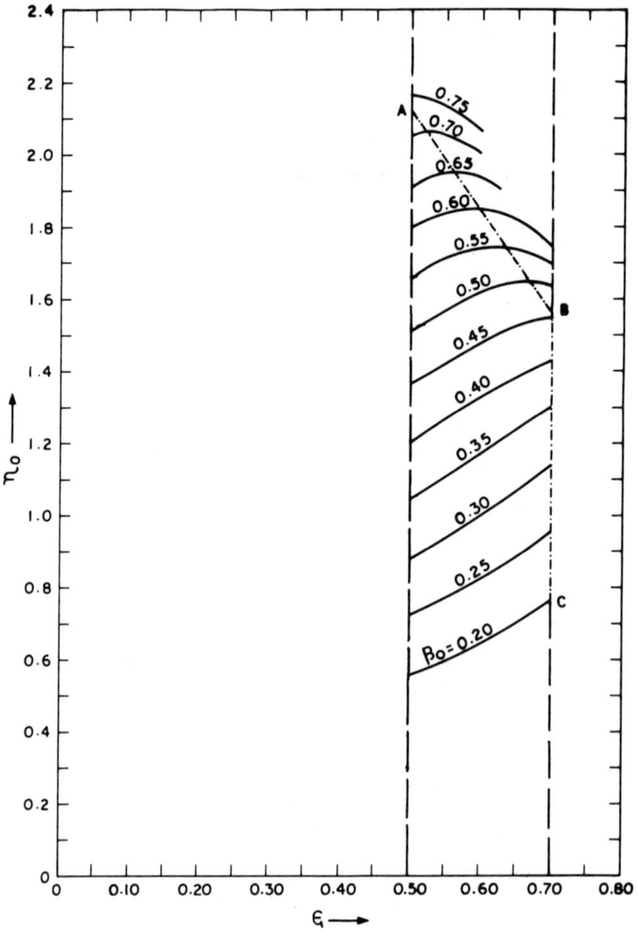

FIGURE 9. Variation of η_0 with ϵ for different β_0 values (partly suspended).

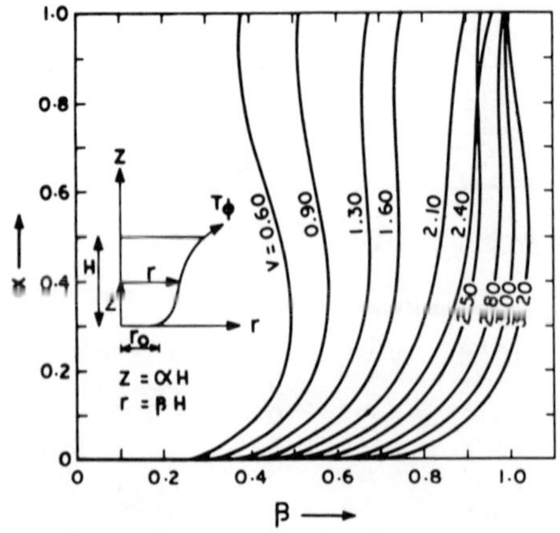

FIGURE 10. Nondimensional tank parameters (partly suspended).

The procedure is now repeated taking other β_0 values, and the corresponding curves connecting ϵ and η_0 are drawn as shown in Figure 9. For each curve the point giving maximum value of η_0 is located within the chosen range of ϵ and these points are joined and approximated to the bilinear curve ABC.

The tank shapes corresponding to these maximum points are reproduced in Figure 10 and the coordinate values given in Table 3.

A numerical example is worked out to illustrate the design of a partly suspended water tank. If $V = 300$ cu. ft. ($8.5m^3$), $\sigma = 1500$ psi (10.3 N/mm^2) and the height of tank is chosen to be twice its base width, from Table 3, $\beta_0 = 0.25$, $C = 5.80$ and $v = 0.60$. The height of tank, its base width and thickness of side wall work out to be 7.94 ft. (2.42m), 3.97 ft. (1.21m) and 0.00314 ft. (0.91mm), respectively.

APPLICATIONS AND CONCLUSIONS

The procedure suggested in this chapter enables generation of nondimensional shapes for stress optimized suspended or partly suspended tanks. These shapes have positive curvature in the lower region and negative curvature in the upper region. In the case of partly suspended tank, the supports are deemed to be horizontal.

Depending on the aesthetics preferred or the material economy needed, a nondimensional tank shape is chosen and the actual tank dimensions worked out from the required tank capacity. These tanks can be made of flexible plastics or thin stainless steel sheets in a corrosive environment. Small capacity plastics water tanks can be used in coastal villages and they can be suspended from trees or poles or supported on stabilized sand dunes. Flexible containers made of plastics or canvas are portable and may be useful in military applications. The use of these tanks for storing grains is a distinct possibility that can be explored further.

NOTATION

C = shape constant
H = height of container
p = intensity of liquid pressure
P_φ, P_Θ = total tensile force acting on elemental arc length of parallel and meridian, respectively
R_Θ, R_φ = radii of curvature of parallel and meridian, respectively
S = surface area of container
s = surface area of nondimensional tank shape
T = total tensile force acting on unit length of arc, same at all points and in all directions
T_φ, T_Θ = total tensile force acting on unit arc length of parallel and meridian, respectively
t = thickness of tank wall
V = volume of container
V_m = volume of material of tank wall
v = volume of nondimensional tank shape
v_1, v_2 = volume transferred to the suspension and support, respectively
w = weight of liquid per unit volume
z, r = coordinates of any point on the meridian
r_0 = radius of tank at the origin
α, β = nondimensional tank parameters
β_0 = radius of the nondimensional tank at the origin
η = $v^{4/3}C/s$
ϱ = volume of container/volume of material
σ = allowable tensile stress in material
φ = slope of meridian at any point
ϵ = v_1/v

REFERENCES

1. Day, C. L., "New Spheroidal Design for Large Oil Tanks," *Engineering News Record,* Vol. 103, pp. 416–419 (1929).
2. Flugge, W., *Stresses in Shells,* Springer-Verlag, New York, N.Y., pp. 39–45 (1967).
3. Frei, O., ed., "Tensile Structures," *Pneumatic Structures,* Vol. 1, The M.I.T. Press, Massachusetts Institute of Technology, Cambridge, Mass., pp. 134–135 (1967).
4. Jnanasekaran, N., and D. S. Sandegren, "Stress Optimized Suspended Liquid Containers," *Journal of Engineering Mechanics,* American Society of Civil Engineers, pp. 1350–1356 (Sept. 1984).
5. Timoshenko, S., and S. Woinowsky-Krieger, *Theory of Plates and Shells,* 2nd ed., McGraw-Hill Book Co., Inc., New York, N.Y., pp. 443–445 (1959).

Influence of Drainage Trenches and Tubular Drains on Slope Stability

Bogdan Stanić* and Ervin Nonveiller*

ABSTRACT

An unfavourable groundwater condition is a frequent cause of slope sliding. Pore pressure decreases the effective stress level and consequently the shear resistance of the soil on the slip surface. Thus the slope safety against sliding is reduced. When the unfavourable groundwater condition causes the sliding, the stabilization methods which favourably change the groundwater condition is the most effective one. A change in the groundwater condition is achieved by drainage system. Drainage trenches are suitable for stabilizing relative shallow slidings and tubular drains for stabilizing the deep ones.

INTRODUCTION

Any slope sliding is a consequence of the equilibrium disturbance between the active and the resisting forces on the most critical slip line within the slope section. The material failure occurs with large deformations which are visible on the slope. The slope sliding will be stopped when the shear stresses creating instability are reduced or when shear resistance of the material in that zone is increased, or when both factors are achieved by the stabilization method. Stabilization methods should relate to the main causes of sliding. An unfavourable groundwater condition is a frequent cause of slope instability. The pore pressure decreases the effective stress level and consequently the shear resistence of the material along the slip surface is decreased. In this way the slope safety against sliding is reduced. The stabilization methods which favourably influence the change of groundwater condition present the most effective way for sliding stabilization, the cause of which is the unfavourable groundwater condition. The change in the groundwater condition is achieved by drainage system. It can be divided into the preventive and the active one.

The purpose of the preventive drainage system is to stop the penetration of precipitation which flows on the surface and, penetrating the soil, raises sometimes the groundwater table, as well as to increase the hydrostatic pressure in the existing cracks on the landslide surface.

The active drainage system introduces lower potential into the groundwater which decreases the pore pressures in the sliding mass. It is achieved by drainages. The result of this acting is the improvement in the ratio between the available and the needed shear stress component, thus increasing the factor of safety. If there is a difference of the total potential among water, in various points, water flows in the pores from the place of higher potential to the place of the lower one.

The total pore pressure force on soil element in the flow field could be divided into the component force caused by the potential *h* (buoyancy) and the component force caused by the excess potential, *dh* (seepage force). The total effective force acting on the sliding mass could be determined by adding the volume integrals of the buoyant weight (γ') and of the seepage force ($i \cdot \gamma_w$) or by adding the volume integral of the total weight (γ) and the surface integral of the pore pressure along the mass boundary. The latter is the simpler one.

The above-mentioned procedures are conceptually explained on a very simple example of unit soil element from the infinite slope with inclination "β" (Figure 1).

The drain effectiveness caused by decreasing the pore pressure along the slip surface or by changing the flow direction on the change of the slope stability is presented on the example of the equilibrium of the circular segment of the slope with the groundwater table on the surface of the slope (Figure 2). The total forces of the weight *W*, buoyancy *U* and the seepage force *S* act in the gravity centre of the seg-

*Faculty of Civil Engineering, University of Zagreb, Yugoslavia

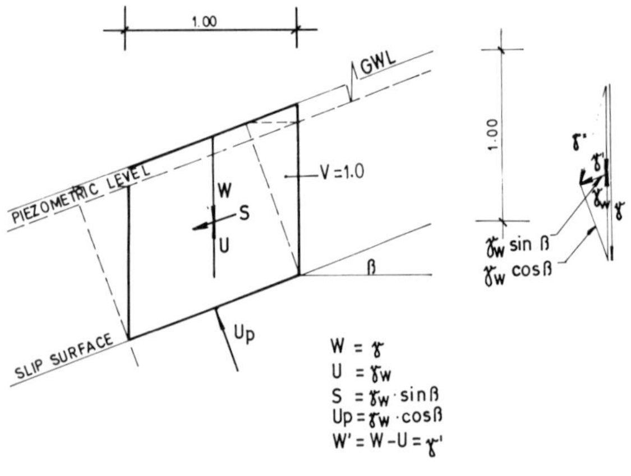

FIGURE 1. Effective force γ'' on the unit soil element in the flow field.

ment. Their value can be calculated if A is the surface of the cross section of the segment thickness 1 m, γ bulk weight of soil and γ_w unit weight of water. The forces W, U, and S_1 give the total force P_1 and N_1 normal component of the soil resistance and T_1 shear stress component provide for its balance.

If we install drains in the slope the direction and the value of the total seepage force on a certain force S_2 will be changed while other two forces W and U will stay unchanged. The total force will be P_2 with the reaction components N_2 and T_2. As it is visible from the polygon of forces, shear stress component $T_2 < T_1$, while normal component $N_2 > N_1$. Because the shear resistance along the failure plane is the function of the normal component N, it is visible that the ratio between the available and the needed shear stress component T is improved by draining. In this way the factor of safety of the segment is increased. It can be achieved if the groundwater table is unchanged, and the drain is effective in the material with low permeability. The water quantities flowing through the drain are insignificant, and the precipitation penetrating the soil in temperate climate region regularly replaces water which has drained out. In the materials with higher pemeability there are periods when drains lower the groundwater table, but there are periods when the groundwater table is equal to the one without drains. These periods are critical for slope stability.

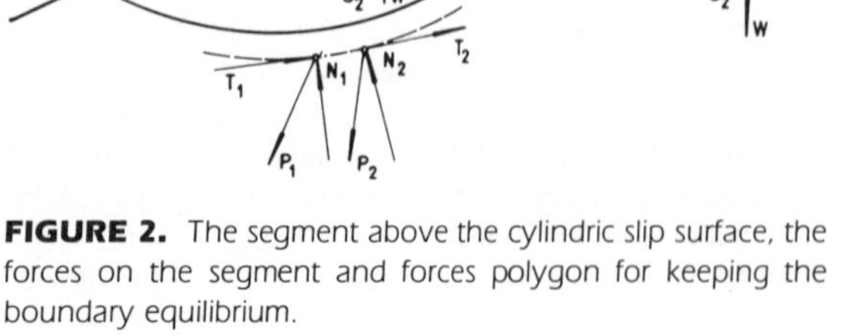

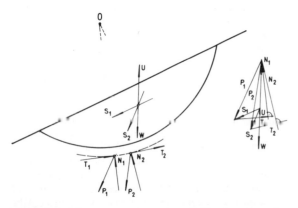

FIGURE 2. The segment above the cylindric slip surface, the forces on the segment and forces polygon for keeping the boundary equilibrium.

Drains installed in the slopes cause the change in the potential field and direct the waterflow more or less crosswise to the slope direction. If it happens with the greatest quantity of water the drains are effective.

The best effect of the drains is achieved by the greatest pore pressure reduction in the failure zone. It will be achieved if the drains are placed near the failure zone when their influence is spread over the largest part of the failure zone and when they are placed in the direction of the maximal slope inclination. For deep landslides the horizontally bored drains represent the most effective way of achieving these conditions. For stabilizing relative shallow slides (to approximately 5.0 m deep) drainage trenches are the most suitable.

Drain galleries are expensive but they can be very useful for stabilizing great and deep slidings, and their efficiency can become greater by boring horizontal and vertical boreholes, which is particularly useful if the material is inhomogenous or bedded.

The perpendicular drains made as shafts or boreholes can be useful in bedded or anisotropic permeable material. In some circumstances they lead water away into the deeper layer with lower groundwater table or the water is occasionally pumped out automatically.

Vegetation can be useful for the decrease of pore pressure evapotranspiration in shallow slidings when the roots contribute to fixing of the material. Grass-covered shroud decreases the precipitation infiltration into the soil and stops the surface erosion of the dug out slopes.

INFLUENCE OF DRAINAGE TRENCHES ON SLOPE STABILITY

For stabilizing relative shallow slides (to approximately 5.0 m deep) which are caused by unfavourable groundwater conditions, the use of drainage trenches is the most suitable stabilizing method. The trenches which are under the slip surface are filled by permeable material; those near the soil by fine-grained material; and those in the middle by coarse-grained material. The drainage trenches must have permanent effectiveness. Streamline flow of water towards the drain must not cause the soil erosion from the boundary zone because it would block the drain. On the contrary, the material which is the filling of the drain must be permeable enough so that water can flow through it with considerable decreased gradient. The drain filling must be adapted to the filter rules: the permeability must be at least ten times greater than that of the surrounding material, and the grain-sizes must be adapted to the surrounding material. The minimal cross section on the lower part of the drain must be great enough to accept the whole flow of the drained water. In respect of the realization of the drains the cross section of the drain will rarely have to be broader than one constructionally needed (60–80 cm). The drains with greater capacity are the drains in inhomogeneous material covering the layers with greater pemeability from which the waterflow can occasionally be great.

The drain ability to lead water away becomes greater if the tubular drains are on the bottom. Such tubular drains are installed in gravel, with grain sizes corresponding to the tube openings and drain fillings. Concrete slab is often put at the bottom of the drain, as the base for the tubular drain. There is no excuse for low permeability of the concrete and such a drain gets water only through lateral planes. The concrete slab can be used if there is sliding of the permeable material along the layer which is essentially water impermeable, i.e., if there is no water flowing into the drain from below.

In the case of the drainage trenches without support the filter layer of the cohesionless material is replaced by the geodrain.

The drain is covered on the surface slope by the layer of clay to stop the sinking of the surface water bringing small particles which would by time decrease its permeability. The drain outlet is achieved by tubular drain. The tube is built in the concrete slab.

On the joint between the drains there is a control shaft. It is very useful to build such shafts to get shorter drains. Figure 3 shows the drainage trench with details. Relatively shallow landslides are frequent in civil engineering. Because there are relatively small slidings and the stabilization methods are mainly urgent, there is no possibility of longlasting investigations. The available data concerning soil is one of the basic reasons for undertaking the parametric study [12]. In most cases the surface layer of the weathered rock slides on the layer of the primary rock which is significantly less permeable (e.g., clay sliding along marl). Because of that the parametric study [12] is made, supposing the sliding plane is on impermeable base. If the impervious layer is deeper the pore pressure reduction in the slip zone is proportionally greater. According to Hutchinson [1] the pore pressures on the sliding plane in a great depth are about 10% lower, in the case of impervious base, than in the case of impermeable sliding plane.

Nonveiller [10] studied the consolidation process after installing the drains in a slope. He took into consideration the possibility that the increase of the stress effectiveness in soft soil increases the pore pressures which are consequences of

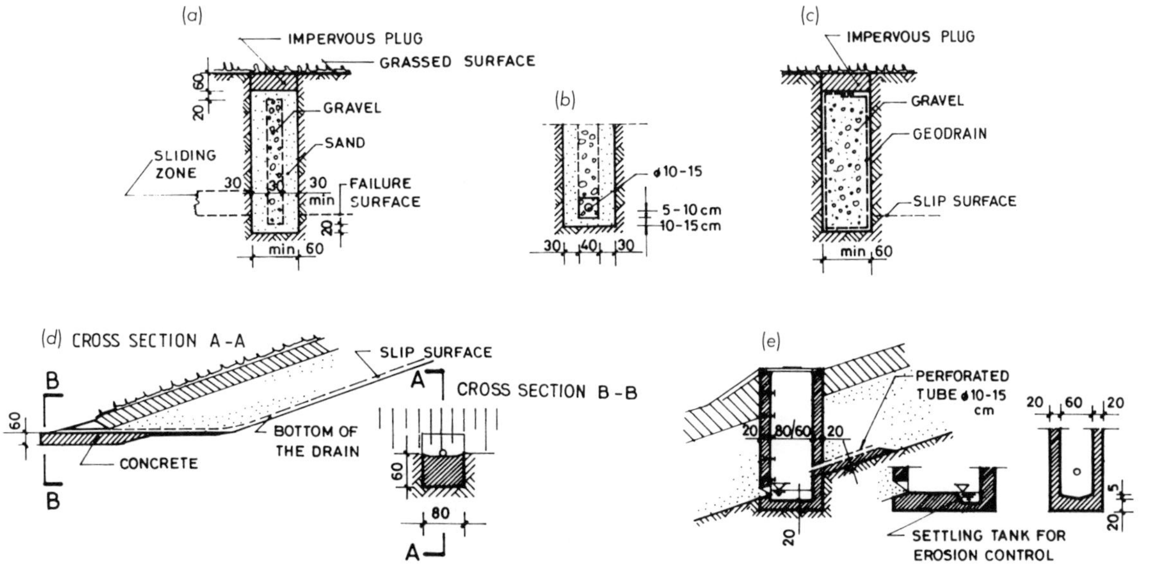

FIGURE 3. Details of the drainage trenches: a) without the tubular drain; b) with the tubular drain; c) with geodrain; d) drain outlet on the slope surface; e) control shaft.

the changed hydrodynamic field and drains become effective in one month's time in silty sand slopes, and in the period of 6 months on clay ones. Because the drainage trenches distance is essentially smaller (it is advisable $S = 20$ m) and the consolidation time is proportional to the square distance among the drains, the drainage trenches become effective in much shorter time. Because of that, in calculating the drainage trenches effectiveness it is not necessary to observe the change of the pore pressures in time, but it is enough to observe the finite stationary flow after consolidation. Consequently, the parametric solution does not depend on coefficient of permeability.

MODELING OF DRAINAGE TRENCHES INFLUENCE

For the parametric study of the spatial influence of drainage trenches on the potential field change in the slope, a typical example is modeled (schematized) in the following way.

1. The slope is infinite in all directions with the inclination β to the horizontal plane.
2. The slip surface is impermeable and parallel to the slope.
3. The groundwater table is parallel to the slope and does not change in time.
4. The soil is of homogenous and isotropic material.
5. The slope will be stabilized by digging the drainage trenches of depth h from the groundwater table to the slip surface and at a spacing s.

The parameters which influence the change of a potential field in the observed area are:

1. Angle β of the groundwater table and slip surface inclination from the horizontal plane
2. Distance s between the drainage trenches' central lines
3. Distance h from the groundwater table to the slip surface

The influence of distance s and height h is taken into account by the normalized distance by the ratio s/h.

MATHEMATICAL SOLUTION OF DRAINAGE TRENCHES INFLUENCE

It is necessary to solve a potential flow problem in the given domain with the prescribed boundary condition. For the seepage through homogenous and isotropic soil the solution is a scalar field of potential $H(x,y,z)$ which satisfies the Laplace differential equation $H(x,y,z) = 0$, in the domain with the given boundary condition.

For the earlier stated model of the drainage trenches influence on the change of the potential field in the slope, the

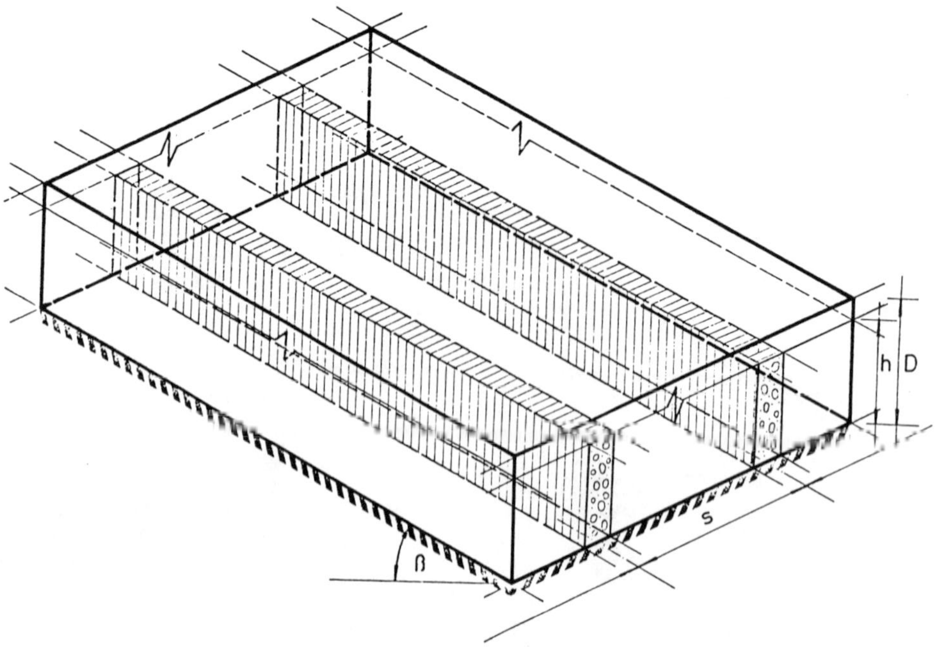

FIGURE 4. Presentation of model for slope drainage trenches.

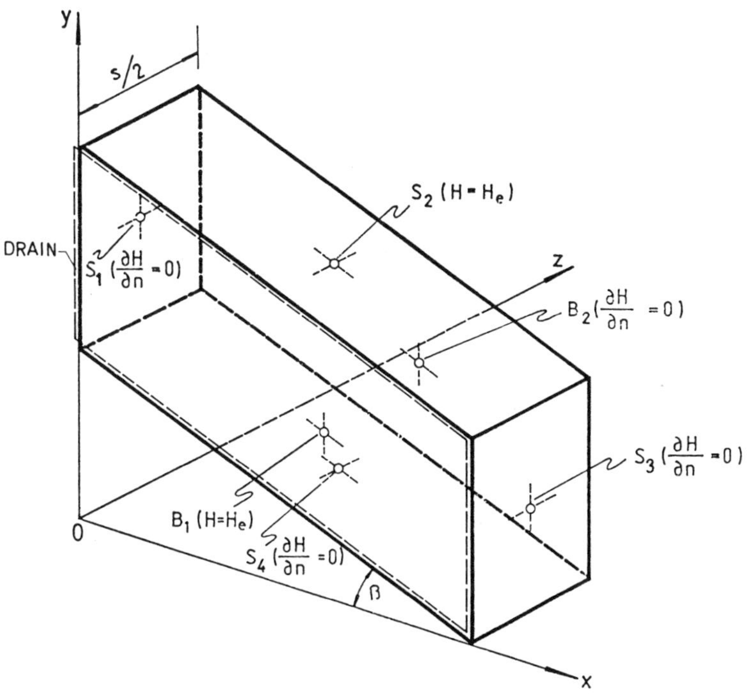

FIGURE 5. Slope with domain boundaries.

mathematical model with the boundaries defined in Figure 5 is specified.

A symmetric part of the slope with the width $s/2$ between two drains is shown. The plane, B_1, in XoY is the drain plane and, therefore, the total potential is equal to the elevation head on this boundary. The plane, B_2, is parallel to B_1 at distance $s/2$ representing the plane of symmetry between the two drains. The plane S_2 is the groundwater table. The total potential is equal to the elevation head on this plane. The plane S_4 is parallel to S_2, and it represents the impervious base. On the boundary planes S_1 and S_3 the values of the potential cannot be determined a priori. If the mathematical model encompasses the sufficiently long portion of the slope the influence of the boundaries on the middle part of the model would be small, especially when the drain distance is such that the dominant water flow is perpendicular to the slope. In this case the assumption that the planes S_1 and S_3 are impervious is acceptable and the potential field in the middle will tend toward the true solution.

Impervious planes and the planes of symmetry are boundary stream planes and the total potential function must satisfy the condition $\partial H/\partial n = 0$ (n is normal to the plane under consideration). The steady state seepage is considered, i.e., the final influence of the drains and not the transient state which occurs immediately upon the drain installation. The stated problem with the given boundary condition is solved by the computer program ELIPTI [2]. The program solves the problems described by the Poisson's eliptic partial differential equation.

$$\frac{\partial}{\partial x_i}\left(k_{ij} \frac{\partial H}{\partial y_i} \right) = f(x_i) \qquad i,j = 1,2,3$$

in which

- H is a scalar function with some physical meaning, e.g., potential
- k_{ij} is symmetric tensor of second order which describes the domain properties, e.g., the coefficient of permeability
- $f(x_i)$ is the field function distributed in the domain

The finite element method is used in order to solve the problem.

INFLUENCE OF DRAINAGE TRENCHES DISTANCE ON WATER POTENTIAL IN DOMAIN

For the described physical model the influence of the drainage trenches distance s and of the trenches height h on the flow pattern has been studied by the stated mathematical model for different slope inclination β.

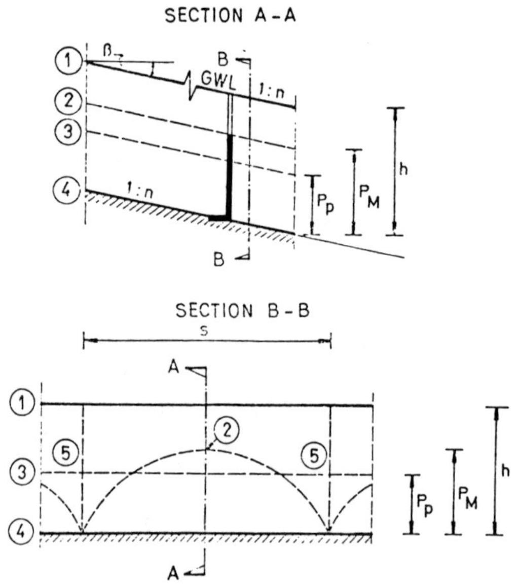

FIGURE 6. Characteristic cross section of slope with drainage trenches.

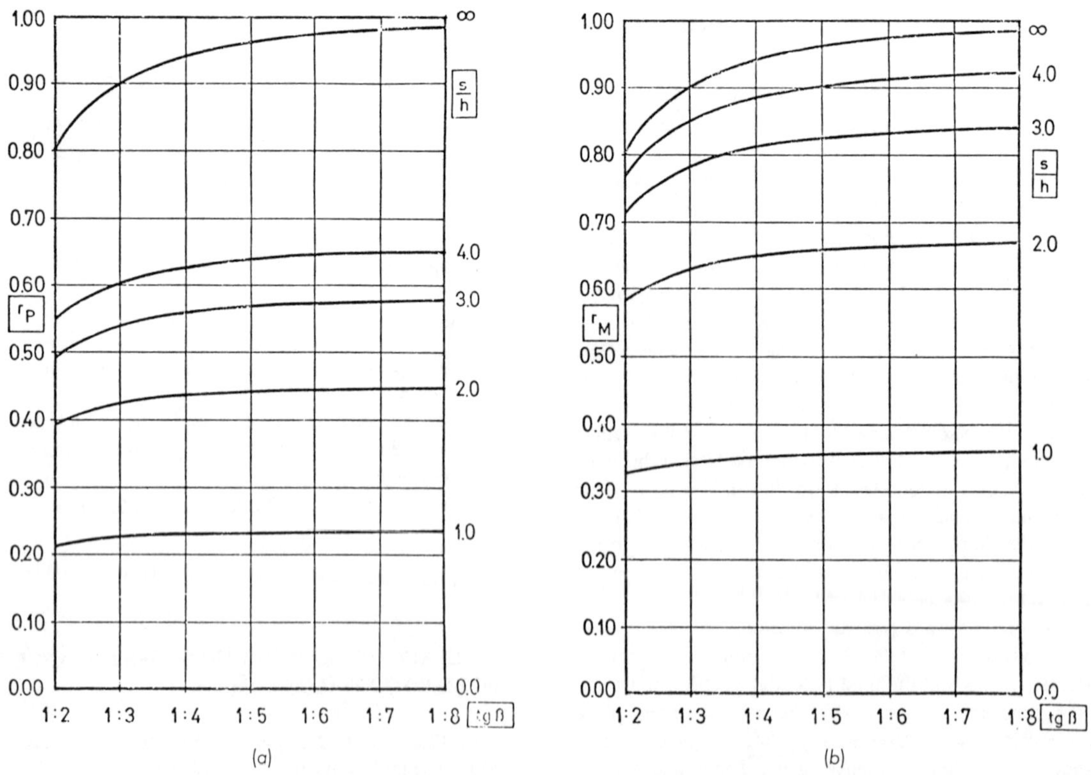

FIGURE 7. Diagrams of normalized peizometric level at the slip surface: a) average; b) maximal.

Figure 6 shows the characteristic cross section of slope with the designation as used. The groundwater table is defined by line 1, maximal piezometric level between two drains by line 2, average piezometric level by line 3, slip surface by line 4, and the drain plane by line 5. The slope inclination is defined by the ratio $1{:}n = tg\,\beta$. The slopes from 1:2 to 1:8 with the drain distance from $s = 1.0\,h$ to $s = 4.0\,h$ and a slope without drain (designation $s/h = \infty$ with the piezometric level at the slip surface $h_p = h \cdot \cos^2 \beta$) are considered.

The maximal and average piezometric levels at the slip surface are designed by P_M and P_p, respectively.

The diagrams in Figure 7 determining the average and maximal piezometric levels on the slip surface are constructed by the parametric study with the following parameters.

$tg\,\beta = 1{:}n$ slope inclination

s/h normalized drain distance s by height h

$r_p = \dfrac{P_p}{h}$ normalized average piezometric level at the slip surface by height h

$r_M = \dfrac{P_M}{h}$ normalized maximal piezometric level at the slip surface by height h

By using these diagrams and the appropriate interpolation and exploration, the piezometric level at the slip surface could be determined with the sufficient accuracy for any slope inclination from 1:2 to 1:8 and for any normalized drain distance s/h.

From the diagram of the maximal normalized piezometric level [Figure 7(a)] it could be seen that the maximal piezometric level has decreased approximately 5% for the case of $s/h = 4.0$ compared to the case $s/h = \infty$ (the slope without drain). Experience has shown that local sliding between drainage trenches is possible when the drain distance is larger. Therefore it is advisable to use the average piezometric level obtained from the diagram in Figure 7(b) only for the values $s/h < 4.0$. Using the influence of the drainage trenches on the potential field according to the stated procedure it is possible to analyze the slope stability by any appropriate calculation method.

THE INFLUENCE OF ANISOTROPY ON THE DECREASE OF PIEZOMETRIC LEVEL

The anisotropy of the coefficient of permeability is taken into consideration where $k_{xx} = k_{zz} \neq k_{yy}$. The normalized piezometric level for various ratios of vertical and horizontal permeability was calculated, with the slope inclination to $tg\beta = 1{:}2$, without drain, with a drain distance $s = h$ and $s = 4\,h$. The results are shown in Figure 8. By the ratio increase, $k_{xx}/k_{yy} = k_{zz}/k_{yy}$, i.e., the increase of horizontal permeability in relation to the vertical one, the piezometric level of the slope without drain decreases. For $k_{xx}/k_{yy} = 10$ it is 50% from the piezometric level of the isotropic soil slope ($k_{xx}/k_{yy} = 1$), and for $k_{xx}/k_{yy} = 100$ about 10% from the piezometric level of the slope.

It is probable that in the conditions of essentially greater horizontal permeability ($k_{xx}/k_{yy} > 10$) sliding caused by unfavourable groundwater condition will not appear.

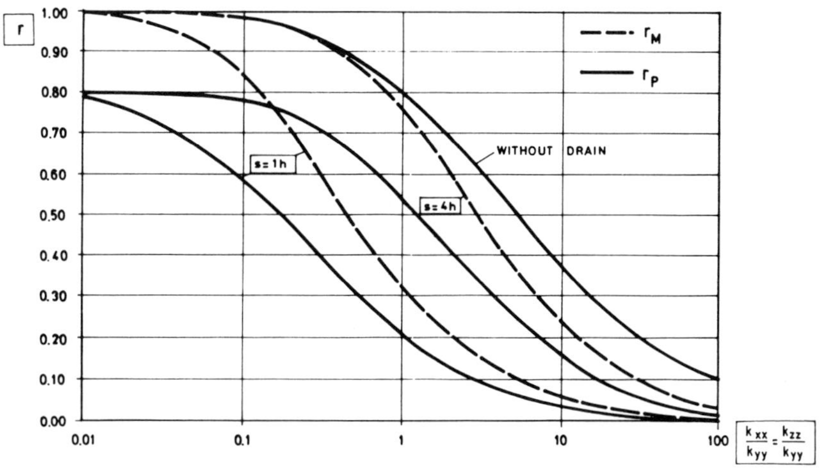

FIGURE 8. Diagrams of normalized piezometric level at slope inclination 1:2 for various ratios of horizontal and vertical permeability.

The effectiveness of the drainage trenches is increased by the increase of horizontal permeability in relation to the vertical one, as is shown in Figure 8.

The piezometric level of the slope without drain is increased by the decrease of the relations $k_{xx}/k_{yy} = k_{zz}/k_{yy}$, i.e., the increase of the vertical permeability in relation to the horizontal one. For $k_{xx}/k_{yy} = 0.1$ it is 122% of the piezometric level of the slope of the isotropic soil, and for $k_{xx}/k_{yy} = 0.1$, 125% of the piezometric level of the isotropic soil.

The effect of the drainage trenches decreases by the increase of the vertical permeability in relation to the horizontal one, as is quantitatively shown in Figure 8. In such cases the drainage "carpets" or horizontally bored drains on a small distance are effective.

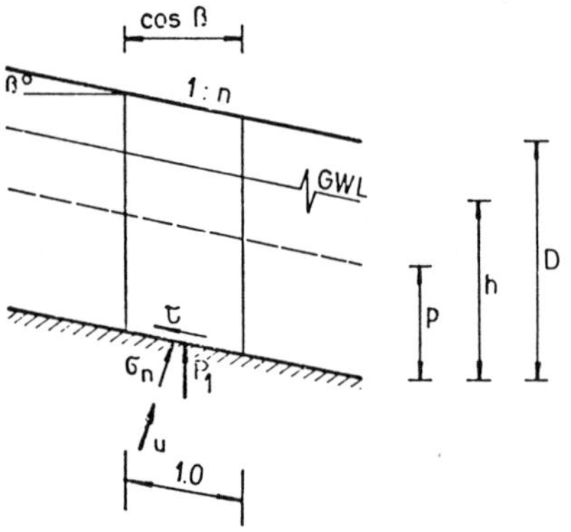

FIGURE 9. Model of infinite slope with used designation.

INFLUENCE OF DRAINAGE TRENCHES DISTANCE ON FACTOR OF SAFETY

The determination of the factor of safety by diagrams for the circular slip surfaces has been developed by many authors [3,9,10,14].

The landslides described in this paper are frequent and of small dimensions; the sliding mass is stretched and shallow; and the data on soil characteristics are often insufficient. Therefore parametric study for the influence of the drainage trenches distance on the factor of safety of an infinite slope is performed. The model with the designations used is shown in Figure 9. A slice with the unit width is considered.

According to Figure 9,

reaction on the slip surface:

$$P_1 = \gamma \cdot D \cdot \cos\beta \tag{1}$$

shear stress on the slip surface:

$$\tau = \gamma \cdot D \cdot \cos\beta \cdot \sin\beta \tag{2}$$

total normal stress on the slip surface:

$$\sigma_n = \gamma \cdot D \cdot \cos^2\beta \tag{3}$$

pore pressure on the slip surface:

$$u = P \cdot \gamma_w = r \cdot h \cdot \gamma_w \tag{4}$$

r is the specific piezometric level on the slip surface which is obtained from diagrams in Figure 7. The shear strength defined by the Mohr-Coulomb failure criterion is as follows:

$$\tau_f = c' + (\sigma_n - u) \cdot tg\phi' \tag{5}$$

Factor of safety is defined as

$$F_s = \tau_f/\tau \tag{6}$$

By substituting Equations (2), (3), (4), and (5) into Equation (6), Equation (7) is obtained:

$$F_s = \frac{c'}{\gamma \cdot D} \cdot \frac{1}{\sin\beta\cos\beta} + \frac{\cos^2\beta \cdot tg\phi'}{\sin\beta\cos\beta} - \frac{h \cdot \gamma_w \cdot r \cdot tg\phi'}{\gamma \cdot D \cdot \sin\beta\cos\beta} \tag{7}$$

By introducing

$$N = \frac{c'}{\gamma \cdot D} \tag{8}$$

and

$$M = \frac{\gamma_w \cdot h}{\gamma \cdot D} \tag{9}$$

Equation (10) is obtained:

$$F_s = \frac{N}{\sin\beta\cos\beta} + \frac{tg\phi'}{tg\beta} - r\,\frac{M \cdot tg\phi'}{\sin\beta\cos\beta} \tag{10}$$

Using

$$n = \frac{N}{\sin\beta\cos\beta} + \frac{tg\phi'}{tg\beta} \tag{11}$$

and

$$m = \frac{M \cdot tg\phi'}{\sin\beta\cos\beta} \tag{12}$$

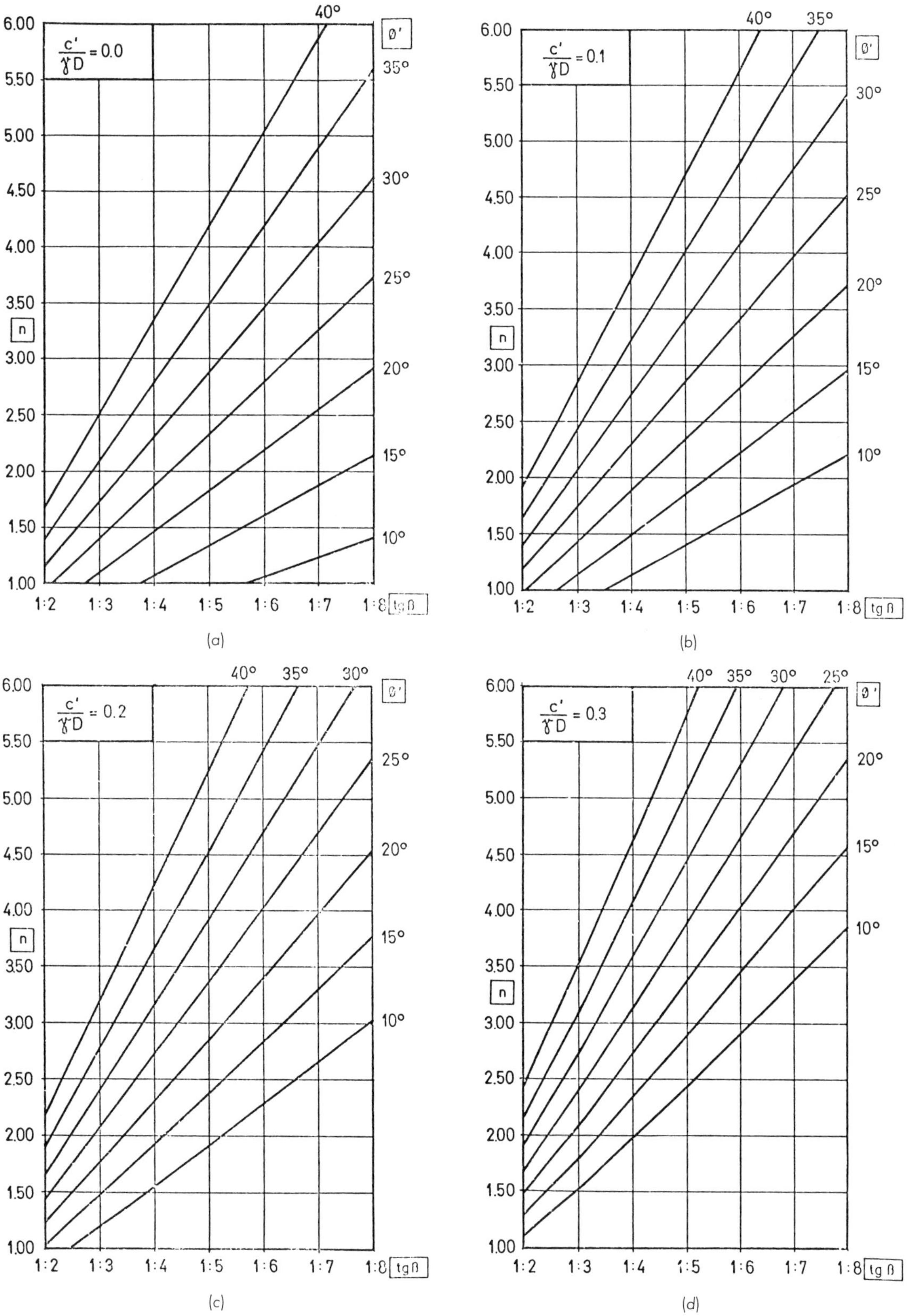

FIGURE 10. Diagrams for n = f (β, ϕ′, N) for a) N = 0.0, b) N = 0.1, c) N = 0.2, d) N = 0.3.

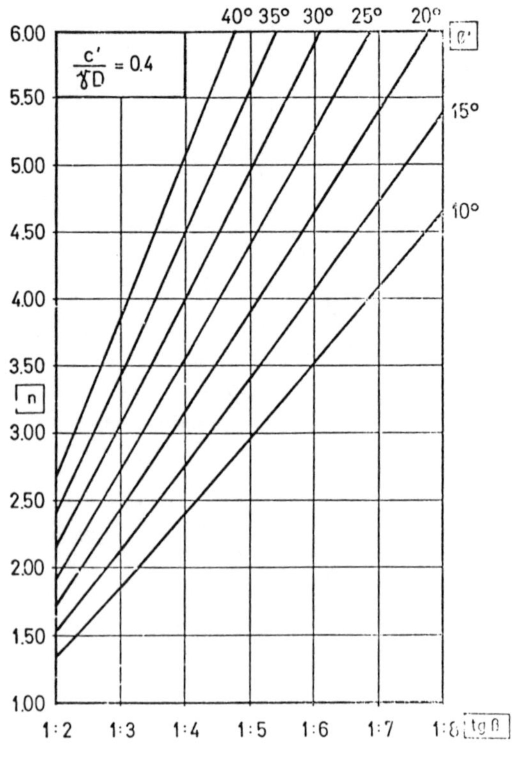

(e)

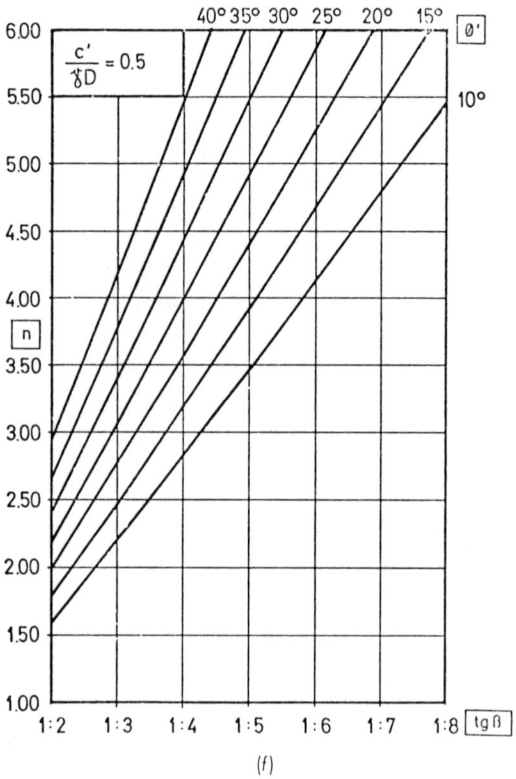

(f)

FIGURE 10 (continued). e) N = 0.4, f) N = 0.5.

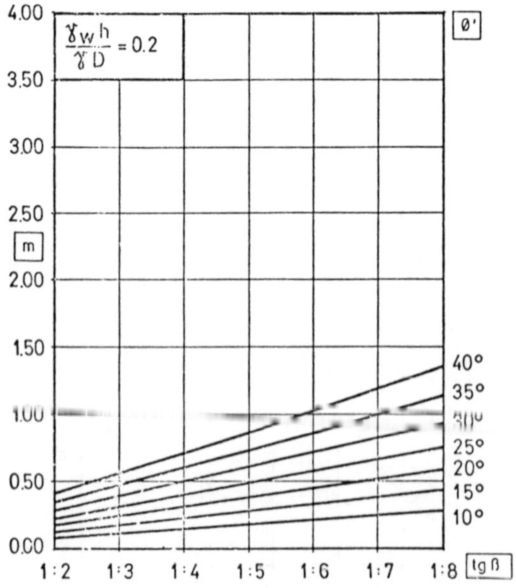

(a)

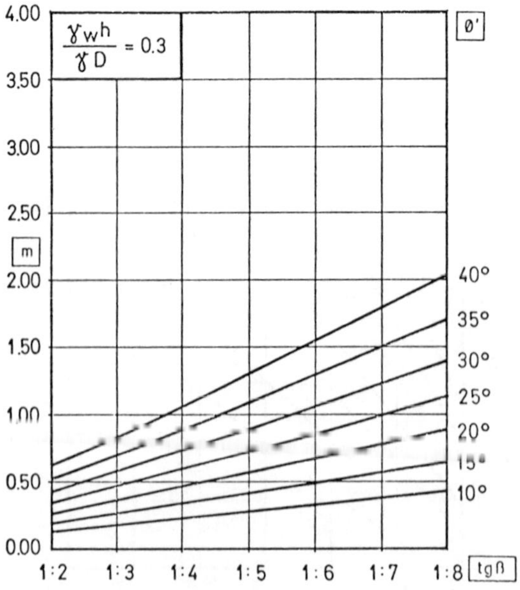

(b)

FIGURE 11. Diagrams for m = f (β, φ′, M): a) M = 0.2, b) M = 0.3.

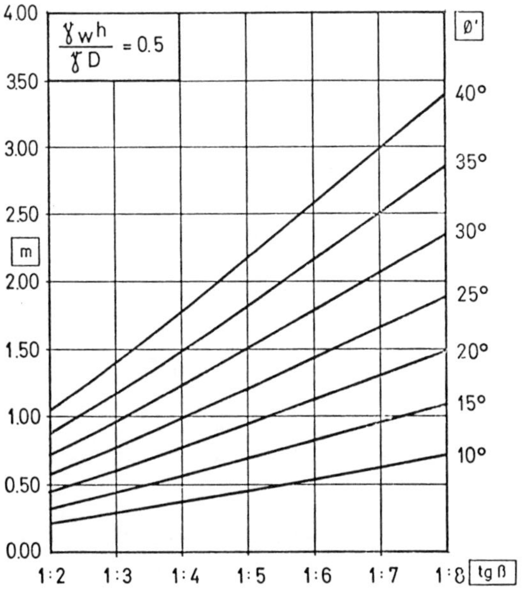

(c)

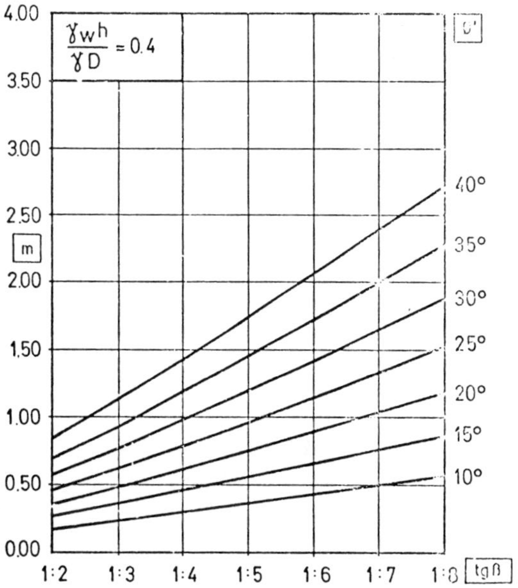

(d)

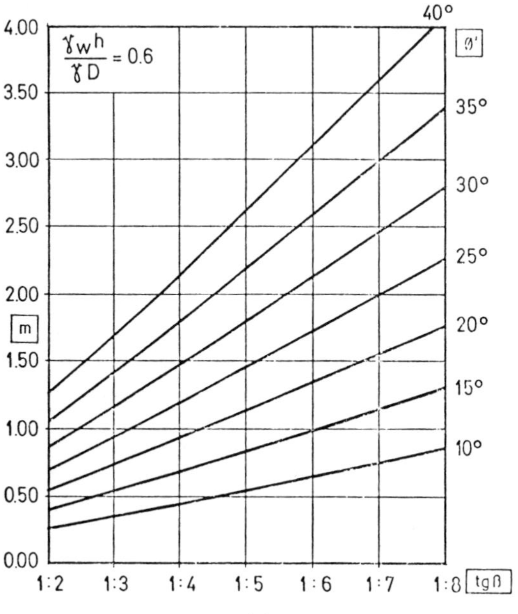

(e)

FIGURE 11 (continued). c) M = 0.4, d) M = 0.5, e) M = 0.6.

Equation (10) becomes

$$F_s = n - r \cdot m \qquad (13)$$

n represents the factor of safety for the slope without water while $r \cdot m$ represents the factor of safety decrease caused by the pore pressure on the slip surface.

For slopes with $tg\beta = 1{:}n$ $(n = 2{,}3 \ldots 8)$, relative values N $(N = 0.0–0.5)$ and M $(M = 0.2–0.6)$ and for the range of friction angle φ' from 10° to 40° the diagrams for determining n and m values have been constructed. The diagrams for n values are shown in Figure 10 and and m values in Figure 11, respectively.

Thus using diagrams for r (Figure 7) all data necessary for calculating the factor of safety by Equation (13) are known.

EXAMPLE OF THE CALCULATED DRAINAGE TRENCHES

It is necessary to increase the factor of safety against sliding by a minimum of 20% for a 1:4 slope. Impervious base depth $D = 5.0$ m; groundwater table level $h = 4.0$ above the slip surface. Soil parameters are $c = 10$ kN/m², $\varphi = 15°$ and $\gamma = 20$ kN/m³.

One calculates:

$$N = \frac{c'}{\gamma \cdot D} = \frac{10.0}{20.0 \cdot 5.0} = 0.100$$

From the diagram in Figure 10b, $n = 1.50$ is obtained.

$$M = \frac{\gamma_w \cdot h}{\gamma \cdot D} = \frac{10.0 \cdot 4.0}{20.0 \cdot 5.0} = 0.400$$

From the diagram in Figure 11c, $m = 0.46$ is obtained.

From the diagram for $s/h = \propto$ in Figure 7, $r_\propto = 0.94$ is obtained.

The factor of safety before installing the drains is calculated by Equation (13).

$$F_s = n - r_\propto \cdot m = 1.5 - 0.94 - 0.46 = 1.07$$

It is required that an increase of 20% for the factor of safety is achieved by the drains. Thus $F_s \geq 1.2 \cdot 1.07 = 1.28$. From Equation (13) the normalized piezometric level for this case is obtained:

$$r \leq \frac{n - F_s}{m} = \frac{1.5 - 1.28}{0.46} = 0.48$$

and the corresponding drain distance is taken from the diagram in Figure 7a $s \leq 2.3h = 9.2\ m$.

Therefore in order to increase the factor of safety by 20%, it is necessary to install the drainage trenches 5.0 m deep at a distance of $s = 10$ m each.

INFLUENCE OF BORED HORIZONTAL DRAINS ON SLOPE STABILITY

For stabilizing deep slidings caused by unfavourable groundwater condition the most economical stabilization methods are drilled horizontal tubular drains. In such cases deep drainage trenches are expensive and very difficult and dangerous to be performed. Horizontal drains have following advantages:

a. The work is performed from the stable domain out of sliding area.
b. Work velocity is great (~ 30 m daily).
c. There is the possibility of drain washing out in case of filter silting.

Tubular drains are installed in already drilled boreholes of proper length and diameter. If the drain length is over 20 m rotary drilling equipment with bit and core barrel is used. Boreholes shorter than 20 m can be drilled by auger of corresponding diameter, which is much cheaper and quicker than drilling with bit and core barrel. In both cases the more effective way is drilling with special drilling equipment which installs casing during drilling if the soil is disturbed before putting the tubular drain. The borehole diameter should be somewhat greater than that of the drain so that it can be installed in the whole length without difficulty.

The drain consists of perforated tube with internal diameter of about 50 mm. The external side of it is covered by porous filter permeable enough not to prevent the flow of water, and with enough small pores to prevent erosion and carrying of the material from the surrounding soil into the drainage tube which could be blocked by time. We installed the granular filter at distances of 60–100 cm made from unified fine sand by bakelite or other artificial resins (Figure 12).

Drain goes out of the slope through a small concrete or stone wall. A broad tube 4–6 m long is installed up to first ring of granular filter which encloses it. The outer part of it is closed by a flange with only a tubular drain with water flowing, going through it. If it is necessary the flange can be taken off and the drain can be washed by means of a pipe with a nozzle or by a packer from which water is injected into the drain under pressure. The water flows around filter and at the end of the drain it comes out through broader pipe in the wall.

The choice of the protection system of tubular drain filters from silting has great importance for efficiency and lasting of the system. The dangers are different depending on

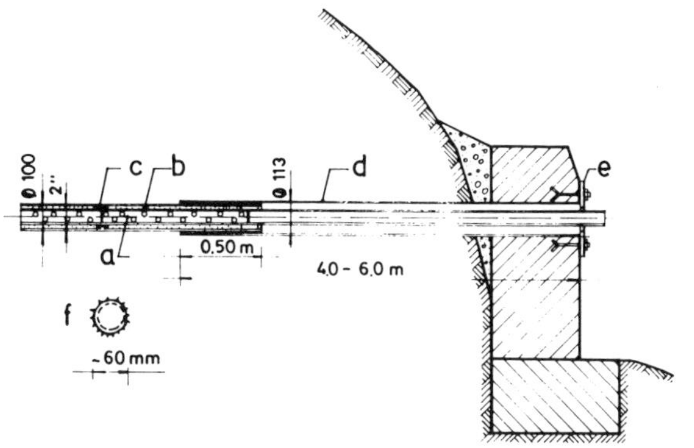

FIGURE 12. Details of tubular drain: a) perforated tube Ø 50 mm; b) granular filter; c) rubber chain; d) casing Ø 113 mm; e) rubber seal with flange and screws; f) alternative protection by geodrain.

materials. Granular filter around the tubular drain must be permanently more permeable than the surrounding material. If the grain size distribution of the material is such that the internal scour and transportation of fine particles are possible, they can silt the filter and decrease its permeability. This unfavourably influences the hydrodynamic field and decreases the efficiency of the system. This danger is smaller if the filter around drains is permeable for the fine particles. After some time a zone of permeable material develops around it and stabilizes further from the drain where the gradient is so small that it does not cause internal scour.

Coarse grained filter around the drain is enough to prevent erosion. It is advisable that the granular filter around the drain should be permeable for particles finer than coarse silt (< 0.06 mm), and its thickness should not be great (< 20 mm).

If there is the possibility of the hydraulic erosion of the surrounding area, it is necessary to control the water leaking out of the tubular drains because of suspended matter and the elutrition in the tubes at the beginning of the working period of the drainage system.

The other processes such as geochemical and biochemical ones can cause the blocking of tubular drains. The sedimentation of iron compounds and manganese ones as well as calcium carbonate from the underground water can in certain conditions block the drains after a longer period. Therefore it is useful to test the chemical composition of the drained water and present documentary evidence in the project.

Various kinds of plastic textile can be used as filter protection of tubular drains (polyester, etc.), but it is necessary to test the duration and filtrating ability of such plastic materials, because there is not much experience with such filter protection of drains.

MODELING OF BORED DRAINS INFLUENCE

The bored drains influence [10] and the change of factor of safety of the slope is modeled in the following way for the parametric study of the spacial influence of bored drains.

1. The slope has inclination from the horizontal, finite length and infinite width.
2. Maximal difference of the total potential H_i is defined.
3. Groundwater table is parallel to the slope and it is not changed in time.
4. Soil is homogenous isotropic material with the coefficient of consolidation C_v.
5. The slope will be stabilized by installing the horizontal bored drains in the slope direction, with length L at distance s.

Figure 13 shows the slope with used designation. The following are the parameters causing the potential field change in the observed model:

1. Angle β of the slope inclination and groundwater table from the horizontal plane
2. Drain lengths L
3. Drain distance s
4. Maximal difference of the total potential H_i
5. Field potential change in compressible soil is time depended on coefficient of consolidation C_v.

MATHEMATICAL SOLUTION OF DRILLED DRAINS INFLUENCE ON THE POTENTIAL FIELD CHANGE

Slope with boundaries, according to Figure 14, is mathematically solved for the defined model of spacial influence of horizontally drilled drains on the potential field change.

The symmetrical part of the slope from the plane B_1 in XoY with the drain, with length L is presented up to its parallel plane B_2 in the middle of the distance between the drains. The plane S_4 is the boundary flow plane, i.e., impermeable boundary. There is no waterflow perpendicular to the impermeable planes and symmetry planes, and the function of the total potential must satisfy the condition $\partial H/\partial n = 0$ (n is normal line at the observed plane).

Plane S_2 is the plane of the groundwater table. The total potential is equal to the elevation head. There is a waterflow perpendicular to the planes S_1 and S_3, and they represent 100% and 0% equipotential planes. Across the drain with length L the total potential is equal to the elevation head.

The stationary flow is observed, i.e., the final drain influence. The described problem with the stated boundary conditions is programmed in TROFIL. It solves the problems described by the Laplace differential equation.

$$\frac{\partial^2 \cdot H}{\partial x_i^2} = 0 \qquad i = 1,2,3$$

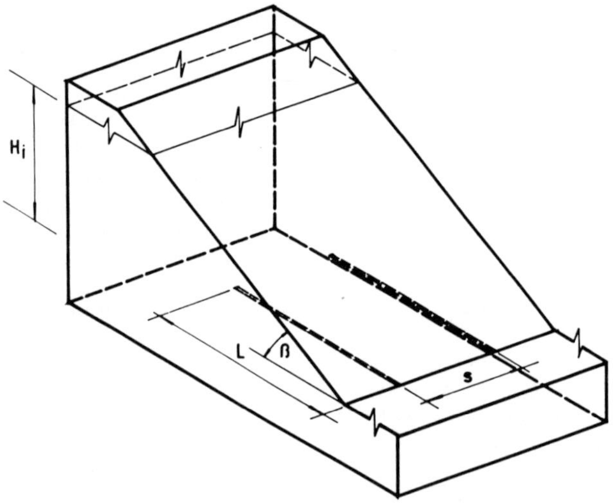

FIGURE 13. Model of the slope with bored drains.

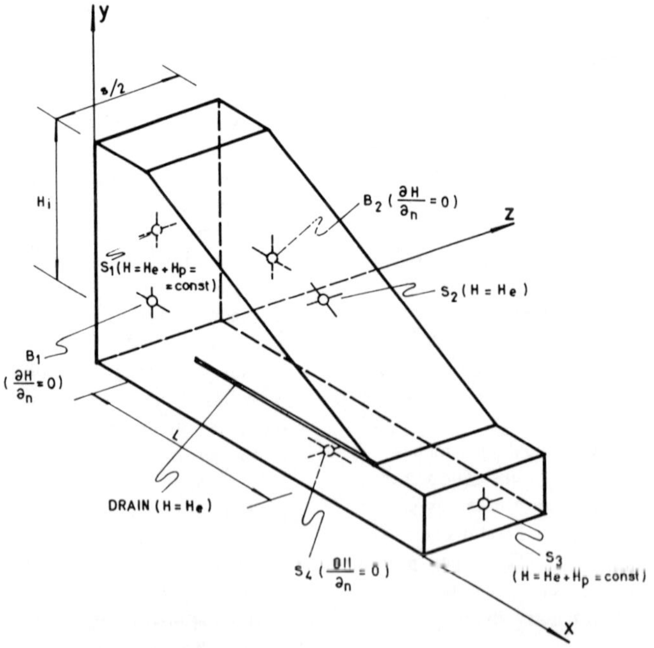

FIGURE 14. Slope with boundaries.

The used numerical technique of solving problems is the finite element method with the iterative process of solving equations according to Krylov [4].

DRAIN LENGTH AND DRAIN DISTANCE INFLUENCE ON FACTOR OF SAFETY

Finite drain length and drain distance influence on the increase of the factor of safety, time depended, is observed on the slopes with inclination 1:2 and 1:3 with groundwater table about 50 m above the foot of the slope. The form of the slope, net of nodes and approximate equipotentials in field among drains for the drain length $L = 100$ m and distances $S = 20$ m and 100 m are presented in Figure 15.

Slope safety against sliding is calculated by means of the program SSTAB1 [15] for circular sliding bodies, for slope without drain (the belonging factor of safety F_o) and for slope with finite influence of the installed drains (the belonging factor of safety F_d).

The results are shown in Figure 16. The increase of factor of safety is defined by the relation $F = F_d/F_o$ for slopes with inclination 1:2 and 1:3 with parameters:

$s_i = S/H_i$ normalized drain distance according to the

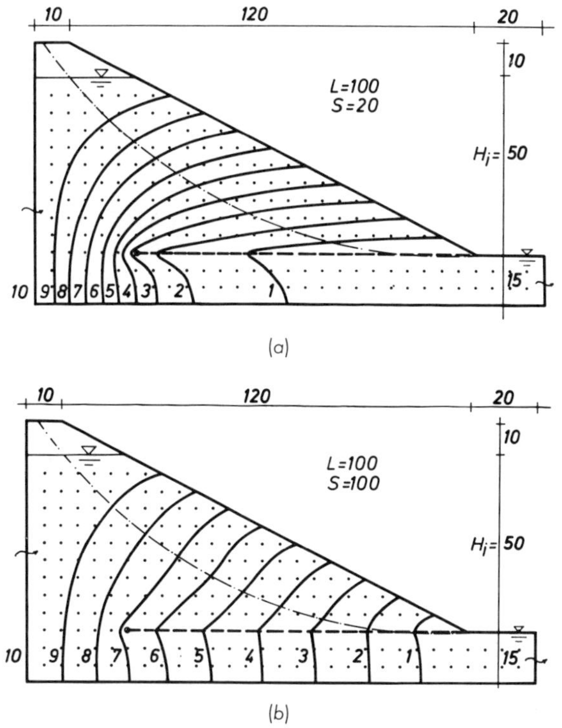

FIGURE 15. Finite approximate equipotentials in field between two drains: a) S = 20 m; b) S = 100 m.

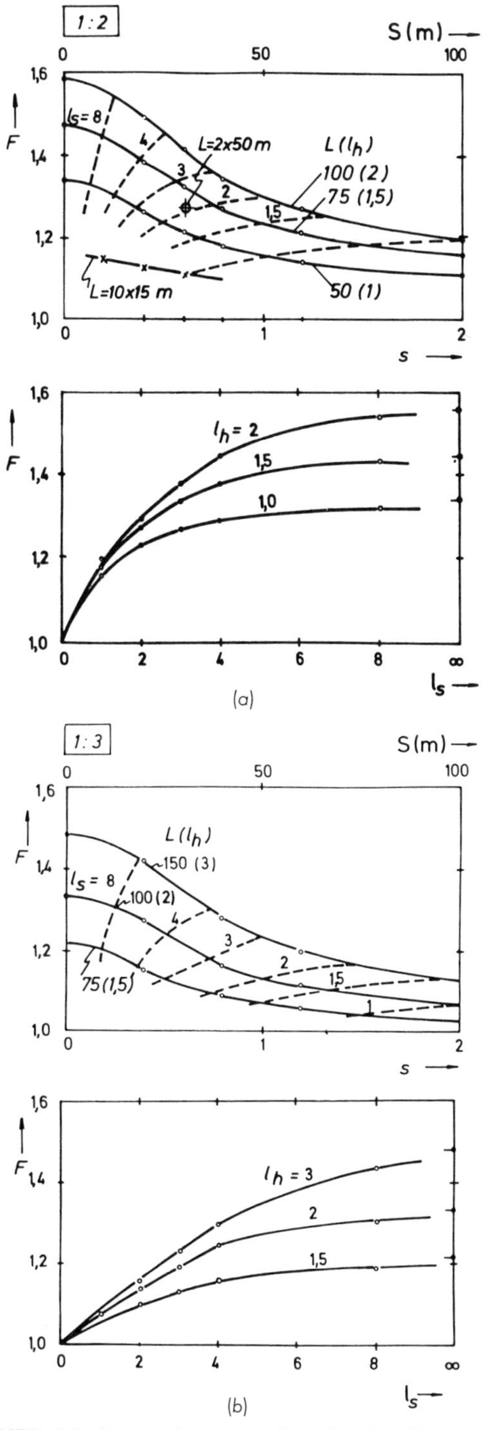

FIGURE 16. Parametric presentation of drain influences of various lengths L and distances S at the increase of the factor of safety $F = F_d/F_o$: a) slope inclination 1:2; b) slope inclination 1:3.

maximal difference of the total potential

$l_s = L/S$ normalized drain length according to the distance

$l_h = L/H_i$ normalized drain length according to the maximal difference of the total potential

In order to get the required increase of the factor of safety the drain distance and drain length can be chosen with these diagrams.

CONSOLIDATION OF INDUCED PORE PRESSURE

The field of initial pore pressure after installing drain in the slope is

$$U(xyz) = \gamma_w\{H\,(xyz)_o - H\,(xyz)_d\}$$

Index (o) denotes the potential in the slope without drain and (d) with the drain (Figure 17).

The change of the effective stress can appear after the corresponding change of pore volume, caused by water flowing towards outlets. There appears the process of consolidation described by the diffuse differential equation:

$$C_a = \frac{\partial^2 \cdot U}{\partial\, a^2} = \frac{\partial U}{\partial t} \qquad a = x,y,z$$

in which C_a is the coefficient of consolidation.

$$C_a = \frac{K_a \cdot M_a}{\gamma_\omega}$$

K_a = Darcy's coefficient in direction a
M_a = coefficient of compressibility in direction a

The numerical solution of this problem must satisfy the boundary conditions.

1. $U_t = 0$ on the boundaries S_1,S_2,S_3 and L (Figure 14)

2. $\dfrac{\partial U_t}{\partial z} = 0$ on the boundaries B_1 and B_2

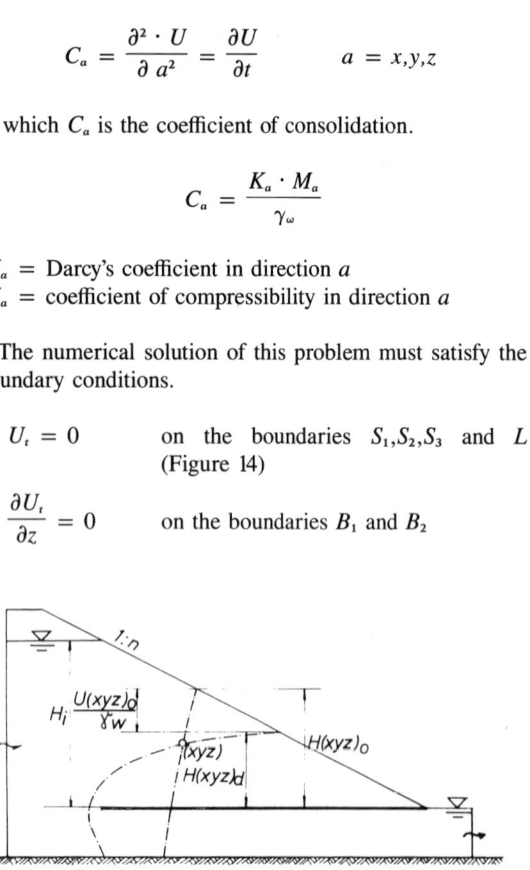

FIGURE 17.

3. $\dfrac{\partial U_t}{\partial y} = 0$ on the boundary S_4

The described problem with the stated boundary conditions is solved by the program SPACON which calculates from the results of the program TROFIL the change of potential from H_o to H_d in the nods (xyz) in chosen periods.

The output of the program SPACON is combined over program for slope stability analysis SSTABI [15]. The factor of safety against sliding is calculated by means of Spencer's method for chosen failure planes in chosen periods of time from the beginning of draining. The results are shown on the diagram F_g/θ_1 for chosen periods of time after installing the drains. They are:

$$F_g = 1 - \frac{F_d - F_t}{F_d - F_o}$$

Indices (o), (t) and (d) denote time $t = 0$, $t = t$, and time of final drain influence:

$$\theta_1 = t \cdot C_v \cdot H_i^{-2}$$

C_v is coefficient of consolidation, and H_i maximal difference of potentials.

The results of the increase of the factor of safety with time for the slope with the inclination 1:2, $C_v = 10^{-3}$m²/s, $H_i = 50$ m and drain lengths $L = 50$, 75 and 100 m, are shown in Figure 18.

These diagrams can be used for approximation of drain length and distance in order to get the required increase of the factor of safety in stated time.

The results of the increase of factor of safety $F = F_t/F_o$ in relation to the factor of time θ_2 for the slope inclination 1:2 with $H_i = 50$ m are shown in Figure 19. These are:

$$\theta_2 = C_v \cdot L \cdot t \cdot H_i^{-2} \cdot S^{-2}$$

The result grouped in relative narrow area show up to $\theta_2 = 1.6$ that the diagram in Figure 19 can interpolate with the results of real slopes of various heights with drains of various lengths and distances.

One can notice that drains are effective on homogenous slope if $C_v > 10^{-7}$ m²/s and time to achieve the desired effect increases proportionally with $1/C_v$.

EXAMPLES OF TUBULAR DRAINS APPLICATION

Drilled tubular drains have been used for stabilizing slopes for more than 40 years. Stanton [13] described their application in stabilizing slope slidings near motor-ways in Canada. Nonveiller and others [6,11] described their usage for stabilization of slidings in Yugoslavia.

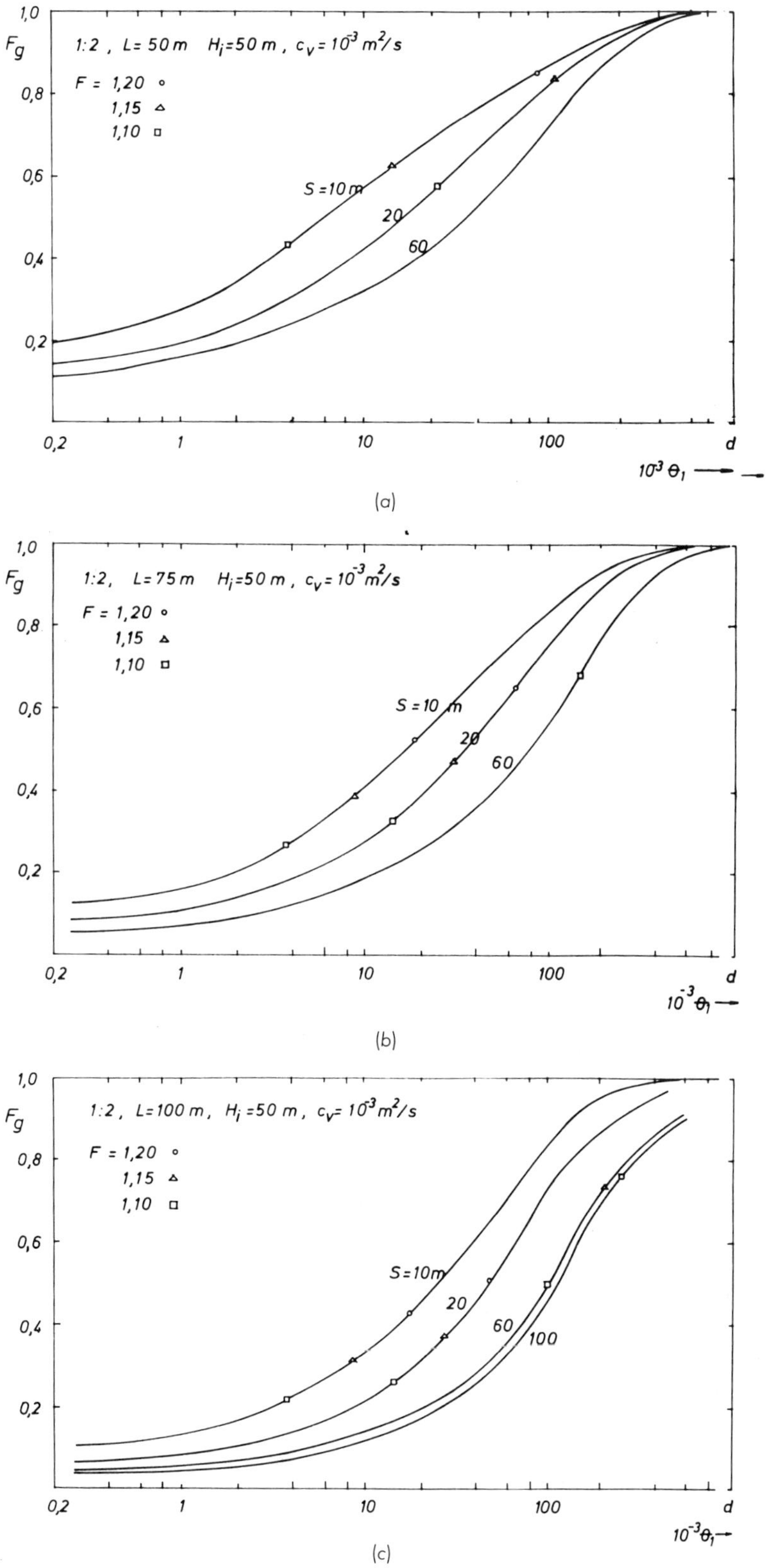

FIGURE 18. Diagrams F_g/Θ_1 for slope inclination 1:2 and drain lengths: a) L = 50 m; b) L = 75 m c) L = 100 m.

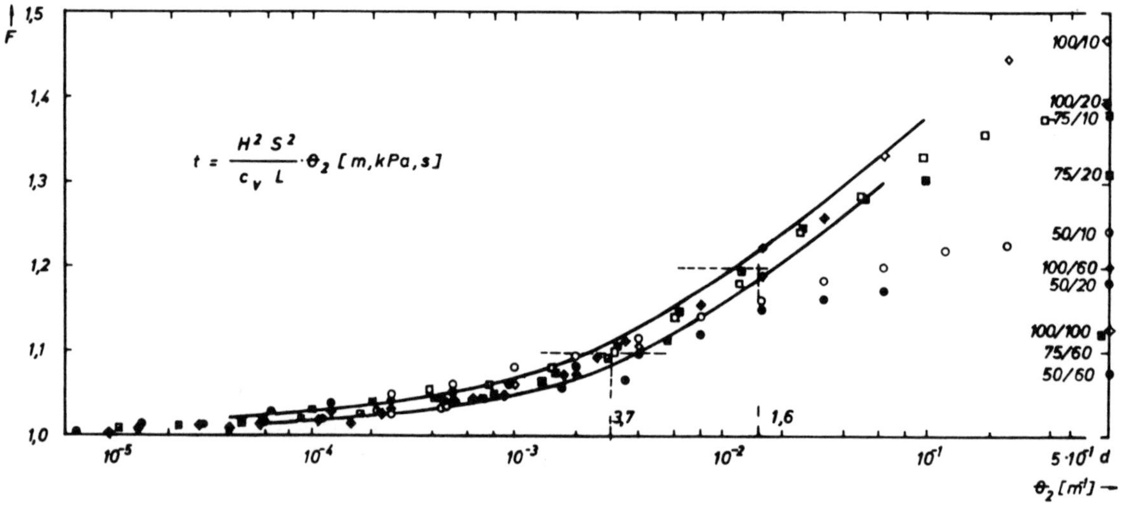

FIGURE 19. Increase of factor of safety F according to factor of time Θ_2 for the slope inclination 1:2.

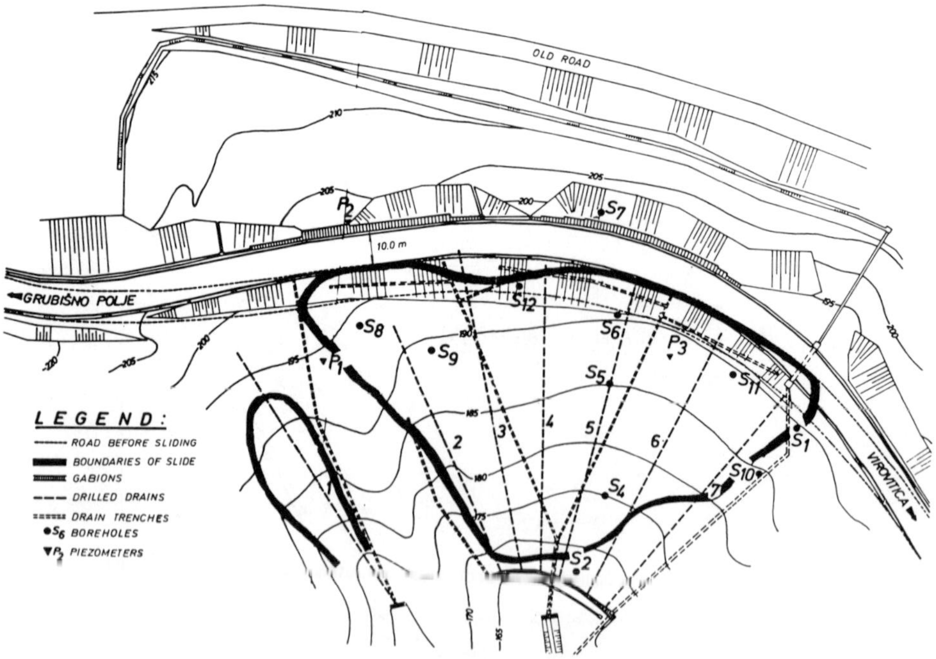

FIGURE 20. Ground-plan of sliding "Lonĉarica" with designed position of drains.

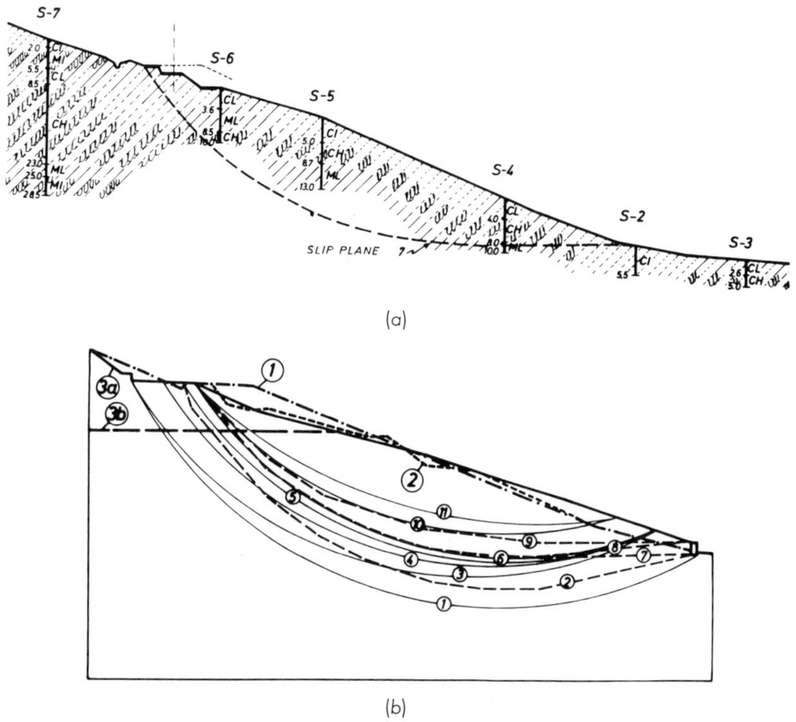

FIGURE 21. a) Characteristic geotechnical profile of the slope; b) cross section: 1) built road; 2) after sliding; 3) after stabilizing.

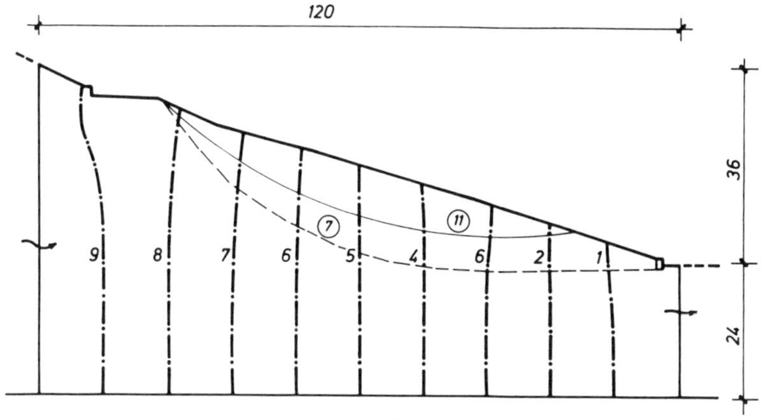

(a)

FIGURE 22. Average potential field: a) without drains.

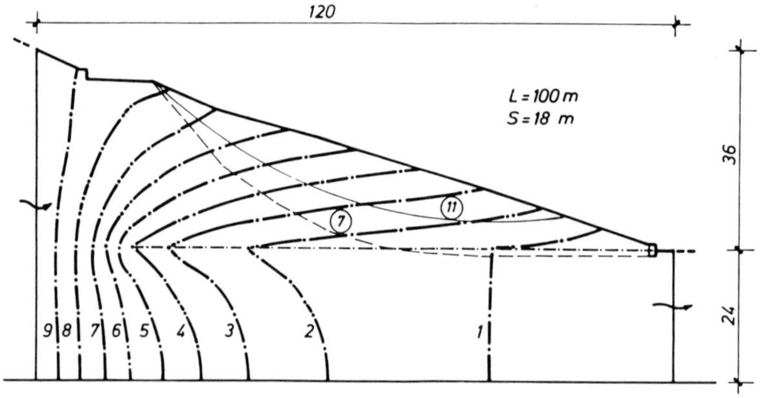

(b)

FIGURE 22 (continued). b) with drains.

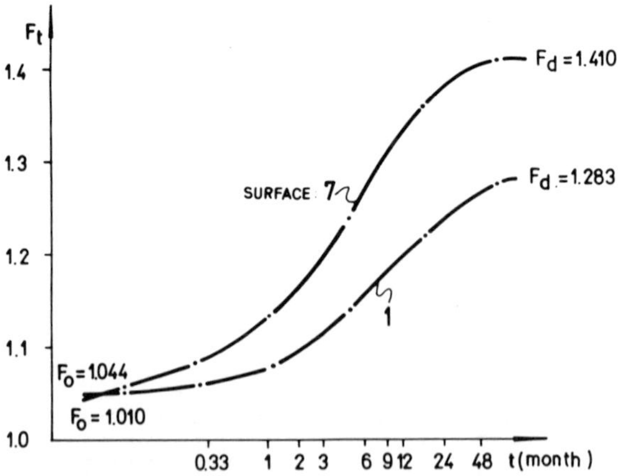

FIGURE 23. Increase of factor of safety in time.

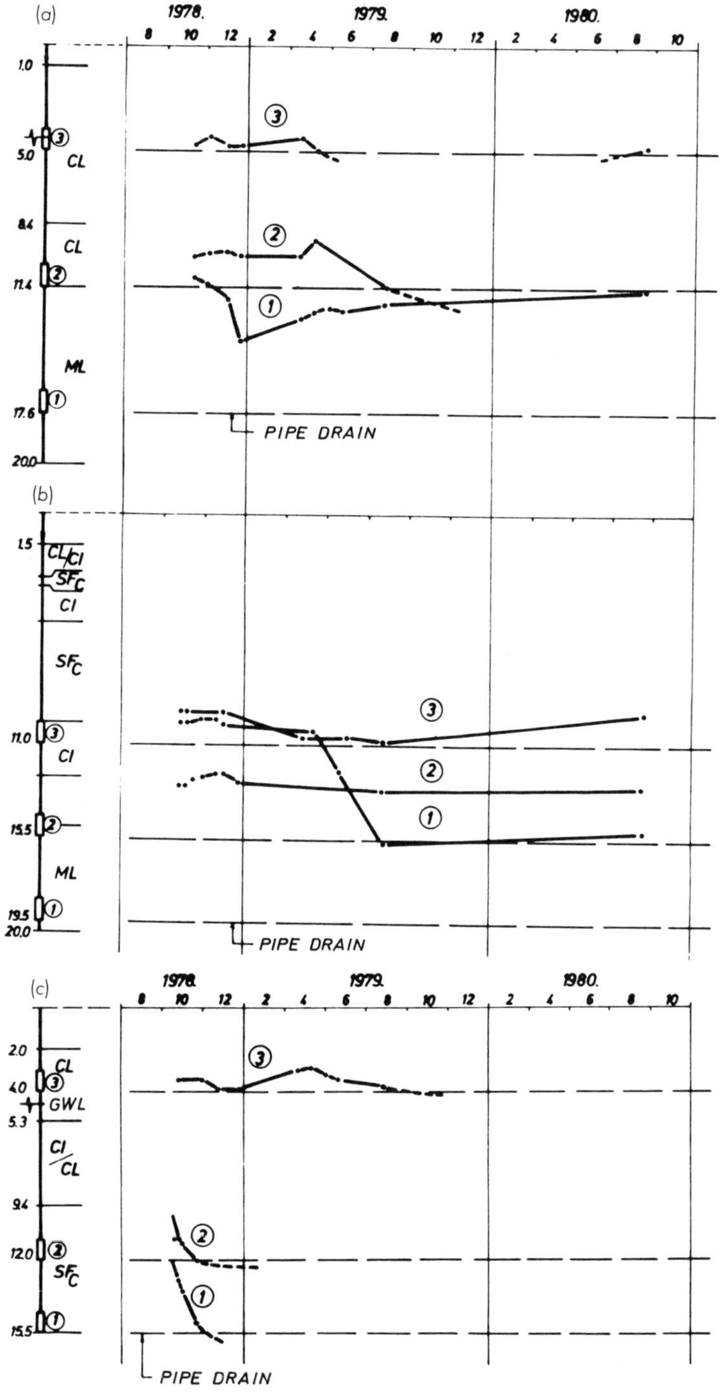

FIGURE 24. Soil composition and groundwater levels: a) piezometer P_1; b) piezometer P_2; c) piezometer P_3.

For stabilizing small slidings which often appear on natural slopes and on cuttings for roads it is enough, after inspecting the place and without large investigations, to drill and install several horizontal tubular drains on their foot to stablize the sliding.

It is useful to set the disturbed slope surface to prevent too great infiltration of precipitation. People maintaining roads can do that if they have small drilling equipment with auger.

After reconstruction of one road in Yugoslavia the sliding "Lonĉarica" became active. It is presented in Figure 20. The soil consists of the layers for Upper Neogene, represented by silty sand and clay of low and middle plasticity. Characteristic geotechnical cross section and model for slope stability analysis are presented in Figure 21. The average field potential is shown in Figure 22. The increase of factor of safety with time characteristic for two sliding planes is presented in Figure 23. To control the efficiency of draining, three piezometers were installed in a borehole on different levels. They showed the drain efficiency. Observation results are shown in Figure 24. The experiences of the application of drainage trenches and tubular drains for stabilization are positive.

NOTATION

i = hydraulic gradient
γ = bulk weight of soil
γ_w = unit weight of water
γ' = buoyant unit weight of soil
β = angle of slope inclination
S = drains distance
h = distance between groundwater table and slip surface
D = depth of slip surface
(xyz) = point coordinates
c' = cohesion
φ' = friction angle
H = total potential head
k = coefficient of permeability
P_p = average piezometric level at slip surface
P_M = maximal piezometric level at slip surface
P_1 = reaction on slip surface
r_M = normalized maximal piezometric level at slip surface
r_P = normalized average piezometric level at slip surface
τ = shear stress
τ_f = Mohr-Coulomb failure criterion
τ_n = total normal stress
u = pore pressure
Δ = Laplace operator
C_v = coefficient of consolidation
H_i = maximal difference of total potential
L = tubular drain strength
F_s = factor of safety
F = increase of factor of safety
F_t = factor of safety in time t
F_o = factor of safety in time o
F = final factor of safety
θ = factor of time
M = coefficient of compressibility

REFERENCES

1. Hutchinson, J. N., "Assessment of the Effectiveness of Corrective Measures in Relation to Geological Conditions and Types of Slope Movement," Bulletin of IAEG, General Report, Theme 3, Krefeld, pp. 131–155 (1977).
2. Jović, V. and T. Radelja. ELIPTI, Program for Solving Steady State Filtration in Saturated Soil and the Seepage Potential (in Croatian) (1981).
3. Kenney, T. C., M. Pazin and W. C. Shoi, "Design of Horizontal Drains for Soil Slopes," *Journal of the Geotechnical Engineering Division, ASCE, 103*(No. GT11), 1311–1323 (1977).
4. Krylov, C. I., V. V. Babkov and P. I. Monastirnyi, "Vychislitelnye metody vshej matematiki," Vyshesjshaja shkola, Minsk (1972).
5. Morgenstern, N. R., "Stability Charts for Earth Slopes During Rapid Drawdown," *Geotechnique, 13*(2), 121–131 (1963).
6. Nonveiller, E., "Sanierung von Hangrutschungen mittels Horizontalbohrungen," *European Civil Engineering,* Bratislava–Praha–Wien, No. 5, pp. 221–228 (1970).
7. Nonveiller, E. and J. Tadić, "Izbor duljine i razmaka cijevnih drenova za stabiliziranje kosina," *Proc. XIVth Congr. Yugoslav SSMFE,* Sarajevo, Vol. II, pp. 313–321 (1978).
8. Nonveiller, E., "Mehanika tla i temeljenje," Ŝkolska knjiga, Zagreb, Chapters 15, 16 (1979).
9. Nonveiller, E and Z. Tuŝar, "Asanacija kliženja ceste kod Lonĉarice pomoću cijevnih drenova," Simpozij istraživanje i sanacija kliziŝta, Bled, Vol. II, pp. 293–302 (1981).
10. Nonveiller, E., "Efficiency of Horizontal Drains on Slope Stability," *X ICSMFE,* Stockholm, Vol. 3, pp. 495–500 (1981).
11. Nonveiller, E., B. Stanić and Z. Tuŝar, "Pipe Drains for Landslide Stabilization," *Gradevinski kalendar,* Beograd, 1983, pp. 443–488 (in Croatian) (1983).
12. Stanić, B., "Influence of Drainage Trenches on Slope Stability," *Journal of the Geotechnical Engineering Division, ASCE, 110*(11), 1624–1636 (1984).
13. Stanton, T. E., "California Experience in Stabilizing Earth Slopes Through the Installation of Horizontal Drains by the Hydrauger Method," *Proc. IInd ICSMFE,* Rotterdam, Vol. III, pp. 256–260 (1948).
14. Taylor, D. W., "Stability on Earth Slopes," *Journal Boston Society of Civil Engineers, 24,* p. 197 (1937).
15. Wright, S. G., "SSTAB-1 Slope Stability Program" (1974).

INDEX